# Polarization Theory of Nuclear Reactions

天香
竞夸天下
独

Qing-Biao Shen

# Polarization Theory of Nuclear Reactions

**Revisers**
Chong-Hai Cai
Ye Tian
Bradley Kemp Wilson

Springer

Qing-Biao Shen
China Institute of Atomic Energy
Beijing, China

ISBN 978-3-031-11877-7 ISBN 978-3-031-11878-4 (eBook)
https://doi.org/10.1007/978-3-031-11878-4



This Springer imprint is published by the registered company Springer Nature Switzerland AG
The registered company address is: Gewerbestrasse 11, 6330 Cham, Switzerland

# Summary

This book develops the polarization theory describing the nuclear reactions of the spin 1/2 and 1 incident particles with unpolarized target nuclei and residual nuclei with arbitrary spin, and studies the polarization theory of the nuclear reactions of two polarized light particles. The unitary transformation relations of the rank-2 and higher-rank tensors between the spherical basis coordinate system and the orthogonal Cartesian coordinate system are found, and the theoretical method for studying the polarization phenomena of the particles with spin equal to or greater than 3/2 is given. The continuum discretized coupled channels (CDCC) theory of the axis-symmetric rotational nuclei describing the incident deuteron breakup reactions is developed. The relativistic nuclear reaction theories are introduced, and the relativistic nuclear reaction Dirac *S* matrix theory is proposed and developed. The polarization phenomena are not considered in the existing transport theory which describes the motion of the nucleons in nuclear medium, and it is considered that all particles are unpolarized, which is an approximate method from the view of microphysics. In order to describe the objective physical process in nature more vividly, this book presents for the first time the concrete scheme and method of establishing the polarization nuclear database. The Monte-Carlo method for describing the transport of the polarized particles and the transport equation of the polarized particles are proposed and developed for the first time. The book provides a technical way for the introduction of the polarization phenomenon in the micro-world into the research and design of the nuclear engineering projects closely related to human life.

# Introduction

When a particle outside of the nucleus moves in the nuclear potential formed by the interactions between the nucleons located inside of the nucleus, not only will its coordinate position and momentum continue to change but the particle and the nucleus itself will also undergo self-rotation, called spin. The spin values of particles and nuclei are quantized and can only take integers or semi-odd numbers. Both experiments and theories have proved that the nuclear potential acting on the spin $\frac{1}{2}$ particle contains spin-orbit coupling potential, and the nuclear potential acting on the spin 1 particle contains not only spin-orbit coupling potential but also tensor potential. In the theory of nuclear structure, the spin-orbit coupling potential causes the nuclear level to be split. Two nucleons with spin $\frac{1}{2}$ can be coupled to two states of total spin $S$ which is 0 or 1, and there is a tensor potential for the state which total spin equals 1. The spin of the deuteron is equal to 1. When studying the deuteron structure, it is found that not only the S state which orbit angular momentum $L = 0$ exists but the D state which orbit angular momentum $L = 2$ exists as well. The above results show that there is a tensor potential between the nucleon-nucleon. In nuclear reaction theory, because the spin-orbital coupling potential and tensor potential related to spin are included in the nuclear potential, the forces felt by particles with different spin magnetic quantum numbers are different, so their motion behavior will also be different. When all the magnetic quantum numbers of spins of the particles are equal and mutually incoherent, the particles are called unpolarized particles. Otherwise, they are polarized particles. In the case with spin-related nuclear potential, the polarized outgoing particles can be obtained through nuclear reaction induced by unpolarized incident particles. This is the polarization phenomenon of nuclear reactions.

The microscopic system of the nucleus is understood through the study of various considerable measurements of nuclear reactions. Nuclear force, nuclear structure, and nuclear reaction mechanism can be studied by studying various cross sections, angular distributions, energy spectra, and double differential sections of nuclear reactions. However, if only the unpolarized states of particles and nuclei are studied, the understanding for the nucleus has limitation. The polarization phenomena greatly enrich the data obtained from scattering and nuclear reactions, provide information on the interaction that can change spin orientation, and provide important verification data for the study of nuclear structures or reaction mechanisms. In addition, the effect of polarization phenomena on nuclear fusion rate is a promising research topic. Nowadays, some peoples are studying to produce the directed neutron sources by using the nuclear reactions $\vec{d}\left(\vec{t}, n\right)\alpha$ or $\vec{d}\left(\vec{d}, n\right)\,^{3}\text{He}$ introduced by two polarized light nuclei, and their civilian and military applications are being explored.

The theory of calculating various polarization quantities from the nuclear reaction amplitude is called polarization theory of nuclear reactions. As early as the 1950s, Wolfinstein et al. conducted theoretical studies on the polarization of nuclear reactions. Later, many papers on polarization phenomena were published, some of them also contained some information on experiments of the polarization nuclear reactions. In 1972, Ohlsen published a paper in *Rep. Prog. Phys.* This paper is a classical article on the polarization theory of nuclear reactions. In 1974, Robson published a book titled *The Theory of Polarization Phenomena* with 119 pages. In their papers, the basic knowledge of polarization theory is introduced. More detailed computational expressions are given for the polarization theory of spin $\frac{1}{2}$ particles with spinless targets. A part of the computational expressions is given for the polarization theory of spin 1 particles with spinless targets. Some specific calculations were also made and compared with experimental data in spinless target case. In Ohlsen's paper, a brief introduction of the polarization formulas of the two polarized light particles is given. In the relativistic optical model, the Dirac equation can be converted into a Schrodinger-like equation, and thus the polarization quantities of the elastic scattering between the nucleon and the target with spinless can also been calculated. For the nuclear reactions in which the target and the residual nucleus with non-zero spin, in 1965, Tamura gave a formula for calculating the polarization rate of the emitted nucleons using the method of averaging the Pauli matrix by the wave functions in his paper on coupling channel calculations. On this basis, some people made effort to study the corresponding analyzing power. In Robson's book, the polarization problem of the nuclear reaction between an incident particle with arbitrary spin and a target nucleus with arbitrary spin is discussed by using the formal theory in the complex spherical basis coordinate, and it does not yet have the practical application value. However, before writing this book no one has ever given a systematic polarization theory and the clear theoretical formulas that can be used to calculate actually the polarization observables for elastic scattering and direct reaction channels of spin $\frac{1}{2}$ and 1 particles with targets with non-zero spin. For example, there are no clear theoretical formulas to calculate the polarization observables for $\vec{1} + A \rightarrow \vec{\frac{1}{2}} + B$ and $\vec{\frac{1}{2}} + A \rightarrow \vec{1} + B$ reactions which target or residual nucleus with non-zero spin certainly, where $\vec{\frac{1}{2}}$ represents the polarized particle with spin $\frac{1}{2}$. Therefore, it is impossible to fully carry out the calculations of various polarization quantities for targets or residual nuclei with non-zero spin. Since the spins of most nuclei (including all odd A nuclei) are not equal to 0, the development of polarization theory of nuclear reactions is constrained.

There are only the theoretical formulas for calculating various polarization quantities of nuclear reactions using the nuclear reaction amplitude, but no specific research is doing on the nuclear reaction amplitude itself, one can only call this kind of theory as the formal theory. We know that the $S$ matrix theory is a basic theory describing the nuclear reactions. The expression of the reaction amplitude given by $S$ matrix theory satisfies the physical properties such as the angular momentum conservation and the parity conservation. When studying the two-body direct reaction that occurs between spin $\frac{1}{2}$ or 1 particle with the unpolarized target with spin $I$ and the unpolarized residual nucleus with spin $I'$, we found that there are some kind of internal relations between the matrix elements of the nuclear reaction amplitude for determined spin magnetic quantum numbers $M_I$ and $M'_I$. Therefore, when making the summation of $M_I$ and $M'_I$ in the polarization formula, some items will cancel each other and some items will be merged due to equality. In this way, some clear expressions for calculating various polarization quantities are obtained. Taking $\overrightarrow{\frac{1}{2}} + \mathrm{A} \rightarrow \overrightarrow{\frac{1}{2}} + \mathrm{B}$ reaction as an example, when the spins of A and B are both 0 and the right hand screw rule is used to select the $y$-axis perpendicular to the reaction plane, the results given by the predecessors have clearly pointed out that only the $y$ components of the polarization analyzing powers $A_i$ $(i = x, y, z)$ and the polarization rates $P^i$ $(i = x, y, z)$ of the outgoing particles for the unpolarized incident particles are not equal to 0, and only 5 of the 9 polarization transfer coefficients $K_i^j$ $(i, j = x, y, z)$ are not equal to 0. This book strictly proves that the above conclusion is still true when the spins of the unpolarized nuclei A and B are not equal to 0. This result has not been given before. We also use the above method to study $\vec{1} + \mathrm{A} \rightarrow \vec{1} + \mathrm{B}$ reactions as well as $\vec{1} + \mathrm{A} \rightarrow \overrightarrow{\frac{1}{2}} + \mathrm{B}$ and $\overrightarrow{\frac{1}{2}} + \mathrm{A} \rightarrow \vec{1} + \mathrm{B}$ reactions. The study for these reactions also found that some components of the polarization quantities are equal to 0 and have certain regularity. The above phenomenon is due to the fact that the nuclear reaction process satisfies the parity conservation.

We know that optical model, coupled channel optical model, distorted wave Born approximation (DWBA), $R$ matrix theory, phase-shift analysis method, and even some microscopic nuclear reaction theory can all be used to calculate $S$ matrix elements of nuclear reactions. These theories belong to the usual nuclear reaction theory. Therefore, it can be said that based on the contents of this book the polarization theory of spin $\frac{1}{2}$ or 1 particles and unpolarized targets and residual nuclei with arbitrary spins has reached the stage that the actual calculations can be made.

This book *Polarization Theory of Nuclear Reactions* comprises six chapters. The first chapter introduces the spin operators, irreducible tensors, polarization operators, density matrices etc., all of them are the basic knowledge of polarization theory of nuclear reactions.

The second chapter introduces the polarization theory of nuclear reactions for spin $\frac{1}{2}$ particles, such as the Pauli matrices, the polarization of spin $\frac{1}{2}$ particle beams, the polarization theory of elastic scattering of unpolarized and polarized spin $\frac{1}{2}$ particles with spinless target, polarization theory of triple elastic scatterings between spin $\frac{1}{2}$ particle and spinless nucleus, two theoretical methods to study the polarization phenomena of nuclear reactions. Then the polarization theory of $\overrightarrow{\frac{1}{2}} + \mathrm{A} \rightarrow \overrightarrow{\frac{1}{2}} + \mathrm{B}$ reactions which can be used for various kinds of two-body direct reactions is introduced, and the spin $\frac{1}{2}$ particle polarization theory is also introduced when both target and residual nucleus are polarized. Finally, the polarization problem of two spin $\frac{1}{2}$ polarized particles is studied theoretically.

The third chapter introduces the polarization theory of nuclear reactions for spin 1 particles, such as the spin operators and polarization operators of spin 1 particles, the vector polarization rate and tensor polarization rate of spin 1 incident particles, the elastic scattering amplitude of spin 1 particles with spinless target, the polarization theory of elastic scattering of spin 1 particle with both vector polarization and tensor polarization from a spinless target, the general form of the nuclear reaction polarization theory of spin 1 particles with non-zero spin target. Then the polarization theories of $\vec{1} + \mathrm{A} \rightarrow \vec{1} + \mathrm{B}$, $\vec{1} + \mathrm{A} \rightarrow \overrightarrow{\frac{1}{2}} + \mathrm{B}$, $\overrightarrow{\frac{1}{2}} + \mathrm{A} \rightarrow \vec{1} + \mathrm{B}$ reactions as well as $\overrightarrow{\frac{1}{2}} + \vec{1}$ and $\vec{1} + \vec{1}$ reactions with unpolarized targets and residual nuclei are introduced, respectively. In addition, the deuteron phenomenological optical potential with tensor term and its corresponding radial equation are introduced, the folding model of the deuteron-nucleus reactions is described, and the spherical nucleus continuum discretized coupled channels (CDCC) theory that only applies to the spinless targets describing the breakup reactions of the loosely bound light complex particles is introduced. This book also develops an axis-symmetric rotational nucleus CDCC theory describing the breakup reaction for loosely bound light complex particle. This theory can be used for targets with arbitrary spin and can consider the excited states of the targets. The above theories belong to the normal nuclear reaction theories of deuterons.

In Chap. 4, we first give the polarization theory of nuclear reactions for spin $\frac{3}{2}$ particles in the spherical basis coordinate system. When we study the five-dimensional rank-2 tensor in spin 1 particle polarization theory, the unitary transformation relation of the rank-2 tensor between the spherical basis coordinate system and the orthogonal Cartesian coordinate system is found. According to the regularity of the unitary transformation relations of the vector (rank-1 tensor) and the rank-2 tensor between the spherical basis coordinate system and the orthogonal Cartesian coordinate system, the unitary transformation relations of the rank-3 and higher-rank tensors between the spherical basis coordinate system and the orthogonal Cartesian coordinate system are found by recursion method. And the generalized Pauli matrix is introduced, thus the theoretical method for studying spin $\geq \frac{3}{2}$ particle polarization phenomena is given. Finally, based on the Maxwell equation of electromagnetic field, the quantization method of electromagnetic field is introduced, and the polarization theory for photon beams is discussed.

The fifth chapter introduces the polarization theory of relativistic nuclear reactions. It introduces the basic theory of relativistic quantum mechanics, transformation of relativistic coordinate systems, relativistic optical model and phenomenological optical potentials, Dirac $S$ matrix theory of relativistic nuclear reactions, Dirac coupling channel theory including elastic scattering and collective inelastic scattering channels, relativistic collective deformation relativistic distorted wave Born approximation (RDWBA) method and calculation of nucleon polarization quantities, relativistic impulse approximation of elastic scattering, relativistic impulse approximation of inelastic scattering, relativistic impulse approximation of (p,n) reactions. Then the relativistic classical field theory and Lagrangian density in quantum hadron dynamics, relativistic Green function theory at zero temperature, real part of nucleon relativistic microscopic optical potential and relativistic nuclear matter properties, imaginary part of nucleon relativistic microscopic optical potential, and contribution to imaginary part of nucleon relativistic microscopic optical potential by fourth order exchange diagrams are introduced. Then it introduces the relativistic Bete-Salpeter equation, the Bonn one boson exchange potential, and the nucleon relativistic microscopic optical potential based on the Dirac-Brueckner-Hartree-Fock (DBHF) theory. Finally, the Proca relativistic dynamics equation of spin 1 particle and its application in elastic scattering calculations, Weinberg relativistic dynamics equation of spin 1 particles and discussion on its application, and the relativistic nuclear reaction theory considering the internal structure of the incident deuterons are introduced. The above theories are the basic relativistic nuclear reaction theories for the study of the high-energy nucleon and deuteron induced nuclear reactions. By solving the corresponding dynamic equations, the reaction amplitude can be obtained and then various microscopic physical quantities including polarization quantities can be calculated.

The sixth chapter introduces the basis of the polarized particle transport theory. The polarized electron transport theory has been studied for a long time. This chapter proposes to carry out the polarized nucleus transport theory research. In order to simplify the theoretical formula, people usually select the $y$-axis along the direction which is perpendicular to the reaction plane, in this way there is the azimuthal angle $\varphi = 0$. If one wants to introduce the polarization process of the nuclear reactions into the particle transport theory, the polarization quantities of any $(\theta, \varphi)$ angles must be known. This chapter introduces the polarization theories of $\overrightarrow{\frac{1}{2}} + \mathrm{A} \rightarrow \overrightarrow{\frac{1}{2}} + \mathrm{B}$, $\vec{1} + \mathrm{A} \rightarrow \vec{1} + \mathrm{B}$, $\vec{1} + \mathrm{A} \rightarrow \overrightarrow{\frac{1}{2}} + \mathrm{B}$, and $\overrightarrow{\frac{1}{2}} + \mathrm{A} \rightarrow \vec{1} + \mathrm{B}$ reactions related to the polar angle $\theta$ and azimuthal angle $\varphi$ at the same time, and puts forward the specific concept of establishing a polarization nuclear database. It is point out what polarization quantities that are only related to the polar angles $\theta$ should be stored in the polarization nuclear database for each considering nuclear reactions. The specific calculation formulas of the polarization quantities related to $(\theta, \varphi)$ angles at the same time by using the data stored in the polarization nuclear database are given. It is easier to deal with $\overrightarrow{\frac{1}{2}} + \mathrm{A} \rightarrow 0 + \mathrm{B}$, $0 + \mathrm{A} \rightarrow \overrightarrow{\frac{1}{2}} + \mathrm{B}$, $\vec{1} + \mathrm{A} \rightarrow 0 + \mathrm{B}$, and $0 + \mathrm{A} \rightarrow \vec{1} + \mathrm{B}$ reactions related to spinless α particles, in which only the incident or outgoing channel is polarized. In addition, the Monte-Carlo methods for simulating the reactions including polarization channels are given and the method of the coordinate system rotation and the transformation of the polarization rates of transported particles are presented. The transport equation of polarized particles with spin $\frac{1}{2}$ or 1 is also proposed and developed. Finally, the prospective problems of the future development of the polarized particle transport theory are discussed.

In the normal particle transport theory, the particle flux function $\psi\left(\vec{r}, \vec{p}, t\right)$ is studied, and in the polarized particle transport theory proposed in this book the particle flux function $\psi\left(\vec{r}, \vec{p}, s, t\right)$ is studied, in which $s = \frac{1}{2}, -\frac{1}{2}$ for nucleons and $s = 1, 0, -1$ for deuterons. In the normal particle transport theory, only the usual nuclear data, which is not related to spin magnetic quantum numbers, are required. In the polarized nucleus transport theory proposed in this book, polarization nuclear data related to spin magnetic quantum numbers are needed for some reaction channels. Due to the existence objectively of spin-orbit forces and tensor forces related to spins of particles, the observable physical quantities of the emitted particles with different spin magnetic quantum numbers are different. Therefore, it can be said that the polarization particle transport theory can more realistically describe the physical processes that exist in nature. As to when it is necessary to consider the polarized particle transport process, this is a subject to be studied in the future.

More than 50 % of the theoretical formulas that appear in this book are new derivation. The main innovations of this book are:

1. Developed a nuclear reaction polarization theory system that can be used for specific calculations for various types of elastic scattering and two-body direct reaction between spin $\frac{1}{2}$ or 1 particle and unpolarized target nucleus with arbitrary spin. It breaks through the bottleneck that previously the people could not fully calculate various polarization quantities when the spins of the target and residual nucleus are not equal to 0.

2. The unitary transformation relations of the rank-2 and higher-rank tensors between the spherical basis coordinate system and the orthogonal Cartesian coordinate system are found, and the theoretical method for studying the spin $\geq \frac{3}{2}$ particle polarization phenomena is given.
3. The axis-symmetric rotational nucleus CDCC theory describing the breakup reaction of loosely bound light complex particles is developed. And the CDCC theory, which was originally applied only to spinless target, is developed to be used for the target with arbitrary spin and to include the inelastic scattering channels of the excited states.
4. The improper introduction of the coulomb potential into the Dirac equation was corrected. The Dirac $S$ matrix theory of relativistic nuclear reactions is proposed and developed, so that when the reaction amplitude of spin $\frac{1}{2}$ particles is solved by the Dirac equation, various microscopic physical quantities including polarization quantities can be calculated using this theory. And when the incident particle energy is quite low, the relativistic Dirac $S$ matrix theory will automatically degenerate into the general non-relativistic $S$ matrix theory of spin $\frac{1}{2}$ particles.
5. For the first time, the idea of establishing a polarization nuclear database was proposed. Which polarization quantities, that are only related to the polar angle $\theta$, are point out for $\overrightarrow{\frac{1}{2}} + \mathrm{A} \rightarrow \overrightarrow{\frac{1}{2}} + \mathrm{B}$, $\vec{1} + \mathrm{A} \rightarrow \vec{1} + \mathrm{B}$, $\vec{1} + \mathrm{A} \rightarrow \overrightarrow{\frac{1}{2}} + \mathrm{B}$, and $\overrightarrow{\frac{1}{2}} + \mathrm{A} \rightarrow \vec{1} + \mathrm{B}$ reactions in the polarization nuclear database, and the methods for obtaining these data from experimental data and theoretical calculations are discussed. The theoretical formulas for calculating the various polarization quantities related to $(\theta, \varphi)$ angles are also given.
6. For the first time, the idea of carrying out research on polarized nucleus transport theory is put forward. The Monte-Carlo method describing the polarized particle transport process and the polarized particle transport equation are proposed and developed. The polarization phenomena of nuclear reactions are objectively existed in the microscopic world. In the existing neutron and other heavy particle transport theories the polarization problem is not included, the above reality can only be considered as an approximation. In order to describe the physical processes that exist in nature more realistically, the polarized nucleus transport theory should be studied, especially when the angular distribution of the emitted particles clearly presents anisotropy. Above circumstances show it is necessary to carry out the research on the polarized nucleus transport theory. Since this book gives a nuclear reaction polarization theory system in which the spin $\frac{1}{2}$ and 1 particles react directly with unpolarized target and residual nucleus with arbitrary spins, and the theoretical method to study the spin $\geq \frac{3}{2}$ particle polarization phenomena is also given, so these make it possible to carry out the

research on polarized nucleus transport theory. That is the polarized nucleus transport theory research becoming feasible. For the first time, the establishment of a polarization nuclear database and the development of the polarized nucleus transport theory are proposed, that will inevitably encounter difficulties and challenges in the future. But after all, this idea is introducing the polarization phenomena of the microscopic world into the research and design of nuclear engineering projects that are closely related to human reality, and showing a way to explore the relationship of the nuclear engineering projects and the nuclear microscopic properties including the polarization phenomena.

Professor Chong-Hai CAI of Nankai University in China made a very careful revision for the Chinese version of this book. During the revision process, most of the theoretical formulas in the manuscript were carefully deduced, many mistakes were discovered and corrected, and many valuable and even key modifications were proposed. When I handed Professor CAI the general formulas for calculating the various polarization quantities of $\overrightarrow{\frac{1}{2}} + \mathrm{A} \rightarrow \overrightarrow{\frac{1}{2}} + \mathrm{B}$ reaction with the target and the residual nucleus with no zero spins, he programmed it into the $R$ matrix program and made the corresponding calculation. In the calculation results, it is found that for the case where the spins of the target and the residual nucleus are not equal to 0, there are also many polarization quantity components are equal to 0. It is also found that there is a certain relationship between certain matrix elements of reaction amplitude. Then, I carefully analyzed and deduced the relevant theoretical formulas, and theoretically proved that the relationship between these matrix elements of reaction amplitude seen in the calculation results does exist. And using these relations, it can be strictly proved that those components of polarization quantities with a calculated result of 0 should indeed be equal to 0. This is a critical step in the polarization theory of nuclear reactions given in this book. Some of the results obtained later in this book are closely related to this step. It can be seen from this issue that Professor CAI has made a very important contribution to this book. I would like to express my heartfelt thanks to Professor CAI, and I feel honored to have such an excellent partner in scientific research.

The Nuclear Data Key Laboratory of China and the China Nuclear Data Center have provided good working conditions and necessary financial support for the writing of this book. Researcher Yin-Lu HAN provided sincere help and support to the writing and publication of this book from beginning to the end. I express my sincere thanks here. Researcher Hai-Rui GUO of the Beijing Institute of Applied Physics and Computational Mathematics and Professor Yong-Li XU of Shanxi Datong University of China read part of the manuscript, and I express my thanks to them for their advice and help. I am very grateful to Mr. Feng-Quan ZHAO for the large number of typing of Chinese manuscript and formulas and making charts for this book.

My wife Ye TIAN, an associate researcher, has made an important contribution to the book. She and I both graduated from the Department of Modern Physics at the University of Science and Technology of China in July 1963 and were also assigned to the China Institute of Atomic Energy, where she worked in reactor physics and then theoretical nuclear physics. Both of us were original members of the China Nuclear Data Center when it was founded in 1975. In the process of writing this book, she assisted the me to do some data collation and participated in the discussion of some related issues. When the Chinese version of the book was finished in July 2018, Ms. Ye TIAN used computer translation software and other means to produce the first English version of the book in 2018.

Then, based on the first English draft, I carried out the revision sentence by paragraph, focusing on the use of special vocabulary and correct expression of the content of the book. At the same time, I also made some amendments and additions to the manuscript. After the draft of the English version of this book was completed, Dr. Bradley Kemp WILSON, a professor at the University of Saint Thomas in Saint Paul, Minnesota, USA, modified and refined most parts of the English version of this book.

In September 2021, I formally submitted the English version of the book to Springer Nature press and applied for publication. I underwent lumbar spine surgery on November 9, 2021, and needed to spend more than 3 months in bed after discharge. Springer Nature press passed the reviews to the manuscript on November 20, 2021, and reviewers gave the book a lot of positive reviews, so there was an urgent need for me to do a lot of work with the publisher via e-mail, such as responding to reviewers' comments, minor amendments, and discussing the content of the publishing agreement, among other things. Since it was difficult for me to get up to work on the computer, in this case, Ms. Ye TIAN discussed and agreed with me on some issues and then communicated with the publisher. This allowed the publication work of the book to proceed without delay. I take comfort in the fact that we 84-year-old couples are still working together professionally.

I would like to express my sincere thanks to Associate Researcher Ye TIAN and Professor Bradley Kemp WILSON for their contributions to the English version of this book.

Only this book commemorates my father Yu-Tian SHEN and my mother Qing-Wen KONG, who were ordinary farmers, they had worked for a lifetime in Shenzhuang Village, Gu'an County, Hebei Province in China.

Beijing, China Qing-Biao Shen
May 1, 2022

# Contents

# About the Author

**Qing-Biao Shen** is a researcher at the China Institute of Atomic Energy (CIAE). He was born on September 1, 1938, in Gu'an County, Hebei Province. In July 1963, he graduated from the University of Science and Technology of China with a degree in physics under the tutelage of Mr. Hong-Yuan ZHU. In the same year, Mr. Shen joined the CIAE, and from September 1963 to January 1965, he participated in a group led by Min YU and Zu-Qia HUANG working on hydrogen bomb theory. Since then, Mr. Shen's career has focused on low- and medium-energy nuclear reaction theory, nuclear many-body theory, and nuclear data theory, some of which he conducted while he was a visiting scholar at the University of Kentucky from 1985 to 1987.

Mr. Shen is the recipient of 12 National or Ministerial Science and Technology Progress Awards, including 5 inaugural prizes, and he has published over 300 academic papers as either a lead author or co-author. He is the author of *Nuclear Reaction Theory of Low and Medium Energies*. In addition, Mr. Shen developed the Skyrme force microscopic optical potential theory, a theory that to this day best fits experimental data.

# Chapter 1
# Basic Knowledge of Polarization Theory of Nuclear Reactions

**Abstract** This chapter introduces the basic knowledge of polarization theory of nuclear reactions, including spin operators, irreducible tensors, polarization operators, density matrices, and so on.

**Keywords** Density matrices · Irreducible tensors · Magnetic quantum numbers · Polarization operators · $S$ matrix elements · Spin operators

## 1.1 Introduction

When a particle outside the nucleus moves under the action of the nuclear potential, its coordinate position and momentum change continuously. At the same time, the particle and the nucleus itself will also undergo self-rotation, called spin. The spin values of particles and nuclei are quantized. For example, the spin of $\alpha$ particles is 0; the spin of neutrons, protons, tritium, and $^3$He is $\frac{1}{2}$; the spin of deuteron and $^6$Li is 1; and the spin of $^5$He, $^5$Li, $^7$Li, $^7$Be, $^9$Be is $\frac{3}{2}$. The unit of spin is ħ, and the spin of a particle is typically represented by $S$. Its projection in a certain spatial direction is called the spin magnetic quantum number, and its value is $S, S-1, \ldots, -S$. The spin of the target is usually represented by $I$.

In the nuclear structure theory, the spin-orbital coupling potential of the nucleons will cause the single particle level to be split. During the nuclear reaction process, the particles with different spin magnetic quantum numbers feel different nuclear forces. Its movement behavior will also be different, so the polarization phenomena will occur. The spin $\vec{S}$ of the particle is a vector in the coordinate space. Its average value $\langle\vec{S}\rangle$ in the nuclear reaction system is called the particle polarization vector. If it is not equal to 0, then the particle is said to be polarized; otherwise it is unpolarized.

Two nucleons with spin $\frac{1}{2}$ can be coupled to two states of total spin, $S = 0$ and 1, and there is a tensor potential for the $S = 1$ state. The deuteron spin $\vec{S}$ is a vector, and the vector $\vec{S}$ can also constitute a rank-2 tensor. Since the tensor potential is non-central, the total orbital angular momentum $L$ is no longer a good quantum

 
Q.-B. Shen, *Polarization Theory of Nuclear Reactions*,
https://doi.org/10.1007/978-3-031-11878-4_1

number, and the coupling between different $L$ waves will occur when solving the deuteron Schrodinger equation. Because of the tensor force existing between the nucleon-nucleon, it is found that when studying the deuterium structure not only the S state with $L = 0$ exists, but the D state with $L = 2$ exists as well. It is known that the spin of the deuteron is equal to 1 and its magnetic quantum numbers have three states: 1, 0, and $-1$. The action intensities of the spin-orbital potential on these three states are different: the action intensities of the tensor potential are also different. Therefore, there is not only vector polarization, but also tensor polarization for deuteron. They are all observables. For particles whose spin is greater than 1, there are more complex polarization observables. The polarization phenomenon greatly enriches the data obtained from scattering and nuclear reaction experiment; provides information on the interaction that can change spin orientation; and provides important verification data for the study of nuclear structures or reaction mechanisms.

The theory of calculating various polarization quantities from the nuclear reaction amplitude is called polarization theory of nuclear reactions. In the 1950s, Wolfenstein et al. conducted theoretical studies on the polarization of nuclear reactions [1–3]. Subsequently, many papers [4–22] on polarization phenomena have been published, with some of them also containing information on experiments of polarization nuclear reactions. In 1972, a paper published in *Rep. Prog. Phys.* by Ohlsen is a classical article [6] on the polarization theory of nuclear reactions. In 1974, Robson published a short book titled *The Theory of Polarization Phenomena* [9]. In these studies, more detailed computational expressions are given for the polarization theory of spin $\frac{1}{2}$ particles with spinless targets, and partial computational expressions are given for the polarization theory of spin 1 particles with spinless targets. Some specific calculations were also made and compared with experimental data in the spinless target case. In Ohlsen's paper, a brief introduction of the polarization formulas of the two polarized light particles was also given. In the relativistic optical model, the Dirac equation can be converted into a Schrodinger-like equation; thus, the polarization quantities of elastic scattering between the nucleon and the target with spinless can be calculated. For nuclear reactions with the non-zero spin target and the residual nucleus, in 1965, Tamura in his paper [23] on coupling channel calculations gave a formula for calculating the polarization rate of the emitted nucleons using the method of averaging the Pauli matrix by the wave functions. On this basis, effort was devoted by some people to studying the corresponding analyzing power [24, 25]. In Robson's book [9], only in the complex spherical basis coordinate system, the polarization problem of the nuclear reaction between the incident particle with arbitrary spin and the target nucleus with arbitrary spin has been studied briefly by the formal theory, which has no yet the practical value. However, before this book, no one ever gave a systematic polarization theory with clear theoretical formulas that can be used to calculate the polarization observables for elastic scattering and direct reaction channels of spin $\frac{1}{2}$ and 1 particles with targets with non-zero spins. For example, there is no clear theoretical formula to calculate the polarization observables for $\vec{1} + \mathrm{A} \rightarrow \overrightarrow{\frac{1}{2}} + \mathrm{B}$ and $\overrightarrow{\frac{1}{2}} + \mathrm{A} \rightarrow \vec{1} + \mathrm{B}$

reactions with non-zero spin target or residual nucleus, where $\frac{\overrightarrow{1}}{2}$ represents the polarized particle with spin $\frac{1}{2}$. Therefore, it is impossible to carry out fully the calculations of various polarization quantities for targets or residual nuclei with non-zero spins. Since the spins of most nuclei (including all odd $A$ nuclei) are not equal to 0, the development of polarization theory of nuclear reactions is constrained.

In order to perform a specific calculation of the polarization quantity, the nuclear reaction theory [26, 27] must be used to study how to get the corresponding reaction amplitude. We know that the $S$ matrix theory with $j$–$j$ coupling [26] or $S$-$L$ coupling [27] can give expressions for scattering amplitude, or reaction amplitude of the two-body nuclear reactions, including the C-G coefficients, the spherical harmonic functions, the Legendre functions, and the $S$ matrix elements. But the key issue is how to calculate the $S$ matrix elements. In nuclear reaction theory, $S$ matrix elements can be calculated by the phase shift analysis method for nucleon-nucleon or nucleon-deuteron elastic scattering, and it is also possible to extend this method to other types of nuclear reactions. For the light nucleus resonance energy region, the $R$ matrix theory can be used to calculate the $S$ matrix elements. The $S$ matrix elements in the case of elastic scattering of a particle with a spinless target can be calculated using a spherical optical model. The polarization problem of elastic scattering of the spin $\frac{1}{2}$ particles and spinless targets is introduced in Ref. [26], in which the $S$ matrix elements are calculated by the spherical optical model. At present, two groups of authors have given the universal deuteron optical potentials in the energy region below 200 MeV [28, 29]. However, neither group considered the deuteron polarization data when determining the parameters of the two sets of deuteron optical potentials. Moreover, in another set of deuteron universal optical potential suitable for 12–90 MeV energy rangy [30], it only considered the vector polarization data. In order to calculate the deuteron observables that can fit vector and tensor polarization experimental data simultaneously, the tensor potentials should be included in the deuteron optical potentials. The $S$ matrix elements of the inelastic scattering can be calculated by using the coupled channel optical model. The $T$ matrix elements of the two-body direct reactions can be obtained by using the DWBA method and can be converted into the $S$ matrix elements [26]. With the development of computers, coupled reaction channel theory considering elastic scattering, inelastic scattering, and other two-body direct reactions can simultaneously give the $S$ matrix elements of these reaction channels. In addition, the three-body reaction theory starting from realistic nuclear forces can also give scattering amplitude or reaction amplitude, so it can also be considered to perform a specific calculation of the polarization observables. Since statistical average assumptions are made in the study of equilibrium and pre-equilibrium reaction theories, the polarization problem does not need to be considered in these theories.

We know that the quantum theory of angular momentum is the main theoretical tool for studying how to obtain various polarization quantities from the amplitude of the nuclear reactions. In this respect, a very detailed and practical Ref. [31] can be found. Given this, it is possible now to develop a more widely applicable and practical polarization theory of nuclear reactions based on the existing polarization theory of nuclear reactions.

This book will study the polarization theory of nuclear reactions in which spin $\frac{1}{2}$ or 1 particle reacts directly with the unpolarized or polarized targets and residual nuclei with arbitrary spins, including $\vec{1} + \mathrm{A} \rightarrow \vec{\frac{1}{2}} + \mathrm{B}$ and $\vec{\frac{1}{2}} + \mathrm{A} \rightarrow \vec{1} + \mathrm{B}$ reactions. This book will also give the polarization theory of nuclear reactions in which the two particles in the incident channel are polarized; and make study on the polarization theory of nuclear reactions of the spin $\frac{3}{2}$ particles and the polarization theory of the photon beams; and give the theory of relativistic nuclear reactions and the corresponding polarization theory. We know that in some energy regions, even if the target, the residual nucleus, and the incident particle are all unpolarized, the emitted particles are generally polarized. For example, for the neutron induced reaction, the nuclear forces felt by neutrons with spin up and spin down are different, so their motion behaviors are different. Although the polarization phenomena exist objectively in nuclear reactions, at present people only use unpolarized nuclear reaction data when studying the nuclear transport process, and do not consider the polarization effect. Obviously, this approach is only an approximation in some energy regions. Therefore, in the last chapter of this book, we will discuss how to establish polarization nuclear database and make studying and discussion on the development of polarized nucleus transport theory.

## 1.2 Spin Operators

In general, we adopt the natural unit system, that is, letting $c = \hbar = 1$. In the actual calculation, the dimensionless spin and spin projection should all be multiplied by $\hbar$.

For coordinate space, the Cartesian coordinate system is generally used, that is, the rectangular coordinate system, and the coordinate position vector $\vec{r}$ can be represented as $(x, y, z)$. Ohlsen, an authority on polarization theory, pointed out that formally constructing another coordinate system represented by complex numbers, in which the coordinate position vector $\vec{r}$ can be represented as $(r_1, r_0, r_{-1})$, has many advantages in theoretical derivation [7]. In the monograph *Quantum Theory of Angular Momentum* [31], the coordinate system represented by $(r, \theta, \varphi)$ is called the polar coordinate system, and the coordinate system represented by $(r_1, r_0, r_{-1})$ is called the spherical coordinate system. In an article on polarization theory published in *Phys. Rev. C* [32], the coordinate system represented by $(r_1, r_0, r_{-1})$ is also referred to as the spherical coordinate system. In Rose's book *Elementary Theory of Angular Momentum* [33], the coordinate system represented by $(r_1, r_0, r_{-1})$ is called the "spherical coordinate representation", and the relations between the Cartesian basis functions and the spherical basis functions are given. In order to avoid confusion with the spherical coordinate system represented by $(r, \theta, \varphi)$, this book will refer to the coordinate system represented by $(r_1, r_0, r_{-1})$ as the spherical basis coordinate system. Note that in the spin $S$ space, there are $2S + 1$ spherical basis coordinate axes.

If the spin of a particle is $S$, then its spin operator $\hat{S}$ is a vector with three components and typically represented by a set of three square $(2S + 1) \times (2S + 1)$ matrices. The spin operator $\hat{S}$ is Hermitian

$$\hat{S}^{+} = \hat{S} \tag{1.1}$$

For the Cartesian coordinate components of $\hat{S}$, Eq. (1.1) representing the Hermitian property has the form

$$(\hat{S}_i)^{+} = \hat{S}_i, \qquad i = x,y,z \tag{1.2}$$

and for spherical basis coordinate components there are

$$(\hat{S}_\mu)^{+} = \hat{S}^{\mu} = (-1)^{\mu}\hat{S}_{-\mu}, \qquad \mu = \pm 1,0 \tag{1.3}$$

Spin operator $\hat{S}$ satisfies the following relation

$$\hat{S} \times \hat{S} = \mathrm{i}\hat{S}, \tag{1.4}$$

and also satisfies the following commutation relation

$$[(\vec{a} \cdot \hat{S}),(\vec{b} \cdot \hat{S})] = \mathrm{i}(\vec{a} \times \vec{b}) \cdot \hat{S} \tag{1.5}$$

where $\vec{a}$ and $\vec{b}$ are any constant vectors. When letting $\vec{a} = \vec{\mathrm{e}}_i$, $\vec{b} = \vec{\mathrm{e}}_k$ $(i, k = x, y, z)$, where $\vec{\mathrm{e}}_i$ and $\vec{\mathrm{e}}_k$ are the basis vectors of the Cartesian coordinate system, the following commutation relations can be obtained from Eq. (1.5)

$$[\hat{S}_i, \hat{S}_k] = \mathrm{i}\varepsilon_{ikl}\hat{S}_l, \quad [\hat{S}^2, \hat{S}_i] = 0, \quad i,k,l = x,y,z \tag{1.6}$$

where

$$\varepsilon_{ikl} = \begin{cases} 1, & \text{if the combinaton of } ikl \text{ is obtained by an even permutation of } xyz \\ -1 & \text{if the combinaton of } ikl \text{ is obtained by an odd permutation of } xyz \\ 0 & \text{if at least two indices among } ikl \text{ are equal} \end{cases} \tag{1.7}$$

The square of the spin operator $\hat{S}^2$ may be expressed in terms of Cartesian coordinate components $\hat{S}_i$ $(i = x, y, z)$ as

$$\hat{S}^2 = \sum_i \hat{S}_i^2 = \hat{S}_x^2 + \hat{S}_y^2 + \hat{S}_z^2 \tag{1.8}$$

In Ref. [26], Eq. (3.4.48) gives the definition of the spherical basis vectors, and Eqs. (3.4.51) and (3.4.52) give the spherical basis coordinate representations of any three-dimensional vector $\vec{A}$ as follows:

$$\vec{A} = \sum_{\mu=-1}^{1} (-1)^\mu A_\mu \vec{e}_{-\mu} \tag{1.9}$$

$$A_1 = -\frac{1}{\sqrt{2}}(A_x + iA_y), \quad A_0 = A_z, \quad A_{-1} = \frac{1}{\sqrt{2}}(A_x - iA_y) \tag{1.10}$$

And we can find out

$$\hat{S}^2 = \sum_\mu (-1)^\mu \hat{S}_{-\mu} \hat{S}_\mu = \sum_\mu \hat{S}^\mu \hat{S}_\mu \tag{1.11}$$

The spin function of spin $S$ particles can be expressed as $\chi(\sigma)$, where $\sigma$ is the projection of the spin on the $z$-axis, $\sigma = S, S-1, \cdots, -S$. The spin functions $\chi(\sigma)$ are commonly written as column matrices of $2S+1$ elements

$$\hat{\chi} = \begin{pmatrix} \chi(S) \\ \chi(S-1) \\ \vdots \\ \chi(-S) \end{pmatrix} \tag{1.12}$$

The corresponding Hermitian conjugate function $\hat{\chi}^+$ has the form of a row matrix

$$\hat{\chi}^+ = (\chi^*(S) \;\; \chi^*(S-1) \;\; \cdots \;\; \chi^*(-S)) \tag{1.13}$$

Its normalization condition is

$$\hat{\chi}^+ \hat{\chi} = \sum_{\sigma=-S}^{S} |\chi(\sigma)|^2 = 1 \tag{1.14}$$

The basis spin function is a spin state wave function that has definite spin and spin projection on the $z$-axis. The basis spin functions $\chi_{Sm}$ are the common eigenfunctions of spin operators $\hat{S}^2$ and $\hat{S}_z$, that is,

$$\hat{S}^2 \chi_{Sm} = S(S+1)\chi_{Sm}, \quad \hat{S}_z \chi_{Sm} = m\chi_{Sm} \tag{1.15}$$

In the spherical basis representation, the dependence of the basis spin functions $\chi_{Sm}(\sigma)$ on the spin variable $\sigma$ is given by

$$\chi_{Sm}(\sigma) = \delta_{m\sigma} \tag{1.16}$$

The basis spin functions of the spin $S$ particles can be written as column matrices as follows:

$$\chi_{SS} = \begin{pmatrix} 1 \\ 0 \\ \vdots \\ 0 \\ 0 \end{pmatrix}, \quad \chi_{S\ S-1} = \begin{pmatrix} 0 \\ 1 \\ \vdots \\ 0 \\ 0 \end{pmatrix}, \quad \cdots, \quad \chi_{S\ -S} = \begin{pmatrix} 0 \\ 0 \\ \vdots \\ 0 \\ 1 \end{pmatrix} \tag{1.17}$$

The collection of $2S+1$ basis spin functions $\chi_{Sm}$ $(m = S, S-1, \cdots, -S)$ constitutes a complete orthonormal set of functions, and its orthonormality condition is

$$\overset{+}{\chi}_{Sm} \chi_{Sm'} = \delta_{mm'} \tag{1.18}$$

Its completeness condition may be written in matrix form

$$\sum_{m=-S}^{S} \chi_{Sm} \overset{+}{\chi}_{Sm} = \hat{I}, \tag{1.19}$$

where $\hat{I}$ is the unit $(2S+1) \times (2S+1)$ matrix.

The spherical basis components of the spin operator $\hat{S}$ may be expressed in terms of the basis spin functions $\chi_{Sm}$ as [31]

$$\hat{S}_\mu = \sqrt{S(S+1)} \sum_{mm'} C^{S\ m'}_{Sm\ 1\mu} \chi_{Sm'} \overset{+}{\chi}_{Sm}, \qquad \mu = \pm 1{,}0 \tag{1.20}$$

From Eq. (1.20) and Eq. (1.16), the matrix elements of the spherical basis component of the spin matrix in the spherical basis representation can be obtained as

$$\left(\hat{S}_\mu\right)_{\sigma'\sigma} = \sqrt{S(S+1)}\ C^{S\sigma'}_{S\sigma\ 1\mu}, \qquad \sigma,\sigma' = S, S-1, \cdots, -S \tag{1.21}$$

The components of the spin operator $\hat{S}$ in the Cartesian coordinate system can be obtained from its components in the spherical basis coordinate system given by Eq. (1.20) by use of the following formulas:

$$\hat{S}_x = \frac{1}{\sqrt{2}}(\hat{S}_{-1} - \hat{S}_1), \quad \hat{S}_y = \frac{i}{\sqrt{2}}(\hat{S}_{-1} + \hat{S}_1), \quad \hat{S}_z = \hat{S}_0 \tag{1.22}$$

Its inverse transformation relations are

$$\hat{S}_1 = -\frac{1}{\sqrt{2}}\left(\hat{S}_x + i\hat{S}_y\right), \quad \hat{S}_0 = \hat{S}_z, \quad \hat{S}_{-1} = \frac{1}{\sqrt{2}}\left(\hat{S}_x - i\hat{S}_y\right) \tag{1.23}$$

In the Cartesian coordinate system, there are the following trace (tr) formulas for the products of spin matrices for an arbitrary spin $S$ [31]:

$$\begin{aligned}
&\mathrm{tr}\{\hat{S}_i\} = 0, \quad \mathrm{tr}\{\hat{S}_i\hat{S}_k\} = \frac{S(S+1)(2S+1)}{3}\delta_{ik}, \\
&\mathrm{tr}\{\hat{S}_i\hat{S}_k\hat{S}_l\} = i\frac{S(S+1)(2S+1)}{6}\varepsilon_{ikl}, \\
&\mathrm{tr}\left\{\hat{S}_i\hat{S}_k\hat{S}_l\hat{S}_j\right\} = \frac{S(S+1)(2S+1)}{15}\left\{\left[S(S+1) + \frac{1}{2}\right]\left(\delta_{ik}\delta_{lj} + \delta_{ij}\delta_{kl}\right)\right. \\
&\qquad \left. + \left[S(S+1) - 2\right]\delta_{il}\delta_{kj}\right\}, \quad i, k, l, j = x, y, z
\end{aligned} \tag{1.24}$$

The first symbol i on the right side of the third equation of the above expressions represents the imaginary number. In the spherical basis coordinate system, there are the following trace formulas for the products of spin matrices [31]

$$\begin{aligned}
&\mathrm{tr}\left\{\hat{S}_\mu\right\} = 0, \quad \mathrm{tr}\left\{\hat{S}_\mu\hat{S}_\nu\right\} = \frac{S(S+1)(2S+1)}{3}(-1)^\mu\delta_{\mu\ -\nu}, \\
&\mathrm{tr}\left\{\hat{S}_\mu\hat{S}_\nu\hat{S}_\lambda\right\} = -\frac{S(S+1)(2S+1)}{\sqrt{6}}\begin{pmatrix} 1 & 1 & 1 \\ \mu & \nu & \lambda \end{pmatrix} = \frac{S(S+1)(2S+1)}{3\sqrt{2}}(-1)^{1+\lambda}C_{1\mu\ 1\nu}^{1\ -\lambda}, \\
&\mathrm{tr}\left\{\hat{S}_\mu\hat{S}_\nu\hat{S}_\lambda\hat{S}_\rho\right\} = \frac{S(S+1)(2S+1)}{15}\left\{\left[S(S+1) + \frac{1}{2}\right](-1)^{\mu+\lambda}\right. \\
&\qquad \left(\delta_{\mu\ -\nu}\delta_{\lambda\ -\rho} + \delta_{\mu\ -\rho}\delta_{\nu\ -\lambda}\right) + \left.[S(S+1) - 2]\,(-1)^{\mu+\nu}\delta_{\mu\ -\lambda}\delta_{\nu\ -\rho}\right\}, \\
&\qquad \mu, \nu, \lambda, \rho = \pm 1, 0
\end{aligned} \tag{1.25}$$

where $\begin{pmatrix} 1 & 1 & 1 \\ \mu & \nu & \lambda \end{pmatrix}$ is the $3j$ symbol.

## 1.3 Irreducible Tensors

The Wigner $D$ function $D^J_{MM'}(\alpha, \beta, \gamma)$ is defined as the matrix elements of the rotation operator $\hat{D}(\alpha, \beta, \gamma)$ with a definite angular momentum $J$ and magnetic quantum number $M$

$$\langle JM|\hat{D}(\alpha, \beta, \gamma)|J'M'\rangle = \delta_{JJ'}D^J_{MM'}(\alpha, \beta, \gamma) \tag{1.26}$$

where $\alpha$, $\beta$, $\gamma$ are Euler angles. The wave function $\Psi_{JM}$ of a quantum system with angular momentum $J$ and projection $M$ rotates in coordinate space in the following way:

$$\Psi_{JM'}(\theta', \varphi') = \sum_{M=-J}^{J} \Psi_{JM}(\theta, \varphi)D^J_{MM'}(\alpha, \beta, \gamma) \tag{1.27}$$

where $\theta$, $\varphi$ and $\theta'$, $\varphi'$ are the polar and azimuthal angles of the initial and rotated coordinate systems. $\theta$, $\varphi$ and $\theta'$, $\varphi'$ satisfy the following relations [31]:

$$\begin{aligned} \cos\theta' &= \cos\theta\cos\beta + \sin\theta\sin\beta\cos(\varphi - \alpha), \\ \cot(\varphi' + \gamma) &= \cot(\varphi - \alpha)\cos\beta - \frac{\cot\theta\sin\beta}{\sin(\varphi - \alpha)} \end{aligned} \tag{1.28}$$

Its inverse relations are

$$\begin{aligned} \cos\theta &= \cos\theta'\cos\beta - \sin\theta'\sin\beta\cos(\varphi' + \gamma), \\ \cot(\varphi - \alpha) &= \cot(\varphi' + \gamma)\cos\beta + \frac{\cot\theta'\sin\beta}{\sin(\varphi' + \gamma)} \end{aligned} \tag{1.29}$$

The spin projection on the linear momentum direction of a particle is called the helicity, and the helicity of the spin $S$ particle is taken as $\lambda = S, S-1, \cdots, -S$. Let $\vec{n} = \vec{p}/p$ denote the unit vector of the particle momentum direction, where $\theta$ and $\varphi$ are the polar and azimuthal angles of $\vec{n}$. The helicity basis functions $\chi_{S\lambda}(\theta, \varphi)$ are the common eigenfunctions of the operators $\hat{\vec{S}}^2$ and $\hat{\vec{S}} \cdot \vec{n}$, that is,

$$\hat{\vec{S}}^2\chi_{S\lambda}(\theta, \varphi) = S(S+1)\chi_{S\lambda}(\theta, \varphi), \quad \hat{\vec{S}} \cdot \vec{n}\chi_{S\lambda}(\theta, \varphi) = \lambda\chi_{S\lambda}(\theta, \varphi) \tag{1.30}$$

The helicity basis functions can be obtained by rotating the general basis spin functions, which is defined to the $z$-axis [31]

$$
\begin{aligned}
\chi_{S\lambda}(\theta,\varphi) &= \sum_m D^S_{m\lambda}(\varphi,\theta,0)\chi_{Sm}, \\
\chi_{Sm} &= \sum_\lambda D^S_{-\lambda\ -m}(0,\theta,\varphi)\chi_{S\lambda}(\theta,\varphi), \\
\chi^+_{S\lambda}(\theta,\varphi) &= \sum_m (-1)^{\lambda-m} D^S_{-m\ -\lambda}(\varphi,\theta,0)\chi^+_{Sm}, \\
\chi^+_{Sm} &= \sum_\lambda (-1)^{\lambda-m} D^S_{\lambda\ m}(0,\theta,\varphi)\chi^+_{S\lambda}(\theta,\varphi)
\end{aligned}
\tag{1.31}
$$

The orthonormality condition for the helicity basis functions is

$$
\chi^+_{S\lambda'}(\theta,\varphi)\chi_{S\lambda}(\theta,\varphi) = \delta_{\lambda'\lambda} \tag{1.32}
$$

The completeness condition for the helicity basis functions has the following matrix form

$$
\sum_\lambda \chi_{S\lambda}(\theta,\varphi)\chi^+_{S\lambda}(\theta,\varphi) = \hat{I} \tag{1.33}
$$

where $\hat{I}$ is a unit $(2S+1)\times(2S+1)$ matrix.

Irreducible tensors mean that under rotations of coordinate systems these tensors transform in the same manner as eigenfunctions of the angular momentum operator. Assuming that $\mathcal{M}_J$ is an irreducible tensor of rank $J$, where $J$ can be a non-negative integer or semi-odd number, $\mathcal{M}_J$ has $2J+1$ components $\mathcal{M}_{JM}$, $M = -J, -J+1, \cdots, J$. When the coordinate system rotates, a linear transformation of $\mathcal{M}_{JM}$ is performed as follows:

$$
\begin{aligned}
\mathcal{M}_{JM'}(X') &= \hat{D}(\alpha,\beta,\gamma)\mathcal{M}_{JM'}(X)\left[\hat{D}(\alpha,\beta,\gamma)\right]^{-1} \\
&= \sum_{M=-J}^{J} \mathcal{M}_{JM}(X)D^J_{MM'}(\alpha,\beta,\gamma)
\end{aligned}
\tag{1.34}
$$

where $X$ and $X'$ denote sets of all arguments of the tensor in the initial and final coordinate systems, respectively. It can be seen that the linear transformation coefficient of this rotation is the Wigner $D$ function. In fact, the angular momentum operators $\hat{J}$,$\hat{L}$,$\hat{S}$ are the irreducible tensors of rank 1; the spherical harmonic function $Y_{lm}$ is an irreducible tensor of rank $l$; and the spin $S$ wave function is an irreducible tensor of rank $S$.

The spherical basis components of the angular momentum operator $\hat{J}$ and the irreducible tensor $\mathcal{M}_{JM}$ satisfy the following commutation relations:

$$\begin{aligned}\left[\hat{J}_{\pm 1}, \mathcal{M}_{JM}\right] &= \mp \frac{1}{\sqrt{2}} e^{\pm i\delta} \sqrt{J(J+1) - M(M \pm 1)} \mathcal{M}_{J\ M\pm 1}, \\ \left[\hat{J}_0, \mathcal{M}_{JM}\right] &= M \mathcal{M}_{JM}\end{aligned} \tag{1.35}$$

The above formulas can be written in compact form as

$$\left[\hat{J}_\mu, \mathcal{M}_{JM}\right] = e^{i\mu\delta} \sqrt{J(J+1)} C_{J\ M\ 1\mu}^{J\ M+\mu} \mathcal{M}_{J\ M+\mu}, \quad \mu = \pm 1{,}0 \tag{1.36}$$

From these relations it follows that

$$\left[\hat{J}^2, \mathcal{M}_{JM}\right] = J(J+1)\mathcal{M}_{JM} \tag{1.37}$$

Let us adopt the phase $\delta = 0$, that is, $e^{\pm i\delta} = 1$. For the integer rank $J$, the overall phase of the $\mathcal{M}_{JM}$ components is usually defined in such manner that

$$(\mathcal{M}_{JM})^* = (-1)^{-M} \mathcal{M}_{J\ -M} \tag{1.38}$$

This choice of phase coincides with that for spherical harmonics. You can also define another irreducible tensor as

$$\widetilde{\mathcal{M}}_{JM} = i^J \mathcal{M}_{JM} \tag{1.39}$$

Then one has

$$\left(\widetilde{\mathcal{M}}_{JM}\right)^* = (-1)^{J-M} \widetilde{\mathcal{M}}_{J\ -M} \tag{1.40}$$

The irreducible tensor defined in this way can be used in the case where J is an integer or a semi-odd number; $J - M$ is always an integer. According to Eq. (1.3), $\mathcal{M}_{JM}$ and $\mathcal{M}_J^M$ respectively represent covariant and contravariant components of the irreducible tensor $\mathcal{M}_J$, and we have the following relations:

$$\begin{aligned}\mathcal{M}_J^M &= (\mathcal{M}_{JM})^* = (-1)^{-M} \mathcal{M}_{J\ -M}, \\ \widetilde{\mathcal{M}}_J^M &= \left(\widetilde{\mathcal{M}}_{JM}\right)^* = (-1)^{J-M} \widetilde{\mathcal{M}}_{J\ -M}\end{aligned} \tag{1.41}$$

The scalar product of two irreducible tensors $\mathcal{M}_J$ and $\mathcal{N}_J$ of the same rank is defined as

$$(\mathcal{M}_J \cdot \mathcal{N}_J) = \sum_M (-1)^{-M} \mathcal{M}_{JM} \mathcal{N}_{J\ -M} = \sum_M \mathcal{M}_{JM} \mathcal{N}_{JM}^* = \sum_M \mathcal{M}_{JM} \mathcal{N}_J^M \tag{1.42}$$

$$\left(\widetilde{\mathcal{M}}_J \cdot \widetilde{\mathcal{N}}_J\right) = \sum_M (-1)^{J-M} \widetilde{\mathcal{M}}_{JM} \widetilde{\mathcal{N}}_{J\ -M} = \sum_M \widetilde{\mathcal{M}}_{JM} \widetilde{\mathcal{N}}_{JM}^* = \sum_M \widetilde{\mathcal{M}}_{JM} \widetilde{\mathcal{N}}_J^M \quad (1.43)$$

The components of an irreducible tensor product $\mathcal{L}_J$ of two irreducible tensors $\mathcal{M}_{J_1}$ and $\mathcal{N}_{J_2}$ of ranks $J_1$ and $J_2$ are defined as follows:

$$\mathcal{L}_{JM} = \sum_{M_1 M_2} C_{J_1 M_1\ J_2 M_2}^{J\ M} \mathcal{M}_{J_1 M_1} \mathcal{N}_{J_2 M_2} \quad (1.44)$$

The irreducible tensor product is also denoted as

$$\mathcal{L}_J \equiv \left\{\mathcal{M}_{J_1} \otimes \mathcal{N}_{J_2}\right\}_J \quad (1.45)$$

where $\mathcal{M}_{J_1 M_1} \mathcal{N}_{J_2 M_2}$ is the direct product of a set of $(2J_1 + 1)(2J_2 + 1)$ components of two irreducible tensors $\mathcal{M}_{J_1}$ and $\mathcal{N}_{J_2}$. According to Eq. (1.44) one can obtain

$$\mathcal{M}_{J_1 M_1} \mathcal{N}_{J_2 M_2} = \sum_{J=|J_1 - J_2|}^{J_1 + J_2} C_{J_1 M_1\ J_2 M_2}^{J\ M} \mathcal{L}_{JM} \quad (1.46)$$

This formula shows that the direct product $\mathcal{M}_{J_1 M_1} \mathcal{N}_{J_2 M_2}$ is reducible and can be expanded with irreducible tensor $\mathcal{L}_{JM}$ with different rank $J$. If one use $\widetilde{\mathcal{M}}_{J_1 M_1}$ and $\widetilde{\mathcal{N}}_{J_2 M_2}$ defined by Eq. (1.39) to get $\widetilde{\mathcal{L}}_{JM}$, one can prove

$$\left(\widetilde{\mathcal{L}}_{JM}\right)^* = (-1)^{J-M} \widetilde{\mathcal{L}}_{J\ -M} \quad (1.47)$$

This formula shows that $\widetilde{\mathcal{M}}_{J_1 M_1}$, $\widetilde{\mathcal{N}}_{J_2 M_2}$, and $\widetilde{\mathcal{L}}_{JM}$ all satisfy the relation (1.40). Although $\mathcal{M}_{J_1 M_1}$ and $\mathcal{N}_{J_2 M_2}$ satisfy relation (1.38), $\mathcal{L}_{JM}$ does not except $J_1 + J_2 - J =$ even. From the above discussion, it is understandable why $\mathrm{i}^l \mathrm{Y}_{l\ m_l}$ is used instead of using only $\mathrm{Y}_{l\ m_l}$ in the coupling wave function of the nuclear reactions.

## 1.4 Polarization Operators

The spin observables are different for particles with different spins. Table 1.1 gives the types of the spin observables corresponding to various spin particles.

The polarization operator is used to describe the polarization state of particles. The polarization operator $\hat{T}_{LM}(S)(M = -L, -L + 1, \cdots, L$ and $L = 0, 1, \ldots, 2S$; both $L$ and $M$ being integers) are square $(2S + 1) \times (2S + 1)$ matrices which act on spin functions. When the coordinate system is rotated, this operator is linearly transformed with the Wigner $D$ function in the way given by Eq. (1.34), that is,

**Table 1.1** Types of the spin observables corresponding to various spin particles

| Spin observables | Spin $S$ | | | | |
|---|---|---|---|---|---|
| | 0 | $\frac{1}{2}$ | 1 | $\frac{3}{2}$ | ...... |
| Unpolarized, scalar, rank-0 tensor | √ | √ | √ | √ | ...... |
| Vector, rank-1 tensor | | √ | √ | √ | ...... |
| Rank-2 tensor | | | √ | √ | ...... |
| Rank-3 tensor | | | | √ | ...... |
| ...... | | | | | ...... |

$\hat{T}_{LM}(S)$ are irreducible tensors of rank $L$. According to Eq. (1.36), the commutation relations between the spherical basis components of the spin operator $\hat{S}_\mu$ and the polarization operators $\hat{T}_{LM}(S)$ can be written as

$$\left[\hat{S}_\mu, \hat{T}_{LM}(S)\right] = \sqrt{L(L+1)}C_{LM\ 1\mu}^{L\ \ M+\mu}\,\hat{T}_{L\ M+\mu}(S) \tag{1.48}$$

Let us normalize the polarization operators by the condition

$$\mathrm{tr}\left\{\hat{T}_{LM}(S)\hat{T}^{+}_{L'M'}(S)\right\} = (2S+1)\,\delta_{LL'}\delta_{MM'} \tag{1.49}$$

and choose the phase factors to satisfy the relations

$$\hat{T}^{+}_{LM}(S) = (-1)^M\hat{T}_{L\ -M}(S) \tag{1.50}$$

The above three conditions completely determine the polarization operator $\hat{T}_{LM}(S)$. The above definition of $\hat{T}_{LM}(S)$ in this book is consistent with that in the polarization theory articles [6, 34] in $S$=1 case. But the defined polarization operator $\hat{T}_{LM}(S)$ in this book compared with that in the monograph titled *Quantum Theory of Angular Momentum* [31] has a more factor $\sqrt{2S+1}$ . Based on the trace formula $\mathrm{tr}\left\{\hat{\sigma}_\mu\hat{\sigma}^{+}_{\mu'}\right\} = 2\delta_{\mu\mu'}$ and Eq. (1.49), the defined $T_{1M}\left(\frac{1}{2}\right)$ in this book is equal to the Pauli matrix $\hat{\sigma}_M$. The polarization operator $\hat{T}_{LM}(S)$ has $\sum\limits_{L=0}^{2S}(2L+1) = (2S+1)^2$ linear independent components, and each component is a square $(2S+1)\times(2S+1)$ matrix. These $(2S+1)^2$ matrices constitute a set of complete linear independent matrices.

The polarization operator $\hat{T}_{LM}(S)$ can be represented by a basis spin function $\chi_{Sm}$

$$\hat{T}_{LM}(S) = \sqrt{2L+1}\sum_{m\ m'}C_{Sm\ LM}^{S\ m'}\chi_{Sm'}\chi^{+}_{Sm} \tag{1.51}$$

Its inverse relation is

$$\chi_{Sm'}\chi_{Sm}^{+} = \sum_{LM} \frac{\sqrt{2L+1}}{2S+1} C_{Sm\ LM}^{S\ m'} \hat{T}_{LM}(S) \tag{1.52}$$

Using Eq. (1.16) in the spherical basis representation, the matrix element of $\hat{T}_{LM}(S)$ in the spherical basis representation can be obtained as

$$\left[\hat{T}_{LM}(S)\right]_{\sigma'\sigma} = \sqrt{2L+1} C_{S\sigma\ LM}^{S\sigma'}, \quad \sigma,\sigma' = -S, -S+1,\cdots,S \tag{1.53}$$

When $L = 0$ from the above relation, one can obtain

$$\hat{T}_{00}(S) = \hat{I} \tag{1.54}$$

where $\hat{I}$ is a unit $(2S + 1) \times (2S + 1)$ matrix. Comparing Eq. (1.20) with Eq. (1.51) in $L = 1$ case, the following relation can be obtained:

$$\hat{T}_{1M}(S) = \sqrt{\frac{3}{S(S+1)}} \hat{S}_M, \quad M = \pm 1,0 \tag{1.55}$$

According to Eq. (1.49), it can be seen that $\frac{1}{\sqrt{(2S+1)}} \hat{T}_{LM}(S)$ constitutes a set of complete linear independent matrices. Therefore, any square $(2S + 1) \times (2S + 1)$ matrix $\hat{A}$ can be expanded with a series of polarization operators $\frac{1}{\sqrt{(2S+1)}} \hat{T}_{LM}(S)$ as

$$\hat{A} = \frac{1}{\sqrt{2S+1}} \sum_{L=0}^{2S} \sum_{M=-L}^{L} A_{LM} \hat{T}_{LM}(S) \tag{1.56}$$

By use of Eq. (1.49) and the above formula, we can get

$$\mathrm{tr}\left\{\hat{A}\hat{T}_{LM}^{+}(S)\right\} = \sqrt{2S+1}\, A_{LM} \tag{1.57}$$

If $\hat{A}$ is Hermitian, that is, $\hat{A}^{+} = \hat{A}$, then by using Eqs. (1.50) and (1.56) one can obtain

$$\begin{aligned} \hat{A} = \hat{A}^{+} &= \frac{1}{\sqrt{2S+1}} \sum_{LM} A_{LM}^{*} (-1)^{M} \hat{T}_{L\ -M}(S) \\ &= \frac{1}{\sqrt{2S+1}} \sum_{LM} A_{L\ -M}^{*} (-1)^{M} \hat{T}_{LM}(S) \end{aligned} \tag{1.58}$$

Comparing Eq. (1.56) and Eq. (1.58), the following formulas can be obtained:

$$A_{LM} = (-1)^M A^*_{L\ -M}, \quad A^*_{LM} = (-1)^M A_{L\ -M} \tag{1.59}$$

Products of two polarization operators $\hat{T}_{L_1M_1}(S)$ and $\hat{T}_{L_2M_2}(S)$ may be written as [31]

$$\hat{T}_{L_1M_1}(S)\hat{T}_{L_2M_2}(S) = \hat{L}_1\hat{L}_2\hat{S}\sum_{LM}(-1)^{L_1+L_2+L}W(L_1L_2SS;LS)C^{LM}_{L_1M_1\ L_2M_2}\hat{T}_{LM}(S) \tag{1.60}$$

where $\hat{L} \equiv \sqrt{2L+1}$, $\hat{S} \equiv \sqrt{2S+1}$, and Racah coefficient $W$ is used. Substituting Eq. (1.51) into the right end of Eq. (1.60), one can obtain

$$I_R \equiv \hat{L}_1\hat{L}_2\hat{S}\sum_{LM}(-1)^{L_1+L_2+L}\hat{L}W(L_1L_2SS;LS)C^{LM}_{L_1M_1\ L_2M_2}\sum_{mm'}C^{Sm'}_{Sm\ LM}\chi_{Sm'}\chi^+_{Sm} \tag{1.61}$$

Substituting Eq. (1.51) into the left end of Eq. (1.60) and using Eq. (1.18), one can obtain

$$\begin{aligned} I_L &\equiv \hat{L}_1\hat{L}_2\left(\sum_{m_1m_1'}C^{S\ m_1'}_{Sm_1\ L_1M_1}\chi_{Sm_1'}\chi^+_{Sm_1}\right)\left(\sum_{m_2m_2'}C^{S\ m_2'}_{Sm_2\ L_2M_2}\chi_{Sm_2'}\chi^+_{Sm_2}\right) \\ &= \hat{L}_1\hat{L}_2\sum_{m_1m'm}C^{S\ m'}_{Sm_1\ L_1M_1}C^{S\ m_1}_{Sm\ L_2M_2}\chi_{Sm'}\chi^+_{Sm} \end{aligned} \tag{1.62}$$

Using the following formula including the C–G coefficient and the Racah coefficient

$$\sum_{\varepsilon}C^{e\varepsilon}_{a\alpha\ b\beta}C^{c\gamma}_{e\varepsilon\ d\delta} = \sum_{f\varphi}\hat{e}\hat{f}\ W(abcd;ef)C^{f\varphi}_{b\beta\ d\delta}C^{c\gamma}_{a\alpha\ f\varphi} \tag{1.63}$$

one can obtain

$$\begin{aligned} \sum_{m_1}C^{Sm'}_{Sm_1\ L_1M_1}C^{Sm_1}_{Sm\ L_2M_2} &= \sum_{LM}\hat{S}\hat{L}W(SL_2SL_1;SL)C^{LM}_{L_2M_2\ L_1M_1}C^{Sm'}_{Sm\ LM} \\ &= \sum_{LM}(-1)^{L_1+L_2+L}\hat{S}\hat{L}W(L_1L_2SS;LS)C^{LM}_{L_1M_1\ L_2M_2}C^{Sm'}_{Sm\ LM} \end{aligned} \tag{1.64}$$

Substituting the above formula into Eq. (1.62), it is proved that the two ends of Eq. (1.60) are equal.

The polarization operators satisfy the following commutation relations [31]:

$$\begin{aligned} \left[\hat{T}_{L_1M_1}(S), \hat{T}_{L_2M_2}(S)\right] &= \hat{L}_1\hat{L}_2\hat{S}\sum_{L_3}(-1)^{L_1+L_2+L_3}\left[1-(-1)^{L_1+L_2+L_3}\right] \\ &\quad W(L_1L_2SS;L_3S)C^{L_3M_3}_{L_1M_1\ L_2M_2}\hat{T}_{L_3M_3}(S) \end{aligned} \tag{1.65}$$

Based on Eq. (1.55), we know that Eq. (1.48) is only a special case of Eq. (1.65) in the case of $L_1 = 1$. The anti-commutation relations of the polarization operators are

$$\begin{aligned}\{\hat{T}_{L_1M_1}(S), \hat{T}_{L_2M_2}(S)\} = \hat{L}_1\hat{L}_2\hat{S}\sum_{L_3}(-1)^{L_1+L_2+L_3}\left[1+(-1)^{L_1+L_2+L_3}\right] \\ W(L_1L_2SS;L_3S)C_{L_1M_1\ L_2M_2}^{L_3M_3}\hat{T}_{L_3M_3}(S)\end{aligned} \tag{1.66}$$

Here note the introduced symbol $\hat{S} \equiv \sqrt{2S+1}$. The following formulas can be obtained using Eqs. (1.49), (1.50), and (1.54)

$$\mathrm{tr}\left\{\hat{T}_{LM}(S)\right\} = \hat{S}^2\delta_{L0}\delta_{M0} \tag{1.67}$$

$$\mathrm{tr}\left\{\hat{T}_{L_1M_1}(S)\hat{T}_{L_2M_2}(S)\right\} = (-1)^{M_1}\hat{S}^2\delta_{L_1L_2}\delta_{M_1\ -M_2} \tag{1.68}$$

Using Eq. (1.60) and Eq. (1.68), one can get

$$\begin{aligned}&\mathrm{tr}\{\hat{T}_{L_1M_1}(S)\hat{T}_{L_2M_2}(S)\hat{T}_{L_3M_3}(S)\} \\ &= (-1)^{L_1+L_2+L_3+M_3}\hat{L}_1\hat{L}_2\hat{S}^3 C_{L_1M_1\ L_2M_2}^{L_3\ -M_3} W(L_1L_2SS;L_3S)\end{aligned} \tag{1.69}$$

Using Eq. (1.60) and Eq. (1.69), one can also get

$$\begin{aligned}&\mathrm{tr}\{\hat{T}_{L_1M_1}(S)\hat{T}_{L_2M_2}(S)\hat{T}_{L_3M_3}(S)\hat{T}_{L_4M_4}(S)\} \\ &\quad = \sum_{L\le 2S}(-1)^{L_1+L_2+L}\hat{L}_1\hat{L}_2\hat{S}W(L_1L_2SS;LS)C_{L_1M_1\ L_2M_2}^{L\ M_1+M_2} \\ &\quad (-1)^{L+L_3+L_4+M_4}\hat{L}\hat{L}_3\hat{S}^3 C_{L\ M_1+M_2\ L_3M_3}^{L_4\ -M_4} W(LL_3SS;L_4S) \\ &\quad = (-1)^{L_1+L_2+L_3+L_4+M_4}\hat{L}_1\hat{L}_2\hat{L}_3\hat{S}^4 \\ &\quad \sum_{L\le 2S}\hat{L}C_{L_1M_1\ L_2M_2}^{L\ M_1+M_2}C_{L\ M_1+M_2\ L_3M_3}^{L_4\ -M_4} W(L_1L_2SS;LS)W(LL_3SS;L_4S)\end{aligned} \tag{1.70}$$

## 1.5 Density Matrices

The density matrix is defined as [31]

$$\hat{\rho} = \sum_{\alpha} w_\alpha|\psi_\alpha\rangle\langle\psi_\alpha| \tag{1.71}$$

where $|\psi_\alpha\rangle$ is the $\alpha$ sub-state of the system, $w_\alpha$ is the weight and the normalization condition is given by

$$\sum_{\alpha} w_\alpha = 1 \tag{1.72}$$

Obviously the density matrix defined by Eq. (1.71) is Hermitian.

Set $\{|\varphi_i\rangle\}$ as a set of orthogonal complete basis, the matrix element of the density matrix $\hat{\rho}$ in the orthogonal basis is

$$\rho_{ik} = \langle\varphi_i|\hat{\rho}|\varphi_k\rangle = \sum_\alpha \langle\varphi_i|\psi_\alpha\rangle w_\alpha \langle\psi_\alpha|\varphi_k\rangle \tag{1.73}$$

In general $\hat{\rho}$ is not a diagonal matrix. The definition of the trace of the observable $\hat{O}$ is

$$\mathrm{tr}\{\hat{O}\} = \sum_i \langle\varphi_i|\hat{O}|\varphi_i\rangle \tag{1.74}$$

The density matrix can be traced, using Eqs. (1.71), (1.72), and (1.74) as well as the orthonormal properties of the wave functions, as follows:

$$\mathrm{tr}\{\hat{\rho}\} = \sum_i \langle\varphi_i|\sum_\alpha w_\alpha|\psi_\alpha\rangle\langle\psi_\alpha|\varphi_i\rangle = \sum_\alpha w_\alpha \sum_i \langle\psi_\alpha|\varphi_i\rangle\langle\varphi_i|\psi_\alpha\rangle = \sum_\alpha w_\alpha = 1 \tag{1.75}$$

The average value of any physical quantity $\hat{O}$ is

$$\langle\hat{O}\rangle = \frac{\sum_\alpha w_\alpha \langle\psi_\alpha|\hat{O}|\psi_\alpha\rangle}{\sum_\alpha w_\alpha \langle\psi_\alpha|\psi_\alpha\rangle} \tag{1.76}$$

The complete condition of orthogonal basis $|\varphi_i\rangle$ is

$$\sum_i |\varphi_i\rangle\langle\varphi_i| = \hat{I} \tag{1.77}$$

Here $\hat{I}$ is the unit matrix. Note that the matrix element is a value that can be exchanged with other physical quantities, so using Eqs. (1.77) and (1.73) one can rewrite Eq. (1.76) as

$$\begin{aligned}\langle\hat{O}\rangle &= \frac{\sum_{\alpha ik} w_\alpha \langle\psi_\alpha|\varphi_k\rangle\langle\varphi_k|\hat{O}|\varphi_i\rangle\langle\varphi_i|\psi_\alpha\rangle}{\sum_{\alpha i} w_\alpha \langle\psi_\alpha|\varphi_i\rangle\langle\varphi_i|\psi_\alpha\rangle} = \frac{\sum_{ik}\left\{\sum_\alpha \langle\varphi_i|\psi_\alpha\rangle w_\alpha \langle\psi_\alpha|\varphi_k\rangle\right\}\langle\varphi_k|\hat{O}|\varphi_i\rangle}{\sum_i\left\{\sum_\alpha \langle\varphi_i|\psi_\alpha\rangle w_\alpha \langle\psi_\alpha|\varphi_i\rangle\right\}} \\ &= \frac{\sum_{ik}\rho_{ik}O_{ki}}{\sum_i \rho_{ii}} = \frac{\mathrm{tr}\{\hat{\rho}\hat{O}\}}{\mathrm{tr}\{\hat{\rho}\}} = \frac{\mathrm{tr}\{\hat{O}\hat{\rho}\}}{\mathrm{tr}\{\hat{\rho}\}}\end{aligned} \tag{1.78}$$

The polarization density matrix $\hat{\rho}$ of the spin $S$ particles is a square $(2S + 1) \times (2S + 1)$ matrix defined by the spin wave functions $\chi(\sigma)$ in the spin space in the following way

$$\rho_{\sigma\sigma'} = \langle \chi(\sigma)\chi^*(\sigma') \rangle_\xi \tag{1.79}$$

or in matrix form by

$$\hat{\rho} = \langle \hat{\chi}\hat{\chi}^+ \rangle_\xi \tag{1.80}$$

where $\langle \, \rangle_\xi$ denotes the statistical average. In particular, for pure states, one gets

$$\hat{\rho} = \hat{\chi}\hat{\chi}^+ = |\chi\rangle\langle\chi| \tag{1.81}$$

From the above definition, it can be seen that the density matrix is Hermitian, that is,

$$\hat{\rho}^+ = \hat{\rho}, \qquad \hat{\rho}^*_{\sigma\sigma'} = \hat{\rho}_{\sigma'\sigma} \tag{1.82}$$

and normalized, that is,

$$\mathrm{tr}\{\hat{\rho}\} = 1, \qquad \sum_\sigma \rho_{\sigma\sigma} = 1 \tag{1.83}$$

For pure states, from Eq. (1.81) and Eq. (1.14), one can prove

$$\hat{\rho}^2 = \hat{\chi}\hat{\chi}^+\hat{\chi}\hat{\chi}^+ = \hat{\chi}\hat{\chi}^+ = \hat{\rho} \tag{1.84}$$

We use $|\chi_\mathrm{i}\rangle$ and $|\chi_\mathrm{f}\rangle$ to represent the initial and final states of the spin space, respectively. The reaction amplitude is defined by the following formula:

$$|\chi_\mathrm{f}\rangle = \hat{F}|\chi_\mathrm{i}\rangle \tag{1.85}$$

Using Eq. (1.81), the density matrix of the initial state and the final state can be expressed as

$$\hat{\rho}_\mathrm{in} = |\chi_\mathrm{i}\rangle\langle\chi_\mathrm{i}| \tag{1.86}$$

$$\hat{\rho}_\mathrm{out} = |\chi_\mathrm{f}\rangle\langle\chi_\mathrm{f}| = \hat{F}|\chi_\mathrm{i}\rangle\langle\chi_\mathrm{i}|\hat{F}^+ = \hat{F}\hat{\rho}_\mathrm{in}\hat{F}^+ \tag{1.87}$$

The average value of any physical quantity $\hat{O}$ in the final state can be obtained by using the complete condition of the basis spin functions $|\chi_{Sm}\rangle$

$$\langle \hat{O} \rangle = \frac{\langle \chi_{\mathrm{f}} | \hat{O} | \chi_{\mathrm{f}} \rangle}{\langle \chi_{\mathrm{f}} | \chi_{\mathrm{f}} \rangle} = \frac{\langle \chi_{\mathrm{i}} | \hat{F}^{+} \hat{O} \hat{F} | \chi_{\mathrm{i}} \rangle}{\langle \chi_{\mathrm{i}} | \hat{F}^{+} \hat{F} | \chi_{\mathrm{i}} \rangle} = \frac{\sum\limits_{mm'} \langle \chi_{\mathrm{i}} | \hat{F}^{+} | \chi_{Sm} \rangle \langle \chi_{Sm} | \hat{O} | \chi_{Sm'} \rangle \langle \chi_{Sm'} | \hat{F} | \chi_{\mathrm{i}} \rangle}{\sum\limits_{m} \langle \chi_{\mathrm{i}} | \hat{F}^{+} | \chi_{Sm} \rangle \langle \chi_{Sm} | \hat{F} | \chi_{\mathrm{i}} \rangle} \tag{1.88}$$

Since the matrix element $\langle \ldots \rangle$ is a numerical value, the matrix element can exchange positions with other physical quantities, so the above formula can be rewritten as

$$\begin{aligned} \langle \hat{O} \rangle &= \frac{\sum\limits_{m\,m'} \langle \chi_{Sm'} | \hat{F} | \chi_{\mathrm{i}} \rangle \langle \chi_{\mathrm{i}} | \hat{F}^{+} | \chi_{Sm} \rangle \langle \chi_{Sm} | \hat{O} | \chi_{Sm'} \rangle}{\sum\limits_{m} \langle \chi_{Sm} | \hat{F} | \chi_{\mathrm{i}} \rangle \langle \chi_{\mathrm{i}} | \hat{F}^{+} | \chi_{Sm} \rangle} \\ &= \frac{\sum\limits_{m'} \langle \chi_{Sm'} | \hat{F} | \chi_{\mathrm{i}} \rangle \langle \chi_{\mathrm{i}} | \hat{F}^{+} \hat{O} | \chi_{Sm'} \rangle}{\sum\limits_{m} \langle \chi_{Sm} | \hat{F} | \chi_{\mathrm{i}} \rangle \langle \chi_{\mathrm{i}} | \hat{F}^{+} | \chi_{Sm} \rangle} = \frac{\mathrm{tr}\{\hat{\rho}_{\mathrm{out}} \hat{O}\}}{\mathrm{tr}\{\hat{\rho}_{\mathrm{out}}\}} = \frac{\mathrm{tr}\{\hat{O} \hat{\rho}_{\mathrm{out}}\}}{\mathrm{tr}\{\hat{\rho}_{\mathrm{out}}\}} \end{aligned} \tag{1.89}$$

Equation (1.89) obtained in the spin space is the same as Eq. (1.78). If the polarization problem is studied by using the first equation of Eq. (1.88), it is called the expectation value method. If the polarization problem is studied by using the last right end of Eq. (1.89), it is called the tracing method. These two methods are equivalent.

If the density matrix is normalized, an expectation value of any polarization operator $\hat{T}$ in a state described by the density matrix $\hat{\rho}$ may be evaluated by

$$\langle \hat{T} \rangle = \mathrm{tr}\{\hat{T} \hat{\rho}\} = \mathrm{tr}\{\hat{\rho} \hat{T}\} \tag{1.90}$$

The normalized density matrix $\hat{\rho}$ of spin $S$ particles can be expanded by a set of complete linear independent matrices $\frac{1}{\sqrt{2S+1}} \hat{T}_{LM}(S)$ that have been formed before, and the corresponding expansion coefficient can be represented by $\frac{1}{\sqrt{2S+1}} t_{LM}(S)$. Given this, we have the following formula:

$$\hat{\rho} = \frac{1}{2S+1} \sum_{L=0}^{2S} t_L(S) \cdot \hat{T}_L(S) \tag{1.91}$$

where $\hat{T}_L$ is the polarization operator discussed earlier and $t_L$ is called the statistical tensor, which is a spatial coordinate function. This book does not use the polarization operator defined in Ref. [31], but uses the polarization operator normalized by Eq. (1.49). It can be seen from the above formula that there are some advantages by using our approach. When the density matrix given by Eq. (1.91) is used to find the differential cross section, the factor $\frac{1}{2S+1}$, which should be multiplied when the average of the initial states is carrying out, will appear automatically at the front of the expression. The point multiplication in Eq. (1.91) represents the scalar product of two irreducible tensors defined by Eq. (1.41), so Eq. (1.91) can be rewritten as

$$\begin{aligned}\hat{\rho} &= \frac{1}{2S+1}\sum_{L=0}^{2S}\sum_{M=-L}^{L}(-1)^M t_{L\ -M}(S)\hat{T}_{LM}(S)\\ &= \frac{1}{2S+1}\sum_{L=0}^{2S}\sum_{M=-L}^{L} t_{LM}(S)(-1)^M\hat{T}_{L\ -M}(S)\end{aligned} \tag{1.92}$$

We can obtain the following relation according to Eqs. (1.90), (1.92), and (1.68):

$$\left\langle \hat{T}_{LM}(S)\right\rangle = \mathrm{tr}\left\{\hat{\rho}\,\hat{T}_{LM}(S)\right\} = t_{LM}(S) \tag{1.93}$$

The above formula shows that $t_{LM}(S)$ is the expectation value of the polarization operator $\hat{T}_{LM}(S)$ in the state described by the normalized density matrix $\hat{\rho}$.

In the spherical basis representation, using Eq. (1.53) we can get

$$(-1)^M\left(\hat{T}_{L\ -M}(S)\right)_{\sigma\sigma'} = (-1)^M\sqrt{2L+1}C_{S\sigma'\ L\ -M}^{S\sigma} = \sqrt{2L+1}C_{S\sigma\ LM}^{S\sigma'} \tag{1.94}$$

Substituting Eq. (1.94) into Eq. (1.92) we have

$$\rho_{\sigma\sigma'} = \sum_{LM}\frac{\sqrt{2L+1}}{2S+1}C_{S\sigma\ LM}^{S\sigma'}\,t_{LM}(S) \tag{1.95}$$

Using the formula of the C-G coefficient

$$C_{S\sigma\ LM}^{S\sigma'} = (-1)^{S-\sigma}\frac{\sqrt{2S+1}}{\sqrt{2L+1}}C_{S\sigma\ S\ -\sigma'}^{L\ -M} \tag{1.96}$$

and Eq. (1.95) we may obtain

$$\begin{aligned}\sum_{\sigma\sigma'}C_{S\sigma\ LM}^{S\sigma'}\rho_{\sigma\sigma'} &= \frac{\sqrt{2S+1}}{\sqrt{2L+1}}\sum_{\sigma\sigma'}\sum_{L'M'}\frac{\sqrt{2L'+1}}{2S+1}\frac{\sqrt{2S+1}}{\sqrt{2L'+1}}\\ &C_{S\sigma\ S\ -\sigma'}^{L\ -M}C_{S\sigma\ S\ -\sigma'}^{L'\ -M'}t_{L'M'}(S) = \frac{1}{\sqrt{2L+1}}t_{LM}(S)\end{aligned} \tag{1.97}$$

which yields

$$t_{LM}(S) = \sqrt{2L+1}\sum_{\sigma\sigma'}C_{S\sigma\ LM}^{S\sigma'}\rho_{\sigma\sigma'} \tag{1.98}$$

By use of Eq. (1.83) the following relation is easily obtained from Eq. (1.98):

$$t_{00}(S) = 1 \tag{1.99}$$

It can easily be proven that the density matrix $\hat{\rho}$ of spin space defined by Eq. (1.91) or Eq. (1.92) is normalized using Eqs. (1.67), (1.99), and (1.54). According to Eq. (1.82) and Eq. (1.98), we can prove

$$t_{LM}^{*}(S) = \sqrt{2L+1}\sum_{\sigma\sigma'}(-1)^{M}C_{S\sigma'\ L\ -M}^{S\sigma}\,\rho_{\sigma'\sigma} = (-1)^{M}t_{L\ -M}(S) \tag{1.100}$$

## References


1. Wolfenstein L, Ashkin J. Invariance conditions on the scattering amplitudes for spin 1/2 particles. Phys Rev. 1952;85:947.
2. Wolfenstein L. Possible triple-scattering experimrnts. Phys Rev. 1954;96:1654.
3. Wolfenstein L. Polarization of fast nucleons. Ann Rev Nucl Sci. 1956;6:43.
4. Hoshizaki N. Appendix: Formalism of nucleon-nucleon scattering. Prog Theor Phys Suppl. 1968;42:107.
5. Binstock J, Bryan R. Nucleon-nucleon scattering near 50 MeV. II. Sensitivity of various n-p observables to the phase parameters. Phys Rev. 1974;D9:2528.
6. Ohlsen GG. Polarization transfer and spin correlation experiments in nuclear physics. Rep Prog Phys. 1972;35:717.
7. Ohlsen G G, Gammel J L, and Keaton P W. Description $^4$He(d,d)$^4$He polarization-transfer experiments. Phys Rev., 1972, C5:1205
8. Salzman GC, Mitchell CK, Ohlsen GG. Techniques for polarization transfer coefficient determination. Nucl Inst Meth. 1973;109:61.
9. Robson BA. The theory of polarization phenomena. Oxford: Clarendon Press; 1974.
10. Sperisen F, Gruebler W, Konig V. A general formalism for polarization transfer measurements. Nucl Inst Meth. 1983;204:491.
11. Seiler F, Darden SE, Mcintyre LC, Weitkamp WG. Tensor polarization of deuterons scattered from He$^4$ between 4 and 7.5 MeV. Nucl Phys. 1964;53:65.
12. Schwandt P, Haeberli W. Elastic scattering of polarized deuterons from $^{27}$Al, Si and $^{60}$Ni between 7 and 11 MeV. Nucl Phys. 1968;A110:585.
13. Cords H, Din GU, Ivanovich M, Robson BA. Tensor polarization of deuterons from $^{12}$C-d elastic scattering. Nucl Phys. 1968;A113:608.
14. Schwandt P, Haeberli W. Optical-model analysis of d-Ca polarization and cross-section measurements from 5 to 34 MeV. Nucl Phys. 1969;A123:401.
15. Gruebler W, Konig V, Schmelzbach PA, Mamier P. Elastic scatteriong of vector polarized deutrons on $^4$He. Nucl Phys. 1969;A134:686.
16. Djaloeis A, Nurzynski J. Tensor polarization of deuterons from the Mg(d,d)Mg elastic scattering at 7.0 MeV. Nucl Phys. 1971;A163:113.
17. Djaloeis A, Nurzynski J. Tensor polarization and differential cross sections for the elastic scattering of deuterons by Si at low energies. Nucl Phys. 1972;A181:280.
18. Irshad M, Robson BA. Elastic scattering of 15 MeV deuterons. Nucl Phys. 1974;A218:504.
19. Goddard RP, Haeberli W. The optical model for elastic scattering of 10 to 15 MeV polarized deuterons from medium-weight nuclei. Nucl Phys. 1979;A316:116.

20. Matsuoka N, Sakai H, Saito T, et al. Optical model and folding model potential for elastic scattering of 56 MeV polarized deuterons. Nucl Phys, 1986, A455:413
21. Takei M, Aoki Y, Tagishi Y, Yagi K. Tensor interaction in elastic scattering of polarized deutrons from medium-weight nuclei near $E_d$=22 MeV. Nucl Phys. 1987;A472:41.
22. Iseri Y, Kameyama H, Kamimura M, Yahiro M, Tanifuji M. Virtual breakup effects in elastic scattering of polarized deutrons. Nucl Phys. 1988;A490:383.
23. Tamura T. Analyses of the scattering of nuclear particles by collective nuclei in terms of the coupled-channel calculation. Rev Mod Phys. 1965;37:679.
24. Li R, Sun WL, Soukhovitskii ES, et al. Dispersive coupled-channels optical-model potential with soft-rotator couplings for Cr, Fe, and Ni isotopes. Phys Rev. 2013;C87:054611.
25. Sun WL, Li R, Soukhovitskii ES, et al. A fully Lane-consistent dispersive optical model potential for even Fe isotopes based on a soft-rotator model. Nucl Data Sheets. 2014;118:191.
26. Shen QB. Nuclear reaction theory of low and medium energy (upper volume). Beijing: Science Press; 2005. (in Chinese)
27. Shen QB. Nuclear reaction theory of low and medium energy (middle volume). Beijing: Science Press; 2012. (in Chinese)
28. An HX, Cai CH. Global deuteron optical model potential for the energy range up to 183 MeV. Phys Rev. 2006;C73:054605.
29. Han YL, Shi YY, Shen QB. Deuteron global optical model potential for energies up to 200 MeV. Phys Rev. 2006;C74:044615.
30. Daehnick WW, Childs JD, Vrcelj Z. Global optical model potential for elastic deuteron scattering from 12 to 90 MeV. Phys Rev. 1980;C21:2253.
31. Varshalovich DA, Moskalev AN, Khersonskii VK. Quantum theory of angular momentum. Singapore/New Jersey/Hong Kong: World Scientific; 1988.
32. Zhang JS, Liu KF, Shuy GW. Neutron suppression in polarized dd fusion reaction. Phys Rev. 1999;C60:054614.
33. Rose ME. Elementary theory of angular momentum. New York: Wiley; 1957.
34. Lakin W. Spin polarization of the deuteron. Phys Rev. 1955;98:139.

# Chapter 2
# Polarization Theory of Nuclear Reactions for Spin $\frac{1}{2}$ Particles

**Abstract** This chapter will introduce the polarization theory of nuclear reactions for spin $\frac{1}{2}$ particles, such as the polarization of spin $\frac{1}{2}$ particle beams, and the polarization theory of elastic scattering of unpolarized and polarized spin $\frac{1}{2}$ particles with spinless target. The polarization theory of $\overrightarrow{\frac{1}{2}} + \text{A} \rightarrow \overrightarrow{\frac{1}{2}} + \text{B}$ reactions which can be used for various kinds of two-body direct reactions will be introduced. Finally, the polarization problem of two spin $\frac{1}{2}$ polarized particles will be studied theoretically.

**Keywords** Elastic scattering · Nuclear reactions · Pauli matrices · Polarization phenomena · Residual nucleus · Spin $\frac{1}{2}$ particle · Spinless targets

## 2.1 Pauli Matrices

According to Eq. (1.17), the basis spin functions of a spin $\frac{1}{2}$ particle may be written as

$$\chi_{\frac{1}{2}\ \frac{1}{2}} = \begin{pmatrix} 1 \\ 0 \end{pmatrix}, \quad \chi_{\frac{1}{2}\ -\frac{1}{2}} = \begin{pmatrix} 0 \\ 1 \end{pmatrix} \tag{2.1}$$

Using Eq. (2.1) and Table 2.1, the spherical basis components of the spin operator $\hat{S}$ of a spin $\frac{1}{2}$ particle is obtained from Eq. (1.20) as

$$\hat{S}_1 = -\frac{1}{\sqrt{2}}\begin{pmatrix} 0 & 1 \\ 0 & 0 \end{pmatrix}, \quad \hat{S}_0 = \frac{1}{2}\begin{pmatrix} 1 & 0 \\ 0 & -1 \end{pmatrix}, \quad \hat{S}_{-1} = \frac{1}{\sqrt{2}}\begin{pmatrix} 0 & 0 \\ 1 & 0 \end{pmatrix} \tag{2.2}$$



Q.-B. Shen, *Polarization Theory of Nuclear Reactions*,
https://doi.org/10.1007/978-3-031-11878-4_2

**Table 2.1** Clebsch-Gordan coefficient $C_{l\ m-\mu\ \ 1\mu}^{j\ m}$ [1]

| | $\mu = 1$ | $\mu = 0$ | $\mu = -1$ |
|---|---|---|---|
| $j = l + 1$ | $\sqrt{\frac{(l+m)(l+m+1)}{2(l+1)(2l+1)}}$ | $\sqrt{\frac{(l-m+1)(l+m+1)}{(l+1)(2l+1)}}$ | $\sqrt{\frac{(l-m)(l-m+1)}{2(l+1)(2l+1)}}$ |
| $j = l$ | $-\sqrt{\frac{(l+m)(l-m+1)}{2l(l+1)}}$ | $\frac{m}{\sqrt{l(l+1)}}$ | $\sqrt{\frac{(l-m)(l+m+1)}{2l(l+1)}}$ |
| $j = l - 1$ | $\sqrt{\frac{(l-m)(l-m+1)}{2l(2l+1)}}$ | $-\sqrt{\frac{(l-m)(l+m)}{l(2l+1)}}$ | $\sqrt{\frac{(l+m)(l+m+1)}{2l(2l+1)}}$ |

Using Eq. (1.22) and the formulas above, the Cartesian basis components of the spin operator $\hat{S}$ of a spin $\frac{1}{2}$ particle are

$$\hat{S}_x = \frac{1}{2}\begin{pmatrix} 0 & 1 \\ 1 & 0 \end{pmatrix}, \quad \hat{S}_y = \frac{1}{2}\begin{pmatrix} 0 & -\mathrm{i} \\ \mathrm{i} & 0 \end{pmatrix}, \quad \hat{S}_z = \frac{1}{2}\begin{pmatrix} 1 & 0 \\ 0 & -1 \end{pmatrix} \tag{2.3}$$

For spin $\frac{1}{2}$ particles, let

$$\hat{S} = \frac{1}{2}\hat{\sigma} \tag{2.4}$$

where $\hat{\sigma}$ is called the Pauli matrix. The spherical and Cartesian basis components of the Pauli matrix $\hat{\sigma}$ can be obtained by Eqs. (2.2) and (2.3), respectively

$$\hat{\sigma}_1 = \begin{pmatrix} 0 & -\sqrt{2} \\ 0 & 0 \end{pmatrix}, \quad \hat{\sigma}_0 = \begin{pmatrix} 1 & 0 \\ 0 & -1 \end{pmatrix}, \quad \hat{\sigma}_{-1} = \begin{pmatrix} 0 & 0 \\ \sqrt{2} & 0 \end{pmatrix} \tag{2.5}$$

$$\hat{\sigma}_x = \begin{pmatrix} 0 & 1 \\ 1 & 0 \end{pmatrix}, \quad \hat{\sigma}_y = \begin{pmatrix} 0 & -\mathrm{i} \\ \mathrm{i} & 0 \end{pmatrix}, \quad \hat{\sigma}_z = \begin{pmatrix} 1 & 0 \\ 0 & -1 \end{pmatrix} \tag{2.6}$$

The Pauli matrix $\hat{\sigma}$ has the following properties:

$$\hat{\sigma}_i^+ = \hat{\sigma}_i, \qquad i = x, y, z \tag{2.7}$$

$$\hat{\sigma}_\mu^+ = (-1)^\mu \hat{\sigma}_{-\mu}, \qquad \mu = \pm 1, 0 \tag{2.8}$$

$$\hat{\sigma}_i^2 = \hat{I}, \qquad i = x, y, z \tag{2.9}$$

$$\hat{\sigma}^2 = \hat{\sigma}_x^2 + \hat{\sigma}_y^2 + \hat{\sigma}_z^2 = 3\hat{I} \tag{2.10}$$

$$\hat{\sigma}_x\hat{\sigma}_y = -\hat{\sigma}_y\hat{\sigma}_x = \mathrm{i}\hat{\sigma}_z, \quad \hat{\sigma}_y\hat{\sigma}_z = -\hat{\sigma}_z\hat{\sigma}_y = \mathrm{i}\hat{\sigma}_x, \quad \hat{\sigma}_z\hat{\sigma}_x = -\hat{\sigma}_x\hat{\sigma}_z = \mathrm{i}\hat{\sigma}_y \tag{2.11}$$

where $\hat{I}$ is the unit $2 \times 2$ matrix. The Eq. (2.9) can be combined with Eq. (2.11) as

$$\hat{\sigma}_i\hat{\sigma}_k = \delta_{ik}\hat{I} + \mathrm{i}\sum_l \varepsilon_{ikl}\hat{\sigma}_l \tag{2.12}$$

The definition of $\varepsilon_{ikl}$ is given by Eq. (1.7). The Pauli matrix satisfies the following commutation and anti-commutation relations:

$$\hat{\sigma}\times\hat{\sigma} = 2\mathrm{i}\hat{\sigma}, \quad [\hat{\sigma}_i, \hat{\sigma}_k] = 2i\varepsilon_{ikl}\hat{\sigma}_l, \quad \{\hat{\sigma}_i, \hat{\sigma}_k\} = 2\delta_{ik}\hat{I}, \qquad i = x, y, z \tag{2.13}$$

$$[\hat{\sigma}_\mu, \hat{\sigma}_\nu] = -2\sqrt{2}C_{1\mu\ 1\nu}^{1\lambda}\hat{\sigma}_\lambda, \quad \{\hat{\sigma}_\mu, \hat{\sigma}_\nu\} = 2(-1)^\mu\delta_{\mu\ -\nu}\hat{I}, \qquad \mu, \nu, \lambda = \pm 1, 0 \tag{2.14}$$

From Eq. (2.8) and Eq. (2.14), one can also obtain

$$[\hat{\sigma}_\mu^+, \hat{\sigma}_\nu] = 2\sqrt{2}(-1)^{1+\mu}C_{1\ -\mu\ 1\nu}^{1\ \nu-\mu}\hat{\sigma}_{\nu-\mu}, \qquad \mu, \nu = \pm 1, 0 \tag{2.15}$$

Letting $\vec{a}$ be a constant vector, the following formula can be proved using the anti-commutation relation (2.13):

$$(\vec{a}\cdot\hat{\sigma})(\vec{a}\cdot\hat{\sigma}) = \left(a_x^2 + a_y^2 + a_z^2\right)\hat{I} = a^2\hat{I} \tag{2.16}$$

By means of Eq. (2.12), it can also be proved that [2]

$$\hat{\sigma}(\vec{a}\cdot\hat{\sigma}) = \vec{a} + \mathrm{i}(\vec{a}\times\hat{\sigma}), \quad (\vec{a}\cdot\hat{\sigma})\hat{\sigma} = \vec{a} - \mathrm{i}(\vec{a}\times\hat{\sigma}) \tag{2.17}$$

In addition, there is a vector formula

$$\vec{a}\times(\vec{b}\times\vec{c}) = \vec{b}(\vec{a}\cdot\vec{c}) - \vec{c}(\vec{a}\cdot\vec{b}) \tag{2.18}$$

Based on the above formulas, we have

$$(\hat{\sigma}\times\vec{a})\times\vec{a} = -a^2\hat{\sigma} + (\vec{a}\cdot\hat{\sigma})\vec{a} \tag{2.19}$$

Using Eq. (2.12) and the properties of $\varepsilon_{ijk}$ defined by Eq. (1.7), we can prove the following relation:

$$(\vec{a}\cdot\hat{\sigma})\hat{\sigma}(\vec{a}\cdot\hat{\sigma}) = 2(\vec{a}\cdot\hat{\sigma})\vec{a} - a^2\hat{\sigma} \tag{2.20}$$

When the spherical basis components of the Pauli matrix $\hat{\sigma}$ act on the basis spin functions, we have the following relations:

$$\hat{\sigma}_{\mu}\chi_{\frac{1}{2}m} = -\sqrt{3}C_{1\mu\ \frac{1}{2}m}^{\frac{1}{2}\ m+\mu}\chi_{\frac{1}{2}\ m+\mu} = \sqrt{2}(-1)^{\frac{1}{2}-m}C_{\frac{1}{2}\ m+\mu\ \frac{1}{2}\ -m}^{1\ \mu}\chi_{\frac{1}{2}\ m+\mu} \tag{2.21}$$

and we can obtain

$$\langle m|\hat{\sigma}|m\rangle = \sum_{\mu}(-1)^{\mu}\langle m|\hat{\sigma}_{\mu}|m\rangle \vec{\mathrm{e}}_{-\mu} = \sqrt{2}\sum_{\mu}(-1)^{\mu+\frac{1}{2}-m}C_{\frac{1}{2}\ m+\mu\ \frac{1}{2}\ -m}^{1\ \mu}\vec{\mathrm{e}}_{-\mu} \tag{2.22}$$

The following tracing relations of products of the Pauli matrix Cartesian basis components are obtained, by use of Eqs. (2.6), (2.9), (2.11), and (2.12), as follows:

$$\begin{aligned} &\mathrm{tr}\{\hat{\sigma}_i\} = 0, \quad \mathrm{tr}\{\hat{\sigma}_i\hat{\sigma}_k\} = 2\delta_{ik}, \quad \mathrm{tr}\{\hat{\sigma}_i\hat{\sigma}_k\hat{\sigma}_l\} = 2i\varepsilon_{ikl}, \\ &\mathrm{tr}\{\hat{\sigma}_i\hat{\sigma}_k\hat{\sigma}_l\hat{\sigma}_m\} = 2(\delta_{ik}\delta_{lm} - \delta_{il}\delta_{km} + \delta_{im}\delta_{kl}), \qquad i,k,l,m = x,y,z \end{aligned} \tag{2.23}$$

For the Pauli matrix spherical basis components, the following tracing relations hold [1]

$$\begin{aligned} &\mathrm{tr}\{\hat{\sigma}_{\mu}\} = 0, \quad \mathrm{tr}\{\hat{\sigma}_{\mu}\hat{\sigma}_{\nu}\} = 2(-1)^{\mu}\delta_{\mu\ -\nu}, \\ &\mathrm{tr}\{\hat{\sigma}_{\mu}\hat{\sigma}_{\nu}\hat{\sigma}_{\lambda}\} = -2\sqrt{6}\begin{pmatrix} 1 & 1 & 1 \\ \mu & \nu & \lambda \end{pmatrix} = 2\sqrt{2}(-1)^{1+\lambda}C_{1\mu\ 1\nu}^{1\ -\lambda}, \\ &\mathrm{tr}\{\hat{\sigma}_{\mu}\hat{\sigma}_{\nu}\hat{\sigma}_{\lambda}\hat{\sigma}_{\rho}\} = 2(-1)^{\mu}\Big\{(-1)^{\lambda}\delta_{\mu\ -\nu}\delta_{\lambda\ -\rho} - (-1)^{\nu}\delta_{\mu\ -\lambda}\delta_{\nu\ -\rho} \\ &\qquad + (-1)^{\nu}\delta_{\mu\ -\rho}\delta_{\nu\ -\lambda}\Big\}, \qquad \mu,\nu,\lambda,\rho = \pm 1, 0 \end{aligned} \tag{2.24}$$

## 2.2 Polarization of the Spin $\frac{1}{2}$ Particle Beams

It is a commonly used experimental method to obtain the polarized particle beams through the previous scattering or reaction. Moreover, when the unpolarized particle beams pass through a uniform strong magnetic field, the particles with different magnetic quantum numbers are separated. It is also an experimental method to obtain the polarized particle beams.

Any spin wave function $X_{\frac{1}{2}}$ of a spin $\frac{1}{2}$ particle can be expanded using the basis spin functions

$$X_{\frac{1}{2}} = \sum_{m=-\frac{1}{2}}^{\frac{1}{2}} a^{m}\chi_{\frac{1}{2}m} = \begin{pmatrix} a^{\frac{1}{2}} \\ a^{-\frac{1}{2}} \end{pmatrix} \tag{2.25}$$

where $a^m$ is the contravariant component of the spin wave function $X_{\frac{1}{2}}$. The Hermitian conjugate wave function corresponding to Eq. (2.25) is

$$X_{\frac{1}{2}}^{+} = \sum_{m=-\frac{1}{2}}^{\frac{1}{2}} a^{m*}\chi_{\frac{1}{2}m}^{+} = \begin{pmatrix} a^{\frac{1}{2}*} & a^{-\frac{1}{2}*} \end{pmatrix} \tag{2.26}$$

From the normalization condition

$$X_{\frac{1}{2}}^{+}X_{\frac{1}{2}} = 1 \tag{2.27}$$

the following expression can be obtained:

$$\left|a^{\frac{1}{2}}\right|^2 + \left|a^{-\frac{1}{2}}\right|^2 = 1 \tag{2.28}$$

Therefore, the coefficient $a^m$ may be interpreted as the probability amplitude when the particle spin projection on the $z$-axis in a given state $X_{\frac{1}{2}}$ is equal to $m$. If the particle spin is directed along the $z$-axis, then

$$a^{\frac{1}{2}} = 1, \quad a^{-\frac{1}{2}} = 0, \quad X_{\frac{1}{2}} = \chi_{\frac{1}{2}\,\frac{1}{2}} = \begin{pmatrix} 1 \\ 0 \end{pmatrix} \tag{2.29}$$

In this case, the particles are completely polarized along the $z$-axis. If the particle spin is directed along the negative $z$-axis, then

$$a^{\frac{1}{2}} = 0, \quad a^{-\frac{1}{2}} = 1, \quad X_{\frac{1}{2}} = \chi_{\frac{1}{2}\,-\frac{1}{2}} = \begin{pmatrix} 0 \\ 1 \end{pmatrix} \tag{2.30}$$

In this case, the particles are completely polarized along the negative $z$-axis.

The polarization density matrix for pure states of the spin $\frac{1}{2}$ particles can be written as

$$\hat{\rho} = X_{\frac{1}{2}}X_{\frac{1}{2}}^{+} = \left(\sum_{m} a^m \chi_{\frac{1}{2}m}\right)\left(\sum_{m'} a^{m'*}\chi_{\frac{1}{2}m'}^{+}\right) \tag{2.31}$$

There are no requirements for the normalization of $X_{\frac{1}{2}}$ and $\hat{\rho}$. The matrix element of $\hat{\rho}$ is

$$\rho_{ik} = \chi_{\frac{1}{2}\,i}^{+}\hat{\rho}\chi_{\frac{1}{2}\,k} = a^i a^{k*} \tag{2.32}$$

When writing the above formula in matrix form, it becomes

$$\hat{\rho} = X_{\frac{1}{2}} X_{\frac{1}{2}}^{+} = \begin{pmatrix} \left| a^{\frac{1}{2}} \right|^2 & a^{\frac{1}{2}} a^{-\frac{1}{2}*} \\ a^{-\frac{1}{2}} a^{\frac{1}{2}*} & \left| a^{-\frac{1}{2}} \right|^2 \end{pmatrix} \tag{2.33}$$

First we can obtain the relation

$$I \equiv \mathrm{tr}\{\hat{\rho}\} = \left| a^{\frac{1}{2}} \right|^2 + \left| a^{-\frac{1}{2}} \right|^2 \tag{2.34}$$

Then we define the polarization rate of the spin $\frac{1}{2}$ particles as

$$Ip_i = \langle \sigma_i \rangle, \qquad i = x, y, z \tag{2.35}$$

The following polarization rates can be obtained by using the expectation value method given by Eq. (1.89) as well as Eqs. (2.35), (2.6), and (2.33):

$$\begin{aligned} Ip_x &= a^{\frac{1}{2}} a^{-\frac{1}{2}*} + a^{-\frac{1}{2}} a^{\frac{1}{2}*} = 2\,\mathrm{Re}\left( a^{\frac{1}{2}} a^{-\frac{1}{2}*} \right), \\ Ip_y &= \mathrm{i}\left( a^{\frac{1}{2}} a^{-\frac{1}{2}*} - a^{-\frac{1}{2}} a^{\frac{1}{2}*} \right) = -2\mathrm{Im}\left( a^{\frac{1}{2}} a^{-\frac{1}{2}*} \right), \\ Ip_z &= \left| a^{\frac{1}{2}} \right|^2 - \left| a^{-\frac{1}{2}} \right|^2 \end{aligned} \tag{2.36}$$

where Re and Im denote taking the real part and the imaginary part, respectively. The above results show that only the beam particle wave functions with different spin magnetic quantum numbers yield the same probabilities and there is no coherence between the different beam particle waves; the beam is unpolarized. Then we obtain the following relation from Eqs. (2.33) and (2.36):

$$\hat{\rho} = \frac{1}{2} I \begin{pmatrix} 1 + p_z & p_x - \mathrm{i}p_y \\ p_x + \mathrm{i}p_y & 1 - p_z \end{pmatrix} \tag{2.37}$$

Assume that the spin $\frac{1}{2}$ particle coming from the polarized ion source has a polarization vector $\vec{p}$ along the direction of the external magnetic field; $\vec{p}$ is just the particle polarization direction. Note that the particle polarization direction and the motion direction of the particle beam are two different things. Here let us choose the particle polarization direction as the $z$-axis. In the $xyz$ coordinate system with the particle polarization direction as the $z$-axis, the components of the particle polarization vector must satisfy $p_x = p_y = 0$, so it can be seen from Eq. (2.37) that in this case the particle density matrix is a diagonal matrix. We use $N_{\frac{1}{2}}$ and $N_{-\frac{1}{2}}$ to represent the relative particle numbers corresponding to the spin-up and the spin-down, respectively, and require them to satisfy the normalization condition

$$N_{\frac{1}{2}} + N_{-\frac{1}{2}} = 1 \tag{2.38}$$

In this *xyz* coordinate system, the spin $\frac{1}{2}$ particle density matrix is

$$\hat{\rho}_0 = \begin{pmatrix} \sqrt{N_{\frac{1}{2}}} \\ 0 \end{pmatrix} \begin{pmatrix} \sqrt{N_{\frac{1}{2}}} & 0 \end{pmatrix} + \begin{pmatrix} 0 \\ \sqrt{N_{-\frac{1}{2}}} \end{pmatrix} \begin{pmatrix} 0 & \sqrt{N_{-\frac{1}{2}}} \end{pmatrix} = \begin{pmatrix} N_{\frac{1}{2}} & 0 \\ 0 & N_{-\frac{1}{2}} \end{pmatrix} \tag{2.39}$$

Thus, we can find out the following results according to Eqs. (2.38), (2.39), and (2.6):

$$I = \mathrm{tr}\{\rho_0\} = N_{\frac{1}{2}} + N_{-\frac{1}{2}} = 1, \quad Ip_z = \mathrm{tr}\{\hat{\sigma}_z \rho_0\} = N_{\frac{1}{2}} - N_{-\frac{1}{2}} \tag{2.40}$$

$$Ip_i = = \mathrm{tr}\{\hat{\sigma}_i \rho_0\} = 0, \quad i = x,y \tag{2.41}$$

This is a special polarized particle state, which different spin magnetic quantum number states are incoherent, but may have different shares.

The above results are given in the *xyz* coordinate system where the *z*-axis is along the polarization direction of the incident particles. Now we choose a new $x'y'z'$ coordinate system where the $z'$-axis is along the move direction of the incident particle beam, and assume that the *z*-axis of the original coordinate system in the new $x'y'z'$ coordinate system has the polar angle $\theta$ and the azimuthal angle $\varphi$. The $I = N_{\frac{1}{2}} + N_{-\frac{1}{2}}$ representing the number of particles does not change in the two coordinate systems, but the component $Ip_z = N_{\frac{1}{2}} - N_{-\frac{1}{2}}$ belonging to the original coordinate system becomes three components in the new coordinate system as

$$Ip_{x'} = \left(N_{\frac{1}{2}} - N_{-\frac{1}{2}}\right)\sin\theta\cos\varphi, \; Ip_{y'} = \left(N_{\frac{1}{2}} - N_{-\frac{1}{2}}\right)\sin\theta\sin\varphi, \; Ip_{z'} = \left(N_{\frac{1}{2}} - N_{-\frac{1}{2}}\right)\cos\theta \tag{2.42}$$

Using Eq. (2.42) and Eq. (2.37), the polarization density matrix in $x'y'z'$ coordinate system becomes

$$\hat{\rho}' = \frac{1}{2}\begin{pmatrix} \left(N_{\frac{1}{2}} + N_{-\frac{1}{2}}\right) + \left(N_{\frac{1}{2}} - N_{-\frac{1}{2}}\right)\cos\theta & \left(N_{\frac{1}{2}} - N_{-\frac{1}{2}}\right)\sin\theta\, \mathrm{e}^{-\mathrm{i}\varphi} \\ \left(N_{\frac{1}{2}} - N_{-\frac{1}{2}}\right)\sin\theta\, \mathrm{e}^{\mathrm{i}\varphi} & \left(N_{\frac{1}{2}} + N_{-\frac{1}{2}}\right) - \left(N_{\frac{1}{2}} - N_{-\frac{1}{2}}\right)\cos\theta \end{pmatrix} \tag{2.43}$$

Let $\chi'_{\frac{1}{2}\frac{1}{2}}$ and $\chi'_{\frac{1}{2}\,-\frac{1}{2}}$ represent the components of the particle spin wave function in the new $x'y'z'$ coordinate system; the corresponding weights of the particle number with spin up and spin down remain the $N_{\frac{1}{2}}$ and $N_{-\frac{1}{2}}$, respectively. And assume

$$\chi'_{\frac{1}{2}\frac{1}{2}}=\begin{pmatrix} u \\ v \end{pmatrix}, \quad \chi'_{\frac{1}{2}\,-\frac{1}{2}}=\begin{pmatrix} t \\ s \end{pmatrix} \tag{2.44}$$

Then the following relation can be obtained:

$$\hat{\rho}'=\sum_m N_m \chi'_{\frac{1}{2}m}\chi'^{+}_{\frac{1}{2}m}=\begin{pmatrix} N_{\frac{1}{2}}|u|^2+N_{-\frac{1}{2}}|t|^2 & N_{\frac{1}{2}}uv^*+N_{-\frac{1}{2}}ts^* \\ N_{\frac{1}{2}}u^*v+N_{-\frac{1}{2}}t^*s & N_{\frac{1}{2}}|v|^2+N_{-\frac{1}{2}}|s|^2 \end{pmatrix} \tag{2.45}$$

Comparing Eq. (2.45) with Eq. (2.43), one can get

$$\begin{aligned} uu^* &= \frac{1}{2}(1+\cos\theta), & vv^* &= \frac{1}{2}(1-\cos\theta), \\ uv^* &= \frac{1}{2}\sin\theta\, \mathrm{e}^{-\mathrm{i}\varphi}, & u^*v &= \frac{1}{2}\sin\theta\, \mathrm{e}^{\mathrm{i}\varphi}, \\ tt^* &= \frac{1}{2}(1-\cos\theta), & ss^* &= \frac{1}{2}(1+\cos\theta), \\ ts^* &= -\frac{1}{2}\sin\theta\, \mathrm{e}^{-\mathrm{i}\varphi}, & t^*s &= -\frac{1}{2}\sin\theta\, \mathrm{e}^{\mathrm{i}\varphi} \end{aligned} \tag{2.46}$$

By using the triangular formulas, one can formalize them into

$$\begin{aligned} uu^* &= \cos^2\frac{\theta}{2}, & vv^* &= \sin^2\frac{\theta}{2}, \\ uv^* &= \sin\frac{\theta}{2}\cos\frac{\theta}{2}\mathrm{e}^{-\mathrm{i}\varphi}, & u^*v &= \sin\frac{\theta}{2}\cos\frac{\theta}{2}\mathrm{e}^{\mathrm{i}\varphi}, \\ tt^* &= \sin^2\frac{\theta}{2}, & ss^* &= \cos^2\frac{\theta}{2}, \\ ts^* &= -\sin\frac{\theta}{2}\cos\frac{\theta}{2}\mathrm{e}^{-\mathrm{i}\varphi}, & t^*s &= -\sin\frac{\theta}{2}\cos\frac{\theta}{2}\mathrm{e}^{\mathrm{i}\varphi} \end{aligned} \tag{2.47}$$

So the following results can be obtained:

$$\begin{aligned} u &= \cos\frac{\theta}{2}\mathrm{e}^{-\mathrm{i}\frac{\varphi}{2}}, & v &= \sin\frac{\theta}{2}\mathrm{e}^{\mathrm{i}\frac{\varphi}{2}}, \\ t &= -\sin\frac{\theta}{2}\mathrm{e}^{-\mathrm{i}\frac{\varphi}{2}}, & s &= \cos\frac{\theta}{2}\mathrm{e}^{\mathrm{i}\frac{\varphi}{2}} \end{aligned} \tag{2.48}$$

If $u$ and $v$ or $t$ and $s$ change the sign at the same time, Eq. (2.47) is still satisfied. This is only a phase extraction problem, whereas the square of the wave function has a physical meaning of probability. It follows that Eq. (2.44) can be rewritten as

$$\chi'_{\frac{1}{2}\frac{1}{2}} = \begin{pmatrix} \cos\frac{\theta}{2}\mathrm{e}^{-\mathrm{i}\frac{\varphi}{2}} \\ \sin\frac{\theta}{2}\mathrm{e}^{\mathrm{i}\frac{\varphi}{2}} \end{pmatrix}, \quad \chi'_{\frac{1}{2}\,-\frac{1}{2}} = \begin{pmatrix} -\sin\frac{\theta}{2}\mathrm{e}^{-\mathrm{i}\frac{\varphi}{2}} \\ \cos\frac{\theta}{2}\mathrm{e}^{\mathrm{i}\frac{\varphi}{2}} \end{pmatrix} \tag{2.49}$$

The spin projection on the linear momentum direction of a particle is called the helicity. The helicity basis functions $\chi_{\frac{1}{2}\lambda}(\theta, \varphi)$ $\left(\lambda = \pm\frac{1}{2}\right)$ of a spin $\frac{1}{2}$ particle describe the spin states with spin projections $\lambda$ along its linear momentum direction $\vec{n}(\theta, \varphi) = \vec{p}/p$. These functions are the common eigenfunctions of operators $\hat{S}^2$ and $\hat{S} \cdot \vec{n}$, that is,

$$\hat{S}^2\chi_{\frac{1}{2}\lambda}(\theta, \varphi) = \frac{3}{4}\chi_{\frac{1}{2}\lambda}(\theta, \varphi), \quad (\hat{S} \cdot \vec{n})\chi_{\frac{1}{2}\lambda}(\theta, \varphi) = \lambda\chi_{\frac{1}{2}\lambda}(\theta, \varphi) \tag{2.50}$$

According to Eq. (1.31) the corresponding helicity basis functions can be obtained from the usual basis spin functions as follows:

$$\chi_{\frac{1}{2}\lambda}(\theta, \varphi) = \sum_m D^{\frac{1}{2}}_{m\lambda}(\varphi, \theta, 0)\chi_{\frac{1}{2}m} \tag{2.51}$$

$$\chi_{\frac{1}{2}m} = \sum_\lambda D^{\frac{1}{2}}_{-\lambda\ -m}(0, \theta, \varphi)\chi_{\frac{1}{2}\lambda}(\theta, \varphi) \tag{2.52}$$

The corresponding Hermitian conjugate expressions are

$$\chi^+_{\frac{1}{2}\lambda}(\theta, \varphi) = \sum_m (-1)^{\lambda-m} D^{\frac{1}{2}}_{-m\ -\lambda}(\varphi, \theta, 0)\chi^+_{\frac{1}{2}m} \tag{2.53}$$

$$\chi^+_{\frac{1}{2}m} = \sum_m (-1)^{\lambda-m} D^{\frac{1}{2}}_{\lambda\ m}(0, \theta, \varphi)\chi^+_{\frac{1}{2}\lambda}(\theta, \varphi) \tag{2.54}$$

Its orthogonal normalization condition is

$$\chi^+_{\frac{1}{2}\lambda'}(\theta, \varphi)\chi_{\frac{1}{2}\lambda}(\theta, \varphi) = \delta_{\lambda'\lambda} \tag{2.55}$$

And its completeness condition is

$$\sum_\lambda \chi_{\frac{1}{2}\lambda}(\theta, \varphi)\chi^+_{\frac{1}{2}\lambda}(\theta, \varphi) = \hat{I} \tag{2.56}$$

The $D$ function has the following expression:

$$D^J_{MM'}(\alpha, \beta, \gamma) = \mathrm{e}^{-\mathrm{i}M\alpha} d^J_{MM'}(\beta)\,\mathrm{e}^{-\mathrm{i}M'\gamma} \tag{2.57}$$

**Table 2.2** Expressions of $d^{\frac{1}{2}}_{MM'}(\beta)$ [1]

| $M$ | $M'$: $\frac{1}{2}$ | $M'$: $-\frac{1}{2}$ |
|---|---|---|
| $\frac{1}{2}$ | $\cos\frac{\beta}{2}$ | $-\sin\frac{\beta}{2}$ |
| $-\frac{1}{2}$ | $\sin\frac{\beta}{2}$ | $\cos\frac{\beta}{2}$ |

Table 2.2 gives the expressions of $d^{\frac{1}{2}}_{MM'}(\beta)$. The specific expressions of the helicity basis functions can be obtained based on Eqs. (2.51) and (2.53) [1]:

$$\chi_{\frac{1}{2}\frac{1}{2}}(\theta,\varphi)=\begin{pmatrix}\cos\frac{\theta}{2}\mathrm{e}^{-\mathrm{i}\frac{\varphi}{2}}\\ \sin\frac{\theta}{2}\mathrm{e}^{\mathrm{i}\frac{\varphi}{2}}\end{pmatrix},\quad \chi_{\frac{1}{2}\,-\frac{1}{2}}(\theta,\varphi)=\begin{pmatrix}-\sin\frac{\theta}{2}\mathrm{e}^{-\mathrm{i}\frac{\varphi}{2}}\\ \cos\frac{\theta}{2}\mathrm{e}^{\mathrm{i}\frac{\varphi}{2}}\end{pmatrix} \tag{2.58}$$

$$\begin{aligned}\chi^{+}_{\frac{1}{2}\frac{1}{2}}(\theta,\varphi)&=\left(\cos\frac{\theta}{2}\mathrm{e}^{\mathrm{i}\frac{\varphi}{2}}\quad \sin\frac{\theta}{2}\mathrm{e}^{-\mathrm{i}\frac{\varphi}{2}}\right),\\ \chi^{+}_{\frac{1}{2}\,-\frac{1}{2}}(\theta,\varphi)&=\left(-\sin\frac{\theta}{2}\mathrm{e}^{\mathrm{i}\frac{\varphi}{2}}\quad \cos\frac{\theta}{2}\mathrm{e}^{-\mathrm{i}\frac{\varphi}{2}}\right)\end{aligned} \tag{2.59}$$

The results given by Eq. (2.58) and Eq. (2.49) are exactly the same.

Based on Eq. (1.91) we can write the corresponding normalized polarization density matrix for spin $\frac{1}{2}$ particles as

$$\hat{\rho}=\frac{1}{2}\left[t_{00}\hat{T}_{00}\left(\frac{1}{2}\right)+\vec{t}_1\cdot\hat{T}_1\left(\frac{1}{2}\right)\right] \tag{2.60}$$

From Eq. (1.55) we obtain

$$\hat{T}_1\left(\frac{1}{2}\right)=2\hat{S}=\hat{\sigma} \tag{2.61}$$

It can be seen that $\hat{T}_1\left(\frac{1}{2}\right)$ is equal to the Pauli matrix $\hat{\sigma}$. Letting

$$\vec{p}=\vec{t}_1 \tag{2.62}$$

and using Eqs. (1.54) and (1.99), from Eq. (2.60), we can obtain

$$\hat{\rho}=\frac{1}{2}\left(\hat{I}+\vec{p}\cdot\hat{\sigma}\right) \tag{2.63}$$

The expression of the above formula in the Cartesian coordinate system is

$$\hat{\rho} = \frac{1}{2}\left(\hat{I} + p_x\hat{\sigma}_x + p_y\hat{\sigma}_y + p_z\hat{\sigma}_z\right) \tag{2.64}$$

Note that the density matrix $\hat{\rho}$ of the incident particles is normalized. Based on Eq. (1.89) and the tracing formula (2.23), the following relation can be obtained:

$$\langle\vec{\sigma}\rangle = \mathrm{tr}\{\vec{\sigma}\hat{\rho}\} = \vec{p} \tag{2.65}$$

It is clear that $\vec{p}$ is the polarization vector of the system, and its absolute value $p$ is called polarization degree, and there is $0 \le p \le 1$. Based on Eq. (2.62) and Eq. (1.98), one has

$$p_\mu = \sqrt{3}\sum_{\sigma\sigma'} C^{\frac{1}{2}\sigma'}_{\frac{1}{2}\sigma\ 1\mu}\,\rho_{\sigma\sigma'} \tag{2.66}$$

For $\vec{p} = 0$ unpolarized state, the following result for spin $\frac{1}{2}$ particles can be obtained from Eq. (2.63)

$$\hat{\rho}_{\mathrm{unpol}} = \frac{1}{2}\hat{I} \tag{2.67}$$

## 2.3 Polarization Theory of Elastic Scattering of Unpolarized Spin $\frac{1}{2}$ Particles with Spinless Targets

Equation (5.5.14) in Ref. [2] gives the differential cross section of the spin $\frac{1}{2}$ particles with the spinless target in the unpolarization case as follows:

$$\frac{\mathrm{d}\sigma^0}{\mathrm{d}\Omega} = \frac{1}{2}\sum_{\mu\mu'}\left|F_{\mu'\mu}(\theta,\varphi)\right|^2 \tag{2.68}$$

where the scattering amplitude is

$$F_{\mu'\mu}(\theta,\varphi) = f_{\mathrm{C}}(\theta)\,\delta_{\mu'\mu} + \frac{\mathrm{i}\sqrt{\pi}}{k}\sum_{lj}\sqrt{2l+1}\,\mathrm{e}^{2\mathrm{i}\sigma_l}\left(1 - S_l^j\right) C^{j\,\mu}_{l0\ \frac{1}{2}\mu}\,C^{j\,\mu}_{lm_l\ \frac{1}{2}\mu'}\,Y_{lm_l}(\theta,\varphi) \tag{2.69}$$

and $k$ is the particle wave number. $S_l^j$ is the $S$ matrix element obtained by using the spherical optical model. Equation (3.3.41) in Ref. [2] presents the Coulomb scattering amplitude $f_C(\theta)$ as follows:

$$f_C(\theta) = -\frac{\eta}{2k\sin^2\frac{\theta}{2}} e^{-i\eta\ln\left(\sin^2\frac{\theta}{2}\right)+2i\sigma_0} \tag{2.70}$$

where

$$\eta = \frac{\mu z Z e^2}{\hbar^2 k} \tag{2.71}$$

In above formula, $z$ and $Z$ are incident particle and target charges, respectively, $\mu$ is reduced mass. Phase shift factor $\sigma_l$ satisfies the relation

$$e^{2i\sigma_l} = \frac{\Gamma(l+1+i\eta)}{\Gamma(l+1-i\eta)}, \qquad \sigma_0 = \frac{1}{2i}\ln\frac{\Gamma(1+i\eta)}{\Gamma(1-i\eta)} \tag{2.72}$$

where $\Gamma$ denotes the gamma function.

There are the following spherical harmonic function and associated Legendre function expressions:

$$Y_{lm_l}(\theta,\varphi) = (-1)^{m_l}\sqrt{\frac{2l+1}{4\pi}}\sqrt{\frac{(l-m_l)!}{(l+m_l)!}}P_l^{m_l}(\cos\theta)\, e^{im_l\varphi} \tag{2.73}$$

$$P_l^{-m}(x) = (-1)^m\frac{(l-m)!}{(l+m)!}P_l^m(x) \tag{2.74}$$

yielding

$$Y_{l1} = -\sqrt{\frac{2l+1}{4\pi}}\sqrt{\frac{1}{l(l+1)}}P_l^1 e^{i\varphi}, \quad Y_{l\,-1} = \sqrt{\frac{2l+1}{4\pi}}\sqrt{\frac{1}{l(l+1)}}P_l^1 e^{-i\varphi} \tag{2.75}$$

The following C–G coefficients can be obtained from Table 2.3:

$$\begin{aligned} C_{l0\,\frac{1}{2}\frac{1}{2}}^{l+\frac{1}{2}\,\frac{1}{2}} &= \sqrt{\frac{l+1}{2l+1}}, & C_{l0\,\frac{1}{2}\frac{1}{2}}^{l-\frac{1}{2}\,\frac{1}{2}} &= -\sqrt{\frac{l}{2l+1}} \\ C_{l0\,\frac{1}{2}\,-\frac{1}{2}}^{l+\frac{1}{2}\,-\frac{1}{2}} &= \sqrt{\frac{l+1}{2l+1}}, & C_{l0\,\frac{1}{2}\,-\frac{1}{2}}^{l-\frac{1}{2}\,-\frac{1}{2}} &= \sqrt{\frac{l}{2l+1}} \end{aligned} \tag{2.76}$$

$$\begin{aligned} C_{l1\,\frac{1}{2}\,-\frac{1}{2}}^{l+\frac{1}{2}\,\frac{1}{2}} &= \sqrt{\frac{l}{2l+1}}, & C_{l1\,\frac{1}{2}\,-\frac{1}{2}}^{l-\frac{1}{2}\,\frac{1}{2}} &= \sqrt{\frac{l+1}{2l+1}} \\ C_{l\,-1\,\frac{1}{2}\frac{1}{2}}^{l+\frac{1}{2}\,-\frac{1}{2}} &= \sqrt{\frac{l}{2l+1}}, & C_{l\,-1\,\frac{1}{2}\frac{1}{2}}^{l-\frac{1}{2}\,-\frac{1}{2}} &= -\sqrt{\frac{l+1}{2l+1}} \end{aligned} \tag{2.77}$$

**Table 2.3** Expressions of Clebsch-Gordan coefficient $C_{l\ m-\mu\ \frac{1}{2}\ \mu}^{j\ m}$

| | $\mu=\frac{1}{2}$ | $\mu=-\frac{1}{2}$ |
|---|---|---|
| $j=l+\frac{1}{2}$ | $\sqrt{\frac{l+m+\frac{1}{2}}{2l+1}}$ | $\sqrt{\frac{l-m+\frac{1}{2}}{2l+1}}$ |
| $j=l-\frac{1}{2}$ | $-\sqrt{\frac{l-m+\frac{1}{2}}{2l+1}}$ | $\sqrt{\frac{l+m+\frac{1}{2}}{2l+1}}$ |

The following results can be obtained substituting Eqs. (2.75) – (2.77) into Eq. (2.69):

$$F_{\frac{1}{2}\ \frac{1}{2}}(\theta,\varphi)=F_{-\frac{1}{2}\ -\frac{1}{2}}(\theta,\varphi)=A(\theta) \tag{2.78}$$

$$F_{\frac{1}{2}\ -\frac{1}{2}}(\theta,\varphi)=-\mathrm{i}B(\theta)\,\mathrm{e}^{-\mathrm{i}\varphi} \tag{2.79}$$

$$F_{-\frac{1}{2}\ \frac{1}{2}}(\theta,\varphi)=\mathrm{i}B(\theta)\,\mathrm{e}^{\mathrm{i}\varphi} \tag{2.80}$$

Writing the above three formulas in matrix form, one obtains

$$\hat{F}(\theta,\varphi)=\begin{pmatrix} A(\theta) & -\mathrm{i}B(\theta)\,\mathrm{e}^{-\mathrm{i}\varphi} \\ \mathrm{i}B(\theta)\,\mathrm{e}^{\mathrm{i}\varphi} & A(\theta) \end{pmatrix},\ \hat{F}^{+}(\theta,\varphi)=\begin{pmatrix} A^{*}(\theta) & -\mathrm{i}B^{*}(\theta)\,\mathrm{e}^{-\mathrm{i}\varphi} \\ \mathrm{i}B^{*}(\theta)\,\mathrm{e}^{\mathrm{i}\varphi} & A^{*}(\theta) \end{pmatrix} \tag{2.81}$$

where

$$A(\theta)=f_C(\theta)+\frac{\mathrm{i}}{2k}\sum_l\left[(l+1)\left(1-S_l^{l+\frac{1}{2}}\right)+l\left(1-S_l^{l-\frac{1}{2}}\right)\right]\mathrm{e}^{2\mathrm{i}\sigma_l}\mathrm{P}_l(\cos\theta) \tag{2.82}$$

$$B(\theta)=\frac{1}{2k}\sum_l\left[S_l^{l+\frac{1}{2}}-S_l^{l-\frac{1}{2}}\right]\mathrm{e}^{2\mathrm{i}\sigma_l}\mathrm{P}_l^1(\cos\theta) \tag{2.83}$$

Based on Eq. (2.63), the polarized incident particle density matrix can be written as

$$\hat{\rho}_{\mathrm{in}}=\frac{1}{2}\left(\hat{I}+\vec{p}_{\mathrm{in}}\cdot\hat{\sigma}\right) \tag{2.84}$$

According to Eq. (2.65), $\vec{p}_{in}$ is the polarization vector of incident particles. The physical significance of the scattering amplitude is that the final state can be obtained when letting the scattering amplitude act on the initial state. Therefore, when the target spin is 0, the polarization density matrix of the elastic scattering outgoing particles can be written as

$$\hat{\rho}_{\text{out}} = \hat{F}\hat{\rho}_{\text{in}}\hat{F}^{+} \tag{2.85}$$

Assuming the incident particles are unpolarized, that is, $\vec{p}_{in} = 0$, Eq. (2.84) is simplified to

$$\hat{\rho}_{\text{in}}^{0} = \frac{1}{2}\hat{I} \tag{2.86}$$

and Eq. (2.85) is simplified to

$$\hat{\rho}_{\text{out}}^{0} = \frac{1}{2}\hat{F}\hat{F}^{+} \tag{2.87}$$

From Eq. (2.81) we can find that the differential cross section of the outgoing particles is

$$\frac{\mathrm{d}\sigma^0}{\mathrm{d}\Omega} = \mathrm{tr}\{\hat{\rho}_{\text{out}}^{0}\} = |A(\theta)|^2 + |B(\theta)|^2 \tag{2.88}$$

In the center-of-mass (c.m.) coordinate system, the incident particle wave vector is $\vec{k}_{\mathrm{i}}$; the outgoing particle wave vector is $\vec{k}_{\mathrm{f}}$; the reaction plane is formed by $\vec{k}_{\mathrm{i}}$ and $\vec{k}_{\mathrm{f}}$; and the following unit vectors are introduced:

$$\vec{I} = \frac{\vec{k}_{\mathrm{i}}}{\left|\vec{k}_{\mathrm{i}}\right|}, \qquad \vec{F} = \frac{\vec{k}_{\mathrm{f}}}{\left|\vec{k}_{\mathrm{f}}\right|} \tag{2.89}$$

$$\vec{n} = \frac{\vec{k}_{\mathrm{i}} \times \vec{k}_{\mathrm{f}}}{\left|\vec{k}_{\mathrm{i}} \times \vec{k}_{\mathrm{f}}\right|} = \frac{\vec{I} \times \vec{F}}{\left|\vec{I} \times \vec{F}\right|} \tag{2.90}$$

$$\vec{P} = \frac{\vec{I} + \vec{F}}{\left|\vec{I} + \vec{F}\right|}, \qquad \vec{K} = \frac{\vec{F} - \vec{I}}{\left|\vec{F} - \vec{I}\right|} \tag{2.91}$$

It is easy to see

$$\vec{n} \cdot \vec{P} = 0, \qquad \vec{n} \cdot \vec{K} = 0 \tag{2.92}$$

And we can prove

$$\vec{P}\cdot\vec{K}=\frac{1}{\left|\vec{I}+\vec{F}\right|\left|\vec{F}-\vec{I}\right|}\left(\vec{I}\cdot\vec{F}-1+1-\vec{I}\cdot\vec{F}\right)=0 \tag{2.93}$$

Then it can be seen that the unit vectors $\vec{K},\vec{n},\vec{P}$ form a Cartesian coordinate system.

Since the Pauli matrix $\hat{\sigma}$ and the two-dimensional unit matrix $\hat{I}$ constitute a completeness independent two-dimensional matrix system, the two-dimensional scattering amplitude matrix $\hat{F}(\theta,\varphi)$ given by Eq. (2.81) can be expanded as follows:

$$\hat{F}(\theta,\varphi)=a(\theta,\varphi)\hat{I}+b(\theta,\varphi)\vec{h}\cdot\hat{\sigma} \tag{2.94}$$

We take the unit vectors of the $x$, $y$, and $z$ axes of the Cartesian coordinate system as $\vec{K},\vec{n},\vec{P}$, respectively, so that we can write

$$\vec{h}\cdot\hat{\sigma}=h_x\left(\vec{K}\cdot\hat{\sigma}\right)+h_y\left(\vec{n}\cdot\hat{\sigma}\right)+h_z\left(\vec{P}\cdot\hat{\sigma}\right) \tag{2.95}$$

In the case of spinless target, the scattering amplitude $\hat{F}(\theta,\varphi)$ must satisfy the parity conservation, that is, it must remain unchanged when space reflection is performed. It can be seen from Eqs. (2.89) – (2.91) that after substituting Eq. (2.95) into Eq. (2.94) and the space reflections $\vec{K}\to-\vec{K},\ \vec{n}\to\vec{n},\ \vec{P}\to-\vec{P}$ are performed, $h_x=h_z=0$ should hold and $h_y=h=1$ can be taken since the parity conservation. So $\vec{h}=\vec{n}$ is the unit vector in the polarization direction of the outgoing particles. Then Eq. (2.94) can be rewritten as

$$\hat{F}(\theta,\varphi)=a(\theta,\varphi)\hat{I}+b(\theta,\varphi)\,\vec{n}\cdot\hat{\sigma} \tag{2.96}$$

Now let us reselect a new coordinate system. The direction of the incident particle (i.e., the $\vec{k}_{\rm i}$ direction) is the $z$-axis, the $xy$ plane is vertical to $\vec{k}_{\rm i}$, and the projection of $\vec{k}_{\rm f}$ on the $xy$ plane is $\vec{p}_{\rm f}$, which is located on the intersection of the reaction plane and the $xy$ plane. The unit vector $\vec{n}$ given by Eq. (2.90) is perpendicular to the reaction plane. Since $\vec{n}$ is perpendicular to the $z$-axis and passes through the coordinate origin, it is also in the $xy$ plane. Figure 2.1 shows the position relationship of $\vec{p}_{\rm f}$ and $\vec{n}$ in the $xy$ plane.

**Fig. 2.1** Vector relationship in the $xy$ plane

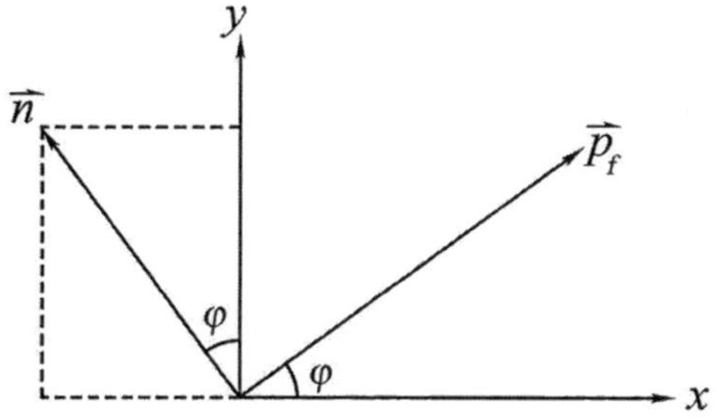

In this figure, $\varphi$ is the azimuthal angle of the emitted particle wave vector $\vec{k}_f$. According to Fig. 2.1 we can write

$$\vec{n}\cdot\hat{\sigma} = -\hat{\sigma}_x \sin\varphi + \hat{\sigma}_y \cos\varphi \tag{2.97}$$

Using Eq. (2.6) the above formula can be written as a matrix

$$\vec{n}\cdot\hat{\sigma} = \begin{pmatrix} 0 & -\mathrm{i}e^{-\mathrm{i}\varphi} \\ \mathrm{i}e^{\mathrm{i}\varphi} & 0 \end{pmatrix} \tag{2.98}$$

Substituting the first equation of Eq. (2.81) and Eq. (2.98) into the Eq. (2.96), one can obtain

$$a(\theta,\varphi) = A(\theta), \qquad b(\theta,\varphi) = B(\theta) \tag{2.99}$$

Then Eq. (2.96) can be rewritten as

$$\hat{F}(\theta,\varphi) = A(\theta)\hat{I} + B(\theta)\,\vec{n}\cdot\hat{\sigma} \tag{2.100}$$

When choosing the $y$-axis in the direction along $\vec{n}$, Eqs. (2.100) and (2.81) become

$$\hat{F} = A\hat{I} + B\hat{\sigma}_y \tag{2.101}$$

$$\hat{F} = \begin{pmatrix} A & -\mathrm{i}B \\ \mathrm{i}B & A \end{pmatrix}, \quad \hat{F}^+ = \begin{pmatrix} A^* & -\mathrm{i}B^* \\ \mathrm{i}B^* & A^* \end{pmatrix} \tag{2.102}$$

Another approach to find $\hat{F}$ is discussed below. By use of $\hat{I}$ and $\hat{\sigma}_i$ $(i = x, y, z)$, $\hat{F}$ can be expanded as

$$\hat{F} = a\hat{I} + u\hat{\sigma}_x + b\hat{\sigma}_y + v\hat{\sigma}_z \tag{2.103}$$

By means of Eq. (2.102) and Eq. (2.6), the above formula becomes

$$\begin{pmatrix} A & -\mathrm{i}B \\ \mathrm{i}B & \mathrm{A} \end{pmatrix} = a\begin{pmatrix} 1 & 0 \\ 0 & 1 \end{pmatrix} + u\begin{pmatrix} 0 & 1 \\ 1 & 0 \end{pmatrix} + b\begin{pmatrix} 0 & -\mathrm{i} \\ \mathrm{i} & 0 \end{pmatrix} + v\begin{pmatrix} 1 & 0 \\ 0 & -1 \end{pmatrix} \tag{2.104}$$

Based on the four matrix elements of Eq. (2.104), one may obtain

$$a + v = A, \quad u - \mathrm{i}b = -\mathrm{i}B, \quad u + \mathrm{i}b = \mathrm{i}B, \quad a - v = A \tag{2.105}$$

Thus, the following results can be obtained from the above formulas:

$$v = 0, \quad u = 0, \quad a = A, \quad b = B \tag{2.106}$$

That is, Eq. (2.101) can be obtained from Eq. (2.103) immediately. The above results show that the scattering amplitude obtained from the nuclear reaction theory has automatically satisfied various physical symmetry requirements and does not need to be additionally restricted.

Substituting Eq. (2.100) into Eq. (2.87), for which the incident particle is also unpolarized, one obtains

$$\hat{\rho}^0_{\text{out}} = \frac{1}{2}\left(A\hat{I} + B\vec{n}\cdot\hat{\sigma}\right)\left(A^*\hat{I} + B^*\vec{n}\cdot\hat{\sigma}\right) \tag{2.107}$$

Using Eq. (2.16) the above formula becomes

$$\hat{\rho}^0_{\text{out}} = \frac{1}{2}\left[\left(|A|^2 + |B|^2\right)\hat{I} + (AB^* + A^*B)\left(\vec{n}\cdot\hat{\sigma}\right)\right] \tag{2.108}$$

Since $\hat{\rho}^0_{\text{out}}$ given by the above formula is not normalized, from Eqs. (1.78) and (2.17) as well as the tracing formula (2.23), the differential cross section and polarization vectors of the emitted particles can be obtained as follows:

$$I^0 = \text{tr}\{\hat{\rho}^0_{\text{out}}\} = |A|^2 + |B|^2 \tag{2.109}$$

$$\vec{P}^0_{\text{out}} = \langle\hat{\sigma}\rangle^0_{\text{out}} = \frac{\text{tr}\{\hat{\sigma}\hat{\rho}^0_{\text{out}}\}}{\text{tr}\{\hat{\rho}^0_{\text{out}}\}} = \frac{AB^* + A^*B}{|A|^2 + |B|^2}\vec{n} = \frac{2\text{Re}(AB^*)}{|A|^2 + |B|^2}\vec{n} \tag{2.110}$$

where Re($A$) denotes taking the real part of $A$.

It can be seen from Fig. 2.1 that if the $y$-axis is selected along the direction $\vec{n}$, then there is an azimuthal angle $\varphi = 0$. In this case, the polarization vector of the emitted particle can be said to be in the $y$-axis direction. The value $P^0_{\text{out}}(\theta)$ of the polarization vector of the emitted particles is also called the emitted particle polarization degree.

## 2.4 Polarization Theory of Elastic Scattering of Polarized Spin $\frac{1}{2}$ Particles with Spinless Targets

If the incident particles are polarized, the polarization density matrix of the incident particles given by Eq. (2.84) should be used. Substituting Eq. (2.84) and Eq. (2.100) into Eq. (2.85), and noting that $\hat{\sigma}$ is a Hermitian matrix, one can obtain

$$\hat{\rho}_{\text{out}} = \frac{1}{2}\left(A\hat{I} + B\vec{n}\cdot\hat{\sigma}\right)\left(\hat{I} + \vec{p}_{\text{in}}\cdot\hat{\sigma}\right)\left(A^*\hat{I} + B^*\vec{n}\cdot\hat{\sigma}\right) = \hat{\rho}^0_{\text{out}} + \hat{\rho}^1_{\text{out}} \tag{2.111}$$

The item $\hat{\rho}^0_{\text{out}}$ corresponding with $\vec{p}_{\text{in}} = 0$ has been given by Eq. (2.108). Using Eq. (2.17) and Eq. (2.20), the following formula can be obtained:

$$
\begin{aligned}
&\left(A\hat{I}+B\vec{n}\cdot\hat{\sigma}\right)\hat{\sigma}\left(A^{*}\hat{I}+B^{*}\vec{n}\cdot\hat{\sigma}\right)\\
&\quad=|A|^{2}\hat{\sigma}+AB^{*}\hat{\sigma}\left(\vec{n}\cdot\hat{\sigma}\right)+A^{*}B\left(\vec{n}\cdot\hat{\sigma}\right)\hat{\sigma}+|B|^{2}\left(-\hat{\sigma}+2\left(\vec{n}\cdot\hat{\sigma}\right)\vec{n}\right)\\
&\quad=\left(|A|^{2}-|B|^{2}\right)\hat{\sigma}+\left[(AB^{*}+A^{*}B)\hat{I}+2|B|^{2}\left(\vec{n}\cdot\hat{\sigma}\right)\right]\vec{n}+\mathrm{i}(AB^{*}-A^{*}B)\left(\vec{n}\times\hat{\sigma}\right)
\end{aligned}
\tag{2.112}
$$

Subsequently yielding

$$
\begin{aligned}
\hat{\rho}_{\text{out}}^{1}=\frac{1}{2}\Big\{&\left(|A|^{2}-|B|^{2}\right)\hat{\sigma}+\left[(AB^{*}+A^{*}B)\hat{I}+2|B|^{2}\left(\vec{n}\cdot\hat{\sigma}\right)\right]\vec{n}\\
&+\mathrm{i}(AB^{*}-A^{*}B)\left(\vec{n}\times\hat{\sigma}\right)\Big\}\cdot\vec{p}_{\text{in}}
\end{aligned}
\tag{2.113}
$$

Substituting Eq. (2.108) and Eq. (2.113) into Eq. (2.111), one obtains

$$
\begin{aligned}
\hat{\rho}_{\text{out}}=\frac{1}{2}\Big\{&\left(|A|^{2}+|B|^{2}\right)\hat{I}+(AB^{*}+A^{*}B)\left(\vec{n}\cdot\hat{\sigma}\right)\\
&+\Big[\left(|A|^{2}-|B|^{2}\right)\hat{\sigma}+\left((AB^{*}+A^{*}B)\hat{I}+2|B|^{2}\left(\vec{n}\cdot\hat{\sigma}\right)\right)\vec{n}\\
&+\mathrm{i}(AB^{*}-A^{*}B)\left(\vec{n}\times\hat{\sigma}\right)\Big]\cdot\vec{p}_{\text{in}}\Big\}
\end{aligned}
\tag{2.114}
$$

Using the tracing formula (2.23), the angular distribution of the emitted particles in the case of polarized incident particles can be obtained by Eq. (2.114).

$$
\frac{\mathrm{d}\sigma}{\mathrm{d}\Omega}=\mathrm{tr}\{\hat{\rho}_{\text{out}}\}=|A|^{2}+|B|^{2}+2\mathrm{Re}(AB^{*})\left(\vec{n}\cdot\vec{p}_{\text{in}}\right)
\tag{2.115}
$$

From this formula it can be seen that only when the projection of $\vec{p}_{\text{in}}$ in the direction $\vec{n}$ is not 0, $\vec{p}_{\text{in}}$ has contributions to the angular distribution of the emitted particles. According to Eq. (2.88) we can rewrite Eq. (2.115) as [3]

$$
\frac{\mathrm{d}\sigma}{\mathrm{d}\Omega}=\frac{\mathrm{d}\sigma^{0}}{\mathrm{d}\Omega}\left[1+\left(\vec{n}\cdot\vec{p}_{\text{in}}\right)A_{y}(\theta)\right]
\tag{2.116}
$$

where $A_y(\theta)$ is called the analyzing power and $y$ refers to selecting the $y$-axis in the direction $\vec{n}$. Comparing Eq. (2.116) with Eq. (2.115), it can be seen that in the case of target with spinless $A_y(\theta)$ is exactly equal to the polarization vector value $P_{\text{out}}^{0}(\theta)$ of the emitted particles corresponding to the unpolarized incident particles given by Eq. (2.110). Now assume that the components of the incident particle polarization vector $\vec{p}_{\text{in}}$ with spin up and down in the direction $\vec{n}$ are $p_+$ and $p_-$, respectively; that is,

$$\vec{p}_{\text{in}} = (p_+ - p_-)\,\vec{n} \tag{2.117}$$

and define

$$\frac{\mathrm{d}\sigma_\pm}{\mathrm{d}\Omega} = \frac{\mathrm{d}\sigma^0}{\mathrm{d}\Omega}\left[1 \pm p_\pm A_y(\theta)\right] \tag{2.118}$$

From the above formulas, one can obtain

$$\begin{aligned} p_+\frac{\mathrm{d}\sigma_-}{\mathrm{d}\Omega} &= p_+\frac{\mathrm{d}\sigma^0}{\mathrm{d}\Omega}\left(1 - p_- A_y\right) = \frac{\mathrm{d}\sigma^0}{\mathrm{d}\Omega}\left(p_+ - p_+ p_- A_y\right), \\ p_-\frac{\mathrm{d}\sigma_+}{\mathrm{d}\Omega} &= p_-\frac{\mathrm{d}\sigma^0}{\mathrm{d}\Omega}\left(1 + p_+ A_y\right) = \frac{\mathrm{d}\sigma^0}{\mathrm{d}\Omega}\left(p_- + p_- p_+ A_y\right) \end{aligned} \tag{2.119}$$

Using Eq. (2.118) and adding the two formulas in Eq. (2.119), we obtain

$$A_y(\theta) = \frac{\frac{\mathrm{d}\sigma_+}{\mathrm{d}\Omega} - \frac{\mathrm{d}\sigma_-}{\mathrm{d}\Omega}}{\left(p_+ + p_-\right)\frac{\mathrm{d}\sigma^0}{\mathrm{d}\Omega}} = \frac{\frac{\mathrm{d}\sigma_+}{\mathrm{d}\Omega} - \frac{\mathrm{d}\sigma_-}{\mathrm{d}\Omega}}{p_+\frac{\mathrm{d}\sigma_-}{\mathrm{d}\Omega} + p_-\frac{\mathrm{d}\sigma_+}{\mathrm{d}\Omega}} \tag{2.120}$$

Letting $\frac{\mathrm{d}\sigma_\pm}{\mathrm{d}\Omega}$ represent the complete polarization differential cross sections of the spin up ($p_+ = 1,\ p_- = 0$) and spin down ($p_+ = 0,\ p_- = 1$) of the incident particle, respectively, Eq. (2.118) can be rewritten as

$$\frac{\mathrm{d}\sigma_\pm}{\mathrm{d}\Omega} = \frac{\mathrm{d}\sigma^0}{\mathrm{d}\Omega}\left[1 \pm A_y(\theta)\right] \tag{2.121}$$

then we can obtain

$$A_y(\theta) = \frac{\frac{\mathrm{d}\sigma_+}{\mathrm{d}\Omega} - \frac{\mathrm{d}\sigma_-}{\mathrm{d}\Omega}}{\frac{\mathrm{d}\sigma_+}{\mathrm{d}\Omega} + \frac{\mathrm{d}\sigma_-}{\mathrm{d}\Omega}} \tag{2.122}$$

This is the source of the analyzing power name.

Using the vector expression

$$\vec{a}\cdot\left(\vec{b}\times\vec{c}\right) = \left(\vec{a}\times\vec{b}\right)\cdot\vec{c} \tag{2.123}$$

we can derive the following formula:

$$(\vec{n}\times\hat{\sigma})\cdot\vec{p}_{\text{in}} = -\hat{\sigma}\cdot\left(\vec{n}\times\vec{p}_{\text{in}}\right) \tag{2.124}$$

Using Eq. (2.22) and Eq. (2.124), the following is obtained from Eq. (2.114):

$$\begin{aligned}\mathrm{tr}\{\hat{\sigma}\hat{\rho}_{\mathrm{out}}\} = {} & \left[AB^* + A^*B + 2|B|^2\left(\vec{n}\cdot\vec{p}_{\mathrm{in}}\right)\right]\vec{n} + \left(|A|^2 - |B|^2\right)\vec{p}_{\mathrm{in}} \\ & - \mathrm{i}(AB^* - A^*B)\left(\vec{n}\times\vec{p}_{\mathrm{in}}\right)\end{aligned} \tag{2.125}$$

From Eq. (2.115) and Eq. (2.125), we can find that the polarization vector of the emitted particles in the case of the polarized incident particles is

$$\begin{aligned}\vec{P}_{\mathrm{out}} &= \frac{\mathrm{tr}\{\hat{\sigma}\hat{\rho}_{\mathrm{out}}\}}{\mathrm{tr}\{\hat{\rho}_{\mathrm{out}}\}} \\ &= \frac{2\left[\mathrm{Re}(AB^*) + |B|^2(\vec{n}\cdot\vec{p}_{\mathrm{in}})\right]\vec{n} + \left(|A|^2 - |B|^2\right)\vec{p}_{\mathrm{in}} + 2\mathrm{Im}(AB^*)(\vec{n}\times\vec{p}_{\mathrm{in}})}{|A|^2 + |B|^2 + 2\mathrm{Re}(AB^*)(\vec{n}\cdot\vec{p}_{\mathrm{in}})}\end{aligned} \tag{2.126}$$

where Im($A$) denotes taking the imaginary part of $A$. If $\vec{p}_{\mathrm{in}}$ is known, the value and direction of $\vec{P}_{\mathrm{out}}$ can be obtained from the above expression, in which $P_{\mathrm{out}}$ is called the polarization degree of the emitted particles in the case of polarized incident particles. When $\vec{p}_{\mathrm{in}} = 0$, Eq. (2.126) degenerates to Eq. (2.110).

The following relation may be obtained by point multiplying Eq. (2.126) by $\vec{n}$

$$\vec{n}\cdot\vec{P}_{\mathrm{out}} = \frac{2\,\mathrm{Re}\,(AB^*) + \left(|A|^2 + |B|^2\right)\left(\vec{n}\cdot\vec{p}_{\mathrm{in}}\right)}{|A|^2 + |B|^2 + 2\,\mathrm{Re}\,(AB^*)\left(\vec{n}\cdot\vec{p}_{\mathrm{in}}\right)} \tag{2.127}$$

And above relation can be rewritten as

$$\vec{n}\cdot\vec{p}_{\mathrm{in}} = \frac{\left(|A|^2 + |B|^2\right)\left(\vec{n}\cdot\vec{P}_{\mathrm{out}}\right) - 2\,\mathrm{Re}\,(AB^*)}{|A|^2 + |B|^2 - 2\,\mathrm{Re}\,(AB^*)\left(\vec{n}\cdot\vec{P}_{\mathrm{out}}\right)} \tag{2.128}$$

This formula shows that if the projection of $\vec{P}_{\mathrm{out}}$ in the direction $\vec{n}$ is known, the projection of $\vec{p}_{\mathrm{in}}$ on the direction $\vec{n}$ can be derived out. When the emitted particle polarization vector $\vec{P}_{\mathrm{out}} = 0$, the formula above yields the following relation:

$$\vec{n}\cdot\vec{p}_{\mathrm{in}} = -\frac{2\,\mathrm{Re}\,(AB^*)}{|A|^2 + |B|^2} \tag{2.129}$$

For the differential cross section and the polarization vector value of the emitted particles corresponding to the spin $\frac{1}{2}$ unpolarized incident particles with the spinless targets given by Eqs. (2.88) and (2.110), the following symbols are introduced, respectively:

$$I^0 = |A|^2 + |B|^2 \tag{2.130}$$

$$P = \frac{2\,\mathrm{Re}\,(AB^*)}{|A|^2 + |B|^2} \tag{2.131}$$

Additionally, we introduce two symbols

$$Q = \frac{2\mathrm{Im}(AB^*)}{|A|^2 + |B|^2} \tag{2.132}$$

$$W = \frac{|A|^2 - |B|^2}{|A|^2 + |B|^2} \tag{2.133}$$

And we can derive the following formula:

$$\begin{aligned}&\left(|A|^2 - |B|^2\right)^2 + [2\mathrm{Im}(AB^*)]^2 + [2\,\mathrm{Re}\,(AB^*)]^2 \\ &= |A|^4 + |B|^4 - 2|A|^2|B|^2 + 4|A|^2|B|^2 = \left(|A|^2 + |B|^2\right)^2\end{aligned} \tag{2.134}$$

Thus, we can prove the following relation:

$$W^2 + Q^2 + P^2 = 1 \tag{2.135}$$

If $P$ and $Q$ are considered to be independent polarization functions, then $W$ is not considered to be an independent polarization function. Eqs. (2.126) – (2.128) can also be rewritten as

$$\vec{P}_{\text{out}} = \frac{\left[P + (1-W)\left(\vec{n}\cdot\vec{p}_{\text{in}}\right)\right]\vec{n} + W\vec{p}_{\text{in}} + Q\left(\vec{n}\times\vec{p}_{\text{in}}\right)}{1 + P\left(\vec{n}\cdot\vec{p}_{\text{in}}\right)} \tag{2.136}$$

$$\vec{n}\cdot\vec{P}_{\text{out}} = \frac{P + \left(\vec{n}\cdot\vec{p}_{\text{in}}\right)}{1 + P\left(\vec{n}\cdot\vec{p}_{\text{in}}\right)} \tag{2.137}$$

$$\vec{n}\cdot\vec{p}_{\text{in}} = \frac{\left(\vec{n}\cdot\vec{P}_{\text{out}}\right) - P}{1 - P\left(\vec{n}\cdot\vec{P}_{\text{out}}\right)} \tag{2.138}$$

We choose the direction of the incident particle as the $z$-axis; the direction of $\vec{n}$ as the $y$-axis; and the $x$-axis is chosen according to the right-handed screw rule. From Fig. 2.1 it can be seen that $\varphi = 0$ in the chosen coordinate system. This coordinate system is also a helicity coordinate system for incident particles. This section studies

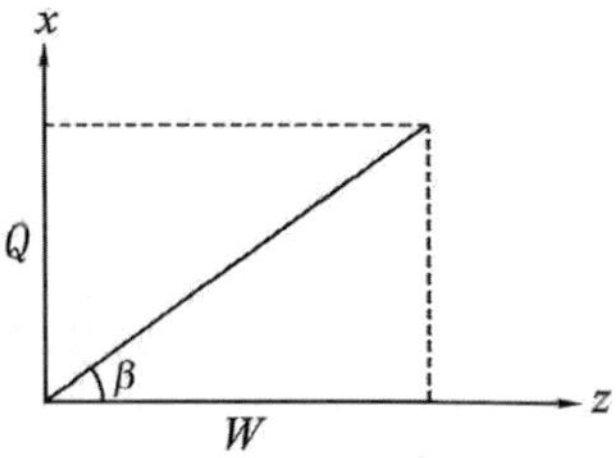


**Fig. 2.2** Polarization vector relationship diagram in scattering plane when $\vec{p}_{\text{in}} = p_z \vec{e}_z$

the first reaction process in which incident and outgoing particles can be studied in the same coordinate system. We use $\vec{e}_x, \vec{e}_y, \vec{e}_z$ to represent the unit vectors of the $x$, $y$, $z$ axes, and there is $\vec{n} = \vec{e}_y$. $\vec{p}_{\text{in}}$ is the polarization vector of the incident particles and belongs to the nature of the incident particle beam before the reaction. The three components $p_x$, $p_y$, $p_z$ of $\vec{p}_{\text{in}}$ can be determined by Eq. (2.36). When taking $\vec{p}_{\text{in}} = p_z \vec{e}_z$, based on Eqs. (2.115), (2.136), (2.130) – (2.133), the following relations can be obtained:

$$I = I^0 = |A|^2 + |B|^2, \quad \vec{P}_{\text{out}} = P\vec{e}_y + p_z W\,\vec{e}_z + p_z Q\,\vec{e}_x \tag{2.139}$$

Figure 2.2 presents the polarization vector relationship diagram in scattering plane when $\vec{P}_{\text{in}} = p_z \vec{e}_z$. It can be seen from Fig. 2.2 that

$$\tan\beta = \frac{Q}{W} = \frac{2\text{Im}(AB^*)}{|A|^2 - |B|^2} = \frac{2\text{Im}(AB^*)}{I^0 W} \tag{2.140}$$

where $\beta$ is the angle in the center-of-mass system. The above results show that when the incident particles have a polarization rate $p_z$ only in the $z$-axis direction, the outgoing particles have a polarization rate $p_z W$ in the $z$-axis direction and have a polarization rate $p_z Q$ in the $x$-axis direction. When taking $\vec{p}_{\text{in}} = p_x \vec{e}_x$, from Eqs. (2.115), (2.136), (2.130) – (2.133), one can obtain

$$I = I^0 = |A|^2 + |B|^2, \quad \vec{P}_{\text{out}} = P\vec{e}_y + p_x W\vec{e}_x - p_x Q\vec{e}_z \tag{2.141}$$

The above formulas show that when the incident particle has a polarization rate $p_x$ only in the $x$-axis direction, the outgoing particles have a polarization rate $p_x W$ in the $x$-axis direction and have a polarization rate $-p_x Q$ in the $z$-axis direction. In view of the above results, we call Q a spin rotation function, and W (Wolfenstein) a spin continuation function [4, 5]. The above results also show that when the incident particles are polarized only in the $x$-axis direction and/or in the $z$-axis direction, the differential cross section of the outgoing particles is equal to that of the unpolarized incident particles. When taking $\vec{p}_{\text{in}} = p_y \vec{e}_y$, from Eqs. (2.115), (2.136), (2.130) – (2.133) one can obtain

$$I = |A|^2 + |B|^2 + 2\,\mathrm{Re}\,(AB^*)p_y = I^0(1 + p_y P), \quad \vec{P}_{\text{out}} = \frac{P + p_y}{1 + p_y P}\vec{e}_y \tag{2.142}$$

It can be seen that when $\vec{p}_{in} = p_y \vec{e}_y$, the differential cross section is different from that of the unpolarized incident particle. And the polarization vector of the outgoing particles only appears in the $\vec{e}_y$ direction.

When the three components of $\vec{P}_{\text{out}}$ are represented by $P^x$, $P^y$, $P^z$, from Eq. (2.126), one can obtain

$$\begin{aligned} P^x &= \frac{\left(|A|^2 - |B|^2\right)p_x + 2\mathrm{Im}(AB^*)p_z}{|A|^2 + |B|^2 + 2\,\mathrm{Re}\,(AB^*)p_y} = \frac{p_x W + p_z Q}{1 + p_y P}, \\ P^y &= \frac{2\,\mathrm{Re}\,(AB^*) + \left(|A|^2 + |B|^2\right)p_y}{|A|^2 + |B|^2 + 2\,\mathrm{Re}\,(AB^*)p_y} = \frac{P + p_y}{1 + p_y P}, \\ P^z &= \frac{\left(|A|^2 - |B|^2\right)p_z - 2\mathrm{Im}(AB^*)p_x}{|A|^2 + |B|^2 + 2\,\mathrm{Re}\,(AB^*)p_y} = \frac{p_z W - p_x Q}{1 + p_y P} \end{aligned} \tag{2.143}$$

In the coordinate system selected above, from Eqs. (2.102), (2.23), and (2.131), one can get

$$P^i(\theta) \equiv \frac{\mathrm{tr}\{\hat{\sigma}_i \hat{F} \hat{F}^+\}}{\mathrm{tr}\{\hat{F} \hat{F}^+\}} = \delta_{iy} P \tag{2.144}$$

where $P^i(\theta)$ represents the value of the $i$th component of the polarization vector obtained from the unpolarized incident particles. At the same time, one can get

$$A_i(\theta) \equiv \frac{\mathrm{tr}\{\hat{F}\,\hat{\sigma}_i \hat{F}^+\}}{\mathrm{tr}\{\hat{F} \hat{F}^+\}} = \delta_{iy} P, \quad A_y(\theta) = P \tag{2.145}$$

where $A_i(\theta)$ represents the $i$th component of the analyzing power polarization vector of the incident particles. Obviously, for spinless target there is

$$P^i(\theta) = A_i(\theta) = \delta_{iy} P \tag{2.146}$$

It means that both $P^i(\theta)$ and $A_i(\theta)$ are not 0 only in the y-axis direction. We define the following symbol to represent the polarization transfer coefficient from the $i$th initial polarization component to the $k$th final polarization component

**Table 2.4** Polarization transfer coefficient $K_i^k(\theta)$ of spin $\frac{1}{2}$ particles with spinless targets

| Out $k$ | In $i$ | | |
|---|---|---|---|
| | $x$ | $y$ | $z$ |
| $x$ | $W$ | | $Q$ |
| $y$ | | $1$ | |
| $z$ | $-Q$ | | $W$ |

$$K_i^k(\theta) = \frac{\mathrm{tr}\left\{\hat{\sigma}_k \hat{F}\, \hat{\sigma}_i \hat{F}^+\right\}}{\mathrm{tr}\left\{\hat{F}\hat{F}^+\right\}}, \qquad i,k = x,y,z \tag{2.147}$$

Table 2.4 lists the specific results of $K_i^k(\theta)$ for spin $\frac{1}{2}$ particles with spinless targets derived from Eq. (2.147) by using Eqs. (2.102), (2.6), and (2.23). In this table, the empty cells are to be filled with 0.

The spin $\frac{1}{2}$ particle differential data including the differential cross section $I$ and the polarization vector $\vec{p}$ can be represented by the following Stokes vector [6]:

$$S_c = I \begin{pmatrix} 1 \\ p_x \\ p_y \\ p_z \end{pmatrix}_c \tag{2.148}$$

where $c$ represents the selected coordinate system. Below we use $S_c^{(\mathrm{i})}$ and $S_c^{(\mathrm{f})}$ to represent the Stokes vectors for the incident particles and outgoing particles, respectively. In this section they are all given in the same coordinate system $c$. By use of the previous symbols, one can write the following formulas:

$$S_c^{(\mathrm{i})} = I^0 \begin{pmatrix} 1 \\ p_x \\ p_y \\ p_z \end{pmatrix}_c, \qquad S_c^{(\mathrm{f})} = I \begin{pmatrix} 1 \\ P^x \\ P^y \\ P^z \end{pmatrix}_c \tag{2.149}$$

Using the simplified symbols introduced in this section, the physical quantities appeared in Eq. (2.116) are $I = \frac{\mathrm{d}\sigma}{\mathrm{d}\Omega}$, $I^0 = \frac{\mathrm{d}\sigma^0}{\mathrm{d}\Omega}$, $p_y = \vec{n} \cdot \vec{p}_{\mathrm{in}}, P = A_y(\theta)$, then according to Eqs. (2.142) and (2.143) the following relation can be obtained from Eq. (2.149):

$$S_c^{(\mathrm{f})} = \hat{Z}_{cc} S_c^{(\mathrm{i})} \tag{2.150}$$

where $\hat{Z}_{cc}$ is a $4 \times 4$ matrix, called the Mueller matrix [6] , in the form of

$$\hat{Z}_{cc} = \begin{pmatrix} 1 & 0 & P & 0 \\ 0 & W & 0 & Q \\ P & 0 & 1 & 0 \\ 0 & -Q & 0 & W \end{pmatrix}_{cc} \tag{2.151}$$

## 2.5 Polarization Theory of Triple Elastic Scatterings Between Spin $\frac{1}{2}$ Particles and Spinless Nuclei

This section studies the polarization theory of triple elastic scatterings between spin $\frac{1}{2}$ particles and spinless nuclei. For the sake of simplicity, we constrain the incident particles to be the same, but the targets can be the same or different. Considering that the calculations of the differential cross sections and the polarization physical quantities must be carried out in the center-of-mass ($C$) system, and the tracking and sampling of particles should be carried out in the laboratory ($L$) system, the coordinate system transformation problem should be considered. Since the incident direction of the particles is generally selected as the $z$-axis, the transformation between $C$ system and $L$ system is also carried out along the $z$-axis direction. Therefore, only the polar angle $\theta$ changes under this transformation; the azimuthal angle $\varphi$ is not affected.

Let $I_C(\theta_C, \varphi)$ and $I_L(\theta_L, \varphi)$ denote the emitted particle angular distributions in $C$ system and $L$ system, respectively. By using Eq. (2.59) and Eq. (2.2.56) given in Ref. [2], the coordinate transformation relations between them can be obtained as

$$I_L(\theta_L, \varphi) = X(\theta_C) I_C(\theta_C, \varphi), \quad I_C(\theta_C, \varphi) = Y(\theta_L) I_L(\theta_L, \varphi) \tag{2.152}$$

where

$$X(\theta_C) = \frac{\left(1 + \gamma^2 + 2\gamma \cos\theta_C\right)^{3/2}}{|1 + \gamma \cos\theta_C|} \tag{2.153}$$

$$Y(\theta_L) = \frac{\sqrt{1 - \gamma^2 \sin^2\theta_L}}{\left(\sqrt{1 - \gamma^2 \sin^2\theta_L} + \gamma \cos\theta_L\right)^2} \tag{2.154}$$

For elastic scattering

$$\gamma = \frac{m}{M} \le 1 \tag{2.155}$$

where $m$ and $M$ are masses of incident particle and target, respectively. For the polarization observables $P$, $Q$, and $W$ introduced in Sect. 2.4, because the molecules and the denominators of their expressions require coordinate system transformation at the same time, so we have

$$P_C(\theta_C, \varphi) = P_L(\theta_L, \varphi), \quad Q_C(\theta_C, \varphi) = Q_L(\theta_L, \varphi), \quad W_C(\theta_C, \varphi) = W_L(\theta_L, \varphi) \tag{2.156}$$

The angular relations between the two coordinate systems are

$$\begin{aligned} \cos\theta_C &= \cos\theta_L\sqrt{1-\gamma^2\sin^2\theta_L} - \gamma\sin^2\theta_L, \\ \cos\theta_L &= \frac{\gamma + \cos\theta_C}{(1+\gamma^2+2\gamma\cos\theta_C)^{1/2}} \end{aligned} \tag{2.157}$$

For the elastic scattering, there are the following energy relations:

$$E_C = \frac{M}{m+M}E_L \tag{2.158}$$

$$\varepsilon_C = \frac{M}{m+M}E_C \tag{2.159}$$

$$\varepsilon_L = \varepsilon_C\left(1+\gamma^2+2\gamma\cos\theta_C\right) \tag{2.160}$$

where $E_C$ and $E_L$ are the total kinetic energies of incident particle relative to the target in $C$ system and $L$ system, respectively, and $\varepsilon_C$ and $\varepsilon_L$ are the kinetic energies of the emitted particle in $C$ system and $L$ system, respectively. Substituting the first equation of Eq. (2.157) into Eq. (2.160), one can obtain

$$\varepsilon_L = \varepsilon_C\left[1+\gamma^2+2\gamma\left(\cos\theta_L\sqrt{1-\gamma^2\sin^2\theta_L} - \gamma\sin^2\theta_L\right)\right] \tag{2.161}$$

It is assumed that the incident particle beam of the first scattering is along the $z$-axis and is unpolarized. In order to make the polarization direction after the first scattering along the $y$-axis, we choose $\varphi_1 = 0$. Thus, according to the discussion in Sect. 2.4, we can write that the Stokes vector in $C_1$ system after the first scattering as

$$S_{C_1}^{(f)} = I_{C_1}\begin{pmatrix} 1 \\ 0 \\ P_{C_1} \\ 0 \end{pmatrix} \tag{2.162}$$

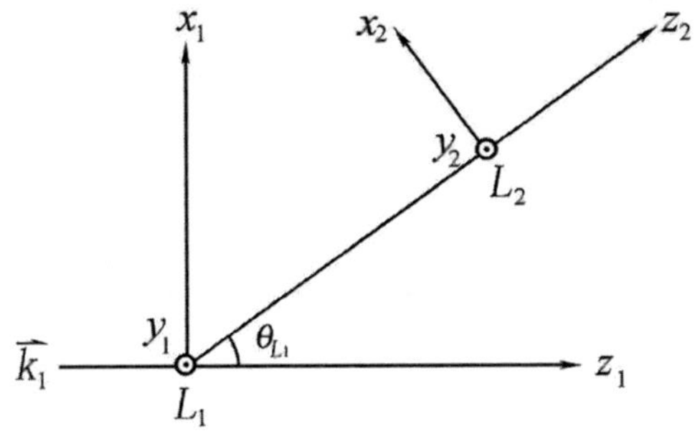


**Fig. 2.3** Coordinates in the laboratory system for the particle first twice scatterings

Using Eq. (2.109) and Eq. (2.110), we can show

$$I_{C_1} = |A_1|^2 + |B_1|^2 \tag{2.163}$$

$$P_{C_1} = \frac{2\,\mathrm{Re}\left(A_1 B_1^*\right)}{|A_1|^2 + |B_1|^2} \tag{2.164}$$

The expressions of $A_i$ and $B_i$ have been given by Eq. (2.82) and Eq. (2.83). By use of Eq. (2.152) the Stokes vector in $L_1$ system after the first scattering can be obtained as

$$S_{L_1}^{(\mathrm{f})} = X(\theta_{C_1})\, I_{C_1} \begin{pmatrix} 1 \\ 0 \\ P_{C_1} \\ 0 \end{pmatrix} \tag{2.165}$$

By use of Eq. (2.157) the angle $\theta_{C_1}$ at the right end of the above formula can be changed to angle $\theta_{L_1}$ in $L_1$ system.

Figure 2.3 shows coordinates in the laboratory system for the particle first twice scatterings. The axes $y_1$ and $y_2$ are parallel to each other and taken the vertical page upward direction. The direction of the second scattering incident particle is taken as $z_2$-axis, and the angle between the $z_2$-axis and the $z_1$-axis is $\theta_{L_1}$. The wave number of the first scattering incident particle beam in $C_1$ system is

$$k_1 = \frac{\sqrt{2\mu_1 E_{C_1}}}{\hbar} \tag{2.166}$$

where $\mu_1$ is the reduced mass

$$\mu_1 = \frac{m}{m + M_1} \tag{2.167}$$

Since the $y$-axes of $L_1$ and $L_2$ systems are parallel to each other, the Stokes vector of the incident particles in $L_2$ system can be obtained directly from Eq. (2.165):

$$S_{L_2}^{(i)} = X(\theta_{C_1})\, I_{C_1} \begin{pmatrix} 1 \\ 0 \\ P_{C_1} \\ 0 \end{pmatrix} \tag{2.168}$$

Transforming the above formula into $C_2$ system yields

$$S_{C_2}^{(i)} = Y(\theta_{L_1})X(\theta_{C_1})\, I_{C_1} \begin{pmatrix} 1 \\ 0 \\ P_{C_1} \\ 0 \end{pmatrix} \equiv I_{C_2}^{(i)} \begin{pmatrix} 1 \\ 0 \\ P_{C_2}^{(i)} \\ 0 \end{pmatrix} \tag{2.169}$$

The above formula shows that the incident particle of the second scattering has no polarization component in $x_2$ and $z_2$ axis directions.

The final state Stokes vector after the second scattering can be expressed as

$$S_{C_2}^{(f)} = I_{C_2}^{(f)} \begin{pmatrix} 1 \\ P_{x_2} \\ P_{y_2} \\ P_{z_2} \end{pmatrix} \tag{2.170}$$

The wave number of the incident particle steam of the second scattering in $C_2$ system is

$$k_2 = \frac{\sqrt{2\mu_2 \varepsilon_{C_2}}}{\hbar} \tag{2.171}$$

And we can write out

$$\varepsilon_{C_2} = \frac{M_2}{m + M_2} \varepsilon_{L_1} \tag{2.172}$$

$$\varepsilon_{L_1} = \varepsilon_{C_1} \left[ 1 + \gamma_1^2 + 2\gamma_1 \left( \cos\theta_{L_1} \sqrt{1 - \gamma^2 \sin^2\theta_{L_1}} - \gamma_1 \sin^2\theta_{L_1} \right) \right] \tag{2.173}$$

$$\varepsilon_{C_1} = \frac{M_1}{m + M_1} E_{C_1} \tag{2.174}$$

$$E_{C_1} = \frac{M_1}{m + M_1} E_{L_1} \tag{2.175}$$

It can be seen that $k_2$ is related to the angle $\theta_{L_1}$.

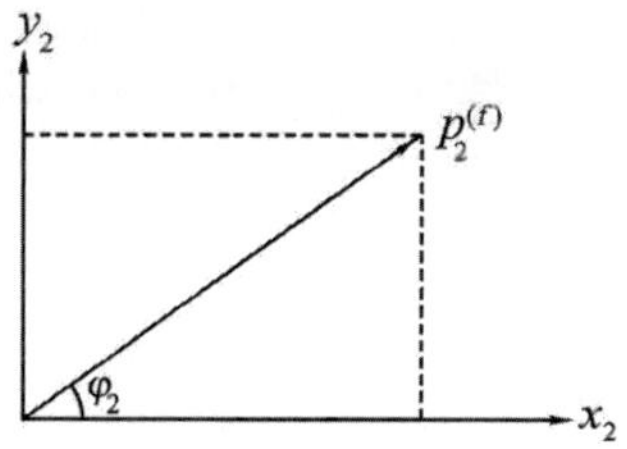


**Fig. 2.4** Projection of the wave vector of the emitted particle in $C_2$ system on the $x_2y_2$ plane

The polar angle of the wave vector $\vec{k}_2^{(f)}$ of the emitted particles of the second scattering in $C_2$ system is $\theta_{C2}$; its azimuthal angle is $\varphi_2$; and its projection on the $x_2y_2$ plane is $p_2^{(f)}$ (see Fig. 2.4). According to Eq. (2.81) the scattering amplitude of the second scattering can be written as

$$\begin{aligned}\hat{F}_2(\theta_{C_2},\varphi_2) &= \begin{pmatrix} A_2(\theta_{C_2}) & -\mathrm{i}B_2(\theta_{C_2})\,\mathrm{e}^{-\mathrm{i}\varphi_2} \\ \mathrm{i}B_2(\theta_{C_2})\,\mathrm{e}^{\mathrm{i}\varphi_2} & A_2(\theta_{C_2}) \end{pmatrix}, \\ \hat{F}_2^+(\theta_{C_2},\varphi_2) &= \begin{pmatrix} A_2^*(\theta_{C_2}) & -\mathrm{i}B_2^*(\theta_{C_2})\,\mathrm{e}^{-\mathrm{i}\varphi_2} \\ \mathrm{i}B_2^*(\theta_{C_2})\,\mathrm{e}^{\mathrm{i}\varphi_2} & A_2^*(\theta_{C_2}) \end{pmatrix}\end{aligned} \tag{2.176}$$

and the following expression can be obtained:

$$\hat{F}_2\hat{F}_2^+ = \begin{pmatrix} |A_2|^2+|B_2|^2 & -\mathrm{i}\left(A_2B_2^*+B_2A_2^*\right)\mathrm{e}^{-\mathrm{i}\varphi_2} \\ \mathrm{i}\left(A_2B_2^*+B_2A_2^*\right)\mathrm{e}^{\mathrm{i}\varphi_2} & |A_2|^2+|B_2|^2 \end{pmatrix} \tag{2.177}$$

By using Eq. (2.177) and Eq. (2.6), the contributions of the unpolarized part of the second incident particle can be obtained as

$$I_{C_2}^0 = \frac{1}{2}\mathrm{tr}\left\{\hat{F}_2\hat{F}_2^+\right\} = |A_2|^2+|B_2|^2 \tag{2.178}$$

$$I_{C_2}^0P_{x_2}^0 = \frac{1}{2}\mathrm{tr}\left\{\hat{\sigma}_x\hat{F}_2\hat{F}_2^+\right\} = -2\,\mathrm{Re}\left(A_2B_2^*\right)\,\sin\varphi_2 = -I_{C_2}^0P_{C_2}\sin\varphi_2 \tag{2.179}$$

$$I_{C_2}^0P_{y_2}^0 = \frac{1}{2}\mathrm{tr}\left\{\hat{\sigma}_y\hat{F}_2\hat{F}_2^+\right\} = 2\,\mathrm{Re}\left(A_2B_2^*\right)\,\cos\varphi_2 = I_{C_2}^0P_{C_2}\cos\varphi_2 \tag{2.180}$$

$$I_{C_2}^0P_{z_2}^0 = \frac{1}{2}\mathrm{tr}\left\{\hat{\sigma}_z\hat{F}_2\hat{F}_2^+\right\} = 0 \tag{2.181}$$

where

$$P_{C_2} = \frac{2\,\mathrm{Re}\left(A_2B_2^*\right)}{|A_2|^2+|B_2|^2} \tag{2.182}$$

When considering the initial state of $C_2$ system given by Eq. (2.169), one first finds the following expression based on Eqs. (2.176) and (2.6):

$$\hat{F}_2\sigma_y\hat{F}_2^+ = \begin{pmatrix} A_2 & -\mathrm{i}B_2\mathrm{e}^{-\mathrm{i}\varphi_2} \\ \mathrm{i}B_2\mathrm{e}^{\mathrm{i}\varphi_2} & A_2 \end{pmatrix}\begin{pmatrix} B_2^*\mathrm{e}^{\mathrm{i}\varphi_2} & -\mathrm{i}A_2^* \\ \mathrm{i}A_2^* & B_2^*\mathrm{e}^{-\mathrm{i}\varphi_2} \end{pmatrix}$$
$$= \begin{pmatrix} A_2B_2^*\mathrm{e}^{\mathrm{i}\varphi_2} + B_2A_2^*\mathrm{e}^{-\mathrm{i}\varphi_2} & -\mathrm{i}\left(|A_2|^2 + |B_2|^2\mathrm{e}^{-2\mathrm{i}\varphi_2}\right) \\ \mathrm{i}\left(|A_2|^2 + |B_2|^2\mathrm{e}^{2\mathrm{i}\varphi_2}\right) & B_2A_2^*\mathrm{e}^{\mathrm{i}\varphi_2} + A_2B_2^*\mathrm{e}^{-\mathrm{i}\varphi_2} \end{pmatrix} \tag{2.183}$$

It can also be found that the contributions of the polarized part of the second incident particles are

$$I_{C_2}^1 = \frac{1}{2}\mathrm{tr}\left\{\hat{F}_2\hat{\sigma}_y\hat{F}_2^+\right\} = I_{C_2}^0 P_{C_2}\cos\varphi_2 \tag{2.184}$$

$$I_{C_2}^0 P_{x_2}^1 = \frac{1}{2}\mathrm{tr}\left\{\hat{\sigma}_x\hat{F}_2\hat{\sigma}_y\hat{F}_2^+\right\} = -I_{C_2}^0\frac{1}{2}(1 - W_{C_2})\sin(2\varphi_2) \tag{2.185}$$

$$I_{C_2}^0 P_{y_2}^1 = \frac{1}{2}\mathrm{tr}\left\{\hat{\sigma}_y\hat{F}_2\hat{\sigma}_y\hat{F}_2^+\right\} = |A_2|^2 + |B_2|^2\cos(2\varphi_2)$$
$$= I_{C_2}^0\left[1 - \frac{1}{2}(1 - W_{C_2})\,(1 - \cos(2\varphi_2))\right] \tag{2.186}$$

$$I_{C_2}^0 P_{z_2}^1 = \frac{1}{2}\mathrm{tr}\left\{\hat{\sigma}_z\hat{F}_2\hat{\sigma}_y\hat{F}_2^+\right\} = -2\mathrm{Im}\left(A_2B_2^*\right)\sin\varphi_2 = -I_{C_2}^0 Q_{C_2}\sin\varphi_2 \tag{2.187}$$

where

$$Q_{C_2} = \frac{2\mathrm{Im}\left(A_2B_2^*\right)}{|A_2|^2 + |B_2|^2} \tag{2.188}$$

$$W_{C_2} = \frac{|A_2|^2 - |B_2|^2}{|A_2|^2 + |B_2|^2} \tag{2.189}$$

By using the above results, according to Eq. (2.169) one can obtain

$$I_{C_2}^{(\mathrm{f})} = I_{C_2}^{(\mathrm{i})} I_{C_2}^0\left(1 + p_{C_2}^{(\mathrm{i})} P_{C_2}\cos\varphi_2\right) \tag{2.190}$$

$$P_{x_2} = P_{x_2}^0 + p_{C_2}^{(\mathrm{i})} P_{x_2}^1 = -P_{C_2}\sin\varphi_2 - p_{C_2}^{(\mathrm{i})}\frac{1}{2}(1 - W_{C_2})\sin(2\varphi_2) \tag{2.191}$$

$$P_{y_2} = P_{y_2}^0 + p_{C_2}^{(\mathrm{i})} P_{y_2}^1 = P_{C_2}\cos\varphi_2 + p_{C_2}^{(\mathrm{i})}\left[1 - \frac{1}{2}(1 - W_{C_2})\,(1 - \cos(2\varphi_2))\right] \tag{2.192}$$

$$P_{z_2} = P_{z_2}^0 + p_{C_2}^{(\mathrm{i})} P_{z_2}^1 = -p_{C_2}^{(\mathrm{i})} Q_{C_2}\sin\varphi_2 \tag{2.193}$$

In this way the final state Stokes vector after the second scattering given by Eq. (2.170) is obtained. Then transforming it to $L_2$ system, we can get

$$S_{L_2}^{(f)} = X(\theta_{C_2}) S_{C_2}^{(f)} = X(\theta_{C_2})\, I_{C_2}^{(f)} \begin{pmatrix} 1 \\ P_{x_2} \\ P_{y_2} \\ P_{z_2} \end{pmatrix} \tag{2.194}$$

Figure 2.5 shows laboratory system for the third scattering. The angle between the $z_3$-axis and the $z_2$-axis is $\theta_{L_2}$; the polar angle of the $z_3$-axis in $L_2$ system is $\theta_{L_2}$; and its azimuthal angle is $\varphi_2$. From Eq. (2.194) we can see that the polarization vector in $L_2$ system is

$$\vec{P}^{L_2} \equiv P_{x_2} \vec{e}_x + P_{y_2} \vec{e}_y + P_{z_2} \vec{e}_z \tag{2.195}$$

Based on Eq. (2.57) and Table 2.5, we can obtain

$$\hat{D}^1(\alpha, \beta, \gamma) = \begin{pmatrix} \frac{1+\cos\beta}{2} e^{-i(\alpha+\gamma)} & -\frac{\sin\beta}{\sqrt{2}} e^{-i\alpha} & \frac{1-\cos\beta}{2} e^{i(\gamma-\alpha)} \\ \frac{\sin\beta}{\sqrt{2}} e^{-i\gamma} & \cos\beta & -\frac{\sin\beta}{\sqrt{2}} e^{i\gamma} \\ \frac{1-\cos\beta}{2} e^{i(\alpha-\gamma)} & \frac{\sin\beta}{\sqrt{2}} e^{i\alpha} & \frac{1+\cos\beta}{2} e^{i(\gamma+\alpha)} \end{pmatrix} \tag{2.196}$$

According to Eq. (1.27) we can write

$$A'_{1M'} = \sum_M A_{1M} D^1_{MM'}(\alpha, \beta, \gamma) \tag{2.197}$$

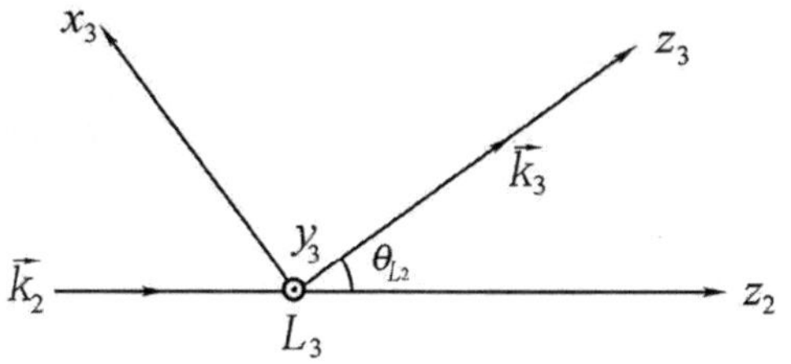


**Fig. 2.5** Laboratory system for the third scattering

**Table 2.5** Expressions of $d^1_{MM'}(\beta)$[1]

| | $M'$ | | |
|---|---|---|---|
| $M$ | 1 | 0 | −1 |
| 1 | $\frac{1+\cos\beta}{2}$ | $-\frac{\sin\beta}{\sqrt{2}}$ | $\frac{1-\cos\beta}{2}$ |
| 0 | $\frac{\sin\beta}{\sqrt{2}}$ | $\cos\beta$ | $-\frac{\sin\beta}{\sqrt{2}}$ |
| −1 | $\frac{1-\cos\beta}{2}$ | $\frac{\sin\beta}{\sqrt{2}}$ | $\frac{1+\cos\beta}{2}$ |

The following relations can be obtained by using Eqs. (2.197), (2.196), and (1.23):

$$
\begin{aligned}
A'_{11} &= -\frac{1+\cos\beta}{2}\mathrm{e}^{-\mathrm{i}(\alpha+\gamma)}\frac{1}{\sqrt{2}}(A_x+\mathrm{i}A_y)+\frac{\sin\beta}{\sqrt{2}}\mathrm{e}^{-\mathrm{i}\gamma}A_z \\
&\quad+\frac{1-\cos\beta}{2}\mathrm{e}^{\mathrm{i}(\alpha-\gamma)}\frac{1}{\sqrt{2}}(A_x-\mathrm{i}A_y) \\
A'_{10} &= \frac{\sin\beta}{\sqrt{2}}\mathrm{e}^{-\mathrm{i}\alpha}\frac{1}{\sqrt{2}}(A_x+\mathrm{i}A_y)+\cos\beta A_z+\frac{\sin\beta}{\sqrt{2}}\mathrm{e}^{\mathrm{i}\alpha}\frac{1}{\sqrt{2}}(A_x-\mathrm{i}A_y) \\
A'_{1\,-1} &= -\frac{1-\cos\beta}{2}\mathrm{e}^{\mathrm{i}(\gamma-\alpha)}\frac{1}{\sqrt{2}}(A_x+\mathrm{i}A_y)-\frac{\sin\beta}{\sqrt{2}}\mathrm{e}^{\mathrm{i}\gamma}A_z \\
&\quad+\frac{1+\cos\beta}{2}\mathrm{e}^{\mathrm{i}(\gamma+\alpha)}\frac{1}{\sqrt{2}}(A_x-\mathrm{i}A_y)
\end{aligned}
\tag{2.198}
$$

Then by use of Eq. (1.22) the following relations can be obtained from Eq. (2.198):

$$
\begin{aligned}
A'_x &= \frac{1}{\sqrt{2}}(A'_{1\,-1}-A'_{11}) = \left[-\frac{1-\cos\beta}{4}\mathrm{e}^{\mathrm{i}(\gamma-\alpha)}+\frac{1+\cos\beta}{4}\mathrm{e}^{-\mathrm{i}(\alpha+\gamma)}\right](A_x+\mathrm{i}A_y) \\
&\quad+\left[\frac{1+\cos\beta}{4}\mathrm{e}^{\mathrm{i}(\gamma+\alpha)}-\frac{1-\cos\beta}{4}\mathrm{e}^{\mathrm{i}(\alpha-\gamma)}\right](A_x-\mathrm{i}A_y)-\left[\frac{\sin\beta}{2}\mathrm{e}^{\mathrm{i}\gamma}+\frac{\sin\beta}{2}\mathrm{e}^{-\mathrm{i}\gamma}\right]A_z \\
&= \left[\frac{1+\cos\beta}{2}\cos(\alpha+\gamma)-\frac{1-\cos\beta}{2}\cos(\alpha-\gamma)\right]A_x \\
&\quad+\left[\frac{1+\cos\beta}{2}\sin(\alpha+\gamma)-\frac{1-\cos\beta}{2}\sin(\alpha-\gamma)\right]A_y-\sin\beta\cos\gamma A_z \\
&= \frac{1}{2}[(1+\cos\beta)(\cos\alpha\cos\gamma-\sin\alpha\sin\gamma)-(1-\cos\beta)(\cos\alpha\cos\gamma+\sin\alpha\sin\gamma)]A_x \\
&\quad+\frac{1}{2}[(1+\cos\beta)(\sin\alpha\cos\gamma+\cos\alpha\sin\gamma)-(1-\cos\beta)(\sin\alpha\cos\gamma-\cos\alpha\sin\gamma)]A_y \\
&\quad-\sin\beta\cos\gamma A_z \\
&= (\cos\beta\cos\alpha\cos\gamma-\sin\alpha\sin\gamma)A_x+(\cos\beta\sin\alpha\cos\gamma+\cos\alpha\sin\gamma)A_y-\sin\beta\cos\gamma A_z
\end{aligned}
\tag{2.199}
$$

In the same way, the following relations can also be obtained:

$$
\begin{aligned}
A'_y &= -(\cos\beta\cos\alpha\sin\gamma+\sin\alpha\cos\gamma)A_x \\
&\quad-(\cos\beta\sin\alpha\sin\gamma-\cos\alpha\cos\gamma)A_y+\sin\beta\sin\gamma A_z \\
A'_z &= \sin\beta\cos\alpha A_x+\sin\beta\sin\alpha A_y+\cos\beta A_z
\end{aligned}
\tag{2.200}
$$

So that we can subsequently yield

$$A_i' = \sum_{k=x,y,z} A_k a_{ki}, \qquad i = x, y, z \tag{2.201}$$

The rotation matrix composed by matrix elements $a_{ki}$ is [1]

$$\hat{a}(\alpha, \beta, \gamma) = \begin{pmatrix} \cos\beta\cos\alpha\cos\gamma - \sin\alpha\sin\gamma & -\cos\beta\cos\alpha\sin\gamma - \sin\alpha\cos\gamma & \sin\beta\cos\alpha \\ \cos\beta\sin\alpha\cos\gamma + \cos\alpha\sin\gamma & -\cos\beta\sin\alpha\sin\gamma + \cos\alpha\cos\gamma & \sin\beta\sin\alpha \\ -\sin\beta\cos\gamma & \sin\beta\sin\gamma & \cos\beta \end{pmatrix} \tag{2.202}$$

If we take $\alpha = \varphi, \beta = \theta, \gamma = 0$, Eq. (2.202) degenerates into

$$\hat{B}(\theta, \varphi) = \begin{pmatrix} \cos\theta\cos\varphi & -\sin\varphi & \sin\theta\cos\varphi \\ \cos\theta\sin\varphi & \cos\varphi & \sin\theta\sin\varphi \\ -\sin\theta & 0 & \cos\theta \end{pmatrix} \tag{2.203}$$

The inverse of a rotation matrix $\hat{D}^J(\alpha, \beta, \gamma)$ is $\left[\hat{D}^J(\alpha, \beta, \gamma)\right]^{-1} = \left[\hat{D}^J(\alpha, \beta, \gamma)\right]^{+}$. Since the matrix $\hat{B}(\theta, \varphi)$ given by Eq. (2.203) is a real number matrix, its inverse is its transpose matrix, that is,

$$\hat{b}(\theta, \varphi) = \widetilde{B}(\theta, \varphi) = \begin{pmatrix} \cos\theta\cos\varphi & \cos\theta\sin\varphi & -\sin\theta \\ -\sin\varphi & \cos\varphi & 0 \\ \sin\theta\cos\varphi & \sin\theta\sin\varphi & \cos\theta \end{pmatrix} \tag{2.204}$$

We can see that at $\theta = \varphi = 0$, $\hat{B}$ and $\hat{b}$ are all unit matrix.

When the object and the coordinate system rotate at the same time, the relative relationship between the object and the coordinate system will not change. Let $\hat{D}(\alpha, \beta, \gamma)$ be the rotation Euler angles operator; then we get

$$\begin{aligned} &\{\hat{D}(\alpha, \beta, \gamma)\}_{\text{object}} \cdot \{\hat{D}(\alpha, \beta, \gamma)\}_{\text{coord. system}} = 1 \\ &\{D^J_{MM'}(\alpha, \beta, \gamma)\}_{\text{object}} = \left\{ \left[D^{-1}(\alpha, \beta, \gamma)\right]^J_{MM'} \right\}_{\text{coord. system}} \end{aligned} \tag{2.205}$$

In the density matrix given by Eq. (1.91), $t_L(S) \cdot \hat{T}_L(S)$ is a scalar with rotational invariance. We think of the particle vector spin operator or tensor polarization operator $\hat{T}_L(S)$ as a coordinate system, and the particle vector or tensor polarization rate $t_L(S)$ as a physical quantity of the object. Equation (2.201) gives the rotation relation of the coordinate system, while the particle polarization vector represents the property of the object, so its corresponding rotation relation is

$$p_i' = \sum_{k=x,y,z} p_k b_{ki} = \sum_{k=x,y,z} B_{ik} p_k, \qquad i = x, y, z \tag{2.206}$$

For example, assuming that the components of the polarization vector before rotation are $p_x = p_y = 0$, $p_z = p_Z$ and the particle polarization vector is rotated by angles $(\theta, \varphi)$, then we can obtain $p_x' = \sin\theta \cos\varphi \, p_Z$, $p_y' = \sin\theta \sin\varphi \, p_Z$, $p_z' = \cos\theta \, p_Z$ from Eqs. (2.206) and (2.203). These formulas are the familiar results. It is also understood that in the Cartesian coordinate system *XYZ* taking the polarization vector as *Z*-axis, only the component $p_Z$ is not equal to 0; then the coordinate system is rotated reversely by angles $(\theta, \varphi)$, and the same results can also be obtained.

$\vec{P}^{L_2}$ given by Eq. (2.195) is the polarization vector in $L_2$ system. $L_3$ system is rotated by angles $(\theta_{L_2}, \varphi_2)$ relative to $L_2$ system; $\vec{p}^{L_3}$ is $\vec{P}^{L_2}$ described in $L_3$ system. This situation is equivalent to only rotating the coordinate system and the object is not rotated, so there is

$$\vec{p}^{L_3} = \hat{b}(\theta_{L_2}, \varphi_2) \vec{P}^{L_2} \tag{2.207}$$

Using Eq. (2.204) we can obtain

$$\begin{aligned} p_{x_3} &= \cos\theta_{L_2} \cos\varphi_2 P_{x_2} + \cos\theta_{L_2} \sin\varphi_2 P_{y_2} - \sin\theta_{L_2} P_{z_2} \\ p_{y_3} &= -\sin\varphi_2 P_{x_2} + \cos\varphi_2 P_{y_2} \\ p_{z_3} &= \sin\theta_{L_2} \cos\varphi_2 P_{x_2} + \sin\theta_{L_2} \sin\varphi_2 P_{y_2} + \cos\theta_{L_2} P_{z_2} \end{aligned} \tag{2.208}$$

After the transformation of the polarization vector given by Eq. (2.208), from the final state of $L_2$ system given by Eq. (2.194) we can obtain the initial state in $L_3$ system as

$$S_{L_3}^{(\mathrm{i})} = X(\theta_{C_2}) \, I_{C_2}^{(\mathrm{f})} \begin{pmatrix} 1 \\ p_{x_3} \\ p_{y_3} \\ p_{z_3} \end{pmatrix} \tag{2.209}$$

Then transforming $S_{L_3}^{(\mathrm{i})}$ to $C_3$ system we have

$$S_{C_3}^{(\mathrm{i})} = Y(\theta_{L_2}) X(\theta_{C_2}) I_{C_2}^{(\mathrm{f})} \begin{pmatrix} 1 \\ p_{x_3} \\ p_{y_3} \\ p_{z_3} \end{pmatrix} \equiv I_{C_3}^{(\mathrm{i})} \begin{pmatrix} 1 \\ p_{x_3}^{(\mathrm{i})} \\ p_{y_3}^{(\mathrm{i})} \\ p_{z_3}^{(\mathrm{i})} \end{pmatrix} \tag{2.210}$$

The final state Stokes vector after the third scattering can be expressed as

$$S_{C_3}^{(\mathrm{f})} = I_{C_3}^{(\mathrm{f})} \begin{pmatrix} 1 \\ P_{x_3} \\ P_{y_3} \\ P_{z_3} \end{pmatrix} \tag{2.211}$$

The wave number of the incident particle beam of the third scattering in $C_3$ system is

$$k_3 = \frac{\sqrt{2\mu_3 \varepsilon_{C_3}}}{\hbar} \tag{2.212}$$

We can also write out

$$\varepsilon_{C_3} = \frac{M_3}{m + M_3} \varepsilon_{L_2} \tag{2.213}$$

$$\varepsilon_{L_2} = \varepsilon_{C_2} \left[ 1 + \gamma_2^2 + 2\gamma_2 \left( \cos\theta_{L_2} \sqrt{1 - \gamma_2^2 \sin^2\theta_{L_2}} - \gamma_2 \sin^2\theta_{L_2} \right) \right] \tag{2.214}$$

$$\gamma_2 = \frac{m}{M_2} \tag{2.215}$$

where $\varepsilon_{C_2}$ has been given by Eq. (2.172). It can be seen that $k_3$ is the function of $\theta_{L_1}$ and $\theta_{L_2}$.

The polar angle of the emitted particle wave vector $\vec{k}_3^{(\mathrm{f})}$ of the third scattering in $C_3$ system is $\theta_{C_3}$; its azimuthal angle is $\varphi_3$; and its projection on the $x_3 y_3$ plane is $p_3^{(\mathrm{f})}$ (see Fig. 2.6). The contributions of the unpolarized part of the third scattering are, according to Eqs. (2.178)–(2.182), as follows:

$$I_{C_3}^0 = \frac{1}{2} \mathrm{tr}\left\{ \hat{F}_3 \hat{F}_3^+ \right\} = |A_3|^2 + |B_3|^2 \tag{2.216}$$

$$I_{C_3}^0 P_{x_3}^0 = \frac{1}{2} \mathrm{tr}\left\{ \hat{\sigma}_x \hat{F}_3 \hat{F}_3^+ \right\} = -I_{C_3}^0 P_{C_3} \sin\varphi_3 \tag{2.217}$$

$$I_{C_3}^0 P_{y_3}^0 = \frac{1}{2} \mathrm{tr}\left\{ \hat{\sigma}_y \hat{F}_3 \hat{F}_3^+ \right\} = I_{C_3}^0 P_{C_3} \cos\varphi_3 \tag{2.218}$$

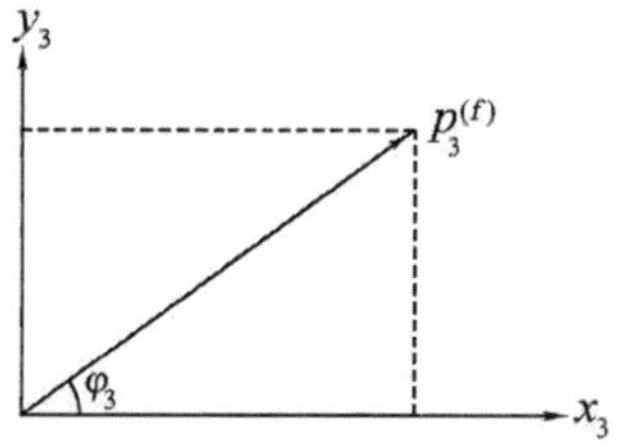


**Fig. 2.6** Projection of the emitted particle wave vector on the $x_3 y_3$ plane in $C_3$ system

$$I_{C_3}^0 P_{z_3}^0 = \frac{1}{2}\mathrm{tr}\left\{\hat{\sigma}_z \hat{F}_3 \hat{F}_3^+\right\} = 0 \tag{2.219}$$

where

$$P_{C_3} = \frac{2\,\mathrm{Re}\left(A_3 B_3^*\right)}{|A_3|^2 + |B_3|^2} \tag{2.220}$$

Comparing Eq. (2.210) with Eq. (2.169), we should note that the incident particle beams of the second scattering only in the y direction are polarized, i.e., $p_{y_2}^{(\mathrm{i})} = p_{C_2}^{(\mathrm{i})}$, $p_{x_2}^{(\mathrm{i})} = p_{z_2}^{(\mathrm{i})} = 0$, but the incident particle beams of the third scattering in $x$, $y$, and $z$ directions are all polarized, i.e., $p_{x_3}^{(\mathrm{i})}, p_{y_3}^{(\mathrm{i})}, p_{z_3}^{(\mathrm{i})}$ are all not equal to 0. We can obtain the following results according to Eqs. (2.184)–(2.189):

$$I_{C_3}^{1y} = \frac{1}{2}\mathrm{tr}\left\{\hat{F}_3 \hat{\sigma}_y \hat{F}_3^+\right\} = I_{C_3}^0 P_{C_3} \cos\varphi_3 \tag{2.221}$$

$$I_{C_3}^0 P_{x_3}^{1y} = \frac{1}{2}\mathrm{tr}\left\{\hat{\sigma}_x \hat{F}_3 \hat{\sigma}_y \hat{F}_3^+\right\} = -I_{C_3}^0 \frac{1}{2}(1 - W_{C_3}) \sin(2\varphi_3) \tag{2.222}$$

$$I_{C_3}^0 P_{y_3}^{1y} = \frac{1}{2}\mathrm{tr}\left\{\hat{\sigma}_y \hat{F}_3 \hat{\sigma}_y \hat{F}_3^+\right\} = I_{C_3}^0 \left[1 - \frac{1}{2}(1 - W_{C_3})(1 - \cos(2\varphi_3))\right] \tag{2.223}$$

$$I_{C_3}^0 P_{z_3}^{1y} = \frac{1}{2}\mathrm{tr}\left\{\hat{\sigma}_z \hat{F}_3 \hat{\sigma}_y \hat{F}_3^+\right\} = -I_{C_3}^0 Q_{C_3} \sin\varphi_3 \tag{2.224}$$

where

$$Q_{C_3} = \frac{2\mathrm{Im}\left(A_3 B_3^*\right)}{|A_3|^2 + |B_3|^2} \tag{2.225}$$

$$W_{C_3} = \frac{|A_3|^2 - |B_3|^2}{|A_3|^2 + |B_3|^2} \tag{2.226}$$

According to Eq. (2.183) we can also obtain the following formulas from Eqs. (2.176) and (2.6):

$$\begin{aligned}
\hat{F}_3 \hat{\sigma}_x \hat{F}_3^+ &= \begin{pmatrix} A_3 & -\mathrm{i}B_3 \mathrm{e}^{-\mathrm{i}\varphi_3} \\ \mathrm{i}B_3 \mathrm{e}^{\mathrm{i}\varphi_3} & A_3 \end{pmatrix} \begin{pmatrix} \mathrm{i}B_3^* \mathrm{e}^{\mathrm{i}\varphi_3} & A_3^* \\ A_3^* & -\mathrm{i}B_3^* \mathrm{e}^{-\mathrm{i}\varphi_3} \end{pmatrix} \\
&= \begin{pmatrix} \mathrm{i}\left(A_3 B_3^* \mathrm{e}^{\mathrm{i}\varphi_3} - B_3 A_3^* \mathrm{e}^{-\mathrm{i}\varphi_3}\right) & |A_3|^2 - |B_3|^2 \mathrm{e}^{-2\mathrm{i}\varphi_3} \\ |A_3|^2 - |B_3|^2 \mathrm{e}^{2\mathrm{i}\varphi_3} & \mathrm{i}\left(B_3 A_3^* \mathrm{e}^{\mathrm{i}\varphi_3} - A_3 B_3^* \mathrm{e}^{-\mathrm{i}\varphi_3}\right) \end{pmatrix}
\end{aligned} \tag{2.227}$$

$$
\begin{aligned}
\hat{F}_3\hat{\sigma}_z\hat{F}_3^+ &= \begin{pmatrix} A_3 & -\mathrm{i}B_3\mathrm{e}^{-\mathrm{i}\varphi_3} \\ \mathrm{i}B_3\mathrm{e}^{\mathrm{i}\varphi_3} & A_3 \end{pmatrix}\begin{pmatrix} A_3^* & -\mathrm{i}B_3^*\mathrm{e}^{-\mathrm{i}\varphi_3} \\ -\mathrm{i}B_3^*\mathrm{e}^{\mathrm{i}\varphi_3} & -A_3^* \end{pmatrix} \\
&= \begin{pmatrix} |A_3|^2-|B_3|^2 & -\mathrm{i}\left(A_3B_3^*-B_3A_3^*\right)\mathrm{e}^{-\mathrm{i}\varphi_3} \\ \mathrm{i}\left(B_3A_3^*-A_3B_3^*\right)\mathrm{e}^{\mathrm{i}\varphi_3} & -\left(|A_3|^2-|B_3|^2\right) \end{pmatrix}
\end{aligned} \tag{2.228}
$$

Given these, it can be found that the contributions of the polarization part of the incident particle of the third scattering in the $x$ and $z$ directions are

$$
I_{C_3}^{1x}=\frac{1}{2}\mathrm{tr}\left\{\hat{F}_3\hat{\sigma}_x\hat{F}_3^+\right\}=-I_{C_3}^0P_{C_3}\sin\varphi_3 \tag{2.229}
$$

$$
I_{C_3}^0P_{x_3}^{1x}=\frac{1}{2}\mathrm{tr}\left\{\hat{\sigma}_x\hat{F}_3\hat{\sigma}_x\hat{F}_3^+\right\}=I_{C_3}^0\left[1-\frac{1}{2}(1-W_{C_3})(1+\cos(2\varphi_3))\right] \tag{2.230}
$$

$$
I_{C_3}^0P_{y_3}^{1x}=\frac{1}{2}\mathrm{tr}\left\{\hat{\sigma}_y\hat{F}_3\hat{\sigma}_x\hat{F}_3^+\right\}=-I_{C_3}^0\frac{1}{2}(1-W_{C_3})\sin(2\varphi_3) \tag{2.231}
$$

$$
I_{C_3}^0P_{z_3}^{1x}=\frac{1}{2}\mathrm{tr}\left\{\hat{\sigma}_z\hat{F}_3\hat{\sigma}_x\hat{F}_3^+\right\}=-I_{C_3}^0Q_{C_3}\cos\varphi_3 \tag{2.232}
$$

$$
I_{C_3}^{1z}=\frac{1}{2}\mathrm{tr}\left\{\hat{F}_3\hat{\sigma}_z\hat{F}_3^+\right\}=0 \tag{2.233}
$$

$$
I_{C_3}^0P_{x_3}^{1z}=\frac{1}{2}\mathrm{tr}\left\{\hat{\sigma}_x\hat{F}_3\hat{\sigma}_z\hat{F}_3^+\right\}=I_{C_3}^0Q_{C_3}\cos\varphi_3 \tag{2.234}
$$

$$
I_{C_3}^0P_{y_3}^{1z}=\frac{1}{2}\mathrm{tr}\left\{\hat{\sigma}_y\hat{F}_3\hat{\sigma}_z\hat{F}_3^+\right\}=I_{C_3}^0Q_{C_3}\sin\varphi_3 \tag{2.235}
$$

$$
I_{C_3}^0P_{z_3}^{1z}=\frac{1}{2}\mathrm{tr}\left\{\hat{\sigma}_z\hat{F}_3\hat{\sigma}_z\hat{F}_3^+\right\}=I_{C_3}^0W_{C_3} \tag{2.236}
$$

The following physical quantities appearing in Eq. (2.211) can be obtained, by using Eqs. (2.216) – (2.219), (2.221) – (2.224), and (2.229) – (2.236):

$$
I_{C_3}^{(\mathrm{f})}=I_{C_3}^{(\mathrm{i})}I_{C_3}^0\left[1-P_{C_3}\left(p_{x_3}^{(\mathrm{i})}\sin\varphi_3-p_{y_3}^{(\mathrm{i})}\cos\varphi_3\right)\right] \tag{2.237}
$$

$$
\begin{aligned}
P_{x_3}=&-P_{C_3}\sin\varphi_3+p_{x_3}^{(\mathrm{i})}\left[1-\frac{1}{2}(1-W_{C_3})(1+\cos(2\varphi_3))\right] \\
&-p_{y_3}^{(\mathrm{i})}\frac{1}{2}(1-W_{C_3})\sin(2\varphi_3)+p_{z_3}^{(\mathrm{i})}Q_{C_3}\cos\varphi_3
\end{aligned} \tag{2.238}
$$

$$
\begin{aligned}
P_{y_3}=&P_{C_3}\cos\varphi_3-p_{x_3}^{(\mathrm{i})}\frac{1}{2}(1-W_{C_3})\sin(2\varphi_3) \\
&+p_{y_3}^{(\mathrm{i})}\left[1-\frac{1}{2}(1-W_{C_3})\,(1-\cos(2\varphi_3))\right]+p_{z_3}^{(\mathrm{i})}Q_{C_3}\sin\varphi_3
\end{aligned} \tag{2.239}
$$

$$P_{z_3} = -p_{x_3}^{(\mathrm{i})} Q_{C_3} \cos\varphi_3 - p_{y_3}^{(\mathrm{i})} Q_{C_3} \sin\varphi_3 + p_{z_3}^{(\mathrm{i})} W_{C_3} \tag{2.240}$$

In this way, the final state Stokes vector after the third scattering given by Eq. (2.211) can be obtained. And then transforming it to $L_3$ system we have:

$$S_{L_3}^{(\mathrm{f})} = X(\theta_{C_3}) S_{C_3}^{(\mathrm{f})} = X(\theta_{C_3})\, I_{C_3}^{(\mathrm{f})} \begin{pmatrix} 1 \\ P_{x_3} \\ P_{y_3} \\ P_{z_3} \end{pmatrix} \tag{2.241}$$

The method described in this section can be used to study the Monte-Carlo transport process when spin $\frac{1}{2}$ particles and spinless targets are only undergoing elastic scattering. Of course, the position of the emitted particle must be recorded at any time. For the first scattering, we select the $x$-axis on the scattering plane, which is equivalent to $\varphi_1 = 0$, and for the second and third scatterings we can choose any azimuthal angles $\varphi_2$ and $\varphi_3$; so the process is more complicated to a certain extent. When the spin $\frac{1}{2}$ particle elastic scattering with spinless target is considered only, if the required physical quantities $I^0$, $P$, $Q$, $W$ can be obtained directly from the existing nuclear databases and the transport process calculation is only carried out in the laboratory system, the problem discussed above will be simplified. However, there are still many types of reactions in the actual particle transport process. The problems of the polarization nuclear databases and polarized nucleus transport theory will be studied in the final chapter of this book.

## 2.6 Two Theoretical Methods to Study the Polarization Phenomena of Nuclear Reactions

The polarization theory of the nuclear reactions described above is based on the density matrix and the tracing method. In fact, the study of polarization theory of nuclear reactions can also use the expectation value method based on the orthonormal properties of the spin wave function [2, 7, 8]. This section will introduce the second method of directly calculating the polarization observables.

Equations (2.81) – (2.83) give the scattering amplitude of the spin $\frac{1}{2}$ projectile with the spinless target. Utilizing Eq. (2.100) and Eq. (2.16), one can obtain

$$\hat{F}^{+}\hat{F} = \left(|A|^2 + |B|^2\right)\hat{I} + 2\,\mathrm{Re}\,(AB^*)\left(\vec{n}\cdot\hat{\sigma}\right) \tag{2.242}$$

The direction of the incident particle is usually taken as the $z$-axis; thus $\vec{n}$ is the unit vector located in the $xy$ plane (see Fig. 2.1). If the incident particles are unpolarized and there is no coherence between the spin magnetic quantum numbers of the incident particles, then the differential cross sections can be obtained by use of Eqs. (2.1) and (2.242) as follows:

$$I^0 \equiv \frac{\mathrm{d}\sigma^0}{\mathrm{d}\Omega} = \frac{1}{2}\sum_{\mu}\left\langle \chi_{\frac{1}{2}\mu}\middle|\hat{F}^+\hat{F}\middle|\chi_{\frac{1}{2}\mu}\right\rangle = |A|^2 + |B|^2 \tag{2.243}$$

The following expression can be obtained utilizing Eqs. (2.100), (2.17), and (2.20):

$$\begin{aligned}\hat{F}^+\hat{\sigma}\hat{F} &= \left(A^*\hat{I} + B^*\vec{n}\cdot\hat{\sigma}\right)\hat{\sigma}\left(A\hat{I} + B\vec{n}\cdot\hat{\sigma}\right)\\ &= \left(|A|^2 - |B|^2\right)\hat{\sigma} + 2\left[\mathrm{Re}\,(AB^*) + |B|^2(\vec{n}\cdot\hat{\sigma})\right]\vec{n} - \mathrm{i}(AB^* - BA^*)\,(\vec{n}\times\hat{\sigma})\end{aligned} \tag{2.244}$$

For the unpolarized incident particles, the polarization vector of the outgoing particles can be obtained by using Eqs. (2.1), (2.6), (2.98), and (2.244):

$$\vec{P}^0_{\mathrm{out}} = \langle\hat{\sigma}\rangle^0_{\mathrm{out}} = \frac{1}{I^0}\frac{1}{2}\sum_{\mu}\left\langle \chi_{\frac{1}{2}\mu}\middle|\hat{F}^+\hat{\sigma}\hat{F}\middle|\chi_{\frac{1}{2}\mu}\right\rangle = \frac{2\,\mathrm{Re}\,(AB^*)}{|A|^2 + |B|^2}\vec{n} \tag{2.245}$$

Equation (2.243) and Eq. (2.245) are exactly the same as Eq. (2.88) and Eq. (2.110), respectively.

Equation (2.25) gives the arbitrary spin wave function $X_{\frac{1}{2}}$ of the spin $\frac{1}{2}$ particles, which is a mixed state of the basis spin functions $\chi_{\frac{1}{2}\frac{1}{2}}$ and $\chi_{\frac{1}{2}\,-\frac{1}{2}}$. It indicates that in the spin wave function $X_{\frac{1}{2}}$ there are either couplings between the states with different spin magnetic quantum numbers, or they are coherent. In this case, it can be seen from Eq. (2.36) that the incident particle beam described by $X_{\frac{1}{2}}$ is polarized. The orthogonal normalization condition requires

$$\left\langle X_{\frac{1}{2}}|X_{\frac{1}{2}}\right\rangle = 1 \tag{2.246}$$

The polarization vector of the spin $\frac{1}{2}$ incident particles is defined as

$$\vec{p}_{\mathrm{in}} = \left\langle X_{\frac{1}{2}}|\hat{\sigma}|X_{\frac{1}{2}}\right\rangle \tag{2.247}$$

The following result can be obtained using Eqs. (2.242), (2.246), and (2.247):

$$I \equiv \frac{\mathrm{d}\sigma}{\mathrm{d}\Omega} = \left\langle X_{\frac{1}{2}}\middle|\hat{F}^+\hat{F}\middle|X_{\frac{1}{2}}\right\rangle = |A|^2 + |B|^2 + 2\,\mathrm{Re}\,(AB^*)\left(\vec{n}\cdot\vec{p}_{\mathrm{in}}\right) \tag{2.248}$$

This expression is exactly the same as Eq. (2.115). It can be found that the polarization vector of the outgoing particles corresponding to the polarized incident particles is, by using Eqs. (2.244), (2.246), and (2.247), as follows:

$$\vec{P}_{\mathrm{out}} = \frac{1}{I}\left\langle X_{\frac{1}{2}} \middle| \hat{F}^{+}\hat{\sigma}\,\hat{F} \middle| X_{\frac{1}{2}} \right\rangle$$
$$= \frac{2\left[\mathrm{Re}(AB^{*}) + |B|^{2}\left(\vec{n}\cdot\vec{p}_{\mathrm{in}}\right)\right]\vec{n} + \left(|A|^{2} - |B|^{2}\right)\vec{p}_{\mathrm{in}} + 2\mathrm{Im}(AB^{*})\left(\vec{n}\times\vec{p}_{\mathrm{in}}\right)}{|A|^{2} + |B|^{2} + 2\mathrm{Re}(AB^{*})\left(\vec{n}\cdot\vec{p}_{\mathrm{in}}\right)} \tag{2.249}$$

This expression is exactly the same as Eq. (2.126).

The above results show that for spinless target, the polarization theoretical formulas of the spin $\frac{1}{2}$ particles, obtained by use of the first equation of Eq. (1.88) belonging to the expectation value method and the last two equations of Eq.(1.89) belonging to the tracing method, are exactly the same.

## 2.7 Polarization Theory for $\overrightarrow{\frac{1}{2}} + \mathrm{A} \rightarrow \overrightarrow{\frac{1}{2}} + \mathrm{B}$ Reactions

For the target and the residual nucleus with non-zero spin, Eq. (3.8.20) in Ref. [2] gives the differential cross section of the *j*–*j* coupling mode without considering the polarization as

$$\frac{\mathrm{d}\sigma_{\alpha' n', \alpha n}}{\mathrm{d}\Omega} = \frac{1}{\hat{i}^{2}\,\hat{I}^{2}} \sum_{\mu M_I \mu' M_I'} \left| f_{\alpha' n' \mu' M_I', \, \alpha n \mu M_I}(\Omega) \right|^{2} \tag{2.250}$$

It can be seen from the above equation that in the case of the unpolarized incident particle, target nucleus, and residual nucleus, the outgoing particles belong to a special polarized particle state, in which all the outgoing particle states with different spin quantum numbers of the incident particle, the outgoing particle, the target nucleus, and the residual nucleus are incoherent, but the shares of the outgoing particle states with different spin quantum numbers of the incident particle, the outgoing particle, the target nucleus, and the residual nucleus may be different. However, in Eq. (2.250) after summing the final state particle spin magnetic quantum numbers and averaging the initial state particle spin magnetic quantum numbers, it becomes the total differential cross section of the outgoing particles. The reaction amplitude appearing in Eq. (2.250) has been given by Eq. (3.8.12) in Ref. [2] as follows:

$$
\begin{aligned}
f_{\alpha'n'\mu'M'_I,\,\alpha n\mu M_I}(\Omega) = {} & f_C(\theta)\,\delta_{\alpha'n',\,\alpha n}\delta_{\mu'M'_I,\,\mu M_I} \\
& + \frac{\mathrm{i}\sqrt{\pi}}{k}\sum_{ljl'j'J}\hat{l}\,\mathrm{e}^{\mathrm{i}(\sigma_l+\sigma_{l'})}\left(\delta_{\alpha'n'l'j',\,\alpha nlj} - S^J_{\alpha'n'l'j',\,\alpha nlj}\right) \\
& C^{j\mu}_{l0\ i\mu} C^{J\ M}_{j\mu\ IM_I} C^{j'\ m'_j}_{l'm'_l\ i'\mu'} C^{J\ M}_{j'm'_j\ I'M'_I} \mathrm{Y}_{l'm'_l}(\theta,\varphi)
\end{aligned}
\tag{2.251}
$$

The simplification symbol $\hat{i} \equiv \sqrt{2i+1}$ is introduced in the above expressions. $\alpha$ and $\alpha'$ represent the types of incident particles and outgoing particles; $n$ and $n'$ represent the level labels of the target and the residual nucleus, respectively. $S^J_{\alpha'n'l'j',\alpha nlj}$ is a $S$ matrix element of the nuclear reactions. For the $S$–$L$ coupling mode, Eq. (2.250) can still be used. However its reaction amplitude needs to be changed to the form given by Eq. (13.8.13) in Ref. [9] as follows:

$$
\begin{aligned}
f_{\alpha'n'\mu'M'_I,\,\alpha n\mu M_I}(\Omega) = {} & f_C(\theta)\,\delta_{\alpha'n',\alpha n}\delta_{\mu'M'_I,\,\mu M_I} \\
& + \frac{\mathrm{i}\sqrt{\pi}}{k}\sum_{lS\,l'S'J}\hat{l}\mathrm{e}^{\mathrm{i}\,(\sigma_l+\sigma_{l'})}\left(\delta_{\alpha'n'l'S',\alpha nlS} - S^J_{\alpha'n'l'S',\alpha nlS}\right) \\
& C^{J\ M}_{l0\ SM} C^{S\ M}_{i\mu\ IM_I} C^{J\ M}_{l'm'_l\ S'M'_S} C^{S'\ M'_S}_{i'\mu'\ I'M'_I} \mathrm{Y}_{l'm'_l}(\theta,\varphi)
\end{aligned}
\tag{2.252}
$$

Nucleon-nucleon elastic scattering is only a special case of the nuclear reaction described by Eq. (2.252), which is equivalent to taking $\alpha' = \alpha$, $n' = n$, $i = i' = I = I' = \frac{1}{2}$ in the above formula.

Eq. (2.73) has given

$$
\mathrm{Y}_{l'm'_l}(\theta,\varphi) = (-1)^{m'_l}\sqrt{\frac{2l'+1}{4\pi}}\sqrt{\frac{(l'-m'_l)!}{(l'+m'_l)!}}\mathrm{P}^{m'_l}_{l'}(\cos\theta)\,\mathrm{e}^{\mathrm{i}m'_l\varphi}
\tag{2.253}
$$

From the C–G coefficients in Eq. (2.251) or Eq. (2.252), the following relation can be obtained:

$$
m'_l = \mu + M_I - \mu' - M'_I
\tag{2.254}
$$

In the case of both incident particles and outgoing particles with spin $\frac{1}{2}$, the following symbols are introduced for the $j$–$j$ coupling mode and $S$–$L$ coupling mode, respectively:

$$f^{ll'J}_{\mu'\mu}(m'_l) = \hat{l}\,\hat{l}'\mathrm{e}^{\mathrm{i}(\sigma_l+\sigma_{l'})}\sqrt{\frac{(l'-m'_l)!}{(l'+m'_l)!}}\sum_{jj'} C^{j\ \mu}_{l0\ \frac{1}{2}\mu} C^{J\ M}_{j\mu\ IM_I} C^{j'\ m'_j}_{l'm'_l\ \frac{1}{2}\mu'} C^{J\ M}_{j'm'_j\ I'M'_I}\left(\delta_{\alpha' n' l' j', \alpha n l j} - S^J_{\alpha' n' l' j', \alpha n l j}\right) \tag{2.255}$$

$$f^{ll'J}_{\mu'\mu}(m'_l) = \hat{l}\,\hat{l}'\mathrm{e}^{\mathrm{i}(\sigma_l+\sigma_{l'})}\sqrt{\frac{(l'-m'_l)!}{(l'+m'_l)!}}\sum_{SS'} C^{J\ M}_{l0\ SM} C^{S\ M}_{\frac{1}{2}\mu\ IM_I} C^{J\ M}_{l'm'_l\ S'M'_S} C^{S'\ M'_S}_{\frac{1}{2}\mu'\ I'M'_I}\left(\delta_{\alpha' n' l' S', \alpha n l S} - S^J_{\alpha' n' l' S', \alpha n l S}\right) \tag{2.256}$$

Therefore, the reaction amplitude can be uniformly written in the following form under both angular momentum coupling modes:

$$\begin{aligned} f_{\alpha' n'\mu' M'_I, \alpha n \mu M_I}(\Omega) = {} & f_{\mathrm{C}}(\theta)\, \delta_{\alpha' n', \alpha n}\delta_{\mu' M'_I, \mu M_I} \\ & + \frac{\mathrm{i}}{2k}(-1)^{\mu+M_I-\mu'-M'_I}\sum_{ll'J} f^{ll'J}_{\mu'\mu}(\mu+M_I-\mu'-M'_I) \\ & \mathrm{P}^{\mu+M_I-\mu'-M'_I}_{l'}(\cos\theta)\, \mathrm{e}^{\mathrm{i}(\mu+M_I-\mu'-M'_I)\varphi} \end{aligned} \tag{2.257}$$

Note that $\delta_{\mu' M'_I,\ \mu M_I}$ is contained in the first item on right end of the above formula, so that it can be multiplied by a factor $e^{\mathrm{i}(M_I - M'_I)\varphi}$ which is equal to 1. We define

$$\begin{aligned} A_{M'_I M_I}(\theta) = {} & f_{\mathrm{C}}(\theta)\, \delta_{\alpha' n', \alpha n}\delta_{M'_I,\, M_I} + \frac{\mathrm{i}}{2k}(-1)^{M_I-M'_I} \\ & \sum_{ll'J} f^{ll'J}_{\frac{1}{2}\ \frac{1}{2}}(M_I-M'_I)\ \mathrm{P}^{M_I-M'_I}_{l'}(\cos\theta) \end{aligned} \tag{2.258}$$

$$B_{M'_I M_I}(\theta) = -\frac{1}{2k}(-1)^{M_I-M'_I-1}\sum_{ll'J} f^{ll'J}_{\frac{1}{2}\ -\frac{1}{2}}(M_I-M'_I-1)\ \mathrm{P}^{M_I-M'_I-1}_{l'}(\cos\theta) \tag{2.259}$$

$$C_{M'_I M_I}(\theta) = \frac{1}{2k}(-1)^{M_I-M'_I+1}\sum_{ll'J} f^{ll'J}_{-\frac{1}{2}\ \frac{1}{2}}(M_I-M'_I+1)\ \mathrm{P}^{M_I-M'_I+1}_{l'}(\cos\theta) \tag{2.260}$$

$$\begin{aligned} D_{M'_I M_I}(\theta) = {} & f_{\mathrm{C}}(\theta)\, \delta_{\alpha' n', \alpha n}\delta_{M'_I,\, M_I} + \frac{\mathrm{i}}{2k}(-1)^{M_I-M'_I} \\ & \sum_{ll'J} f^{ll'J}_{-\frac{1}{2}\ -\frac{1}{2}}(M_I-M'_I)\mathrm{P}^{M_I-M'_I}_{l'}(\cos\theta) \end{aligned} \tag{2.261}$$

Now introducing the following simplification symbols

$$F_{\mu'\mu}(\theta, \varphi) \equiv f_{\alpha' n'\mu' M'_I,\ \alpha n\mu M_I}(\Omega) \tag{2.262}$$

then we obtain

$$F_{\frac{1}{2}\,\frac{1}{2}}(\theta,\varphi)=A_{M_I'M_I}(\theta)\,\mathrm{e}^{\mathrm{i}(M_I-M_I')\varphi} \tag{2.263}$$

$$F_{\frac{1}{2}\,-\frac{1}{2}}(\theta,\varphi)=-\mathrm{i}B_{M_I'M_I}(\theta)\,\mathrm{e}^{\mathrm{i}(M_I-M_I')\varphi}\mathrm{e}^{-\mathrm{i}\varphi} \tag{2.264}$$

$$F_{-\frac{1}{2}\,\frac{1}{2}}(\theta,\varphi)=\mathrm{i}C_{M_I'M_I}(\theta)\,\mathrm{e}^{\mathrm{i}(M_I-M_I')\varphi}\mathrm{e}^{\mathrm{i}\varphi} \tag{2.265}$$

$$F_{-\frac{1}{2}\,-\frac{1}{2}}(\theta,\varphi)=D_{M_I'M_I}(\theta)\,\mathrm{e}^{\mathrm{i}(M_I-M_I')\varphi} \tag{2.266}$$

When the foot note $M_I'M_I$ and argument $(\theta)$ of $A$, $B$, $C$, and $D$ are omitted, the reaction amplitude with definite $M_I'M_I$, in which arguments $(\theta,\varphi)$ are emitted, can be written as

$$\begin{aligned}\hat{F}&=\begin{pmatrix}A & -\mathrm{i}B\mathrm{e}^{-\mathrm{i}\varphi}\\ \mathrm{i}C\mathrm{e}^{\mathrm{i}\varphi} & D\end{pmatrix}\mathrm{e}^{\mathrm{i}(M_I-M_I')\varphi},\\ \hat{F}^{+}&=\begin{pmatrix}A^* & -\mathrm{i}C^*\mathrm{e}^{-\mathrm{i}\varphi}\\ \mathrm{i}B^*\mathrm{e}^{\mathrm{i}\varphi} & D^*\end{pmatrix}\mathrm{e}^{-\mathrm{i}(M_I-M_I')\varphi}\end{aligned} \tag{2.267}$$

We first study the case where the target and the residual nucleus are all unpolarized. In this case the different $M_I$ components are incoherent and the different $M_I'$ components are also incoherent. For the components with definite $M_I'M_I$, from Eq. (2.267) one can obtain

$$\begin{aligned}\hat{F}\hat{F}^{+}&=\begin{pmatrix}A & -\mathrm{i}B\mathrm{e}^{-\mathrm{i}\varphi}\\ \mathrm{i}C\mathrm{e}^{\mathrm{i}\varphi} & D\end{pmatrix}\begin{pmatrix}A^* & -\mathrm{i}C^*\mathrm{e}^{-\mathrm{i}\varphi}\\ \mathrm{i}B^*\mathrm{e}^{\mathrm{i}\varphi} & D^*\end{pmatrix}\\ &=\begin{pmatrix}|A|^2+|B|^2 & -\mathrm{i}(AC^*+BD^*)\,\mathrm{e}^{-\mathrm{i}\varphi}\\ \mathrm{i}(CA^*+DB^*)\,\mathrm{e}^{\mathrm{i}\varphi} & |C|^2+|D|^2\end{pmatrix}\end{aligned} \tag{2.268}$$

For the sake of simplicity, we select the $y$-axis in the direction $\vec{n}$ perpendicular to the reaction plane. In this case there is $\varphi=0$; thus Eqs. (2.267) and (2.268) are simplified, respectively, as

$$\hat{F}=\begin{pmatrix}A & -\mathrm{i}B\\ \mathrm{i}C & D\end{pmatrix},\quad \hat{F}^{+}=\begin{pmatrix}A^* & -\mathrm{i}C^*\\ \mathrm{i}B^* & D^*\end{pmatrix} \tag{2.269}$$

$$\hat{F}\hat{F}^{+}=\begin{pmatrix}|A|^2+|B|^2 & -\mathrm{i}(AC^*+BD^*)\\ \mathrm{i}(CA^*+DB^*) & |C|^2+|D|^2\end{pmatrix} \tag{2.270}$$

From Eq. (2.270) we can obtain

$$I^0 = \frac{1}{2}\mathrm{tr}\{\hat{F}\hat{F}^+\} = \frac{1}{2}\left(|A|^2 + |B|^2 + |C|^2 + |D|^2\right) \tag{2.271}$$

From Eq. (2.269) and Eq. (2.6) we can also obtain

$$\hat{\sigma}_x\hat{F}^+ = \begin{pmatrix} \mathrm{i}B^* & D^* \\ A^* & -\mathrm{i}C^* \end{pmatrix}, \quad \hat{\sigma}_y\hat{F}^+ = \begin{pmatrix} B^* & -\mathrm{i}D^* \\ \mathrm{i}A^* & C^* \end{pmatrix}, \quad \hat{\sigma}_z\hat{F}^+ = \begin{pmatrix} A^* & -\mathrm{i}C^* \\ -\mathrm{i}B^* & -D^* \end{pmatrix} \tag{2.272}$$

$$\begin{aligned} \hat{F}\,\hat{\sigma}_x\hat{F}^+ &= \begin{pmatrix} \mathrm{i}(AB^* - BA^*) & AD^* - BC^* \\ -CB^* + DA^* & \mathrm{i}(CD^* - DC^*) \end{pmatrix} \\ \hat{F}\,\hat{\sigma}_y\hat{F}^+ &= \begin{pmatrix} AB^* + BA^* & -\mathrm{i}(AD^* + BC^*) \\ \mathrm{i}(CB^* + DA^*) & CD^* + DC^* \end{pmatrix} \\ \hat{F}\,\hat{\sigma}_z\hat{F}^+ &= \begin{pmatrix} |A|^2 - |B|^2 & -\mathrm{i}(AC^* - BD^*) \\ \mathrm{i}(CA^* - DB^*) & |C|^2 - |D|^2 \end{pmatrix} \end{aligned} \tag{2.273}$$

Then from Eq. (2.273) the following results can be obtained:

$$\begin{aligned} A_x &= \frac{1}{2I^0}\mathrm{tr}\left\{\hat{F}\,\hat{\sigma}_x\hat{F}^+\right\} = -\frac{1}{I^0}\mathrm{Im}(AB^* + CD^*), \\ A_y &= \frac{1}{2I^0}\mathrm{tr}\left\{\hat{F}\,\hat{\sigma}_y\hat{F}^+\right\} = \frac{1}{I^0}\,\mathrm{Re}\,(AB^* + CD^*), \\ A_z &= \frac{1}{2I^0}\mathrm{tr}\left\{\hat{F}\,\hat{\sigma}_z\hat{F}^+\right\} = \frac{1}{2I^0}\left(|A|^2 - |B|^2 + |C|^2 - |D|^2\right) \end{aligned} \tag{2.274}$$

From Eq. (2.270) and Eq. (2.6) we can also obtain

$$\begin{aligned} P^{0x} &= \frac{1}{2I^0}\mathrm{tr}\left\{\hat{\sigma}_x\hat{F}\hat{F}^+\right\} = \frac{\mathrm{i}}{2I^0}(CA^* + DB^* - AC^* - BD^*) = \frac{1}{I^0}\mathrm{Im}(AC^* + BD^*), \\ P^{0y} &= \frac{1}{2I^0}\mathrm{tr}\left\{\hat{\sigma}_y\hat{F}\hat{F}^+\right\} = \frac{1}{2I^0}(CA^* + DB^* + AC^* + BD^*) = \frac{1}{I^0}\,\mathrm{Re}\,(AC^* + BD^*), \\ P^{0z} &= \frac{1}{2I^0}\mathrm{tr}\left\{\hat{\sigma}_z\hat{F}\hat{F}^+\right\} = \frac{1}{2I^0}\left(|A|^2 + |B|^2 - |C|^2 - |D|^2\right) \end{aligned} \tag{2.275}$$

We define the polarization transfer coefficient as

$$K_i^k = \frac{1}{2I^0}\mathrm{tr}\left\{\hat{\sigma}_k\hat{F}\,\hat{\sigma}_i\hat{F}^+\right\}, \qquad i,k = x,y,z \tag{2.276}$$

The following results can be obtained from Eqs. (2.276), (2.273), and (2.6):

$$
\begin{aligned}
K_x^x &= \frac{1}{2I^0}(-CB^* + DA^* + AD^* - BC^*) = \frac{1}{I^0}\operatorname{Re}(AD^* - BC^*),\\
K_x^y &= \frac{\mathrm{i}}{2I^0}(CB^* - DA^* + AD^* - BC^*) = -\frac{1}{I^0}\operatorname{Im}(AD^* - BC^*),\\
K_x^z &= \frac{\mathrm{i}}{2I^0}(AB^* - BA^* - CD^* + DC^*) = -\frac{1}{I^0}\operatorname{Im}(AB^* - CD^*),\\
K_y^x &= \frac{\mathrm{i}}{2I^0}(CB^* + DA^* - AD^* - BC^*) = \frac{1}{I^0}\operatorname{Im}(AD^* + BC^*),\\
K_y^y &= \frac{1}{2I^0}(CB^* + DA^* + AD^* + BC^*) = \frac{1}{I^0}\operatorname{Re}(AD^* + BC^*),\\
K_y^z &= \frac{1}{2I^0}(AB^* + BA^* - CD^* - DC^*) = \frac{1}{I^0}\operatorname{Re}(AB^* - CD^*),\\
K_z^x &= \frac{\mathrm{i}}{2I^0}(CA^* - DB^* - AC^* + BD^*) = \frac{1}{I^0}\operatorname{Im}(AC^* - BD^*),\\
K_z^y &= \frac{1}{2I^0}(CA^* - DB^* + AC^* - BD^*) = \frac{1}{I^0}\operatorname{Re}(AC^* - BD^*),\\
K_z^z &= \frac{1}{2I^0}\left(|A|^2 - |B|^2 - |C|^2 + |D|^2\right)
\end{aligned}
\tag{2.277}
$$

Since $A$, $B$, $C$, and $D$ introduced by Eqs. (2.258) –(2.261) have foot note $M_I'M_I$, it follows that $I^0$, $A_i$, $P^{0k}$, $K_i^k$ $(i, k = x, y, z)$ obtained previously also have foot note $M_I'M_I$. In the case that both the target and residual nucleus are unpolarized, the simplification summation symbol for averaging to $M_I$ and summing to $M_I'$ concurrently is introduced as

$$
\widetilde{\Sigma} \equiv \frac{1}{2I+1}\sum_{M_I\, M_I'} \tag{2.278}
$$

The differential cross section of the unpolarized incident particle can be obtained by Eq. (2.271) as

$$
\overline{I}^0 = \widetilde{\Sigma}\, I^0 \tag{2.279}
$$

And the analyzing power of the incident particles can be obtained by Eq. (2.274) as

$$
\overline{A}_i = \frac{1}{\overline{I}^0}\widetilde{\Sigma}\, I^0 A_i, \qquad i = x, y, z \tag{2.280}
$$

According to Eq. (2.63) and Eq. (2.34), the differential cross section of the polarized incident particles can be obtained as

$$\bar{I}=\bar{I}^0\left(1+\sum_{i=x,y,z} p_i\bar{A}_i\right) \tag{2.281}$$

where $p_i$ $(i = x, y, z)$ are the polarization rates of the incident particles. The polarization rates of the outgoing particles corresponding to the unpolarized incident particles can be obtained by Eq. (2.275) as

$$\overline{P}^{0k} = \frac{1}{\overline{I}^0}\tilde{\Sigma}\, I^0 P^{0k}, \qquad k=x,y,z \tag{2.282}$$

The polarization transfer coefficient can be obtained by Eq. (2.277):

$$\overline{K}_i^k = \frac{1}{\overline{I}^0}\tilde{\Sigma}\, I^0 K_i^k, \qquad i,k=x,y,z \tag{2.283}$$

From the above three formulas and Eq. (2.63), the polarization rates of the outgoing particles corresponding to the polarized incident particles can be obtained as

$$\overline{P}^k = \frac{\overline{I}^0}{\overline{I}}\left(\overline{P}^{0k}+\sum_{i=x,y,z} p_i\overline{K}_i^k\right), \qquad k=x,y,z \tag{2.284}$$

According to Eq. (3.2.8) and Eq. (3.2.12) given in Ref. [2], when $\varphi = 0$ one can obtain

$$\mathrm{Y}_{lm}(\theta,0)=(-1)^m\sqrt{\frac{(2l+1)(l-m)!}{4\pi(l+m)!}}\mathrm{P}_l^m(\cos\theta) \tag{2.285}$$

$$\mathrm{Y}_{l\ -m}(\theta,0)=(-1)^m\sqrt{\frac{(2l+1)(l+m)!}{4\pi(l-m)!}}\mathrm{P}_l^{-m}(\cos\theta) \tag{2.286}$$

Based on Eq. (3.2.10) given in Ref. [2], one can also obtain

$$\mathrm{P}_l^m(\cos\theta)=(-1)^m\frac{(l+m)!}{(l-m)!}\mathrm{P}_l^{-m}(\cos\theta) \tag{2.287}$$

Substituting Eq. (2.287) into Eq. (2.285) and using Eq. (2.286), the following relation can be obtained:

$$\mathrm{Y}_{lm}(\theta,0)=(-1)^m\mathrm{Y}_{l\ -m}(\theta,0) \tag{2.288}$$

In the next analyses we will use Eq. (2.252) belonging to the *S*–*L* coupling mode. First we have the following relations based on the properties of the C–G coefficient:

$$C_{\frac{1}{2}\mu\ \ IM_I}^{S\ \ \mu+M_I} = (-1)^{I+\frac{1}{2}-S} C_{\frac{1}{2}\ -\mu\ \ I\ -M_I}^{S\ \ -\mu-M_I}, \quad C_{\frac{1}{2}\mu'\ \ I'M'_I}^{S'\ \ \mu'+M'_I} = (-1)^{I'+\frac{1}{2}-S'} C_{\frac{1}{2}\ -\mu'\ \ I'\ -M'_I}^{S'\ \ -\mu'-M'_I},$$
$$C_{l0\ \ SM}^{J\ M} = (-1)^{l+S-J} C_{l0\ \ S\ -M}^{J\ -M}, \quad C_{l'm'_l\ \ S'M'_S}^{J\ M} = (-1)^{l'+S'-J} C_{l'\ -m'_l\ \ S'\ -M'_S}^{J\ -M} \tag{2.289}$$

We divide Eq. (2.252) into two items and study the second item first. Using Eq. (2.288) and Eq. (2.289), for the part which is not related with angle $\varphi$, the following phase factor comes out when all spin magnetic quantum numbers change their signs

$$(-1)^{l+S-J+l'+S'-J+I+\frac{1}{2}-S+I'+\frac{1}{2}-S'+\mu+M_I-\mu'-M'_I}$$
$$=(-1)^{2J-1+l+I+l'+I'+\mu+M_I-\mu'-M'_I}$$

Let $\pi_i$,$\pi'_i$,$\pi_I$, and $\pi'_I$ represent the parities of the incident particle, outgoing particle, target, and residual nucleus, respectively. If $\pi_i\pi_I=\pi'_i\pi'_I$, $l$ and $l'$ should all be even or odd, and in this case let $\Pi=0$. On the other hand, if $\pi_i\pi_I=-\pi'_i\pi'_I$, $l$ and $l'$ should be that for which one is even and another is odd, and in this case let $\Pi=1$. Moreover we can express the above results in a unified form as

$$\Pi = \begin{cases} 0 & \text{when} \quad \pi_i\pi_I = \pi'_i\pi'_I \\ 1 & \text{when} \quad \pi_i\pi_I = -\pi'_i\pi'_I \end{cases} \tag{2.290}$$

In the case that the spins of both the incident particle and the outgoing particle are $\frac{1}{2}$, the target spin $I$ and the residual nucleus spin $I'$ must be the integers or the semi-odd numbers at the same time. If $I$ and $I'$ are all integers, then $J$ is a semi-odd number, so $(-1)^{2J-1}=1$, and in this case let $\Lambda=0$. If $I$ and $I'$ are all semi-odd numbers, then $J$ is an integer, so $(-1)^{2J-1}=-1$, and in this case let $\Lambda=1$. Moreover we can express the above results in a unified form as

$$\Lambda = \begin{cases} 0 & \text{when } I \text{ and } I' \text{ are all integers} \\ 1 & \text{when } I \text{ and } I' \text{ are all semi-odd numbers} \end{cases} \tag{2.291}$$

Thus we can get the following relation.

$$f^{(2)}_{\alpha' n'\mu' M'_I, \alpha n\mu M_I}(\theta) = (-1)^{\Pi+\Lambda+I+I'+\mu+M_I-\mu'-M'_I} f^{(2)}_{\alpha' n'-\mu'-M'_I, \alpha n-\mu-M_I}(\theta) \tag{2.292}$$

Now we come back to study the first item on the right end of Eq. (2.252). For elastic scattering, we have $\Pi=0$, $I'=I$, $(-1)^{\Lambda+2I}=1$, and the δ symbol requires $\mu=\mu'$, $M_I=M'_I$, so $f^{(1)}$ also satisfies the relation given by Eq. (2.292). Thus, for the independent part with angle $\varphi$ of the total reaction amplitude $f$, there is the following relation:

$$f_{\alpha' n' \mu' M_I', \alpha n \mu M_I}(\theta) = (-1)^{\Pi + \Lambda + I + I' + \mu + M_I - \mu' - M_I'} f_{\alpha' n' - \mu' - M_I', \alpha n - \mu - M_I}(\theta) \tag{2.293}$$

The following relations can be obtained from Eqs. (2.263)–(2.266):

$$\begin{aligned} &A_{M_I' M_I}(\theta) = f_{\alpha' n' \frac{1}{2} M_I', \alpha n \frac{1}{2} M_I}(\theta), \quad B_{M_I' M_I}(\theta) = \mathrm{i} f_{\alpha' n' \frac{1}{2} M_I', \alpha n - \frac{1}{2} M_I}(\theta), \\ &C_{M_I' M_I}(\theta) = -\mathrm{i} f_{\alpha' n' - \frac{1}{2} M_I', \alpha n \frac{1}{2} M_I}(\theta), \quad D_{M_I' M_I}(\theta) = f_{\alpha' n' - \frac{1}{2} M_I', \alpha n - \frac{1}{2} M_I}(\theta) \end{aligned} \tag{2.294}$$

According to Eq. (2.293) and Eq. (2.294) we have

$$D_{M_I' M_I}(\theta) = (-1)^{\Pi + \Lambda + I + I' + M_I - M_I'} A_{-M_I' \; -M_I}(\theta) \tag{2.295}$$

$$C_{M_I' M_I}(\theta) = (-1)^{\Pi + \Lambda + I + I' + M_I - M_I'} B_{-M_I' \; -M_I}(\theta) \tag{2.296}$$

The phase factor $L \equiv \Pi + \Lambda + I + I' + M_I - M_I'$ appearing in Eq. (2.295) and Eq. (2.296) must be an integer, so there must be $(-1)^{2L} = 1$. In the summation symbol $\widetilde{\Sigma}$ given by Eq. (2.278), the sum of $M_I$ and the sum of $-M_I$ are all from $-I$ to $I$, and the sum of $M'_I$ and the sum of $-M_I'$ are all from $-I'$ to $I'$, so using Eqs. (2.295) and (2.296) the following results can be proved:

$$\begin{aligned} &\widetilde{\Sigma}\mathrm{Im}(AB^* + CD^*) = \widetilde{\Sigma}\mathrm{Im}(AB^* + BA^*) = 0 \\ &\widetilde{\Sigma}\,\mathrm{Re}\,(AB^* + CD^*) = \widetilde{\Sigma}\,\mathrm{Re}\,(AB^* + BA^*) = 2\widetilde{\Sigma}\,\mathrm{Re}\,(AB^*) \\ &\widetilde{\Sigma}\mathrm{Im}(AC^* + BD^*) = \widetilde{\Sigma}\mathrm{Im}(AC^* + CA^*) = 0 \\ &\widetilde{\Sigma}\,\mathrm{Re}\,(AC^* + BD^*) = \widetilde{\Sigma}\,\mathrm{Re}\,(AC^* + CA^*) = 2\widetilde{\Sigma}\,\mathrm{Re}\,(AC^*) \\ &\widetilde{\Sigma}\mathrm{Im}(AB^* - CD^*) = \widetilde{\Sigma}\mathrm{Im}(AB^* - BA^*) = 2\widetilde{\Sigma}\mathrm{Im}(AB^*) \\ &\widetilde{\Sigma}\,\mathrm{Re}\,(AB^* - CD^*) = \widetilde{\Sigma}\,\mathrm{Re}\,(AB^* - BA^*) = 0 \\ &\widetilde{\Sigma}\mathrm{Im}(AC^* - BD^*) = \widetilde{\Sigma}\mathrm{Im}(AC^* - CA^*) = 2\widetilde{\Sigma}\mathrm{Im}(AC^*) \\ &\widetilde{\Sigma}\,\mathrm{Re}\,(AC^* - BD^*) = \widetilde{\Sigma}\,\mathrm{Re}\,(AC^* - CA^*) = 0 \\ &\widetilde{\Sigma}|D|^2 = \widetilde{\Sigma}|A|^2, \qquad \widetilde{\Sigma}|C|^2 = \widetilde{\Sigma}|B|^2 \end{aligned} \tag{2.297}$$

In addition, the following expression can also be proved:

$$\begin{aligned} \sum_{M_I M_I'} A_{M_I' M I} D^*_{M_I' M_I} &= \frac{1}{2} \sum_{M_I M_I'} \left( A_{M_I' M I} D^*_{M_I' M_I} + A_{-M_I' \; -M} D^*_{-M_I' \; -M_I} \right) \\ &= \frac{1}{2} \sum_{M_I M_I'} \left( A_{M_I' M I} D^*_{M_I' M_I} + D_{M_I' M} A^*_{M_I' M_I} \right) \end{aligned} \tag{2.298}$$

From the above formula we can see that $AD^*$ is a real number. Similarly, $BC^*$ should be a real number. Given these, we can get

$$\tilde{\Sigma}\mathrm{Im}(AD^* \pm BC^*) = 0 \tag{2.299}$$

The following relation can be obtained utilizing Eqs. (2.279), (2.271), and (2.297):

$$\bar{I}^0 = \tilde{\Sigma}\left(|A|^2 + |B|^2\right) \tag{2.300}$$

The following relations can also be obtained by use of Eqs. (2.280), (2.274), and (2.297):

$$\bar{A}_x = \bar{A}_z = 0, \qquad \bar{A}_y = \frac{2}{\bar{I}^0}\tilde{\Sigma}\ \mathrm{Re}\,(AB^*) \tag{2.301}$$

It follows that Eq. (2.281) can be rewritten as

$$\bar{I} = \bar{I}^0\left(1 + p_y\bar{A}_y\right) \tag{2.302}$$

We can also obtain the following relations by using Eqs. (2.282), (2.275), and (2.297):

$$\bar{P}^{0x} = \bar{P}^{0z} = 0, \qquad \bar{P}^{0y} = \frac{2}{\bar{I}^0}\tilde{\Sigma}\ \mathrm{Re}\,(AC^*) \tag{2.303}$$

And the following results can be obtained utilizing Eqs. (2.283), (2.277), (2.297), and (2.299):

$$\begin{aligned}
&\bar{K}_x^y = \bar{K}_y^x = \bar{K}_y^z = \bar{K}_z^y = 0, \quad \bar{K}_x^x = \frac{1}{\bar{I}^0}\tilde{\Sigma}\ \mathrm{Re}\,(AD^* - BC^*),\\
&\bar{K}_x^z = -\frac{2}{\bar{I}^0}\tilde{\Sigma}\,\mathrm{Im}(AB^*), \quad \bar{K}_y^y = \frac{1}{\bar{I}^0}\tilde{\Sigma}\ \mathrm{Re}\,(AD^* + BC^*),\\
&\bar{K}_z^x = \frac{2}{\bar{I}^0}\tilde{\Sigma}\,\mathrm{Im}(AC^*), \quad \bar{K}_z^z = \frac{1}{\bar{I}^0}\tilde{\Sigma}\left(|A|^2 - |B|^2\right)
\end{aligned} \tag{2.304}$$

In addition Eq. (2.284) can be rewritten as

$$\begin{aligned}
&\bar{P}^x = \frac{\bar{I}^0}{\bar{I}}\left(p_x\bar{K}_x^x + p_z\bar{K}_z^x\right), \quad \bar{P}^y = \frac{\bar{I}^0}{\bar{I}}\left(\bar{P}^{0y} + p_y\bar{K}_y^y\right),\\
&\bar{P}^z = \frac{\bar{I}^0}{\bar{I}}\left(p_x\bar{K}_x^z + p_z\bar{K}_z^z\right)
\end{aligned} \tag{2.305}$$

For a definite component $M_I'M_I$, from Eq. (2.271) one can get

$$\begin{aligned}(I^0)^2 = &\frac{1}{4}\Big[|A|^4 + |B|^4 + |C|^4 + |D|^4 + 2\Big(|A|^2|B|^2 + |A|^2|C|^2 + |A|^2|D|^2 \\ &+ |B|^2|C|^2 + |B|^2|D|^2 + |C|^2|D|^2\Big)\Big]\end{aligned} \tag{2.306}$$

From Eq. (2.277) the following relations can be obtained:

$$\begin{aligned}(K_z^z)^2 = &\frac{1}{4(I^0)^2}\Big[|A|^4 + |B|^4 + |C|^4 + |D|^4 + 2\Big(|A|^2|D|^2 + |B|^2|C|^2 \\ &- |A|^2|B|^2 - |A|^2|C|^2 - |B|^2|D|^2 - |C|^2|D|^2\Big)\Big]\end{aligned} \tag{2.307}$$

$$\begin{aligned}(K_x^x)^2 &= \frac{1}{4(I^0)^2}(AD^* + A^*D - CB^* - C^*B)^2 \\ &= \frac{1}{4(I^0)^2}\Big[(AD^* + A^*D)^2 + (CB^* + C^*B)^2 \\ &- 2AD^*CB^* - 2AD^*C^*B - 2A^*DCB^* - 2A^*DC^*B\Big]\end{aligned} \tag{2.308}$$

$$\begin{aligned}\left(K_y^y\right)^2 &= \frac{1}{4(I^0)^2}(AD^* + A^*D + CB^* + C^*B)^2 \\ &= \frac{1}{4(I^0)^2}\Big[(AD^* + A^*D)^2 + (CB^* + C^*B)^2 \\ &+2AD^*CB^* + 2AD^*C^*B + 2A^*DCB^* + 2A^*DC^*B\Big]\end{aligned} \tag{2.309}$$

Notice

$$\begin{aligned}&\mathrm{Re}\,(AB^* + CD^*) = \mathrm{Re}\,(AB^* + C^*D), \\ &\mathrm{Im}(AB^* - CD^*) = \mathrm{Im}(AB^* + C^*D)\end{aligned} \tag{2.310}$$

From Eq. (2.274) and Eq. (2.277), one can obtain

$$\begin{aligned}(A_y)^2 + (K_x^z)^2 &= \frac{1}{(I^0)^2}|AB^* + DC^*|^2 \\ &= \frac{1}{(I^0)^2}\Big( |A|^2|B|^2 + |D|^2|C|^2 + AB^*D^*C + A^*BDC^*\Big)\end{aligned} \tag{2.311}$$

And note

$$\begin{aligned}&\mathrm{Re}\,(AC^* + BD^*) = \mathrm{Re}\,(AC^* + B^*D), \\ &\mathrm{Im}(AC^* - BD^*) = \mathrm{Im}(AC^* + B^*D)\end{aligned} \tag{2.312}$$

Thus from Eq. (2.275) and Eq. (2.277), one can also obtain

$$\begin{aligned}\left(P^{0y}\right)^2+\left(K_z^x\right)^2&=\frac{1}{\left(I^0\right)^2}\left|AC^*+DB^*\right|^2\\&=\frac{1}{\left(I^0\right)^2}\left(|A|^2|C|^2+|D|^2|B|^2+AC^*D^*B+A^*CDB^*\right)\end{aligned} \tag{2.313}$$

Then the following relation is obtained when $M_I$ and $M_I'$ are taking the definite values

$$\left(A_y\right)^2+\left(K_x^z\right)^2+\left(K_z^z\right)^2+\left(P^{0y}\right)^2+\left(K_z^x\right)^2+\left(K_x^x\right)^2-\left(K_y^y\right)^2=1 \tag{2.314}$$

Section 2.4 has already pointed out that when the spins of both the target and the residual nucleus are equal to 0 and the *y*-axis is selected to be perpendicular to the reaction plane, only their y components of the polarization analyzing powers $A_i(i=x,y,z)$ and the polarization rate $P^{0i}(i{=}x,\ y,\ z)$ corresponding to the unpolarized incident particles are not equal to 0; only 5 of the 9 polarization transfer coefficient components $K_i^j(i,\ j{=}x,\ y,\ z)$ are not equal to 0. The theoretical derivation of this section has proved that the conclusions mentioned above are still true when the spins of the unpolarized target and residual nucleus are not equal to 0. This is due to the fact that the nuclear reaction process satisfies the conservation of parity. If let $D=A,\ C=B$ in Eqs. (2.300) – (2.305) and remove the summation symbol $\widetilde{\Sigma}$, they automatically degenerate into the result given in Sect. 2.4 in which the spins of the target and residual nucleus are equal to 0. Simultaneously, Eq. (2.314) will also automatically degenerate into Eq. (2.135) under the condition mentioned above.

## 2.8 Polarization Theory of Nuclear Reactions of Spin $\frac{1}{2}$ Particle with Polarized Target and Residual Nucleus

Setting $\Phi_{IM_I}$ as the basis spin function of the target along the *z*-axis, then according to Eq. (2.25) the arbitrary spin wave function $\Phi_I$ of the target with spin *I* can be expressed as [7]

$$\Phi_I=\sum_{M_I}a^{M_I}\Phi_{IM_I} \tag{2.315}$$

where $a^{M_I}$ is a contravariant component and can also be called the target polarization rate. The corresponding Hermitian conjugate function is

$$\Phi_I^+=\sum_{M_I}a^{M_I\,*}\,\Phi_{IM_I}^+ \tag{2.316}$$

The normalization condition requires

$$\sum_{M_I} \left| a^{M_I} \right|^2 = 1 \tag{2.317}$$

If the target spin wave function $\Phi_I$ can be expanded with the basis spin functions $\Phi_{IN_I}$ along the $z'$-axis (e.g., the direction of the target motion along the $z'$ axis), then we have

$$\Phi_I = \sum_{N_I} a^{N_I} \Phi_{IN_I} \tag{2.318}$$

If the expansion coefficient $a^{N_I}$ is known, and the $z'$-axis is not parallel to the $z$-axis, thus by use of the $D$ function we can get the following relation:

$$a^{M_I} = \sum_{N_I} a^{N_I} D^I_{N_I M_I}(\varphi_A, \theta_A, 0) \tag{2.319}$$

where $\theta_A$ and $\varphi_A$ are the polar and azimuthal angles of the $z$-axis in the $z'$-axis coordinate system. The residual nucleus spin wave function $\Phi_{I'}$ can be expressed as

$$\Phi_{I'} = \sum_{M'_I} b^{M'_I} \Phi_{I'M'_I} \tag{2.320}$$

where $b^{M'_I}$ is a contravariant component and can also be called the residual nucleus polarization rate. The normalization condition is

$$\sum_{M'_I} \left| b^{M'_I} \right|^2 = 1 \tag{2.321}$$

In the 2.7 section discussions, we assume that both the target and the residual nucleus are unpolarized, that is, the probabilities of the target and the residual nucleus at different spin magnetic quantum numbers are equal and they are incoherent to each other. However, the contributions of the different $M_I$ and $M'_I$ states of the target and the residual nucleus to the reaction amplitude, after the interactions with the incident particles and the outgoing particles, may not be the same, but they still remain incoherent. In this section we assume that the target and the residual nucleus are in the states of the arbitrary spin wave functions described by Eqs. (2.315) and (2.320), that is, in the mixed states of different $M_I$ and $M'_I$. Thus the coherence effect between different $M_I$ and different $M'_I$ should be considered, that is, it is assumed that both the target and the residual nucleus are in the polarized states.

If we select the $y$-axis along the direction of the unit vector $\vec{n}$ vertical to the reaction plane in Fig. 2.1, then the azimuthal angle $\varphi = 0$. According to Eq. (2.315) and Eq. (2.320) and refer to Eq. (2.269), we express the amplitude of the reaction as

$$\hat{F}=\sum_{M_I M_I'} a^{M_I} b^{M_I'} \begin{pmatrix} A_{M_I' M_I} & -\mathrm{i}B_{M_I' M_I} \\ \mathrm{i}C_{M_I' M_I} & D_{M_I' M_I} \end{pmatrix}, \quad \hat{F}^+=\sum_{\overline{M}_I \overline{M}_I'} a^{\overline{M}_I^*} b^{\overline{M}_I'^*} \begin{pmatrix} A^*_{\overline{M}_I' \overline{M}_I} & -\mathrm{i}C^*_{\overline{M}_I' \overline{M}_I} \\ \mathrm{i}B^*_{\overline{M}_I' \overline{M}_I} & D^*_{\overline{M}_I' \overline{M}_I} \end{pmatrix} \tag{2.322}$$

Based on Eq. (2.271) one can write

$$I^0_{\overline{M}_I' \overline{M}_I M_I' M_I}=\frac{1}{2}\left(A^*_{\overline{M}_I' \overline{M}_I} A_{M_I' M_I}+B^*_{\overline{M}_I' \overline{M}_I} B_{M_I' M_I}+C^*_{\overline{M}_I' \overline{M}_I} C_{M_I' M_I}+D^*_{\overline{M}_I' \overline{M}_I} D_{M_I' M_I}\right) \tag{2.323}$$

Based on Eq. (2.274) one can also write

$$\begin{aligned}
A_{x,\overline{M}_I' \overline{M}_I M_I' M_I} &= -\frac{1}{I^0_{\overline{M}_I' \overline{M}_I M_I' M_I}} \mathrm{Im}\left(A_{M_I' M_I} B^*_{\overline{M}_I' \overline{M}_I}+C_{M_I' M_I} D^*_{\overline{M}_I' \overline{M}_I}\right), \\
A_{y,\overline{M}_I' \overline{M}_I M_I' M_I} &= \frac{1}{I^0_{\overline{M}_I' \overline{M}_I M_I' M_I}} \mathrm{Re}\left(A_{M_I' M_I} B^*_{\overline{M}_I' \overline{M}_I}+C_{M_I' M_I} D^*_{\overline{M}_I' \overline{M}_I}\right), \\
A_{z,\overline{M}_I' \overline{M}_I M_I' M_I} &= \frac{1}{2I^0_{\overline{M}_I' \overline{M}_I M_I' M_I}} \left(A^*_{\overline{M}_I' \overline{M}_I} A_{M_I' M_I}-B^*_{\overline{M}_I' \overline{M}_I} B_{M_I' M_I}+C^*_{\overline{M}_I' \overline{M}_I} C_{M_I' M_I}-D^*_{\overline{M}_I' \overline{M}_I} D_{M_I' M_I}\right)
\end{aligned} \tag{2.324}$$

Based on Eq. (2.275) one can obtain

$$\begin{aligned}
P^{0x}_{\overline{M}_I' \overline{M}_I M_I' M_I} &= \frac{1}{I^0_{\overline{M}_I' \overline{M}_I M_I' M_I}} \mathrm{Im}\left(A_{M_I' M_I} C^*_{\overline{M}_I' \overline{M}_I}+B_{M_I' M_I} D^*_{\overline{M}_I' \overline{M}_I}\right), \\
P^{0y}_{\overline{M}_I' \overline{M}_I M_I' M_I} &= \frac{1}{I^0_{\overline{M}_I' \overline{M}_I M_I' M_I}} \mathrm{Re}\left(A_{M_I' M_I} C^*_{\overline{M}_I' \overline{M}_I}+B_{M_I' M_I} D^*_{\overline{M}_I' \overline{M}_I}\right), \\
P^{0z}_{\overline{M}_I' \overline{M}_I M_I' M_I} &= \frac{1}{2I^0_{\overline{M}_I' \overline{M}_I M_I' M_I}} \left(A^*_{\overline{M}_I' \overline{M}_I} A_{M_I' M_I}+B^*_{\overline{M}_I' \overline{M}_I} B_{M_I' M_I}-C^*_{\overline{M}_I' \overline{M}_I} C_{M_I' M_I}-D^*_{\overline{M}_I' \overline{M}_I} D_{M_I' M_I}\right)
\end{aligned} \tag{2.325}$$

Based on Eq. (2.277) one can get

$$
\begin{aligned}
K^x_{x,\overline{M}_I'\overline{M}_I M_I' M_I} &= \frac{1}{I^0_{\overline{M}_I'\overline{M}_I M_I' M_I}} \mathrm{Re}\left(A_{M_I' M_I} D^*_{\overline{M}_I'\overline{M}_I} - B_{M_I' M_I} C^*_{\overline{M}_I'\overline{M}_I}\right), \\
K^y_{x,\overline{M}_I'\overline{M}_I M_I' M_I} &= -\frac{1}{I^0_{\overline{M}_I'\overline{M}_I M_I' M_I}} \mathrm{Im}\left(A_{M_I' M_I} D^*_{\overline{M}_I'\overline{M}_I} - B_{M_I' M_I} C^*_{\overline{M}_I'\overline{M}_I}\right), \\
K^z_{x,\overline{M}_I'\overline{M}_I M_I' M_I} &= -\frac{1}{I^0_{\overline{M}_I'\overline{M}_I M_I' M_I}} \mathrm{Im}\left(A_{M_I' M_I} B^*_{\overline{M}_I'\overline{M}_I} - C_{M_I' M_I} D^*_{\overline{M}_I'\overline{M}_I}\right), \\
K^x_{y,\overline{M}_I'\overline{M}_I M_I' M_I} &= \frac{1}{I^0_{\overline{M}_I'\overline{M}_I M_I' M_I}} \mathrm{Im}\left(A_{M_I' M_I} D^*_{\overline{M}_I'\overline{M}_I} + B_{M_I' M_I} C^*_{\overline{M}_I'\overline{M}_I}\right), \\
K^y_{y,\overline{M}_I'\overline{M}_I M_I' M_I} &= \frac{1}{I^0_{\overline{M}_I'\overline{M}_I M_I' M_I}} \mathrm{Re}\left(A_{M_I' M_I} D^*_{\overline{M}_I'\overline{M}_I} + B_{M_I' M_I} C^*_{\overline{M}_I'\overline{M}_I}\right), \\
K^z_{y,\overline{M}_I'\overline{M}_I M_I' M_I} &= \frac{1}{I^0_{\overline{M}_I'\overline{M}_I M_I' M_I}} \mathrm{Re}\left(A_{M_I' M_I} B^*_{\overline{M}_I'\overline{M}_I} - C_{M_I' M_I} D^*_{\overline{M}_I'\overline{M}_I}\right), \\
K^x_{z,\overline{M}_I'\overline{M}_I M_I' M_I} &= \frac{1}{I^0_{\overline{M}_I'\overline{M}_I M_I' M_I}} \mathrm{Im}\left(A_{M_I' M_I} C^*_{\overline{M}_I'\overline{M}_I} - B_{M_I' M_I} D^*_{\overline{M}_I'\overline{M}_I}\right), \\
K^y_{z,\overline{M}_I'\overline{M}_I M_I' M_I} &= \frac{1}{I^0_{\overline{M}_I'\overline{M}_I M_I' M_I}} \mathrm{Re}\left(A_{M_I' M_I} C^*_{\overline{M}_I'\overline{M}_I} - B_{M_I' M_I} D^*_{\overline{M}_I'\overline{M}_I}\right), \\
K^z_{z,\overline{M}_I'\overline{M}_I M_I' M_I} &= \frac{1}{2I^0_{\overline{M}_I'\overline{M}_I M_I' M_I}} \left(A^*_{\overline{M}_I'\overline{M}_I} A_{M_I' M_I} - B^*_{\overline{M}_I'\overline{M}_I} B_{M_I' M_I}\right. \\
&\quad \left. - C^*_{\overline{M}_I'\overline{M}_I} C_{M_I' M_I} + D^*_{\overline{M}_I'\overline{M}_I} D_{M_I' M_I}\right)
\end{aligned}
\tag{2.326}
$$

Introduce the following simplification summation symbol

$$
\overline{\Sigma} \equiv \frac{1}{2I+1} \sum_{M_I \overline{M}_I M_I' \overline{M}_I'} a^{\overline{M}_I *} a^{M_I} b^{\overline{M}_I' *} b^{M_I'}
\tag{2.327}
$$

Then based on Eq. (2.279) the following result can be obtained:

$$
\overline{I}^0 = \overline{\Sigma}\, I^0_{\overline{M}_I'\overline{M}_I M_I' M_I}
\tag{2.328}
$$

Based on Eq. (2.280) we can obtain

$$
\overline{A}_i = \frac{1}{\overline{I}^0} \overline{\Sigma}\, I^0_{\overline{M}_I'\overline{M}_I M_I' M_I} A_{i,\overline{M}_I'\overline{M}_I M_I' M_I}, \qquad i = x, y, z
\tag{2.329}
$$

Equation (2.281), which is still applicable here, is given again below:

$$
\overline{I} = \overline{I}^0 \left(1 + \sum_{i=x,y,z} p_i \overline{A}_i\right)
\tag{2.330}
$$

Based on Eq. (2.282) we have

$$\overline{P}^{0k}=\frac{1}{\overline{I}^0}\Sigma\, I^0_{\overline{M}_I'\overline{M}_I M_I' M_I} P^{0k}_{\overline{M}_I'\overline{M}_I M_I' M_I}, \qquad k=x,y,z \tag{2.331}$$

Based on Eq. (2.283) we can obtain

$$\overline{K}_i^k=\frac{1}{\overline{I}^0}\Sigma\, I^0_{\overline{M}_I'\overline{M}_I M_I' M_I} K^k_{i,\overline{M}_I'\overline{M}_I M_I' M_I}, \qquad i,k=x,y,z \tag{2.332}$$

Equation (2.284), which is still applicable here, is given again below:

$$\overline{P}^k=\frac{\overline{I}^0}{\overline{I}}\left(\overline{P}^{0k}+\sum_{i=x,y,z} p_i\overline{K}_i^k\right), \qquad k=x,y,z \tag{2.333}$$

In the expressions given in this section, only if the polarization rate $a^{M_I}$ and/or $b^{M_I'}$ of the target and/or the residual nucleus are known, can the calculations of the polarization physical quantities of the nuclear reactions with spin $\frac{1}{2}$ particles be done for the polarized target and/or the polarized residual nucleus. The unpolarization conditions of the target and the residual nucleus are, respectively,

$$a^{\overline{M}_I^*}a^{M_I}=\delta_{\overline{M}_I M_I} \tag{2.334}$$

$$b^{\overline{M}_I'^*}b^{M_I'}=\delta_{\overline{M}_I' M_I'} \tag{2.335}$$

If substituting Eq. (2.334) and Eq. (2.335) into Eqs. (2.323) – (2.333), then these expressions mentioned above automatically degenerate into the expressions calculating the polarization physical quantities of the nuclear reactions for the unpolarized target and residual nucleus with spin $\frac{1}{2}$ particles given by Eqs. (2.271), (2.274), (2.275), (2.277) –(2.284), respectively.

## 2.9 Polarization Theory for $\frac{\vec{1}}{2}+\frac{\vec{1}}{2}$ Reactions

This section will study the polarization theory of the nuclear reactions between two different polarized particles with spin $\frac{1}{2}$, and take the n–p elastic scattering as an example.

The *S–L* angular momentum coupling mode is used for nucleon- nucleon interactions. The reaction amplitude of the n–p elastic scattering can be written from Eq. (2.252) as

$$f_{\mu'\nu',\mu\,\nu}(\Omega)=\frac{\mathrm{i}\sqrt{\pi}}{k}\sum_{lS\,l'S'J}\hat{l}\mathrm{e}^{\mathrm{i}(\sigma_l+\sigma_{l'})}\left(\delta_{l'S',lS}-S^J_{l'S',lS}\right)$$
$$C^{J\ \ M}_{l0\ \ SM}C^{S\ \ M}_{\frac{1}{2}\mu\ \ \frac{1}{2}\nu}C^{J\ \ M}_{l'm'_l\ \ S'M'_S}C^{S'\ \ M'_S}_{\frac{1}{2}\mu'\ \ \frac{1}{2}\nu'}\mathrm{Y}_{l'm'_l}(\theta,\varphi) \tag{2.336}$$

The $S$ matrix element $S^J_{l'S',lS}$ can be calculated by the phase shift analysis method of nucleon-nucleon elastic scattering. From the C–G coefficients in the above formula one can also see

$$m'_l=\mu+\nu-\mu'-\nu' \tag{2.337}$$

where $\mu$ and $\mu'$, $\nu$ and $\nu'$ represent the spin magnetic quantum numbers of the incident neutron and the outgoing neutron, the target proton and the residual nucleus proton, respectively. Introducing the following symbols

$$f^{ll'J}_{\mu'\nu',\mu\nu}=\hat{l}\hat{l}'\mathrm{e}^{\mathrm{i}(\sigma_l+\sigma_{l'})}\sqrt{\frac{(l'-m'_l)!}{(l'+m'_l)!}}\sum_{SS'}C^{J\ \ M}_{l0\ \ SM}C^{S\ \ M}_{\frac{1}{2}\mu\ \ \frac{1}{2}\nu}C^{J\ \ M}_{l'm'_l\ \ S'M'_S}C^{S'\ \ M'_S}_{\frac{1}{2}\mu'\ \ \frac{1}{2}\nu'}\left(\delta_{l'S',lS}-S^J_{l'S',lS}\right) \tag{2.338}$$

and using Eq. (2.253), we can obtain the following expressions from Eq. (2.336):

$$f_{\mu'\nu',\mu\nu}(\Omega)=F_{\mu'\nu',\mu\nu}(\theta)\mathrm{e}^{\mathrm{i}(\mu+\nu-\mu'-\nu')\varphi} \tag{2.339}$$

$$F_{\mu'\nu',\mu\nu}(\theta)=\frac{\mathrm{i}}{2k}(-1)^{\mu+\nu-\mu'-\nu'}\sum_{ll'J}f^{ll'J}_{\mu'\nu',\mu\,\nu}\mathrm{P}^{\mu+\nu-\mu'-\nu'}_{l'}(\cos\theta) \tag{2.340}$$

Considering parity conservation, $l$ and $l'$ must be all even or odd numbers, so the following C–G coefficient relations can be obtained:

$$C^{J\ \ \mu+\nu}_{l0\ \ S\ \ \mu+\nu}C^{S\ \ \mu+\nu}_{\frac{1}{2}\mu\ \ \frac{1}{2}\nu}C^{J\ \ \mu+\nu}_{l'\ \ \mu+\nu-\mu'-\nu'\ \ S'\ \ \mu'+\nu'}C^{S'\ \ \mu'+\nu'}_{\frac{1}{2}\mu'\ \ \frac{1}{2}\nu'}$$
$$=C^{J\ \ -\mu-\nu}_{l0\ \ S\ \ -\mu-\nu}C^{S\ \ -\mu-\nu}_{\frac{1}{2}\ -\mu\ \ \frac{1}{2}\ -\nu}C^{J\ \ -\mu-\nu}_{l'\ \ -\mu-\nu+\mu'+\nu'\ \ S'\ \ -\mu'-\nu'}C^{S'\ \ -\mu'-\nu'}_{\frac{1}{2}\ -\mu'\ \ \frac{1}{2}\ -\nu'} \tag{2.341}$$

By using Eq. (2.287) the following relation can be obtained from Eqs. (2.338) and (2.340):

$$F_{-\mu'-\nu',-\mu-\nu}(\theta)=(-1)^{\mu+\nu-\mu'-\nu'}F_{\mu'\nu',\mu\,\nu}(\theta) \tag{2.342}$$

Taking the $y$-axis in the direction of the unit vector $\vec{n}$ perpendicular to the reaction plane, there is $\varphi=0$. According to Eq. (2.342) we define the following matrix elements:

$$
\begin{aligned}
&A(\theta)=F_{\frac{1}{2}\frac{1}{2},\frac{1}{2}\frac{1}{2}}(\theta)=F_{-\frac{1}{2}\,-\frac{1}{2},-\frac{1}{2}\,-\frac{1}{2}}(\theta), && B(\theta)=\mathrm{i}F_{\frac{1}{2}\frac{1}{2},\frac{1}{2}\,-\frac{1}{2}}(\theta)=-\mathrm{i}F_{-\frac{1}{2}\,-\frac{1}{2},-\frac{1}{2}\frac{1}{2}}(\theta),\\
&C(\theta)=\mathrm{i}F_{\frac{1}{2}\frac{1}{2},-\frac{1}{2}\frac{1}{2}}(\theta)=-\mathrm{i}F_{-\frac{1}{2}\,-\frac{1}{2},\frac{1}{2}\,-\frac{1}{2}}(\theta), && D(\theta)=-F_{\frac{1}{2}\frac{1}{2},-\frac{1}{2}\,-\frac{1}{2}}(\theta)=-F_{-\frac{1}{2}\,-\frac{1}{2},\frac{1}{2}\frac{1}{2}}(\theta),\\
&E(\theta)=\mathrm{i}F_{\frac{1}{2}\,-\frac{1}{2},-\frac{1}{2}\,-\frac{1}{2}}(\theta)=-\mathrm{i}F_{-\frac{1}{2}\frac{1}{2},\frac{1}{2}\frac{1}{2}}(\theta), && F(\theta)=F_{\frac{1}{2}\,-\frac{1}{2},-\frac{1}{2}\frac{1}{2}}(\theta)=F_{-\frac{1}{2}\frac{1}{2},\frac{1}{2}\,-\frac{1}{2}}(\theta),\\
&G(\theta)=F_{\frac{1}{2}\,-\frac{1}{2},\frac{1}{2}\,-\frac{1}{2}}(\theta)=F_{-\frac{1}{2}\frac{1}{2},-\frac{1}{2}\frac{1}{2}}(\theta), && H(\theta)=\mathrm{i}F_{-\frac{1}{2}\frac{1}{2},-\frac{1}{2}\,-\frac{1}{2}}(\theta)=-\mathrm{i}F_{\frac{1}{2}\,-\frac{1}{2},\frac{1}{2}\frac{1}{2}}(\theta)
\end{aligned}
\tag{2.343}
$$

Then the 4 × 4 reaction matrix given by Eq. (2.340) can be expressed as

$$
\hat{F}=\begin{pmatrix} A & -\mathrm{i}B & -\mathrm{i}C & -D \\ \mathrm{i}H & G & F & -\mathrm{i}E \\ \mathrm{i}E & F & G & -\mathrm{i}H \\ -D & \mathrm{i}C & \mathrm{i}B & A \end{pmatrix},\quad \hat{F}^{+}=\begin{pmatrix} A^{*} & -\mathrm{i}H^{*} & -\mathrm{i}E^{*} & -D^{*} \\ \mathrm{i}B^{*} & G^{*} & F^{*} & -\mathrm{i}C^{*} \\ \mathrm{i}C^{*} & F^{*} & G^{*} & -\mathrm{i}B^{*} \\ -D^{*} & \mathrm{i}E^{*} & \mathrm{i}H^{*} & A^{*} \end{pmatrix}
\tag{2.344}
$$

The matrix elements $F_{\mu'\nu',\mu\nu}(\theta)$ of the matrix $\hat{F}$ have double sets of subscripts: $\mu'$ and $\nu'$ correspond to the final states of neutron and proton, respectively, and $\mu$ and $\nu$ correspond to the initial states of neutron and proton, respectively. $\mu(\mu')$ and $\nu(\nu')$ take on $\frac{1}{2}$ or $-\frac{1}{2}$. If the 4 × 4 matrix $\hat{F}$ is divided into 4 small 2 × 2 matrices, then we will find that for the upper left 2 × 2 matrix $\mu'=\mu=\frac{1}{2}$; for the lower right matrix $\mu'=\mu=-\frac{1}{2}$; for the upper right matrix $\mu'=\frac{1}{2}$ and $\mu=-\frac{1}{2}$; and for the lower left matrix $\mu'=-\frac{1}{2}$ and $\mu=\frac{1}{2}$. The second subscripts $\nu'$ and $\nu$ of the four small matrices are all arranged in the same way: in the upper left corner they are $\frac{1}{2}\,\frac{1}{2}$; in the lower right corner they are $-\frac{1}{2}\,-\frac{1}{2}$; in the upper right corner they are $\frac{1}{2}\,-\frac{1}{2}$; and in the lower left corner they are $-\frac{1}{2}\,\frac{1}{2}$. To facilitate the next operation, we introduce the following 2 × 2 submatrices:

$$
\hat{t}_{\mu'\mu}=\begin{pmatrix} F_{\mu'\frac{1}{2},\,\mu\frac{1}{2}} & F_{\mu'\frac{1}{2},\,\mu\,-\frac{1}{2}} \\ F_{\mu'\,-\frac{1}{2},\,\mu\frac{1}{2}} & F_{\mu'\,-\frac{1}{2},\,\mu\,-\frac{1}{2}} \end{pmatrix}
\tag{2.345}
$$

For the sake of simplicity, in the following the foot note $\frac{1}{2}$ is replaced by + and the foot note $-\frac{1}{2}$ is replaced by –; then according to Eq. (2.343) or Eq. (2.344) we can write

$$\hat{t}_{++}=\begin{pmatrix} A & -\mathrm{i}B \\ \mathrm{i}H & G \end{pmatrix},\ \hat{t}_{++}^{+}=\begin{pmatrix} A^* & -\mathrm{i}H^* \\ \mathrm{i}B^* & G^* \end{pmatrix},\ \hat{t}_{+-}=\begin{pmatrix} -\mathrm{i}C & -D \\ F & -\mathrm{i}E \end{pmatrix},\ \hat{t}_{+-}^{+}=\begin{pmatrix} \mathrm{i}C^* & F^* \\ -D^* & \mathrm{i}E^* \end{pmatrix}$$
$$\hat{t}_{-+}=\begin{pmatrix} \mathrm{i}E & F \\ -D & \mathrm{i}C \end{pmatrix},\ \hat{t}_{-+}^{+}=\begin{pmatrix} -\mathrm{i}E^* & -D^* \\ F^* & -\mathrm{i}C^* \end{pmatrix},\ \hat{t}_{--}=\begin{pmatrix} G & -\mathrm{i}H \\ \mathrm{i}B & A \end{pmatrix},\ \hat{t}_{--}^{+}=\begin{pmatrix} G^* & -\mathrm{i}B^* \\ \mathrm{i}H^* & A^* \end{pmatrix} \tag{2.346}$$

and the matrix $\hat{F}$ given by Eq. (2.344) can be expressed as

$$\hat{F}=\begin{pmatrix} \hat{t}_{++} & \hat{t}_{+-} \\ \hat{t}_{-+} & \hat{t}_{--} \end{pmatrix},\quad \hat{F}^{+}=\begin{pmatrix} \hat{t}_{++}^{+} & \hat{t}_{-+}^{+} \\ \hat{t}_{+-}^{+} & \hat{t}_{--}^{+} \end{pmatrix} \tag{2.347}$$

According to Eq. (2.64), the polarization density matrix of the normalized incident particle channel can be written as

$$\hat{\rho}_{\mathrm{in}}=\frac{1}{4}\left(\hat{I}+p_{\mathrm{n}x}\hat{\sigma}_x^{\mathrm{n}}+p_{\mathrm{n}y}\hat{\sigma}_y^{\mathrm{n}}+p_{\mathrm{n}z}\hat{\sigma}_z^{\mathrm{n}}\right)\left(\hat{I}+p_{\mathrm{p}x}\hat{\sigma}_x^{\mathrm{p}}+p_{\mathrm{p}y}\hat{\sigma}_y^{\mathrm{p}}+p_{\mathrm{p}z}\hat{\sigma}_z^{\mathrm{p}}\right) \tag{2.348}$$

where $p_{\mathrm{n}i}$ and $p_{\mathrm{p}j}$ $(i,j=x,y,z)$ are the polarization rates of the neutrons and protons in the incident channel, respectively. $\hat{\sigma}_i^{\mathrm{n}}$ $(i=x,\ y,\ z)$ act on the first set of subscripts $\mu'$, $\mu$ corresponding to the neutron states only, and $\hat{\sigma}_j^{\mathrm{p}}$ $(j=x,\ y,\ z)$ act on the second set of subscripts $\nu'$, $\nu$ corresponding to the proton states only.

Let $\hat{A}$ and $\hat{B}$ be the n- and m-dimensional square matrices such that their direct product is defined as

$$\left(\hat{A}\times\hat{B}\right)_{ia,jb}\equiv A_{ij}B_{ab} \tag{2.349}$$

Let $\hat{I}$ be a 2 × 2 unit matrix and $\hat{0}$ be a 2 × 2 zero matrix. Using the above direct product definition, we can obtain

$$\hat{\Sigma}_i^{\mathrm{n}}\equiv\hat{\sigma}_i^{\mathrm{n}}\times\hat{I}=\begin{pmatrix} \sigma_{i,++}^{\mathrm{n}}\hat{I} & \sigma_{i,+-}^{\mathrm{n}}\hat{I} \\ \sigma_{i,-+}^{\mathrm{n}}\hat{I} & \sigma_{i,--}^{\mathrm{n}}\hat{I} \end{pmatrix} \tag{2.350}$$

$$\hat{\Sigma}_j^{\mathrm{p}}\equiv\hat{I}\times\hat{\sigma}_j^{\mathrm{p}}=\begin{pmatrix} \hat{\sigma}_j^{\mathrm{p}} & \hat{0} \\ \hat{0} & \hat{\sigma}_j^{\mathrm{p}} \end{pmatrix} \tag{2.351}$$

It can be seen that the resulting 4 × 4 matrix $\hat{\Sigma}_i^{\mathrm{n}}$ only acts on the subscript $\mu'$, $\mu$ of neutrons, and $\hat{\Sigma}_j^{\mathrm{p}}$ only acts on the subscript $\nu'$, $\nu$ of protons. $\hat{\Sigma}_i^{\mathrm{n}}$ and $\hat{\Sigma}_j^{\mathrm{p}}$ are commutative of each other. Letting $\hat{I}_4$ be the unit 4 × 4 matrix, then Eq. (2.348) can be rewritten as

$$\hat{\rho}_{\mathrm{in}} = \frac{1}{4}\left(\hat{I}_4 + p_{\mathrm{nx}}\hat{\Sigma}_x^{\mathrm{n}} + p_{\mathrm{ny}}\hat{\Sigma}_y^{\mathrm{n}} + p_{\mathrm{nz}}\hat{\Sigma}_z^{\mathrm{n}}\right)\left(\hat{I}_4 + p_{\mathrm{px}}\hat{\Sigma}_x^{\mathrm{p}} + p_{\mathrm{py}}\hat{\Sigma}_y^{\mathrm{p}} + p_{\mathrm{pz}}\hat{\Sigma}_z^{\mathrm{p}}\right) \quad (2.352)$$

When both n and p in the incident channel are unpolarized, the differential cross sections of the emitted particles are

$$I_0 \equiv \frac{\mathrm{d}\sigma^0}{\mathrm{d}\Omega} = \frac{1}{4}\mathrm{tr}\left\{\hat{F}\hat{F}^+\right\} \quad (2.353)$$

Substituting Eq. (2.344) into the above formula, the following result can be obtained immediately:

$$I_0 = \frac{1}{2}\left(|A|^2 + |B|^2 + |C|^2 + |D|^2 + |E|^2 + |F|^2 + |G|^2 + |H|^2\right) \quad (2.354)$$

The polarization vector components of the outgoing particle n and p corresponding to the unpolarized incident channel are, respectively,

$$P_0^{ki'} = \frac{\mathrm{tr}\left\{\hat{\Sigma}_{i'}^k \hat{F}\hat{F}^+\right\}}{\mathrm{tr}\left\{\hat{F}\hat{F}^+\right\}}, \quad k = \mathrm{n,p}; \quad i' = x', y', z' \quad (2.355)$$

From Eq. (2.347) we can obtain

$$\hat{F}\hat{F}^+ = \begin{pmatrix} \hat{t}_{++}\hat{t}_{++}^+ + \hat{t}_{+-}\hat{t}_{+-}^+ & \hat{t}_{++}\hat{t}_{-+}^+ + \hat{t}_{+-}\hat{t}_{--}^+ \\ \hat{t}_{-+}\hat{t}_{++}^+ + \hat{t}_{--}\hat{t}_{+-}^+ & \hat{t}_{-+}\hat{t}_{-+}^+ + \hat{t}_{--}\hat{t}_{--}^+ \end{pmatrix} \equiv \begin{pmatrix} \hat{g}_{++} & \hat{g}_{+-} \\ \hat{g}_{-+} & \hat{g}_{--} \end{pmatrix} \quad (2.356)$$

Moreover using Eq. (2.356) and Eq. (2.346), we obtain

$$\begin{aligned} \hat{g}_{++} &= \begin{pmatrix} A & -\mathrm{i}B \\ \mathrm{i}H & G \end{pmatrix}\begin{pmatrix} A^* & -\mathrm{i}H^* \\ \mathrm{i}B^* & G^* \end{pmatrix} + \begin{pmatrix} -\mathrm{i}C & -D \\ F & -\mathrm{i}E \end{pmatrix}\begin{pmatrix} \mathrm{i}C^* & F^* \\ -D^* & \mathrm{i}E^* \end{pmatrix} \\ &= \begin{pmatrix} |A|^2 + |B|^2 + |C|^2 + |D|^2 & -\mathrm{i}(AH^* + BG^* + CF^* + DE^*) \\ \mathrm{i}(HA^* + GB^* + FC^* + ED^*) & |H|^2 + |G|^2 + |F|^2 + |E|^2 \end{pmatrix} \end{aligned} \quad (2.357)$$

$$\begin{aligned} \hat{g}_{+-} &= \begin{pmatrix} A & -\mathrm{i}B \\ \mathrm{i}H & G \end{pmatrix}\begin{pmatrix} -\mathrm{i}E^* & -D^* \\ F^* & -\mathrm{i}C^* \end{pmatrix} + \begin{pmatrix} -\mathrm{i}C & -D \\ F & -\mathrm{i}E \end{pmatrix}\begin{pmatrix} G^* & -\mathrm{i}B^* \\ \mathrm{i}H^* & A^* \end{pmatrix} \\ &= \begin{pmatrix} -\mathrm{i}(AE^* + BF^* + CG^* + DH^*) & -(AD^* + BC^* + CB^* + DA^*) \\ HE^* + GF^* + FG^* + EH^* & -\mathrm{i}(HD^* + GC^* + FB^* + EA^*) \end{pmatrix} \end{aligned} \quad (2.358)$$

$$\hat{g}_{-+} = \begin{pmatrix} \mathrm{i}E & F \\ -D & \mathrm{i}C \end{pmatrix} \begin{pmatrix} A^* & -\mathrm{i}H^* \\ \mathrm{i}B^* & G^* \end{pmatrix} + \begin{pmatrix} G & -\mathrm{i}H \\ \mathrm{i}B & A \end{pmatrix} \begin{pmatrix} \mathrm{i}C^* & F^* \\ -D^* & \mathrm{i}E^* \end{pmatrix}$$
$$= \begin{pmatrix} \mathrm{i}(EA^* + FB^* + GC^* + HD^*) & EH^* + FG^* + GF^* + HE^* \\ -(DA^* + CB^* + BC^* + AD^*) & \mathrm{i}(DH^* + CG^* + BF^* + AE^*) \end{pmatrix} \tag{2.359}$$

$$\hat{g}_{--} = \begin{pmatrix} \mathrm{i}E & F \\ -D & \mathrm{i}C \end{pmatrix} \begin{pmatrix} -\mathrm{i}E^* & -D^* \\ F^* & -\mathrm{i}C^* \end{pmatrix} + \begin{pmatrix} G & -\mathrm{i}H \\ \mathrm{i}B & A \end{pmatrix} \begin{pmatrix} G^* & -\mathrm{i}B^* \\ \mathrm{i}H^* & A^* \end{pmatrix}$$
$$= \begin{pmatrix} |E|^2 + |F|^2 + |G|^2 + |H|^2 & -\mathrm{i}(ED^* + FC^* + GB^* + HA^*) \\ \mathrm{i}(DE^* + CF^* + BG^* + AH^*) & |D|^2 + |C|^2 + |B|^2 + |A|^2 \end{pmatrix} \tag{2.360}$$

It can be seen that substituting Eqs. (2.356), (2.357), and (2.360) into Eq. (2.353) we get Eq. (2.354). First we obtain the following relations by using Eqs. (2.6), (2.350), (2.355), and (2.356) –(2.360):

$$\hat{\Sigma}_x^{\mathrm{n}} \hat{F} \hat{F}^+ = \begin{pmatrix} \hat{0} & \hat{I} \\ \hat{I} & \hat{0} \end{pmatrix} \begin{pmatrix} \hat{g}_{++} & \hat{g}_{+-} \\ \hat{g}_{-+} & \hat{g}_{--} \end{pmatrix} = \begin{pmatrix} \hat{g}_{-+} & \hat{g}_{--} \\ \hat{g}_{++} & \hat{g}_{+-} \end{pmatrix} \tag{2.361}$$

$$P_0^{\mathrm{n}x'} = 0 \tag{2.362}$$

$$\hat{\Sigma}_y^{\mathrm{n}} \hat{F} \hat{F}^+ = \begin{pmatrix} \hat{0} & -\mathrm{i}\hat{I} \\ \mathrm{i}\hat{I} & \hat{0} \end{pmatrix} \begin{pmatrix} \hat{g}_{++} & \hat{g}_{+-} \\ \hat{g}_{-+} & \hat{g}_{--} \end{pmatrix} = -\mathrm{i} \begin{pmatrix} \hat{g}_{-+} & \hat{g}_{--} \\ -\hat{g}_{++} & -\hat{g}_{+-} \end{pmatrix} \tag{2.363}$$

$$P_0^{\mathrm{n}y'} = \frac{1}{I_o} \mathrm{Re}\,(AE^* + BF^* + CG^* + DH^*) \tag{2.364}$$

$$\hat{\Sigma}_z^{\mathrm{n}} \hat{F} \hat{F}^+ = \begin{pmatrix} \hat{I} & \hat{0} \\ \hat{0} & -\hat{I} \end{pmatrix} \begin{pmatrix} \hat{g}_{++} & \hat{g}_{+-} \\ \hat{g}_{-+} & \hat{g}_{--} \end{pmatrix} = \begin{pmatrix} \hat{g}_{++} & \hat{g}_{+-} \\ -\hat{g}_{-+} & -\hat{g}_{--} \end{pmatrix} \tag{2.365}$$

$$P_0^{\mathrm{n}z'} = 0 \tag{2.366}$$

Because

$$\hat{\Sigma}_j^{\mathrm{p}} \hat{F} \hat{F}^+ = \begin{pmatrix} \hat{\sigma}_j^{\mathrm{p}} & \hat{0} \\ \hat{0} & \hat{\sigma}_j^{\mathrm{p}} \end{pmatrix} \begin{pmatrix} \hat{g}_{++} & \hat{g}_{+-} \\ \hat{g}_{-+} & \hat{g}_{--} \end{pmatrix} = \begin{pmatrix} \hat{\sigma}_j^{\mathrm{p}} \hat{g}_{++} & \hat{\sigma}_j^{\mathrm{p}} \hat{g}_{+-} \\ \hat{\sigma}_j^{\mathrm{p}} \hat{g}_{-+} & \hat{\sigma}_j^{\mathrm{p}} \hat{g}_{--} \end{pmatrix} \tag{2.367}$$

therefore $\hat{\Sigma}_j^{\mathrm{p}}$ acting on the $4 \times 4$ matrix is equivalent to $\hat{\sigma}_j^{\mathrm{p}}$ acting on the $2 \times 2$ matrices of its blocks. In order to find $P^{\mathrm{p}i'}$, first let us calculate the following terms:

$$\hat{\sigma}_x^{\mathrm{p}}\hat{g}_{++}=\begin{pmatrix}0&1\\1&0\end{pmatrix}\begin{pmatrix}|A|^2+|B|^2+|C|^2+|D|^2 & -\mathrm{i}(AH^*+BG^*+CF^*+DE^*)\\ \mathrm{i}(HA^*+GB^*+FC^*+ED^*) & |H|^2+|G|^2+|F|^2+|E|^2\end{pmatrix}$$
$$=\begin{pmatrix}\mathrm{i}(HA^*+GB^*+FC^*+ED^*) & |H|^2+|G|^2+|F|^2+|E|^2\\ |A|^2+|B|^2+|C|^2+|D|^2 & -\mathrm{i}(AH^*+BG^*+CF^*+DE^*)\end{pmatrix},$$
$$\hat{\sigma}_x^{\mathrm{p}}\hat{g}_{--}=\begin{pmatrix}0&1\\1&0\end{pmatrix}\begin{pmatrix}|E|^2+|F|^2+|G|^2+|H|^2 & -\mathrm{i}(ED^*+FC^*+GB^*+HA^*)\\ \mathrm{i}(DE^*+CF^*+BG^*+AH^*) & |D|^2+|C|^2+|B|^2+|A|^2\end{pmatrix}$$
$$=\begin{pmatrix}\mathrm{i}(DE^*+CF^*+BG^*+AH^*) & |D|^2+|C|^2+|B|^2+|A|^2\\ |E|^2+|F|^2+|G|^2+|H|^2 & -\mathrm{i}(ED^*+FC^*+GB^*+HA^*)\end{pmatrix} \tag{2.368}$$

Substituting Eq. (2.368) into Eq. (2.355), we get

$$P_0^{\mathrm{p}x'}=0 \tag{2.369}$$

According to Eq. (2.368) we have

$$\hat{\sigma}_y^{\mathrm{p}}\hat{g}_{++}=\begin{pmatrix}HA^*+GB^*+FC^*+ED^* & -\mathrm{i}\left(|H|^2+|G|^2+|F|^2+|E|^2\right)\\ \mathrm{i}\left(|A|^2+|B|^2+|C|^2+|D|^2\right) & AH^*+BG^*+CF^*+DE^*\end{pmatrix},$$
$$\hat{\sigma}_y^{\mathrm{p}}\hat{g}_{--}=\begin{pmatrix}DE^*+CF^*+BG^*+AH^* & -\mathrm{i}\left(|D|^2+|C|^2+|B|^2+|A|^2\right)\\ \mathrm{i}\left(|E|^2+|F|^2+|G|^2+|H|^2\right) & ED^*+FC^*+GB^*+HA^*\end{pmatrix} \tag{2.370}$$

Substituting Eq. (2.370) into Eq. (2.355), the following result is obtained:

$$P_0^{\mathrm{p}y'}=\frac{1}{I_0}\,\mathrm{Re}\,(AH^*+BG^*+CF^*+DE^*) \tag{2.371}$$

According to Eq. (2.368) we have

$$\hat{\sigma}_z^{\mathrm{p}}\hat{g}_{++}=\begin{pmatrix}|A|^2+|B|^2+|C|^2+|D|^2 & -\mathrm{i}(AH^*+BG^*+CF^*+DE^*)\\ -\mathrm{i}(HA^*+GB^*+FC^*+ED^*) & -\left(|H|^2+|G|^2+|F|^2+|E|^2\right)\end{pmatrix},$$
$$\hat{\sigma}_z^{\mathrm{p}}\hat{g}_{--}=\begin{pmatrix}|E|^2+|F|^2+|G|^2+|H|^2 & -\mathrm{i}(ED^*+FC^*+GB^*+HA^*)\\ -\mathrm{i}(DE^*+CF^*+BG^*+AH^*) & -\left(|D|^2+|C|^2+|B|^2+|A|^2\right)\end{pmatrix} \tag{2.372}$$

Substituting Eq. (2.372) into Eq. (2.355), we get

$$P_0^{\mathrm{p}z'} = 0 \tag{2.373}$$

The above results show that $P_0^{\mathrm{n}x'} = P_0^{\mathrm{n}z'} = P_0^{\mathrm{p}x'} = P_0^{\mathrm{p}z'} = 0,\ P_0^{\mathrm{n}y'} \neq 0,\ P_0^{\mathrm{p}y'} \neq 0.$

The polarization analyzing power for n − p elastic scattering is defined as

$$A_i^k = \frac{\mathrm{tr}\left\{\hat{F}\hat{\Sigma}_i^k\hat{F}^+\right\}}{\mathrm{tr}\left\{\hat{F}\hat{F}^+\right\}}, \quad k = \mathrm{n,p}; \quad i = x, y, z \tag{2.374}$$

$$A_{ij}^{\mathrm{np}} = \frac{\mathrm{tr}\left\{\hat{F}\hat{\Sigma}_i^{\mathrm{n}}\hat{\Sigma}_j^{\mathrm{p}}\hat{F}^+\right\}}{\mathrm{tr}\left\{\hat{F}\hat{F}^+\right\}}, \quad i,j = x, y, z \tag{2.375}$$

where $A_{ij}^{\mathrm{np}}$ can also be called the correlation analyzing power [10]. We can obtain the following expression from Eqs. (2.6), (2.347), and (2.350):

$$\begin{aligned}
\hat{F}\hat{\Sigma}_x^{\mathrm{n}}\hat{F}^+ &= \begin{pmatrix} \hat{t}_{++} & \hat{t}_{+-} \\ \hat{t}_{-+} & \hat{t}_{--} \end{pmatrix}\begin{pmatrix} \hat{0} & \hat{I} \\ \hat{I} & \hat{0} \end{pmatrix}\begin{pmatrix} \hat{t}_{++}^+ & \hat{t}_{-+}^+ \\ \hat{t}_{+-}^+ & \hat{t}_{--}^+ \end{pmatrix} \\
&= \begin{pmatrix} \hat{t}_{++} & \hat{t}_{+-} \\ \hat{t}_{-+} & \hat{t}_{--} \end{pmatrix}\begin{pmatrix} \hat{t}_{+-}^+ & \hat{t}_{--}^+ \\ \hat{t}_{++}^+ & \hat{t}_{-+}^+ \end{pmatrix} \\
&= \begin{pmatrix} \hat{t}_{++}\hat{t}_{+-}^+ + \hat{t}_{+-}\hat{t}_{++}^+ & \hat{t}_{++}\hat{t}_{--}^+ + \hat{t}_{+-}\hat{t}_{-+}^+ \\ \hat{t}_{-+}\hat{t}_{+-}^+ + \hat{t}_{--}\hat{t}_{++}^+ & \hat{t}_{-+}\hat{t}_{--}^+ + \hat{t}_{--}\hat{t}_{-+}^+ \end{pmatrix} \equiv \begin{pmatrix} \hat{X}_{++}^n & \hat{X}_{+-}^n \\ \hat{X}_{-+}^n & \hat{X}_{--}^n \end{pmatrix}
\end{aligned} \tag{2.376}$$

We can get the following results using Eqs. (2.376), (2.346), and (2.374):

$$\begin{aligned}
\hat{X}_{++}^{\mathrm{n}} &= \begin{pmatrix} A & -\mathrm{i}B \\ \mathrm{i}H & G \end{pmatrix}\begin{pmatrix} \mathrm{i}C^* & F^* \\ -D^* & \mathrm{i}E^* \end{pmatrix} + \begin{pmatrix} -\mathrm{i}C & -D \\ F & -\mathrm{i}E \end{pmatrix}\begin{pmatrix} A^* & -\mathrm{i}H^* \\ \mathrm{i}B^* & G^* \end{pmatrix} \\
&= \begin{pmatrix} \mathrm{i}(AC^* + BD^* - CA^* - DB^*) & AF^* + BE^* - CH^* - DG^* \\ -(HC^* + GD^* - FA^* - EB^*) & \mathrm{i}(HF^* + GE^* - FH^* - EG^*) \end{pmatrix}
\end{aligned} \tag{2.377}$$

$$\begin{aligned}
\hat{X}_{+-}^{\mathrm{n}} &= \begin{pmatrix} A & -\mathrm{i}B \\ \mathrm{i}H & G \end{pmatrix}\begin{pmatrix} G^* & -\mathrm{i}B^* \\ \mathrm{i}H^* & A^* \end{pmatrix} + \begin{pmatrix} -\mathrm{i}C & -D \\ F & -\mathrm{i}E \end{pmatrix}\begin{pmatrix} -\mathrm{i}E^* & -D^* \\ F^* & -\mathrm{i}C^* \end{pmatrix} \\
&= \begin{pmatrix} AG^* + BH^* - CE^* - DF^* & -\mathrm{i}(AB^* + BA^* - CD^* - DC^*) \\ \mathrm{i}(HG^* + GH^* - FE^* - EF^*) & HB^* + GA^* - FD^* - EC^* \end{pmatrix}
\end{aligned} \tag{2.378}$$

$$\hat{X}^{\mathrm{n}}_{-+} = \begin{pmatrix} \mathrm{i}E & F \\ -D & \mathrm{i}C \end{pmatrix}\begin{pmatrix} \mathrm{i}C^* & F^* \\ -D^* & \mathrm{i}E^* \end{pmatrix} + \begin{pmatrix} G & -\mathrm{i}H \\ \mathrm{i}B & A \end{pmatrix}\begin{pmatrix} A^* & -\mathrm{i}H^* \\ \mathrm{i}B^* & G^* \end{pmatrix}$$
$$= \begin{pmatrix} -(EC^* + FD^* - GA^* - HB^*) & \mathrm{i}(EF^* + FE^* - GH^* - HG^*) \\ -\mathrm{i}(DC^* + CD^* - BA^* - AB^*) & -(DF^* + CE^* - BH^* - AG^*) \end{pmatrix} \tag{2.379}$$

$$\hat{X}^{\mathrm{n}}_{--} = \begin{pmatrix} \mathrm{i}E & F \\ -D & \mathrm{i}C \end{pmatrix}\begin{pmatrix} G^* & -\mathrm{i}B^* \\ \mathrm{i}H^* & A^* \end{pmatrix} + \begin{pmatrix} G & -\mathrm{i}H \\ \mathrm{i}B & A \end{pmatrix}\begin{pmatrix} -\mathrm{i}E^* & -D^* \\ F^* & -\mathrm{i}C^* \end{pmatrix}$$
$$= \begin{pmatrix} \mathrm{i}(EG^* + FH^* - GE^* - HF^*) & EB^* + FA^* - GD^* - HC^* \\ -(DG^* + CH^* - BE^* - AF^*) & \mathrm{i}(DB^* + CA^* - BD^* - AC^*) \end{pmatrix} \tag{2.380}$$

$$A^{\mathrm{n}}_x = 0 \tag{2.381}$$

$$\hat{F}\hat{\Sigma}^{\mathrm{n}}_y\hat{F}^+ = \begin{pmatrix} \hat{t}_{++} & \hat{t}_{+-} \\ \hat{t}_{-+} & \hat{t}_{--} \end{pmatrix}\begin{pmatrix} \hat{0} & -\mathrm{i}\hat{I} \\ \mathrm{i}\hat{I} & \hat{0} \end{pmatrix}\begin{pmatrix} \hat{t}^+_{++} & \hat{t}^+_{-+} \\ \hat{t}^+_{+-} & \hat{t}^+_{--} \end{pmatrix}$$
$$= \mathrm{i}\begin{pmatrix} \hat{t}_{++} & \hat{t}_{+-} \\ \hat{t}_{-+} & \hat{t}_{--} \end{pmatrix}\begin{pmatrix} -\hat{t}^+_{+-} & -\hat{t}^+_{--} \\ \hat{t}^+_{++} & \hat{t}^+_{-+} \end{pmatrix}$$
$$= -\mathrm{i}\begin{pmatrix} \hat{t}_{++}\hat{t}^+_{+-} - \hat{t}_{+-}\hat{t}^+_{++} & \hat{t}_{++}\hat{t}^+_{--} - \hat{t}_{+-}\hat{t}^+_{-+} \\ \hat{t}_{-+}\hat{t}^+_{+-} - \hat{t}_{--}\hat{t}^+_{++} & \hat{t}_{-+}\hat{t}^+_{--} - \hat{t}_{--}\hat{t}^+_{-+} \end{pmatrix} \equiv \begin{pmatrix} \hat{Y}^n_{++} & \hat{Y}^n_{+-} \\ \hat{Y}^n_{-+} & \hat{Y}^n_{--} \end{pmatrix} \tag{2.382}$$

We can also get the following results using Eqs. (2.382), (2.346), and (2.374):

$$\hat{Y}^{\mathrm{n}}_{++} = -\mathrm{i}\begin{pmatrix} A & -\mathrm{i}B \\ \mathrm{i}H & G \end{pmatrix}\begin{pmatrix} \mathrm{i}C^* & F^* \\ -D^* & \mathrm{i}E^* \end{pmatrix} + \mathrm{i}\begin{pmatrix} -\mathrm{i}C & -D \\ F & -\mathrm{i}E \end{pmatrix}\begin{pmatrix} A^* & -\mathrm{i}H^* \\ \mathrm{i}B^* & G^* \end{pmatrix}$$
$$= \begin{pmatrix} AC^* + BD^* + CA^* + DB^* & -\mathrm{i}(AF^* + BE^* + CH^* + DG^*) \\ \mathrm{i}(HC^* + GD^* + FA^* + EB^*) & HF^* + GE^* + FH^* + EG^* \end{pmatrix} \tag{2.383}$$

$$\hat{Y}^{\mathrm{n}}_{+-} = -\mathrm{i}\begin{pmatrix} A & -\mathrm{i}B \\ \mathrm{i}H & G \end{pmatrix}\begin{pmatrix} G^* & -\mathrm{i}B^* \\ \mathrm{i}H^* & A^* \end{pmatrix} + \mathrm{i}\begin{pmatrix} -\mathrm{i}C & -D \\ F & -\mathrm{i}E \end{pmatrix}\begin{pmatrix} -\mathrm{i}E^* & -D^* \\ F^* & -\mathrm{i}C^* \end{pmatrix}$$
$$= \begin{pmatrix} -\mathrm{i}(AG^* + BH^* + CE^* + DF^*) & -(AB^* + BA^* + CD^* + DC^*) \\ HG^* + GH^* + FE^* + EF^* & -\mathrm{i}(HB^* + GA^* + FD^* + EC^*) \end{pmatrix} \tag{2.384}$$

$$\hat{Y}^{\rm n}_{-+} = -\mathrm{i}\begin{pmatrix} \mathrm{i}E & F \\ -D & \mathrm{i}C \end{pmatrix}\begin{pmatrix} \mathrm{i}C^* & F^* \\ -D^* & \mathrm{i}E^* \end{pmatrix} + \mathrm{i}\begin{pmatrix} G & -\mathrm{i}H \\ \mathrm{i}B & A \end{pmatrix}\begin{pmatrix} A^* & -\mathrm{i}H^* \\ \mathrm{i}B^* & G^* \end{pmatrix}$$
$$= \begin{pmatrix} \mathrm{i}(EC^* + FD^* + GA^* + HB^*) & EF^* + FE^* + GH^* + HG^* \\ -(DC^* + CD^* + BA^* + AB^*) & \mathrm{i}(DF^* + CE^* + BH^* + AG^*) \end{pmatrix} \tag{2.385}$$

$$\hat{Y}^{\rm n}_{--} = -\mathrm{i}\begin{pmatrix} \mathrm{i}E & F \\ -D & \mathrm{i}C \end{pmatrix}\begin{pmatrix} G^* & -\mathrm{i}B^* \\ \mathrm{i}H^* & A^* \end{pmatrix} + \mathrm{i}\begin{pmatrix} G & -\mathrm{i}H \\ \mathrm{i}B & A \end{pmatrix}\begin{pmatrix} -\mathrm{i}E^* & -D^* \\ F^* & -\mathrm{i}C^* \end{pmatrix}$$
$$= \begin{pmatrix} EG^* + FH^* + GE^* + HF^* & -\mathrm{i}(EB^* + FA^* + GD^* + HC^*) \\ \mathrm{i}(DG^* + CH^* + BE^* + AF^*) & DB^* + CA^* + BD^* + AC^* \end{pmatrix} \tag{2.386}$$

$$A_y^{\rm n} = \frac{1}{I_0}\mathrm{Re}(AC^* + BD^* + EG^* + FH^*) \tag{2.387}$$

$$\hat{F}\hat{\Sigma}_z^{\rm n}\hat{F}^+ = \begin{pmatrix} \hat{t}_{++} & \hat{t}_{+-} \\ \hat{t}_{-+} & \hat{t}_{--} \end{pmatrix}\begin{pmatrix} \hat{I} & \hat{0} \\ \hat{0} & -\hat{I} \end{pmatrix}\begin{pmatrix} \hat{t}^+_{++} & \hat{t}^+_{-+} \\ \hat{t}^+_{+-} & \hat{t}^+_{--} \end{pmatrix}$$
$$= \begin{pmatrix} \hat{t}_{++} & \hat{t}_{+-} \\ \hat{t}_{-+} & \hat{t}_{--} \end{pmatrix}\begin{pmatrix} \hat{t}^+_{++} & \hat{t}^+_{-+} \\ -\hat{t}^+_{+-} & -\hat{t}^+_{--} \end{pmatrix}$$
$$= \begin{pmatrix} \hat{t}_{++}\hat{t}^+_{++} - \hat{t}_{+-}\hat{t}^+_{+-} & \hat{t}_{++}\hat{t}^+_{-+} - \hat{t}_{+-}\hat{t}^+_{--} \\ \hat{t}_{-+}\hat{t}^+_{++} - \hat{t}_{--}\hat{t}^+_{+-} & \hat{t}_{-+}\hat{t}^+_{-+} - \hat{t}_{--}\hat{t}^+_{-+} \end{pmatrix} \equiv \begin{pmatrix} \hat{Z}^n_{++} & \hat{Z}^n_{+-} \\ \hat{Z}^n_{-+} & \hat{Z}^n_{--} \end{pmatrix} \tag{2.388}$$

The following results can be got using Eqs. (2.388), (2.346), and (2.374):

$$\hat{Z}^{\rm n}_{++} = \begin{pmatrix} A & -\mathrm{i}B \\ \mathrm{i}H & G \end{pmatrix}\begin{pmatrix} A^* & -\mathrm{i}H^* \\ \mathrm{i}B^* & G^* \end{pmatrix} - \begin{pmatrix} -\mathrm{i}C & -D \\ F & -\mathrm{i}E \end{pmatrix}\begin{pmatrix} \mathrm{i}C^* & F^* \\ -D^* & \mathrm{i}E^* \end{pmatrix}$$
$$= \begin{pmatrix} |A|^2 + |B|^2 - |C|^2 - |D|^2 & -\mathrm{i}(AH^* + BG^* - CF^* - DE^*) \\ \mathrm{i}(HA^* + GB^* - FC^* - ED^*) & |H|^2 + |G|^2 - |F|^2 - |E|^2 \end{pmatrix} \tag{2.389}$$

$$\hat{Z}^{\rm n}_{+-} = \begin{pmatrix} A & -\mathrm{i}B \\ \mathrm{i}H & G \end{pmatrix}\begin{pmatrix} -\mathrm{i}E^* & -D^* \\ F^* & -\mathrm{i}C^* \end{pmatrix} - \begin{pmatrix} -\mathrm{i}C & -D \\ F & -\mathrm{i}E \end{pmatrix}\begin{pmatrix} G^* & -\mathrm{i}B^* \\ \mathrm{i}H^* & A^* \end{pmatrix}$$
$$= \begin{pmatrix} -\mathrm{i}(AE^* + BF^* - CG^* - DH^*) & -(AD^* + BC^* - CB^* - DA^*) \\ HE^* + GF^* - FG^* - EH^* & -\mathrm{i}(HD^* + GC^* - FB^* - EA^*) \end{pmatrix} \tag{2.390}$$

$$\hat{Z}^{\rm n}_{-+} = \begin{pmatrix} \mathrm{i}E & F \\ -D & \mathrm{i}C \end{pmatrix}\begin{pmatrix} A^* & -\mathrm{i}H^* \\ \mathrm{i}B^* & G^* \end{pmatrix} - \begin{pmatrix} G & -\mathrm{i}H \\ \mathrm{i}B & A \end{pmatrix}\begin{pmatrix} \mathrm{i}C^* & F^* \\ -D^* & \mathrm{i}E^* \end{pmatrix}$$
$$= \begin{pmatrix} \mathrm{i}(EA^* + FB^* - GC^* - HD^*) & EH^* + FG^* - GF^* - HE^* \\ -(DA^* + CB^* - BC^* - AD^*) & \mathrm{i}(DH^* + CG^* - BF^* - AE^*) \end{pmatrix} \tag{2.391}$$

$$\hat{Z}^{\mathrm{n}}_{--}=\begin{pmatrix} \mathrm{i}E & F \\ -D & \mathrm{i}C \end{pmatrix}\begin{pmatrix} -\mathrm{i}E^* & -D^* \\ F^* & -\mathrm{i}C^* \end{pmatrix}-\begin{pmatrix} G & -\mathrm{i}H \\ \mathrm{i}B & A \end{pmatrix}\begin{pmatrix} G^* & -\mathrm{i}B^* \\ \mathrm{i}H^* & A^* \end{pmatrix}$$
$$=\begin{pmatrix} |E|^2+|F|^2-|G|^2-|H|^2 & -\mathrm{i}(ED^*+FC^*-GB^*-HA^*) \\ \mathrm{i}(DE^*+CF^*-BG^*-AH^*) & |D|^2+|C|^2-|B|^2-|A|^2 \end{pmatrix} \tag{2.392}$$

$$A^{\mathrm{n}}_{z}=0 \tag{2.393}$$

We can write the following expressions according to Eqs. (2.356), (2.351), (2.346), and (2.374):

$$\hat{F}\hat{\Sigma}^{\mathrm{p}}_{x}\hat{F}^{+}=\begin{pmatrix} \hat{t}_{++}\hat{\sigma}^{\mathrm{p}}_{x}\hat{t}^{+}_{++}+\hat{t}_{+-}\hat{\sigma}^{\mathrm{p}}_{x}\hat{t}^{+}_{+-} & \hat{t}_{++}\hat{\sigma}^{\mathrm{p}}_{x}\hat{t}^{+}_{-+}+\hat{t}_{+-}\hat{\sigma}^{\mathrm{p}}_{x}\hat{t}^{+}_{--} \\ \hat{t}_{-+}\hat{\sigma}^{\mathrm{p}}_{x}\hat{t}^{+}_{++}+\hat{t}_{--}\hat{\sigma}^{\mathrm{p}}_{x}\hat{t}^{+}_{+-} & \hat{t}_{-+}\hat{\sigma}^{\mathrm{p}}_{x}\hat{t}^{+}_{-+}+\hat{t}_{--}\hat{\sigma}^{\mathrm{p}}_{x}\hat{t}^{+}_{--} \end{pmatrix}$$
$$\equiv\begin{pmatrix} \hat{X}^{\mathrm{p}}_{++} & \hat{X}^{\mathrm{p}}_{+-} \\ \hat{X}^{\mathrm{p}}_{-+} & \hat{X}^{\mathrm{p}}_{--} \end{pmatrix} \tag{2.394}$$

$$\hat{X}^{\mathrm{p}}_{++}=\begin{pmatrix} A & -\mathrm{i}B \\ \mathrm{i}H & G \end{pmatrix}\begin{pmatrix} 0 & 1 \\ 1 & 0 \end{pmatrix}\begin{pmatrix} A^* & -\mathrm{i}H^* \\ \mathrm{i}B^* & G^* \end{pmatrix}+\begin{pmatrix} -\mathrm{i}C & -D \\ F & -\mathrm{i}E \end{pmatrix}\begin{pmatrix} 0 & 1 \\ 1 & 0 \end{pmatrix}\begin{pmatrix} \mathrm{i}C^* & F^* \\ -D^* & \mathrm{i}E^* \end{pmatrix}$$
$$=\begin{pmatrix} A & -\mathrm{i}B \\ \mathrm{i}H & G \end{pmatrix}\begin{pmatrix} \mathrm{i}B^* & G^* \\ A^* & -\mathrm{i}H^* \end{pmatrix}+\begin{pmatrix} -\mathrm{i}C & -D \\ F & -\mathrm{i}E \end{pmatrix}\begin{pmatrix} -D^* & \mathrm{i}E^* \\ \mathrm{i}C^* & F^* \end{pmatrix}$$
$$=\begin{pmatrix} \mathrm{i}(AB^*-BA^*+CD^*-DC^*) & AG^*-BH^*+CE^*-DF^* \\ -(HB^*-GA^*+FD^*-EC^*) & \mathrm{i}(HG^*-GH^*+FE^*-EF^*) \end{pmatrix} \tag{2.395}$$

$$\hat{X}^{\mathrm{p}}_{+-}=\begin{pmatrix} A & -\mathrm{i}B \\ \mathrm{i}H & G \end{pmatrix}\begin{pmatrix} 0 & 1 \\ 1 & 0 \end{pmatrix}\begin{pmatrix} -\mathrm{i}E^* & -D^* \\ F^* & -\mathrm{i}C^* \end{pmatrix}+\begin{pmatrix} -\mathrm{i}C & -D \\ F & -\mathrm{i}E \end{pmatrix}\begin{pmatrix} 0 & 1 \\ 1 & 0 \end{pmatrix}\begin{pmatrix} G^* & -\mathrm{i}B^* \\ \mathrm{i}H^* & A^* \end{pmatrix}$$
$$=\begin{pmatrix} A & -\mathrm{i}B \\ \mathrm{i}H & G \end{pmatrix}\begin{pmatrix} F^* & -\mathrm{i}C^* \\ -\mathrm{i}E^* & -D^* \end{pmatrix}+\begin{pmatrix} -\mathrm{i}C & -D \\ F & -\mathrm{i}E \end{pmatrix}\begin{pmatrix} \mathrm{i}H^* & A^* \\ G^* & -\mathrm{i}B^* \end{pmatrix}$$
$$=\begin{pmatrix} AF^*-BE^*+CH^*-DG^* & -\mathrm{i}(AC^*-BD^*+CA^*-DB^*) \\ \mathrm{i}(HF^*-GE^*+FH^*-EG^*) & HC^*-GD^*+FA^*-EB^* \end{pmatrix} \tag{2.396}$$

$$\hat{X}^{\mathrm{p}}_{-+}=\begin{pmatrix} \mathrm{i}E & F \\ -D & \mathrm{i}C \end{pmatrix}\begin{pmatrix} 0 & 1 \\ 1 & 0 \end{pmatrix}\begin{pmatrix} A^* & -\mathrm{i}H^* \\ \mathrm{i}B^* & G^* \end{pmatrix}+\begin{pmatrix} G & -\mathrm{i}H \\ \mathrm{i}B & A \end{pmatrix}\begin{pmatrix} 0 & 1 \\ 1 & 0 \end{pmatrix}\begin{pmatrix} \mathrm{i}C^* & F^* \\ -D^* & \mathrm{i}E^* \end{pmatrix}$$
$$=\begin{pmatrix} \mathrm{i}E & F \\ -D & \mathrm{i}C \end{pmatrix}\begin{pmatrix} \mathrm{i}B^* & G^* \\ A^* & -\mathrm{i}H^* \end{pmatrix}+\begin{pmatrix} G & -\mathrm{i}H \\ \mathrm{i}B & A \end{pmatrix}\begin{pmatrix} -D^* & \mathrm{i}E^* \\ \mathrm{i}C^* & F^* \end{pmatrix}$$
$$=\begin{pmatrix} -(EB^*-FA^*+GD^*-HC^*) & \mathrm{i}(EG^*-FH^*+GE^*-HF^*) \\ -\mathrm{i}(DB^*-CA^*+BD^*-AC^*) & -(DG^*-CH^*+BE^*-AF^*) \end{pmatrix} \tag{2.397}$$

$$\hat{X}^{\mathrm{p}}_{--} = \begin{pmatrix} \mathrm{i}E & F \\ -D & \mathrm{i}C \end{pmatrix} \begin{pmatrix} 0 & 1 \\ 1 & 0 \end{pmatrix} \begin{pmatrix} -\mathrm{i}E^* & -D^* \\ F^* & -\mathrm{i}C^* \end{pmatrix} + \begin{pmatrix} G & -\mathrm{i}H \\ \mathrm{i}B & A \end{pmatrix} \begin{pmatrix} 0 & 1 \\ 1 & 0 \end{pmatrix} \begin{pmatrix} G^* & -\mathrm{i}B^* \\ \mathrm{i}H^* & A^* \end{pmatrix}$$
$$= \begin{pmatrix} \mathrm{i}E & F \\ -D & \mathrm{i}C \end{pmatrix} \begin{pmatrix} F^* & -\mathrm{i}C^* \\ -\mathrm{i}E^* & -D^* \end{pmatrix} + \begin{pmatrix} G & -\mathrm{i}H \\ \mathrm{i}B & A \end{pmatrix} \begin{pmatrix} \mathrm{i}H^* & A^* \\ G^* & -\mathrm{i}B^* \end{pmatrix}$$
$$= \begin{pmatrix} \mathrm{i}(EF^* - FE^* + GH^* - HG^*) & EC^* - FD^* + GA^* - HB^* \\ -(DF^* - CE^* + BH^* - AG^*) & \mathrm{i}(DC^* - CD^* + BA^* - AB^*) \end{pmatrix} \tag{2.398}$$

$$A^{\mathrm{p}}_x = 0 \tag{2.399}$$

We can also obtain the following expressions according to Eqs. (2.356), (2.351), (2.346), and (2.374):

$$\hat{F}\hat{\Sigma}^{\mathrm{p}}_y\hat{F}^+ = \begin{pmatrix} \hat{t}_{++}\hat{\sigma}^{\mathrm{p}}_y\hat{t}^+_{++} + \hat{t}_{+-}\hat{\sigma}^{\mathrm{p}}_y\hat{t}^+_{+-} & \hat{t}_{++}\hat{\sigma}^{\mathrm{p}}_y\hat{t}^+_{-+} + \hat{t}_{+-}\hat{\sigma}^{\mathrm{p}}_y\hat{t}^+_{--} \\ \hat{t}_{-+}\hat{\sigma}^{\mathrm{p}}_y\hat{t}^+_{++} + \hat{t}_{--}\hat{\sigma}^{\mathrm{p}}_y\hat{t}^+_{+-} & \hat{t}_{-+}\hat{\sigma}^{\mathrm{p}}_y\hat{t}^+_{-+} + \hat{t}_{--}\hat{\sigma}^{\mathrm{p}}_y\hat{t}^+_{--} \end{pmatrix} \equiv \begin{pmatrix} \hat{Y}^{\mathrm{p}}_{++} & \hat{Y}^{\mathrm{p}}_{+-} \\ \hat{Y}^{\mathrm{p}}_{-+} & \hat{Y}^{\mathrm{p}}_{--} \end{pmatrix} \tag{2.400}$$

$$\hat{Y}^{\mathrm{p}}_{++} = \begin{pmatrix} A & -\mathrm{i}B \\ \mathrm{i}H & G \end{pmatrix} \begin{pmatrix} 0 & -\mathrm{i} \\ \mathrm{i} & 0 \end{pmatrix} \begin{pmatrix} A^* & -\mathrm{i}H^* \\ \mathrm{i}B^* & G^* \end{pmatrix} + \begin{pmatrix} -\mathrm{i}C & -D \\ F & -\mathrm{i}E \end{pmatrix} \begin{pmatrix} 0 & -\mathrm{i} \\ \mathrm{i} & 0 \end{pmatrix} \begin{pmatrix} \mathrm{i}C^* & F^* \\ -D^* & \mathrm{i}E^* \end{pmatrix}$$
$$= \begin{pmatrix} A & -\mathrm{i}B \\ \mathrm{i}H & G \end{pmatrix} \begin{pmatrix} B^* & -\mathrm{i}G^* \\ \mathrm{i}A^* & H^* \end{pmatrix} + \begin{pmatrix} -\mathrm{i}C & -D \\ F & -\mathrm{i}E \end{pmatrix} \begin{pmatrix} \mathrm{i}D^* & E^* \\ -C^* & \mathrm{i}F^* \end{pmatrix}$$
$$= \begin{pmatrix} AB^* + BA^* + CD^* + DC^* & -\mathrm{i}(AG^* + BH^* + CE^* + DF^*) \\ \mathrm{i}(HB^* + GA^* + FD^* + EC^*) & HG^* + GH^* + FE^* + EF^* \end{pmatrix} \tag{2.401}$$

$$\hat{Y}^{\mathrm{p}}_{+-} = \begin{pmatrix} A & -\mathrm{i}B \\ \mathrm{i}H & G \end{pmatrix} \begin{pmatrix} 0 & -\mathrm{i} \\ \mathrm{i} & 0 \end{pmatrix} \begin{pmatrix} -\mathrm{i}E^* & -D^* \\ F^* & -\mathrm{i}C^* \end{pmatrix} + \begin{pmatrix} -\mathrm{i}C & -D \\ F & -\mathrm{i}E \end{pmatrix} \begin{pmatrix} 0 & -\mathrm{i} \\ \mathrm{i} & 0 \end{pmatrix} \begin{pmatrix} G^* & -\mathrm{i}B^* \\ \mathrm{i}H^* & A^* \end{pmatrix}$$
$$= \begin{pmatrix} A & -\mathrm{i}B \\ \mathrm{i}H & G \end{pmatrix} \begin{pmatrix} -\mathrm{i}F^* & -C^* \\ E^* & -\mathrm{i}D^* \end{pmatrix} + \begin{pmatrix} -\mathrm{i}C & -D \\ F & -\mathrm{i}E \end{pmatrix} \begin{pmatrix} H^* & -\mathrm{i}A^* \\ \mathrm{i}G^* & B^* \end{pmatrix}$$
$$= \begin{pmatrix} -\mathrm{i}(AF^* + BE^* + CH^* + DG^*) & -(AC^* + BD^* + CA^* + DB^*) \\ HF^* + GE^* + FH^* + EG^* & -\mathrm{i}(HC^* + GD^* + FA^* + EB^*) \end{pmatrix} \tag{2.402}$$

$$\hat{Y}^{\mathrm{p}}_{-+} = \begin{pmatrix} \mathrm{i}E & F \\ -D & \mathrm{i}C \end{pmatrix} \begin{pmatrix} 0 & -\mathrm{i} \\ \mathrm{i} & 0 \end{pmatrix} \begin{pmatrix} A^* & -\mathrm{i}H^* \\ \mathrm{i}B^* & G^* \end{pmatrix} + \begin{pmatrix} G & -\mathrm{i}H \\ \mathrm{i}B & A \end{pmatrix} \begin{pmatrix} 0 & -\mathrm{i} \\ \mathrm{i} & 0 \end{pmatrix} \begin{pmatrix} \mathrm{i}C^* & F^* \\ -D^* & \mathrm{i}E^* \end{pmatrix}$$
$$= \begin{pmatrix} \mathrm{i}E & F \\ -D & \mathrm{i}C \end{pmatrix} \begin{pmatrix} B^* & -\mathrm{i}G^* \\ \mathrm{i}A^* & H^* \end{pmatrix} + \begin{pmatrix} G & -\mathrm{i}H \\ \mathrm{i}B & A \end{pmatrix} \begin{pmatrix} \mathrm{i}D^* & E^* \\ -C^* & \mathrm{i}F^* \end{pmatrix}$$
$$= \begin{pmatrix} \mathrm{i}(EB^* + FA^* + GD^* + HC^*) & EG^* + FH^* + GE^* + HF^* \\ -(DB^* + CA^* + BD^* + AC^*) & \mathrm{i}(DG^* + CH^* + BE^* + AF^*) \end{pmatrix} \tag{2.403}$$

$$\hat{Y}^{\mathrm{p}}_{--}=\begin{pmatrix} \mathrm{i}E & F \\ -D & \mathrm{i}C \end{pmatrix}\begin{pmatrix} 0 & -\mathrm{i} \\ \mathrm{i} & 0 \end{pmatrix}\begin{pmatrix} -\mathrm{i}E^* & -D^* \\ F^* & -\mathrm{i}C^* \end{pmatrix}+\begin{pmatrix} G & -\mathrm{i}H \\ \mathrm{i}B & A \end{pmatrix}\begin{pmatrix} 0 & -\mathrm{i} \\ \mathrm{i} & 0 \end{pmatrix}\begin{pmatrix} G^* & -\mathrm{i}B^* \\ \mathrm{i}H^* & A^* \end{pmatrix}$$
$$=\begin{pmatrix} \mathrm{i}E & F \\ -D & \mathrm{i}C \end{pmatrix}\begin{pmatrix} -\mathrm{i}F^* & -C^* \\ E^* & -\mathrm{i}D^* \end{pmatrix}+\begin{pmatrix} G & -\mathrm{i}H \\ \mathrm{i}B & A \end{pmatrix}\begin{pmatrix} H^* & -\mathrm{i}A^* \\ \mathrm{i}G^* & B^* \end{pmatrix}$$
$$=\begin{pmatrix} EF^*+FE^*+GH^*+HG^* & -\mathrm{i}(EC^*+FD^*+GA^*+HB^*) \\ \mathrm{i}(DF^*+CE^*+BH^*+AG^*) & DC^*+CD^*+BA^*+AB^* \end{pmatrix} \tag{2.404}$$

$$A_y^{\mathrm{p}}=\frac{1}{I_0}\,\mathrm{Re}\,(AB^*+CD^*+EF^*+GH^*) \tag{2.405}$$

And we can also obtain the following expressions according to Eqs. (2.356), (2.351), (2.346), and (2.374):

$$\hat{F}\hat{\Sigma}_z^{\mathrm{p}}\hat{F}^+=\begin{pmatrix} \hat{t}_{++}\hat{\sigma}_z^{\mathrm{p}}\hat{t}^+_{++}+\hat{t}_{+-}\hat{\sigma}_z^{\mathrm{p}}\hat{t}^+_{+-} & \hat{t}_{++}\hat{\sigma}_z^{\mathrm{p}}\hat{t}^+_{-+}+\hat{t}_{+-}\hat{\sigma}_z^{\mathrm{p}}\hat{t}^+_{--} \\ \hat{t}_{-+}\hat{\sigma}_z^{\mathrm{p}}\hat{t}^+_{++}+\hat{t}_{--}\hat{\sigma}_z^{\mathrm{p}}\hat{t}^+_{+-} & \hat{t}_{-+}\hat{\sigma}_z^{\mathrm{p}}\hat{t}^+_{-+}+\hat{t}_{--}\hat{\sigma}_z^{\mathrm{p}}\hat{t}^+_{--} \end{pmatrix}\equiv\begin{pmatrix} \hat{Z}^{\mathrm{p}}_{++} & \hat{Z}^{\mathrm{p}}_{+-} \\ \hat{Z}^{\mathrm{p}}_{-+} & \hat{Z}^{\mathrm{p}}_{--} \end{pmatrix} \tag{2.406}$$

$$\hat{Z}^{\mathrm{p}}_{++}=\begin{pmatrix} A & -\mathrm{i}B \\ \mathrm{i}H & G \end{pmatrix}\begin{pmatrix} 1 & 0 \\ 0 & -1 \end{pmatrix}\begin{pmatrix} A^* & -\mathrm{i}H^* \\ \mathrm{i}B^* & G^* \end{pmatrix}+\begin{pmatrix} -\mathrm{i}C & -D \\ F & -\mathrm{i}E \end{pmatrix}\begin{pmatrix} 1 & 0 \\ 0 & -1 \end{pmatrix}\begin{pmatrix} \mathrm{i}C^* & F^* \\ -D^* & \mathrm{i}E^* \end{pmatrix}$$
$$=\begin{pmatrix} A & -\mathrm{i}B \\ \mathrm{i}H & G \end{pmatrix}\begin{pmatrix} A^* & -\mathrm{i}H^* \\ -\mathrm{i}B^* & -G^* \end{pmatrix}+\begin{pmatrix} -\mathrm{i}C & -D \\ F & -\mathrm{i}E \end{pmatrix}\begin{pmatrix} \mathrm{i}C^* & F^* \\ D^* & -\mathrm{i}E^* \end{pmatrix}$$
$$=\begin{pmatrix} |A|^2-|B|^2+|C|^2-|D|^2 & -\mathrm{i}(AH^*-BG^*+CF^*-DE^*) \\ \mathrm{i}(HA^*-GB^*+FC^*-ED^*) & |H|^2-|G|^2+|F|^2-|E|^2 \end{pmatrix} \tag{2.407}$$

$$\hat{Z}^{\mathrm{p}}_{+-}=\begin{pmatrix} A & -\mathrm{i}B \\ \mathrm{i}H & G \end{pmatrix}\begin{pmatrix} 1 & 0 \\ 0 & -1 \end{pmatrix}\begin{pmatrix} -\mathrm{i}E^* & -D^* \\ F^* & -\mathrm{i}C^* \end{pmatrix}+\begin{pmatrix} -\mathrm{i}C & -D \\ F & -\mathrm{i}E \end{pmatrix}\begin{pmatrix} 1 & 0 \\ 0 & -1 \end{pmatrix}\begin{pmatrix} G^* & -\mathrm{i}B^* \\ \mathrm{i}H^* & A^* \end{pmatrix}$$
$$=\begin{pmatrix} A & -\mathrm{i}B \\ \mathrm{i}H & G \end{pmatrix}\begin{pmatrix} -\mathrm{i}E^* & -D^* \\ -F^* & \mathrm{i}C^* \end{pmatrix}+\begin{pmatrix} -\mathrm{i}C & -D \\ F & -\mathrm{i}E \end{pmatrix}\begin{pmatrix} G^* & -\mathrm{i}B^* \\ -\mathrm{i}H^* & -A^* \end{pmatrix}$$
$$=\begin{pmatrix} -\mathrm{i}(AE^*-BF^*+CG^*-DH^*) & -(AD^*-BC^*+CB^*-DA^*) \\ HE^*-GF^*+FG^*-EH^* & -\mathrm{i}(HD^*-GC^*+FB^*-EA^*) \end{pmatrix} \tag{2.408}$$

$$\hat{Z}^{\mathrm{p}}_{-+}=\begin{pmatrix} \mathrm{i}E & F \\ -D & \mathrm{i}C \end{pmatrix}\begin{pmatrix} 1 & 0 \\ 0 & -1 \end{pmatrix}\begin{pmatrix} A^* & -\mathrm{i}H^* \\ \mathrm{i}B^* & G^* \end{pmatrix}+\begin{pmatrix} G & -\mathrm{i}H \\ \mathrm{i}B & A \end{pmatrix}\begin{pmatrix} 1 & 0 \\ 0 & -1 \end{pmatrix}\begin{pmatrix} \mathrm{i}C^* & F^* \\ -D^* & \mathrm{i}E^* \end{pmatrix}$$
$$=\begin{pmatrix} \mathrm{i}E & F \\ -D & \mathrm{i}C \end{pmatrix}\begin{pmatrix} A^* & -\mathrm{i}H^* \\ -\mathrm{i}B^* & -G^* \end{pmatrix}+\begin{pmatrix} G & -\mathrm{i}H \\ \mathrm{i}B & A \end{pmatrix}\begin{pmatrix} \mathrm{i}C^* & F^* \\ D^* & -\mathrm{i}E^* \end{pmatrix}$$
$$=\begin{pmatrix} \mathrm{i}(EA^*-FB^*+GC^*-HD^*) & EH^*-FG^*+GF^*-HE^* \\ -(DA^*-CB^*+BC^*-AD^*) & \mathrm{i}(DH^*-CG^*+BF^*-AE^*) \end{pmatrix} \tag{2.409}$$

$$\hat{Z}^{\mathrm{p}}_{--}=\begin{pmatrix} \mathrm{i}E & F \\ -D & \mathrm{i}C \end{pmatrix}\begin{pmatrix} 1 & 0 \\ 0 & -1 \end{pmatrix}\begin{pmatrix} -\mathrm{i}E^* & -D^* \\ F^* & -\mathrm{i}C^* \end{pmatrix}+\begin{pmatrix} G & -\mathrm{i}H \\ \mathrm{i}B & A \end{pmatrix}\begin{pmatrix} 1 & 0 \\ 0 & -1 \end{pmatrix}\begin{pmatrix} G^* & -\mathrm{i}B^* \\ \mathrm{i}H^* & A^* \end{pmatrix}$$
$$=\begin{pmatrix} \mathrm{i}E & F \\ -D & \mathrm{i}C \end{pmatrix}\begin{pmatrix} -\mathrm{i}E^* & -D^* \\ -F^* & \mathrm{i}C^* \end{pmatrix}+\begin{pmatrix} G & -\mathrm{i}H \\ \mathrm{i}B & A \end{pmatrix}\begin{pmatrix} G^* & -\mathrm{i}B^* \\ -\mathrm{i}H^* & -A^* \end{pmatrix}$$
$$=\begin{pmatrix} |E|^2-|F|^2+|G|^2-|H|^2 & -\mathrm{i}(ED^*-FC^*+GB^*-HA^*) \\ \mathrm{i}(DE^*-CF^*+BG^*-AH^*) & |D|^2-|C|^2+|B|^2-|A|^2 \end{pmatrix} \tag{2.410}$$

$$A^{\mathrm{p}}_z=0 \tag{2.411}$$

Refer to Eq. (2.376) we can write

$$\hat{F}\hat{\Sigma}^{\mathrm{n}}_x\hat{\Sigma}^{\mathrm{p}}_i\hat{F}^+=\begin{pmatrix} \hat{t}_{++}\hat{\sigma}^{\mathrm{p}}_i\hat{t}^+_{+-}+\hat{t}_{+-}\hat{\sigma}^{\mathrm{p}}_i\hat{t}^+_{++} & \hat{t}_{++}\hat{\sigma}^{\mathrm{p}}_i\hat{t}^+_{--}+\hat{t}_{+-}\hat{\sigma}^{\mathrm{p}}_i\hat{t}^+_{-+} \\ \hat{t}_{-+}\hat{\sigma}^{\mathrm{p}}_i\hat{t}^+_{+-}+\hat{t}_{--}\hat{\sigma}^{\mathrm{p}}_i\hat{t}^+_{++} & \hat{t}_{-+}\hat{\sigma}^{\mathrm{p}}_i\hat{t}^+_{--}+\hat{t}_{--}\hat{\sigma}^{\mathrm{p}}_i\hat{t}^+_{-+} \end{pmatrix}$$
$$\equiv\begin{pmatrix} \hat{X}^i_{++} & \hat{X}^i_{+-} \\ \hat{X}^i_{-+} & \hat{X}^i_{--} \end{pmatrix},\quad i=x,y,z \tag{2.412}$$

$$\hat{X}^x_{++}=\begin{pmatrix} A & -\mathrm{i}B \\ \mathrm{i}H & G \end{pmatrix}\begin{pmatrix} 0 & 1 \\ 1 & 0 \end{pmatrix}\begin{pmatrix} \mathrm{i}C^* & F^* \\ -D^* & \mathrm{i}E^* \end{pmatrix}+\begin{pmatrix} -\mathrm{i}C & -D \\ F & -\mathrm{i}E \end{pmatrix}\begin{pmatrix} 0 & 1 \\ 1 & 0 \end{pmatrix}\begin{pmatrix} A^* & -\mathrm{i}H^* \\ \mathrm{i}B^* & G^* \end{pmatrix}$$
$$=\begin{pmatrix} A & -\mathrm{i}B \\ \mathrm{i}H & G \end{pmatrix}\begin{pmatrix} -D^* & \mathrm{i}E^* \\ \mathrm{i}C^* & F^* \end{pmatrix}+\begin{pmatrix} -\mathrm{i}C & -D \\ F & -\mathrm{i}E \end{pmatrix}\begin{pmatrix} \mathrm{i}B^* & G^* \\ A^* & -\mathrm{i}H^* \end{pmatrix}$$
$$=\begin{pmatrix} -(AD^*-BC^*-CB^*+DA^*) & \mathrm{i}(AE^*-BF^*-CG^*+DH^*) \\ -\mathrm{i}(HD^*-GC^*-FB^*+EA^*) & -(HE^*-GF^*-FG^*+EH^*) \end{pmatrix} \tag{2.413}$$

$$\hat{X}^x_{+-}=\begin{pmatrix} A & -\mathrm{i}B \\ \mathrm{i}H & G \end{pmatrix}\begin{pmatrix} 0 & 1 \\ 1 & 0 \end{pmatrix}\begin{pmatrix} G^* & -\mathrm{i}B^* \\ \mathrm{i}H^* & A^* \end{pmatrix}+\begin{pmatrix} -\mathrm{i}C & -D \\ F & -\mathrm{i}E \end{pmatrix}\begin{pmatrix} 0 & 1 \\ 1 & 0 \end{pmatrix}\begin{pmatrix} -\mathrm{i}E^* & -D^* \\ F^* & -\mathrm{i}C^* \end{pmatrix}$$
$$=\begin{pmatrix} A & -\mathrm{i}B \\ \mathrm{i}H & G \end{pmatrix}\begin{pmatrix} \mathrm{i}H^* & A^* \\ G^* & -\mathrm{i}B^* \end{pmatrix}+\begin{pmatrix} -\mathrm{i}C & -D \\ F & -\mathrm{i}E \end{pmatrix}\begin{pmatrix} F^* & -\mathrm{i}C^* \\ -\mathrm{i}E^* & -D^* \end{pmatrix}$$
$$=\begin{pmatrix} \mathrm{i}(AH^*-BG^*-CF^*+DE^*) & |A|^2-|B|^2-|C|^2+|D|^2 \\ -\left(|H|^2-|G|^2-|F|^2+|E|^2\right) & \mathrm{i}(HA^*-GB^*-FC^*+ED^*) \end{pmatrix} \tag{2.414}$$

$$\hat{X}^x_{-+}=\begin{pmatrix} \mathrm{i}E & F \\ -D & \mathrm{i}C \end{pmatrix}\begin{pmatrix} 0 & 1 \\ 1 & 0 \end{pmatrix}\begin{pmatrix} \mathrm{i}C^* & F^* \\ -D^* & \mathrm{i}E^* \end{pmatrix}+\begin{pmatrix} G & -\mathrm{i}H \\ \mathrm{i}B & A \end{pmatrix}\begin{pmatrix} 0 & 1 \\ 1 & 0 \end{pmatrix}\begin{pmatrix} A^* & -\mathrm{i}H^* \\ \mathrm{i}B^* & G^* \end{pmatrix}$$
$$=\begin{pmatrix} \mathrm{i}E & F \\ -D & \mathrm{i}C \end{pmatrix}\begin{pmatrix} -D^* & \mathrm{i}E^* \\ \mathrm{i}C^* & F^* \end{pmatrix}+\begin{pmatrix} G & -\mathrm{i}H \\ \mathrm{i}B & A \end{pmatrix}\begin{pmatrix} \mathrm{i}B^* & G^* \\ A^* & -\mathrm{i}H^* \end{pmatrix}$$
$$=\begin{pmatrix} -\mathrm{i}(ED^*-FC^*-GB^*+HA^*) & -\left(|E|^2-|F|^2-|G|^2+|H|^2\right) \\ |D|^2-|C|^2-|B|^2+|A|^2 & -\mathrm{i}(DE^*-CF^*-BG^*+AH^*) \end{pmatrix} \tag{2.415}$$

$$\hat{X}^x_{--}=\begin{pmatrix} iE & F\\ -D & iC\end{pmatrix}\begin{pmatrix}0 & 1\\ 1 & 0\end{pmatrix}\begin{pmatrix}G^* & -iB^*\\ iH^* & A^*\end{pmatrix}+\begin{pmatrix}G & -iH\\ iB & A\end{pmatrix}\begin{pmatrix}0 & 1\\ 1 & 0\end{pmatrix}\begin{pmatrix}-iE^* & -D^*\\ F^* & -iC^*\end{pmatrix}$$
$$=\begin{pmatrix} iE & F\\ -D & iC\end{pmatrix}\begin{pmatrix}iH^* & A^*\\ G^* & -iB^*\end{pmatrix}+\begin{pmatrix}G & -iH\\ iB & A\end{pmatrix}\begin{pmatrix}F^* & -iC^*\\ -iE^* & -D^*\end{pmatrix}$$
$$=\begin{pmatrix}-(EH^*-FG^*-GF^*+HE^*) & i(EA^*-FB^*-GC^*+HD^*)\\ -i(DH^*-CG^*-BF^*+AE^*) & -(DA^*-CB^*-BC^*+AD^*)\end{pmatrix} \tag{2.416}$$

$$A^{np}_{xx}=-\frac{1}{I_0}\mathrm{Re}(AD^*-BC^*+EH^*-FG^*) \tag{2.417}$$

$$\hat{X}^y_{++}=\begin{pmatrix} A & -iB\\ iH & G\end{pmatrix}\begin{pmatrix}0 & -i\\ i & 0\end{pmatrix}\begin{pmatrix}iC^* & F^*\\ -D^* & iE^*\end{pmatrix}+\begin{pmatrix}-iC & -D\\ F & -iE\end{pmatrix}\begin{pmatrix}0 & -i\\ i & 0\end{pmatrix}\begin{pmatrix}A^* & -iH^*\\ iB^* & G^*\end{pmatrix}$$
$$=\begin{pmatrix} A & -iB\\ iH & G\end{pmatrix}\begin{pmatrix}iD^* & E^*\\ -C^* & iF^*\end{pmatrix}+\begin{pmatrix}-iC & -D\\ F & -iE\end{pmatrix}\begin{pmatrix}B^* & -iG^*\\ iA^* & H^*\end{pmatrix}$$
$$=\begin{pmatrix}i(AD^*+BC^*-CB^*-DA^*) & AE^*+BF^*-CG^*-DH^*\\ -(HD^*+GC^*-FB^*-EA^*) & i(HE^*+GF^*-FG^*-EH^*)\end{pmatrix} \tag{2.418}$$

$$\hat{X}^y_{+-}=\begin{pmatrix} A & -iB\\ iH & G\end{pmatrix}\begin{pmatrix}0 & -i\\ i & 0\end{pmatrix}\begin{pmatrix}G^* & -iB^*\\ iH^* & A^*\end{pmatrix}+\begin{pmatrix}-iC & -D\\ F & -iE\end{pmatrix}\begin{pmatrix}0 & -i\\ i & 0\end{pmatrix}\begin{pmatrix}-iE^* & -D^*\\ F^* & -iC^*\end{pmatrix}$$
$$=\begin{pmatrix} A & -iB\\ iH & G\end{pmatrix}\begin{pmatrix}H^* & -iA^*\\ iG^* & B^*\end{pmatrix}+\begin{pmatrix}-iC & -D\\ F & -iE\end{pmatrix}\begin{pmatrix}-iF^* & -C^*\\ E^* & -iD^*\end{pmatrix}$$
$$=\begin{pmatrix}AH^*+BG^*-CF^*-DE^* & -i\left(|A|^2+|B|^2-|C|^2-|D|^2\right)\\ i\left(|H|^2+|G|^2-|F|^2-|E|^2\right) & HA^*+GB^*-FC^*-ED^*\end{pmatrix} \tag{2.419}$$

$$\hat{X}^y_{-+}=\begin{pmatrix} iE & F\\ -D & iC\end{pmatrix}\begin{pmatrix}0 & -i\\ i & 0\end{pmatrix}\begin{pmatrix}iC^* & F^*\\ -D^* & iE^*\end{pmatrix}+\begin{pmatrix}G & -iH\\ iB & A\end{pmatrix}\begin{pmatrix}0 & -i\\ i & 0\end{pmatrix}\begin{pmatrix}A^* & -iH^*\\ iB^* & G^*\end{pmatrix}$$
$$=\begin{pmatrix} iE & F\\ -D & iC\end{pmatrix}\begin{pmatrix}iD^* & E^*\\ -C^* & iF^*\end{pmatrix}+\begin{pmatrix}G & -iH\\ iB & A\end{pmatrix}\begin{pmatrix}B^* & -iG^*\\ iA^* & H^*\end{pmatrix}$$
$$=\begin{pmatrix}-(ED^*+FC^*-GB^*-HA^*) & i\left(|E|^2+|F|^2-|G|^2-|H|^2\right)\\ -i\left(|D|^2+|C|^2-|B|^2-|A|^2\right) & -(DE^*+CF^*-BG^*-AH^*)\end{pmatrix} \tag{2.420}$$

$$\hat{X}^y_{--}=\begin{pmatrix} iE & F\\ -D & iC\end{pmatrix}\begin{pmatrix}0 & -i\\ i & 0\end{pmatrix}\begin{pmatrix}G^* & -iB^*\\ iH^* & A^*\end{pmatrix}+\begin{pmatrix}G & -iH\\ iB & A\end{pmatrix}\begin{pmatrix}0 & -i\\ i & 0\end{pmatrix}\begin{pmatrix}-iE^* & -D^*\\ F^* & -iC^*\end{pmatrix}$$
$$=\begin{pmatrix} iE & F\\ -D & iC\end{pmatrix}\begin{pmatrix}H^* & -iA^*\\ iG^* & B^*\end{pmatrix}+\begin{pmatrix}G & -iH\\ iB & A\end{pmatrix}\begin{pmatrix}-iF^* & -C^*\\ E^* & -iD^*\end{pmatrix}$$
$$=\begin{pmatrix}i(EH^*+FG^*-GF^*-HE^*) & EA^*+FB^*-GC^*-HD^*\\ -(DH^*+CG^*-BF^*-AE^*) & i(DA^*+CB^*-BC^*-AD^*)\end{pmatrix} \tag{2.421}$$

$$A_{xy}^{\mathrm{np}}=0 \tag{2.422}$$

$$\begin{aligned}
\hat{X}_{++}^{z}&=\begin{pmatrix} A & -\mathrm{i}B \\ \mathrm{i}H & G \end{pmatrix}\begin{pmatrix} 1 & 0 \\ 0 & -1 \end{pmatrix}\begin{pmatrix} \mathrm{i}C^{*} & F^{*} \\ -D^{*} & \mathrm{i}E^{*} \end{pmatrix}+\begin{pmatrix} -\mathrm{i}C & -D \\ F & -\mathrm{i}E \end{pmatrix}\begin{pmatrix} 1 & 0 \\ 0 & -1 \end{pmatrix}\begin{pmatrix} A^{*} & -\mathrm{i}H^{*} \\ \mathrm{i}B^{*} & G^{*} \end{pmatrix}\\
&=\begin{pmatrix} A & -\mathrm{i}B \\ \mathrm{i}H & G \end{pmatrix}\begin{pmatrix} \mathrm{i}C^{*} & F^{*} \\ D^{*} & -\mathrm{i}E^{*} \end{pmatrix}+\begin{pmatrix} -\mathrm{i}C & -D \\ F & -\mathrm{i}E \end{pmatrix}\begin{pmatrix} A^{*} & -\mathrm{i}H^{*} \\ -\mathrm{i}B^{*} & G^{*} \end{pmatrix}\\
&=\begin{pmatrix} \mathrm{i}(AC^{*}-BD^{*}-CA^{*}+DB^{*}) & AF^{*}-BE^{*}-CH^{*}+DG^{*} \\ -(HC^{*}-GD^{*}-FA^{*}+EB^{*}) & \mathrm{i}(HF^{*}-GE^{*}-FH^{*}+EG^{*}) \end{pmatrix}
\end{aligned} \tag{2.423}$$

$$\begin{aligned}
\hat{X}_{+-}^{z}&=\begin{pmatrix} A & -\mathrm{i}B \\ \mathrm{i}H & G \end{pmatrix}\begin{pmatrix} 1 & 0 \\ 0 & -1 \end{pmatrix}\begin{pmatrix} G^{*} & -\mathrm{i}B^{*} \\ \mathrm{i}H^{*} & A^{*} \end{pmatrix}+\begin{pmatrix} -\mathrm{i}C & -D \\ F & -\mathrm{i}E \end{pmatrix}\begin{pmatrix} 1 & 0 \\ 0 & -1 \end{pmatrix}\begin{pmatrix} -\mathrm{i}E^{*} & -D^{*} \\ F^{*} & -\mathrm{i}C^{*} \end{pmatrix}\\
&=\begin{pmatrix} A & -\mathrm{i}B \\ \mathrm{i}H & G \end{pmatrix}\begin{pmatrix} G^{*} & -\mathrm{i}B^{*} \\ -\mathrm{i}H^{*} & -A^{*} \end{pmatrix}+\begin{pmatrix} -\mathrm{i}C & -D \\ F & -\mathrm{i}E \end{pmatrix}\begin{pmatrix} -\mathrm{i}E^{*} & -D^{*} \\ -F^{*} & \mathrm{i}C^{*} \end{pmatrix}\\
&=\begin{pmatrix} AG^{*}-BH^{*}-CE^{*}+DF^{*} & -\mathrm{i}(AB^{*}-BA^{*}-CD^{*}+DC^{*}) \\ \mathrm{i}(HG^{*}-GH^{*}-FE^{*}+EF^{*}) & HB^{*}-GA^{*}-FD^{*}-EC^{*} \end{pmatrix}
\end{aligned} \tag{2.424}$$

$$\begin{aligned}
\hat{X}_{-+}^{z}&=\begin{pmatrix} \mathrm{i}E & F \\ -D & \mathrm{i}C \end{pmatrix}\begin{pmatrix} 1 & 0 \\ 0 & -1 \end{pmatrix}\begin{pmatrix} \mathrm{i}C^{*} & F^{*} \\ -D^{*} & \mathrm{i}E^{*} \end{pmatrix}+\begin{pmatrix} G & -\mathrm{i}H \\ \mathrm{i}B & A \end{pmatrix}\begin{pmatrix} 1 & 0 \\ 0 & -1 \end{pmatrix}\begin{pmatrix} A^{*} & -\mathrm{i}H^{*} \\ \mathrm{i}B^{*} & G^{*} \end{pmatrix}\\
&=\begin{pmatrix} \mathrm{i}E & F \\ -D & \mathrm{i}C \end{pmatrix}\begin{pmatrix} \mathrm{i}C^{*} & F^{*} \\ D^{*} & -\mathrm{i}E^{*} \end{pmatrix}+\begin{pmatrix} G & -\mathrm{i}H \\ \mathrm{i}B & A \end{pmatrix}\begin{pmatrix} A^{*} & -\mathrm{i}H^{*} \\ -\mathrm{i}B^{*} & -G^{*} \end{pmatrix}\\
&=\begin{pmatrix} -(EC^{*}-FD^{*}-GA^{*}+HB^{*}) & \mathrm{i}(EF^{*}-FE^{*}-GH^{*}+HG^{*}) \\ -\mathrm{i}(DC^{*}-CD^{*}-BA^{*}+AB^{*}) & -(DF^{*}-CE^{*}-BH^{*}+AG^{*}) \end{pmatrix}
\end{aligned} \tag{2.425}$$

$$\begin{aligned}
\hat{X}_{--}^{z}&=\begin{pmatrix} \mathrm{i}E & F \\ -D & \mathrm{i}C \end{pmatrix}\begin{pmatrix} 1 & 0 \\ 0 & -1 \end{pmatrix}\begin{pmatrix} G^{*} & -\mathrm{i}B^{*} \\ \mathrm{i}H^{*} & A^{*} \end{pmatrix}+\begin{pmatrix} G & -\mathrm{i}H \\ \mathrm{i}B & A \end{pmatrix}\begin{pmatrix} 1 & 0 \\ 0 & -1 \end{pmatrix}\begin{pmatrix} -\mathrm{i}E^{*} & -D^{*} \\ F^{*} & -\mathrm{i}C^{*} \end{pmatrix}\\
&=\begin{pmatrix} \mathrm{i}E & F \\ -D & \mathrm{i}C \end{pmatrix}\begin{pmatrix} G^{*} & -\mathrm{i}B^{*} \\ -\mathrm{i}H^{*} & -A^{*} \end{pmatrix}+\begin{pmatrix} G & -\mathrm{i}H \\ \mathrm{i}B & A \end{pmatrix}\begin{pmatrix} -\mathrm{i}E^{*} & -D^{*} \\ -F^{*} & \mathrm{i}C^{*} \end{pmatrix}\\
&=\begin{pmatrix} \mathrm{i}(EG^{*}-FH^{*}-GE^{*}+HF^{*}) & EB^{*}-FA^{*}-GD^{*}+HC^{*} \\ -(DG^{*}-CH^{*}-BE^{*}+AF^{*}) & \mathrm{i}(DB^{*}-CA^{*}-BD^{*}+AC^{*}) \end{pmatrix}
\end{aligned} \tag{2.426}$$

$$A_{xz}^{\mathrm{np}}=-\frac{1}{I_0}\mathrm{Im}(AC^{*}-BD^{*}+EG^{*}-FH^{*}) \tag{2.427}$$

We can write the following results according to Eqs. (2.382), (2.346), and (2.375):

$$\hat{F}\hat{\Sigma}_y^{\mathrm{n}}\hat{\Sigma}_i^{\mathrm{p}}\hat{F}^+ = -\mathrm{i}\begin{pmatrix} \hat{t}_{++}\hat{\sigma}_i^{\mathrm{p}}\hat{t}_{+-}^+ - \hat{t}_{+-}\hat{\sigma}_i^{\mathrm{p}}\hat{t}_{++}^+ & \hat{t}_{++}\hat{\sigma}_i^{\mathrm{p}}\hat{t}_{--}^+ - \hat{t}_{+-}\hat{\sigma}_i^{\mathrm{p}}\hat{t}_{-+}^+ \\ \hat{t}_{-+}\hat{\sigma}_i^{\mathrm{p}}\hat{t}_{+-}^+ - \hat{t}_{--}\hat{\sigma}_i^{\mathrm{p}}\hat{t}_{++}^+ & \hat{t}_{-+}\hat{\sigma}_i^{\mathrm{p}}\hat{t}_{--}^+ - \hat{t}_{--}\hat{\sigma}_i^{\mathrm{p}}\hat{t}_{-+}^+ \end{pmatrix} \equiv \begin{pmatrix} \hat{Y}_{++}^i & \hat{Y}_{+-}^i \\ \hat{Y}_{-+}^i & \hat{Y}_{--}^i \end{pmatrix},\ i=x,y,z \tag{2.428}$$

$$\begin{aligned}\hat{Y}_{++}^x &= -\mathrm{i}\begin{pmatrix} A & -\mathrm{i}B \\ \mathrm{i}H & G \end{pmatrix}\begin{pmatrix} 0 & 1 \\ 1 & 0 \end{pmatrix}\begin{pmatrix} \mathrm{i}C^* & F^* \\ -D^* & \mathrm{i}E^* \end{pmatrix} + \mathrm{i}\begin{pmatrix} -\mathrm{i}C & -D \\ F & -\mathrm{i}E \end{pmatrix}\begin{pmatrix} 0 & 1 \\ 1 & 0 \end{pmatrix}\begin{pmatrix} A^* & -\mathrm{i}H^* \\ \mathrm{i}B^* & G^* \end{pmatrix} \\ &= -\mathrm{i}\begin{pmatrix} A & -\mathrm{i}B \\ \mathrm{i}H & G \end{pmatrix}\begin{pmatrix} -D^* & \mathrm{i}E^* \\ \mathrm{i}C^* & F^* \end{pmatrix} + \mathrm{i}\begin{pmatrix} -\mathrm{i}C & -D \\ F & -\mathrm{i}E \end{pmatrix}\begin{pmatrix} \mathrm{i}B^* & G^* \\ A^* & -\mathrm{i}H^* \end{pmatrix} \\ &= \begin{pmatrix} \mathrm{i}(AD^* - BC^* + CB^* - DA^*) & AE^* - BF^* + CG^* - DH^* \\ -(HD^* - GC^* + FB^* - EA^*) & \mathrm{i}(HE^* - GF^* + FG^* - EH^*) \end{pmatrix}\end{aligned} \tag{2.429}$$

$$\begin{aligned}\hat{Y}_{+-}^x &= -\mathrm{i}\begin{pmatrix} A & -\mathrm{i}B \\ \mathrm{i}H & G \end{pmatrix}\begin{pmatrix} 0 & 1 \\ 1 & 0 \end{pmatrix}\begin{pmatrix} G^* & -\mathrm{i}B^* \\ \mathrm{i}H^* & A^* \end{pmatrix} + \mathrm{i}\begin{pmatrix} -\mathrm{i}C & -D \\ F & -\mathrm{i}E \end{pmatrix}\begin{pmatrix} 0 & 1 \\ 1 & 0 \end{pmatrix}\begin{pmatrix} -\mathrm{i}E^* & -D^* \\ F^* & -\mathrm{i}C^* \end{pmatrix} \\ &= -\mathrm{i}\begin{pmatrix} A & -\mathrm{i}B \\ \mathrm{i}H & G \end{pmatrix}\begin{pmatrix} \mathrm{i}H^* & A^* \\ G^* & -\mathrm{i}B^* \end{pmatrix} + \mathrm{i}\begin{pmatrix} -\mathrm{i}C & -D \\ F & -\mathrm{i}E \end{pmatrix}\begin{pmatrix} F^* & -\mathrm{i}C^* \\ -\mathrm{i}E^* & -D^* \end{pmatrix} \\ &= \begin{pmatrix} AH^* - BG^* + CF^* - DE^* & -\mathrm{i}\left(|A|^2 - |B|^2 + |C|^2 - |D|^2\right) \\ \mathrm{i}\left(|H|^2 - |G|^2 + |F|^2 - |E|^2\right) & HA^* - GB^* + FC^* - ED^* \end{pmatrix},\end{aligned} \tag{2.430}$$

$$\begin{aligned}\hat{Y}_{-+}^x &= -\mathrm{i}\begin{pmatrix} \mathrm{i}E & F \\ -D & \mathrm{i}C \end{pmatrix}\begin{pmatrix} 0 & 1 \\ 1 & 0 \end{pmatrix}\begin{pmatrix} \mathrm{i}C^* & F^* \\ -D^* & \mathrm{i}E^* \end{pmatrix} + \mathrm{i}\begin{pmatrix} G & -\mathrm{i}H \\ \mathrm{i}B & A \end{pmatrix}\begin{pmatrix} 0 & 1 \\ 1 & 0 \end{pmatrix}\begin{pmatrix} A^* & -\mathrm{i}H^* \\ \mathrm{i}B^* & G^* \end{pmatrix} \\ &= -\mathrm{i}\begin{pmatrix} \mathrm{i}E & F \\ -D & \mathrm{i}C \end{pmatrix}\begin{pmatrix} -D^* & \mathrm{i}E^* \\ \mathrm{i}C^* & F^* \end{pmatrix} + \mathrm{i}\begin{pmatrix} G & -\mathrm{i}H \\ \mathrm{i}B & A \end{pmatrix}\begin{pmatrix} \mathrm{i}B^* & G^* \\ A^* & -\mathrm{i}H^* \end{pmatrix} \\ &= \begin{pmatrix} -(ED^* - FC^* + GB^* - HA^*) & \mathrm{i}\left(|E|^2 - |F|^2 + |G|^2 - |H|^2\right) \\ -\mathrm{i}\left(|D|^2 - |C|^2 + |B|^2 - |A|^2\right) & -(DE^* - CF^* + BG^* - AH^*) \end{pmatrix}\end{aligned} \tag{2.431}$$

$$\begin{aligned}\hat{Y}_{--}^x &= -\mathrm{i}\begin{pmatrix} \mathrm{i}E & F \\ -D & \mathrm{i}C \end{pmatrix}\begin{pmatrix} 0 & 1 \\ 1 & 0 \end{pmatrix}\begin{pmatrix} G^* & -\mathrm{i}B^* \\ \mathrm{i}H^* & A^* \end{pmatrix} + \mathrm{i}\begin{pmatrix} G & -\mathrm{i}H \\ \mathrm{i}B & A \end{pmatrix}\begin{pmatrix} 0 & 1 \\ 1 & 0 \end{pmatrix}\begin{pmatrix} -\mathrm{i}E^* & -D^* \\ F^* & -\mathrm{i}C^* \end{pmatrix} \\ &= -\mathrm{i}\begin{pmatrix} \mathrm{i}E & F \\ -D & \mathrm{i}C \end{pmatrix}\begin{pmatrix} \mathrm{i}H^* & A^* \\ G^* & -\mathrm{i}B^* \end{pmatrix} + \mathrm{i}\begin{pmatrix} G & -\mathrm{i}H \\ \mathrm{i}B & A \end{pmatrix}\begin{pmatrix} F^* & -\mathrm{i}C^* \\ -\mathrm{i}E^* & -D^* \end{pmatrix} \\ &= \begin{pmatrix} \mathrm{i}(EH^* - FG^* + GF^* - HE^*) & EA^* - FB^* + GC^* - HD^* \\ -(DH^* - CG^* + BF^* - AE^*) & \mathrm{i}(DA^* - CB^* + BC^* - AD^*) \end{pmatrix}\end{aligned} \tag{2.432}$$

$$A_{yx}^{\mathrm{np}} = 0 \tag{2.433}$$

$$\hat{Y}^{y}_{++}=-\mathrm{i}\begin{pmatrix} A & -\mathrm{i}B \\ \mathrm{i}H & G \end{pmatrix}\begin{pmatrix} 0 & -\mathrm{i} \\ \mathrm{i} & 0 \end{pmatrix}\begin{pmatrix} \mathrm{i}C^* & F^* \\ -D^* & \mathrm{i}E^* \end{pmatrix}+\mathrm{i}\begin{pmatrix} -\mathrm{i}C & -D \\ F & -\mathrm{i}E \end{pmatrix}\begin{pmatrix} 0 & -\mathrm{i} \\ \mathrm{i} & 0 \end{pmatrix}\begin{pmatrix} A^* & -\mathrm{i}H^* \\ \mathrm{i}B^* & G^* \end{pmatrix}$$
$$=-\mathrm{i}\begin{pmatrix} A & -\mathrm{i}B \\ \mathrm{i}H & G \end{pmatrix}\begin{pmatrix} \mathrm{i}D^* & E^* \\ -C^* & \mathrm{i}F^* \end{pmatrix}+\mathrm{i}\begin{pmatrix} -\mathrm{i}C & -D \\ F & -\mathrm{i}E \end{pmatrix}\begin{pmatrix} B^* & -\mathrm{i}G^* \\ \mathrm{i}A^* & H^* \end{pmatrix}$$
$$=\begin{pmatrix} AD^*+BC^*+CB^*+DA^* & -\mathrm{i}(AE^*+BF^*+CG^*+DH^*) \\ \mathrm{i}(HD^*+GC^*+FB^*+EA^*) & HE^*+GF^*+FG^*+EH^* \end{pmatrix} \tag{2.434}$$

$$\hat{Y}^{y}_{+-}=-\mathrm{i}\begin{pmatrix} A & -\mathrm{i}B \\ \mathrm{i}H & G \end{pmatrix}\begin{pmatrix} 0 & -\mathrm{i} \\ \mathrm{i} & 0 \end{pmatrix}\begin{pmatrix} G^* & -\mathrm{i}B^* \\ \mathrm{i}H^* & A^* \end{pmatrix}+\mathrm{i}\begin{pmatrix} -\mathrm{i}C & -D \\ F & -\mathrm{i}E \end{pmatrix}\begin{pmatrix} 0 & -\mathrm{i} \\ \mathrm{i} & 0 \end{pmatrix}\begin{pmatrix} -\mathrm{i}E^* & -D^* \\ F^* & -\mathrm{i}C^* \end{pmatrix}$$
$$=-\mathrm{i}\begin{pmatrix} A & -\mathrm{i}B \\ \mathrm{i}H & G \end{pmatrix}\begin{pmatrix} H^* & -\mathrm{i}A^* \\ \mathrm{i}G^* & B^* \end{pmatrix}+\mathrm{i}\begin{pmatrix} -\mathrm{i}C & -D \\ F & -\mathrm{i}E \end{pmatrix}\begin{pmatrix} -\mathrm{i}F^* & -C^* \\ E^* & -\mathrm{i}D^* \end{pmatrix}$$
$$=\begin{pmatrix} -\mathrm{i}(AH^*+BG^*+CF^*+DE^*) & -\left(|A|^2+|B|^2+|C|^2+|D|^2\right) \\ |H|^2+|G|^2+|F|^2+|E|^2 & -\mathrm{i}(HA^*+GB^*+FC^*+ED^*) \end{pmatrix} \tag{2.435}$$

$$\hat{Y}^{y}_{-+}=-\mathrm{i}\begin{pmatrix} \mathrm{i}E & F \\ -D & \mathrm{i}C \end{pmatrix}\begin{pmatrix} 0 & -\mathrm{i} \\ \mathrm{i} & 0 \end{pmatrix}\begin{pmatrix} \mathrm{i}C^* & F^* \\ -D^* & \mathrm{i}E^* \end{pmatrix}+\mathrm{i}\begin{pmatrix} G & -\mathrm{i}H \\ \mathrm{i}B & A \end{pmatrix}\begin{pmatrix} 0 & -\mathrm{i} \\ \mathrm{i} & 0 \end{pmatrix}\begin{pmatrix} A^* & -\mathrm{i}H^* \\ \mathrm{i}B^* & G^* \end{pmatrix}$$
$$=-\mathrm{i}\begin{pmatrix} \mathrm{i}E & F \\ -D & \mathrm{i}C \end{pmatrix}\begin{pmatrix} \mathrm{i}D^* & E^* \\ -C^* & \mathrm{i}F^* \end{pmatrix}+\mathrm{i}\begin{pmatrix} G & -\mathrm{i}H \\ \mathrm{i}B & A \end{pmatrix}\begin{pmatrix} B^* & -\mathrm{i}G^* \\ \mathrm{i}A^* & H^* \end{pmatrix}$$
$$=\begin{pmatrix} \mathrm{i}(ED^*+FC^*+GB^*+HA^*) & |E|^2+|F|^2+|G|^2+|H|^2 \\ -\left(|D|^2+|C|^2+|B|^2+|A|^2\right) & \mathrm{i}(DE^*+CF^*+BG^*+AH^*) \end{pmatrix} \tag{2.436}$$

$$\hat{Y}^{y}_{--}=-\mathrm{i}\begin{pmatrix} \mathrm{i}E & F \\ -D & \mathrm{i}C \end{pmatrix}\begin{pmatrix} 0 & -\mathrm{i} \\ \mathrm{i} & 0 \end{pmatrix}\begin{pmatrix} G^* & -\mathrm{i}B^* \\ \mathrm{i}H^* & A^* \end{pmatrix}+\mathrm{i}\begin{pmatrix} G & -\mathrm{i}H \\ \mathrm{i}B & A \end{pmatrix}\begin{pmatrix} 0 & -\mathrm{i} \\ \mathrm{i} & 0 \end{pmatrix}\begin{pmatrix} -\mathrm{i}E^* & -D^* \\ F^* & -\mathrm{i}C^* \end{pmatrix}$$
$$=-\mathrm{i}\begin{pmatrix} \mathrm{i}E & F \\ -D & \mathrm{i}C \end{pmatrix}\begin{pmatrix} H^* & -\mathrm{i}A^* \\ \mathrm{i}G^* & B^* \end{pmatrix}+\mathrm{i}\begin{pmatrix} G & -\mathrm{i}H \\ \mathrm{i}B & A \end{pmatrix}\begin{pmatrix} -\mathrm{i}F^* & -C^* \\ E^* & -\mathrm{i}D^* \end{pmatrix}$$
$$=\begin{pmatrix} EH^*+FG^*+GF^*+HE^* & -\mathrm{i}(EA^*+FB^*+GC^*+HD^*) \\ \mathrm{i}(DH^*+CG^*+BF^*+AE^*) & DA^*+CB^*+BC^*+AD^* \end{pmatrix} \tag{2.437}$$

$$A_{yy}^{\mathrm{np}}=\frac{1}{I_0}\mathrm{Re}(AD^*+BC^*+EH^*+FG^*) \tag{2.438}$$

$$\hat{Y}^z_{++}=-\mathrm{i}\begin{pmatrix} A & -\mathrm{i}B \\ \mathrm{i}H & G \end{pmatrix}\begin{pmatrix} 1 & 0 \\ 0 & -1 \end{pmatrix}\begin{pmatrix} \mathrm{i}C^* & F^* \\ -D^* & \mathrm{i}E^* \end{pmatrix}+\mathrm{i}\begin{pmatrix} -\mathrm{i}C & -D \\ F & -\mathrm{i}E \end{pmatrix}\begin{pmatrix} 1 & 0 \\ 0 & -1 \end{pmatrix}\begin{pmatrix} A^* & -\mathrm{i}H^* \\ \mathrm{i}B^* & G^* \end{pmatrix}$$
$$=-\mathrm{i}\begin{pmatrix} A & -\mathrm{i}B \\ \mathrm{i}H & G \end{pmatrix}\begin{pmatrix} \mathrm{i}C^* & F^* \\ D^* & -\mathrm{i}E^* \end{pmatrix}+\mathrm{i}\begin{pmatrix} -\mathrm{i}C & -D \\ F & -\mathrm{i}E \end{pmatrix}\begin{pmatrix} A^* & -\mathrm{i}H^* \\ -\mathrm{i}B^* & -G^* \end{pmatrix}$$
$$=\begin{pmatrix} AC^*-BD^*+CA^*-DB^* & -\mathrm{i}(AF^*-BE^*+CH^*-DG^*) \\ \mathrm{i}(HC^*-GD^*+FA^*-EB^*) & HF^*-GE^*+FH^*-EG^* \end{pmatrix} \tag{2.439}$$

$$\hat{Y}^z_{+-}=-\mathrm{i}\begin{pmatrix} A & -\mathrm{i}B \\ \mathrm{i}H & G \end{pmatrix}\begin{pmatrix} 1 & 0 \\ 0 & -1 \end{pmatrix}\begin{pmatrix} G^* & -\mathrm{i}B^* \\ \mathrm{i}H^* & A^* \end{pmatrix}+\mathrm{i}\begin{pmatrix} -\mathrm{i}C & -D \\ F & -\mathrm{i}E \end{pmatrix}\begin{pmatrix} 1 & 0 \\ 0 & -1 \end{pmatrix}\begin{pmatrix} -\mathrm{i}E^* & -D^* \\ F^* & -\mathrm{i}C^* \end{pmatrix}$$
$$=-\mathrm{i}\begin{pmatrix} A & -\mathrm{i}B \\ \mathrm{i}H & G \end{pmatrix}\begin{pmatrix} G^* & -\mathrm{i}B^* \\ -\mathrm{i}H^* & -A^* \end{pmatrix}+\mathrm{i}\begin{pmatrix} -\mathrm{i}C & -D \\ F & -\mathrm{i}E \end{pmatrix}\begin{pmatrix} -\mathrm{i}E^* & -D^* \\ -F^* & \mathrm{i}C^* \end{pmatrix}$$
$$=\begin{pmatrix} -\mathrm{i}(AG^*-BH^*+CE^*-DF^*) & -(AB^*-BA^*+CD^*-DC^*) \\ HG^*-GH^*+FE^*-EF^* & -\mathrm{i}(HB^*-GA^*+FD^*-EC^*) \end{pmatrix} \tag{2.440}$$

$$\hat{Y}^z_{-+}=-\mathrm{i}\begin{pmatrix} \mathrm{i}E & F \\ -D & \mathrm{i}C \end{pmatrix}\begin{pmatrix} 1 & 0 \\ 0 & -1 \end{pmatrix}\begin{pmatrix} \mathrm{i}C^* & F^* \\ -D^* & \mathrm{i}E^* \end{pmatrix}+\mathrm{i}\begin{pmatrix} G & -\mathrm{i}H \\ \mathrm{i}B & A \end{pmatrix}\begin{pmatrix} 1 & 0 \\ 0 & -1 \end{pmatrix}\begin{pmatrix} A^* & -\mathrm{i}H^* \\ \mathrm{i}B^* & G^* \end{pmatrix}$$
$$=-\mathrm{i}\begin{pmatrix} \mathrm{i}E & F \\ -D & \mathrm{i}C \end{pmatrix}\begin{pmatrix} \mathrm{i}C^* & F^* \\ D^* & -\mathrm{i}E^* \end{pmatrix}+\mathrm{i}\begin{pmatrix} G & -\mathrm{i}H \\ \mathrm{i}B & A \end{pmatrix}\begin{pmatrix} A^* & -\mathrm{i}H^* \\ -\mathrm{i}B^* & -G^* \end{pmatrix}$$
$$=\begin{pmatrix} \mathrm{i}(EC^*-FD^*+GA^*-HB^*) & EF^*-FE^*+GH^*-HG^* \\ -(DC^*-CD^*+BA^*-AB^*) & \mathrm{i}(DF^*-CE^*+BH^*-AG^*) \end{pmatrix} \tag{2.441}$$

$$\hat{Y}^z_{--}=-\mathrm{i}\begin{pmatrix} \mathrm{i}E & F \\ -D & \mathrm{i}C \end{pmatrix}\begin{pmatrix} 1 & 0 \\ 0 & -1 \end{pmatrix}\begin{pmatrix} G^* & -\mathrm{i}B^* \\ \mathrm{i}H^* & A^* \end{pmatrix}+\mathrm{i}\begin{pmatrix} G & -\mathrm{i}H \\ \mathrm{i}B & A \end{pmatrix}\begin{pmatrix} 1 & 0 \\ 0 & -1 \end{pmatrix}\begin{pmatrix} -\mathrm{i}E^* & -D^* \\ F^* & -\mathrm{i}C^* \end{pmatrix}$$
$$=-\mathrm{i}\begin{pmatrix} \mathrm{i}E & F \\ -D & \mathrm{i}C \end{pmatrix}\begin{pmatrix} G^* & -\mathrm{i}B^* \\ -\mathrm{i}H^* & -A^* \end{pmatrix}+\mathrm{i}\begin{pmatrix} G & -\mathrm{i}H \\ \mathrm{i}B & A \end{pmatrix}\begin{pmatrix} -\mathrm{i}E^* & -D^* \\ -F^* & \mathrm{i}C^* \end{pmatrix}$$
$$=\begin{pmatrix} EG^*-FH^*+GE^*-HF^* & -\mathrm{i}(EB^*-FA^*+GD^*-HC^*) \\ \mathrm{i}(DG^*-CH^*+BE^*-AF^*) & DB^*-CA^*+BD^*-AC^* \end{pmatrix} \tag{2.442}$$

$$A^{\mathrm{np}}_{yz}=0 \tag{2.443}$$

And we can obtain the following results according to Eqs. (2.388), (2.346), and (2.375):

$$\hat{F}\hat{\Sigma}_z^{\mathrm{n}}\hat{\Sigma}_i^{\mathrm{p}}\hat{F}^+ = \begin{pmatrix} \hat{t}_{++}\hat{\sigma}_i^{\mathrm{p}}\hat{t}_{++}^+ - \hat{t}_{+-}\hat{\sigma}_i^{\mathrm{p}}\hat{t}_{+-}^+ & \hat{t}_{++}\hat{\sigma}_i^{\mathrm{p}}\hat{t}_{-+}^+ - \hat{t}_{+-}\hat{\sigma}_i^{\mathrm{p}}\hat{t}_{--}^+ \\ \hat{t}_{-+}\hat{\sigma}_i^{\mathrm{p}}\hat{t}_{++}^+ - \hat{t}_{--}\hat{\sigma}_i^{\mathrm{p}}\hat{t}_{+-}^+ & \hat{t}_{-+}\hat{\sigma}_i^{\mathrm{p}}\hat{t}_{-+}^+ - \hat{t}_{--}\hat{\sigma}_i^{\mathrm{p}}\hat{t}_{--}^+ \end{pmatrix}$$
$$\equiv \begin{pmatrix} \hat{Z}_{++}^i & \hat{Z}_{+-}^i \\ \hat{Z}_{-+}^i & \hat{Z}_{--}^i \end{pmatrix}, \ i = x, y, z \tag{2.444}$$

$$\hat{Z}_{++}^x = \begin{pmatrix} A & -\mathrm{i}B \\ \mathrm{i}H & G \end{pmatrix}\begin{pmatrix} 0 & 1 \\ 1 & 0 \end{pmatrix}\begin{pmatrix} A^* & -\mathrm{i}H^* \\ \mathrm{i}B^* & G^* \end{pmatrix} - \begin{pmatrix} -\mathrm{i}C & -D \\ F & -\mathrm{i}E \end{pmatrix}\begin{pmatrix} 0 & 1 \\ 1 & 0 \end{pmatrix}\begin{pmatrix} \mathrm{i}C^* & F^* \\ -D^* & \mathrm{i}E^* \end{pmatrix}$$
$$= \begin{pmatrix} A & -\mathrm{i}B \\ \mathrm{i}H & G \end{pmatrix}\begin{pmatrix} \mathrm{i}B^* & G^* \\ A^* & -\mathrm{i}H^* \end{pmatrix} - \begin{pmatrix} -\mathrm{i}C & -D \\ F & -\mathrm{i}E \end{pmatrix}\begin{pmatrix} -D^* & \mathrm{i}E^* \\ \mathrm{i}C^* & F^* \end{pmatrix}$$
$$= \begin{pmatrix} \mathrm{i}(AB^* - BA^* - CD^* + DC^*) & AG^* - BH^* - CE^* + DF^* \\ -(HB^* - GA^* - FD^* + EC^*) & \mathrm{i}(HG^* - GH^* - FE^* + EF^*) \end{pmatrix} \tag{2.445}$$

$$\hat{Z}_{+-}^x = \begin{pmatrix} A & -\mathrm{i}B \\ \mathrm{i}H & G \end{pmatrix}\begin{pmatrix} 0 & 1 \\ 1 & 0 \end{pmatrix}\begin{pmatrix} -\mathrm{i}E^* & -D^* \\ F^* & -\mathrm{i}C^* \end{pmatrix} - \begin{pmatrix} -\mathrm{i}C & -D \\ F & -\mathrm{i}E \end{pmatrix}\begin{pmatrix} 0 & 1 \\ 1 & 0 \end{pmatrix}\begin{pmatrix} G^* & -\mathrm{i}B^* \\ \mathrm{i}H^* & A^* \end{pmatrix}$$
$$= \begin{pmatrix} A & -\mathrm{i}B \\ \mathrm{i}H & G \end{pmatrix}\begin{pmatrix} F^* & -\mathrm{i}C^* \\ -\mathrm{i}E^* & -D^* \end{pmatrix} - \begin{pmatrix} -\mathrm{i}C & -D \\ F & -\mathrm{i}E \end{pmatrix}\begin{pmatrix} \mathrm{i}H^* & A^* \\ G^* & -\mathrm{i}B^* \end{pmatrix}$$
$$= \begin{pmatrix} AF^* - BE^* - CH^* + DG^* & -\mathrm{i}(AC^* - BD^* - CA^* + DB^*) \\ \mathrm{i}(HF^* - GE^* - FH^* + EG^*) & HC^* - GD^* - FA^* + EB^* \end{pmatrix} \tag{2.446}$$

$$\hat{Z}_{-+}^x = \begin{pmatrix} \mathrm{i}E & F \\ -D & \mathrm{i}C \end{pmatrix}\begin{pmatrix} 0 & 1 \\ 1 & 0 \end{pmatrix}\begin{pmatrix} A^* & -\mathrm{i}H^* \\ \mathrm{i}B^* & G^* \end{pmatrix} - \begin{pmatrix} G & -\mathrm{i}H \\ \mathrm{i}B & A \end{pmatrix}\begin{pmatrix} 0 & 1 \\ 1 & 0 \end{pmatrix}\begin{pmatrix} \mathrm{i}C^* & F^* \\ -D^* & \mathrm{i}E^* \end{pmatrix}$$
$$= \begin{pmatrix} \mathrm{i}E & F \\ -D & \mathrm{i}C \end{pmatrix}\begin{pmatrix} \mathrm{i}B^* & G^* \\ A^* & -\mathrm{i}H^* \end{pmatrix} - \begin{pmatrix} G & -\mathrm{i}H \\ \mathrm{i}B & A \end{pmatrix}\begin{pmatrix} -D^* & \mathrm{i}E^* \\ \mathrm{i}C^* & F^* \end{pmatrix}$$
$$= \begin{pmatrix} -(EB^* - FA^* - GD^* + HC^*) & \mathrm{i}(EG^* - FH^* - GE^* + HF^*) \\ -\mathrm{i}(DB^* - CA^* - BD^* + AC^*) & -(DG^* - CH^* - BE^* + AF^*) \end{pmatrix} \tag{2.447}$$

$$\hat{Z}_{--}^x = \begin{pmatrix} \mathrm{i}E & F \\ -D & \mathrm{i}C \end{pmatrix}\begin{pmatrix} 0 & 1 \\ 1 & 0 \end{pmatrix}\begin{pmatrix} -\mathrm{i}E^* & -D^* \\ F^* & -\mathrm{i}C^* \end{pmatrix} - \begin{pmatrix} G & -\mathrm{i}H \\ \mathrm{i}B & A \end{pmatrix}\begin{pmatrix} 0 & 1 \\ 1 & 0 \end{pmatrix}\begin{pmatrix} G^* & -\mathrm{i}B^* \\ \mathrm{i}H^* & A^* \end{pmatrix}$$
$$= \begin{pmatrix} \mathrm{i}E & F \\ -D & \mathrm{i}C \end{pmatrix}\begin{pmatrix} F^* & -\mathrm{i}C^* \\ -\mathrm{i}E^* & -D^* \end{pmatrix} - \begin{pmatrix} G & -\mathrm{i}H \\ \mathrm{i}B & A \end{pmatrix}\begin{pmatrix} \mathrm{i}H^* & A^* \\ G^* & -\mathrm{i}B^* \end{pmatrix}$$
$$= \begin{pmatrix} \mathrm{i}(EF^* - FE^* - GH^* + HG^*) & EC^* - FD^* - GA^* + HB^* \\ -(DF^* - CE^* - BH^* + AG^*) & \mathrm{i}(DC^* - CD^* - BA^* + AB^*) \end{pmatrix} \tag{2.448}$$

$$A_{zx}^{\mathrm{np}} = -\frac{1}{I_0}\mathrm{Im}(AB^* - CD^* + EF^* - GH^*) \tag{2.449}$$

$$\hat{Z}^{y}_{++} = \begin{pmatrix} A & -\mathrm{i}B \\ \mathrm{i}H & G \end{pmatrix}\begin{pmatrix} 0 & -\mathrm{i} \\ \mathrm{i} & 0 \end{pmatrix}\begin{pmatrix} A^* & -\mathrm{i}H^* \\ \mathrm{i}B^* & G^* \end{pmatrix} - \begin{pmatrix} -\mathrm{i}C & -D \\ F & -\mathrm{i}E \end{pmatrix}\begin{pmatrix} 0 & -\mathrm{i} \\ \mathrm{i} & 0 \end{pmatrix}\begin{pmatrix} \mathrm{i}C^* & F^* \\ -D^* & \mathrm{i}E^* \end{pmatrix}$$
$$= \begin{pmatrix} A & -\mathrm{i}B \\ \mathrm{i}H & G \end{pmatrix}\begin{pmatrix} B^* & -\mathrm{i}G^* \\ \mathrm{i}A^* & H^* \end{pmatrix} - \begin{pmatrix} -\mathrm{i}C & -D \\ F & -\mathrm{i}E \end{pmatrix}\begin{pmatrix} \mathrm{i}D^* & E^* \\ -C^* & \mathrm{i}F^* \end{pmatrix}$$
$$= \begin{pmatrix} AB^* + BA^* - CD^* - DC^* & -\mathrm{i}(AG^* + BH^* - CE^* - DF^*) \\ \mathrm{i}(HB^* + GA^* - FD^* - EC^*) & HG^* + GH^* - FE^* - EF^* \end{pmatrix} \quad (2.450)$$

$$\hat{Z}^{y}_{+-} = \begin{pmatrix} A & -\mathrm{i}B \\ \mathrm{i}H & G \end{pmatrix}\begin{pmatrix} 0 & -\mathrm{i} \\ \mathrm{i} & 0 \end{pmatrix}\begin{pmatrix} -\mathrm{i}E^* & -D^* \\ F^* & -\mathrm{i}C^* \end{pmatrix} - \begin{pmatrix} -\mathrm{i}C & -D \\ F & -\mathrm{i}E \end{pmatrix}\begin{pmatrix} 0 & -\mathrm{i} \\ \mathrm{i} & 0 \end{pmatrix}\begin{pmatrix} G^* & -\mathrm{i}B^* \\ \mathrm{i}H^* & A^* \end{pmatrix}$$
$$= \begin{pmatrix} A & -\mathrm{i}B \\ \mathrm{i}H & G \end{pmatrix}\begin{pmatrix} -\mathrm{i}F^* & -C^* \\ E^* & -\mathrm{i}D^* \end{pmatrix} - \begin{pmatrix} -\mathrm{i}C & -D \\ F & -\mathrm{i}E \end{pmatrix}\begin{pmatrix} H^* & -\mathrm{i}A^* \\ \mathrm{i}G^* & B^* \end{pmatrix}$$
$$= \begin{pmatrix} -\mathrm{i}(AF^* + BE^* - CH^* - DG^*) & -(AC^* + BD^* - CA^* - DB^*) \\ HF^* + GE^* - FH^* - EG^* & -\mathrm{i}(HC^* + GD^* - FA^* - EB^*) \end{pmatrix} \quad (2.451)$$

$$\hat{Z}^{y}_{-+} = \begin{pmatrix} \mathrm{i}E & F \\ -D & \mathrm{i}C \end{pmatrix}\begin{pmatrix} 0 & -\mathrm{i} \\ \mathrm{i} & 0 \end{pmatrix}\begin{pmatrix} A^* & -\mathrm{i}H^* \\ \mathrm{i}B^* & G^* \end{pmatrix} - \begin{pmatrix} G & -\mathrm{i}H \\ \mathrm{i}B & A \end{pmatrix}\begin{pmatrix} 0 & -\mathrm{i} \\ \mathrm{i} & 0 \end{pmatrix}\begin{pmatrix} \mathrm{i}C^* & F^* \\ -D^* & \mathrm{i}E^* \end{pmatrix}$$
$$= \begin{pmatrix} \mathrm{i}E & F \\ -D & \mathrm{i}C \end{pmatrix}\begin{pmatrix} B^* & -\mathrm{i}G^* \\ \mathrm{i}A^* & H^* \end{pmatrix} - \begin{pmatrix} G & -\mathrm{i}H \\ \mathrm{i}B & A \end{pmatrix}\begin{pmatrix} \mathrm{i}D^* & E^* \\ -C^* & \mathrm{i}F^* \end{pmatrix}$$
$$= \begin{pmatrix} \mathrm{i}(EB^* + FA^* - GD^* - HC^*) & EG^* + FH^* - GE^* - HF^* \\ -(DB^* + CA^* - BD^* - AC^*) & \mathrm{i}(DG^* + CH^* - BE^* - AF^*) \end{pmatrix} \quad (2.452)$$

$$\hat{Z}^{y}_{--} = \begin{pmatrix} \mathrm{i}E & F \\ -D & \mathrm{i}C \end{pmatrix}\begin{pmatrix} 0 & -\mathrm{i} \\ \mathrm{i} & 0 \end{pmatrix}\begin{pmatrix} -\mathrm{i}E^* & -D^* \\ F^* & -\mathrm{i}C^* \end{pmatrix} - \begin{pmatrix} G & -\mathrm{i}H \\ \mathrm{i}B & A \end{pmatrix}\begin{pmatrix} 0 & -\mathrm{i} \\ \mathrm{i} & 0 \end{pmatrix}\begin{pmatrix} G^* & -\mathrm{i}B^* \\ \mathrm{i}H^* & A^* \end{pmatrix}$$
$$= \begin{pmatrix} \mathrm{i}E & F \\ -D & \mathrm{i}C \end{pmatrix}\begin{pmatrix} -\mathrm{i}F^* & -C^* \\ E^* & -\mathrm{i}D^* \end{pmatrix} - \begin{pmatrix} G & -\mathrm{i}H \\ \mathrm{i}B & A \end{pmatrix}\begin{pmatrix} H^* & -\mathrm{i}A^* \\ \mathrm{i}G^* & B^* \end{pmatrix}$$
$$= \begin{pmatrix} EF^* + FE^* - GH^* - HG^* & -\mathrm{i}(EC^* + FD^* - GA^* - HB^*) \\ \mathrm{i}(DF^* + CE^* - BH^* - AG^*) & DC^* + CD^* - BA^* - AB^* \end{pmatrix} \quad (2.453)$$

$$A^{\mathrm{np}}_{zy} = 0 \quad (2.454)$$

$$\hat{Z}^{z}_{++} = \begin{pmatrix} A & -\mathrm{i}B \\ \mathrm{i}H & G \end{pmatrix}\begin{pmatrix} 1 & 0 \\ 0 & -1 \end{pmatrix}\begin{pmatrix} A^* & -\mathrm{i}H^* \\ \mathrm{i}B^* & G^* \end{pmatrix} - \begin{pmatrix} -\mathrm{i}C & -D \\ F & -\mathrm{i}E \end{pmatrix}\begin{pmatrix} 1 & 0 \\ 0 & -1 \end{pmatrix}\begin{pmatrix} \mathrm{i}C^* & F^* \\ -D^* & \mathrm{i}E^* \end{pmatrix}$$
$$= \begin{pmatrix} A & -\mathrm{i}B \\ \mathrm{i}H & G \end{pmatrix}\begin{pmatrix} A^* & -\mathrm{i}H^* \\ -\mathrm{i}B^* & -G^* \end{pmatrix} - \begin{pmatrix} -\mathrm{i}C & -D \\ F & -\mathrm{i}E \end{pmatrix}\begin{pmatrix} \mathrm{i}C^* & F^* \\ D^* & -\mathrm{i}E^* \end{pmatrix}$$
$$= \begin{pmatrix} |A|^2 - |B|^2 - |C|^2 + |D|^2 & -\mathrm{i}(AH^* - BG^* - CF^* + DE^*) \\ \mathrm{i}(HA^* - GB^* - FC^* + ED^*) & |H|^2 - |G|^2 - |F|^2 + |E|^2 \end{pmatrix} \quad (2.455)$$

$$\hat{Z}^z_{+-}=\begin{pmatrix} A & -\mathrm{i}B \\ \mathrm{i}H & G \end{pmatrix}\begin{pmatrix} 1 & 0 \\ 0 & -1 \end{pmatrix}\begin{pmatrix} -\mathrm{i}E^* & -D^* \\ F^* & -\mathrm{i}C^* \end{pmatrix}-\begin{pmatrix} -\mathrm{i}C & -D \\ F & -\mathrm{i}E \end{pmatrix}\begin{pmatrix} 1 & 0 \\ 0 & -1 \end{pmatrix}\begin{pmatrix} G^* & -\mathrm{i}B^* \\ \mathrm{i}H^* & A^* \end{pmatrix}$$
$$=\begin{pmatrix} A & -\mathrm{i}B \\ \mathrm{i}H & G \end{pmatrix}\begin{pmatrix} -\mathrm{i}E^* & -D^* \\ -F^* & \mathrm{i}C^* \end{pmatrix}-\begin{pmatrix} -\mathrm{i}C & -D \\ F & -\mathrm{i}E \end{pmatrix}\begin{pmatrix} G^* & -\mathrm{i}B^* \\ -\mathrm{i}H^* & -A^* \end{pmatrix}$$
$$=\begin{pmatrix} -\mathrm{i}(AE^*-BF^*-CG^*+DH^*) & -(AD^*-BC^*-CB^*+DA^*) \\ HE^*-GF^*-FG^*+EH^* & -\mathrm{i}(HD^*-GC^*-FB^*+EA^*) \end{pmatrix} \tag{2.456}$$

$$\hat{Z}^z_{-+}=\begin{pmatrix} \mathrm{i}E & F \\ -D & \mathrm{i}C \end{pmatrix}\begin{pmatrix} 1 & 0 \\ 0 & -1 \end{pmatrix}\begin{pmatrix} A^* & -\mathrm{i}H^* \\ \mathrm{i}B^* & G^* \end{pmatrix}-\begin{pmatrix} G & -\mathrm{i}H \\ \mathrm{i}B & A \end{pmatrix}\begin{pmatrix} 1 & 0 \\ 0 & -1 \end{pmatrix}\begin{pmatrix} \mathrm{i}C^* & F^* \\ -D^* & \mathrm{i}E^* \end{pmatrix}$$
$$=\begin{pmatrix} \mathrm{i}E & F \\ -D & \mathrm{i}C \end{pmatrix}\begin{pmatrix} A^* & -\mathrm{i}H^* \\ -\mathrm{i}B^* & -G^* \end{pmatrix}-\begin{pmatrix} G & -\mathrm{i}H \\ \mathrm{i}B & A \end{pmatrix}\begin{pmatrix} \mathrm{i}C^* & F^* \\ D^* & -\mathrm{i}E^* \end{pmatrix}$$
$$=\begin{pmatrix} \mathrm{i}(EA^*-FB^*-GC^*+HD^*) & EH^*-FG^*-GF^*+HE^* \\ -(DA^*-CB^*-BC^*+AD^*) & \mathrm{i}(DH^*-CG^*-BF^*+AE^*) \end{pmatrix} \tag{2.457}$$

$$\hat{Z}^z_{--}=\begin{pmatrix} \mathrm{i}E & F \\ -D & \mathrm{i}C \end{pmatrix}\begin{pmatrix} 1 & 0 \\ 0 & -1 \end{pmatrix}\begin{pmatrix} -\mathrm{i}E^* & -D^* \\ F^* & -\mathrm{i}C^* \end{pmatrix}-\begin{pmatrix} G & -\mathrm{i}H \\ \mathrm{i}B & A \end{pmatrix}\begin{pmatrix} 1 & 0 \\ 0 & -1 \end{pmatrix}\begin{pmatrix} G^* & -\mathrm{i}B^* \\ \mathrm{i}H^* & A^* \end{pmatrix}$$
$$=\begin{pmatrix} \mathrm{i}E & F \\ -D & \mathrm{i}C \end{pmatrix}\begin{pmatrix} -\mathrm{i}E^* & -D^* \\ -F^* & \mathrm{i}C^* \end{pmatrix}-\begin{pmatrix} G & -\mathrm{i}H \\ \mathrm{i}B & A \end{pmatrix}\begin{pmatrix} G^* & -\mathrm{i}B^* \\ -\mathrm{i}H^* & -A^* \end{pmatrix}$$
$$=\begin{pmatrix} |E|^2-|F|^2-|G|^2+|H|^2 & -\mathrm{i}(ED^*-FC^*-GB^*+HA^*) \\ \mathrm{i}(DE^*-CF^*-BG^*+AH^*) & |D|^2-|C|^2-|B|^2+|A|^2 \end{pmatrix} \tag{2.458}$$

$$A^{\mathrm{np}}_{zz}=\frac{1}{2I_0}\left(|A|^2-|B|^2-|C|^2+|D|^2+|E|^2-|F|^2-|G|^2+|H|^2\right) \tag{2.459}$$

Based on the non-zero analyzing powers obtained above, the differential cross section of the emitted particles can be obtained by using Eq. (2.352) in the case of the polarized incident particle channel as

$$I\equiv\frac{\mathrm{d}\sigma}{\mathrm{d}\Omega}=\mathrm{tr}\{\hat{F}\hat{\rho}_{\mathrm{in}}\hat{F}^+\}=I_0\left(1+\sum_{k=\mathrm{n,p}}p_{ky}A^k_y+p_{\mathrm{n}y}p_{\mathrm{p}y}A^{\mathrm{np}}_{yy}+\sum_{i,j=x,z}p_{\mathrm{n}i}p_{\mathrm{p}j}A^{\mathrm{np}}_{ij}\right) \tag{2.460}$$

where $A^{\mathrm{np}}_{yy}$ and $A^{\mathrm{np}}_{ij}$ $(i,j=x,z)$ are the correlation analyzing powers [10].

Define the following polarization transfer coefficients

$$K^{k\,i'}_{m\,i}\equiv\frac{\mathrm{tr}\{\hat{\Sigma}^k_{i'}\hat{F}\hat{\Sigma}^m_i\hat{F}^+\}}{\mathrm{tr}\{\hat{F}\hat{F}^+\}},\quad k,m=\mathrm{n,p};\ \ i=x,\ y,\ z;\ \ i'=x',y',z' \tag{2.461}$$

$$K^{k\,i'}_{\mathrm{n}i\mathrm{p}j}\equiv\frac{\mathrm{tr}\left\{\hat{\Sigma}^k_{i'}\hat{F}\hat{\Sigma}^{\mathrm{n}}_i\hat{\Sigma}^{\mathrm{p}}_j\hat{F}^+\right\}}{\mathrm{tr}\left\{\hat{F}\hat{F}^+\right\}},\quad k=\mathrm{n,p};\ \ i,j=x,\ y,\ z;\ \ i'=x',y',z' \tag{2.462}$$

where $K_{nipj}^{ki'}$ can also be called the polarization transfer coefficient of the correlated incident particles. For the outgoing particles, the correlation problem may not be considered. Using the matrix expressions of $\hat{F}\hat{\Sigma}_i^m\hat{F}^+$ $(m = \text{n}, \text{p};\ i = x, y, z)$ and $\hat{F}\hat{\Sigma}_i^{\text{n}}\hat{\Sigma}_j^{\text{p}}\hat{F}^+$ $(i, j = x,\ y,\ z)$ obtained previously, the following polarization transfer coefficients can be obtained:

$$
\begin{aligned}
&I_0K_{\text{nx}}^{\text{nx}'} = \text{Re}(AG^* + BH^* - CE^* - DF^*)\\
&I_0K_{\text{nx}}^{\text{nz}'} = -\text{Im}(AC^* + BD^* - EG^* - FH^*), \quad I_0K_{\text{nx}}^{\text{px}'} = \text{Re}(AF^* + BE^* - CH^* - DG^*)\\
&I_0K_{\text{nx}}^{\text{pz}'} = -\text{Im}(AC^* + BD^* + EG^* + FH^*), \quad I_0K_{\text{ny}}^{\text{ny}'} = \text{Re}(AG^* + BH^* + CE^* + DF^*)\\
&I_0K_{\text{ny}}^{\text{py}'} = \text{Re}(AF^* + BE^* + CH^* + DG^*) \quad I_0K_{\text{nz}}^{\text{nx}'} = \text{Im}(AE^* + BF^* - CG^* - DH^*)\\
&I_0K_{\text{nz}}^{\text{nz}'} = \frac{1}{2}\left(|A|^2 + |B|^2 - |C|^2 - |D|^2 - |E|^2 - |F|^2 + |G|^2 + |H|^2\right)\\
&I_0K_{\text{nz}}^{\text{px}'} = \text{Im}(AH^* + BG^* - CF^* - DE^*)\\
&I_0K_{\text{nz}}^{\text{pz}'} = \frac{1}{2}\left(|A|^2 + |B|^2 - |C|^2 - |D|^2 + |E|^2 + |F|^2 - |G|^2 - |H|^2\right)\\
&I_0K_{\text{px}}^{\text{nx}'} = \text{Re}(AF^* - BE^* + CH^* - DG^*), \quad I_0K_{\text{px}}^{\text{nz}'} = -\text{Im}(AB^* + CD^* - EF^* - GH^*)\\
&I_0K_{\text{px}}^{\text{px}'} = \text{Re}(AG^* - BH^* + CE^* - DF^*), \quad I_0K_{\text{px}}^{\text{pz}'} = -\text{Im}(AB^* + CD^* + EF^* + GH^*)\\
&I_0K_{\text{py}}^{\text{ny}'} = \text{Re}(AF^* + BE^* + CH^* + DG^*), \quad I_0K_{\text{py}}^{\text{py}'} = \text{Re}(AG^* + BH^* + CE^* + DF^*)\\
&I_0K_{\text{pz}}^{\text{nx}'} = \text{Im}(AE^* - BF^* + CG^* - DH^*)\\
&I_0K_{\text{pz}}^{\text{nz}'} = \frac{1}{2}\left(|A|^2 - |B|^2 + |C|^2 - |D|^2 - |E|^2 + |F|^2 - |G|^2 + |H|^2\right)\\
&I_0K_{\text{pz}}^{\text{px}'} = \text{Im}(AH^* - BG^* + CF^* - DE^*)\\
&I_0K_{\text{pz}}^{\text{pz}'} = \frac{1}{2}\left(|A|^2 - |B|^2 + |C|^2 - |D|^2 + |E|^2 - |F|^2 + |G|^2 - |H|^2\right)\\
&I_0K_{\text{nxpx}}^{\text{ny}'} = -\text{Re}(AH^* - BG^* - CF^* + DE^*), \quad I_0K_{\text{nxpx}}^{\text{py}'} = -\text{Re}(AE^* - BF^* - CG^* + DH^*)\\
&I_0K_{\text{nxpy}}^{\text{nx}'} = \text{Re}(AH^* + BG^* - CF^* - DE^*), \quad I_0K_{\text{nxpy}}^{\text{nz}'} = -\text{Im}(AD^* + BC^* - EH^* - FG^*)\\
&I_0K_{\text{nxpy}}^{\text{px}'} = \text{Re}(AE^* + BF^* - CG^* - DH^*), \quad I_0K_{\text{nxpy}}^{\text{pz}'} = -\text{Im}(AD^* + BC^* + EH^* + FG^*)\\
&I_0K_{\text{nxpz}}^{\text{ny}'} = -\text{Im}(AG^* - BH^* - CE^* + DF^*), \quad I_0K_{\text{nxpz}}^{\text{py}'} = -\text{Im}(AF^* - BE^* - CH^* + DG^*)\\
&I_0K_{\text{nypx}}^{\text{nx}'} = \text{Re}(AH^* - BG^* + CF^* - DE^*), \quad I_0K_{\text{nypx}}^{\text{nz}'} = -\text{Im}(AD^* - BC^* - EH^* + FG^*)\\
&I_0K_{\text{nypx}}^{\text{px}'} = \text{Re}(AE^* - BF^* + CG^* - DH^*), \quad I_0K_{\text{nypx}}^{\text{pz}'} = -\text{Im}(AD^* - BC^* + EH^* - FG^*)\\
&I_0K_{\text{nypy}}^{\text{ny}'} = \text{Re}(AH^* + BG^* + CF^* + DE^*), \quad I_0K_{\text{nypy}}^{\text{py}'} = \text{Re}(AE^* + BF^* + CG^* + DH^*)\\
&I_0K_{\text{nypz}}^{\text{nx}'} = \text{Im}(AG^* - BH^* + CE^* - DF^*), \quad I_0K_{\text{nypz}}^{\text{nz}'} = \text{Re}(AC^* - BD^* - EG^* + FH^*)\\
&I_0K_{\text{nypz}}^{\text{px}'} = \text{Im}(AF^* - BE^* + CH^* - DG^*), \quad I_0K_{\text{nypz}}^{\text{pz}'} = \text{Re}(AC^* - BD^* + EG^* - FH^*)\\
&I_0K_{\text{nzpx}}^{\text{ny}'} = -\text{Im}(AF^* - BE^* - CH^* + DG^*), \quad I_0K_{\text{nzpx}}^{\text{py}'} = -\text{Im}(AG^* - BH^* - CE^* + DF^*)\\
&I_0K_{\text{nzpy}}^{\text{nx}'} = \text{Im}(AF^* + BE^* - CH^* - DG^*), \quad I_0K_{\text{nzpy}}^{\text{nz}'} = \text{Re}(AB^* - CD^* - EF^* + GH^*)\\
&I_0K_{\text{nzpy}}^{\text{px}'} = \text{Re}(AG^* + BH^* - CE^* - DF^*), \quad I_0K_{\text{nzpy}}^{\text{pz}'} = \text{Re}(AB^* - CD^* + EF^* - GH^*)\\
&I_0K_{\text{nzpz}}^{\text{ny}'} = \text{Re}(AE^* - BF^* - CG^* + DH^*), \quad I_0K_{\text{nzpz}}^{\text{py}'} = \text{Re}(AH^* - BG^* - CF^* + DE^*)
\end{aligned}
\tag{2.463}
$$

The other polarization transfer coefficient components not given above are all equal to 0.

According to the non-zero polarization transfer coefficients obtained above, the components of the polarization vector of the emitted particles can be obtained by use of Eq. (2.352) in the polarized incident particle channel as

$$P^{ky'} = \frac{I_0}{I}\left(P_0^{ky'} + \sum_{m=\mathrm{n,p}} p_{my}K_{my}^{ky'} + p_{\mathrm{ny}}p_{\mathrm{py}}K_{\mathrm{nypy}}^{ky'} + \sum_{i,j=x,z} p_{\mathrm{n}i}p_{\mathrm{p}j}K_{\mathrm{n}i\mathrm{p}j}^{ky'}\right), \quad k=\mathrm{n,p} \quad (2.464)$$

$$P^{ki'} = \frac{I_0}{I}\left(\sum_{\substack{m=\mathrm{n,\,p}\\ i=x,\,z}} p_{mi}K_{mi}^{ki'} + p_{\mathrm{ny}}\sum_{i=x,\,z} p_{\mathrm{p}i}K_{\mathrm{nyp}i}^{ki'} + p_{\mathrm{py}}\sum_{i=x,\,z} p_{\mathrm{n}i}K_{\mathrm{n}i\mathrm{py}}^{ki'}\right), \quad k=\mathrm{n,p}; \quad i'=x',z' \quad (2.465)$$

If the $y$-axis is not selected in the direction $\vec{n}$ perpendicular to the reaction plane, then similar derivations can be made using Eq. (2.339) containing the azimuthal angle $\varphi$, and the formulas for the calculations of the various polarization physical quantities containing the $\theta$ and $\varphi$ at the same time can also be obtained, but they are more complex.

## References


1. Varshalovich DA, Moskalev AN, Khersonskii VK. Quantum theory of angular momentum. Singapore/New Jersey/Hong Kong: World Scientific; 1988.
2. Shen QB. Nuclear reaction theory of low and medium energy (upper volume). Beijing: Science Press; 2005. (in Chinese)
3. Kocher DC, Bertrand FE, Gross EE, Newman E. $^{58}$Ni($\vec{\mathrm{p}}$, p') reaction at 60 MeV: study of the analyzing power for inelastic excitation of the giant resonance of the nuclear continuum and of low-lying bound states. Phys Rev. 1976; C14:1392
4. Wolfenstein L. Possible triple-scattering experimrnts. Phys Rev. 1954;96:1654.
5. Wolfenstein L. Polarization of fast nucleons. Ann Rev Nucl Sci. 1956;6:43.
6. Robson BA. The theory of polarization phenomena. Oxford: Clarendon Press; 1974.
7. Tamura T. Analyses of the scattering of nuclear particles by collective nuclei in terms of the coupled-channel calculation. Rev Mod Phys. 1965;37:679.
8. Joachain CJ. Quantum collision theory. North-Holland Publishing Company. 1975:488.
9. Shen QB. Nuclear reaction theory of low and medium energy (middle volume). Beijing: Science Press; 2012. (in Chinese)
10. Ohlsen GG. Polarization transfer and spin correlation experiments in nuclear physics. Rep Prog Phys. 1972;35:717.

# Chapter 3
# Polarization Theory of Nuclear Reactions for Spin 1 Particles

**Abstract** This chapter will introduce the polarization theory of nuclear reactions for spin 1 particles, such as the spin operators and polarization operators of spin 1 particles. The polarization theories of $\vec{1} + \mathrm{A} \rightarrow \vec{1} + \mathrm{B}$, $\vec{1} + \mathrm{A} \rightarrow \overrightarrow{\frac{1}{2}} + \mathrm{B}$, $\overrightarrow{\frac{1}{2}} + \mathrm{A} \rightarrow \vec{1} + \mathrm{B}$ reactions with unpolarized targets and residual nuclei as well as $\overrightarrow{\frac{1}{2}} + \vec{1}$ and $\vec{1} + \vec{1}$ reactions will be introduced, respectively. This chapter will develop an axis-symmetric rotational nucleus CDCC theory describing the breakup reaction for loosely bound light complex particle.

**Keywords** Axisymmetric rotational nucleus · Breakup reaction · Cartesian coordinate systems · Deuteron folding model · Elastic scattering amplitude · Polarization operators · Polarization rate · Spherical basis coordinate systems · Spin 1 particle

## 3.1 Spin Operators and Polarization Operators of Spin 1 Particles

### *3.1.1 General Expressions of S = 1 Spin Operators and Polarization Operators [1–3]*

The basis spin functions $\chi_m(m = \pm 1, 0)$ are the common eigenfunctions of the operators $\hat{S}^2$ and $\hat{S}_z$, that is,

$$\hat{S}^2 \chi_m = S(S+1)\chi_m = 2\chi_m, \qquad \hat{S}_z \chi_m = m\chi_m \tag{3.1}$$

It can be seen that in $S = 1$ case there is

 
Q.-B. Shen, *Polarization Theory of Nuclear Reactions*,
https://doi.org/10.1007/978-3-031-11878-4_3

$$\hat{S}^2 = 2\hat{I} \tag{3.2}$$

where

$$\hat{I} = \begin{pmatrix} 1 & 0 & 0 \\ 0 & 1 & 0 \\ 0 & 0 & 1 \end{pmatrix} \tag{3.3}$$

The transformation relation between the basis spin functions $\chi_m(m = \pm 1, 0)$ in spherical basis coordinate system and the basis spin functions $\chi_i(i = x, y, z)$ in Cartesian coordinate system is as follows:

$$\chi_x = \frac{1}{\sqrt{2}}(\chi_{-1} - \chi_1), \quad \chi_y = \frac{\mathrm{i}}{\sqrt{2}}(\chi_{-1} + \chi_1), \quad \chi_z = \chi_0 \tag{3.4}$$

$$\chi_1 = -\frac{1}{\sqrt{2}}(\chi_x + \mathrm{i}\chi_y), \quad \chi_0 = \chi_z, \quad \chi_{-1} = \frac{1}{\sqrt{2}}(\chi_x - \mathrm{i}\chi_y) \tag{3.5}$$

Their orthogonal normalization conditions are

$$\chi_{m'}^{+}\chi_m = \delta_{m'm}, \qquad \chi_i^{+}\chi_k = \delta_{ik} \tag{3.6}$$

Their completeness conditions are

$$\sum_{m=\pm 1,0} \chi_m \chi_m^{+} = \hat{I}, \qquad \sum_{i=x,y,z} \chi_i \chi_i^{+} = \hat{I} \tag{3.7}$$

When the particle spin $S = 1$, the spin operator $\hat{S}$ and the polarization operator $\hat{T}_{LM}(L=0,\ 1,\ 2;\ -L \le M \le L)$ are all square $3 \times 3$ matrices. In this case from Eq. (1.54) we can get

$$\hat{T}_{00} = \hat{I} \tag{3.8}$$

and from Eq. (1.55) we can also get

$$\hat{T}_{1M} = \sqrt{\frac{3}{2}}\hat{S}_M, \qquad M = \pm 1, 0 \tag{3.9}$$

It follows that the following expressions can be obtained from Eqs. (3.9) and (1.23):

$$\hat{T}_{1\ \pm 1} = \mp \frac{\sqrt{3}}{2}(\hat{S}_x \pm \mathrm{i}\hat{S}_y), \quad \hat{T}_{10} = \sqrt{\frac{3}{2}}\hat{S}_z \tag{3.10}$$

Moreover from Eq. (1.51) we can obtain

$$\hat{T}_{2M} = \sqrt{5}\sum_{mm'} C_{1m\ \ 2M}^{1\ \ m'} \chi_{m'}\chi_m^+ \tag{3.11}$$

and from Eq. (1.20) we can also obtain

$$\hat{S}_\mu = \sqrt{2}\sum_{mm'} C_{1m\ \ 1\mu}^{1\ \ m'} \chi_{m'}\chi_m^+ \tag{3.12}$$

Let

$$\hat{X} = \sqrt{3}\sum_{\mu\nu} C_{1\mu\ \ 1\nu}^{2\ M} \hat{S}_\mu \hat{S}_\nu \tag{3.13}$$

Substituting Eq. (3.12) into Eq. (3.13), using Eq. (1.18) and the following relation

$$\sum_{\beta\delta\varepsilon} C_{a\alpha\ b\beta}^{e\varepsilon} C_{e\varepsilon\ d\delta}^{c\gamma} C_{b\beta\ d\delta}^{f\varphi} = \hat{e}\hat{f}\ C_{a\alpha\ f\phi}^{c\gamma} W(abcd;ef) \tag{3.14}$$

as well as noting $W(1111;12) = \frac{1}{6}$, then comparing the obtained result with Eq. (3.11), we can get $\hat{X} = \hat{T}_{2M}$, that is,

$$\hat{T}_{2M} = \sqrt{3}\sum_{\mu\nu} C_{1\mu\ \ 1\nu}^{2\ M} \hat{S}_\mu \hat{S}_\nu, \quad M = \pm 2, \pm 1, 0 \tag{3.15}$$

The above formula is an expression of $\hat{T}_{2M}$ in the spherical basis coordinate system. It has a total of five linear independent components, each of which is a square 3 × 3 matrix.

The expressions of the C-G coefficient $C_{l\ \ m-\mu\ 1\mu}^{j\ \ m}$ with definite $m$ value can be obtained from Table 2.1 and are listed in Table 3.1.

The specific results of the C-G coefficient $C_{1\ \ m-\mu\ 1\mu}^{1\ \ m}$ can be extracted from Table 3.1 and are listed in Table 3.2.

And the specific results of the C-G coefficient $C_{1\mu\ \ 1\nu}^{2\ \ \mu+\nu}$ can also be extracted from Table 3.1 and are listed in Table 3.3.

The following results can be obtained utilizing Eqs. (3.15), (1.23), and (3.2) and Table 3.3:

**Table 3.1** Expressions of C-G coefficient $C_{l\ m-\mu\ \ 1\mu}^{j\ \ m}$

| $j$ | $m-\mu$ | $\mu=1$ | $\mu=0$ | $\mu=-1$ |
|---|---|---|---|---|
| $l+1$ | 0 | $\sqrt{\frac{l+2}{2(2l+1)}}$ | $\sqrt{\frac{l+1}{2l+1}}$ | $\sqrt{\frac{l+2}{2(2l+1)}}$ |
| $l$ | 0 | $-\frac{1}{\sqrt{2}}$ | 0 | $\frac{1}{\sqrt{2}}$ |
| $l-1$ | 0 | $\sqrt{\frac{l-1}{2(2l+1)}}$ | $-\sqrt{\frac{l}{2l+1}}$ | $\sqrt{\frac{l-1}{2(2l+1)}}$ |
| $l+1$ | 1 | $\sqrt{\frac{(l+2)(l+3)}{2(l+1)(2l+1)}}$ | $\sqrt{\frac{l(l+2)}{(l+1)(2l+1)}}$ | $\sqrt{\frac{l}{2(2l+1)}}$ |
| $l$ | 1 | $-\sqrt{\frac{(l-1)(l+2)}{2l(l+1)}}$ | $\frac{1}{\sqrt{l(l+1)}}$ | $\frac{1}{\sqrt{2}}$ |
| $l-1$ | 1 | $\sqrt{\frac{(l-2)(l-1)}{2l(2l+1)}}$ | $-\sqrt{\frac{(l-1)(l+1)}{l(2l+1)}}$ | $\sqrt{\frac{l+1}{2(2l+1)}}$ |
| $l+1$ | −1 | $\sqrt{\frac{l}{2(2l+1)}}$ | $\sqrt{\frac{l(l+2)}{(l+1)(2l+1)}}$ | $\sqrt{\frac{(l+2)(l+3)}{2(l+1)(2l+1)}}$ |
| $l$ | −1 | $-\frac{1}{\sqrt{2}}$ | $-\frac{1}{\sqrt{l(l+1)}}$ | $\sqrt{\frac{(l+2)(l-1)}{2l(l+1)}}$ |
| $l-1$ | −1 | $\sqrt{\frac{l+1}{2(2l+1)}}$ | $-\sqrt{\frac{(l-1)(l+1)}{l(2l+1)}}$ | $\sqrt{\frac{(l-2)(l-1)}{2l(2l+1)}}$ |
| $l+1$ | 2 | $\sqrt{\frac{(l+3)(l+4)}{2(l+1)(2l+1)}}$ | $\sqrt{\frac{(l-1)(l+3)}{(l+1)(2l+1)}}$ | $\sqrt{\frac{(l-1)l}{2(l+1)(2l+1)}}$ |
| $l$ | 2 | $-\sqrt{\frac{(l-2)(l+3)}{2l(l+1)}}$ | $\frac{2}{\sqrt{l(l+1)}}$ | $\sqrt{\frac{(l-1)(l+2)}{2l(l+1)}}$ |
| $l-1$ | 2 | $\sqrt{\frac{(l-3)(l-2)}{2l(2l+1)}}$ | $-\sqrt{\frac{(l-2)(l+2)}{l(2l+1)}}$ | $\sqrt{\frac{(l+1)(l+2)}{2(2l+1)}}$ |
| $l+1$ | −2 | $\sqrt{\frac{(l-1)\,l}{2(l+1)(2l+1)}}$ | $\sqrt{\frac{(l-1)(l+3)}{(l+1)(2l+1)}}$ | $\sqrt{\frac{(l+3)(l+4)}{2(l+1)(2l+1)}}$ |
| $l$ | −2 | $-\sqrt{\frac{(l-1)(l+2)}{2l(l+1)}}$ | $-\frac{2}{\sqrt{l(l+1)}}$ | $\sqrt{\frac{(l-2)(l+3)}{2l(l+1)}}$ |
| $l-1$ | −2 | $\sqrt{\frac{(l+1)(l+2)}{2l(2l+1)}}$ | $-\sqrt{\frac{(l-2)(l+2)}{l(2l+1)}}$ | $\sqrt{\frac{(l-3)(l-2)}{2l(2l+1)}}$ |

**Table 3.2** C-G coefficients $C_{1\ m-\mu\ 1\mu}^{1\ m}$

| | $m$ | | |
|---|---|---|---|
| $\mu$ | 1 | 0 | −1 |
| 1 | $-\frac{1}{\sqrt{2}}$ | $-\frac{1}{\sqrt{2}}$ | 0 |
| 0 | $\frac{1}{\sqrt{2}}$ | 0 | $-\frac{1}{\sqrt{2}}$ |
| −1 | 0 | $\frac{1}{\sqrt{2}}$ | $\frac{1}{\sqrt{2}}$ |

**Table 3.3** C-G coefficients $C_{1\mu\ 1\nu}^{2\ \mu+\nu}$

| | $\nu$ | | |
|---|---|---|---|
| $\mu$ | 1 | 0 | −1 |
| 1 | 1 | $\frac{1}{\sqrt{2}}$ | $\frac{1}{\sqrt{6}}$ |
| 0 | $\frac{1}{\sqrt{2}}$ | $\frac{2}{\sqrt{6}}$ | $\frac{1}{\sqrt{2}}$ |
| −1 | $\frac{1}{\sqrt{6}}$ | $\frac{1}{\sqrt{2}}$ | 1 |

$$\begin{aligned}\hat{T}_{2\ 0} &= \sqrt{2}\hat{S}_z^2 - \frac{1}{2\sqrt{2}}\left[\left(\hat{S}_x + \mathrm{i}\hat{S}_y\right)\left(\hat{S}_x - \mathrm{i}\hat{S}_y\right) + \left(\hat{S}_x - \mathrm{i}\hat{S}_y\right)\left(\hat{S}_x + \mathrm{i}\hat{S}_y\right)\right] \\ &= \sqrt{2}\hat{S}_z^2 - \frac{1}{2\sqrt{2}}\left(2\hat{S}_x^2 + 2\hat{S}_y^2\right) = \frac{1}{\sqrt{2}}\left[3\hat{S}_z^2 - \left(\hat{S}_x^2 + \hat{S}_y^2 + \hat{S}_z^2\right)\right] \\ &= \frac{1}{\sqrt{2}}\left(3\hat{S}_z^2 - 2\right)\end{aligned} \tag{3.16}$$

$$\hat{T}_{2\ \pm1} = \mp\frac{\sqrt{3}}{2}\left[\left(\hat{S}_x \pm \mathrm{i}\hat{S}_y\right)\hat{S}_z + \hat{S}_z\left(\hat{S}_x \pm \mathrm{i}\hat{S}_y\right)\right] \tag{3.17}$$

$$\hat{T}_{2\ \pm2} = \frac{\sqrt{3}}{2}\left(\hat{S}_x \pm \mathrm{i}\hat{S}_y\right)^2 \tag{3.18}$$

Equations (3.10) and (3.16)–(3.18) have been given in References [4] and [5], so it can be seen that $T_{LM}$ determined by Eq. (1.51) is a common form in the polarization theory of nuclear reactions.

We can use nine square 3 × 3 matrices $\hat{Q}_{ik}(i,\ k = x,\ y,\ z)$ in Cartesian coordinate systems equivalent to $\hat{T}_{2M}(M = \pm 2, \pm 1, 0)$ in spherical basis coordinate system, but we require the following conditions to be satisfied:

$$\sum_{i=x,y,z} \hat{Q}_{ii} = 0 \tag{3.19}$$

$$\hat{Q}_{ik} = \hat{Q}_{ki}, \quad i \neq k; \quad i, k = x, y, z \tag{3.20}$$

Equations (3.19) and (3.20) require that in spatial coordinate $\hat{Q}$ is not a trace tensor, but is a symmetric tensor, respectively. The two formulas give one and three constraints to tensor components, respectively, and therefore only five $\hat{Q}_{ik}$ are linearly independent. If we select $\hat{Q}_{ik}$ according to the following relations [2]

$$
\begin{aligned}
&\hat{Q}_{xx}=\frac{\sqrt{3}}{2}\left(\hat{T}_{2\,2}+\hat{T}_{2\,-2}\right)-\frac{1}{\sqrt{2}}\hat{T}_{2\,0}, &&\hat{Q}_{xy}=\hat{Q}_{yx}=\frac{\mathrm{i}\sqrt{3}}{2}\left(\hat{T}_{2\,-2}-\hat{T}_{2\,2}\right),\\
&\hat{Q}_{yy}=-\frac{\sqrt{3}}{2}\left(\hat{T}_{2\,2}+\hat{T}_{2\,-2}\right)-\frac{1}{\sqrt{2}}\hat{T}_{2\,0}, &&\hat{Q}_{xz}=\hat{Q}_{zx}=\frac{\sqrt{3}}{2}\left(\hat{T}_{2\,-1}-\hat{T}_{2\,1}\right),\\
&\hat{Q}_{zz}=\sqrt{2}\hat{T}_{2\,0}, &&\hat{Q}_{yz}=\hat{Q}_{zy}=\frac{\mathrm{i}\sqrt{3}}{2}\left(\hat{T}_{2\,-1}+\hat{T}_{2\,1}\right).\\
&\hat{Q}_{xx-yy}\equiv\hat{Q}_{xx}-\hat{Q}_{yy}=\sqrt{3}\left(\hat{T}_{2\,-2}+\hat{T}_{2\,2}\right)
\end{aligned}
\tag{3.21}
$$

then the obtained $\hat{Q}_{ik}$ satisfies Eq. (3.19) and Eq. (3.20), and their inverse relations can be obtained as follows:

$$
\hat{T}_{2\,\pm2}=\frac{1}{2\sqrt{3}}\left(\hat{Q}_{xx-yy}\pm2\mathrm{i}\hat{Q}_{xy}\right),\quad \hat{T}_{2\,\pm1}=\mp\frac{1}{\sqrt{3}}\left(\hat{Q}_{xz}\pm\mathrm{i}\hat{Q}_{yz}\right),\quad \hat{T}_{2\,0}=\frac{1}{\sqrt{2}}\hat{Q}_{zz}
\tag{3.22}
$$

From the above results it can be seen that $\hat{Q}_{xy}, \hat{Q}_{xz}, \hat{Q}_{yz}, \hat{Q}_{xx-yy}, \hat{Q}_{zz}$ form five linearly independent tensor operators in the Cartesian coordinate system.

We choose [2]

$$
\hat{Q}_{ik}=\frac{3}{2}\left(\hat{S}_i\hat{S}_k+\hat{S}_k\hat{S}_i-\frac{4}{3}\delta_{ik}\hat{I}\right),\quad i,k=x,y,z
\tag{3.23}
$$

This formula was published as early as 1958 in a paper in *Nucl. Phys.*, 7,622 by Goldfarb. It can be seen directly that $\hat{Q}_{ik}$ given by the above formula satisfies Eq. (3.20), and using Eq. (3.2) we can prove that $\hat{Q}_{ik}$ also satisfies Eq. (3.19). Substituting Eq. (3.15) into the first equation of Eq. (3.21) we can obtain

$$
\begin{aligned}
\hat{Q}_{xx}&=\sum_{\mu\nu}\left[\frac{3}{2}\left(C_{1\mu\,1\nu}^{2\,2}+C_{1\mu\,1\nu}^{2\,-2}\right)-\sqrt{\frac{3}{2}}C_{1\mu\,1\nu}^{2\,0}\right]\hat{S}_\mu\hat{S}_\nu\\
&=\frac{3}{2}\left(C_{11\,11}^{2\,2}\hat{S}_1\hat{S}_1+C_{1\,-1\,1\,-1}^{2\,-2}\hat{S}_{-1}\hat{S}_{-1}\right)\\
&\quad-\sqrt{\frac{3}{2}}\left(C_{11\,1\,-1}^{2\,0}\hat{S}_1\hat{S}_{-1}+C_{1\,0\,1\,0}^{2\,0}\hat{S}_0\hat{S}_0+C_{1\,-1\,11}^{2\,0}\hat{S}_{-1}\hat{S}_1\right)
\end{aligned}
\tag{3.24}
$$

Using Table 3.3, or Ref. [1], there are

$$C_{11\ 11}^{2\ 2} = C_{1-1\ 1-1}^{2\ -2} = 1, \quad C_{11\ 1-1}^{2\ 0} = C_{1-1\ 11}^{2\ 0} = \frac{1}{\sqrt{6}}, \quad C_{10\ 10}^{2\ 0} = \frac{2}{\sqrt{6}} \tag{3.25}$$

From Eq. (3.24) we can obtain the following expression utilizing Eqs. (3.25), (1.23), and (3.2):

$$\begin{aligned}
\hat{Q}_{xx} &= \frac{3}{4}\left[\left(\hat{S}_x + \mathrm{i}\hat{S}_y\right)^2 + \left(\hat{S}_x - \mathrm{i}\hat{S}_y\right)^2\right] \\
&\quad + \frac{1}{4}\left[\left(\hat{S}_x + \mathrm{i}\hat{S}_y\right)\left(\hat{S}_x - \mathrm{i}\hat{S}_y\right) - 4\hat{S}_z^2 + \left(\hat{S}_x - \mathrm{i}\hat{S}_y\right)\left(\hat{S}_x + \mathrm{i}\hat{S}_y\right)\right] \\
&= \frac{3}{2}\left(\hat{S}_x^2 - \hat{S}_y^2\right) + \frac{1}{4}\left(\hat{S}_x^2 - \mathrm{i}\hat{S}_x\hat{S}_y + \mathrm{i}\hat{S}_y\hat{S}_x + \hat{S}_y^2 - 4\hat{S}_z^2 + \hat{S}_x^2 + \mathrm{i}\hat{S}_x\hat{S}_y - \mathrm{i}\hat{S}_y\hat{S}_x + \hat{S}_y^2\right) \\
&= \frac{3}{2}\hat{S}_x^2 - \frac{3}{2}\hat{S}_y^2 + \frac{1}{2}\hat{S}_x^2 + \frac{1}{2}\hat{S}_y^2 - \hat{S}_z^2 = 3\hat{S}_x^2 - \left(\hat{S}_x^2 + \hat{S}_y^2 + \hat{S}_z^2\right) = 3\hat{S}_x^2 - 2\hat{I}
\end{aligned} \tag{3.26}$$

It can be seen that $\hat{Q}_{xx}$ obtained by Eq. (3.21) is exactly the same as $\hat{Q}_{xx}$ obtained by Eq. (3.23). The other components of $\hat{Q}$ can also be proved by a similar method, so that $\hat{Q}_{ik}$ given by Eq. (3.21) and Eq. (3.23) are exactly the same.

In $S = 1$ case the nine matrices $\hat{T}_{LM}(L = 0, 1, 2; \ -L \leq M \leq L)$ in the spherical basis coordinate system are equivalent to the nine linear independent matrices of $\hat{I}, \hat{S}, \hat{Q}_{ik}$ in the Cartesian coordinate system.

Since in $S = 1$ case the basis spin functions may take $\chi_{1m}(m = \pm 1, 0)$ or $\chi_i(i = x, y, z)$, the spin operators and polarization operators can be presented in the spherical basis representation or the Cartesian basis representation. It is noted that the expressions given earlier in this section, including Eq. (3.23), are applicable to both representations.

### *3.1.2 Specific Expressions in the Spherical Basis Representation*

The component of the basis spin functions of the spherical basis coordinate in the spherical basis representation is defined as

$$\chi_{Sm}(\sigma) = \delta_{m\sigma}, \quad m, \sigma = S, S-1, \ \ldots, -S+1, -S \tag{3.27}$$

In $S = 1$ case, there are

$$\chi_1 = \begin{pmatrix} 1 \\ 0 \\ 0 \end{pmatrix}, \quad \chi_0 = \begin{pmatrix} 0 \\ 1 \\ 0 \end{pmatrix}, \quad \chi_{-1} = \begin{pmatrix} 0 \\ 0 \\ 1 \end{pmatrix} \tag{3.28}$$

Given these, the following expression is satisfied:

$$\chi_m^* = \chi_m, \qquad m = \pm 1,0 \tag{3.29}$$

Using Eq. (1.22) or Eq. (3.4), from Eq. (3.28) we can find that the $S = 1$ particle basis spin functions of the Cartesian coordinate system in the spherical basis representation are as follows:

$$\chi_x = \frac{1}{\sqrt{2}}\begin{pmatrix} -1 \\ 0 \\ 1 \end{pmatrix}, \quad \chi_y = \frac{i}{\sqrt{2}}\begin{pmatrix} 1 \\ 0 \\ 1 \end{pmatrix}, \quad \chi_z = \begin{pmatrix} 0 \\ 1 \\ 0 \end{pmatrix} \tag{3.30}$$

Equation (1.21) gives the matrix element of the spin operator component in the spherical basis coordinate system and the spherical basis representation. In $S = 1$ case, this formula becomes

$$\left(\hat{S}_\mu\right)_{\sigma'\sigma} = \sqrt{2}C_{1\sigma\ 1\mu}^{1\sigma'}, \qquad \mu, \sigma, \sigma' = \pm 1, 0 \tag{3.31}$$

In the spherical basis representation, using Table 3.2 and Eq. (3.31), the expressions of the $S = 1$ particle spin operator component in the spherical basis coordinate system can be obtained as follows:

$$\hat{S}_1 = -\begin{pmatrix} 0 & 1 & 0 \\ 0 & 0 & 1 \\ 0 & 0 & 0 \end{pmatrix}, \quad \hat{S}_0 = \begin{pmatrix} 1 & 0 & 0 \\ 0 & 0 & 0 \\ 0 & 0 & -1 \end{pmatrix}, \quad \hat{S}_{-1} = \begin{pmatrix} 0 & 0 & 0 \\ 1 & 0 & 0 \\ 0 & 1 & 0 \end{pmatrix} \tag{3.32}$$

and using Eqs. (1.22) and (3.32) the expressions of the $S = 1$ particle spin operator components in the Cartesian spherical basis coordinate system can be obtained as follows:

$$\hat{S}_x = \frac{1}{\sqrt{2}}\begin{pmatrix} 0 & 1 & 0 \\ 1 & 0 & 1 \\ 0 & 1 & 0 \end{pmatrix}, \quad \hat{S}_y = \frac{i}{\sqrt{2}}\begin{pmatrix} 0 & -1 & 0 \\ 1 & 0 & -1 \\ 0 & 1 & 0 \end{pmatrix}, \quad \hat{S}_z = \begin{pmatrix} 1 & 0 & 0 \\ 0 & 0 & 0 \\ 0 & 0 & -1 \end{pmatrix} \tag{3.33}$$

The above spin operator matrices satisfy the following relations:

$$\hat{S}_i^+ = \hat{S}_i, \qquad i = x, y, z \tag{3.34}$$

$$\hat{S}_\mu^+ = (-1)^\mu \hat{S}_{-\mu}, \qquad \mu = \pm 1, 0 \tag{3.35}$$

Using Eq. (3.33) the specific expressions of $\hat{Q}_{ik}$ in the spherical basis representation can be obtained by Eq. (3.23) as [3]

$$\hat{Q}_{xx}=\frac{1}{2}\begin{pmatrix}-1 & 0 & 3\\ 0 & 2 & 0\\ 3 & 0 & -1\end{pmatrix},\ \hat{Q}_{yy}=\frac{1}{2}\begin{pmatrix}-1 & 0 & -3\\ 0 & 2 & 0\\ -3 & 0 & -1\end{pmatrix},\hat{Q}_{zz}=\begin{pmatrix}1 & 0 & 0\\ 0 & -2 & 0\\ 0 & 0 & 1\end{pmatrix},$$

$$\hat{Q}_{xy}=\hat{Q}_{yx}=\frac{3i}{2}\begin{pmatrix}0 & 0 & -1\\ 0 & 0 & 0\\ 1 & 0 & 0\end{pmatrix},\ \hat{Q}_{xz}=\hat{Q}_{zx}=\frac{3}{2\sqrt{2}}\begin{pmatrix}0 & 1 & 0\\ 1 & 0 & -1\\ 0 & -1 & 0\end{pmatrix},$$

$$\hat{Q}_{yz}=\hat{Q}_{zy}=\frac{3i}{2\sqrt{2}}\begin{pmatrix}0 & -1 & 0\\ 1 & 0 & 1\\ 0 & -1 & 0\end{pmatrix},\ \hat{Q}_{xx-yy}\equiv\hat{Q}_{xx}-\hat{Q}_{yy}=3\begin{pmatrix}0 & 0 & 1\\ 0 & 0 & 0\\ 1 & 0 & 0\end{pmatrix}\tag{3.36}$$

From the above formulas it can be seen that $\hat{Q}_{ik}$ is a Hermitian matrix

$$\hat{Q}_{ik}^{+}=\hat{Q}_{ik}\tag{3.37}$$

And $\hat{Q}_{xx}$, $\hat{Q}_{yy}$, $\hat{Q}_{zz}$, $\hat{Q}_{xz}$, $\hat{Q}_{zx}$, $\hat{Q}_{xx-yy}$ are real numbers, but $\hat{Q}_{xy}$, $\hat{Q}_{yx}$, $\hat{Q}_{yz}$, $\hat{Q}_{zy}$ are pure imaginary numbers.

The specific expressions of the polarization operators $\hat{T}_{LM}$ in the spherical basis representation can be obtained by using Eqs. (3.8), (3.9), (3.32), (3.22), and (3.36):

$$\hat{T}_{00}=\begin{pmatrix}1 & 0 & 0\\ 0 & 1 & 0\\ 0 & 0 & 1\end{pmatrix},\ \hat{T}_{11}=-\sqrt{\frac{3}{2}}\begin{pmatrix}0 & 1 & 0\\ 0 & 0 & 1\\ 0 & 0 & 0\end{pmatrix},\ \hat{T}_{10}=\sqrt{\frac{3}{2}}\begin{pmatrix}1 & 0 & 0\\ 0 & 0 & 0\\ 0 & 0 & -1\end{pmatrix},$$

$$\hat{T}_{1\ -1}=\sqrt{\frac{3}{2}}\begin{pmatrix}0 & 0 & 0\\ 1 & 0 & 0\\ 0 & 1 & 0\end{pmatrix},\ \hat{T}_{22}=\sqrt{3}\begin{pmatrix}0 & 0 & 1\\ 0 & 0 & 0\\ 0 & 0 & 0\end{pmatrix},\ \hat{T}_{21}=\sqrt{\frac{3}{2}}\begin{pmatrix}0 & -1 & 0\\ 0 & 0 & 1\\ 0 & 0 & 0\end{pmatrix},$$

$$\hat{T}_{20}=\frac{1}{\sqrt{2}}\begin{pmatrix}1 & 0 & 0\\ 0 & -2 & 0\\ 0 & 0 & 1\end{pmatrix},\ \hat{T}_{2\ -1}=\sqrt{\frac{3}{2}}\begin{pmatrix}0 & 0 & 0\\ 1 & 0 & 0\\ 0 & -1 & 0\end{pmatrix},\ \hat{T}_{2\ -2}=\sqrt{3}\begin{pmatrix}0 & 0 & 0\\ 0 & 0 & 0\\ 1 & 0 & 0\end{pmatrix}\tag{3.38}$$

From the above formulas it can be seen that all $\hat{T}_{LM}(L=0,\ 1,\ 2;\quad -L\le M\le L)$ are real numbers and satisfy

$$\hat{T}^{+}_{L\,M} = (-1)^M \hat{T}_{L\ -M} \tag{3.39}$$

From Eq. (1.53) we can see that the matrix element of the $S = 1$ particle polarization operator in the spherical basis representation is

$$\left(\hat{T}_{LM}(1)\right)_{\sigma'\sigma} = \sqrt{2L+1}C^{1\sigma'}_{1\sigma\ LM}, \quad L = 0,\ 1,\ 2; \quad -L \le M \le L; \quad \sigma, \sigma' = \pm 1, 0 \tag{3.40}$$

### 3.1.3 Specific Expressions in the Cartesian Basis Representation

In the Cartesian basis representation, the component of the $S = 1$ basis spin function in the Cartesian coordinate system is defined as

$$\chi_i(\sigma) = \delta_{i\sigma}, \qquad i, \sigma = x, y, z \tag{3.41}$$

The corresponding basis spin functions are

$$\chi_x = \begin{pmatrix} 1 \\ 0 \\ 0 \end{pmatrix}, \quad \chi_y = \begin{pmatrix} 0 \\ 1 \\ 0 \end{pmatrix}, \quad \chi_z = \begin{pmatrix} 0 \\ 0 \\ 1 \end{pmatrix} \tag{3.42}$$

From above formulas it can be seen that the following relation is satisfied:

$$\chi_i^* = \chi_i, \qquad i = x, y, z \tag{3.43}$$

Using Eq. (3.5) and Eq. (3.42) the expressions of the basis spin functions of the $S = 1$ particles in the spherical basis coordinate system and Cartesian basis representation can be obtained as follows:

$$\chi_1 = -\frac{1}{\sqrt{2}}\begin{pmatrix} 1 \\ \mathrm{i} \\ 0 \end{pmatrix}, \quad \chi_0 = \begin{pmatrix} 0 \\ 0 \\ 1 \end{pmatrix}, \quad \chi_{-1} = \frac{1}{\sqrt{2}}\begin{pmatrix} 1 \\ -\mathrm{i} \\ 0 \end{pmatrix}, \tag{3.44}$$

And they satisfy the following relation:

$$\chi_m^* = (-1)^m \chi_{-m}, \qquad m = \pm 1, 0 \tag{3.45}$$

Using Eq. (3.44) and the C-G coefficients given in Table 3.2, we can obtain the following expressions by Eq. (3.12):

$$\hat{S}_1 = \frac{1}{\sqrt{2}} \begin{pmatrix} 0 & 0 & 1 \\ 0 & 0 & \mathrm{i} \\ -1 & -\mathrm{i} & 0 \end{pmatrix}, \quad \hat{S}_0 = \begin{pmatrix} 0 & -\mathrm{i} & 0 \\ \mathrm{i} & 0 & 0 \\ 0 & 0 & 0 \end{pmatrix}, \quad \hat{S}_{-1} = \frac{1}{\sqrt{2}} \begin{pmatrix} 0 & 0 & 1 \\ 0 & 0 & -\mathrm{i} \\ -1 & \mathrm{i} & 0 \end{pmatrix} \tag{3.46}$$

According to Eq. (1.22) we can also obtain the following expressions by Eq. (3.46):

$$\hat{S}_x = \begin{pmatrix} 0 & 0 & 0 \\ 0 & 0 & -\mathrm{i} \\ 0 & \mathrm{i} & 0 \end{pmatrix}, \quad \hat{S}_y = \begin{pmatrix} 0 & 0 & \mathrm{i} \\ 0 & 0 & 0 \\ -\mathrm{i} & 0 & 0 \end{pmatrix}, \quad \hat{S}_z = \begin{pmatrix} 0 & -\mathrm{i} & 0 \\ \mathrm{i} & 0 & 0 \\ 0 & 0 & 0 \end{pmatrix} \tag{3.47}$$

Equations (3.46) and (3.47) give the specific expressions of the spin operator components of the $S = 1$ particles in the spherical basis coordinate system and the Cartesian coordinate system in the Cartesian basis representation, respectively. The matrix element of $\hat{S}_i$ can be written as

$$\left(\hat{S}_i\right)_{kl} = -\mathrm{i}\varepsilon_{ikl}, \quad i, k, l = x, y, z \tag{3.48}$$

where $\varepsilon_{ikl}$ is given by Eq. (1.7). In addition, there are the following relations:

$$\left(\hat{S}_i\right)^+ = \hat{S}_i, \quad i = x, y, z \tag{3.49}$$

$$\left(\hat{S}_\mu\right)^+ = (-1)^\mu \, \hat{S}_{-\mu}, \quad \mu = \pm 1, 0 \tag{3.50}$$

Substituting Eq. (3.47) into Eq. (3.23), we can get the specific expressions of $\hat{Q}_{ik}$ in the Cartesian basis representation as follows:

$$\begin{gathered}
\hat{Q}_{xx} = \begin{pmatrix} -2 & 0 & 0 \\ 0 & 1 & 0 \\ 0 & 0 & 1 \end{pmatrix}, \quad \hat{Q}_{yy} = \begin{pmatrix} 1 & 0 & 0 \\ 0 & -2 & 0 \\ 0 & 0 & 1 \end{pmatrix}, \quad \hat{Q}_{zz} = \begin{pmatrix} 1 & 0 & 0 \\ 0 & 1 & 0 \\ 0 & 0 & -2 \end{pmatrix}, \\
\hat{Q}_{xy} = \hat{Q}_{yz} = \frac{3}{2} \begin{pmatrix} 0 & -1 & 0 \\ -1 & 0 & 0 \\ 0 & 0 & 0 \end{pmatrix}, \quad \hat{Q}_{xz} = \hat{Q}_{zx} = \frac{3}{2} \begin{pmatrix} 0 & 0 & -1 \\ 0 & 0 & 0 \\ -1 & 0 & 0 \end{pmatrix}, \\
\hat{Q}_{yz} = \hat{Q}_{zy} = \frac{3}{2} \begin{pmatrix} 0 & 0 & 0 \\ 0 & 0 & -1 \\ 0 & -1 & 0 \end{pmatrix}, \quad \hat{Q}_{xx-yy} \equiv \hat{Q}_{xx} - \hat{Q}_{yy} = 3 \begin{pmatrix} -1 & 0 & 0 \\ 0 & 1 & 0 \\ 0 & 0 & 0 \end{pmatrix}
\end{gathered} \tag{3.51}$$

Its matrix elements can be summarized as

$$(\hat{Q}_{ik})_{lm} = -\frac{3}{2}(\delta_{il}\delta_{km} + \delta_{im}\delta_{kl} - \frac{2}{3}\delta_{ik}\delta_{lm}), \quad i, k, l, m = x, y, z \tag{3.52}$$

This formula is only applied in the Cartesian coordinate system and the Cartesian basis representation such that

$$\hat{Q}^{+}_{ik} = \hat{Q}_{ik}, \qquad \hat{Q}^{*}_{ik} = \hat{Q}_{ik} \tag{3.53}$$

The specific expressions of the polarization operators $\hat{T}_{LM}(1)$ in the Cartesian basis representation can be obtained, by using Eqs. (3.8), (3.9), (3.46), (3.22), and (3.51), as follows:

$$\hat{T}_{00} = \begin{pmatrix} 1 & 0 & 0 \\ 0 & 1 & 0 \\ 0 & 0 & 1 \end{pmatrix}, \ \hat{T}_{11} = \frac{\sqrt{3}}{2}\begin{pmatrix} 0 & 0 & 1 \\ 0 & 0 & \mathrm{i} \\ -1 & -\mathrm{i} & 0 \end{pmatrix}, \ \hat{T}_{10} = \sqrt{\frac{3}{2}}\begin{pmatrix} 0 & -\mathrm{i} & 0 \\ \mathrm{i} & 0 & 0 \\ 0 & 0 & 0 \end{pmatrix},$$
$$\hat{T}_{1\ -1} = \frac{\sqrt{3}}{2}\begin{pmatrix} 0 & 0 & 1 \\ 0 & 0 & -\mathrm{i} \\ -1 & \mathrm{i} & 0 \end{pmatrix}, \ \hat{T}_{22} = \frac{\sqrt{3}}{2}\begin{pmatrix} -1 & -\mathrm{i} & 0 \\ -\mathrm{i} & 1 & 0 \\ 0 & 0 & 0 \end{pmatrix}, \ \hat{T}_{21} = \frac{\sqrt{3}}{2}\begin{pmatrix} 0 & 0 & 1 \\ 0 & 0 & \mathrm{i} \\ 1 & \mathrm{i} & 0 \end{pmatrix},$$
$$\hat{T}_{20} = \frac{1}{\sqrt{2}}\begin{pmatrix} 1 & 0 & 0 \\ 0 & 1 & 0 \\ 0 & 0 & -2 \end{pmatrix}, \ \hat{T}_{2\ -1} = \frac{\sqrt{3}}{2}\begin{pmatrix} 0 & 0 & -1 \\ 0 & 0 & \mathrm{i} \\ -1 & \mathrm{i} & 0 \end{pmatrix}, \ \hat{T}_{2\ -2} = \frac{\sqrt{3}}{2}\begin{pmatrix} -1 & \mathrm{i} & 0 \\ \mathrm{i} & 1 & 0 \\ 0 & 0 & 0 \end{pmatrix} \tag{3.54}$$

where $\hat{T}_{LM}$ satisfies the following relations:

$$\hat{T}^{+}_{LM} = (-1)^{M}\hat{T}_{L\ -M}, \ \hat{T}^{*}_{LM} = (-1)^{L+M}\hat{T}_{L\ -M}, \ L = 0,1,2; \quad -L \le M \le L \tag{3.55}$$

### 3.1.4 Transformation Relations Between the Spherical Basis and Cartesian Coordinate Systems

The coordinate system transformation here refers to the transformation relations of the components of the polarization quantities between the different coordinate systems.

The following unitary matrix $\hat{U}$ can be used to transform the components $A_x$, $A_y$, $A_z$ of the vector $\vec{A}$ in the Cartesian coordinate system, and transform the contravariant components $A^1, A^0, A^{-1}$ and the covariant components $A_1, A_0, A_{-1}$ of the vector $\vec{A}$ in the spherical basis coordinate system

$$\hat{U}=\begin{pmatrix} -\frac{1}{\sqrt{2}} & \frac{\mathrm{i}}{\sqrt{2}} & 0 \\ 0 & 0 & 1 \\ \frac{1}{\sqrt{2}} & \frac{\mathrm{i}}{\sqrt{2}} & 0 \end{pmatrix}, \quad \hat{U}^{+}=\hat{U}^{-1}=\begin{pmatrix} -\frac{1}{\sqrt{2}} & 0 & \frac{1}{\sqrt{2}} \\ -\frac{\mathrm{i}}{\sqrt{2}} & 0 & -\frac{\mathrm{i}}{\sqrt{2}} \\ 0 & 1 & 0 \end{pmatrix} \tag{3.56}$$

Their transformation relations are as follows:

$$\begin{pmatrix} A^1 \\ A^0 \\ A^{-1} \end{pmatrix}=\hat{U}\begin{pmatrix} A_x \\ A_y \\ A_z \end{pmatrix}, \quad \begin{pmatrix} A_x \\ A_y \\ A_z \end{pmatrix}=\hat{U}^{-1}\begin{pmatrix} A^1 \\ A^0 \\ A^{-1} \end{pmatrix} \tag{3.57}$$

$$\begin{pmatrix} A_1 \\ A_0 \\ A_{-1} \end{pmatrix}=\hat{U}^{*}\begin{pmatrix} A_x \\ A_y \\ A_z \end{pmatrix}, \quad \begin{pmatrix} A_x \\ A_y \\ A_z \end{pmatrix}=\hat{U}^{-1*}\begin{pmatrix} A_1 \\ A_0 \\ A_{-1} \end{pmatrix} \tag{3.58}$$

For example, using

$$\begin{pmatrix} \hat{S}_1 \\ \hat{S}_0 \\ \hat{S}_{-1} \end{pmatrix}=\begin{pmatrix} -\frac{1}{\sqrt{2}} & -\frac{\mathrm{i}}{\sqrt{2}} & 0 \\ 0 & 0 & 1 \\ \frac{1}{\sqrt{2}} & -\frac{\mathrm{i}}{\sqrt{2}} & 0 \end{pmatrix}\begin{pmatrix} \hat{S}_x \\ \hat{S}_y \\ \hat{S}_z \end{pmatrix} \tag{3.59}$$

one can get Eq. (1.23), and using

$$\begin{pmatrix} \hat{S}_x \\ \hat{S}_y \\ \hat{S}_z \end{pmatrix}=\begin{pmatrix} -\frac{1}{\sqrt{2}} & 0 & \frac{1}{\sqrt{2}} \\ \frac{\mathrm{i}}{\sqrt{2}} & 0 & \frac{\mathrm{i}}{\sqrt{2}} \\ 0 & 1 & 0 \end{pmatrix}\begin{pmatrix} \hat{S}_1 \\ \hat{S}_0 \\ \hat{S}_{-1} \end{pmatrix} \tag{3.60}$$

one can obtain Eq. (1.22). In addition, the transformation relations given by Eq. (3.4) and Eq. (3.5) can also be obtained by the above method.

The transformation relations between the polarization tensor components $\hat{T}_{2M}$ and $\hat{Q}_{ij}$ have been given by Eq. (3.21) and Eq. (3.22). If the spherical basis coordinate system is used, then both the spin operator and the polarization operator can be uniformly represented by irreducible tensors $\hat{T}_{LM}(L=1, 2; M=-L, \cdots, L)$. Each $\hat{T}_{LM}$ is independent, easy to process, and suitable for theoretical calculations. Its disadvantage is that only $\hat{T}_{L0}$ is observable; $\hat{T}_{LM}(M\neq 0)$ is not observable. In the common eigenstates of $\hat{S}^2$ and $\hat{S}_z$, the obtained expectation value is not a real number (it becomes a real number in the special case of azimuthal angle $\varphi = 0$). If the Cartesian coordinate system is used, the spin operator and the polarization operator cannot be represented in a unified mathematical form. They can only be expressed as vectors $\hat{S}_i(i=x, y, z)$ and rank-2 second reducible tensors $\hat{Q}_{ij}(i, j=x, y, z)$, respectively. Only five of the nine $\hat{Q}_{ij}$ are independent, but all expectation values of the above components are real numbers. All of them are observable physical quantities.

### 3.1.5 Transformation Relations Between the Spherical Basis and Cartesian Basis Representations

The representation transformation refers to the transformation relations between the $3 \times 3$ matrix expressions of the components of the same polarized physical quantity in the spherical basis coordinate system or Cartesian coordinate system in different representations.

Let $\hat{A}_S$ and $\hat{A}_C$ be the $3 \times 3$ matrix expressions of the same polarization physical quantity components in the spherical basis representation and the Cartesian basis representation, respectively. The following relations should be satisfied between them:

$$\hat{A}_S\hat{U}=\hat{U}\hat{A}_C, \quad \hat{A}_S=\hat{U}\hat{A}_C\hat{U}^{-1}, \quad \hat{A}_C=\hat{U}^{-1}\hat{A}_S\hat{U} \tag{3.61}$$

Equations (3.32) and (3.46) yield, respectively

$$\hat{S}_{1,S}=-\begin{pmatrix} 0 & 1 & 0 \\ 0 & 0 & 1 \\ 0 & 0 & 0 \end{pmatrix}, \quad \hat{S}_{1,C}=\frac{1}{\sqrt{2}}\begin{pmatrix} 0 & 0 & 1 \\ 0 & 0 & \mathrm{i} \\ -1 & -\mathrm{i} & 0 \end{pmatrix} \tag{3.62}$$

The following relations can be proved using matrix $\hat{U}$ given by Eq. (3.56):

$$\hat{S}_{1,S}\hat{U} = -\begin{pmatrix} 0 & 1 & 0 \\ 0 & 0 & 1 \\ 0 & 0 & 0 \end{pmatrix}\begin{pmatrix} -\frac{1}{\sqrt{2}} & \frac{i}{\sqrt{2}} & 0 \\ 0 & 0 & 1 \\ \frac{1}{\sqrt{2}} & \frac{i}{\sqrt{2}} & 0 \end{pmatrix} = -\begin{pmatrix} 0 & 0 & 1 \\ \frac{1}{\sqrt{2}} & \frac{i}{\sqrt{2}} & 0 \\ 0 & 0 & 0 \end{pmatrix},$$

$$\hat{U}\hat{S}_{1,C} = \begin{pmatrix} -\frac{1}{\sqrt{2}} & \frac{i}{\sqrt{2}} & 0 \\ 0 & 0 & 1 \\ \frac{1}{\sqrt{2}} & \frac{i}{\sqrt{2}} & 0 \end{pmatrix}\begin{pmatrix} 0 & 0 & \frac{1}{\sqrt{2}} \\ 0 & 0 & \frac{i}{\sqrt{2}} \\ -\frac{1}{\sqrt{2}} & -\frac{i}{\sqrt{2}} & 0 \end{pmatrix} = -\begin{pmatrix} 0 & 0 & 1 \\ \frac{1}{\sqrt{2}} & \frac{i}{\sqrt{2}} & 0 \\ 0 & 0 & 0 \end{pmatrix} \tag{3.63}$$

Equation (3.63) indicates that Eq. (3.61) is valid for $\hat{S}_1$. For another example, Eqs. (3.36) and (3.51) yield, respectively

$$\hat{Q}_{xx,S} = \frac{1}{2}\begin{pmatrix} -1 & 0 & 3 \\ 0 & 2 & 0 \\ 3 & 0 & -1 \end{pmatrix}, \quad \hat{Q}_{xx,C} = \begin{pmatrix} -2 & 0 & 0 \\ 0 & 1 & 0 \\ 0 & 0 & 1 \end{pmatrix} \tag{3.64}$$

Using operator $\hat{U}$ given by Eq. (3.56) one can also prove

$$\hat{Q}_{xx,S}\hat{U} = \frac{1}{2}\begin{pmatrix} -1 & 0 & 3 \\ 0 & 2 & 0 \\ 3 & 0 & -1 \end{pmatrix}\begin{pmatrix} -\frac{1}{\sqrt{2}} & \frac{i}{\sqrt{2}} & 0 \\ 0 & 0 & 1 \\ \frac{1}{\sqrt{2}} & \frac{i}{\sqrt{2}} & 0 \end{pmatrix} = \frac{1}{2}\begin{pmatrix} 2\sqrt{2} & \sqrt{2}i & 0 \\ 0 & 0 & 2 \\ -2\sqrt{2} & \sqrt{2}i & 0 \end{pmatrix},$$

$$\hat{U}\hat{Q}_{xx,C} = \begin{pmatrix} -\frac{1}{\sqrt{2}} & \frac{i}{\sqrt{2}} & 0 \\ 0 & 0 & 1 \\ \frac{1}{\sqrt{2}} & \frac{i}{\sqrt{2}} & 0 \end{pmatrix}\begin{pmatrix} -2 & 0 & 0 \\ 0 & 1 & 0 \\ 0 & 0 & 1 \end{pmatrix} = \frac{1}{2}\begin{pmatrix} 2\sqrt{2} & \sqrt{2}i & 0 \\ 0 & 0 & 2 \\ -2\sqrt{2} & \sqrt{2}i & 0 \end{pmatrix} \tag{3.65}$$

Equation (3.65) indicates that Eq. (3.61) is also valid for $\hat{Q}_{xx}$. In fact Eq. (3.61) can be used for the representation transformation of any polarization physical quantity component.

## 3.2 Some Mathematical Formulas Related to Spin Operators and Polarization Operators of Spin 1 Particles

### 3.2.1 *Coordinate Rotation of S = 1 Spin Wave Function and Spin Operator*

The $S = 1$ spin wave function $\chi_m(m = \pm 1, 0)$ in the spherical basis coordinate can be rotated by Wigner $D$ function $\hat{D}^1(\alpha, \beta, \gamma)$, that is,

$$\chi'_{m'} = \sum_{m=\pm 1,0} \chi_m \hat{D}^1_{m\,m'}(\alpha, \beta, \gamma) \tag{3.66}$$

Based on Eq. (2.57) and Table 2.5, Eq. (2.196) has given

$$\hat{D}^1(\alpha, \beta, \gamma) = \begin{pmatrix} \frac{1+\cos\beta}{2}e^{-i(\alpha+\gamma)} & -\frac{\sin\beta}{\sqrt{2}}e^{-i\alpha} & \frac{1-\cos\beta}{2}e^{i(\gamma-\alpha)} \\ \frac{\sin\beta}{\sqrt{2}}e^{-i\gamma} & \cos\beta & -\frac{\sin\beta}{\sqrt{2}}e^{i\gamma} \\ \frac{1-\cos\beta}{2}e^{i(\alpha-\gamma)} & \frac{\sin\beta}{\sqrt{2}}e^{i\alpha} & \frac{1+\cos\beta}{2}e^{i(\gamma+\alpha)} \end{pmatrix} \tag{3.67}$$

The $S = 1$ spin wave function $\chi_i(i = x, y, z)$ in a Cartesian coordinate system can be rotated as follows:

$$\chi'_i = \sum_{k=x,\,y,\,z} \chi_k a_{ki} \tag{3.68}$$

The rotation matrix composed of matrix elements $a_{ki}$ has been given by Eq. (2.202) as [1]

$$\hat{a}(\alpha, \beta, \gamma) = \begin{pmatrix} \cos\beta\cos\alpha\cos\gamma - \sin\alpha\sin\gamma & -\cos\beta\cos\alpha\sin\gamma - \sin\alpha\cos\gamma & \sin\beta\cos\alpha \\ \cos\beta\sin\alpha\cos\gamma + \cos\alpha\sin\gamma & -\cos\beta\sin\alpha\sin\gamma + \cos\alpha\cos\gamma & \sin\beta\sin\alpha \\ -\sin\beta\cos\gamma & \sin\beta\sin\gamma & \cos\beta \end{pmatrix} \tag{3.69}$$

The corresponding inverse rotation operator of the coordinate system is [1]

$$\left[\hat{D}^1(\alpha,\ \beta,\ \gamma)\right]^{-1} = \left[\hat{D}^1(\alpha,\ \beta,\ \gamma)\right]^{+} = \hat{D}^1(-\gamma,\ -\beta,\ -\alpha) = \hat{D}^1(\pi-\gamma, \beta, -\pi-\alpha) \tag{3.70}$$

In addition, in $S = 1$ case the corresponding relations of the helicity basis functions can be obtained by Eqs. (1.30)–(1.33).

### 3.2.2 Action of the S = 1 Spin Operators and Polarization Operators on Basis Spin Functions

In the spherical basis representation, when the spin operator $\hat{S}_\mu(\mu = \pm 1, 0)$ and polarization operator $\hat{T}_{2M}(M = \pm 2, \pm 1, 0)$ in the spherical basis coordinate system act on the basis spin function $\chi_m(m = \pm 1, 0)$, we can obtain [1]

$$\hat{S}_\mu \chi_m = \sqrt{2} C_{1m\ 1\mu}^{1\ m+\mu} \chi_{m+\mu} \tag{3.71}$$

$$\hat{T}_{2M} \chi_m = \sqrt{5} C_{1m\ 2M}^{1\ m+M} \chi_{m+M} \tag{3.72}$$

In the Cartesian basis representation, when the spin operator $\hat{S}_i(i = x, y, z)$ and the polarization operator $\hat{Q}_{ik}(i, k = x, y, z)$ in the Cartesian coordinate system act on the basis spin function $\chi_i(i = x, y, z)$, we can obtain [1]

$$\hat{S}_i \chi_k = \mathrm{i}\varepsilon_{ikl} \chi_l \tag{3.73}$$

$$\hat{Q}_{ik} \chi_l = \frac{3}{2}\left(\frac{2}{3}\delta_{ik}\chi_l - \delta_{il}\chi_k - \delta_{kl}\chi_i\right) \tag{3.74}$$

### 3.2.3 Products of S = 1 Spin Operators and Polarization Operators

1. **Products of the operator component $\hat{S}_i$ and $\hat{Q}_{ik}$ in Cartesian coordinate system (*i*, *k*, *l*, et al. taking *x*, *y*, *z*)**

Equations (1.4) and (1.6) have given

$$\hat{S} \times \hat{S} = \mathrm{i}\hat{S}, \quad \left[\hat{S}_i,\ \hat{S}_k\right] \equiv \hat{S}_i\hat{S}_k - \hat{S}_k\hat{S}_i = \mathrm{i}\varepsilon_{ikl}\hat{S}_l \tag{3.75}$$

From Eq. (3.23) and the second formula of Eq. (3.75), we can get

$$\hat{S}_i\hat{S}_k = \frac{2}{3}\delta_{ik}\hat{I} + \frac{\mathrm{i}}{2}\varepsilon_{ikl}\hat{S}_l + \frac{1}{3}\hat{Q}_{ik} \tag{3.76}$$

Using $\sum_i \hat{Q}_{ii} = 0$ and the other equations, the following relations can be obtained from the above formula:

$$\hat{S}^2 = \hat{S}_x^2 + \hat{S}_y^2 + \hat{S}_z^2 = 2\hat{I} \tag{3.77}$$

$$\{\hat{S}_i, \hat{S}_k\} \equiv \hat{S}_i\hat{S}_k + \hat{S}_k\hat{S}_i = \frac{4}{3}\delta_{ik}\hat{I} + \frac{2}{3}\hat{Q}_{ik} \tag{3.78}$$

and

$$\begin{aligned} \hat{S}_i\hat{S}_k\hat{S}_l &= \frac{\mathrm{i}}{3}\varepsilon_{ikl}\hat{I} + \frac{1}{2}\left(\delta_{ik}\hat{S}_l + \delta_{kl}\hat{S}_i\right) + \frac{\mathrm{i}}{3}\sum_m \varepsilon_{ilm}\hat{Q}_{km} \\ &= \frac{\mathrm{i}}{3}\varepsilon_{ikl}\hat{I} + \frac{1}{2}\left(\delta_{ik}\hat{S}_l + \delta_{kl}\hat{S}_i\right) + \frac{\mathrm{i}}{6}\sum_m \left(\varepsilon_{ikm}\hat{Q}_{lm} + \varepsilon_{ilm}\hat{Q}_{km} + \varepsilon_{klm}\hat{Q}_{im}\right) \end{aligned} \tag{3.79}$$

From the above formula we can also obtain

$$\hat{S}_i\hat{S}_k\hat{S}_l + \hat{S}_l\hat{S}_k\hat{S}_i = \delta_{ik}\hat{S}_l + \delta_{kl}\hat{S}_i \tag{3.80}$$

$$\hat{S}_i\hat{S}_k\hat{S}_i = \delta_{ik}\hat{S}_i \ (\text{no sum to } i); \qquad \sum_i \hat{S}_i\hat{S}_k\hat{S}_i = \hat{S}_k \tag{3.81}$$

$$\hat{S}_i\hat{S}_k\hat{S}_k + \hat{S}_k\hat{S}_k\hat{S}_i = \hat{S}_i + \delta_{ik}\hat{S}_k \ (\text{no sum to } k) \tag{3.82}$$

$$\hat{S}_x\hat{S}_y\hat{S}_z + \hat{S}_y\hat{S}_z\hat{S}_x + \hat{S}_z\hat{S}_x\hat{S}_y = \mathrm{i}\hat{I}, \quad \hat{S}_x\hat{S}_z\hat{S}_y + \hat{S}_z\hat{S}_y\hat{S}_x + \hat{S}_y\hat{S}_x\hat{S}_z = -\mathrm{i}\hat{I} \tag{3.83}$$

$$\hat{S}_x\hat{S}_z\hat{S}_z = \hat{S}_y\hat{S}_y\hat{S}_x = \frac{1}{2}\hat{S}_x - \frac{\mathrm{i}}{3}\hat{Q}_{yz}, \quad \hat{S}_x\hat{S}_y\hat{S}_y = \hat{S}_z\hat{S}_z\hat{S}_x = \frac{1}{2}\hat{S}_x + \frac{\mathrm{i}}{3}\hat{Q}_{yz} \tag{3.84}$$

$$\hat{S}_y\hat{S}_x\hat{S}_x = \hat{S}_z\hat{S}_z\hat{S}_y = \frac{1}{2}\hat{S}_y - \frac{\mathrm{i}}{3}\hat{Q}_{xz}, \quad \hat{S}_y\hat{S}_z\hat{S}_z = \hat{S}_x\hat{S}_x\hat{S}_y = \frac{1}{2}\hat{S}_y + \frac{\mathrm{i}}{3}\hat{Q}_{xz} \tag{3.85}$$

$$\hat{S}_z\hat{S}_y\hat{S}_y = \hat{S}_x\hat{S}_x\hat{S}_z = \frac{1}{2}\hat{S}_z - \frac{\mathrm{i}}{3}\hat{Q}_{xy}, \quad \hat{S}_z\hat{S}_x\hat{S}_x = \hat{S}_y\hat{S}_y\hat{S}_z = \frac{1}{2}\hat{S}_z + \frac{\mathrm{i}}{3}\hat{Q}_{xy} \tag{3.86}$$

Now, there are

$$\hat{S}_i^{2n} = \frac{2}{3}\hat{I} + \frac{\hat{Q}_{ii}}{3}, \ n = 1, 2\cdots; \qquad \hat{S}_i^{2n+1} = \hat{S}_i, \ n = 0, 1, 2\cdots \tag{3.87}$$

If $\vec{n}$ is a unit vector, then there are

$$\begin{aligned}\left(\hat{S}\cdot\vec{n}\right)^{2k} &= \frac{2}{3}\hat{I}+\frac{1}{3}\sum_{il}n_i n_l \hat{Q}_{il},\ k=1,2,3\cdots;\\ \left(\hat{S}\cdot\vec{n}\right)^{2k+1} &= \left(\hat{S}\cdot\vec{n}\right),\ k=0,1,2\cdots\end{aligned}\tag{3.88}$$

And there are

$$\hat{Q}_{ik}\hat{S}_l=\frac{3}{4}\left(\delta_{il}\hat{S}_k+\delta_{kl}\hat{S}_i-\frac{2}{3}\delta_{ik}\hat{S}_l\right)+\frac{\mathrm{i}}{2}\sum_m\left(\varepsilon_{ilm}\hat{Q}_{km}+\varepsilon_{klm}\hat{Q}_{im}\right)\tag{3.89}$$

$$\hat{S}_i\hat{Q}_{kl}=\frac{3}{4}\left(\delta_{ik}\hat{S}_l+\delta_{il}\hat{S}_k-\frac{2}{3}\delta_{kl}\hat{S}_i\right)+\frac{\mathrm{i}}{2}\sum_m\left(\varepsilon_{ikm}\hat{Q}_{lm}+\varepsilon_{ilm}\hat{Q}_{km}\right)\tag{3.90}$$

$$\begin{aligned}\hat{Q}_{ik}\hat{Q}_{lm}=&\ \frac{3}{2}\left(\delta_{il}\delta_{km}+\delta_{im}\delta_{kl}-\frac{2}{3}\delta_{ik}\delta_{lm}\right)\hat{I}\\ &-\frac{3}{4}\left(\delta_{il}\hat{Q}_{km}+\delta_{im}\hat{Q}_{kl}+\delta_{km}\hat{Q}_{il}+\delta_{kl}\hat{Q}_{im}-\frac{4}{3}\delta_{ik}\hat{Q}_{lm}-\frac{4}{3}\delta_{lm}\hat{Q}_{ik}\right)\\ &+\frac{9\mathrm{i}}{8}\sum_p\left(\delta_{il}\varepsilon_{kmp}+\delta_{im}\varepsilon_{klp}+\delta_{kl}\varepsilon_{imp}+\delta_{km}\varepsilon_{ilp}\right)\hat{S}_p\end{aligned}\tag{3.91}$$

2. **The products of the components of the spin operator $\hat{S}$ and the polarization operator $\hat{T}_{LM}$ in the spherical basis coordinate system ($\mu, \nu = \pm 1, 0$)**

We have the following formula: [1]

$$\hat{S}_\mu\hat{S}_\nu=\frac{2}{3}(-1)^\mu\delta_{\mu\ -\nu}\hat{I}-\frac{1}{\sqrt{2}}C^{1\lambda}_{1\mu\ 1\nu}\hat{S}_\lambda+\frac{1}{\sqrt{3}}C^{2M}_{1\mu\ 1\nu}\hat{T}_{2M}\tag{3.92}$$

From the above formula we can obtain

$$\hat{S}^2=-\hat{S}_1\hat{S}_{-1}+\hat{S}_0\hat{S}_0-\hat{S}_{-1}\hat{S}_1=2\hat{I}\tag{3.93}$$

$$\left[\hat{S}_\mu,\hat{S}_\nu\right]\equiv\hat{S}_\mu\hat{S}_\nu-\hat{S}_\nu\hat{S}_\mu=-\sqrt{2}C^{1\lambda}_{1\mu\ 1\nu}\hat{S}_\lambda\tag{3.94}$$

$$\left\{\hat{S}_\mu,\hat{S}_\nu\right\}\equiv\hat{S}_\mu\hat{S}_\nu+\hat{S}_\nu\hat{S}_\mu=\frac{4}{3}(-1)^\mu\delta_{\mu\ -\nu}\hat{I}+\frac{2}{\sqrt{3}}C^{2M}_{1\mu\ 1\nu}\hat{T}_{2M}\tag{3.95}$$

and

$$\hat{S}^2_{\pm1}=\frac{1}{\sqrt{3}}\hat{T}_{2\ \pm2},\quad \hat{S}^3_{\pm1}=\hat{S}^4_{\pm1}=\hat{S}^5_{\pm1}=\ldots=0\tag{3.96}$$

$$\hat{S}_0^2 = \frac{2}{3}\hat{I} + \frac{\sqrt{2}}{3}\hat{T}_{20} \tag{3.97}$$

$$\hat{S}_0^{2n} = \hat{S}_0^2, \quad n = 1, 2, \ldots ; \qquad \hat{S}_0^{2n+1} = \hat{S}_0, \quad n = 0, 1, 2, \ldots \tag{3.98}$$

$$\hat{T}_{L_1M_1}\hat{T}_{L_2M_2} = \sqrt{3}\hat{L}_1\hat{L}_2\sum_L(-1)^{L_1+L_2+L}W(L_1L_2 11; L1)C^{LM}_{L_1M_1\ L_2M_2}\hat{T}_{LM} \tag{3.99}$$

From Eq. (3.99) we can also obtain

$$\hat{T}_{LM}\hat{T}_{00} = \hat{T}_{00}\hat{T}_{LM} = \hat{T}_{LM}, \quad L = 0, 1, 2; \quad -L \le M \le L \tag{3.100}$$

$$\hat{S}_\mu\hat{T}_{2M} = -\frac{\sqrt{5}}{2}C^{1\nu}_{1\mu\,2M}\hat{S}_\nu - \sqrt{\frac{3}{2}}C^{2N}_{1\mu\,2M}\hat{T}_{2N} \tag{3.101}$$

$$\hat{T}_{2M}\hat{S}_\mu = -\frac{\sqrt{5}}{2}C^{1\nu}_{1\mu\,2M}\hat{S}_\nu + \sqrt{\frac{3}{2}}C^{2N}_{1\mu\,2M}\hat{T}_{2N} \tag{3.102}$$

$$\hat{T}_{2M}\hat{T}_{2N} = (-1)^M\delta_{M\ -N}\hat{I} + \frac{3}{2}\sqrt{\frac{5}{2}}C^{1\mu}_{2M\,2N}\hat{S}_\mu + \frac{\sqrt{7}}{2}C^{2\Lambda}_{2M\,2N}\hat{T}_{2\Lambda} \tag{3.103}$$

### 3.2.4 Traces of the $S = 1$ Spin Operators and Polarization Operators

When $S = 1$ one can obtain the following results from Eq. (1.24):

$$\begin{gathered} \mathrm{tr}\{\hat{S}_i\} = 0, \quad \mathrm{tr}\{\hat{S}_i\hat{S}_k\} = 2\delta_{ik}, \quad \mathrm{tr}\{\hat{S}_i\hat{S}_k\hat{S}_l\} = \mathrm{i}\varepsilon_{ikl}, \\ \mathrm{tr}\{\hat{S}_i\hat{S}_k\hat{S}_l\hat{S}_m\} = \delta_{ik}\delta_{lm} + \delta_{im}\delta_{kl} \end{gathered} \tag{3.104}$$

For $\hat{Q}_{ik}$ there are the following trace formulas: [1]

$$\begin{aligned} &\mathrm{tr}\{\hat{Q}_{ik}\} = 0, \quad \mathrm{tr}\{\hat{S}_i\hat{Q}_{kl}\} = 0, \quad \mathrm{tr}\{\hat{Q}_{ik}\hat{S}_l\} = 0, \\ &\mathrm{tr}\{\hat{Q}_{ik}\hat{Q}_{lm}\} = \frac{9}{2}\left(\delta_{il}\delta_{km} + \delta_{kl}\delta_{im} - \frac{2}{3}\delta_{ik}\delta_{lm}\right), \\ &\mathrm{tr}\{\hat{Q}_{ii}\hat{Q}_{ii}\} = 6, \quad \mathrm{tr}\{\hat{Q}_{ii}\hat{Q}_{kk}\}_{i\neq k} = -3, \quad \mathrm{tr}\{\hat{Q}_{ik}\hat{Q}_{ik}\}_{i\neq k} = \frac{9}{2} \end{aligned} \tag{3.105}$$

In the above two formulas, take $i, k, l, m = x, y, z$.

Let $a_{ij}$ and $b_{ij}$ represent the matrix elements of the matrix $A$ and $B$, respectively, so that we can prove

$$\mathrm{tr}\{AB\}=\sum_i\sum_j a_{ij}b_{ji}=\sum_j\sum_i b_{ji}a_{ij}=\mathrm{tr}\{BA\} \tag{3.106}$$

Naturally we can prove the following relation:

$$\mathrm{tr}\{ABC\}=\mathrm{tr}\{CAB\}=\mathrm{tr}\{BCA\} \tag{3.107}$$

The following relation can be obtained utilizing Eqs. (3.107), (3.105), (3.75), and (3.76) and $Q_{lm} = Q_{ml}$

$$\begin{aligned}\mathrm{tr}\{\hat{S}_m\hat{Q}_{ik}\hat{S}_l\} &=\mathrm{tr}\{\hat{Q}_{ik}\hat{S}_l\hat{S}_m\} = \frac{1}{3}\mathrm{tr}\{\hat{Q}_{ik}\hat{Q}_{lm}\} \\ &= \frac{3}{2}\left(\delta_{il}\delta_{km}+\delta_{kl}\delta_{im}-\frac{2}{3}\delta_{ik}\delta_{lm}\right)\end{aligned} \tag{3.108}$$

And the following relation can also be obtained utilizing Eqs. (3.107), (3.105), (3.104), (3.89) (or (3.90), or (3.91)):

$$\begin{aligned}\mathrm{tr}\{\hat{S}_j\hat{Q}_{ik}\hat{Q}_{lm}\} &=\mathrm{tr}\{\hat{Q}_{ik}\hat{Q}_{lm}\hat{S}_j\} \\ &= \frac{9\mathrm{i}}{4}\left(\delta_{il}\varepsilon_{kmj}+\delta_{im}\varepsilon_{klj}+\delta_{kl}\varepsilon_{imj}+\delta_{km}\varepsilon_{ilj}\right)\end{aligned} \tag{3.109}$$

Equation (3.89) can be rewritten as

$$\hat{Q}_{ik}\hat{S}_j=\frac{3}{4}\left(\delta_{ij}\hat{S}_k+\delta_{kj}\hat{S}_i-\frac{2}{3}\delta_{ik}\hat{S}_j\right)+\frac{\mathrm{i}}{2}\sum_n\left(\varepsilon_{ijn}\hat{Q}_{kn}+\varepsilon_{kjn}\hat{Q}_{in}\right) \tag{3.110}$$

Using Eq. (3.105) and Eq. (3.109) we can obtain

$$\begin{aligned}\mathrm{tr}\{\hat{Q}_{ik}\hat{S}_j\hat{Q}_{lm}\} &= \frac{9\mathrm{i}}{4}\sum_n\Bigg[\varepsilon_{ijn}\left(\delta_{kl}\delta_{nm}+\delta_{nl}\delta_{km}-\frac{2}{3}\delta_{kn}\delta_{lm}\right) \\ &\quad +\varepsilon_{kjn}\left(\delta_{il}\delta_{nm}+\delta_{nl}\delta_{im}-\frac{2}{3}\delta_{in}\delta_{lm}\right)\Bigg] \\ &= \frac{9\mathrm{i}}{4}\left(\varepsilon_{ijm}\delta_{kl}+\varepsilon_{ijl}\delta_{km}-\frac{2}{3}\varepsilon_{ijk}\delta_{lm}+\varepsilon_{kjm}\delta_{il}+\varepsilon_{kjl}\delta_{im}-\frac{2}{3}\varepsilon_{kji}\delta_{lm}\right) \\ &= -\frac{9\mathrm{i}}{4}\left(\delta_{il}\varepsilon_{kmj}+\delta_{im}\varepsilon_{klj}+\delta_{kl}\varepsilon_{imj}+\delta_{km}\varepsilon_{ilj}\right) \\ &= -\mathrm{tr}\left(\hat{S}_j\hat{Q}_{ik}\hat{Q}_{lm}\right)\end{aligned} \tag{3.111}$$

When $S = 1$ we can obtain the following results from Eq. (1.25):

$$\begin{aligned}
&\mathrm{tr}\{\hat{S}_\mu\}=0, \quad tr\{\hat{S}_\mu\hat{S}_\nu\}=2(-1)^\mu\delta_{\mu\ -\nu},\\
&\mathrm{tr}\{\hat{S}_\mu\hat{S}_\nu\hat{S}_\lambda\}=-\sqrt{6}\begin{pmatrix}1 & 1 & 1\\ \mu & \nu & \lambda\end{pmatrix}=\sqrt{2}(-1)^{1+\lambda}C^{1\ -\lambda}_{1\mu\ 1\nu},\\
&\mathrm{tr}\{\hat{S}_\mu\hat{S}_\nu\hat{S}_\lambda\hat{S}_\rho\}=(-1)^{\mu+\lambda}\left(\delta_{\mu\ -\nu}\delta_{\lambda\ -\rho}+\delta_{\mu\ -\rho}\delta_{\nu\ -\lambda}\right)
\end{aligned}\tag{3.112}$$

where $\mu$, $\nu$, $\lambda$, $\rho=\pm 1, 0$. We can also obtain the following trace formulas from Eqs. (1.67)–(1.70):

$$\begin{aligned}
&\mathrm{tr}\{T_{LM}(1)\}=3\delta_{L0}\delta_{M0}, \quad \mathrm{tr}\{T_{L_1M_1}(1)T_{L_2M_2}(1)\}=3(-1)^{M_1}\delta_{L_1L_2}\delta_{M_1\ -M_2},\\
&\mathrm{tr}\{T_{L_1M_1}(1)T_{L_2M_2}(1)T_{L_3M_3}(1)\}\\
&=3\sqrt{3}(-1)^{L_1+L_2+L_3+M_3}\hat{L}_1\hat{L}_2C^{L_3\ -M_3}_{L_1M_1\ L_2M_2}W(L_1L_211;L_31),\\
&\mathrm{tr}\{\hat{T}_{L_1M_1}(1)\hat{T}_{L_2M_2}(1)\hat{T}_{L_3M_3}(1)\hat{T}_{L_4M_4}(1)\}=9(-1)^{L_1+L_2+L_3+L_4+M_4}\hat{L}_1\hat{L}_2\hat{L}_3\\
&\sum_{L\le 2}\hat{L}C^{L\ M_1+M_2}_{L_1M_1\ L_2M_2}C^{L_4\ -M_4}_{L\ M_1+M_2\ L_3M_3}W(L_1L_211;L1)W(LL_311;L_41)
\end{aligned}\tag{3.113}$$

Utilizing Eq. (3.9) we can further obtain the following relations from Eq. (3.113):

$$\mathrm{tr}\{\hat{S}_{M_1}\hat{T}_{L_2M_2}(1)\}=\mathrm{tr}\{\hat{T}_{L_2M_2}(1)\hat{S}_{M_1}\}=\sqrt{6}(-1)^{M_1}\delta_{1L_2}\delta_{M_1\ -M_2}\tag{3.114}$$

$$\mathrm{tr}\{\hat{S}_{M_1}\hat{T}_{L_2M_2}(1)\hat{T}_{L_3M_3}(1)\}=-3\sqrt{6}(-1)^{L_2+L_3+M_3}\hat{L}_2C^{L_3\ -M_3}_{1M_1\ L_2M_2}W(1L_211;L_31),\tag{3.115}$$

$$\begin{aligned}
\mathrm{tr}\{\hat{T}_{L_2M_2}(1)\hat{S}_{M_1}\hat{T}_{L_3M_3}(1)\}&=-3\sqrt{6}(-1)^{L_2+L_3+M_3}\hat{L}_2C^{L_3\ -M_3}_{L_2M_2\ 1M_1}W(L_2111;L_31)\\
&=3\sqrt{6}(-1)^{M_3}\hat{L}_2C^{L_3\ -M_3}_{1M_1\ L_2M_2}W(1L_211;L_31)\\
&=-(-1)^{L_2+L_3}\mathrm{tr}\{\hat{S}_{M_1}\hat{T}_{L_2M_2}(1)\hat{T}_{L_3M_3}(1)\}
\end{aligned}\tag{3.116}$$

Based on Eq. (3.107) there is

$$\mathrm{tr}\{\hat{T}_{L_2M_2}(1)\hat{T}_{L_3M_3}(1)\hat{S}_{M_1}\}=\mathrm{tr}\{\hat{S}_{M_1}\hat{T}_{L_2M_2}(1)\hat{T}_{L_3M_3}(1)\}\tag{3.117}$$

We can also obtain

$$\mathrm{tr}\{\hat{S}_{M_1}\hat{S}_{M_2}\hat{T}_{L_3M_3}(1)\}=6\sqrt{3}(-1)^{L_3+M_3}C^{L_3\ -M_3}_{1M_1\ 1M_2}W(1111;L_31)\tag{3.118}$$

$$\begin{aligned}&\mathrm{tr}\left\{\hat{S}_{M_1}\hat{T}_{L_3M_3}(1)\hat{S}_{M_2}\right\}=6(-1)^{L_3+M_3}\hat{L}_3C_{1M_1\ L_3M_3}^{1\ \ -M_2}W(1L_311;11)\\&=6\sqrt{3}(-1)^{M_3}C_{1M_1\ 1M_2}^{L_3\ -M_3}W(1111;L_31)=(-1)^{L_3}\mathrm{tr}\left\{\hat{S}_{M_1}\hat{S}_{M_2}\hat{T}_{L_3M_3}(1)\right\}\end{aligned} \tag{3.119}$$

Based on Eq. (3.107) the following relations are satisfied:

$$\mathrm{tr}\left\{\hat{T}_{L_3M_3}(1)\hat{S}_{M_1}\hat{S}_{M_2}\right\}=\mathrm{tr}\left\{\hat{S}_{M_1}\hat{S}_{M_2}\hat{T}_{L_3M_3}(1)\right\}=\mathrm{tr}\left\{\hat{S}_{M_2}\hat{T}_{L_3M_3}(1)\hat{S}_{M_1}\right\} \tag{3.120}$$

### 3.2.5 *General Spin Wave Functions and Density Matrices for S = 1 Particles*

Any spin wave function of the spin 1 particles can be expanded as follows:

$$\chi=\sum_{m=\pm1,0}a^m\chi_m=\sum_{m=\pm1,0}(-1)^m a_{-m}\chi_m=\sum_{i=x,\,y,\,z}a_i\chi_i \tag{3.121}$$

Its Hermitian conjugate wave function $\chi^+$ is

$$\chi^+=\sum_{m=\pm1,0}a^{m*}\chi_m^+=\sum_{m=\pm1,0}(-1)^m a_{-m}^*\chi_m^+=\sum_{i=x,\,y,\,z}a_i^*\chi_i^+ \tag{3.122}$$

Usually one calls $\vec{a}$ the polarization vector of spin 1 particles. Its normalization condition $\chi^+\chi=1$ requires

$$\left|\vec{a}\right|^2=\vec{a}^*\cdot\vec{a}=1 \tag{3.123}$$

Let $\vec{a}=\vec{a}_\mathrm{r}+\mathrm{i}\vec{a}_\mathrm{i}$, so we can obtain

$$\vec{a}\times\vec{a}^*=-\mathrm{i}\left(\vec{a}_\mathrm{r}\times\vec{a}_\mathrm{i}-\vec{a}_\mathrm{i}\times\vec{a}_\mathrm{r}\right)=-2\mathrm{i}\left(\vec{a}_\mathrm{r}\times\vec{a}_\mathrm{i}\right) \tag{3.124}$$

Then we can define a real vector $\vec{p}$ with a polarization vector $\vec{a}$

$$\vec{p}\equiv\mathrm{i}\left[\vec{a}\times\vec{a}^*\right] \tag{3.125}$$

It can be seen that vectors $\vec{p}$ is perpendicular to vector $\vec{a}$.

Using the spherical basis coordinate expansion formula

$$\vec{A} = \sum_{\mu=-1}^{1} (-1)^{\mu} A_{\mu} \vec{e}_{-\mu} = \sum_{\mu=-1}^{1} A^{-\mu} \vec{e}_{-\mu} \tag{3.126}$$

we can write the following expression from Eq. (3.125):

$$\vec{p} = \mathrm{i} \sum_{mn} a^{-m} a^{-n*} \left( \vec{e}_{-m} \times \vec{e}^{\,*}_{-n} \right) \tag{3.127}$$

Using expression

$$\vec{e}_m \times \vec{e}_n = \mathrm{i}\sqrt{2} C^{1\ m+n}_{1m\ 1n}\ \vec{e}_{m+n} \tag{3.128}$$

we can obtain

$$\begin{aligned} \vec{e}_{-m} \times \vec{e}^{\,*}_{-n} &= (-1)^{-n} \vec{e}_{-m} \times \vec{e}_n = \mathrm{i}\sqrt{2}(-1)^{-n} C^{1\ -m+n}_{1\ -m\ 1n} \vec{e}_{-m+n} \\ &= -\mathrm{i}\sqrt{2}(-1)^{m-n} C^{1\ -n}_{1\ -m\ \ 1\,m-n} \vec{e}_{-m+n} \end{aligned} \tag{3.129}$$

Substituting Eq. (3.129) into Eq. (3.127), we can get

$$\begin{aligned} \vec{p} &= \sqrt{2} \sum_{mn} a^{-m} a^{-n*} (-1)^{m-n} C^{1\ -n}_{1\ -m\ \ 1\,m-n} \vec{e}_{-m+n} \\ &= \sqrt{2} \sum_{mn} a^{m} a^{n*} (-1)^{-m+n} C^{1\ n}_{1\ m\ 1\,-m+n} \vec{e}_{m-n} \end{aligned} \tag{3.130}$$

Comparing this formula with Eq. (3.126), we have

$$p_{\mu} = \sqrt{2} \sum_{mn} C^{1\,n}_{1\,m\ 1\,\mu} a^{m} a^{n*}, \qquad m, n, \mu = \pm 1, 0 \tag{3.131}$$

Refer to Eqs. (1.98) and (1.79), we obtain

$$p_{\mu} = \sqrt{\frac{2}{3}} t_{1\mu}, \qquad \mu = \pm 1, 0 \tag{3.132}$$

In fact, according to Eq. (1.93), the transformation from $t_{1\mu}$ to $p_{\mu}$ is the same as the transformation from $\hat{T}_{1\mu}$ to $\hat{S}_{\mu}$, so Eq. (3.132) can be obtained directly from Eq. (3.9).

When $S = 1$ the normalized density matrix $\hat{\rho}$ can be divided into three terms according to the $L$ value based on Eqs. (1.81) and (1.92):

$$\chi\chi^{+} = \hat{\rho} = \hat{\rho}_0 + \hat{\rho}_1 + \hat{\rho}_2 \tag{3.133}$$

We can obtain the following expression from Eqs. (1.92), (1.99), and (3.8):

$$\hat{\rho}_0 = \frac{1}{3} t_{00} \hat{T}_{00} = \frac{1}{3} \hat{I} \tag{3.134}$$

The following expression can be obtained from Eqs. (1.92), (3.9), and (3.132):

$$\hat{\rho}_1 = \frac{1}{3} \sum_{\mu=-1}^{1} (-1)^{\mu} t_{1\ -\mu} \hat{T}_{1\mu} = \frac{1}{2} \sum_{\mu=-1}^{1} (-1)^{\mu} p_{-\mu} \hat{S}_{\mu} = \frac{1}{2} \vec{p} \cdot \hat{S} \tag{3.135}$$

Next, order the rank-2 irreducible tensor

$$p_2 = t_2 \tag{3.136}$$

usually $t_2$ or $p_2$ is referred to as the statistical tensor or polarization tensor, and we can write

$$\hat{\rho}_2 = \frac{1}{3} \sum_{\mu=-2}^{2} (-1)^{M} t_{2\ -M} \hat{T}_{2M} = \frac{1}{3} p_2 \cdot \hat{T}_2 \tag{3.137}$$

where $p_2 \cdot \hat{T}_2$ in the above formula represents the scalar product of two rank-2 irreducible tensors defined by Eq. (1.41). We can get the following expression from Eqs. (1.98), (1.79), and (3.136):

$$p_{2M} = t_{2M} = \sqrt{5} \sum_{mn} C_{1m\ 2M}^{1n} a^{m} a^{n*}, \qquad m, n = \pm 1, 0; \quad M = \pm 2, \pm 1, 0 \tag{3.138}$$

Then in $S = 1$ case the density matrix can be expressed in compact form and in spherical basis coordinate system according to Eqs. (3.133)–(3.135) and Eq. (3.137), respectively:

$$\chi\chi^{+} = \hat{\rho} = \frac{1}{3} \left( \hat{I} + \frac{3}{2} \vec{p} \cdot \hat{S} + p_2 \cdot \hat{T}_2 \right) \tag{3.139}$$

$$\chi\chi^{+} = \hat{\rho} = \frac{1}{3} \left( \hat{I} + \frac{3}{2} \sum_{\mu=-1}^{1} (-1)^{\mu} p_{-\mu} \hat{S}_{\mu} + \sum_{M=-2}^{2} (-1)^{M} p_{2\ -M} \hat{T}_{2M} \right) \tag{3.140}$$

In the following context, Eq. (3.139) will be discussed in the Cartesian coordinate system. Each term in Eq. (3.139) is a square $3 \times 3$ matrix in the spin space. $\vec{p}$ is a vector in the spatial coordinate composed of three real components. $\hat{S}$ is a vector in the spatial coordinate composed of three square $3 \times 3$ matrices in spin space. $p_2$ is a rank-2 tensor in spatial coordinate composed of nine numerical components. $\hat{T}_2$ is a rank-2 tensor in spatial coordinate composed of nine rank-2 tensors presented by square 3 × 3 matrices in spin space. The three components of the vector $\vec{p}$ and $\hat{S}$ in the Cartesian coordinate system are $p_i$ and $\hat{S}_i$ $(i = x, y, z)$. The scalar product of two rank-2 tensors in a Cartesian coordinate system is defined as

$$T \cdot Q = \left( \sum_{ik} T_{ik} \vec{e}_k \vec{e}_i \right) : \left( \sum_{lm} Q_{lm} \vec{e}_l \vec{e}_m \right) = \sum_{ik} T_{ik} Q_{ik} \tag{3.141}$$

where the symbol ":" represents taking the scalar product of two vectors, that are close to each other, twice. The 3.1 section has pointed out that nine of $\hat{Q}_{ik}$ in the Cartesian coordinate system are equivalent to five of $\hat{T}_{2M}$ in the spherical basis coordinate systems, and the nine components of the rank-2 tensor $p_2$ and $\hat{T}_2$ in the Cartesian coordinate system are $p_{ik}$ and $\hat{Q}_{ik}(i, k = x, y, z)$, respectively. According to Eq. (3.21) it can be seen that the transformation relation of the five independent polarization tensor operators from a spherical basis coordinate system to a Cartesian coordinate system is

$$\begin{pmatrix} \hat{H}_{xx-yy} \\ \hat{H}_{xy} \\ \hat{H}_{xz} \\ \hat{H}_{yz} \\ \hat{H}_{zz} \end{pmatrix} \equiv \begin{pmatrix} \sqrt{\frac{1}{6}} \hat{Q}_{xx-yy} \\ \sqrt{\frac{2}{3}} \hat{Q}_{xy} \\ \sqrt{\frac{2}{3}} \hat{Q}_{xz} \\ \sqrt{\frac{2}{3}} \hat{Q}_{yz} \\ \sqrt{\frac{1}{2}} \hat{Q}_{zz} \end{pmatrix} = \begin{pmatrix} \frac{1}{\sqrt{2}} & 0 & 0 & 0 & \frac{1}{\sqrt{2}} \\ -\mathrm{i}\frac{1}{\sqrt{2}} & 0 & 0 & 0 & \mathrm{i}\frac{1}{\sqrt{2}} \\ 0 & -\frac{1}{\sqrt{2}} & 0 & \frac{1}{\sqrt{2}} & 0 \\ 0 & \mathrm{i}\frac{1}{\sqrt{2}} & 0 & \mathrm{i}\frac{1}{\sqrt{2}} & 0 \\ 0 & 0 & 1 & 0 & 0 \end{pmatrix} \begin{pmatrix} \hat{T}_{2\,2} \\ \hat{T}_{2\,1} \\ \hat{T}_{2\,0} \\ \hat{T}_{2\,-1} \\ \hat{T}_{2\,-2} \end{pmatrix} \tag{3.142}$$

In the above formula the 5 × 5 matrix can be denoted by $\hat{U}$ such that

$$
\hat{U}\hat{U}^{+}=\begin{pmatrix} \frac{1}{\sqrt{2}} & 0 & 0 & 0 & \frac{1}{\sqrt{2}} \\ -\mathrm{i}\frac{1}{\sqrt{2}} & 0 & 0 & 0 & \mathrm{i}\frac{1}{\sqrt{2}} \\ 0 & -\frac{1}{\sqrt{2}} & 0 & \frac{1}{\sqrt{2}} & 0 \\ 0 & \mathrm{i}\frac{1}{\sqrt{2}} & 0 & \mathrm{i}\frac{1}{\sqrt{2}} & 0 \\ 0 & 0 & 1 & 0 & 0 \end{pmatrix}\begin{pmatrix} \frac{1}{\sqrt{2}} & \mathrm{i}\frac{1}{\sqrt{2}} & 0 & 0 & 0 \\ 0 & 0 & -\frac{1}{\sqrt{2}} & -\mathrm{i}\frac{1}{\sqrt{2}} & 0 \\ 0 & 0 & 0 & 0 & 1 \\ 0 & 0 & \frac{1}{\sqrt{2}} & -\mathrm{i}\frac{1}{\sqrt{2}} & 0 \\ \frac{1}{\sqrt{2}} & -\mathrm{i}\frac{1}{\sqrt{2}} & 0 & 0 & 0 \end{pmatrix}
$$
$$
=\begin{pmatrix} 1 & 0 & 0 & 0 & 0 \\ 0 & 1 & 0 & 0 & 0 \\ 0 & 0 & 1 & 0 & 0 \\ 0 & 0 & 0 & 1 & 0 \\ 0 & 0 & 0 & 0 & 1 \end{pmatrix} \tag{3.143}
$$

Apparently the transformation from $\hat{T}_{2M}$ to $\hat{H}_i$ given by Eq. (3.142) is the unitary transformation, then the basis functions $\left(\hat{H}_{xx-yy}, \hat{H}_{xy}, \hat{H}_{xz}, \hat{H}_{yz}, \hat{H}_{zz}\right)$ of the polarization tensor form the unit vectors of the coordinate axes of the five-dimensional orthogonal Cartesian coordinate system, and the corresponding

$$
\left(h_{xx-yy}, h_{xy}, h_{xz}, h_{yz}, h_{zz}\right) \equiv \left(\sqrt{\frac{1}{6}}p_{xx-yy}, \sqrt{\frac{2}{3}}p_{xy}, \sqrt{\frac{2}{3}}p_{xz}, \sqrt{\frac{2}{3}}p_{yz}, \sqrt{\frac{1}{2}}p_{zz}\right)
$$

are the particle polarization tensor components in the five-dimensional orthogonal Cartesian coordinate system. From Eq. (1.93) we can see that $p_{ij}$ should be the expectation value of $\hat{Q}_{ij}$, so that the transformation relation from $p_{2M}$ to $p_{ij}$ is the same as the transformation relation from $\hat{T}_{2M}$ to $\hat{Q}_{ij}$. Therefore we can get

$$
\begin{aligned} \sum_M(-1)^M p_{2\,-M}\hat{T}_{2M} &= \frac{2}{3}\left(p_{xy}\hat{Q}_{xy}+p_{xz}\hat{Q}_{xz}+p_{yz}\hat{Q}_{yz}\right) \\ &\quad +\frac{1}{6}p_{xx-yy}\hat{Q}_{xx-yy}+\frac{1}{2}p_{zz}\hat{Q}_{zz} \end{aligned} \tag{3.144}
$$

where $p_{xx-yy}\equiv p_{xx}-p_{yy}$. From $\hat{Q}_{xx}+\hat{Q}_{yy}+\hat{Q}_{zz}=0$ and $p_{xx}+p_{yy}+p_{zz}=0$ we can obtain

$$\frac{1}{6}\left(p_{xx}-p_{yy}\right)\left(\hat{Q}_{xx}-\hat{Q}_{yy}\right)+\frac{1}{2}p_{zz}\hat{Q}_{zz}=\frac{1}{3}\left(p_{xx}\hat{Q}_{xx}+p_{yy}\hat{Q}_{yy}+p_{zz}\hat{Q}_{zz}\right) \tag{3.145}$$

So using Eq. (3.20) we can obtain the following result from Eq. (3.144):

$$\begin{aligned}\sum_{M}(-1)^{M}p_{2\ -M}\hat{T}_{2M}&=\frac{2}{3}\left(p_{xy}\hat{Q}_{xy}+p_{xz}\hat{Q}_{xz}+p_{yz}\hat{Q}_{yz}\right)\\&+\frac{1}{3}\left(p_{xx}\hat{Q}_{xx}+p_{yy}\hat{Q}_{yy}+p_{zz}\hat{Q}_{zz}\right)=\frac{1}{3}\sum_{ij}p_{ij}\hat{Q}_{ij}\end{aligned} \tag{3.146}$$

If using the method given in the monograph *Quantum Theory of Angular Momentum* [1], the transformation coefficients in the right end of Eq. (3.21) are all divided by $\sqrt{3}$, and the transformation coefficients in the right end of Eq. (3.22) are all multiplied by $\sqrt{3}$, in this case the corresponding results obtained by the transformation relation Eq. (3.142) are $\hat{Q}'_{ij}$, and we can get

$$\sum_{M}(-1)^{M}p_{2\ -M}\hat{T}_{2M}=\sum_{ij}p'_{ij}\hat{Q}'_{ij} \tag{3.147}$$

However, it should be noted that there are five items in $\hat{T}_{2M}$ and nine items in $\hat{Q}'_{ij}$. Therefore the average value of $p'_{ij}\hat{Q}'_{ij}$ is 1.8 times smaller than the average value of $(-1)^{M}p_{2-M}\hat{T}_{2M}$, and the average value of $\hat{Q}'_{ij}$ is $\sqrt{1.8}\approx 1.34$ times smaller than the average value of $\hat{T}_{2M}$. When using Eq. (3.146), the average value of $\hat{Q}_{ij}$ is $\sqrt{\frac{3}{1.8}}\approx 1.29$ times larger than the average value of $\hat{T}_{2M}$. It can be seen that by using any of the two methods, the range of the expectation value of $\hat{Q}_{ij}$ is not much different from the range of the expectation value of $\hat{T}_{2M}$. Although the transformation method given by Eq. (3.147) based on Reference [1] is better in mathematical form, however the transformation method given by Eq. (3.146), Eq. (3.21), and Eq. (3.22) has been widely used by people engaged in polarization theory research [2, 3, 6]. Therefore, this book also uses the transformation method represented by Eq. (3.146), that is, using transformation formulas given by Eq. (3.21) and Eq. (3.22). According to the above choice in the Cartesian coordinate system, Eq. (3.139) can be rewritten as

$$\chi\chi^{+}=\hat{\rho}=\frac{1}{3}\left(\hat{I}+\frac{3}{2}\sum_{i}p_{i}\hat{S}_{i}+\frac{1}{3}\sum_{ik}p_{ik}\hat{Q}_{ik}\right) \tag{3.148}$$

where $\hat{\rho}$, $\hat{I}$, $\hat{S}_i$, $\hat{Q}_{ik}$ are all square $3\times 3$ matrices in spin space.

In the Cartesian coordinate system there is

$$\vec{e}_k \times \vec{e}_l = \varepsilon_{ikl} \vec{e}_i, \qquad i, k, l = x, y, z \tag{3.149}$$

It follows that Eq. (3.125) can be rewritten as

$$\vec{p} = \mathrm{i} \sum_{ikl} \varepsilon_{ik\ell} a_k a_\ell^* \vec{e}_i \tag{3.150}$$

that is,

$$p_i = \mathrm{i} \sum_{k\ell} \varepsilon_{ik\ell} a_k a_l^*, \qquad i, k, \ell = x, y, z \tag{3.151}$$

In the following context, we will find the matrix element of the $l$th row and $m$th column in the spin space for each term of Eq. (3.148) in the Cartesian coordinate system and the Cartesian basis representation. First we can write the following expressions according to Eqs. (3.121) and (3.122):

$$\rho_{\ell m} = a_\ell a_m^* \tag{3.152}$$

$$\left(\frac{1}{3}\hat{I}\right)_{\ell m} = \frac{1}{3}\delta_{\ell m} \tag{3.153}$$

Using Eq. (3.151) and the expression for the matrix element of $\hat{S}_i$ in the Cartesian basis representation given by Eq. (3.48), we can obtain

$$\frac{1}{2}\sum_i p_i \left(\hat{S}_i\right)_{\ell m} = \frac{1}{2}\sum_i \sum_{jn} \varepsilon_{ijn} a_j a_n^* \varepsilon_{i\ell m} = \frac{1}{2}\left(a_\ell a_m^* - a_m a_\ell^*\right) \tag{3.154}$$

Then using the expression for the matrix element of $\hat{Q}_{ik}$ in the Cartesian basis representation given by Eq. (3.52), we can obtain

$$\begin{aligned} \frac{1}{9}\sum_{ik} p_{ik} \left(\hat{Q}_{ik}\right)_{\ell m} &= -\frac{1}{6}\sum_{ik} p_{ik} \left(\delta_{i\ell}\delta_{km} + \delta_{im}\delta_{k\ell} - \frac{2}{3}\delta_{ik}\delta_{\ell m}\right) \\ &= -\frac{1}{6}\left[p_{\ell m} + p_{m\ell} - \frac{2}{3}\left(\sum_i p_{ii}\right)\delta_{\ell m}\right] \end{aligned} \tag{3.155}$$

From Eq. (3.148) the following formula can be obtained using Eqs. (3.152)–(3.155):

$$0 = \frac{1}{3}\delta_{\ell m} - \frac{1}{2}\left(a_\ell a_m^* + a_m a_\ell^*\right) - \frac{1}{6}\left(p_{\ell m} + p_{m\ell}\right) + \frac{1}{9}\left(\sum_i p_{ii}\right)\delta_{\ell m} \tag{3.156}$$

From the above formula one can immediately see

$$p_{\ell m} = p_{m\ell}, \quad \ell \neq m; \quad \ell, m = x, y, z \tag{3.157}$$

which means that $p_2$ is a symmetric tensor in spatial coordinate. Let

$$\sum_{i=x,\,y,\,z} p_{ii} = 0 \tag{3.158}$$

that is, require that $p_2$ is not a trace tensor in spatial coordinate. So from Eq. (3.156) we can obtain

$$p_{\ell m} = -\frac{3}{2}\left(a_\ell a_m^* + a_m a_\ell^* - \frac{2}{3}\delta_{\ell m}\right) \tag{3.159}$$

From this formula we can see that $p_{\ell m}$ is a real number. And only five components $p_{\ell m}$ are independent. Here Eq. (3.159) describing $p_{\ell m}$ was proved in the Cartesian basis representation.

We can obtain the following results according to Eq. (1.89) and using Eqs. (3.140), (3.112)–(3.114)

$$\langle \hat{S} \rangle = \chi^+ \hat{S} \chi = \mathrm{tr}\{\hat{\rho}\hat{S}\} = \vec{p} \tag{3.160}$$

$$\langle \hat{T}_{2M} \rangle = \chi^+ \hat{T}_{2M} \chi = \mathrm{tr}\{\hat{\rho}\hat{T}_{2M}\} = p_{2M} \tag{3.161}$$

We can also obtain Eq. (3.160) using Eq. (3.148), Eq. (3.104), and Eq. (3.105). The following relation can also be obtained:

$$\langle \hat{Q}_{\ell m} \rangle = \chi^+ \hat{Q}_{\ell m} \chi = \mathrm{tr}\{\hat{\rho}\, \hat{Q}_{\ell m}\} = \frac{1}{9}\sum_{ik} p_{ik} \frac{9}{2}\left(\delta_{i\ell}\delta_{km} + \delta_{k\ell}\delta_{im} - \frac{2}{3}\delta_{ik}\delta_{\ell m}\right) = p_{\ell m} \tag{3.162}$$

Note that Eqs. (3.157) and (3.158) are needed when deriving the above formula.

## 3.3 Vector Polarization Rate and Tensor Polarization Rate of Spin 1 Incident Particles

Select the direction $\vec{k}_{\text{in}}$ of the incident particles as the $z$-axis, where the angle between the direction $\vec{k}_{\text{out}}$ of the outgoing particles and the z axis is $\theta$. Next, define the unit vector $\vec{n} = \left(\vec{k}_{\text{in}} \times \vec{k}_{\text{out}}\right) / \left|\vec{k}_{\text{in}} \times \vec{k}_{\text{out}}\right|$, where $\hat{n}$ is perpendicular to the reaction plane formed by $\vec{k}_{\text{in}}$ and $\vec{k}_{\text{out}}$. If the $y$-axis is selected in the direction $\vec{n}$, and $\vec{k}_{\text{out}}$ is on the $xz$ plane, then the azimuthal angle of the outgoing particle $\varphi = 0$. The results of the second chapter of this book have pointed out that in the $xyz$ coordinate system mentioned above, for the problem of spin $\frac{1}{2}$ particle scattering, if the incident particles are unpolarized, then only the polarization rate of the outgoing particles in the $y$-axis direction is not equal to 0, that is, the outgoing particles are polarized along the direction $\vec{n}$. Therefore, $\vec{n}$ is sometimes referred to as the polarization direction of the outgoing particles. In fact, if the incident particles are polarized, then the polarization rates of the outgoing particles may not be equal to 0 in $x$, $y$, and $z$ directions simultaneously. The previous discussion was carried out in the $xyz$ coordinate system with the incident particle direction along the $z$-axis, that is, it is the helicity coordinate system of the incident particles.

Let us now discuss the polarization rate of spin $S = 1$ incident particles. Polarization particle beam is provided by the polarization ion sources, and the quantum axis of incident particles is consistent with the magnetic field direction of the experimental device. The direction of the quantum axis of the particles is just the polarization direction of the particles. We select the $Z$ axis along the quantum axis of the particles, and the $X$ axis and the $Y$ axis are arbitrary in $XYZ$ Cartesian coordinate system. The $xyz$ and $XYZ$ systems are both different spatial coordinate system. First, we will carry out studying in the $XYZ$ coordinate system. We use $N_\mu$ to represent the share of spin 1 particles in the spin magnetic quantum number $\mu\ (= \pm 1, 0)$ subspace, and obviously it should be satisfied that

$$N_1 + N_0 + N_{-1} = 1. \tag{3.163}$$

Refer to the discussion on spin $\frac{1}{2}$ particles in Sect. 2.2, in $XYZ$ coordinate system taking spin $S = 1$ particle polarization direction as the $Z$-axis, the three components of the polarization vector and the five independent components of the polarization tensor of the incident particles should all be equal to 0 except $p_Z$ and $p_{ZZ}$, and in this case the particle density matrix should be a diagonal matrix. Refer to Eq. (2.39) in this case the density matrix should be

$$\begin{aligned}\hat{\rho}_0 &= \begin{pmatrix}\sqrt{N_1}\\ 0\\ 0\end{pmatrix}\begin{pmatrix}\sqrt{N_1} & 0 & 0\end{pmatrix} + \begin{pmatrix}0\\ \sqrt{N_0}\\ 0\end{pmatrix}\begin{pmatrix}0 & \sqrt{N_0} & 0\end{pmatrix}\\ &+ \begin{pmatrix}0\\ 0\\ \sqrt{N_{-1}}\end{pmatrix}\begin{pmatrix}0 & 0 & \sqrt{N_{-1}}\end{pmatrix} = \begin{pmatrix}N_1 & 0 & 0\\ 0 & N_0 & 0\\ 0 & 0 & N_{-1}\end{pmatrix}\end{aligned} \tag{3.164}$$

The following results can be obtained according to Eqs. (3.163), (3.164), and (3.33):

$$
\begin{aligned}
I = \mathrm{tr}\{\hat{\rho}_0\} = N_1 + N_0 + N_{-1} = 1, \quad p_Z = \mathrm{tr}\{\hat{S}_Z\hat{\rho}_0\} = N_1 - N_{-1}, \\
p_i = \mathrm{tr}\{\hat{S}_i\hat{\rho}_0\} = 0, \quad i = X, Y
\end{aligned}
\tag{3.165}
$$

We can also obtain the following results according to Eqs. (3.163), (3.164), and (3.36):

$$
\begin{aligned}
&p_{ZZ} = \mathrm{tr}\{\hat{Q}_{ZZ}\hat{\rho}_0\} = N_1 - 2N_0 + N_{-1}, \\
&p_i = \mathrm{tr}\{\hat{Q}_i\hat{\rho}_0\} = -\frac{1}{2}(N_1 - 2N_0 + N_{-1}) = -\frac{1}{2}p_{ZZ}, \quad i = XX, YY \\
&p_j = \mathrm{tr}\{\hat{Q}_j\hat{\rho}_0\} = 0, \quad j = XY, XZ, YZ, XX - YY
\end{aligned}
\tag{3.166}
$$

This case also belongs to the special polarized particle states, for which the states with different spin magnetic quantum numbers are incoherent, but they may have different shares. It can be seen that the result given by Eq. (3.166) satisfies $p_{XX} + p_{YY} + p_{ZZ} = 0$ given by Eq. (3.158). We can also see that $1 \geq p_Z \geq -1$ and $1 \geq p_{ZZ} \geq -2$. When $N_1 = N_0 = N_{-1} = \frac{1}{3}$, there are $p_Z = 0$ and $p_{ZZ} = 0$; in this case the particles are unpolarized. And notice that as long as there is $N_1 = N_{-1}$, there is $p_Z = 0$; as long as there is $N_0 = \frac{1}{3}$, there is $p_{ZZ} = 0$

It is also noted that if the coordinate transformation method represented by Eq. (3.147) is used, since $\hat{Q}'_{zz} = \frac{1}{\sqrt{3}}\hat{Q}_{zz}$ and based on Eq. (3.166) there is $p'_{ZZ} = \frac{1}{\sqrt{3}}p_{ZZ}$, then it can be ensured that the results given by the right end of the second equal sign of Eq. (3.146) and the right end of Eq. (3.147) are equal to each other. Assuming that the angle between the $Z$ axis and the $z$ axis is $\beta$, the angle between the projection of the $Z$ axis on the $xy$ plane of the $xyz$ coordinate system and the $x$ axis is $\psi$. This is the usual spherical coordinate selection method. It is easy to write

$$
p_x = p_Z \sin\beta\cos\psi, \quad p_y = p_Z \sin\beta\sin\psi, \quad p_z = p_Z \cos\beta \tag{3.167}
$$

The above equations can be written in a matrix form

$$
\begin{pmatrix} p_x \\ p_y \\ p_z \end{pmatrix} = \begin{pmatrix} U_{xx} & U_{xy} & \sin\beta\cos\psi \\ U_{yx} & U_{yy} & \sin\beta\sin\psi \\ U_{zx} & U_{zy} & \cos\beta \end{pmatrix} \begin{pmatrix} 0 \\ 0 \\ p_Z \end{pmatrix} \tag{3.168}
$$

And it can be rewritten by the vector and matrix form as

$$
\vec{p}_{xyz} = \hat{U}\vec{p}_{XYZ} \tag{3.169}
$$

where $\hat{U}$ is a real $3 \times 3$ matrix. It can be seen from Eq. (3.168) that only $U_{xz}$, $U_{yz}$, $U_{zz}$ are known. Let us set $\tilde{U}$ as the transpose matrix of $\hat{U}$ , and require that $\hat{U}$ is a unitary matrix, that is, $\hat{U}\tilde{U} = I$.

For vector $\vec{A}$ and $\vec{B}$, we define the outer product

$$(AB) \equiv \vec{A}\widetilde{\vec{B}} = \begin{pmatrix} A_x \\ A_y \\ A_z \end{pmatrix} \begin{pmatrix} B_x & B_y & B_z \end{pmatrix} = \begin{pmatrix} A_xB_x & A_xB_y & A_xB_z \\ A_yB_x & A_yB_y & A_yB_z \\ A_zB_x & A_zB_y & A_zB_z \end{pmatrix} \tag{3.170}$$

Then, letting $\vec{A}' = \hat{U}\vec{A}$, we have $\widetilde{\vec{A}'} = \widetilde{\vec{A}}\,\widetilde{U}$, so we can obtain

$$(AB)' = \hat{U}(AB)\widetilde{U} \tag{3.171}$$

From Eq. (3.166) we can see that the tensor polarization rate in the coordinate system *XYZ* is

$$\vec{p}\vec{p}_{XYZ} = \begin{pmatrix} -\frac{1}{2}p_{ZZ} & 0 & 0 \\ 0 & -\frac{1}{2}p_{ZZ} & 0 \\ 0 & 0 & p_{ZZ} \end{pmatrix} \tag{3.172}$$

And it can be rewritten as [2]

$$\vec{p}\vec{p}_{XYZ} = \frac{3}{2}p_{ZZ}\begin{pmatrix} 0 & 0 & 0 \\ 0 & 0 & 0 \\ 0 & 0 & 1 \end{pmatrix} - \frac{1}{2}p_{ZZ}\begin{pmatrix} 1 & 0 & 0 \\ 0 & 1 & 0 \\ 0 & 0 & 1 \end{pmatrix} \tag{3.173}$$

Transforming the above formula into the *xyz* coordinate system with Eq. (3.171), we can obtain

$$\vec{p}\vec{p}_{xyz} = \frac{3}{2}p_{ZZ}\hat{U}\begin{pmatrix} 0 & 0 & 0 \\ 0 & 0 & 0 \\ 0 & 0 & 1 \end{pmatrix}\widetilde{U} - \frac{1}{2}p_{ZZ}\begin{pmatrix} 1 & 0 & 0 \\ 0 & 1 & 0 \\ 0 & 0 & 1 \end{pmatrix} \tag{3.174}$$

where $\hat{U}\widetilde{U} = I$ is used in the above formula. The following relation can be obtained by Eqs. (3.174) and (3.168):

$$\vec{p}\vec{p}_{xyz} = \frac{3}{2}p_{ZZ}\hat{U}\begin{pmatrix} 0 & 0 & 0 \\ 0 & 0 & 0 \\ \sin\beta\cos\psi & \sin\beta\sin\psi & \cos\beta \end{pmatrix} - \frac{1}{2}p_{ZZ}\begin{pmatrix} 1 & 0 & 0 \\ 0 & 1 & 0 \\ 0 & 0 & 1 \end{pmatrix} \tag{3.175}$$

Further we can get the following results from the above formula and Eq. (3.168):

$$
\begin{aligned}
p_{xx} &= \frac{1}{2}\left(3\sin^2\beta\cos^2\psi - 1\right)p_{ZZ}, & p_{yy} &= \frac{1}{2}\left(3\sin^2\beta\sin^2\psi - 1\right)p_{ZZ}, \\
p_{zz} &= \frac{1}{2}\left(3\cos^2\beta - 1\right)p_{ZZ}, & p_{xy} &= \frac{3}{2}\sin^2\beta\cos\psi\sin\psi p_{ZZ}, \\
p_{xz} &= \frac{3}{2}\sin\beta\cos\beta\cos\psi p_{ZZ}, & p_{yz} &= \frac{3}{2}\sin\beta\cos\beta\sin\psi p_{ZZ}
\end{aligned}
\tag{3.176}
$$

From the above formula it can be seen that $p_{xx} + p_{yy} + p_{zz} = 0$ and we have

$$
p_{xx-yy} \equiv p_{xx} - p_{yy} = \frac{3}{2}\sin^2\beta\cos(2\psi)p_{ZZ} \tag{3.177}
$$

We can find that the tensor polarization rates of the incident particles in the spherical basis coordinate system are, using Eqs. (3.22), (3.176), and (3.177), as follows:

$$
\begin{aligned}
&p_{2\,\pm2} = \frac{\sqrt{3}}{4}\sin^2\beta\,\mathrm{e}^{\pm 2\mathrm{i}\psi}p_{ZZ}, \quad p_{2\,\pm1} = \mp\frac{\sqrt{3}}{4}\sin(2\beta)\,\mathrm{e}^{\pm\mathrm{i}\psi}p_{ZZ}, \\
&p_{20} = \frac{1}{2\sqrt{2}}\left(3\cos^2\beta - 1\right)p_{ZZ}
\end{aligned}
\tag{3.178}
$$

## 3.4 Elastic Scattering Amplitude of Spin 1 Particles with Spinless Targets

Based on Eq. (5.5.51) in Ref. [7], the scattering amplitude of an incident particle with spin 1 and a spinless target can be written as

$$
F_{\mu'\mu}(\theta,\varphi) = f_C(\theta)\delta_{\mu'\mu} + \frac{\mathrm{i}\sqrt{\pi}}{k}\sum_{\ell j} C^{j\,\mu}_{\ell 0\,1\mu} C^{j\,\mu}_{\ell\,\mu-\mu'\,1\mu'}\left(1 - S^j_\ell\right)\mathrm{e}^{2\mathrm{i}\sigma_\ell}Y_{\ell\,\mu-\mu'}(\theta,\varphi) \tag{3.179}
$$

where $S^j_\ell$ is the $S$ matrix element. Let

$$
f^\ell_{\mu'\mu} = \hat{\ell}^2\sum_j C^{j\,\mu}_{\ell 0\,1\mu} C^{j\,\mu}_{\ell\,\mu-\mu'\,1\mu'}\left(1 - S^j_\ell\right), \tag{3.180}
$$

here $\hat{\ell} \equiv \sqrt{2\ell + 1}$. Next we introduce the following symbols:

$$
\alpha^+_\ell = 1 - S^{\ell+1}_\ell, \quad \alpha^0_\ell = 1 - S^\ell_\ell, \quad \alpha^-_\ell = 1 - S^{\ell-1}_\ell \tag{3.181}
$$

Using the C-G coefficients given in Table 3.1, we can obtain the following expressions by Eqs. (3.180) and (3.181):

$$f_{00}^{\ell} = (\ell+1)\alpha_{\ell}^{+} + \ell\alpha_{\ell}^{-} \tag{3.182}$$

$$f_{11}^{\ell} = f_{-1\ -1}^{\ell} = \frac{1}{2}\left[(\ell+2)\alpha_{\ell}^{+} + (2\ell+1)\alpha_{\ell}^{0} + (\ell-1)\alpha_{\ell}^{-}\right] \tag{3.183}$$

$$f_{10}^{\ell} = f_{-1\ 0}^{\ell} = \sqrt{\frac{\ell(\ell+1)}{2}}\left(\alpha_{\ell}^{+} - \alpha_{\ell}^{-}\right) \tag{3.184}$$

$$f_{01}^{\ell} = f_{0\ -1}^{\ell} = \frac{1}{\sqrt{2\ell(\ell+1)}}\left[\ell(\ell+2)\alpha_{\ell}^{+} - (2\ell+1)\alpha_{\ell}^{0} - (\ell-1)(\ell+1)\alpha_{\ell}^{-}\right] \tag{3.185}$$

$$f_{1\ -1}^{\ell} = f_{-1\ 1}^{\ell} = \frac{1}{2}\sqrt{\frac{(\ell-1)(\ell+2)}{\ell(\ell+1)}}\left[\ell\alpha_{\ell}^{+} - (2\ell+1)\alpha_{\ell}^{0} + (\ell+1)\alpha_{\ell}^{-}\right] \tag{3.186}$$

We can get the following relations according to Eqs. (3.73), (3.75), and (3.77) in Ref. [7]:

$$\begin{gathered}\mathrm{Y}_{\ell 0} = \sqrt{\frac{2\ell+1}{4\pi}}\mathrm{P}_{\ell}, \quad \mathrm{Y}_{\ell\ \pm 1} = \mp\sqrt{\frac{2\ell+1}{4\pi}}\sqrt{\frac{1}{\ell(\ell+1)}}\ \mathrm{P}_{\ell}^{1}\mathrm{e}^{\pm \mathrm{i}\varphi}, \\ \mathrm{Y}_{\ell\ \pm 2} = \sqrt{\frac{2\ell+1}{4\pi}}\sqrt{\frac{1}{(\ell-1)\ell(\ell+1)(\ell+2)}}\mathrm{P}_{\ell}^{2}\mathrm{e}^{\pm 2\mathrm{i}\varphi}\end{gathered} \tag{3.187}$$

Next we introduce the following symbols according to Eqs. (3.179)–(3.187):

$$A(\theta) = f_{\mathrm{C}}(\theta) + \frac{\mathrm{i}}{2k}\sum_{\ell}\left[(\ell+1)\alpha_{\ell}^{+} + \ell\alpha_{\ell}^{-}\right]\mathrm{e}^{2\mathrm{i}\sigma_{\ell}}\mathrm{P}_{\ell}(\cos\theta) \tag{3.188}$$

$$\begin{aligned}B(\theta) = f_{\mathrm{C}}(\theta) + \frac{\mathrm{i}}{2k}\sum_{\ell}\frac{1}{2}&\left[(\ell+2)\alpha_{\ell}^{+} + (2\ell+1)\alpha_{\ell}^{0} + (\ell-1)\alpha_{\ell}^{-}\right] \\ &\mathrm{e}^{2\mathrm{i}\sigma_{\ell}}\mathrm{P}_{\ell}(\cos\theta)\end{aligned} \tag{3.189}$$

$$C(\theta) = -\frac{1}{2k}\sum_{\ell}\frac{1}{\sqrt{2}}\left(\alpha_{\ell}^{+} - \alpha_{\ell}^{-}\right)\mathrm{e}^{2\mathrm{i}\sigma_{\ell}}\mathrm{P}_{\ell}^{1}(\cos\theta) \tag{3.190}$$

$$\begin{aligned}D(\theta) = -\frac{1}{2k}\sum_{\ell}\frac{1}{\sqrt{2}}\frac{1}{\ell(\ell+1)}&\left[\ell(\ell+2)\alpha_{\ell}^{+} - (2\ell+1)\alpha_{\ell}^{0} - (\ell-1)(\ell+1)\alpha_{\ell}^{-}\right] \\ &\mathrm{e}^{2\mathrm{i}\sigma_{\ell}}\mathrm{P}_{\ell}^{1}(\cos\theta)\end{aligned} \tag{3.191}$$

$$E(\theta) = -\frac{\mathrm{i}}{2k}\sum_{\ell}\frac{1}{2\ell(\ell+1)}\left[\ell\alpha_{\ell}^{+} - (2\ell+1)\alpha_{\ell}^{0} + (\ell+1)\alpha_{\ell}^{-}\right]\mathrm{e}^{2\mathrm{i}\sigma_{\ell}}\mathrm{P}_{\ell}^{2}(\cos\theta) \tag{3.192}$$

Given these, the following relations can be obtained by Eq. (3.179):

$$
\begin{aligned}
&F_{00}(\theta,\varphi)=A(\theta),\ F_{11}(\theta,\varphi)=F_{-1-1}(\theta,\varphi)=B(\theta),\ F_{10}(\theta,\varphi)=-\mathrm{i}C(\theta)\mathrm{e}^{-\mathrm{i}\varphi},\\
&F_{-1\,0}(\theta,\varphi)=\mathrm{i}C(\theta)\mathrm{e}^{\mathrm{i}\varphi},\ F_{0\,1}(\theta,\varphi)=\mathrm{i}D(\theta)\mathrm{e}^{\mathrm{i}\varphi},\ F_{0\,-1}(\theta,\varphi)=-\mathrm{i}D(\theta)\mathrm{e}^{-\mathrm{i}\varphi},\\
&F_{1\,-1}(\theta,\varphi)=-E(\theta)\mathrm{e}^{-2\mathrm{i}\varphi},\ F_{-1\,1}(\theta,\varphi)=-E(\theta)\mathrm{e}^{2\mathrm{i}\varphi}
\end{aligned}
\tag{3.193}
$$

In this way, the matrix elements given by Eq. (3.179) can form the following scattering matrix in the spherical basis representation:

$$
\begin{aligned}
\hat{F}(\theta,\varphi)&=\begin{pmatrix} B(\theta) & -\mathrm{i}C(\theta)\mathrm{e}^{-\mathrm{i}\varphi} & -E(\theta)\mathrm{e}^{-2\mathrm{i}\varphi}\\ \mathrm{i}D(\theta)\mathrm{e}^{\mathrm{i}\varphi} & A(\theta) & -\mathrm{i}D(\theta)\mathrm{e}^{-\mathrm{i}\varphi}\\ -E(\theta)\mathrm{e}^{2\mathrm{i}\varphi} & \mathrm{i}C(\theta)\mathrm{e}^{\mathrm{i}\varphi} & B(\theta)\end{pmatrix},\\
\hat{F}^{+}(\theta,\varphi)&=\begin{pmatrix} B^*(\theta) & -\mathrm{i}D^*(\theta)\mathrm{e}^{-\mathrm{i}\varphi} & -E^*(\theta)\mathrm{e}^{-2\mathrm{i}\varphi}\\ \mathrm{i}C^*(\theta)\mathrm{e}^{\mathrm{i}\varphi} & A^*(\theta) & -\mathrm{i}C^*(\theta)\mathrm{e}^{-\mathrm{i}\varphi}\\ -E^*(\theta)\mathrm{e}^{2\mathrm{i}\varphi} & \mathrm{i}D^*(\theta)\mathrm{e}^{\mathrm{i}\varphi} & B^*(\theta)\end{pmatrix}
\end{aligned}
\tag{3.194}
$$

Skipping the argument $(\theta)$ of $A$, $B$, $C$, $D$, and $E$, the following expression can be obtained by Eq. (3.194):

$$
\begin{aligned}
\hat{F}\hat{F}^{+}&=\begin{pmatrix} B & -\mathrm{i}C\mathrm{e}^{-\mathrm{i}\varphi} & -E\mathrm{e}^{-2\mathrm{i}\varphi}\\ \mathrm{i}D\mathrm{e}^{\mathrm{i}\varphi} & A & -\mathrm{i}D\mathrm{e}^{-\mathrm{i}\varphi}\\ -E\mathrm{e}^{2\mathrm{i}\varphi} & \mathrm{i}C\mathrm{e}^{\mathrm{i}\varphi} & B\end{pmatrix}\begin{pmatrix} B^* & -\mathrm{i}D^*\mathrm{e}^{-\mathrm{i}\varphi} & -E^*\mathrm{e}^{-2\mathrm{i}\varphi}\\ \mathrm{i}C^*\mathrm{e}^{\mathrm{i}\varphi} & A^* & -\mathrm{i}C^*\mathrm{e}^{-\mathrm{i}\varphi}\\ -E^*\mathrm{e}^{2\mathrm{i}\varphi} & \mathrm{i}D^*\mathrm{e}^{\mathrm{i}\varphi} & B^*\end{pmatrix}\\
&=\begin{pmatrix} |B|^2+|C|^2+|E|^2 & -\mathrm{i}(BD^*+CA^*+ED^*)\mathrm{e}^{-\mathrm{i}\varphi} & -\left(BE^*+|C|^2+EB^*\right)\mathrm{e}^{-2\mathrm{i}\varphi}\\ \mathrm{i}(DB^*+AC^*+DE^*)\mathrm{e}^{\mathrm{i}\varphi} & 2|D|^2+|A|^2 & -\mathrm{i}(DB^*+AC^*+DE^*)\mathrm{e}^{-\mathrm{i}\varphi}\\ -\left(EB^*+|C|^2+BE^*\right)\mathrm{e}^{2\mathrm{i}\varphi} & \mathrm{i}(BD^*+CA^*+ED^*)\mathrm{e}^{\mathrm{i}\varphi} & |B|^2+|C|^2+|E|^2\end{pmatrix}
\end{aligned}
\tag{3.195}
$$

And it can be found that the differential cross section of the unpolarized incident particles is

$$
I_0=\frac{1}{3}\mathrm{tr}\left\{\hat{F}\hat{F}^{+}\right\}=\frac{1}{3}\left[\,|A|^2+2\left(\,|B|^2+|C|^2+|D|^2+|E|^2\right)\,\right]
\tag{3.196}
$$

According to Eq. (3.140) the scattering matrix given by Eq. (3.194) can be expanded in the spherical basis coordinate system

$$
\hat{F}=\frac{1}{3}u\,\hat{I}+\frac{1}{2}\sum_{\mu}(-1)^{\mu}v_{-\mu}\hat{S}_{\mu}+\frac{1}{3}\sum_{M}(-1)^{M}w_{-M}\hat{T}_{2M}
\tag{3.197}
$$

From Eq. (3.194) and Eq. (3.197), the following relations can be obtained by using $\hat{S}_{\mu}$ given by Eq. (3.32) and $\hat{T}_{2M}$ given by Eq. (3.38) in the spherical basis representation

$$
\begin{aligned}
&B=\frac{1}{3}u+\frac{1}{2}v_0+\frac{1}{3\sqrt{2}}w_0, && -\mathrm{i}C\mathrm{e}^{-\mathrm{i}\varphi}=\frac{1}{2}v_{-1}+\frac{1}{\sqrt{6}}w_{-1},\\
&-E\mathrm{e}^{-2\mathrm{i}\varphi}=\frac{1}{\sqrt{3}}w_{-2}, && \mathrm{i}D\mathrm{e}^{\mathrm{i}\varphi}=-\frac{1}{2}v_1-\frac{1}{\sqrt{6}}w_1,\\
&A=\frac{1}{3}u-\frac{\sqrt{2}}{3}w_0, && -\mathrm{i}D\mathrm{e}^{-\mathrm{i}\varphi}=\frac{1}{2}v_{-1}-\frac{1}{\sqrt{6}}w_{-1},\\
&-E\mathrm{e}^{2\mathrm{i}\varphi}=\frac{1}{\sqrt{3}}w_2, && \mathrm{i}C\mathrm{e}^{\mathrm{i}\varphi}=-\frac{1}{2}v_1+\frac{1}{\sqrt{6}}w_1,\\
&B=\frac{1}{3}u-\frac{1}{2}v_0+\frac{1}{3\sqrt{2}}w_0
\end{aligned} \tag{3.198}
$$

From the above formulas we can obtain

$$
\begin{aligned}
&u=A+2B,\ \ v_{-1}=-\mathrm{i}(D+C)\mathrm{e}^{-\mathrm{i}\varphi},\ \ v_0=0,\ \ v_1=-\mathrm{i}(D+C)\mathrm{e}^{\mathrm{i}\varphi},\\
&w_{-2}=-\sqrt{3}E\mathrm{e}^{-2\mathrm{i}\varphi},\ \ w_{-1}=\mathrm{i}\sqrt{\frac{3}{2}}(D-C)\mathrm{e}^{-\mathrm{i}\varphi},\ \ w_0=-\sqrt{2}(A-B),\\
&w_1=-\mathrm{i}\sqrt{\frac{3}{2}}(D-C)\mathrm{e}^{\mathrm{i}\varphi},\ \ w_2=-\sqrt{3}E\mathrm{e}^{2\mathrm{i}\varphi}
\end{aligned} \tag{3.199}
$$

The above results show that $\hat{F}$ contains the vector polarization terms $\hat{S}_1$ and $\hat{S}_{-1}$ as well as tensor polarization terms $\hat{T}_{2M}(M=\pm 2, \pm 1, 0)$. If $\varphi=0$ we have that $v_{-1}=v_1$, $w_{-2}=w_2$, $w_{-1}=-w_1$, then there is only one term $\hat{S}_1$ for the vector polarization and there are only three terms $\hat{T}_{20}$, $\hat{T}_{21}$, $\hat{T}_{22}$ for the tensor polarization.

We can get the following expressions utilizing Eqs. (1.23) and (3.199):

$$
\begin{aligned}
&\frac{1}{2}\sum_{\mu}(-1)^{\mu}v_{-\mu}\hat{S}_{\mu}\\
&=\frac{1}{2\sqrt{2}}v_{-1}\left(\hat{S}_x+\mathrm{i}\hat{S}_y\right)+\frac{1}{2}v_0\hat{S}_z-\frac{1}{2\sqrt{2}}v_1\left(\hat{S}_x-\mathrm{i}\hat{S}_y\right)\\
&=\frac{1}{2\sqrt{2}}(v_{-1}-v_1)\hat{S}_x+\frac{1}{2}v_0\hat{S}_z+\frac{\mathrm{i}}{2\sqrt{2}}(v_{-1}+v_1)\hat{S}_y\\
&=-\frac{\mathrm{i}}{2\sqrt{2}}(D+C)\left(\mathrm{e}^{-\mathrm{i}\varphi}-\mathrm{e}^{\mathrm{i}\varphi}\right)\hat{S}_x+\frac{1}{2\sqrt{2}}(D+C)\left(\mathrm{e}^{-\mathrm{i}\varphi}+\mathrm{e}^{\mathrm{i}\varphi}\right)\hat{S}_y\\
&=-\frac{1}{\sqrt{2}}(D+C)\sin\varphi\,\hat{S}_x+\frac{1}{\sqrt{2}}(D+C)\cos\varphi\,\hat{S}_y\\
&=\frac{1}{\sqrt{2}}(D+C)\left(\cos\varphi\,\hat{S}_y-\sin\varphi\,\hat{S}_x\right)
\end{aligned} \tag{3.200}
$$

We can also get the following expressions utilizing Eqs. (3.22) and (3.199):

$$
\begin{aligned}
&\frac{1}{3}\sum_{M}(-1)^{M}w_{-M}\hat{T}_{2M}\\
&=\frac{1}{6\sqrt{3}}w_2\big(\hat{Q}_{xx-yy}-2\mathrm{i}\hat{Q}_{xy}\big)-\frac{1}{3\sqrt{3}}w_1\big(\hat{Q}_{xz}-\mathrm{i}\hat{Q}_{yz}\big)+\frac{1}{3\sqrt{2}}w_0\hat{Q}_{zz}\\
&\quad+\frac{1}{3\sqrt{3}}w_{-1}\big(\hat{Q}_{xz}+\mathrm{i}\hat{Q}_{yz}\big)+\frac{1}{6\sqrt{3}}w_{-2}\big(\hat{Q}_{xx-yy}+2\mathrm{i}\hat{Q}_{xy}\big)\\
&=-\frac{1}{3}(A-B)\hat{Q}_{zz}+\frac{\mathrm{i}}{3\sqrt{2}}(D-C)\big(\mathrm{e}^{\mathrm{i}\varphi}+\mathrm{e}^{-\mathrm{i}\varphi}\big)\hat{Q}_{xz}\\
&\quad+\frac{1}{3\sqrt{2}}(D-C)\big(\mathrm{e}^{\mathrm{i}\varphi}-\mathrm{e}^{-\mathrm{i}\varphi}\big)\hat{Q}_{yz}-\frac{1}{6}E\big(\mathrm{e}^{2\mathrm{i}\varphi}+\mathrm{e}^{-2\mathrm{i}\varphi}\big)\hat{Q}_{xx-yy}\\
&\quad+\frac{\mathrm{i}}{3}E\big(\mathrm{e}^{2\mathrm{i}\varphi}-\mathrm{e}^{-2\mathrm{i}\varphi}\big)\hat{Q}_{xy}\\
&=-\frac{1}{3}(A-B)\hat{Q}_{zz}+\mathrm{i}\frac{\sqrt{2}}{3}(D-C)\big(\cos\varphi\,\hat{Q}_{xz}+\sin\varphi\,\hat{Q}_{yz}\big)\\
&\quad-\frac{1}{3}E\big(\cos(2\varphi)\hat{Q}_{xx-yy}+2\sin(2\varphi)\hat{Q}_{xy}\big)
\end{aligned}
\tag{3.201}
$$

Substituting the first equation of Eq. (3.199), Eq. (3.200), and Eq. (3.201) into Eq. (3.197), we can obtain

$$
\begin{aligned}
\hat{F}=&\frac{1}{3}(A+2B)\hat{I}+\frac{1}{\sqrt{2}}(D+C)\big(\cos\varphi\hat{S}_y-\sin\varphi\hat{S}_x\big)\\
&+\mathrm{i}\frac{\sqrt{2}}{3}(D-C)\big(\cos\varphi\hat{Q}_{xz}+\sin\varphi\hat{Q}_{yz}\big)\\
&-\frac{1}{3}E\big[\cos(2\varphi)\hat{Q}_{xx-yy}+2\sin(2\varphi)\hat{Q}_{xy}\big]-\frac{1}{3}(A-B)\hat{Q}_{zz}
\end{aligned}
\tag{3.202}
$$

If one takes the $y$-axis in the direction $\vec{n}$, in this case there is $\varphi=0$, then Eq. (3.202) is simplified to

$$
\begin{aligned}
\hat{F}=&\frac{1}{3}(A+2B)\hat{I}+\frac{1}{\sqrt{2}}(D+C)\hat{S}_y+\mathrm{i}\frac{\sqrt{2}}{3}(D-C)\hat{Q}_{xz}\\
&-\frac{1}{3}E\hat{Q}_{xx-yy}-\frac{1}{3}(A-B)\hat{Q}_{zz}
\end{aligned}
\tag{3.203}
$$

This formula is consistent with the included terms in the result given in Reference [3]. From Eq. (3.203) we can obtain the following result by using the tracing formulas Eqs. (3.104) and (3.105):

$$
\begin{aligned}
I_0 &= \frac{1}{3}\mathrm{tr}\left\{\hat{F}\hat{F}^+\right\} = \frac{1}{3}\left[\frac{1}{3}(A+2B)(A^*+2B^*)+(D+C)(D^*+C^*)\right. \\
&\quad \left. +(D-C)(D^*-C^*)+\frac{2}{3}(A-B)(A^*-B^*)+2|E|^2\right] \\
&= \frac{1}{3}\left[\,|A|^2+2\left(|B|^2+|C|^2+|D|^2+|E|^2\right)\right]
\end{aligned}
\tag{3.204}
$$

This formula is exactly the same as Eq. (3.196).

The scattering matrix $\hat{F}$ is given by Eq. (3.194) in the spherical basis representation, and it can be expanded in the spherical basis coordinate system using Eq. (3.197). Their expansion coefficients have been given by Eq. (3.199). And Eq. (3.202) gives the scattering matrix $\hat{F}$ that has been transformed into the Cartesian coordinate system. We use $F_{\mu'\mu}(\mu,\,\mu' = \pm 1,\,0)$ to represent the matrix elements in the spherical basis coordinate system and the spherical basis representation. We can obtain the following results according to Eq. (3.202) as well as Eqs. (3.33) and (3.36) in the spherical basis representation:

$$
\begin{aligned}
&F_{1\,1} = \frac{1}{3}(A+2B) - \frac{1}{3}(A-B) = B, \\
&F_{1\,0} = -\frac{1}{2}(D+C)(\mathrm{i}\cos\varphi + \sin\varphi) + \frac{1}{2}(D-C)(\mathrm{i}\cos\varphi + \sin\varphi) = -\mathrm{i}C\mathrm{e}^{-\mathrm{i}\varphi}, \\
&F_{1\,-1} = -E[\cos(2\varphi) - \mathrm{i}\sin(2\varphi)] = -E\mathrm{e}^{-2\mathrm{i}\varphi}, \\
&F_{0\,1} = \frac{1}{2}(D+C)(\mathrm{i}\cos\varphi - \sin\varphi) + \frac{1}{2}(D-C)(\mathrm{i}\cos\varphi - \sin\varphi) = \mathrm{i}D\mathrm{e}^{\mathrm{i}\varphi}, \\
&F_{0\,0} = \frac{1}{3}(A+2B) + \frac{2}{3}(A-B) = A, \\
&F_{0\,-1} = \frac{1}{2}(D+C)(-\mathrm{i}\cos\varphi - \sin\varphi) + \frac{1}{2}(D-C)(-\mathrm{i}\cos\varphi - \sin\varphi) = -\mathrm{i}D\mathrm{e}^{-\mathrm{i}\varphi}, \\
&F_{-1\,1} = -E[\cos(2\varphi) + \mathrm{i}\sin(2\varphi)] = -E\mathrm{e}^{2\mathrm{i}\varphi}, \\
&F_{-1\,0} = \frac{1}{2}(D+C)(\mathrm{i}\cos\varphi - \sin\varphi) - \frac{1}{2}(D-C)(\mathrm{i}\cos\varphi - \sin\varphi) = \mathrm{i}C\mathrm{e}^{\mathrm{i}\varphi}, \\
&F_{-1-1} = \frac{1}{3}(A+2B) - \frac{1}{3}(A-B) = B
\end{aligned}
\tag{3.205}
$$

The above results are consistent with Eq. (3.194). And the above discussions are carried out in the spherical basis representation.

## 3.5 Polarization Theory of Elastic Scattering of Unpolarized Spin 1 Particles with Spinless Targets

According to Eq. (3.148) we will write out the density matrix of the incident particles in the Cartesian coordinate system. Because $\hat{Q}_{ik} = \hat{Q}_{ki}$ in $i \neq k$ case, in the density matrix expression we only keep one of them but multiply it by 2. It follows that

$$\hat{\rho}_{\text{in}} = \frac{1}{3}\Big[\hat{I} + \frac{3}{2}\left(p_x\hat{S}_x + p_y\hat{S}_y + p_z\hat{S}_z\right) + \frac{2}{3}\left(p_{xy}\hat{Q}_{xy} + p_{xz}\hat{Q}_{xz} + p_{yz}\hat{Q}_{yz}\right) + \frac{1}{3}\left(p_{xx}\hat{Q}_{xx} + p_{yy}\hat{Q}_{yy} + p_{zz}\hat{Q}_{zz}\right)\Big] \tag{3.206}$$

where $p_i$ is the component of the incident particle polarization vector and $p_{ik}$ is the component of the incident particle polarization tensor. Eq. (3.19) requires

$$\hat{Q}_{xx} + \hat{Q}_{yy} + \hat{Q}_{zz} = \begin{pmatrix} 0 & 0 & 0 \\ 0 & 0 & 0 \\ 0 & 0 & 0 \end{pmatrix} \tag{3.207}$$

and Eq. (3.158) requires

$$p_{xx} + p_{yy} + p_{zz} = 0 \tag{3.208}$$

By using Eq. (3.104) and Eq. (3.105), the following results are easily demonstrated from Eqs. (3.206) and (3.208):

$$p_i = \langle \hat{S}_i \rangle = \text{tr}\{\hat{S}_i\hat{\rho}_{\text{in}}\}, \qquad i = x,\, y,\, z \tag{3.209}$$

$$p_{ik} = \langle \hat{Q}_{ik} \rangle = \text{tr}\{\hat{Q}_{ik}\hat{\rho}_{\text{in}}\}, \qquad i,k = x,\, y,\, z \tag{3.210}$$

We can obtain the following relations according to Eqs. (3.207) and (3.208):

$$\frac{1}{2}\left(p_{xx} - p_{yy}\right)\left(\hat{Q}_{xx} - \hat{Q}_{yy}\right) + \frac{3}{2}p_{zz}\hat{Q}_{zz} = p_{xx}\hat{Q}_{xx} + p_{yy}\hat{Q}_{yy} + p_{zz}\hat{Q}_{zz} \tag{3.211}$$

$$\frac{1}{2}\left(p_{xx} - p_{zz}\right)\left(\hat{Q}_{xx} - \hat{Q}_{zz}\right) + \frac{3}{2}p_{yy}\hat{Q}_{yy} = p_{xx}\hat{Q}_{xx} + p_{yy}\hat{Q}_{yy} + p_{zz}\hat{Q}_{zz} \tag{3.212}$$

$$\frac{1}{2}\left(p_{yy} - p_{zz}\right)\left(\hat{Q}_{yy} - \hat{Q}_{zz}\right) + \frac{3}{2}p_{xx}\hat{Q}_{xx} = p_{xx}\hat{Q}_{xx} + p_{yy}\hat{Q}_{yy} + p_{zz}\hat{Q}_{zz} \tag{3.213}$$

Noting $\hat{Q}_{xx-yy}$ appearing in the matrix $\hat{F}$ given by Eq. (3.203) and the transformation formula Eq. (3.22), using Eq. (3.211) we can rewrite Eq. (3.206) as

$$\hat{\rho}_{\text{in}} = \frac{1}{3}\Big[\hat{I} + \frac{3}{2}\left(p_x\hat{S}_x + p_y\hat{S}_y + p_z\hat{S}_z\right) + \frac{2}{3}\left(p_{xy}\hat{Q}_{xy} + p_{xz}\hat{Q}_{xz} + p_{yz}\hat{Q}_{yz}\right) + \frac{1}{6}\left(p_{xx} - p_{yy}\right)\left(\hat{Q}_{xx} - \hat{Q}_{yy}\right) + \frac{1}{2}p_{zz}\hat{Q}_{zz}\Big] \tag{3.214}$$

From Eq. (3.214) we can still obtain Eq. (3.209) and Eq. (3.210) by using Eq. (3.104) and Eq. (3.105). We can also obtain

$$p_{xx-yy} \equiv p_{xx} - p_{yy} = \langle \hat{Q}_{xx} - \hat{Q}_{yy} \rangle = \mathrm{tr}\{ (\hat{Q}_{xx} - \hat{Q}_{yy}) \hat{\rho}_{in} \} \tag{3.215}$$

In this case, $\hat{Q}_{xx-yy}(\equiv \hat{Q}_{xx} - \hat{Q}_{yy})$, $\hat{Q}_{xy}$, $\hat{Q}_{xz}$, $\hat{Q}_{yz}$, $\hat{Q}_{zz}$ form 5 independent tensor operators. In the practical applications, Eq. (3.206) or Eq. (3.214) can all be used. This book generally uses Eq. (3.214) which contains only 5 independent tensor operators.

If the incident particle beam is polarized only in the $y$-axis direction, that is, $p_x = p_z = 0$, $p_{xy} = p_{xz} = p_{yz} = 0$, $p_{xx} = p_{zz} = -\frac{1}{2} p_{yy}$, $\hat{Q}_{xx} = \hat{Q}_{zz} = -\frac{1}{2} \hat{Q}_{yy}$, then according to Eqs. (3.206) and (3.212), we can write

$$\hat{\rho}_{\mathrm{in}} = \frac{1}{3} \left( \hat{I} + \frac{3}{2} p_y \hat{S}_y + \frac{1}{2} p_{yy} \hat{Q}_{yy} \right) \tag{3.216}$$

Notice Eq. (3.216) is only a special case.

If the incident particles are unpolarized, then we have

$$\hat{\rho}^0_{\mathrm{in}} = \frac{1}{3} \hat{I} \tag{3.217}$$

$$\hat{\rho}^0_{\mathrm{out}} = \frac{1}{3} \hat{F} \hat{F}^+ \tag{3.218}$$

In the 3.4 section, Eq. (3.196) has given

$$\frac{\mathrm{d}\sigma^0}{\mathrm{d}\Omega} = I_0 = \mathrm{tr}\left(\hat{\rho}^0_{\mathrm{out}}\right) = \frac{1}{3} \mathrm{tr}\left\{ \hat{F} \hat{F}^+ \right\} = \frac{1}{3} \left[ |A|^2 + 2\left( |B|^2 + |C|^2 + |D|^2 + |E|^2 \right) \right] \tag{3.219}$$

When the $y$-axis is taken along the direction $\vec{n}$, there is $\varphi = 0$, then Eqs. (3.194) and (3.195) can be simplified, respectively, as

$$\hat{F} = \begin{pmatrix} B & -\mathrm{i}C & -E \\ \mathrm{i}D & A & -\mathrm{i}D \\ -E & \mathrm{i}C & B \end{pmatrix}, \quad \hat{F}^+ = \begin{pmatrix} B^* & -\mathrm{i}D^* & -E^* \\ \mathrm{i}C^* & A^* & -\mathrm{i}C^* \\ -E^* & \mathrm{i}D^* & B^* \end{pmatrix} \tag{3.220}$$

$$\hat{F}\hat{F}^+ = \begin{pmatrix} |B|^2 + |C|^2 + |E|^2 & -\mathrm{i}(BD^* + CA^* + ED^*) & -\left( BE^* + |C|^2 + EB^* \right) \\ \mathrm{i}(DB^* + AC^* + DE^*) & 2|D|^2 + |A|^2 & -\mathrm{i}(DB^* + AC^* + DE^*) \\ -\left( EB^* + |C|^2 + BE^* \right) & \mathrm{i}(BD^* + CA^* + ED^*) & |B|^2 + |C|^2 + |E|^2 \end{pmatrix} \tag{3.221}$$

Next we define the following as the component of the polarization vector of the outgoing particles corresponding to the unpolarized incident particles:

$$P_0^{i'} = \frac{\mathrm{tr}\left\{\hat{S}_{i'}\hat{F}\hat{F}^+\right\}}{\mathrm{tr}\left\{\hat{F}\hat{F}^+\right\}}, \qquad i' = x', y', z' \tag{3.222}$$

We also define the following as the component of the polarization tensor of the outgoing particles corresponding to the unpolarized incident particles:

$$P_0^{i'} = \frac{\mathrm{tr}\left\{\hat{Q}_{i'}\hat{F}\hat{F}^+\right\}}{\mathrm{tr}\left\{\hat{F}\hat{F}^+\right\}}, \qquad i' = x'y', x'z', y'z', x'x' - y'y', z'z' \tag{3.223}$$

From Eq. (3.222) and Eq. (3.223), we can obtain the following results by using Eqs. (3.33), (3.36), (3.219), and (3.221):

$$P_0^{x'} = \frac{1}{3\sqrt{2}I_0}\left[\mathrm{i}(DB^* + AC^* + DE^*) - \mathrm{i}(DB^* + AC^* + DE^*)\right] = 0 \tag{3.224}$$

$$\begin{aligned} P_0^{y'} &= \frac{\mathrm{i}}{3\sqrt{2}I_0}\Big[-\mathrm{i}(DB^* + AC^* + DE^*) - 2\mathrm{i}(BD^* + CA^* + ED^*) \\ &\quad - \mathrm{i}(DB^* + AC^* + DE^*)\Big] = \frac{2\sqrt{2}}{3I_0}\,\mathrm{Re}\,(DB^* + AC^* + DE^*) \end{aligned} \tag{3.225}$$

$$P_0^{z'} = \frac{1}{3I_0}\left[\left(|B|^2 + |C|^2 + |E|^2\right) - \left(|B|^2 + |C|^2 + |E|^2\right)\right] = 0 \tag{3.226}$$

$$P_0^{x'y'} = \frac{\mathrm{i}}{2I_0}\left[\left(EB^* + |C|^2 + BE^*\right) - \left(BE^* + |C|^2 + EB^*\right)\right] = 0 \tag{3.227}$$

$$\begin{aligned} P_0^{x'z'} &= \frac{1}{2\sqrt{2}I_0}\Big[\mathrm{i}(DB^* + AC^* + DE^*) - 2\mathrm{i}(BD^* + CA^* + ED^*) \\ &\quad + \mathrm{i}(DB^* + AC^* + DE^*)\Big] = -\frac{\sqrt{2}}{I_0}\,\mathrm{Im}(DB^* + AC^* + DE^*) \end{aligned} \tag{3.228}$$

$$P_0^{y'z'} = \frac{\mathrm{i}}{2\sqrt{2}I_0}\left[-\mathrm{i}(DB^* + AC^* + DE^*) + \mathrm{i}(DB^* + AC^* + DE^*)\right] = 0 \tag{3.229}$$

$$\begin{aligned} P_0^{x'x'-y'y'} &= \frac{1}{I_0}\left[-\left(EB^* + |C|^2 + BE^*\right) - \left(BE^* + |C|^2 + EB^*\right)\right] \\ &= -\frac{2}{I_0}\left[|C|^2 + 2\,\mathrm{Re}\,(EB^*)\right] \end{aligned} \tag{3.230}$$

$$P_0^{z'z'} = \frac{1}{3I_0}\left[\left(|B|^2+|C|^2+|E|^2\right) - 2\left(2|D|^2+|A|^2\right) + \left(|B|^2+|C|^2+|E|^2\right)\right]$$
$$= \frac{2}{3I_0}\left(|B|^2+|C|^2+|E|^2-|A|^2-2|D|^2\right) \tag{3.231}$$

From the above results we can see that only $P_0^{i'}$ $(i' = y', x'z', x'x' - y'y', z'z')$ are not equal to 0. The above results also show that even if the spin 1 incident particle is unpolarized, when the spins of the target and the residual nucleus are equal to 0, the vector polarization rate and tensor polarization rate of the spin 1 outgoing particle are generally not equal to 0. That is, the outgoing particles will also be polarized.

## 3.6 Polarization Theory of Elastic Scattering of Spin 1 Particle with both Vector Polarization and Tensor Polarization from a Spinless Target

If the spin 1 incident particle contains both vector and tensor polarization components, then the items following the first item in the incident particle density matrix given by Eq. (3.214) should be considered.

We define the following as the vector polarization analyzing power

$$A_i = \frac{\text{tr}\left\{\hat{F}\hat{S}_i\hat{F}^+\right\}}{\text{tr}\left\{\hat{F}\hat{F}^+\right\}}, \qquad i = x, y, z \tag{3.232}$$

and define the tensor polarization analyzing power as

$$A_i = \frac{\text{tr}\left\{\hat{F}\hat{Q}_i\hat{F}^+\right\}}{\text{tr}\left\{\hat{F}\hat{F}^+\right\}}, \qquad i = xy, xz, yz, xx - yy, zz \tag{3.233}$$

From Eq. (3.232) and Eq. (3.233), we can obtain the following expressions using Eqs. (3.33), (3.36), (3.219), and (3.220):

$$\hat{S}_x\hat{F}^+ = \frac{1}{\sqrt{2}}\begin{pmatrix} \mathrm{i}C^* & A^* & -\mathrm{i}C^* \\ B^*-E^* & 0 & B^*-E^* \\ \mathrm{i}C^* & A^* & -\mathrm{i}C^* \end{pmatrix},$$

$$\hat{F}\hat{S}_x\hat{F}^+ = \frac{1}{\sqrt{2}}$$

$$\begin{pmatrix} \mathrm{i}(BC^* - CB^* + CE^* - EC^*) & BA^* - EA^* & -\mathrm{i}(BC^* + CB^* - CE^* - EC^*) \\ AB^* - AE^* & 0 & AB^* - AE^* \\ -\mathrm{i}(EC^* - CB^* + CE^* - BC^*) & -EA^* + BA^* & \mathrm{i}(EC^* + CB^* - CE^* - BC^*) \end{pmatrix} \tag{3.234}$$

$$A_x = \frac{1}{3\sqrt{2}I_0}[\mathrm{i}(BC^* - CB^* + CE^* - EC^*) + \mathrm{i}(EC^* + CB^* - CE^* - BC^*)] = 0 \tag{3.235}$$

$$\hat{S}_y\hat{F}^+ = \frac{\mathrm{i}}{\sqrt{2}}\begin{pmatrix} -\mathrm{i}C^* & -A^* & \mathrm{i}C^* \\ B^* + E^* & -2\mathrm{i}D^* & -(E^* + B^*) \\ \mathrm{i}C^* & A^* & -\mathrm{i}C^* \end{pmatrix},$$

$$\hat{F}\hat{S}_y\hat{F}^+ = \frac{\mathrm{i}}{\sqrt{2}}$$

$$\begin{pmatrix} -\mathrm{i}(BC^* + CB^* + CE^* + EC^*) & -(BA^* + 2CD^* + EA^*) & \mathrm{i}(BC^* + CE^* + CB^* + EC^*) \\ 2DC^* + AB^* + AE^* & -2\mathrm{i}(DA^* + AD^*) & -(2DC^* + AE^* + AB^*) \\ \mathrm{i}(EC^* + CB^* + CE^* + BC^*) & EA^* + 2CD^* + BA^* & -\mathrm{i}(EC^* + CE^* + CB^* + BC^*) \end{pmatrix} \tag{3.236}$$

$$\begin{aligned} A_y = {} & \frac{\mathrm{i}}{3\sqrt{2}I_0}[-\mathrm{i}(BC^* + CB^* + CE^* + EC^*) - 2\mathrm{i}(DA^* + AD^*) \\ & -\mathrm{i}(EC^* + CE^* + CB^* + BC^*)] = \frac{2\sqrt{2}}{3I_0}\,\mathrm{Re}\,(BC^* + CE^* + DA^*) \end{aligned} \tag{3.237}$$

$$\begin{aligned} \hat{S}_z\hat{F}^+ &= \begin{pmatrix} B^* & -\mathrm{i}D^* & -E^* \\ 0 & 0 & 0 \\ E^* & -\mathrm{i}D^* & -B^* \end{pmatrix}, \\ \hat{F}\hat{S}_z\hat{F}^+ &= \begin{pmatrix} |B|^2 - |E|^2 & -\mathrm{i}(BD^* - ED^*) & -(BE^* - EB^*) \\ \mathrm{i}(DB^* - DE^*) & 0 & -\mathrm{i}(DE^* - DB^*) \\ -(EB^* - BE^*) & \mathrm{i}(ED^* - BD^*) & |E|^2 - |B|^2 \end{pmatrix} \end{aligned} \tag{3.238}$$

$$A_z = \frac{1}{3I_0}\left[\left(|B|^2 - |E|^2\right) + \left(|E|^2 - |B|^2\right)\right] = 0 \tag{3.239}$$

$$\hat{Q}_{xy}\hat{F}^{+} = \frac{3i}{2}\begin{pmatrix} E^{*} & -iD^{*} & -B^{*} \\ 0 & 0 & 0 \\ B^{*} & -iD^{*} & -E^{*} \end{pmatrix},$$

$$\hat{F}\hat{Q}_{xy}\hat{F}^{+} = \frac{3i}{2}\begin{pmatrix} BE^{*}-EB^{*} & -i(BD^{*}-ED^{*}) & -|B|^{2}+|E|^{2} \\ i(DE^{*}-DB^{*}) & 0 & -i(DB^{*}-DE^{*}) \\ -|E|^{2}+|B|^{2} & i(ED^{*}-BD^{*}) & EB^{*}-BE^{*} \end{pmatrix} \tag{3.240}$$

$$A_{xy} = \frac{i}{2I_0}[(BE^{*}-EB^{*})+(EB^{*}-BE^{*})]=0 \tag{3.241}$$

$$\hat{Q}_{xz}\hat{F}^{+} = \frac{3}{2\sqrt{2}}\begin{pmatrix} iC^{*} & A^{*} & -iC^{*} \\ B^{*}+E^{*} & -2iD^{*} & -(E^{*}+B^{*}) \\ -iC^{*} & -A^{*} & iC^{*} \end{pmatrix},$$

$$\hat{F}\hat{Q}_{xz}\hat{F}^{+} = \frac{3}{2\sqrt{2}}$$

$$\begin{pmatrix} i(BC^{*}-CB^{*}-CE^{*}+EC^{*}) & BA^{*}-2CD^{*}+EA^{*} & -i(BC^{*}-CE^{*}-CB^{*}+EC^{*}) \\ -2DC^{*}+AB^{*}+AE^{*} & 2i(DA^{*}-AD^{*}) & 2DC^{*}-AE^{*}-AB^{*} \\ -i(EC^{*}-CB^{*}-CE^{*}+BC^{*}) & -EA^{*}+2CD^{*}-BA^{*} & i(EC^{*}-CE^{*}-CB^{*}+BC^{*}) \end{pmatrix} \tag{3.242}$$

$$\begin{aligned} A_{xz} &= \frac{1}{2\sqrt{2}I_0}[i(BC^{*}-CB^{*}-CE^{*}+EC^{*})+2i(DA^{*}-AD^{*}) \\ &\quad +i(EC^{*}-CE^{*}-CB^{*}+BC^{*})] = -\frac{\sqrt{2}}{I_0}\mathrm{Im}(BC^{*}+DA^{*}+EC^{*}) \end{aligned} \tag{3.243}$$

$$\hat{Q}_{yz}\hat{F}^{+} = \frac{3i}{2\sqrt{2}}\begin{pmatrix} -iC^{*} & -A^{*} & iC^{*} \\ B^{*}-E^{*} & 0 & -E^{*}+B^{*} \\ -iC^{*} & -A^{*} & iC^{*} \end{pmatrix},$$

$$\hat{F}\hat{Q}_{yz}\hat{F}^{+} = \frac{3i}{2\sqrt{2}}$$

$$\begin{pmatrix} -i(BC^{*}+CB^{*}-CE^{*}-EC^{*}) & -BA^{*}+EA^{*} & i(BC^{*}+CE^{*}-CB^{*}-EC^{*}) \\ AB^{*}-AE^{*} & 0 & -AE^{*}+AB^{*} \\ i(EC^{*}+CB^{*}-CE^{*}-BC^{*}) & EA^{*}-BA^{*} & -i(EC^{*}+CE^{*}-CB^{*}-BC^{*}) \end{pmatrix} \tag{3.244}$$

$$A_{yz} = \frac{\mathrm{i}}{2\sqrt{2}I_0}[-\mathrm{i}(BC^* + CB^* - CE^* - EC^*) - \mathrm{i}(EC^* + CE^* - CB^* - BC^*)] = 0 \tag{3.245}$$

$$\hat{Q}_{xx-yy}\hat{F}^+ = 3\begin{pmatrix} -E^* & \mathrm{i}D^* & B^* \\ 0 & 0 & 0 \\ B^* & -\mathrm{i}D^* & -E^* \end{pmatrix},$$

$$\hat{F}\hat{Q}_{xx-yy}\hat{F}^+ = 3\begin{pmatrix} -BE^* - EB^* & \mathrm{i}(BD^* + ED^*) & |B|^2 + |E|^2 \\ -\mathrm{i}(DE^* + DB^*) & -2|D|^2 & \mathrm{i}(DB^* + DE^*) \\ |E|^2 + |B|^2 & -\mathrm{i}(ED^* + BD^*) & -EB^* - BE^* \end{pmatrix} \tag{3.246}$$

$$A_{xx-yy} = \frac{1}{I_0}\left[-2(BE^* + EB^*) - 2|D|^2\right] = -\frac{2}{I_0}\left[2\,\mathrm{Re}\,(BE^*) + |D|^2\right] \tag{3.247}$$

$$\hat{Q}_{zz}\hat{F}^+ = \begin{pmatrix} B^* & -\mathrm{i}D^* & -E^* \\ -2\mathrm{i}C^* & -2A^* & 2\mathrm{i}C^* \\ -E^* & \mathrm{i}D^* & B^* \end{pmatrix},$$

$$\hat{F}\hat{Q}_{zz}\hat{F}^+ =$$

$$\begin{pmatrix} |B|^2 - 2|C|^2 + |E|^2 & -\mathrm{i}(BD^* - 2CA^* + ED^*) & -BE^* + 2|C|^2 - EB^* \\ \mathrm{i}(DB^* - 2AC^* + DE^*) & 2|D|^2 - 2|A|^2 & -\mathrm{i}(DE^* - 2AC^* + DB^*) \\ -EB^* + 2|C|^2 - BE^* & \mathrm{i}(ED^* - 2CA^* + BD^*) & |E|^2 - 2|C|^2 + |B|^2 \end{pmatrix} \tag{3.248}$$

$$\begin{aligned} A_{zz} &= \frac{1}{3I_0}\left[2\left(|B|^2 - 2|C|^2 + |E|^2\right) + 2|D|^2 - 2|A|^2\right] \\ &= \frac{2}{3I_0}\left(|B|^2 + |E|^2 + |D|^2 - |A|^2 - 2|C|^2\right) \end{aligned} \tag{3.249}$$

From the above results we can see that only $A_i$ ($i = y, xz, xx - yy, zz$) are not equal to 0.

According to Eq. (3.214) it can be written that the differential cross section of the outgoing particle corresponding to the polarized incident particle is

$$\frac{\mathrm{d}\sigma}{\mathrm{d}\Omega} = I = I_0\left(1 + \frac{3}{2}p_y A_y + \frac{2}{3}p_{xz}A_{xz} + \frac{1}{6}p_{xx-yy}A_{xx-yy} + \frac{1}{2}p_{zz}A_{zz}\right) \tag{3.250}$$

The polarization transfer coefficient from the $i$th initial polarization vector component of the incident particles to the $i'$th final polarization vector component of the outgoing particles is defined as

$$K_i^{i'} = \frac{\mathrm{tr}\left\{\hat{S}_{i'}\hat{F}\hat{S}_i\hat{F}^+\right\}}{\mathrm{tr}\left\{\hat{F}\hat{F}^+\right\}}, \quad i = x, y, z; \quad i' = x', y', z' \tag{3.251}$$

The polarization transfer coefficient from the $i$th initial polarization vector component to the $i'$th final polarization tensor component is defined as

$$K_i^{i'} = \frac{\mathrm{tr}\left\{\hat{Q}_{i'}\hat{F}\hat{S}_i\hat{F}^+\right\}}{\mathrm{tr}\left\{\hat{F}\hat{F}^+\right\}}, \quad i = x, y, z; \quad i' = x'y', x'z', y'z', x'x' - y'y', z'z' \tag{3.252}$$

The polarization transfer coefficient from the $i$th initial polarization tensor component to the $i'$th final polarization vector component is defined as

$$K_i^{i'} = \frac{\mathrm{tr}\left\{\hat{S}_{i'}\hat{F}\hat{Q}_i\hat{F}^+\right\}}{\mathrm{tr}\left\{\hat{F}\hat{F}^+\right\}}, \quad i = xy, xz, yz, xx - yy, zz; \quad i' = x', y', z' \tag{3.253}$$

Finally, the polarization transfer coefficient from the $i$th initial polarization tensor component to the $i'$th final polarization tensor component is defined as

$$\begin{gathered} K_i^{i'} = \frac{\mathrm{tr}\left\{\hat{Q}_{i'}\hat{F}\hat{Q}_i\hat{F}^+\right\}}{\mathrm{tr}\left\{\hat{F}\hat{F}^+\right\}}, \quad i = xy, xz, yz, xx - yy, zz; \\ i' = x'y', x'z', y'z', x'x' - y'y', z'z' \end{gathered} \tag{3.254}$$

The following results can be obtained from Eqs. (3.251)–(3.254) and utilizing Eqs. (3.33), (3.36), and (3.219) as well as the matrices $\hat{F}\hat{S}_i\hat{F}^+$ and $\hat{F}\hat{Q}_i\hat{F}^+$ obtained early in this section

$$\begin{aligned} K_x^{x'} &= \frac{1}{6I_0}\left[(AB^* - AE^*) + 2(BA^* - EA^*) + (AB^* - AE^*)\right] \\ &= \frac{2}{3I_0}\,\mathrm{Re}\,(AB^* - AE^*) \end{aligned} \tag{3.255}$$

$$K_x^{y'} = \frac{\mathrm{i}}{6I_0}\left[(-AB^* + AE^*) + (AB^* - AE^*)\right] = 0 \tag{3.256}$$

$$K_x^{z'} = \frac{1}{3\sqrt{2}I_0}[\mathrm{i}(BC^* - CB^* + CE^* - EC^*) + \mathrm{i}(BC^* - CB^* + CE^* - EC^*)]$$
$$= -\frac{2\sqrt{2}}{3I_0}\mathrm{Im}(BC^* + CE^*) \tag{3.257}$$

$$K_x^{x'y'} = \frac{\mathrm{i}}{2\sqrt{2}I_0}[-\mathrm{i}(BC^* + CB^* - CE^* - EC^*) - \mathrm{i}(BC^* + CB^* - CE^* - EC^*)]$$
$$= \frac{\sqrt{2}}{I_0}\mathrm{Re}\,(CB^* - EC^*) \tag{3.258}$$

$$K_x^{x'z'} = \frac{1}{4I_0}[(AB^* - AE^*) - (AB^* - AE^*)] = 0 \tag{3.259}$$

$$K_x^{y'z'} = \frac{\mathrm{i}}{4I_0}[-(AB^* - AE^*) + 2(BA^* - EA^*) - (AB^* - AE^*)]$$
$$= \frac{1}{I_0}\mathrm{Im}(AB^* - AE^*) \tag{3.260}$$

$$K_x^{x'x'-y'y'} = \frac{1}{\sqrt{2}I_0}[-\mathrm{i}(EC^* - CB^* + CE^* - BC^*) - \mathrm{i}(BC^* + CB^* - CE^* - EC^*)] = 0 \tag{3.261}$$

$$K_x^{z'z'} = \frac{1}{3\sqrt{2}I_0}[\mathrm{i}(BC^* - CB^* + CE^* - EC^*) - \mathrm{i}(BC^* - CB^* + CE^* - EC^*)] = 0 \tag{3.262}$$

$$K_y^{x'} = \frac{\mathrm{i}}{6I_0}[(2DC^* + AB^* + AE^*) - (2DC^* + AB^* + AE^*)] = 0 \tag{3.263}$$

$$K_y^{y'} = -\frac{1}{6I_0}[(-2DC^* - AB^* - AE^*) + (-2BA^* - 4CD^* - 2EA^*)$$
$$+(-2DC^* - AB^* - AE^*)] = \frac{2}{3I_0}\mathrm{Re}\,(AB^* + AE^* + 2DC^*) \tag{3.264}$$

$$K_y^{z'} = \frac{\mathrm{i}}{3\sqrt{2}I_0}[-\mathrm{i}(BC^* + CB^* + CE^* + EC^*) + \mathrm{i}(BC^* + CB^* + CE^* + EC^*)] = 0 \tag{3.265}$$

$$K_y^{x'y'} = -\frac{1}{2\sqrt{2}I_0}[-\mathrm{i}(BC^* + CB^* + CE^* + EC^*) + \mathrm{i}(BC^* + CB^* + CE^* + EC^*)] = 0 \tag{3.266}$$

$$K_y^{x'z'} = \frac{\mathrm{i}}{4I_0}[(AB^* + AE^* + 2DC^*) - 2(BA^* + EA^* + 2CD^*) + (AB^* + AE^* + 2DC^*)] = -\frac{1}{I_0}\mathrm{Im}(AB^* + AE^* + 2DC^*) \quad (3.267)$$

$$K_y^{y'z'} = -\frac{1}{4I_0}[-(AB^* + AE^* + 2DC^*) + (AB^* + AE^* + 2DC^*)] = 0 \quad (3.268)$$

$$K_y^{x'x'-y'y'} = \frac{\mathrm{i}}{\sqrt{2}I_0}[\mathrm{i}(EC^* + CB^* + CE^* + BC^*) + \mathrm{i}(BC^* + CE^* + CB^* + EC^*)] = -\frac{2\sqrt{2}}{I_0}\mathrm{Re}\,(CB^* + EC^*) \quad (3.269)$$

$$K_y^{z'z'} = \frac{\mathrm{i}}{3\sqrt{2}I_0}[-\mathrm{i}(BC^* + CB^* + CE^* + EC^*) + 4\mathrm{i}(AD^* + DA^*) - \mathrm{i}(BC^* + CB^* + CE^* + EC^*)] = \frac{2\sqrt{2}}{3I_0}\mathrm{Re}\,(BC^* + CE^* - 2DA^*) \quad (3.270)$$

$$K_z^{x'} = \frac{1}{3\sqrt{2}I_0}[\mathrm{i}(DB^* - DE^*) - 2\mathrm{i}(BD^* - ED^*) + \mathrm{i}(DB^* - DE^*)] = -\frac{2\sqrt{2}}{3I_0}\mathrm{Im}(DB^* - DE^*) \quad (3.271)$$

$$K_z^{y'} = \frac{\mathrm{i}}{3\sqrt{2}I_0}[-\mathrm{i}(DB^* - DE^*) + \mathrm{i}(DB^* - DE^*)] = 0 \quad (3.272)$$

$$K_z^{z'} = \frac{1}{3I_0}\left[\left(|B|^2 - |E|^2\right) + \left(|B|^2 - |E|^2\right)\right] = \frac{2}{3I_0}\left(|B|^2 - |E|^2\right) \quad (3.273)$$

$$K_z^{x'y'} = \frac{\mathrm{i}}{2I_0}[(EB^* - BE^*) - (BE^* - EB^*)] = -\frac{2}{I_0}\mathrm{Im}(EB^*) \quad (3.274)$$

$$K_z^{x'z'} = \frac{1}{2\sqrt{2}I_0}[\mathrm{i}(DB^* - DE^*) - \mathrm{i}(DB^* - DE^*)] = 0 \quad (3.275)$$

$$K_z^{y'z'} = \frac{\mathrm{i}}{2\sqrt{2}I_0}[-\mathrm{i}(DB^* - DE^*) - 2\mathrm{i}(BD^* - ED^*) - \mathrm{i}(DB^* - DE^*)] = \frac{\sqrt{2}}{I_0}\mathrm{Re}\,(DB^* - DE^*) \quad (3.276)$$

$$K_z^{x'x'-y'y'} = \frac{1}{I_0}[-(EB^* - BE^*) - (BE^* - EB^*)] = 0 \quad (3.277)$$

$$K_z^{z'z'} = \frac{1}{3I_0}\left[\left(|B|^2 - |E|^2\right) + \left(|E|^2 - |B|^2\right)\right] = 0 \quad (3.278)$$

$$\begin{aligned} K_{xy}^{x'} &= \frac{\mathrm{i}}{2\sqrt{2}I_0}[-\mathrm{i}(DB^* - DE^*) - 2\mathrm{i}(BD^* - ED^*) - \mathrm{i}(DB^* - DE^*)] \\ &= \frac{\sqrt{2}}{I_0}\,\mathrm{Re}\,(BD^* - DE^*) \end{aligned} \tag{3.279}$$

$$K_{xy}^{y'} = -\frac{1}{2\sqrt{2}I_0}[\mathrm{i}(DB^* - DE^*) - \mathrm{i}(DB^* - DE^*)] = 0 \tag{3.280}$$

$$K_{xy}^{z'} = \frac{\mathrm{i}}{2I_0}[(BE^* - EB^*) - (EB^* - BE^*)] = -\frac{2}{I_0}\mathrm{Im}(BE^*) \tag{3.281}$$

$$K_{xy}^{x'y'} = -\frac{3}{4I_0}\left[\left(|E|^2 - |B|^2\right) + \left(-|B|^2 + |E|^2\right)\right] = \frac{3}{2I_0}\left(|B|^2 - |E|^2\right) \tag{3.282}$$

$$K_{xy}^{x'z'} = \frac{3\mathrm{i}}{4\sqrt{2}I_0}[\mathrm{i}(DE^* - DB^*) + \mathrm{i}(DB^* - DE^*)] = 0 \tag{3.283}$$

$$\begin{aligned} K_{xy}^{y'z'} &= -\frac{3}{4\sqrt{2}I_0}[-\mathrm{i}(DE^* - DB^*) - 2\mathrm{i}(BD^* - ED^*) + \mathrm{i}(DB^* - DE^*)] \\ &= -\frac{3}{\sqrt{2}I_0}\mathrm{Im}(BD^* + DE^*) \end{aligned} \tag{3.284}$$

$$K_{xy}^{x'x'-y'y''} = \frac{3\mathrm{i}}{2I_0}\left(-|E|^2 + |B|^2 - |B|^2 + |E|^2\right) = 0 \tag{3.285}$$

$$K_{xy}^{z'z'} = \frac{\mathrm{i}}{2I_0}[(BE^* - EB^*) + (EB^* - BE^*)] = 0 \tag{3.286}$$

$$K_{xz}^{x'} = \frac{1}{4I_0}[(-2DC^* + AB^* + AE^*) + (2DC^* - AE^* - AB^*)] = 0 \tag{3.287}$$

$$\begin{aligned} K_{xz}^{y'} = \frac{\mathrm{i}}{4I_0}&[(2DC^* - AB^* - AE^*) + 2(BA^* - 2CD^* + EA^*) \\ &+(2DC^* - AE^* - AB^*)] = \frac{1}{I_0}\mathrm{Im}(AB^* + AE^* - 2DC^*) \end{aligned} \tag{3.288}$$

$$K_{xz}^{z'} = \frac{1}{2\sqrt{2}I_0}[\mathrm{i}(BC^* - CB^* - CE^* + EC^*) - \mathrm{i}(BC^* - CB^* - CE^* + EC^*)] = 0 \tag{3.289}$$

$$K_{xz}^{x'y'} = \frac{3\,\mathrm{i}}{4\sqrt{2}I_0}[\mathrm{i}(EC^* - CB^* - CE^* + BC^*) - \mathrm{i}(BC^* - CE^* - CB^* + EC^*)] = 0 \tag{3.290}$$

$$\begin{aligned} K_{xz}^{x'z'} &= \frac{3}{8I_0}[(AB^* + AE^* - 2DC^*) + 2(BA^* + EA^* - 2CD^*) \\ &\quad +(AB^* + AE^* - 2DC^*)] = \frac{3}{2I_0}\,\mathrm{Re}\,(AB^* + AE^* - 2DC^*) \end{aligned} \tag{3.291}$$

$$K_{xz}^{y'z'} = \frac{3\,\mathrm{i}}{8I_0}[(2DC^* - AB^* - AE^*) - (2DC^* - AE^* - AB^*)] = 0 \tag{3.292}$$

$$K_{xz}^{x'x'-y'y'} = \frac{3}{2\sqrt{2}I_0}[-\mathrm{i}(EC^* - CB^* - CE^* + BC^*) - \mathrm{i}(BC^* - CE^* - CB^* + EC^*)] = -\frac{3\sqrt{2}}{I_0}\mathrm{Im}(CB^* - EC^*) \tag{3.293}$$

$$K_{xz}^{z'z'} = \frac{1}{2\sqrt{2}I_0}[-\mathrm{i}(CB^* - BC^* + CE^* - EC^*) + 4\mathrm{i}(AD^* - DA^*) - \mathrm{i}(CB^* - BC^* + CE^* - EC^*)] = -\frac{\sqrt{2}}{I_0}\mathrm{Im}(BC^* - CE^* - 2DA^*) \tag{3.294}$$

$$K_{yz}^{x'} = \frac{\mathrm{i}}{4I_0}[(AB^* - AE^*) - 2(BA^* - EA^*) + (AB^* - AE^*)] = -\frac{1}{I_0}\mathrm{Im}(AB^* - AE^*) \tag{3.295}$$

$$K_{yz}^{y'} = -\frac{1}{4I_0}[-(AB^* - AE^*) + (AB^* - AE^*)] = 0 \tag{3.296}$$

$$K_{yz}^{z'} = \frac{\mathrm{i}}{2\sqrt{2}I_0}[-\mathrm{i}(BC^* + CB^* - CE^* - EC^*) + \mathrm{i}(EC^* + CE^* - CB^* - BC^*)] = \frac{\sqrt{2}}{I_0}\mathrm{Re}\,(BC^* - CE^*) \tag{3.297}$$

$$K_{yz}^{x'y'} = -\frac{3}{4\sqrt{2}I_0}[-\mathrm{i}(EC^* + CB^* - CE^* - BC^*) + \mathrm{i}(BC^* + CE^* - CB^* - EC^*)] = -\frac{3}{\sqrt{2}I_0}\mathrm{Im}(CB^* + EC^*) \tag{3.298}$$

$$K_{yz}^{x'z'} = \frac{3\mathrm{i}}{8I_0}[(AB^* - AE^*) - (AB^* - AE^*)] = 0 \tag{3.299}$$

$$K_{yz}^{y'z'} = -\frac{3}{8I_0}[-(AB^* - AE^*) - 2(BA^* - EA^*) - (AB^* - AE^*)] = \frac{3}{2I_0}\mathrm{Re}\,(AB^* - AE^*) \tag{3.300}$$

$$K_{yz}^{x'x'-y'y'} = \frac{3\mathrm{i}}{2\sqrt{2}I_0}[\mathrm{i}(EC^* + CB^* - CE^* - BC^*) + \mathrm{i}(BC^* + CE^* - CB^* - EC^*)] = 0 \tag{3.301}$$

$$K_{yz}^{z'z'} = \frac{\mathrm{i}}{2\sqrt{2}I_0}[-\mathrm{i}(BC^* + CB^* - CE^* - EC^*) - \mathrm{i}(EC^* + CE^* - CB^* - BC^*)] = 0 \tag{3.302}$$

$$K_{xx-yy}^{x'}=\frac{1}{\sqrt{2}I_0}[-\mathrm{i}(DE^*+DB^*)+\mathrm{i}(BD^*+ED^*)-\mathrm{i}(ED^*+BD^*)+\mathrm{i}(DB^*+DE^*)]=0 \tag{3.303}$$

$$\begin{aligned} K_{xx-yy}^{y'} &= \frac{\mathrm{i}}{\sqrt{2}I_0}[\mathrm{i}(DE^*+DB^*)+\mathrm{i}(BD^*+ED^*)+\mathrm{i}(ED^*+BD^*)+\mathrm{i}(DB^*+DE^*)] \\ &= -\frac{2\sqrt{2}}{I_0}\mathrm{Re}\,(BD^*+DE^*) \end{aligned} \tag{3.304}$$

$$K_{xx-yy}^{z'}=\frac{1}{I_0}[-BE^*-EB^*+EB^*+BE^*]=0 \tag{3.305}$$

$$K_{xx-yy}^{x'y'}=\frac{3\mathrm{i}}{2I_0}\left[-|E|^2-|B|^2+|B|^2+|E|^2\right]=0 \tag{3.306}$$

$$\begin{aligned} K_{xx-yy}^{x'z'} &= \frac{3}{2\sqrt{2}I_0}[-\mathrm{i}(DE^*+DB^*)+\mathrm{i}(BD^*+ED^*)+\mathrm{i}(ED^*+BD^*) \\ &\quad -\mathrm{i}(DB^*+DE^*)]=-\frac{3\sqrt{2}}{I_0}\mathrm{Im}(BD^*-DE^*) \end{aligned} \tag{3.307}$$

$$K_{xx-yy}^{y'z'}=\frac{3\,\mathrm{i}}{2\sqrt{2}I_0}[\mathrm{i}(DE^*+DB^*)+\mathrm{i}(BD^*+ED^*)-\mathrm{i}(ED^*+BD^*)-\mathrm{i}(DB^*+DE^*)]=0 \tag{3.308}$$

$$K_{xx-yy}^{x'x'-y'y'}=\frac{3}{I_0}\left(|E|^2+|B|^2+|B|^2+|E|^2\right)=\frac{6}{I_0}\left(|B|^2+|E|^2\right) \tag{3.309}$$

$$K_{xx-yy}^{z'z'}=\frac{1}{I_0}\left(-BE^*-EB^*+4|D|^2-EB^*-BE^*\right)=-\frac{4}{I_0}\left[\mathrm{Re}\,(BE^*)-|D|^2\right] \tag{3.310}$$

$$K_{zz}^{x'}=\frac{1}{3\sqrt{2}I_0}[\mathrm{i}(DB^*-2AC^*+DE^*)-\mathrm{i}(DE^*-2AC^*+DB^*)]=0 \tag{3.311}$$

$$\begin{aligned} K_{zz}^{y'} &= \frac{\mathrm{i}}{3\sqrt{2}I_0}[-\mathrm{i}(DB^*-2AC^*+DE^*)-2\mathrm{i}(BD^*-2CA^*+ED^*) \\ &\quad -\mathrm{i}(DB^*-2AC^*+DE^*)]=\frac{2\sqrt{2}}{3I_0}\mathrm{Re}\,(DB^*+DE^*-2AC^*) \end{aligned} \tag{3.312}$$

$$K_{zz}^{z'}=\frac{1}{3I_0}\left[\left(|B|^2-2|C|^2+|E|^2\right)+\left(-|E|^2+2|C|^2-|B|^2\right)\right]=0 \tag{3.313}$$

$$K_{zz}^{x'y'}=\frac{\mathrm{i}}{2I_0}\left[\left(EB^*-2|C|^2+BE^*\right)-\left(BE^*-2|C|^2+EB^*\right)\right]=0 \tag{3.314}$$

$$K_{zz}^{x'z'} = \frac{1}{2\sqrt{2}I_0}[\mathrm{i}(DB^* - 2AC^* + DE^*) - 2\mathrm{i}(BD^* - 2CA^* + ED^*) \\ + \mathrm{i}(DB^* - 2AC^* + DE^*)] = -\frac{\sqrt{2}}{I_0}\mathrm{Im}(DB^* + DE^* - 2AC^*) \tag{3.315}$$

$$K_{zz}^{y'z'} = \frac{\mathrm{i}}{2\sqrt{2}I_0}[-\mathrm{i}(DB^* - 2AC^* + DE^*) + \mathrm{i}(DB^* - 2AC^* + DE^*)] = 0 \tag{3.316}$$

$$K_{zz}^{x'x'-y'y'} = \frac{1}{I_0}\left(-EB^* + 2|C|^2 - BE^* - BE^* + 2|C|^2 - EB^*\right) \\ = -\frac{4}{I_0}\left[\mathrm{Re}\,(EB^*) - |C|^2\right] \tag{3.317}$$

$$K_{zz}^{z'z'} = \frac{1}{3I_0}\left[\left(|B|^2 - 2|C|^2 + |E|^2\right) - 4\left(|D|^2 - |A|^2\right) + \left(|E|^2 - 2|C|^2 + |B|^2\right)\right] \\ = \frac{2}{3I_0}\left(|B|^2 + |E|^2 + 2|A|^2 - 2|C|^2 - 2|D|^2\right) \tag{3.318}$$

From the above results we can see

$$K_{xy}^{x'} = K_z^{y'z'},\quad K_{xy}^{z'} = -K_z^{x'y'},\quad K_{xy}^{x'y'} = \frac{9}{4}K_z^{z'},\quad K_{xy}^{y'z'} = -\frac{9}{4}K_z^{x'}, \\ K_{yz}^{x'} = -K_x^{y'z'},\quad K_{yz}^{z'} = K_x^{x'y'},\quad K_{yz}^{x'y'} = -\frac{9}{4}K_x^{z'},\quad K_{yz}^{y'z'} = \frac{9}{4}K_x^{x'} \tag{3.319}$$

We divide the Cartesian coordinate indexes of spin 1 particles into two groups

$$\varepsilon_1 = y,\ xz,\ xx - yy,\ zz; \qquad \varepsilon_2 = x,\ z,\ xy,\ yz \tag{3.320}$$

For the polarization transfer coefficient $K_i^{i'}$, due to the conservation of parity, $K_i^{i'}$ is not equal to 0 only when $i$ and $i'$ are in the same group; otherwise $K_i^{i'}$ must be equal to 0. There are a total of 16 $K_i^{i'}$ of $\varepsilon_1$ which are not equal to 0, and 16 $K_i^{i'}$ of $\varepsilon_2$ that are not equal to 0; however, taking into account Eq. (3.319), only 8 non-zero $K_i^{i'}$ of $\varepsilon_2$ are independent, that is, only a total of 24 non-zero $K_i^{i'}$ are independent. Also note that $P_0^{i'}$ and $A_i$ given above are not equal to 0 only in the first group $\varepsilon_1$ of the coordinate index.

According to Eq. (3.214) the polarization vector and polarization tensor components of the outgoing particles corresponding to the polarized incident particles can be written as follows:

$$P^{i'} = (I_0/I)\left(P_0^{i'} + \frac{3}{2}p_y K_y^{i'} + \frac{2}{3}p_{xz}K_{xz}^{i'} + \frac{1}{6}p_{xx-yy}K_{xx-yy}^{i'} + \frac{1}{2}p_{zz}K_{zz}^{i'}\right), \\ i' = \varepsilon_1' = y',\ x'z',\ x'x' - y'y',\ z'z' \tag{3.321}$$

$$P^{i'} = (I_0/I)\left(\frac{3}{2}p_x K_x^{i'} + \frac{3}{2}p_z K_z^{i'} + \frac{2}{3}p_{xy} K_{xy}^{i'} + \frac{2}{3}p_{yz} K_{yz}^{i'}\right), \quad i' = \varepsilon_2' = x', z', x'y', y'z' \tag{3.322}$$

Using Eq. (1.99) and Eq. (3.8), in $S = 1$ case the following normalized incident particle density matrix can be written as from Eq. (1.92):

$$\hat{\rho}_{\text{in}} = \frac{1}{3}\left[\hat{I} + \sum_{\mu}(-1)^{\mu}\left(-\mathrm{i}t_{1\,-\mu}\right)\left(\mathrm{i}\hat{T}_{1\mu}\right) + \sum_{M}(-1)^{M} t_{2\,-M}\hat{T}_{2M}\right] \tag{3.323}$$

Utilizing Eq. (3.113) we can prove

$$\left\langle \mathrm{i}\hat{T}_{1\mu}\right\rangle = \mathrm{tr}\left\{\mathrm{i}\hat{T}_{1\mu}\hat{\rho}_{\text{in}}\right\} = \mathrm{i}t_{1\mu}, \qquad \left\langle \hat{T}_{2\mu}\right\rangle = \mathrm{tr}\left\{\hat{T}_{2M}\hat{\rho}_{\text{in}}\right\} = t_{2M} \tag{3.324}$$

We define

$$\mathrm{i}T_{1\mu} = \frac{\mathrm{tr}\left\{\hat{F}\,\mathrm{i}\hat{T}_{1\mu}F^{+}\right\}}{\mathrm{tr}\left\{\hat{F}F^{+}\right\}}, \quad \mu = \pm 1, 0; \quad T_{2M} = \frac{\mathrm{tr}\left\{\hat{F}\hat{T}_{2M}F^{+}\right\}}{\mathrm{tr}\left\{\hat{F}F^{+}\right\}}, \quad M = \pm 2, \pm 1, 0 \tag{3.325}$$

Here $\mathrm{i}T_{1\mu}$ and $T_{2M}$ are no longer the operators; referring to Eqs. (3.232) and (3.233), $\mathrm{i}T_{1\mu}$ and $T_{2M}$ can be called the polarization analyzing powers in the spherical basis coordinate system.

When the $y$-axis is not parallel to the unit vector $\vec{n}$, which is perpendicular to the reaction plan, the corresponding scattering amplitude $\hat{F}$ has been given by Eq. (3.194), and $\hat{F}$ and $\hat{F}^{+}$ can be simplified as

$$\hat{F} = \begin{pmatrix} B & -\mathrm{i}C\mathrm{e}^{-\mathrm{i}\varphi} & -E\mathrm{e}^{-2\mathrm{i}\varphi} \\ \mathrm{i}D\mathrm{e}^{\mathrm{i}\varphi} & A & -\mathrm{i}D\mathrm{e}^{-\mathrm{i}\varphi} \\ -E\mathrm{e}^{2\mathrm{i}\varphi} & \mathrm{i}C\mathrm{e}^{\mathrm{i}\varphi} & B \end{pmatrix}, \quad \hat{F}^{+} = \begin{pmatrix} B^{*} & -\mathrm{i}D^{*}\mathrm{e}^{-\mathrm{i}\varphi} & -E^{*}\mathrm{e}^{-2\mathrm{i}\varphi} \\ \mathrm{i}C^{*}\mathrm{e}^{\mathrm{i}\varphi} & A^{*} & -\mathrm{i}C^{*}\mathrm{e}^{-\mathrm{i}\varphi} \\ -E^{*}\mathrm{e}^{2\mathrm{i}\varphi} & \mathrm{i}D^{*}\mathrm{e}^{\mathrm{i}\varphi} & B^{*} \end{pmatrix} \tag{3.326}$$

We can find the following expressions from Eqs. (3.38), (3.326), (3.196), and (3.325):

$$
\begin{aligned}
\mathrm{i}\hat{T}_{11}\hat{F}^{+} &= -\sqrt{\frac{3}{2}}\,\mathrm{i}\begin{pmatrix} \mathrm{i}C^{*}\mathrm{e}^{\mathrm{i}\varphi} & A^{*} & -\mathrm{i}C^{*}\mathrm{e}^{-\mathrm{i}\varphi} \\ -E^{*}\mathrm{e}^{2\mathrm{i}\varphi} & \mathrm{i}D^{*}\mathrm{e}^{\mathrm{i}\varphi} & B^{*} \\ 0 & 0 & 0 \end{pmatrix}, \\
\mathrm{i}T_{11} &= -\frac{\sqrt{3}\,\mathrm{i}}{3\sqrt{2}I_0}\left[\mathrm{i}(BC^{*}+CE^{*})\mathrm{e}^{\mathrm{i}\varphi}+\mathrm{i}(DA^{*}+AD^{*})\mathrm{e}^{\mathrm{i}\varphi}+\mathrm{i}(EC^{*}+CB^{*})\mathrm{e}^{\mathrm{i}\varphi}\right] \\
&= \frac{\sqrt{2}}{\sqrt{3}I_0}\,\mathrm{Re}\,(CB^{*}+CE^{*}+AD^{*})\mathrm{e}^{\mathrm{i}\varphi}
\end{aligned}
\tag{3.327}
$$

$$
\begin{aligned}
\mathrm{i}\hat{T}_{10}\hat{F}^{+} &= \sqrt{\frac{3}{2}}\,\mathrm{i}\begin{pmatrix} B^{*} & -\mathrm{i}D^{*}\mathrm{e}^{-\mathrm{i}\varphi} & -E^{*}\mathrm{e}^{-2\mathrm{i}\varphi} \\ 0 & 0 & 0 \\ E^{*}\mathrm{e}^{2\mathrm{i}\varphi} & -\mathrm{i}D^{*}\mathrm{e}^{\mathrm{i}\varphi} & -B^{*} \end{pmatrix}, \\
\mathrm{i}T_{10} &= \frac{\sqrt{3}\,\mathrm{i}}{3\sqrt{2}I_0}\left[\left(|B|^2-|E|^2\right)+\left(|E|^2-|B|^2\right)\right]=0
\end{aligned}
\tag{3.328}
$$

$$
\begin{aligned}
\mathrm{i}\hat{T}_{1\,-1}\hat{F}^{+} &= \sqrt{\frac{3}{2}}\,\mathrm{i}\begin{pmatrix} 0 & 0 & 0 \\ B^{*} & -\mathrm{i}D^{*}\mathrm{e}^{-\mathrm{i}\varphi} & -E^{*}\mathrm{e}^{-2\mathrm{i}\varphi} \\ \mathrm{i}C^{*}\mathrm{e}^{\mathrm{i}\varphi} & A^{*} & -\mathrm{i}C^{*}\mathrm{e}^{-\mathrm{i}\varphi} \end{pmatrix}, \\
\mathrm{i}T_{1\,-1} &= \frac{\sqrt{3}\,\mathrm{i}}{3\sqrt{2}I_0}\left[-\mathrm{i}(CB^{*}+EC^{*})\,\mathrm{e}^{-\mathrm{i}\varphi}-\mathrm{i}(AD^{*}+DA^{*})\,\mathrm{e}^{-\mathrm{i}\varphi}-\mathrm{i}(CE^{*}+BC^{*})\,\mathrm{e}^{-\mathrm{i}\varphi}\right] \\
&= \frac{\sqrt{2}}{\sqrt{3}I_0}\mathrm{Re}\,(CB^{*}+CE^{*}+AD^{*})\,\mathrm{e}^{-\mathrm{i}\varphi}
\end{aligned}
\tag{3.329}
$$

$$
\begin{aligned}
\hat{T}_{22}\hat{F}^{+} &= \sqrt{3}\begin{pmatrix} -E^{*}\mathrm{e}^{2\mathrm{i}\varphi} & \mathrm{i}D^{*}\mathrm{e}^{\mathrm{i}\varphi} & B^{*} \\ 0 & 0 & 0 \\ 0 & 0 & 0 \end{pmatrix}, \\
T_{22} &= -\frac{1}{\sqrt{3}I_0}\left(BE^{*}+|D|^2+EB^{*}\right)\mathrm{e}^{2\mathrm{i}\varphi} = -\frac{1}{\sqrt{3}I_0}\left[2\,\mathrm{Re}\,(BE^{*})+|D|^2\right]\mathrm{e}^{2\mathrm{i}\varphi}
\end{aligned}
\tag{3.330}
$$

$$
\begin{aligned}
\hat{T}_{21}\hat{F}^{+} &= \sqrt{\frac{3}{2}}\begin{pmatrix} -\mathrm{i}C^{*}\mathrm{e}^{\mathrm{i}\varphi} & -A^{*} & \mathrm{i}C^{*}\mathrm{e}^{-\mathrm{i}\varphi} \\ -E^{*}\mathrm{e}^{2\mathrm{i}\varphi} & \mathrm{i}D^{*}\mathrm{e}^{\mathrm{i}\varphi} & B^{*} \\ 0 & 0 & 0 \end{pmatrix}, \\
T_{21} &= \frac{1}{\sqrt{6}I_0}\left[-\mathrm{i}(BC^{*}-CE^{*})\mathrm{e}^{\mathrm{i}\varphi}-\mathrm{i}(DA^{*}-AD^{*})\mathrm{e}^{\mathrm{i}\varphi}-\mathrm{i}(EC^{*}-CB^{*})\mathrm{e}^{\mathrm{i}\varphi}\right] \\
&= -\frac{\sqrt{2}}{\sqrt{3}I_0}\mathrm{Im}(CB^{*}+CE^{*}+AD^{*})\,\mathrm{e}^{\mathrm{i}\varphi}
\end{aligned}
\tag{3.331}
$$

$$\hat{T}_{20}\hat{F}^{+}=\frac{1}{\sqrt{2}}\begin{pmatrix} B^{*} & -\mathrm{i}D^{*}\mathrm{e}^{-\mathrm{i}\varphi} & -E^{*}\mathrm{e}^{-2\mathrm{i}\varphi} \\ -2\mathrm{i}C^{*}\mathrm{e}^{\mathrm{i}\varphi} & -2A^{*} & 2\mathrm{i}C^{*}\mathrm{e}^{-\mathrm{i}\varphi} \\ -E^{*}\mathrm{e}^{2\mathrm{i}\varphi} & \mathrm{i}D^{*}\mathrm{e}^{\mathrm{i}\varphi} & B^{*} \end{pmatrix},$$

$$\begin{aligned} T_{20} &= \frac{1}{3\sqrt{2}I_0}\left[\left(|B|^2-2|C|^2+|E|^2\right)+\left(|D|^2-2|A|^2+|D|^2\right)+\left(|E|^2-2|C|^2+|B|^2\right)\right] \\ &= \frac{\sqrt{2}}{3I_0}\left(|B|^2+|E|^2-|A|^2+|D|^2-2|C|^2\right) \end{aligned} \tag{3.332}$$

$$\hat{T}_{2\,-1}\hat{F}^{+}=\sqrt{\frac{3}{2}}\begin{pmatrix} 0 & 0 & 0 \\ B^{*} & -\mathrm{i}D^{*}\mathrm{e}^{-\mathrm{i}\varphi} & -E^{*}\mathrm{e}^{-2\mathrm{i}\varphi} \\ -\mathrm{i}C^{*}\mathrm{e}^{\mathrm{i}\varphi} & -A^{*} & \mathrm{i}C^{*}\mathrm{e}^{-\mathrm{i}\varphi} \end{pmatrix},$$

$$\begin{aligned} T_{2\,-1} &= \frac{\sqrt{3}}{3\sqrt{2}I_0}\left[-\mathrm{i}(CB^{*}-EC^{*})\mathrm{e}^{-\mathrm{i}\varphi}-\mathrm{i}(AD^{*}-DA^{*})\mathrm{e}^{-\mathrm{i}\varphi}-\mathrm{i}(CE^{*}-BC^{*})\mathrm{e}^{-\mathrm{i}\varphi}\right] \\ &= \frac{\sqrt{2}}{\sqrt{3}I_0}\mathrm{Im}(CB^{*}+CE^{*}+AD^{*})\,\mathrm{e}^{-\mathrm{i}\varphi} \end{aligned} \tag{3.333}$$

$$\hat{T}_{2\,-2}\hat{F}^{+}=\sqrt{3}\begin{pmatrix} 0 & 0 & 0 \\ 0 & 0 & 0 \\ B^{*} & -\mathrm{i}D^{*}\mathrm{e}^{-\mathrm{i}\varphi} & -E^{*}\mathrm{e}^{-2\mathrm{i}\varphi} \end{pmatrix},$$

$$T_{2\,-2} = -\frac{1}{\sqrt{3}I_0}\left(EB^{*}+|D|^2+BE^{*}\right)\mathrm{e}^{-2\mathrm{i}\varphi} = -\frac{1}{\sqrt{3}I_0}\left[2\mathrm{Re}\,(BE^{*})+|D|^2\right]\mathrm{e}^{-2\mathrm{i}\varphi} \tag{3.334}$$

Therefore, according to Eq. (3.323), the differential cross section of the elastic scattering of the polarized spin 1 incident particle beam with the spinless targets can be written as

$$\begin{aligned} \frac{\mathrm{d}\sigma}{\mathrm{d}\Omega} &= I(\theta,\varphi)=\mathrm{tr}\left\{\hat{F}\hat{\rho}_{\mathrm{in}}\hat{F}^{+}\right\}=I_0(\theta)\,[1+\mathrm{i}t_{1\,-1}(\mathrm{i}T_{11})+\mathrm{i}t_{11}(\mathrm{i}T_{1\,-1}) \\ &\quad +t_{2\,-2}T_{22}-t_{2\,-1}T_{21}+t_{20}T_{20}-t_{21}T_{2\,-1}+t_{22}T_{2\,-2}] \end{aligned} \tag{3.335}$$

The analyzing powers in the spherical basis coordinate system appearing in the above formula have been given by Eqs. (3.327) and (3.329)–(3.334).

We can obtain the following results from Eqs. (3.327)–(3.329) and using Eqs. (3.232), (1.22), and (3.9):

$$A_x=\frac{-\mathrm{i}}{\sqrt{3}}(\mathrm{i}T_{1\,-1}-\mathrm{i}T_{11})=-\frac{2\sqrt{2}}{3I_0}\mathrm{Re}\,(CB^{*}+CE^{*}+AD^{*})\sin\varphi \tag{3.336}$$

$$A_y = \frac{1}{\sqrt{3}}(\mathrm{i}T_{1\ -1} + \mathrm{i}T_{11}) = \frac{2\sqrt{2}}{3I_0}\,\mathrm{Re}\,(CB^* + CE^* + AD^*)\cos\varphi \tag{3.337}$$

$$A_z = \sqrt{\frac{2}{3}}(-\mathrm{i})\,\mathrm{i}T_{10} = 0 \tag{3.338}$$

Noting Eq. (3.324) we can obtain the following results utilizing Eqs. (1.22) and (3.132):

$$p_x = \frac{1}{\sqrt{3}}(t_{1\ -1} - t_{11}),\ \ p_y = \frac{1}{\sqrt{3}}(\mathrm{i}t_{1\ -1} + \mathrm{i}t_{11}),\ \ p_z = \sqrt{\frac{2}{3}}t_{10} \tag{3.339}$$

and their inverse relations are

$$\mathrm{i}t_{11} = \frac{\sqrt{3}}{2}(p_y - \mathrm{i}p_x),\ \ \mathrm{i}t_{1\ -1} = \frac{\sqrt{3}}{2}(p_y + \mathrm{i}p_x),\ \ t_{10} = \sqrt{\frac{3}{2}}p_z \tag{3.340}$$

In $\varphi = 0$ case Eqs. (3.336), (3.337), and (3.338) are reduced to Eqs. (3.235), (3.237), and (3.239), respectively. In this case, because we have $A_x = 0$, then $\mathrm{i}T_{1\ -1} = \mathrm{i}T_{11}$, so from Eq. (3.337) we can obtain

$$A_y = \frac{2}{\sqrt{3}}\mathrm{i}T_{11}, \quad \mathrm{i}T_{11} = \frac{\sqrt{3}}{2}A_y \tag{3.341}$$

Using Eq. (3.233) and Eq. (3.21) we can obtain the following results from Eqs. (3.330)–(3.334):

$$A_{xy} = \frac{\mathrm{i}\sqrt{3}}{2}(T_{2\ -2} - T_{22}) = -\frac{1}{I_0}\left[2\,\mathrm{Re}\,(BE^*) + |D|^2\right]\sin(2\varphi) \tag{3.342}$$

$$A_{xz} = \frac{\sqrt{3}}{2}(T_{2\ -1} - T_{21}) = \frac{\sqrt{2}}{I_0}\mathrm{Im}(CB^* + CE^* + AD^*)\cos\varphi \tag{3.343}$$

$$A_{yz} = \frac{\mathrm{i}\sqrt{3}}{2}(T_{2\ -1} + T_{21}) = \frac{\sqrt{2}}{I_0}\mathrm{Im}(CB^* + CE^* + AD^*)\sin\varphi \tag{3.344}$$

$$A_{xx-yy} = \sqrt{3}(T_{2\ -2} + T_{22}) = -\frac{2}{I_0}\left[2\,\mathrm{Re}\,(BE^*) + |D|^2\right]\cos(2\varphi) \tag{3.345}$$

$$A_{zz} = \sqrt{2}T_{20} = \frac{2}{3I_0}\left(|B|^2 + |E|^2 - |A|^2 + |D|^2 - 2|C|^2\right) \tag{3.346}$$

And we can find the following expressions based on Eqs. (3.21), (3.136), and (3.324):

$$p_{xy} = \frac{i\sqrt{3}}{2}(t_{2\,-2} - t_{22}),\ \ p_{xz} = \frac{\sqrt{3}}{2}(t_{2\,-1} - t_{21}),\ \ p_{yz} = \frac{i\sqrt{3}}{2}(t_{2\,-1} + t_{21}),$$
$$p_{xx-yy} = \sqrt{3}(t_{2\,-2} + t_{22}),\ \ p_{zz} = \sqrt{2}t_{20} \tag{3.347}$$

Using Eq. (3.22) we can also obtain

$$t_{2\,\pm2} = \frac{1}{2\sqrt{3}}\left(p_{xx-yy} \pm 2\mathrm{i}p_{xy}\right),\ \ t_{2\,\pm1} = \mp\frac{1}{\sqrt{3}}\left(p_{xz} \pm \mathrm{i}p_{yz}\right),\ \ t_{20} = \frac{1}{\sqrt{2}}p_{zz} \tag{3.348}$$

If $\varphi = 0$, Eqs. (3.342)–(3.346) will degenerate to Eqs. (3.241), (3.243), (3.245), (3.247), and (3.249), respectively, and in this case because $A_{xy} = 0,\ A_{yz} = 0$, we can see $T_{2\,-2} = T_{22}$, $T_{2\,-1} = -\ T_{21}$ from Eq. (3.21), and we can obtain

$$A_{xz} = -\sqrt{3}T_{21},\quad A_{xx-yy} = 2\sqrt{3}T_{22},\quad A_{zz} = \sqrt{2}T_{20} \tag{3.349}$$

From the above formulas we can also get

$$T_{21} = -\frac{1}{\sqrt{3}}A_{xz},\quad T_{20} = \frac{1}{\sqrt{2}}A_{zz},\quad T_{22} = \frac{1}{2\sqrt{3}}A_{xx-yy} \tag{3.350}$$

In References [5, 8–18] the measurement and theoretical calculation results of deuteron elastic scattering vector polarization and tensor polarization analyzing powers of some spinless targets are given, and most of them are given in the form of $\mathrm{i}T_{11}$, $T_{20}$, $T_{21}$, $T_{22}$.

If the incident particle beam is polarized only in the $y$-axis direction, then the initial density matrix should be in the form given by Eq. (3.216), and the corresponding differential cross section of the outgoing particles is

$$\frac{\mathrm{d}\sigma}{\mathrm{d}\Omega} = I(\theta) = \mathrm{tr}\left\{\hat{F}\hat{\rho}_{\mathrm{in}}\hat{F}^{+}\right\} = I_0(\theta)\left[1 + \frac{3}{2}P_yA_y(\theta) + \frac{1}{2}P_{yy}A_{yy}(\theta)\right] \tag{3.351}$$

References [19–21] give the measurement and calculation results of deuteron elastic scattering polarization analyzing power $A_y(\theta)$ and $A_{yy}(\theta)$ of some targets in this type of polarization case.

If we use the spherical optical model or the $R$ matrix theory to calculate the $S$ matrix elements of elastic scattering between the deuteron and the spinless target, the polarization data can be calculated using the theory described in this section.

## 3.7 General Form of Nuclear Reaction Polarization Theory of Spin 1 Particles with Non-zero Spin Target

Equations (2.251) and (2.252) give the reaction amplitude expressions of the *j*-*j* coupling [7] and *S*-*L* coupling [22], respectively. In the case where the spins of the incident particles and the outgoing particles are both 1, the following symbols are introduced for the *j*-*j* coupling and the *S*-*L* coupling, respectively:

$$f_{\mu'\mu}^{ll'J}(m_l') = \hat{l}\hat{l}' \mathrm{e}^{\mathrm{i}(\sigma_l+\sigma_{l'})} \sqrt{\frac{(l'-m_l')!}{(l'+m_l')!}} \sum_{jj'} C_{l0\ 1\mu}^{j\ \mu} C_{j\ \mu\ IM_I}^{J\ M} C_{l'm_l'\ 1\mu'}^{j'\ m_j'} C_{j'm_j'\ I'M_I'}^{J\ M} \left(\delta_{\alpha'n'l'j',\alpha nlj} - S_{\alpha'n'l'j',\alpha nlj}^{J}\right) \tag{3.352}$$

$$f_{\mu'\mu}^{ll'J}(m_l') = \hat{l}\hat{l}' \mathrm{e}^{\mathrm{i}(\sigma_l+\sigma_{l'})} \sqrt{\frac{(l'-m_l')!}{(l'+m_l')!}} \sum_{SS'} C_{l0\ SM}^{J\ M} C_{1\mu\ IM_I}^{S\ M} C_{l'm_l'\ S'M_S'}^{J\ M} C_{1\mu'\ I'M_I'}^{S'\ M_S'} \left(\delta_{\alpha'n'l'S',\alpha nlS} - S_{\alpha'n'l'S',\alpha nlS}^{J}\right) \tag{3.353}$$

Given these, the reaction amplitudes can be uniformly written in the following form under the two angular momentum coupling modes according to Eqs. (2.251)–(2.254):

$$f_{\alpha'n'\mu'M_I',\alpha n\mu M_I}(\Omega) = f_{\mathrm{C}}(\theta)\ \delta_{\alpha'n',\alpha n}\delta_{\mu'M_I',\mu M_I} + \frac{\mathrm{i}}{2k}(-1)^{\mu+M_I-\mu'-M_I'} \sum_{ll'J} f_{\mu'\mu}^{ll'J}(\mu+M_I-\mu'-M_I')\ \mathrm{P}_{l'}^{\mu+M_I-\mu'-M_I'}(\cos\theta)\ \mathrm{e}^{\mathrm{i}(\mu+M_I-\mu'-M_I')\varphi} \tag{3.354}$$

Note that $\delta_{M_I',M_I}$ is contained in the first item of the above formula so that this item can be multiplied by $\mathrm{e}^{\mathrm{i}(M_I-M_I')\varphi}$, which is equal to 1 since $\delta_{M_I',M_I}$. Let

$$A_{M_I'M_I}(\theta) = f_{\mathrm{C}}(\theta)\ \delta_{\alpha'n',\alpha n}\delta_{M_I',M_I} + \frac{\mathrm{i}}{2k}(-1)^{M_I-M_I'} \sum_{ll'J} f_{0\,0}^{ll'J}(M_I-M_I')\ \mathrm{P}_{l'}^{M_I-M_I'}(\cos\theta) \tag{3.355}$$

$$B_{M_I'M_I}(\theta) = f_{\mathrm{C}}(\theta)\ \delta_{\alpha'n',\alpha n}\delta_{M_I',M_I} + \frac{\mathrm{i}}{2k}(-1)^{M_I-M_I'} \sum_{ll'J} f_{1\,1}^{ll'J}(M_I-M_I')\ \mathrm{P}_{l'}^{M_I-M_I'}(\cos\theta) \tag{3.356}$$

$$H_{M_I' M_I}(\theta) = f_C(\theta)\,\delta_{\alpha' n', \alpha n}\delta_{M_I', M_I} + \frac{i}{2k}(-1)^{M_I - M_I'}\sum_{ll'J} f_{-1\,-1}^{ll'J}(M_I - M_I')\,P_{l'}^{M_I - M_I'}(\cos\theta) \tag{3.357}$$

$$C_{M_I' M_I}(\theta) = -\frac{1}{2k}(-1)^{M_I - M_I' - 1}\sum_{ll'J} f_{1\,0}^{ll'J}(M_I - M_I' - 1)\,P_{l'}^{M_I - M_I' - 1}(\cos\theta) \tag{3.358}$$

$$G_{M_I' M_I}(\theta) = \frac{1}{2k}(-1)^{M_I - M_I' + 1}\sum_{ll'J} f_{-1\,0}^{ll'J}(M_I - M_I' + 1)\,P_{l'}^{M_I - M_I' + 1}(\cos\theta) \tag{3.359}$$

$$D_{M_I' M_I}(\theta) = -\frac{1}{2k}(-1)^{M_I - M_I' - 1}\sum_{ll'J} f_{0\,-1}^{ll'J}(M_I - M_I' - 1)\,P_{l'}^{M_I - M_I' - 1}(\cos\theta) \tag{3.360}$$

$$O_{M_I' M_I}(\theta) = \frac{1}{2k}(-1)^{M_I - M_I' + 1}\sum_{ll'J} f_{0\,1}^{ll'J}(M_I - M_I' + 1)\,P_{l'}^{M_I - M_I' + 1}(\cos\theta) \tag{3.361}$$

$$E_{M_I' M_I}(\theta) = -\frac{i}{2k}(-1)^{M_I - M_I' - 2}\sum_{ll'J} f_{1\,-1}^{ll'J}(M_I - M_I' - 2)\,P_{l'}^{M_I - M_I' - 2}(\cos\theta) \tag{3.362}$$

$$F_{M_I' M_I}(\theta) = -\frac{i}{2k}(-1)^{M_I - M_I' + 2}\sum_{ll'J} f_{-1\,1}^{ll'J}(M_I - M_I' + 2)\,P_{l'}^{M_I - M_I' + 2}(\cos\theta) \tag{3.363}$$

Introducing the following symbol

$$F_{\mu'\mu}(\theta, \varphi) = f_{\alpha' n' \mu' M_I', \alpha n \mu M_I}(\Omega) \tag{3.364}$$

then one can get

$$F_{0\,0}(\theta, \varphi) = A_{M_I' M_I}(\theta) e^{i(M_I - M_I')\varphi} \tag{3.365}$$

$$F_{1\,1}(\theta, \varphi) = B_{M_I' M_I}(\theta) e^{i(M_I - M_I')\varphi} \tag{3.366}$$

$$F_{-1\,-1}(\theta, \varphi) = H_{M_I' M_I}(\theta) e^{i(M_I - M_I')\varphi} \tag{3.367}$$

$$F_{1\,0}(\theta, \varphi) = -iC_{M_I' M_I}(\theta) e^{i(M_I - M_I')\varphi} e^{-i\varphi} \tag{3.368}$$

$$F_{-1\,0}(\theta, \varphi) = iG_{M_I' M_I}(\theta) e^{i(M_I - M_I')\varphi} e^{i\varphi} \tag{3.369}$$

$$F_{0\,-1}(\theta, \varphi) = -iD_{M_I' M_I}(\theta) e^{i(M_I - M_I')\varphi} e^{-i\varphi} \tag{3.370}$$

$$F_{0\,1}(\theta, \varphi) = iO_{M_I' M_I}(\theta) e^{i(M_I - M_I')} e^{i\varphi} \tag{3.371}$$

$$F_{1\,-1}(\theta, \varphi) = -E_{M_I' M_I}(\theta) e^{i(M_I - M_I')} e^{-2i\varphi} \tag{3.372}$$

$$F_{-1\,1}(\theta,\varphi) = -F_{M_I'M_I}(\theta)\mathrm{e}^{\mathrm{i}\left(M_I - M_I'\right)}\mathrm{e}^{2\mathrm{i}\varphi} \tag{3.373}$$

If the footnote $M_I'M_I$ and argument ($\theta$) of $A, B, H, C, G, D, O, E, F$ are omitted, then the reaction amplitude $\hat{F}$ with definite footnote $M_I'M_I$, but which footnote $M_I'M_I$ and argument $(\theta, \varphi)$ are omitted, can be written as

$$\hat{F} = \begin{pmatrix} B & -\mathrm{i}C\mathrm{e}^{-\mathrm{i}\varphi} & -E\mathrm{e}^{-2\mathrm{i}\varphi} \\ \mathrm{i}O\mathrm{e}^{\mathrm{i}\varphi} & A & -\mathrm{i}D\mathrm{e}^{-\mathrm{i}\varphi} \\ -F\mathrm{e}^{2\mathrm{i}\varphi} & \mathrm{i}G\mathrm{e}^{\mathrm{i}\varphi} & H \end{pmatrix} \mathrm{e}^{\mathrm{i}\left(M_I - M_I'\right)\varphi} \tag{3.374}$$

Comparing this formula with Eq. (3.194), it shows that when the target spin $I = 0$ and the residual nucleus spin $I' = 0$ there are

$$\begin{gathered} A_{M_I'M_I} = A, \quad H_{M_I'M_I} = B_{M_I'M_I} = B, \quad G_{M_I'M_I} = C_{M_I'M_I} = C, \\ O_{M_I'M_I} = D_{M_I'M_I} = D, \quad F_{M_I'M_I} = E_{M_I'M_I} = E \end{gathered} \tag{3.375}$$

When the target and the residual nucleus are both unpolarized, the magnetic quantum number footnotes of the matrix elements of $\hat{F}$ and $\hat{F}^+$ appearing in the polarization theory formulas are the same, i.e., they are incoherent. So from Eq. (3.374) we can obtain

$$\hat{F}^+ = \begin{pmatrix} B^* & -\mathrm{i}O^*\mathrm{e}^{-\mathrm{i}\varphi} & -F^*\mathrm{e}^{-2\mathrm{i}\varphi} \\ \mathrm{i}C^*\mathrm{e}^{\mathrm{i}\varphi} & A^* & -\mathrm{i}G^*\mathrm{e}^{-\mathrm{i}\varphi} \\ -E^*\mathrm{e}^{2\mathrm{i}\varphi} & \mathrm{i}D^*\mathrm{e}^{\mathrm{i}\varphi} & H^* \end{pmatrix} \mathrm{e}^{-\mathrm{i}\left(M_I - M_I'\right)\varphi} \tag{3.376}$$

$$\hat{F}\hat{F}^+ = \begin{pmatrix} |B|^2+|C|^2+|E|^2 & -\mathrm{i}(BO^*+CA^*+ED^*)\,\mathrm{e}^{-\mathrm{i}\varphi} & -(BF^*+CG^*+EH^*)\,\mathrm{e}^{-2\mathrm{i}\varphi} \\ \mathrm{i}(OB^*+AC^*+DE^*)\,\mathrm{e}^{\mathrm{i}\varphi} & |O|^2+|A|^2+|D|^2 & -\mathrm{i}(OF^*+AG^*+DH^*)\,\mathrm{e}^{-\mathrm{i}\varphi} \\ -(FB^*+GC^*+HE^*)\,\mathrm{e}^{2\mathrm{i}\varphi} & \mathrm{i}(FO^*+GA^*+HD^*)\,\mathrm{e}^{\mathrm{i}\varphi} & |F|^2+|G|^2+|H|^2 \end{pmatrix} \tag{3.377}$$

For the definite $M'_IM_I$ from Eq. (3.377) we can obtain

$$I_0 = \frac{1}{3}\mathrm{tr}\left\{\hat{F}\hat{F}^+\right\} = \frac{1}{3}\left(|A|^2+|B|^2+|H|^2+|C|^2+|G|^2+|D|^2+|O|^2+|E|^2+|F|^2\right) \tag{3.378}$$

For the sake of simplicity, choose the $y$-axis along the direction $\vec{n}$. In this case there is $\varphi = 0$. Then the following formulas can be obtained from Eqs. (3.374), (3.376), and (3.377):

$$\hat{F}=\begin{pmatrix} B & -\mathrm{i}C & -E \\ \mathrm{i}O & A & -\mathrm{i}D \\ -F & \mathrm{i}G & H \end{pmatrix}, \quad \hat{F}^{+}=\begin{pmatrix} B^{*} & -\mathrm{i}O^{*} & -F^{*} \\ \mathrm{i}C^{*} & A^{*} & -\mathrm{i}G^{*} \\ -E^{*} & \mathrm{i}D^{*} & H^{*} \end{pmatrix} \tag{3.379}$$

$$\hat{F}\hat{F}^{+}=\begin{pmatrix} |B|^2+|C|^2+|E|^2 & -\mathrm{i}(BO^{*}+CA^{*}+ED^{*}) & -(BF^{*}+CG^{*}+EH^{*}) \\ \mathrm{i}(OB^{*}+AC^{*}+DE^{*}) & |O|^2+|A|^2+|D|^2 & -\mathrm{i}(OF^{*}+AG^{*}+DH^{*}) \\ -(FB^{*}+GC^{*}+HE^{*}) & \mathrm{i}(FO^{*}+GA^{*}+HD^{*}) & |F|^2+|G|^2+|H|^2 \end{pmatrix} \tag{3.380}$$

For definite $M_I'M_I$ the component of the outgoing particle polarization vector corresponding to the unpolarized incident particles is defined as

$$P_0^{i'}=\frac{\mathrm{tr}\{\hat{S}_{i'}\hat{F}\hat{F}^{+}\}}{\mathrm{tr}\{\hat{F}\hat{F}^{+}\}}, \quad i'=x', y', z' \tag{3.381}$$

And the component of the outgoing particle polarization tensor corresponding to the unpolarized incident particles is defined as

$$P_0^{i'}=\frac{\mathrm{tr}\left\{\hat{Q}_{i'}\hat{F}\hat{F}^{+}\right\}}{\mathrm{tr}\{\hat{F}\hat{F}^{+}\}}, \quad i'=x'y', x'z', y'z', x'x'-y'y', z'z' \tag{3.382}$$

From Eq. (3.381) and Eq. (3.382), the following results can be obtained utilizing Eqs. (3.33), (3.36), (3.378), and (3.380):

$$\begin{aligned} P_0^{x'} &= \frac{1}{3\sqrt{2}I_0}[\mathrm{i}(OB^{*}+AC^{*}+DE^{*})-\mathrm{i}(BO^{*}+CA^{*}+ED^{*}) \\ &\quad +\mathrm{i}(FO^{*}+GA^{*}+HD^{*})-\mathrm{i}(OF^{*}+AG^{*}+DH^{*})] \\ &= -\frac{\sqrt{2}}{3I_0}\mathrm{Im}[(OB^{*}+AC^{*}+DE^{*})+(FO^{*}+GA^{*}+HD^{*})] \end{aligned} \tag{3.383}$$

$$\begin{aligned} P_0^{y'} &= \frac{\mathrm{i}}{3\sqrt{2}I_0}[-\mathrm{i}(OB^{*}+AC^{*}+DE^{*})-\mathrm{i}(BO^{*}+CA^{*}+ED^{*}) \\ &\quad -\mathrm{i}(FO^{*}+GA^{*}+HD^{*})-\mathrm{i}(OF^{*}+AG^{*}+DH^{*})] \\ &= \frac{\sqrt{2}}{3I_0}\mathrm{Re}\,[(OB^{*}+AC^{*}+DE^{*})+(FO^{*}+GA^{*}+HD^{*})] \end{aligned} \tag{3.384}$$

$$P_0^{z'}=\frac{1}{3I_0}\left[\left(|B|^2+|C|^2+|E|^2\right)-\left(|H|^2+|G|^2+|F|^2\right)\right] \tag{3.385}$$

$$
\begin{aligned}
P_0^{x'y'} &= \frac{\mathrm{i}}{2I_0}[(FB^* + GC^* + HE^*) - (BF^* + CG^* + EH^*)] \\
&= -\frac{1}{I_0}\mathrm{Im}(FB^* + GC^* + HE^*)
\end{aligned}
\tag{3.386}
$$

$$
\begin{aligned}
P_0^{x'z'} = &\frac{1}{2\sqrt{2}I_0}[\mathrm{i}(OB^* + AC^* + DE^*) - \mathrm{i}(BO^* + CA^* + ED^*) \\
&- \mathrm{i}(FO^* + GA^* + HD^*) + \mathrm{i}(OF^* + AG^* + DH^*)] \\
= &-\frac{1}{\sqrt{2}I_0}\mathrm{Im}[(OB^* + AC^* + DE^*) - (FO^* + GA^* + HD^*)]
\end{aligned}
\tag{3.387}
$$

$$
\begin{aligned}
P_0^{y'z'} = &\frac{\mathrm{i}}{2\sqrt{2}I_0}[-\mathrm{i}(OB^* + AC^* + DE^*) - \mathrm{i}(BO^* + CA^* + ED^*) \\
&+\mathrm{i}(FO^* + GA^* + HD^*) + \mathrm{i}(OF^* + AG^* + DH^*)] \\
= &\frac{1}{\sqrt{2}I_0}\mathrm{Re}\,[(OB^* + AC^* + DE^*) - (FO^* + GA^* + HD^*)]
\end{aligned}
\tag{3.388}
$$

$$
\begin{aligned}
P_0^{x'x'-y'y'} &= \frac{1}{I_0}[-(FB^* + GC^* + HE^*) - (BF^* + CG^* + EH^*)] \\
&= -\frac{2}{I_0}\mathrm{Re}\,(FB^* + GC^* + HE^*)
\end{aligned}
\tag{3.389}
$$

$$
P_0^{z'z'} = \frac{1}{3I_0}\left[|B|^2 + |C|^2 + |E|^2 + |F|^2 + |G|^2 + |H|^2 - 2\left(|O|^2 + |A|^2 + |D|^2\right)\right]
\tag{3.390}
$$

And in $I = I' = 0$ case, using Eq. (3.375) we can see that Eqs. (3.383)–(3.390) will automatically degenerate to Eqs. (3.224)–(3.231).

For definite $M_I'M_I$ the polarization vector analyzing power and polarization tensor analyzing power are defined as, respectively

$$
A_i = \frac{\mathrm{tr}\{\hat{F}\hat{S}_i\hat{F}^+\}}{\mathrm{tr}\{\hat{F}\hat{F}^+\}}, \qquad i = x,\ y,\ z
\tag{3.391}
$$

$$
A_i = \frac{\mathrm{tr}\{\hat{F}\hat{Q}_i\hat{F}^+\}}{\mathrm{tr}\{\hat{F}\hat{F}^+\}}, \qquad i = xy,\ xz,\ yz,\ xx - yy,\ zz
\tag{3.392}
$$

From Eq. (3.391) and Eq. (3.392), we can obtain the following expressions utilizing Eqs. (3.33), (3.36), (3.378), and (3.379):

$$\hat{S}_x\hat{F}^+ = \frac{1}{\sqrt{2}}\begin{pmatrix} iC^* & A^* & -iG^* \\ B^*-E^* & i(D^*-O^*) & H^*-F^* \\ iC^* & A^* & -iG^* \end{pmatrix},$$

$$\hat{F}\hat{S}_x\hat{F}^+ = \frac{1}{\sqrt{2}}$$

$$\begin{pmatrix} i(BC^*-CB^*+CE^*-EC^*) & BA^*-EA^*+CD^*-CO^* & -i(BG^*-EG^*+CH^*-CF^*) \\ AB^*-AE^*-OC^*+DC^* & i(OA^*-DA^*+AD^*-AO^*) & OG^*-DG^*+AH^*-AF^* \\ -i(FC^*-GB^*+GE^*-HC^*) & -FA^*+HA^*-GD^*+GO^* & i(FG^*-HG^*+GH^*-GF^*) \end{pmatrix} \tag{3.393}$$

$$A_x = -\frac{\sqrt{2}}{3I_0}\mathrm{Im}(BC^* + CE^* + OA^* + AD^* + FG^* + GH^*) \tag{3.394}$$

$$\hat{S}_y\hat{F}^+ = \frac{i}{\sqrt{2}}\begin{pmatrix} -iC^* & -A^* & iG^* \\ B^*+E^* & -i(D^*+O^*) & -(F^*+H^*) \\ iC^* & A^* & -iG^* \end{pmatrix},$$

$$\hat{F}\hat{S}_y\hat{F}^+ = \frac{i}{\sqrt{2}}$$

$$\begin{pmatrix} -i(BC^*+CB^*+CE^*+EC^*) & -(BA^*+EA^*+CD^*+CO^*) & i(BG^*+EG^*+CF^*+CH^*) \\ DC^*+OC^*+AB^*+AE^* & -i(OA^*+DA^*+AD^*+AO^*) & -(OG^*+DG^*+AF^*+AH^*) \\ i(FC^*+HC^*+GB^*+GE^*) & FA^*+HA^*+GD^*+GO^* & -i(FG^*+HG^*+GF^*+GH^*) \end{pmatrix} \tag{3.395}$$

$$A_y = \frac{\sqrt{2}}{3I_0}\,\mathrm{Re}\,(BC^* + CE^* + OA^* + AD^* + FG^* + GH^*) \tag{3.396}$$

$$\hat{S}_z\hat{F}^+ = \begin{pmatrix} B^* & -iO^* & -F^* \\ 0 & 0 & 0 \\ E^* & -iD^* & -H^* \end{pmatrix},$$
$$\hat{F}\hat{S}_z\hat{F}^+ = \begin{pmatrix} |B|^2-|E|^2 & -i(BO^*-ED^*) & -(BF^*-EH^*) \\ i(OB^*-DE^*) & |O|^2-|D|^2 & -i(OF^*-DH^*) \\ -(FB^*-HE^*) & i(FO^*-HD^*) & |F|^2-|H|^2 \end{pmatrix} \tag{3.397}$$

$$A_z = \frac{1}{3I_0}\left(|B|^2 - |H|^2 + |F|^2 - |E|^2 + |O|^2 - |D|^2\right) \tag{3.398}$$

$$\hat{Q}_{xy}\hat{F}^+ = \frac{3i}{2}\begin{pmatrix} E^* & -iD^* & -H^* \\ 0 & 0 & 0 \\ B^* & -iO^* & -F^* \end{pmatrix},$$

$$\hat{F}\hat{Q}_{xy}\hat{F}^+ = \frac{3i}{2}\begin{pmatrix} BE^* - EB^* & -i(BD^* - EO^*) & -BH^* + EF^* \\ i(OE^* - DB^*) & OD^* - DO^* & -i(OH^* - DF^*) \\ -FE^* + HB^* & i(FD^* - HO^*) & FH^* - HF^* \end{pmatrix} \tag{3.399}$$

$$A_{xy} = -\frac{1}{I_0}\mathrm{Im}(BE^* + OD^* + FH^*) \tag{3.400}$$

$$\hat{Q}_{xz}\hat{F}^+ = \frac{3}{2\sqrt{2}}\begin{pmatrix} iC^* & A^* & -iG^* \\ B^* + E^* & -i(O^* + D^*) & -(F^* + H^*) \\ -iC^* & -A^* & iG^* \end{pmatrix},$$

$$\hat{F}\hat{Q}_{xz}\hat{F}^+ = \frac{3}{2\sqrt{2}}$$

$$\begin{pmatrix} i(BC^* - CB^* - CE^* + EC^*) & BA^* - CO^* - CD^* + EA^* & -i(BG^* - CF^* - CH^* + EG^*) \\ -OC^* - DC^* + AB^* + AE^* & i(OA^* + DA^* - AO^* - AD^*) & OG^* + DG^* - AF^* - AH^* \\ -i(FC^* - GB^* - GE^* + HC^*) & -FA^* + GO^* + GD^* - HA^* & i(FG^* - GF^* - GH^* + HG^*) \end{pmatrix} \tag{3.401}$$

$$A_{xz} = -\frac{1}{\sqrt{2}I_0}\mathrm{Im}(BC^* - CE^* + OA^* - AD^* + FG^* - GH^*) \tag{3.402}$$

$$\hat{Q}_{yz}\hat{F}^+ = \frac{3i}{2\sqrt{2}}\begin{pmatrix} -iC^* & -A^* & iG^* \\ B^* - E^* & i(D^* - O^*) & -(F^* - H^*) \\ -iC^* & -A^* & iG^* \end{pmatrix},$$

$$\hat{F}\hat{Q}_{yz}\hat{F}^+ = \frac{3i}{2\sqrt{2}}$$

$$\begin{pmatrix} -i(BC^* + CB^* - CE^* - EC^*) & -BA^* + EA^* + CD^* - CO^* & i(BG^* - EG^* + CF^* - CH^*) \\ AB^* - AE^* + OC^* - DC^* & -i(OA^* - DA^* - AD^* + AO^*) & -AF^* + AH^* - OG^* + DG^* \\ i(FC^* - HC^* + GB^* - GE^*) & FA^* - HA^* - GD^* + GO^* & -i(FG^* - HG^* + GF^* - GH^*) \end{pmatrix} \tag{3.403}$$

$$A_{yz} = \frac{1}{\sqrt{2}I_0}\mathrm{Re}\,(BC^* - CE^* + OA^* - AD^* + FG^* - GH^*) \tag{3.404}$$

$$\hat{Q}_{xx-yy}\hat{F}^{+} = 3\begin{pmatrix} -E^* & \mathrm{i}D^* & H^* \\ 0 & 0 & 0 \\ B^* & -\mathrm{i}O^* & -F^* \end{pmatrix},$$

$$\hat{F}\hat{Q}_{xx-yy}\hat{F}^{+} = 3\begin{pmatrix} -BE^* - EB^* & \mathrm{i}(BD^* + EO^*) & BH^* + EF^* \\ -\mathrm{i}(OE^* + DB^*) & -(OD^* + DO^*) & \mathrm{i}(OH^* + DF^*) \\ FE^* + HB^* & -\mathrm{i}(FD^* + HO^*) & -(FH^* + HF^*) \end{pmatrix} \tag{3.405}$$

$$A_{xx-yy} = -\frac{2}{I_0}\,\mathrm{Re}\,(BE^* + OD^* + FH^*) \tag{3.406}$$

$$\hat{Q}_{zz}\hat{F}^{+} = \begin{pmatrix} B^* & -\mathrm{i}O^* & -F^* \\ -2\mathrm{i}C^* & -2A^* & 2\mathrm{i}G^* \\ -E^* & \mathrm{i}D^* & H^* \end{pmatrix},$$

$$\hat{F}\hat{Q}_{zz}\hat{F}^{+} =$$

$$\begin{pmatrix} |B|^2 - 2|C|^2 + |E|^2 & -\mathrm{i}(BO^* - 2CA^* + ED^*) & -BF^* + 2CG^* - EH^* \\ \mathrm{i}(OB^* - 2AC^* + DE^*) & |O|^2 - 2|A|^2 + |D|^2 & -\mathrm{i}(OF^* - 2AG^* + DH^*) \\ -FB^* + 2GC^* - HE^* & \mathrm{i}(FO^* - 2GA^* + HD^*) & |F|^2 - 2|G|^2 + |H|^2 \end{pmatrix} \tag{3.407}$$

$$A_{zz} = \frac{1}{3I_0}\left[|B|^2 + |H|^2 + |E|^2 + |F|^2 + |O|^2 + |D|^2 - 2\left(|C|^2 + |A|^2 + |G|^2\right)\right] \tag{3.408}$$

In $I = I' = 0$ case, using Eq. (3.375) we can see that Eqs. (3.393)–(3.408) will automatically degenerate to Eqs. (3.234)–(3.249).

For the definite $M_I' M_I$, we still define the polarization transfer coefficients in the form of the expressions given by Eqs. (3.251)–(3.254). The following results can be obtained from Eqs. (3.251)–(3.254) and utilizing Eqs. (3.33), (3.36), and (3.378) as well as the matrices $\hat{F}\hat{S}_i\hat{F}^{+}$ and $\hat{F}\hat{Q}_i\hat{F}^{+}$ obtained earlier in this section

$$K_x^{x'} = \frac{1}{3I_0}\,\mathrm{Re}\,(AB^* + DC^* + GO^* + HA^* - OC^* - AE^* - FA^* - GD^*) \tag{3.409}$$

$$K_x^{y'} = \frac{1}{3I_0}\,\mathrm{Im}(AB^* + DC^* + GO^* + HA^* - OC^* - AE^* - FA^* - GD^*) \tag{3.410}$$

$$K_x^{z'} = -\frac{\sqrt{2}}{3I_0}\,\mathrm{Im}(BC^* + CE^* - FG^* - GH^*) \tag{3.411}$$

$$K_x^{x'y'} = \frac{1}{\sqrt{2}I_0} \mathrm{Re}\,(GB^* + HC^* - FC^* - GE^*) \tag{3.412}$$

$$K_x^{x'z'} = \frac{1}{2I_0} \mathrm{Re}\,(AB^* + DC^* - GO^* - HA^* - OC^* - AE^* + FA^* + GD^*) \tag{3.413}$$

$$K_x^{y'z'} = \frac{1}{2I_0} \mathrm{Im}(AB^* + DC^* - GO^* - HA^* - OC^* - AE^* + FA^* + GD^*) \tag{3.414}$$

$$K_x^{x'x'-y'y'} = -\frac{\sqrt{2}}{I_0} \mathrm{Im}(GB^* + HC^* - FC^* - GE^*) \tag{3.415}$$

$$K_x^{z'z'} = -\frac{\sqrt{2}}{3I_0} \mathrm{Im}[BC^* + CE^* + FG^* + GH^* - 2(OA^* + AD^*)] \tag{3.416}$$

$$K_y^{x'} = -\frac{1}{3I_0} \mathrm{Im}(AB^* + DC^* + GO^* + HA^* + OC^* + AE^* + FA^* + GD^*) \tag{3.417}$$

$$K_y^{y'} = \frac{1}{3I_0} \mathrm{Re}\,(AB^* + DC^* + GO^* + HA^* + OC^* + AE^* + FA^* + GD^*) \tag{3.418}$$

$$K_y^{z'} = \frac{\sqrt{2}}{3I_0} \mathrm{Re}\,(BC^* + CE^* - FG^* - GH^*) \tag{3.419}$$

$$K_y^{x'y'} = -\frac{1}{\sqrt{2}I_0} \mathrm{Im}(GB^* + HC^* + FC^* + GE^*) \tag{3.420}$$

$$K_y^{x'z'} = -\frac{1}{2I_0} \mathrm{Im}(AB^* + DC^* - GO^* - HA^* + OC^* + AE^* - FA^* - GD^*) \tag{3.421}$$

$$K_y^{y'z'} = \frac{1}{2I_0} \mathrm{Re}\,(AB^* + DC^* - GO^* - HA^* + OC^* + AE^* - FA^* - GD^*) \tag{3.422}$$

$$K_y^{x'x'-y'y'} = -\frac{\sqrt{2}}{I_0} \mathrm{Re}\,(GB^* + HC^* + FC^* + GE^*) \tag{3.423}$$

$$K_y^{z'z'} = \frac{\sqrt{2}}{3I_0} \mathrm{Re}\,[BC^* + CE^* + FG^* + GH^* - 2(OA^* + AD^*)] \tag{3.424}$$

$$K_z^{x'} = -\frac{\sqrt{2}}{3I_0} \mathrm{Im}(OB^* - DE^* + FO^* - HD^*) \tag{3.425}$$

$$K_z^{y'} = \frac{\sqrt{2}}{3I_0} \mathrm{Re}\,(OB^* - DE^* + FO^* - HD^*) \tag{3.426}$$

$$K_z^{z'} = \frac{1}{3I_0} \left( |B|^2 - |E|^2 + |H|^2 - |F|^2 \right) \tag{3.427}$$

$$K_z^{x'y'} = -\frac{1}{I_0} \mathrm{Im}(FB^* - HE^*) \tag{3.428}$$

$$K_z^{x'z'} = -\frac{1}{\sqrt{2}I_0}\mathrm{Im}(OB^* - DE^* - FO^* + HD^*) \tag{3.429}$$

$$K_z^{y'z'} = \frac{1}{\sqrt{2}I_0}\mathrm{Re}\,(OB^* - DE^* - FO^* + HD^*) \tag{3.430}$$

$$K_z^{x'x'-y'y'} = -\frac{2}{I_0}\mathrm{Re}\,(FB^* - HE^*) \tag{3.431}$$

$$K_z^{z'z'} = \frac{1}{3I_0}\left[\,|B|^2 - |E|^2 - |H|^2 + |F|^2 - 2\left(\,|O|^2 - |D|^2\right)\right] \tag{3.432}$$

$$K_{xy}^{x'} = \frac{1}{\sqrt{2}I_0}\mathrm{Re}\,(BD^* + OH^* - OE^* - FD^*) \tag{3.433}$$

$$K_{xy}^{y'} = -\frac{1}{\sqrt{2}I_0}\mathrm{Im}(BD^* + OH^* + OE^* + FD^*) \tag{3.434}$$

$$K_{xy}^{z'} = -\frac{1}{I_0}\mathrm{Im}(BE^* - FH^*) \tag{3.435}$$

$$K_{xy}^{x'y'} = \frac{3}{2I_0}\mathrm{Re}\,(HB^* - FE^*) \tag{3.436}$$

$$K_{xy}^{x'z'} = \frac{3}{2\sqrt{2}I_0}\mathrm{Re}\,(BD^* - OH^* - OE^* + FD^*) \tag{3.437}$$

$$K_{xy}^{y'z'} = -\frac{3}{2\sqrt{2}I_0}\mathrm{Im}(BD^* - OH^* + OE^* - FD^*) \tag{3.438}$$

$$K_{xy}^{x'x'-y'y'} = -\frac{3}{I_0}\mathrm{Im}(HB^* - FE^*) \tag{3.439}$$

$$K_{xy}^{z'z'} = -\frac{1}{I_0}\mathrm{Im}(BE^* + FH^* - 2OD^*) \tag{3.440}$$

$$K_{xz}^{x'} = \frac{1}{2I_0}\mathrm{Re}\,(AB^* - DC^* + GO^* - HA^* - OC^* + AE^* - FA^* + GD^*) \tag{3.441}$$

$$K_{xz}^{y'} = \frac{1}{2I_0}\mathrm{Im}(AB^* - DC^* + GO^* - HA^* - OC^* + AE^* - FA^* + GD^*) \tag{3.442}$$

$$K_{xz}^{z'} = -\frac{1}{\sqrt{2}I_0}\mathrm{Im}(BC^* - CE^* - FG^* + GH^*) \tag{3.443}$$

$$K_{xz}^{x'y'} = \frac{3}{2\sqrt{2}I_0}\mathrm{Re}\,(GB^* - HC^* - FC^* + GE^*) \tag{3.444}$$

$$K_{xz}^{x'z'} = \frac{3}{4I_0}\mathrm{Re}\,(AB^* - DC^* - GO^* + HA^* - OC^* + AE^* + FA^* - GD^*) \tag{3.445}$$

$$K_{xz}^{y'z'} = \frac{3}{4I_0}\mathrm{Im}(AB^* - DC^* - GO^* + HA^* - OC^* + AE^* + FA^* - GD^*) \quad (3.446)$$

$$K_{xz}^{x'x'-y'y'} = -\frac{3}{\sqrt{2}I_0}\mathrm{Im}(GB^* - HC^* - FC^* + GE^*) \quad (3.447)$$

$$K_{xz}^{z'z'} = -\frac{1}{\sqrt{2}I_0}\mathrm{Im}[BC^* - CE^* + FG^* - GH^* - 2(OA^* - AD^*)] \quad (3.448)$$

$$K_{yz}^{x'} = -\frac{1}{2I_0}\mathrm{Im}(AB^* - DC^* + GO^* - HA^* + OC^* - AE^* + FA^* - GD^*) \quad (3.449)$$

$$K_{yz}^{y'} = \frac{1}{2I_0}\mathrm{Re}\,(AB^* - DC^* + GO^* - HA^* + OC^* - AE^* + FA^* - GD^*) \quad (3.450)$$

$$K_{yz}^{z'} = \frac{1}{\sqrt{2}I_0}\mathrm{Re}\,(BC^* - CE^* - FG^* + GH^*) \quad (3.451)$$

$$K_{yz}^{x'y'} = -\frac{3}{2\sqrt{2}I_0}\mathrm{Im}(GB^* - HC^* + FC^* - GE^*) \quad (3.452)$$

$$K_{yz}^{x'z'} = -\frac{3}{4I_0}\mathrm{Im}(AB^* - DC^* - GO^* + HA^* + OC^* - AE^* - FA^* + GD^*) \quad (3.453)$$

$$K_{yz}^{y'z'} = \frac{3}{4I_0}\mathrm{Re}\,(AB^* - DC^* - GO^* + HA^* + OC^* - AE^* - FA^* + GD^*) \quad (3.454)$$

$$K_{yz}^{x'x'-y'y'} = -\frac{3}{\sqrt{2}I_0}\mathrm{Re}\,(GB^* - HC^* + FC^* - GE^*) \quad (3.455)$$

$$K_{yz}^{z'z'} = \frac{1}{\sqrt{2}I_0}\mathrm{Re}\,[BC^* - CE^* + FG^* - GH^* - 2(OA^* - AD^*)] \quad (3.456)$$

$$K_{xx-yy}^{x'} = -\frac{\sqrt{2}}{I_0}\mathrm{Im}(BD^* + OH^* - OE^* - FD^*) \quad (3.457)$$

$$K_{xx-yy}^{y'} = -\frac{\sqrt{2}}{I_0}\mathrm{Re}\,(BD^* + OH^* + OE^* + FD^*) \quad (3.458)$$

$$K_{xx-yy}^{z'} = -\frac{2}{I_0}\mathrm{Re}\,(BE^* - FH^*) \quad (3.459)$$

$$K_{xx-yy}^{x'y'} = \frac{3}{I_0}\mathrm{Im}(HB^* + FE^*) \quad (3.460)$$

$$K_{xx-yy}^{x'z'} = -\frac{3}{\sqrt{2}I_0}\mathrm{Im}(BD^* - OH^* - OE^* + FD^*) \quad (3.461)$$

$$K_{xx-yy}^{y'z'} = -\frac{3}{\sqrt{2}I_0}\mathrm{Re}\,(BD^* - OH^* + OE^* - FD^*) \quad (3.462)$$

$$K_{xx-yy}^{x'x'-y'y'} = \frac{6}{I_0}\,\mathrm{Re}\,(HB^* + FE^*) \tag{3.463}$$

$$K_{xx-yy}^{z'z'} = -\frac{2}{I_0}\,\mathrm{Re}\,(BE^* + FH^* - 2OD^*) \tag{3.464}$$

$$K_{zz}^{x'} = -\frac{\sqrt{2}}{3I_0}\mathrm{Im}(OB^* + DE^* - 2AC^* + FO^* + HD^* - 2GA^*) \tag{3.465}$$

$$K_{zz}^{y'} = \frac{\sqrt{2}}{3I_0}\,\mathrm{Re}\,(OB^* + DE^* - 2AC^* + FO^* + HD^* - 2GA^*) \tag{3.466}$$

$$K_{zz}^{z'} = -\frac{1}{3I_0}\left[\,|B|^2 - |H|^2 + |E|^2 - |F|^2 - 2\left(\,|C|^2 - |G|^2\right)\right] \tag{3.467}$$

$$K_{zz}^{x'y'} = -\frac{1}{I_0}\mathrm{Im}(FB^* + HE^* - 2GC^*) \tag{3.468}$$

$$K_{zz}^{x'z'} = -\frac{1}{\sqrt{2}I_0}\mathrm{Im}(OB^* + DE^* - 2AC^* - FO^* - HD^* + 2GA^*) \tag{3.469}$$

$$K_{zz}^{y'z'} = \frac{1}{\sqrt{2}I_0}\,\mathrm{Re}\,(OB^* + DE^* - 2AC^* - FO^* - HD^* + 2GA^*) \tag{3.470}$$

$$K_{zz}^{x'x'-y'y'} = -\frac{2}{I_0}\,\mathrm{Re}\,(FB^* + HE^* - 2GC^*) \tag{3.471}$$

$$K_{zz}^{z'z'} = \frac{1}{3I_0}\left(|B|^2 + |H|^2 + |E|^2 + |F|^2 - 2|C|^2 - 2|G|^2 + 4|A|^2 - 2|O|^2 - 2|D|^2\right) \tag{3.472}$$

In $I = I' = 0$ case, using Eq. (3.375) we can see that Eqs. (3.409)–(3.472) will automatically degenerate to Eqs. (3.255)–(3.318).

In the case where both the target and the residual nucleus are polarized and the $y$-axis along to direction $\vec{n}$ is chosen, so there is $\varphi = 0$, the footnote of the matrix element of $\hat{F}$ in Eq. (3.379) is marked by $M_I' M_I$, and the footnote of the matrix element of complex conjugate matrix $\hat{F}^+$ in Eq. (3.379) is marked by $\overline{M_I'}\,\overline{M}_I$. And the following changes for the previous results are made:

$$|X|^2 \rightarrow X^*_{\overline{M_I'}\,\overline{M}_I} X_{M_I' M_I}, \qquad X = A,B,C,D,E,F,G,H,O \tag{3.473}$$

So according to Eq. (3.378) we can write

$$\begin{aligned} I_{0,M_I'\overline{M_I'}M_I\overline{M}_I} = \frac{1}{3}\Big(&A^*_{\overline{M_I'}\,\overline{M}_I}A_{M_I'M_I} + B^*_{\overline{M_I'}\,\overline{M}_I}B_{M_I'M_I} + H^*_{\overline{M_I'}\,\overline{M}_I}H_{M_I'M_I} \\ &+ C^*_{\overline{M_I'}\,\overline{M}_I}C_{M_I'M_I} + G^*_{\overline{M_I'}\,\overline{M}_I}G_{M_I'M_I} + D^*_{\overline{M_I'}\,\overline{M}_I}D_{M_I'M_I} \\ &+ O^*_{\overline{M_I'}\,\overline{M}_I}O_{M_I'M_I} + E^*_{\overline{M_I'}\,\overline{M}_I}E_{M_I'M_I} + F^*_{\overline{M_I'}\,\overline{M}_I}F_{M_I'M_I}\Big) \end{aligned} \tag{3.474}$$

Using the form of the target and the residual nucleus polarization wave functions introduced by Eq. (2.315) and Eq. (2.320), while still using the simplification summation symbol $\overline{\Sigma}$ for the polarized target and residual nucleus introduced by Eq. (2.327), the angular distribution of the outgoing particles corresponding to the unpolarized incident particles can be expressed as

$$\overline{I}^0 \equiv I^0_{\alpha' n', \alpha n} = \frac{\mathrm{d}\sigma^0_{\alpha' n', \alpha n}}{\mathrm{d}\Omega} = \overline{\Sigma}\, I_{0, M_I' \overline{M}_I' M_I \overline{M}_I} \tag{3.475}$$

Next introduce the following Cartesian coordinate index set

$$\varepsilon = x,\ y,\ z,\ xy,\ xz,\ yz,\ xx - yy,\ zz \tag{3.476}$$

The components of the polarization vector and polarization tensor corresponding to the unpolarized incident particles are

$$\overline{P}^{0i'} \equiv P^{0i'}_{\alpha' n', \alpha n} = \frac{1}{\overline{I}^0} \overline{\Sigma}\, I_{0, M_I' \overline{M}_I' M_I \overline{M}_I} P^{i'}_{0, M_I' \overline{M}_I' M_I \overline{M}_I}, \quad i' = \varepsilon' \tag{3.477}$$

When the normal variables are added with the footnote $M_I' M_I$ and the complex conjugate variables are added with the footnote $\overline{M}_I' \overline{M}_I$ in Eqs. (3.383)–(3.390), and using Eq. (3.473), the various $P^{i'}_{0, M_I' \overline{M}_I' M_I \overline{M}_I}$ in Eq. (3.477) can be obtained.

When the incident particles, targets, and residual nuclei are all polarized, the components of the vector polarization analyzing power and tensor polarization analyzing power of the nuclear reaction are

$$\overline{A}_i \equiv A_{i, \alpha' n', \alpha n} = \frac{1}{\overline{I}^0} \overline{\Sigma}\, I_{0, M_I' \overline{M}_I' M_I \overline{M}_I} A_{i, M_I' \overline{M}_I' M_I \overline{M}_I}, \quad i = \varepsilon \tag{3.478}$$

where $A_{i, M_I' \overline{M}_I' M_I \overline{M}_I}$ has been given by Eqs. (3.394)–(3.408). Referring to Eq. (3.214), we can write the corresponding outgoing particle differential cross section as

$$\begin{aligned} \overline{I} \equiv I_{\alpha' n', \alpha n} = \frac{\mathrm{d}\sigma_{\alpha' n', \alpha n}}{\mathrm{d}\Omega} &= \overline{I}^0 \Big[ 1 + \frac{3}{2} \left( p_x \overline{A}_x + p_y \overline{A}_y + p_z \overline{A}_z \right) \\ &+ \frac{2}{3} \left( p_{xy} \overline{A}_{xy} + p_{xz} \overline{A}_{xz} + p_{yz} \overline{A}_{yz} \right) + \frac{1}{6} p_{xx-yy} \overline{A}_{xx-yy} + \frac{1}{2} p_{zz} \overline{A}_{zz} \Big] \end{aligned} \tag{3.479}$$

The corresponding polarization transfer coefficient is

$$\overline{K}_i^{i'} \equiv K^{i'}_{i, \alpha' n', \alpha n} = \frac{1}{\overline{I}^0} \overline{\Sigma} I_{0, M_I' \overline{M}_I' M_I \overline{M}_I} K^{i'}_{i, M_I' \overline{M}_I' M_I \overline{M}_I}, \quad i = \varepsilon; \quad i' = \varepsilon' \tag{3.480}$$

where $K^{i'}_{i,M_I'\overline{M}_I'M_I\overline{M}_I}$ has been given by Eqs. (3.409)–(3.472). Referring to Eq. (3.214), the corresponding components of the polarization vector and polarization tensor of the outgoing particles can be written as

$$\begin{aligned}\overline{P}^{i'} \equiv P^{i'}_{\alpha' n',\alpha n} = \frac{\overline{I}^0}{\overline{I}}\Big[1 + \frac{3}{2}\left(p_x\overline{K}^{i'}_x + p_y\overline{K}^{i'}_y + p_z\overline{K}^{i'}_z\right) \\ + \frac{2}{3}\left(p_{xy}\overline{K}^{i'}_{xy} + p_{xz}\overline{K}^{i'}_{xz} + p_{yz}\overline{K}^{i'}_{yz}\right) + \frac{1}{6}p_{xx-yy}\overline{K}^{i'}_{xx-yy} + \frac{1}{2}p_{zz}\overline{K}^{i'}_{zz}\Big], \quad i' = \varepsilon'\end{aligned} \tag{3.481}$$

## 3.8 Polarization Theory for $\vec{1} + A \rightarrow \vec{1} + B$ Reactions

Equations (2.334) and (2.335) give the conditions for the target and the residual nucleus to be unpolarized, respectively. Using the abovementioned two conditions and the simplification summation symbol $\widetilde{\Sigma}$ introduced in Eq. (2.278) for the unpolarized target and residual nucleus, Eqs. (3.474), (3.475), (3.477), (3.478), and (3.480) can be simplified as follows, respectively:

$$\begin{aligned}I_{0,M_I'M_I} = \frac{1}{3}\Big(\left|A_{M_I'M_I}\right|^2 + \left|B_{M_I'M_I}\right|^2 + \left|H_{M_I'M_I}\right|^2 + \left|C_{M_I'M_I}\right|^2 + \left|G_{M_I'M_I}\right|^2 \\ + \left|D_{M_I'M_I}\right|^2 + \left|O_{M_I'M_I}\right|^2 + \left|E_{M_I'M_I}\right|^2 + \left|F_{M_I'M_I}\right|^2\Big)\end{aligned} \tag{3.482}$$

$$\overline{I}^0 = \widetilde{\Sigma} I_{0,M_I'M_I} \tag{3.483}$$

$$\overline{P}^{0\,i'} = \frac{1}{\overline{I}^0}\widetilde{\Sigma} I_{0,M_I'M_I} P^{i'}_{0,M_I'M_I}, \qquad i' = \varepsilon' \tag{3.484}$$

$$\overline{A}_i = \frac{1}{\overline{I}^0}\widetilde{\Sigma} I_{0,M_I'M_I} A_{i,M_I'M_I}, \qquad i = \varepsilon \tag{3.485}$$

$$\overline{K}^{i'}_i = \frac{1}{\overline{I}^0}\widetilde{\Sigma} I_{0,M_I'M_I} K^{i'}_{i,M_I'M_I}, \qquad i = \varepsilon; \qquad i' = \varepsilon' \tag{3.486}$$

Next we analyze Eq. (3.353) which belongs to the $S$-$L$ coupling mode. The following relations can be obtained according to the C-G coefficient properties:

$$\begin{aligned}&C^{S\,\mu+M_I}_{1\mu\,IM_I} = (-1)^{I+1-S} C^{S\,-\mu-M_I}_{1\,-\mu\,I\,-M_I}, \quad C^{S'\,\mu'+M_I'}_{1\mu'\,I'M_I'} = (-1)^{I'+1-S'} C^{S'\,-\mu'\,-M_I'}_{1\,-\mu'\,I'\,-M_I'} \\ &C^{J\,M}_{l0\,\,SM} = (-1)^{l+S-J} C^{J\,-M}_{l0\,S\,-M}, \quad C^{J\,M}_{l'm_l'\,S'M_S'} = (-1)^{l'+S-J'} C^{J\,-M}_{l'\,-m_l'\,S'\,-M_S'}\end{aligned} \tag{3.487}$$

Dividing Eq. (2.252) with $i = i' = 1$ into two items, we study first the second item. For the part which is not related to angle $\varphi$, the following phase factor comes out using Eqs. (2.288) and (3.487) when all spin magnetic quantum numbers change their signs

$$(-1)^{l+S-J+l'+S'-J+I+1-S+I'+1-S'+\mu+M_I-\mu'-M'_I} = (-1)^{2J+l+I+l'+I'+\mu+M_I-\mu'-M'_I}$$

Using the same procedure to get Eq. (2.290), for $\vec{1} + A \rightarrow \vec{1} + B$ reaction we can obtain the following similar expression:

$$\Pi = \begin{cases} 0 & \text{when } \pi_i \pi_I = \pi'_i \pi'_I \\ 1 & \text{when } \pi_i \pi_I = -\pi'_i \pi'_I \end{cases} \tag{3.488}$$

In the case that the spins of both the incident particle and the outgoing particle are 1, the target spin $I$ and the residual nucleus spin $I'$ must be either the integers or the semi-odd numbers at the same time. If $I$ and $I'$ are all integers, $J$ is an integer, so $(-1)^{2J} = 1$, and in this case let $\Lambda = 0$; if $I$ and $I'$ are all semi-odd numbers, $J$ is a semi-odd number, so $(-1)^{2J} = -1$, and in this case let $\Lambda = 1$. Moreover, we can express the above results in a unified form as

$$\Lambda = \begin{cases} 0 & \text{when } I \text{ and } I' \text{ are all integers} \\ 1 & \text{when } I \text{ and } I' \text{ are all semi-odd numbers} \end{cases} \tag{3.489}$$

This last expression (3.489) is similar to Eq. (2.290). Given this, we can get the following relation:

$$f^{(2)}_{\alpha' n' \mu' M'_I, \alpha n \mu M_I}(\theta) = (-1)^{\Pi+\Lambda+I+I'+\mu+M_I-\mu'-M'_I} f^{(2)}_{\alpha' n' -\mu' -M'_I, \alpha n -\mu -M_I}(\theta) \tag{3.490}$$

We can now study the first item of Eq. (2.252) with $i = i' = 1$. For elastic scattering, we have $\Pi = 0$, $I' = I$, $(-1)^{\Lambda+2I} = 1$, and the δ symbol requires $\mu = \mu'$, $M_I = M'_I$, so $f^{(1)}$ also satisfies the relation given by Eq. (3.490). Thus, for the independent part with angle $\varphi$ of the total reaction amplitude $f$, we have

$$f_{\alpha' n' \mu' M'_I, \alpha n \mu M_I}(\theta) = (-1)^{\Pi+\Lambda+I+I'+\mu+M_I-\mu'-M'_I} f_{\alpha' n' -\mu' -M'_I, \alpha n -\mu -M_I}(\theta) \tag{3.491}$$

The following relations can be obtained from Eqs. (3.354)–(3.363):

$$\begin{aligned}
&A_{M'_I M_I}(\theta) = f_{\alpha' n' 0 M'_I, \alpha n 0 M_I}(\theta), \\
&B_{M'_I M_I}(\theta) = f_{\alpha' n' 1 M'_I, \alpha n 1 M_I}(\theta), && H_{M'_I M_I}(\theta) = f_{\alpha' n' -1 M'_I, \alpha n -1 M_I}(\theta), \\
&C_{M'_I M_I}(\theta) = \mathrm{i} f_{\alpha' n' 1 M'_I, \alpha n 0 M_I}(\theta), && G_{M'_I M_I}(\theta) = -\mathrm{i} f_{\alpha' n' -1 M'_I, \alpha n 0 M_I}(\theta), \\
&D_{M'_I M_I}(\theta) = \mathrm{i} f_{\alpha' n' 0 M'_I, \alpha n -1 M_I}(\theta), && O_{M'_I M_I}(\theta) = -\mathrm{i} f_{\alpha' n' 0 M'_I, \alpha n 1 M_I}(\theta), \\
&E_{M'_I M_I}(\theta) = -f_{\alpha' n' 1 M'_I, \alpha n -1 M_I}(\theta), && F_{M'_I M_I}(\theta) = -f_{\alpha' n' -1 M'_I, \alpha n 1 M_I}(\theta)
\end{aligned} \tag{3.492}$$

Given this, we have the following relations according to Eqs. (3.491) and (3.492):

$$A_{M_I'M_I}(\theta)=(-1)^{\Pi+\Lambda+I+I'+M_I'-M_I}A_{-M_I'\ -M_I}(\theta) \tag{3.493}$$

$$H_{M_I'M_I}(\theta)=(-1)^{\Pi+\Lambda+I+I'+M_I-M_I'}B_{-M_I'\ -M_I}(\theta) \tag{3.494}$$

$$G_{M_I'M_I}(\theta)=(-1)^{\Pi+\Lambda+I+I'+M_I-M_I'}C_{-M_I'\ -M_I}(\theta) \tag{3.495}$$

$$O_{M_I'M_I}(\theta)=(-1)^{\Pi+\Lambda+I+I'+M_I-M_I'}D_{-M_I'\ -M_I}(\theta) \tag{3.496}$$

$$F_{M_I'M_I}(\theta)=(-1)^{\Pi+\Lambda+I+I'+M_I-M_I'}E_{-M_I'\ -M_I}(\theta) \tag{3.497}$$

By using the method to prove Eq. (2.297), and using once again the simplification summation symbol $\widetilde{\Sigma}$ introduced by Eq. (2.278), the following relations can be proved based on Eqs. (3.493)–(3.497):

$$\begin{aligned}
&\widetilde{\Sigma}\mathrm{Im}(BC^*+GH^*)=0,\ \ \widetilde{\Sigma}\mathrm{Im}(CE^*+FG^*)=0,\ \ \widetilde{\Sigma}\mathrm{Im}(OA^*+AD^*)=0,\\
&\widetilde{\Sigma}\,\mathrm{Re}\,(BC^*+GH^*)=2\widetilde{\Sigma}\,\mathrm{Re}\,(BC^*),\ \ \widetilde{\Sigma}\,\mathrm{Re}\,(CE^*+FG^*)=2\widetilde{\Sigma}\,\mathrm{Re}\,(CE^*),\\
&\widetilde{\Sigma}\,\mathrm{Re}\,(OA^*+AD^*)=2\widetilde{\Sigma}\,\mathrm{Re}\,(OA^*),\ \ \widetilde{\Sigma}\mathrm{Im}(BC^*-GH^*)=2\widetilde{\Sigma}\mathrm{Im}(BC^*),\\
&\widetilde{\Sigma}\mathrm{Im}(CE^*-FG^*)=2\widetilde{\Sigma}\mathrm{Im}(CE^*),\ \ \widetilde{\Sigma}\mathrm{Im}(OA^*-AD^*)=2\widetilde{\Sigma}\mathrm{Im}(OA^*),\\
&\widetilde{\Sigma}\,\mathrm{Re}\,(BC^*-GH^*)=0,\ \ \widetilde{\Sigma}\,\mathrm{Re}\,(CE^*-FG^*)=0,\ \ \widetilde{\Sigma}\,\mathrm{Re}\,(OA^*-AD^*)=0,\\
&\widetilde{\Sigma}|H|^2=\widetilde{\Sigma}|B|^2,\ \ \widetilde{\Sigma}|G|^2=\widetilde{\Sigma}|C|^2,\ \ \widetilde{\Sigma}|O|^2=\widetilde{\Sigma}|D|^2,\ \ \widetilde{\Sigma}|F|^2=\widetilde{\Sigma}|E|^2
\end{aligned} \tag{3.498}$$

In addition, other expressions similar to the above formulas can also be obtained. By using the method used to prove Eq. (2.298), we can prove

$$\begin{aligned}
&\widetilde{\Sigma}HB^*,\ \ \widetilde{\Sigma}GC^*,\ \ \widetilde{\Sigma}OD^*,\ \ \widetilde{\Sigma}FE^*,\ \ \widetilde{\Sigma}(BC^*+GH^*),\ \ \widetilde{\Sigma}(BD^*+OH^*),\\
&\widetilde{\Sigma}(BE^*+FH^*),\ \ \widetilde{\Sigma}(CD^*+OG^*),\ \ \widetilde{\Sigma}(CE^*+FG^*),\ \ \widetilde{\Sigma}(DE^*+FO^*)
\end{aligned} \tag{3.499}$$

such that their corresponding complex conjugate terms are all real numbers.

Now, defining

$$\widetilde{I}_{0,\,M_I'M_I}=\frac{1}{3}\left[\left|A_{M_I'M_I}\right|^2+2\left(\left|B_{M_I'M_I}\right|^2+\left|C_{M_I'M_I}\right|^2+\left|D_{M_I'M_I}\right|^2+\left|E_{M_I'M_I}\right|^2\right)\right] \tag{3.500}$$

we can prove the following expression from Eqs. (3.482), (3.483), and (3.498):

$$\widetilde{I}^0=\widetilde{\Sigma}\,\widetilde{I}_{0,M_I'M_I} \tag{3.501}$$

From Eq. (3.484) we can obtain the following results according to Eqs. (3.383)–(3.390) and by using Eqs. (3.493)–(3.499):

$$\overline{P}^{0\, x'} = \overline{P}^{0\, z'} = \overline{P}^{0\, x'y'} = \overline{P}^{0\, y'z'} = 0 \tag{3.502}$$

$$\overline{P}^{0\, y'} = \frac{2\sqrt{2}}{3\overline{I}^0} \widetilde{\Sigma} \operatorname{Re}(OB^* + AC^* + DE^*) \tag{3.503}$$

$$\overline{P}^{0\, x'z'} = -\frac{\sqrt{2}}{\overline{I}^0} \widetilde{\Sigma} \operatorname{Im}(OB^* + AC^* + DE^*) \tag{3.504}$$

$$\overline{P}^{0\ x'x'-y'y'} = -\frac{2}{\overline{I}^0} \widetilde{\Sigma} \operatorname{Re}(2FB^* + GC^*) \tag{3.505}$$

$$\overline{P}^{0\, z'z'} = \frac{2}{3\overline{I}^0} \widetilde{\Sigma}\left(|B|^2 + |C|^2 + |E|^2 - |A|^2 - 2|D|^2\right) \tag{3.506}$$

And from Eq. (3.485) the following results can be obtained according to the Eqs. (3.394)–(3.408):

$$\overline{A}_x = \overline{A}_z = \overline{A}_{xy} = \overline{A}_{yz} = 0 \tag{3.507}$$

$$\overline{A}_y = \frac{2\sqrt{2}}{3\overline{I}^0} \widetilde{\Sigma} \operatorname{Re}(BC^* + CE^* + OA^*) \tag{3.508}$$

$$\overline{A}_{xz} = -\frac{\sqrt{2}}{\overline{I}^0} \widetilde{\Sigma} \operatorname{Im}(BC^* + OA^* + FG^*) \tag{3.509}$$

$$\overline{A}_{xx-yy} = -\frac{2}{\overline{I}^0} \widetilde{\Sigma} \operatorname{Re}(2BE^* + OD^*) \tag{3.510}$$

$$A_{zz} = \frac{2}{3\overline{I}^0} \widetilde{\Sigma}\left(|B|^2 + |E|^2 + |D|^2 - |A|^2 - 2|C|^2\right) \tag{3.511}$$

From Eq. (3.486), $\overline{K}_i^{i'}$ can be found according to Eqs. (3.409)–(3.472). First, we can see

$$\overline{K}_i^{y'} = \overline{K}_i^{x'z'} = \overline{K}_i^{x'x'-y'y'} = \overline{K}_i^{z'z'} = 0, \qquad i = x,\, z,\, xy,\, yz \tag{3.512}$$

$$\overline{K}_i^{x'} = \overline{K}_i^{z'} = \overline{K}_i^{x'y'} = \overline{K}_i^{y'z'} = 0, \qquad i = y,\, xz,\, xx - yy,\, zz \tag{3.513}$$

Then we can still get

$$\overline{K}_x^{x'} = \frac{2}{3\overline{I}^0} \widetilde{\Sigma} \operatorname{Re}(AB^* - AE^* + DC^* - OC^*) \tag{3.514}$$

$$\overline{K}_x^{z'} = -\frac{2\sqrt{2}}{3\overline{I}^0} \widetilde{\Sigma} \operatorname{Im}(BC^* + CE^*) \tag{3.515}$$

$$\overline{K}_x^{x'y'} = \frac{\sqrt{2}}{\overline{I}^0}\widetilde{\Sigma}\,\mathrm{Re}\,(GB^* - FC^*) \tag{3.516}$$

$$\overline{K}_x^{y'z'} = \frac{1}{\overline{I}^0}\widetilde{\Sigma}\mathrm{Im}(AB^* - AE^* + DC^* - OC^*) \tag{3.517}$$

$$\overline{K}_y^{y'} = \frac{2}{3\overline{I}^0}\widetilde{\Sigma}\,\mathrm{Re}\,(AB^* + AE^* + DC^* + OC^*) \tag{3.518}$$

$$\overline{K}_y^{x'z'} = -\frac{1}{\overline{I}^0}\widetilde{\Sigma}\mathrm{Im}(AB^* + AE^* + DC^* + OC^*) \tag{3.519}$$

$$\overline{K}_y^{x'x'-y'y'} = -\frac{2\sqrt{2}}{\overline{I}^0}\widetilde{\Sigma}\,\mathrm{Re}\,(GB^* + FC^*) \tag{3.520}$$

$$\overline{K}_y^{z'z'} = \frac{2\sqrt{2}}{3\overline{I}^0}\widetilde{\Sigma}\,\mathrm{Re}\,(BC^* + CE^* - 2OA^*) \tag{3.521}$$

$$\overline{K}_z^{x'} = -\frac{2\sqrt{2}}{3\overline{I}^0}\widetilde{\Sigma}\mathrm{Im}(OB^* - DE^*) \tag{3.522}$$

$$\overline{K}_z^{z'} = \frac{2}{3\overline{I}^0}\widetilde{\Sigma}\left(|B|^2 - |E|^2\right) \tag{3.523}$$

$$\overline{K}_z^{x'y'} = -\frac{2}{\overline{I}^0}\widetilde{\Sigma}\mathrm{Im}(FB^*) \tag{3.524}$$

$$\overline{K}_z^{y'z'} = \frac{\sqrt{2}}{\overline{I}^0}\widetilde{\Sigma}\,\mathrm{Re}\,(OB^* - DE^*) \tag{3.525}$$

$$\overline{K}_{xy}^{x'} = \frac{\sqrt{2}}{\overline{I}^0}\widetilde{\Sigma}\,\mathrm{Re}\,(BD^* - OE^*) \tag{3.526}$$

$$\overline{K}_{xy}^{z'} = -\frac{2}{\overline{I}^0}\widetilde{\Sigma}\mathrm{Im}(BE^*) \tag{3.527}$$

$$\overline{K}_{xy}^{x'y'} = \frac{3}{2\overline{I}^0}\widetilde{\Sigma}\,\mathrm{Re}\,(HB^* - FE^*) \tag{3.528}$$

$$\overline{K}_{xy}^{y'z'} = -\frac{3}{\sqrt{2}\overline{I}^0}\widetilde{\Sigma}\mathrm{Im}(BD^* + OE^*) \tag{3.529}$$

$$\overline{K}_{xz}^{y'} = \frac{1}{\overline{I}^0}\widetilde{\Sigma}\mathrm{Im}(AB^* + AE^* - DC^* - OC^*) \tag{3.530}$$

$$\overline{K}_{xz}^{x'z'} = \frac{3}{2\overline{I}^0}\widetilde{\Sigma}\,\mathrm{Re}\,(AB^* + AE^* - DC^* - OC^*) \tag{3.531}$$

$$\overline{K}_{xz}^{x'x'-y'y'} = -\frac{3\sqrt{2}}{\overline{I}^0}\widetilde{\Sigma}\mathrm{Im}(GB^* - FC^*) \tag{3.532}$$

$$\overline{K}_{xz}^{z'z'} = -\frac{\sqrt{2}}{\overline{I}^0}\widetilde{\Sigma}\mathrm{Im}(BC^* - CE^* - 2OA^*) \tag{3.533}$$

$$\overline{K}_{yz}^{x'} = -\frac{1}{\overline{I}^0}\widetilde{\Sigma}\mathrm{Im}(AB^* - AE^* - DC^* + OC^*) \tag{3.534}$$

$$\overline{K}_{yz}^{z'} = \frac{\sqrt{2}}{\overline{I}^0}\widetilde{\Sigma}\,\mathrm{Re}\,(BC^* - CE^*) \tag{3.535}$$

$$\overline{K}_{yz}^{x'y'} = -\frac{3}{\sqrt{2}\overline{I}^0}\widetilde{\Sigma}\mathrm{Im}(GB^* + FC^*) \tag{3.536}$$

$$\overline{K}_{yz}^{y'z'} = \frac{3}{2\overline{I}^0}\widetilde{\Sigma}\,\mathrm{Re}\,(AB^* - AE^* - DC^* + OC^*) \tag{3.537}$$

$$\overline{K}_{xx-yy}^{y'} = -\frac{2\sqrt{2}}{\overline{I}^0}\widetilde{\Sigma}\,\mathrm{Re}\,(BD^* + OE^*) \tag{3.538}$$

$$\overline{K}_{xx-yy}^{x'z'} = -\frac{3\sqrt{2}}{\overline{I}^0}\widetilde{\Sigma}\mathrm{Im}(BD^* - OE^*) \tag{3.539}$$

$$\overline{K}_{xx-yy}^{x'x'-y'y'} = \frac{6}{\overline{I}^0}\widetilde{\Sigma}\,\mathrm{Re}\,(HB^* + FE^*) \tag{3.540}$$

$$\overline{K}_{xx-yy}^{z'z'} = -\frac{4}{\overline{I}^0}\widetilde{\Sigma}\,\mathrm{Re}\,(BE^* - OD^*) \tag{3.541}$$

$$\overline{K}_{zz}^{y'} = \frac{2\sqrt{2}}{3\overline{I}^0}\widetilde{\Sigma}\,\mathrm{Re}\,(OB^* + DE^* - 2AC^*) \tag{3.542}$$

$$\overline{K}_{zz}^{x'z'} = -\frac{\sqrt{2}}{\overline{I}^0}\widetilde{\Sigma}\mathrm{Im}(OB^* + DE^* - 2AC^*) \tag{3.543}$$

$$\overline{K}_{zz}^{x'x'-y'y'} = -\frac{4}{\overline{I}^0}\widetilde{\Sigma}\,\mathrm{Re}\,(FB^* - GC^*) \tag{3.544}$$

$$\overline{K}_{zz}^{z'z'} = \frac{2}{3\overline{I}^0}\widetilde{\Sigma}\left(|B|^2 + |E|^2 + 2|A|^2 - 2|C|^2 - 2|D|^2\right) \tag{3.545}$$

Here, and still according to Eq. (3.320), the component labels of the Cartesian coordinate of the spin-1 particle are divided into two groups $\varepsilon_1 = y,\ xz,\ xx - yy,\ zz$ and $\varepsilon_2 = x,\ z,\ xy,\ yz$. From the results obtained previously in this section, we can see

$$\overline{P}^{0\ i'}\begin{cases} \neq 0 & i' = \varepsilon_1' \\ = 0 & i' = \varepsilon_2' \end{cases} \tag{3.546}$$

$$\overline{A}_i\begin{cases} \neq 0 & i = \varepsilon_1 \\ = 0 & i = \varepsilon_2 \end{cases} \tag{3.547}$$

$$\overline{K}_i^{i'}\begin{cases}\neq 0 & \left(i=\varepsilon_1' \text{ and } i'=\varepsilon_1'\right) \text{ or } \left(i=\varepsilon_2 \text{ and } i'=\varepsilon_2'\right)\\ =0 & \left(i=\varepsilon_1 \text{ and } i'=\varepsilon_2'\right) \text{ or } \left(i=\varepsilon_2 \text{ and } i'=\varepsilon_1'\right)\end{cases} \tag{3.548}$$

Now when the spins of the unpolarized target and the unpolarized residual nucleus are not equal to 0, it can be seen that for $\vec{1}+A\rightarrow\vec{1}+B$ reaction the presence circumstance of the non-zero polarization physical quantities is consistent with the case where the spins of the target and the residual nucleus are equal to 0. It shows that the influence of parity conservation exists in both cases, and we can obtain from Eqs. (3.479) and (3.481)

$$\overline{I}=\overline{I}^0\left(1+\frac{3}{2}p_y\overline{A}_y+\frac{2}{3}p_{xz}\overline{A}_{xz}+\frac{1}{6}p_{xx-yy}\overline{A}_{xx-yy}+\frac{1}{2}p_{zz}\overline{A}_{zz}\right) \tag{3.549}$$

$$\overline{P}^{i'}=\frac{\overline{I}^0}{\overline{I}}\left(\overline{P}^{0\,i'}+\frac{3}{2}p_y\overline{K}_y^{i'}+\frac{2}{3}p_{xz}\overline{K}_{xz}^{i'}+\frac{1}{6}p_{xx-yy}\overline{K}_{xx-yy}^{i'}+\frac{1}{2}p_{zz}\overline{K}_{zz}^{i'}\right),\quad i'=\varepsilon_1{}';$$
$$\overline{P}^{i'}=\frac{\overline{I}^0}{\overline{I}}\left[\frac{3}{2}\left(p_x\overline{K}_x^{i'}+\overline{p}_z\overline{K}_z^{i'}\right)+\frac{2}{3}\left(p_{xy}\overline{K}_{xy}^{i'}+p_{yz}\overline{K}_{yz}^{i'}\right)\right],\quad i'=\varepsilon_2{}' \tag{3.550}$$

If we let $H=B$, $G=C$, $O=D$, $F=E$ and we remove the simplification summation symbol $\widetilde{\Sigma}$, the expressions obtained in this section will automatically degenerate to the corresponding expressions given in Sects. 3.5 and 3.6 where the spins of the target and residual nucleus are all equal to 0.

## 3.9 Polarization Theory for $\vec{1}+\mathrm{A}\rightarrow\vec{\frac{1}{2}}+\mathrm{B}$ Reactions

In this section, we study the polarization theory for $\vec{1}+\mathrm{A}\rightarrow\vec{\frac{1}{2}}+\mathrm{B}$ reactions such that the spins of the target and the residual nucleus are not equal to 0, but both are unpolarized. The expressions for the reaction amplitude of the *j-j* coupling and the *S-L* coupling are given in Eq. (2.251) and Eq. (2.252), respectively. In the case where the incident particle spin is 1 and the outgoing particle spin is $\frac{1}{2}$, the following symbols are introduced for j-j coupling and S-L coupling, respectively:

$$f_{\mu'\mu}^{ll'J}(m_l')=\hat{l}\hat{l}'\mathrm{e}^{\mathrm{i}(\sigma_l+\sigma_{l'})}\sqrt{\frac{(l'-m_l')!}{(l'+m_l')!}}$$
$$\sum_{jj'}C_{l0\ 1\mu}^{j\ \mu}C_{j\mu\ IM_I}^{J\ M}C_{l'm_l'\ \frac{1}{2}\mu'}^{j'\ m_j'}C_{j'm_j'\ I'M_I'}^{J\ M}\left(\delta_{\alpha'n'l'j',\alpha nlj}-S_{\alpha'n'l'j',\alpha nlj}^{J}\right) \tag{3.551}$$

$$f_{\mu'\mu}^{ll'J}(m_l') = \hat{l}\hat{l}' e^{i(\sigma_l+\sigma_{l'})} \sqrt{\frac{(l'-m_l')!}{(l'+m_l')!}} \sum_{SS'} C_{l0\ SM}^{J\ M} C_{1\mu\ IM_I}^{S\ M} C_{l'm_l'\ S'M_S'}^{J\ M} C_{\frac{1}{2}\mu'\ I'M_I'}^{S'M_S'} \left(\delta_{\alpha'n'l'S',\alpha nlS} - S_{\alpha'n'l'S',\alpha nlS}^{J}\right) \tag{3.552}$$

where $\hat{l} \equiv \sqrt{2l+1}$. According to Eqs. (2.251)–(2.254), it follows that the reaction amplitude can be uniformly written in the following form under the two angular momentum coupling schemes:

$$f_{\alpha'n'\mu'M_I',\alpha n\mu M_I}(\Omega) = f_C(\theta)\delta_{\alpha'n',\alpha n}\delta_{\mu'M_I',\mu M_I} + \frac{i}{2k}(-1)^{(\mu+M_I-\mu'-M_I')} \sum_{ll'J} f_{\mu'\mu}^{ll'J}(\mu+M_I-\mu'-M_I') P_{l'}^{\mu+M_I-\mu'-M_I'}(\cos\theta)\, e^{i(\mu+M_I-\mu'-M_I')\varphi} \tag{3.553}$$

Because there can be no elastic scattering channel for $\vec{1} + A \to \vec{\frac{1}{2}} + B$ reactions, so the first term representing Coulomb scattering in the above formula does not appear. Let

$$A_{M_I'M_I}(\theta) = \frac{i}{2k}(-1)^{M_I-M_I'+\frac{1}{2}} \sum_{ll'J} f_{\frac{1}{2}\ 1}^{ll'J}\left(M_I - M_I' + \frac{1}{2}\right) P_{l'}^{M_I-M_I'+\frac{1}{2}}(\cos\theta) \tag{3.554}$$

$$B_{M_I'M_I}(\theta) = \frac{i}{2k}(-1)^{M_I-M_I'-\frac{1}{2}} \sum_{ll'J} f_{-\frac{1}{2}\ -1}^{ll'J}\left(M_I - M_I' - \frac{1}{2}\right) P_{l'}^{M_I-M_I'-\frac{1}{2}}(\cos\theta) \tag{3.555}$$

$$C_{M_I'M_I}(\theta) = -\frac{1}{2k}(-1)^{M_I-M_I'-\frac{1}{2}} \sum_{ll'J} f_{\frac{1}{2}\ 0}^{ll'J}\left(M_I - M_I' - \frac{1}{2}\right) P_{l'}^{M_I-M_I'-\frac{1}{2}}(\cos\theta) \tag{3.556}$$

$$D_{M_I'M_I}(\theta) = \frac{1}{2k}(-1)^{M_I-M_I'+\frac{1}{2}} \sum_{ll'J} f_{-\frac{1}{2}\ 0}^{ll'J}\left(M_I - M_I' + \frac{1}{2}\right) P_{l'}^{M_I-M_I'+\frac{1}{2}}(\cos\theta) \tag{3.557}$$

$$E_{M_I'M_I}(\theta) = -\frac{i}{2k}(-1)^{M_I-M_I'-\frac{3}{2}} \sum_{ll'J} f_{\frac{1}{2}\ -1}^{ll'J}\left(M_I - M_I' - \frac{3}{2}\right) P_{l'}^{M_I-M_I'-\frac{3}{2}}(\cos\theta) \tag{3.558}$$

$$F_{M_I'M_I}(\theta) = -\frac{i}{2k}(-1)^{M_I-M_I'+\frac{3}{2}} \sum_{ll'J} f_{-\frac{1}{2}\ 1}^{ll'J}\left(M_I - M_I' + \frac{3}{2}\right) P_{l'}^{M_I-M_I'+\frac{3}{2}}(\cos\theta) \tag{3.559}$$

And introduce the following simplification symbols

$$F_{\mu'\mu}(\theta,\varphi)=f_{\alpha' n'\mu' M_I',\alpha n\mu M_I}(\Omega) \tag{3.560}$$

then we can get

$$F_{\frac{1}{2}\ 1}(\theta,\varphi)=A_{M_I'M_I}(\theta)\mathrm{e}^{\mathrm{i}\left(M_I-M_I'+\frac{1}{2}\right)\varphi} \tag{3.561}$$

$$F_{\frac{1}{2}\ 0}(\theta,\varphi)=-\mathrm{i}C_{M_I'M_I}(\theta)\mathrm{e}^{\mathrm{i}\left(M_I-M_I'-\frac{1}{2}\right)\varphi} \tag{3.562}$$

$$F_{\frac{1}{2}\ -1}(\theta,\varphi)=-E_{M_I'M_I}(\theta)\mathrm{e}^{\mathrm{i}\left(M_I-M_I'-\frac{3}{2}\right)\varphi} \tag{3.563}$$

$$F_{-\frac{1}{2}\ 1}(\theta,\varphi)=-F_{M_I'M_I}(\theta)\mathrm{e}^{\mathrm{i}\left(M_I-M_I'+\frac{3}{2}\right)\varphi} \tag{3.564}$$

$$F_{-\frac{1}{2}\ 0}(\theta,\varphi)=\mathrm{i}D_{M_I'M_I}(\theta)\mathrm{e}^{\mathrm{i}\left(M_I-M_I'+\frac{1}{2}\right)\varphi} \tag{3.565}$$

$$F_{-\frac{1}{2}\ -1}(\theta,\varphi)=B_{M_I'M_I}(\theta)\mathrm{e}^{\mathrm{i}\left(M_I-M_I'-\frac{1}{2}\right)\varphi} \tag{3.566}$$

When the foot note $M_I'M_I$ and argument ($\theta$) of $A$, $B$, $C$, $D$, $E$, and $F$ are omitted, the reaction amplitude $\hat{F}$, in which arguments $(\theta,\varphi)$ are emitted, with definite $M_I'M_I$ can be written as

$$\hat{F}=\begin{pmatrix} A\mathrm{e}^{\frac{\mathrm{i}}{2}\varphi} & -\mathrm{i}C\mathrm{e}^{-\frac{\mathrm{i}}{2}\varphi} & -E\mathrm{e}^{-\frac{3\mathrm{i}}{2}\varphi} \\ -F\mathrm{e}^{\frac{3\mathrm{i}}{2}\varphi} & \mathrm{i}D\mathrm{e}^{\frac{\mathrm{i}}{2}\varphi} & B\mathrm{e}^{-\frac{\mathrm{i}}{2}\varphi} \end{pmatrix}\mathrm{e}^{\mathrm{i}(M_I-M_I')\varphi} \tag{3.567}$$

From here we can see that $\hat{F}$ is no longer a square Matrix, but rather a rectangular matrix with an unequal numbers of rows and columns. According to Eqs. (2.334) and (2.335), it can be seen that the magnetic quantum numbers as the footnotes of the matrix elements of $\hat{F}$ and $\hat{F}^+$ are the same for each term in the case where both the target and the residual nucleus are unpolarized, and so from Eq. (3.567) we can obtain

$$\hat{F}^+=\begin{pmatrix} A^*\mathrm{e}^{-\frac{\mathrm{i}}{2}\varphi} & -F^*\mathrm{e}^{-\frac{3\mathrm{i}}{2}\varphi} \\ \mathrm{i}C^*\mathrm{e}^{\frac{\mathrm{i}}{2}\varphi} & -\mathrm{i}D^*\mathrm{e}^{-\frac{\mathrm{i}}{2}\varphi} \\ -E^*\mathrm{e}^{\frac{3\mathrm{i}}{2}\varphi} & B^*\mathrm{e}^{\frac{\mathrm{i}}{2}\varphi} \end{pmatrix}\mathrm{e}^{-\mathrm{i}(M_I-M_I')\varphi} \tag{3.568}$$

$$\hat{F}\hat{F}^+=\begin{pmatrix} |A|^2+|C|^2+|E|^2 & -(AF^*+CD^*+EB^*)\,\mathrm{e}^{-\mathrm{i}\varphi} \\ -(FA^*+DC^*+BE^*)\,\mathrm{e}^{\mathrm{i}\varphi} & |B|^2+|D|^2+|F|^2 \end{pmatrix} \tag{3.569}$$

For the definite $M_I'M_I$, the following result can be obtained from Eq. (3.569):

$$I_0 = \frac{1}{3}\mathrm{tr}\left\{\hat{F}\hat{F}^+\right\} = \frac{1}{3}\left( |A|^2 + |B|^2 + |C|^2 + |D|^2 + |E|^2 + |F|^2\right) \tag{3.570}$$

For the sake of simplicity, we select the $y$-axis along the direction $\vec{n}$; in this case $\varphi = 0$. Then we can obtain the following expressions from Eqs. (3.567)–(3.569):

$$\hat{F} = \begin{pmatrix} A & -\mathrm{i}C & -E \\ -F & \mathrm{i}D & B \end{pmatrix}, \quad \hat{F}^+ = \begin{pmatrix} A^* & -F^* \\ \mathrm{i}C^* & -\mathrm{i}D^* \\ -E^* & B^* \end{pmatrix} \tag{3.571}$$

$$\hat{F}\hat{F}^+ = \begin{pmatrix} |A|^2 + |C|^2 + |E|^2 & -(AF^* + CD^* + EB^*) \\ -(FA^* + DC^* + BE^*) & |B|^2 + |D|^2 + |F|^2 \end{pmatrix} \tag{3.572}$$

For the definite $M_I'M_I$ we define the following as the components of the polarization vector of the outgoing particles corresponding to the unpolarized incident particles:

$$P_0^{i'} = \frac{\mathrm{tr}\left\{\hat{\sigma}_{i'}\hat{F}\hat{F}^+\right\}}{\mathrm{tr}\left\{\hat{F}\hat{F}^+\right\}}, \qquad i' = x', y', z' \tag{3.573}$$

From Eq. (3.573) we can obtain the following results using Eqs. (2.6), and (3.570)–(3.572):

$$\begin{aligned} P_0^{x'} &= \frac{1}{3I_0}\left[-(FA^* + DC^* + BE^*) - (AF^* + CD^* + EB^*)\right] \\ &= -\frac{2}{3I_0}\,\mathrm{Re}\,(FA^* + DC^* + BE^*) \end{aligned} \tag{3.574}$$

$$\begin{aligned} P_0^{y'} &= \frac{1}{3I_0}\left[\mathrm{i}(FA^* + DC^* + BE^*) - \mathrm{i}(AF^* + CD^* + EB^*)\right] \\ &= -\frac{2}{3I_0}\mathrm{Im}(FA^* + DC^* + BE^*) \end{aligned} \tag{3.575}$$

$$P_0^{z'} = \frac{1}{3I_0}\left(|A|^2 + |C|^2 + |E|^2 - |B|^2 - |D|^2 - |F|^2\right) \tag{3.576}$$

For the definite $M_I'M_I$ we also define

$$A_i = \frac{\mathrm{tr}\left\{\hat{F}\hat{S}_i\hat{F}^+\right\}}{\mathrm{tr}\left\{\hat{F}\hat{F}^+\right\}}, \qquad i = x, y, z \tag{3.577}$$

and

$$A_i = \frac{\mathrm{tr}\left\{\hat{F}\hat{Q}_i\hat{F}^+\right\}}{\mathrm{tr}\left\{\hat{F}\hat{F}^+\right\}}, \qquad i = xy, xz, yz, xx - yy, zz \tag{3.578}$$

as the polarization vector analyzing powers and the polarization tensor analyzing powers, respectively. By means of Eqs. (3.33), (3.36), (3.570), and (3.571), we can obtain the following expressions from Eqs. (3.577) and (3.578):

$$\hat{S}_x\hat{F}^+ = \frac{1}{\sqrt{2}}\begin{pmatrix} \mathrm{i}C^* & -\mathrm{i}D^* \\ A^* - E^* & -F^* + B^* \\ \mathrm{i}C^* & -\mathrm{i}D^* \end{pmatrix}$$

$$\hat{F}\hat{S}_x\hat{F}^+ = \frac{1}{\sqrt{2}}\begin{pmatrix} \mathrm{i}(AC^* - CA^* + CE^* - EC^*) & -\mathrm{i}(AD^* - CF^* + CB^* - ED^*) \\ -\mathrm{i}(FC^* - DA^* + DE^* - BC^*) & \mathrm{i}(FD^* - DF^* + DB^* - BD^*) \end{pmatrix} \tag{3.579}$$

$$A_x = -\frac{\sqrt{2}}{3I_0}\mathrm{Im}(AC^* + CE^* + FD^* + DB^*) \tag{3.580}$$

$$\hat{S}_y\hat{F}^+ = \frac{\mathrm{i}}{\sqrt{2}}\begin{pmatrix} -\mathrm{i}C^* & \mathrm{i}D^* \\ A^* + E^* & -(F^* + B^*) \\ \mathrm{i}C^* & -\mathrm{i}D^* \end{pmatrix}$$

$$\hat{F}\hat{S}_y\hat{F}^+ = \frac{1}{\sqrt{2}}\begin{pmatrix} AC^* + CA^* + CE^* + EC^* & -(AD^* + CF^* + CB^* + ED^*) \\ -(FC^* + DA^* + DE^* + BC^*) & FD^* + DF^* + DB^* + BD^* \end{pmatrix} \tag{3.581}$$

$$A_y = \frac{\sqrt{2}}{3I_0}\mathrm{Re}\,(AC^* + CE^* + FD^* + DB^*) \tag{3.582}$$

$$S_z\hat{F}^+ = \begin{pmatrix} A^* & -F^* \\ 0 & 0 \\ E^* & -B^* \end{pmatrix}$$

$$\hat{F}S_z\hat{F}^+ = \begin{pmatrix} |A|^2 - |E|^2 & -AF^* + EB^* \\ -FA^* + BE^* & |F|^2 - |B|^2 \end{pmatrix} \tag{3.583}$$

$$A_z = \frac{1}{3I_0}\left(|A|^2 - |B|^2 - |E|^2 + |F|^2\right) \tag{3.584}$$

$$\hat{Q}_{xy}\hat{F}^{+} = \frac{3\mathrm{i}}{2}\begin{pmatrix} E^* & -B^* \\ 0 & 0 \\ A^* & -F^* \end{pmatrix},$$

$$\hat{F}\hat{Q}_{xy}\hat{F}^{+} = \frac{3\mathrm{i}}{2}\begin{pmatrix} AE^* - EA^* & -AB^* + EF^* \\ -FE^* + BA^* & FB^* - BF^* \end{pmatrix} \tag{3.585}$$

$$A_{xy} = -\frac{1}{I_0}\mathrm{Im}(AE^* + FB^*) \tag{3.586}$$

$$\hat{Q}_{xz}\hat{F}^{+} = \frac{3}{2\sqrt{2}}\begin{pmatrix} \mathrm{i}C^* & -\mathrm{i}D^* \\ A^* + E^* & -(F^* + B^*) \\ -\mathrm{i}C^* & \mathrm{i}D^* \end{pmatrix},$$

$$\hat{F}\hat{Q}_{xz}\hat{F}^{+} = \frac{3}{2\sqrt{2}}\begin{pmatrix} \mathrm{i}(AC^* - CA^* - CE^* + EC^*) & -\mathrm{i}(AD^* - CF^* - CB^* + ED^*) \\ -\mathrm{i}(FC^* - DA^* - DE^* + BC^*) & \mathrm{i}(FD^* - DF^* - DB^* + BD^*) \end{pmatrix} \tag{3.587}$$

$$A_{xz} = -\frac{1}{\sqrt{2}I_0}\mathrm{Im}(AC^* - CE^* + FD^* - DB^*) \tag{3.588}$$

$$\hat{Q}_{yz}\hat{F}^{+} = \frac{3\mathrm{i}}{2\sqrt{2}}\begin{pmatrix} -\mathrm{i}C^* & \mathrm{i}D^* \\ A* - E^* & -F^* + B^* \\ -\mathrm{i}C^* & \mathrm{i}D^* \end{pmatrix},$$

$$\hat{F}\hat{Q}_{yz}\hat{F}^{+} = \frac{3}{2\sqrt{2}}\begin{pmatrix} AC^* + CA^* - CE^* - EC^* & -(AD^* + CF^* - CB^* - ED^*) \\ -(FC^* + DA^* - DE^* - BC^*) & FD^* + DF^* - DB^* - BD^* \end{pmatrix} \tag{3.589}$$

$$A_{yz} = \frac{1}{\sqrt{2}I_0}\,\mathrm{Re}\,(AC^* - CE^* + FD^* - DB^*) \tag{3.590}$$

$$\hat{Q}_{xx-yy}\hat{F}^{+} = 3\begin{pmatrix} -E^* & B^* \\ 0 & 0 \\ A^* & -F^* \end{pmatrix},$$

$$\hat{F}\hat{Q}_{xx-yy}\hat{F}^{+} = 3\begin{pmatrix} -AE^* - EA^* & AB^* + EF^* \\ FE^* + BA^* & -FB^* - BF^* \end{pmatrix} \tag{3.591}$$

$$A_{xx-yy} = -\frac{2}{I_0}\,\mathrm{Re}\,(AE^* + FB^*) \tag{3.592}$$

$$\hat{Q}_{zz}\hat{F}^{+} = \begin{pmatrix} A^* & -F^* \\ -2iC^* & 2iD^* \\ -E^* & B^* \end{pmatrix} \tag{3.593}$$

$$\hat{F}\hat{Q}_{zz}\hat{F}^{+} = \begin{pmatrix} |A|^2 - 2|C|^2 + |E|^2 & -AF^* + 2CD^* - EB^* \\ -FA^* + 2DC^* - BE^* & |F|^2 - 2|D|^2 + |B|^2 \end{pmatrix}$$

$$A_{zz} = \frac{1}{3I_0}\left[\,|A|^2 + |B|^2 + |E|^2 + |F|^2 - 2|C|^2 - 2|D|^2\right] \tag{3.594}$$

Now define

$$K_i^{i'} = \frac{\mathrm{tr}\left\{\hat{\sigma}_{i'}\hat{F}\hat{S}_i\hat{F}^{+}\right\}}{\mathrm{tr}\left\{\hat{F}\hat{F}^{+}\right\}}, \quad i = x, y, z; \quad i' = x', y', z' \tag{3.595}$$

as the components of the polarization transfer coefficients from the incident particle polarization vector to the outgoing particle polarization vector, and define

$$K_i^{i'} = \frac{\mathrm{tr}\left\{\hat{\sigma}_{i'}\hat{F}\hat{Q}_i\hat{F}^{+}\right\}}{\mathrm{tr}\left\{\hat{F}\hat{F}^{+}\right\}}, \quad i = xy, xz, yz, xx-yy, zz; \quad i' = x', y', z' \tag{3.596}$$

as the components of the polarization transfer coefficients from the incident particle polarization tensor to the outgoing particle polarization vector. Using Eqs. (2.6) and (3.570) as well as the matrices $\hat{F}\hat{S}_i\hat{F}^{+}$ and $\hat{F}\hat{Q}_i\hat{F}^{+}$ obtained in this section previously, we can obtain from Eqs. (3.595) and (3.596)

$$K_x^{x'} = \frac{\sqrt{2}}{3I_0}\mathrm{Im}(AD^* + CB^* + FC^* + DE^*) \tag{3.597}$$

$$K_x^{y'} = \frac{\sqrt{2}}{3I_0}\mathrm{Re}\,(AD^* + CB^* - FC^* - DE^*) \tag{3.598}$$

$$K_x^{z'} = -\frac{\sqrt{2}}{3I_0}\mathrm{Im}(AC^* + CE^* - FD^* - DB^*) \tag{3.599}$$

$$K_y^{x'} = -\frac{\sqrt{2}}{3I_0}\mathrm{Re}\,(AD^* + CB^* + FC^* + DE^*) \tag{3.600}$$

$$K_y^{y'} = \frac{\sqrt{2}}{3I_0}\mathrm{Im}(AD^* + CB^* - FC^* - DE^*) \tag{3.601}$$

$$K_y^{z'} = \frac{\sqrt{2}}{3I_0} \operatorname{Re}(AC^* + CE^* - FD^* - DB^*) \tag{3.602}$$

$$K_z^{x'} = -\frac{2}{3I_0} \operatorname{Re}(FA^* - BE^*) \tag{3.603}$$

$$K_z^{y'} = -\frac{2}{3I_0} \operatorname{Im}(FA^* - BE^*) \tag{3.604}$$

$$K_z^{z'} = \frac{1}{3I_0} \left( |A|^2 + |B|^2 - |E|^2 - |F|^2 \right) \tag{3.605}$$

$$K_{xy}^{x'} = \frac{1}{I_0} \operatorname{Im}(AB^* + FE^*) \tag{3.606}$$

$$K_{xy}^{y'} = \frac{1}{I_0} \operatorname{Re}(AB^* - FE^*) \tag{3.607}$$

$$K_{xy}^{z'} = -\frac{1}{I_0} \operatorname{Im}(AE^* - FB^*) \tag{3.608}$$

$$K_{xz}^{x'} = \frac{1}{\sqrt{2}I_0} \operatorname{Im}(AD^* - CB^* + FC^* - DE^*) \tag{3.609}$$

$$K_{xz}^{y'} = \frac{1}{\sqrt{2}I_0} \operatorname{Re}(AD^* - CB^* - FC^* + DE^*) \tag{3.610}$$

$$K_{xz}^{z'} = -\frac{1}{\sqrt{2}I_0} \operatorname{Im}(AC^* - CE^* - FD^* + DB^*) \tag{3.611}$$

$$K_{yz}^{x'} = -\frac{1}{\sqrt{2}I_0} \operatorname{Re}(AD^* - CB^* + FC^* - DE^*) \tag{3.612}$$

$$K_{yz}^{y'} = \frac{1}{\sqrt{2}I_0} \operatorname{Im}(AD^* - CB^* - FC^* + DE^*) \tag{3.613}$$

$$K_{yz}^{z'} = \frac{1}{\sqrt{2}I_0} \operatorname{Re}(AC^* - CE^* - FD^* + DB^*) \tag{3.614}$$

$$K_{xx-yy}^{x'} = \frac{2}{I_0} \operatorname{Re}(AB^* + FE^*) \tag{3.615}$$

$$K_{xx-yy}^{y'} = -\frac{2}{I_0} \operatorname{Im}(AB^* - FE^*) \tag{3.616}$$

$$K_{xx-yy}^{z'} = -\frac{2}{I_0} \operatorname{Re}(AE^* - FB^*) \tag{3.617}$$

$$K_{zz}^{x'} = -\frac{2}{3I_0} \operatorname{Re}(FA^* - 2DC^* + BE^*) \tag{3.618}$$

$$K_{zz}^{y'} = -\frac{2}{3I_0}\,\mathrm{Im}(FA^* - 2DC^* + BE^*) \tag{3.619}$$

$$K_{zz}^{z'} = \frac{1}{3I_0}\left(|A|^2 - |B|^2 + |E|^2 - |F|^2 - 2|C|^2 + 2|D|^2\right) \tag{3.620}$$

In the case where both the target and the residual nucleus are polarized, and continuing to choose the $y$-axis along the direction $\vec{n}$ corresponding to $\vec{n}$, where, in Eq. (3.571), the footnote of the matrix element of $\hat{F}$ is $M_I{'}M_I$ and the foot note of the matrix element of $\hat{F}^+$ is taken as $\overline{M}_I{'}\overline{M}_I$, and making the following changes to the previous results

$$|X|^2 \to X^*_{\overline{M}_I{'}\overline{M}_I} X_{M_I{'}M_I}, \quad X = A,\, B,\, C,\, D,\, E,\, F \tag{3.621}$$

Then according to Eq. (3.570) we can write

$$\begin{aligned} I_{0,M_I{'}\overline{M}_I{'}M_I\overline{M}_I} = \frac{1}{3}\big(& A^*_{\overline{M}_I{'}\overline{M}_I} A_{M_I{'}M_I} + B^*_{\overline{M}_I{'}\overline{M}_I} B_{M_I{'}M_I} + C^*_{\overline{M}_I{'}\overline{M}_I} C_{M_I{'}M_I} \\ &+ D^*_{\overline{M}_I{'}\overline{M}_I} D_{M_I{'}M_I} + E^*_{\overline{M}_I{'}\overline{M}_I} E_{M_I{'}M_I} + F^*_{\overline{M}_I{'}\overline{M}_I} F_{M_I{'}M_I}\big) \end{aligned} \tag{3.622}$$

By means of the form of the polarization wave functions of the target and the residual nucleus introduced by Eqs. (2.315) and (2.320) and continuing to use the simplification summation symbol $\overline{\Sigma}$ for the polarized target and residual nucleus introduced by Eq. (2.327), the angular distribution of the emitted particles corresponding to the unpolarized incident particles can be expressed as

$$\overline{I}^0 \equiv I^0_{\alpha' n',\, \alpha n} = \frac{d\sigma^0_{\alpha' n', \alpha n}}{d\Omega} = \overline{\Sigma}\, I_{0,\, M_I{'}\overline{M}_I{'}M_I\overline{M}_I} \tag{3.623}$$

The component of the emitted particle polarization vector corresponding to the unpolarized incident particle is

$$\overline{P}^{0\, i'} \equiv P^{0\, i'}_{\alpha' n',\, \alpha n} = \frac{1}{\overline{I}^0}\overline{\Sigma}\, I_{0,\, M_I{'}\overline{M}_I{'}M_I\overline{M}_I} P^{i'}_{0,\, M_I{'}\overline{M}_I{'}M_I\overline{M}_I}, \quad i' = x',\, y',\, z' \tag{3.624}$$

When the normal variable in the Eqs. (3.574)–(3.576) is added with the footnote $M_I'M_I$, and the complex conjugate variable is added with the footnote $\overline{M}_I'\overline{M}_I$, and using Eq. (3.621), then every $P^{i'}_{0,\, M_I{'}\overline{M}_I{'}M_I\overline{M}_I}$ in Eq. (3.624) can be obtained.

When the incident particles, targets, and residual nuclei are all polarized, the components of the vector polarization analyzing power and the tensor polarization analyzing power of the nuclear reactions are

$$\overline{A}_i \equiv A_{i,\ \alpha' n',\ \alpha n} = \frac{1}{\overline{I}^0} \overline{\Sigma} I_{0,\ M_I'\overline{M}_I' M_I \overline{M}_I} A_{i,\ M_I'\overline{M}_I' M_I \overline{M}_I}, \quad i = \varepsilon \tag{3.625}$$

where $\varepsilon$ is the symbol of the label set of the components of the Cartesian coordinate introduced by Eq. (3.476). Here $A_{i, M_I'\overline{M}_I' M_I \overline{M}_I}$ have been given by Eqs. (3.580), (3.582), (3.584), (3.586), (3.588), (3.590), (3.592), and (3.594). Referring to $\hat{\rho}_{\text{in}}$ given by Eq. (3.214) and $\hat{\rho}_{\text{out}}$ given by Eq. (2.85), the corresponding differential cross section of the emitted particles can be written as

$$\begin{aligned} \overline{I} \equiv I_{\alpha' n',\ \alpha n} = \frac{\mathrm{d}\sigma_{\alpha' n',\ \alpha n}}{\mathrm{d}\Omega} = \operatorname{tr}\{\hat{\rho}_{\text{out}}\} = \overline{I}^0 \Big[ 1 + \frac{3}{2} (p_x \overline{A}_x + p_y \overline{A}_y + p_z \overline{A}_z) \\ + \frac{2}{3} (p_{xy} \overline{A}_{xy} + p_{xz} \overline{A}_{xz} + p_{yz} \overline{A}_{yz}) + \frac{1}{6} p_{xx-yy} \overline{A}_{xx-yy} + \frac{1}{2} p_{zz} \overline{A}_{zz} \Big] \end{aligned} \tag{3.626}$$

The corresponding polarization transfer coefficients are

$$\overline{K}_i^{i'} \equiv K_{i,\ \alpha' n',\ \alpha n}^{i'} = \frac{1}{\overline{I}^0} \overline{\Sigma} I_{0,\ M_I' M_I' M_I \overline{M}_I} K_{i,\ M_I'\overline{M}_I' M_I \overline{M}_I}^{i'}, \quad i = \varepsilon, \quad i' = x', y', z' \tag{3.627}$$

where $K_{i,\ M_I'\overline{M}_I' M_I \overline{M}_I}^{i'}$ have been given by Eqs. (3.597)–(3.620). Referring to Eqs. (3.214) and (2.85), the corresponding components of the polarization vector of the emitted particles can be written as follows:

$$\begin{aligned} \overline{P}^{i'} \equiv P_{\alpha' n',\ \alpha n}^{i'} = \frac{\operatorname{tr}\{\hat{\sigma}_{i'} \hat{\rho}_{\text{out}}\}}{\operatorname{tr}\{\hat{\rho}_{\text{out}}\}} = \frac{\overline{I}^0}{\overline{I}} \Big[ \overline{P}^{0\ i'} + \frac{3}{2} \left( p_x \overline{K}_x^{i'} + p_y \overline{K}_y^{i'} + p_z \overline{K}_z^{i'} \right) \\ + \frac{2}{3} \left( p_{xy} \overline{K}_{xy}^{i'} + p_{xz} \overline{K}_{xz}^{i'} + p_{yz} \overline{K}_{yz}^{i'} \right) + \frac{1}{6} p_{xx-yy} \overline{K}_{xx-yy}^{i'} + \frac{1}{2} p_{zz} \overline{K}_{zz}^{i'} \Big], \ i' = x', y', z' \end{aligned} \tag{3.628}$$

In the case where both the target and the residual nucleus are unpolarized, and using Eqs. (2.334) and (2.335) as well as the simplification summation symbols $\widetilde{\Sigma}$ given by Eq. (2.278), the Eqs. (3.622)–(3.625), and (3.627) can be simplified, respectively, as follows:

$$I_{0,\ M_I' M_I} = \frac{1}{3} \left( |A_{M_I' M_I}|^2 + |B_{M_I' M_I}|^2 + |C_{M_I' M_I}|^2 + |D_{M_I' M_I}|^2 + |E_{M_I' M_I}|^2 + |F_{M_I' M_I}|^2 \right) \tag{3.629}$$

$$\overline{I}^0 = \widetilde{\Sigma} I_{0,\ M_I' M_I} \tag{3.630}$$

$$\overline{P}^{0 i'} = \frac{1}{\overline{I}^0} \widetilde{\Sigma} I_{0,\ M_I' M_I} P_{0,\ M_I' M_I}^{i'}, \quad i' = x', y', z' \tag{3.631}$$

$$\overline{A}_i = \frac{1}{\overline{I}^0} \widetilde{\Sigma} I_{0,\ M_I{}' M_I} A_{i,\ M_I{}' M_I}, \quad i = \varepsilon \tag{3.632}$$

$$\overline{K}_i^{i'} = \frac{1}{\overline{I}^0} \widetilde{\Sigma} I_{0,\ M_I{}' M_I} K_{i,\ M_I{}' M_I}^{i'}, \quad i = \varepsilon; \quad i' = x', y', z' \tag{3.633}$$

Next, we will analyze Eq. (3.552) which belongs to *S*-*L* coupling scheme. Based on the properties of the C-G coefficient, there are the following relations:

$$\begin{aligned}
&C_{1\mu\ IM_I}^{S\ \mu+M_I} = (-1)^{I+1-S} C_{1\ -\mu\ I\ -M_I}^{S\ -\mu-M_I}, \quad C_{\frac{1}{2}\mu'\ I'M_I'}^{S'\ \mu'+M_I'} = (-1)^{I'+\frac{1}{2}-S'} C_{\frac{1}{2}\ -\mu'\ I'\ -M_I'}^{S'\ -\mu'-M_I'}, \\
&C_{\ell 0\ SM}^{J\ M} = (-1)^{\ell+S-J} C_{\ell 0\ S\ -M}^{J\ -M}, \quad C_{\ell' m_{\ell'}\ S' M_{S'}}^{J\ M} = (-1)^{\ell'+S'-J} C_{\ell'\ -m_{\ell'}\ S'\ -M_{S'}}^{J\ -M}
\end{aligned} \tag{3.634}$$

Now we only need to study the second item of Eq. (3.553) with $i = 1$ and $i' = \frac{1}{2}$. Using Eqs. (2.288) and (3.634), for the part which is not associated with angle $\varphi$, the following phase factor comes out when all spin magnetic quantum numbers change signs:

$$(-1)^{\ell+S-J+\ell'+S'-J+I+1-S+I'+\frac{1}{2}-S'+\mu+M_I-\mu'-M_I'} = (-1)^{2J+\ell+I+\ell'+I'+\frac{3}{2}+\mu+M_I-\mu'-M_I'}$$

The symbol $\Pi$ introduced by Eq. (3.488) can also be used here, so $(-1)^{\ell+\ell'} = (-1)^{\Pi}$. For $\vec{1} + \mathrm{A} \to \overrightarrow{\frac{1}{2}} + \mathrm{B}$ reaction, *I* and *I′* must be that one is an integer and another is a semi-odd number. When *I* is an integer and *I′* is a semi-odd number, *J* is an integer; when *I* is a semi-odd number and *I′* is an integer, *J* is a semi-odd number. Let

$$\Lambda = \begin{cases} 0 & \text{when } I \text{ is an integer and } I' \text{ is a semi-odd number} \\ 1 & \text{when } I \text{ is a semi-odd number and } I' \text{ is an integer} \end{cases} \tag{3.635}$$

thus $(-1)^{2J}$ can be replaced by $(-1)^{\Lambda}$, and we can get

$$f_{\alpha' n' \mu' M_I', \alpha n \mu M_I}(\theta) = (-1)^{\Pi+\Lambda+I+I'+\frac{3}{2}+\mu+M_I-\mu'-M_I'} f_{\alpha' n'\ -\mu'\ -M_I', \alpha n\ -\mu\ -M_I}(\theta) \tag{3.636}$$

The following relations can be seen from Eqs. (3.553)–(3.559):

$$\begin{aligned}
&A_{M_I' M_I}(\theta) = f_{\alpha' n' \frac{1}{2} M_I', \alpha n 1 M_I}(\theta), && B_{M_I' M_I}(\theta) = f_{\alpha' n'\ -\frac{1}{2} M_I', \alpha n\ -1 M_I}(\theta), \\
&C_{M_I' M_I}(\theta) = \mathrm{i} f_{\alpha' n' \frac{1}{2} M_I', \alpha n 0 M_I}(\theta), && D_{M_I' M_I}(\theta) = -\mathrm{i} f_{\alpha' n'\ -\frac{1}{2} M_I', \alpha n 0 M_I}(\theta), \\
&E_{M_I' M_I}(\theta) = -f_{\alpha' n' \frac{1}{2} M_I', \alpha n\ -1 M_I}(\theta), && F_{M_I' M_I}(\theta) = -f_{\alpha' n'\ -\frac{1}{2} M_I', \alpha n 1 M_I}(\theta)
\end{aligned} \tag{3.637}$$

And the following relations can be obtained according to Eqs. (3.636) and (3.637):

$$B_{M_I'M_I}(\theta) = (-1)^{\Pi+\Lambda+I+I'+M_I-M_I'+1} A_{-M_I'\ -M_I}(\theta) \tag{3.638}$$

$$D_{M_I'M_I}(\theta) = (-1)^{\Pi+\Lambda+I+I'+M_I-M_I'+1} C_{-M_I'\ -M_I}(\theta) \tag{3.639}$$

$$F_{M_I'M_I}(\theta) = (-1)^{\Pi+\Lambda+I+I'+M_I-M_I'+1} E_{-M_I'\ -M_I}(\theta) \tag{3.640}$$

Here we continue to use the simplification summation symbol $\tilde{\Sigma}$ given by Eq. (2.278). Note that it must be $(-1)^{2X} = 1$ when $X$ is an integer, and also note that the sum of $M_I$ and the sum of $-M_I$ are all from $-I$ to $I$, and the sum of $M_I'$ and the sum of $-M_I'$ are all from $-I'$ to $I'$, so using the method to prove Eq. (2.297), from Eqs. (2.423)–(2.425) the following results can be proved:

$$\begin{aligned}
&\tilde{\Sigma}\,\mathrm{Im}(AC^* + DB^*) = 0,\ \tilde{\Sigma}\,\mathrm{Im}(AE^* + FB^*) = 0,\ \tilde{\Sigma}\,\mathrm{Im}(CE^* + FD^*) = 0,\\
&\tilde{\Sigma}\,\mathrm{Re}(AC^* + DB^*) = 2\tilde{\Sigma}\,\mathrm{Re}(AC^*),\ \tilde{\Sigma}\,\mathrm{Re}(AE^* + FB^*) = 2\tilde{\Sigma}\,\mathrm{Re}(AE^*),\\
&\tilde{\Sigma}\,\mathrm{Re}(CE^* + FD^*) = 2\tilde{\Sigma}\,\mathrm{Re}(CE^*),\ \tilde{\Sigma}\,\mathrm{Im}(AC^* - DB^*) = 2\tilde{\Sigma}\,\mathrm{Im}(AC^*),\\
&\tilde{\Sigma}\,\mathrm{Im}(AE^* - FB^*) = 2\tilde{\Sigma}\,\mathrm{Im}(AE^*),\ \tilde{\Sigma}\,\mathrm{Im}(CE^* - FD^*) = 2\tilde{\Sigma}\,\mathrm{Im}(CE^*),\\
&\tilde{\Sigma}\,\mathrm{Re}(AC^* - DB^*) = 0,\ \tilde{\Sigma}\,\mathrm{Re}(AE^* - FB^*) = 0,\ \tilde{\Sigma}\,\mathrm{Re}(CE^* - FD^*) = 0,\\
&\tilde{\Sigma}\,|B|^2 = \tilde{\Sigma}\,|A|^2,\quad \tilde{\Sigma}\,|D|^2 = \tilde{\Sigma}\,|C|^2,\quad \tilde{\Sigma}\,|F|^2 = \tilde{\Sigma}\,|E|^2
\end{aligned} \tag{3.641}$$

And by using the method to prove Eq. (2.298), we can prove

$$\tilde{\Sigma}BA^*,\ \tilde{\Sigma}DC^*,\ \tilde{\Sigma}FE^*,\ \tilde{\Sigma}(AC^* + DB^*),\ \tilde{\Sigma}(AE^* + FB^*),\ \tilde{\Sigma}(CE^* + FD^*) \tag{3.642}$$

such that their corresponding complex conjugate terms are all real numbers. There are other items that also satisfy the above two expressions.

Defining

$$\tilde{I}_{0,M_I'M_I} = \frac{2}{3}\left(\left|A_{M_I'M_I}\right|^2 + \left|C_{M_I'M_I}\right|^2 + \left|E_{M_I'M_I}\right|^2\right) \tag{3.643}$$

then from Eqs. (3.629), (3.630), and (3.641), we can prove

$$\bar{I}^0 = \tilde{\Sigma}\,\tilde{I}_{0,M_I'M_I} \tag{3.644}$$

From Eq. (3.631), and according to Eqs. (3.574)–(3.576), we can obtain the following results using Eqs. (3.638)–(3.642):

$$\bar{P}^{0\,x'} = -\frac{2}{3\bar{I}_0}\tilde{\Sigma}\,\mathrm{Re}\,(2FA^* + DC^*) \tag{3.645}$$

$$\bar{P}^{0\,y'} = \bar{P}^{0\,z'} = 0 \tag{3.646}$$

And from Eq. (3.632) we can obtain the following results according to Eqs. (3.580), (3.582), (3.584), (3.586), (3.588), (3.590), (3.592), and (3.594):

$$\overline{A}_x=\overline{A}_z=\overline{A}_{xy}=\overline{A}_{yz}=0 \tag{3.647}$$

$$\overline{A}_y=\frac{2\sqrt{2}}{3\overline{I}_0}\widetilde{\Sigma}\,\mathrm{Re}\,(AC^*+CE^*) \tag{3.648}$$

$$\overline{A}_{xz}=-\frac{\sqrt{2}}{\overline{I}_0}\widetilde{\Sigma}\mathrm{Im}(AC^*-CE^*) \tag{3.649}$$

$$\overline{A}_{xx-yy}=-\frac{4}{\overline{I}_0}\widetilde{\Sigma}\,\mathrm{Re}\,(AE^*) \tag{3.650}$$

$$\overline{A}_{zz}=\frac{2}{3\overline{I}_0}\widetilde{\Sigma}\left(|A|^2+|E|^2-2|C|^2\right) \tag{3.651}$$

Further from Eq. (3.633) we can obtain the following results according to Eqs. (3.597)–(3.620):

$$\overline{K}_i^{i'}=0,\qquad (i=\varepsilon_1 \text{ and } i'=y',\,z') \text{ or } (i=\varepsilon_2 \text{ and } i'=x') \tag{3.652}$$

such that the definition of the subscripts $\varepsilon_1$ and $\varepsilon_2$ are given by Eq. (3.320). Given this, we can obtain

$$\overline{K}_x^{y'}=\frac{2\sqrt{2}}{3\overline{I}_0}\widetilde{\Sigma}\,\mathrm{Re}\,(AD^*-FC^*) \tag{3.653}$$

$$\overline{K}_x^{z'}=-\frac{2\sqrt{2}}{3\overline{I}_0}\widetilde{\Sigma}\mathrm{Im}(AC^*+CE^*) \tag{3.654}$$

$$\overline{K}_y^{x'}=-\frac{2\sqrt{2}}{3\overline{I}_0}\widetilde{\Sigma}\,\mathrm{Re}\,(AD^*+FC^*) \tag{3.655}$$

$$\overline{K}_z^{y'}=-\frac{4}{3\overline{I}_0}\widetilde{\Sigma}\mathrm{Im}(FA^*) \tag{3.656}$$

$$\overline{K}_z^{z'}=\frac{2}{3\overline{I}_0}\widetilde{\Sigma}\left(|A|^2-|E|^2\right) \tag{3.657}$$

$$\overline{K}_{xy}^{y'}=\frac{1}{\overline{I}_0}\widetilde{\Sigma}\,\mathrm{Re}\,(AB^*-FE^*) \tag{3.658}$$

$$\overline{K}_{xy}^{z'}=-\frac{2}{\overline{I}_0}\widetilde{\Sigma}\mathrm{Im}(AE^*) \tag{3.659}$$

$$\overline{K}_{xz}^{x'} = \frac{\sqrt{2}}{\overline{I}_0} \widetilde{\Sigma} \mathrm{Im}(AD^* + FC^*) \tag{3.660}$$

$$\overline{K}_{yz}^{y'} = \frac{\sqrt{2}}{\overline{I}_0} \widetilde{\Sigma} \mathrm{Im}(AD^* - FC^*) \tag{3.661}$$

$$\overline{K}_{yz}^{z'} = \frac{\sqrt{2}}{\overline{I}_0} \widetilde{\Sigma} \mathrm{Re}\,(AC^* - CE^*) \tag{3.662}$$

$$\overline{K}_{xx-yy}^{x'} = \frac{2}{\overline{I}_0} \widetilde{\Sigma} \mathrm{Re}\,(AB^* + FE^*) \tag{3.663}$$

$$\overline{K}_{zz}^{x'} = -\frac{4}{3\overline{I}_0} \widetilde{\Sigma} \mathrm{Re}\,(FA^* - DC^*) \tag{3.664}$$

From the results obtained above, it can be seen that

$$\overline{P}^{0\,i'} \begin{cases} \neq 0 & i' = x' \\ = 0 & i' = y',\, z' \end{cases} \tag{3.665}$$

$$\overline{A}_i \begin{cases} \neq 0 & i = \varepsilon_1 \\ = 0 & i = \varepsilon_2 \end{cases} \tag{3.666}$$

$$\overline{K}_i^{i'} \begin{cases} \neq 0 & (i = \varepsilon_1 \text{ and } i' = x') \text{ or } (i = \varepsilon_2 \text{ and } i' = y',\ z') \\ = 0 & (i = \varepsilon_1 \text{ and } i' = y',\ z') \text{ or } (i = \varepsilon_2 \text{ and } i' = x') \end{cases} \tag{3.667}$$

And we can obtain the following results from Eqs. (3.626) and (3.628):

$$\overline{I} = \overline{I}^0 \left[ 1 + \frac{3}{2} p_y \overline{A}_y + \frac{2}{3} p_{xz} \overline{A}_{xz} + \frac{1}{6} p_{xx-yy} \overline{A}_{xx-yy} + \frac{1}{2} p_{zz} \overline{A}_{zz} \right] \tag{3.668}$$

$$\overline{P}^{i'} = \frac{\overline{I}^0}{\overline{I}} \left[ \overline{P}^{0i'} + \frac{3}{2} p_y \overline{K}_y^{i'} + \frac{2}{3} p_{xz} \overline{K}_{xz}^{i'} + \frac{1}{6} p_{xx-yy} \overline{K}_{xx-yy}^{i'} + \frac{1}{2} p_{zz} \overline{K}_{zz}^{i'} \right], \quad i' = x' \tag{3.669}$$

$$\overline{P}^{i'} = \frac{\overline{I}^0}{\overline{I}} \left[ \frac{3}{2} \left( p_x \overline{K}_x^{i'} + p_z \overline{K}_z^{i'} \right) + \frac{2}{3} \left( p_{xy} \overline{K}_{xy}^{i'} + p_{yz} \overline{K}_{yz}^{i'} \right) \right], \quad i' = y', z' \tag{3.670}$$

## 3.10 Polarization Theory for $\vec{\frac{1}{2}} + \mathrm{A} \to \vec{1} + \mathrm{B}$ Reactions

This section studies the polarization theory for $\vec{\frac{1}{2}} + \mathrm{A} \to \vec{1} + \mathrm{B}$ reactions, where target and residual nucleus spins are not equal to 0, but are both unpolarized.

In the case of a spin $\frac{1}{2}$ incident particle and a spin 1 outgoing particle, and referring to Eqs. (2.251) and (2.252), we introduce the following symbols for *j*-*j* coupling and *S*-*L* coupling, respectively:

$$f_{\mu'\mu}^{ll'J}(m_l') = \hat{l}\hat{l}' \mathrm{e}^{\mathrm{i}(\sigma_l+\sigma_{l'})} \sqrt{\frac{(l'-m_l')!}{(l'+m_l')!}} \sum_{jj'} C_{l0\,\frac{1}{2}\mu}^{j\,\mu} C_{j\mu\,IM_I}^{J\,M} C_{l'm_l'\,1\mu'}^{j'm_j'} C_{j'm_j'\,I'M_I'}^{J\,M} \left(\delta_{\alpha' n' l' j', \alpha n l j} - S_{\alpha' n' l' j',\,\alpha n l j}^{J}\right) \tag{3.671}$$

$$f_{\mu'\mu}^{ll'J}(m_l') = \hat{l}\hat{l}' \mathrm{e}^{\mathrm{i}(\sigma_l+\sigma_{l'})} \sqrt{\frac{(l'-m_l')!}{(l'+m_l')!}} \sum_{SS'} C_{l0\,SM}^{J\,M} C_{\frac{1}{2}\mu\,IM_I}^{S\,M} C_{l'm_l'\,S'M'_S}^{J\,M} C_{1\mu'\,I'M_I'}^{S'\,M'_S} \left(\delta_{\alpha' n' l' S',\,\alpha n l S} - S_{\alpha' n' l' S',\,\alpha n l S}^{J}\right) \tag{3.672}$$

Given this, the reaction amplitude can be uniformly written in the form of Eq. (3.553) under two angular momentum coupling schemes according to Eqs. (2.251)–(2.254).

Because there is no elastic scattering channel for $\vec{\frac{1}{2}} + \mathrm{A} \to \vec{1} + \mathrm{B}$ reactions, the first term representing Coulomb scattering in Eq. (3.553) does not appear. Let

$$A_{M_I'M_I}(\theta) = \frac{\mathrm{i}}{2k}(-1)^{M_I - M_I' - \frac{1}{2}} \sum_{ll'J} f_{1\,\frac{1}{2}}^{ll'J}\left(M_I - M_I' - \frac{1}{2}\right) \mathrm{P}_{l'}^{M_I - M_I' - \frac{1}{2}}(\cos\theta) \tag{3.673}$$

$$B_{M_I'M_I}(\theta) = \frac{\mathrm{i}}{2k}(-1)^{M_I - M_I' + \frac{1}{2}} \sum_{ll'J} f_{-1\,-\frac{1}{2}}^{ll'J}\left(M_I - M_I' + \frac{1}{2}\right) \mathrm{P}_{l'}^{M_I - M_I' + \frac{1}{2}}(\cos\theta) \tag{3.674}$$

$$C_{M_I'M_I}(\theta) = -\frac{1}{2k}(-1)^{M_I - M_I' + \frac{1}{2}} \sum_{ll'J} f_{0\,\frac{1}{2}}^{ll'J}\left(M_I - M_I' + \frac{1}{2}\right) \mathrm{P}_{l'}^{M_I - M_I' + \frac{1}{2}}(\cos\theta) \tag{3.675}$$

$$D_{M_I'M_I}(\theta) = \frac{1}{2k}(-1)^{M_I - M_I' - \frac{1}{2}} \sum_{ll'J} f_{0\,-\frac{1}{2}}^{ll'J}\left(M_I - M_I' - \frac{1}{2}\right) \mathrm{P}_{l'}^{M_I - M_I' - \frac{1}{2}}(\cos\theta) \tag{3.676}$$

$$E_{M_I'M_I}(\theta) = -\frac{\mathrm{i}}{2k}(-1)^{M_I - M_I' + \frac{3}{2}} \sum_{ll'J} f_{-1\,\frac{1}{2}}^{ll'J}\left(M_I - M_I' + \frac{3}{2}\right) \mathrm{P}_{l'}^{M_I - M_I' + \frac{3}{2}}(\cos\theta) \tag{3.677}$$

$$F_{M_I'M_I}(\theta) = -\frac{\mathrm{i}}{2k}(-1)^{M_I - M_I' - \frac{3}{2}} \sum_{ll'J} f_{1\ -\frac{1}{2}}^{ll'J}\left(M_I - M_I' - \frac{3}{2}\right) \mathrm{P}_{l'}^{M_I - M_I' - \frac{3}{2}}(\cos\theta) \tag{3.678}$$

Using the simplified symbols introduced in Eq. (3.560), we can get

$$F_{1\ \frac{1}{2}}(\theta, \varphi) = A_{M_I'M_I}(\theta) \mathrm{e}^{\mathrm{i}\left(M_I - M_I' - \frac{1}{2}\right)\varphi} \tag{3.679}$$

$$F_{1\ -\frac{1}{2}}(\theta, \varphi) = -F_{M_I'M_I}(\theta) \mathrm{e}^{\mathrm{i}\left(M_I - M_I' - \frac{3}{2}\right)\varphi} \tag{3.680}$$

$$F_{0\ \frac{1}{2}}(\theta, \varphi) = -\mathrm{i}C_{M_I'M_I}(\theta) \mathrm{e}^{\mathrm{i}\left(M_I - M_I' + \frac{1}{2}\right)\varphi} \tag{3.681}$$

$$F_{0\ -\frac{1}{2}}(\theta, \varphi) = \mathrm{i}D_{M_I'M_I}(\theta) \mathrm{e}^{\mathrm{i}\left(M_I - M_I' - \frac{1}{2}\right)\varphi} \tag{3.682}$$

$$F_{-1\ \frac{1}{2}}(\theta, \varphi) = -E_{M_I'M_I}(\theta) \mathrm{e}^{\mathrm{i}\left(M_I - M_I' + \frac{3}{2}\right)\varphi} \tag{3.683}$$

$$F_{-1\ -\frac{1}{2}}(\theta, \varphi) = B_{M_I'M_I}(\theta) \mathrm{e}^{\mathrm{i}\left(M_I - M_I' + \frac{1}{2}\right)\varphi} \tag{3.684}$$

When the footnote $M_I'M_I$ and argument $(\theta)$ of $A$, $B$, $C$, $D$, $E$, and $F$ are omitted, the reaction amplitude $\hat{F}$, in which foot note $M_I'M_I$ and arguments $(\theta, \varphi)$ are emitted, too, can be written as

$$\hat{F} = \begin{pmatrix} A\mathrm{e}^{-\frac{\mathrm{i}}{2}\varphi} & -F\mathrm{e}^{-\frac{3\mathrm{i}}{2}\varphi} \\ -\mathrm{i}C\mathrm{e}^{\frac{\mathrm{i}}{2}\varphi} & \mathrm{i}D\mathrm{e}^{-\frac{\mathrm{i}}{2}\varphi} \\ -E\mathrm{e}^{\frac{3\mathrm{i}}{2}\varphi} & B\mathrm{e}^{\frac{\mathrm{i}}{2}\varphi} \end{pmatrix} \mathrm{e}^{\mathrm{i}\left(M_I - M_I'\right)\varphi} \tag{3.685}$$

In the case of that both the target and the residual nucleus are unpolarized, from Eqs. (2.334) and (2.335), we can see that the magnetic quantum number footnotes of the matrix element of $\hat{F}$ and $\hat{F}^+$ are always the same in the polarization theory formulas, so from Eq. (3.685) we can obtain

$$\hat{F}^+ = \begin{pmatrix} A^*\mathrm{e}^{\frac{\mathrm{i}}{2}\varphi} & \mathrm{i}C^*\mathrm{e}^{-\frac{\mathrm{i}}{2}\varphi} & -E^*\mathrm{e}^{-\frac{3\mathrm{i}}{2}\varphi} \\ -F^*\mathrm{e}^{\frac{3\mathrm{i}}{2}\varphi} & -iD^*\mathrm{e}^{\frac{\mathrm{i}}{2}\varphi} & B\mathrm{e}^{-\frac{\mathrm{i}}{2}\varphi} \end{pmatrix} \mathrm{e}^{-\mathrm{i}(M_I - M_I')\varphi} \tag{3.686}$$

$$\hat{F}\hat{F}^+ = \begin{pmatrix} |A|^2 + |F|^2 & \mathrm{i}(AC^* + FD^*)\mathrm{e}^{-\mathrm{i}\varphi} & -(AE^* + FB^*)\mathrm{e}^{-2\mathrm{i}\varphi} \\ -\mathrm{i}(CA^* + DF^*)\,\mathrm{e}^{\mathrm{i}\varphi} & |C|^2 + |D|^2 & \mathrm{i}(CE^* + DB^*)\,\mathrm{e}^{-\mathrm{i}\varphi} \\ -(EA^* + BF^*)\,\mathrm{e}^{2\mathrm{i}\varphi} & -\mathrm{i}(EC^* + BD^*)\,\mathrm{e}^{\mathrm{i}\varphi} & |E|^2 + |B|^2 \end{pmatrix} \tag{3.687}$$

For the definite $M_I'M_I$, from Eq. (3.687) we can obtain

$$I_0 = \frac{1}{2}\text{tr}\left\{\hat{F}\hat{F}^+\right\} = \frac{1}{2}\left(|A|^2 + |B|^2 + |C|^2 + |D|^2 + |E|^2 + |F|^2\right) \tag{3.688}$$

For the sake of simplicity, select the $y$-axis along the direction $\vec{n}$, such that $\varphi = 0$. Given this, the following expressions can be obtained from Eqs. (3.685)–(3.687):

$$\hat{F} = \begin{pmatrix} A & -F \\ -\text{i}C & \text{i}D \\ -E & B \end{pmatrix}, \qquad \hat{F}^+ = \begin{pmatrix} A^* & \text{i}C^* & -E^* \\ -F^* & -\text{i}D^* & B^* \end{pmatrix} \tag{3.689}$$

$$\hat{F}\hat{F}^+ = \begin{pmatrix} |A|^2 + |F|^2 & \text{i}(AC^* + FD^*) & -(AE^* + FB^*) \\ -\text{i}(CA^* + DF^*) & |C|^2 + |D|^2 & \text{i}(CE^* + DB^*) \\ -(EA^* + BF^*) & -\text{i}(EC^* + BD^*) & |E|^2 + |B|^2 \end{pmatrix} \tag{3.690}$$

For the definite $M_I'M_I$, we define the following as the component of the outgoing particle polarization vector corresponding to the unpolarized incident particle,

$$P_0^{i'} = \frac{\text{tr}\left\{\hat{S}_{i'}\hat{F}\hat{F}^+\right\}}{\text{tr}\left\{\hat{F}\hat{F}^+\right\}}, \qquad i' = x', y', z' \tag{3.691}$$

and we define

$$P_0^{i'} = \frac{\text{tr}\left\{\hat{Q}_{i'}\hat{F}\hat{F}^+\right\}}{\text{tr}\left\{\hat{F}\hat{F}^+\right\}}, \qquad i' = x'y', x'z', y'z', x'x' - y'y', z'z' \tag{3.692}$$

as the component of the outgoing particle polarization tensor corresponding to the unpolarized incident particle. Utilizing Eqs. (3.33), (3.36), (3.688), and (3.690), the following results can be obtained from Eqs. (3.691) and (3.692):

$$\begin{aligned} P_0^{x'} &= \frac{1}{2\sqrt{2}I_0}[-\text{i}(CA^* + DF^*) + \text{i}(AC^* + FD^*) \\ &\quad - \text{i}(EC^* + BD^*) + \text{i}(CE^* + DB^*)] \\ &= \frac{\sqrt{2}}{2I_0}\text{Im}(CA^* + DF^* + EC^* + BD^*) \end{aligned} \tag{3.693}$$

$$
\begin{aligned}
P_0^{y'} &= \frac{\mathrm{i}}{2\sqrt{2}I_0}[\mathrm{i}(CA^* + DF^*) + \mathrm{i}(AC^* + FD^*) \\
&\quad + \mathrm{i}(EC^* + BD^*) + \mathrm{i}(CE^* + DB^*)] \\
&= -\frac{\sqrt{2}}{2I_0}\,\mathrm{Re}\,(CA^* + DF^* + EC^* + BD^*)
\end{aligned}
\tag{3.694}
$$

$$
P_0^{z'} = \frac{1}{2I_0}\left[\left(|A|^2 + |F|^2\right) - \left(|E|^2 + |B|^2\right)\right]
\tag{3.695}
$$

$$
\begin{aligned}
P_0^{x'y'} &= \frac{3\mathrm{i}}{4I_0}[(EA^* + BF^*) - (AE^* + FB^*)] \\
&= -\frac{3}{2I_0}\mathrm{Im}(EA^* + BF^*)
\end{aligned}
\tag{3.696}
$$

$$
\begin{aligned}
P_0^{x'z'} &= \frac{3}{4\sqrt{2}I_0}[-\mathrm{i}(CA^* + DF^*) + \mathrm{i}(AC^* + FD^*) \\
&\quad + \mathrm{i}(EC^* + BD^*) - \mathrm{i}(CE^* + DB^*)] \\
&= \frac{3}{2\sqrt{2}I_0}\mathrm{Im}(CA^* + DF^* - EC^* - BD^*)
\end{aligned}
\tag{3.697}
$$

$$
\begin{aligned}
P_0^{y'z'} &= \frac{3i}{4\sqrt{2}I_0}[\mathrm{i}(CA^* + DF^*) + \mathrm{i}(AC^* + FD^*) \\
&\quad - \mathrm{i}(EC^* + BD^*) - \mathrm{i}(CE^* + DB^*)] \\
&= -\frac{3}{2\sqrt{2}I_0}\,\mathrm{Re}\,(CA^* + DF^* - EC^* - BD^*)
\end{aligned}
\tag{3.698}
$$

$$
\begin{aligned}
P_0^{x'x'-y'y'} &= \frac{3}{2I_0}[-(EA^* + BF^*) - (AE^* + FB^*)] \\
&= -\frac{3}{I_0}\,\mathrm{Re}\,(EA^* + BF^*)
\end{aligned}
\tag{3.699}
$$

$$
P_0^{z'z'} = \frac{1}{2I_0}\left(|A|^2 + |B|^2 + |E|^2 + |F|^2 - 2|C|^2 - 2|D|^2\right)
\tag{3.700}
$$

For the definite $M_I'M_I$, we define the polarization vector analyzing power as

$$
A_i = \frac{\mathrm{tr}\left\{\hat{F}\hat{\sigma}_i\hat{F}^+\right\}}{\mathrm{tr}\left\{\hat{F}\hat{F}^+\right\}}, \qquad i = x, y, z
\tag{3.701}
$$

The following expressions can be obtained from Eq. (3.701) and utilizing Eqs. (2.6), (3.688), and Eq. (3.689)

$$\hat{\sigma}_x\hat{F}^+ = \begin{pmatrix} -F^* & -\mathrm{i}D^* & B^* \\ A^* & \mathrm{i}C^* & -E^* \end{pmatrix},$$

$$\hat{F}\hat{\sigma}_x\hat{F}^+ = \begin{pmatrix} -(AF^*+FA^*) & -\mathrm{i}(AD^*+FC^*) & AB^*+FE^* \\ \mathrm{i}(CF^*+DA^*) & -(CD^*+DC^*) & -\mathrm{i}(CB^*+DE^*) \\ EF^*+BA^* & \mathrm{i}(ED^*+BC^*) & -(EB^*+BE^*) \end{pmatrix} \tag{3.702}$$

$$A_x = -\frac{1}{I_0}\,\mathrm{Re}\,(AF^*+CD^*+EB^*) \tag{3.703}$$

$$\hat{\sigma}_y\hat{F}^+ = \begin{pmatrix} \mathrm{i}F^* - D^* - \mathrm{i}B^* \\ \mathrm{i}A^* - C^* - \mathrm{i}E^* \end{pmatrix},$$

$$\hat{F}\hat{\sigma}_y\hat{F}^+ = \begin{pmatrix} \mathrm{i}(AF^*-FA^*) & -AD^*+FC^* & -\mathrm{i}(AB^*-FE^*) \\ CF^*-DA^* & \mathrm{i}(CD^*-DC^*) & -CB^*+DE^* \\ -\mathrm{i}(EF^*-BA^*) & ED^*-BC^* & \mathrm{i}(EB^*-BE^*) \end{pmatrix} \tag{3.704}$$

$$A_y = -\frac{1}{I_0}\mathrm{Im}(AF^*+CD^*+EB^*) \tag{3.705}$$

$$\hat{\sigma}_z\hat{F}^+ = \begin{pmatrix} A^* & \mathrm{i}C^* & -E^* \\ F^* & \mathrm{i}D^* & -B^* \end{pmatrix},$$

$$\hat{F}\hat{\sigma}_z\hat{F}^+ = \begin{pmatrix} |A|^2-|F|^2 & \mathrm{i}(AC^*-FD^*) & -AE^*+FB^* \\ -\mathrm{i}(CA^*-DF^*) & |C|^2-|D|^2 & \mathrm{i}(CE^*-DB^*) \\ -EA^*+BF^* & -\mathrm{i}(EC^*-BD^*) & |E|^2-|B|^2 \end{pmatrix} \tag{3.706}$$

$$A_z = \frac{1}{2I_0}\left[\,|A|^2+|C|^2+|E|^2-|F|^2-|D|^2-|B|^2\right] \tag{3.707}$$

Next we define the polarization transfer coefficient from the incident particle polarization vector to the outgoing particle polarization vector as

$$K_i^{i'} = \frac{\mathrm{tr}\left\{\hat{S}_{i'}\hat{F}\hat{\sigma}_i\hat{F}^+\right\}}{\mathrm{tr}\left\{\hat{F}\hat{F}^+\right\}}, \qquad i=x,y,z; \qquad i'=x',y',z' \tag{3.708}$$

and define.

$$K_i^{i'} = \frac{\mathrm{tr}\left\{\hat{Q}_{i'}\hat{F}\hat{\sigma}_i\hat{F}^+\right\}}{\mathrm{tr}\left\{\hat{F}\hat{F}^+\right\}}, \quad i=x,y,z; \quad i'=x'y',x'z',y'z',x'x'-y'y',z'z' \tag{3.709}$$

as the polarization transfer coefficient from the incident particle polarization vector to the outgoing particle polarization tensor. Using Eqs. (3.33), (3.36), and (3.688), as well as the matrices $\hat{F}\hat{\sigma}_i\hat{F}^+ (i = x, y, z)$ obtained previously in this section, the following results can be obtained by Eqs. (3.708) and (3.709):

$$K_x^{x'} = -\frac{1}{\sqrt{2}I_0}\mathrm{Im}(DA^* + BC^* + CF^* + ED^*) \tag{3.710}$$

$$K_y^{x'} = -\frac{1}{\sqrt{2}I_0}\mathrm{Re}\,(DA^* + BC^* - CF^* - ED^*) \tag{3.711}$$

$$K_z^{x'} = \frac{1}{\sqrt{2}I_0}\mathrm{Im}(CA^* - DF^* + EC^* - BD^*) \tag{3.712}$$

$$K_x^{y'} = \frac{1}{\sqrt{2}I_0}\mathrm{Re}\,(DA^* + BC^* + CF^* + ED^*) \tag{3.713}$$

$$K_y^{y'} = -\frac{1}{\sqrt{2}I_0}\mathrm{Im}(DA^* + BC^* - CF^* - ED^*) \tag{3.714}$$

$$K_z^{y'} = -\frac{1}{\sqrt{2}I_0}\mathrm{Re}\,(CA^* - DF^* + EC^* - BD^*) \tag{3.715}$$

$$K_x^{z'} = -\frac{1}{I_0}\mathrm{Re}\,(AF^* - EB^*) \tag{3.716}$$

$$K_y^{z'} = -\frac{1}{I_0}\mathrm{Im}(AF^* - EB^*) \tag{3.717}$$

$$K_z^{z'} = \frac{1}{2I_0}\left(|A|^2 - |F|^2 - |E|^2 + |B|^2\right) \tag{3.718}$$

$$K_x^{x'y'} = \frac{3}{2I_0}\mathrm{Im}(BA^* + EF^*) \tag{3.719}$$

$$K_y^{x'y'} = \frac{3}{2I_0}\mathrm{Re}\,(BA^* - EF^*) \tag{3.720}$$

$$K_z^{x'y'} = -\frac{3}{2I_0}\mathrm{Im}(EA^* - BF^*) \tag{3.721}$$

$$K_x^{x'z'} = -\frac{3}{2\sqrt{2}I_0}\mathrm{Im}(DA^* - BC^* + CF^* - ED^*) \tag{3.722}$$

$$K_y^{x'z'} = -\frac{3}{2\sqrt{2}I_0}\mathrm{Re}\,(DA^* - BC^* - CF^* + ED^*) \tag{3.723}$$

$$K_z^{x'z'} = \frac{3}{2\sqrt{2}I_0}\mathrm{Im}(CA^* - DF^* - EC^* + BD^*) \tag{3.724}$$

$$K_x^{y'z'} = \frac{3}{2\sqrt{2}I_0} \mathrm{Re}\,(DA^* - BC^* + CF^* - ED^*) \tag{3.725}$$

$$K_y^{y'z'} = -\frac{3}{2\sqrt{2}I_0} \mathrm{Im}(DA^* - BC^* - CF^* + ED^*) \tag{3.726}$$

$$K_z^{y'z'} = -\frac{3}{2\sqrt{2}I_0} \mathrm{Re}\,(CA^* - DF^* - EC^* + BD^*) \tag{3.727}$$

$$K_x^{x'x'-y'y'} = \frac{3}{I_0} \mathrm{Re}\,(BA^* + EF^*) \tag{3.728}$$

$$K_y^{x'x'-y'y'} = -\frac{3}{I_0} \mathrm{Im}(BA^* - EF^*) \tag{3.729}$$

$$K_z^{x'x'-y'y'} = -\frac{3}{I_0} \mathrm{Re}\,(EA^* - BF^*) \tag{3.730}$$

$$K_x^{z'z'} = -\frac{1}{I_0} \mathrm{Re}\,(AF^* + EB^* - 2CD^*) \tag{3.731}$$

$$K_y^{z'z'} = -\frac{1}{I_0} \mathrm{Im}(AF^* + EB^* - 2CD^*) \tag{3.732}$$

$$K_z^{z'z'} = \frac{1}{2I_0}\left[\,|A|^2 - |F|^2 + |E|^2 - |B|^2 + 2|D|^2 - 2|C|^2\right] \tag{3.733}$$

In the case where both the target and the residual nucleus are polarized, and continuing to choose the y-axis along to direction $\vec{n}$, that is, $\varphi = 0$, the footnote of the matrix element of $\hat{F}$ in Eq. (3.689) is marked by $M_I' M_I$, and the footnote of the matrix element of the complex conjugate matrix $\hat{F}^+$ in Eq. (3.689) is marked by $\overline{M}_I' \overline{M}_I$. Also, we make the following changes to the previous results:

$$|X|^2 \to X^*_{\overline{M}_I' \overline{M}_I} X_{M_I' M_I}, \qquad X = A,\, B,\, C,\, D,\, E,\, F \tag{3.734}$$

Given this, and according to Eq. (3.688), we can write

$$\begin{aligned} I_{0,\, M_I' \overline{M}_I' M_I \overline{M}_I} = \frac{1}{2}\Big( & A^*_{\overline{M}_I' \overline{M}_I} A_{M_I' M_I} + B^*_{\overline{M}_I' \overline{M}_I} B_{M_I' M_I} + C^*_{\overline{M}_I' \overline{M}_I} C_{M_I' M_I} \\ & + D^*_{\overline{M}_I' \overline{M}_I} D_{M_I' M_I} + E^*_{\overline{M}_I' \overline{M}_I} E_{M_I' M_I} + F^*_{\overline{M}_I' \overline{M}_I} F_{M_I' M_I} \Big) \end{aligned} \tag{3.735}$$

By means of the form of the polarization wave functions of the target and the residual nucleus introduced by Eqs. (2.315) and (2.320), and continuing to use the simplification summation symbol $\overline{\Sigma}$ for the polarized target and residual nucleus introduced by Eq. (2.327), the angular distribution of the emitted particles corresponding to the unpolarized incident particles can be expressed as

$$\overline{I}^0 \equiv I^0_{\alpha' n',\, \alpha n} = \frac{d\sigma^0_{\alpha' n',\, \alpha n}}{d\Omega} = \Sigma I_{0,\, M_I' \overline{M}_I' M_I \overline{M}_I} \tag{3.736}$$

The components of the polarization vector and polarization tensor corresponding to the unpolarized incident particles are

$$\overline{P}^{0\, i'} \equiv P^{0\, i'}_{\alpha' n',\, \alpha n} = \frac{1}{\overline{I}^0} \Sigma I_{0,\, M_I' \overline{M}_I' M_I \overline{M}_I} P^{i'}_{0,\, M_I' \overline{M}_I' M_I \overline{M}_I}, \qquad i' = \varepsilon' \tag{3.737}$$

where $\varepsilon'$ is the Cartesian coordinate index set introduced by Eq. (3.476). When the normal variables are added with the footnote $M_I' M_I$ and the complex conjugate variables are added with the footnote $\overline{M}_I' \overline{M}_I$ in Eqs. (3.693)–(3.700) and using Eq. (3.734), the various $P^{i'}_{0,\, M_I' \overline{M}_I' M_I \overline{M}_I}$ in Eq. (3.737) can be obtained.

When the incident particles, targets, and residual nuclei are all polarized, the component of the nuclear reaction polarization vector analyzing power is

$$\overline{A}_i \equiv A_{i, \alpha' n', \alpha n} = \frac{1}{\overline{I}^0} \Sigma I_{0, M_I' \overline{M}_I' M_I \overline{M}_I} A_{i, M_I' \overline{M}_I' M_I \overline{M}_I}, \qquad i = x, y, z \tag{3.738}$$

where $A_{i, M_I' \overline{M}_I' M_I \overline{M}_I}$ have been given by Eqs. (3.703), (3.705), and (3.707). Referring to $\hat{\rho}_{in}$ given by Eq. (3.214) and $\hat{\rho}_{out}$ given by Eq. (2.85), then the differential cross section of the emitted particles corresponding to polarized incident particles can be written as

$$\overline{I} \equiv I_{\alpha' n', \alpha n} = \frac{d\sigma_{\alpha' n', \alpha n}}{d\Omega} = \mathrm{tr}\{\hat{\rho}_{\mathrm{out}}\} = \overline{I}_0 \left[1 + \frac{3}{2}\left(p_x \overline{A}_x + p_y \overline{A}_y + p_z \overline{A}_z\right)\right] \tag{3.739}$$

The corresponding polarization transfer coefficient is

$$\overline{K}^{i'}_i \equiv K^{i'}_{i, \alpha' n', \alpha n} = \frac{1}{\overline{I}^0} \Sigma I_{0, M_I' \overline{M}_I' M_I \overline{M}_I} K^{i'}_{i, M_I' \overline{M}_I' M_I \overline{M}_I}, \qquad i = x, y, z; \quad i' = \varepsilon' \tag{3.740}$$

where $K^{i'}_{i, M_I' \overline{M}_I' M_I \overline{M}_I}$ have been given by Eqs. (3.710)–(3.733). Referring to Eqs. (3.214) and (2.85) we can write the components of the polarization vector and polarization tensor of the outgoing particles as

$$\overline{P}^{i'} \equiv P^{i'}_{\alpha' n', \alpha n} = \frac{\overline{I}^0}{\overline{I}} \left[\overline{P}^{0\, i'} + \frac{3}{2}\left(p_x \overline{K}^{i'}_x + p_y \overline{K}^{i'}_y + p_z \overline{K}^{i'}_z\right)\right], \quad i' = \varepsilon' \tag{3.741}$$

Now, using Eqs. (2.334) and (2.335), as well as the simplification summation symbols $\widetilde{\Sigma}$ given by Eq. (2.278), Eqs. (3.735)–(3.738), and (3.740) can be simplified, respectively, in the case where both the target and the residual nucleus are unpolarized, as follows:

$$I_{0,M_I'M_I}=\frac{1}{2}\left(\left|A_{M_I'M_I}\right|^2+\left|B_{M_I'M_I}\right|^2+\left|C_{M_I'M_I}\right|^2+\left|D_{M_I'M_I}\right|^2+\left|E_{M_I'M_I}\right|^2+\left|F_{M_I'M_I}\right|^2\right) \tag{3.742}$$

$$\overline{I}^0=\widetilde{\Sigma}I_{0,M_I'M_I} \tag{3.743}$$

$$\overline{P}^{0\ i'}=\frac{1}{\overline{I}^0}\widetilde{\Sigma}I_{0,M_I'M_I}P^{i'}_{0,M_I'M_I},\quad i'=\varepsilon' \tag{3.744}$$

$$\overline{A}_i=\frac{1}{\overline{I}^0}\widetilde{\Sigma}I_{0,M_I'M_I}A_{i,M_I'M_I},\quad i=x,\,y,\,z \tag{3.745}$$

$$K_i^{i'}=\frac{1}{\overline{I}^0}\widetilde{\Sigma}I_{0,M_I'M_I}K^{i'}_{i,M_I'M_I},\quad i=x,\,y,\,z;\quad i'=\varepsilon' \tag{3.746}$$

Next, we will analyze Eq. (3.672) which belongs to S-L coupling scheme. Based on the properties of the C-G coefficient, there are the following relations:

$$\begin{aligned}&C^{S\ \mu+M_I}_{\frac{1}{2}\mu\ IM_I}=(-1)^{I+\frac{1}{2}-S}C^{S\ -\mu-M_I}_{\frac{1}{2}\ -\mu\ I\ -M_I},\ C^{S'\ \mu'+M_I'}_{1\mu'\ I'M_I'}=(-1)^{I'+1-S'}C^{S'\ -\mu'-M_I'}_{1\ -\mu'\ I'\ -M_I'}\\&C^{J\ M}_{l0\ SM}=(-1)^{l+S-J}C^{J\ -M}_{l0\ S\ -M},\ C^{J\ M}_{l'm_l'\ S'M_S'}=(-1)^{l'+S'-J}C^{J\ -M}_{l'\ -m_l'\ S'\ -M_S'}\end{aligned} \tag{3.747}$$

Now we only need to study the second item of Eq. (3.553) with $i=\frac{1}{2}$ and $i'=1$. Using Eqs. (2.288) and (3.747), for the part which is not associated with angle $\varphi$, the following phase factor comes out when all spin magnetic quantum numbers change signs:

$$(-1)^{l+S-J+l'+S'-J+I+\frac{1}{2}-S+I'+1-S'+\mu+M_I-\mu'-M_I'}=(-1)^{2J+l+I+l'+I'+\frac{3}{2}+\mu+M_I-\mu'-M_I'}$$

The symbol $\Pi$ introduced in Eq. (3.488) can also be used here, so we have $(-1)^{l+l'}=(-1)^{\Pi}$. In $\overrightarrow{\frac{1}{2}}+\text{A}\rightarrow\vec{1}+\text{B}$ reactions, $I$ and $I'$ must be that one is an integer and another is a semi-odd number. When $I$ is a semi-odd number and $I'$ is an integer, J is an integer. When $I$ is an integer and $I'$ is a semi-odd number, J is a semi-odd number. Letting

$$\Lambda=\begin{cases}0 & \text{when } I \text{ is a semi-odd number and } I' \text{ is an integer}\\ 1 & \text{when } I \text{ is an integer and } I' \text{ is a semi-odd number}\end{cases} \tag{3.748}$$

then one can use $(-1)^{\Lambda}$ to replace $(-1)^{2J}$ in this case. Note that the definition of $\Lambda$ given by Eq. (3.748) and the definition of $\Lambda$ given by Eq. (3.635) are not the same; thus, we can get

$$f_{\alpha' n' \mu' M_I', \alpha n \mu M_I}(\theta) = (-1)^{\Pi + \Lambda + I + I' + \frac{3}{2} + \mu + M_I - \mu' - M_I'} f_{\alpha' n' - \mu' - M_I', \alpha n - \mu - M_I}(\theta) \tag{3.749}$$

Now, we can obtain the following relations from Eqs. (3.553) and (3.673)–(3.678):

$$\begin{aligned} &A_{M_I' M_I}(\theta) = f_{\alpha' n' 1 M_I', \alpha n \frac{1}{2} M_I}(\theta), && B_{M_I' M_I}(\theta) = f_{\alpha' n' -1 M_I', \alpha n - \frac{1}{2} M_I}(\theta), \\ &C_{M_I' M_I}(\theta) = \mathrm{i} f_{\alpha' n' 0 M_I', \alpha n \frac{1}{2} M_I}(\theta), && D_{M_I' M_I}(\theta) = -\mathrm{i} f_{\alpha' n' 0 M_I', \alpha n - \frac{1}{2} M_I}(\theta), \\ &E_{M_I' M_I}(\theta) = -f_{\alpha' n' -1 M_I', \alpha n \frac{1}{2} M_I}(\theta), && F_{M_I' M_I}(\theta) = -f_{\alpha' n' 1 M_I', \alpha n - \frac{1}{2} M_I}(\theta) \end{aligned} \tag{3.750}$$

According to Eqs. (3.749) and (3.750), we have

$$B_{M_I' M_I}(\theta) = (-1)^{\Pi + \Lambda + I + I' + M_I - M_I'} A_{-M_I' \, -M_I}(\theta) \tag{3.751}$$

$$D_{M_I' M_I}(\theta) = (-1)^{\Pi + \Lambda + I + I' + M_I - M_I'} C_{-M_I' \, -M_I}(\theta) \tag{3.752}$$

$$F_{M_I' M_I}(\theta) = (-1)^{\Pi + \Lambda + I + I' + M_I - M_I'} E_{-M_I' \, -M_I}(\theta) \tag{3.753}$$

One can see that Eqs. (3.641) and (3.642) are still applicable in this section by comparing Eqs. (3.751)–(3.753) with Eqs. (3.638)–(3.640). Letting

$$\widetilde{I}_{0, M_I' M_I}(\theta) = |A_{M_I' M_I}|^2 + |C_{M_I' M_I}|^2 + |E_{M_I' M_I}|^2 \tag{3.754}$$

then according to Eq. (3.641) the following relation can be obtained from Eqs. (3.742) and (3.743):

$$\overline{I}^0 = \widetilde{\Sigma} \widetilde{I}_{0, M_I' M_I} \tag{3.755}$$

Given this, the following results can be obtained from Eq. (3.744), according to Eqs. (3.693)–(3.700), and using Eqs. (3.751)–(3.753), (3.641), and (3.642):

$$\overline{P}^{0 \, x'} = \overline{P}^{0 \, z'} = \overline{P}^{0 \, x'y'} = \overline{P}^{0 \, y'z'} = 0 \tag{3.756}$$

$$\overline{P}^{0 \, y'} = -\frac{\sqrt{2}}{\overline{I}^0} \widetilde{\Sigma} \, \mathrm{Re}\,(CA^* + DF^*) \tag{3.757}$$

$$\overline{P}^{0 \, x'z'} = \frac{3}{\sqrt{2} \overline{I}^0} \widetilde{\Sigma} \mathrm{Im}(CA^* + DF^*) \tag{3.758}$$

$$\overline{P}^{0\ x'x'-y'y'} = -\frac{6}{\overline{I}^0}\widetilde{\Sigma}\,\mathrm{Re}\,(EA^*) \tag{3.759}$$

$$\overline{P}^{0\ z'z'} = \frac{1}{\overline{I}^0}\widetilde{\Sigma}\left(|A|^2 + |E|^2 - 2|C|^2\right) \tag{3.760}$$

Further, according to Eqs. (3.703), (3.705), and (3.707), we can obtain the following results by Eq. (3.745):

$$\overline{A}_x = -\frac{1}{\overline{I}^0}\widetilde{\Sigma}\,\mathrm{Re}\,(2AF^* + CD^*) \tag{3.761}$$

$$\overline{A}_y = \overline{A}_z = 0 \tag{3.762}$$

It follows that from Eq. (3.746) we can obtain the following results according to Eqs. (3.710)–(3.733):

$$\overline{K}_i^{i'} = 0, \qquad (i = x \text{ and } i' = \varepsilon_2') \text{ or } (i = y, z \text{ and } i' = \varepsilon_1') \tag{3.763}$$

where the subscript $\varepsilon_1$ and $\varepsilon_2$ are given by Eq. (3.320). Further we can obtain

$$\overline{K}_y^{x'} = -\frac{\sqrt{2}}{\overline{I}^0}\widetilde{\Sigma}\,\mathrm{Re}\,(DA^* - CF^*) \tag{3.764}$$

$$\overline{K}_z^{x'} = \frac{\sqrt{2}}{\overline{I}^0}\widetilde{\Sigma}\mathrm{Im}(CA^* - DF^*) \tag{3.765}$$

$$\overline{K}_x^{y'} = \frac{\sqrt{2}}{\overline{I}^0}\widetilde{\Sigma}\,\mathrm{Re}\,(DA^* + CF^*) \tag{3.766}$$

$$\overline{K}_y^{z'} = -\frac{2}{\overline{I}^0}\widetilde{\Sigma}\mathrm{Im}(AF^*) \tag{3.767}$$

$$\overline{K}_z^{z'} = \frac{1}{\overline{I}^0}\widetilde{\Sigma}\left(|A|^2 - |F|^2\right) \tag{3.768}$$

$$\overline{K}_y^{x'y'} = \frac{3}{2\overline{I}^0}\widetilde{\Sigma}\,\mathrm{Re}\,(BA^* - EF^*) \tag{3.769}$$

$$\overline{K}_z^{x'y'} = -\frac{3}{\overline{I}^0}\widetilde{\Sigma}\mathrm{Im}(EA^*) \tag{3.770}$$

$$\overline{K}_x^{x'z'} = -\frac{3}{\sqrt{2}\overline{I}^0}\widetilde{\Sigma}\mathrm{Im}(DA^* + CF^*) \tag{3.771}$$

$$\overline{K}_y^{y'z'} = -\frac{3}{\sqrt{2}\overline{I}^0} \widetilde{\Sigma} \mathrm{Im}(DA^* - CF^*) \tag{3.772}$$

$$\overline{K}_z^{y'z'} = -\frac{3}{\sqrt{2}\overline{I}^0} \widetilde{\Sigma} \mathrm{Re}\,(CA^* - DF^*) \tag{3.773}$$

$$\overline{K}_x^{x'x' - y'y'} = \frac{3}{\overline{I}^0} \widetilde{\Sigma} \mathrm{Re}\,(BA^* + EF^*) \tag{3.774}$$

$$\overline{K}_x^{z'z'} = -\frac{2}{\overline{I}^0} \widetilde{\Sigma} \mathrm{Re}\,(AF^* - CD^*) \tag{3.775}$$

Now from the results obtained above, it can be seen that

$$\overline{P}^{0i'} \begin{cases} \neq 0 & i' = \varepsilon_1{}' \\ = 0 & i' = \varepsilon_2{}' \end{cases} \tag{3.776}$$

$$\overline{A}_i \begin{cases} \neq 0 & i = x \\ = 0 & i = y,\, z \end{cases} \tag{3.777}$$

$$\overline{K}_i^{i'} \begin{cases} \neq 0 & (i = x \text{ and } i' = \varepsilon_1{}') \text{ or } (i = y,\ z \text{ and } i' = \varepsilon_2{}') \\ = 0 & (i = x \text{ and } i' = \varepsilon_2{}') \text{ or } (i = y,\ z \text{ and } i' = \varepsilon_1{}') \end{cases} \tag{3.778}$$

And we can obtain the following results from Eqs. (3.739) and (3.741):

$$\overline{I} = \overline{I}^0 \left(1 + \frac{3}{2} p_x \overline{A}_x\right) \tag{3.779}$$

$$\overline{P}^{i'} = \frac{\overline{I}^0}{\overline{I}} \left(\overline{P}^{0\,i'} + \frac{3}{2} p_x \overline{K}_x^{i'}\right), \qquad i' = \varepsilon_1{}' \tag{3.780}$$

$$\overline{P}^{i'} = \frac{\overline{I}^0}{\overline{I}} \frac{3}{2} \left(p_y \overline{K}_y^{i'} + p_z K_z^{i'}\right), \qquad i' = \varepsilon_2{}' \tag{3.781}$$

References [23–29] gave the experiment data of the polarization analyzing powers for the (p, d) reaction with the targets $^{12}$C,$^{13}$C,$^{28}$Si, $^{24}$Mg, $^{208}$Pb, separately. The transition matrix element $T$ can be calculated by using the DWBA method, and it can be converted into the $S$ matrix element. It follows that the theoretical analyses of the corresponding polarization physical quantities can be carried out using the methods described in this section.

## 3.11 Polarization Theory for $\vec{\frac{1}{2}} + \vec{1}$ Reactions

In this section we will study the polarization theory of the reactions between polarized spin $\frac{1}{2}$ particles and polarized spin 1 particles, and take $\vec{\mathrm{N}} + \vec{\mathrm{d}}$ elastic scattering as an example, where N may be neutron or proton. In order to simplify this problem, people generally only study either the $\vec{\mathrm{N}} + \mathrm{d}$ reactions of polarized nucleon and unpolarized deuteron [30–35], or the $\vec{\mathrm{d}} + \mathrm{N}$ reactions of polarized deuteron and unpolarized nucleon [32–35] . With the theory described in this book, these two cases can be treated together.

We know that for the few-body reactions, the angular distribution anisotropy is obvious even in the keV energy region, and its polarization effect should not be ignored. In particular, some people are currently studying whether fusion reactions using polarized charged particles as fuel are likely to increase the fusion reaction rate [6, 36–40], and there are also people studying $\vec{\mathrm{d}} + \vec{\mathrm{t}} \rightarrow \mathrm{n} + \alpha$ and $\vec{\mathrm{d}} + \overrightarrow{{}^3\mathrm{He}} \rightarrow \mathrm{p} + \alpha$ reactions [2, 39, 40], both of which belong to $\vec{1} + \vec{\frac{1}{2}}$ reaction, where the spin of the emitted particles is 0 or $\frac{1}{2}$, and their reaction amplitude constitutes a $2 \times 6$ matrix.

The elastic scattering of the polarized nucleons (deuterons) and deuterons (nucleons) that are the secondary particles of the radioactive nuclear beam, or both as the secondary particles in the device, belongs to $\vec{\frac{1}{2}} + \vec{1}$ reaction. Of course, the polarization experiments can also be done directly using already polarized nucleons and polarized deuterons.

The N + d elastic scattering can be studied by using the phenomenological phase shift analysis method to obtain the corresponding $S$ matrix element, or using the microscopic three-body theory [41–45]; in this way the corresponding $T$ matrix elements are first obtained, and then the corresponding $S$ matrix elements are obtained from the $T$ matrix elements.

For the nucleon-deuteron interactions, the $S$-$L$ angular momentum coupling mode is used. The reaction amplitude of the N − d elastic scattering can be written by Eq. (2.252) as

$$\begin{aligned} f_{\mu'\nu',\mu\nu}(\Omega) = & f_{\mathrm{C}}(\theta)\delta_{\mu'\nu',\mu\nu} + \frac{\mathrm{i}\sqrt{\pi}}{k}\sum_{lSl'S'J} \hat{l}\mathrm{e}^{\mathrm{i}(\sigma_l+\sigma_{l'})}\left(\delta_{l'S',lS} - S^J_{l'S',lS}\right) \\ & C^{J\ M}_{l0\ SM} C^{S\ M}_{\frac{1}{2}\mu\ 1\nu} C^{J\ M}_{l'm'_l\ S'M'_S} C^{S'\ M'_S}_{\frac{1}{2}\mu'\ 1\nu'} Y_{l'm'_l}(\theta, \varphi) \end{aligned} \tag{3.782}$$

One can see that

$$m_l' = \mu + \nu - \mu' - \nu' \tag{3.783}$$

where $\mu$ and $\mu'$, $\nu$ and $\nu'$ represent the spin magnetic quantum numbers of the incident nucleon N and the target deuteron d in incident channel and the outgoing channel, respectively. If N is a proton, the first item on the right side of Eq. (3.782) representing the Coulomb interaction cannot be ignored. Now, introduce the following symbols

$$f^{ll'J}_{\mu'\nu',\mu\nu} = \hat{l}\hat{l}' \mathrm{e}^{\mathrm{i}(\sigma_l+\sigma_{l'})} \sqrt{\frac{(l'-m_l')!}{(l'+m_l')!}} \sum_{SS'} C^{J\ M}_{l0\ SM} C^{S\ M}_{\frac{1}{2}\mu\ 1\nu} C^{J\ M}_{l'm_l'\ S'M_S'} C^{S'\ M_S'}_{\frac{1}{2}\mu'\ 1\nu'} \left(\delta_{l'S',lS} - S^J_{l'S',lS}\right) \tag{3.784}$$

Using Eq. (2.253), one can obtain the following from Eq. (3.782):

$$f_{\mu'\nu',\mu\nu}(\Omega) = F_{\mu'\nu',\mu\nu}(\theta)\ \mathrm{e}^{\mathrm{i}(\mu+\nu-\mu'-\nu')\varphi} \tag{3.785}$$

$$F_{\mu'\nu',\mu\nu}(\theta) = f_C(\theta)\ \delta_{\mu'\nu',\mu\nu} + \frac{\mathrm{i}}{2k}(-1)^{\mu+\nu-\mu'-\nu'} \sum_{ll'J} f^{ll'J}_{\mu'\nu',\mu\nu} \mathrm{P}^{\mu+\nu-\mu'-\nu'}_{l'}(\cos\theta) \tag{3.786}$$

Considering parity conservation, $l$ and $l'$ must simultaneously be even or odd, and because 2J is odd, the following C-G coefficient relations can be obtained:

$$\begin{aligned} & C^{J\ \mu+\nu}_{l0\ S\ \mu+\nu} C^{S\ \mu+\nu}_{\frac{1}{2}\mu\ 1\nu} C^{J\ \mu+\nu}_{l'\ \mu+\nu-\mu'-\nu'\ S'\ \mu'+\nu'} C^{S'\ \mu'+\nu'}_{\frac{1}{2}\mu'\ 1\nu'} \\ & = C^{J\ -\mu-\nu}_{l0\ S\ -\mu-\nu} C^{S\ -\mu-\nu}_{\frac{1}{2}\ -\mu\ 1\ -\nu} C^{J\ -\mu-\nu}_{l'\ -\mu-\nu+\mu'+\nu'\ S'\ -\mu'-\nu'} C^{S'\ -\mu'-\nu'}_{\frac{1}{2}\ -\mu'\ 1\ -\nu'} \end{aligned} \tag{3.787}$$

Noting the relation expression (2.288), we can obtain the following from Eqs. (3.782)–(3.787):

$$F_{\mu'\nu',\mu\nu}(\theta) = (-1)^{\mu+\nu-\mu'-\nu'} F_{-\mu'\ -\nu',\ -\mu\ -\nu}(\theta) \tag{3.788}$$

Next, take the $y$-axis along the direction $\vec{n}$, so that $\varphi = 0$, and introduce the following matrix element symbols according to Eq. (3.788):

$$
\begin{aligned}
&A(\theta)=F_{\frac{1}{2}1,\,\frac{1}{2}1}(\theta)=F_{-\frac{1}{2}\,-1,\,-\frac{1}{2}\,-1}(\theta), && B(\theta)=\mathrm{i}F_{\frac{1}{2}1,\,\frac{1}{2}0}(\theta)=-\mathrm{i}F_{-\frac{1}{2}\,-1,\,-\frac{1}{2}\,0}(\theta),\\
&C(\theta)=-F_{\frac{1}{2}1,\,\frac{1}{2}\,-1}(\theta)=-F_{-\frac{1}{2}\,-1,\,-\frac{1}{2}\,1}(\theta), && D(\theta)=-\mathrm{i}F_{\frac{1}{2}1,\,-\frac{1}{2}1}(\theta)=\mathrm{i}F_{-\frac{1}{2}\,-1,\,\frac{1}{2}\,-1}(\theta),\\
&E(\theta)=F_{\frac{1}{2}1,\,-\frac{1}{2}0}(\theta)=F_{-\frac{1}{2}\,-1,\,\frac{1}{2}\,0}(\theta), && F(\theta)=\mathrm{i}F_{\frac{1}{2}1,\,-\frac{1}{2}\,-1}(\theta)=-\mathrm{i}F_{-\frac{1}{2}\,-1,\,\frac{1}{2}\,1}(\theta),\\
&G(\theta)=F_{\frac{1}{2}0,\,-\frac{1}{2}\,-1}(\theta)=F_{-\frac{1}{2}\,0,\,\frac{1}{2}\,1}(\theta), && H(\theta)=-\mathrm{i}F_{\frac{1}{2}0,\,-\frac{1}{2}\,0}(\theta)=\mathrm{i}F_{-\frac{1}{2}\,0,\,\frac{1}{2}\,0}(\theta),\\
&J(\theta)=-F_{\frac{1}{2}0,\,-\frac{1}{2}\,1}(\theta)=-F_{-\frac{1}{2}\,0,\,\frac{1}{2}\,-1}(\theta), && K(\theta)=\mathrm{i}F_{\frac{1}{2}0,\,\frac{1}{2}\,-1}(\theta)=-\mathrm{i}F_{-\frac{1}{2}\,0,\,-\frac{1}{2}\,1}(\theta),\\
&L(\theta)=F_{\frac{1}{2}0,\,\frac{1}{2}0}(\theta)=F_{-\frac{1}{2}\,0,\,-\frac{1}{2}\,0}(\theta), && M(\theta)=-\mathrm{i}F_{\frac{1}{2}0,\,\frac{1}{2}1}(\theta)=\mathrm{i}F_{-\frac{1}{2}\,0,\,-\frac{1}{2}\,-1}(\theta),\\
&N(\theta)=-F_{\frac{1}{2}\,-1,\,\frac{1}{2}1}(\theta)=-F_{-\frac{1}{2}\,1,\,-\frac{1}{2}\,-1}(\theta), && O(\theta)=-\mathrm{i}F_{\frac{1}{2}\,-1,\,\frac{1}{2}0}(\theta)=\mathrm{i}F_{-\frac{1}{2}\,1,\,-\frac{1}{2}0}(\theta),\\
&Q(\theta)=F_{\frac{1}{2}\,-1,\,\frac{1}{2}\,-1}(\theta)=F_{-\frac{1}{2}\,1,\,-\frac{1}{2}\,1}(\theta), && R(\theta)=\mathrm{i}F_{\frac{1}{2}\,-1,\,-\frac{1}{2}1}(\theta)=-\mathrm{i}F_{-\frac{1}{2}\,1,\,\frac{1}{2}\,-1}(\theta),\\
&S(\theta)=-F_{\frac{1}{2}\,-1,\,-\frac{1}{2}0}(\theta)=-F_{-\frac{1}{2}\,1,\,\frac{1}{2}\,0}(\theta), && T(\theta)=-\mathrm{i}F_{\frac{1}{2}\,-1,\,-\frac{1}{2}\,-1}(\theta)=\mathrm{i}F_{-\frac{1}{2}\,1,\,\frac{1}{2}\,1}(\theta)
\end{aligned}
\tag{3.789}
$$

Then the 6 × 6 reaction matrix given by Eq. (3.786) can be expressed as

$$
\hat{F}=\begin{pmatrix}
A & -\mathrm{i}B & -C & \mathrm{i}D & E & -\mathrm{i}F\\
\mathrm{i}M & L & -\mathrm{i}K & -J & \mathrm{i}H & G\\
-N & \mathrm{i}O & Q & -\mathrm{i}R & -S & \mathrm{i}T\\
-\mathrm{i}T & -S & \mathrm{i}R & Q & -\mathrm{i}O & -N\\
G & -\mathrm{i}H & -J & \mathrm{i}K & L & -\mathrm{i}M\\
\mathrm{i}F & E & -\mathrm{i}D & -C & \mathrm{i}B & A
\end{pmatrix},
\quad
\hat{F}^{+}=\begin{pmatrix}
A^{*} & -\mathrm{i}M^{*} & -N^{*} & \mathrm{i}T^{*} & G^{*} & -\mathrm{i}F^{*}\\
\mathrm{i}B^{*} & L^{*} & -\mathrm{i}O^{*} & -S^{*} & \mathrm{i}H^{*} & E^{*}\\
-C^{*} & \mathrm{i}K^{*} & Q^{*} & -\mathrm{i}R^{*} & -J^{*} & \mathrm{i}D^{*}\\
-\mathrm{i}D^{*} & -J^{*} & \mathrm{i}R^{*} & Q^{*} & -\mathrm{i}K^{*} & -C^{*}\\
E^{*} & -\mathrm{i}H^{*} & -S^{*} & \mathrm{i}O^{*} & L^{*} & -\mathrm{i}B^{*}\\
\mathrm{i}F^{*} & G^{*} & -\mathrm{i}T^{*} & -N^{*} & \mathrm{i}M^{*} & A^{*}
\end{pmatrix}
\tag{3.790}
$$

Now introduce the following 3 × 3 submatrices:

$$
\hat{t}_{\mu'\mu}=\begin{pmatrix}
F_{\mu'1,\,\mu 1} & F_{\mu'1,\,\mu 0} & F_{\mu'1,\,\mu\,-1}\\
F_{\mu'0,\,\mu 1} & F_{\mu'0,\,\mu 0} & F_{\mu'0,\,\mu\,-1}\\
F_{\mu'\,-1,\,\mu 1} & F_{\mu'\,-1,\,\mu 0} & F_{\mu'\,-1,\,\mu\,-1}
\end{pmatrix}
\tag{3.791}
$$

For the sake of simplicity, the footnote $\frac{1}{2}$ is represented by the footnote +, and the footnote $-\frac{1}{2}$ is represented by the footnote −, so that the following expressions can be written according to Eq. (3.790) as follows:

$$\begin{aligned}
\hat{t}_{++} &= \begin{pmatrix} A & -\mathrm{i}B & -C \\ \mathrm{i}M & L & -\mathrm{i}K \\ -N & \mathrm{i}O & Q \end{pmatrix}, \quad \hat{t}^{+}_{++} = \begin{pmatrix} A^* & -\mathrm{i}M^* & -N^* \\ \mathrm{i}B^* & L^* & -\mathrm{i}O^* \\ -C^* & \mathrm{i}K^* & Q^* \end{pmatrix} \\
\hat{t}_{+-} &= \begin{pmatrix} \mathrm{i}D & E & -\mathrm{i}F \\ -J & \mathrm{i}H & G \\ -\mathrm{i}R & -S & \mathrm{i}T \end{pmatrix}, \quad \hat{t}^{+}_{+-} = \begin{pmatrix} -\mathrm{i}D^* & -J^* & \mathrm{i}R^* \\ E^* & -\mathrm{i}H^* & -S^* \\ \mathrm{i}F^* & G^* & -\mathrm{i}T^* \end{pmatrix} \\
\hat{t}_{-+} &= \begin{pmatrix} -\mathrm{i}T & -S & \mathrm{i}R \\ G & -\mathrm{i}H & -J \\ \mathrm{i}F & E & -\mathrm{i}D \end{pmatrix}, \quad \hat{t}^{+}_{-+} = \begin{pmatrix} \mathrm{i}T^* & G^* & -\mathrm{i}F^* \\ -S^* & \mathrm{i}H^* & E^* \\ -\mathrm{i}R^* & -J^* & \mathrm{i}D^* \end{pmatrix} \\
\hat{t}_{--} &= \begin{pmatrix} Q & -\mathrm{i}O & -N \\ \mathrm{i}K & L & -\mathrm{i}M \\ -C & \mathrm{i}B & A \end{pmatrix}, \quad \hat{t}^{+}_{--} = \begin{pmatrix} Q^* & -\mathrm{i}K^* & -C^* \\ \mathrm{i}O^* & L^* & -\mathrm{i}B^* \\ -N^* & \mathrm{i}M^* & A^* \end{pmatrix}
\end{aligned} \tag{3.792}$$

and the matrix $\hat{F}$ and $\hat{F}^{+}$ given by Eq. (3.790) can be expressed as

$$\hat{F} = \begin{pmatrix} \hat{t}_{++} & \hat{t}_{+-} \\ \hat{t}_{-+} & \hat{t}_{--} \end{pmatrix}, \quad \hat{F}^{+} = \begin{pmatrix} \hat{t}^{+}_{++} & \hat{t}^{+}_{-+} \\ \hat{t}^{+}_{+-} & \hat{t}^{+}_{--} \end{pmatrix} \tag{3.793}$$

The polarization density matrix of the normalized incident channel can be written according to Eqs. (2.64) and (3.214) as follows:

$$\begin{aligned}
\hat{\rho}_{\mathrm{in}} = \frac{1}{6}\left(\hat{I}_2 + p^{\mathrm{N}}_x\hat{\sigma}_x + p^{\mathrm{N}}_y\hat{\sigma}_y + p^{\mathrm{N}}_z\hat{\sigma}_z\right)\Big[\hat{I}_3 + \frac{3}{2}\left(p^{\mathrm{d}}_x\hat{S}_x + p^{\mathrm{d}}_y\hat{S}_y + p^{\mathrm{d}}_z\hat{S}_z\right) \\
+\frac{2}{3}\left(p^{\mathrm{d}}_{xy}\hat{Q}_{xy} + p^{\mathrm{d}}_{xz}\hat{Q}_{xz} + p^{\mathrm{d}}_{yz}\hat{Q}_{yz}\right) + \frac{1}{6}p^{\mathrm{d}}_{xx-yy}\hat{Q}_{xx-yy} + \frac{1}{2}p^{\mathrm{d}}_{zz}\hat{Q}_{zz}\Big]
\end{aligned} \tag{3.794}$$

where $\hat{I}_2$ and $\hat{I}_3$ are the unit $2 \times 2$ and $3 \times 3$ matrices. $p^{\mathrm{N}}_i$ and $p^{\mathrm{d}}_i$ $(i = x, y, z)$ are the vector polarization rates of the incident nucleons and deuterons, respectively, and $p^{\mathrm{d}}_i$ $(i = xy, xz, yz, xx - yy, zz)$ are the tensor polarization rates of the incident deuterons.

Now, let $\hat{0}_2$ and $\hat{0}_3$ be the zero $2 \times 2$ and $3 \times 3$ matrices, respectively. According to the definition of the matrix direct product given by Eq. (2.349), we have

$$\hat{\Sigma}_i \equiv \hat{\sigma}_i \times \hat{I}_3 = \begin{pmatrix} \sigma_{i,++}\hat{I}_3 & \sigma_{i,+-}\hat{I}_3 \\ \sigma_{i,-+}\hat{I}_3 & \sigma_{i,--}\hat{I}_3 \end{pmatrix}, \quad i = x, y, z \tag{3.795}$$

$$\hat{\Gamma}_i \equiv \hat{I}_2 \times \hat{S}_i = \begin{pmatrix} \hat{S}_i & \hat{0}_3 \\ \hat{0}_3 & \hat{S}_i \end{pmatrix}, \quad i = x, y, z \tag{3.796}$$

$$\hat{\Omega}_i \equiv \hat{I}_2 \times \hat{Q}_i = \begin{pmatrix} \hat{Q}_i & \hat{0}_3 \\ \hat{0}_3 & \hat{Q}_i \end{pmatrix}, \quad i = xy, xz, yz, xx - yy, zz \tag{3.797}$$

Also, setting $\hat{I}_6$ to be the unit $6 \times 6$ matrix, Eq. (3.794) can be rewritten as

$$\begin{aligned} \hat{\rho}_{\text{in}} = \frac{1}{6}\left(\hat{I}_6 + p_x^{\text{N}}\hat{\Sigma}_x + p_y^{\text{N}}\hat{\Sigma}_y + p_z^{\text{N}}\hat{\Sigma}_z\right)\Big[\hat{I}_6 + \frac{3}{2}\left(p_x^{\text{d}}\hat{\Gamma}_x + p_y^{\text{d}}\hat{\Gamma}_y + p_z^{\text{d}}\hat{\Gamma}_z\right) \\ + \frac{2}{3}\left(p_{xy}^{\text{d}}\hat{\Omega}_{xy} + p_{xz}^{\text{d}}\hat{\Omega}_{xz} + p_{yz}^{\text{d}}\hat{\Omega}_{yz}\right) + \frac{1}{6}p_{xx-yy}^{\text{d}}\hat{\Omega}_{xx-yy} + \frac{1}{2}p_{zz}^{\text{d}}\hat{\Omega}_{zz}\Big] \end{aligned} \tag{3.798}$$

From Eqs. (3.790) and (3.798), when N and d in the incident channel are both unpolarized, we can obtain

$$\begin{aligned} I_0 \equiv \frac{\text{d}\sigma^0}{\text{d}\Omega} = \frac{1}{6}\text{tr}\left\{\hat{F}\hat{F}^+\right\} = \frac{1}{3}\Big(|A|^2 + |B|^2 + |C|^2 + |D|^2 + |E|^2 + |F|^2 + |G|^2 + |H|^2 \\ + |J|^2 + |K|^2 + |L|^2 + |M|^2 + |N|^2 + |O|^2 + |Q|^2 + |R|^2 + |S|^2 + |T|^2\Big) \end{aligned} \tag{3.799}$$

The polarization vector component of the outgoing nucleon N corresponding to the unpolarized incident channel is

$$P_0^{\text{N}\,i'} = \frac{\text{tr}\{\hat{\Sigma}_{i'}\hat{F}\hat{F}^+\}}{\text{tr}\{\hat{F}\hat{F}^+\}}, \quad i' = x', y', z' \tag{3.800}$$

The corresponding polarization vector component of the outgoing d is

$$P_0^{\text{d}\,i'} = \frac{\text{tr}\left\{\hat{\Gamma}_{i'}\hat{F}\hat{F}^+\right\}}{\text{tr}\left\{\hat{F}\hat{F}^+\right\}}, \quad i' = x', y', z' \tag{3.801}$$

The corresponding polarization tensor component of the outgoing d is

$$P_0^{\text{d}\,i'} = \frac{\text{tr}\{\hat{\Omega}_{i'}\hat{F}\hat{F}^+\}}{\text{tr}\{\hat{F}\hat{F}^+\}}, \quad i' = x'y', x'z', y'z', x'x' - y'y', z'z' \tag{3.802}$$

From Eq. (3.793) we can obtain

$$\hat{F}\hat{F}^+ = \begin{pmatrix} \hat{t}_{++}\hat{t}^+_{++} + \hat{t}_{+-}\ \hat{t}^+_{+-} & \hat{t}_{++}\hat{t}^+_{-+} + \hat{t}_{+-}\ \hat{t}^+_{--} \\ \hat{t}_{-+}\hat{t}^+_{++} + \hat{t}_{--}\ \hat{t}^+_{+-} & \hat{t}_{-+}\hat{t}^+_{-+} + \hat{t}_{--}\ \hat{t}^+_{--} \end{pmatrix} \equiv \begin{pmatrix} \hat{g}_{++} & \hat{g}_{+-} \\ \hat{g}_{-+} & \hat{g}_{--} \end{pmatrix} \tag{3.803}$$

Using Eq. (3.792), we can further obtain

$$\hat{g}_{++} = \begin{pmatrix} A & -\mathrm{i}B & -C \\ \mathrm{i}M & L & -\mathrm{i}K \\ -N & \mathrm{i}O & Q \end{pmatrix} \begin{pmatrix} A^* & -\mathrm{i}M^* & -N^* \\ \mathrm{i}B^* & L^* & -\mathrm{i}O^* \\ -C^* & \mathrm{i}K^* & Q^* \end{pmatrix} + \begin{pmatrix} \mathrm{i}D & E & -\mathrm{i}F \\ -J & \mathrm{i}H & G \\ -\mathrm{i}R & -S & \mathrm{i}T \end{pmatrix} \begin{pmatrix} -\mathrm{i}D^* & -J^* & \mathrm{i}R^* \\ E^* & -\mathrm{i}H^* & -S^* \\ \mathrm{i}F^* & G^* & -\mathrm{i}T^* \end{pmatrix}$$

$$= \begin{pmatrix} |A|^2+|B|^2+|C|^2+|D|^2+|E|^2+|F|^2 & -\mathrm{i}(AM^*+BL^*+CK^*+DJ^*+EH^*+FG^*) & -(AN^*+BO^*+CQ^*+DR^*+ES^*+FT^*) \\ \mathrm{i}(MA^*+LB^*+KC^*+JD^*+HE^*+GF^*) & |M|^2+|L|^2+|K|^2+|J|^2+|H|^2+|G|^2 & -\mathrm{i}(MN^*+LO^*+KQ^*+JR^*+HS^*+GT^*) \\ -(NA^*+OB^*+QC^*+RD^*+SE^*+TF^*) & \mathrm{i}(NM^*+OL^*+QK^*+RJ^*+SH^*+TG^*) & |N|^2+|O|^2+|Q|^2+|R|^2+|S|^2+|T|^2 \end{pmatrix} \tag{3.804}$$

$$\hat{g}_{+-} = \begin{pmatrix} A & -\mathrm{i}B & -C \\ \mathrm{i}M & L & -\mathrm{i}K \\ -N & \mathrm{i}O & Q \end{pmatrix} \begin{pmatrix} \mathrm{i}T^* & G^* & -\mathrm{i}F^* \\ -S^* & \mathrm{i}H^* & E^* \\ -\mathrm{i}R^* & -J^* & \mathrm{i}D^* \end{pmatrix} + \begin{pmatrix} \mathrm{i}D & E & -\mathrm{i}F \\ -J & \mathrm{i}H & G \\ -\mathrm{i}R & -S & \mathrm{i}T \end{pmatrix} \begin{pmatrix} Q^* & -\mathrm{i}K^* & -C^* \\ \mathrm{i}O^* & L^* & -\mathrm{i}B^* \\ -N^* & \mathrm{i}M^* & A^* \end{pmatrix}$$

$$= \begin{pmatrix} \mathrm{i}(AT^*+BS^*+CR^*+DQ^*+EO^*+FN^*) & AG^*+BH^*+CJ^*+DK^*+EL^*+FM^* & -\mathrm{i}(AF^*+BE^*+CD^*+DC^*+EB^*+FA^*) \\ -(MT^*+LS^*+KR^*+JQ^*+HO^*+GN^*) & \mathrm{i}(MG^*+LH^*+KJ^*+JK^*+HL^*+GM^*) & MF^*+LE^*+KD^*+JC^*+HB^*+GA^* \\ -\mathrm{i}(NT^*+OS^*+QR^*+RQ^*+SO^*+TN^*) & -(NG^*+OH^*+QJ^*+RK^*+SL^*+TM^*) & \mathrm{i}(NF^*+OE^*+QD^*+RC^*+SB^*+TA^*) \end{pmatrix} \tag{3.805}$$

$$\hat{g}_{-+}=\begin{pmatrix}-\mathrm{i}T & -S & \mathrm{i}R\\ G & -\mathrm{i}H & -J\\ \mathrm{i}F & E & -\mathrm{i}D\end{pmatrix}\begin{pmatrix}A^* & -\mathrm{i}M^* & -N^*\\ \mathrm{i}B^* & L^* & -\mathrm{i}O^*\\ -C^* & \mathrm{i}K^* & Q^*\end{pmatrix}+\begin{pmatrix}Q & -\mathrm{i}O & -N\\ \mathrm{i}K & L & -\mathrm{i}M\\ -C & \mathrm{i}B & A\end{pmatrix}\begin{pmatrix}-\mathrm{i}D^* & -J^* & \mathrm{i}R^*\\ E^* & -\mathrm{i}H^* & -S^*\\ \mathrm{i}F^* & G^* & -\mathrm{i}T^*\end{pmatrix}$$

$$=\begin{pmatrix}-\mathrm{i}(TA^*+SB^*+RC^*+QD^*+OE^*+NF^*) & -(TM^*+SL^*+RK^*+QJ^*+OH^*+NG^*) & \mathrm{i}(TN^*+SO^*+RQ^*+QR^*+OS^*+NT^*)\\ GA^*+HB^*+JC^*+KD^*+LE^*+MF^* & -\mathrm{i}(GM^*+HL^*+JK^*+KJ^*+LH^*+MG^*) & -(GN^*+HO^*+JQ^*+KR^*+LS^*+MT^*)\\ \mathrm{i}(FA^*+EB^*+DC^*+CD^*+BE^*+AF^*) & FM^*+EL^*+DK^*+CJ^*+BH^*+AG^* & -\mathrm{i}(FN^*+EO^*+DQ^*+CR^*+BS^*+AT^*)\end{pmatrix} \tag{3.806}$$

$$\hat{g}_{--}=\begin{pmatrix}-\mathrm{i}T & -S & \mathrm{i}R\\ G & -\mathrm{i}H & -J\\ \mathrm{i}F & E & -\mathrm{i}D\end{pmatrix}\begin{pmatrix}\mathrm{i}T^* & G^* & -\mathrm{i}F^*\\ -S^* & \mathrm{i}H^* & E^*\\ -\mathrm{i}R^* & -J^* & \mathrm{i}D^*\end{pmatrix}+\begin{pmatrix}Q & -\mathrm{i}O & -N\\ \mathrm{i}K & L & -\mathrm{i}M\\ -C & \mathrm{i}B & A\end{pmatrix}\begin{pmatrix}Q^* & -\mathrm{i}K^* & -C^*\\ \mathrm{i}O^* & L^* & -\mathrm{i}B^*\\ -N^* & \mathrm{i}M^* & A^*\end{pmatrix}$$

$$=\begin{pmatrix}|T|^2+|S|^2+|R|^2+|Q|^2+|O|^2+|N|^2 & -\mathrm{i}(TG^*+SH^*+RJ^*+QK^*+OL^*+NM^*) & -(TF^*+SE^*+RD^*+QC^*+OB^*+NA^*)\\ \mathrm{i}(GT^*+HS^*+JR^*+KQ^*+LO^*+MN^*) & |G|^2+|H|^2+|J|^2+|K|^2+|L|^2+|M|^2 & -\mathrm{i}(GF^*+HE^*+JD^*+KC^*+LB^*+MA^*)\\ -(FT^*+ES^*+DR^*+CQ^*+BO^*+AN^*) & \mathrm{i}(FG^*+EH^*+DJ^*+CK^*+BL^*+AM^*) & |F|^2+|E|^2+|D|^2+|C|^2+|B|^2+|A|^2\end{pmatrix} \tag{3.807}$$

Next, we can obtain first the following expression from Eqs. (2.6), (3.795), and (3.803):

$$\hat{\Sigma}_x\hat{F}\hat{F}^+=\begin{pmatrix}\hat{0}_3 & \hat{I}_3\\ \hat{I}_3 & \hat{0}_3\end{pmatrix}\begin{pmatrix}\hat{g}_{++} & \hat{g}_{+-}\\ \hat{g}_{-+} & \hat{g}_{--}\end{pmatrix}=\begin{pmatrix}\hat{g}_{-+} & \hat{g}_{--}\\ \hat{g}_{++} & \hat{g}_{+-}\end{pmatrix} \tag{3.808}$$

Then we can obtain the following result from Eqs. (3.800), (3.805), and (3.806):

$$P_0^{\mathrm{N}\,x'}=0 \tag{3.809}$$

Using the same method we can obtain

$$\hat{\Sigma}_y\hat{F}\hat{F}^+=\begin{pmatrix}\hat{0}_3 & -\mathrm{i}\hat{I}_3\\ \mathrm{i}\hat{I}_3 & \hat{0}_3\end{pmatrix}\begin{pmatrix}\hat{g}_{++} & \hat{g}_{+-}\\ \hat{g}_{-+} & \hat{g}_{--}\end{pmatrix}=-\mathrm{i}\begin{pmatrix}\hat{g}_{-+} & \hat{g}_{--}\\ -\hat{g}_{++} & -\hat{g}_{+-}\end{pmatrix} \tag{3.810}$$

$$P_0^{\mathrm{N}\,y'}=-\frac{2}{3I_0}\,\mathrm{Re}\,\left(AT^*+BS^*+CR^*+DQ^*+EO^*+FN^*+GM^*+HL^*+JK^*\right) \tag{3.811}$$

$$\hat{\Sigma}_z\hat{F}\hat{F}^+=\begin{pmatrix}\hat{I}_3 & \hat{0}_3\\ \hat{0}_3 & -\hat{I}_3\end{pmatrix}\begin{pmatrix}\hat{g}_{++} & \hat{g}_{+-}\\ \hat{g}_{-+} & \hat{g}_{--}\end{pmatrix}=\begin{pmatrix}\hat{g}_{++} & \hat{g}_{+-}\\ -\hat{g}_{-+} & -\hat{g}_{--}\end{pmatrix} \tag{3.812}$$

$$P_0^{\mathrm{N}\,z'}=0 \tag{3.813}$$

Now in order to calculate $P^{\mathrm{d}\,i'}\,(i'=x',\,y',\,z')$, first we find the following expressions from Eqs. (3.33), (3.796), (3.804), and (3.807):

$$\begin{aligned}\mathrm{tr}\{\hat{S}_x\hat{g}_{++}\}=&\frac{1}{\sqrt{2}}[\mathrm{i}(MA^*+LB^*+KC^*+JD^*+HE^*+GF^*)\\&-\mathrm{i}(AM^*+BL^*+CK^*+DJ^*+EH^*+FG^*)\\&+\mathrm{i}(NM^*+OL^*+QK^*+RJ^*+SH^*+TG^*)\\&-\mathrm{i}(MN^*+LO^*+KQ^*+JR^*+HS^*+GT^*)]\\=&\sqrt{2}\mathrm{Im}(AM^*+BL^*+CK^*+DJ^*+EH^*+FG^*\\&+GT^*+HS^*+JR^*+KQ^*+LO^*+MN^*)\end{aligned} \tag{3.814}$$

$$\begin{aligned}\mathrm{tr}\{\hat{S}_x\hat{g}_{--}\}=&\frac{1}{\sqrt{2}}[\mathrm{i}(GT^*+HS^*+JR^*+KQ^*+LO^*+MN^*)\\&-\mathrm{i}(TG^*+SH^*+RJ^*+QK^*+OL^*+NM^*)\\&+\mathrm{i}(FG^*+EH^*+DJ^*+CK^*+BL^*+AM^*)\\&-\mathrm{i}(GF^*+HE^*+JD^*+KC^*+LB^*+MA^*)]\\=&-\sqrt{2}\mathrm{Im}(AM^*+BL^*+CK^*+DJ^*+EH^*+FG^*\\&+GT^*+HS^*+JR^*+KQ^*+LO^*+MN^*)\end{aligned} \tag{3.815}$$

Then from Eq. (3.801) we find

$$P_0^{\mathrm{d}\,x'}=0 \tag{3.816}$$

Using the same method we can obtain

$$\begin{aligned}P_0^{\mathrm{d}\,y'}=&\frac{1}{6I_0}\mathrm{tr}\{\hat{S}_y\hat{g}_{++}+\hat{S}_y\hat{g}_{--}\}=\frac{\sqrt{2}}{3I_0}\,\mathrm{Re}\,(AM^*+BL^*+CK^*+DJ^*\\&+EH^*+FG^*+GT^*+HS^*+JR^*+KQ^*+LO^*+MN^*)\end{aligned} \tag{3.817}$$

$$P_0^{\mathrm{d}\,z'}=\frac{1}{6I_0}\mathrm{tr}\{\hat{S}_z\hat{g}_{++}+\hat{S}_z\hat{g}_{--}\}=0 \tag{3.818}$$

And using the same method we can obtain the following results according to Eqs. (3.36) and (3.797):

$$P_0^{\mathrm{d}\,x'y'} = \frac{1}{6I_0}\mathrm{tr}\{\hat{Q}_{x'y'}\hat{g}_{++} + \hat{Q}_{x'y'}\hat{g}_{--}\} = 0 \tag{3.819}$$

$$\begin{aligned} P_0^{\mathrm{d}\,x'z'} &= \frac{1}{6I_0}\mathrm{tr}\{\hat{Q}_{x'z'}\hat{g}_{++} + \hat{Q}_{x'z'}\hat{g}_{--}\} \\ &= \frac{1}{\sqrt{2}I_0}\mathrm{Im}[(AM^* + BL^* + CK^* + DJ^* + EH^* + FG^*) \\ &\quad - (GT^* + HS^* + JR^* + KQ^* + LO^* + MN^*)] \end{aligned} \tag{3.820}$$

$$P_0^{\mathrm{d}\,y'z'} = \frac{1}{6I_0}\mathrm{tr}\{\hat{Q}_{y'z'}\hat{g}_{++} + \hat{Q}_{y'z'}\hat{g}_{--}\} = 0 \tag{3.821}$$

$$\begin{aligned} P_0^{\mathrm{d}\,x'x'-y'y'} &= \frac{1}{6I_0}\mathrm{tr}\{\hat{Q}_{x'x'-y'y'}\hat{g}_{++} + \hat{Q}_{x'x'-y'y'}\hat{g}_{--}\} \\ &= -\frac{2}{3I_0}\mathrm{Re}\,(AN^* + BO^* + CQ^* + DR^* + ES^* + FT^*) \end{aligned} \tag{3.822}$$

$$\begin{aligned} P_0^{\mathrm{d}\,z'z'} = \frac{1}{3I_0}\Big[ & |A|^2 + |B|^2 + |C|^2 + |D|^2 + |E|^2 + |F|^2 + |N|^2 + |O|^2 + |Q|^2 \\ & + |R|^2 + |S|^2 + |T|^2 - 2\left(|G|^2 + |H|^2 + |J|^2 + |K|^2 + |L|^2 + |M|^2\right)\Big] \end{aligned} \tag{3.823}$$

Next, introduce a simplification symbol of the coordinate index set:

$$\gamma \equiv xy,\ xz,\ yz,\ xx - yy,\ zz \tag{3.824}$$

Given this, the following polarization analyzing powers can be defined for $\vec{\mathrm{N}}\left(\vec{\mathrm{d}}, \vec{\mathrm{d}}\right)\vec{\mathrm{N}}$ reaction:

$$A_i^{\mathrm{N}} = \frac{\mathrm{tr}\left\{\hat{F}\hat{\Sigma}_i\hat{F}^+\right\}}{\mathrm{tr}\left\{\hat{F}\hat{F}^+\right\}}, \quad i = x, y, z \tag{3.825}$$

$$\begin{aligned} A_i^{\mathrm{d}} &= \frac{\mathrm{tr}\left\{\hat{F}\hat{\Gamma}_i\hat{F}^+\right\}}{\mathrm{tr}\left\{\hat{F}\hat{F}^+\right\}}, \quad i = x, y, z \\ &= \frac{\mathrm{tr}\left\{\hat{F}\hat{\Omega}_i\hat{F}^+\right\}}{\mathrm{tr}\left\{\hat{F}\hat{F}^+\right\}}, \quad i = \gamma \end{aligned} \tag{3.826}$$

$$A_{i,j}^{\mathrm{N\,d}} = \frac{\mathrm{tr}\left\{\hat{F}\hat{\Sigma}_i\hat{\Gamma}_j\hat{F}^+\right\}}{\mathrm{tr}\left\{\hat{F}\hat{F}^+\right\}}, \qquad i,j = x, y, z$$
$$= \frac{\mathrm{tr}\left\{\hat{F}\hat{\Sigma}_i\hat{\Omega}_j\hat{F}^+\right\}}{\mathrm{tr}\left\{\hat{F}\hat{F}^+\right\}}, \qquad i = x, y, z; \quad j = \gamma \tag{3.827}$$

where $A_{i,j}^{N\,d}$ can also be called a correlation analyzing power. The following polarization transfer coefficients can also be defined:

$$K_{\mathrm{N}\,i}^{\mathrm{N}\,i'} = \frac{\mathrm{tr}\left\{\hat{\Sigma}_{i'}\hat{F}\hat{\Sigma}_i\hat{F}^+\right\}}{\mathrm{tr}\left\{\hat{F}\hat{F}^+\right\}}, \qquad i = x, y, z; \quad i' = x', y', z' \tag{3.828}$$

$$K_{\mathrm{N}\,i}^{\mathrm{d}\,i'} = \frac{\mathrm{tr}\left\{\hat{\Gamma}_{i'}\hat{F}\hat{\Sigma}_i\hat{F}^+\right\}}{tr\left\{\hat{F}\hat{F}^+\right\}}, \qquad i = x, y, z; \quad i' = x', y', z'$$
$$= \frac{\mathrm{tr}\left\{\hat{\Omega}_{i'}\hat{F}\hat{\Sigma}_i\hat{F}^+\right\}}{\mathrm{tr}\left\{\hat{F}\hat{F}^+\right\}}, \qquad i = x, y, z; \quad i' = \gamma' \tag{3.829}$$

$$K_{\mathrm{d}\,i}^{\mathrm{N}\,i'} = \frac{\mathrm{tr}\left\{\hat{\Sigma}_{i'}\hat{F}\hat{\Gamma}_i\hat{F}^+\right\}}{\mathrm{tr}\left\{\hat{F}\hat{F}^+\right\}}, \qquad i = x, y, z; \quad i' = x', y', z'$$
$$= \frac{\mathrm{tr}\left\{\hat{\Sigma}_{i'}\hat{F}\hat{\Omega}_i\hat{F}^+\right\}}{\mathrm{tr}\left\{\hat{F}\hat{F}^+\right\}}, \qquad i = \gamma; \quad i' = x', y', z' \tag{3.830}$$

$$K_{\mathrm{d}\,i}^{\mathrm{d}\,i'} = \frac{\mathrm{tr}\left\{\hat{\Gamma}_{i'}\hat{F}\hat{\Gamma}_i\hat{F}^+\right\}}{\mathrm{tr}\left\{\hat{F}\hat{F}^+\right\}}, \quad i = x, y, z; \quad i' = x', y', z'$$
$$= \frac{\mathrm{tr}\left\{\hat{\Omega}_{i'}\hat{F}\hat{\Gamma}_i\hat{F}^+\right\}}{\mathrm{tr}\left\{\hat{F}\hat{F}^+\right\}}, \quad i = x, y, z; \quad i' = \gamma'$$
$$= \frac{\mathrm{tr}\left\{\hat{\Gamma}_{i'}\hat{F}\hat{\Omega}_i\hat{F}^+\right\}}{\mathrm{tr}\left\{\hat{F}\hat{F}^+\right\}}, \quad i = \gamma; \quad i' = x', y', z'$$
$$= \frac{\mathrm{tr}\left\{\hat{\Omega}_{i'}\hat{F}\hat{\Omega}_i\hat{F}^+\right\}}{\mathrm{tr}\left\{\hat{F}\hat{F}^+\right\}}, \quad i = \gamma; \quad i' = \gamma' \tag{3.831}$$

Now the polarization transfer coefficients defined by the above four formulas are all introduced assuming that only one particle in the incident channel is polarized. If both particles in the incident channel are polarized, then a correlation polarization transfer coefficient needs to be introduced. For the outgoing particles, the correlation problem need not be considered.

If it is not necessary to study the polarization rate of the outgoing particles corresponding to the polarized incident particles, then there is no need to find the polarization transfer coefficient; therefore, this section will not deduce the polarization transfer coefficient. It follows that when deriving the polarization analyzing power, it will no longer be necessary to prepare the matrices for deriving the polarization transfer coefficients. Instead, we first obtain the following relation from Eqs. (2.6), (3.795), and (3.793):

$$\hat{\Sigma}_x\hat{F}^+ = \begin{pmatrix} \hat{0}_3 & \hat{I}_3 \\ \hat{I}_3 & \hat{0}_3 \end{pmatrix}\begin{pmatrix} \hat{t}^+_{++} & \hat{t}^+_{-+} \\ \hat{t}^+_{+-} & \hat{t}^+_{--} \end{pmatrix} = \begin{pmatrix} \hat{t}^+_{+-} & \hat{t}^+_{--} \\ \hat{t}^+_{++} & \hat{t}^+_{-+} \end{pmatrix} \tag{3.832}$$

Now we can obtain the following result from Eqs. (3.792), (3.793), and (3.825):

$$\begin{aligned} A_x^{\mathrm{N}} &= \frac{1}{6I_0}\mathrm{tr}\left\{\hat{F}\hat{\Sigma}_x\hat{F}^+\right\} = \frac{1}{6I_0}\mathrm{tr}\left(\hat{t}_{++}\hat{t}^+_{+-} + \hat{t}_{+-}\hat{t}^+_{++} + \hat{t}_{-+}\hat{t}^+_{--} + \hat{t}_{--}\hat{t}^+_{-+}\right) \\ &= \frac{1}{6I_0}[-\mathrm{i}(AD^* + BE^* + CF^*) - \mathrm{i}(MJ^* + LH^* + KG^*) - \mathrm{i}(NR^* + OS^* + QT^*) \\ &+\mathrm{i}(DA^* + EB^* + FC^*) + \mathrm{i}(JM^* + HL^* + GK^*) + \mathrm{i}(RN^* + SO^* + TQ^*) \\ &-\mathrm{i}(TQ^* + SO^* + RN^*) - \mathrm{i}(GK^* + HL^* + JM^*) - \mathrm{i}(FC^* + EB^* + DA^*) \\ &+\mathrm{i}(QT^* + OS^* + NR^*) + \mathrm{i}(KG^* + LH^* + MJ^*) + \mathrm{i}(CF^* + BE^* + AD^*)] = 0 \end{aligned} \tag{3.833}$$

Using the same method we can obtain

$$\hat{\Sigma}_y\hat{F}^+ = \begin{pmatrix} \hat{0}_3 & -\mathrm{i}\hat{I}_3 \\ \mathrm{i}\hat{I}_3 & \hat{0}_3 \end{pmatrix}\begin{pmatrix} \hat{t}^+_{++} & \hat{t}^+_{-+} \\ \hat{t}^+_{+-} & \hat{t}^+_{--} \end{pmatrix} = -\mathrm{i}\begin{pmatrix} \hat{t}^+_{+-} & \hat{t}^+_{--} \\ -\hat{t}^+_{++} & -\hat{t}^+_{-+} \end{pmatrix} \tag{3.834}$$

Next, referring to the results given by Eq. (3.833), we can obtain

$$\begin{aligned} A_y^{\mathrm{N}} &= \frac{1}{6I_0}\mathrm{tr}\left\{\hat{F}\hat{\Sigma}_y\hat{F}^+\right\} = -\frac{\mathrm{i}}{6I_0}\mathrm{tr}\left(\hat{t}_{++}\hat{t}^+_{+-} - \hat{t}_{+-}\hat{t}^+_{++} + \hat{t}_{-+}\hat{t}^+_{--} - \hat{t}_{--}\hat{t}^+_{-+}\right) \\ &= -\frac{2}{3I_0}\mathrm{Re}\left(AD^* + BE^* + CF^* + MJ^* + LH^* + KG^* + NR^* + OS^* + QT^*\right) \end{aligned} \tag{3.835}$$

We can also obtain

$$\hat{\Sigma}_z\hat{F}^+ = \begin{pmatrix} \hat{I}_3 & \hat{0}_3 \\ \hat{0}_3 & -\hat{I}_3 \end{pmatrix}\begin{pmatrix} \hat{t}^+_{++} & \hat{t}^+_{-+} \\ \hat{t}^+_{+-} & \hat{t}^+_{--} \end{pmatrix} = \begin{pmatrix} \hat{t}^+_{++} & \hat{t}^+_{-+} \\ -\hat{t}^+_{+-} & -\hat{t}^+_{--} \end{pmatrix} \tag{3.836}$$

Now from Eq. (3.792) we can directly see

$$A_z^{\mathrm{N}} = \frac{1}{6I_0}\mathrm{tr}\left\{\hat{F}\hat{\Sigma}_z\hat{F}^+\right\} = \frac{1}{6I_0}\mathrm{tr}\{\hat{t}_{++}\hat{t}^+_{++} - \hat{t}_{+-}\hat{t}^+_{+-} + \hat{t}_{-+}\hat{t}^+_{-+} - \hat{t}_{--}\hat{t}^+_{--}\} = 0 \tag{3.837}$$

and we can write the following expression according to Eqs. (3.803) and (3.796):

$$\mathrm{tr}\left\{\hat{F}\hat{\Gamma}_i\hat{F}^+\right\} = \mathrm{tr}\{\hat{t}_{++}\hat{S}_i\hat{t}^+_{++} + \hat{t}_{+-}\hat{S}_i\hat{t}^+_{+-} + \hat{t}_{-+}\hat{S}_i\hat{t}^+_{-+} + \hat{t}_{--}\hat{S}_i\hat{t}^+_{--}\}, \quad i = x, y, z \tag{3.838}$$

Also, we can obtain the following expressions from Eqs. (3.33) and (3.792):

$$\begin{aligned}
\hat{S}_x\hat{t}^+_{++} &= \frac{1}{\sqrt{2}}\begin{pmatrix} \mathrm{i}B^* & L^* & -\mathrm{i}O^* \\ A^* - C^* & -\mathrm{i}(M^* - K^*) & -(N^* - Q^*) \\ \mathrm{i}B^* & L^* & -\mathrm{i}O^* \end{pmatrix}, \\
\hat{S}_x\hat{t}^+_{+-} &= \frac{1}{\sqrt{2}}\begin{pmatrix} E^* & -\mathrm{i}H^* & -S^* \\ -\mathrm{i}(D^* - F^*) & -(J^* - G^*) & \mathrm{i}(R^* - T^*) \\ E^* & -\mathrm{i}H^* & -S^* \end{pmatrix}, \\
\hat{S}_x\hat{t}^+_{-+} &= \frac{1}{\sqrt{2}}\begin{pmatrix} -S^* & \mathrm{i}H^* & E^* \\ \mathrm{i}(T^* - R^*) & G^* - J^* & -\mathrm{i}(F^* - D^*) \\ -S^* & \mathrm{i}H^* & E^* \end{pmatrix}, \\
\hat{S}_x\hat{t}^+_{--} &= \frac{1}{\sqrt{2}}\begin{pmatrix} \mathrm{i}O^* & L^* & -\mathrm{i}B^* \\ Q^* - N^* & -\mathrm{i}(K^* - M^*) & -(C^* - A^*) \\ \mathrm{i}O^* & L^* & -\mathrm{i}B^* \end{pmatrix}
\end{aligned} \tag{3.839}$$

$$
\begin{aligned}
A_x^{\mathrm{d}} &= \frac{1}{6I_0}\mathrm{tr}\left\{\hat{F}\hat{\Gamma}_x\hat{F}^+\right\} \\
&= \frac{1}{6\sqrt{2}I_0}[\mathrm{i}(AB^* - BA^* + BC - CB^*) + \mathrm{i}(ML^* - LM^* + LK^* - KL^*) \\
&+\mathrm{i}(NO^* - ON^* + OQ^* - QO^*) + \mathrm{i}(DE^* - ED^* + EF^* - FE^*) \\
&+\mathrm{i}(JH^* - HJ^* + HG^* - GH^*) + \mathrm{i}(RS^* - SR^* + ST^* - TS^*) \\
&+\mathrm{i}(TS^* - ST^* + SR^* - RS^*) + \mathrm{i}(GH^* - HG^* + HJ^* - JH^*) \\
&+\mathrm{i}(FE^* - EF^* + ED^* - DE^*) + \mathrm{i}(QO^* - OQ^* + ON^* - NO^*) \\
&+\mathrm{i}(KL^* - LK^* + LM^* - ML^*) + \mathrm{i}(CB^* - BC^* + BA^* - AB^*)] = 0
\end{aligned} \tag{3.840}
$$

From Eq. (3.839), a comparison of $\hat{S}_x$ and $\hat{S}_y$ in Eq. (3.33) permits us to directly write

$$
\begin{aligned}
\hat{S}_y\hat{t}^+_{++} &= \frac{\mathrm{i}}{\sqrt{2}}\begin{pmatrix} -\mathrm{i}B^* & -L^* & \mathrm{i}O^* \\ A^* + C^* & -\mathrm{i}(M^* + K^*) & -(N^* + Q^*) \\ \mathrm{i}B^* & L^* & -\mathrm{i}O^* \end{pmatrix}, \\
\hat{S}_y\hat{t}^+_{+-} &= \frac{\mathrm{i}}{\sqrt{2}}\begin{pmatrix} -E^* & \mathrm{i}H^* & S^* \\ -\mathrm{i}(D^* + F^*) & -(J^* + G^*) & \mathrm{i}(R^* + T^*) \\ E^* & -\mathrm{i}H^* & -S^* \end{pmatrix}, \\
\hat{S}_y\hat{t}^+_{-+} &= \frac{\mathrm{i}}{\sqrt{2}}\begin{pmatrix} S^* & -\mathrm{i}H^* & -E^* \\ \mathrm{i}(T^* + R^*) & G^* + J^* & -\mathrm{i}(F^* + D^*) \\ -S^* & -\mathrm{i}H^* & E^* \end{pmatrix}, \\
\hat{S}_y\hat{t}^+_{--} &= \frac{\mathrm{i}}{\sqrt{2}}\begin{pmatrix} -\mathrm{i}O^* & -L^* & \mathrm{i}B^* \\ Q^* + N^* & -\mathrm{i}(K^* + M^*) & -(C^* + A^*) \\ \mathrm{i}O^* & L^* & -\mathrm{i}B^* \end{pmatrix}
\end{aligned} \tag{3.841}
$$

and referring to Eq. (3.840), we can immediately write

$$
\begin{aligned}
A_y^{\mathrm{d}} &= \frac{1}{6I_0}\mathrm{tr}\left\{\hat{F}\hat{\Gamma}_y\hat{F}^+\right\} \\
&= \frac{\mathrm{i}}{6\sqrt{2}I_0}[-\mathrm{i}(AB^* + BA^* + BC + CB^*) - \mathrm{i}(ML^* + LM^* + LK^* + KL^*) \\
&-\mathrm{i}(NO^* + ON^* + OQ^* + QO^*) - \mathrm{i}(DE^* + ED^* + EF^* + FE^*) \\
&-\mathrm{i}(JH^* + HJ^* + HG^* + GH^*) - \mathrm{i}(RS^* + SR^* + ST^* + TS^*) \\
&-\mathrm{i}(TS^* + ST^* + SR^* + RS^*) - \mathrm{i}(GH^* + HG^* + HJ^* + JH^*) \\
&-\mathrm{i}(FE^* + EF^* + ED^* + DE^*) - \mathrm{i}(QO^* + OQ^* + ON^* + NO^*) \\
&-\mathrm{i}(KL^* + LK^* + LM^* + ML^*) - \mathrm{i}(CB^* + BC^* + BA^* + AB^*)] \\
&= \frac{\sqrt{2}}{3I_0}\mathrm{Re}\,(AB^* + BC^* + DE^* + EF^* + GH^* + HJ^* \\
&+KL^* + LM^* + NO^* + OQ^* + RS^* + ST^*)
\end{aligned} \tag{3.842}
$$

Further we can also obtain the following expressions from Eqs. (3.33) and (3.792):

$$\hat{S}_z\hat{t}^+_{++} = \begin{pmatrix} A^* & -iM^* & -N^* \\ 0 & 0 & 0 \\ C^* & -iK^* & -Q^* \end{pmatrix}, \quad \hat{S}_z\hat{t}^+_{+-} = \begin{pmatrix} -iD^* & -J^* & iR^* \\ 0 & 0 & 0 \\ -iF^* & -G^* & iT^* \end{pmatrix},$$
$$\hat{S}_z\hat{t}^+_{-+} = \begin{pmatrix} iT^* & G^* & -iF^* \\ 0 & 0 & 0 \\ iR^* & J^* & -iD^* \end{pmatrix}, \quad \hat{S}_z\hat{t}^+_{--} = \begin{pmatrix} Q^* & -iK^* & -C^* \\ 0 & 0 & 0 \\ N^* & -iM^* & -A^* \end{pmatrix} \tag{3.843}$$

Now the following results can be obtained based on Eqs. (3.792) and (3.838):

$$\begin{aligned} A^{\mathrm{d}}_z = \frac{1}{6I_0}\mathrm{tr}\left\{\hat{F}\hat{\Gamma}_z\hat{F}^+\right\} = \frac{1}{6I_0}(\ &|A|^2 - |C|^2 + |M|^2 - |K|^2 + |N|^2 - |Q|^2 \\ &+|D|^2 - |F|^2 + |J|^2 - |G|^2 + |R|^2 - |T|^2 + |T|^2 - |R|^2 + |G|^2 - |J|^2 \\ &+|F|^2 - |D|^2 + |Q|^2 - |N|^2 + |K|^2 - |M|^2 + |C|^2 - |A|^2) = 0 \end{aligned} \tag{3.844}$$

And we can also write the following expression according to Eqs. (3.797) and (3.803):

$$\mathrm{tr}\left\{\hat{F}\hat{\Omega}_i\hat{F}^+\right\} = \mathrm{tr}\{\hat{t}_{++}\hat{Q}_i\hat{t}^+_{++} + \hat{t}_{+-}\hat{Q}_i\hat{t}^+_{+-} + \hat{t}_{-+}\hat{Q}_i\hat{t}^+_{-+} + \hat{t}_{--}\hat{Q}_i\hat{t}^+_{--}\}, \quad i = \gamma \tag{3.845}$$

where the definition of the index set $\gamma$ has been given by Eq. (3.824). Moreover the following expressions can be obtained from Eqs. (3.36) and (3.792):

$$\hat{Q}_{xy}\hat{t}^+_{++} = \frac{3i}{2}\begin{pmatrix} C^* & -iK^* & -Q^* \\ 0 & 0 & 0 \\ A^* & -iM^* & -N^* \end{pmatrix}, \quad \hat{Q}_{xy}\hat{t}^+_{+-} = \frac{3i}{2}\begin{pmatrix} -iF^* & -G^* & iT^* \\ 0 & 0 & 0 \\ -iD^* & -J^* & iR^* \end{pmatrix},$$
$$\hat{Q}_{xy}\hat{t}^+_{-+} = \frac{3i}{2}\begin{pmatrix} iR^* & J^* & -iD^* \\ 0 & 0 & 0 \\ iT^* & G^* & -iF^* \end{pmatrix}, \hat{Q}_{xy}\hat{t}^+_{--} = \frac{3i}{2}\begin{pmatrix} N^* & -iM^* & -A^* \\ 0 & 0 & 0 \\ Q^* & -iK^* & -C^* \end{pmatrix} \tag{3.846}$$

$$\begin{aligned} A^{\mathrm{d}}_{xy} = \frac{1}{6I_0}\mathrm{tr}\left\{\hat{F}\hat{\Omega}_{xy}\hat{F}^+\right\} = \frac{i}{4I_0}(&AC^* - CA^* + MK^* - KM^* + NQ^* - QN^* \\ &+DF^* - FD^* + JG^* - GJ^* + RT^* - TR^* + TR^* - RT^* + GJ^* - JG^* \\ &+FD^* - DF^* + QN^* - NQ^* + KM^* - MK^* + CA^* - AC^*) = 0 \end{aligned} \tag{3.847}$$

We can also obtain

$$
\begin{aligned}
\hat{Q}_{xz}\hat{t}^{+}_{++} &= \frac{3}{2\sqrt{2}}\begin{pmatrix} \mathrm{i}B^* & L^* & -\mathrm{i}O^* \\ A^*+C^* & -\mathrm{i}(M^*+K^*) & -(N^*+Q^*) \\ -\mathrm{i}B^* & -L^* & \mathrm{i}O^* \end{pmatrix}, \\
\hat{Q}_{xz}\hat{t}^{+}_{+-} &= \frac{3}{2\sqrt{2}}\begin{pmatrix} E^* & -\mathrm{i}H^* & -S^* \\ -\mathrm{i}(D^*+F^*) & -(J^*+G^*) & \mathrm{i}(R^*+T^*) \\ -E^* & \mathrm{i}H^* & S^* \end{pmatrix}, \\
\hat{Q}_{xz}\hat{t}^{+}_{-+} &= \frac{3}{2\sqrt{2}}\begin{pmatrix} -S^* & \mathrm{i}H^* & E^* \\ \mathrm{i}(T^*+R^*) & G^*+J^* & -\mathrm{i}(F^*+D^*) \\ S^* & -\mathrm{i}H^* & -E^* \end{pmatrix}, \\
\hat{Q}_{xz}\hat{t}^{+}_{--} &= \frac{3}{2\sqrt{2}}\begin{pmatrix} \mathrm{i}O^* & L^* & -\mathrm{i}B^* \\ Q^*+N^* & -\mathrm{i}(K^*+M^*) & -(C^*+A^*) \\ -\mathrm{i}O^* & -L^* & \mathrm{i}B^* \end{pmatrix}
\end{aligned}
\tag{3.848}
$$

and we can obtain the following result utilizing Eqs. (3.845), (3.848), and (3.792):

$$
\begin{aligned}
A^{\mathrm{d}}_{xz} &= \frac{1}{6I_0}\mathrm{tr}\left\{\hat{F}\hat{\Omega}_{xz}\hat{F}^+\right\} \\
&= \frac{1}{4\sqrt{2}I_0}[\mathrm{i}(AB^*-BA^*-BC^*+CB^*)+\mathrm{i}(ML^*-LM^*-LK^*+KL^*) \\
&+\mathrm{i}(NO^*-ON^*-OQ^*+QO^*)+\mathrm{i}(DE^*-ED^*-EF^*+FE^*) \\
&+\mathrm{i}(JH^*-HJ^*-HG^*+GH^*)+\mathrm{i}(RS^*-SR^*-ST^*+TS^*) \\
&+\mathrm{i}(TS^*-ST^*-SR^*+RS^*)+\mathrm{i}(GH^*-HG^*-HJ^*+JH^*) \\
&+\mathrm{i}(FE^*-EF^*-ED^*+DE^*)+\mathrm{i}(QO^*-OQ^*-ON^*+NO^*) \\
&+\mathrm{i}(KL^*-LK^*-LM^*+ML^*)+\mathrm{i}(CB^*-BC^*-BA^*+AB^*)] \\
&= -\frac{1}{\sqrt{2}I_0}\mathrm{Im}(AB^*-BC^*+DE^*-EF^*+GH^*-HJ^* \\
&+KL^*-LM^*+NO^*-OQ^*+RS^*-ST^*)
\end{aligned}
\tag{3.849}
$$

Comparing $\hat{Q}_{xz}$ and $\hat{Q}_{yz}$ in Eq. (3.36), from Eq. (3.848), we can directly write

$$
\begin{aligned}
\hat{Q}_{yz}\hat{t}^{+}_{++} &= \frac{3\mathrm{i}}{2\sqrt{2}}\begin{pmatrix} -\mathrm{i}B^* & -L^* & \mathrm{i}O^* \\ A^*-C^* & -\mathrm{i}(M^*-K^*) & -(N^*-Q^*) \\ -\mathrm{i}B^* & -L^* & \mathrm{i}O^* \end{pmatrix}, \\
\hat{Q}_{yz}\hat{t}^{+}_{+-} &= \frac{3\mathrm{i}}{2\sqrt{2}}\begin{pmatrix} -E^* & \mathrm{i}H^* & S^* \\ -\mathrm{i}(D^*-F^*) & -(J^*-G^*) & \mathrm{i}(R^*-T^*) \\ -E^* & \mathrm{i}H^* & S^* \end{pmatrix}, \\
\hat{Q}_{yz}\hat{t}^{+}_{-+} &= \frac{3\mathrm{i}}{2\sqrt{2}}\begin{pmatrix} S^* & -\mathrm{i}H^* & -E^* \\ \mathrm{i}(T^*-R^*) & G^*-J^* & -\mathrm{i}(F^*-D^*) \\ S^* & -\mathrm{i}H^* & -E^* \end{pmatrix}, \\
\hat{Q}_{yz}\hat{t}^{+}_{--} &= \frac{3\mathrm{i}}{2\sqrt{2}}\begin{pmatrix} -\mathrm{i}O^* & -L^* & \mathrm{i}B^* \\ Q^*-N^* & -\mathrm{i}(K^*-M^*) & -(C^*-A^*) \\ -\mathrm{i}O^* & -L^* & \mathrm{i}B^* \end{pmatrix}
\end{aligned}
\tag{3.850}
$$

Next, comparing Eq. (3.850) with Eq. (3.848), from Eq. (3.849), we can directly write

$$
\begin{aligned}
A^{\mathrm{d}}_{yz} &= \frac{1}{6I_0}\mathrm{tr}\left\{\hat{F}\hat{\Omega}_{yz}\hat{F}^+\right\} \\
&= \frac{\mathrm{i}}{4\sqrt{2}I_0}[-\mathrm{i}(AB^*+BA^*-BC-CB^*)-\mathrm{i}(ML^*+LM^*-LK^*-KL^*) \\
&\quad -\mathrm{i}(NO^*+ON^*-OQ^*-QO^*)-\mathrm{i}(DE^*+ED^*-EF^*-FE^*) \\
&\quad -\mathrm{i}(JH^*+HJ^*-HG^*-GH^*)-\mathrm{i}(RS^*+SR^*-ST^*-TS^*) \\
&\quad -\mathrm{i}(TS^*+ST^*-SR^*-RS^*)-\mathrm{i}(GH^*+HG^*-HJ^*-JH^*) \\
&\quad -\mathrm{i}(FE^*+EF^*-ED^*-DE^*)-\mathrm{i}(QO^*+OQ^*-ON^*-NQ^*) \\
&\quad -\mathrm{i}(KL^*+LK^*-LM^*-ML^*)-\mathrm{i}(CB^*+BC^*-BA^*-AB^*)]=0
\end{aligned}
\tag{3.851}
$$

We can also obtain the following expressions from Eqs. (3.36) and (3.792):

$$
\begin{aligned}
&\hat{Q}_{xx-yy}\hat{t}^{+}_{++}=3\begin{pmatrix} -C^* & \mathrm{i}K^* & Q^* \\ 0 & 0 & 0 \\ A^* & -\mathrm{i}M^* & -N^* \end{pmatrix}, \quad \hat{Q}_{xx-yy}\hat{t}^{+}_{+-}=3\begin{pmatrix} \mathrm{i}F^* & G^* & -\mathrm{i}T^* \\ 0 & 0 & 0 \\ -\mathrm{i}D^* & -J^* & \mathrm{i}R^* \end{pmatrix}, \\
&\hat{Q}_{xx-yy}\hat{t}^{+}_{-+}=3\begin{pmatrix} -\mathrm{i}R^* & -J^* & \mathrm{i}D^* \\ 0 & 0 & 0 \\ \mathrm{i}T^* & G^* & -\mathrm{i}F^* \end{pmatrix}, \quad \hat{Q}_{xx-yy}\hat{t}^{+}_{--}=3\begin{pmatrix} -N^* & \mathrm{i}M^* & A^* \\ 0 & 0 & 0 \\ Q^* & -\mathrm{i}K^* & -C^* \end{pmatrix}
\end{aligned}
\tag{3.852}
$$

$$A^{\mathrm{d}}_{xx-yy}=\frac{1}{6I_0}\mathrm{tr}\left\{\hat{F}\hat{\Omega}_{xx-yy}\hat{F}^{+}\right\}=\frac{1}{2I_0}(-AC^*-CA^*-MK^*-KM^*-NQ^*$$
$$-QN^*-DF^*-FD^*-JG^*-GJ^*-RT^*-TR^*-TR^*-RT^*-GJ^*$$
$$-JG^*-FD^*-DF^*-QN^*-NQ^*-KM^*-MK^*-CA^*-AC^*)$$
$$=-\frac{2}{I_0}\mathrm{Re}\,(AC^*+MK^*+NQ^*+DF^*+JG^*+RT^*) \tag{3.853}$$

The following expressions can also be obtained from Eqs. (3.36) and (3.792):

$$\hat{Q}_{zz}\hat{t}^{+}_{++}=\begin{pmatrix} A^* & -\mathrm{i}M^* & -N^* \\ -2\mathrm{i}B^* & -2L^* & 2\mathrm{i}O^* \\ -C^* & \mathrm{i}K^* & Q^* \end{pmatrix},\quad \hat{Q}_{zz}\hat{t}^{+}_{+-}=\begin{pmatrix} -\mathrm{i}D^* & -J^* & \mathrm{i}R^* \\ -2E^* & 2\mathrm{i}H^* & 2S^* \\ \mathrm{i}F^* & G^* & -\mathrm{i}T^* \end{pmatrix},$$
$$\hat{Q}_{zz}\hat{t}^{+}_{-+}=\begin{pmatrix} \mathrm{i}T^* & G^* & -\mathrm{i}F^* \\ 2S^* & -2\mathrm{i}H^* & -2E^* \\ -\mathrm{i}R^* & -J^* & \mathrm{i}D^* \end{pmatrix},\quad \hat{Q}_{zz}\hat{t}^{+}_{--}=\begin{pmatrix} Q^* & -\mathrm{i}K^* & -C^* \\ -2\mathrm{i}O^* & -2L^* & 2\mathrm{i}B^* \\ -N^* & \mathrm{i}M^* & A^* \end{pmatrix} \tag{3.854}$$

$$A^{\mathrm{d}}_{zz}=\frac{1}{6I_0}\mathrm{tr}\left\{\hat{F}\hat{\Omega}_{zz}\hat{F}^{+}\right\}=\frac{1}{6I_0}(|A|^2-2|B|^2+|C|^2+|M|^2-2|L|^2+|K|^2+|N|^2$$
$$-2|O|^2+|Q|^2+|D|^2-2|E|^2+|F|^2+|J|^2-2|H|^2+|G|^2+|R|^2-2|S|^2+|T|^2$$
$$+|T|^2-2|S|^2+|R|^2+|G|^2-2|H|^2+|J|^2+|F|^2-2|E|^2+|D|^2+|Q|^2$$
$$-2|O|^2+|N|^2+|K|^2-2|L|^2+|M|^2+|C|^2-2|B|^2+|A|^2)$$
$$=\frac{1}{3I_0}[|A|^2+|C|^2+|D|^2+|F|^2+|G|^2+|J|^2+|K|^2+|M|^2+|N|^2+|Q|^2$$
$$+|R|^2+|T|^2-2\left(|B|^2+|E|^2+|H|^2+|L|^2+|O|^2+|S|^2\right)] \tag{3.855}$$

There are also the following expressions according to Eqs. (3.833), (3.835), (3.837), (3.796), and (3.797):

$$\mathrm{tr}\left\{\hat{F}\hat{\Sigma}_x\hat{\Gamma}_i\hat{F}^{+}\right\}=\mathrm{tr}(\hat{t}_{++}\hat{S}_i\hat{t}^{+}_{+-}+\hat{t}_{+-}\hat{S}_i\hat{t}^{+}_{++}+\hat{t}_{-+}\hat{S}_i\hat{t}^{+}_{--}+\hat{t}_{--}\hat{S}_i\hat{t}^{+}_{-+}),$$
$$\mathrm{tr}\left\{\hat{F}\hat{\Sigma}_y\hat{\Gamma}_i\hat{F}^{+}\right\}=-\mathrm{i}\,\mathrm{tr}(\hat{t}_{++}\hat{S}_i\hat{t}^{+}_{+-}-\hat{t}_{+-}\hat{S}_i\hat{t}^{+}_{++}+\hat{t}_{-+}\hat{S}_i\hat{t}^{+}_{--}-\hat{t}_{--}\hat{S}_i\hat{t}^{+}_{-+}),$$
$$\mathrm{tr}\left\{\hat{F}\hat{\Sigma}_z\hat{\Gamma}_i\hat{F}^{+}\right\}=\mathrm{tr}(\hat{t}_{++}\hat{S}_i\hat{t}^{+}_{++}-\hat{t}_{+-}\hat{S}_i\hat{t}^{+}_{+-}+\hat{t}_{-+}\hat{S}_i\hat{t}^{+}_{-+}-\hat{t}_{--}\hat{S}_i\hat{t}^{+}_{--}),$$
$$i=x,y,z \tag{3.856}$$

$$
\begin{aligned}
&\mathrm{tr}\left\{\hat{F}\hat{\Sigma}_x\hat{\Omega}_i\hat{F}^+\right\} = \mathrm{tr}\left(\hat{t}_{++}\hat{Q}_i\hat{t}^+_{+-} + \hat{t}_{+-}\hat{Q}_i\hat{t}^+_{++} + \hat{t}_{-+}\hat{Q}_i\hat{t}^+_{--} + \hat{t}_{--}\hat{Q}_i\hat{t}^+_{-+}\right),\\
&\mathrm{tr}\left\{\hat{F}\hat{\Sigma}_y\hat{\Omega}_i\hat{F}^+\right\} = -\mathrm{i}\,\mathrm{tr}\left(\hat{t}_{++}\hat{Q}_i\hat{t}^+_{+-} - \hat{t}_{+-}\hat{Q}_i\hat{t}^+_{++} + \hat{t}_{-+}\hat{Q}_i\hat{t}^+_{--} - \hat{t}_{--}\hat{Q}_i\hat{t}^+_{-+}\right),\\
&\mathrm{tr}\left\{\hat{F}\hat{\Sigma}_z\hat{\Omega}_i\hat{F}^+\right\} = \mathrm{tr}\left(\hat{t}_{++}\hat{Q}_i\hat{t}^+_{++} - \hat{t}_{+-}\hat{Q}_i\hat{t}^+_{+-} + \hat{t}_{-+}\hat{Q}_i\hat{t}^+_{-+} - \hat{t}_{--}\hat{Q}_i\hat{t}^+_{--}\right),\\
&\qquad\qquad\qquad\qquad i = \gamma
\end{aligned}
\tag{3.857}
$$

Utilizing Eqs. (3.839) and (3.792), we can find the following result from Eqs. (3.827), (3.795), (3.796), and (3.856):

$$
\begin{aligned}
A^{\mathrm{Nd}}_{x,x} &= \frac{1}{6I_0}\mathrm{tr}\left\{\hat{F}\hat{\Sigma}_x\hat{\Gamma}_x\hat{F}^+\right\} = \frac{1}{6\sqrt{2}I_0}(AE^* - BD^* + BF^* - CE^* + MH^* - LJ^*\\
&\quad + LG^* - KH^* + NS^* - OR^* + OT^* - QS^* - DB^* + EA^* - EC^* + FB^* - JL^*\\
&\quad + HM^* - HK^* + GL^* - RO^* + SN^* - SQ^* + TO^* + TO^* - SQ^* + SN^* - RO^*\\
&\quad + GL^* - HK^* + HM^* - JL^* + FB^* - EC^* + EA^* - DB^* - QS^* + OT^* - OR^*\\
&\quad + NS^* - KH^* + LG^* - LJ^* + MH^* - CE^* + BF^* - BD^* + AE^*)\\
&= \frac{\sqrt{2}}{3I_0}\,\mathrm{Re}\,(AE^* - BD^* + BF^* - CE^* + MH^* - LJ^* + LG^* - KH^*\\
&\quad + NS^* - OR^* + OT^* - QS^*)
\end{aligned}
\tag{3.858}
$$

Now, according to Eqs. (3.856) and (3.858), we get the following result:

$$
A^{\mathrm{Nd}}_{y,x} = 0 \tag{3.859}
$$

Comparing the third equation of Eq. (3.856) with Eq. (3.838), from Eq. (3.840), we can obtain

$$
\begin{aligned}
A^{\mathrm{Nd}}_{z,x} = \frac{1}{6I_0}\mathrm{tr}\left\{\hat{F}\hat{\Sigma}_z\hat{\Gamma}_x\hat{F}^+\right\} = -\frac{\sqrt{2}}{3I_0}\mathrm{Im}(AB^* + BC^* - DE^* - EF^*\\
+ GH^* + HJ^* - KL^* - LM^* + NO^* + OQ^* - RS^* - ST^*)
\end{aligned}
\tag{3.860}
$$

Using the same method, we can obtain the following result via Eqs. (3.792), (3.795), (3.796), (3.827), (3.841), and (3.856):

$$
\begin{aligned}
A_{x,y}^{\mathrm{Nd}} = & \frac{\mathrm{i}}{6\sqrt{2}I_0}(-AE^* - BD^* - BF^* - CE^* - MH^* - LJ^* - LG^* \\
& - KH^* - NS^* - OR^* - OT^* - QS^* + DB^* + EA^* + EC^* + FB^* \\
& + JL^* + HM^* + HK^* + GL^* + RO^* + SN^* + SQ^* + TO^* - TO^* \\
& - SQ^* - SN^* - RO^* - GL^* - HK^* - HM^* - JL^* - FB^* - EC^* \\
& - EA^* - DB^* + QS^* + OT^* + OR^* + NS^* + KH^* + LG^* + LJ^* \\
& + MH^* + CE^* + BF^* + BD^* + AE^*) = 0
\end{aligned}
\tag{3.861}
$$

We can also obtain the following result according to Eqs. (3.856) and (3.861):

$$
\begin{aligned}
A_{y,y}^{\mathrm{Nd}} = & -\frac{\sqrt{2}}{3I_0}\mathrm{Re}\,(AE^* + BD^* + BF^* + CE^* + MH^* + LJ^* \\
& + LG^* + KH^* + NS^* + OR^* + OT^* + QS^*)
\end{aligned}
\tag{3.862}
$$

Comparing the third equation of Eq. (3.856) with Eq. (3.838), from Eq. (3.842), we can obtain

$$
A_{z,y}^{\mathrm{Nd}} = 0 \tag{3.863}
$$

Next, the following result can be obtained utilizing Eqs. (3.792), (3.795), (3.796), (3.827), (3.843), and (3.856):

$$
\begin{aligned}
A_{x,z}^{\mathrm{Nd}} = & \frac{1}{6I_0}(-\mathrm{i}AD^* + \mathrm{i}CF^* - \mathrm{i}MJ^* + \mathrm{i}KG^* - \mathrm{i}NR^* + \mathrm{i}QT^* + \mathrm{i}DA^* - \mathrm{i}FC^* \\
& + \mathrm{i}JM^* - \mathrm{i}GK^* + \mathrm{i}RN^* - \mathrm{i}TQ^* - \mathrm{i}TQ^* + \mathrm{i}RN^* - \mathrm{i}GK^* + \mathrm{i}JM^* - \mathrm{i}FC^* \\
& + \mathrm{i}DA^* + \mathrm{i}QT^* - \mathrm{i}NR^* + \mathrm{i}KG^* - \mathrm{i}MJ^* + \mathrm{i}CF^* - \mathrm{i}AD^*) \\
= & \frac{2}{3I_0}\mathrm{Im}(AD^* - CF^* + MJ^* - KG^* + NR^* - QT^*)
\end{aligned}
\tag{3.864}
$$

and we can get the following according to Eqs. (3.856) and (3.864):

$$
A_{y,z}^{\mathrm{Nd}} = 0 \tag{3.865}
$$

Comparing the third equation of Eq. (3.856) with Eq. (3.838), from Eq. (3.844) we can obtain

$$
\begin{aligned}
A_{z,z}^{\mathrm{Nd}} = & \frac{1}{3I_0}\Big(|A|^2 - |C|^2 + |F|^2 - |D|^2 + |G|^2 - |J|^2 + |M|^2 - |K|^2 \\
& + |N|^2 - |Q|^2 + |T|^2 - |R|^2\Big)
\end{aligned}
\tag{3.866}
$$

and the following results can be obtained by using Eqs. (3.846), (3.857), (3.792), and (3.797):

$$\begin{aligned} A^{\mathrm{Nd}}_{x,xy} &= \frac{\mathrm{i}}{4I_0}( -\mathrm{i}AF^* + \mathrm{i}CD^* - \mathrm{i}MG^* + \mathrm{i}KJ^* - \mathrm{i}NT^* + \mathrm{i}QR^* + \mathrm{i}DC^* \\ &\quad - \mathrm{i}FA^* + \mathrm{i}JK^* - \mathrm{i}GM^* + \mathrm{i}RQ^* - \mathrm{i}TN^* - \mathrm{i}TN^* + \mathrm{i}RQ^* - \mathrm{i}GM^* + \mathrm{i}JK^* \\ &\quad - \mathrm{i}FA^* + \mathrm{i}DC^* + \mathrm{i}QR^* - \mathrm{i}NT^* + \mathrm{i}KJ^* - \mathrm{i}MG^* + \mathrm{i}CD^* - \mathrm{i}AF^*) \\ &= \frac{1}{I_0}\,\mathrm{Re}\,(AF^* - CD^* + MG^* - KJ^* + NT^* - QR^*) \end{aligned} \tag{3.867}$$

$$A^{\mathrm{Nd}}_{y,xy} = 0 \tag{3.868}$$

Using the third equation of Eq. (3.857) and Eq. (3.847), we can obtain

$$A^{\mathrm{Nd}}_{z,xy} = -\frac{1}{I_0}\mathrm{Im}(AC^* - DF^* + MK^* - JG^* + NQ^* - RT^*) \tag{3.869}$$

The following results can also be obtained by using Eqs. (3.848), (3.857), (3.792), and (3.797):

$$\begin{aligned} A^{\mathrm{Nd}}_{x,xz} = \frac{1}{4\sqrt{2}I_0}&(AE^* - BD^* - BF^* + CE^* + MH^* - LJ^* - LG^* + KH^* \\ &+NS^* - OR^* - OT^* + QS^* - DB^* + EA^* + EC^* - FB^* \\ &- JL^* + HM^* + HK^* - GL^* - RO^* + SN^* + SQ^* - TO^* \\ &+TO^* - SQ^* - SN^* + RO^* + GL^* - HK^* - HM^* + JL^* \\ &+FB^* - EC^* - EA^* + DB^* - QS^* + OT^* + OR^* - NS^* \\ &- KH^* + LG^* + LJ^* - MH^* - CE^* + BF^* + BD^* - AE^*) = 0 \end{aligned} \tag{3.870}$$

$$\begin{aligned} A^{\mathrm{Nd}}_{y,xz} = \frac{1}{\sqrt{2}I_0}&\mathrm{Im}(AE^* - BD^* - BF^* + CE^* + MH^* - LJ^* \\ &- LG^* + KH^* + NS^* - OR^* - OT^* + QS^*) \end{aligned} \tag{3.871}$$

Next, using the third equation of Eq. (3.857) and Eq. (3.849), we can obtain

$$A^{\mathrm{Nd}}_{z,xz} = 0 \tag{3.872}$$

We can also obtain the following results by using Eqs. (3.850), (3.857), (3.792), and (3.797):

$$\begin{aligned}A^{\mathrm{Nd}}_{x,yz} &= \frac{\mathrm{i}}{4\sqrt{2}I_0}(-AE^* - BD^* + BF^* + CE^* - MH^* - LJ^* + LG^* + KH^* \\ &\quad - NS^* - OR^* + OT^* + QS^* + DB^* + EA^* - EC^* - FB^* \\ &\quad + JL^* + HM^* - HK^* - GL^* + RO^* + SN^* - SQ^* - TO^* \\ &\quad - TO^* - SQ^* + SN^* + RO^* - GL^* - HK^* + HM^* + JL^* \\ &\quad - FB^* - EC^* + EA^* + DB^* + QS^* + OT^* - OR^* - NS^* \\ &\quad + KH^* + LG^* - LJ^* - MH^* + CE^* + BF^* - BD^* - AE^*) \\ &= \frac{1}{\sqrt{2}I_0}\mathrm{Im}(AE^* + BD^* - BF^* - CE^* + MH^* + LJ^* \\ &\quad - LG^* - KH^* + NS^* + OR^* - OT^* - QS^*)\end{aligned} \tag{3.873}$$

$$A^{\mathrm{Nd}}_{y,yz} = 0 \tag{3.874}$$

Now, using the third equation of Eq. (3.857) together with Eq. (3.851), we can obtain

$$\begin{aligned}A^{\mathrm{Nd}}_{z,yz} &= \frac{1}{\sqrt{2}I_0}\mathrm{Re}\,(AB^* - BC^* - DE^* + EF^* + ML^* - LK^* \\ &\quad - JH^* + HG^* + NO^* - OQ^* - RS^* + ST^*)\end{aligned} \tag{3.875}$$

Next, we can obtain the following results by using Eqs. (3.852), (3.857), (3.792), and (3.797):

$$\begin{aligned}A^{\mathrm{Nd}}_{x,xx-yy} &= \frac{1}{2I_0}(-\mathrm{i}DC^* - \mathrm{i}FA^* - \mathrm{i}JK^* - \mathrm{i}GM^* - \mathrm{i}RQ^* - \mathrm{i}TN^* \\ &\quad + \mathrm{i}AF^* + \mathrm{i}CD^* + \mathrm{i}MG^* + \mathrm{i}KJ^* + \mathrm{i}NT^* + \mathrm{i}QR^* \\ &\quad - \mathrm{i}QR^* - \mathrm{i}NT^* - \mathrm{i}KJ^* - \mathrm{i}MG^* - \mathrm{i}CD^* - \mathrm{i}AF^* \\ &\quad + \mathrm{i}TN^* + \mathrm{i}RQ^* + \mathrm{i}GM^* + \mathrm{i}JK^* + \mathrm{i}FA^* + \mathrm{i}DC^*) = 0\end{aligned} \tag{3.876}$$

$$A^{\mathrm{Nd}}_{y,xx-yy} = \frac{2}{I_0}\mathrm{Re}\,(AF^* + CD^* + MG^* + KJ^* + NT^* + QR^*) \tag{3.877}$$

$$\begin{aligned}A^{\mathrm{Nd}}_{z,xx-yy} &= \frac{1}{2I_0}[-AC^* - CA^* - MK^* - KM^* - NQ^* - QN^* \\ &\quad - (-DF^* - FD^* - JG^* - GJ^* - RT^* - TR^*) \\ &\quad - TR^* - RT^* - GJ^* - JG^* - FD^* - DF^* \\ &\quad - (-QN^* - NQ^* - KM^* - MK^* - CA^* - AC^*)] = 0\end{aligned} \tag{3.878}$$

The following results can also be obtained by using Eqs. (3.854), (3.857), (3.792), and (3.797):

$$\begin{aligned} A^{\mathrm{Nd}}_{x,zz} = \frac{1}{6I_0}(&\mathrm{i}DA^* - 2\mathrm{i}EB^* + \mathrm{i}FC^* + \mathrm{i}JM^* - 2\mathrm{i}HL^* + \mathrm{i}GK^* \\ &+\mathrm{i}RN^* - 2\mathrm{i}SO^* + \mathrm{i}TQ^* - \mathrm{i}AD^* + 2\mathrm{i}BE^* - \mathrm{i}CF^* \\ &-\mathrm{i}MJ^* + 2\mathrm{i}LH^* - \mathrm{i}KG^* - \mathrm{i}NR^* + 2\mathrm{i}OS^* - \mathrm{i}QT^* \\ &+\mathrm{i}QT^* - 2\mathrm{i}OS^* + \mathrm{i}NR^* + \mathrm{i}KG^* - 2\mathrm{i}LH^* + \mathrm{i}MJ^* \\ &+\mathrm{i}CF^* - 2\mathrm{i}BE^* + \mathrm{i}AD^* - \mathrm{i}TQ^* + 2\mathrm{i}SO^* - \mathrm{i}RN^* \\ &-\mathrm{i}GK^* + 2\mathrm{i}HL^* - \mathrm{i}JM^* - \mathrm{i}FC^* + 2\mathrm{i}EB^* - \mathrm{i}DA^*) = 0 \end{aligned} \tag{3.879}$$

$$\begin{aligned} A^{\mathrm{Nd}}_{y,zz} = -\frac{2}{3I_0}\,\mathrm{Re}\,[&AD^* + CF^* + MJ^* + KG^* + NR^* + QT^* \\ &-2(BE^* + LH^* + OS^*)] \end{aligned} \tag{3.880}$$

Now referring to Eq. (3.855), we have

$$A^{\mathrm{Nd}}_{z,zz} = 0 \tag{3.881}$$

According to the non-zero analyzing powers obtained above and $\hat{\rho}_{\mathrm{in}}$ given by Eq. (3.798), it can be seen that when N and d of the incident channel are polarized the differential cross section is

$$\begin{aligned} I \equiv \frac{\mathrm{d}\sigma}{\mathrm{d}\Omega} &= \mathrm{tr}\left\{\hat{F}\hat{\rho}_{\mathrm{in}}\hat{F}^+\right\} \\ &= I_0\Big[1 + p^{\mathrm{N}}_y A^{\mathrm{N}}_y + \frac{3}{2}p^{\mathrm{d}}_y A^{\mathrm{d}}_y + \frac{2}{3}p^{\mathrm{d}}_{xz} A^{\mathrm{d}}_{xz} + \frac{1}{6}p^{\mathrm{d}}_{xx-yy} A^{\mathrm{d}}_{xx-yy} + \frac{1}{2}p^{\mathrm{d}}_{zz} A^{\mathrm{d}}_{zz} \\ &\quad + p^{\mathrm{N}}_y\left(\frac{3}{2}p^{\mathrm{d}}_y A^{\mathrm{Nd}}_{y,y} + \frac{2}{3}p^{d}_{xz} A^{\mathrm{Nd}}_{y,xz} + \frac{1}{6}p^{\mathrm{d}}_{xx-yy} A^{\mathrm{Nd}}_{y,xx-yy} + \frac{1}{2}p^{\mathrm{d}}_{zz} A^{\mathrm{Nd}}_{y,zz}\right) \\ &\quad + \sum_{i=x,\,z} p^{\mathrm{N}}_i \left(\frac{3}{2}\sum_{j=x,\,z} p^{\mathrm{d}}_j A^{\mathrm{Nd}}_{i,j} + \frac{2}{3}\sum_{j=xy,\,yz} p^{\mathrm{d}}_j A^{\mathrm{Nd}}_{i,j}\right)\Big] \end{aligned} \tag{3.882}$$

where $A^{\mathrm{Nd}}_{i,j}$ is called a correlation analyzing power.

## 3.12 Polarization Theory for $\vec{1} + \vec{1}$ Reactions

The deuteron fusion reaction d(d, p)t and d(d, n)$^3$He belong to the typical rearrangement collision process, and are the representative reactions to be used to discuss whether using the polarized charged particles as fuel can increase the fusion reaction rate. Some individuals have conducted theoretical research [2, 6, 39, 40] on the reactions $\vec{\mathrm{d}}\left(\vec{\mathrm{d}},\mathrm{p}\right)\mathrm{t}$ and $\vec{\mathrm{d}}\left(\vec{\mathrm{d}},\mathrm{n}\right){}^{3}\mathrm{He}$ that belong to the $\vec{1}+\vec{1}$ reaction. This section will take the $\vec{\mathrm{d}}\left(\vec{\mathrm{d}},\mathrm{p}\right)\mathrm{t}$ reaction as an example for theoretical derivation.

Using $\mu$, $\nu$, $\mu'$, $\nu'$ to represent the spin magnetic quantum numbers for the first $d$, second $d$, $p$, and $t$ in $\mathrm{d}+\mathrm{d}\rightarrow\mathrm{p}+\mathrm{t}$ reaction, respectively, and noting that the types of particles before and after the reaction are different, then according to Eq. (2.252) the reaction amplitude of $\mathrm{d}+\mathrm{d}\rightarrow\mathrm{p}+\mathrm{t}$ reaction can be written as

$$\begin{aligned} f_{\mu'\nu',\mu\nu}(\Omega)=&f_{\mathrm{C}}(\theta)\delta_{\mu'\nu',\mu\nu}+\frac{\mathrm{i}\sqrt{\pi}}{k}\sum_{lSl'S'J}\hat{l}\mathrm{e}^{\mathrm{i}(\sigma_l+\sigma_{l'})}\left(\delta_{l'S',lS}-S^J_{l'S',lS}\right)\\ &C^{J\ M}_{l0\ SM}C^{S\ M}_{1\mu\ 1\nu}C^{J\ M}_{l'm'_l\ S'M_{S'}}C^{S'\ M'_S}_{\frac{1}{2}\mu'\ \frac{1}{2}\nu'}\mathrm{Y}_{l'm_{l'}}(\theta,\varphi)\end{aligned} \tag{3.883}$$

From this formula we can see

$$m'_l=\mu+\nu-\mu'-\nu' \tag{3.884}$$

Now, introduce the following symbol

$$f^{ll'J}_{\mu'\nu',\mu\nu}=\hat{l}\hat{l}'\mathrm{e}^{\mathrm{i}(\sigma_l+\sigma_{l'})}\sqrt{\frac{(l'-m'_l)!}{(l'+m'_l)!}}\sum_{SS'}C^{J\ M}_{l0\ SM}C^{S\ M}_{1\mu\ 1\nu}C^{J\ M}_{l'm'_l\ S'M'_S}C^{S'M'_S}_{\frac{1}{2}\mu'\ \frac{1}{2}\nu'}\left(\delta_{l'S',lS}-S^J_{l'S',lS}\right) \tag{3.885}$$

Using Eq. (2.253), we can obtain the following expressions from Eq. (3.883)

$$f_{\mu'\nu',\mu\nu}(\Omega)=F_{\mu'\nu',\mu\nu}(\theta)\ \mathrm{e}^{\mathrm{i}(\mu+\nu-\mu'-\nu')\varphi} \tag{3.886}$$

$$F_{\mu'\nu',\mu\nu}=f_{\mathrm{C}}(\theta)\delta_{\mu'\nu',\mu\nu}+\frac{\mathrm{i}}{2k}(-1)^{\mu+\nu-\mu'-\nu'}\sum_{ll'J}f^{ll'J}_{\mu'\nu',\mu\nu}\mathrm{P}^{\mu+\nu-\mu'-\nu'}_{l'}(\cos\theta) \tag{3.887}$$

Now for parity conservation, $l$ and $l'$ must be even or odd simultaneously, and because $S$ and $S'$ are all integers, $2J$ must be even. Given this, the following C-G coefficient relation can be obtained:

$$\begin{aligned} &C^{J\ \mu+\nu}_{l0\ S\ \mu+\nu}C^{S\ \mu+\nu}_{1\mu\ 1\nu}C^{J\ \mu+\nu}_{l'\ \mu+\nu-\mu'-\nu'\ S'\ \mu'+\nu'}C^{S'\ \mu'+\nu'}_{\frac{1}{2}\mu'\ \frac{1}{2}\nu'}\\ &=-C^{J\ -\mu-\nu}_{l0\ S\ -\mu-\nu}C^{S\ -\mu-\nu}_{1\ -\mu\ 1-\nu}C^{J\ -\mu-\nu}_{l'\ -\mu-\nu+\mu'+\nu'\ S'\ -\mu'-\nu'}C^{S'\ -\mu'-\nu'}_{\frac{1}{2}\ -\mu'\ \frac{1}{2}\ -\nu'}\end{aligned} \tag{3.888}$$

Noting Eq. (2.287), we can obtain the following relation from Eqs. (3.885) and (3.887):

$$F_{\mu'\nu',\mu\nu}(\theta) = (-1)^{\mu+\nu-\mu'-\nu'+1} F_{-\mu'-\nu',-\mu-\nu}(\theta) \tag{3.889}$$

Once again, if we take the $y$-axis along the direction $\vec{n}$, then $\varphi = 0$. Next, introduce the following matrix element symbols according to Eq. (3.889):

$$\begin{array}{ll}
A(\theta) = F_{\frac{1}{2}\frac{1}{2},11}(\theta) = F_{-\frac{1}{2}-\frac{1}{2},-1-1}(\theta), & B(\theta) = \mathrm{i}F_{\frac{1}{2}\frac{1}{2},10}(\theta) = -\mathrm{i}F_{-\frac{1}{2}-\frac{1}{2},-10}(\theta), \\
C(\theta) = -F_{\frac{1}{2}\frac{1}{2},1-1}(\theta) = -F_{-\frac{1}{2}-\frac{1}{2},-11}(\theta), & D(\theta) = \mathrm{i}F_{\frac{1}{2}\frac{1}{2},01}(\theta) = -\mathrm{i}F_{-\frac{1}{2}-\frac{1}{2},0-1}(\theta), \\
E(\theta) = -F_{\frac{1}{2}\frac{1}{2},00}(\theta) = -F_{-\frac{1}{2}-\frac{1}{2},00}(\theta), & F(\theta) = -\mathrm{i}F_{\frac{1}{2}\frac{1}{2},0-1}(\theta) = \mathrm{i}F_{-\frac{1}{2}-\frac{1}{2},0\,1}(\theta), \\
G(\theta) = -F_{\frac{1}{2}\frac{1}{2},-11}(\theta) = -F_{-\frac{1}{2}-\frac{1}{2},1-1}(\theta), & H(\theta) = -\mathrm{i}F_{\frac{1}{2}\frac{1}{2},-10}(\theta) = \mathrm{i}F_{-\frac{1}{2}-\frac{1}{2},10}(\theta), \\
J(\theta) = F_{\frac{1}{2}\frac{1}{2},-1-1}(\theta) = F_{-\frac{1}{2}-\frac{1}{2},11}(\theta), & K(\theta) = \mathrm{i}F_{\frac{1}{2}-\frac{1}{2},11}(\theta) = -\mathrm{i}F_{-\frac{1}{2}\frac{1}{2},-1-1}(\theta), \\
L(\theta) = -F_{\frac{1}{2}-\frac{1}{2},10}(\theta) = -F_{-\frac{1}{2}\frac{1}{2},-10}(\theta), & M(\theta) = -\mathrm{i}F_{\frac{1}{2}-\frac{1}{2},1-1}(\theta) = \mathrm{i}F_{-\frac{1}{2}\frac{1}{2},-11}(\theta), \\
N(\theta) = -F_{\frac{1}{2}-\frac{1}{2},01}(\theta) = -F_{-\frac{1}{2}\frac{1}{2},0-1}(\theta), & O(\theta) = -\mathrm{i}F_{\frac{1}{2}-\frac{1}{2},00}(\theta) = \mathrm{i}F_{-\frac{1}{2}\frac{1}{2},00}(\theta), \\
Q(\theta) = F_{\frac{1}{2}-\frac{1}{2},0-1}(\theta) = F_{-\frac{1}{2}\frac{1}{2},01}(\theta), & R(\theta) = -\mathrm{i}F_{\frac{1}{2}-\frac{1}{2},-11}(\theta) = \mathrm{i}F_{-\frac{1}{2}\frac{1}{2},1-1}(\theta), \\
S(\theta) = F_{\frac{1}{2}-\frac{1}{2},-10}(\theta) = F_{-\frac{1}{2}\frac{1}{2},10}(\theta), & T(\theta) = \mathrm{i}F_{\frac{1}{2}-\frac{1}{2},-1-1}(\theta) = -\mathrm{i}F_{-\frac{1}{2}\frac{1}{2},11}(\theta)
\end{array} \tag{3.890}$$

In the following, the footnote $\frac{1}{2}$ is represented by the footnote $+$, and the footnote $-\frac{1}{2}$ is represented by the footnote $-$, so the $4 \times 9$ reaction amplitude matrix given by Eq. (3.887) can be expressed as

$$\begin{aligned}
\hat{F} &= \begin{pmatrix}
F_{++,11} & F_{++,10} & F_{++,1-1} & F_{++,01} & F_{++,00} & F_{++,0-1} & F_{++,-11} & F_{++,-10} & F_{++,-1-1} \\
F_{+-,11} & F_{+-,10} & F_{+-,1-1} & F_{+-,01} & F_{+-,00} & F_{+-,0-1} & F_{+-,-11} & F_{+-,-10} & F_{+-,-1-1} \\
F_{-+,11} & F_{-+,10} & F_{-+,1-1} & F_{-+,01} & F_{-+,00} & F_{-+,0-1} & F_{-+,-11} & F_{-+,-10} & F_{-+,-1-1} \\
F_{--,11} & F_{--,10} & F_{--,1-1} & F_{--,01} & F_{--,00} & F_{--,0-1} & F_{--,-11} & F_{--,-10} & F_{--,-1-1}
\end{pmatrix} \\
&= \begin{pmatrix}
A & -\mathrm{i}B & -C & -\mathrm{i}D & -E & \mathrm{i}F & -G & \mathrm{i}H & J \\
-\mathrm{i}K & -L & \mathrm{i}M & -N & \mathrm{i}O & Q & \mathrm{i}R & S & -\mathrm{i}T \\
\mathrm{i}T & S & -\mathrm{i}R & Q & -\mathrm{i}O & -N & -\mathrm{i}M & -L & \mathrm{i}K \\
J & -\mathrm{i}H & -G & -\mathrm{i}F & -E & \mathrm{i}D & -C & \mathrm{i}B & A
\end{pmatrix}
\end{aligned} \tag{3.891}$$

$$\hat{F}^+ = \begin{pmatrix} F^*_{++,11} & F^*_{+-,11} & F^*_{-+,11} & F^*_{--,11} \\ F^*_{++,10} & F^*_{+-,10} & F^*_{-+,10} & F^*_{--,10} \\ F^*_{++,1-1} & F^*_{+-,1-1} & F^*_{-+,1-1} & F^*_{--,1-1} \\ F^*_{++,01} & F^*_{+-,01} & F^*_{-+,01} & F^*_{--,01} \\ F^*_{++,00} & F^*_{+-,00} & F^*_{-+,00} & F^*_{--,00} \\ F^*_{++,0-1} & F^*_{+-,0-1} & F^*_{-+,0-1} & F^*_{--,0-1} \\ F^*_{++,-11} & F^*_{+-,-11} & F^*_{-+,-11} & F^*_{--,-11} \\ F^*_{++,-10} & F^*_{+-,-10} & F^*_{-+,-10} & F^*_{--,-10} \\ F^*_{++,-1-1} & F^*_{+-,-1-1} & F^*_{-+,-1-1} & F^*_{--,-1-1} \end{pmatrix}$$
$$= \begin{pmatrix} A^* & \mathrm{i}K^* & -\mathrm{i}T^* & J^* \\ \mathrm{i}B^* & -L^* & S^* & \mathrm{i}H^* \\ -C^* & -\mathrm{i}M^* & \mathrm{i}R^* & -G^* \\ \mathrm{i}D^* & -N^* & Q^* & \mathrm{i}F^* \\ -E^* & -\mathrm{i}O^* & \mathrm{i}O^* & -E^* \\ -\mathrm{i}F^* & Q^* & -N^* & -\mathrm{i}D^* \\ -G^* & -\mathrm{i}R^* & \mathrm{i}M^* & -C^* \\ -\mathrm{i}H^* & S^* & -L^* & -\mathrm{i}B^* \\ J^* & \mathrm{i}T^* & -\mathrm{i}K^* & A^* \end{pmatrix} \quad (3.892)$$

Now, from Eqs. (3.891) and (3.892) we can find that the differential cross section is as follows when the two deuterons of the incident channel are unpolarized:

$$\begin{aligned} I_0 \equiv \frac{\mathrm{d}\sigma^0}{\mathrm{d}\Omega} &= \frac{1}{9}\mathrm{tr}\{FF^+\} \\ &= \frac{2}{9}\Big( |A|^2 + |B|^2 + |C|^2 + |D|^2 + |E|^2 + |F|^2 + |G|^2 + |H|^2 + |J|^2 \\ &\quad + |K|^2 + |L|^2 + |M|^2 + |N|^2 + |O|^2 + |Q|^2 + |R|^2 + |S|^2 + |T|^2 \Big) \end{aligned} \quad (3.893)$$

According to Eq. (3.214) the polarization density matrix of the normalized incident channel can be written as

$$\begin{aligned} \hat{\rho}_{\mathrm{in}} = \frac{1}{9} \prod_{k=1,2} \Big[ \hat{I}_3 &+ \frac{3}{2}\left( p_x^k \hat{S}_x^{(k)} + p_y^k \hat{S}_y^{(k)} + p_z^k \hat{S}_z^{(k)} \right) \\ &+ \frac{2}{3}\left( p_{xy}^k \hat{Q}_{xy}^{(k)} + p_{xz}^k \hat{Q}_{xz}^{(k)} + p_{yz}^k \hat{Q}_{yz}^{(k)} \right) + \frac{1}{6} p_{xx-yy}^k Q_{xx-yy}^{(k)} + \frac{1}{2} p_{zz}^k Q_{zz}^{(k)} \Big] \end{aligned} \quad (3.894)$$

where $k = 1, 2$ such that the two multiplied items represent the first deuteron and the second deuteron of the incident channel, respectively.

Next, define the following polarization analyzing powers:

$$A_i^{(k)} = \begin{cases} \dfrac{\mathrm{tr}\left\{\hat{F}\hat{S}_i^{(k)}\hat{F}^+\right\}}{\mathrm{tr}\left\{\hat{F}\hat{F}^+\right\}}, & k=1,2; \quad i=x,y,z \\ \dfrac{\mathrm{tr}\left\{\hat{F}\hat{Q}_i^{(k)}\hat{F}^+\right\}}{\mathrm{tr}\left\{\hat{F}\hat{F}^+\right\}}, & k=1,2; \quad i=\gamma \end{cases} \tag{3.895}$$

$$A_{i,j}^{(12)} = \begin{cases} \dfrac{\mathrm{tr}\left\{\hat{F}\hat{S}_i^{(1)}\hat{S}_j^{(2)}\hat{F}^+\right\}}{\mathrm{tr}\left\{\hat{F}\hat{F}^+\right\}}, & i,j=x,y,z \\ \dfrac{\mathrm{tr}\left\{\hat{F}\hat{S}_i^{(1)}\hat{Q}_j^{(2)}\hat{F}^+\right\}}{\mathrm{tr}\left\{\hat{F}\hat{F}^+\right\}}, & i=x,y,z; \quad j=\gamma \\ \dfrac{\mathrm{tr}\left\{\hat{F}\hat{Q}_i^{(1)}\hat{S}_j^{(2)}\hat{F}^+\right\}}{\mathrm{tr}\left\{\hat{F}\hat{F}^+\right\}}, & i=\gamma; \quad j=x,y,z \\ \dfrac{\mathrm{tr}\left\{\hat{F}\hat{Q}_i^{(1)}\hat{Q}_j^{(2)}\hat{F}^+\right\}}{\mathrm{tr}\left\{\hat{F}\hat{F}^+\right\}}, & i,j=\gamma \end{cases} \tag{3.896}$$

where the definition of the footnote set $\gamma$ has been given by Eq. (3.824), that is, $\gamma = xy, xz, yz, xx-yy, zz$. $A_{i,j}^{(12)}$ is also known as correlation polarization analyzing power. It follows that according to Eq. (3.894) the general form of the differential cross section for $\vec{\mathrm{d}}\left(\vec{\mathrm{d}},\mathrm{p}\right)$ t reaction can be written as

$$I \equiv \frac{\mathrm{d}\sigma}{\mathrm{d}\Omega} = \mathrm{tr}\{F\hat{\rho}_{\mathrm{in}}F^+\} = I_0\left(1 + \sum_{k=1,2}\ \sum_{i=x,y,z,\gamma} \overline{p}_i^k A_i^{(k)} + \sum_{i,j=x,y,z,\gamma} \overline{p}_i^1\overline{p}_j^2 A_{i,j}^{(12)}\right) \tag{3.897}$$

where

$$\overline{p}_i^k = \begin{cases} \frac{3}{2}p_i^k, & i=x,\ y,\ z \\ \frac{2}{3}p_i^k, & i=xy,\ xz,\ yz \\ \frac{1}{6}p_i^k, & i=xx-yy \\ \frac{1}{2}p_i^k, & i=zz \end{cases} \tag{3.898}$$

If the expectation value method given by Eq. (1.88) is used to find the mean value of the physical quantity for $\vec{\mathrm{d}} + \vec{\mathrm{d}} \rightarrow \mathrm{p} + \mathrm{t}$ reaction, the operators $\hat{S}_i^{(1)}$ and $\hat{Q}_i^{(1)}$ in this section can only act on the spin wave function of the first deuteron, and $\hat{S}_i^{(2)}$ and $\hat{Q}_i^{(2)}$ act on the spin wave function of the second deuteron. When we use the tracing method, $\hat{S}_i^{(1)}$ and $\hat{Q}_i^{(1)}$ only works on the first deuteron magnetic quantum number $\mu$ of $F^+_{\mu'\nu',\mu\nu}$ in the same $\mu'\nu'\nu$ case, but $\hat{S}_i^{(2)}$ and $\hat{Q}_i^{(2)}$ only works on the second deuteron magnetic quantum number $\nu$ of $F^+_{\mu'\nu',\mu\nu}$ in the same $\mu'\nu'\mu$ case.

Next, define the following vectors:

$$\begin{aligned} &\vec{U}^{(2)}_{i,lmn} = \hat{S}_i^{(2)} \begin{pmatrix} F^*_{lm,n1} \\ F^*_{lm,n0} \\ F^*_{lm,n-1} \end{pmatrix}, \quad \vec{U}^{(1)}_{i,lmn} = \hat{S}_i^{(1)} \begin{pmatrix} F^*_{lm,1n} \\ F^*_{lm,0n} \\ F^*_{lm,-1\,n} \end{pmatrix}, \quad i=x,y,z \\ &\vec{V}^{(2)}_{i,lmn} = \hat{Q}_i^{(2)} \begin{pmatrix} F^*_{lm,n1} \\ F^*_{lm,n0} \\ F^*_{lm,n-1} \end{pmatrix}, \quad \vec{V}^{(1)}_{i,lmn} = \hat{Q}_i^{(1)} \begin{pmatrix} F^*_{lm,1n} \\ F^*_{lm,0n} \\ F^*_{lm,-1\,n} \end{pmatrix}, \quad i=\gamma \end{aligned} \tag{3.899}$$

The $b$ $(=1,0,-1)$ components of the vectors $\vec{U}^{(k)}_{i,lma}$ and $\vec{V}^{(k)}_{i,lma}$ are represented by $U^{(k)}_{i,lma,b}$ and $V^{(k)}_{i,lma,b}$ separately below. Now redefine

$$\vec{W}^{(12)}_{ij,lmn} = \begin{cases} \hat{S}_i^{(1)} \begin{pmatrix} U^{(2)}_{j,lm1,n} \\ U^{(2)}_{j,lm0,n} \\ U^{(2)}_{j,lm-1,n} \end{pmatrix}, & i,j=x,y,z \\ \hat{S}_i^{(1)} \begin{pmatrix} V^{(2)}_{j,lm1,n} \\ V^{(2)}_{j,lm0,n} \\ V^{(2)}_{j,lm-1,n} \end{pmatrix}, & i=x,y,z; \quad j=\gamma \end{cases} \tag{3.900}$$

$$\vec{Z}^{(12)}_{ij,lmn}=\begin{cases}\hat{Q}^{(1)}_i\begin{pmatrix}U^{(2)}_{j,lm1,n}\\ U^{(2)}_{j,lm0,n}\\ U^{(2)}_{j,lm-1,n}\end{pmatrix}, & i=\gamma; \quad j=x,y,z\\ \hat{Q}^{(1)}_i\begin{pmatrix}V^{(2)}_{j,lm1,n}\\ V^{(2)}_{j,lm0,n}\\ V^{(2)}_{j,lm-1,n}\end{pmatrix}, & i,j=\gamma\end{cases} \tag{3.901}$$

where in the above three formulas $l, m=+,-$; $n=1,0,-1$.

For the second deuteron, one can obtain the following results from Eqs. (3.33), (3.892), and (3.899):

$$\vec{U}^{(2)}_{x,++1}=\hat{S}^{(2)}_x\begin{pmatrix}F^*_{++,11}\\ F^*_{++,10}\\ F^*_{++,1-1}\end{pmatrix}=\frac{1}{\sqrt{2}}\begin{pmatrix}0&1&0\\1&0&1\\0&1&0\end{pmatrix}\begin{pmatrix}A^*\\ \mathrm{i}B^*\\ -C^*\end{pmatrix}=\frac{1}{\sqrt{2}}\begin{pmatrix}\mathrm{i}B^*\\ A^*-C^*\\ \mathrm{i}B^*\end{pmatrix},$$

$$\vec{U}^{(2)}_{x,++0}=\frac{1}{\sqrt{2}}\begin{pmatrix}-E^*\\ \mathrm{i}(D^*-F^*)\\ -E^*\end{pmatrix},\qquad \vec{U}^{(2)}_{x,++-1}=\frac{1}{\sqrt{2}}\begin{pmatrix}-\mathrm{i}H^*\\ J^*-G^*\\ -\mathrm{i}H^*\end{pmatrix},$$

$$\vec{U}^{(2)}_{x,+-1}=\frac{1}{\sqrt{2}}\begin{pmatrix}-L^*\\ \mathrm{i}(K^*-M^*)\\ -L^*\end{pmatrix},\qquad \vec{U}^{(2)}_{x,+-0}=\frac{1}{\sqrt{2}}\begin{pmatrix}-\mathrm{i}O^*\\ Q^*-N^*\\ -\mathrm{i}O^*\end{pmatrix},$$

$$\vec{U}^{(2)}_{x,+-\ -1}=\frac{1}{\sqrt{2}}\begin{pmatrix}S^*\\ \mathrm{i}(T^*-R^*)\\ S^*\end{pmatrix},\qquad \vec{U}^{(2)}_{x,-+1}=\frac{1}{\sqrt{2}}\begin{pmatrix}S^*\\ \mathrm{i}(R^*-T^*)\\ S^*\end{pmatrix},$$

$$\vec{U}^{(2)}_{x,-+0}=\frac{1}{\sqrt{2}}\begin{pmatrix}\mathrm{i}O^*\\ Q^*-N^*\\ \mathrm{i}O^*\end{pmatrix},\qquad \vec{U}^{(2)}_{x,-+\ -1}=\frac{1}{\sqrt{2}}\begin{pmatrix}-L^*\\ \mathrm{i}(M^*-K^*)\\ -L^*\end{pmatrix},$$

$$\vec{U}^{(2)}_{x,--1}=\frac{1}{\sqrt{2}}\begin{pmatrix}\mathrm{i}H^*\\ J^*-G^*\\ \mathrm{i}H^*\end{pmatrix},\qquad \vec{U}^{(2)}_{x,--0}=\frac{1}{\sqrt{2}}\begin{pmatrix}-E^*\\ \mathrm{i}(F^*-D^*)\\ -E^*\end{pmatrix},$$

$$\vec{U}^{(2)}_{x,--\ -1}=\frac{1}{\sqrt{2}}\begin{pmatrix}-\mathrm{i}B^*\\ A^*-C^*\\ -\mathrm{i}B^*\end{pmatrix} \tag{3.902}$$

such that one can now write

$$\hat{S}_x^{(2)}\hat{F}^+ = \frac{1}{\sqrt{2}}\begin{pmatrix} \mathrm{i}B^* & -L^* & S^* & \mathrm{i}H^* \\ A^*-C^* & \mathrm{i}(K^*-M^*) & \mathrm{i}(R^*-T^*) & J^*-G^* \\ \mathrm{i}B^* & -L^* & S^* & \mathrm{i}H^* \\ -E^* & -\mathrm{i}O^* & \mathrm{i}O^* & -E^* \\ \mathrm{i}(D^*-F^*) & Q^*-N^* & Q^*-N^* & \mathrm{i}(F^*-D^*) \\ -E^* & -\mathrm{i}O^* & \mathrm{i}O^* & -E^* \\ -\mathrm{i}H^* & S^* & -L^* & -\mathrm{i}B^* \\ J^*-G^* & \mathrm{i}(T^*-R^*) & \mathrm{i}(M^*-K^*) & A^*-C^* \\ -\mathrm{i}H^* & S^* & -L^* & -\mathrm{i}B^* \end{pmatrix} \tag{3.903}$$

Substituting Eqs. (3.891) and (3.903) into Eq. (3.895), one can obtain the following result:

$$\begin{aligned} A_x^{(2)} = \frac{1}{9I_0}\mathrm{tr}\left\{F\hat{S}_x^{(2)}F^+\right\} = \frac{\mathrm{i}}{9\sqrt{2}I_0}(&AB^* - BA^* + BC^* - CB^* + DE^* + EF^* \\ &- ED^* - FE^* + GH^* - HG^* + HJ^* - JH^* + KL^* - LK^* + LM^* - ML^* \\ &+NO^* + OQ^* - ON^* - QO^* + RS^* - SR^* + ST^* - TS^* + TS^* - ST^* \\ &+SR^* - RS^* + QO^* - OQ^* + ON^* - NO^* + ML^* - LM^* + LK^* - KL^* \\ &+JH^* + HG^* - HJ^* - GH^* + FE^* + ED^* - EF^* - DE^* + CB^* + BA^* \\ &- BC^* - AB^*) = 0 \end{aligned} \tag{3.904}$$

One can also obtain

$$\vec{U}_{y,++1}^{(2)} = \hat{S}_y^{(2)}\begin{pmatrix} F_{++,11}^* \\ F_{++,10}^* \\ F_{++,1-1}^* \end{pmatrix} = \frac{\mathrm{i}}{\sqrt{2}}\begin{pmatrix} 0 & -1 & 0 \\ 1 & 0 & -1 \\ 0 & 1 & 0 \end{pmatrix}\begin{pmatrix} A^* \\ \mathrm{i}B^* \\ -C^* \end{pmatrix} = \frac{1}{\sqrt{2}}\begin{pmatrix} B^* \\ \mathrm{i}(A^*+C^*) \\ -B^* \end{pmatrix},$$

$$\vec{U}_{y,++0}^{(2)} = \frac{1}{\sqrt{2}}\begin{pmatrix} \mathrm{i}E^* \\ -(D^*+F^*) \\ -\mathrm{i}E^* \end{pmatrix}, \quad \vec{U}_{y,++\ -1}^{(2)} = \frac{1}{\sqrt{2}}\begin{pmatrix} -H^* \\ -\mathrm{i}(G^*+J^*) \\ H^* \end{pmatrix},$$

$$\vec{U}_{y,+-1}^{(2)} = \frac{1}{\sqrt{2}}\begin{pmatrix} \mathrm{i}L^* \\ -(K^*+M^*) \\ -\mathrm{i}L^* \end{pmatrix}, \quad \vec{U}_{y,+-0}^{(2)} = \frac{1}{\sqrt{2}}\begin{pmatrix} -O^* \\ -\mathrm{i}(N^*+Q^*) \\ O^* \end{pmatrix},$$

$$\vec{U}_{y,+-\ -1}^{(2)} = \frac{1}{\sqrt{2}}\begin{pmatrix} -\mathrm{i}S^* \\ R^*+T^* \\ \mathrm{i}S^* \end{pmatrix}, \quad \vec{U}_{y,-+1}^{(2)} = \frac{1}{\sqrt{2}}\begin{pmatrix} -\mathrm{i}S^* \\ R^*+T^* \\ \mathrm{i}S^* \end{pmatrix},$$

$$\vec{U}^{(2)}_{y,-+0} = \frac{1}{\sqrt{2}}\begin{pmatrix} O^* \\ \mathrm{i}(Q^* + N^*) \\ -O^* \end{pmatrix}, \qquad \vec{U}^{(2)}_{y,-+\ -1} = \frac{1}{\sqrt{2}}\begin{pmatrix} \mathrm{i}L^* \\ -(K^* + M^*) \\ -\mathrm{i}L^* \end{pmatrix},$$

$$\vec{U}^{(2)}_{y,--1} = \frac{1}{\sqrt{2}}\begin{pmatrix} H^* \\ \mathrm{i}(G^* + J^*) \\ -H^* \end{pmatrix}, \qquad \vec{U}^{(2)}_{y,--0} = \frac{1}{\sqrt{2}}\begin{pmatrix} \mathrm{i}E^* \\ -(D^* + F^*) \\ -\mathrm{i}E^* \end{pmatrix},$$

$$\vec{U}^{(2)}_{y,--\ -1} = \frac{1}{\sqrt{2}}\begin{pmatrix} -B^* \\ -\mathrm{i}(A^* + C^*) \\ B^* \end{pmatrix} \tag{3.905}$$

such that one can now write

$$\hat{S}^{(2)}_y \hat{F}^+ = \frac{1}{\sqrt{2}}\begin{pmatrix} B^* & \mathrm{i}L^* & -\mathrm{i}S^* & H^* \\ \mathrm{i}(A^*+C^*) & -(K^*+M^*) & R^*+T^* & \mathrm{i}(G^*+J^*) \\ -B^* & -\mathrm{i}L^* & \mathrm{i}S^* & -H^* \\ \mathrm{i}E^* & -O^* & O^* & \mathrm{i}E^* \\ -(D^*+F^*) & -\mathrm{i}(N^*+Q^*) & \mathrm{i}(Q^*+N^*) & -(D^*+F^*) \\ -\mathrm{i}E^* & O^* & -O^* & -\mathrm{i}E^* \\ -H^* & -\mathrm{i}S^* & \mathrm{i}L^* & -B^* \\ -\mathrm{i}(G^*+J^*) & R^*+T^* & -(K^*+M^*) & -\mathrm{i}(A^*+C^*) \\ H^* & \mathrm{i}S^* & -\mathrm{i}L^* & B^* \end{pmatrix} \tag{3.906}$$

Substituting Eqs. (3.891) and (3.906) into Eq. (3.895), one can obtain the following result:

$$\begin{aligned} A^{(2)}_y &= \frac{1}{9I_0}\mathrm{tr}\left\{F\hat{S}^{(2)}_y F^+\right\} = \frac{1}{9\sqrt{2}I_0}(AB^* + BA^* + BC^* + CB^* + DE^* + ED^* \\ &\quad +EF^* + FE^* + GH^* + HG^* + HJ^* + JH^* + KL^* + LK^* + LM^* + ML^* \\ &\quad +NO^* + ON^* + OQ^* + QO^* + RS^* + SR^* + ST^* + TS^* + TS^* + ST^* \\ &\quad +SR^* + RS^* + QO^* + OQ^* + ON^* + NO^* + ML^* + LM^* + LK^* + KL^* \\ &\quad +JH^* + HJ^* + HG^* + GH^* + FE^* + EF^* + ED^* + DE^* + CB^* + BC^* \\ &\quad +BA^* + AB^*) \\ &= \frac{2\sqrt{2}}{9I_0}(AB^* + BC^* + DE^* + EF^* + GH^* + HJ^* + KL^* + LM^* + NO^* \\ &\quad +OQ^* + RS^* + ST^*) \end{aligned} \tag{3.907}$$

and one can still get

$$\begin{aligned}
&\vec{U}^{(2)}_{z,++1}=\hat{S}^{(2)}_z\begin{pmatrix} F^*_{++,11}\\ F^*_{++,10}\\ F^*_{++,1-1}\end{pmatrix}=\begin{pmatrix}1&0&0\\0&0&0\\0&0&-1\end{pmatrix}\begin{pmatrix}A^*\\ \mathrm{i}B^*\\ -C^*\end{pmatrix}=\begin{pmatrix}A^*\\0\\C^*\end{pmatrix},\\
&\vec{U}^{(2)}_{z,++0}=\begin{pmatrix}\mathrm{i}D^*\\0\\ \mathrm{i}F^*\end{pmatrix},\vec{U}^{(2)}_{z,++-1}=\begin{pmatrix}-G^*\\0\\-J^*\end{pmatrix},\vec{U}^{(2)}_{z,+-1}=\begin{pmatrix}\mathrm{i}K^*\\0\\ \mathrm{i}M^*\end{pmatrix},\\
&\vec{U}^{(2)}_{z,+-0}=\begin{pmatrix}-N^*\\0\\-Q^*\end{pmatrix},\vec{U}^{(2)}_{z,+--1}=\begin{pmatrix}-\mathrm{i}R^*\\0\\-\mathrm{i}T^*\end{pmatrix},\vec{U}^{(2)}_{z,-+1}=\begin{pmatrix}-\mathrm{i}T^*\\0\\-\mathrm{i}R^*\end{pmatrix},\\
&\vec{U}^{(2)}_{z,-+0}=\begin{pmatrix}Q^*\\0\\N^*\end{pmatrix},\vec{U}^{(2)}_{z,-+-1}=\begin{pmatrix}\mathrm{i}M^*\\0\\ \mathrm{i}K^*\end{pmatrix},\vec{U}^{(2)}_{z,--1}=\begin{pmatrix}J^*\\0\\G^*\end{pmatrix},\\
&\vec{U}^{(2)}_{z,--0}=\begin{pmatrix}\mathrm{i}F^*\\0\\ \mathrm{i}D^*\end{pmatrix},\vec{U}^{(2)}_{z,---1}=\begin{pmatrix}-C^*\\0\\-A^*\end{pmatrix}
\end{aligned}\tag{3.908}$$

It follows that

$$\hat{S}^{(2)}_z\hat{F}^+=\begin{pmatrix}
A^* & \mathrm{i}K^* & -\mathrm{i}T^* & J^*\\
0&0&0&0\\
C^* & \mathrm{i}M^* & -\mathrm{i}R^* & G^*\\
\mathrm{i}D^* & -N^* & Q^* & \mathrm{i}F^*\\
0&0&0&0\\
\mathrm{i}F^* & -Q^* & N^* & \mathrm{i}D^*\\
-G^* & -\mathrm{i}R^* & \mathrm{i}M^* & -C^*\\
0&0&0&0\\
-J^* & -\mathrm{i}T^* & \mathrm{i}K^* & -A^*
\end{pmatrix}\tag{3.909}$$

Substituting Eqs. (3.891) and (3.909) into Eq. (3.895), one can obtain the following result:

$$\begin{aligned}
A^{(2)}_z=\frac{1}{9I_0}\mathrm{tr}\left\{F\hat{S}^{(2)}_zF^+\right\}=\frac{1}{9I_0}(&|A|^2-|C|^2+|D|^2-|F|^2+|G|^2\\
&-|J|^2+|K|^2-|M|^2+|N|^2-|Q|^2+|R|^2-|T|^2+|T|^2-|R|^2+|Q|^2\\
&-|N|^2+|M|^2-|K|^2+|J|^2-|G|^2+|F|^2-|D|^2+|C|^2-|A|^2)=0
\end{aligned}\tag{3.910}$$

Now for the first deuteron, one can obtain the following results from Eqs. (3.33), (3.892), and (3.899):

$$\vec{U}^{(1)}_{x,++1}=\hat{S}^{(1)}_x\begin{pmatrix}F^*_{++,11}\\F^*_{++,01}\\F^*_{++,-11}\end{pmatrix}=\frac{1}{\sqrt{2}}\begin{pmatrix}0&1&0\\1&0&1\\0&1&0\end{pmatrix}\begin{pmatrix}A^*\\iD^*\\-G^*\end{pmatrix}=\frac{1}{\sqrt{2}}\begin{pmatrix}iD^*\\A^*-G^*\\iD^*\end{pmatrix},$$
$$\vec{U}^{(1)}_{x,++0}=\frac{1}{\sqrt{2}}\begin{pmatrix}-E^*\\i(B^*-H^*)\\-E^*\end{pmatrix},\qquad \vec{U}^{(1)}_{x,++\ -1}=\frac{1}{\sqrt{2}}\begin{pmatrix}-iF^*\\-(C^*-J^*)\\-iF^*\end{pmatrix},$$
$$\vec{U}^{(1)}_{x,+-1}=\frac{1}{\sqrt{2}}\begin{pmatrix}-N^*\\i(K^*-R^*)\\-N^*\end{pmatrix},\qquad \vec{U}^{(1)}_{x,+-0}=\frac{1}{\sqrt{2}}\begin{pmatrix}-iO^*\\-(L^*-S^*)\\-iO^*\end{pmatrix},$$
$$\vec{U}^{(1)}_{x,+-\ -1}=\frac{1}{\sqrt{2}}\begin{pmatrix}Q^*\\-i(M^*-T^*)\\Q^*\end{pmatrix},\qquad \vec{U}^{(1)}_{x,-+1}=\frac{1}{\sqrt{2}}\begin{pmatrix}Q^*\\i(M^*-T^*)\\Q^*\end{pmatrix},\tag{3.911}$$
$$\vec{U}^{(1)}_{x,-+0}=\frac{1}{\sqrt{2}}\begin{pmatrix}iO^*\\-(L^*-S^*)\\iO^*\end{pmatrix},\qquad \vec{U}^{(1)}_{x,-+\ -1}=\frac{1}{\sqrt{2}}\begin{pmatrix}-N^*\\-i(K^*-R^*)\\-N^*\end{pmatrix},$$
$$\vec{U}^{(1)}_{x,--1}=\frac{1}{\sqrt{2}}\begin{pmatrix}iF^*\\-(C^*-J^*)\\iF^*\end{pmatrix},\qquad \vec{U}^{(1)}_{x,--0}=\frac{1}{\sqrt{2}}\begin{pmatrix}-E^*\\-i(B^*-H^*)\\-E^*\end{pmatrix},$$
$$\vec{U}^{(1)}_{x,--\ -1}=\frac{1}{\sqrt{2}}\begin{pmatrix}-iD^*\\A^*-G^*\\-iD^*\end{pmatrix}$$

and the following can be written:

$$\hat{S}^{(1)}_x\hat{F}^+=\frac{1}{\sqrt{2}}\begin{pmatrix}iD^* & -N^* & Q^* & iF^*\\ -E^* & -iO^* & iO^* & -E^*\\ -iF^* & Q^* & -N^* & -iD^*\\ A^*-G^* & i(K^*-R^*) & i(M^*-T^*) & J^*-C^*\\ i(B^*-H^*) & S^*-L^* & S^*-L^* & i(H^*-B^*)\\ J^*-C^* & -i(M^*-T^*) & i(R^*-K^*) & A^*-G^*\\ iD^* & -N^* & Q^* & iF^*\\ -E^* & -iO^* & iO^* & -E^*\\ -iF^* & Q^* & -N^* & -iD^*\end{pmatrix}\tag{3.912}$$

Substituting Eqs. (3.891) and (3.912) into Eq. (3.895), one can obtain the following result:

$$A_x^{(1)} = \frac{1}{9I_0}\mathrm{tr}\left\{F\hat{S}_x^{(1)}F^+\right\} = \frac{\mathrm{i}}{9\sqrt{2}I_0}(AD^* + BE^* + CF^* - DA^* + DG^* - EB^*$$
$$+EH^* - FC^* + FJ^* - GD^* - HE^* - JF^* + KN^* + LO^* + MQ^*$$
$$-NK^* + NR^* - OL^* + OS^* - QM^* + QT^* - RN^* - SO^* - TQ^* + TQ^*$$
$$+SO^* + RN^* - QT^* + QM^* - OS^* + OL^* - NR^* + NK^* - MQ^* - LO^*$$
$$-KN^* + JF^* + HE^* + GD^* - FJ^* + FC^* - EH^* + EB^* - DG^* + DA^*$$
$$-CF^* - BE^* - AD^*) = 0 \tag{3.913}$$

One can also obtain

$$\vec{U}^{(1)}_{y,++1} = \hat{S}_y^{(1)}\begin{pmatrix} F^*_{++,11} \\ F^*_{++,01} \\ F^*_{++,-11}\end{pmatrix} = \frac{\mathrm{i}}{\sqrt{2}}\begin{pmatrix} 0 & -1 & 0 \\ 1 & 0 & -1 \\ 0 & 1 & 0\end{pmatrix}\begin{pmatrix} A^* \\ \mathrm{i}D^* \\ -G^*\end{pmatrix} = \frac{1}{\sqrt{2}}\begin{pmatrix} D^* \\ \mathrm{i}(A^*+G^*) \\ -D^*\end{pmatrix},$$

$$\vec{U}^{(1)}_{y,++0} = \frac{1}{\sqrt{2}}\begin{pmatrix} \mathrm{i}E^* \\ -(B^*+H^*) \\ -\mathrm{i}E^*\end{pmatrix},\quad \vec{U}^{(1)}_{y,++\,-1} = \frac{1}{\sqrt{2}}\begin{pmatrix} -F^* \\ -\mathrm{i}(C^*+J^*) \\ F^*\end{pmatrix},$$

$$\vec{U}^{(1)}_{y,+-1} = \frac{1}{\sqrt{2}}\begin{pmatrix} \mathrm{i}N^* \\ -(K^*+R^*) \\ -\mathrm{i}N^*\end{pmatrix},\quad \vec{U}^{(1)}_{y,+-0} = \frac{1}{\sqrt{2}}\begin{pmatrix} -O^* \\ -\mathrm{i}(L^*+S^*) \\ O^*\end{pmatrix},$$

$$\vec{U}^{(1)}_{y,+-\,-1} = \frac{1}{\sqrt{2}}\begin{pmatrix} -\mathrm{i}Q^* \\ M^*+T^* \\ \mathrm{i}Q^*\end{pmatrix},\quad \vec{U}^{(1)}_{y,-+1} = \frac{1}{\sqrt{2}}\begin{pmatrix} -\mathrm{i}Q^* \\ T^*+M^* \\ \mathrm{i}Q^*\end{pmatrix},$$

$$\vec{U}^{(1)}_{y,-+0} = \frac{1}{\sqrt{2}}\begin{pmatrix} O^* \\ \mathrm{i}(S^*+L^*) \\ -O^*\end{pmatrix},\quad \vec{U}^{(1)}_{y,-+\,-1} = \frac{1}{\sqrt{2}}\begin{pmatrix} \mathrm{i}N^* \\ -(R^*+K^*) \\ -\mathrm{i}N^*\end{pmatrix},$$

$$\vec{U}^{(1)}_{y,--1} = \frac{1}{\sqrt{2}}\begin{pmatrix} F^* \\ \mathrm{i}(J^*+C^*) \\ -F^*\end{pmatrix},\quad \vec{U}^{(1)}_{y,--0} = \frac{1}{\sqrt{2}}\begin{pmatrix} \mathrm{i}E^* \\ -(H^*+B^*) \\ -\mathrm{i}E^*\end{pmatrix},$$

$$\vec{U}^{(1)}_{y,--\,-1} = \frac{1}{\sqrt{2}}\begin{pmatrix} -D^* \\ -\mathrm{i}(G^*+A^*) \\ D^*\end{pmatrix} \tag{3.914}$$

therefore, the following can be written:

$$\hat{S}_y^{(1)}\hat{F}^+ = \frac{1}{\sqrt{2}}\begin{pmatrix} D^* & iN^* & -iQ^* & F^* \\ iE^* & -O^* & O^* & iE^* \\ -F^* & -iQ^* & iN^* & -D^* \\ i(A^*+G^*) & -(K^*+R^*) & T^*+M^* & i(J^*+C^*) \\ -(B^*+H^*) & -i(L^*+S^*) & i(S^*+L^*) & -(H^*+B^*) \\ -i(C^*+J^*) & M^*+T^* & -(R^*+K^*) & -i(G^*+A^*) \\ -D^* & -iN^* & iQ^* & -F^* \\ -iE^* & O^* & -O^* & -iE^* \\ F^* & iQ^* & -iN^* & D^* \end{pmatrix} \tag{3.915}$$

Substituting Eqs. (3.891) and (3.915) into Eq. (3.895), one can obtain the following result:

$$\begin{aligned} A_y^{(1)} &= \frac{1}{9I_0}\mathrm{tr}\left\{F\hat{S}_y^{(1)}F^+\right\} = \frac{1}{9\sqrt{2}I_0}(AD^* + BE^* + CF^* + DA^* + DG^* + EB^* \\ &\quad +EH^* + FC^* + FJ^* + GD^* + HE^* + JF^* + KN^* + LO^* + MQ^* + NK^* \\ &\quad +NR^* + OL^* + OS^* + QM^* + QT^* + RN^* + SO^* + TQ^* + TQ^* + SO^* \\ &\quad +RN^* + QT^* + QM^* + OS^* + OL^* + NR^* + NK^* + MQ^* + LO^* + KN^* \\ &\quad +JF^* + HE^* + GD^* + FJ^* + FC^* + EH^* + EB^* + DG^* + DA^* + CF^* \\ &\quad +BE^* + AD^*) \\ &= \frac{2\sqrt{2}}{9I_0}\,\mathrm{Re}\,(AD^* + BE^* + CF^* + DG^* + EH^* + FJ^* + KN^* + LO^* + MQ^* \\ &\quad +NR^* + OS^* + QT^*) \end{aligned} \tag{3.916}$$

and we can still get

$$
\begin{aligned}
&\vec{U}^{(1)}_{z,++1}=\hat{S}^{(1)}_z\begin{pmatrix}F^*_{++,11}\\F^*_{++,01}\\F^*_{++,-11}\end{pmatrix}=\begin{pmatrix}1&0&0\\0&0&0\\0&0&-1\end{pmatrix}\begin{pmatrix}A^*\\\mathrm{i}D^*\\-G^*\end{pmatrix}=\begin{pmatrix}A^*\\0\\G^*\end{pmatrix},\\
&\vec{U}^{(1)}_{z,++0}=\begin{pmatrix}\mathrm{i}B^*\\0\\\mathrm{i}H^*\end{pmatrix},\quad \vec{U}^{(1)}_{z,++\,-1}=\begin{pmatrix}-C^*\\0\\-J^*\end{pmatrix},\quad \vec{U}^{(1)}_{z,+-1}=\begin{pmatrix}\mathrm{i}K^*\\0\\\mathrm{i}R^*\end{pmatrix},\\
&\vec{U}^{(1)}_{z,+-0}=\begin{pmatrix}-L^*\\0\\-S^*\end{pmatrix},\quad \vec{U}^{(1)}_{z,+-\,-1}=\begin{pmatrix}-\mathrm{i}M^*\\0\\-\mathrm{i}T^*\end{pmatrix},\quad \vec{U}^{(1)}_{z,-+1}=\begin{pmatrix}-\mathrm{i}T^*\\0\\-\mathrm{i}M^*\end{pmatrix},\\
&\vec{U}^{(1)}_{z,-+0}=\begin{pmatrix}S^*\\0\\L^*\end{pmatrix},\quad \vec{U}^{(1)}_{z,-+\,-1}=\begin{pmatrix}\mathrm{i}R^*\\0\\\mathrm{i}K^*\end{pmatrix},\quad \vec{U}^{(1)}_{z,--1}=\begin{pmatrix}J^*\\0\\C^*\end{pmatrix},\\
&\vec{U}^{(1)}_{z,--0}=\begin{pmatrix}\mathrm{i}H^*\\0\\\mathrm{i}B^*\end{pmatrix},\quad \vec{U}^{(1)}_{z,--\,-1}=\begin{pmatrix}-G^*\\0\\-A^*\end{pmatrix}
\end{aligned}
\tag{3.917}
$$

So one can write

$$
\hat{S}^{(1)}_z\hat{F}^+=\begin{pmatrix}
A^* & \mathrm{i}K^* & -\mathrm{i}T^* & J^*\\
\mathrm{i}B^* & -L^* & S^* & \mathrm{i}H^*\\
-C^* & -\mathrm{i}M^* & \mathrm{i}R^* & -G^*\\
0&0&0&0\\
0&0&0&0\\
0&0&0&0\\
G^* & \mathrm{i}R^* & -\mathrm{i}M^* & C^*\\
\mathrm{i}H^* & -S^* & L^* & \mathrm{i}B^*\\
-J^* & -\mathrm{i}T^* & \mathrm{i}K^* & -A^*
\end{pmatrix}
\tag{3.918}
$$

Substituting Eqs. (3.891) and (3.918) into Eq. (3.895), one can obtain the following result:

$$
\begin{aligned}
A^{(1)}_z=\frac{1}{9I_0}\mathrm{tr}\left\{F\hat{S}^{(1)}_zF^+\right\}=\frac{1}{9I_0}(&|A|^2+|B|^2+|C|^2-|G|^2-|H|^2-|J|^2\\
&+|K|^2+|L|^2+|M|^2-|R|^2-|S|^2-|T|^2+|T|^2+|S|^2+|R|^2-|M|^2\\
&-|L|^2-|K|^2+|J|^2+|H|^2+|G|^2-|C|^2-|B|^2-|A|^2)=0
\end{aligned}
\tag{3.919}
$$

Also, one can obtain the following expressions from Eqs. (3.900) and (3.902):

$$\vec{W}^{(12)}_{xx,++1} = \frac{1}{2}\begin{pmatrix} 0 & 1 & 0 \\ 1 & 0 & 1 \\ 0 & 1 & 0 \end{pmatrix}\begin{pmatrix} iB^* \\ -E^* \\ -iH^* \end{pmatrix} = \frac{1}{2}\begin{pmatrix} -E^* \\ i(B^* - H^*) \\ -E^* \end{pmatrix},$$

$$\vec{W}^{(12)}_{xx,++0} = \frac{1}{2}\begin{pmatrix} i(D^* - F^*) \\ A^* - C^* + J^* - G^* \\ i(D^* - F^*) \end{pmatrix}, \quad \vec{W}^{(12)}_{xx,++\ -1} = \frac{1}{2}\begin{pmatrix} -E^* \\ i(B^* - H^*) \\ -E^* \end{pmatrix},$$

$$\vec{W}^{(12)}_{xx,+-1} = \frac{1}{2}\begin{pmatrix} -iO^* \\ -L^* + S^* \\ -iO^* \end{pmatrix}, \quad \vec{W}^{(12)}_{xx,+-0} = \frac{1}{2}\begin{pmatrix} Q^* - N^* \\ i(K^* - M^* + T^* - R^*) \\ Q^* - N^* \end{pmatrix},$$

$$\vec{W}^{(12)}_{xx,+-\ -1} = \frac{1}{2}\begin{pmatrix} -iO^* \\ -L^* + S^* \\ -iO^* \end{pmatrix}, \quad \vec{W}^{(12)}_{xx,-+1} = \frac{1}{2}\begin{pmatrix} iO^* \\ S^* - L^* \\ iO^* \end{pmatrix},$$

$$\vec{W}^{(12)}_{xx,-+0} = \frac{1}{2}\begin{pmatrix} Q^* - N^* \\ i(R^* - T^* + M^* - K^*) \\ Q^* - N^* \end{pmatrix}, \quad \vec{W}^{(12)}_{xx,-+\ -1} = \frac{1}{2}\begin{pmatrix} iO^* \\ S^* - L^* \\ iO^* \end{pmatrix},$$

$$\vec{W}^{(12)}_{xx,--1} = \frac{1}{2}\begin{pmatrix} -E^* \\ -i(B^* - H^*) \\ -E^* \end{pmatrix}, \quad \vec{W}^{(12)}_{xx,--0} = \frac{1}{2}\begin{pmatrix} i(F^* - D^*) \\ J^* - G^* + A^* - C^* \\ i(F^* - D^*) \end{pmatrix},$$

$$\vec{W}^{(12)}_{xx,--\ -1} = \frac{1}{2}\begin{pmatrix} -E^* \\ -i(B^* - H^*) \\ -E^* \end{pmatrix} \tag{3.920}$$

Therefore, one can write

$$
\hat{S}_x^{(1)}\hat{S}_x^{(2)}\hat{F}^{+}=\frac{1}{2}
\begin{pmatrix}
-E^* & -\mathrm{i}O^* & \mathrm{i}O^* & -E^* \\
\mathrm{i}(D^*-F^*) & Q^*-N & Q^*-N^* & -\mathrm{i}(D^*-F^*) \\
-E^*-\mathrm{i}O^* & \mathrm{i}O^*-E^* & & \\
\mathrm{i}(B^*-H^*) & S^*-L & S^*-L^* & -\mathrm{i}(B^*-H^*) \\
A^*-C^*+J^*-G^* & \mathrm{i}(K^*-M^*+T^*-R^*) & -\mathrm{i}(K^*-M^*+T^*-R^*) & A^*-C^*+J^*-G^* \\
\mathrm{i}(B^*-H^*) & S^*-L^* & S^*-L^* & -\mathrm{i}(B^*-H^*) \\
-E^* & -\mathrm{i}O^* & \mathrm{i}O^* & -E^* \\
\mathrm{i}(D^*-F^*) & Q^*-N^* & Q^*-N^* & -\mathrm{i}(D^*-F^*) \\
-E^* & -\mathrm{i}O^* & \mathrm{i}O^* & -E^*
\end{pmatrix}
\tag{3.921}
$$

Substituting Eqs. (3.891) and (3.921) into Eq. (3.896), one can obtain the following results:

$$
\begin{aligned}
A_{x,x}^{(12)}=&\frac{1}{9I_0}\mathrm{tr}\left\{F\hat{S}_x^{(1)}\hat{S}_x^{(2)}F^{+}\right\}=\frac{1}{18I_0}[-AE^*+B(D^*-F^*)+CE^*+D(B^*-H^*)\\
&-E(A^*-C^*+J^*-G^*)-F(B^*-H^*)+GE^*-H(D^*-F^*)-JE^*\\
&-KO^*-L(Q^*-N^*)+MO^*-N(S^*-L^*)-O(K^*-M^*+T^*-R^*)\\
&+Q(S^*-L^*)+RO^*+S(Q^*-N^*)-TO^*-TO^*+S(Q^*-N^*)+RO^*\\
&+Q(S^*-L^*)-O(K^*-M^*+T^*-R^*)-N(S^*-L^*)+MO^*-L(Q^*-N^*)\\
&-KO^*-JE^*-H(D^*-F^*)+GE^*-F(B^*-H^*)-E(A^*-C^*+J^*-G^*)\\
&+D(B^*-H^*)+CE^*+B(D^*-F^*)-AE^*]\\
=&\frac{1}{9I_0}[-E(A^*-C^*+J^*-G^*)-(A-C+J-G)E^*\\
&-O(K^*-M^*+T^*-R^*)-(K-M+T-R)O^*+(B-H)(D^*-F^*)\\
&+(D-F)(B^*-H^*)+(S-L)(Q^*-N^*)+(Q-N)(S^*-L^*)]\\
=&-\frac{2}{9I_0}\mathrm{Re}[(A-C+J-G)E^*+(K-M+T-R)O^*-(B-H)(D^*-F^*)\\
&-(L-S)(N^*-Q^*)]
\end{aligned}
\tag{3.922}
$$

Other items of $A_{i,j}^{(12)}(i,j=x,\,y,\,z)$ can be derived by the above method. By using Eq. (3.36) further, $A_i^{(k)}(k=1,\,2,\;\;i=\gamma)$ and all remaining items of $A_{i,j}^{(12)}(i,j=x,\,y,\,z,\,\gamma)$ can be derived. Due to parity conservation, there must be some analyzing powers with a value of 0.

Referring to Eqs. (3.795)–(3.797), the following 9 × 9 matrices can be introduced:

$$\hat{S}_{1i} \equiv \begin{pmatrix} S_{i,1\,1}\hat{I}_3 & S_{i,1\,0}\hat{I}_3 & S_{i,1\,-1}\hat{I}_3 \\ S_{i,0\,1}\hat{I}_3 & S_{i,0\,0}\hat{I}_3 & S_{i,0\,-1}\hat{I}_3 \\ S_{i,-1\,1}\hat{I}_3 & S_{i,-1\,0}\hat{I}_3 & S_{i,-1\,-1}\hat{I}_3 \end{pmatrix}, \quad i=x,\,y,\,z \tag{3.923}$$

$$\hat{G}_{1i} \equiv \begin{pmatrix} G_{i,1\,1}\hat{I}_3 & G_{i,1\,0}\hat{I}_3 & G_{i,1\,-1}\hat{I}_3 \\ G_{i,0\,1}\hat{I}_3 & G_{i,0\,0}\hat{I}_3 & G_{i,0\,-1}\hat{I}_3 \\ G_{i,-1\,1}\hat{I}_3 & G_{i,-1\,0}\hat{I}_3 & G_{i,-1\,-1}\hat{I}_3 \end{pmatrix}, \quad i=xy,xz,yz,xx-yy,zz \tag{3.924}$$

$$\hat{S}_{2i} \equiv \begin{pmatrix} \hat{S}_i & \hat{0}_3 & \hat{0}_3 \\ \hat{0}_3 & \hat{S}_i & \hat{0}_3 \\ \hat{0}_3 & \hat{0}_3 & \hat{S}_i \end{pmatrix}, \quad i=x,\,y,\,z \tag{3.925}$$

$$\hat{G}_{2i} \equiv \begin{pmatrix} \hat{G}_i & \hat{0}_3 & \hat{0}_3 \\ \hat{0}_3 & \hat{G}_i & \hat{0}_3 \\ \hat{0}_3 & \hat{0}_3 & \hat{G}_i \end{pmatrix}, \quad i=xy,xz,yz,xx-yy,zz \tag{3.926}$$

The matrices $\hat{S}_{1i}$ and $\hat{G}_{1i}$ act on the first deuteron functions, and the matrices $\hat{S}_{2i}$ and $\hat{G}_{2i}$ act on the second deuteron functions. It follows that the $\vec{1}+\vec{1}$ reaction can be processed according to the method described in Sect. 3.11.

The previous method used for the arrangement of reaction amplitude footnotes belongs to the usual method. In order to facilitate the operation, the reaction amplitude matrix element $F_{\mu'\nu',\mu\nu}$ given by Eq. (3.887) can also be rewritten as $F_{\mu'\mu,\nu'\nu}$ [6], such that for the $\vec{\mathrm{d}}+\vec{\mathrm{d}}\to\mathrm{p}+\mathrm{t}$ reaction the footnote $\mu'$ is the line number of p, the footnote $\mu$ is the column number of the first d, the footnote $\nu'$ is the line number of t, and the footnote $\nu$ is the column number of the second d. For the definite $\nu'\nu$, the footnotes $\mu'\mu$ constitute a 2 × 3 submatrix $\hat{F}^{(1)}_{\nu'\nu}$; for the definite $\mu'\mu$, the footnotes $\nu'\nu$ constitute a 2 × 3 submatrix $\hat{F}^{(2)}_{\mu'\mu}$; $\hat{S}^{(1)}_i$ and $\hat{Q}^{(1)}_i$ act on $\hat{F}^{(1)+}_{\nu'\nu}$; $\hat{S}^{(2)}_i$ and $\hat{Q}^{(2)}_i$ act on $\hat{F}^{(2)+}_{\mu'\mu}$. In order to find the correlation analyzing powers, one must first form a new 9 × 4 matrix $\hat{X}^+_j$ after $\hat{S}^{(2)}_j$ or $\hat{Q}^{(2)}_j$ acting on all $\hat{F}^{(2)}_{\mu'\mu}$, and then calculate $\hat{S}^{(1)}_i$ or $\hat{Q}^{(1)}_i$ acting on the $\hat{X}^{(1)+}_{j,\nu'\nu}$.

In fact, because the matrix operations of $\vec{1}+\vec{1}$ reaction are too large, the above complex formula derivation cannot be performed; only the numerical calculation program including the tracing needs to be compiled. As for how to calculate the $S$ matrix elements of a few-body reaction, there are several options which can be employed, including, but not limited to, direct reaction theory, $R$ matrix theory, the phase-shift analysis method, resonance group theory, Faddeev equation, and CDCC theory.

The theoretical approach described in this section can also be applied to polarization nuclear reaction $\vec{\mathrm{d}}\left(\vec{\mathrm{d}},\mathrm{n}\right)^3\mathrm{He}$.

## 3.13 Deuteron Phenomenological Optical Potentials with Tensor Terms and Corresponding Radial Equations

### 3.13.1 *Deuteron Phenomenological Optical Potentials Containing Tensor Terms*

If the optical potential only contains the spin-independent central potential, the incident particle will not induce any spin polarization phenomenon. In order to consider the particle polarization phenomenon, only the spin-orbital coupling potential needs to be introduced for the reactions of $S = \frac{1}{2}$ particles with spinless targets. For $S = 1$ particles, not only does the spin-orbital coupling potential of two vectors need to be considered, it is also necessary to consider the tensor potential coupled by two rank 2 tensors of spin space and coordinate space. The tensor potential to be included in the deuteron optical potential must be a scalar, that is, it satisfies the invariance of rotation, and it should also satisfy the parity conservation condition.

In the coordinate space, we have three vectors: coordinate $\vec{r}$, momentum $\vec{p} = -\mathrm{i}\vec{\nabla}$, and angular momentum $\vec{L} = -\mathrm{i}\vec{r} \times \vec{\nabla}$. Here $\vec{p}$ is the wave vector $\vec{k}$. Now, we have the following expression: [22, 46]

$$(\vec{a} \cdot \vec{c})(\vec{b} \cdot \vec{d}) = \frac{1}{3}(\vec{a} \cdot \vec{b})\,(\vec{c} \cdot \vec{d}) + \frac{1}{2}(\vec{a} \times \vec{b}) \cdot (\vec{c} \times \vec{d}) + T_2(\vec{a}, \vec{b}) \cdot T_2(\vec{c}, \vec{d}) \tag{3.927}$$

If in the above formula we let $\vec{c} = \vec{d} = \hat{S}$ and $\vec{a} = \vec{b} = \vec{r}$, or $\vec{a} = \vec{b} = \vec{p}$, or $\vec{a} = \vec{b} = \vec{L}$, and note that

$$\vec{r} \times \vec{r} = \vec{p} \times \vec{p} = 0, \quad \vec{L} \times \vec{L} = \mathrm{i}\vec{L}, \quad \hat{S} \times \hat{S} = \mathrm{i}\hat{S}, \quad \hat{S}^2 = 2\hat{I} \tag{3.928}$$

then the following tensor potentials of the optical potential can be obtained: [46]

$$T_r \equiv \frac{1}{r^2} T_2(\vec{r}, \vec{r}) \cdot T_2(\hat{S}, \hat{S}) = (\hat{S} \cdot \vec{r})^2 / r^2 - \frac{2}{3} \tag{3.929}$$

$$T_p \equiv T_2(\vec{p}, \vec{p}) \cdot T_2(\hat{S}, \hat{S}) = (\hat{S} \cdot \vec{p})^2 - \frac{2}{3} p^2 \tag{3.930}$$

$$T_L \equiv T_2(\vec{L}, \vec{L}) \cdot T_2(\hat{S}, \hat{S}) = (\vec{L} \cdot \hat{S})^2 + \frac{1}{2} \vec{L} \cdot \hat{S} - \frac{2}{3} L^2 \tag{3.931}$$

The component of $T_2(\hat{S}, \hat{S})$ appearing in the above three formulas is

$$T_{2M}(\hat{S}, \hat{S}) = \sum_{\mu\nu} C_{1\mu\ 1\nu}^{2\ M} \hat{S}_\mu \hat{S}_\nu \tag{3.932}$$

Comparing the above formula with Eq. (3.15) we can see

$$T_{2M}(\hat{S}, \hat{S}) = \frac{1}{\sqrt{3}} \hat{T}_{2M}(1) \tag{3.933}$$

Using Eqs. (3.16)–(3.18) and (1.23), we can obtain the following expressions:

$$T_{20}(\hat{S}, \hat{S}) = \frac{1}{\sqrt{6}} (3\hat{S}_0^2 - 2) \tag{3.934}$$

$$T_{2\ \pm1}(\hat{S}, \hat{S}) = \frac{1}{\sqrt{2}} (\hat{S}_{\pm1}\hat{S}_0 + \hat{S}_0\hat{S}_{\pm1}) \tag{3.935}$$

$$T_{2\ \pm2}(\hat{S}, \hat{S}) = \hat{S}_{\pm1}^2 \tag{3.936}$$

The component of $T_2(\vec{r}, \vec{r})$ appearing in Eq. (3.929) is

$$T_{2M}(\vec{r}, \vec{r}) = \sum_{\mu\nu} C_{1\mu\ 1\nu}^{2\ M} r_\mu r_\nu \tag{3.937}$$

Using Table 3.3 the following results can be obtained from Eq. (3.937):

$$T_{20}(\vec{r}, \vec{r}) = \sqrt{\frac{2}{3}} (r_0^2 + r_{+1} r_{-1}) = \frac{1}{\sqrt{6}} (3r_0^2 - r^2) \tag{3.938}$$

$$T_{2\ \pm1}(\vec{r}, \vec{r}) = \sqrt{2} r_0 r_{\pm1} \tag{3.939}$$

$$T_{2\ \pm2}(\vec{r}, \vec{r}) = r_{\pm1}^2 \tag{3.940}$$

Now the three-dimensional spherical harmonic function is defined as

$$\mathrm{Y}_{lm}\left(\vec{r}\right) = r^l \mathrm{Y}_{lm}(\hat{r}) \tag{3.941}$$

Also, there is the following relation

$$\mathrm{Y}_{1\mu}\left(\vec{r}\right) = r \mathrm{Y}_{1\mu}(\hat{r}) = \sqrt{\frac{3}{4\pi}} r_\mu \tag{3.942}$$

For the spherical harmonic function, there is the following relation

$$\mathrm{Y}_{l_1 m_1}\mathrm{Y}_{l_2 m_2} = \sum_l \frac{\hat{l}_1 \hat{l}_2}{\sqrt{4\pi}\,\hat{l}} C^{l\ m_1+m_2}_{l_1 m_1\ l_2 m_2} C^{l\ 0}_{l_1 0\ l_2 0} \mathrm{Y}_{l\ m_1+m_2} \tag{3.943}$$

where $\hat{l} \equiv \sqrt{2l+1}$. Using the C-G coefficient properties and Eq. (3.943) we have

$$\begin{aligned}
\sum_{m_1 m_2} C^{l\ m}_{l_1 m_1\ l_2 m_2} \mathrm{Y}_{l_1 m_1} \mathrm{Y}_{l_2 m_2} &= \frac{\hat{l}_1 \hat{l}_2}{\sqrt{4\pi}\,\hat{l}} C^{l\ 0}_{l_1 0\ l_2 0} \mathrm{Y}_{lm}, \\
\mathrm{Y}_{lm} &= \frac{\sqrt{4\pi}\,\hat{l}}{\hat{l}_1 \hat{l}_2 C^{l\ 0}_{l_1 0\ l_2 0}} \sum_{m_1 m_2} C^{l\ m}_{l_1 m_1\ l_2 m_2} \mathrm{Y}_{l_1 m_1} \mathrm{Y}_{l_2 m_2}
\end{aligned} \tag{3.944}$$

Checking Table 3.3 we can get

$$\begin{aligned}
\mathrm{Y}_{2M}(\hat{r}) &= \sqrt{\frac{10\pi}{3}} \sum_{\mu\nu} C^{2M}_{1\mu\ 1\nu} \mathrm{Y}_{1\mu}(\hat{r}) \mathrm{Y}_{1\nu}(\hat{r}), \\
\mathrm{Y}_{2M}(\vec{r}) &= \sqrt{\frac{10\pi}{3}} \sum_{\mu\nu} C^{2M}_{1\mu\ 1\nu} \mathrm{Y}_{1\mu}(\vec{r}) \mathrm{Y}_{1\nu}(\vec{r})
\end{aligned} \tag{3.945}$$

Using Eqs. (3.937), (3.942), and (3.945), we can then obtain the following relation:

$$T_{2M}(\vec{r}, \vec{r}) = \sqrt{\frac{8\pi}{15}} \mathrm{Y}_{2M}(\vec{r}) \tag{3.946}$$

Further, we can obtain the following relation according to Eqs. (3.929), (3.933), and (3.946):

$$T_r = \sqrt{\frac{8\pi}{45}} \mathrm{Y}_2(\hat{r}) \cdot \hat{T}_2(1) \tag{3.947}$$

Now the general form of the deuteron spherical phenomenological optical potential can be written as

$$\begin{aligned}
V(\vec{r}) = {} & V_C(r) + V_c(r) + \mathrm{i}W_c(r) + [V_{LS}(r) + \mathrm{i}W_{LS}(r)]\vec{L} \cdot \hat{S} \\
& + [V_{TR}(r) + \mathrm{i}W_{TR}(r)]T_r + [V_{TL}(r) + \mathrm{i}W_{TL}(r)]T_L \\
& + \frac{1}{2}\{T_p[V_{TP}(r) + \mathrm{i}W_{TP}(r)] + [V_{TP}(r) + \mathrm{i}W_{TP}(r)]T_p\}
\end{aligned} \tag{3.948}$$

where $V_C(r)$ is the Coulomb potential, $V_c(r)$ is the real part of the center potential, and $W_c(r)$ is the imaginary part of the center potential which includes both the volume absorption and the surface absorption parts. Nuclear forces include two-body force and three-body force, whereas the optical potential is equivalent to the average field, and is a one-body potential. Let $U_r(r)$ be the coefficient of the $T_r$ term in the optical potential, and we have

$$[U_r(r)T_{2M}(\vec{r},\ \vec{r})]^+ = (-1)^M T_{2\ -M}(\vec{r},\ \vec{r})U_r^*(r) \tag{3.949}$$

Because $\vec{r}$ and $U_r(r)$ is commutative, $U_r^*(r)$ can be moved to the front of $T_{2\ -M}(\vec{r},\ \vec{r})$ in the above formula. It can be seen that the real part $V_{TR}(r)T_r$ is Hermitian, and $\mathrm{i}W_{TR}(r)T_r$ is anti-Hermitian. Because $T_L$ and the corresponding $U_L(r)$ is also commutative, it follows that $V_{TL}(r)T_L$ is Hermitian, and $\mathrm{i}W_{TL}(r)T_L$ is anti-Hermitian. However, because $\vec{p}$ and corresponding $U_p(r)$ is not commutative, it is impossible to move $U_p^*(r)$ to the front of $T_p^+$ for the momentum term. Nevertheless, using the form of the $T_p$ term in Eq. (3.948) can ensure which real part is Hermitian and which imaginary part is anti-Hermitian. In fact, the central potential and spin-orbit coupling potential also have the above properties.

In References [47–49], the deuteron universal phenomenological optical potentials are given separately, but none of them contain tensor potential. Reference [49] gives the calculations of the vector polarization quantities. The deuteron optical potentials given in References [13–17, 50] all contain the $T_r$ term, and in References [8–10, 12] the deuteron optical potentials contain the $T_r$ term and the $T_L$ term simultaneously. In Reference [51], the necessity of the inclusion of the $T_p$ term in the deuteron optical potential is discussed, and in Reference [17] there is also some discussion on the $T_p$ term, but at present there is a question about whether or not the $T_p$ term needs to be included; besides, the handling of the $T_p$ term in the deuteron optical potential has yet to be further studied, so we will no longer include the $T_p$ term in subsequent discussions. This noted, it is generally believed that the $T_r$ term in the three tensor potentials is the most important, and it plays an important role in fitting experimental data of $S = 1$ particle polarization analyzing powers.

### 3.13.2 Deuteron Radial Equations for the Spherical Nuclei when Containing Tensor Potentials

The stationary Schrodinger equation satisfied by the deuteron wave function $\Phi_{JM}(\vec{r})$ with total angular momentum $J$ is

$$\left[-\frac{\hbar^2}{2\mu}\vec{\nabla}^2 + V\left(\vec{r}\right)\right]\Phi_{JM}(\vec{r}) = E\Phi_{JM}(\vec{r}) \tag{3.950}$$

where $\mu$ is the deuteron reduced mass. The optical potential $V(\vec{r})$ takes the form given by Eq. (3.948), but it does not include the last $T_p$ term. When the target spin is equal to 0, $\Phi_{JM}(\vec{r})$ can be written as

$$\Phi_{JM}(\vec{r}) = \sum_{L} \frac{u_L^J(r)}{r} \mathbb{Y}_{LJM} \tag{3.951}$$

$$\mathbb{Y}_{LJM} = \sum_{M_L\nu} C_{LM_L\ 1\nu}^{J\ M} \mathrm{i}^L Y_{LM_L}(\hat{r}) \chi_{1\nu} \tag{3.952}$$

The coordinate tensor operator between the two nucleons is defined as

$$S_{12} = \frac{3}{r^2} (\hat{\sigma}_1 \cdot \vec{r})\,(\hat{\sigma}_2 \cdot \vec{r}) - (\hat{\sigma}_1 \cdot \hat{\sigma}_2) \tag{3.953}$$

and let

$$\hat{S} = \frac{1}{2} (\hat{\sigma}_1 + \hat{\sigma}_2) \tag{3.954}$$

Now we can obtain the following expression from Eqs. (3.953) and (3.929): [22]

$$S_{12} = 6\left[\left(\hat{S} \cdot \vec{r}\right)^2 / r^2 - \frac{1}{3}\hat{S}^2\right] = 6\left[\left(\hat{S} \cdot \vec{r}\right)^2 / r^2 - \frac{2}{3}\right] = 6T_r \tag{3.955}$$

In Reference [22], the matrix elements $\left\langle \mathbb{Y}_{L'JM} \middle| \vec{L} \cdot \hat{S} \middle| \mathbb{Y}_{LJM} \right\rangle$ and $\langle \mathbb{Y}_{L'JM} | S_{12} | \mathbb{Y}_{LJM} \rangle$ have been obtained, and we list them in Tables 3.4 and 3.5, respectively:

Utilizing Eq. (3.931) we can obtain

$$T_L \mathbb{Y}_{LJM} = \left\{ \frac{1}{4} [J(J+1) - L(L+1) - 2]^2 + \frac{1}{4} [J(J+1) - L(L+1) - 2] - \frac{2}{3} L(L+1) \right\} \mathbb{Y}_{LJM} \tag{3.956}$$

Note that $\mathbb{Y}_{LJM}$ is an orthogonal normalized wave function, and it follows that the matrix elements $\langle \mathbb{Y}_{L'JM} | T_L | \mathbb{Y}_{LJM} \rangle$ can be obtained and are listed in Table 3.6 [46].

The above results show that $\vec{L} \cdot \hat{S}$ and $T_L$ only have the non-zero diagonal elements for the orbital angular momentum $L$, but $T_r$ has non-zero off-diagonal matrix elements with a difference of 2 for $L$, so in this case $L$ is no longer a good quantum number. For a definite $J$, $L = J \pm 1$, and two states will be coupled, but due to the parity conservation, the $L = J$ state and the $L = J \pm 1$ states are not coupled, and Table 3.5 clearly shows the above results. Since the deuteron spin $S = 1$, then it must be that for a definite total angular momentum $J$, $L = J \pm 1$ and $L = J$, and the three states of $J$ must be solved at the same time. When $J = 0$, there are only two states of $L$, namely, $L = 0, 1$.

Substituting Eq. (3.951) into Eq. (3.950), multiplying $\mathbb{Y}_{L'JM}^{+}$ on both sides of the equation from the left concurrently, and integrating the direction angles of the coordinate space to the obtained equation, the following three equations can be obtained.

**Table 3.4** Matrix elements $\left\langle \mathbb{Y}_{L'JM} \middle| \vec{L} \cdot \widehat{S} \middle| \mathbb{Y}_{LJM} \right\rangle$

| | $L'$ | | |
|---|---|---|---|
| $L$ | $J+1$ | $J$ | $J-1$ |
| $J+1$ | $-(J+2)$ | 0 | 0 |
| $J$ | 0 | $-1$ | 0 |
| $J-1$ | 0 | 0 | $J-1$ |

**Table 3.5** Matrix elements $\langle \mathbb{Y}_{L'JM} | S_{12} | \mathbb{Y}_{LJM} \rangle$

| | $L'$ | | |
|---|---|---|---|
| $L$ | $J+1$ | $J$ | $J-1$ |
| $J+1$ | $-\frac{2(J+2)}{2J+1}$ | 0 | $-\frac{6\sqrt{J(J+1)}}{2J+1}$ |
| $J$ | 0 | 2 | 0 |
| $J-1$ | $-\frac{6\sqrt{J(J+1)}}{2J+1}$ | 0 | $-\frac{2(J-1)}{2J+1}$ |

**Table 3.6** Matrix elements $\langle \mathbb{Y}_{L'JM} | T_L | \mathbb{Y}_{LJM} \rangle$

| | $L'$ | | |
|---|---|---|---|
| $L$ | $J+1$ | $J$ | $J-1$ |
| $J+1$ | $\frac{1}{6}(J+2)(2J+5)$ | 0 | 0 |
| $J$ | 0 | $-\frac{1}{6}(2J-1)(2J+3)$ | 0 |
| $J-1$ | 0 | 0 | $\frac{1}{6}(J-1)(2J-3)$ |

$$\left[\frac{\mathrm{d}^2}{\mathrm{d}r^2} - \frac{J(J-1)}{r^2} + k^2 - U_0(r) - U_{LS}(r)C^J_{J-1} - U_{TL}(r)B^J_{J-1} - U_{TR}(r)A^J_{J-1,J-1}\right]u^J_{J-1}$$
$$= U_{TR}(r)A^J_{J-1,J+1}u^J_{J+1} \tag{3.957}$$

$$\left[\frac{\mathrm{d}^2}{\mathrm{d}r^2} - \frac{J(J+1)}{r^2} + k^2 - U_0(r) - U_{LS}(r)C^J_J - U_{TL}(r)B^J_J - U_{TR}(r)A^J_{J,J}\right]u^J_J = 0 \tag{3.958}$$

$$\left[\frac{\mathrm{d}^2}{\mathrm{d}r^2} - \frac{(J+1)(J+2)}{r^2} + k^2 - U_0(r) - U_{LS}(r)C^J_{J+1} - U_{TL}(r)B^J_{J+1} - U_{TR}(r)A^J_{J+1,J+1}\right]u^J_{J+1}$$
$$= U_{TR}(r)A^J_{J+1,J-1}u^J_{J-1} \tag{3.959}$$

where

$$k = \frac{\sqrt{2\mu E}}{\hbar} \tag{3.960}$$

$$U_0(r) = \frac{2\mu}{\hbar^2}\left[V_C(r) + V_c(r) + iW_c(r)\right] \tag{3.961}$$

$$U_x(r) = \frac{2\mu}{\hbar^2}\left[V_x(r) + iW_x(r)\right], \quad x = LS, TL, TR \tag{3.962}$$

$$C_L^J = \begin{cases} J-1 & L = J-1 \\ -1 & L = J \\ -(J+2) & L = J+1 \end{cases} \tag{3.963}$$

$$B_L^J = \begin{cases} \frac{1}{6}(J-1)(2J-3) & L = J-1 \\ -\frac{1}{6}(2J-1)(2J+3) & L = J \\ \frac{1}{6}(J+2)(2J+5) & L = J+1 \end{cases} \tag{3.964}$$

$$\begin{gathered} A^J_{J-1,\,J-1} = -\frac{J-1}{3(2J+1)}, \quad A^J_{J,J} = \frac{1}{3}, \quad A^J_{J+1,J+1} = -\frac{J+2}{3(2J+1)}, \\ A^J_{J-1,J+1} = A^J_{J+1,J-1} = -\frac{\sqrt{J(J+1)}}{2J+1} \end{gathered} \tag{3.965}$$

Equations (3.957) and (3.959) are simultaneous, and Eq. (3.958) is independent. For each total angular momentum $J$, the above three equations are solved concurrently, and the boundary conditions are linked to the incident channel, so the required S matrix elements can be obtained. It is worth noting that it is not uncommon for the off-diagonal term of $T_r$ to be neglected, as it is only an approximate method [8].

## 3.14 Folding Model Describing the Reaction Between Deuteron and Nucleus

### *3.14.1 Deuteron Folding Model Without Breakup Channel [46, 52, 53]*

The basic assumption of the deuteron single folding model is that the interaction between the deuteron and the target nucleus can be approximately described by a potential field $U_0(\vec{R})$, where $U_0(\vec{R})$ is obtained based on the interactions of the neutron and proton in the deuteron with the target after being averaged by the deuteron internal wave function $\varphi(\vec{r})$. The effective Hamiltonian of the system composed by the neutron and proton in the deuteron and target nucleus is

$$H = -\frac{\hbar^2}{2m}\left(\vec{\nabla}_n^2 + \vec{\nabla}_p^2\right) + V_n\left(\vec{r}_n\right) + V_p\left(\vec{r}_p\right) + v\left(\vec{r}_n - \vec{r}_p\right) \tag{3.966}$$

where $V_n(\vec{r}_n)$ and $V_p(\vec{r}_p)$ are the potential fields felt by the neutron and proton in the deuteron and are usually regarded as the optical potentials of the neutron and proton in the deuteron, respectively. Also, $v\left(\vec{r}_n - \vec{r}_p\right)$ are the n − p interactions within the deuteron that determine the deuteron binding energy; this potential belongs to nuclear force. Next we define

$$\vec{R} = \frac{1}{2}\left(\vec{r}_n + \vec{r}_p\right), \qquad \vec{r} = \vec{r}_n - \vec{r}_p \tag{3.967}$$

where the two expressions represent the center of mass coordinate of the deuteron and the relative coordinate between the neutron and proton, respectively (see Fig. 3.1). Now we can obtain

$$\vec{\nabla}_n^2 + \vec{\nabla}_p^2 = \frac{1}{2}\vec{\nabla}_R^2 + 2\vec{\nabla}_r^2 \tag{3.968}$$

Letting $\Psi(\vec{R}, \vec{r})$ be the system wave function, we can write the stationary Schrodinger equation of the whole system as

$$\left[-\frac{\hbar^2}{4m}\vec{\nabla}_R^2 - \frac{\hbar^2}{m}\vec{\nabla}_r^2 + V_n\left(\vec{R} + \frac{1}{2}\vec{r}\right) + V_p\left(\vec{R} - \frac{1}{2}\vec{r}\right) + v(\vec{r})\right]\Psi(\vec{R}, \vec{r}) = E\Psi(\vec{R}, \vec{r}) \tag{3.969}$$

where $E$ represents the total energy of the system. Next, let

$$V(\vec{R}, \vec{r}) = V_n\left(\vec{R} + \frac{1}{2}\vec{r}\right) + V_p\left(\vec{R} - \frac{1}{2}\vec{r}\right) \tag{3.970}$$

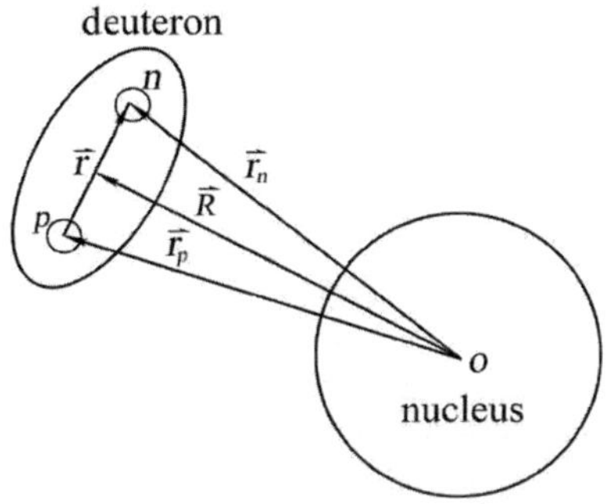


**Fig. 3.1** Single folding model coordinate system of deuteron

Now, when $\vec{R}$ is relatively large, it is possible that the variables of $\Psi(\vec{R}, \vec{r})$ can be separated and becomes

$$\Psi(\vec{R}, \vec{r}) = \varphi(\vec{r})\Phi(\vec{R}) \tag{3.971}$$

where $\Phi(\vec{R})$ is a movement wave function of the center of mass of the deuteron with respect to the target nucleus, and $\varphi(\vec{r})$ is a relative motion wave function of the neutron and proton inside the deuteron. Now it may be that the deuteron is first absorbed by the target, followed by breakage in the target, with the separated neutron and proton again synthesizing a deuteron, subsequently emitted from the target. This process corresponds to the influence of the non-local potential. Here we ignore this process, for this process can be described by the imaginary part of the optical potential. Therefore, in Eq. (3.971), the components, for which the variables cannot be separated, can be approximately ignored. At the same time, we also assume that the target spin is equal to 0. It follows that Eq. (3.969) can be rewritten as

$$\left[-\frac{\hbar^2}{4m}\vec{\nabla}_R^2 + V(\vec{R}, \vec{r}) - \frac{\hbar^2}{m}\vec{\nabla}_r^2 + v(\vec{r})\right]\varphi(\vec{r})\Phi(\vec{R}) = E\varphi(\vec{r})\Phi(\vec{R}) \tag{3.972}$$

Now the equation satisfied by the internal motion wave function $\varphi(\vec{r})$ of the deuteron is

$$\left[-\frac{\hbar^2}{m}\vec{\nabla}_r^2 + v(\vec{r})\right]\varphi(\vec{r}) = \varepsilon\varphi(\vec{r}) \tag{3.973}$$

The deuteron has a ground bound state $\varphi_0(\vec{r})$, $\varepsilon_0 = -2.225$ MeV. The other solutions of Eq. (3.973) are the deuteron breakup states $\varphi_k(\vec{r})$ with continuous positive energy $\varepsilon_k$, where $\vec{k}$ is a n-p relative motion wave vector such that

$$\varepsilon_k = \frac{\hbar^2 k^2}{2\mu}, \quad \mu = \frac{m}{2} \tag{3.974}$$

In this case Eq. (3.971) can be rewritten as

$$\Psi(\vec{R}, \vec{r}) = \varphi_0(\vec{r})\Phi_0(\vec{R}) + \int d\vec{k}\,\varphi_k(\vec{r})\Phi_k(\vec{R}) \tag{3.975}$$

where $\varphi_0(\vec{r})$ and $\Phi_0(\vec{R})$ all have dimensions fm$^{-3/2}$, though $\varphi_k(\vec{r})$ and $\Phi_k(\vec{R})$ are all dimensionless.

It is only meaningful to calculate the optical potential when there is no deuteron breakage, but there is indeed the possibility of such a breakup when studying deuteron scattering. Therefore, it is necessary to study the influence of the deuteron breakup on the deuteron optical potential. Note that when the energy of the incident deuteron is relatively low, the probability of deuteron breakup is very small. Therefore, we first study the case that Eq. (3.975) does not include a breakup channel.

Substituting $\Psi(\vec{R}, \vec{r}) = \varphi_0(\vec{r})\Phi_0(\vec{R})$ into Eq. (3.972), multiplying $\varphi_0^+(\vec{r})$ from the left on both sides of the equation, and integrating to $\vec{r}$, we obtain

$$\left[-\frac{\hbar^2}{4m}\vec{\nabla}_R^2 + V_0(\vec{R})\right]\Phi_0(\vec{R}) = (E - \varepsilon_0)\Phi_0(\vec{R}) \tag{3.976}$$

where

$$V_0(\vec{R}) = \int \varphi_0^+(\vec{r})\, V(\vec{R}, \vec{r})\varphi_0(\vec{r})\mathrm{d}\vec{r} \tag{3.977}$$

is the deuteron folding optical potential and $E_0 \equiv E - \varepsilon_0$ represents the deuteron incident energy in the c.m. system. Next, we can prove [22]

$$\frac{1}{4\pi}\int \left(\hat{\sigma}_1 \cdot \vec{r}\right)\left(\hat{\sigma}_2 \cdot \vec{r}\right) \mathrm{d}\Omega = \frac{r^2}{3}(\hat{\sigma}_1 \cdot \hat{\sigma}_2) \tag{3.978}$$

Now, there is the following relation for $S_{12}$ given by Eq. (3.953)

$$\int S_{12}\mathrm{d}\Omega = 0 \tag{3.979}$$

The n-p interaction potential containing the tensor term in the deuteron can be written as

$$v = v_0(r) + v_\sigma(r)\hat{\sigma}_1 \cdot \hat{\sigma}_2 + v_T(r)S_{12} \tag{3.980}$$

Since $v$ is a two-body nuclear force, tensor terms can be included. The Schrodinger equation satisfied by the deuteron ground state wave function is

$$\left[-\frac{\hbar^2}{2\mu}\vec{\nabla}^2 + v_0(r) + v_\sigma(r)\hat{\sigma}_1 \cdot \hat{\sigma}_2 + v_T(r)S_{12}\right]\varphi_0(\vec{r}) = \varepsilon_0\varphi_0(\vec{r}) \tag{3.981}$$

For the deuteron, we have that spin $S = 1$, total angular momentum $J = 1$, orbital angular momentum $l = 0$ (S state), and $l = 2$ (D state); the deuteron ground state wave function can be written as

$$\varphi_0(\vec{r}) = \frac{u(r)}{r}\mathbb{Y}_{01M} + \frac{\mathrm{w}(r)}{r}\mathbb{Y}_{21M} \tag{3.982}$$

$$\mathbb{Y}_{l1M} = \sum_{m_l \nu} C_{lm_l\ 1\nu}^{1\ \ M} \mathrm{i}^l \mathrm{Y}_{lm_l}(\Omega)\chi_{1\nu} \tag{3.983}$$

and note

$$(\hat{\sigma}_1 \cdot \hat{\sigma}_2)\chi_{1\nu} = \chi_{1\nu}, \quad (\hat{\sigma}_1 \cdot \hat{\sigma}_2)\chi_{00} = -3\chi_{00} \tag{3.984}$$

Now, substitute Eq. (3.982) into Eq. (3.981), multiply $\mathbb{Y}_{01M}^+$ and $\mathbb{Y}_{21M}^+$ from the left of the two sides of the equation, respectively, and integrate to the direction angles of the spherical coordinate system. Next, use the results given by Eq. (3.984) and Table 3.5, and note that $\vec{\nabla}^2 = \frac{1}{r}\frac{\mathrm{d}^2}{\mathrm{d}r^2}r - \frac{l(l+1)}{r^2}$. Now when $l = 0$ there is $l(l+1) = 0$, and when $l = 2$ there is $l(l+1) = 6$; in this way we can obtain

$$-\frac{\hbar^2}{2\mu}\frac{\mathrm{d}^2 u}{\mathrm{d}r^2} + [v_0(r) + v_\sigma(r)]u - \varepsilon_0 u = \sqrt{8}v_T(r)w \tag{3.985}$$

$$-\frac{\hbar^2}{2\mu}\left(\frac{\mathrm{d}^2 w}{\mathrm{d}r^2} - \frac{6}{r^2}w\right) + [v_0(r) + v_\sigma(r)]w - 2v_T(r)w - \varepsilon_0 w = \sqrt{8}v_T(r)\,u \tag{3.986}$$

From Eq. (3.983) we can see

$$\mathbb{Y}_{01M} = \frac{1}{\sqrt{4\pi}}\chi_{1M} \tag{3.987}$$

Assuming that Eq. (3.982) can be rewritten as [16, 54]

$$\varphi_0(\vec{r}) = \frac{1}{\sqrt{4\pi}}\left[\frac{u(r)}{r} + \frac{w(r)}{\sqrt{8}r}S_{12}\right]\chi_{1M} \tag{3.988}$$

and noting that the square of any component of the Pauli matrix in Cartesian coordinate system is 1 and satisfies the anti-commutation relation, it can be proved that

$$\frac{\left(\hat{\sigma} \cdot \vec{r}\right)^2}{r^2} = 1 \tag{3.989}$$

For $\chi_{1M}$ we can get

$$S_{12}^2 = 9 - 6\frac{\left(\hat{\sigma}_1 \cdot \vec{r}\right)\left(\hat{\sigma}_2 \cdot \vec{r}\right)}{r^2} + 1 = 8 - 2S_{12} \tag{3.990}$$

Also, we can obtain the following relations utilizing Eqs. (3.953), (3.979), and (3.990):

$$\frac{1}{4\pi}\int (S_{12})^2 \mathrm{d}\Omega = 8 \tag{3.991}$$

$$\frac{1}{4\pi}\int (S_{12})^3 \mathrm{d}\Omega = \frac{1}{4\pi}\int (8-2S_{12})S_{12}\mathrm{d}\Omega = -16 \tag{3.992}$$

In the process of substituting Eq. (3.988) into Eq. (3.981) and multiplying $\frac{1}{\sqrt{4\pi}}\chi_{1M}^{+}$ or $\frac{1}{\sqrt{4\pi}}\frac{S_{12}}{\sqrt{8}}\chi_{1M}^{+}$ from the left of the above mentioned equations, then integrating the direction angles of the spherical coordinate system and utilizing Eqs. (3.984), (3.979), (3.991), and (3.992), we can obtain Eqs. (3.985) and (3.986), respectively. It can be seen that Eq. (3.988) is the deuteron ground state wave function and satisfies the following normalization condition:

$$\int_0^{\infty} \left[u^2(r) + w^2(r)\right] \mathrm{d}r = 1 \tag{3.993}$$

The neutron optical potential $V_{\mathrm{n}}(\vec{r}_{\mathrm{n}})$ and proton optical potential $V_{\mathrm{p}}(\vec{r}_{\mathrm{p}})$ in Eq. (3.966) are the one-body potentials of spin $\frac{1}{2}$ particles. They can contain spin-orbital coupling potentials and cannot contain tensor potentials. For spherical optical potentials, they are taken as the following form:

$$V_{\mathrm{n}}(r_{\mathrm{n}}) = V_{\mathrm{n}}^{\mathrm{c}}(r_{\mathrm{n}}) + V_{\mathrm{n}}^{ls}(r_{\mathrm{n}})\,\vec{l}_{\mathrm{n}}\cdot\hat{s}_{\mathrm{n}} \tag{3.994}$$

$$V_{\mathrm{p}}(r_{\mathrm{p}}) = V_{\mathrm{p}}^{\mathrm{C}}(r_{\mathrm{p}}) + V_{\mathrm{p}}^{\mathrm{c}}(r_{\mathrm{p}}) + V_{\mathrm{p}}^{ls}(r_{\mathrm{p}})\,\vec{l}_{\mathrm{p}}\cdot\hat{s}_{\mathrm{p}} \tag{3.995}$$

where $V_{\mathrm{p}}^{\mathrm{C}}(r_{\mathrm{p}})$ is the proton Coulomb potential and $V_{\mathrm{p}}^{\mathrm{c}}(r_{\mathrm{p}})$ is the proton central potential.

Now we have the following relations:

$$\begin{aligned} r_{\mathrm{n}} &= \left|\vec{R} + \frac{1}{2}\vec{r}\right| = \left(R^2 + \frac{r^2}{4} + Rr\cos\theta\right)^{1/2}, \\ r_{\mathrm{p}} &= \left|\vec{R} - \frac{1}{2}\vec{r}\right| = \left(R^2 + \frac{r^2}{4} - Rr\cos\theta\right)^{1/2} \end{aligned} \tag{3.996}$$

and perform the following expansion:

$$V_i^x(r_i) = \sum_l (2l+1)\, V_{i,l}^x(R, r)\, \mathrm{P}_l(\cos\theta), \quad x = \mathrm{c}, ls, \quad i = \mathrm{n}; \quad x = \mathrm{C}, \mathrm{c}, ls, \quad i = \mathrm{p} \tag{3.997}$$

where

$$V_{i,l}^{x}(R,r)=\frac{1}{2}\int V_{i}^{x}(r_i)\mathrm{P}_l(\cos\theta)\mathrm{d}\cos\theta \tag{3.998}$$

such that Eq. (3.997) can be rewritten as

$$V_{i}^{x}(r_i)=4\pi\sum_{lm}V_{i,l}^{x}(R,r)\mathrm{Y}_{lm}(\hat{R})\mathrm{Y}_{lm}^{*}(\hat{r}) \tag{3.999}$$

According to Eqs. (3.955), (3.929), and (3.946) and the definition of the scalar product of the two irreducible tensors given by Eq. (1.41), we can get the following relation:

$$S_{12}=4\sqrt{\frac{6\pi}{5}}\sum_{M}T_{2M}^{*}(\hat{S},\hat{S})\ \mathrm{Y}_{2M}(\hat{r}) \tag{3.1000}$$

where $\hat{S}$ represents an operator, $\hat{r}$ represents the direction angle in spherical coordinate system, and $T_{2M}(\hat{S},\hat{S})$ has been given by Eq. (3.932).

Firstly, we can obtain the following expression from Eqs. (3.988) and (3.990):

$$\varphi_0^{+}(\vec{r})\varphi_0(\vec{r})=\frac{1}{4\pi r^2}\left[u^2+w^2+\frac{w}{\sqrt{2}}\left(u-\frac{w}{\sqrt{8}}S_{12}\right)\right] \tag{3.1001}$$

For $V_{\mathrm{n}}^{\mathrm{c}}(r_{\mathrm{n}})$, $V_{\mathrm{p}}^{\mathrm{C}}(r_{\mathrm{p}})$, and $V_{\mathrm{p}}^{\mathrm{c}}(r_{\mathrm{p}})$ we can obtain the following relations using Eqs. (3.999)–(3.1001) and (3.955):

$$\begin{aligned}V_{i}^{x}(\vec{R})=&\int\varphi_0^{+}(\vec{r})\ V_{i}^{x}(r_i)\varphi_0(\vec{r})\mathrm{d}\vec{r}=\int_0^{\infty}\left(u^2+w^2\right)\ V_{i,0}^{x}(R,r)\mathrm{d}r\\&+3\sqrt{2}\left(\int_0^{\infty}V_{i,2}^{x}(R,\ r)w\left(u-\frac{w}{\sqrt{8}}\right)\mathrm{d}r\right)T_R\end{aligned} \tag{3.1002}$$

where

$$T_R=\frac{\left(\hat{S}\cdot\vec{R}\right)^2}{R^2}-\frac{2}{3} \tag{3.1003}$$

For spin-orbit coupling potential, Eq. (3.1002) cannot be used because $\vec{l}_i\cdot\hat{s}_i$ and $S_{12}$ are not commutable. Set $\vec{L}$ as the orbital angular momentum of the deuteron center of mass

$$\vec{L} = -\mathrm{i}\vec{R} \times \vec{\nabla}_R \tag{3.1004}$$

where $\vec{l}_n, \vec{l}_p, \vec{L}$ are all orbital angular momentum relative to the target center of mass. Let $\vec{l}$ be the orbital angular momentum of the relative motion between n-p

$$\vec{l} = -\mathrm{i}\vec{r} \times \vec{\nabla}_r \tag{3.1005}$$

Now we know that the nucleon spin-orbital potential $V^{ls}$ is a small amount relative to the central potential $V^{c}$, and the deuteron D state is a small amount relative to the deuteron S state. Therefore, when we use Eq. (3.977) to find the deuteron folding optical potential, we ignore the influence on $V^{ls}$ by the $w$ part representing the D state in $\varphi_0(\vec{r})$. That is, when finding the deuteron spin-orbital coupling potential $V^{LS}$ only the effect on $V^{ls}$ by the $u$ part representing the S state in $\varphi_0(\vec{r})$ is considered. When only the S-state in the deuteron is considered, $l = 0$. At this time, for the most extreme situation where $r = 0$, $\vec{r}_n = \vec{r}_p = \vec{R}$, it follows that when $l = 0$ we should have

$$\vec{l}_\mathrm{n} = \vec{l}_\mathrm{p} = \vec{L} \tag{3.1006}$$

Now, with some loss of generality, though not unreasonable, let us assume the following special provision:

$$V_\mathrm{n}^{ls}(r_\mathrm{n}) = V_\mathrm{p}^{ls}(r_\mathrm{p}) \tag{3.1007}$$

At this time we can obtain

$$\vec{l}_\mathrm{n} \cdot \hat{s}_\mathrm{n} + \vec{l}_\mathrm{p} \cdot \hat{s}_\mathrm{p} = \vec{L} \cdot \hat{S} \tag{3.1008}$$

where $\hat{S}$ has been given by Eq. (3.954). From here we can get [46]

$$V^{LS}(R) = \int \varphi_0^{+}(\vec{r})\, V_\mathrm{n}^{ls}(r_\mathrm{n})\varphi_0(\vec{r})\mathrm{d}\vec{r} \cong \int_0^{\infty} u^2 V_{\mathrm{n},0}^{ls}(R, r)\mathrm{d}r \tag{3.1009}$$

When taking into account only the contribution of the deuteron ground state, according to Eq. (3.977) the folding optical potential of the deuteron is obtained in the following form:

$$V_0(\vec{R}) = V^{\mathrm{C}}(R) + V^{\mathrm{c}}(R) + V^{LS}(R)\vec{L} \cdot \hat{S} + V^{\mathrm{T}}(R)T_R \tag{3.1010}$$

Substituting Eq. (3.998) into the results obtained above, we can obtain

$$V^{\mathrm{C}}(R) = \frac{1}{4\pi}\int_0^{\infty} \frac{u^2 + w^2}{r^2} V_{\mathrm{p}}^{\mathrm{C}}\left(\left|\vec{R} - \frac{1}{2}\vec{r}\right|\right)\mathrm{d}\vec{r} \tag{3.1011}$$

$$V^{\mathrm{c}}(R) = \frac{1}{4\pi}\int_0^{\infty} \frac{u^2 + w^2}{r^2}\left[V_{\mathrm{n}}^{\mathrm{c}}\left(\left|\vec{R} + \frac{1}{2}\vec{r}\right|\right) + V_{\mathrm{p}}^{\mathrm{c}}\left(\left|\vec{R} - \frac{1}{2}\vec{r}\right|\right)\right]\mathrm{d}\vec{r} \tag{3.1012}$$

$$V^{LS}(R) = \frac{1}{4\pi}\int_0^{\infty} \frac{u^2}{r^2} V_{\mathrm{n}}^{ls}\left(\left|\vec{R} + \frac{1}{2}\vec{r}\right|\right)\mathrm{d}\vec{r} \tag{3.1013}$$

$$\begin{aligned} V^{\mathrm{T}}(R) = {} & \frac{6}{4\pi\sqrt{2}}\int_0^{\infty} \frac{w}{r^2}\left(u - \frac{w}{\sqrt{8}}\right)\left[V_{\mathrm{p}}^{\mathrm{C}}\left(\left|\vec{R} - \frac{1}{2}\vec{r}\right|\right) + V_{\mathrm{n}}^{\mathrm{c}}\left(\left|\vec{R} + \frac{1}{2}\vec{r}\right|\right)\right. \\ & \left. + V_{\mathrm{p}}^{\mathrm{c}}\left(\left|\vec{R} - \frac{1}{2}\vec{r}\right|\right)\right] \mathrm{P}_2(\cos\theta)\mathrm{d}\vec{r} \end{aligned} \tag{3.1014}$$

References [16, 53] present the contribution of the nucleon spin-orbit coupling potential $V^{ls}$ to deuteron optical potential in the presence of deuteron D state, but there are still some problems to be examined.

When the $l = 2$ D-state exists in the deuteron ground state structure, only the tensor force is contained in the nucleon-nucleon interaction, and in this case the experimental data of the deuteron magnetic moment can be explained by theory. Because the sum of the spins of the two nucleons can be 1, there can be tensor force in the two-body forces between the nucleon-nucleon. The optical potentials for the neutrons and protons in the deuteron with target are one-body potential; the tensor potential cannot exist in them. But since the deuteron wave function Eq. (3.988) contains the contribution of tensor force, the tensor potential $V^{\mathrm{T}}(R)T_R$ appears in the deuteron optical potential given by Eq. (3.1010) obtained through folding the deuteron ground state wave functions by using Eq. (3.977). The discussion above gave out the theoretical basis for the introduction of the tensor potential $T_R$ in the deuteron phenomenological optical potential.

### *3.14.2 Deuteron Folding Optical Potentials Containing Break Channel*

When the break channel is included, then substituting Eq. (3.975) into Eq. (3.972) and multiplying $\varphi_0^+(\vec{r})$ from the left on both sides of the equation, or multiplying $\varphi_k^+(\vec{r})$ from the left on both sides of the equation, and then integrating on $\vec{r}$, one can yield the following equations, respectively:

$$\left[-\frac{\hbar^2}{4m}\vec{\nabla}_R^2 + \langle 0|V|0\rangle - (E - \varepsilon_0)\right]\Phi_0(\vec{R}) = -\int \mathrm{d}\vec{k}\,\langle 0|V|k\rangle \Phi_k(\vec{R}) \tag{3.1015}$$

$$\left[-\frac{\hbar^2}{4m}\vec{\nabla}_R^2+\langle k|V|k\rangle-(E-\varepsilon_k)\right]\Phi_k(\vec{R})=-\langle k|V|0\rangle\Phi_0(\vec{R})-\int d\vec{k}'\langle k|V|k'\rangle\Phi_{k'}(\vec{R}) \tag{3.1016}$$

where

$$\langle 0|V|0\rangle=\int\varphi_0^+(\vec{r})\,V(\vec{R},\vec{r})\varphi_0(\vec{r})d\vec{r}=V_0(\vec{R}) \tag{3.1017}$$

$$\langle k|V|k'\rangle=\int\varphi_k^+(\vec{r})\,V(\vec{R},\vec{r})\varphi_{k'}(\vec{r})d\vec{r} \tag{3.1018}$$

When the contribution of the break channel is ignored, Eq. (3.1015) will be reduced to Eq. (3.976).

It should be clear that the greatest contribution of the break channel likely comes from the area where $k\approx 0$ and $k'\approx 0$. Now, if we omit the $\langle k|V|k'\rangle$ $(k'\neq k)$ term in Eq. (3.1016), then we get

$$\left[-\frac{\hbar^2}{4m}\vec{\nabla}_R^2+\langle k|V|k\rangle-(E-\varepsilon_k)\right]\Phi_k(\vec{R})=-\langle k|V|0\rangle\Phi_0(\vec{R}) \tag{3.1019}$$

Referring to Eq. (3.976) again and assuming that the following formula is also approximately true

$$\left[-\frac{\hbar^2}{4m}\vec{\nabla}_R^2+\langle 0|V|0\rangle-(E-\varepsilon_0)\right]\Phi_k(\vec{R})=0 \tag{3.1020}$$

then from the above two equations we can obtain

$$\Phi_k(\vec{R})=\frac{\langle k|V|0\rangle\Phi_0(\vec{R})}{\langle 0|V|0\rangle-\langle k|V|k\rangle-\varepsilon_k+\varepsilon_0} \tag{3.1021}$$

Substituting the above equation into Eq. (3.1015), we can get

$$\left[-\frac{\hbar^2}{4m}\vec{\nabla}_R^2+V(\vec{R})-(E-\varepsilon_0)\right]\Phi_0(\vec{R})=0 \tag{3.1022}$$

$$V(\vec{R})=V_0(\vec{R})+V_1(\vec{R}) \tag{3.1023}$$

where $V_0(\vec{R})$ has been given by Eq. (3.977) or Eq. (3.1017), $V_0(\vec{R})$ represents the contribution of the deuteron bound state, and

$$V_1(\vec{R}) = \int \mathrm{d}\vec{k} \frac{\langle 0|V|k\rangle\langle k|V|0\rangle}{\langle 0|V|0\rangle - \langle k|V|k\rangle - \varepsilon_k + \varepsilon_0} \tag{3.1024}$$

It can be seen that $V_1(\vec{R})$ represents the contribution of the deuteron break channel to the deuteron folding optical potential.

Referring to Eqs. (3.994), (3.995), and (3.1010), $\langle 0|V|k\rangle$ or $\langle k|V|0\rangle$ all have potential to contain the $\vec{l} \cdot \hat{S}$ terms; in Eq. (3.1024) they appear by multiplication. Therefore, in a certain sense, it provides a theoretical basis for including the $\left(\vec{L} \cdot \hat{S}\right)^2$ terms, or angular momentum tensor $T_L$ terms, in the deuteron optical potential.

## 3.15 The Spherical Nucleus CDCC Theory Describing the Breakup Reaction for Incident Loosely Bound Light Complex Particles

### *3.15.1 Development of CDCC Theory*

The microscopic three-body theory usually describes the elastic scattering or three-body breakup reaction between the nucleon and the deuteron. The nucleon-nucleon interaction uses the realistic nuclear force. Since the nucleons are fermions, the total wave function of the three-nucleon system must be completely asymmetric. The continuous discretization coupling channel (CDCC) theory [18, 55–72] is usually used to describe the reaction process of the deuteron (or other loosely bound light complex particles) elastic scattering or breakup reaction with a heavier nucleus A. After the deuteron breaking, it becomes a n + p + A three-body problem. n − A and p − A interactions are usually described with an optical potential, and the n − p interaction is usually described by nuclear force. In CDCC theory, the asymmetry problem is not considered generally. This kind of method can be one considered as an approximation, and the excitation of the target is not considered for spherical nuclei. Therefore, it can be said that CDCC theory only describes the direct reaction process. Since the $S$-matrix elements of deuteron elastic scattering can be calculated in CDCC theory, the polarization problem of the deuteron can also be studied using CDCC theory [18, 21, 60].

In References [63–65], the elastic scattering and three-body breakup reaction caused by the deuteron are studied. If a loosely bound light complex particle that is heavier than a deuteron is used as an incident particle, the complex particle may split into two groups during the reaction process, and the complex particle in the elastic scattering outgoing channel can be in the ground state or in the excited state. For example, the incident particle $^6$He can be regarded as the (2n + α) two-body weakly bound state of the double neutron nucleus and α particle; thus they can be described by CDCC theory [66]. It is also possible to regard the $^6$He as the (n + n +α) three-body loosely bound state of two neutrons and an α particle; then the CDCC theory of four-body breakup [67–69] is developed in which $^6$He can be a ground state $0^+$ or an excited state $2^+$. Some studies have also introduced the internal freedom of the target in the CDCC theory, using it to study inelastic scattering and two-step reaction processes [70]. In References [71] and [72], the target $^6$Li($^7$Li) is considered as the $\alpha$ + d(t) loosely bound state, and the ground state and excited state of the target $^6$Li ($^7$Li) can be solved with the internal interaction potential between the $\alpha$ and d(t). One then uses the three-body CDCC theory to solve the $^6$Li($^7$Li)+n(p) reaction, in which outgoing channel $^6$Li($^7$Li) may be in the ground state (elastic scattering), or excited state (inelastic scattering), or breakup. However, this book only introduces the CDCC theory of the deuteron induced reaction, as the deuteron does not have any excited bound state. This section will introduce the CDCC theory of the spherical nuclei, and the next section will give the CDCC theory of the axis-symmetric rotating nuclei.

### *3.15.2 Spherical Nucleus CDCC Equation*

In the center of mass system, the spatial coordinate positions of the neutron and proton of the deuteron relative to the target center of mass are $\vec{r}_n$ and $\vec{r}_p$, respectively. Let

$$\vec{R} = \frac{1}{2}\left(\vec{r}_{\mathrm{n}} + \vec{r}_{\mathrm{p}}\right), \qquad \vec{r} = \vec{r}_{\mathrm{n}} - \vec{r}_{\mathrm{p}} \tag{3.1025}$$

Since the masses of neutrons and protons are basically equal, $\vec{R}$ represents the coordinate of the deuteron center of mass, and $\vec{r}$ is the relative coordinate between n and p. The total Hamiltonian of the system is

$$H = -\frac{\hbar^2}{2\mu_R}\vec{\nabla}_R^2 + U_{\mathrm{n}}(r_{\mathrm{n}}) + U_{\mathrm{p}}\left(r_{\mathrm{p}}\right) + U_{\mathrm{C}}\left(r_{\mathrm{p}}\right) + H_{\mathrm{np}} \tag{3.1026}$$

$$H_{\mathrm{np}} = -\frac{\hbar^2}{2\mu_r}\vec{\nabla}_r^2 + V_{\mathrm{np}}(\vec{r}) \tag{3.1027}$$

where the reduced masses are

$$\mu_R = \frac{m_{\mathrm{d}} m_{\mathrm{A}}}{m_{\mathrm{d}} + m_{\mathrm{A}}}, \qquad \mu_r = \frac{m_{\mathrm{n}} m_{\mathrm{p}}}{m_{\mathrm{n}} + m_{\mathrm{p}}} \tag{3.1028}$$

and $V_{\mathrm{np}}(\vec{r})$ is the n–p interaction potential, and $U_{\mathrm{n}}(r_{\mathrm{n}})$ and $U_{\mathrm{p}}(r_{\mathrm{p}})$ are the nucleon potentials of n and p relative to the target center of mass; for the deuteron, they can be replaced by the nucleon optical potential of half the kinetic energy of the deuteron. $U_{\mathrm{C}}(r_{\mathrm{p}})$ is the Coulomb potential of the proton relative to the target center of mass; there is no Coulomb interaction between n–p.

Since the CDCC theory describing the reaction between the deuteron and the spherical nucleus does not include the physical mechanism of the target excited by the internal degree of freedom of the target, that is, there is no operator that can change the target state in the interaction potential, the target spin is assumed to be 0 [18, 60] in the spherical nucleus CDCC theory, as in the spherical optical model. If the relative orbital angular momentum $\vec{l}$ inside the deuteron is coupled with the spin $\vec{S}$ of the deuteron, then it is coupled with the deuteron orbital angular momentum $\vec{L}$ relative to the target center of mass, and finally it is coupled with the target spin $\vec{I}_A$, it will be found that taking different value for $I_A$ has no effect to the last results. That is, $\vec{L}+\vec{l}+\vec{S}$ is the total angular momentum of the system. This shows that target spin $I_{\rm A}=0$ should be selected in the spherical nucleus CDCC theory. Therefore, using the spherical nucleus CDCC theory for odd A nucleus is only an approximation.

We stipulate that $\vec{l}$ and $\vec{L}$ are the deuteron orbital angular momentum relative to $\vec{r}$ and $\vec{R}$, respectively, the spins of n and p are expressed by $\vec{s}_{\rm n}$ and $\vec{s}_{\rm p}$, and we define

$$\vec{S}=\vec{s}_{\rm n}+\vec{s}_{\rm p}, \qquad \vec{j}=\vec{l}+\vec{S} \tag{3.1029}$$

where $\vec{S}$ represents the deuteron spin, and $\vec{j}$ is the total angular momentum of the deuteron. Now, redefine

$$\vec{J}=\vec{L}+\vec{j} \tag{3.1030}$$

where $\vec{J}$ is the total angular momentum of the system, and $\nu_{\rm n}$, $\nu_{\rm p}$, $\nu$, $m$, $\mu$, $M_L$, $M$ are used to represent the $z$ components of the angular momentum $\vec{s}_{\rm n},\vec{s}_{\rm p},\vec{S},\vec{l},\vec{j},\vec{L},\vec{J}$.

Let $\varepsilon_0$ be the deuteron bound energy such that $\varepsilon_0=-2.225$ MeV. Using $\varepsilon_0$ to represent the n−p continuous state relative motion energy after the deuteron breaking, the total energy conservation of the system in the system center of mass frame requires

$$E=E_0+\varepsilon_0=\in+\varepsilon, \qquad \in=\frac{\hbar^2K^2}{2\mu_R} \tag{3.1031}$$

where the incident deuteron kinetic energy is

$$E_0=\frac{\hbar^2K_0^2}{2\mu_R} \tag{3.1032}$$

and $K_0$ and $K$ are the deuteron momentum wave number of the bound state and the continuous state in the deuteron center of mass system, respectively. Next, let

$$\varepsilon = \frac{\hbar^2 k^2}{2\mu_r} \tag{3.1033}$$

where $k$ represents the n–p relative motion wave number of the deuteron continuous state. We can see from Eq. (3.1031) that the maximum value of $\varepsilon$ is

$$\varepsilon_{\max} = E_0 + \varepsilon_0 = \frac{\hbar^2 k_{\max}^2}{2\mu_r} \tag{3.1034}$$

We know that deuteron spin $S = 1$ and total angular momentum $j = 1$. Let $\psi^0_{l11\mu}(\vec{r})$ be the deuteron binding state wave function, its subscript $l1$ represent the internal orbital angular momentum $l$ and spin $S = 1$, the subscript $1\mu$ are the deuteron total angular momentum $j = 1$ and the corresponding $z$ component $\mu$. $\psi^0_{l11\mu}(\vec{r})$ satisfies the following Schrodinger equation:

$$\left(H_{\rm np} - \varepsilon_0\right) \psi^0_{l11\mu}(\vec{r}) = 0 \tag{3.1035}$$

The method to obtain the detailed solution of the above equation is given in Section 13.5 of Reference [22]. Its solution $\psi^0_{l11\mu}(\vec{r})$ can be written formally as

$$\psi^0_{l11\mu}(\vec{r}) = \frac{\phi^0_l(r)}{r} \mathbb{Y}_{l11\mu}(\Omega_r) \tag{3.1036}$$

$$\mathbb{Y}_{lSj\mu}(\Omega_r) = \sum_{m\nu} C^{j\ \mu}_{lm\ S\nu}\, {\rm i}^l Y_{lm}(\Omega_r)\chi_{S\nu} \tag{3.1037}$$

and $\phi^0_l(r)$ is the deuteron bound state normalized radial wave function, and $\chi_{S\nu}$ is the deuteron spin wave function. The deuteron continuous state equation with n–p relative motion wave number $k$ can be written as

$$\left(H_{\rm np} - \varepsilon\right) \psi^k_{lSj\mu}(\vec{r}) = 0 \tag{3.1038}$$

where $j\mu$ are the total angular momentum and its $z$ component of the deuteron continuous state, and its solution can be formally written as

$$\psi^k_{lSj\mu}(\vec{r}) = \frac{\phi^k_{lSj}(r)}{r} \mathbb{Y}_{lSj\mu}(\Omega_r) \tag{3.1039}$$

where $\phi^k_{lSj}(r)$ is the normalization radial wave function of the deuteron continuous state. For the deuteron bound state, if only the central potential is included in $V_{np}(\vec{r})$, then only $l = 0$ S wave is obtained in its solution. If the tensor force belonging to the non-central potential is also contained in $V_{np}(\vec{r})$, then the $l = 2$ D wave will be included in its solution in addition to the S wave. Now the key question is how to select $V_{\mathrm{np}}(\vec{r})$. Some studies have chosen the Reid soft core potential for the deuteron bound state [18] , with the following Gaussian nuclear force $V_{\mathrm{np}}(r)$ for the deuteron continuum state adopted

$$V_{\mathrm{np}}(r) = v_{lSj}\, \mathrm{e}^{-\left(\frac{r}{a_{lSj}}\right)^2} \tag{3.1040}$$

In Reference [18], the specific values for the used $v_{lSj}$ and $a_{lSj}$ are given. The orbital angular momentum wave function of the deuteron relative to the target center of mass, coupled further with $\mathbb{Y}_{lSj\mu}$ given by Eq.(3.1037), becomes

$$\varphi_{lSjLJM}(\Omega_R, \Omega_r) = \sum_{M_L\mu} C_{LM_L\ j\mu}^{J\ M}\, \mathrm{i}^L Y_{LM_L}(\Omega_R) \mathbb{Y}_{lSj\mu}(\Omega_r) \tag{3.1041}$$

The total wave function of a system with total angular momentum *JM* satisfies the following Schrodinger equation:

$$(H - E)\, \Psi_{JM}(\vec{R}, \vec{r}) = 0 \tag{3.1042}$$

For the bound state, the *JM* component of the total wave function of the system is

$$\Psi_{JM}^0(\vec{R}, \vec{r}) = \sum_{lL} \frac{u_{0l11L}^J(R)}{R} \frac{\phi_l^0(r)}{r} \varphi_{l11LJM}(\Omega_R, \Omega_r) \tag{3.1043}$$

where $u_{0l11L}^J(R)$ is the radial wave function of the bound state deuteron center of mass moving relative to the target center of mass. For the continuous states, the *JM* component of the total wave function of the system is

$$\Psi_{JM}^k(\vec{R}, \vec{r}) = \sum_{lSjL} \frac{u_{klSjL}^J(R)}{R} \frac{\phi_{lSj}^k(r)}{r} \varphi_{lSjLJM}(\Omega_R, \Omega_r) \tag{3.1044}$$

where $u_{klSjL}^J(R)$ is the radial wave function of the continuous state deuteron center of mass moving relative to the target center of mass. It follows that the *JM* component of the total wave function of the system can be written as

$$\Psi_{JM}(\vec{R}, \vec{r}) = \Psi_{JM}^0(\vec{R}, \vec{r}) + \int_0^{k_{\max}} \Psi_{JM}^k(\vec{R}, \vec{r})\, \mathrm{d}k \tag{3.1045}$$

For the deuteron bound state, there are $S = 1$ and $j = 1$, and according to parity conservation, $l = 0, 2$ can only be taken. For the deuteron continuous state, there are $S = 0, 1$, $l = 0, 2, 4, \cdots, l_{\max}$, and $j = |l - S|, \ldots, l + S$. The larger the incident deuteron energy $E_0$, the larger the $l_{\max}$; $l_{\max}$ is also related with the weight of the target. When $E_0$ is only few tens of MeV, $l_{\max} = 4$ is generally taken for the light target. We then dissociate the integral of $k$ appearing in Eq. (3.1045) and divide 0–$k_{\max}$ into $N$ equal parts. Let

$$\Delta = \frac{k_{\max}}{N}, \quad k_i = i\Delta, \quad i = 0,1,2,\cdots,N \tag{3.1046}$$

thus the continuously changing $k$ is divided into $[0, k_1]$, $[k_1, k_2]$, $\cdots$, $[k_{i-1}, k_i]$ , $\cdots$, $[k_{N-1}, k_N]$ $N$ equal parts. Next, introduce the following symbols for the deuteron continuous state wave functions:

$$\phi^i_{lSj}(r) = \frac{1}{\sqrt{\Delta}} \int_{k_{i-1}}^{k_i} \phi^k_{lSj}(r)\, \mathrm{d}k \tag{3.1047}$$

$$u^J_{ilSjL}(R) = \frac{1}{\sqrt{\Delta}} \int_{k_{i-1}}^{k_i} u^J_{klSjL}(R)\, \mathrm{d}k \tag{3.1048}$$

where Eqs. (3.1039) and (3.1044) can be now rewritten as

$$\psi^i_{lSj\mu}(\vec{r}) = \frac{\phi^i_{lSj}(r)}{r} \mathbb{Y}_{lSj\mu}(\Omega_r) \tag{3.1049}$$

$$\Psi^i_{JM}(\vec{R}, \vec{r}) = \sum_{lSjL} \frac{u^J_{ilSjL}(R)}{R} \frac{\phi^i_{lSj}(r)}{r} \varphi_{lSjLJM}(\Omega_R, \Omega_r) \tag{3.1050}$$

Now, concurrently introduce the following symbols for the deuteron bound state wave functions:

$$\phi^0_{lSj}(r) = \begin{cases} \phi^0_l(r), & \text{when } S = j = 1, \quad l = 0, 2 \\ 0, & \text{otherwise} \end{cases} \tag{3.1051}$$

$$u^J_{0lSjL}(R) = \begin{cases} u^J_{0l11L}(R), & \text{when } S = j = 1, \quad l = 0, 2 \\ 0, & \text{otherwise} \end{cases} \tag{3.1052}$$

Thus Eqs. (3.1036) and (3.1043) can be written in the form of Eqs. (3.1049) and (3.1050) only needing to let $i = 0$. In this way, Eq. (3.1045) can be rewritten as

$$\Psi_{JM}(\vec{R}, \vec{r}) = \sum_{i=0}^{N} \Psi^i_{JM}(\vec{R}, \vec{r}) \tag{3.1053}$$

where $i = 0$ corresponds to the deuteron bound state and $i = 1, 2, \ldots, N$ correspond to the deuteron continuous states. In addition, we define

$$\varepsilon_i = \frac{\hbar^2}{4\mu_r}\left(k_{i-1}^2 + k_i^2\right), \qquad i = 1,2,\cdots,N \tag{3.1054}$$

and assume that the radial wave functions after average processing with Eqs. (3.1047) and (3.1048) still satisfy the orthogonal normalization condition, that is,

$$\int \phi_{lSj}^{i*}(r)\phi_{l'S'j'}^{i'}(r)\mathrm{d}r = \delta_{ilSj,i'l'S'j'} \tag{3.1055}$$

$$\int u_{ilSjL}^{J*}(R)u_{i'l'S'j'L'}^{J}(R)\mathrm{d}R = \delta_{ilSjL,i'l'S'j'L'} \tag{3.1056}$$

Let $V_C(R)$ be the spherical symmetric Coulomb potential felt by the deuteron in the center of mass when assuming that the deuteron charge is at its center of mass. Then substituting Eq. (3.1026) into Eq. (3.1042), we can obtain the following equation for the spherical target:

$$\begin{aligned}&\left[\frac{\hbar^2}{2\mu_R}\left(\frac{1}{R}\frac{\mathrm{d}^2}{\mathrm{d}R^2}R - \frac{L(L+1)}{R^2}\right) - V_\mathrm{C}(R) - H_\mathrm{np} + E\right]\Psi_{JM}(\vec{R},\vec{r})\\&= \left[U_\mathrm{n}(r_\mathrm{n}) + U_\mathrm{p}(r_\mathrm{p}) + U_\mathrm{C}(r_p) - V_\mathrm{C}(R)\right]\Psi_{JM}(\vec{R},\vec{r})\end{aligned} \tag{3.1057}$$

In the above equation, we do not ignore the Coulomb distortion effect caused by $(U_\mathrm{C}(r_\mathrm{p}) - V_\mathrm{C}(R))$. Substituting $\Psi_{JM}(\vec{R},\vec{r})$ given by Eq. (3.1053) into Eq. (3.1057); referring to Eq. (3.1050) and multiplying $R\frac{\phi_{lSj}^{i*}(r)}{r}\varphi_{lSjLJM}^{+}(\Omega_R,\Omega_r)$ on both sides of the above equation from the left; integrating over $\mathrm{d}\vec{r}\mathrm{d}\Omega_R$ for the obtained equation; and using the orthogonal normalization relation given by Eq. (3.1055) and other relations, we can obtain

$$\begin{aligned}&\left[\frac{\hbar^2}{2\mu_R}\left(\frac{\mathrm{d}^2}{\mathrm{d}R^2} - \frac{L(L+1)}{R^2}\right) - V_\mathrm{C}(R) + (E-\varepsilon_i)\right]u_{ilSjL}^{J}(R)\\&= \sum_{i'l'S'j'L'} F_{ilSjL,\ i'l'S'j'L'}^{J}(R)\,u_{i'l'S'j'L'}^{J}(R)\end{aligned} \tag{3.1058}$$

where

$$F_{ilSjL,i'l'S'j'L'}^{J}(R) = \int \frac{\phi_{lSj}^{i*}(r)}{r}\varphi_{ilSjLJM}^{+}(\Omega_R,\Omega_r)U(\vec{R},\vec{r})\frac{\phi_{l'S'j'}^{i'}(r)}{r}\varphi_{l'S'j'L'JM}(\Omega_R,\Omega_r)\mathrm{d}\vec{r}\,\mathrm{d}\Omega_R \tag{3.1059}$$

$$U(\vec{R}, \vec{r}) = U_{\mathrm{n}}(r_{\mathrm{n}}) + U_{\mathrm{p}}(r_{\mathrm{p}}) + U_{\mathrm{C}}(r_{\mathrm{p}}) - V_{\mathrm{C}}(R) \tag{3.1060}$$

From here, we can see that Eq. (3.1058) is a coupled channel equation that separates the continuous momentum wave number $k$ of the deuteron breakup channel. An appropriate method can be chosen for finding its numerical solution [73].

Nucleon optical potential can usually be expressed as

$$U_{\mathrm{N}}(r_{\mathrm{N}}) = U_{\mathrm{NA}}^{\mathrm{CE}}(r_{\mathrm{N}}) + U_{\mathrm{NA}}^{\mathrm{SO}}(r_{\mathrm{N}})\ \vec{l}_{\mathrm{N}} \cdot \vec{s}_{\mathrm{N}}, \quad \mathrm{N} = \mathrm{n,p} \tag{3.1061}$$

Therefore, $U(\vec{R}, \vec{r})$ given by Eq. (3.1060) can be divided into two parts.

$$U(\vec{R}, \vec{r}) = U_1(\vec{R}, \vec{r}) + U_2(\vec{R}, \vec{r}) \tag{3.1062}$$

$$U_1(\vec{R}, \vec{r}) = U_{\mathrm{nA}}^{\mathrm{CE}}(r_{\mathrm{n}}) + U_{\mathrm{pA}}^{\mathrm{CE}}(r_{\mathrm{p}}) + U_{\mathrm{C}}(r_{\mathrm{p}}) - V_{\mathrm{C}}(R) \tag{3.1063}$$

$$U_2(\vec{R}, \vec{r}) = U_{\mathrm{nA}}^{\mathrm{SO}}(r_{\mathrm{n}})\ \vec{l}_{\mathrm{n}} \cdot \vec{s}_{\mathrm{n}} + U_{\mathrm{pA}}^{\mathrm{SO}}(r_{\mathrm{p}})\ \vec{l}_{\mathrm{p}} \cdot \vec{s}_{\mathrm{p}} \tag{3.1064}$$

Let $\theta$ be an angle between $\vec{r}$ and $\vec{R}$. Then we have

$$\begin{aligned} r_{\mathrm{n}} &= \left|\vec{R} + \frac{1}{2}\vec{r}\right| = \left(R^2 + \frac{r^2}{4} + Rr\cos\theta\right)^{1/2}, \\ r_{\mathrm{p}} &= \left|\vec{R} - \frac{1}{2}\vec{r}\right| = \left(R^2 + \frac{r^2}{4} - Rr\cos\theta\right)^{1/2} \end{aligned} \tag{3.1065}$$

For $U_{\mathrm{nA}}^{\mathrm{CE}}(r_{\mathrm{n}})$ we can do the following expansion:

$$U_{\mathrm{nA}}^{\mathrm{CE}}(r_{\mathrm{n}}) = \sum_{\lambda}(2\lambda + 1)\ U_{\mathrm{nA},\lambda}^{\mathrm{CE}}(R, r)\ \mathrm{P}_{\lambda}(\cos\theta) \tag{3.1066}$$

$$U_{\mathrm{nA},\lambda}^{\mathrm{CE}}(R, r) = \frac{1}{2}\int U_{\mathrm{nA}}^{\mathrm{CE}}(r_{\mathrm{n}})\ \mathrm{P}_{\lambda}(\cos\theta)\ \mathrm{d}\cos\theta \tag{3.1067}$$

such that Eq. (3.1066) can be rewritten as

$$U_{\mathrm{nA}}^{\mathrm{CE}}(r_{\mathrm{n}}) = 4\pi\sum_{\lambda\ m_\lambda} U_{\mathrm{nA},\lambda}^{\mathrm{CE}}(R, r)\mathrm{Y}_{\lambda\ m_\lambda}(\Omega_R)\mathrm{Y}_{\lambda\ m_\lambda}^{*}(\Omega_r) \tag{3.1068}$$

Also, $U_{\mathrm{pA}}^{\mathrm{CE}}(r_{\mathrm{p}})$ and $U_{\mathrm{C}}(r_{\mathrm{p}})$ can be expanded in the same way. Now Eq. (3.1059) can be divided into two parts according to Eq. (3.1062):

$$F^{J}_{ilSjL,i'l'S'j'L'} = F^{J(1)}_{ilSjL,i'l'S'j'L'} + F^{J(2)}_{ilSjL,i'l'S'j'L'} \tag{3.1069}$$

Since there is no spin operator in $U_1(\vec{R}, \vec{r})$, it follows that according to Eqs. (3.1041) and (3.1037) we can write the following expression:

$$\begin{aligned} F^{J(1)}_{ilSjL,i'l'S'j'L'}(R) = \delta_{SS'} \sum_{\nu} \int \Bigg( & \sum_{M_L \mu} C^{J\ M}_{LM_L\ j\mu} \mathrm{i}^{-L} \mathrm{Y}^*_{LM_L}(\Omega_R) \frac{\phi^{i*}_{lSj}(r)}{r} \sum_{m} C^{j\ \mu}_{lm\ S\nu} \mathrm{i}^{-l} \mathrm{Y}^*_{lm}(\Omega_r) \Bigg) U_1(\vec{R}, \vec{r}) \\ & \left( \sum_{M'_L \mu'} C^{J\ M}_{L'M'_L\ j'\mu'} \mathrm{i}^{L'} \mathrm{Y}_{L'M'_L}(\Omega_R) \frac{\phi^{i'}_{l'S\,j'}(r)}{r} \sum_{m'} C^{j'\mu'}_{l'm'\ S\nu}\, \mathrm{i}^{l'} \mathrm{Y}_{l'm'}(\Omega_r) \right) \mathrm{d}\vec{r}\, \mathrm{d}\Omega_R \end{aligned} \tag{3.1070}$$

Next, we first find the contribution to the above formula by the first item of $U_1(\vec{R}, \vec{r})$ given by Eq. (3.1063). Using Eq. (3.1068) and paying attention to the following relations

$$\int \mathrm{Y}^*_{lm}(\Omega_r) \mathrm{Y}^*_{\lambda m_\lambda}(\Omega_r) \mathrm{Y}_{l'm'}(\Omega_r) \mathrm{d}\Omega_r = (-1)^{m_\lambda} \frac{\hat{l}'\hat{\lambda}}{\sqrt{4\pi}\,\hat{l}} C^{l\ m}_{l'm'\ \lambda\ -m_\lambda} C^{l\ 0}_{l'0\ \lambda 0} \tag{3.1071}$$

$$\int \mathrm{Y}^*_{LM_L}(\Omega_R) \mathrm{Y}_{\lambda m_\lambda}(\Omega_R) \mathrm{Y}_{L'M'_L}(\Omega_R) \mathrm{d}\Omega_R = \frac{\hat{L}'\hat{\lambda}}{\sqrt{4\pi}\,\hat{L}} C^{L\ M_L}_{L'M'_L\ \lambda m_\lambda} C^{L\ 0}_{L'0\ \lambda 0} \tag{3.1072}$$

and ordering

$$B^{S\lambda(1,\,1)}_{ilj,i'l'j'}(R) = 4\pi \int_0^\infty \phi^{i*}_{lSj}(r) U^{CE}_{nA,\lambda}(R, r) \phi^{i'}_{l'Sj'}(r) \mathrm{d}r \tag{3.1073}$$

we can obtain the following expression according to Eq. (3.1070):

$$\begin{aligned} F^{J(1,\,1)}_{ilSjL,i'l'S'j'L'}(R) = \delta_{SS'} \mathrm{i}^{L'-L+l'-l} \sum_{\lambda} B^{S\lambda(1,\,1)}_{ilj,i'l'j'}(R) \frac{\hat{l}'\hat{L}'\hat{\lambda}^2}{\hat{l}\hat{L}} C^{l0}_{l'0\ \lambda 0} C^{L0}_{L'0\ \lambda 0} \\ \sum_{\substack{\nu\ M_L \mu\ m \\ m_\lambda M'_L \mu' m'}} C^{J\ M}_{LM_L\ j\mu} C^{j\ \mu}_{lm\ S\nu} (-1)^{m_\lambda} C^{l\ m}_{l'm'\ \lambda\ -m_\lambda} C^{L\ M_L}_{L'M'_L\ \lambda m_\lambda} C^{J\ M}_{L'M'_L\ j'\mu'} C^{j'\ \mu'}_{l'm'\ S\nu} \end{aligned} \tag{3.1074}$$

Because Eq. (3.1074) contains the factor $C^{l0}_{l'0\ \lambda 0}$, $\lambda$ is an integer and there is $l + l' + \lambda =$ even; thus, using Eq. (3.14) we can obtain

$$\sum_{mm'\nu} C_{lm\ S\nu}^{j\ \mu} C_{l'm'\ \lambda\ -m_\lambda}^{l\ m} C_{l'm'\ S\nu}^{j'\ \mu'} = \widehat{l'j} C_{\lambda\ -m_\lambda\ j'\mu'}^{j\ \mu} W(\lambda l' jS; lj') \tag{3.1075}$$

And there is the following C-G coefficient formula

$$\sum_{\substack{\alpha\beta \\ \delta\varepsilon\varphi}} C_{a\alpha\ b\beta}^{e\varepsilon} C_{e\varepsilon\ d\delta}^{c\gamma} C_{b\beta\ d\delta}^{f\varphi} C_{a\alpha\ f\varphi}^{c\gamma} = \hat{e}\hat{f}\ W(abcd; ef) \tag{3.1076}$$

using this formula we can get

$$\sum_{\substack{M_L \mu M'_L \\ \mu' m_\lambda}} C_{LM_L\ j\mu}^{J\ M} (-1)^{m_\lambda} C_{L'M'_L\ \lambda\ m_\lambda}^{L\ M_L} C_{L'M'_L\ j'\mu'}^{J\ M} C_{\lambda\ -m_\lambda\ j'\mu'}^{j\ \mu} = (-1)^{j-j'} \hat{L}\hat{j} W(L'\lambda Jj; lj') \tag{3.1077}$$

Next, the following expression can be obtained substituting Eqs. (3.1075) and (3.1077) into Eq. (3.1074):

$$\begin{aligned} F_{ilSjL,i'l'S'j'L'}^{J(1,\ 1)}(R) &= \delta_{SS'} \mathrm{i}^{L'-L+l'-l} (-1)^{j-j'} \sum_\lambda \widehat{l'j'} \hat{L}' \hat{\lambda}^2 \hat{j} \hat{j}\, C_{l'0\ \lambda 0}^{l0} C_{L'0\ \lambda 0}^{L0} \\ &\quad W(\lambda l' jS; lj') W(L'\lambda Jj; Lj') B_{ilj,i'l'j'}^{S\lambda(1,\ 1)}(R) \\ &= \delta_{SS'} \mathrm{i}^{L'-L+l'-l} (-1)^{j-j'-S-J} \hat{L}\hat{L}' \hat{l}\hat{l}' \widehat{jj'} \sum_\lambda C_{l0\ \ l'0}^{\lambda 0} C_{L0\ L'0}^{\lambda 0} \\ &\quad W(ll'jj'; \lambda S) W(LL'jj'; \lambda J) B_{ilj,i'l'j'}^{S\lambda(1,\ 1)}(R) \end{aligned} \tag{3.1078}$$

Similar derivations can be done for the second and third items $U_{\mathrm{pA}}^{\mathrm{CE}}(r_{\mathrm{p}})$ and $U_{\mathrm{C}}(r_{\mathrm{p}})$ of Eq. (3.1063). Based on Eq. (3.1070) the fourth item $-V_{\mathrm{C}}(R)$ can be easily obtained as

$$F_{ilSjL,i'l'S'j'L'}^{J(1,\ 4)}(R) = -V_{\mathrm{C}}(R) \delta_{ilSjL,i'l'S'j'L'} \tag{3.1079}$$

Now, we can obtain the following results using Eqs. (3.1078) and (3.1079):

$$\begin{aligned} F_{ilSjL,i'l'S'j'L'}^{J(1)}(R) &= \delta_{SS'} \mathrm{i}^{L'-L+l'-l} (-1)^{j-j'-S-J} \hat{L}\hat{L}' \hat{l}\hat{l}' \widehat{jj'} \sum_\lambda C_{l0\ l'0}^{\lambda 0} C_{L0\ L'0}^{\lambda 0} \\ &\quad W(ll'jj'; \lambda S) W(LL'jj'; \lambda J) B_{ilj,i'l'j'}^{S\lambda(1)}(R) - V_{\mathrm{C}}(R) \delta_{ilSjL,i'l'S'j'L'} \end{aligned} \tag{3.1080}$$

where

$$B_{ilj,i'l'j'}^{S\lambda(1)}(R)=4\pi\int_0^{\infty}\phi_{lSj}^{i*}(r)\left[U_{\mathrm{nA},\lambda}^{\mathrm{CE}}(R,\ r)+U_{\mathrm{pA},\lambda}^{\mathrm{CE}}(R,r)+U_{\mathrm{C},\lambda}(R,r)\right]\phi_{l'Sj'}^{i'}(r)\mathrm{d}r \tag{3.1081}$$

where $U_{\mathrm{pA},\lambda}^{\mathrm{CE}}(R, r)$ and $U_{\mathrm{C},\ \lambda}(R, r)$ can be found using the same method used to find Eq. (3.1067).

For the spin-orbit coupling potential given by Eq. (3.1064), Eq. (3.1029) yields

$$\vec{S}=\vec{s}_{\mathrm{n}}+\vec{s}_{\mathrm{p}}=\frac{1}{2}\left(\vec{\sigma}_{\mathrm{n}}+\vec{\sigma}_{\mathrm{p}}\right) \tag{3.1082}$$

In the deuteron bound state, the S wave is the main component. The deuteron continuous state evolves from the deuteron bound state, so the S wave is also the main component. Therefore, for the deuteron spin-orbit coupling potential, the S-wave approximation method given in the previous section can be used. By limiting

$$U_{\mathrm{nA}}^{\mathrm{SO}}(r_{\mathrm{n}})=U_{\mathrm{pA}}^{\mathrm{SO}}(r_{\mathrm{p}}) \tag{3.1083}$$

and assuming that the deuteron has a spin-orbit coupling potential only when $l=0$, then the following relation can be obtained from Eq. (3.1064) using Eqs. (3.1029) and (3.1006):

$$U_2(\vec{R},\ \vec{r})=U_{\mathrm{nA}}^{\mathrm{SO}}(r_{\mathrm{n}})\left(\vec{L}\cdot\vec{S}\right)\delta_{l0} \tag{3.1084}$$

where $\vec{S}$ has been given by Eq. (3.1082). Because $\vec{L}\cdot\vec{S}$ and $U_{\mathrm{nA}}^{\mathrm{SO}}(r_{\mathrm{n}})$ are not commutated, the above equation can be rewritten as

$$U_2(\vec{R},\ \vec{r})=\frac{1}{2}\left[\left(\vec{L}\cdot\vec{S}\right)U_{\mathrm{nA}}^{\mathrm{SO}}(r_{\mathrm{n}})+U_{\mathrm{nA}}^{\mathrm{SO}}(r_{\mathrm{n}})\left(\vec{L}\cdot\vec{S}\right)\right]\delta_{l0} \tag{3.1085}$$

Now we have the following relations according to Eqs. (3.1029) and (3.1030):

$$\vec{J}=\vec{L}+\vec{l}+\vec{S} \tag{3.1086}$$

$$\vec{L}\cdot\vec{S}=\frac{1}{2}\left[\vec{J}^2-\vec{L}^2-\vec{l}^2-\vec{S}^2-2\vec{l}\cdot\left(\vec{L}+\vec{S}\right)\right] \tag{3.1087}$$

Referring to Eq. (3.1080), we can write the following expression using Eqs. (3.1085) and (3.1087):

$$F^{J(2)}_{ilSjL,i'l'S'j'L'}(R) = \delta_{SS'} \mathrm{i}^{L'-L+l'-l}(-1)^{j-j'-S-J}\hat{L}\hat{L}'\hat{l}\hat{l}'\hat{j}\hat{j}'\sum_{\lambda} C^{\lambda\,0}_{l0\,l'0} C^{\lambda\,0}_{L0\,L'0} W(ll'jj';\lambda S)$$
$$W(LL'jj';\lambda J)\frac{1}{2}\left\{J(J+1) - \frac{1}{2}\left[L(L+1) + L'(L'+1) + 2S(S+1)\right]\right\}B^{S\lambda(2)}_{ilj,i'l'j'}\delta_{l0}\delta_{l'0} \tag{3.1088}$$

where

$$B^{S\lambda(2)}_{ilj,i'l'j'}(R) = 4\pi\int_0^{\infty}\phi^{i*}_{lSj}(r)U^{\mathrm{SO}}_{\mathrm{nA},\lambda}(R,\,r)\phi^{i'}_{l'Sj'}(r)\mathrm{d}r \tag{3.1089}$$

where $U^{\mathrm{SO}}_{\mathrm{nA},\lambda}(R,\,r)$ can be obtained using the same method to find Eq. (3.1067). Equations (3.1080) and (3.1088) show that $S$ is a conservation quantity during the entire reaction process. Because $S=1$ for the deuteron bound state, $S=1$ is maintained throughout the reaction process.

### 3.15.3 Angular Distribution of Deuteron Elastic Scattering

The spherical nucleus CDCC equation (3.1058) describes the process of elastic scattering and breakup of the deuteron. When $R$ is large enough, the nuclear potential contained in the term on the right side of this equation will tend to 0, and $U_{\mathrm{C}}(r_{\mathrm{p}}) - V_{\mathrm{C}}(R)$ will also tend to 0. In this case, the solutions of this equation become its asymptotic solutions $F_L(K_iR)$ and $G_L(K_iR)$, where

$$K_i = \frac{\sqrt{2\mu_R(E-\varepsilon_i)}}{\hbar}, \quad i=0,1,\cdots,N \tag{3.1090}$$

and $F_1$ and $G_L$ are the regular and irregular Coulomb wave functions, respectively. The speed of the deuteron center of mass is

$$V_i = \frac{\hbar K_i}{\mu_R} \tag{3.1091}$$

In the center of mass system, the incident wave function of the deuteron relative to the target A is

$$\Psi^{(\mathrm{in})}_{\mu}\left(\vec{R},\,\vec{r}\right) = \frac{1}{\sqrt{V_0}}\mathrm{e}^{\mathrm{i}\vec{K}_0\cdot\vec{R}}\,\psi_{\mu}\left(\vec{r}\right) \tag{3.1092}$$

where $\psi_\mu\left(\vec{r}\right)$ is a normalization bound state wave function of the deuteron with the total angular momentum magnetic quantum number $\mu$. Referring to Eq. (3.1036), we can write

$$\psi_\mu\left(\vec{r}\right) = \sum_l \frac{\phi_l^0(r)}{r} \mathbb{Y}_{l1 1\mu}(\Omega_r) \tag{3.1093}$$

thus Eq. (3.1092) can be rewritten as [7]

$$\Psi_\mu^{(\text{in})}\left(\vec{R}, \vec{r}\right) = \sqrt{\frac{4\pi}{V_0}} \sum_L \hat{L} e^{\mathrm{i}\sigma_L^{(0)}} \frac{F_L(K_0 R)}{K_0 R} \mathrm{i}^L Y_{L0}(\theta_R, 0) \sum_l \frac{\phi_l^0(r)}{r} \mathbb{Y}_{l1 1\mu}(\Omega_r) \tag{3.1094}$$

where $\sigma_L^{(0)}$ is a Coulomb phase shift corresponding to the deuteron bound state. Using Eq. (3.1041) we can write

$$\mathrm{i}^L Y_{l0} \mathbb{Y}_{l11\mu} = \sum_J C_{L0\ 1\mu}^{J\ \mu} \varphi_{l11LJ\mu} \tag{3.1095}$$

and Eq. (3.1094) can be rewritten as

$$\begin{aligned}\Psi_\mu^{(\text{in})} &= \frac{1}{K_0 R} \sqrt{\frac{4\pi}{V_0}} \sum_{lLJ} \hat{L} C_{L0\ 1\mu}^{J\ \mu} e^{\mathrm{i}\sigma_L^{(0)}} F_L(K_0 R) \frac{\phi_l^0(r)}{r} \varphi_{l11LJ\mu} \\ &= \frac{\mathrm{i}\sqrt{\pi}}{K_0 R \sqrt{V_0}} \sum_{lLJ} \hat{L} C_{L0\ 1\mu}^{J\ \mu} e^{\mathrm{i}\sigma_L^{(0)}} \left[(G_L - \mathrm{i}F_L) - (G_L + \mathrm{i}F_L)\right] \frac{\phi_l^0(r)}{r} \varphi_{l11LJ\mu}\end{aligned} \tag{3.1096}$$

In an isolated system where n–p elastic scattering occurs, the total angular momentum $j\mu$ is the conserved quantity. If the n–p interaction potential given by Eq. (3.1040) is taken, for which the tensor force is not included, and in this case, the n–p relative motion orbital angular momentum $l$ is also a good quantum number. For discrete energy $\varepsilon_i$, from Eq. (3.1038) we have

$$\left(H_{\text{np}} - \varepsilon_i\right) \psi_{lj\mu}^i\left(\vec{r}\right) = 0 \tag{3.1097}$$

The form of $\psi_{lj\mu}^i\left(\vec{r}\right)$ has been given by Eq. (3.1049), but $S = 1$ is no longer marked here. Assuming that the n–p collision takes place along the z-axis direction, the initial total wave function for n–p scattering is [7]

$$\Psi^{(\mathrm{in})}_{\nu_n\nu_p}\left(\vec{r}\right)=\sqrt{\frac{4\pi}{v_i}}\left(\sum_l \hat{L}j_l(k_ir)\ \mathrm{i}^l\mathrm{Y}_{l0}(\theta,\ 0)\right)\chi_{\frac{1}{2}\nu_n}\chi_{\frac{1}{2}\nu_p} \tag{3.1098}$$

The n–p relative motion speed is

$$v_i=\frac{\hbar k_i}{\mu_r} \tag{3.1099}$$

where $j_l$ is the spherical Bessel function. Taking the spin coupled wave functions as

$$\chi_{1\nu}=\sum_{\nu_n\nu_p}C^{1\ \ \nu}_{\frac{1}{2}\nu_n\ \frac{1}{2}\nu_p}\chi_{\frac{1}{2}\nu_n}\chi_{\frac{1}{2}\nu_p} \tag{3.1100}$$

$$\chi_{\frac{1}{2}\nu_n}\chi_{\frac{1}{2}\nu_p}=\sum_{\nu}C^{1\ \ \nu}_{\frac{1}{2}\nu_n\ \frac{1}{2}\nu_p}\chi_{1\nu} \tag{3.1101}$$

and utilizing Eq. (3.1037), we can get

$$\mathrm{i}^l\mathrm{Y}_{l0}\chi_{1\nu}=\sum_j C^{j\nu}_{l0\ \ 1\nu}\mathbb{Y}_{l1j\nu} \tag{3.1102}$$

Note that when there is no Coulomb interaction $\frac{F_l(kr)}{kr}\to\mathrm{j}_l$, $\frac{G_l(kr)}{kr}\to-\mathrm{n}_l(kr)$; thus, Eq. (3.1098) can be rewritten as

$$\begin{aligned}\Psi^{(\mathrm{in})}_{\nu_\mathrm{n}\nu_\mathrm{p}}\left(\vec{r}\right)&=\sqrt{\frac{4\pi}{v_i}}\sum_{lj\nu}\hat{l}C^{j\ \nu}_{l0\ 1\nu}C^{1\ \ \nu}_{\frac{1}{2}\nu_\mathrm{n}\ \frac{1}{2}\nu_\mathrm{p}}\mathrm{j}_l(k_rr)\mathbb{Y}_{l1j\nu}=\frac{\mathrm{i}\sqrt{\pi}}{\sqrt{v_i}}\sum_{lj\nu}\hat{l}C^{j\ \nu}_{l0\ 1\nu}C^{1\ \ \nu}_{\frac{1}{2}\nu_\mathrm{n}\ \frac{1}{2}\nu_\mathrm{p}}\\&\quad\{[-\mathrm{n}_l(k_ir)-\mathrm{i}\mathrm{j}_l(k_ir)]-[-\mathrm{n}_l(k_ir)+\mathrm{i}\mathrm{j}_l(k_ir)]\}\mathbb{Y}_{l1j\nu}\end{aligned} \tag{3.1103}$$

If n–p elastic scattering occurs, and its energy $\varepsilon_i$, angular momentum $l$, and $S=1$ have not changed, then, when $r$ is large enough to ignore the nuclear potential, the corresponding total wave function becomes

$$\begin{aligned}\Psi^{(\mathrm{t})}_{\nu_\mathrm{n}\nu_\mathrm{p}}\left(\vec{r}\right)&=\frac{\mathrm{i}\sqrt{\pi}}{\sqrt{v_i}}\sum_{lj\nu}\hat{l}C^{j\ \nu}_{l0\ 1\nu}C^{1\ \ \nu}_{\frac{1}{2}\nu_\mathrm{n}\ \frac{1}{2}\nu_\mathrm{p}}\\&\quad\{[-\mathrm{n}_l(k_ir)-\mathrm{i}\mathrm{j}_l(k_ir)]-s_{ilj}[-\mathrm{n}_l(k_ir)+\mathrm{i}\mathrm{j}_l(k_ir)]\}\mathbb{Y}_{l1j\nu}\\&=\sqrt{\frac{4\pi}{v_i}}\sum_{lj\nu}\hat{l}C^{j\ \nu}_{l0\ 1\nu}C^{1\ \ \nu}_{\frac{1}{2}\nu_\mathrm{n}\ \frac{1}{2}\nu_\mathrm{p}}\left\{\mathrm{j}_l(k_ir)+\frac{\mathrm{i}}{2}\left(1-s_{ilj}\right)\left[-\mathrm{n}_l(k_ir)+\mathrm{i}\mathrm{j}_l(k_ir)\right]\right\}\mathbb{Y}_{l1j\nu}\end{aligned} \tag{3.1104}$$

where $s_{ilj}$ is a n–p elastic scattering $S$-matrix element. Letting the outer boundary of $r$ be $r_m$, then when $r\geq r_m$, the nuclear potential between n–p can be ignored.

We stipulate that the symbol "$'$" represents the derivative for $k_i r$. According to Eq. (3.1049) and the first equation of Eq. (3.1104), the boundary condition can be written as

$$\left.\frac{\left(\phi_{lj}^{i}(r)\right)'}{\phi_{lj}^{i}(r)}\right|_{r_m}=\left.\frac{\left[-(k_i r\,\mathrm{n}_l(k_i r))'-\mathrm{i}(k_i r\,\mathrm{j}_l(k_i r))'\right]-s_{ilj}\left[-(k_i r\,\mathrm{n}_l(k_i r))'+\mathrm{i}(k_i r\,\mathrm{j}_l(k_i r))'\right]}{\left[-(k_i r\,\mathrm{n}_l(k_i r))-\mathrm{i}(k_i r\,\mathrm{j}_l(k_i r))\right]-s_{ilj}\left[-(k_i r\,\mathrm{n}_l(k_i r))+\mathrm{i}(k_i r\,\mathrm{j}_l(k_i r))\right]}\right|_{r_m} \tag{3.1105}$$

where $\phi_{lj}^{i}(r)$ is obtained by numerically solving the Schrodinger equation in the n–p system inner region of the existence of the nuclear potential. The $S$ matrix element $s_{ilj}$ of the n–p elastic scattering can be obtained by using the boundary conditions given by the above equation. Note that

$$-\mathrm{n}_l(k_i r)+\mathrm{i}\,\mathrm{j}_l(k_i r)\xrightarrow{r\to\infty}(-\mathrm{i})^l\frac{\mathrm{e}^{\mathrm{i}k_i r}}{k_i r} \tag{3.1106}$$

thus we can find that the final state total wave function is

$$\begin{aligned}\Psi_{\nu_\mathrm{n}\nu_\mathrm{p}}^{(\mathrm{f})}(\vec{r})&=\Psi_{\nu_\mathrm{n}\nu_\mathrm{p}}^{(\mathrm{t})}(\vec{r})-\Psi_{\nu_\mathrm{n}\nu_\mathrm{p}}^{(\mathrm{in})}(\vec{r})\\&=\frac{\mathrm{i}\sqrt{\pi}}{\sqrt{v_i}}\sum_{lj\nu}\hat{l}C_{l0\,1\nu}^{j\,\nu}C_{\frac{1}{2}\nu_\mathrm{n}\,\frac{1}{2}\nu_\mathrm{p}}^{1\,\nu}\left(1-s_{ilj}\right)\left[-\mathrm{n}_l(k_i r)+\mathrm{i}\mathrm{j}_l(k_i r)\right]\mathbb{Y}_{l1j\nu}\\&\xrightarrow{r\to\infty}\frac{\mathrm{i}\sqrt{\pi}}{\sqrt{v_i}}\sum_{lj\nu}\hat{l}C_{l0\,1\nu}^{j\,\nu}C_{\frac{1}{2}\nu_\mathrm{n}\,\frac{1}{2}\nu_\mathrm{p}}^{1\,\nu}\left(1-s_{ilj}\right)\frac{\mathrm{e}^{\mathrm{i}k_i r}}{k_i r}\\&\quad\sum_{\nu'_\mathrm{n}\nu'_\mathrm{p}}C_{l\,\nu-\nu'_\mathrm{n}-\nu'_\mathrm{p}\,1\,\nu'_\mathrm{n}+\nu'_\mathrm{p}}^{j\,\nu}C_{\frac{1}{2}\nu'_\mathrm{n}\,\frac{1}{2}\nu'_\mathrm{p}}^{1\,\nu'_\mathrm{n}+\nu'_\mathrm{p}}\mathrm{Y}_{l\,\nu-\nu'_\mathrm{n}-\nu'_\mathrm{p}}(\Omega_r)\chi_{\frac{1}{2}\nu'_\mathrm{n}}\chi_{\frac{1}{2}\nu'_\mathrm{p}}\end{aligned} \tag{3.1107}$$

From the above expression, the n–p elastic scattering amplitude can be written as

$$\begin{aligned}f_{\nu_\mathrm{n}'\nu_\mathrm{p}',\nu_\mathrm{n}\nu_\mathrm{p}}^{(i)}(\Omega_r)=&\frac{\mathrm{i}\sqrt{\pi}}{k_i}\sum_{lj\nu}\hat{l}C_{l0\,1\nu}^{j\,\nu}C_{\frac{1}{2}\nu_\mathrm{n}\,\frac{1}{2}\nu_\mathrm{p}}^{1\,\nu}\left(1-s_{ilj}\right)\\&C_{l\,\nu-\nu'_\mathrm{n}-\nu'_\mathrm{p}\,1\,\nu'_\mathrm{n}+\nu'_\mathrm{p}}^{j\,\nu}C_{\frac{1}{2}\nu'_\mathrm{n}\,\frac{1}{2}\nu'_\mathrm{p}}^{1\,\nu'_\mathrm{n}+\nu'_\mathrm{p}}\mathrm{Y}_{l\,\nu-\nu'_\mathrm{n}-\nu'_\mathrm{p}}(\Omega_r)\end{aligned} \tag{3.1108}$$

If the deuteron breaks under the interaction with the target, the outgoing channels may be $i = 1, 2, \ldots, N$. We assume that n–p is an isolated system during the derivation process of Eqs. (3.1097)–(3.1108) of n–p elastic scattering. However, n–p is not an isolated system in the d + A → n + p + A reaction process. In fact, the interactions of n and p with A cause the deuteron to be changed from a bound state to a continuous state after breaking, and $j\mu$ are no longer the conservation quantities; but here we introduce the elastic scattering channel approximation [61, 74, 75]. That is, the radial wave function $\phi_{lj}^{i}(r)$ of the n–p relative motion of the deuteron breakup channel is obtained by the occurrence of n–p elastic scattering. The corresponding $S$ matrix element $s_{ilj}$ is only related to the outgoing channel subscripts, and the solution method has been given by Eq. (3.1105).

If there is only a central potential between the deuteron and the spherical target nucleus, the orbital angular momentum $L$ is a conservation quantity when the deuteron undergoes an elastic scattering reaction. However, in the CDCC theory, the deuteron feels the folding potential. It can be seen from Eq. (3.1058) that this potential is non-central, and the elastic scattering channel is coupled with the breakup channel, so $L$ is no longer a conservation quantity. It follows that, according to Eq. (3.1096), the total wave function of the system after the reaction can be written as

$$\begin{aligned}\Psi_{\mu}^{(\mathrm{t})} = \frac{\mathrm{i}\sqrt{\pi}}{K_0 R}\sum_{lLJ}\hat{L}C_{L0\ 1\mu}^{J\ \mu}\mathrm{e}^{\mathrm{i}\sigma_L^{(0)}}\Bigg[&\frac{1}{\sqrt{V_0}}(G_L(K_0R)-\mathrm{i}F_L(K_0R))\frac{\phi_l^0(r)}{r}\varphi_{l11LJM}\\ &-\sum_{l'L'}\frac{1}{\sqrt{V_0}}S_{0l'11L',0l11L}^{J}(G_{L'}(K_0R)+\mathrm{i}F_{L'}(K_0R))\frac{\phi_{l'}^0(r)}{r}\varphi_{l'11L'JM}\\ &-\sum_{i'=1}^{N}\sum_{l'j'L'}\frac{1}{\sqrt{V_{i'}}}S_{i'l'1j'L',0l11L}^{J}s_{i'l'j'}(G_{L'}(K_{i'}R)+\mathrm{i}F_{L'}(K_{i'}R))\\ &(-\mathrm{n}_{l'}(k_{i'}r)+\mathrm{i}\mathrm{j}_{l'}(k_{i'}r))\ \varphi_{l'1j'L'JM}\Bigg]\end{aligned} \tag{3.1109}$$

The first item in the square brackets of the above expression is the spherical incident wave; the second item is the spherical outgoing wave of the elastic scattering channel; and the third item is the spherical outgoing wave of the breakup channel in which $s_{i'l'j'}$ is used, indicating that the elastic scattering channel approximation is used. The n–p continuous state is actually created because of the breakup of the deuteron, but here it is approximated that when $R$ is large enough the n-p continuous state is obtained by the occurrence of n–p elastic scattering.

Letting the outer boundary of $R$ be $R_m$, then when $R \geq R_m$ the nuclear potential and Coulomb distortion potential between the deuteron and the target can be ignored. We stipulate that the symbol "$'$" represents the derivative for $K_0R$, so the boundary conditions for the elastic scattering process of the deuteron center of mass can be written according to Eqs. (3.1043) and (3.1109) as follows:

$$\left.\frac{(u_{0l'11L'}(R))'}{u_{0l'11L'}(R)}\right|_{R_m} = \left.\frac{(G_L'(K_0R)-\mathrm{i}F_L'(K_0R))\delta_{l'L',lL}-S_{0l'11L',0l11L}^{J}(G_{L'}'(K_0R)+\mathrm{i}F_{L'}'(K_0R))}{(G_L(K_0R)-\mathrm{i}F_L(K_0R))\delta_{l'L',lL}-S_{0l'11L',0l11L}^{J}(G_{L'}(K_0R)+\mathrm{i}F_{L'}(K_0R))}\right|_{R_m} \tag{3.1110}$$

where $u^J_{0l'11L'}(R)$ is obtained by numerically solving the coupling channel equation (3.1058) in the internal region of the existence of the nuclear potential, which includes the influence of the breakup channel on the elastic scattering channel. In the actual calculation, the elastic scattering channel and breakup channel must be solved simultaneously, and Eq. (3.1110) is just the equation to determine the $S$ matrix element $S^J_{0l'11L',0l11L}$.

Now, we can rewrite Eq. (3.1109) as

$$\begin{aligned}\Psi^{(\mathrm{t})}_{\mu} = \frac{\sqrt{4\pi}}{K_0 R}\sum_{lLJ}\hat{L}C^{J\,\mu}_{L0\,1\mu}\mathrm{e}^{\mathrm{i}\sigma^{(0)}_L}\Bigg[&\frac{1}{\sqrt{V_0}}F_L(K_0R)\frac{\phi^0_l(r)}{r}\varphi_{l11LJM}\\ &+\sum_{l'L'}\frac{\mathrm{i}}{2\sqrt{V_0}}\left(\delta_{l'L',lL}-S^J_{0l'11L',0l11L}\right)\left(G_{L'}(K_0R)+\mathrm{i}F_{L'}(K_0R)\right)\frac{\phi^0_{l'}(r)}{r}\varphi_{l'11L'JM}\\ &-\sum_{i'=1}^{N}\sum_{l'j'L'}\frac{\mathrm{i}}{2\sqrt{V_{i'}}}S^J_{i'l'1j'L',0l11L}s_{i'l'j'}\left(G_{L'}(K_{i'}R)+\mathrm{i}F_{L'}(K_{i'}R)\right)\\ &\left(-\mathrm{n}_{l'}(k_{i'}r)+\mathrm{i}\mathrm{j}_{l'}(k_{i'}r)\right)\varphi_{l'1j'L'JM}\Bigg]\end{aligned} \tag{3.1111}$$

and note

$$G_{L'}(K_{i'}R)+\mathrm{i}F_{L'}(K_{i'}R)\xrightarrow{R\to\infty}\mathrm{e}^{\mathrm{i}K_{i'}R}(-\mathrm{i})^{L'}\mathrm{e}^{\mathrm{i}\sigma^{(i')}_{L'}} \tag{3.1112}$$

Recalling Eq. (3.1106), we can get the final state total wave function from Eqs. (3.1111) and (3.1096) as follows:

$$\begin{aligned}\Psi^{(\mathrm{out})}_{\mu} = \Psi^{(\mathrm{t})}_{\mu} - \Psi^{(\mathrm{in})}_{\mu} = &\frac{\mathrm{i}\sqrt{\pi}}{K_0R}\sum_{lLJ}\hat{L}C^{J\,\mu}_{L0\,1\mu}\mathrm{e}^{\mathrm{i}\sigma^{(0)}_L}\\ &\Bigg[\sum_{l'L'}\frac{1}{\sqrt{V_0}}\left(\delta_{l'L',lL}-S^J_{0l'11L',0l11L}\right)\left(G_{L'}(K_0R)+\mathrm{i}F_{L'}(K_0R)\right)\frac{\phi^0_{l'}(r)}{r}\varphi_{l'11L'JM}\\ &-\sum_{i'=1}^{N}\sum_{l'j'L'}\frac{1}{\sqrt{V_{i'}}}S^J_{i'l'1j'L',0l11L}s_{i'l'j'}\left(G_{L'}(K_{i'}R)+\mathrm{i}F_{L'}(K_{i'}R)\right)\\ &\left(-\mathrm{n}_{l'}(k_{i'}r)+\mathrm{i}\mathrm{j}_{l'}(k_{i'}r)\right)\varphi_{l'1j'L'JM}\Bigg]\end{aligned}$$

$$\rightarrow \frac{\mathrm{i}\sqrt{\pi}}{K_0}\sum_{ljJ}\hat{L}C_{l0\ 1\mu}^{J\ \mu}\mathrm{e}^{\mathrm{i}\sigma_L^{(0)}}\left[\sum_{l'L'}\frac{\mathrm{e}^{\mathrm{i}\sigma_{L'}^{(0)}}}{\sqrt{V_0}}\left(\delta_{l'L',lL}-S^J_{0l'11L',0l11L}\right)\right.$$
$$\frac{\mathrm{e}^{\mathrm{i}K_0R}}{R}\frac{\phi^0_{l'}(r)}{r}\sum_{\mu'}C_{L'\ \mu-\mu'\ 1\mu'}^{J\ \mu}\mathrm{Y}_{L'\ \mu-\mu'}(\Omega_R)\mathbb{Y}_{l'11\mu'}(\Omega_r)$$
$$-\sum_{i'=1}^{N}\sum_{l'j'L'}\frac{\mathrm{e}^{\mathrm{i}\sigma_{L'}^{(i')}}}{\sqrt{V_{i'}k_{i'}}}S^J_{i'l'1j'L',0l11L}s_{i'l'j'}\frac{\mathrm{e}^{\mathrm{i}K_{i'}R}}{R}\frac{\mathrm{e}^{\mathrm{i}k_{i'}r}}{r}$$
$$\left.\sum_{\mu'\nu'}C_{L'\ \mu-\mu'\ j'\mu'}^{J\ \mu}C_{l'\ \mu'-\nu'\ 1\nu'}^{j'\ \mu'}\mathrm{Y}_{L'\ \mu-\mu'}(\Omega_R)\mathrm{Y}_{l'\ \mu'-\nu'}(\Omega_r)\chi_{1\nu'}\right] \tag{3.1113}$$

where Eqs. (3.1041) and. (3.1037) were also used in the final step of the derivation of Eq. (3.1113). Now it is known that the outgoing wave of the Coulomb scattering is [7]

$$\Psi_\mu^{(\mathrm{C})} \rightarrow \frac{1}{\sqrt{V_0}R}f_{\mathrm{C}}(\theta_R)\,\mathrm{e}^{\mathrm{i}K_0R}\psi_\mu\left(\vec{r}\right) \tag{3.1114}$$

It follows that from the above two expressions, the final state wave function of the deuteron elastic scattering can be written as

$$\Psi_\mu^{(\mathrm{r})}=\frac{1}{\sqrt{V_0}}\left[f_{\mathrm{C}}(\theta_R)\sum_l\frac{\phi^0_l(r)}{r}\mathbb{Y}_{l11\mu}(\Omega_r)+\frac{\mathrm{i}\sqrt{\pi}}{K_0}\sum_{lLJ}\hat{L}C_{L0\ 1\mu}^{J\ \mu}\mathrm{e}^{\mathrm{i}\sigma_L^{(0)}}\sum_{l'L'}\mathrm{e}^{\mathrm{i}\sigma_{L'}^{(0)}}\right.$$
$$\left.\left(\delta_{l'L',lL}-S^J_{0l'11L',0l11L}\right)\sum_{\mu'}C_{L'\ \mu-\mu'\ 1\mu'}^{J\ \mu}\mathrm{Y}_{L'\ \mu-\mu'}(\Omega_R)\frac{\phi^0_{l'}(r)}{r}\mathbb{Y}_{l'11\mu'}(\Omega_r)\right]\frac{\mathrm{e}^{\mathrm{i}K_0R}}{R} \tag{3.1115}$$

and from this expression, the deuteron elastic scattering amplitude can be written as

$$f_{l'\mu',l\mu}(\Omega_r)=f_{\mathrm{C}}(\theta_R)\,\delta_{l'\mu',l\mu}+\frac{\mathrm{i}\sqrt{\pi}}{K_0}\sum_{LJ}\hat{L}C_{L0\ 1\mu}^{J\ \mu}\mathrm{e}^{\mathrm{i}\sigma_L^{(0)}}$$
$$\sum_{L'}\mathrm{e}^{\mathrm{i}\sigma_{L'}^{(0)}}\left(\delta_{l'L',lL}-S^J_{0l'11L',0l11L}\right)C_{L'\ \mu-\mu'\ 1\mu'}^{J\ \mu}\mathrm{Y}_{L'\ \mu-\mu'}(\Omega_R) \tag{3.1116}$$

such that both incident and outgoing waves can be S waves or D waves. With this elastic scattering amplitude, various corresponding polarization quantities of the deuteron can be studied. In the center of mass system, the deuteron elastic scattering angular distribution is

$$\frac{d\sigma}{d\Omega_R} = \frac{1}{3} \sum_{l\mu\, l'\mu'} |f_{l'\mu',l\mu}(\Omega_R)|^2 \tag{3.1117}$$

where the deuteron elastic scattering angular distribution in the laboratory system can be obtained through the coordinate system transformation for the above expression [7]. The deuteron elastic scattering angular distribution obtained here includes the influence of the breakup channel; that calculated by the ordinary deuteron optical model does not take into account the influence of the breakup channel. The problem of calculating the absorption cross section of the deuteron induced nuclear reaction, using the obtained S-matrix element $S^J_{0l'11L,0l11L}$, can be further studied here.

### 3.15.4 Double Differential Cross Section of the Ejected Nucleons After Deuteron Breaking

Stipulating that the symbol "′" represents the derivative for $K_{i'}R$, the boundary conditions for the deuteron breakup channel $(1 \leq i \leq N)$ in the case of the introduction of elastic scattering channel approximation can be written according to Eqs. (3.1050) and (3.1109) as follows:

$$\left.\frac{\left(u^J_{i'l'1j'L'}(R)\right)'}{u^J_{i'l'1j'L'}(R)}\right|_{R_m} = \left.\frac{(G'_L(K_{i'}R) - \mathrm{i}F'_L(K_{i'}R))\delta_{i'l'j'L',0l1L} - S^J_{i'l'1j'L',0l11L}(G'_{L'}(K_{i'}R) + \mathrm{i}F'_{L'}(K_{i'}R))}{(G_L(K_{i'}R) - \mathrm{i}F_L(K_{i'}R))\delta_{i'l'j'L',0l1L} - S^J_{i'l'1j'L',0l11L}(G_{L'}(K_{i'}R) + \mathrm{i}F_{L'}(K_{i'}R))}\right|_{R_m} \tag{3.1118}$$

It is important to note that the boundary condition Eq. (3.1110) of elastic scattering is only a special case of the above equation, and the boundary condition Eq. (3.1118) can be used uniformly. The above equation is the equation to determine the $S$ matrix element $S^J_{i'l'1j'L',0l11L}$. The reaction amplitude of the breakup channel can be written by Eq. (3.1113) as

$$f_{\nu',\mu}(\Omega_R, \Omega_r) = -\frac{\mathrm{i}\sqrt{\pi}}{K_0} \sum_{lLJ} \hat{L} C^{J\,\mu}_{L0\;1\mu} \mathrm{e}^{\mathrm{i}\sigma_L^{(0)}} \sum_{i'=1}^{N} \sum_{l'j'L'} \frac{\mathrm{e}^{\mathrm{i}\sigma_{L'}^{(i')}}}{k_{i'}} S^J_{i'l'1j'L',0l11L}\, s_{i'l'j'} \sum_{\mu'} C^{J\,\mu}_{L'\,\mu-\mu'\;j'\mu'} C^{j'\,\mu'}_{l'\,\mu'-\nu'\;1\nu'}\, Y_{L'\,\mu-\mu'}(\Omega_R)\, Y_{l'\,\mu'-\nu'}(\Omega_r) \tag{3.1119}$$

Recall that $\varepsilon_i$ is defined by Eq. (3.1054). If $k_{i-1} \le k \le k_i$, then the amount related to k is obtained through interpolation in the interval $[k_{i-1}, k_i]$. Using $c = 0$ to represent the bound state, and $c \neq 0$ to represent the continuous state, Eq. (3.1119) can be rewritten as

$$f_{\nu',\mu}(\Omega_R, \Omega_r) = -\frac{\mathrm{i}\sqrt{\pi}}{K_0}\sum_{lLJ}\hat{L}C_{L0\ 1\mu}^{J\ \mu}\mathrm{e}^{\mathrm{i}\sigma_L^{(0)}}\sum_{l'j'L'\mu'}C_{L'\ \mu-\mu'\ j'\mu'}^{J\ \mu}C_{l'\ \mu'-\nu'\ 1\nu'}^{j'\ \mu'}$$
$$\mathrm{Y}_{L'\ \mu-\mu'}(\Omega_R)\mathrm{Y}_{l'\ \mu'-\nu'}(\Omega_r)\int_0^{k_{\max}}\frac{e^{\mathrm{i}\sigma_{L'}(k)}}{k\Delta}S_{cl'1j'L',0l11L}^{J}(k)s_{cl'j'}(k)\mathrm{d}k \tag{3.1120}$$

and stipulate that

$$S_{cl'1j'L',0l11L}^{J}(0) = S_{cl'1j'L',0l11L}^{J}(k_{\max}) = s_{cl'j'}(0) = s_{cl'j'}(k_{\max}) = 0 \tag{3.1121}$$

Note that $k = 0$ still belongs to the $c \neq 0$ continuous state, and the difference of the $c = 0$ bound state and the $k = 0$ continuous state is the binding energy $\varepsilon_0$. So according to Eq. (3.1121), the five-dimensional reaction amplitude of the breakup channel can be written as

$$f_{\nu',\mu}(k, \Omega_R, \Omega_r) = -\frac{\mathrm{i}\sqrt{\pi}}{K_0}\sum_{lLJ}\hat{L}C_{L0\ 1\mu}^{J\ \mu}\mathrm{e}^{\mathrm{i}\sigma_L^{(0)}}\sum_{l'j'L'\mu'}C_{L'\ \mu-\mu'\ j'\ \mu'}^{J\ \mu}C_{l'\ \mu'-\nu'\ 1\nu'}^{j'\ \mu'}$$
$$\frac{\mathrm{e}^{\mathrm{i}\sigma_{L'}(k)}}{k\Delta}S_{cl'1j'L',0l11L}^{J}(k)s_{cl'j'}(k)\mathrm{Y}_{L'\ \mu-\mu'}(\Omega_R)\mathrm{Y}_{l'\ \mu'-\nu'}(\Omega_r) \tag{3.1122}$$

The corresponding five-differential cross sections in the center of mass system are

$$\frac{\mathrm{d}^5\sigma}{\mathrm{d}k\ \mathrm{d}\Omega_R\mathrm{d}\Omega_r} = \frac{1}{3}\sum_{\mu\ \nu'}|f_{\nu',\mu}(k,\ \Omega_R,\ \Omega_r)\ |^2 \tag{3.1123}$$

Also, note the relation

$$\int \mathrm{Y}_{l''\ \mu''-\nu'}^{*}(\Omega_r)\mathrm{Y}_{l'\ \mu'-\nu'}(\Omega_r)\ \mathrm{d}\Omega_r = \delta_{l''\mu'',l'\mu'} \tag{3.1124}$$

From Eq. (3.1123) we can find that the double differential cross section of the deuteron center of mass in the center of mass system is

$$\frac{d^2\sigma}{dk\, d\Omega_R} = \frac{1}{3}\sum_{\mu\nu' l' \mu'} \left| f_{l'\mu'\nu',\mu}(k,\ \Omega_R) \right|^2 \tag{3.1125}$$

$$f_{l'\mu'\nu',\mu}(k,\ \Omega_R) = -\frac{\mathrm{i}\sqrt{\pi}}{K_0}\sum_{lLJ}\hat{L}C_{L0\ 1\mu}^{J\ \mu}\mathrm{e}^{\mathrm{i}\sigma_L^{(0)}}\sum_{j'L'}C_{L'\ \mu-\mu'\ j'\ \mu'}^{J\ \mu}C_{l'\ \mu'-\nu'\ 1\nu'}^{j'\ \mu'} \frac{\mathrm{e}^{\mathrm{i}\sigma_{L'}(k)}}{k\Delta}S_{cl'1j'L',0l11L}^{J}(k)s_{cl'j'}(k)\mathrm{Y}_{L'\ \mu-\mu'}(\Omega_R) \tag{3.1126}$$

Now because the entire reaction system is axisymmetric, the above two equations will not change with azimuth angle $\varphi_R$. The double differential cross section of the deuteron center of mass in the laboratory system can be obtained from Eq. (3.1125) through the coordinate system transformation.

The momentum wave vectors of neutrons, protons, and target nuclei in the breakup channel in laboratory system are $\vec{k}_{\mathrm{n}}$, $\vec{k}_{\mathrm{p}}$, and $\vec{k}_{\mathrm{A}}$, respectively, and the total momentum wave vector of the system is

$$\vec{k}_{\mathrm{t}} = \vec{k}_{\mathrm{n}} + \vec{k}_{\mathrm{p}} + \vec{k}_{\mathrm{A}} \tag{3.1127}$$

where $\vec{k}_{\mathrm{t}}$ is along the $z$-axis direction, i.e., the incident deuteron direction. The total kinetic energy of the system in the laboratory system after the deuteron breaking is

$$E_{\mathrm{t}} = \frac{\hbar^2}{2}\left(\frac{k_{\mathrm{n}}^2}{m_{\mathrm{n}}} + \frac{k_{\mathrm{p}}^2}{m_{\mathrm{p}}} + \frac{k_{\mathrm{A}}^2}{m_{\mathrm{A}}}\right) \tag{3.1128}$$

In the laboratory system, the total energy of the incident channel is

$$E_{\mathrm{in}} = \frac{m_{\mathrm{d}} + m_{\mathrm{A}}}{m_{\mathrm{A}}}E_0 + \varepsilon_0 = E_L + \varepsilon_0 \tag{3.1129}$$

where $E_L$ is the deuteron incident energy in the laboratory system, and we have

$$E_L = \frac{\hbar^2 k_{\mathrm{in}}^2}{2m_{\mathrm{d}}}, \qquad k_{\mathrm{in}} = \frac{\sqrt{2m_{\mathrm{d}}E_L}}{\hbar} \tag{3.1130}$$

In the laboratory system, the momentum conservation requires

$$\vec{k}_{\mathrm{t}} = \vec{k}_{\mathrm{in}} \tag{3.1131}$$

and it can be seen that $k_t$ is a known amount. It follows that Eq. (3.1129) can be rewritten as

$$E_{\text{in}} = \frac{\hbar^2 k_{\text{t}}^2}{2m_{\text{d}}} + \varepsilon_0 \tag{3.1132}$$

Using the energy conservation condition given by Eq. (3.1031), Eq. (3.1123) can be rewritten as

$$\mathrm{d}^5\sigma = \frac{1}{3}\left(\sum_{\mu\nu'} |f_{\nu l,\mu}(k,\ \Omega_R,\ \Omega_r)|^2\right)\int_\delta \delta(E_0 + \varepsilon_0 - \in - \varepsilon)\,\mathrm{d}\vec{k}\,\mathrm{d}\vec{K} \tag{3.1133}$$

where

$$\in = \frac{\hbar^2 K^2}{2\mu_R} \tag{3.1134}$$

The $\delta$ function indicates that the energy conservation is satisfied in the center of mass system, and $\int_\delta$ indicates that the integral will be done only to the argument of the $\delta$ function. Now introduce the integral amount that appeared in Eq.(3.1133) as

$$\mathrm{d}W \equiv \int_\delta \delta(E_0 + \varepsilon_0 - \in - \varepsilon)\mathrm{d}\vec{k}\mathrm{d}\vec{K} \tag{3.1135}$$

Since $\mathrm{d}W$ is an integral amount that is independent with the coordinate system, so the above equation can be rewritten in the laboratory system as

$$\mathrm{d}W = \int_\delta \delta\left(\vec{k}_{\text{t}} - \vec{k}_{\text{n}} - \vec{k}_{\text{p}} - \vec{k}_{\text{A}}\right)\delta(E_{\text{in}} - E_{\text{t}})\mathrm{d}\vec{k}_{\text{n}}\mathrm{d}\vec{k}_{\text{p}}\mathrm{d}\vec{k}_{\text{A}} \tag{3.1136}$$

The two $\delta$ functions in the above equation represent the momentum conservation and the energy conservation in the laboratory system. From the momentum conservation we have

$$\vec{k}_{\text{A}} = \vec{k}_{\text{t}} - \vec{k}_{\text{n}} - \vec{k}_{\text{p}} \tag{3.1137}$$

and substituting this equation into Eq. (3.1128) we can get

$$E_{\text{t}} = \frac{\hbar^2}{2}\left\{\frac{k_{\text{n}}^2}{m_{\text{n}}} + \frac{k_{\text{p}}^2}{m_{\text{p}}} + \frac{1}{m_{\text{A}}}\left[k_{\text{t}}^2 + k_{\text{n}}^2 + k_{\text{p}}^2 + 2\left(\vec{k}_{\text{n}}\cdot\vec{k}_{\text{p}} - \vec{k}_{\text{n}}\cdot\vec{k}_{\text{t}} - \vec{k}_{\text{p}}\cdot\vec{k}_{\text{t}}\right)\right]\right\} \tag{3.1138}$$

It follows that Eq. (3.1136) can be rewritten as

$$
\begin{aligned}
\mathrm{d}W = & \int_{\delta} \delta\Big(\varepsilon_0 - \frac{\hbar^2}{2}\Big[\left(\frac{1}{m_\mathrm{n}}+\frac{1}{m_\mathrm{A}}\right)k_\mathrm{n}^2 + \left(\frac{1}{m_\mathrm{p}}+\frac{1}{m_\mathrm{A}}\right)k_\mathrm{p}^2 \\
& - \left(\frac{1}{m_\mathrm{d}}-\frac{1}{m_\mathrm{A}}\right)k_\mathrm{t}^2 - \frac{2}{m_\mathrm{A}}\left(\vec{k}_\mathrm{t}-\vec{k}_\mathrm{n}\right)\cdot\vec{k}_\mathrm{p} - \frac{2}{m_\mathrm{A}}\vec{k}_\mathrm{n}\cdot\vec{k}_\mathrm{t}\Big]\Big)\mathrm{d}\vec{k}_\mathrm{n}\mathrm{d}\vec{k}_\mathrm{p} \\
= & \frac{2m_\mathrm{p}m_\mathrm{A}}{(m_\mathrm{p}+m_\mathrm{A})\,\hbar^2}\int \delta\Big(k_\mathrm{p}^2 - \frac{2m_\mathrm{p}}{m_\mathrm{p}+m_\mathrm{A}}\left(\vec{k}_\mathrm{t}-\vec{k}_\mathrm{n}\right)\cdot\vec{k}_\mathrm{p} + \frac{m_\mathrm{p}(m_\mathrm{n}+m_\mathrm{A})}{m_\mathrm{n}(m_\mathrm{p}+m_\mathrm{A})}k_\mathrm{n}^2 \\
& - \frac{2m_\mathrm{p}}{m_\mathrm{p}+m_\mathrm{A}}\vec{k}_\mathrm{n}\cdot\vec{k}_\mathrm{t} - \frac{m_\mathrm{p}(m_\mathrm{A}-m_\mathrm{d})}{m_\mathrm{d}(m_\mathrm{p}+m_\mathrm{A})}k_\mathrm{t}^2 - \frac{2m_\mathrm{p}m_\mathrm{A}}{(m_\mathrm{p}+m_\mathrm{A})\,\hbar^2}\varepsilon_0\Big)\mathrm{d}\vec{k}_\mathrm{n}\mathrm{d}\vec{k}_\mathrm{p}
\end{aligned}
\tag{3.1139}
$$

When the $\delta$ function is integrated in the above equation, $k_p$ is taken as the integral variable. It is known that $\vec{k}_\mathrm{t}$ is along the $z$-axis direction, and let

$$\vec{k}_\mathrm{n}\cdot\vec{k}_\mathrm{p} = k_\mathrm{n}k_\mathrm{p}\cos\theta_\mathrm{np} \tag{3.1140}$$

$$\cos\theta_\mathrm{np} = \cos\theta_\mathrm{n}\cos\theta_\mathrm{p} + \sin\theta_\mathrm{n}\sin\theta_\mathrm{p}\cos\left(\varphi_\mathrm{n}-\varphi_\mathrm{p}\right) \tag{3.1141}$$

The following equation can be obtained from the δ function in Eq. (3.1139):

$$
\begin{aligned}
& k_\mathrm{p}^2 - \frac{2m_\mathrm{p}}{m_\mathrm{p}+m_\mathrm{A}}\left(k_\mathrm{t}\cos\theta_\mathrm{p} - k_\mathrm{n}\cos\theta_\mathrm{np}\right)k_\mathrm{p} \\
& + \frac{m_\mathrm{p}}{m_\mathrm{p}+m_\mathrm{A}}\left(\frac{m_\mathrm{n}+m_\mathrm{A}}{m_\mathrm{n}}k_\mathrm{n}^2 - 2k_\mathrm{n}k_\mathrm{t}\cos\theta_\mathrm{n} - \frac{m_\mathrm{A}-m_\mathrm{d}}{m_\mathrm{d}}k_\mathrm{t}^2 - \frac{2m_\mathrm{A}}{\hbar^2}\varepsilon_0\right) = 0
\end{aligned}
\tag{3.1142}
$$

and its solution is

$$
\begin{aligned}
k_{\mathrm{p}\pm} = & \frac{m_\mathrm{p}}{m_\mathrm{p}+m_\mathrm{A}}\Big\{\left(k_\mathrm{t}\cos\theta_\mathrm{p} - k_\mathrm{n}\cos\theta_\mathrm{np}\right) \pm \Big[\left(k_\mathrm{t}\cos\theta_\mathrm{p} - k_\mathrm{n}\cos\theta_\mathrm{np}\right)^2 \\
& - \frac{m_\mathrm{p}+m_\mathrm{A}}{m_\mathrm{p}}\left(\frac{m_\mathrm{n}+m_\mathrm{A}}{m_\mathrm{n}}k_\mathrm{n}^2 - 2k_\mathrm{n}k_\mathrm{t}\cos\theta_\mathrm{n} - \frac{m_\mathrm{A}-m_\mathrm{d}}{m_\mathrm{d}}k_\mathrm{t}^2 - \frac{2m_\mathrm{A}}{\hbar^2}\varepsilon_0\right)\Big]^{1/2}\Big\}
\end{aligned}
\tag{3.1143}
$$

The condition that the above equation has a solution is

$$
\begin{aligned}
& \left(k_\mathrm{t}\cos\theta_\mathrm{p} - k_\mathrm{n}\cos\theta_\mathrm{np}\right)^2 - \frac{m_\mathrm{p}+m_\mathrm{A}}{m_\mathrm{p}}\left(\frac{m_\mathrm{n}+m_\mathrm{A}}{m_\mathrm{n}}k_\mathrm{n}^2\right. \\
& \left. - 2k_\mathrm{n}k_\mathrm{t}\cos\theta_\mathrm{n} - \frac{m_\mathrm{A}-m_\mathrm{d}}{m_\mathrm{d}}k_\mathrm{t}^2 - \frac{2m_\mathrm{A}}{\hbar^2}\varepsilon_0\right) \geq 0
\end{aligned}
\tag{3.1144}
$$

and areas that do not meet this condition are forbidden. Only the $k_p \geq 0$ solutions are taken in the above equation. If $k_{p-}<0$, then $k_{p-}$ is an unphysical solution, and it is not selected. Now the $\delta$ function has the following properties:

$$\delta(f(x)) = \sum_i \frac{\delta(x - x_i)}{|f'(x_i)|} \tag{3.1145}$$

Therefore, from Eq. (3.1139), we can obtain

$$\begin{aligned} dW = & \frac{m_p m_A}{(m_p + m_A)\hbar^2} \left[ \frac{k_{p+}^2}{|k_{p+} - \frac{m_p}{m_p + m_A}(k_t \cos\theta_p - k_n \cos\theta_{np})|} \right. \\ & \left. + \frac{k_{p-}^2 \eta(k_{p-})}{|k_{p-} - \frac{m_p}{m_p + m_A}(k_t \cos\theta_p - k_n \cos\theta_{np})|} \right] k_n^2 dk_n d\Omega_n d\Omega_p \end{aligned} \tag{3.1146}$$

where

$$\eta(k_{p-}) = \begin{cases} 1 & \text{when } k_{p-} > 0 \\ 0 & \text{when } k_{p-} \leq 0 \end{cases} \tag{3.1147}$$

Next let

$$E_{p\pm} = \frac{\hbar^2 k_{p\pm}^2}{2m_p} \tag{3.1148}$$

then we can get

$$dW = \rho(E_n, \Omega_n, \Omega_p)\, dE_n d\Omega_n d\Omega_p \tag{3.1149}$$

where

$$\begin{aligned} \rho(E_n, \Omega_n, \Omega_p) = & \frac{2m_n m_p^2 m_A}{(m_p + m_A)\hbar^6} \sqrt{m_n E_n} \\ & \left[ \frac{E_{p+}}{|\sqrt{m_p E_{p+}} - \frac{m_p}{m_p + m_A}(\sqrt{m_d E_L} \cos\theta_p - \sqrt{m_n E_n} \cos\theta_{np})|} \right. \\ & \left. + \frac{E_{p-} \eta(k_{p-})}{|\sqrt{m_p E_{p-}} - \frac{m_p}{m_p + m_A}(\sqrt{m_d E_L} \cos\theta_p - \sqrt{m_n E_n} \cos\theta_{np})|} \right] \end{aligned} \tag{3.1150}$$

It follows that we can obtain the five-differential cross section, according to Eqs. (3.1133), (3.1135), and (3.1149), as follows:

$$\frac{d^5\sigma}{dE_n d\Omega_n d\Omega_p} = \frac{1}{3}\left(\sum_{\mu\nu'} |f_{\nu\prime,\mu}(k, \quad \Omega_R, \quad \Omega_r)|^2\right)\rho(E_n, \Omega_n, \Omega_p) \tag{3.1151}$$

From the this equation we can find that the double differential cross section of the emitted neutrons is

$$\frac{d^2\sigma}{dE_n d\Omega_n} = \int \frac{d^5\sigma}{dE_n d\Omega_n d\Omega_p} d\cos\theta_p d\varphi_p \tag{3.1152}$$

and the same method can be used to obtain $\frac{d^2\sigma}{dE_p d\Omega_p}$.

The remaining question is how to replace $(k, \Omega_R, \Omega_r)$ by $(E_n, \Omega_n, \Omega_p)$ in Eq. (3.1151).

The speed vectors of the neutron, proton, and target nucleus in laboratory system are, respectively,

$$\vec{v}_n = \frac{\hbar \vec{k}_n}{m_n}, \quad \vec{v}_p = \frac{\hbar \vec{k}_p}{m_p}, \quad \vec{v}_A = \frac{\hbar \vec{k}_A}{m_A} \tag{3.1153}$$

The speed vector of the incident deuteron in laboratory system is

$$\vec{v}_d = \frac{\hbar \vec{k}_t}{m_d} \tag{3.1154}$$

Also, the center of mass velocity of the incident channel system is

$$\vec{v}_c^{(in)} = \frac{\hbar \vec{k}_t}{m_d + m_A} \tag{3.1155}$$

where the total mass of the system before the deuteron breaking is $m_d + m_A$, and that after the deuteron breaking becomes

$$m_t = m_n + m_p + m_A \tag{3.1156}$$

In the exit channel system, due to the deuteron breaking the total mass of the system has increased a bit, but it has consumed a part of the kinetic energy, so the center of mass of the system will also become a little slower. According to Eq. (3.1155), the center of mass velocity of the exit channel system can be written as

$$\vec{v}_c^{\,(\text{out})} = \frac{\hbar \vec{k}_\text{t}}{m_\text{n} + m_\text{p} + m_\text{A}} = \frac{\hbar \vec{k}_\text{t}}{m_\text{t}} \tag{3.1157}$$

Therefore it can be found that the momentum wave vectors of neutron, proton, and target nucleus in the exit channel center of mass system are, respectively,

$$\vec{p}_\text{n} = \frac{m_\text{n}}{\hbar}\left(\vec{v}_\text{n} - \vec{v}_c^{\,(\text{out})}\right) = \vec{k}_\text{n} - \frac{m_\text{n}}{m_\text{t}}\vec{k}_\text{t} \tag{3.1158}$$

$$\vec{p}_\text{p} = \frac{m_\text{p}}{\hbar}\left(\vec{v}_\text{p} - \vec{v}_c^{\,(\text{out})}\right) = \vec{k}_\text{p} - \frac{m_\text{p}}{m_\text{t}}\vec{k}_\text{t} \tag{3.1159}$$

$$\vec{p}_\text{A} = \frac{m_\text{A}}{\hbar}\left(\vec{v}_\text{A} - \vec{v}_c^{\,(\text{out})}\right) = \vec{k}_\text{A} - \frac{m_\text{A}}{m_\text{t}}\vec{k}_\text{t} = \frac{m_\text{n} + m_\text{p}}{m_\text{t}}\vec{k}_\text{t} - \vec{k}_\text{n} - \vec{k}_\text{p} \tag{3.1160}$$

such that the following relation is satisfied:

$$\vec{p}_\text{n} + \vec{p}_\text{p} + \vec{p}_\text{A} = 0 \tag{3.1161}$$

The relative motion wave vector of neutron n and proton p is [22]

$$\vec{k} = \frac{1}{m_\text{n} + m_\text{p}}\left(m_\text{p}\vec{p}_\text{n} - m_\text{n}\vec{p}_\text{p}\right) \tag{3.1162}$$

and the relative motion wave vector of the center of mass of neutron and proton two-nucleons system relative to the remaining nucleus A is

$$\begin{aligned}\vec{K} &= \frac{1}{m_\text{t}}\left[(m_\text{n} + m_\text{p})\vec{p}_\text{A} - m_\text{A}\left(\vec{p}_\text{n} + \vec{p}_\text{p}\right)\right] \\ &= -\frac{1}{m_\text{t}}\left[(m_\text{n} + m_\text{p})\left(\vec{p}_\text{n} + \vec{p}_\text{p}\right) + m_\text{A}\left(\vec{p}_\text{n} + \vec{p}_\text{p}\right)\right] = -\left(\vec{p}_\text{n} + \vec{p}_\text{p}\right)\end{aligned} \tag{3.1163}$$

where it can be seen that $\vec{K}$ is just $\vec{p}_\text{A}$. From the above two expressions we can obtain

$$k^2 = \frac{1}{(m_\text{n} + m_\text{p})^2}\left(m_\text{p}^2 p_\text{n}^2 + m_\text{n}^2 p_\text{p}^2 - 2 m_\text{n} m_\text{p} p_\text{n} p_\text{p} \cos\theta_\text{np}^c\right) \tag{3.1164}$$

$$K^2 = p_\text{n}^2 + p_\text{p}^2 + 2 p_\text{n} p_\text{p} \cos\theta_\text{np}^c \tag{3.1165}$$

and making point multiplications using $\frac{1}{k_\text{t}}\vec{k}_\text{t}$ on Eqs. (3.1162) and (3.1163), we can get

$$k\cos\theta_r = \frac{1}{m_{\rm n}+m_{\rm p}}\left(m_{\rm p}p_{\rm n}\cos\theta_{\rm n}^c - m_{\rm n}p_{\rm p}\cos\theta_{\rm p}^c\right) \tag{3.1166}$$

$$K\cos\theta_R = -\left(p_{\rm n}\cos\theta_{\rm n}^c + p_{\rm p}\cos\theta_{\rm p}^c\right) \tag{3.1167}$$

where

$$\cos\theta_{\rm np}^c = \cos\theta_{\rm n}^c\cos\theta_{\rm p}^c + \sin\theta_{\rm n}^c\sin\theta_{\rm p}^c\cos\left(\varphi_{\rm n}^c - \varphi_{\rm p}^c\right) \tag{3.1168}$$

and $\varphi_R$ is symmetrical for the $z$ axis. From the five equations Eqs. (3.1164)–(3.1167) and Eq. (3.1031), $\left(p_{\rm n}, p_{\rm p}, \cos\theta_{\rm n}^{\rm c}, \cos\theta_{\rm p}^{\rm c}, \left(\varphi_{\rm n}^{\rm c} - \varphi_{\rm p}^{\rm c}\right)\right)$ can be solved to be expressed by $(k, K, \cos\theta_r, \cos\theta_R, \varphi_r)$. Squaring Eqs. (3.1158) and (3.1159), respectively, we can get the following two equations:

$$p_{\rm n}^2 = k_{\rm n}^2 + \frac{m_{\rm n}^2}{m_{\rm t}^2}k_{\rm t}^2 - \frac{2m_{\rm n}}{m_{\rm t}}k_{\rm n}k_{\rm t}\cos\theta_{\rm n} \tag{3.1169}$$

$$p_{\rm p}^2 = k_{\rm p}^2 + \frac{m_{\rm p}^2}{m_{\rm t}^2}k_{\rm t}^2 - \frac{2m_{\rm p}}{m_{\rm t}}k_{\rm p}k_{\rm t}\cos\theta_{\rm p} \tag{3.1170}$$

and making point multiplications using $\frac{1}{k_{\rm t}}\vec{k}_{\rm t}$ on Eq. (3.1158) and Eq. (3.1159), the following two equations can also be obtained:

$$p_{\rm n}\cos\theta_{\rm n}^c = k_{\rm n}\cos\theta_{\rm n} - \frac{m_{\rm n}}{m_{\rm t}}k_{\rm t} \tag{3.1171}$$

$$p_{\rm p}\cos\theta_{\rm p}^c = k_{\rm p}\cos\theta_{\rm p} - \frac{m_{\rm p}}{m_{\rm t}}k_{\rm t} \tag{3.1172}$$

Also note that when making the coordinate system transformation between the center of mass system and the laboratory system, $\varphi_R$, $\varphi_{\rm n}$, $\varphi_{\rm p}$ are constants, so by using Eqs. (3.1169)–(3.1172) the expressions for $\left(p_{\rm n}, p_{\rm p}, \cos\theta_{\rm n}^{\rm c}, \cos\theta_{\rm p}^{\rm c}\right)$ can be solved to be expressed by $(k_{\rm n}, k_{\rm p}, \cos\theta_{\rm n}, \cos\theta_{\rm p})$; this solves the problem of substituting $(k, \Omega_R, \Omega_r)$ by $(E_n, \Omega_{\rm n}, \Omega_{\rm p})$ in Eq. (3.1151).

If one wishes to use the method described in this section to the nuclear reaction d + n(p), then it must first be assumed that the n(p) spin is 0, which is a large approximation. It must also be noted that the n(p) double differential cross section generated by the deuteron breakup, and the n(p) double differential cross section as the recoil nucleus, must be added before the n(p) double differential cross section can be compared with the experimental data. Also, all interactions should use nuclear force and can no longer use optical potential.

## 3.16 Axis Symmetric Rotational Nucleus CDCC Theory Describing the Breakup Reaction for Incident Loosely Bound Light Complex Particles

### 3.16.1 Axis-Symmetric Rotational Nucleus CDCC Equation

The spherical nucleus CDCC theory only describes the reaction process of the target nucleus with spin 0. This section will provide the CDCC give the CDCC theory of the axis symmetric rotational nucleus. In this case, the spin of target nucleus can be any integer or semi-odd number. At the same time, the various energy levels of the target nucleus in the same ground state rotational band can also be considered.

Referring to Eq. (3.1026), we can write the total Hamiltonian of the system as

$$H = -\frac{\hbar^2}{2\mu_R}\vec{\nabla}_R^2 + U_{\mathrm{n}}(r_{\mathrm{n}}) + U_{\mathrm{p}}(r_{\mathrm{p}}) + U_{\mathrm{C}}(r_{\mathrm{p}}) + H_{\mathrm{T}} + H_{\mathrm{np}} \tag{3.1173}$$

where $H_{\mathrm{T}}$ is the inner motion Hamiltonian of the target nucleus. Let $\Phi_{I_n M_n}$ be the spin wave function of the target; $I_n M_n$ are the total spin and its z component of the nth energy level of the ground state rotational band of the target; and

$$H_{\mathrm{T}}\Phi_{I_n M_n} = e_n \Phi_{I_n M_n} \tag{3.1174}$$

where $e_n$ is the excitation energy of the $n$th energy level. Total energy conservation in the center of mass system requires the following:

$$E = E_0 + \varepsilon_0 + e_{n_0} = \in_n + \varepsilon_0 + e_n = E_n + \varepsilon_n + e_n \tag{3.1175}$$

$$\in_n = E_0 + e_{n_0} - e_n, \qquad E_n = \frac{\hbar^2 K_n^2}{2\mu_R} \tag{3.1176}$$

$$\varepsilon_n = \frac{\hbar^2 k_n^2}{2\mu_r}, \qquad \varepsilon_{n,\max} = E - e_n = E_0 + \varepsilon_0 + e_{n_0} - e_n = \frac{\hbar^2 k_{n,\max}^2}{2\mu_r} \tag{3.1177}$$

where $\varepsilon_0$ is the deuteron bound state energy. It is assumed that the initial state of the target is at the $n_0$ energy level, and when $n_0 = 0$ it is the ground state. The three terms on the right hand side of the first equal sign of Eq. (3.1175) correspond to the incident channel and the elastic scattering channel. If $n \neq n_0$, the three terms on the right hand side of the second equal sign of Eq. (3.1175) correspond to the inelastic scattering channel, and $\in_n$ is the deuteron kinetic energy after inelastic scattering occurs. The three terms on the right hand side of the third equal sign of Eq. (3.1175) correspond to the breakup channel, allowing $n = n_0$ or $n \neq n_0$, that is, the residual nucleus after the deuteron breaking can be on the ground state or the excited state. $E_n$ is the n−p system center of mass kinetic energy after deuteron breaking.

The angular momentum coupling method given in Eq. (3.1029) is still valid here, but Eq. (3.1030) needs to be changed to

$$\vec{I} = \vec{L} + \vec{j}, \qquad \vec{J} = \vec{I} + \vec{I}_n \tag{3.1178}$$

where $\vec{L}$ represents the orbital angular momentum of deuteron center of mass. We use $M_I$ to represent the $z$ component of the angular momentum operator $\vec{I}$. When elastic or inelastic scattering occurs, the deuteron is still in the bound state, and its wave function can still be written as

$$\psi^0_{l11\mu}\left(\vec{r}\right) = \frac{\phi^0_l(r)}{r} \mathbb{Y}_{l11\mu}(\Omega_r) \tag{3.1179}$$

The deuteron continuous state wave function can be written formally as

$$\psi^{k_n}_{nlSj\mu}\left(\vec{r}\right) = \frac{\phi^{k_n}_{nlSj}(r)}{r} \mathbb{Y}_{lSj\mu}(\Omega_r) \tag{3.1180}$$

where $\phi^{k_n}_{nlSj}(r)$ is the deuteron continuous state radial wave function corresponding to $E_n$ and $\varepsilon_n$. Now introduce the following coupling wave function

$$Z_{lSjLIM_I}(\Omega_R, \Omega_r) = \sum_{M_L \mu} C^{I\ M_I}_{LM_L\ j\mu} \mathrm{i}^L Y_{LM_L}(\Omega_R) \mathbb{Y}_{lSj\mu}(\Omega_r) \tag{3.1181}$$

$$\varphi_{nlSjLIJM}(\Omega_R, \Omega_r) = \sum_{M_I M_n} C^{J\ M}_{IM_I\ I_n M_n} Z_{lSjLIM_I}(\Omega_R, \Omega_r)\ \Phi_{I_n M_n} \tag{3.1182}$$

The *nJM* component of the system total wave function for elastic and inelastic scattering channels is

$$\Psi^0_{nJM}\left(\vec{R}, \vec{r}\right) = \sum_{lLI} \frac{u^J_{0nl11LI}(R)}{R} \frac{\phi^0_l(r)}{r} \varphi_{nl11LIJM}(\Omega_R, \Omega_r) \tag{3.1183}$$

For the breakup channel, there is

$$\Psi^{k_n}_{nJM}\left(\vec{R}, \vec{r}\right) = \sum_{lSjLI} \frac{u^J_{k_n nlSjLI}(R)}{R} \frac{\phi^{k_n}_{nlSj}(r)}{r} \varphi_{nlSjLIJM}(\Omega_R, \Omega_r) \tag{3.1184}$$

It follows that the *JM* component of the system total wave function can be written as

$$\Psi_{JM}\left(\vec{R}, \vec{r}\right) = \sum_{n=0}^{\overline{N}} \left[\Psi^0_{nJM}\left(\vec{R}, \vec{r}\right) + \int_0^{k_{n,\max}} \Psi^{k_n}_{nJM}\left(\vec{R}, \vec{r}\right) \mathrm{d}k_n\right] \tag{3.1185}$$

where $\overline{N}$ is the number of target nucleus excited states to be considered. Next divide $0 - k_{n,\ \max}$ into $N$ equal parts and let

$$\Delta_n = \frac{k_{n,\max}}{N}, \quad k_{ni} = i\Delta_n, \qquad i = 0,1,2,\cdots,N \tag{3.1186}$$

Thus, the continuously changed $k_n$ can be divided into $[0, k_{n1}]$, $[k_{n1}, k_{n2}]$, $\cdots$, $[k_{n\ i-1}, k_{n\ i}]$, $\cdots$, $[k_{n\ N-1}, k_{n\ N}]$ $N$ equal parts. Now introduce the following symbols for the continuous state wave functions

$$\phi^i_{nlSj}(r) = \frac{1}{\sqrt{\Delta_n}} \int_{k_{n\ i-1}}^{k_{ni}} \phi^{k_n}_{nlSj}(r) \mathrm{d}k_n \tag{3.1187}$$

$$u^J_{inlSjLI}(R) = \frac{1}{\sqrt{\Delta_n}} \int_{k_{n\ i-1}}^{k_{ni}} u^J_{k_n nlSjLI}(R) \mathrm{d}k_n \tag{3.1188}$$

such that Eqs. (3.1180) and (3.1184) can be rewritten as

$$\psi^i_{nlSj\mu}\left(\vec{r}\right) = \frac{\phi^i_{nlSj}(r)}{r} \mathcal{Y}_{lSj\mu}(\Omega_r) \tag{3.1189}$$

$$\Psi^i_{nJM}\left(\vec{R}, \vec{r}\right) = \sum_{lSjLI} \frac{u^J_{inlSjLI}(R)}{R} \frac{\phi^i_{nlSj}(r)}{r} \varphi_{nlSjLIJM}(\Omega_R, \Omega_r) \tag{3.1190}$$

Next introduce the following symbols for the deuteron bound state

$$\phi^0_{nlSj}(r) = \begin{cases} \phi^0_l(r) & \text{when } S = j = 1, \ \ l = 0,\ 2; \ \ n \text{ may take any value} \\ 0, & \text{otherwise} \end{cases} \tag{3.1191}$$

Note that the right end of the above formula is independent of n, indicating that the target nucleus does not affect the internal state of the deuteron. Now introduce the symbols

$$u^J_{0nlSjLI}(R) = \begin{cases} u^J_{0nl11LI}(R) & \text{when } S = j = 1, \ \ l = 0,\ 2 \\ 0, & \text{otherwise} \end{cases} \tag{3.1192}$$

such that Eq. (3.1185) can be rewritten as

$$\Psi_{JM}\left(\vec{R}, \vec{r}\right) = \sum_{n=0}^{\overline{N}} \sum_{i=0}^{N} \Psi^i_{nJM}\left(\vec{R}, \vec{r}\right) \tag{3.1193}$$

Next, we redefine

$$\varepsilon_{ni} = \frac{\hbar^2}{4\mu_r}\left(k_{n\ i-1}^2 + k_{n\ i}^2\right), \qquad i = 1, 2, \cdots, N \tag{3.1194}$$

and assume that the radial wave function after average processing by Eqs. (3.1187) and (3.1188) still satisfies the orthogonal normalization condition, that is,

$$\int \phi_{nlSj}^{i*}(r)\phi_{n'l'S'j'}^{i'}(r)\mathrm{d}r = \delta_{inlSj,i'n'l'S'j'} \tag{3.1195}$$

$$\int u_{inlSjLI}^{J*}(R)u_{i'n'l'S'j'L'I'}^{J}(R)\mathrm{d}R = \delta_{inlSjLI,i'n'l'S'j'L'I'} \tag{3.1196}$$

We assume that the target nucleus is an axisymmetric deformational nucleus, so the nuclear potential and Coulomb potential felt by the nucleons around the target nucleus should also be axisymmetric deformational potentials. Letting $R$ be the distance from the deuteron center of mass to the target center of mass, the Coulomb potential can be expressed by $V_c(R)$ independently of the angle after averaging on the $4\pi$ angle. Substituting Eqs. (3.1173), (3.1193), and (3.1190) into Eq. (3.1042), and multiplying the following expression on both sides of the obtained equation from the left, respectively,

$$R\frac{\phi_{nlSj}^{i*}(r)}{r}\varphi_{nlSjLIJM}^{+}(\Omega_R, \Omega_r)$$

and making the integral for $\mathrm{d}\vec{r}\,\mathrm{d}\Omega_R\mathrm{d}\xi$, where $\xi$ represents the internal degree of freedom of the target nucleus, as well as using the orthogonal normalization property of the wave function, we can obtain

$$\begin{aligned}&\left[\frac{\hbar^2}{2\mu_R}\left(\frac{\mathrm{d}^2}{\mathrm{d}R^2} - \frac{L(L+1)}{R^2}\right) - V_{\mathrm{C}}(R) + (E - e_n - \varepsilon_{ni})\right]u_{inlSjLI}^{J}(R)\\ &\quad = \sum_{i'n'l'S'j'L'I'} F_{inlSjLI,i'n'l'S'j'L'I'}^{J}(R)\, u_{i'n'l'S'j'L'I'}^{J}(R)\end{aligned} \tag{3.1197}$$

where

$$\begin{aligned}F_{inlSjLI,i'n'l'S'j'L'I'}^{J}(R) = &\int \frac{\phi_{nlSj}^{i*}(r)}{r}\varphi_{nlSjLIJM}^{+}(\Omega_R, \Omega_r)U\left(\vec{R}', \vec{r}\right)\\ &\frac{\phi_{n'l'S'j'}^{i'}(r)}{r}\varphi_{n'l'S'j'L'I'JM}(\Omega_R, \Omega_r)\mathrm{d}\vec{r}\,\mathrm{d}\Omega_R\mathrm{d}\xi\end{aligned} \tag{3.1198}$$

$$U\left(\vec{R}', \vec{r}\right) = U_{\mathrm{n}}(r_{\mathrm{n}}) + U_{\mathrm{p}}(r_{\mathrm{p}}) + U_{\mathrm{C}}(r_{\mathrm{p}}) - V_{\mathrm{C}}(R) \tag{3.1199}$$

The coordinate positions of the neutron and proton relative to the target center of mass are $\vec{r}_n$ and $\vec{r}_p$, respectively, and let

$$\vec{R}' = \frac{1}{2}\left(\vec{r}_\text{n} + \vec{r}_\text{p}\right), \qquad \vec{r} = \vec{r}_\text{n} - \vec{r}_\text{p} \tag{3.1200}$$

For axisymmetric targets, the axis of symmetry of the target nucleus is selected as the $z'$ axis, and the optical potential is also axisymmetric in the $z'$ coordinate system. If the $z'$ coordinate system is fixed on the target nucleus, then the radius $R'$ in the optical potential can be expanded as follows:

$$R' \equiv R'(\theta') = R\left[1 + \sum_{\Lambda} \beta_\Lambda \mathrm{Y}_{\Lambda 0}(\theta', 0)\right] \tag{3.1201}$$

For axis-symmetric target also with left and right symmetry, only $\beta_2$ and $\beta_4$ two items need to be taken generally. The nucleon optical potential is still in the form given by Eq. (3.1061), so the optical potential given by Eq. (3.1199) can be divided into two terms:

$$U\left(\vec{R}', \vec{r}\right) = U_1\left(\vec{R}', \vec{r}\right) + U_2\left(\vec{R}', \vec{r}\right) \tag{3.1202}$$

where

$$U_1\left(\vec{R}', \vec{r}\right) = U_\text{nA}^\text{CE}(r_n) + U_\text{pA}^\text{CE}(r_\text{p}) + U_\text{C}(r_\text{p}) - V_\text{C}(R) \tag{3.1203}$$

$$U_2\left(\vec{R}', \vec{r}\right) = U_\text{nA}^\text{SO}(r_\text{n})\, \vec{l}_\text{n} \cdot \vec{s}_\text{n} + U_\text{pA}^\text{SO}(r_\text{p})\, \vec{l}_\text{p} \cdot \vec{s}_\text{p} \tag{3.1204}$$

Letting $\theta$ be the angle between $\vec{r}$ and $\vec{R}'$, then we have the following:

$$\begin{aligned} r_\text{n} &= \left|\vec{R}' + \frac{1}{2}\vec{r}\right| = \left(R'^2 + \frac{r^2}{4} + R' r \cos\theta\right)^{1/2}, \\ r_\text{p} &= \left|\vec{R}' - \frac{1}{2}\vec{r}\right| = \left(R'^2 + \frac{r^2}{4} - R' r \cos\theta\right)^{1/2} \end{aligned} \tag{3.1205}$$

where $U_\text{nA}^\text{CE}(r_\text{n})$ can be expanded as follows:

$$\begin{aligned} U_\text{nA}^\text{CE}(r_\text{n}) &= \sum_\lambda (2\lambda + 1)\, U_{\text{nA},\lambda}^\text{CE}(R', r)\, \mathrm{P}_\lambda(\cos\theta) \\ &= 4\pi \sum_{\lambda m_\lambda} U_{\text{nA},\lambda}^\text{CE}(R', r) \mathrm{Y}_{\lambda m_\lambda}(\Omega_R) \mathrm{Y}^*_{\lambda m_\lambda}(\Omega_r) \end{aligned} \tag{3.1206}$$

$$U^{\mathrm{CE}}_{\mathrm{nA},\lambda}(R', r) = \frac{1}{2}\int U^{\mathrm{CE}}_{\mathrm{nA}}(r_{\mathrm{n}})\ \mathrm{P}_{\lambda}(\cos\theta)\mathrm{d}\cos\theta \tag{3.1207}$$

and $U^{\mathrm{CE}}_{\mathrm{pA}}(r_{\mathrm{p}})$ and $U_C(r_{\mathrm{p}})$ can be expanded in the same way. Since the spin-orbit coupling potential is a small item, we also assume, according to convention, that the spin-orbit coupling potential is spherical [7, 76]. Given this, we can obtain the following relations from Eqs. (3.1029) and (3.1178):

$$\vec{J} = \vec{L} + \vec{l} + \vec{S} + \vec{I}_n \tag{3.1208}$$

$$\vec{L}\cdot\vec{S} = \frac{1}{2}\left[\vec{J}^2 - \vec{L}^2 - \vec{l}^2 - \vec{S}^2 - \vec{I}_n^2 - 2\vec{l}\cdot\left(\vec{L}+\vec{S}+\vec{I}_n\right) - 2\left(\vec{L}+\vec{S}\right)\cdot\vec{I}_n\right] \tag{3.1209}$$

Now in the following derivation we ignore the item $\left(\vec{L}+\vec{S}\right)\cdot\vec{I}_n$ that appears in the above formula. According to the results of the previous section, and noting that the spin-orbit coupling potential is the spherical and that the S wave approximation has been used, we make

$$\begin{aligned} U^{\lambda J}_{nlSL,n'l'S'L'}(R', r) &= U^{\mathrm{CE}}_{\mathrm{nA},\lambda}(R', r) + U^{\mathrm{CE}}_{\mathrm{pA},\lambda}(R', r) + U_{\mathrm{C},\lambda}(R', r) \\ &+\frac{1}{2}\{J(J+1) - \frac{1}{2}[L(L+1) + L'(L'+1) + S(S+1) + S'(S'+1) \\ &+I_n(I_n+1) + I'_n(I'_n+1)]\ \}\ U^{\mathrm{SO}}_{\mathrm{nA},\lambda}(R', r)\ \delta_{l0}\delta_{l'0} \end{aligned} \tag{3.1210}$$

with the following expansion

$$U^{\lambda J}_{nlSL,n'l'S'L'}(R'(\theta'), r) = \sqrt{4\pi}\sum_{\Lambda} V^{\lambda\Lambda J}_{nlSL,n'l'S'L'}(R', r)\mathrm{Y}_{\Lambda 0}(\theta', 0) \tag{3.1211}$$

$$U^{\lambda\Lambda J}_{nlSL,n'l'S'L'}(R', r) = \frac{1}{\sqrt{4\pi}}\int U^{\lambda J}_{nlSL,n'l'S'L'}(R'(\theta'), r)\ \mathrm{Y}_{\Lambda 0}(\theta', 0)\ \mathrm{d}\Omega' \tag{3.1212}$$

According to Eq. (1.27), $\mathrm{Y}_{lm}(\theta,\varphi)$ in the z coordinate system and $\mathrm{Y}_{lm'}(\theta',\varphi')$ in the $z'$ coordinate system satisfy the following relation:

$$\mathrm{Y}_{lm'}(\theta',\varphi') = \sum_m \mathrm{Y}_{lm}(\theta,\varphi)D^l_{m,m'}(\alpha,\beta,\gamma) \tag{3.1213}$$

where $\alpha, \beta, \gamma$ are Euler angles, $D^l_{m,m'}$ is a matrix element of the $D$ function, such that

$$D^{l*}_{m,m'} = (-1)^{m-m'}D^l_{-m,-m'} \tag{3.1214}$$

It follows that

$$\begin{aligned} \mathrm{Y}_{\Lambda 0}(\theta', 0) &= \sum_{M_\Lambda} \mathrm{Y}_{\Lambda M_\Lambda}(\Omega_R) D^{\Lambda}_{M_\Lambda,0}(\alpha, \beta, \gamma) \\ &= \sum_{M_\Lambda} (-1)^{M_\Lambda} \mathrm{Y}_{\Lambda\ -M_\Lambda}(\Omega_R) D^{\Lambda *}_{M_\Lambda,0}(\alpha, \beta, \gamma) \end{aligned} \tag{3.1215}$$

Now the normalized eigenfunction of the rotational nucleus is

$$\Phi_{I_n M_n} = \sqrt{\frac{2I_n + 1}{8\pi^2}} D^{I_n *}_{M_n,K} \tag{3.1216}$$

where $M_n$ and $K$ are the projections of the angular momentum $\vec{I}_n$ on $z$ axis and $z'$ axis, respectively, and $K$ is the mark of the nuclear energy level rotational band. Here we take $K = 0$, which indicates that only the ground state rotational band is considered, so the theory given here only applies to even $A$ nuclei. When we only consider the rotation transition dynamics for the internal degree of freedom $\xi$ of the target nucleus, we have

$$\mathrm{d}\xi = \sin\beta \mathrm{d}\beta \mathrm{d}\alpha \mathrm{d}\gamma \tag{3.1217}$$

and

$$\int D^{j_3 *}_{\mu_3, m_3} D^{j_2}_{\mu_2, m_2} D^{j_1}_{\mu_1, m_1} \sin\beta \mathrm{d}\beta \mathrm{d}\alpha \mathrm{d}\gamma = \frac{8\pi^2}{\hat{j}_3^2} C^{j_3 \mu_3}_{j_1 \mu_1\ j_2 \mu_2} C^{j_3 m_3}_{j_1 m_1\ j_2 m_2} \tag{3.1218}$$

It follows that we can obtain

$$\int D^{I_n}_{M_n,0} D^{\Lambda *}_{M_\Lambda,0} D^{I'_n *}_{M'_n,0} \sin\beta \mathrm{d}\beta \mathrm{d}\alpha \mathrm{d}\gamma = \frac{8\pi^2}{\hat{I}_n^2} C^{I_n M_n}_{I'_n M'_n\ \Lambda M_\Lambda} C^{I_n 0}_{I'_n 0\ \Lambda 0} \tag{3.1219}$$

and

$$\int \Phi^*_{I_n M_n} D^{\Lambda *}_{M_\Lambda,0} \Phi_{I'_n M'_n} \mathrm{d}\xi = \frac{\hat{I}'_n}{\hat{I}_n} C^{I_n M_n}_{I'_n M'_n\ \Lambda M_\Lambda} C^{I_n 0}_{I'_n 0\ \Lambda 0} \tag{3.1220}$$

where $C^{I_n 0}_{I'_n 0\ \Lambda 0}$ requires $I_n + I'_n + \Lambda = \text{even}$.

Now, we have the following formulas:

$$\mathrm{Y}_{\lambda m_\lambda} \mathrm{Y}_{\Lambda\ -M_\Lambda} = \sum_{\Gamma M_\Gamma} \frac{\hat{\lambda}\hat{\Lambda}}{\sqrt{4\pi}\hat{\Gamma}} C^{\Gamma M_\Gamma}_{\lambda m_\lambda\ \Lambda\ -M_\Lambda} C^{\Gamma 0}_{\lambda 0\ \Lambda 0} \mathrm{Y}_{\Gamma M_\Gamma} \tag{3.1221}$$

$$\int \mathrm{Y}^*_{L M_L} \mathrm{Y}_{\Gamma M_\Gamma} \mathrm{Y}_{L' M'_L} \mathrm{d}\Omega_R = \frac{\hat{L}'\hat{\Gamma}}{\sqrt{4\pi}\hat{L}} C^{L M_L}_{L' M'_L\ \Gamma M_\Gamma} C^{L 0}_{L' 0\ \Gamma 0} \tag{3.1222}$$

and let

$$B^{\lambda\Lambda J}_{inlSjL,i'n'l'S'j'L'}(R)=4\pi\int_0^\infty \phi^{i*}_{nlSj}(r)\,V^{\lambda\Lambda J}_{nlSL,n'l'S'L'}(R,r)\,\phi^{i'}_{n'l'S'j'}(r)\mathrm{d}r \tag{3.1223}$$

From Eq. (3.1198) we can obtain the following results by using Eqs. (3.1182), (3.1181), (3.1037), (3.1202)–(3.1223), and (3.1071):

$$\begin{aligned}
&F^J_{inlSjLI,i'n'l'S'j'L'I'}(R)=\int\Bigg(\sum_{M_IM_n}C^{J\ M}_{IM_I\ I_nM_n}\sum_{M_L\mu}C^{I\ M_I}_{LM_L\ j\mu}\mathrm{i}^{-L}\mathrm{Y}^*_{LM_L}(\Omega_R)\\
&\sum_{m\nu}C^{j\ \mu}_{lm\ S\nu}\mathrm{i}^{-l}\mathrm{Y}^*_{lm}(\Omega_r)\chi^+_{S\nu}\Phi^*_{I_nM_n}\Bigg)\frac{\phi^{i*}_{nlSj}(r)}{r}(4\pi)^{3/2}\\
&\left(\sum_{\lambda\ m_\lambda}\mathrm{Y}_{\lambda\ m_\lambda}(\Omega_R)\mathrm{Y}^*_{\lambda\ m_\lambda}(\Omega_r)\sum_{\Lambda}V^{\lambda\Lambda J}_{nlSL,n'l'S'L'}(R,\ r)\sum_{M_\Lambda}(-1)^{M_\Lambda}D^{\Lambda^*}_{M_\Lambda,0}\mathrm{Y}_{\Lambda\ -M_\Lambda}(\Omega_R)\right)\\
&\Bigg(\sum_{M_I'M_n'}C^{J\ M}_{I'M_I'\ I_n'M_n'}\sum_{M_L'\mu'}C^{I'\ M_I'}_{L'M_L'\ j'\mu'}\mathrm{i}^{L'}\mathrm{Y}_{L'M_L'}(\Omega_R)\\
&\sum_{\mu'\nu'}C^{j'\ \mu'}_{l'm'\ S'\nu'}\mathrm{i}^{l'}\mathrm{Y}_{l'm'}(\Omega_r)\chi_{S'\nu'}\Phi_{I_n'M_n'}\Bigg)\frac{\phi^{i'}_{n'l'S'j'}(r)}{r}\mathrm{d}\vec{r}\,\mathrm{d}\Omega_R\mathrm{d}\xi\\
&=\delta_{SS'}\sum_{\lambda\Lambda\Gamma}B^{\lambda\Lambda J}_{inlSjL,i'n'l'S'j'L'}(R)\,\mathrm{i}^{L'-L+l'-l}\frac{\hat{l'}\hat{\lambda}^2\hat{\Lambda}\hat{L'}\hat{I'}_n}{\hat{l}\hat{L}\hat{I}_n}C^{l0}_{l'0\ \lambda0}C^{\Gamma0}_{\lambda0\ \Lambda0}C^{L0}_{L'0\ \Gamma0}C^{I_n0}_{I_n'0\ \Lambda0}\\
&\sum_{\substack{M_IM_nM_L\mu\ m\ \nu\ m_\lambda\\ M_\Lambda M_\Gamma M_I'M_n'M_L'\mu'\ m'}}C^{J\ M}_{IM_I\ I_nM_n}C^{I\ M_I}_{LM_L\ j\mu}C^{j\ \mu}_{lm\ S\nu}(-1)C^{I_nM_n}_{I_n'M_n'\ \Lambda M_\Lambda}C^{\Gamma M_\Gamma}_{\lambda m_\lambda\ \Lambda-M_\Lambda}C^{L\ M_L}_{L'M_L'\ \Gamma M_\Gamma}\\
&(-1)C^{l\ m}_{l'm'\ \lambda\ -m_\lambda}C^{J\ M}_{I'M_I'\ I_n'M_n'}C^{I'\ M_I'}_{L'M_L'\ j'\mu'}C^{j'\ \mu'}_{l'm'\ S\nu}-V_C(R)\,\delta_{inlSjLI,i'n'l'S'j'L'I'}
\end{aligned} \tag{3.1224}$$

Next, we have the following $9j$ symbol formula:

$$C^{c\ \gamma}_{a\alpha\ b\beta}\begin{Bmatrix} a & b & c\\ d & e & f\\ g & h & i\end{Bmatrix}=\frac{\hat{c}}{\hat{h}\,\hat{i}^2\hat{g}\hat{f}}\sum_{\varepsilon\eta\varphi\nu\delta\rho}C^{h\ \eta}_{b\beta\ e\varepsilon}C^{i\ \nu}_{c\gamma\ f\varphi}C^{g\ \rho}_{a\alpha\ d\delta}C^{f\ \varphi}_{d\delta\ e\varepsilon}C^{i\ \nu}_{g\rho\ h\eta} \tag{3.1225}$$

thus, by using above formula and $9j$ symbol symmetry, we can obtain

$$\sum_{\substack{M_L\mu M_\lambda \\ M_\Gamma M'_L\mu'}} C^{I\ M_I}_{LM_L\ j\mu} C^{\Gamma M_\Gamma}_{\lambda m_\lambda\ \Lambda\ -M_\Lambda} C^{L\ M_L}_{L'M'_L\ \Gamma M_\Gamma} (-1)^{m_\lambda} C^{I'\ M'_I}_{L'M'_L\ j'\mu'} C^{j\ \mu}_{\lambda\ -m_\lambda\ j'\mu'}$$

$$= (-1)^{-M'_I}(-1)^{I-L+\lambda+\Lambda-\Gamma+L'} \frac{\hat{I}\hat{I}'\hat{j}\hat{\Gamma}\hat{L}}{\hat{\Lambda}} C^{\Lambda\ -M_\Lambda}_{IM_I\ I'\ -M'_I} \begin{Bmatrix} I & I' & \Lambda \\ j & j' & \lambda \\ L & L' & \Gamma \end{Bmatrix} \tag{3.1226}$$

$$= (-1)^{-M'_I}(-1)^{j+j'+I'} \frac{\hat{I}\hat{I}'\hat{j}\hat{\Gamma}\hat{L}}{\hat{\Lambda}} C^{\Lambda\ -M_\Lambda}_{IM_I\ I'-M'_I} \begin{Bmatrix} L & L' & \Gamma \\ j & j' & \lambda \\ I & I' & \Lambda \end{Bmatrix}$$

Using Eq. (3.1076) we can also get

$$\sum_{\substack{M_I M_n M_\Lambda \\ M'_I M'_n}} (-1)^{M_\Lambda - M'_I} C^{J\ M}_{IM_I\ I_nM_n} C^{I_n\ M_n}_{I'_nM'_n\ \Lambda M_\Lambda} C^{J\ M}_{I'M'_I\ I'_nM'_n} C^{\Lambda\ -M_\Lambda}_{IM_I\ I'\ -M'_I}$$

$$= (-1)^{I'_n - I_n + I + \Lambda} \hat{\Lambda}\hat{I}_n W(I\Lambda\, JI'_n; I'I_n) = (-1)^{I'+I+I'_n-J} \hat{\Lambda}\hat{I}_n W(II'I_nI'_n; \Lambda J) \tag{3.1227}$$

From Eq. (3.1208), it can be shown that $J$ and $I_n$ (or $I'_n$) are integer or semi-odd number concurrently, so we have

$$(-1)^{2I'_n - J} = (-1)^{2I'_n - J + 2(J + I'_n)} = (-1)^J$$

It follows that we can obtain the following results substituting Eqs. (3.1223), (3.1226), and (3.1227) into Eq. (3.1224):

$$\begin{aligned} F^J_{inlSjLI,i'n'l'S'j'L'I'}(R) &= \delta_{SS'} \mathrm{i}^{L'-L+l'-l} \sum_{\lambda\Lambda\Gamma} B^{\lambda\Lambda J}_{inlSjL,i'n'l'S'j'L'}(R)(-1)^{j+j'+I+I'_n-J} \\ &\quad \hat{l}'\hat{j}\hat{j}'\hat{I}\hat{I}'\hat{L}'\hat{I}'_n\hat{\lambda}^2\hat{\Lambda}\hat{\Gamma} C^{l\ 0}_{l'0\ \lambda 0} C^{\Gamma\ 0}_{\lambda 0\ \Lambda 0} C^{L\ 0}_{L'0\ \Gamma 0} C^{I_n\ 0}_{I'_n 0\ \Lambda 0} W(\lambda l' jS; lj') \\ &\quad \begin{Bmatrix} L & L' & \Gamma \\ j & j' & \lambda \\ I & I' & \Lambda \end{Bmatrix} W(II'I_nI'_n; \Lambda J) - V_{\mathrm{C}}(R)\delta_{inlSjLI,i'n'l'S'j'L'I'} \\ &= \delta_{SS'} \mathrm{i}^{L'-L+l'-l}(-1)^{S+j+L'+I+J} \hat{l}\hat{l}'\hat{j}\hat{j}'\hat{L}\hat{L}'\hat{I}\hat{I}'\hat{I}_n\hat{I}'_n \sum_{\lambda\Lambda\Gamma} \hat{\lambda} C^{\lambda 0}_{l0\ l'0} C^{\Gamma 0}_{\lambda 0\ \Lambda 0} C^{\Gamma 0}_{L0\ L'0} \\ &\quad C^{\Lambda 0}_{I_n 0\ I'_n 0} W(ll'jj'; \lambda S) \begin{Bmatrix} L & L' & \Gamma \\ j & j' & \lambda \\ I & I' & \Lambda \end{Bmatrix} W(II'I_nI'_n; \Lambda J) B^{\lambda\Lambda J}_{inlSjL,i'n'l'S'j'L'}(R) \\ &\quad - V_{\mathrm{C}}(R)\delta_{inlSjLI,i'n'l'S'j'L'I'} \end{aligned} \tag{3.1228}$$

Due to the deuteron total spin $S = 1$, it can be seen from the above formula that $S = 1$ has been maintained throughout the reaction process.

### 3.16.2 Angular Distribution of Deuteron Elastic and Inelastic Scattering

When $R$ is large enough, the potential given by Eq. (3.1199) on the right side of Eq. (3.1197) tends to 0, and in this case the solutions of Eq. (3.1197) are $F_L(K_{ni}R)$ and $G_L(K_{ni}R)$, where

$$K_{ni} = \frac{\sqrt{2\mu_R(E - e_n - \varepsilon_{ni})}}{\hbar} \tag{3.1229}$$

the speed of the deuteron center of mass is

$$V_{ni} = \frac{\hbar K_{ni}}{\mu_R} \tag{3.1230}$$

and the system incident wave function is

$$\begin{aligned}
\Psi^{(\text{in})}_{\mu\, M_{n_0}}\left(\vec{R}, \vec{r}\right) &= \frac{1}{\sqrt{V_{n_0 0}}} \mathrm{e}^{\mathrm{i}\vec{K}_{n_0 0}\cdot\vec{R}} \psi_\mu\left(\vec{r}\right) \Phi_{I_{n_0} M_{n_0}} \\
&= \sqrt{\frac{4\pi}{V_{n_0 0}}} \sum_L \hat{L} \mathrm{e}^{\mathrm{i}\sigma_L^{(n_0 0)}} \frac{F_L(K_{n_0 0}R)}{K_{n_0 0}R} \mathrm{i}^L Y_{L0}(\theta_R, 0) \left( \sum_l \frac{\phi_l^0(r)}{r} \mathbb{Y}_{l1 1\mu}(\Omega_R) \right) \Phi_{I_{n_0} M_{n_0}} \\
&= \frac{1}{K_{n_0 0}R} \sqrt{\frac{4\pi}{V_{n_0 0}}} \sum_{lLIJ} \hat{L} C^{J\ \mu+M_{n_0}}_{I\ \mu\ \ I_{n_0} M_{n_0}} C^{I\ \mu}_{L0\ 1\mu} \mathrm{e}^{\mathrm{i}\sigma_L^{(n_0 0)}} F_L(K_{n_0 0}R) \frac{\phi_l^0(r)}{r} \varphi_{n_0 l 1 1 I J\ \mu+M_{n_0}} \\
&= \frac{\mathrm{i}\sqrt{\pi}}{K_{n_0 0}R\sqrt{V_{n_0 0}}} \sum_{lLIJ} \hat{L} C^{J\ \mu+M_{n_0}}_{I\ \mu\ \ I_{n_0} M_{n_0}} C^{I\ \mu}_{L0\ 1\mu} \mathrm{e}^{\mathrm{i}\sigma_L^{(n_0 0)}} \\
&\quad \left[ (G_L - \mathrm{i}F_L) - (G_L + \mathrm{i}F_L) \right] \frac{\phi_l^0(r)}{r} \varphi_{n_0 l 1 1 I J\ \mu+M_{n_0}}
\end{aligned} \tag{3.1231}$$

where $n_0$ in $n_0 0$ represents the energy level label of the target nucleus, and 0 in $n_0 0$ is the energy point label of the emitted nucleon. For the n - p elastic scattering with relative motion energy $\varepsilon_{ni}$, the corresponding $S$ matrix element $s_{inlj}$ can be obtained by using the method given by Eqs. (3.1106)–(3.1117). So according to Eq. (3.1231), the total wave function of the system after the reaction can be written as

$$
\begin{aligned}
\Psi_{\mu M_{n_0}}^{(t)} = & \frac{\mathrm{i}\sqrt{\pi}}{K_{n_0 0}R}\sum_{lLIJ}\hat{L}C_{I\ \mu\ I_{n_0}M_{n_0}}^{J\ \mu+M_{n_0}}C_{L0\ 1\mu}^{I\ \mu}\mathrm{e}^{\mathrm{i}\sigma_L^{(n_0 0)}}\left[\frac{1}{\sqrt{V_{n_0 0}}}(G_L-\mathrm{i}F_L)\frac{\phi_l^0(r)}{r}\varphi_{n_0 l11LIJ\ \mu+M_{n_0}}\right. \\
& -\sum_{n'l'L'I'}\frac{1}{\sqrt{V_{n'0}}}S_{0n'l'11L'I',0n_0l11LI}^{J}(G_{L'}+\mathrm{i}F_{L'})\frac{\phi_{l'}^0(r)}{r}\varphi_{n'l'11L'I'J\ \mu+M_{n_0}} \\
& \left.-\sum_{i'=1}^{N}\sum_{n'l'j'L'I'}\frac{1}{\sqrt{V_{n'i'}}}S_{i'n'l'1j'L'I',0n_0l11LI}^{J}s_{i'n'l'j'}(G_{L'}+\mathrm{i}F_{L'})(-\mathrm{n}_{l'}+\mathrm{i}\mathrm{j}_{l'})\varphi_{n'l'1j'L'I'J\ \mu+M_{n_0}}\right]
\end{aligned}
\tag{3.1232}
$$

The first item in the square brackets of the above equation is the spherical wave of the incident channel, the second item is the spherical wave of the elastic and inelastic scattering exit channels, and the third item is the spherical wave of the breakup exit channel. The elastic scattering channel approximation is also used for the n-p particle pair in the breakup channel.

For the elastic and inelastic scattering processes, the boundary conditions can be written, based on Eqs. (3.1183) and (3.1232), as follows:

$$
\begin{aligned}
&\left.\frac{\left(u_{0n'l'11L'I'}^{J}(R)\right)'}{u_{0n'l'11L'I'}^{J}(R)}\right|_{R_m} = \\
&\left.\frac{\left(G_L'(k_{n'0}R)-\mathrm{i}F_L'(k_{n'0}R)\right)\delta_{n'l'L'I',n_0lLI}-S_{0n'l'11L'I',0n_0l11LI}^{J}\left(G_{L'}'(k_{n'0}R)+\mathrm{i}F_{L'}'(k_{n'0}R)\right)}{\left(G_L(k_{n'0}R)-\mathrm{i}F_L(k_{n'0}R)\right)\delta_{n'l'L'I',n_0lLI}-S_{0n'l'11L'I',0n_0l11LI}^{J}\left(G_{L'}(k_{n'0}R)+\mathrm{i}F_{L'}(k_{n'0}R)\right)}\right|_{R_m}
\end{aligned}
\tag{3.1233}
$$

In the above formula, "$'$" represents the derivative for $k_{n'0}R$. $u_{0n'l'11L'I'}^{J}(R)$ is obtained by numerically solving Eq. (3.1197), which contains the influence of the breakup channel on the elastic and inelastic scattering channels. Now we can also rewrite Eq. (3.1232) as

$$
\begin{aligned}
\Psi_{\mu M_{n_0}}^{(t)} = & \frac{\sqrt{4\pi}}{K_{n_0 0}R}\sum_{lLIJ}\hat{L}C_{I\ \mu\ I_{n_0}M_{n_0}}^{J\ \mu+M_{n_0}}C_{L0\ 1\mu}^{I\ \mu}\mathrm{e}^{\mathrm{i}\sigma_L^{(n_0 0)}}\left[\frac{1}{\sqrt{V_{n_0 0}}}F_L\frac{\phi_l^0(r)}{r}\varphi_{n_0 l11LIJ\ \mu+M_{n_0}}\right. \\
& +\sum_{n'l'L'I'}\frac{\mathrm{i}}{2\sqrt{V_{n'0}}}\left(\delta_{n'l'L'I',n_0lLI}-S_{0n'l'11L'I',0n_0l11LI}^{J}\right)(G_{L'}+\mathrm{i}F_{L'})\frac{\phi_{l'}(r)}{r}\varphi_{n'l'11L'I'J\ \mu+M_{n_0}} \\
& \left.-\sum_{i'=1}^{N}\sum_{n'l'j'L'I'}\frac{\mathrm{i}}{2\sqrt{V_{n'i'}}}S_{i'n'l'1j'L'I',0n_0l11LI}^{J}s_{i'n'l'j'}(G_{L'}+\mathrm{i}F_{L'})(-\mathrm{n}_{l'}+\mathrm{i}\mathrm{j}_{l'})\varphi_{n'l'1j'L'I'J\ \mu+M_{n_0}}\right]
\end{aligned}
\tag{3.1234}
$$

From Eqs. (3.1234) and (3.1231), and using Eqs. (3.1121) and (3.1115), the final state total wave function can be obtained as follows:

$$\Psi^{(\mathrm{out})}_{\mu M_{n_0}} = \Psi^{(\mathrm{t})}_{\mu M_{n_0}} - \Psi^{(\mathrm{in})}_{\mu M_{n_0}} = \frac{\mathrm{i}\sqrt{\pi}}{K_{n_0 0}R}\sum_{lLIJ}\hat{L}\, C^{J\ \mu+M_{n_0}}_{I\ \mu\ \ I_{n_0}M_{n_0}} C^{I\ \mu}_{L0\ \ 1\mu}\mathrm{e}^{\mathrm{i}\sigma_L^{(n_0 0)}}$$
$$\left[\sum_{n'l'L'I'}\frac{1}{\sqrt{V_{n'0}}}\left(\delta_{n'l'L'I',\,n_0 lLI} - S^J_{0n'l'11L'I',0n_0 l11LI}\right)(G_{L'}+\mathrm{i}F_{L'})\frac{\phi_{l'}(r)}{r}\varphi_{n'l'11L'I'J\ \ \mu+M_{n_0}}\right.$$
$$\left.-\sum_{i'=1}^{N}\sum_{n'l'j'L'I'}\frac{1}{\sqrt{V_{n'i'}}}S^J_{i'n'l'1j'L'I',0n_0 l11LI}s_{i'n'l'j'}(G_{L'}+\mathrm{i}F_{L'})\left(-\mathrm{n}_{l'}+\mathrm{i}\mathrm{j}_{l'}\right)\varphi_{n'l'1j'L'I'J\ \ \mu+M_{n_0}}\right]$$
$$\rightarrow \frac{\mathrm{i}\sqrt{\pi}}{K_{n_0 0}}\sum_{lLIJ}\hat{L}C^{J\ \mu+M_{n_0}}_{I\mu\ \ I_{n_0}M_{n_0}} C^{I\ \mu}_{L0\ \ 1\mu}\mathrm{e}^{\mathrm{i}\sigma_L^{(n_0 0)}}\left[\sum_{n'l'L'I'}\frac{\mathrm{e}^{\mathrm{i}\sigma_{L'}^{(n'0)}}}{\sqrt{V_{n'0}}}\left(\delta_{n'l'L'I',n_0 lLI} - S^J_{0n'l'11L'I',0n_0 l11LI}\right)\right.$$
$$\frac{\mathrm{e}^{\mathrm{i}K_{n_0 0}R}}{R}\frac{\phi^0_{l'}(r)}{r}\sum_{\mu'}C^{J\ \mu+M_{n_0}}_{I'\ \mu+M_{n_0}-M'_n\ \ I'_n M'_n}C^{I'\ \mu+M_{n_0}-M'_n}_{L'\ \mu-\mu'+M_{n_0}-M'_n\ \ 1\mu'}$$
$$\mathrm{Y}_{L'\ \mu-\mu'+M_{n_0}-M'_n}(\Omega_R)\mathbb{Y}_{l'11\mu'}(\Omega_r)\Phi_{I'_n M'_n}$$
$$-\sum_{i'=1}^{N}\sum_{n'l'j'L'I'}\frac{\mathrm{e}^{\mathrm{i}\sigma_{L'}^{(n'i')}}}{\sqrt{V_{n'i'}}k_{n'i'}}S^J_{i'n'l'1j'L'I',0n_0 l11LI}s_{i'n'l'j'}\frac{\mathrm{e}^{\mathrm{i}K_{n'i'}R}}{R}\frac{\mathrm{e}^{\mathrm{i}k_{n'i'}r}}{r}\sum_{\mu'\nu'}C^{J\ \mu+M_{n_0}}_{I'\ \mu+M_{n_0}-M'_n\ \ I'_n M'_n}$$
$$\left.C^{I'\ \mu+M_{n_0}-M'_n}_{L'\ \mu-\mu'+M_{n_0}-M'_n\ \ j'\mu'}C^{j'\ \mu'}_{l'\ \mu'-\nu'\ \ 1\nu'}\mathrm{Y}_{L'\ \mu-\mu'+M_{n_0}-M'_n}(\Omega_R)\mathrm{Y}_{l'\ \mu'-\nu'}(\Omega_r)\chi_{1\nu'}\Phi_{I'_n M'_n}\right] \tag{3.1235}$$

Referring to Eq. (3.1114) we can write the Coulomb scattering outgoing wave as

$$\Psi^{(\mathrm{C})}_{\mu M_{n_0}} \rightarrow \frac{1}{\sqrt{V_{n_0 0}}R}f_{\mathrm{C}}(\theta_R)\,\mathrm{e}^{\mathrm{i}K_{n_0 0}R}\psi_\mu\left(\vec{r}\right)\,\Phi_{I_{n_0}M_{n_0}} \tag{3.1236}$$

where from the above two formulas the final state wave functions of the deuteron elastic and inelastic scattering can be written as

$$\Psi^{(\mathrm{r})}_{\mu M_{n_0}} = \frac{1}{\sqrt{V_{n_0 0}}}f_{\mathrm{C}}(\theta_R)\left(\sum_l\frac{\phi^0_l(r)}{r}\mathbb{Y}_{l11\mu}(\Omega_r)\right)\Phi_{I_{n_0}M_{n_0}}\frac{\mathrm{e}^{\mathrm{i}K_{n_0 0}R}}{R}$$
$$+\frac{\mathrm{i}\sqrt{\pi}}{K_{n_0 0}}\sum_{lLIJ}\hat{L}C^{J\ \mu+M_{n_0}}_{I\ \mu\ \ I_{n_0}M_{n_0}}C^{I\ \mu}_{L0\ \ 1\mu}\mathrm{e}^{\mathrm{i}\sigma_L^{(n_0 0)}}\sum_{n'l'L'I'}\frac{\mathrm{e}^{\mathrm{i}\sigma_{L'}^{(n'0)}}}{\sqrt{V_{n'0}}}$$
$$\left(\delta_{n'l'L'I',n_0 lLI} - S^J_{0n'l'11L'I',0n_0 l11LI}\right)\sum_{\mu'}C^{J\ \mu+M_{n_0}}_{I'\ \mu+M_{n_0}-M'_n\ \ I'_n M'_n}C^{I'\ \mu+M_{n_0}-M'_n}_{L'\ \mu-\mu'+M_{n_0}-M'_n\ \ 1\mu'}$$
$$\mathrm{Y}_{L'\ \mu-\mu'+M_{n_0}-M'_n}(\Omega_R)\frac{\phi^0_{l'}(r)}{r}\mathbb{Y}_{l'11\mu'}(\Omega_r)\,\Phi_{I'_n M'_n}\frac{\mathrm{e}^{\mathrm{i}K_{n'0}R}}{R} \tag{3.1237}$$

From this, the deuteron elastic and inelastic scattering amplitude can be written as

$$f_{n'I'\mu'M'_n,n_0I\mu M_{n_0}}(\Omega_R)=f_C(\theta_R)\,\delta_{n'I'\mu'M'_n,n_0I\mu M_{n_0}}+\frac{\mathrm{i}\sqrt{\pi}}{K_{n_00}}\sum_{LIJ}\hat{L}$$
$$C^{J\ \mu+M_{n_0}}_{I\mu\ \ I_{n_0}M_{n_0}}C^{I\ \mu}_{L0\ 1\mu}\mathrm{e}^{\mathrm{i}\sigma_L^{(n_00)}}\sum_{L'I'}\mathrm{e}^{\mathrm{i}\sigma_{L'}^{(n'0)}}\left(\delta_{n'I'L'I',n_0ILI}-S^J_{0n'I'11L'I',0n_0I11LI}\right) \tag{3.1238}$$
$$C^{J\ \mu+M_{n_0}}_{I'\ \mu+M_{n_0}-M'_n\ I'_nM'_n}C^{I'\ \mu+M_{n_0}-M'_n}_{L'\ \mu-\mu'+M_{n_0}-M'_n\ 1\mu'}Y_{L'\ \mu-\mu'+M_{n_0}-M'_n}(\Omega_R)$$

Then using the above elastic and inelastic scattering amplitude, the corresponding polarization quantities of the deuteron can be studied. In the center of mass system, the total elastic and inelastic scattering angular distributions of the deuterons are as follows:

$$\frac{\mathrm{d}\sigma_{n'}}{\mathrm{d}\sigma_R}=\frac{1}{3(2I_{n_0}+1)}\sum_{\substack{I\mu\ M_{n_0}\\ I'\mu'M'_n}}\left|f_{n'I'\mu'M'_n,n_0I\mu\ M_{n_0}}(\Omega_R)\right|^2 \tag{3.1239}$$

If $n'=n_0$, then it is an elastic scattering angular distribution, and if $n'\neq n_0$, then it is an inelastic scattering angular distribution. By carrying out the coordinate transformation, the angular distributions of the deuteron elastic and inelastic scattering in the laboratory system can also be obtained by Eq. (3.1239) [7]. The angular distributions of the elastic and inelastic scattering of the deuterons obtained here contain the influence of the deuteron breakup channel, and the angular distributions of the elastic and inelastic scattering of the deuterons obtained by the usual axis-symmetric rotational nucleus coupling channel optical model [7, 76] do not take into account the effects of the deuteron breakup.

### 3.16.3 Double Differential Cross Section of the Ejected Nucleons After Deuteron Breaking

For the breakup channel ($1\le i\le N$), in the case of the introduction of the elastic scattering channel approximation, the boundary conditions can be written, according to Eqs. (3.1190) and (3.1232), as follows:

$$\left.\frac{\left(u^J_{i'n'I'1j'L'I'}(R)\right)'}{u^J_{i'n'I'1j'L'I'}(R)}\right|_{R_m}=$$
$$\left.\frac{\left(G'_L(k_{n'i'}R)-\mathrm{i}F'_L(k_{n'i'}R)\right)\delta_{i'n'I'j'L'I',0n_0I1LI}-S^J_{i'n'I'j'L'I',0n_0I1LI}\left(G'_{L'}(k_{n'i'}R)+\mathrm{i}F'_{L'}(k_{n'i'}R)\right)}{\left(G_L(k_{n'i'}R)-\mathrm{i}F_L(k_{n'i'}R)\right)\delta_{i'n'I'j'L'I',0n_0I1LI}-S^J_{i'n'I'j'L'I',0n_0I1LI}\left(G_{L'}(k_{n'i'}R)+\mathrm{i}F_{L'}(k_{n'i'}R)\right)}\right|_{R_m} \tag{3.1240}$$

In this formula, "′" represents the derivative for $K_{n'i'}R$. Moreover, the elastic and inelastic scattering boundary condition Eq. (3.1233) is only a special case of this formula. From Eq. (3.1235), the reaction amplitude of the breakup channel can be written as

$$\begin{aligned} f_{n'\nu'M_n',n_0\mu\,M_{n_0}}(\Omega_R,\Omega_r) = & -\frac{\mathrm{i}\sqrt{\pi}}{K_{n_00}}\sum_{lLIJ}\hat{L}C_{I\mu\ I_{n_0}M_{n_0}}^{J\ \mu+M_{n_0}}C_{L0\ 1\mu}^{I\ \mu}\mathrm{e}^{\mathrm{i}\sigma_L^{(n_00)}} \\ & \sum_{i'=1}^{N}\sum_{l'j'L'I'}\frac{\mathrm{e}^{\mathrm{i}\sigma_{L'}^{(n'i')}}}{k_{n'i'}}S_{i'n'l'1L'I',0n_0l11LI}^{J}s_{i'n'l'j'}\sum_{\mu'}C_{I'\ \mu+M_{n_0}-M_n'\ I_n'M_n'}^{J\ \mu+M_{n_0}} \\ & C_{L'\ \mu-\mu'+M_{n_0}-M_n'\ j'\mu'}^{I'\ \mu+M_{n_0}-M_n'}C_{l'\ \mu'-\nu'\ 1\nu'}^{j'\mu'}Y_{L'\ \mu-\mu'+M_{n_0}-M_n'}(\Omega_R)Y_{l'\ \mu'-\nu'}(\Omega_r) \end{aligned} \tag{3.1241}$$

Noting $\varepsilon_{ni}$ defined by Eq. (3.1194), if $k_{n\ i-1}\le k\le k_{ni}$, then the amount related to $k$ is obtained by interpolation in the interval $[k_{n\ i-1},k_{ni}]$. We use $c=0$ to represent the bound state, $c\neq 0$ to represent the continuous states, and stipulate that

$$\begin{aligned} S_{cn'l'1j'L'I',0n_0l11LI}^{J}(0) &= S_{cn'l'1j'L'I',0n_0l11LI}^{J}(k_{n',\max}) = s_{cn'l'j'}(0) \\ &= s_{cn'l'j'}(k_{n',\max}) = 0 \end{aligned} \tag{3.1242}$$

So according to Eq. (3.1241), the five-dimensional reaction amplitude of the breakup channel can be written as

$$\begin{aligned} f_{n'\nu'M_n',n_0\mu\,M_{n_0}}(k_{n'},\Omega_R,\Omega_r) = & -\frac{\mathrm{i}\sqrt{\pi}}{K_{n_00}}\sum_{lLIJ}\hat{L}C_{I\mu\ I_{n_0}M_{n_0}}^{J\ \mu+M_{n_0}}C_{L0\ 1\mu}^{I\ \mu}\mathrm{e}^{\mathrm{i}\sigma_L^{(n_00)}} \\ & \sum_{l'j'L'I'\mu'}C_{I'\ \mu+M_{n_0}-M_n'\ I_n'M_n'}^{J\ \mu+M_{n_0}}C_{L'\ \mu-\mu'+M_{n_0}-M_n'\ j'\mu'}^{I'\ \mu+M_{n_0}-M_n'}C_{l'\mu'-\nu'\ 1\nu'}^{j'\mu'}\frac{\mathrm{e}^{\mathrm{i}\sigma_{L'}^{(n'i')}}}{k_{n'}\Delta_{n'}} \\ & S_{cn'l'1j'L'I',0n_0l11LI}^{J}(k_{n'})s_{cn'l'j'}(k_{n'})Y_{L'\ \mu-\mu'+M_{n_0}-M_n'}(\Omega_R)Y_{l'\ \mu'-\nu'}(\Omega_r) \end{aligned} \tag{3.1243}$$

and we can write the five-differential cross sections in the center of mass system as follows:

$$\frac{\mathrm{d}^5\sigma_{n'}}{\mathrm{d}k_{n'}\mathrm{d}\Omega_R\mathrm{d}\Omega_r} = \frac{1}{3(2I_{n_0}+1)}\sum_{\mu\,M_{n_0}\nu'M_n'}\left|f_{n'\nu'M_n',n_0\mu\,M_{n_0}}(k_{n'},\Omega_R,\Omega_r)\right|^2 \tag{3.1244}$$

Noting Eq. (3.1124), the double differential cross section of the deuteron center of mass in c.m. system can be obtained as

$$\frac{\mathrm{d}^2\sigma_{n'}}{\mathrm{d}k_{n'}\mathrm{d}\Omega_R} = \frac{1}{3(2I_{n_0}+1)}\sum_{\substack{\mu\,M_{n_0}\nu' \\ M_n'l'\mu'}}\left|f_{l'\mu',n'\nu'M_n',n_0\mu M_{n_0}}(k_{n'},\Omega_R)\right|^2 \tag{3.1245}$$

where

$$
\begin{aligned}
f_{l'\mu',n'\nu'M'_n,n_0\mu\,M_{n_0}}(k_{n'},\Omega_R) = & -\frac{\mathrm{i}\sqrt{\pi}}{K_{n_0 0}}\sum_{lLIJ}\hat{L}C^{J\ \mu+M_{n_0}}_{I\mu\ \ I_{n_0}M_{n_0}}C^{I\ \mu}_{L0\ 1\mu}\mathrm{e}^{\mathrm{i}\sigma_L^{(n_0 0)}} \\
& \sum_{j'L'I'}C^{J\ \mu+M_{n_0}}_{I'\ \mu+M_{n_0}-M'_n\ \ I'_n M'_n}C^{I'\ \mu+M_{n_0}-M'_n}_{L'\ \mu-\mu'+M_{n_0}-M'_n\ \ j'\mu'}C^{j'\mu'}_{l'\ \mu'-\nu'\ \ 1\nu'}\frac{\mathrm{e}^{\mathrm{i}\sigma_{L'}^{(n'l')}}}{k_{n'}\Delta_{n'}} \\
& S^J_{cn'l'1j'L'I',0n_0l11LI}(k_{n'})s_{cn'l'j'}(k_{n'})\mathrm{Y}_{L'\ \mu-\mu'+M_{n_0}-M'_n}(\Omega_R)
\end{aligned}
\tag{3.1246}
$$

On the question regarding how to transform the five-differential cross section given by Eq. (3.1244) into the laboratory system, Eqs. (3.1127) and (3.1128) can still be applied, but Eq. (3.1129) needs to be changed to

$$
E_{\mathrm{in}} = \frac{m_{\mathrm{d}} + M_{\mathrm{A}}}{M_{\mathrm{A}}}E_0 + \varepsilon_0 + e_{n_0} = E_L + \varepsilon_0 + e_{n_0}
\tag{3.1247}
$$

Equations (3.1130) and (3.1131) can still be applied here, too, while Eq. (3.1132) needs to be changed to

$$
E_{\mathrm{in}} = \frac{\hbar^2 k_{\mathrm{t}}^2}{2m_{\mathrm{d}}} + \varepsilon_0 + e_{n_0}
\tag{3.1248}
$$

By using the energy conservation condition given by Eq. (3.1175), Eq. (3.1244) can be rewritten as

$$
\begin{aligned}
\mathrm{d}^5\sigma_{n'} = & \frac{1}{3(2I_{n_0}+1)}\left(\sum_{\mu\ M_{n_0}\nu' M'_n}\left|f_{n'\nu'M'_n,n_0\mu M_{n_0}}(k_{n'},\Omega_R,\Omega_r)\right|^2\right) \\
& \int_{\delta}\delta(E_0+\varepsilon_0+e_{n_0}-E_{n'}-\varepsilon_{n'}-e_{n'})\,\mathrm{d}\vec{k}_{n'}\mathrm{d}\vec{K}_{n'}
\end{aligned}
\tag{3.1249}
$$

Next, we can write the following expressions:

$$
\begin{aligned}
\mathrm{d}W \equiv & \int_{\delta}\delta(E_0+\varepsilon_0+e_{n_0}-E_{n'}-\varepsilon_{n'}-e_{n'})\mathrm{d}\vec{k}_{n'}\mathrm{d}\vec{K}_{n'} \\
= & \int_{\delta}\delta\left(\vec{k}_{\mathrm{t}}-\vec{k}_{\mathrm{n}}-\vec{k}_{\mathrm{p}}-\vec{k}_{\mathrm{A}}\right)\delta(E_{\mathrm{in}}-E_{\mathrm{t}})\mathrm{d}\vec{k}_{\mathrm{n}}\mathrm{d}\vec{k}_{\mathrm{p}}\mathrm{d}\vec{k}_{\mathrm{A}}
\end{aligned}
\tag{3.1250}
$$

Now note that $k_{n'}$ on the right hand side of the first equation of Eq. (3.1250) is the n-p relative motion wave number in the center of mass system, $n'$ represents the target nucleus energy level label, $k_{\mathrm{n}}$ on the right hand side of the second equation of Eq. (3.1250) is the neutron wave number in the laboratory system, and n represents neutrons. The five-differential cross sections in the laboratory system can also be found as follows based on the derivation given by Eqs. (3.1137)–(3.1150):

$$\frac{d^5\sigma_{n'}}{dE_n d\sigma_n d\sigma_p} = \frac{1}{3(2I_{n_0}+1)}\left(\sum_{\mu\, M_{n_0} \nu' M'_n} \left|f_{n'\nu' M'_n, n_0\mu\, M_{n_0}}(k_{n'}, \Omega_R, \Omega_r)\right|^2\right)\rho(E_n, \Omega_n, \Omega_p) \tag{3.1251}$$

where $\rho(E_n, \Omega_n, \Omega_p)$ in above formula has been given by Eq. (3.1150), and $n'$ in the same expression indicates that the residual nucleus is in the $n'$ state after the deuteron breaking. For the $n'$ state, using a similar expression with Eq. (3.1152), a double differential cross section of the emitted neutrons can be obtained. The proton emission can also be studied in a similar way. The method introduced by Eqs. (3.1153)–(3.1172) can solve the problem of conversion from $(k_{n'}, \Omega_R, \Omega_r)$ to $(E_n, \Omega_n, \Omega_p)$ in Eq. (3.1251).

The CDCC theory of axisymmetric rotational nucleus described in this section allows for the use of the true spin of the target nucleus, and the excited states of the target nucleus have also be considered.

## References

1. Varshalovich DA, Moskalev AN, Khersonskii VK. Quantum theory of angular momentum. Singapore/New Jersey/Hong Kong: World Scientific; 1988.
2. Ohlsen GG. Polarization transfer and spin correlation experiments in nuclear physics. Rep Prog Phys. 1972;35:717.
3. Ohlsen G G, Gammel J L, and Keaton P W. Description $^4$He(d,d)$^4$He polarization-transfer experiments. Phys Rev., 1972, C5:1205
4. Lakin W. Spin polarization of the deuteron. Phys Rev. 1955;98:139.
5. Seiler F, Darden SE, Mcintyre LC, Weitkamp WG. Tensor polarization of deuterons scattered from He$^4$ between 4 and 7.5 MeV. Nucl Phys. 1964;53:65.
6. Zhang JS, Liu KF, Shuy GW. Neutron suppression in polarized *dd* fusion reaction. Phys Rev. 1999;C60:054614.
7. Shen QB. Nuclear reaction theory of low and medium energy (upper volume). Beijing: Science Press; 2005. (in Chinese)
8. Schwandt P, Haeberli W. Elastic scattering of polarized deuterons from $^{27}$Al, Si and $^{60}$Ni between 7 and 11 MeV. Nucl Phys. 1968;A110:585.
9. Cords H, Din GU, Ivanovich M, Robson BA. Tensor polarization of deuterons from $^{12}$C-d elastic scattering. Nucl Phys. 1968;A113:608.
10. Schwandt P, Haeberli W. Optical-model analysis of d-Ca polarization and cross-section measurements from 5 to 34 MeV. Nucl Phys. 1969;A123:401.
11. Gruebler W, Konig V, Schmelzbach PA, Mamier P. Elastic scatteriong of vector polarized deutrons on $^4$He. Nucl Phys. 1969;A134:686.
12. Djaloeis A, Nurzynski J. Tensor polarization of deuterons from the Mg(d,d)Mg elastic scattering at 7.0 MeV. Nucl. Phys. 1971;A163:113.
13. Djaloeis A, Nurzynski J. Tensor polarization and differential cross sections for the elastic scattering of deuterons by Si at low energies. Nucl. Phys. 1972;A181:280.

14. Irshad M, Robson BA. Elastic scattering of 15 MeV deuterons. Nucl Phys. 1974;A218:504.
15. Goddard RP, Haeberli W. The optical model for elastic scattering of 10 to 15 MeV polarized deuterons from medium-weight nuclei. Nucl Phys. 1979;A316:116.
16. Matsuoka N, Sakai H, Saito T, et al. Optical model and folding model potential for elastic scattering of 56 MeV polarized deuterons. Nucl Phys, 1986, A455:413
17. Takei M, Aoki Y, Tagishi Y, Yagi K. Tensor interaction in elastic scattering of polarized deutrons from medium-weight nuclei near $E_d$=22 MeV. Nucl Phys. 1987;A472:41.
18. Iseri Y, Kameyama H, Kamimura M, Yahiro M, Tanifuji M. Virtual breakup effects in elastic scattering of polarized deutrons. Nucl Phys. 1988;A490:383.
19. Extermann P. Deuteron polarization in p-d elastic scattering. Nucl Phys. 1967;A95:615.
20. Van Sen N, Arvieux J, Yanlin Y, et al. Elastic scattering of polarized deuterons from $^{40}$Ca and $^{58}$Ni at tntermediate energies. Phys Lett. 1985;156B:185.
21. Al-Khalili JS, Tostevin JA, Johnson RC. Effects of singlet breakup on deuteron elastic scattering at intermediate energies. Phys Rev. 1990;C41:R806.
22. Shen QB. Nuclear reaction theory of low and medium energy (middle volume). Beijing: Science Press; 2012. (in Chinese)
23. Liljestrand R P, Cameron J M, Hutcheon D A, et al. Analyzing power measurements for the $^{13}$C ($\vec{p}$,d)$^{12}$C reaction. Phys Lett, 1981, 99B:311
24. Rost E, Shepard J R, and Murdock D. Dirac wave function in nuclear distorted-wave calculations. Phys Rev Lett, 1982, 49:448
25. Shepard JR, Rost E, Kunz PD. Failure of the distorted-wave Born approximation in analysis of the $^{24}$Mg($\vec{p}$,d)$^{23}$Mg reaction at $T_p$=94 MeV. Phys Rev. 1982; C25:1127.
26. Miller DW, Jacobs WW, Devins DW, Jones WP. $^{24}$Mg($\vec{p}$,d) analyzing- power measurements at 95 MeV. Phys Rev. 1982;C26:1793.
27. Kraushaar JJ, Shepard JR, Miller DW, et al. The $^{13}$C($\vec{p}$,d)$^{12}$C and $^{208}$Pb($\vec{p}$,d)$^{207}$Pb reactions at 123 MeV. Nucl Phys. 1983;A394:118.
28. Hatanaka K, Fujiwara M, Hosono K, et al. $^{24}$Mg($\vec{p}$,d)$^{23}$Mg reaction at 65 and 80 MeV. Phys Rev. 1984;C29:13.
29. Ohnuma H. Cross sections and analyzing powers in the (p,d) reaction around 500 MeV. Phys Lett. 1984;147B:253.
30. Sagara K, Oguri H, Shimizu S, et al. Energy dependence of analyzing power $A_y$ and cross section for p+d scattering below 18 Mev. Phys Rev. 1994;C50:576.
31. Neidel EM, Tornow W, Gonzalez DE, et al. A new twist to the long-standing three-nucleon analyzing power puzzle. Phys Lett. 2003;B552:29.
32. Weisel GJ, Tornow W, Crowe BJ, et al. Neutron-deuteron analyzing power data at 19.0 Mev. Phys. Rev. 2010;C81:024003.
33. Sperisen F, Gruebler W, Konig V, et al. Comparison of a nearly complete pd elastic scattering data set with Faddeev calculations. Nucl Phys. 1984;A422:81.
34. Shimizu S, Sagara K, Nakamura H, et al. Analyzing power of p+d scattering below the deuteron breakup threshold. Phys Rev. 1995;C52:1193.
35. Carlson J, Schiavilla R. Structure and dynamics of few-nucleon systems. Rev Mod Phys. 1998;70(743):821.
36. Kulsrud RM, Furth HP, Valeo EJ, Goldhaber M. Fusion reactor plasmas with polarized nuclei. Phys Rev Lett. 1982;49:1248.
37. More RM. Nuclear spin-polarized fuel in inertial fusion. Phys Rev Lett. 1983;51:396.
38. Fletcher KA, Ayer Z, Black TC, et al. Tensor analyzing power for $^2$H(d,p)$^3$H and $^2$H(d,n)$^3$He at deuteron energies of 25,40, and 80 keV. Phys Rev. 1994;C49:2305.
39. Schieck H, Paetz gen. The status of "polarized fusion". Eur Phys J. 2010;A44:321.
40. Schieck H, Paetz gen. "Polarized fusion": new aspects of an old project. Few-Body Syst. 2013;54:2159.
41. Kievsky A, Viviani M, Rosati S. Cross section, polarization observables, and phase-shift parameters in p-d and n-d elastic scattering. Phys Rev. 1995;C52:R15.

42. Kievsky A, Rosati S, Viviani M. Proton-deuteron elastic scattering above the deuteron breakup. Phys Rev Lett. 1999;82:3759.
43. Kievsky A, Viviani M, Rosati S. Polarization observables in p-d scattering below 30 MeV. Phys Rev. 2001;C64:024002.
44. Kievsky A. Polarization observables in p-d and p-$^{3}$He scattering. Nucl Phys. 2001;A689:361c.
45. Deltuva A, Fonseca AC, Sauer PU. Momentum- space treatment of the Coulomb interaction in three- nucleon reactions with two protons. Phys Rev. 2005;C71:054005.
46. Satchler GR. Spin-orbit coupling for spin-1 particles. Nucl Phys. 1960;21:116.
47. An HX, Cai CH. Global deuteron optical model potential for the energy range up to 183 MeV. Phys Rev. 2006;C73:054605.
48. Han YL, Shi YY, Shen QB. Deuteron global optical model potential for energies up to 200 MeV. Phys Rev. 2006;C74:044615.
49. Daehnick WW, Childs JD, Vrcelj Z. Global optical model potential for elastic deuteron scattering from 12 to 90 Mev. Phys Rev. 1980;C21:2253.
50. Roche R, Sen NV, Perrin G, el al. Evidence for the deuteron-nucleus tensor interaction. Nucl. Phys. 1974:A220:381.
51. Ioannides AA, Johnson RC. Propagation of a deuteron in nuclear matter and the spin dependence of the deuteron optical potential. Phys Rev. 1978;C17:1331.
52. Stamp AP. The strength of the tensor interaction in the deuteron optical potential. Nucl Phys. 1970;A159:399.
53. Keaton PW, Armstrong DD. Deuteron optical potential with a tentor term and breakup. Phys Rev. 1973;C8:1692.
54. Sachs RG. Nuclear theory, vol. Ch.3. Cambridge: Addison-Wesley Publishing Company; 1953.
55. Johnson RC, Soper PJR. Contribution of deuteron breakup channels to deuteron stripping and elastic scattering. Phys Rev. 1970;C1:976.
56. Rawitscher GH. Effect of deuteron breakup on elastic deuteron-nucleus scattering. Phys Rev. 1974;C9:2210.
57. Yahiro M, Nakano M, Iseri Y, Kamimura M. Coupled-discretized-continuum- channels method for deuteron breakup reactions based on three-body model. Prog Theor Phys. 1982;67:1467.
58. Amakawa H, Austern N, Vincent M. Quasiadiabatic three-body dynamics of deuteron stripping and breakup reactions. Phys Rev. 1984;C29:699.
59. Mukherjee SN, Pandey LN, Srivastava DK, Ganguly NK. Deuteron breakup in the field of a heavy target. Phys. Rev. 1984;C29:1095.
60. Yahiro M, Iseri Y, Kameyama H, et al. Effects of deuteron virtual breakup on deuteron elastic and inelastic scattering. Prog Theor Phys Suppl. 1986;89:32.
61. Austern N, Iseri Y, Kamimura M, et al. Continuum-discretized coupled-channels calculations for three-body models of deuteron-nucleus reactions. Phys Rep. 1987;154:125.
62. Piyadasa RAD, Kawai M, Kamimura M, Yahiro M. Convergence of the solution of the continuum discretized coupled channels method. Phys Rev. 1999;C60:044611.
63. Ye T, Watanabe Y, Ogata K, Chiba S. Analysis of deuteron elastic scattering from $^{6,7}$Li using the continuum discretized coupled channels method. Phys Rev. 2008;C78:024611.
64. Ye T, Watanabe Y, Ogata K. Analysis of deuteron breakup reactions on $^{7}$Li for energies up to 100 MeV. Phys Rev. 2009;C80:014604.
65. Ye T, Hashimoto S, Watanabe Y, et al. Analysis of inclusive (d, xp) reactions on nuclei from $^{9}$Be to $^{238}$U at 100 MeV. Phys Rev. 2011;C84:054606.
66. Moro AM, Rusek K, Arias JM, et al. Improved di-neutron cluster model for $^{6}$He scattering. Phys. Rev. 2007;C75:064607.
67. Matsumoto T, Hiyama E, Ogata K, et al. Continuum-discretized coupled-channels method for four-body nuclear breakup in $^{6}$He+$^{12}$C scattering. Phys Rev. 2004;C70:061601(R).
68. Rodriguez-Gallardo M, Arias JM, Gomez-Camacho J, et al. Four-body continuum-discretized coupled-channels calculations using a transformed harmonic oscillator basis. Phys Rev. 2008; C77:064609.

69. Rodriguez-Gallardo M, Arias JM, Gomez-Camacho J, et al. Four-body continuum-discretized coupled-channels calculations. Phys Rev. 2009;C80:051601(R).
70. Otomar DR, Lubian J, Gomes PRS, Correa T. Breakup following neutron transfer for the $^{7}$Li +$^{144}$Sm system. J Phys G Nucl Part Phys. 2013;40:125105.
71. Guo HR, Watanabe Y, Matsumoto T, et al. Systematic analysis of nucleon scattering from $^{6,7}$Li with the continuum discretized coupled channels method. Phys Rev. 2013;C87:024610.
72. Guo HR, Nagaoka K, Watanabe Y, et al. Application of the continuum discretized coupled channels method to nucleon-induced reactions on $^{6,7}$Li for energies up to 150 MeV. Nucl Data Sheets. 2014;118:254.
73. Clarke NM. Improved numerical integration for coupled-reaction-channels calculations. J Phys G Nucl Phys. 1984;10:1535.
74. Iseri Y, Yahiro M, Kamimura M. Coupled-channels approach to deuteron and $^{3}$He breakup reactions. Prog Theor Phys Suppl. 1986;89:84.
75. Baur G, Fosel F, Trautmann D, et al. Fragmentation processes in nuclear reactions. Phys Rep. 1984;111:333.
76. Tamura T. Analyses of the scattering of nuclear particles by collective nuclei in terms of the coupled-channel calculation. Rev Mod Phys. 1965;37:679.

# Chapter 4
# Polarization Theory of Nuclear Reactions for Spin $\frac{3}{2}$ Particles and Polarization Theory for Photon Beams

**Abstract** This chapter will first give the polarization theory of nuclear reactions for spin $\frac{3}{2}$ particles in the spherical basis coordinate system. According to the regularity of the unitary transformation relations of the vector and the rank-2 tensor between the spherical basis coordinate system and the orthogonal Cartesian coordinate system, the unitary transformation relations of the rank-3 and higher-rank tensors between the spherical basis coordinate system and the orthogonal Cartesian coordinate system will be found by recursion method. And the generalized Pauli matrix will be introduced, thus the theoretical method for studying spin $\geq \frac{3}{2}$ particle polarization phenomena can be given. Finally, the polarization theory for photon beams will be discussed.

**Keywords** Classical electromagnetic field theory · Generalized Pauli matrix · Hamilton canonical equation · Lagrange equation · Maxwell equations · Photon beams · Poisson bracket · Recursion method · Spin $\geq \frac{3}{2}$ particle · Unitary transformation relations

## 4.1 Polarization Theory of Nuclear Reactions for Spin $\frac{3}{2}$ Particles

The ground states of the nuclide $^5$He, $^7$He, $^5$Li, $^7$Li, $^9$Li, $^{11}$Li, $^7$Be, $^9$Be, $^7$B, $^9$B, $^{11}$B, $^{13}$B, $^{17}$B, $^9$C, $^{11}$C, $^{13}$O are $\frac{3}{2}^-$ state. However, they are generally not the main transport particles in nuclear devices. Only under special circumstances do the polarization problems of nuclear reactions of spin $\frac{3}{2}$ particles need to be considered.

The $S = \frac{3}{2}$ basis spin function $\chi_{Sm}\left(m = \pm\frac{3}{2}, \pm\frac{1}{2}\right)$ is a common eigenfunction of the spin operators $\hat{S}^2$ and $\hat{S}_z$; based on Eq. (1.15) we have


Q.-B. Shen, *Polarization Theory of Nuclear Reactions*,
https://doi.org/10.1007/978-3-031-11878-4_4

$$\hat{S}^2 = \frac{15}{4}\hat{I}, \qquad \hat{I} = \begin{pmatrix} 1 & 0 & 0 & 0 \\ 0 & 1 & 0 & 0 \\ 0 & 0 & 1 & 0 \\ 0 & 0 & 0 & 1 \end{pmatrix} \tag{4.1}$$

where $\hat{I}$ is a $4 \times 4$ unit matrix. The components of the basis spin function in the spherical basis coordinate are defined as

$$\chi_{Sm}(\sigma) = \delta_{m\sigma}, \qquad m,\sigma = S, S-1, \ldots, -S+1, S \tag{4.2}$$

Thus, we can write the following:

$$\chi_{\frac{3}{2}\frac{3}{2}} = \begin{pmatrix} 1 \\ 0 \\ 0 \\ 0 \end{pmatrix}, \quad \chi_{\frac{3}{2}\frac{1}{2}} = \begin{pmatrix} 0 \\ 1 \\ 0 \\ 0 \end{pmatrix}, \quad \chi_{\frac{3}{2}\,-\frac{1}{2}} = \begin{pmatrix} 0 \\ 0 \\ 1 \\ 0 \end{pmatrix}, \quad \chi_{\frac{3}{2}\,-\frac{3}{2}} = \begin{pmatrix} 0 \\ 0 \\ 0 \\ 1 \end{pmatrix} \tag{4.3}$$

From Eq. (1.21) the spherical basis coordinate component of the spin operator $\hat{S}$ can be obtained as

$$\left(\hat{S}_\mu\right)_{\sigma'\sigma} = \frac{\sqrt{15}}{2} C_{\frac{3}{2}\sigma\ 1\mu}^{\frac{3}{2}\sigma'} = \sqrt{5}(-1)^{\frac{3}{2}-\sigma} C_{\frac{3}{2}\sigma'\ \frac{3}{2}\,-\sigma}^{1\mu}, \quad \sigma',\sigma = \pm\frac{3}{2}, \pm\frac{1}{2};\ \mu = \pm 1,0 \tag{4.4}$$

where $\sigma'$ represents a row label and $\sigma$ indicates a column label.

Utilizing Eq. (4.4) and Table 4.1, we can obtain

**Table 4.1** C–G coefficient $C_{\frac{3}{2}\alpha\ \frac{3}{2}\beta}^{1\ \alpha+\beta}$ table [1]

| $\alpha$ | $\beta$ | | | |
|---|---|---|---|---|
| | $\frac{3}{2}$ | $\frac{1}{2}$ | $-\frac{1}{2}$ | $-\frac{3}{2}$ |
| $\frac{3}{2}$ | | | $\sqrt{\frac{3}{10}}$ | $\frac{3}{2\sqrt{5}}$ |
| $\frac{1}{2}$ | | $-\sqrt{\frac{2}{5}}$ | $-\frac{1}{2\sqrt{5}}$ | $\sqrt{\frac{3}{10}}$ |
| $-\frac{1}{2}$ | $\sqrt{\frac{3}{10}}$ | $-\frac{1}{2\sqrt{5}}$ | $-\sqrt{\frac{2}{5}}$ | |
| $-\frac{3}{2}$ | $\frac{3}{2\sqrt{5}}$ | $\sqrt{\frac{3}{10}}$ | | |

$$\hat{S}_1 = -\begin{pmatrix} 0 & \sqrt{\frac{3}{2}} & 0 & 0 \\ 0 & 0 & \sqrt{2} & 0 \\ 0 & 0 & 0 & \sqrt{\frac{3}{2}} \\ 0 & 0 & 0 & 0 \end{pmatrix}, \quad \hat{S}_0 = \frac{1}{2}\begin{pmatrix} 3 & 0 & 0 & 0 \\ 0 & 1 & 0 & 0 \\ 0 & 0 & -1 & 0 \\ 0 & 0 & 0 & -3 \end{pmatrix},$$
$$\hat{S}_{-1} = \begin{pmatrix} 0 & 0 & 0 & 0 \\ \sqrt{\frac{3}{2}} & 0 & 0 & 0 \\ 0 & \sqrt{2} & 0 & 0 \\ 0 & 0 & \sqrt{\frac{3}{2}} & 0 \end{pmatrix} \tag{4.5}$$

Using Eq. (1.22), the matrix expressions of the Cartesian coordinate system component of the $S = \frac{3}{2}$ particle spin operator can be obtained from Eq. (4.5) as follows:

$$\hat{S}_x = \frac{1}{\sqrt{2}}\begin{pmatrix} 0 & \sqrt{\frac{3}{2}} & 0 & 0 \\ \sqrt{\frac{3}{2}} & 0 & \sqrt{2} & 0 \\ 0 & \sqrt{2} & 0 & \sqrt{\frac{3}{2}} \\ 0 & 0 & \sqrt{\frac{3}{2}} & 0 \end{pmatrix}, \quad \hat{S}_y = \frac{\mathrm{i}}{\sqrt{2}}\begin{pmatrix} 0 & -\sqrt{\frac{3}{2}} & 0 & 0 \\ \sqrt{\frac{3}{2}} & 0 & -\sqrt{2} & 0 \\ 0 & \sqrt{2} & 0 & -\sqrt{\frac{3}{2}} \\ 0 & 0 & \sqrt{\frac{3}{2}} & 0 \end{pmatrix},$$
$$\hat{S}_z = \frac{1}{2}\begin{pmatrix} 3 & 0 & 0 & 0 \\ 0 & 1 & 0 & 0 \\ 0 & 0 & -1 & 0 \\ 0 & 0 & 0 & -3 \end{pmatrix} \tag{4.6}$$

where the above spin operator matrices satisfy the following relations:

$$\hat{S}_i^+ = \hat{S}_i, \qquad i = x, y, z \tag{4.7}$$

$$\hat{S}_\mu^+ = (-1)^\mu \hat{S}_{-\mu}, \qquad \mu = \pm 1, 0 \tag{4.8}$$

Now the matrix element of the polarization operator $\hat{T}_{LM}\left(\frac{3}{2}\right)$ can be written according to Eq. (1.53) as follows:

$$\left[\hat{T}_{LM}\left(\frac{3}{2}\right)\right]_{\sigma'\sigma} = \sqrt{2L+1}C^{\frac{3}{2}\sigma'}_{\frac{3}{2}\sigma \ LM} = 2(-1)^{\frac{3}{2}-\sigma}C^{LM}_{\frac{3}{2}\sigma' \ \frac{3}{2} \ -\sigma},$$
$$\sigma', \sigma = \pm\frac{3}{2}, \pm\frac{1}{2}; \qquad L = 0, 1, 2, 3; \qquad -L \le M \le L \tag{4.9}$$

such that we know the following relations from Eqs. (1.54) and (1.55)

$$\hat{T}_{00}\left(\frac{3}{2}\right) = \hat{I} \tag{4.10}$$

$$\hat{T}_{1M}\left(\frac{3}{2}\right) = \frac{2}{\sqrt{5}}\hat{S}_M, \qquad M = \pm 1{,}0 \tag{4.11}$$

where the specific expression of the operator $\hat{S}_M$ has been given by Eq. (4.5).

From Eq. (1.51) we can obtain

$$\hat{T}_{2M}\left(\frac{3}{2}\right) = \sqrt{5}\sum_{mm'} C^{\frac{3}{2}m'}_{\frac{3}{2}m \ 2M}\chi_{\frac{3}{2}m'}\chi^{+}_{\frac{3}{2}m} \tag{4.12}$$

and according to Eq. (1.20) we also have

$$\hat{S}_\mu = \frac{\sqrt{15}}{2}\sum_{mm'} C^{\frac{3}{2}m'}_{\frac{3}{2}m \ 1\mu}\chi_{\frac{3}{2}m'}\chi_{\frac{3}{2}m} \tag{4.13}$$

letting

$$\hat{X} = \sum_{\mu\nu} C^{2M}_{1\mu \ 1\nu}\hat{S}_\mu\hat{S}_\nu \tag{4.14}$$

then substituting Eq. (4.13) into Eq. (4.14) and using Eqs. (1.18), (3.14) and $W\left(21\frac{3}{2}\frac{3}{2};1\frac{3}{2}\right) = -\frac{\sqrt{2}}{5\sqrt{3}}$, we can obtain

$$\begin{aligned}
\hat{X} &= \frac{15}{4}\sum_{\mu\nu} C^{2M}_{1\mu \ 1\nu}\sum_{mm'} C^{\frac{3}{2}m'}_{\frac{3}{2}m \ 1\mu}\sum_{\overline{m}\,\overline{m}'} C^{\frac{3}{2}\overline{m}'}_{\frac{3}{2}\overline{m} \ 1\nu}\chi_{\frac{3}{2}m'}\chi^{+}_{\frac{3}{2}m}\chi_{\frac{3}{2}\overline{m}'}\chi^{+}_{\frac{3}{2}\overline{m}} \\
&= \frac{15}{4}\sum_{\mu\nu mm'\overline{m}} C^{2M}_{1\mu \ 1\nu}C^{\frac{3}{2}m'}_{\frac{3}{2}m \ 1\mu}C^{\frac{3}{2}m}_{\frac{3}{2}\overline{m} \ 1\nu}\chi_{\frac{3}{2}m'}\chi^{+}_{\frac{3}{2}\overline{m}} = \frac{15}{4}2\sqrt{5}\sum_{m'\overline{m}} C^{\frac{3}{2}m'}_{\frac{3}{2}\overline{m} \ 2M}W\left(\frac{3}{2}1\frac{3}{2}1;\frac{3}{2}2\right)\chi_{\frac{3}{2}m'}\chi^{+}_{\frac{3}{2}\overline{m}} \\
&= -\frac{15\sqrt{5}}{2}\sum_{m'm} C^{\frac{3}{2}m'}_{\frac{3}{2}m \ 2M}W\left(21\frac{3}{2}\frac{3}{2};1\frac{3}{2}\right)\chi_{\frac{3}{2}m'}\chi^{+}_{\frac{3}{2}m} = \sqrt{\frac{15}{2}}\sum_{m'm} C^{\frac{3}{2}m'}_{\frac{3}{2}m \ 2M}\chi_{\frac{3}{2}m'}\chi^{+}_{\frac{3}{2}m}
\end{aligned} \tag{4.15}$$

Comparing Eq. (4.15) with Eq. (4.12), we have

**Table 4.2** C–G coefficient $C^{2\ \ \alpha+\beta}_{\frac{3}{2}\alpha\ \frac{3}{2}\beta}$ table [1]

| $\alpha$ | $\beta$: $\frac{3}{2}$ | $\frac{1}{2}$ | $-\frac{1}{2}$ | $-\frac{3}{2}$ |
|---|---|---|---|---|
| $\frac{3}{2}$ | | $\frac{1}{\sqrt{2}}$ | $\frac{1}{\sqrt{2}}$ | $\frac{1}{2}$ |
| $\frac{1}{2}$ | $-\frac{1}{\sqrt{2}}$ | 0 | $\frac{1}{2}$ | $\frac{1}{\sqrt{2}}$ |
| $-\frac{1}{2}$ | $-\frac{1}{\sqrt{2}}$ | $-\frac{1}{2}$ | 0 | $\frac{1}{\sqrt{2}}$ |
| $-\frac{3}{2}$ | $-\frac{1}{2}$ | $-\frac{1}{\sqrt{2}}$ | $-\frac{1}{\sqrt{2}}$ | |

$$\hat{T}_{2M}\left(\frac{3}{2}\right) = \sqrt{\frac{2}{3}}\sum_{\mu\nu} C^{2M}_{1\mu\ 1\nu}\hat{S}_\mu\hat{S}_\nu, \qquad M = \pm 2, \pm 1, 0 \tag{4.16}$$

Now it can be seen that $\hat{T}_{2M}\left(\frac{3}{2}\right)$ has 5 linear independent components such that each component is a square $4 \times 4$ matrix in spin space.

From Eq. (4.9) and Table 4.2, we can obtain

$$\hat{T}_{22}\left(\frac{3}{2}\right) = \sqrt{2}\begin{pmatrix} 0 & 0 & 1 & 0 \\ 0 & 0 & 0 & 1 \\ 0 & 0 & 0 & 0 \\ 0 & 0 & 0 & 0 \end{pmatrix}, \quad \hat{T}_{21}\left(\frac{3}{2}\right) = \sqrt{2}\begin{pmatrix} 0 & -1 & 0 & 0 \\ 0 & 0 & 0 & 0 \\ 0 & 0 & 0 & 1 \\ 0 & 0 & 0 & 0 \end{pmatrix},$$

$$\hat{T}_{20}\left(\frac{3}{2}\right) = \begin{pmatrix} 1 & 0 & 0 & 0 \\ 0 & -1 & 0 & 0 \\ 0 & 0 & -1 & 0 \\ 0 & 0 & 0 & 1 \end{pmatrix}, \quad \hat{T}_{2\ -1}\left(\frac{3}{2}\right) = \sqrt{2}\begin{pmatrix} 0 & 0 & 0 & 0 \\ 1 & 0 & 0 & 0 \\ 0 & 0 & 0 & 0 \\ 0 & 0 & -1 & 0 \end{pmatrix}, \tag{4.17}$$

$$\hat{T}_{2\ -2}\left(\frac{3}{2}\right) = \sqrt{2}\begin{pmatrix} 0 & 0 & 0 & 0 \\ 0 & 0 & 0 & 0 \\ 1 & 0 & 0 & 0 \\ 0 & 1 & 0 & 0 \end{pmatrix}$$

where the above matrices satisfy the following relation:

**Table 4.3** C–G coefficient $C^{3\ \alpha+\beta}_{\frac{3}{2}\alpha\ \frac{3}{2}\beta}$ table [1]

| $\alpha$ | $\beta$ | | | |
|---|---|---|---|---|
| | $\frac{3}{2}$ | $\frac{1}{2}$ | $-\frac{1}{2}$ | $-\frac{3}{2}$ |
| $\frac{3}{2}$ | 1 | $\frac{1}{\sqrt{2}}$ | $\frac{1}{\sqrt{5}}$ | $\frac{1}{2\sqrt{5}}$ |
| $\frac{1}{2}$ | $\frac{1}{\sqrt{2}}$ | $\sqrt{\frac{3}{5}}$ | $\frac{3}{2\sqrt{5}}$ | $\frac{1}{\sqrt{5}}$ |
| $-\frac{1}{2}$ | $\frac{1}{\sqrt{5}}$ | $\frac{3}{2\sqrt{5}}$ | $\sqrt{\frac{3}{5}}$ | $\frac{1}{\sqrt{2}}$ |
| $-\frac{3}{2}$ | $\frac{1}{2\sqrt{5}}$ | $\frac{1}{\sqrt{5}}$ | $\frac{1}{\sqrt{2}}$ | 1 |

$$\hat{T}^{+}_{2M}\left(\frac{3}{2}\right) = (-1)^M \hat{T}_{2\ -M}\left(\frac{3}{2}\right) \tag{4.18}$$

Also, from Eq. (4.9) and Table 4.3, we can obtain

$$\hat{T}_{33}\left(\frac{3}{2}\right) = 2\begin{pmatrix} 0 & 0 & 0 & -1 \\ 0 & 0 & 0 & 0 \\ 0 & 0 & 0 & 0 \\ 0 & 0 & 0 & 0 \end{pmatrix}, \quad \hat{T}_{32}\left(\frac{3}{2}\right) = \sqrt{2}\begin{pmatrix} 0 & 0 & 1 & 0 \\ 0 & 0 & 0 & -1 \\ 0 & 0 & 0 & 0 \\ 0 & 0 & 0 & 0 \end{pmatrix},$$
$$\hat{T}_{31}\left(\frac{3}{2}\right) = \frac{2}{\sqrt{5}}\begin{pmatrix} 0 & -1 & 0 & 0 \\ 0 & 0 & \sqrt{3} & 0 \\ 0 & 0 & 0 & -1 \\ 0 & 0 & 0 & 0 \end{pmatrix}, \quad \hat{T}_{30}\left(\frac{3}{2}\right) = \frac{1}{\sqrt{5}}\begin{pmatrix} 1 & 0 & 0 & 0 \\ 0 & -3 & 0 & 0 \\ 0 & 0 & 3 & 0 \\ 0 & 0 & 0 & -1 \end{pmatrix},$$
$$\hat{T}_{3\ -1}\left(\frac{3}{2}\right) = \frac{2}{\sqrt{5}}\begin{pmatrix} 0 & 0 & 0 & 0 \\ 1 & 0 & 0 & 0 \\ 0 & -\sqrt{3} & 0 & 0 \\ 0 & 0 & 1 & 0 \end{pmatrix}, \quad \hat{T}_{3\ -2}\left(\frac{3}{2}\right) = \sqrt{2}\begin{pmatrix} 0 & 0 & 0 & 0 \\ 0 & 0 & 0 & 0 \\ 1 & 0 & 0 & 0 \\ 0 & -1 & 0 & 0 \end{pmatrix},$$
$$\hat{T}_{3\ -3}\left(\frac{3}{2}\right) = 2\begin{pmatrix} 0 & 0 & 0 & 0 \\ 0 & 0 & 0 & 0 \\ 0 & 0 & 0 & 0 \\ 1 & 0 & 0 & 0 \end{pmatrix} \tag{4.19}$$

where these matrices satisfy the relation

$$\hat{T}^{+}_{3M}\left(\frac{3}{2}\right) = (-1)^M \hat{T}_{3\ -M}\left(\frac{3}{2}\right) \tag{4.20}$$

Referring to Eq. (3.323), the normalized incident particle density matrix for $S = \frac{3}{2}$ particles can be written, according to Eqs. (1.92), (1.99), and (4.10), as follows:

$$\begin{aligned}\hat{\rho}_{\text{in}} = \frac{1}{4}\Bigg[\hat{I} &+ \sum_M (-1)^M (-\text{i}t_{1\ -M})\left(\text{i}\hat{T}_{1M}\right) + \sum_M (-1)^M t_{2\ -M}\hat{T}_{2M} \\ &+ \sum_M (-1)^M (-\text{i}t_{3\ -M})\left(\text{i}\hat{T}_{3M}\right)\Bigg]\end{aligned} \tag{4.21}$$

where in this formula, $\text{i}t_{1M}$, $t_{2M}$, and $\text{i}t_{3M}$ are the first-order tensor (vector), second-order tensor, and third-order tensor initial polarization rates of the $S = \frac{3}{2}$ incident particles, respectively.

From Eq. (5.408) or Eq. (5.5.51) in Ref. [2], the elastic scattering amplitude of the spin $\frac{3}{2}$ incident particles with the spinless target nucleus can be written as

$$f_{\mu',\mu}(\theta,\varphi) = f_{\text{C}}(\theta)\delta_{\mu'\mu} + \frac{\text{i}\sqrt{\pi}}{k}\sum_{lj}\hat{l}C^{j\ \mu}_{l0\ \frac{3}{2}\mu}C^{j\ \mu}_{l\ \mu-\mu'\ \frac{3}{2}\mu'}\left(1 - S^j_l\right)\text{e}^{2\text{i}\sigma_l}\text{Y}_{l\ \mu-\mu'}(\theta,\varphi) \tag{4.22}$$

where $S^j_l$ are the elastic scattering $S$ matrix elements. Using Eq. (2.253) we can rewrite the above formula as

$$f_{\mu',\mu}(\theta,\varphi) = F_{\mu',\mu}(\theta)\text{e}^{\text{i}(\mu-\mu')\varphi} \tag{4.23}$$

$$F_{\mu',\mu}(\theta) = f_{\text{C}}(\theta)\delta_{\mu'\mu} + \frac{\text{i}}{2k}(-1)^{\mu-\mu'}\sum_{lj} f^{lj}_{\mu',\mu}\text{P}^{\mu-\mu'}_l(\cos\theta) \tag{4.24}$$

$$f^{lj}_{\mu',\mu} = \hat{l}^2\text{e}^{2\text{i}\sigma_l}\sqrt{\frac{(l-\mu+\mu')!}{(l+\mu-\mu')!}}C^{j\ \mu}_{l0\ \frac{3}{2}\mu}C^{j\ \mu}_{l\ \mu-\mu'\ \frac{3}{2}\mu'}\left(1 - S^j_l\right) \tag{4.25}$$

where it is straightforward to prove

$$C^{j\ \mu}_{l0\ \frac{3}{2}\mu}C^{j\ \mu}_{l\ \mu-\mu'\ \frac{3}{2}\mu'} = C^{j\ -\mu}_{l0\ \frac{3}{2}\ -\mu}C^{j\ -\mu}_{l\ -\mu+\mu'\ \frac{3}{2}\ -\mu'} \tag{4.26}$$

Further noting the relation expression (2.287), from Eqs. (4.25) and (4.26) we can get

$$F_{\mu',\mu}(\theta) = (-1)^{\mu-\mu'} F_{-\mu',-\mu}(\theta) \tag{4.27}$$

Now, we select the unit vector $\vec{n}$ perpendicular to the reaction plane as the $y$-axis, and based upon this rectangular coordinate system, we define the spherical basis coordinate so that the azimuthal angle $\varphi = 0$ can be made. Next, we introduce the following matrix element symbols according to Eq. (4.27):

$$\begin{aligned}
A(\theta) &= F_{\frac{3}{2},\frac{3}{2}}(\theta) = F_{-\frac{3}{2},-\frac{3}{2}}(\theta), & B(\theta) &= \mathrm{i}F_{\frac{3}{2},\frac{1}{2}}(\theta) = -\mathrm{i}F_{-\frac{3}{2},-\frac{1}{2}}(\theta), \\
C(\theta) &= -F_{\frac{3}{2},-\frac{1}{2}}(\theta) = -F_{-\frac{3}{2},\frac{1}{2}}(\theta), & D(\theta) &= -\mathrm{i}F_{\frac{3}{2},-\frac{3}{2}}(\theta) = \mathrm{i}F_{-\frac{3}{2},\frac{3}{2}}(\theta), \\
E(\theta) &= -F_{\frac{1}{2},-\frac{3}{2}}(\theta) = -F_{-\frac{1}{2},\frac{3}{2}}(\theta), & F(\theta) &= \mathrm{i}F_{\frac{1}{2},-\frac{1}{2}}(\theta) = -\mathrm{i}F_{-\frac{1}{2},\frac{1}{2}}(\theta), \\
G(\theta) &= F_{\frac{1}{2},\frac{1}{2}}(\theta) = F_{-\frac{1}{2},-\frac{1}{2}}(\theta), & H(\theta) &= -\mathrm{i}F_{\frac{1}{2},\frac{3}{2}}(\theta) = \mathrm{i}F_{-\frac{1}{2},-\frac{3}{2}}(\theta)
\end{aligned} \tag{4.28}$$

It follows that the scattering amplitude $4 \times 4$ matrix given by Eq. (4.24) can be expressed as

$$\hat{F} = \begin{pmatrix} A & -\mathrm{i}B & -C & \mathrm{i}D \\ \mathrm{i}H & G & -\mathrm{i}F & -E \\ -E & \mathrm{i}F & G & -\mathrm{i}H \\ -\mathrm{i}D & -C & \mathrm{i}B & A \end{pmatrix}, \quad \hat{F}^+ = \begin{pmatrix} A^* & -\mathrm{i}H^* & -E^* & \mathrm{i}D^* \\ \mathrm{i}B^* & G^* & -\mathrm{i}F^* & -C^* \\ -C^* & \mathrm{i}F^* & G^* & -\mathrm{i}B^* \\ -\mathrm{i}D^* & -E^* & \mathrm{i}H^* & A^* \end{pmatrix} \tag{4.29}$$

$$\hat{F}\hat{F}^+ = \begin{pmatrix}
|A|^2+|B|^2+|C|^2+|D|^2 & -\mathrm{i}(AH^*+BG^*+CF^*+DE^*) & -(AE^*+BF^*+CG^*+DH^*) & \mathrm{i}(AD^*+BC^*+CB^*+DA^*) \\
\mathrm{i}(HA^*+GB^*+FC^*+ED^*) & |H|^2+|G|^2+|F|^2+|E|^2 & -\mathrm{i}(HE^*+GF^*+FG^*+EH^*) & -(HD^*+GC^*+FB^*+EA^*) \\
-(EA^*+FB^*+GC^*+HD^*) & \mathrm{i}(EH^*+FG^*+GF^*+HE^*) & |E|^2+|F|^2+|G|^2+|H|^2 & -\mathrm{i}(ED^*+FC^*+GB^*+HA^*) \\
-\mathrm{i}(DA^*+CB^*+BC^*+AD^*) & -(DH^*+CG^*+BF^*+AE^*) & \mathrm{i}(DE^*+CF^*+BG^*+AH^*) & |D|^2+|C|^2+|B|^2+|A|^2
\end{pmatrix} \tag{4.30}$$

Now if the incident particles are unpolarized, then the differential cross section of the incident particles can be found from Eq. (4.30) as

$$I_0 = \frac{1}{4}\mathrm{tr}\left\{\hat{F}\hat{F}^+\right\} = \frac{1}{2}\left(|A|^2+|B|^2+|C|^2+|D|^2+|E|^2+|F|^2+|G|^2+|H|^2\right) \tag{4.31}$$

When the incident particles are unpolarized, the components of the outgoing particle polarization vector are

$$\mathrm{i}P_0^{M'} = \frac{1}{4I_0}\mathrm{tr}\left\{\mathrm{i}\hat{S}_{M'}\hat{F}\hat{F}^+\right\}, \qquad M' = \pm 1, 0 \tag{4.32}$$

Thus, we can obtain

$$\begin{aligned}\mathrm{i}P_0^1 &= -\frac{\mathrm{i}}{4I_0}\left[\mathrm{i}\sqrt{\frac{3}{2}}(HA^* + GB^* + FC^* + ED^*) + \mathrm{i}\sqrt{2}(EH^* + FG^*\right.\\ &\left.+GF^* + HE^*) + \mathrm{i}\sqrt{\frac{3}{2}}(DE^* + CF^* + BG^* + AH^*)\right]\\ &= \frac{1}{2I_0}\,\mathrm{Re}\left[\sqrt{\frac{3}{2}}(AH^* + BG^* + CF^* + DE^*) + \sqrt{2}(EH^* + FG^*)\right]\end{aligned} \tag{4.33}$$

$$\mathrm{i}P_0^0 = 0 \tag{4.34}$$

$$\begin{aligned}\mathrm{i}P_0^{-1} &= \frac{\mathrm{i}}{4I_0}\left[-\mathrm{i}\sqrt{\frac{3}{2}}(AH^* + BG^* + CF^* + DE^*) - \mathrm{i}\sqrt{2}(HE^* + GF^*\right.\\ &\left.+FG^* + EH^*) - \mathrm{i}\sqrt{\frac{3}{2}}(ED^* + FC^* + GB^* + HA^*)\right]\\ &= \frac{1}{2I_0}\,\mathrm{Re}\left[\sqrt{\frac{3}{2}}(AH^* + BG^* + CF^* + DE^*) + \sqrt{2}(EH^* + FG^*)\right] = \mathrm{i}P_0^1\end{aligned} \tag{4.35}$$

Moreover, according to Eq. (4.11) we have

$$\mathrm{i}P_0^{1M'} = \frac{1}{4I_0}\mathrm{tr}\left\{\mathrm{i}\hat{T}_{1M'}\hat{F}\hat{F}^+\right\} = \frac{2}{\sqrt{5}}\mathrm{i}P_0^{M'}, \qquad M' = \pm 1, 0 \tag{4.36}$$

Now, the components of the corresponding polarization rank-2 tensor are

$$P_0^{2M'} = \frac{1}{4I_0}\mathrm{tr}\left\{\hat{T}_{2M'}\hat{F}\hat{F}^+\right\}, \qquad M' = \pm 2, \pm 1, 0 \tag{4.37}$$

thus we can obtain

$$\begin{aligned}P_0^{22} &= \frac{\sqrt{2}}{4I_0}[-(EA^* + FB^* + GC^* + HD^*) - (DH^* + CG^* + BF^* + AE^*)]\\ &= -\frac{1}{\sqrt{2}I_0}\,\mathrm{Re}\,(AE^* + BF^* + CG^* + DH^*)\end{aligned} \tag{4.38}$$

$$
\begin{aligned}
P_0^{21} &= \frac{\sqrt{2}}{4I_0}[-\mathrm{i}(HA^* + GB^* + FC^* + ED^*) + \mathrm{i}(DE^* + CF^* + BG^* + AH^*)] \\
&= -\frac{1}{\sqrt{2}I_0}\mathrm{Im}(AH^* + BG^* + CF^* + DE^*)
\end{aligned}
\tag{4.39}
$$

$$
P_0^{20} = \frac{1}{2I_0}\left[\left(|A|^2 + |B|^2 + |C|^2 + |D|^2\right) - \left(|E|^2 + |F|^2 + |G|^2 + |H|^2\right)\right] \tag{4.40}
$$

$$
\begin{aligned}
P_0^{2\ -1} &= \frac{\sqrt{2}}{4I_0}[-\mathrm{i}(AH^* + BG^* + CF^* + DE^*) + \mathrm{i}(ED^* + FC^* + GB^* + HA^*)] \\
&= \frac{1}{\sqrt{2}I_0}\mathrm{Im}(AH^* + BG^* + CF^* + DE^*) = -P_0^{21}
\end{aligned}
\tag{4.41}
$$

$$
\begin{aligned}
P_2^{2\ -2} &= \frac{\sqrt{2}}{4I_0}[-(AE^* + BF^* + CG^* + DH^*) - (HD^* + GC^* + FB^* + EA^*)] \\
&= -\frac{1}{\sqrt{2}I_0}\mathrm{Re}\,(AE^* + BF^* + CG^* + DH^*) = P_0^{22}
\end{aligned}
\tag{4.42}
$$

The corresponding polarization rank-3 tensor components are

$$
\mathrm{i}P_0^{3M'} = \frac{1}{4I_0}\mathrm{tr}\left\{\mathrm{i}\hat{T}_{3M'}\hat{F}\hat{F}^+\right\}, \qquad M' = \pm 3, \pm 2, \pm 1, 0 \tag{4.43}
$$

yielding

$$
\mathrm{i}P_0^{33} = \frac{\mathrm{i}(2\mathrm{i})}{4I_0}(DA^* + CB^* + BC^* + AD^*) = -\frac{1}{I_0}\mathrm{Re}\,(AD^* + BC^*) \tag{4.44}
$$

$$
\begin{aligned}
\mathrm{i}P_0^{32} &= \frac{\mathrm{i}\sqrt{2}}{4I_0}[-(EA^* + FB^* + GC^* + HD^*) + (DH^* + CG^* + BF^* + AE^*)] \\
&= -\frac{1}{\sqrt{2}I_0}\mathrm{Im}(AE^* + BF^* + CG^* + DH^*)
\end{aligned}
\tag{4.45}
$$

$$
\begin{aligned}
\mathrm{i}P_0^{31} &= \frac{2\mathrm{i}}{4\sqrt{5}I_0}\Big[-\mathrm{i}(HA^* + GB^* + FC^* + ED^*) + \mathrm{i}\sqrt{3}(EH^* + FG^* + GF^* + HE^*) \\
&\quad -\mathrm{i}(DE^* + CF^* + BG^* + AH^*)] \\
&= \frac{1}{\sqrt{5}I_0}\mathrm{Re}\left[(AH^* + BG^* + CF^* + DE^*) - \sqrt{3}(EH^* + FG^*)\right]
\end{aligned}
\tag{4.46}
$$

$$\mathrm{i}P_0^{30} = 0 \tag{4.47}$$

$$\begin{aligned}\mathrm{i}P_0^{3\ -1} &= \frac{2\mathrm{i}}{4\sqrt{5}I_0}\left[-\mathrm{i}(AH^* + BG^* + CF^* + DE^*)\right.\\ &\quad \left.+\mathrm{i}\sqrt{3}(HE^* + GF^* + FG^* + EH^*) - \mathrm{i}(ED^* + FC^* + GB^* + HA^*)\right]\\ &= \frac{1}{\sqrt{5}I_0}\,\mathrm{Re}\left[(AH^* + BG^* + CF^* + DE^*) - \sqrt{3}(EH^* + FG^*)\right] = \mathrm{i}P_0^{31}\end{aligned} \tag{4.48}$$

$$\begin{aligned}\mathrm{i}P_0^{3\ -2} &= \frac{\mathrm{i}\sqrt{2}}{4I_0}\left[-(AE^* + BF^* + CG^* + DH^*) + (HD^* + GC^* + FB^* + EA^*)\right]\\ &= \frac{1}{\sqrt{2}I_0}\mathrm{Im}(AE^* + BF^* + CG^* + DH^*) = -\mathrm{i}P_0^{32}\end{aligned} \tag{4.49}$$

$$\mathrm{i}P_0^{3\ -3} = \frac{\mathrm{i}(2\mathrm{i})}{4I_0}(AD^* + BC^* + CB^* + DA^*) = -\frac{1}{I_0}\,\mathrm{Re}\,(AD^* + BC^*) = \mathrm{i}P_0^{33} \tag{4.50}$$

From these results, it can be seen that the outgoing particle non-zero components and independent components of the polarized rank-1 tensors (vectors), rank-2 tensors, and rank-3 tensors corresponding to the $S = \frac{3}{2}$ unpolarized incident particles are $\mathrm{i}P_0^{11}$; $P_0^{22}$, $P_0^{21}$, $P_0^{20}$; $\mathrm{i}P_0^{33}$, $\mathrm{i}P_0^{32}$, $\mathrm{i}P_0^{31}$.

Next, the corresponding polarization analyzing powers will be studied below. For the polarization vectors, we have

$$\mathrm{i}A_M = \frac{1}{4I_0}\mathrm{tr}\{\hat{F}\,\mathrm{i}\hat{S}_M\hat{F}^+\}, \qquad M = \pm 1, 0 \tag{4.51}$$

from which we can obtain

$$\mathrm{i}\hat{S}_1\hat{F}^+ = -\mathrm{i}\begin{pmatrix} \mathrm{i}\sqrt{\frac{3}{2}}B^* & \sqrt{\frac{3}{2}}G^* & -\mathrm{i}\sqrt{\frac{3}{2}}F^* & -\sqrt{\frac{3}{2}}C^* \\ -\sqrt{2}C^* & \mathrm{i}\sqrt{2}F^* & \sqrt{2}G^* & -\mathrm{i}\sqrt{2}B^* \\ -\mathrm{i}\sqrt{\frac{3}{2}}D^* & -\sqrt{\frac{3}{2}}E^* & \mathrm{i}\sqrt{\frac{3}{2}}H^* & \sqrt{\frac{3}{2}}A^* \\ 0 & 0 & 0 & 0 \end{pmatrix}$$

$$
\begin{aligned}
\mathrm{i}A_1 &= \frac{-\mathrm{i}}{4I_0}\left[\mathrm{i}\left(\sqrt{\frac{3}{2}}AB^* + \sqrt{2}BC^* + \sqrt{\frac{3}{2}}CD^*\right) + \mathrm{i}\left(\sqrt{\frac{3}{2}}HG^* + \sqrt{2}GF^* + \sqrt{\frac{3}{2}}FE^*\right)\right. \\
&\quad \left. +\mathrm{i}\left(\sqrt{\frac{3}{2}}EF^* + \sqrt{2}FG^* + \sqrt{\frac{3}{2}}GH^*\right) + \mathrm{i}\left(\sqrt{\frac{3}{2}}DC^* + \sqrt{2}CB^* + \sqrt{\frac{3}{2}}BA^*\right)\right] \\
&= \frac{1}{2I_0}\,\mathrm{Re}\left[\sqrt{\frac{3}{2}}(AB^* + CD^* + EF^* + GH^*) + \sqrt{2}(BC^* + FG^*)\right]
\end{aligned}
\tag{4.52}
$$

$$
\mathrm{i}\hat{S}_0\hat{F}^+ = \frac{\mathrm{i}}{2}\begin{pmatrix} 3A^* & -3\mathrm{i}H^* & -3E^* & 3\mathrm{i}D^* \\ \mathrm{i}B^* & G^* & -\mathrm{i}F^* & -C^* \\ C^* & -\mathrm{i}F^* & -G^* & \mathrm{i}B^* \\ 3\mathrm{i}D^* & 3E^* & -3\mathrm{i}H^* & -3A^* \end{pmatrix}
$$

$$
\begin{aligned}
\mathrm{i}A_0 = \frac{\mathrm{i}}{8I_0}\Big(&3|A|^2 + |B|^2 - |C|^2 - 3|D|^2 + 3|H|^2 + |G|^2 - |F|^2 - 3|E|^2 \\
&+3|E|^2 + |F|^2 - |G|^2 - 3|H|^2 + 3|D|^2 + |C|^2 - |B|^2 - 3|A|^2\Big) = 0
\end{aligned}
$$

$$
\mathrm{i}\hat{S}_{-1}\hat{F}^+ = \mathrm{i}\begin{pmatrix} 0 & 0 & 0 & 0 \\ \sqrt{\frac{3}{2}}A^* & -\mathrm{i}\sqrt{\frac{3}{2}}H^* & -\sqrt{\frac{3}{2}}E^* & \mathrm{i}\sqrt{\frac{3}{2}}D^* \\ \mathrm{i}\sqrt{2}B^* & \sqrt{2}G^* & -\mathrm{i}\sqrt{2}F^* & -\sqrt{2}C^* \\ -\sqrt{\frac{3}{2}}C^* & \mathrm{i}\sqrt{\frac{3}{2}}F^* & \sqrt{\frac{3}{2}}G^* & -\mathrm{i}\sqrt{\frac{3}{2}}B^* \end{pmatrix}
\tag{4.53}
$$

$$
\begin{aligned}
\mathrm{i}A_{-1} &= \frac{\mathrm{i}}{4I_0}\left[-\mathrm{i}\left(\sqrt{\frac{3}{2}}BA^* + \sqrt{2}CB^* + \sqrt{\frac{3}{2}}DC^*\right) - \mathrm{i}\left(\sqrt{\frac{3}{2}}GH^* + \sqrt{2}FG^* + \sqrt{\frac{3}{2}}EF^*\right)\right. \\
&\quad \left. -\mathrm{i}\left(\sqrt{\frac{3}{2}}FE^* + \sqrt{2}GF^* + \sqrt{\frac{3}{2}}HG^*\right) - \mathrm{i}\left(\sqrt{\frac{3}{2}}CD^* + \sqrt{2}BC^* + \sqrt{\frac{3}{2}}AB^*\right)\right] \\
&= \frac{1}{2I_0}\,\mathrm{Re}\left[\sqrt{\frac{3}{2}}(AB^* + CD^* + EF^* + GH^*) + \sqrt{2}(BC^* + FG^*)\right] = \mathrm{i}A_1
\end{aligned}
\tag{4.54}
$$

such that according to Eq. (4.11) we have

$$
\mathrm{i}A_{1M} = \frac{1}{4I_0}\mathrm{tr}\left\{\hat{F}\,\mathrm{i}\hat{T}_{1M}\hat{F}^+\right\} = \frac{2}{\sqrt{5}}\mathrm{i}A_M, \qquad M = \pm 1{,}0 \tag{4.55}
$$

Now, the rank-2 tensor analyzing powers are

$$A_{2M} = \frac{1}{4I_0}\text{tr}\left\{\hat{F}\hat{T}_{2M}\hat{F}^+\right\}, \qquad M = \pm 2, \pm 1, 0 \tag{4.56}$$

such that we can obtain

$$\hat{T}_{22}\hat{F}^+ = \sqrt{2}\begin{pmatrix} -C^* & \mathrm{i}F^* & G^* & -\mathrm{i}B^* \\ -\mathrm{i}D^* & -E^* & \mathrm{i}H^* & A^* \\ 0 & 0 & 0 & 0 \\ 0 & 0 & 0 & 0 \end{pmatrix}$$

$$\begin{aligned} A_{22} &= \frac{\sqrt{2}}{4I_0}[-(AC^* + BD^* + HF^* + GE^* + EG^* + FH^* + DB^* + CA^*)] \\ &= -\frac{1}{\sqrt{2}I_0}\,\text{Re}\,(AC^* + BD^* + EG^* + FH^*) \end{aligned} \tag{4.57}$$

$$\hat{T}_{21}\hat{F}^+ = \sqrt{2}\begin{pmatrix} -\mathrm{i}B^* & -G^* & \mathrm{i}F^* & C^* \\ 0 & 0 & 0 & 0 \\ -\mathrm{i}D^* & -E^* & \mathrm{i}H^* & A^* \\ 0 & 0 & 0 & 0 \end{pmatrix}$$

$$\begin{aligned} A_{21} &= \frac{\sqrt{2}}{4I_0}[-\mathrm{i}(AB^* - CD^* + HG^* - FE^* + EF^* - GH^* + DC^* - BA^*)] \\ &= \frac{1}{\sqrt{2}I_0}\text{Im}(AB^* - CD^* - GH^* + EF^*) \end{aligned} \tag{4.58}$$

$$\hat{T}_{20}\hat{F}^+ = \begin{pmatrix} A^* & -\mathrm{i}H^* & -E^* & \mathrm{i}D^* \\ -\mathrm{i}B^* & -G^* & \mathrm{i}F^* & C^* \\ C^* & -\mathrm{i}F^* & -G^* & \mathrm{i}B^* \\ -\mathrm{i}D^* & -E^* & \mathrm{i}H^* & A^* \end{pmatrix}$$

$$\begin{aligned} A_{20} &= \frac{1}{4I_0}\left(|A|^2 - |B|^2 - |C|^2 + |D|^2 + |H|^2 - |G|^2 - |F|^2 + |E|^2 \right. \\ &\quad \left. + |E|^2 - |F|^2 - |G|^2 + |H|^2 + |D|^2 - |C|^2 - |B|^2 + |A|^2\right) \\ &= \frac{1}{2I_0}\left(|A|^2 - |B|^2 - |C|^2 + |D|^2 + |E|^2 - |F|^2 - |G|^2 + |H|^2\right) \end{aligned}$$

$$\hat{T}_{2\ -1}\hat{F}^+ = \sqrt{2}\begin{pmatrix} 0 & 0 & 0 & 0 \\ A^* & -\mathrm{i}H^* & -E^* & \mathrm{i}D^* \\ 0 & 0 & 0 & 0 \\ C^* & -\mathrm{i}F^* & -G^* & \mathrm{i}B^* \end{pmatrix} \tag{4.59}$$

$$\begin{aligned} A_{2\ -1} &= \frac{\sqrt{2}}{4I_0}[-\mathrm{i}(BA^* - DC^* + GH^* - EF^* + FE^* - HG^* + CD^* - AB^*)] \\ &= -\frac{1}{\sqrt{2}I_0}\mathrm{Im}(AB^* - CD^* - GH^* + EF^*) = -A_{21} \end{aligned} \tag{4.60}$$

$$\hat{T}_{2\ -2}\hat{F}^+ = \sqrt{2}\begin{pmatrix} 0 & 0 & 0 & 0 \\ 0 & 0 & 0 & 0 \\ A^* & -\mathrm{i}H^* & -E^* & \mathrm{i}D^* \\ \mathrm{i}B^* & G^* & -\mathrm{i}F^* & -C^* \end{pmatrix}$$

$$\begin{aligned} A_{2\ -2} &= \frac{\sqrt{2}}{4I_0}[-(CA^* + DB^* + FH^* + EG^* + GE^* + HF^* + BD^* + AC^*)] \\ &= -\frac{1}{\sqrt{2}I_0}\mathrm{Re}\,(AC^* + BD^* + EG^* + FH^*) = A_{22} \end{aligned} \tag{4.61}$$

In addition, the rank-3 tensor analyzing powers are

$$\mathrm{i}A_{3M} = \frac{1}{4I_0}\mathrm{tr}\left\{\hat{F}\,\mathrm{i}\hat{T}_{3M}\hat{F}^+\right\}, \qquad M = \pm 3, \pm 2, \pm 1{,}0 \tag{4.62}$$

yielding

$$\mathrm{i}\hat{T}_{33}\hat{F}^+ = 2\mathrm{i}\begin{pmatrix} \mathrm{i}D^* & E^* & -\mathrm{i}H^* & -A^* \\ 0 & 0 & 0 & 0 \\ 0 & 0 & 0 & 0 \\ 0 & 0 & 0 & 0 \end{pmatrix}$$

$$\mathrm{i}A_{33} = \frac{2\mathrm{i}}{4I_0}[\mathrm{i}(AD^* + HE^* + EH^* + DA^*)] = -\frac{1}{I_0}\mathrm{Re}\,(AD^* + EH^*)$$

$$\mathrm{i}\hat{T}_{32}\hat{F}^+ = \sqrt{2}\mathrm{i}\begin{pmatrix} -C^* & \mathrm{i}F^* & G^* & -\mathrm{i}B^* \\ \mathrm{i}D^* & E^* & -\mathrm{i}H^* & -A^* \\ 0 & 0 & 0 & 0 \\ 0 & 0 & 0 & 0 \end{pmatrix} \tag{4.63}$$

$$\begin{aligned} \mathrm{i}A_{32} &= \frac{\sqrt{2}\mathrm{i}}{4I_0}[-(AC^* - BD^* + HF^* - GE^* + EG^* - FH^* + DB^* - CA^*)] \\ &= \frac{1}{\sqrt{2}I_0}\mathrm{Im}(AC^* - BD^* + EG^* - FH^*) \end{aligned} \tag{4.64}$$

$$
\mathrm{i}\hat{T}_{31}\hat{F}^{+} = \frac{2i}{\sqrt{5}}\begin{pmatrix} -\mathrm{i}B^* & -G^* & \mathrm{i}F^* & C^* \\ -\sqrt{3}C^* & \mathrm{i}\sqrt{3}F^* & \sqrt{3}G^* & -\mathrm{i}\sqrt{3}B^* \\ \mathrm{i}D^* & E^* & -\mathrm{i}H^* & -A^* \\ 0 & 0 & 0 & 0 \end{pmatrix}
$$

$$
\begin{aligned}
\mathrm{i}A_{31} &= \frac{2i}{4\sqrt{5}I_0}\Big[-\mathrm{i}\Big(AB^* - \sqrt{3}BC^* + CD^* + HG^* - \sqrt{3}GF^* + FE^* \\
&\quad + EF^* - \sqrt{3}FG^* + GH^* + DC^* - \sqrt{3}CB^* + BA^*\Big)\Big] \\
&= \frac{1}{\sqrt{5}I_0}\,\mathrm{Re}\Big[(AB^* + CD^* + EF^* + GH^*) - \sqrt{3}(BC^* + FG^*)\Big]
\end{aligned}
$$

$$
\mathrm{i}\hat{T}_{30}\hat{F}^{+} = \frac{\mathrm{i}}{\sqrt{5}}\begin{pmatrix} A^* & -\mathrm{i}H^* & -E^* & \mathrm{i}D^* \\ -3\mathrm{i}B^* & -3G^* & 3\mathrm{i}F^* & 3C^* \\ -3C^* & 3\mathrm{i}F^* & 3G^* & -3\mathrm{i}B^* \\ \mathrm{i}D^* & E^* & -\mathrm{i}H^* & -A^* \end{pmatrix} \tag{4.65}
$$

$$
\begin{aligned}
\mathrm{i}A_{30} &= \frac{\mathrm{i}}{4\sqrt{5}I_0}\Big(|A|^2 - 3|B|^2 + 3|C|^2 - |D|^2 + |H|^2 - 3|G|^2 + 3|F|^2 - |E|^2 \\
&\quad + |E|^2 - 3|F|^2 + 3|G|^2 - |H|^2 + |D|^2 - 3|C|^2 + 3|B|^2 - |A|^2\Big) = 0
\end{aligned} \tag{4.66}
$$

$$
\mathrm{i}\hat{T}_{3\;-1}\hat{F}^{+} = \frac{2\mathrm{i}}{\sqrt{5}}\begin{pmatrix} 0 & 0 & 0 & 0 \\ A^* & -\mathrm{i}H^* & -E^* & \mathrm{i}D^* \\ -\mathrm{i}\sqrt{3}B^* & -\sqrt{3}G^* & \mathrm{i}\sqrt{3}F^* & \sqrt{3}C^* \\ -C^* & \mathrm{i}F^* & G^* & -\mathrm{i}B^* \end{pmatrix}
$$

$$
\begin{aligned}
\mathrm{i}A_{3\;-1} &= \frac{2\mathrm{i}}{4\sqrt{5}I_0}\Big[-\mathrm{i}\Big(BA^* - \sqrt{3}CB^* + DC^* + GH^* - \sqrt{3}FG^* + EF^* \\
&\quad + FE^* - \sqrt{3}GF^* + HG^* + CD^* - \sqrt{3}BC^* + AB^*\Big)\Big] \\
&= \frac{1}{\sqrt{5}I_0}\,\mathrm{Re}\Big[(AB^* + CD^* + EF^* + GH^*) - \sqrt{3}(BC^* + FG^*)\Big] = \mathrm{i}A_{31}
\end{aligned} \tag{4.67}
$$

$$
\mathrm{i}\hat{T}_{3\;-2}\hat{F}^{+} = \sqrt{2}\mathrm{i}\begin{pmatrix} 0 & 0 & 0 & 0 \\ 0 & 0 & 0 & 0 \\ A^* & -\mathrm{i}H^* & -E^* & \mathrm{i}D^* \\ -\mathrm{i}B^* & -G^* & \mathrm{i}F^* & C^* \end{pmatrix}
$$

$$\begin{aligned} \mathrm{i}A_{3\ -2} &= \frac{\sqrt{2}\mathrm{i}}{4I_0}\left[-(CA^* - DB^* + FH^* - EG^* + GE^* - HF^* + BD^* - AC^*)\right] \\ &= -\frac{1}{\sqrt{2}I_0}\mathrm{Im}(AC^* - BD^* + EG^* - FH^*) = -\mathrm{i}A_{32} \end{aligned} \tag{4.68}$$

$$\mathrm{i}\hat{T}_{3\ -3}\hat{F}^+ = 2\mathrm{i}\begin{pmatrix} 0 & 0 & 0 & 0 \\ 0 & 0 & 0 & 0 \\ 0 & 0 & 0 & 0 \\ A^* & -\mathrm{i}H^* & -E^* & \mathrm{i}D^* \end{pmatrix}$$

$$\mathrm{i}A_{3\ -3} = \frac{2\mathrm{i}}{4I_0}\left[\mathrm{i}(DA^* + EH^* + HE^* + AD^*)\right] = -\frac{1}{I_0}\,\mathrm{Re}\,(AD^* + EH^*) = \mathrm{i}A_{33} \tag{4.69}$$

Now from these results, it can be seen that the non-zero and independent polarization analyzing power components of rank-1 tensors (vectors), rank-2 tensors, and rank-3 tensors corresponding to the $S = \frac{3}{2}$ incident particles are $\mathrm{i}A_{11}$; $A_{22}$, $A_{21}$, $A_{20}$; $\mathrm{i}A_{33}$, $\mathrm{i}A_{32}$, $\mathrm{i}A_{31}$.

Using the polarization analyzing power components obtained above, the differential cross section corresponding to the $S = \frac{3}{2}$ polarized incident particles can be written according to Eq. (4.21) as

$$\begin{aligned} I = \mathrm{tr}\left\{\hat{F}\hat{\rho}_{\mathrm{in}}\hat{F}^+\right\} &= I_0[1 + (\mathrm{i}t_{1\ -1} + \mathrm{i}t_{11})\,\mathrm{i}A_{11} + (t_{2\ -2} + t_{22})A_{22} - (t_{2\ -1} - t_{21})A_{21} \\ &+ t_{20}A_{20} + (\mathrm{i}t_{3\ -3} + \mathrm{i}t_{33})\,\mathrm{i}A_{33} - (\mathrm{i}t_{3\ -2} - \mathrm{i}t_{32})\,\mathrm{i}A_{32} + (\mathrm{i}t_{3\ -1} + \mathrm{i}t_{31})\,\mathrm{i}A_{31}] \end{aligned} \tag{4.70}$$

Next, we define the polarization transfer coefficients of the $S = \frac{3}{2}$ particles as

$$\begin{aligned} \mathrm{i}^{\delta_{L'1}+\delta_{L'3}+\delta_{L1}+\delta_{L3}}K_{LM}^{L'M'} &= \frac{1}{\mathrm{tr}\left\{\hat{F}\hat{F}^+\right\}}\mathrm{tr}\left\{\mathrm{i}^{\delta_{L'1}+\delta_{L'3}}\hat{T}_{L'M'}\hat{F}\,\mathrm{i}^{\delta_{L1}+\delta_{L3}}\hat{T}_{LM}\hat{F}^+\right\}, \\ L',L &= 1,2,3; \quad -L' \le M' \le L'; \quad -L \le M \le L \end{aligned} \tag{4.71}$$

Using Eqs. (4.29), (4.31), (4.5), (4.11), (4.17), and (4.19), the corresponding various polarization transfer coefficients can be obtained from Eq. (4.71). It follows that, according to Eq. (4.21), the various components of the polarization rank-1 tensor (vector), rank-2 tensor, and rank-3 tensor of the outgoing particles can be obtained.

Using the methods described in Chap. 2 and 3, the polarization theory of spin $S = \frac{3}{2}$ particle nuclear reactions can also be studied when target and residual nucleus spins are not 0.

Equation (1.91) gives the general expression for the density matrix of the initial state

$$\hat{\rho}_{\text{in}}(S) = \frac{1}{2S+1}\sum_{L=0}^{2S} t_L(S)\cdot\hat{T}_L(S) \tag{4.72}$$

Here the following generalized Pauli matrix is introduced according to Eq. (1.55):

$$\hat{\Sigma}_M(S) = \hat{T}_{1M}(S) = \sqrt{\frac{3}{S(S+1)}}\,\hat{S}_M(S), \quad M = \pm 1,\ 0 \tag{4.73}$$

When spin $S = \frac{1}{2}$, from Eq. (4.73) we can obtain

$$\hat{\Sigma}_M\left(\frac{1}{2}\right) = 2\,\hat{S}_M\left(\frac{1}{2}\right) = \hat{\sigma}_M, \quad M = \pm 1,\ 0 \tag{4.74}$$

It means that when spin $S = \frac{1}{2}$, the generalized Pauli matrix $\hat{\Sigma}$ is just the Pauli matrix $\hat{\sigma}$. When spin $S = 1$, from Eq. (4.73) we can obtain

$$\hat{\Sigma}_M(1) = \sqrt{\frac{3}{2}}\,\hat{S}_M(1), \quad M = \pm 1,\ 0 \tag{4.75}$$

When spin $S = \frac{3}{2}$, from Eq. (4.73) we can obtain

$$\hat{\Sigma}_M\left(\frac{3}{2}\right) = \frac{2}{\sqrt{5}}\hat{S}_M\left(\frac{3}{2}\right), \quad M = \pm 1,\ 0 \tag{4.76}$$

Note that $\hat{T}_{00}(S) = \hat{I}$ given in Eq. (1.54) and $t_{00}(S) = 1$ given in Eq. (1.99), when spin $S = \frac{1}{2}$ the following expression can be obtained from Eq. (4.72):

$$\hat{\rho}_{\text{in}}\left(\frac{1}{2}\right) = \frac{1}{2}\sum_{L=0}^{1} t_L\left(\frac{1}{2}\right)\cdot\hat{T}_L\left(\frac{1}{2}\right) = \frac{1}{2}\left[\hat{I} + \sum_{M=-1}^{1}(-1)^M(-\mathrm{i}t_{1\ -M})\left(\mathrm{i}\hat{T}_{1M}\right)\right] \tag{4.77}$$

The transformation relations between $\hat{H}_{1i}(i = x,\ y,\ z = 1,\ 2,\ 3)$ and $\hat{T}_{1M}(M = \pm 1, 0)$ can be written, from Eqs. (1.23) and (1.22), as follows:

$$\hat{T}_{1\ \pm 1} = \mp\frac{1}{\sqrt{2}}\left(\hat{H}_{1x} \pm \mathrm{i}\hat{H}_{1y}\right), \quad \hat{T}_{10} = \hat{H}_{1z} \tag{4.78}$$

$$\hat{H}_{1x} = \frac{1}{\sqrt{2}}\left(\hat{T}_{1\ -1} - \hat{T}_{1\ 1}\right), \quad \hat{H}_{1y} = \frac{\mathrm{i}}{\sqrt{2}}\left(\hat{T}_{1\ -1} + \hat{T}_{1\ 1}\right), \quad \hat{H}_{1z} = \hat{T}_{10} \tag{4.79}$$

It is also noted that $h_{1i}(i = x, y, z = 1, 2, 3)$ and $t_{1M}(M = \pm 1, 0)$ are transformed each other by the same transformation relations as the above two formulas. It is easy to find

$$\begin{aligned} &-\left[(-\mathrm{i}t_{1\ -1})\mathrm{i}\hat{T}_{1\ 1} + (-\mathrm{i}t_{1\ 1})\ \mathrm{i}\hat{T}_{1\ -1}\right] \\ &= \frac{1}{2}\left[\left(h_{1x} - \mathrm{i}h_{1y}\right)\left(\hat{H}_{1x} + \mathrm{i}\hat{H}_{1y}\right) + \left(h_{1x} + \mathrm{i}h_{1y}\right)\left(\hat{H}_{1x} - \mathrm{i}\hat{H}_{1y}\right)\right] = h_{1x}\hat{H}_{1x} + h_{1y}\hat{H}_{1y} \end{aligned} \tag{4.80}$$

Then Eq. (4.77) can be rewritten as

$$\hat{\rho}_{\text{in}}\left(\frac{1}{2}\right) = \frac{1}{2}\left[\hat{I} + \sum_{i=x,\ y,\ z} h_{1i}\hat{H}_{1i}\right] = \frac{1}{2}\left[\hat{I} + \sum_{i=1}^{3} h_{1i}\hat{H}_{1i}\right] \tag{4.81}$$

where $\hat{H}_{1i}$ is just the Pauli matrix $\hat{\sigma}_i$ and $h_{1i}$ is the component $p_i$ of the polarization vector. $\hat{\rho}_{\text{in}}, \hat{I}, \hat{H}_{1i}$ are all the square $2 \times 2$ matrix in spin space.

When spin $S = 1$, the following expression can be obtained from Eq. (4.72):

$$\begin{aligned} \hat{\rho}_{\text{in}}(1) &= \frac{1}{3}\sum_{L=0}^{2} t_L(1) \cdot \hat{T}_L(1) \\ &= \frac{1}{3}\left[\hat{I} + \sum_{M=-1}^{1} (-1)^M(-\mathrm{i}t_{1\ -M})\left(\mathrm{i}\hat{T}_{1M}\right) + \sum_{M=-2}^{2} (-1)^M t_{2\ -M}\hat{T}_{2M}\right] \end{aligned} \tag{4.82}$$

The spin 1 particle polarization tensor basis functions in the five-dimensional orthogonal Cartesian coordinate system given by Eq. (3.142) are

$$\begin{aligned} &\left(\hat{H}_{21}, \hat{H}_{22}, \hat{H}_{23}, \hat{H}_{24}, \hat{H}_{25}\right) = \left(\hat{H}_{2\ xx-yy}, \hat{H}_{2xy}, \hat{H}_{2xz}, \hat{H}_{2yz}, \hat{H}_{2zz}\right) \\ &= \left(\sqrt{\frac{1}{6}}\hat{Q}_{xx-yy}, \sqrt{\frac{2}{3}}\hat{Q}_{xy}, \sqrt{\frac{2}{3}}\hat{Q}_{xz}, \sqrt{\frac{2}{3}}\hat{Q}_{yz}, \sqrt{\frac{1}{2}}\hat{Q}_{zz}\right) \end{aligned} \tag{4.83}$$

Using above formula, Eq. (3.22) can be rewritten as

$$\hat{T}_{2\ \pm2} = \frac{1}{\sqrt{2}}\left(\hat{H}_{2\ xx-yy} \pm \mathrm{i}\hat{H}_{2xy}\right), \quad \hat{T}_{2\ \pm1} = \mp\frac{1}{\sqrt{2}}\left(\hat{H}_{2xz} \pm \mathrm{i}\hat{H}_{2yz}\right), \quad \hat{T}_{2\ 0} = \hat{H}_{2zz} \tag{4.84}$$

The last two relations in the above formulas are consistent with the three- dimensional vector transformation relation Eq. (4.78). Using Eq. (4.83) we can rewrite Eq. (3.21) as

$$\begin{aligned}
&\hat{H}_{2\ xx-yy} = \frac{1}{\sqrt{2}}\left(\hat{T}_{2\ -2} + \hat{T}_{2\ 2}\right), \quad \hat{H}_{2xy} = \frac{\mathrm{i}}{\sqrt{2}}\left(\hat{T}_{2\ -2} - \hat{T}_{2\ 2}\right), \\
&\hat{H}_{2xz} = \frac{1}{\sqrt{2}}\left(\hat{T}_{2\ -1} - \hat{T}_{2\ 1}\right), \qquad \hat{H}_{2yz} = \frac{\mathrm{i}}{\sqrt{2}}\left(\hat{T}_{2\ -1} + \hat{T}_{2\ 1}\right). \\
&\hat{H}_{2zz} = \hat{T}_{2\ 0}
\end{aligned} \tag{4.85}$$

These formulas can also be derived from Eq. (4.84) or directly obtained from Eq. (3.142). Using the method of finding Eq. (4.80), we can find

$$\begin{aligned}
-\left(t_{2\ -1}\hat{T}_{2\ 1} + t_{2\ 1}\hat{T}_{2\ -1}\right) &= h_{2xz}\hat{H}_{2xz} + h_{2yz}\hat{H}_{2yz} \\
t_{2\ -2}\hat{T}_{2\ 2} + t_{2\ 2}\hat{T}_{2\ -2} &= h_{2\ xx-yy}\hat{H}_{2\ xx-yy} + h_{2xy}\hat{H}_{2xy}
\end{aligned} \tag{4.86}$$

So using Eq. (4.80) and Eq. (4.86), from Eq. (4.82) we can get

$$\begin{aligned}
\hat{\rho}_{\text{in}}(1) &= \frac{1}{3}\left[\hat{I} + \sum_{i=x,\,y,\,z} h_{1i}\hat{H}_{1i} + \sum_{\substack{i=xx-yy,\,xy,\\ xz,\,yz,\,zz}} h_{2i}\hat{H}_{2i}\right] \\
&= \frac{1}{3}\left[\hat{I} + \sum_{i=1}^{3} h_{1i}\hat{H}_{1i} + \sum_{i=1}^{5} h_{2i}\hat{H}_{2i}\right]
\end{aligned} \tag{4.87}$$

where $\hat{\rho}_{\text{in}}$, $\hat{I}$, $\hat{H}_{1i}$, $\hat{H}_{2i}$ are all square $3 \times 3$ matrix in spin space. $\hat{H}_{1i}$ is the spin 1 particle generalized Pauli matrix, and from Eq. (4.75) we know

$$\hat{H}_{1i}(1) = \sqrt{\frac{3}{2}}\hat{S}_i, \qquad i = x,\ y,\ z \tag{4.88}$$

In the same way we also know

$$h_{1i}(1) = \sqrt{\frac{3}{2}}p_i, \qquad i = x,\ y,\ z \tag{4.89}$$

Refer to Eq. (4.83) there is

$$\begin{aligned}
&(h_{21}, h_{22}, h_{23}, h_{24}, h_{25}) = \left(h_{2\ xx-yy}, h_{2xy}, h_{2xz}, h_{2yz}, h_{2zz}\right) \\
&= \left(\sqrt{\frac{1}{6}}p_{xx-yy}, \sqrt{\frac{2}{3}}p_{xy}, \sqrt{\frac{2}{3}}p_{xz}, \sqrt{\frac{2}{3}}p_{yz}, \sqrt{\frac{1}{2}}p_{zz}\right)
\end{aligned} \tag{4.90}$$

And we can obtain Eq. (3.214) when substituting Eqs. (4.83) and (4.88) – (4.90) into Eq. (4.87).

When spin $S = \frac{3}{2}$, the following expression can be obtained from Eq. (4.72) or Eq. (4.21):

$$
\begin{aligned}
\hat{\rho}_{\text{in}}\left(\frac{3}{2}\right) = \frac{1}{4}\sum_{L=0}^{3} t_L\left(\frac{3}{2}\right)\cdot\hat{T}_L\left(\frac{3}{2}\right) = \frac{1}{4}\Bigg[\hat{I} + \sum_{M=-1}^{1}(-1)^M(-\mathrm{i}t_{1\ -M})\left(\mathrm{i}\hat{T}_{1M}\right) \\
+ \sum_{M=-2}^{2}(-1)^M t_{2\ -M}\hat{T}_{2M} + \sum_{M=-3}^{3}(-1)^M(-\mathrm{i}t_{3\ -M})\left(\mathrm{i}\hat{T}_{3M}\right)\Bigg]
\end{aligned}
\tag{4.91}
$$

From Eq. (4.84) the following unitary transformation relations of the spin $\frac{3}{2}$ particle seven-dimensional rank-3 polarization tensor basis functions can be derived recursively:

$$
\begin{aligned}
\hat{T}_{3\ \pm3} &= \mp\frac{1}{\sqrt{2}}\left(\hat{H}_{31} \pm \mathrm{i}\hat{H}_{32}\right),\quad \hat{T}_{3\ \pm2} = \frac{1}{\sqrt{2}}\left(\hat{H}_{33} \pm \mathrm{i}\hat{H}_{34}\right), \\
\hat{T}_{3\ \pm1} &= \mp\frac{1}{\sqrt{2}}\left(\hat{H}_{35} \pm \mathrm{i}\hat{H}_{36}\right),\quad \hat{T}_{3\,0} = \hat{H}_{37}
\end{aligned}
\tag{4.92}
$$

where $\hat{H}_{3i}$ $(i = 1 \sim 7)$ are the spin $\frac{3}{2}$ particle rank-3 tenser basis functions in the orthogonal Cartesian coordinate system. Their inverse transformation relations can be obtained from Eq. (4.92) as

$$
\begin{aligned}
\hat{H}_{31} &= \frac{1}{\sqrt{2}}\left(\hat{T}_{3\ -3} - \hat{T}_{3\,3}\right),\quad \hat{H}_{32} = \frac{\mathrm{i}}{\sqrt{2}}\left(\hat{T}_{3\ -3} + \hat{T}_{3\,3}\right), \\
\hat{H}_{33} &= \frac{1}{\sqrt{2}}\left(\hat{T}_{3\ -2} + \hat{T}_{3\,2}\right),\quad \hat{H}_{34} = \frac{\mathrm{i}}{\sqrt{2}}\left(\hat{T}_{3\ -2} - \hat{T}_{3\,2}\right), \\
\hat{H}_{35} &= \frac{1}{\sqrt{2}}\left(\hat{T}_{3\ -1} - \hat{T}_{3\,1}\right),\quad \hat{H}_{36} = \frac{\mathrm{i}}{\sqrt{2}}\left(\hat{T}_{3\ -1} + \hat{T}_{3\,1}\right), \\
\hat{H}_{37} &= \hat{T}_{3\,0}
\end{aligned}
\tag{4.93}
$$

The above formula can also be obtained from Eq. (4.85) by recursion method. The expression (4.93) describing the seven-dimensional rank-3 tensors can also be rewritten in an obvious unitary transformation form by reference to Eqs. (3.142) and (3.143) describing the five-dimensional rank-2 tensors. Using the method of finding Eq. (4.80) we can find

$$
\begin{aligned}
&-\left[(-\mathrm{i}t_{3\ -1})\mathrm{i}\hat{T}_{3\,1} + (-\mathrm{i}t_{3\,1})\,\mathrm{i}\hat{T}_{3\ -1}\right] = h_{35}\hat{H}_{35} + h_{36}\hat{H}_{36} \\
&(-\mathrm{i}t_{3\ -2})\,\mathrm{i}\hat{T}_{3\,2} + (-\mathrm{i}t_{3\,2})\,\mathrm{i}\hat{T}_{3\ -2} = h_{33}\hat{H}_{33} + h_{34}\hat{H}_{34} \\
&-\left[(-\mathrm{i}t_{3\ -3})\mathrm{i}\hat{T}_{3\,3} + (-\mathrm{i}t_{3\,3})\,\mathrm{i}\hat{T}_{3\ -3}\right] = h_{31}\hat{H}_{31} + h_{32}\hat{H}_{32}
\end{aligned}
\tag{4.94}
$$

Then we can obtain the following expression from Eq. (4.91) and using Eqs. (4.80), (4.86), and (4.94)

$$\hat{\rho}_{\rm in}\left(\frac{3}{2}\right) = \frac{1}{4}\left[\hat{I} + \sum_{i=1}^{3} h_{1i}\hat{H}_{1i} + \sum_{i=1}^{5} h_{2i}\hat{H}_{2i} + \sum_{i=1}^{7} h_{3i}\hat{H}_{3i}\right] \tag{4.95}$$

where $\hat{\rho}_{\rm in}, \hat{I}, \hat{H}_{1i}, \hat{H}_{2i}, \hat{H}_{3i}$ are all square $4 \times 4$ matrix in spin space. The specific expressions of the spin $\frac{3}{2}$ particle polarization operator $\hat{T}_{2M}$ $(M = \pm 2, \pm 1, 0)$ and $\hat{T}_{3M}$ $(M = \pm 3, \pm 2, \pm 1, 0)$ in the spherical basis coordinate system have been given by Eq. (4.17) and Eq. (4.19), respectively. And the specific expressions of the spin $\frac{3}{2}$ particle polarization operator $\hat{H}_{2i}$ $(i = 1 \sim 5)$ and $\hat{H}_{3i}$ $(i = 1 \sim 7)$ in the orthogonal Cartesian coordinate system can be obtained by Eq. (4.85) and Eq. (4.93), respectively. $\hat{H}_{1i}$ $(i = 1 \sim 3)$ is the spin $\frac{3}{2}$ particle generalized Pauli matrix, and from Eq. (4.76) we know

$$\hat{H}_{1i}\left(\frac{3}{2}\right) = \frac{2}{\sqrt{5}}\hat{S}_i\left(\frac{3}{2}\right), \quad i = 1 \sim 3 \tag{4.96}$$

The spin $\frac{3}{2}$ particle spin operator $\hat{S}_i\left(\frac{3}{2}\right)$, $i = 1 \sim 3$ in Cartesian coordinate system has been given by Eq. (4.6), and the corresponding spin operator $\hat{S}_M\left(\frac{3}{2}\right)$, $M = \pm 1{,}0$ in spherical basis coordinate system has been given by Eq. (4.5).

By using the recursion method, the coordinate transformation relation Eq. (4.92) and Eq. (4.93) can be extended to higher-rank tensors, and thus the relation Eq. (4.95) can be rewritten into a general form suitable for any spin $S$

$$\hat{\rho}_{\rm in}(S) = \frac{1}{2S+1}\sum_{L=0}^{2S}\sum_{i=1}^{2L+1} h_{Li}(S)\hat{H}_{Li}(S) \tag{4.97}$$

Using the spin $\frac{3}{2}$ particle polarization theory given above and the expressions shown in Table 4.4 of the rotation function $d^3_{MM'}(\beta)$ given in Ref. [1], as well as the relation

$$d^J_{MM'}(\beta) = (-1)^{M-M'} d^J_{M'M}(\beta) \tag{4.98}$$

the spin $\frac{3}{2}$ particle polarization theory can be carried out a comprehensive and systematic study, and can reach a practical level. In accordance with the above approach, the spin $S > \frac{3}{2}$ particle polarization theory can also be carried out the specific studying work. However, the higher the spin value of the particle, the more complex the polarization theory is, and the practical value will become smaller. Therefore, under normal circumstances, there is no need to study the polarization problem that the particle spin is greater than $\frac{3}{2}$.

**Table 4.4** Expressions of $d^3_{MM'}(\beta)$ [1]

| $M'$ | $M = 3$ | $M'$ | $M = 2$ |
|---|---|---|---|
| 3 | $\frac{1}{8}(1 + \cos\beta)^3$ | 0 | $\frac{\sqrt{30}}{4}\sin^2\beta\cos\beta$ |
| 2 | $-\frac{\sqrt{6}}{8}\sin\beta(1 + \cos\beta)^2$ | −1 | $-\frac{\sqrt{10}}{8}\sin\beta(1 + 2\cos\beta - 3\cos^2\beta)$ |
| 1 | $\frac{\sqrt{15}}{8}\sin^2\beta(1 + \cos\beta)$ | −2 | $\frac{1}{4}(1 - \cos\beta)^2(2 + 3\cos\beta)$ |
| 0 | $-\frac{\sqrt{5}}{4}\sin^3\beta$ | $M'$ | $M = 1$ |
| −1 | $\frac{\sqrt{15}}{8}\sin^2\beta(1 - \cos\beta)$ | 1 | $-\frac{1}{8}(1 + \cos\beta)(1 + 10\cos\beta - 15\cos^2\beta)$ |
| −2 | $-\frac{\sqrt{6}}{8}\sin\beta(1 - \cos\beta)^2$ | 0 | $\frac{\sqrt{3}}{4}\sin\beta(1 - 5\cos^2\beta)$ |
| −3 | $\frac{1}{8}(1 - \cos\beta)^3$ | −1 | $-\frac{1}{8}(1 - \cos\beta)(1 - 10\cos\beta - 15\cos^2\beta)$ |
| $M'$ | $M = 2$ | $M'$ | $M = 0$ |
| 2 | $-\frac{1}{4}(1 + \cos\beta)^2(2 - 3\cos\beta)$ | 0 | $-\frac{1}{2}\cos\beta(3 - 5\cos^2\beta)$ |
| 1 | $\frac{\sqrt{10}}{8}\sin\beta(1 - 2\cos\beta - 3\cos^2\beta)$ | | |

## 4.2 Polarization Theory for Photon Beams

This section will consider electromagnetic fields in a vacuum and use Gaussian units.

### *4.2.1 Classic Electromagnetic Field Theory [3]*

The force of the electromagnetic field felt by a particle with charge q (Lorentz force) is given by the following formula:

$$\vec{F}(\vec{r}, t) = q\left[\vec{E}(\vec{r}, t) + \frac{1}{c}\vec{v} \times \vec{B}(\vec{r}, t)\right] \tag{4.99}$$

where $\vec{v}$ is particle velocity, c is the speed of light, and $\vec{E}(\vec{r}, t)$ and $\vec{B}(\vec{r}, t)$ are the electric field strength and magnetic induction strength, respectively, which satisfy the following Maxwell equations:

$$\vec{\nabla} \cdot \vec{E} = 4\pi\rho \tag{4.100}$$

$$\vec{\nabla} \times \vec{E} = -\frac{1}{c}\frac{\partial \vec{B}}{\partial t} \tag{4.101}$$

$$\vec{\nabla} \cdot \vec{B} = 0 \tag{4.102}$$

$$\vec{\nabla} \times \vec{B} = \frac{4\pi}{c}\vec{j} + \frac{1}{c}\frac{\partial \vec{E}}{\partial t} \tag{4.103}$$

where $\rho(\vec{r}, t)$ and $\vec{j}(\vec{r}, t)$ are the charge density and electric current density, respectively. Now, because of the charge conservation, we have the following continuous equation:

$$\frac{\partial \rho}{\partial t} + \vec{\nabla} \cdot \vec{j} = 0 \tag{4.104}$$

The divergence of $\vec{B}$ given in Eq. (4.102) is 0, that is, $\vec{B}$ is the transverse field, so it can be expressed as a kind of curl of a vector potential $\vec{A}(\vec{r}, t)$

$$\vec{B} = \vec{\nabla} \times \vec{A} \tag{4.105}$$

Substituting this formula into Eq. (4.101), we can get

$$\vec{\nabla} \times \left(\vec{E} + \frac{1}{c}\frac{\partial \vec{A}}{\partial t}\right) = 0 \tag{4.106}$$

Therefore, the amount in the brackets above must be expressed as a gradient of a scalar potential $\varphi(\vec{r}, t)$, that is,

$$\vec{E} + \frac{1}{c}\frac{\partial \vec{A}}{\partial t} = -\vec{\nabla}\varphi \tag{4.107}$$

which can be rewritten as

$$\vec{E} = -\frac{1}{c}\frac{\partial \vec{A}}{\partial t} - \vec{\nabla}\varphi \tag{4.108}$$

Substituting this formula into Eq. (4.100), we can get

$$\vec{\nabla}^2 \varphi + \frac{1}{c}\frac{\partial}{\partial t}\left(\vec{\nabla} \cdot \vec{A}\right) = -4\pi\rho \tag{4.109}$$

We can also obtain the following relation substituting Eqs. (4.105) and (4.108) into Eq. (4.103)

$$\vec{\nabla} \times \left(\vec{\nabla} \times \vec{A}\right) = \frac{4\pi}{c}\vec{j} - \frac{1}{c}\frac{\partial}{\partial t}\left(\vec{\nabla}\varphi + \frac{1}{c}\frac{\partial \vec{A}}{\partial t}\right) \tag{4.110}$$

Now using the vector formula

$$\vec{\nabla} \times \left(\vec{\nabla} \times \vec{A}\right) = -\vec{\nabla}^2 \vec{A} + \vec{\nabla}\left(\vec{\nabla} \cdot \vec{A}\right) \tag{4.111}$$

Equation (4.110) can be translated into

$$\vec{\nabla}^2 \vec{A} - \frac{1}{c^2}\frac{\partial^2 \vec{A}}{\partial t^2} = -\frac{4\pi}{c}\vec{j} + \vec{\nabla}\left(\vec{\nabla} \cdot \vec{A} + \frac{1}{c}\frac{\partial \varphi}{\partial t}\right) \tag{4.112}$$

Next we will prove that the above equation cannot guarantee that $\varphi$ and $\vec{A}$ have definite solutions, that is, by using different $\varphi$ and $\vec{A}$, we can obtain the same $\vec{E}$ and $\vec{B}$. In order to prove this statement, we introduce a scalar potential $\psi\left(\vec{r}, t\right)$ and define the following:

$$\vec{A}' = \vec{A} + \vec{\nabla}\psi \tag{4.113}$$

$$\varphi' = \varphi - \frac{1}{c}\frac{\partial \psi}{\partial t} \tag{4.114}$$

When we use $\varphi'$ and $\vec{A}'$ given by the above two formulas instead of $\varphi$ and $\vec{A}$ in Eqs. (4.105), (4.108), (4.109), and (4.112), it can be seen that all $\psi$ terms appearing in these four formulas disappear; that is, they all degenerate into the original forms, and Eqs. (4.105), (4.108), (4.109), and (4.112) are equivalent with Eqs. (4.100) – (4.103). The transformation from $\varphi$ and $\vec{A}$ to $\varphi'$ and $\vec{A}'$ given by Eqs. (4.113) and (4.114) is called the gauge transformation. The invariance of the four Maxwell equations is called the gauge invariance, that is, $\vec{E}$ and $\vec{B}$ produced after $\varphi$ and $\vec{A}$ are gauge transformed are still unchanged. Given this, it is permissible to apply certain constraints on the relationship between $\varphi$ and $\vec{A}$. The Lorentz gauge, which is typically applied, is

$$\vec{\nabla} \cdot \vec{A} + \frac{1}{c}\frac{\partial \varphi}{\partial t} = 0 \tag{4.115}$$

With this constraint, we can obtain the following expressions substituting Eq. (4.115) into Eqs. (4.109) and (4.112):

$$\vec{\nabla}^2 \varphi - \frac{1}{c^2}\frac{\partial^2 \varphi}{\partial t^2} = -4\pi\rho \tag{4.116}$$

$$\vec{\nabla}^2 \vec{A} - \frac{1}{c^2}\frac{\partial^2 \vec{A}}{\partial t^2} = -\frac{4\pi}{c}\vec{j} \tag{4.117}$$

Another constraint that has been employed is the Coulomb gauge (or transverse gauge), that is,

$$\vec{\nabla} \cdot \vec{A} = 0 \tag{4.118}$$

where this formula requires that $\vec{A}$ is a transverse potential, and using this formula, Eqs. (4.109) and (4.112) can be converted to

$$\vec{\nabla}^2 \varphi = -4\pi\rho \tag{4.119}$$

$$\vec{\nabla}^2 \vec{A} - \frac{1}{c^2}\frac{\partial^2 \vec{A}}{\partial t^2} = -\frac{4\pi}{c}\vec{j} + \frac{1}{c}\frac{\partial}{\partial t}\left(\vec{\nabla}\varphi\right) \tag{4.120}$$

In the vacuum region $\rho = 0$ and $\vec{j} = 0$, applying the Coulomb gauge condition given by Eq. (4.118), and selecting $\varphi = 0$ according to Eq. (4.119), which is also a gauge condition, then Eq. (4.120) becomes

$$\vec{\nabla}^2 \vec{A} - \frac{1}{c^2}\frac{\partial^2 \vec{A}}{\partial t^2} = 0 \tag{4.121}$$

### 4.2.2 Hamilton Canonical Equation [3]

The Lagrange function of the system is denoted as $L(q_1, \cdots, q_n, \dot{q}_1, \cdots, \dot{q}_n, t)$, or abbreviated as $L(q, \dot{q}, t)$, where $\dot{q}_i$ is a generalized speed. For conservative systems that do not change over time, time $t$ in function $L$ does not appear.

Now set the system to begin at time $t'$, reaching the end point at time $t''$ through a certain track $q(t)$, and define the action quantity

$$S[q(t)] = \int_{t'}^{t''} L(q, \dot{q})\mathrm{d}t \tag{4.122}$$

Note that the principle of minimum action quantity requires that the actual orbit of the particle should let $S$ be minimized. Next, set $q(t) \rightarrow q(t) + \delta q(t)$ in the condition

$$\delta q(t') = \delta q(t'') = 0 \tag{4.123}$$

such that we require

$$\delta S = 0 \tag{4.124}$$

Now, using the partial integral method and noting $\delta \dot{q}_i = \frac{\mathrm{d}}{\mathrm{d}t}\delta q_i$, we can obtain the following expression according to Eqs. (4.122) and (4.123)

$$\begin{aligned}\delta S &= \int_{t'}^{t''} \mathrm{d}t \sum_i \left[\frac{\partial L}{\partial q_i}\delta q_i + \frac{\partial L}{\partial \dot{q}_i}\delta \dot{q}_i\right] \\ &= \sum_i \int_{t'}^{t''} \mathrm{d}t \left[\frac{\partial L}{\partial q_i} - \frac{\mathrm{d}}{\mathrm{d}t}\left(\frac{\partial L}{\partial \dot{q}_i}\right)\right]\delta q_i + \sum_i \left[\frac{\partial L}{\partial \dot{q}_i}\delta q_i\right]_{t'}^{t''} \\ &= \sum_i \int_{t'}^{t''} \mathrm{d}t \left[\frac{\partial L}{\partial q_i} - \frac{\mathrm{d}}{\mathrm{d}t}\left(\frac{\partial L}{\partial \dot{q}_i}\right)\right]\delta q_i = 0\end{aligned} \tag{4.125}$$

where, since $\delta q_i$ is arbitrary, one requires

$$\frac{\partial L}{\partial q_i} - \frac{\mathrm{d}}{\mathrm{d}t}\left(\frac{\partial L}{\partial \dot{q}_i}\right) = 0, \qquad i = 1,2,\cdots,n \tag{4.126}$$

and this formula is the Lagrange equation.

Introducing the generalized momentum

$$p_i = \frac{\partial L}{\partial \dot{q}_i} \tag{4.127}$$

the definition of the Hamilton quantity of the system is

$$H(q,p) = \sum_i p_i \dot{q}_i - L(q,\dot{q}) \tag{4.128}$$

Using Eqs. (4.126) – (4.128), we can prove the following equations:

$$\dot{q}_i = \frac{\partial H}{\partial p_i}, \qquad \dot{p}_i = -\frac{\partial H}{\partial q_i}, \qquad i = 1,2,\ldots,n \tag{4.129}$$

where these equations are the Hamilton canonical equations.

Now, the Poisson bracket of any two mechanical quantities is defined as

$$\{A, B\} \equiv \sum_i \left(\frac{\partial A}{\partial q_i}\frac{\partial B}{\partial p_i} - \frac{\partial A}{\partial p_i}\frac{\partial B}{\partial q_i}\right) \tag{4.130}$$

and from this formula it is easy to prove that

$$\{q_i, p_j\} = \delta_{ij}, \qquad \{q_i, q_j\} = \{p_i, p_j\} = 0 \tag{4.131}$$

Using Poisson bracket, the canonical equation (4.129) can be expressed as

$$\dot{q}_i = \{q_i, H\}, \qquad \dot{p}_i = \{p_i, H\} \tag{4.132}$$

Given the following set of transformations

$$q \to Q(q, p), \qquad p \to P(q, p) \tag{4.133}$$

we can first obtain the following expression:

$$\dot{Q}_j = \sum_i \left( \frac{\partial Q_j}{\partial q_i} \dot{q}_i + \frac{\partial Q_j}{\partial p_i} \dot{p}_i \right) = \sum_i \left( \frac{\partial Q_j}{\partial q_i} \frac{\partial H}{\partial p_i} - \frac{\partial Q_j}{\partial p_i} \frac{\partial H}{\partial q_i} \right) \tag{4.134}$$

Next, transform $H(q,p)$ into $H(Q,P) = H(q,p)$; thus, we can obtain

$$\begin{aligned} \frac{\partial H(q, p)}{\partial p_i} &= \frac{\partial H(Q, P)}{\partial p_i} = \sum_k \left( \frac{\partial H}{\partial Q_k} \frac{\partial Q_k}{\partial p_i} + \frac{\partial H}{\partial P_k} \frac{\partial P_k}{\partial p_i} \right), \\ \frac{\partial H(q, p)}{\partial q_i} &= \frac{\partial H(Q, P)}{\partial q_i} = \sum_k \left( \frac{\partial H}{\partial Q_k} \frac{\partial Q_k}{\partial q_i} + \frac{\partial H}{\partial P_k} \frac{\partial P_k}{\partial q_i} \right) \end{aligned} \tag{4.135}$$

Substituting Eq. (4.135) into Eq. (4.134), and after sorting, we can get

$$\dot{Q}_j = \sum_k \left( \frac{\partial H}{\partial Q_k} \{Q_j, Q_k\} + \frac{\partial H}{\partial P_k} \{Q_j, P_k\} \right) \tag{4.136}$$

In a similar way we can also get

$$\dot{P}_j = \sum_k \left( \frac{\partial H}{\partial Q_k} \{P_j, Q_k\} + \frac{\partial H}{\partial P_k} \{P_j, P_k\} \right) \tag{4.137}$$

Now if the following conditions are met

$$\{Q_j, P_k\} = \delta_{jk}, \quad \{Q_j, Q_k\} = \{P_j, P_k\} = 0 \tag{4.138}$$

then we can obtain the following equations from Eqs. (4.136) and (4.137)

$$\dot{Q}_i = \frac{\partial H}{\partial P_i}, \qquad \dot{P}_i = -\frac{\partial H}{\partial Q_i} \tag{4.139}$$

At this time, $(q, p)$ and $(Q, P)$ are all canonical, and the transformation given by Eq. (4.133) is a canonical transformation.

### 4.2.3 Quantization of Electromagnetic Fields [3–5]

In order to avoid the difficulty of non-normalization in the calculation process, it is assumed that the radiation field is limited to a square box with a volume $V$, and it is also assumed that the electromagnetic field satisfies the periodic boundary conditions; thus $V \to \infty$ is finally ordered. The special solutions of Eq. (4.121) can be obtained using the following separation variable method:

$$\vec{A}(\vec{r}, t) = q(t)\vec{A}(\vec{r}) \tag{4.140}$$

The general solution can be expressed as a linear superposition of these special solutions. Substituting Eq. (4.140) into Eq. (4.121), we can obtain

$$\vec{\nabla}^2 \vec{A}(\vec{r}) + k^2 \vec{A}(\vec{r}) = 0 \tag{4.141}$$

$$\ddot{q}(t) + \omega^2 q(t) = 0 \tag{4.142}$$

where $\ddot{q}(t)$ represents a quadratic derivative of $q(t)$ to $t$; $\omega$ and $\omega = kc$ in the above formulas are constants that do not depend on $\vec{r}$ and $t$; and $c$ is the speed of light.

We study the plane wave light beams that propagate along the $\vec{k}$ direction. Here, $\vec{k}$ is the wave vector of the light beams. For a specific $k$, we use the following unit vectors to compose a right-handed rectangular coordinate system corresponding to $x$, $y$, $z$ axes

$$\vec{\varepsilon}_{k1} = \begin{pmatrix} 1 \\ 0 \\ 0 \end{pmatrix}, \quad \vec{\varepsilon}_{k2} = \begin{pmatrix} 0 \\ 1 \\ 0 \end{pmatrix}, \quad \vec{\varepsilon}_{k3} = \frac{\vec{k}}{k} = \begin{pmatrix} 0 \\ 0 \\ 1 \end{pmatrix} \tag{4.143}$$

such that the following relation is satisfied:

$$\vec{\varepsilon}_{k\mu} \cdot \vec{\varepsilon}_{k\mu'} = \delta_{\mu\mu'} \tag{4.144}$$

Note that Eq. (4.141) can take the following plane wave solutions:

$$\vec{A}_\lambda(\vec{r}) = \frac{1}{\sqrt{V}} \vec{\varepsilon}_\lambda e^{i\vec{k}\cdot\vec{r}} \tag{4.145}$$

where $\lambda = k\mu$ $(\mu = 1, 2, 3)$, and $\vec{\varepsilon}_\lambda$ is a unit vector that describes the direction of photon radiation vibration. We require that $\vec{A}_\lambda$ is periodic with changes in space, and since $\vec{A}_\lambda$ is required to have an entire number of cycles in the square box, the desirable values of $\vec{k}$ are

$$\vec{k} = \frac{2\pi}{V^{1/3}}(l, m, n), \qquad l, m, n = 0, \pm 1, \pm 2, \cdots \quad (\text{except } l = m = n = 0) \tag{4.146}$$

where $l$, $m$, $n$ represent the integer value taken in the $x$, $y$, $z$ direction; thus $\vec{k} \cdot \vec{r}$ is an integer multiple of $2\pi$ at the boundary of the square box of the volume $V$. Each group $l$, $m$, $n$ corresponds to a wave vector, so it follows that $k$ is a separate variable, whereas $k$ becomes a continuous variable only when $V \to \infty$. The following relation is readily seen using the periodic boundary condition Eq. (4.146):

$$\frac{1}{V}\int_V \mathrm{d}\vec{r}\mathrm{e}^{\mathrm{i}\left(\vec{k}-\vec{k}'\right)\cdot\vec{r}} = \delta_{\vec{k}\,\vec{k}'} \tag{4.147}$$

In addition, we have the following $\delta$ function formula:

$$\frac{1}{(2\pi)^3}\int \mathrm{d}\vec{r}\mathrm{e}^{\mathrm{i}\left(\vec{k}-\vec{k}'\right)\cdot\vec{r}} = \delta\left(\vec{k} - \vec{k}'\right) \tag{4.148}$$

Comparing the above two formulas, we can see that there is the following correspondence:

$$\frac{V}{(2\pi)^3}\delta_{\vec{k}\,\vec{k}'} \Leftrightarrow \delta\left(\vec{k} - \vec{k}'\right), \quad \text{when} \quad V \to \infty \tag{4.149}$$

Now, we can obtain the following relations using Eqs. (4.144), (4.145), and (4.147):

$$\int_V \vec{A}_\lambda^* \cdot \vec{A}_{\lambda'}\,\mathrm{d}\vec{r} = \delta_{\lambda\lambda'} \tag{4.150}$$

$$\int_V \vec{A}_\lambda^* \cdot \vec{A}_{\lambda'}^*\,\mathrm{d}\vec{r} = \int_V \vec{A}_\lambda \cdot \vec{A}_{\lambda'}\mathrm{d}\vec{r} = 0 \tag{4.151}$$

Substituting Eq. (4.145) into Eq. (4.140), and then using the vector formula $\vec{\nabla}\left(\vec{a}\cdot\vec{r}\right) = \vec{a}$, from Eq. (4.118) we can obtain

$$\vec{\varepsilon}_\lambda \cdot \vec{k} = 0 \tag{4.152}$$

This is the transverse wave condition, which requires that $\vec{\varepsilon}_\lambda$ is vertical with $\vec{k}$. If we take $\vec{k}$ along the z-axis direction, then the photons have only $\vec{\varepsilon}_{k1}$ and $\vec{\varepsilon}_{k2}$ two polarization waves, and there is no longitudinal polarization wave along the z-axis direction. The solution for Eq. (4.142) is

$$q(t) \propto \mathrm{e}^{\pm \mathrm{i}\omega t} \tag{4.153}$$

where we take $q(t) \sim \mathrm{e}^{-\mathrm{i}\omega t}$. Since the solution of Eq. (4.121) should be a real number, its general solution can be represented by the linear superposition of its special solutions as follows:

$$\begin{aligned} \vec{A}\left(\vec{r}, t\right) &= \sum_{\lambda}\left[q_\lambda(t)\vec{A}_\lambda\left(\vec{r}\right) + q_\lambda^*(t)\vec{A}_\lambda^*\left(\vec{r}\right)\right] \\ &\sim \frac{1}{\sqrt{V}}\sum_{\lambda}\vec{\varepsilon}_\lambda\left[\mathrm{e}^{\mathrm{i}\left(\vec{k}\cdot\vec{r}-\omega t\right)} + \mathrm{e}^{-\mathrm{i}\left(\vec{k}\cdot\vec{r}-\omega t\right)}\right] \end{aligned} \tag{4.154}$$

Using Eq. (4.154) and taking the scalar potential $\varphi = 0$, then from Eqs. (4.108) and (4.105) we can obtain

$$\vec{E} = -\frac{1}{c}\frac{\partial \vec{A}}{\partial t} = \frac{\mathrm{i}}{c}\sum_{\lambda}\omega\left(q_\lambda\vec{A}_\lambda - q_\lambda^*\vec{A}_\lambda^*\right) \tag{4.155}$$

$$\vec{B} = \vec{\nabla}\times\vec{A} = \sum_{\lambda}\left(q_\lambda\vec{\nabla}\times\vec{A}_\lambda + q_\lambda^*\vec{\nabla}\times\vec{A}_\lambda^*\right) \tag{4.156}$$

Utilizing Eq. (4.151) we can obtain the following result from Eq. (4.155):

$$\begin{aligned} \frac{1}{8\pi}\int_V\left|\vec{E}\right|^2 \mathrm{d}\vec{r} &= \frac{1}{8\pi c^2}\sum_{\lambda\lambda'}\omega\omega'\int_V\left(q_\lambda^*\vec{A}_\lambda^* - q_\lambda\vec{A}_\lambda\right)\cdot\left(q_{\lambda'}\vec{A}_{\lambda'} - q_{\lambda'}^*\vec{A}_{\lambda'}^*\right)\mathrm{d}\vec{r} \\ &= \frac{1}{8\pi}\sum_{\lambda\lambda'}kk'\int_V\left(q_\lambda^* q_{\lambda'}\vec{A}_\lambda^*\cdot\vec{A}_{\lambda'} + q_\lambda q_{\lambda'}^*\vec{A}_\lambda\cdot\vec{A}_{\lambda'}^*\right)\mathrm{d}\vec{r} \end{aligned} \tag{4.157}$$

We can also obtain the following result from Eq. (4.156):

$$\begin{aligned} \frac{1}{8\pi}\int_V\left|\vec{B}\right|^2 \mathrm{d}\vec{r} = \frac{1}{8\pi}\sum_{\lambda\lambda'}\int_V\Big[& q_\lambda^* q_{\lambda'}\left(\vec{\nabla}\times\vec{A}_\lambda^*\right)\cdot\left(\vec{\nabla}\times\vec{A}_{\lambda'}\right) + q_\lambda q_{\lambda'}^*\left(\vec{\nabla}\times\vec{A}_\lambda\right)\cdot\left(\vec{\nabla}\times\vec{A}_{\lambda'}^*\right) \\ & + q_\lambda q_{\lambda'}\left(\vec{\nabla}\times\vec{A}_\lambda\right)\cdot\left(\vec{\nabla}\times\vec{A}_{\lambda'}\right) + q_\lambda^* q_{\lambda'}^*\left(\vec{\nabla}\times\vec{A}_\lambda^*\right)\cdot\left(\vec{\nabla}\times\vec{A}_{\lambda'}^*\right)\Big]\mathrm{d}\vec{r} \end{aligned} \tag{4.158}$$

where we have the following vector formula:

$$\vec{\nabla}\cdot\left(\vec{a}\times\vec{b}\right) = \left(\vec{\nabla}\times\vec{a}\right)\cdot\vec{b} - \vec{a}\cdot\left(\vec{\nabla}\times\vec{b}\right) \tag{4.159}$$

It is stipulated that the operator $\vec{\nabla}$ in the parentheses only acts on the quantities in the parentheses. If we take $\vec{b} = \vec{\nabla} \times \vec{c}$, then we have

$$\vec{\nabla} \cdot \left[\vec{a} \times \left(\vec{\nabla} \times \vec{c}\right)\right] = \left(\vec{\nabla} \times \vec{a}\right) \cdot \left(\vec{\nabla} \times \vec{c}\right) - \vec{a} \cdot \left[\vec{\nabla} \times \left(\vec{\nabla} \times \vec{c}\right)\right] \tag{4.160}$$

Utilizing the above formula we can get

$$\left(\vec{\nabla} \times \vec{A}_\lambda^*\right) \cdot \left(\vec{\nabla} \times \vec{A}_{\lambda'}\right) = \vec{\nabla} \cdot \left[\vec{A}_\lambda^* \times \left(\vec{\nabla} \times \vec{A}_{\lambda'}\right)\right] + \vec{A}_\lambda^* \cdot \left[\vec{\nabla} \times \left(\vec{\nabla} \times \vec{A}_{\lambda'}\right)\right] \tag{4.161}$$

Also, from the Gaussian theorem, that is,

$$\oiint \vec{f} \cdot \mathrm{d}\vec{\sigma} = \iiint \left(\vec{\nabla} \cdot \vec{f}\right) \mathrm{d}\vec{r} \tag{4.162}$$

and noting that there is no outflow on the surface of infinite space, the first item on the right-hand side of Eq. (4.161) has no effect. Now, we can obtain the following relation using Eqs. (4.111), (4.118), and (4.141):

$$\vec{\nabla} \times \left(\vec{\nabla} \times \vec{A}_{\lambda'}\right) = k'^2 \vec{A}_{\lambda'} \tag{4.163}$$

Using Eqs. (4.161) – (4.163), as well as noting Eq. (4.151), we can obtain the following relation from Eq. (4.158):

$$\frac{1}{8\pi} \int_V \left|\vec{B}\right|^2 \mathrm{d}\vec{r} = \frac{1}{8\pi} \sum_{\lambda\lambda'} k'^2 \int_V \left(q_\lambda^* q_{\lambda'} \vec{A}_\lambda^* \cdot \vec{A}_{\lambda'} + q_\lambda q_{\lambda'}^* \vec{A}_\lambda \cdot \vec{A}_{\lambda'}^*\right) \mathrm{d}\vec{r} \tag{4.164}$$

Note that the items marked $\lambda'$ as footnote are all behind the items marked $\lambda$ as footnote in Eq. (4.158). Using the orthogonal relation Eq. (4.150), the total energy of the radiation field can be obtained by Eqs. (4.157) and (4.164) as

$$\frac{1}{8\pi} \int_V \left(\left|\vec{E}\right|^2 + \left|\vec{B}\right|^2\right) \mathrm{d}\vec{r} = \frac{1}{4\pi c^2} \sum_\lambda \omega^2 \left(q_\lambda q_\lambda^* + q_\lambda^* q_\lambda\right) \tag{4.165}$$

Since $q_\lambda^*$ and $q_\lambda^*$ are not real variables, the following real variables are defined to facilitate the quantification of the radiation field

$$Q_\lambda = \frac{1}{\sqrt{4\pi c^2}} \left(q_\lambda + q_\lambda^*\right), \quad P_\lambda = \frac{1}{\sqrt{4\pi c^2}} \left(\dot{q}_\lambda + \dot{q}_\lambda^*\right) = -\frac{\mathrm{i}\omega}{\sqrt{4\pi c^2}} \left(q_\lambda - q_\lambda^*\right) \tag{4.166}$$

where their inverse transformations are

$$q_\lambda = \frac{\sqrt{4\pi c^2}}{2}\left(Q_\lambda + \frac{\mathrm{i}}{\omega}P_\lambda\right), \quad q_\lambda^* = \frac{\sqrt{4\pi c^2}}{2}\left(Q_\lambda - \frac{\mathrm{i}}{\omega}P_\lambda\right) \tag{4.167}$$

Substituting Eq. (4.167) into Eq. (4.165) we can obtain

$$H = \frac{1}{2}\sum_\lambda \left(P_\lambda^2 + \omega^2 Q_\lambda^2\right) \tag{4.168}$$

Now it is known that the one-dimensional harmonic oscillator Hamiltonian is

$$H = \frac{p^2}{2m} + \frac{1}{2}m\omega^2 q^2 \tag{4.169}$$

and if we let

$$P = \frac{p}{\sqrt{m}}, \qquad Q = \sqrt{m}q \tag{4.170}$$

then Eq. (4.169) becomes

$$H = \frac{1}{2}\left(P^2 + \omega^2 Q^2\right) \tag{4.171}$$

It can be seen from Eq. (4.168) that the radiation field can be seen as a system consisting of an infinite number of harmonic oscillators; the oscillator frequency $\omega = kc$. In addition, it can be seen from Eq. (4.146) that the system tends to change continuously when $V \to \infty$.

Using the canonical equation (4.139) and Eq. (4.168), we can obtain

$$\begin{aligned}
&\dot{Q}_\lambda = \frac{\partial H}{\partial P_\lambda} = P_\lambda, && \dot{P}_\lambda = -\frac{\partial H}{\partial Q_\lambda} = -\omega^2 Q_\lambda \\
&\ddot{Q}_\lambda = \dot{P}_\lambda = -\omega^2 Q_\lambda, && \ddot{Q}_\lambda + \omega^2 Q_\lambda = 0 \\
&\ddot{P}_\lambda = -\omega^2 \dot{Q}_\lambda = -\omega^2 P_\lambda, && \ddot{P}_\lambda + \omega^2 \dot{P}_\lambda = 0
\end{aligned} \tag{4.172}$$

The above equations are in the same form as Eq. (4.142), where $Q_\lambda$ and $P_\lambda$ can be regarded as the coordinates and momentum that are canonically conjugated to each other. Only after changing $q_i$ and $p_j$ to $Q_\lambda$ and $P_{\lambda'}$ can obtain the following relations using Eqs. (4.130) and (4.131):

$$\{Q_\lambda, P_{\lambda'}\} = \delta_{\lambda\lambda'}, \qquad \{Q_\lambda, Q_{\lambda'}\} = \{P_\lambda, P_{\lambda'}\} = 0 \tag{4.173}$$

The basic idea of field quantification is to find a complete set of canonical "coordinates" and "momentum" describing the classical field and then treat them as corresponding operators to satisfy the commutative relations of canonical coordinates and momentum; thus, they are quantized. Next we establish the following equivalence relation between the Poisson bracket {} in classical electromagnetic field and the commutative relation [ ] in quantum theory.

$$[F, G] = i\hbar\{F, G\} \tag{4.174}$$

From here, and according to Eq. (4.173), the following commutative relations can be obtained in quantum theory.

$$[Q_\lambda, P_{\lambda'}] = i\hbar\delta_{\lambda\lambda'}, \qquad [Q_\lambda, Q_{\lambda'}] = [P_\lambda, P_{\lambda'}] = 0 \tag{4.175}$$

Next, introduce the dimensionless operators in the following way:

$$Q_\lambda = \sqrt{\frac{\hbar}{2\omega}}(a_\lambda + a_\lambda^+), \qquad P_\lambda = -\mathrm{i}\sqrt{\frac{\hbar\omega}{2}}(a_\lambda - a_\lambda^+) \tag{4.176}$$

and then find their inverse solutions. It follows that using Eq. (4.166) together with these solutions, we can obtain

$$\begin{aligned} a_\lambda &= \sqrt{\frac{\omega}{2\hbar}}\left(Q_\lambda + \frac{\mathrm{i}}{\omega}P_\lambda\right) = \frac{1}{\sqrt{4\pi c^2}}\sqrt{\frac{2\omega}{\hbar}}q_\lambda, \\ a_\lambda^+ &= \sqrt{\frac{\omega}{2\hbar}}\left(Q_\lambda - \frac{\mathrm{i}}{\omega}P_\lambda\right) = \frac{1}{\sqrt{4\pi c^2}}\sqrt{\frac{2\omega}{\hbar}}q_\lambda^* \end{aligned} \tag{4.177}$$

Given this, the following relations are straightforward to prove by using Eqs. (4.177) and (4.175):

$$\left[a_\lambda, a_{\lambda'}^+\right] = \delta_{\lambda\lambda'}, \qquad [a_\lambda, a_{\lambda'}] = \left[a_\lambda^+, a_{\lambda'}^+\right] = 0 \tag{4.178}$$

which are precisely the commutative relations satisfied by the generation and annihilation operators of bosons.

Substituting $q_\lambda$ and $q_\lambda^*$ obtained by Eq. (4.177) into the first formula of Eq. (4.154) and writing out the harmonic changing factors in $q_\lambda(t)$ and $q_\lambda^*(t)$ with time by use of Eq. (4.153), thus we can obtain

$$\vec{A}\left(\vec{r}, t\right) = \sqrt{4\pi c^2}\sum_\lambda\sqrt{\frac{\hbar}{2\omega}}\left[\vec{A}_\lambda\left(\vec{r}\right)\mathrm{e}^{-\mathrm{i}\omega t}a_\lambda + \vec{A}_\lambda^*\left(\vec{r}\right)\mathrm{e}^{\mathrm{i}\omega t}a_\lambda^+\right] \tag{4.179}$$

In addition, substituting Eq. (4.145) into the above formula, we can also obtain

$$\vec{A}\left(\vec{r},t\right) = \sqrt{\frac{4\pi\hbar c^2}{V}}\sum_{\lambda}\frac{\vec{\varepsilon}_{\lambda}}{\sqrt{2\omega}}\left[\mathrm{e}^{\mathrm{i}\left(\vec{k}\cdot\vec{r}-\omega t\right)}a_{\lambda} + \mathrm{e}^{-\mathrm{i}\left(\vec{k}\cdot\vec{r}-\omega t\right)}a_{\lambda}^{+}\right] \tag{4.180}$$

where $a_{\lambda}$ and $a_{\lambda}^{+}$ are the Boson annihilation and generation operators that are time invariant. Now, we can obtain the following expressions substituting Eq. (4.180) into Eqs. (4.108) and (4.105) with $\varphi = 0$

$$\vec{E}\left(\vec{r},t\right) = \mathrm{i}\sqrt{\frac{2\pi\hbar ck}{V}}\sum_{\lambda}\vec{\varepsilon}_{\lambda}\left[\mathrm{e}^{\mathrm{i}\left(\vec{k}\cdot\vec{r}-\omega t\right)}a_{\lambda} - \mathrm{e}^{-\mathrm{i}\left(\vec{k}\cdot\vec{r}-\omega t\right)}a_{\lambda}^{+}\right] \tag{4.181}$$

$$\vec{B}\left(\vec{r},t\right) = \mathrm{i}\sqrt{\frac{2\pi\hbar c}{Vk}}\sum_{\lambda}\left(\vec{k}\times\vec{\varepsilon}_{\lambda}\right)\left[\mathrm{e}^{\mathrm{i}\left(\vec{k}\cdot\vec{r}-\omega t\right)}a_{\lambda} - \mathrm{e}^{-\mathrm{i}\left(\vec{k}\cdot\vec{r}-\omega t\right)}a_{\lambda}^{+}\right] \tag{4.182}$$

where $\lambda = k\mu$ such that $\mu = 1, 2$ represents the transverse polarized light in the $x$, $y$ axis directions, respectively.

According to the theory of quadratic quantification, in the occupy number representation we have [3, 6]

$$a_{\lambda}^{+}|n_{\lambda}\rangle = \sqrt{n_{\lambda}+1}\,|n_{\lambda}+1\rangle, \qquad a_{\lambda}|n_{\lambda}\rangle = \sqrt{n_{\lambda}}\,|n_{\lambda}-1\rangle \tag{4.183}$$

such that it is straightforward to verify

$$a_{\lambda}^{+}a_{\lambda}|n_{\lambda}\rangle = n_{\lambda}|n_{\lambda}\rangle \tag{4.184}$$

where $n_{\lambda}$ is the number of photons in $\lambda$ state.

Substituting Eq. (4.176) into Eq. (4.168) and using the commutative relation Eq. (4.178), we can obtain

$$H = \sum_{\lambda}\left(n_{\lambda}+\frac{1}{2}\right)\hbar\omega, \qquad n_{\lambda} = 0, 1, 2, \cdots \tag{4.185}$$

Now, the energy and momentum of each photon on $\lambda$ state are as follows:

$$E = \hbar\omega, \qquad \vec{p} = \hbar\vec{k} \tag{4.186}$$

and because $\omega = kc$, we can see

$$E^2 - p^2c^2 = 0 \tag{4.187}$$

This is the reflection of a photon with a stationary mass of 0.

### 4.2.4 Polarization Theory of Photon Beams

In the particle occupy number representation, according to Eq. (4.183), we have

$$\langle 0|a_\lambda|1\rangle = 1, \qquad \langle 1|a_\lambda^+|0\rangle = 1 \tag{4.188}$$

where the measurement of photon beams belongs to the photon annihilation process. According to Eq. (4.181), we order

$$\vec{E}_\mu^{(-)}\left(k, \vec{r}, t\right) = \mathrm{i}\sqrt{\frac{2\pi\hbar ck}{V}}\vec{\varepsilon}_{k\mu}\mathrm{e}^{\mathrm{i}\left(\vec{k}\cdot\vec{r}-\omega t\right)}a_{k\mu}, \qquad \mu = 1,2 \tag{4.189}$$

Also, according to the Golden rule obtained by perturbation theory, the system transition rate is [2, 3, 5]

$$W = \frac{2\pi}{\hbar}|T|^2\rho_f \tag{4.190}$$

The square of absolute value of the transition matrix element $T$ in above formula is

$$\begin{aligned} |T|^2 &= \frac{1}{2}\left|\sum_\mu \left\langle 0\middle|\vec{E}_\mu^{(-)}(k, \vec{r}, t)\middle|1\right\rangle\right|^2 = \frac{1}{2}\left|\mathrm{i}\sqrt{\frac{2\pi\hbar ck}{V}}\mathrm{e}^{\mathrm{i}\left(\vec{k}\cdot\vec{r}-\omega t\right)}\left(\vec{\varepsilon}_{k1} + \vec{\varepsilon}_{k2}\right)\right|^2 \\ &= \frac{1}{2}\sum_\mu\left|\mathrm{i}\sqrt{\frac{2\pi\hbar ck}{V}}\mathrm{e}^{\mathrm{i}\left(\vec{k}\cdot\vec{r}-\omega t\right)}\vec{\varepsilon}_{k\mu}\right|^2 = \frac{2\pi\hbar ck}{V} \end{aligned} \tag{4.191}$$

where the coefficient $\frac{1}{2}$ represents the averaging of the two transverse vibration directions. According to the principle of phase space, the number of states in the spherical shell, which volume is $V$ and momentum wave number is $k \to k + \mathrm{d}k$, is

$$\mathrm{d}n = \frac{V4\pi\, k^2\mathrm{d}k}{(2\pi)^3} \tag{4.192}$$

For photon $\mathrm{d}E = \hbar\mathrm{d}\omega = \hbar c\mathrm{d}k$, the photon energy state density is

$$\rho_\mathrm{f} = \frac{\mathrm{d}n}{\mathrm{d}E} = \frac{Vk^2}{2\pi^2\hbar c} \tag{4.193}$$

Next, we can obtain the following result substituting Eqs. (4.191) and (4.193) into Eq. (4.190):

$$W = \frac{2k^3}{\hbar} \tag{4.194}$$

Now, we have the following explanation for the above transition rate $W$

$$I_0(k) = \frac{\mathrm{d}M_0\left(k, \vec{r}, t\right)}{\mathrm{d}E\mathrm{d}\vec{r}\,\mathrm{d}t} = W = \frac{2k^3}{\hbar} \tag{4.195}$$

where $M_0\left(k, \vec{r}, t\right)$ represents the number of the measured photons and $I_0(k)$ represents the number of photons with energy $\hbar ck$ measured in unit time, unit volume, and in the unit energy interval; because we are studying the stable plane wave light beam, which has no relationship with position and time, but its intensity is proportional to the third power of photon energy.

Now suppose that we put an ideal plane wave photon detector in the vertical direction with the photon beam. The wavelength of the photon is known as $L(k) = \frac{1}{k}$. When the photon is injected into the detector from the surface of the detector and its depth reaches $L(k)$, then, as an ideal detector, it should be able to obtain all the information of the photons and record the photons; thus, the sensitive detection thickness of the ideal detector to photons is $L(k)$. If the detector has a very high energy resolution and only detects photons with an average energy $\hbar ck$ in a very small energy range $\Delta E = \hbar c\Delta k$, then, using an ideal detector with an effective detection area of S through the detection time $T$, it can be seen that the number of photons recorded should be, according to Eq. (4.195),

$$N_0(k) = I_0(k)SL(k)T\Delta E = 2cSTk^2\Delta k \tag{4.196}$$

It can be seen from Eqs. (4.189) and (4.191) that the amplitudes of the electric field strength $E_\mu^{(-)}\left(k, \vec{r}, t\right)$ in the two directions $\mu = 1$ and $\mu = 2$ are equal, and since $\vec{\varepsilon}_{k1} \cdot \vec{\varepsilon}_{k2} = 0$, there is no coherent term, so that satisfies the unpolarization condition. It means that the original light beam is unpolarized. The normalization density matrix of the unpolarized light beam is

$$\hat{\rho}_0 = \frac{1}{2}\left[\begin{pmatrix}1\\0\\0\end{pmatrix}(1\ \ 0\ \ 0) + \begin{pmatrix}0\\1\\0\end{pmatrix}(0\ \ 1\ \ 0)\right] = \frac{1}{2}\begin{pmatrix}1 & 0 & 0\\0 & 1 & 0\\0 & 0 & 0\end{pmatrix} \tag{4.197}$$

such that, apparently,

$$\frac{1}{2}\sum_\mu \left|\vec{\varepsilon}_{k\mu}\right|^2 = \mathrm{tr}\{\hat{\rho}_0\} \tag{4.198}$$

It follows that Eq. (4.191) can be rewritten as

$$|T|^2 = \frac{2\pi\hbar ck}{V}\mathrm{tr}\{\hat{\rho}_0\} \tag{4.199}$$

and Eq. (4.195) can be rewritten as

$$I_0(k) = \frac{\mathrm{d}M_0\left(k, \vec{r}, t\right)}{\mathrm{d}E\mathrm{d}\vec{r}\,\mathrm{d}t} = \frac{2k^3}{\hbar}\mathrm{tr}\{\hat{\rho}_0\} \tag{4.200}$$

Since photons can be described by three-dimensional vector potential $\vec{A}\left(\vec{r}, t\right)$, the photon spin is equal to 1. Eq. (4.143) is the spin $S = 1$ particle basis spin function in the rectangular basis representation and rectangular coordinate system given by Eq. (3.42). The spin $S = 1$ particle spin operators in the rectangular basis representation and rectangular coordinate system are given by Eq. (3.47) as follows:

$$\hat{S}_x = \begin{pmatrix} 0 & 0 & 0 \\ 0 & 0 & -\mathrm{i} \\ 0 & \mathrm{i} & 0 \end{pmatrix}, \quad \hat{S}_y = \begin{pmatrix} 0 & 0 & \mathrm{i} \\ 0 & 0 & 0 \\ -\mathrm{i} & 0 & 0 \end{pmatrix}, \quad \hat{S}_z = \begin{pmatrix} 0 & -\mathrm{i} & 0 \\ \mathrm{i} & 0 & 0 \\ 0 & 0 & 0 \end{pmatrix} \tag{4.201}$$

Now assume that the photon beam propagated along the $z$-axis passes through both a compensation prism and a polarizer, then we measure the number of photons with a photon detector [7, 8]. Note that at this time the intensity of the light beam may have changed. Although the light beam is still a transverse wave, its amplitude ratio in the $x$ and $y$ directions may have changed, that is, the light beam is polarized, and the phase of the light beam $x$ and $y$ components may also have changed, respectively. According to Eq. (4.189) the total electric field strength of the wave number $k$ in this case can be written as

$$\vec{E}^{(-)}\left(k, \vec{r}, t\right) = \left(a_x \mathrm{e}^{\mathrm{i}\beta_x}\vec{\varepsilon}_{k1}a_{k1} + a_y \mathrm{e}^{\mathrm{i}\beta_y}\vec{\varepsilon}_{k2}a_{k2}\right)\mathrm{i}\sqrt{\frac{2\pi\hbar ck}{V}}\mathrm{e}^{\mathrm{i}\left(\vec{k}\cdot\vec{r}-\omega t\right)} \tag{4.202}$$

$$\langle 0|\vec{E}^{(-)}\left(k, \vec{r}, t\right)|1\rangle = \left(a_x \mathrm{e}^{\mathrm{i}\beta_x}\vec{\varepsilon}_{k1} + a_y \mathrm{e}^{\mathrm{i}\beta_y}\vec{\varepsilon}_{k2}\right)\mathrm{i}\sqrt{\frac{2\pi\hbar ck}{V}}\mathrm{e}^{\mathrm{i}\left(\vec{k}\cdot\vec{r}-\omega t\right)} \tag{4.203}$$

Now introduce the following symbol of the polarized photon wave function

$$|k\rangle = a_x \mathrm{e}^{\mathrm{i}\beta_x}\vec{\varepsilon}_{k1} + a_y \mathrm{e}^{\mathrm{i}\beta_y}\vec{\varepsilon}_{k2} = \begin{pmatrix} a_x \mathrm{e}^{\mathrm{i}\beta_x} \\ a_y \mathrm{e}^{\mathrm{i}\beta_y} \\ 0 \end{pmatrix} \tag{4.204}$$

where it is now possible to define the density matrix of the polarized light beam as

$$\hat{\rho} = |k\rangle\langle k| = \begin{pmatrix} a_x e^{i\beta_x} \\ a_y e^{i\beta_y} \\ 0 \end{pmatrix} \begin{pmatrix} a_x^* e^{-i\beta_x} & a_y^* e^{-i\beta_y} & 0 \end{pmatrix}$$
$$= \begin{pmatrix} |a_x|^2 & a_x a_y^* e^{i(\beta_x - \beta_y)} & 0 \\ a_y a_x^* e^{-i(\beta_x - \beta_y)} & |a_y|^2 & 0 \\ 0 & 0 & 0 \end{pmatrix} \tag{4.205}$$

such that, apparently, we have

$$\mathrm{tr}\{\hat{\rho}\} = |a_x|^2 + |a_y|^2 \tag{4.206}$$

where the density matrix $\hat{\rho}$ is not necessarily normalized. Referring to Eq. (4.200), the transition rate of the polarized light beam can be written as

$$I(k) = \frac{\mathrm{d}M\left(k, \vec{r}, t\right)}{\mathrm{d}E\mathrm{d}\vec{r}\,\mathrm{d}t} = \frac{2k^3}{\hbar}\mathrm{tr}\{\hat{\rho}\} = \left(|a_x|^2 + |a_y|^2\right)\frac{2k^3}{\hbar} \tag{4.207}$$

Using the photon spin operators given by Eq. (4.201) and the density matrix given by Eq. (4.205), the polarization rates of the light beam in $x$, $y$, and $z$ directions can be obtained, respectively, as follows:

$$P_x = \frac{\mathrm{tr}\{\hat{S}_x\hat{\rho}\}}{\mathrm{tr}\{\hat{\rho}\}} = 0, \qquad P_y = \frac{\mathrm{tr}\{\hat{S}_y\hat{\rho}\}}{\mathrm{tr}\{\hat{\rho}\}} = 0$$
$$P_z = \frac{\mathrm{tr}\{\hat{S}_z\hat{\rho}\}}{\mathrm{tr}\{\hat{\rho}\}} = \frac{-ia_y a_x^* e^{-i(\beta_x - \beta_y)} + ia_x a_y^* e^{i(\beta_x - \beta_y)}}{|a_x|^2 + |a_y|^2} = \frac{2\mathrm{Im}\left(a_y a_x^* e^{-i(\beta_x - \beta_y)}\right)}{|a_x|^2 + |a_y|^2} \tag{4.208}$$

such that

$$-1 \le P_z \le 1 \tag{4.209}$$

The above results show that the light beam polarization direction is perpendicular to the vibration direction of the light wave.

From the above discussion we know that the polarization of the particles with mass not equal to 0 is caused by the particle spin, and the photon polarization with mass equal to 0 is caused by the transverse vibration of the photons.In this section, the polarization problem of plane wave light beam is preliminarily discussed, and this subject still needs to be studied from experimental and theoretical aspects. In Refs. [7] and [8] the polarization problem of the light beams is studied theoretically, but photon spin operators are not used in that work. In Reference [9] the polarization problem of the light beams is also discussed.

# References

1. Varshalovich DA, Moskalev AN, Khersonskii VK. Quantum theory of angular momentum. Singapore/New Jersey/Hong Kong: World Scientific; 1988.
2. Shen QB. Nuclear reaction theory of low and medium energy (upper volume). Beijing: Science Press; 2005. (in Chinese)
3. Zeng JY. Quantum mechanics, volume II (third board). Beijing: Science Press; 2003. (in Chinese)
4. Eisenberg JM, Greiner W. Nuclear theory-excitation mechanisms of the nucleus. Amsterdam/ New York/Oxford: North-Holland Publishing Company; 1976. p. 58.
5. Gottfried K, Yan TM. Quantum mechanics: fundamentals, vol. Springer. 2nd ed; 2004. p. 437, 463.
6. Shen QB. Nuclear reaction theory of low and medium energy (middle volume). Beijing: Science Press; 2012. (in Chinese)
7. Glauber RJ. The quantum theory of optical coherence. Phy Rev. 1963;130:2529.
8. Lahiri M, Wolf E. Quantum analysis of polarization properties of optical beams. Phys Rev. 2010; A82:043805.
9. Robson BA. The theory of polarization phenomena. Oxford: Clarendon Press; 1974.

# Chapter 5
# Polarization Theory of Relativistic Nuclear Reactions

**Abstract** This chapter will introduce the polarization theory of relativistic nuclear reactions. A Dirac $S$ matrix theory of relativistic nuclear reactions will be developed. Then the relativistic classical field theory and Lagrangian density in quantum hadron dynamics as well as relativistic Green function theory at zero temperature will be introduced. The nucleon relativistic microscopic optical potentials by relativistic mean field theory and based on DBHF theory will be introduced. Finally, the relativistic dynamics equation of spin 1 particle and its application in elastic scattering calculations will be discussed.

**Keywords** Bjorken-dreel metric · Dirac-Brueckner-Hartree-Fock · Dirac equation · Dirac $S$ matrix theory · Feynman rules · Klein-Gordon equation · Lagrangian density · Lorentz covariance · Pauli metric · Relativistic mean field · Self-energy operator

## 5.1 Basic Theory of Relativistic Quantum Mechanics

### *5.1.1 Klein-Gordon Equation*

The restricted theory of relativity requires that all physical equations remain unchanged under the Lorentz transformation. That is, when we transform the space-time coordinate system $S(x, y, z, t)$ to another space-time coordinate system $S'(x', y', z', t')$, under the condition that the following equation is maintained

$$x^2 + y^2 + z^2 - c^2t^2 = x'^2 + y'^2 + z'^2 - c^2t'^2 \tag{5.1}$$

the physical equation form should remain the same, where $c$ is the speed of light. This is Lorentz covariant [1, 2].

 
Q.-B. Shen, *Polarization Theory of Nuclear Reactions*,
https://doi.org/10.1007/978-3-031-11878-4_5

The Einstein's free particle mass-energy relation is

$$E = mc^2 = \frac{m_0 c^2}{\sqrt{1 - \frac{v^2}{c^2}}} = \frac{m_0 c^2}{\sqrt{1 - \frac{p^2}{m^2 c^2}}} = \frac{m_0 c^2}{\sqrt{1 - \frac{p^2 c^2}{E^2}}} \tag{5.2}$$

where $m_0$ is the particle stationary mass. From the above relation we can obtain

$$E^2 = p^2 c^2 + m_0^2 c^4 \tag{5.3}$$

When $v \ll c$, from Eq. (5.2) we can get

$$E \approx m_0 c^2 + \frac{1}{2} m_0 v^2 \tag{5.4}$$

The second item at the right end of the above equation is the non-relativistic kinetic energy. Next Eq. (5.3) is interpreted using operators as follows:

$$E \to \mathrm{i}\hbar \frac{\partial}{\partial t}, \qquad \vec{p} \to -\mathrm{i}\hbar \vec{\nabla} \tag{5.5}$$

Substituting them into Eq. (5.3) and then acting on the wave function $\psi$, we can obtain

$$\left( \Box + \frac{m_0^2 c^2}{\hbar^2} \right) \psi = 0 \tag{5.6}$$

where

$$\Box = \frac{1}{c^2} \frac{\partial^2}{\partial t^2} - \Delta \tag{5.7}$$

$\Delta = \vec{\nabla}^2$ and $\Box$ are called Laplace operator and D'Alembert operator, respectively. Equation (5.6) is the Klein-Gordon equation of free particles and apparently satisfies the Lorentz covariant. The Klein-Gordon equation is not a relativistic motion equation to describe a single particle, but a field equation. Since the equation has only one component, it describes a field with spinless.

In 1935, Yukawa [3] proposed that the nuclear force between the nucleons is propagated by a field. It can be assumed that this field satisfies the Klein-Gordon equation. The stationary mass $m_0$ in Eq. (5.6) corresponds to the mass of a meson described by this field. The nucleon is the source of this field. Since this source is independent of time, the field produced by it is also independent of time, and the wave function $\psi$ should also be independent of time. Refer to the electromagnetic field scalar potential equation with a stationary mass of 0 given by Eq. (4.116), the static Klein-Gordon equation with source can be written by Eq. (5.6) as

$$\left(\vec{\nabla}^2 - \frac{m_0^2 c^2}{\hbar^2}\right)\psi = -\rho \tag{5.8}$$

Assuming the nucleon is heavy enough to make a classical approximation, and the nucleon is at the origin of the coordinate, then this point source can be expressed as

$$\rho = g\,\delta\left(\vec{r}\right) \tag{5.9}$$

where $g$ is the action constant of the nucleon and this meson field, so that Eq. (5.8) can be rewritten as

$$\left(\vec{\nabla}^2 - \frac{m_0^2 c^2}{\hbar^2}\right)\psi\left(\vec{r}\right) = -g\,\delta\left(\vec{r}\right) \tag{5.10}$$

The following equation has been given in Section 3.11 of the Ref. [4]:

$$\left(\vec{\nabla}^2 + k^2\right) G\left(\vec{r}, \vec{r}'\right) = \delta\left(\vec{r} - \vec{r}'\right) \tag{5.11}$$

Its solution is

$$G\left(\vec{r}, \vec{r}'\right) = -\frac{1}{4\pi}\frac{e^{ik\left|\vec{r}-\vec{r}'\right|}}{\left|\vec{r} - \vec{r}'\right|} \tag{5.12}$$

So we can write the solution of Eq. (5.10) as

$$\psi\left(\vec{r}\right) = \frac{g}{4\pi}\frac{e^{-\frac{m_0 c}{\hbar}r}}{r} \tag{5.13}$$

This formula is called Yukawa potential. If the distance to decay to $\frac{1}{e}$ is defined as the force distance $r_0$, there is

$$r_0 = \frac{\hbar}{m_0 c} = \lambda_C \tag{5.14}$$

here $\lambda_C$ is called the Compton wavelength. If taking nuclear force distance $r_0 = 1.4$ fm, $m_0 \approx 140$ MeV can be obtained by Eq. (5.14). The particles predicted by Yukawa were discovered in cosmic rays in 1947 and are called $\pi$ mesons.

### 5.1.2 Dirac Equation [5]

In order to establish a relativistic motion equation that describes a single particle, the equation should contain only a one order derivative term of the time. In order to satisfy the covariant of the relativistic Lorentz transformation, the equation should also contain only a one order derivative term of the space coordinate. Dirac linearized the relative mass-energy relation given by Eq. (5.3) using operators in the following way:

$$E = c\sqrt{p^2 + m_0^2c^2} = c\sqrt{\sum_i p_i^2 + m_0^2c^2} = c\sum_i \alpha_i p_i + \beta\, m_0 c^2 \tag{5.15}$$

The operators $\alpha_i (i = 1, 2, 3)$ and $\beta$ that are independent of coordinates and momentums are introduced. In order to satisfy the mass-energy relation, taking the square of the above equation and divide it by $c^2$, then we can get

$$\begin{aligned}\sum_i p_i^2 + m_0^2c^2 &= \left(\sum_i \alpha_i p_i + \beta m_0 c\right)\left(\sum_j \alpha_j p_j + \beta m_0 c\right)\\ &= \sum_i \alpha_i^2 p_i^2 + \beta^2 m_0^2 c^2 + \sum_i (\alpha_i\beta + \beta\alpha_i) p_i m_0 c + (\alpha_1\alpha_2 + \alpha_2\alpha_1) p_1 p_2\\ &\quad + (\alpha_1\alpha_3 + \alpha_3\alpha_1) p_1 p_3 + (\alpha_2\alpha_3 + \alpha_3\alpha_2) p_2 p_3\end{aligned} \tag{5.16}$$

In order to make the two sides of the above formula to be equal, the following equations must be established:

$$\alpha_1^2 = \alpha_2^2 = \alpha_3^2 = \beta^2 = 1 \tag{5.17}$$

$$\alpha_i\alpha_j + \alpha_j\alpha_i \equiv \left\{\alpha_i, \alpha_j\right\} = 2\delta_{ij}, \quad \alpha_i\beta + \beta\alpha_i = \{\alpha_i, \beta\} = 0, \quad i, j = 1, 2, 3 \tag{5.18}$$

The above two formulas indicate that the three components of $\vec{\alpha}$ and $\beta$ satisfy the anti-commutative relations to each other. According to Eq. (5.15), the Hamiltonian operator of the free particle can be written as

$$H = c\vec{\alpha}\cdot\vec{p} + \beta\, m_0 c^2 = -ic\hbar\vec{\alpha}\cdot\vec{\nabla} + \beta\, m_0 c^2 \tag{5.19}$$

Let $\psi\left(\vec{r}, t\right)$ be a free particle wave function, and the equation satisfied by it is

$$H\psi = E\psi \tag{5.20}$$

We can obtain the following equations according to Eqs. (5.5), (5.19), and (5.20):

$$\mathrm{i}\hbar\frac{\partial\psi}{\partial t} = H\psi \tag{5.21}$$

$$\left(-\mathrm{i}\hbar\frac{\partial}{\partial t} - \mathrm{i}c\hbar\vec{\alpha}\cdot\vec{\nabla} + \beta\, m_0c^2\right)\psi\left(\vec{r}, t\right) = 0 \tag{5.22}$$

The above two formulas are the Dirac equation. In order to ensure that $H$ is Hermitian, $\vec{\alpha}$ and $\beta$ must also be Hermitian, that is,

$$\vec{\alpha}^{+} = \vec{\alpha}, \qquad \beta^{+} = \beta \tag{5.23}$$

The conjugate equation of Eq. (5.21) is

$$-\mathrm{i}\hbar\frac{\partial\psi^{+}}{\partial t} = \psi^{+}H \tag{5.24}$$

The following relations can be proved utilizing Eqs. (5.21) and (5.24):

$$\frac{\partial}{\partial t}\int \psi^{+}\psi\,\mathrm{d}\vec{r} = \int\left(\frac{\partial\psi^{+}}{\partial t}\psi + \psi^{+}\frac{\partial\psi}{\partial t}\right)\mathrm{d}\vec{r} = \frac{\mathrm{i}}{\hbar}\int\left(\psi^{+}H\psi - \psi^{+}H\psi\right)\mathrm{d}\vec{r} = 0 \tag{5.25}$$

The above formula shows that the Dirac equation satisfies the total spatial probability conservation relationship.

In order to facilitate the representation of the equation, the $\gamma$ matrices are introduced as follows:

$$\gamma_k = -\mathrm{i}\beta\alpha_k \text{ or } \alpha_k = \mathrm{i}\beta\gamma_k, \quad k = 1, 2, 3; \quad \gamma_4 = \beta, \ \gamma_5 = \gamma_1\gamma_2\gamma_3\gamma_4 \tag{5.26}$$

The following relations can be proved using Eqs. (5.17), (5.18), (5.23), and (5.26):

$$\gamma_k^{+} = \mathrm{i}\alpha_k\beta = -\mathrm{i}\beta\alpha_k = \gamma_k, \quad k = 1, 2, 3; \quad \gamma_4^{+} = \gamma_4, \ \gamma_5^{+} = \gamma_5 \tag{5.27}$$

$$\{\gamma_i, \gamma_4\} = 0, \quad \left\{\gamma_i, \gamma_j\right\} = 2\delta_{ij}, \qquad i, j = 1, 2, 3 \tag{5.28}$$

$$\left\{\gamma_\mu, \gamma_\nu\right\} = 2\delta_{\mu\nu}, \quad \mu, \nu = 1, 2, 3, 4 \tag{5.29}$$

And the following equation can be got by multiplying $\beta/\hbar c = \gamma_4/\hbar c$ from the left of Eq. (5.22):

$$\left[\gamma_4\left(-\frac{\mathrm{i}}{c}\frac{\partial}{\partial t}\right) + \sum_i \gamma_i\frac{\partial}{\partial x_i} + \kappa\right]\psi = 0 \tag{5.30}$$

where

$$\kappa = \frac{m_0 c}{\hbar} \tag{5.31}$$

The definitions of the four-dimensional Minkowski coordinate and its derivative are

$$x_\mu = \left(\vec{r}, \mathrm{i}ct\right), \qquad \partial_\mu = \left(\frac{\partial}{\partial x}, \frac{\partial}{\partial y}, \frac{\partial}{\partial z}, -\frac{\mathrm{i}}{c}\frac{\partial}{\partial t}\right) \tag{5.32}$$

Then the Dirac equation (5.30) can be expressed as

$$\left(\gamma_\mu \partial_\mu + \kappa\right) \psi = 0 \tag{5.33}$$

It should be noted that according to Einstein's agreement, two identical Greek letter subscripts represent the summation of four-dimensional coordinates, and two identical English letter subscripts represent the summation of three-dimensional space coordinates. It is known that the two-rank Pauli matrix $\widehat{\sigma}$ satisfies the following anti-commutative relation:

$$\left\{\hat{\sigma}_i, \hat{\sigma}_j\right\} = 2\delta_{ij}, \qquad i, j = 1, 2, 3 \tag{5.34}$$

According to the definition of the square matrix direct product given by Eq. (2.349), we define the two $4 \times 4$ matrices in the following way.

$$\tau = \hat{I} \times \hat{\sigma}, \qquad \rho = \hat{\sigma} \times \hat{I} \tag{5.35}$$

where $\hat{I}$ is the unit $2 \times 2$ matrix. We can further obtain

$$\tau_i = \begin{pmatrix} \hat{\sigma}_i & \hat{0} \\ \hat{0} & \hat{\sigma}_i \end{pmatrix}, \qquad i = 1, 2, 3 \tag{5.36}$$

$$\rho_1 = \begin{pmatrix} \hat{0} & \hat{I} \\ \hat{I} & \hat{0} \end{pmatrix}, \quad \rho_2 = \begin{pmatrix} \hat{0} & -\mathrm{i}\hat{I} \\ \mathrm{i}\hat{I} & \hat{0} \end{pmatrix}, \quad \rho_3 = \begin{pmatrix} \hat{I} & \hat{0} \\ \hat{0} & -\hat{I} \end{pmatrix} \tag{5.37}$$

where $\hat{0}$ is a $2 \times 2$ zero matrix. We define

$$\alpha_i \equiv \rho_1 \tau_i = \begin{pmatrix} \hat{0} & \hat{\sigma}_\mathrm{i} \\ \hat{\sigma}_\mathrm{i} & \hat{0} \end{pmatrix}, \quad \beta = \gamma_4 \equiv \rho_3 = \begin{pmatrix} \hat{I} & \hat{0} \\ \hat{0} & -\hat{I} \end{pmatrix} \tag{5.38}$$

So from Eq. (5.26) we can obtain

$$\gamma_i = \begin{pmatrix} \hat{0} & -\mathrm{i}\hat{\sigma}_i \\ \mathrm{i}\hat{\sigma}_i & \hat{0} \end{pmatrix}, \quad i = 1, 2, 3; \qquad \gamma_5 = \begin{pmatrix} \hat{0} & -\hat{I} \\ -\hat{I} & \hat{0} \end{pmatrix} \tag{5.39}$$

One can verify that Eqs. (5.28) and (5.29) are valid by utilizing Eqs. (5.34), (5.38), and (5.39). And from Eq. (5.39) one can obtain

$$\{\gamma_\mu, \gamma_5\} = 0, \qquad \mu = 1, 2, 3, 4 \tag{5.40}$$

And it can be seen that $\gamma_\mu$ and $\gamma_5$ are Hermitian.

In Eq. (5.33), $\gamma_\mu$ is four-rank matrix, so $\psi$ in the equation must be four-component wave function, which is called Dirac spinor and can be expressed as

$$\psi\left(\vec{r},t\right) = \begin{pmatrix} \psi_1\left(\vec{r},t\right) \\ \psi_2\left(\vec{r},t\right) \\ \psi_3\left(\vec{r},t\right) \\ \psi_4\left(\vec{r},t\right) \end{pmatrix}, \tag{5.41}$$

$$\psi^+\left(\vec{r},t\right) = \left(\psi_1^*\left(\vec{r},t\right) \;\; \psi_2^*\left(\vec{r},t\right) \;\; \psi_3^*\left(\vec{r},t\right) \;\; \psi_4^*\left(\vec{r},t\right)\right)$$

Define a new symbol

$$\overline{\psi} \equiv \psi^+\gamma_4 = \left(\psi_1^*\left(\vec{r},t\right) \;\; \psi_2^*\left(\vec{r},t\right) \;\; -\psi_3^*\left(\vec{r},t\right) \;\; -\psi_4^*\left(\vec{r},t\right)\right) \tag{5.42}$$

Multiplying $\psi^+$ on the left side of the Dirac equation (5.22), there is

$$-i\hbar\psi^+\frac{\partial}{\partial t}\psi - \mathrm{i}c\hbar\psi^+\vec{\alpha}\cdot\vec{\nabla}\psi + \psi^+\beta m_0c^2\psi = 0 \tag{5.43}$$

The conjugate equation of Eq. (5.22) is

$$i\hbar\frac{\partial}{\partial t}\psi^+ + \mathrm{i}c\hbar\vec{\nabla}\psi^+\cdot\vec{\alpha} + \psi^+\beta m_0c^2 = 0 \tag{5.44}$$

Multiplying $\psi$ on the right side of Eq. (5.44), there is

$$\mathrm{i}\hbar\left(\frac{\partial}{\partial t}\psi^+\right)\psi + \mathrm{i}c\hbar\left(\vec{\nabla}\psi^+\cdot\vec{\alpha}\right)\psi + \psi^+\beta m_0c^2\psi = 0 \tag{5.45}$$

Then we subtract Eq. (5.45) from Eq. (5.43), and multiply i/ħ on the obtained result; thus the following equation can be obtained:

$$\frac{\partial}{\partial t}(\psi^+\psi) + \vec{\nabla} \cdot \left(\psi^+ c\vec{\alpha}\psi\right) = 0 \tag{5.46}$$

In the Dirac equation, the probability density and probability flow density are as follows, respectively:

$$\rho\left(\vec{r}, t\right) = \psi^+\psi, \qquad \vec{j}\left(\vec{r}, t\right) = c\psi^+\vec{\alpha}\psi \tag{5.47}$$

So Eq. (5.46) can be rewritten as

$$\frac{\partial\rho\left(\vec{r}, t\right)}{\partial t} + \vec{\nabla} \cdot \vec{j}\left(\vec{r}, t\right) = 0 \tag{5.48}$$

This is the continuity equation. Therefore the Dirac equation is a relativistic motion equation describing a single particle that satisfies Lorentz covariant and guarantees probability conservation. From Eq. (5.47) we can write

$$j_i = c\psi^+\alpha_i\psi = \mathrm{i}c\psi^+\gamma_4\gamma_i\psi = \mathrm{i}c\overline{\psi}\gamma_i\psi \tag{5.49}$$

and according to Eq. (5.32) take

$$j_4 = \mathrm{i}c\rho = \mathrm{i}c\psi^+\psi = \mathrm{i}c\overline{\psi}\gamma_4\psi \tag{5.50}$$

Then the continuity equation (5.48) can be rewritten as

$$\partial_\mu j_\mu = 0 \tag{5.51}$$

The definition of the symbol $\partial_\mu$ has been given by Eq. (5.32).

### 5.1.3 Pauli Metric and Bjorken-Drell Metric

The four-dimensional space-time vectors and $\gamma$ matrices in the Minkowski space discussed above are represented under the Pauli metric; however there is another widely used representation method called Bjorken-Drell metric [6, 7]. The $\hbar = c = 1$ unit system will be adopted below.

In the Bjoken-Drell metric, the time component of the four-dimensional space-time vector in Minkowski space is defined as component of 0, and the space-time vector is expressed by the contravariant components as follows:

$$x = \begin{pmatrix} x^0 \\ x^1 \\ x^2 \\ x^3 \end{pmatrix} = \begin{pmatrix} t \\ x \\ y \\ z \end{pmatrix} \tag{5.52}$$

The Greek letters $\mu$, $\nu$, ... represent the four components of the Minkowski space in the order of 0, 1, 2, and 3. Introduce the Bjoken-Drell metric tensor

$$g = \begin{pmatrix} 1 & 0 & 0 & 0 \\ 0 & -1 & 0 & 0 \\ 0 & 0 & -1 & 0 \\ 0 & 0 & 0 & -1 \end{pmatrix} \tag{5.53}$$

and let its covariant component $g_{\mu\nu}$ equal to its contravariant component $g^{\mu\nu}$, that is,

$$g_{\mu\nu} = g^{\mu\nu} \tag{5.54}$$

The relations between covariant components $x_\mu$ and contravariant components $x^\mu$ of the space-time vector are as follows:

$$x_\mu = g_{\mu\nu} x^\nu, \qquad x^\mu = g^{\mu\nu} x_\nu \tag{5.55}$$

Thus the space-time vector can be expressed as covariant components based on Eqs. (5.52), (5.53), and (5.55) as follows:

$$x = \begin{pmatrix} x_0 \\ x_1 \\ x_2 \\ x_3 \end{pmatrix} = \begin{pmatrix} 1 & 0 & 0 & 0 \\ 0 & -1 & 0 & 0 \\ 0 & 0 & -1 & 0 \\ 0 & 0 & 0 & -1 \end{pmatrix} \begin{pmatrix} t \\ x \\ y \\ z \end{pmatrix} = \begin{pmatrix} t \\ -x \\ -y \\ -z \end{pmatrix} \tag{5.56}$$

In the Bjorken-Drell metric, the $\gamma$ matrices with upper subscripts 0, 1, 2, 3 are used; referring to Eqs. (5.26), (5.38), and (5.39), we define

$$\begin{aligned} &\gamma^0 = \gamma_4 = \beta = \begin{pmatrix} \hat{I} & \hat{0} \\ \hat{0} & -\hat{I} \end{pmatrix}, \quad \gamma^{0+} = \gamma^0, \quad \alpha^i = \alpha_i, \\ &\gamma^i = \gamma^0 \alpha^i = \mathrm{i}\gamma_i = \begin{pmatrix} \hat{0} & \hat{\sigma}_\mathrm{i} \\ -\hat{\sigma}_\mathrm{i} & \hat{0} \end{pmatrix}, \quad \gamma^{i+} = -\gamma^i, \qquad i = 1, 2, 3 \end{aligned} \tag{5.57}$$

$$\gamma^5 = \mathrm{i}\gamma^0\gamma^1\gamma^2\gamma^3 = -\gamma_5 = \begin{pmatrix} \hat{0} & \hat{I} \\ \hat{I} & \hat{0} \end{pmatrix}, \quad \gamma^{5+} = \gamma^5 \tag{5.58}$$

They satisfy the following anti-commutative relation:

$$\{\gamma^\mu, \gamma^\nu\} = 2g^{\mu\nu}, \quad \{\gamma^5, \gamma^\mu\} = 0, \quad \mu, \nu = 0,1,2,3 \tag{5.59}$$

Introduce the following differential operators:

$$\partial_\mu \equiv \frac{\partial}{\partial x^\mu} = \left(\frac{\partial}{\partial t}, \vec{\nabla}\right), \quad \partial^\mu \equiv \frac{\partial}{\partial x_\mu} = \left(\frac{\partial}{\partial t}, -\vec{\nabla}\right) \tag{5.60}$$

The four-dimensional energy-momentum vector can be expressed as contravariant components and covariant components, respectively:

$$p = \begin{pmatrix} p^0 \\ p^1 \\ p^2 \\ p^3 \end{pmatrix} = \begin{pmatrix} E \\ p_x \\ p_y \\ p_z \end{pmatrix}, \quad p = \begin{pmatrix} p_0 \\ p_1 \\ p_2 \\ p_3 \end{pmatrix} = \begin{pmatrix} E \\ -p_x \\ -p_y \\ -p_z \end{pmatrix} \tag{5.61}$$

The following expressions can be obtained according to Eqs. (5.5), (5.56), (5.60), and (5.61):

$$p^\mu = \mathrm{i}\partial^\mu, \quad p_\mu = \mathrm{i}\partial_\mu \tag{5.62}$$

And we can obtain

$$a \cdot b = a_\mu b^\mu = g_{\mu\nu} a^\nu b^\mu = a^0 b^0 - \vec{a} \cdot \vec{b}, \quad \vec{a} \cdot \vec{b} = a^i b^i = a_i b_i \tag{5.63}$$

The square of energy-momentum of the free particles with mass is

$$p^2 = p_\mu p^\mu = E^2 - \vec{p}^2 = m_0^2 \tag{5.64}$$

The D'Alembert operator can be expressed as

$$\square \equiv \partial_\mu \partial^\mu = \partial_0^2 - \vec{\nabla}^2 \tag{5.65}$$

From Eq. (5.57) we can know that $\gamma_4 = \gamma^0$, $\gamma_i = -\mathrm{i}\gamma^i$, so the Dirac equation given by Eq. (5.30) can be rewritten as

$$\left(\mathrm{i}\gamma^\mu \partial_\mu - m_0\right) \psi(x) = 0 \tag{5.66}$$

where the $\hbar = c = 1$ unit system is used.

### 5.1.4 Plane Wave Solution of the Dirac Equation

Equation (5.19) has given that the relativistic Hamiltonian of free particles is

$$H = c\vec{\alpha} \cdot \vec{p} + \beta m_0 c^2 \tag{5.67}$$

The commutative relation between the first component $L_1 = x_2 p_3 - x_3 p_2$ of space angular momentum $\vec{L}$ and Hamiltonian is

$$[H, L_1] = c\vec{\alpha} \cdot \left[\vec{p}, x_2 p_3 - x_3 p_2\right] \tag{5.68}$$

Using the Possion brackets given by Eq. (4.130) and Eq. (4.131) and according the way given by Eq. (4.174), Heisenberg introduced the following commutative relations:

$$\left[q_i, p_j\right] = \mathrm{i}\hbar\delta_{ij}, \quad \left[q_i, q_j\right] = \left[p_i, p_j\right] = 0 \tag{5.69}$$

The dimension of the momentum $p$ is $\frac{\text{MeV}\cdot\text{s}}{\text{fm}}$, where s represents time second, so the dimension of the multiplication of $pq$ is MeV·s; it is the same as the dimension of $\hbar$. So the following result can be obtained from Eq. (5.68):

$$[H, L_1] = -\mathrm{i}\hbar c \sum_i \alpha_i (\delta_{i2} p_3 - \delta_{i3} p_2) = -\mathrm{i}\hbar c(\alpha_2 p_3 - \alpha_3 p_2) = -\mathrm{i}\hbar c\left(\vec{\alpha} \times \vec{p}\right)_1 \neq 0$$

It can be expressed by a three-dimensional vector as follows:

$$\left[H, \vec{L}\right] = \left[c\vec{\alpha} \cdot \vec{p}, \vec{L}\right] = -\mathrm{i}\hbar c\vec{\alpha} \times \vec{p} \neq 0 \tag{5.70}$$

Therefore, each component of orbital angular momentum $\vec{L}$ in the Dirac equation is not a conservation quantity.

In order to ensure the conservation of total angular momentum $\vec{J}$, it is necessary to introduce a four-dimensional spin angular momentum $\vec{S}$ to get the total angular momentum $\vec{J} = \vec{L} + \vec{S}$. Let

$$\vec{S} = \frac{\hbar}{2}\vec{\Sigma} = \frac{\hbar}{2}\begin{pmatrix} \hat{\sigma} & \hat{0} \\ \hat{0} & \hat{\sigma} \end{pmatrix}, \quad \vec{\Sigma} = \begin{pmatrix} \hat{\sigma} & \hat{0} \\ \hat{0} & \hat{\sigma} \end{pmatrix} \tag{5.71}$$

We can prove the following relation:

$$\vec{S} \times \vec{S} = \frac{\hbar^2}{4}\begin{pmatrix} \hat{\sigma} \times \hat{\sigma} & \hat{0} \\ \hat{0} & \hat{\sigma} \times \hat{\sigma} \end{pmatrix} = \frac{2i\hbar^2}{4}\begin{pmatrix} \hat{\sigma} & \hat{0} \\ \hat{0} & \hat{\sigma} \end{pmatrix} = i\hbar\vec{S} \tag{5.72}$$

The following commutative relation can be obtained utilizing Eqs. (5.38) and (5.71):

$$\left[H, \vec{S}\right] = \frac{\hbar}{2}\left[c\vec{\alpha} \cdot \vec{p},\ \vec{\Sigma}\right] = \frac{\hbar c}{2}\begin{pmatrix} \hat{0} & \left[\hat{\sigma} \cdot \vec{p},\ \hat{\sigma}\right] \\ \left[\hat{\sigma} \cdot \vec{p},\ \hat{\sigma}\right] & \hat{0} \end{pmatrix} \tag{5.73a}$$

First of all, the first component of the commutative relation in the above formula is found as follows:

$$\begin{aligned} \left[\hat{\sigma} \cdot \vec{p}, \hat{\sigma}_1\right] &= \left[\hat{\sigma}_1 p_1 + \hat{\sigma}_2 p_2 + \hat{\sigma}_3 p_3, \hat{\sigma}_1\right] = -2i\hat{\sigma}_3 p_2 + 2i\hat{\sigma}_2 p_3 \\ &= 2i\left(\hat{\sigma} \times \vec{p}\right)_1 \end{aligned} \tag{5.73b}$$

So we can get

$$\left[H, \vec{S}\right] = i\hbar c\left(\vec{\alpha} \times \vec{p}\right) \neq 0 \tag{5.74}$$

But the following result can be obtained from Eqs. (5.70) and (5.74):

$$\left[H, \vec{J}\right] = \left[H, \vec{L}\right] + \left[H, \vec{S}\right] = 0 \tag{5.75}$$

So the component $J_z$ of the total angular momentum $\vec{J}$ and $J^2$ are the conserved quantities.

$S_3$ is a diagonal matrix as follows:

$$S_3 = \frac{\hbar}{2}\begin{pmatrix} \hat{\sigma}_3 & \hat{0} \\ \hat{0} & \hat{\sigma}_3 \end{pmatrix} = \frac{\hbar}{2}\begin{pmatrix} 1 & 0 & 0 & 0 \\ 0 & -1 & 0 & 0 \\ 0 & 0 & 1 & 0 \\ 0 & 0 & 0 & -1 \end{pmatrix} \tag{5.76}$$

The four eigenvectors of $S_3$ are

$$
\begin{aligned}
S_3\begin{pmatrix}1\\0\\0\\0\end{pmatrix} &= \frac{\hbar}{2}\begin{pmatrix}1\\0\\0\\0\end{pmatrix}, \quad S_3\begin{pmatrix}0\\1\\0\\0\end{pmatrix} = -\frac{\hbar}{2}\begin{pmatrix}0\\1\\0\\0\end{pmatrix},\\
S_3\begin{pmatrix}0\\0\\1\\0\end{pmatrix} &= \frac{\hbar}{2}\begin{pmatrix}0\\0\\1\\0\end{pmatrix}, \quad S_3\begin{pmatrix}0\\0\\0\\1\end{pmatrix} = -\frac{\hbar}{2}\begin{pmatrix}0\\0\\0\\1\end{pmatrix}
\end{aligned} \tag{5.77}
$$

The corresponding eigenvalues are $\pm\frac{\hbar}{2}$, respectively; they are double degeneracy. It can be seen that the Dirac equation describes the particles (such as electrons and nucleons) that have intrinsic spin $\frac{1}{2}$.

From Eq. (5.21) and Eq. (5.19), it can be seen that the Dirac equation satisfied by the momentum $p = 0$ stationary particles is as follows:

$$
\mathrm{i}\hbar\frac{\partial\psi}{\partial t} = \beta\, m_0 c^2 \psi \tag{5.78}
$$

According to Eq. (5.38), four independent solutions of above formula can be written as

$$
\mathrm{e}^{-\mathrm{i}\frac{m_0c^2}{\hbar}t}\begin{pmatrix}1\\0\\0\\0\end{pmatrix}, \quad \mathrm{e}^{-\mathrm{i}\frac{m_0c^2}{\hbar}t}\begin{pmatrix}0\\1\\0\\0\end{pmatrix}, \quad \mathrm{e}^{\mathrm{i}\frac{m_0c^2}{\hbar}t}\begin{pmatrix}0\\0\\1\\0\end{pmatrix}, \quad \mathrm{e}^{\mathrm{i}\frac{m_0c^2}{\hbar}t}\begin{pmatrix}0\\0\\0\\1\end{pmatrix} \tag{5.79}
$$

Referring to Eq. (5.5) it can be seen that the first two solutions correspond to the $E = m_0c^2$ solutions for spin $\pm\frac{1}{2}\hbar$, and the latter two solutions are the $E = -m_0c^2$ solutions for spin $\pm\frac{1}{2}\hbar$.

When $p \neq 0$ the plane wave solution of the free particle is generally expressed as

$$
\psi = \begin{pmatrix}\psi_{\mathrm{A}}\\ \psi_{\mathrm{B}}\end{pmatrix} = \begin{pmatrix}u_{\mathrm{A}}(p)\\ u_{\mathrm{B}}(p)\end{pmatrix} \mathrm{e}^{\frac{\mathrm{i}}{\hbar}(\vec{p}\cdot\vec{r}-E\,t)} \tag{5.80}
$$

In the above formula $\mathrm{e}^{\frac{\mathrm{i}}{\hbar}(\vec{p}\cdot\vec{r}-E\,t)}$ is the general form of the plane wave, where the energy-related terms are given in Eq. (5.79). The stationary Dirac equation can be written from Eq. (5.20) and using Eqs. (5.67), (5.38), and (5.80) as follows:

$$\left[c\begin{pmatrix}\hat{0} & \hat{\sigma}\\ \hat{\sigma} & \hat{0}\end{pmatrix}\cdot\vec{p}+m_0c^2\begin{pmatrix}\hat{I} & \hat{0}\\ \hat{0} & -\hat{I}\end{pmatrix}\right]\begin{pmatrix}u_{\mathrm{A}}(p)\\ u_{\mathrm{B}}(p)\end{pmatrix}=E\begin{pmatrix}u_{\mathrm{A}}(p)\\ u_{\mathrm{B}}(p)\end{pmatrix},$$
$$c\hat{\sigma}\cdot\vec{p}u_{\mathrm{B}}(p)+m_0c^2u_{\mathrm{A}}(p)=Eu_{\mathrm{A}}(p),$$
$$c\hat{\sigma}\cdot\vec{p}u_{\mathrm{A}}(p)-m_0c^2u_{\mathrm{B}}(p)=Eu_{\mathrm{B}}(p)$$

The above two formulas can be rewritten as

$$u_{\mathrm{A}}(p)=\frac{c}{E-m_0c^2}\hat{\sigma}\cdot\vec{p}\;u_{\mathrm{B}}(p) \tag{5.81}$$

$$u_{\mathrm{B}}(p)=\frac{c}{E+m_0c^2}\hat{\sigma}\cdot\vec{p}\;u_{\mathrm{A}}(p) \tag{5.82}$$

Substituting Eq. (5.82) into Eq. (5.81) and noting that Eq. (2.16) gives $\left(\hat{\sigma}\cdot\vec{p}\right)\left(\hat{\sigma}\cdot\vec{p}\right)=p^2\hat{I}$, so we can get

$$u_{\mathrm{A}}(p)=\frac{p^2c^2}{E^2-m_0^2c^4}u_{\mathrm{A}}(p) \tag{5.83}$$

Above formula gives the energy relations $E^2=p^2c^2+m_0^2c^4$, $E=\pm\sqrt{p^2c^2+m_0^2c^4}$, which can be called the on-shell relation of relativistic energy and momentum. And we can obtain the following relation:

$$\hat{\sigma}\cdot\vec{p}=\sum_i\hat{\sigma}_i p_i=\begin{pmatrix}p_3 & p_1-\mathrm{i}p_2\\ p_1+\mathrm{i}p_2 & -p_3\end{pmatrix} \tag{5.84}$$

When $E>0$ taking $u_{\mathrm{A}}=\begin{pmatrix}1\\0\end{pmatrix}$ and $\begin{pmatrix}0\\1\end{pmatrix}$, and using Eqs. (5.82) and (5.84), we can obtain

$$u_{\mathrm{B}}=\begin{pmatrix}\dfrac{p_3c}{E+m_0c^2}\\ \dfrac{(p_1+\mathrm{i}p_2)c}{E+m_0c^2}\end{pmatrix}\quad\text{and}\quad\begin{pmatrix}\dfrac{(p_1-\mathrm{i}p_2)c}{E+m_0c^2}\\ \dfrac{-p_3c}{E+m_0c^2}\end{pmatrix}$$

Thus the obtained $E>0$ solutions of the two plane waves are

$$u^{(1)}(p)=N\begin{pmatrix}1\\0\\ \dfrac{p_3c}{E+m_0c^2}\\ \dfrac{(p_1+\mathrm{i}p_2)c}{E+m_0c^2}\end{pmatrix},\quad u^{(2)}(p)=N\begin{pmatrix}0\\1\\ \dfrac{(p_1-\mathrm{i}p_2)c}{E+m_0c^2}\\ \dfrac{-p_3c}{E+m_0c^2}\end{pmatrix} \tag{5.85}$$

where $N$ is a normalized coefficient. For negative energy particles, refer to Eq. (5.80), the general representation of free particle plane waves should be changed to

$$\psi = \begin{pmatrix} \psi_A \\ \psi_B \end{pmatrix} = \begin{pmatrix} u_A(p) \\ u_B(p) \end{pmatrix} e^{-\frac{i}{\hbar}(\vec{p}\cdot\vec{r} - Et)} \tag{5.86}$$

That is, the negative signs are added in front of $E$ and $\vec{p}$ of the phase of the plane wave; here $E$ and $\vec{p}$ are still taken positive values. In this case the stationary Dirac equation given by Eqs. (5.20) and (5.67) also becomes

$$\left(-c\vec{\alpha}\cdot\vec{p} + \beta\, m_0c^2\right)\psi = -E\psi \tag{5.87}$$

For negative energy case the relations corresponding to Eqs. (5.81) and (5.82) should be changed to

$$u_A(p) = \frac{c}{E + m_0c^2}\hat{\sigma}\cdot\vec{p}\; u_B(p) \tag{5.88}$$

$$u_B(p) = \frac{c}{E - m_0c^2}\hat{\sigma}\cdot\vec{p}\; u_A(p) \tag{5.89}$$

In this case taking $u_B = \begin{pmatrix} 1 \\ 0 \end{pmatrix}$ and $\begin{pmatrix} 0 \\ 1 \end{pmatrix}$, and using Eqs. (5.88) and (5.84), we can obtain

$$u_A = \begin{pmatrix} \dfrac{p_3c}{E + m_0c^2} \\ \dfrac{(p_1 + ip_2)\,c}{E + m_0c^2} \end{pmatrix} \quad \text{and} \quad \begin{pmatrix} \dfrac{(p_1 - ip_2)c}{E + m_0c^2} \\ \dfrac{-p_3c}{E + m_0c^2} \end{pmatrix}$$

So the obtained solutions of the two plane waves for negative energy case are

$$v^{(1)}(p) = N\begin{pmatrix} \dfrac{p_3c}{E + m_0c^2} \\ \dfrac{(p_1 + \mathrm{i}p_2)c}{E + m_0c^2} \\ 1 \\ 0 \end{pmatrix}, \qquad v^{(2)}(p) = N\begin{pmatrix} \dfrac{(p_1 - \mathrm{i}p_2)c}{E + m_0c^2} \\ \dfrac{-p_3c}{E + m_0c^2} \\ 0 \\ 1 \end{pmatrix} \tag{5.90}$$

Based on Eq. (5.42), let $\overline{u}^{(s)}(p) = u^{(s)+}(p)\gamma^0$ and $\overline{v}^{(s)}(p) = v^{(s)+}(p)\gamma^0$. The following relations can be verified by using Eqs. (5.85) and (5.90):

$$\begin{aligned}\overline{u}^{(s)}(p)u^{(s')}(p) &= \delta_{ss'},\ \ \overline{v}^{(s)}(p)v^{(s')}(p) = -\delta_{ss'},\\ \overline{v}^{(s)}(p)u^{(s')}(p) &= \overline{u}^{(s)}(p)v^{(s')}(p) = 0\end{aligned} \tag{5.91}$$

For example, we can verify the following relations:

$$\overline{u}^{(1)}(p)u^{(1)}(p) = \left[1 - \frac{p^2c^2}{(E+m_0c^2)^2}\right]N^2 = \frac{2m_0c^2}{E+m_0c^2}N^2,$$

$$\overline{v}^{(1)}(p)v^{(1)}(p) = -\left[1 - \frac{p^2c^2}{(E+m_0c^2)^2}\right]N^2 = -\frac{2m_0c^2}{E+m_0c^2}N^2$$

Then the normalization coefficient can be obtained as

$$N = \sqrt{\frac{E+m_0c^2}{2m_0c^2}} \tag{5.92}$$

We know that $u(p)$ is a positive energy plane wave function constituted by $u^{(1)}(p)$ and $u^{(2)}(p)$, and $v(p)$ is a negative energy plane wave function constituted by $v^{(1)}(p)$ and $v^{(2)}(p)$. $\chi$ is a two-dimensional spin wave function; it is an eigenfunction of the helicity operator $\hat{\sigma}\cdot\hat{p} = \hat{\sigma}\cdot\frac{\vec{p}}{|\vec{p}|}$ with eigenvalue $\pm 1$, for which the corresponding basis spin functions are $\begin{pmatrix}1\\0\end{pmatrix}$ and $\begin{pmatrix}0\\1\end{pmatrix}$, respectively. The positive energy solution can be obtained according to Eqs. (5.85), (5.90), and (5.92) as follows:

$$u(p) = \sqrt{\frac{E+m_0c^2}{2m_0c^2}}\begin{pmatrix}\chi\\ \dfrac{c\hat{\sigma}\cdot\vec{p}}{E+m_0c^2}\chi\end{pmatrix} \tag{5.93}$$

The negative energy solution is

$$v(p) = \sqrt{\frac{E+m_0c^2}{2m_0c^2}}\begin{pmatrix}\dfrac{c\hat{\sigma}\cdot\vec{p}}{E+m_0c^2}\chi\\ \chi\end{pmatrix} \tag{5.94}$$

It can be seen that $u(p)$ and $v(p)$ are all four-component vectors.

In order to explain the negative energy solution of the Dirac equation, Dirac proposed the hole theory in 1930, arguing that the hole of the electron is a particle with the same mass as the electron and the opposite charge, i.e., a positron (positive electron). The electron which energy $E > m_0c^2$ is in a free state, i.e., the so-called positive energy state; and the electron which energy $E < -m_0c^2$ is in the Dirac Sea, that is, the so-called negative energy state. Since the electrons belonging to Fermions obey the Pauli incompatibility principle, all the negative states in the Dirac Sea are filled with electrons. At this time, the positive energy electrons cannot fall into the negative states, which can explain the stability of positive energy electrons. The electrons bound by the Coulomb field in the atom are at the energy region $-m_0c^2 < E < m_0c^2$. The so-called vacuum state refers to that the positive energy states are all without electrons, and the negative energy states are all filled with electrons, and this image explains the generation and annihilation of positive and negative electron pairs. In 1932, positrons were observed in cosmic rays; it confirmed the Dirac's prophecy.

### 5.1.5 Lorentz Covariant of the Dirac Equation

In the following discussion the Pauli metric will be used. Linear transformation of the vector $x_\mu \equiv \left(\vec{r}, ict\right)$ in the four-dimensional Minkowski space is as follows:

$$x_\mu \longrightarrow x'_\mu = \alpha_{\mu\nu} x_\nu \tag{5.95}$$

After the transformation it is required that the first three components of $x'_\mu$ are still maintained as real, and the fourth component is imaginary. Therefore, $\alpha_{ik}$ $(i, k = 1, 2, 3)$ is required to be real, $\alpha_{i4}$ and $\alpha_{4i}$ $(i = 1, 2, 3)$ are imaginary, and $\alpha_{44}$ is real. In order to ensure that the speed of light is equal in any coordinate system, the length of the space-time vector after transformation is required to remain unchanged, that is,

$$x_\mu x_\mu \longrightarrow x'_\mu x'_\mu = \alpha_{\mu\nu} x_\nu \alpha_{\mu\rho} x_\rho = x_\rho x_\rho \tag{5.96}$$

Hence from above formula there is the requirement

$$\alpha_{\mu\nu} \alpha_{\mu\rho} = \delta_{\nu\rho} \tag{5.97}$$

The two four-dimensional vectors contraction products (such as $x_\mu x_\mu$, $\partial_\mu \partial_\mu$, $x_\mu \partial_\mu$) are scalar, and Eq. (5.96) indicates that the scalar is unchanged under the Lorentz transformation.

Equation (5.95) is multiplied by $\alpha_{\mu\rho}$ from the left and summed to $\mu$, and using Eq. (5.97) we can obtain

$$\alpha_{\mu\rho} x'_\mu = \alpha_{\mu\rho} \alpha_{\mu\nu} x_\nu = \delta_{\rho\nu} x_\nu = x_\rho \tag{5.98}$$

The inverse transformation of Eq. (5.95) is

$$x_\rho = \alpha^{-1}_{\rho\mu} x'_\mu \tag{5.99}$$

Comparing Eq. (5.99) and Eq. (5.98), the following relation can be obtained:

$$\alpha_{\rho\mu}^{-1} = \alpha_{\mu\rho} \tag{5.100}$$

Therefore the product of positive and inverse matrices satisfies

$$\alpha_{\nu\mu}^{-1}\alpha_{\mu\rho} = \alpha_{\mu\nu}\alpha_{\mu\rho} = \delta_{\nu\rho} \tag{5.101}$$

For the reference systems that move at a uniform speed $v$ along the $z$-axis, let $\beta = v/c$, and then the corresponding Lorentz transformation matrix is

$$\alpha = \begin{pmatrix} 1 & 0 & 0 & 0 \\ 0 & 1 & 0 & 0 \\ 0 & 0 & \dfrac{1}{\sqrt{1-\beta^2}} & \dfrac{i\beta}{\sqrt{1-\beta^2}} \\ 0 & 0 & \dfrac{-i\beta}{\sqrt{1-\beta^2}} & \dfrac{1}{\sqrt{1-\beta^2}} \end{pmatrix} \tag{5.102}$$

Using transformation

$$r' = \alpha\, r, \qquad r = \begin{pmatrix} x \\ y \\ z \\ ict \end{pmatrix} \tag{5.103}$$

we can obtain

$$x' = x,\ \ y' = y,\ \ z' = \frac{z}{\sqrt{1-\beta^2}} - \frac{vt}{\sqrt{1-\beta^2}},\ \ t' = \frac{t}{\sqrt{1-\beta^2}} - \frac{(v/c^2)z}{\sqrt{1-\beta^2}} \tag{5.104}$$

where $(x, y, z, t)$ are the quantities in the fixed coordinate system, and $(x', y', z', t')$ are the quantities in the moving coordinate system. This transformation is called the true Lorentz transformation. The spatial reflection transformation matrix is

$$\alpha = \begin{pmatrix} -1 & 0 & 0 & 0 \\ 0 & -1 & 0 & 0 \\ 0 & 0 & -1 & 0 \\ 0 & 0 & 0 & 1 \end{pmatrix} \tag{5.105}$$

The time inversion transformation matrix is

$$\alpha = \begin{pmatrix} 1 & 0 & 0 & 0 \\ 0 & 1 & 0 & 0 \\ 0 & 0 & 1 & 0 \\ 0 & 0 & 0 & -1 \end{pmatrix} \tag{5.106}$$

It can be seen that the transformation matrices given by Eqs. (5.102), (5.105), and (5.106) all satisfy Eq. (5.97) and Eq. (5.101).

If we write a Dirac equation according to Eq. (5.33)

$$\left(\gamma'_\mu \partial_\mu + \kappa\right) \psi' = 0 \tag{5.107}$$

where $\gamma'_\mu$ satisfies $\left\{ \gamma'_\mu, \gamma'_\nu \right\} = 2\delta_{\mu\nu}$, the Pauli's theorem is that if there are two sets of $4 \times 4$ matrices that satisfy $\{\gamma_\mu, \gamma_\nu\} = 2\delta_{\mu\nu}$ and $\left\{\gamma'_\mu, \gamma'_\nu\right\} = 2\delta_{\mu\nu}$, respectively, there must be a non-singular and unique $4 \times 4$ matrix $S$, so that they satisfy the following transformation relations [1, 6, 7]:

$$S\gamma_\mu S^{-1} = \gamma'_\mu, \qquad S^{-1}\gamma'_\mu S = \gamma_\mu \tag{5.108}$$

where the non-singular means that there must be an inverse matrix of $S$ to satisfy $SS^{-1} = 1$. If $\gamma'_\mu$ is also Hermitian, from Eq. (5.108) we can prove that $S$ is unitary, i.e., $S^+ = S^{-1}$. According to this theorem, Eq. (5.107) is rewritten as

$$\left(S\gamma_\mu S^{-1}\partial_\mu + \kappa\right) SS^{-1}\psi' = 0 \tag{5.109}$$

Multiply $S^{-1}$ from the left of the above formula and note that $S$ is commutable with $\partial_\mu$, and then Eq. (5.109) becomes

$$\left(\gamma_\mu \partial_\mu + \kappa\right) S^{-1}\psi' = 0 \tag{5.110}$$

It can be seen that $S^{-1}\psi'$ is the solution of the Dirac equation, so Eq. (5.107) is equivalent to the Dirac equation (5.33). The relations between $\psi$ and $\psi'$ are as follows:

$$\psi' = S\psi, \qquad \psi'^+ = \psi^+ S^{-1} \tag{5.111}$$

Note that $\alpha_{\nu\mu}$ is only a matrix element; the following expressions can be written according to Eqs. (5.95) and (5.97):

$$\partial'_\nu = \alpha_{\nu\mu}\partial_\mu, \quad \partial_\mu = \alpha_{\nu\mu}\partial'_\nu, \quad \gamma_\mu\partial_\mu = \gamma_\mu\alpha_{\nu\mu}\partial'_\nu = \alpha_{\nu\mu}\gamma_\mu\partial'_\nu \equiv \overline{\gamma}_\nu\partial'_\nu \tag{5.112}$$

where

$$\overline{\gamma}_\nu \equiv \alpha_{\nu\mu}\gamma_\mu \tag{5.113}$$

So the Dirac equation (5.33) can be rewritten according to Eq. (5.112) as follows:

$$\left(\overline{\gamma}_\nu \partial'_\nu + \kappa\right)\psi = 0 \tag{5.114}$$

The following expressions can be proved utilizing Eqs. (5.29) and (5.101):

$$\overline{\gamma}_\mu\overline{\gamma}_\nu + \overline{\gamma}_\nu\overline{\gamma}_\mu = \alpha_{\mu\rho}\gamma_\rho\alpha_{\nu\delta}\gamma_\delta + \alpha_{\nu\delta}\gamma_\delta\alpha_{\mu\rho}\gamma_\rho = \alpha_{\mu\rho}\alpha_{\nu\delta}\left(\gamma_\rho\gamma_\delta + \gamma_\delta\gamma_\rho\right) = 2\delta_{\mu\nu} \tag{5.115}$$

According to the Pauli theorem (5.108), there is a $4 \times 4$ matrix denoted as $L$, and the transformation of the $\gamma$ matrix can be written as follows:

$$\overline{\gamma}_\mu = \alpha_{\mu\nu}\gamma_\nu = L^{-1}\gamma_\mu L \tag{5.116}$$

here $L$ has no relationship with coordinates, and is commutable with $\partial'_\mu$. Then substituting Eq. (5.116) into Eq. (5.114), we can obtain

$$\left(L^{-1}\gamma_\mu L\partial'_\mu + \kappa\right)\psi = 0$$

Multiplying $L$ from the left side of the above equation, the following equation can be got

$$\left(\gamma_\mu\partial'_\mu + \kappa\right)L\psi = 0 \tag{5.117}$$

Therefore under the Lorentz transformation, the change of the wave function and the satisfied equation are, respectively

$$\psi' = L\psi, \qquad \left(\gamma_\mu\partial'_\mu + \kappa\right)\psi' = 0 \tag{5.118}$$

This proves that the Dirac equation has covariance under the Lorentz transformation of four-dimensional space-time coordinates.

The above transformation can also be discussed in another way. Let $L$ be the transformation operator of the matrix element $\alpha_{\mu\nu}$, since this transformation only transforms the four-dimensional coordinates, its matrix elements are all independent of the coordinates, so the Dirac equation given by Eq. (5.33) can be converted into

$$\left(\gamma_\mu\partial'_\mu + \kappa\right)\psi'(x') = 0 \tag{5.119}$$

and there is

$$\psi'(x') = L\psi(x) \tag{5.120}$$

From Eq. (5.95) and Eq. (5.100), we can get

$$\partial'_\mu = \alpha_{\nu\mu}\partial_\nu \tag{5.121}$$

Substituting Eq. (5.120) and Eq. (5.121) into Eq. (5.119) and multiplying $L^{-1}$ from the left side of the obtained formula, the following expression can be obtained:

$$\left(L^{-1}\gamma_\mu L\alpha_{\nu\mu}\partial_\nu + \kappa\right)\psi(x) = 0 \tag{5.122}$$

Only there is

$$L^{-1}\gamma_\mu L = \alpha_{\mu\nu}\gamma_\nu \tag{5.123}$$

Equation (5.122) may become Eq. (5.33).

The conjugate equation of the Dirac equation (5.33) is

$$\partial^*_\mu\psi^+\gamma_\mu + \kappa\psi^+ = 0 \tag{5.124}$$

Multiplying $\gamma_4$ from the right side of the above equation, we can obtain

$$\partial^*_\mu\psi^+\gamma_\mu\gamma_4 + \kappa\psi^+\gamma_4 = 0$$

Noting $\gamma_i\gamma_4 = -\gamma_4\gamma_i$, $\partial^*_i = \partial_i$, $\partial^*_4 = -\partial_4$, based on the definition of Eq. (5.42), the following expression can be obtained from the above formula:

$$\partial_\mu\overline{\psi}\gamma_\mu - \kappa\overline{\psi} = 0 \tag{5.125}$$

And using the inverse transformation $\partial_\mu = \alpha_{\nu\mu}\partial'_\nu$ and Eq. (5.116), from the above equation we can obtain

$$\alpha_{\nu\mu}\partial'_\nu\overline{\psi}\gamma_\mu - \kappa\overline{\psi} = \partial'_\nu\overline{\psi}\alpha_{\nu\mu}\gamma_\mu - \kappa\overline{\psi} = \partial'_\nu\overline{\psi}L^{-1}\gamma_\nu L - \kappa\overline{\psi} = 0 \tag{5.126}$$

Multiplying $L^{-1}$ from the right side of the above equation, and let

$$\overline{\psi'} = \overline{\psi}L^{-1} \tag{5.127}$$

then we get the following equation satisfied by $\overline{\psi'}$ under the Lorentz transformation:

$$\partial'_\nu\overline{\psi'}\gamma_\nu - \kappa\overline{\psi'} = 0 \tag{5.128}$$

It can be seen that $\overline{\psi}$ after the transformation still satisfy the same equation with Eq. (5.125), so under the Lorentz transformation there is

$$\overline{\psi} \to \overline{\psi}' = \overline{\psi} L^{-1} \tag{5.129}$$

Therefore using Eq. (5.129) and Eq. (5.118), we can get

$$\overline{\psi}' \psi' = \overline{\psi} L^{-1} L \psi = \overline{\psi} \psi \tag{5.130}$$

The above formula shows that $\overline{\psi}\psi$ is an invariant under the Lorentz transformation, i.e., $\overline{\psi}\psi$ is a scalar under the Lorentz transformation.

In the following we discuss the $\gamma_5$ behavior under the true Lorentz transformation given by Eq. (5.102). First we can find

$$\begin{aligned}
L^{-1}\gamma_5 L &= L^{-1}\gamma_1 L L^{-1}\gamma_2 L L^{-1}\gamma_3 L L^{-1}\gamma_4 L = (\alpha_{1\mu}\gamma_\mu)(\alpha_{2\nu}\gamma_\nu)(\alpha_{3\rho}\gamma_\rho)(\alpha_{4\lambda}\gamma_\lambda) \\
&= \gamma_1\gamma_2 \left(\frac{1}{\sqrt{1-\beta^2}}\gamma_3 + \frac{\mathrm{i}\beta}{\sqrt{1-\beta^2}}\gamma_4\right)\left(\frac{-\mathrm{i}\beta}{\sqrt{1-\beta^2}}\gamma_3 + \frac{1}{\sqrt{1-\beta^2}}\gamma_4\right) \\
&= \gamma_1\gamma_2\left(\frac{1}{1-\beta^2}\gamma_3\gamma_4 + \frac{\beta^2}{1-\beta^2}\gamma_4\gamma_3\right) = \gamma_1\gamma_2\gamma_3\gamma_4 = \gamma_5
\end{aligned}$$

The above formula indicates that $L$ and $\gamma_5$ are commutable under the true Lorentz transformation, i.e.,

$$L^{-1}\gamma_5 L = \gamma_5, \qquad \gamma_5 L = L\gamma_5 \tag{5.131}$$

Under the spatial reflection transformation, $x_i \to x_i' = -x_i, \ \ t \to t' = t$, it can be seen from Eq. (5.105) that $\alpha_{ik} = -\delta_{ik}, \ \alpha_{44} = 1$. For $\gamma$ matrices there are

$$L^{-1}\gamma_i L = -\gamma_i, \quad \gamma_i L = -L\gamma_i, \quad i = 1, 2, 3; \quad L^{-1}\gamma_4 L = \gamma_4, \quad \gamma_4 L = L\gamma_4 \tag{5.132}$$

The spatial reflection transformation matrix satisfying Eq. (5.132) can be expressed as

$$L = \xi\gamma_4 \tag{5.133}$$

Substituting Eq. (5.133) into (5.132) can verify its validity. Here $\xi$ represents the particle's intrinsic parity, due to requirements $LL = I, \ L^{-1} = L$, so there are $\xi^2 = 1$, $\xi = \pm 1$. The transformations of $\gamma$ matrices during the spatial reflection are

$$\alpha_{\mu\nu}\gamma_\nu = L^{-1}\gamma_\mu L = \xi^2\gamma_4\gamma_\mu\gamma_4 = \gamma_4\gamma_\mu\gamma_4 \tag{5.134}$$

Under the spatial reflection transformation, the transformation of Dirac spinor is

$$\psi(x) \to L\psi(x) = \pm\gamma_4\psi(x) \tag{5.135}$$

The particle's intrinsic parity may be +1 or −1. We can still get

$$L^{-1}\gamma_5 L = \xi^2\gamma_4\gamma_5\gamma_4 = -\gamma_5\gamma_4\gamma_4 = -\gamma_5 \tag{5.136}$$

So for the reflection of space there are

$$\gamma_5 L = -L\gamma_5, \quad L^{-1}\gamma_5 = -\gamma_5 L^{-1} \tag{5.137}$$

For the 4 × 4 matrices, there must be 16 linearly independent matrices, and all 4 × 4 matrices can be composed of their linear combinations. It is possible to select 16 linearly independent matrices, denoted by $\Gamma_A$, composed by four $\gamma$ matrices that satisfy the anti-commutative relations, and then any 4 × 4 matrix can be given with the linear combination of these linearly independent matrices $\Gamma_A$. In the following all the possible multiple products of the $\gamma$ matrices and the number of independent matrices composed by the various multiple products of the $\gamma$ matrices are given.

| | | |
|---|---|---|
| highest product | $\gamma_5 = \gamma_1\gamma_2\gamma_3\gamma_4$ | 1 |
| triple product | $\gamma_\mu\gamma_\nu\gamma_\sigma(\mu \neq \nu \neq \sigma)$ or $\mathrm{i}\gamma_5\gamma_\rho$ | 4 |
| double product | $\gamma_\mu\gamma_\nu$ $(\mu \neq \nu)$ | 6 |
| single product | $\gamma_\mu$ | 4 |
| zero product | $I$ | 1 |

We select 16 linearly independent 4 × 4 matrices $\Gamma_A$ as

$$\Gamma^{\mathrm{S}} = 1, \;\; \Gamma_\mu^{\mathrm{V}} = \mathrm{i}\gamma_\mu, \;\; \Gamma_{\mu\nu}^{\mathrm{T}} = \sigma_{\mu\nu} \equiv \frac{1}{2\mathrm{i}}(\gamma_\mu\gamma_\nu - \gamma_\nu\gamma_\mu), \;\; \Gamma^{\mathrm{P}} = \mathrm{i}\gamma_5, \;\; \Gamma_\mu^{\mathrm{A}} = \mathrm{i}\gamma_5\gamma_\mu \tag{5.138}$$

We can verify, utilizing Eqs. (5.29), (5.39), and (5.40), that they have the following properties:

1. The 16 $\Gamma_A$ matrices all satisfy $|\Gamma_A|^2 = \Gamma_A^+\Gamma_A = I$, so $\Gamma_A^{-1} = \Gamma_A^+$.
2. Except for the unit matrix $\Gamma^{\mathrm{S}}$, for each $\Gamma_A$ there is another $\Gamma_B$ which is anti-commutative with it, i.e., $\Gamma_A\Gamma_B = -\Gamma_B\Gamma_A$.
3. Except for the unit matrix $\Gamma^{\mathrm{S}}$, for each $\Gamma_A$ there is $\mathrm{tr}\{\Gamma_A\} = 0$; when $A \neq B$ there is $\mathrm{tr}\{\Gamma_A\Gamma_B\} = 0$.
4. The determinants of 16 $\Gamma_A$ matrices are all 1. Since the determinant has the following properties: if $A$ and $B$ are the square matrices of the same rank, the determinant $|A{\cdot}B| = |A|{\cdot}|B|$, and after the interchange of rows and columns (reflection about a diagonal) the determinant value does not change. Because $\Gamma_A$ is also a square 4 × 4 matrix, there is determinant $\det(-\Gamma_A) = \det(\Gamma_A)$, and

using Eq. (5.27) it is possible to prove $\det(\Gamma_A^+) = \det(\Gamma_A)$ by $\Gamma_A$ given by Eq. (5.138). So by use of $|\Gamma_A|^2 = I$ we can obtain

$$\det\left(|\Gamma_A|^2\right) = \det(\Gamma_A^+ \Gamma_A) = \det(\Gamma_A^+) \cdot \det(\Gamma_A) = [\det(\Gamma_A)]^2 = 1, \ \det(\Gamma_A) = \pm 1$$

However, in $\Gamma_A$ diagonal representation, except the unit matrix $\Gamma^S$, their traces are still 0. At this time, the numbers of the value $+1$ and $-1$ in the four diagonal elements must be equal, so it can only be that $\det(\Gamma_A) = 1$.

5. The 16 $\Gamma_A$ matrices are linearly independent of each other. This conclusion can be proved as follows: if $\sum_{A=1}^{16} C_A \Gamma_A = 0$, i.e., 16 $\Gamma_A$ are not independent, multiplying $\Gamma_B$ for the front of left side of the above formula one can obtain $C_B + \sum_{A \neq B} C_A \Gamma_B \Gamma_A = 0$; using the preceding 3. clause, tracing the above formula one can find $C_B = 0$. Since $B$ is arbitrary, all of $C_A$ in $\sum_{A=1}^{16} C_A \Gamma_A = 0$ are 0. That is, they are linearly independent.
6. If a matrix is commutative with all 16 $\Gamma_A$ matrices, since the above 16 $\Gamma_A$ matrices are linearly independent of each other, this matrix must be a multiple of the unit matrix.

In the theory of relativistic quantum mechanics, the physical quantity is a Dirac covariable composed by $\overline{\psi}$, $\psi$ and $\gamma_\mu$ matrices. The Dirac covariables of all physical quantities may only have the following five types:

① Scalar $S = \overline{\psi}\psi$ corresponds to zero product.
② Pseudo-scalar $P = \mathrm{i}\overline{\psi}\gamma_5\psi$ corresponds to the highest product.
③ Vector $V_\mu = \mathrm{i}\overline{\psi}\gamma_\mu\psi$ corresponds to single product.
④ Axis-vector $A_\mu = \mathrm{i}\overline{\psi}\gamma_5\gamma_\mu\psi$ corresponds to triple product.
⑤ Anti-symmetric tensor $T_{\mu\nu} = \frac{1}{2\mathrm{i}}\overline{\psi}\left(\gamma_\mu\gamma_\nu - \gamma_\nu\gamma_\mu\right)\psi$ corresponds to double product.

When defining these covariables, except the scalar, the factor i representing imaginary number are introduced in order to make them to meet the properties of vectors and tensors. For example, it is possible to ensure that the first three components of the vector are real, and the fourth component is imaginary, the sign of its Hermitian conjugate real part is invariant, and the sign of its Hermitian conjugate imaginary part is altered. These conclusions can be proved as follows:

$$V_\mu^+ = \left(\mathrm{i}\overline{\psi}\gamma_\mu\psi\right)^+ = -\mathrm{i}\psi^+\gamma_\mu\gamma_4\psi = -\mathrm{i}\overline{\psi}\gamma_4\gamma_\mu\gamma_4\psi = \left(\mathrm{i}\overline{\psi}\gamma_i\psi, \quad -\mathrm{i}\overline{\psi}\gamma_4\psi\right)$$

And keep that the pseudo-scalar is real: $P^+ = -\mathrm{i}\psi^+\gamma_5\gamma_4\psi = \mathrm{i}\overline{\psi}\gamma_5\psi = P$.

Under the true Lorentz transformation, the scalar is invariant:

$$S' = \overline{\psi}\, L^{-1} L\psi = \overline{\psi}\psi = S \tag{5.139}$$

Using Eq. (5.131), under the true Lorentz transformation, the pseudo-scalar is also invariant:

$$P' = \mathrm{i}\overline{\psi}'\gamma_5\psi' = \mathrm{i}\overline{\psi}L^{-1}\gamma_5 L\psi = \mathrm{i}\overline{\psi}\gamma_5\psi = P \tag{5.140}$$

Under the spatial reflection transformation there is still

$$S' = \overline{\psi}\, L^{-1} L\psi = S \tag{5.141}$$

However, due to Eq. (5.137) there is

$$P' = \mathrm{i}\overline{\psi}\, L^{-1}\gamma_5 L\psi = -\mathrm{i}\overline{\psi}\, L^{-1} L\gamma_5\psi = -\mathrm{i}\overline{\psi}\gamma_5\psi = -P \tag{5.142}$$

Therefore, $P$ is called a pseudo-scalar.

Under the true Lorentz transformation, the transformation of the vector $V_\mu$ is

$$V'_\mu = \mathrm{i}\overline{\psi}'\gamma_\mu\psi' = \mathrm{i}\overline{\psi}L^{-1}\gamma_\mu L\psi = \alpha_{\mu\nu}(\mathrm{i}\overline{\psi}\gamma_\nu\psi) = \alpha_{\mu\nu}V_\nu \tag{5.143}$$

and the transformation of the axis vector $A_\mu$ is

$$A'_\mu = \mathrm{i}\overline{\psi}'\gamma_5\gamma_\mu\psi' = \mathrm{i}\overline{\psi}L^{-1}\gamma_5\gamma_\mu L\psi = \mathrm{i}\overline{\psi}\gamma_5 L^{-1}\gamma_\mu L\psi = \alpha_{\mu\nu}A_\nu \tag{5.144}$$

It can be seen that this transformation is the same as the vector $V_\mu$. However, under the spatial reflection transformation, for the vector $V_\mu$ and the axis vector $A_\mu$, using Eqs. (5.105) and (5.137), the following results can be obtained, respectively:

$$V'_\mu = \mathrm{i}\overline{\psi}'\gamma_\mu\psi' = \mathrm{i}\overline{\psi}L^{-1}\gamma_\mu L\psi = \alpha_{\mu\nu}(\mathrm{i}\overline{\psi}\gamma_\nu\psi) = \alpha_{\mu\nu}V_\nu = \begin{cases} -V_i, & i = 1,2,3 \\ V_4, & \mu = 4 \end{cases} \tag{5.145}$$

$$A'_\mu = \mathrm{i}\overline{\psi}'\gamma_5\gamma_\mu\psi' = \mathrm{i}\overline{\psi}L^{-1}\gamma_5\gamma_\mu L\psi = -\alpha_{\mu\nu}(\mathrm{i}\overline{\psi}\gamma_5\gamma_\nu\psi) = \begin{cases} A_i, & i = 1,2,3 \\ -A_4, & \mu = 4 \end{cases} \tag{5.146}$$

It can be seen that the spatial inversion transformation of the axis-vector $A_\mu$ differs by a negative sign from the transformation of the vector $V_\mu$, so $A_\mu$ is called the axis vector.

**Table 5.1** Dirac covariables and their transformation properties under the true Lorentz transformation and the spatial reflection transformation

| | Dirac covariable | True Lorentz transformation | Space reflection transformation | Number |
|---|---|---|---|---|
| Scalar | $S = \overline{\psi}\psi$ | $S' = S$ | $S' = S$ | 1 |
| Pseudo-scalar | $P = \mathrm{i}\overline{\psi}\gamma_5\psi$ | $P' = P$ | $P' = -P$ | 1 |
| Vector | $V_\mu = \mathrm{i}\overline{\psi}\gamma_\mu\psi$ | $V'_\mu = \alpha_{\mu\nu}V_\nu$ | $V'_i = -V_i$, $V'_4 = V_4$ | 4 |
| Axis-vector | $A_\mu = \mathrm{i}\overline{\psi}\gamma_5\gamma_\mu\psi$ | $A'_\mu = \alpha_{\mu\nu}A_\nu$ | $A'_i = A_i$, $A'_4 = -A_4$ | 4 |
| Anti-symmetric tensor | $T_{\mu\nu} = \frac{1}{2\mathrm{i}}\overline{\psi}(\gamma_\mu\gamma_\nu - \gamma_\nu\gamma_\mu)\psi$ | $T'_{\mu\nu} = \alpha_{\mu\rho}\alpha_{\nu\sigma}T_{\rho\sigma}$ | $T'_{ij} = T_{ij}$, $T'_{i4} = -T_{i4}$, $T'_{4j} = -T_{4j}$, $T'_{44} = T_{44}$ | 6 |

Under the true Lorentz transformation, the transformation of the anti-symmetric tensor $T_{\mu\nu}$ is

$$\begin{aligned} T'_{\mu\nu} &= \frac{1}{2\mathrm{i}}\overline{\psi}'(\gamma_\mu\gamma_\nu - \gamma_\nu\gamma_\mu)\psi' = \frac{1}{2\mathrm{i}}\overline{\psi}L^{-1}(\gamma_\mu\gamma_\nu - \gamma_\nu\gamma_\mu)L\psi \\ &= \frac{1}{2\mathrm{i}}\overline{\psi}(\alpha_{\mu\rho}\gamma_\rho\alpha_{\nu\sigma}\gamma_\sigma - \alpha_{\nu\sigma}\gamma_\sigma\alpha_{\mu\rho}\gamma_\rho)\psi \\ &= \alpha_{\mu\rho}\alpha_{\nu\sigma}\frac{1}{2\mathrm{i}}\overline{\psi}(\gamma_\rho\gamma_\sigma - \gamma_\sigma\gamma_\rho)\psi = \alpha_{\mu\rho}\alpha_{\nu\sigma}T_{\rho\sigma} \end{aligned} \tag{5.147}$$

Under the spatial reflection transformation, using Eq. (5.105) the transformations of the anti-symmetric tensor $T_{\mu\nu}$ are

$$\begin{aligned} T'_{ij} &= \alpha_{i\rho}\alpha_{j\sigma}T_{\rho\sigma} = (-\delta_{i\rho})(-\delta_{j\sigma})T_{\rho\sigma} = T_{ij}, \\ T'_{i4} &= \alpha_{i\rho}\alpha_{4\sigma}T_{\rho\sigma} = -\delta_{i\rho}\delta_{4\sigma}T_{\rho\sigma} = -T_{i4}, \quad T'_{4j} = -T_{4j}, \\ T'_{44} &= \alpha_{4\rho}\alpha_{4\sigma}T_{\rho\sigma} = T_{44} \end{aligned} \tag{5.148}$$

Summarizing the above results, all possible Dirac covariables and their transformation properties under the true Lorentz transformation and spatial reflection transformation are listed in Table 5.1. Any other form of Dirac covariables can always be represented by these covariables.

### 5.1.6 The Trace Formulas for $\gamma$ Matrix Products

The trace formulas for the $\gamma$ matrix product of the Bjorken-Drell metric are given below. The trace of an odd number of $\gamma^\mu$ matrix product is zero, and for even number of $\gamma^\mu$ matrix products, the following trace formulas are found [1, 8]:

$$\mathrm{tr}\{\gamma^{\mu}\gamma^{\nu}\} = 4g^{\mu\nu} \tag{5.149}$$

$$\mathrm{tr}\{\gamma^{\mu}\gamma^{\nu}\gamma^{\rho}\gamma^{\sigma}\} = 4(g^{\mu\nu}g^{\rho\sigma} + g^{\mu\sigma}g^{\nu\rho} - g^{\mu\rho}g^{\nu\sigma}) \tag{5.150}$$

$$\mathrm{tr}\{\gamma^{\sigma}\gamma^{\mu}\gamma^{\nu}\gamma^{\rho}\} = \mathrm{tr}\{\gamma^{\mu}\gamma^{\nu}\gamma^{\rho}\gamma^{\sigma}\} \tag{5.151}$$

$$\begin{aligned} \mathrm{tr}\{\gamma^{\mu}\gamma^{\nu}\gamma^{\rho}\gamma^{\sigma}\gamma^{\lambda}\gamma^{\eta}\} &= 4\Big[g^{\mu\nu}\left(g^{\rho\sigma}g^{\lambda\eta} + g^{\rho\eta}g^{\sigma\lambda} - g^{\rho\lambda}g^{\sigma\eta}\right) \\ &-g^{\mu\rho}\left(g^{\nu\sigma}g^{\lambda\eta} + g^{\nu\eta}g^{\sigma\lambda} - g^{\nu\lambda}g^{\sigma\eta}\right) + g^{\mu\sigma}\left(g^{\nu\rho}g^{\lambda\eta} + g^{\nu\eta}g^{\rho\lambda} - g^{\nu\lambda}g^{\rho\eta}\right) \\ &-g^{\mu\lambda}\left(g^{\nu\rho}g^{\sigma\eta} + g^{\nu\eta}g^{\rho\sigma} - g^{\nu\sigma}g^{\rho\eta}\right) + g^{\mu\eta}\left(g^{\nu\rho}g^{\sigma\lambda} + g^{\nu\lambda}g^{\rho\sigma} - g^{\nu\sigma}g^{\rho\lambda}\right)\Big] \end{aligned} \tag{5.152}$$

where $g^{\mu\nu}$ is the matrix element of the metric tensor given by Eq. (5.53).

## 5.2 Transformation of Relativistic Coordinate Systems

In the Pauli metric, the momentum-energy 4 vector composed of momentum $\vec{p}$ and energy $E$ is $\left(\vec{p},\ \mathrm{i}\frac{E}{c}\right)$, and let

$$\beta = \frac{v}{c}, \qquad \gamma = 1/\sqrt{1-\beta^2} \tag{5.153}$$

The Lorentz transformation matrix between the two coordinate systems moving each other at a speed $v$ in the $z$-axis direction has been given by Eq. (5.102) as

$$\alpha = \begin{pmatrix} 1 & 0 & 0 & 0 \\ 0 & 1 & 0 & 0 \\ 0 & 0 & \gamma & \mathrm{i}\beta\gamma \\ 0 & 0 & -\mathrm{i}\beta\gamma & \gamma \end{pmatrix} \tag{5.154}$$

From Eq. (5.100) one can also know

$$\alpha^{-1} = \begin{pmatrix} 1 & 0 & 0 & 0 \\ 0 & 1 & 0 & 0 \\ 0 & 0 & \gamma & -\mathrm{i}\beta\gamma \\ 0 & 0 & \mathrm{i}\beta\gamma & \gamma \end{pmatrix} \tag{5.155}$$

The Lorentz transformation relation of the momentum-energy 4 vector is

$$\begin{pmatrix} p_x{}' \\ p_y{}' \\ p_z{}' \\ \mathrm{i}E'/c \end{pmatrix} = \begin{pmatrix} 1 & 0 & 0 & 0 \\ 0 & 1 & 0 & 0 \\ 0 & 0 & \gamma & \mathrm{i}\beta\gamma \\ 0 & 0 & -\mathrm{i}\beta\gamma & \gamma \end{pmatrix} \begin{pmatrix} p_x \\ p_y \\ p_z \\ \mathrm{i}E/c \end{pmatrix} = \begin{pmatrix} p_x \\ p_y \\ \gamma\left(p_z - \beta E/c\right) \\ -\mathrm{i}\gamma\left(\beta p_z - E/c\right) \end{pmatrix} \tag{5.156}$$

Adding the square of the third row and the square of the fourth row of the right end of the above equation, we can obtain

$$\begin{aligned} &\gamma^2\left[p_z^2 + \beta^2(E/c)^2 - 2\beta p_z E/c\right] - \gamma^2\left[\beta^2 p_z^2 + (E/c)^2 - 2\beta p_z E/c\right] \\ &\quad = \gamma^2(1-\beta^2)p_z^2 - \gamma^2(1-\beta^2)\,(E/c)^2 = p_z^2 - (E/c)^2 \end{aligned} \tag{5.157}$$

It can be seen that the following relation does guarantee

$$p'^2_x + p'^2_y + p'^2_z - \left(\frac{E'}{c}\right)^2 = p_x^2 + p_y^2 + p_z^2 - \left(\frac{E}{c}\right)^2 \tag{5.158}$$

Then the constant $m$ can be introduced, let

$$p^2 - \left(\frac{E}{c}\right)^2 = -m^2c^2 \tag{5.159}$$

And above formula can be rewritten as

$$E^2 = p^2c^2 + m^2c^4 \tag{5.160}$$

where $m$ is the stationary mass of the particle.

In the $S'$ system set $\vec{p}\,' = 0$, which indicates that the particle is stationary in the $S'$ system. Then the following formula can be obtained by Eq. (5.160):

$$E' = mc^2 \tag{5.161}$$

The inverse transformation of momentum-energy 4 vector can be done with Eq. (5.155):

$$\begin{pmatrix} p_x \\ p_y \\ p_z \\ \mathrm{i}E/c \end{pmatrix} = \begin{pmatrix} 1 & 0 & 0 & 0 \\ 0 & 1 & 0 & 0 \\ 0 & 0 & \gamma & -\mathrm{i}\beta\gamma \\ 0 & 0 & \mathrm{i}\beta\gamma & \gamma \end{pmatrix} \begin{pmatrix} 0 \\ 0 \\ 0 \\ \mathrm{i}E'/c \end{pmatrix} = \begin{pmatrix} 0 \\ 0 \\ \gamma m v \\ \mathrm{i}\gamma m c \end{pmatrix} \tag{5.162}$$

So we can get

$$p_z = \gamma m v \tag{5.163}$$

$$E = \gamma m c^2 \tag{5.164}$$

$$v = \frac{p_z}{\gamma m} = \frac{p_z c^2}{E} \tag{5.165}$$

The reference system with a total momentum of zero is called the center of momentum system, referred to as c.m. system. From Eq. (5.163) it can be seen that the momentum is not proportional to speed, so the center of momentum system and the center of mass system are different. Only if they degenerate to satisfy the non-relativistic conditions, they are consistent to each other. For two particle center of momentum system the requirement is

$$\vec{p} = \vec{p}_1 + \vec{p}_2 = 0 \tag{5.166}$$

In the laboratory system, the momentum $\vec{p}_1$ of the incident particle $m_1$ follows the $z$-axis direction, and the target nucleus $m_2$ is stationary. Using the symbol with "′" to represent the physical quantity in the center of momentum system, so we can write the following Lorentz transformations as

$$\begin{pmatrix} 0 \\ 0 \\ p'_{1z} \\ \mathrm{i}E'_1/c \end{pmatrix} = \begin{pmatrix} 1 & 0 & 0 & 0 \\ 0 & 1 & 0 & 0 \\ 0 & 0 & \gamma & \mathrm{i}\beta\gamma \\ 0 & 0 & -\mathrm{i}\beta\gamma & \gamma \end{pmatrix} \begin{pmatrix} 0 \\ 0 \\ p_{1z} \\ \mathrm{i}E_1/c \end{pmatrix} = \begin{pmatrix} 0 \\ 0 \\ \gamma(p_{1z} - \beta E_1/c) \\ -\mathrm{i}\gamma(\beta p_{1z} - E_1/c) \end{pmatrix} \tag{5.167}$$

$$\begin{pmatrix} 0 \\ 0 \\ p'_{2z} \\ \mathrm{i}E'_2/c \end{pmatrix} = \begin{pmatrix} 1 & 0 & 0 & 0 \\ 0 & 1 & 0 & 0 \\ 0 & 0 & \gamma & \mathrm{i}\beta\gamma \\ 0 & 0 & -\mathrm{i}\beta\gamma & \gamma \end{pmatrix} \begin{pmatrix} 0 \\ 0 \\ 0 \\ \mathrm{i}E_2/c \end{pmatrix} = \begin{pmatrix} 0 \\ 0 \\ -\gamma\beta E_2/c \\ \mathrm{i}\gamma E_2/c \end{pmatrix} \tag{5.168}$$

From the above two formulas we may get

$$p'_{1z} = \gamma\left(p_{1z} - \beta\frac{E_1}{c}\right) \tag{5.169}$$

$$E'_1 = -\gamma(\beta c p_{1z} - E_1) \tag{5.170}$$

$$p'_{2z} = -\gamma\beta m_2 c \tag{5.171}$$

$$E'_2 = \gamma m_2 c^2 \tag{5.172}$$

Based on $p'_{1z} + p'_{2z} = 0$ we can find

$$p_{1z} - \beta\frac{E_1}{c} - \beta m_2 c = 0, \quad \beta = \frac{p_{1z}c}{E_1 + m_2 c^2} \tag{5.173}$$

Here $v = \beta c$ is the speed of the motion coordinate system relative to the stationary particle 2. From Eq. (5.160) we get

$$E_1 = \sqrt{p_{1z}^2 c^2 + m_1^2 c^4} \tag{5.174}$$

The total energy of the c.m. system is

$$E' = E'_1 + E'_2 \tag{5.175}$$

The following expression can be obtained from Eqs. (5.170) and (5.172)–(5.175):

$$\begin{aligned} E'^2 &= \gamma^2\left(m_2c^2 + E_1 - \beta c p_{1z}\right)^2 = \frac{1}{1-\beta^2}\left[m_2c^2 + E_1 - \beta^2\left(E_1 + m_2c^2\right)\right]^2 \\ &= \left(1-\beta^2\right)\left(E_1 + m_2c^2\right)^2 = \left(E_1 + m_2c^2\right)^2 - p_{1z}^2c^2 \\ &= E_1^2 + 2E_1m_2c^2 + m_2^2c^4 - E_1^2 + m_1^2c^4 = 2E_1m_2c^2 + m_1^2c^4 + m_2^2c^4 \end{aligned} \tag{5.176}$$

The total energy $E'$ of the c.m. system is also called reaction effective energy.

In $\hbar = c = 1$ unit system, define

$$\sqrt{s} = \left[\left(E_1 + E_2\right)^2 - \left(\vec{p}_1 + \vec{p}_2\right)^2\right]^{1/2} \tag{5.177}$$

Comparing this formula with Eq. (5.3), it can be seen that $\sqrt{s}$ corresponds to a stationary mass term. In the c.m. system of two particles, there is $\vec{p}_1 + \vec{p}_2 = 0$, then we can obtain

$$\sqrt{s} = E_1 + E_2 \tag{5.178}$$

So it can be seen that $\sqrt{s}$ is the total energy of the two particles in the c.m. system, which contains the energy of the stationary mass. It has been assumed that the $S'$ system is uniformly moving with speed $v$ along the $z$-axis direction relative to the

$S$ system, and the physical quantity in the $S'$ system is represented by the symbol with "$'$". Eq. (5.156) has given

$$p'_x = p_x, \quad p'_y = p_y, \quad p'_z = \gamma(p_z - \beta E), \quad E' = -\gamma(\beta p_z - E) \tag{5.179}$$

So we can first find

$$\begin{aligned}(E'_1 + E'_2)^2 &= \gamma^2 [\beta(p_{1z} + p_{2z}) - (E_1 + E_2)]^2 \\ &= \gamma^2 \left[\beta^2 (p_{1z} + p_{2z})^2 + (E_1 + E_2)^2 - 2\beta(p_{1z} + p_{2z})(E_1 + E_2)\right]\end{aligned} \tag{5.180}$$

$$\begin{aligned}\left(\vec{p}'_1 + \vec{p}'_2\right)^2 &= (p_{1x} + p_{2x})^2 + (p_{1y} + p_{2y})^2 + \gamma^2 [p_{1z} + p_{2z} - \beta(E_1 + E_2)]^2 \\ &= (p_{1x} + p_{2x})^2 + (p_{1y} + p_{2y})^2 + \gamma^2 (p_{1z} + p_{2z})^2 + \gamma^2\beta^2 (E_1 + E_2)^2 \\ &\quad -2\gamma^2\beta(p_{1z} + p_{2z})(E_1 + E_2)\end{aligned} \tag{5.181}$$

Using the above two formulas and Eq. (5.177), it can be proved that $\sqrt{s}$ is invariant in different coordinate systems, i.e., $\sqrt{s}$ is Lorentz invariant. In the laboratory system where particle 1 is incident and particle 2 is stationary, from Eqs. (5.177) and (5.3) we can obtain

$$\sqrt{s} = \left(m_1^2 + m_2^2 + 2m_2 E_1\right)^{1/2} \tag{5.182}$$

This formula is consistent with Eq. (5.176).

The wave function equation in the interaction representation is (see Eq. (3.14.6) in Ref. [4])

$$\mathrm{i}\frac{\partial}{\partial t}|t\rangle = V_{\mathrm{I}}(t)\,|t\rangle \tag{5.183}$$

where $V_{\mathrm{I}}(t)$ is the residual interaction. Here we use the $\hbar = c = 1$ unit system. For wave functions we have the relation

$$|t\rangle = u(t, t_0)\,|t_0\rangle \tag{5.184}$$

$S$ matrix element is defined as

$$S \equiv u(\infty, -\infty) \tag{5.185}$$

We discuss the reaction process shown in Fig. 5.1 [9]. Here use 1 and $1'$ to represent incident and outgoing particles, and use 2 and $2'$ to represent the target nucleus and the residual nucleus, respectively. Set $|1, 2\rangle$ as the $t = -\infty$ initial state, $|1', 2'\rangle$ is the $t = \infty$ final state, and its reaction amplitude is

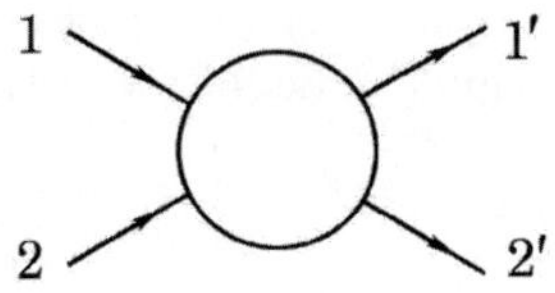


**Fig. 5.1** Reaction process $1 + 2 \to 1' + 2'$

$$\langle 1', 2'|S|1, 2\rangle \tag{5.186}$$

The corresponding probability is

$$W_f = |\langle 1',\ 2'|S|1,\ 2\rangle|^2 \tag{5.187}$$

In the reaction process, the energy conservation and momentum conservation must be satisfied.

From Eq. (5.80) we can see that the free nucleon plane wave function is $\sim e^{i\left(\vec{p}\cdot\vec{r}-Et\right)}$, it is required to have an entire number of cycles in a square box with volume $\Omega$, so the momentum $\vec{p}$ which can be taken is

$$\vec{p} = \frac{2\pi}{\Omega^{1/3}}(l, m, n), \qquad l, m, n = 0, \pm 1, \pm 2, \cdots \quad (\text{except } l = m = n = 0) \tag{5.188}$$

where $l$, $m$, $n$ represent the integer values taken in the $x$, $y$, $z$ directions, respectively, so that $\vec{p}\cdot\vec{r}$ is an integer multiple of $2\pi$ at the square box boundary of the volume $\Omega$. Each group $l$, $m$, $n$ corresponds to a wave vector, i.e., $p$ is a separate variable, whereas $p$ becomes a continuous variable only when $\Omega \to \infty$. Using the periodic boundary condition Eq. (5.188), one can easily obtain

$$\frac{1}{\Omega}\int_\Omega d\vec{r}\, e^{i\left(\vec{p}'-\vec{p}\right)\cdot\vec{r}} = \delta_{\vec{p}',\vec{p}} \tag{5.189}$$

There is also the following δ function formula:

$$\frac{1}{(2\pi)^3}\int d\vec{r}\, e^{i\left(\vec{p}'-\vec{p}\right)\cdot\vec{r}} = \delta\left(\vec{p}'-\vec{p}\right) \tag{5.190}$$

Comparing the above two formulas shows that there is the following correspondence:

$$\frac{\Omega}{(2\pi)^3}\delta_{\vec{p}',\vec{p}} \Leftrightarrow \delta\left(\vec{p}'-\vec{p}\right), \quad \text{when } \Omega \to \infty \tag{5.191}$$

For momentum-energy 4 vectors there is

$$\frac{\Omega}{(2\pi)^4}\delta_{p',p} \Leftrightarrow \delta^4(p'-p), \quad \text{when } \Omega \to \infty \tag{5.192}$$

For the two-particle subsystem, $\Omega$ should become $\Omega^2$, and there is

$$\frac{\Omega^2}{(2\pi)^4}\delta_{p'_1+p'_2,\, p_1+p_2} \Leftrightarrow \delta^4\left(p'_1+p'_2-p_1+p_2\right), \quad \text{when } \Omega \to \infty \tag{5.193}$$

So we can write Eq. (5.186) as [9]

$$\langle 1', 2'|S|1, 2\rangle = \mathcal{M}\frac{(2\pi)^4}{\Omega^2}\delta^4\left(p'_1+p'_2-p_1-p_2\right) \tag{5.194}$$

where $p_1, p_2, p'_1$, and $p'_2$ are the 4 momentum of particles 1, 2, 1′, and 2′, respectively. $\Omega$ represents the system volume. $\mathcal{M}$ is transition matrix element and contains the four exterior lines and the $t$ matrix representing the interaction at the vertex. According to the graphical rules of quantum field theory, $(2\pi)^4\delta^4\left(p'_1+p'_2-p_1-p_2\right)$ is contributed by the vertex.

Substituting Eq. (5.194) into Eq. (5.187), we can obtain

$$W_\mathrm{f} = |\mathcal{M}|^2\left[\frac{(2\pi)^4}{\Omega^2}\delta^4\left(p'_1+p'_2-p_1-p_2\right)\right]^2 \tag{5.195}$$

There are the following relations:

$$\delta^4(p) = \frac{1}{(2\pi)^4}\int \mathrm{d}^4x\mathrm{e}^{\mathrm{i}px}, \qquad (2\pi)^4\delta^4(p) = \int \mathrm{d}^4x\mathrm{e}^{\mathrm{i}px} \tag{5.196}$$

$$(2\pi)^4\delta^4(p)\int \mathrm{d}^4x\mathrm{e}^{\mathrm{i}px} = \left[(2\pi)^4\delta^4(p)\right]^2 \tag{5.197}$$

Since there is $\delta^4(p)$ at the left end of the above formula, one can make $p = 0$ in the integrand, and the following relation can be obtained from the definition of space-time:

$$\int \mathrm{d}^4x = \Omega T \tag{5.198}$$

where $T$ represents the interval of time, and at last one must let $\Omega$ and $T \to \infty$. Then we can obtain the following expression by Eq. (5.197):

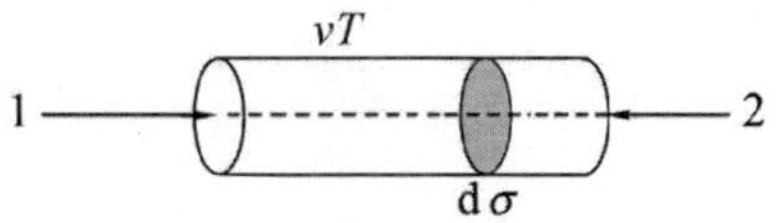


**Fig. 5.2** Two-particle nuclear reaction diagram in center of momentum system, if both particle 1 and particle 2 are in the cylinder $vT\mathrm{d}\sigma$, the reaction $1 + 2 \to 1' + 2'$ will occur within time $T$

$$\left[(2\pi)^4\delta^4(p)\right]^2 = (2\pi)^4\delta^4(p)\Omega T \tag{5.199}$$

Substituting Eq. (5.199) into Eq. (5.195) we get

$$W_{\mathrm{f}} = |\mathscr{M}|^2 \frac{(2\pi)^4}{\Omega^4}\delta^4(p_1' + p_2' - p_1 - p_2)\Omega T \tag{5.200}$$

We study the nuclear reaction between the incident particle labeled with 1 and the target nucleus labeled with 2. The motion direction of the incident particle 1 is selected as the $z$-axis. In the laboratory system, the particle 2 is stationary, and we only consider the coordinate system of inertial motion along the $z$-axis. In the two-particle center of momentum coordinate system, particles 1 and 2 move in the direction of the z-axis at speed $v = \left|\vec{v}_1 - \vec{v}_2\right| = v_1 - v_2$, where $v_1$ and $v_2$ representing the velocities of particles 1 and 2 in the $z$-axis direction, and $v_1 > v_2$ is demanded. We assume that $1 + 2 \to 1' + 2'$ reaction cross section is $\mathrm{d}\sigma$. Now let us consider the cylinder shown in Fig. 5.2. Since the initial state corresponds to the states of only two particles 1 and 2 in the volume $\Omega$, whereas the states, where particles 1 and 2 leave far away, need not to be considered. So the probability that particles 1 and 2 are in the column is

$$\frac{vT\mathrm{d}\sigma}{\Omega} \tag{5.201}$$

If particles 1 and 2 are all in the column, the $1 + 2 \to 1' + 2'$ reaction will occur in time $T$, so its probability is equal to

$$\left|\langle 1',\ 2'|u\left(\frac{T}{2},\ -\frac{T}{2}\right)|1,\ 2\rangle\right|^2 \tag{5.202}$$

The $u$ in the above equation is just the $u$ in Eq. (5.184). At $T \to \infty$ limit, Eq. (5.202) becomes Eq. (5.200), and let Eq. (5.200) and Eq. (5.201) be equivalent to each other, then we can obtain

$$\sigma = \sum \frac{(2\pi)^4}{\Omega^2 v}|\mathscr{M}|^2\delta^4(p_1' + p_2' - p_1 - p_2) \tag{5.203}$$

The summation symbol is added to the right side of the above formula, which indicates to spread throughout various three-dimensional momentum $\vec{p}_1'$ and $\vec{p}_2'$ of the end state, while $\mathrm{d}\sigma$ in the left side of the equation also becomes $\sigma$ representing the integral cross section. According to the phase space theory, we know

$$\sum_{\vec{p}_i} \to \frac{\Omega}{(2\pi)^3} \int \mathrm{d}^3\vec{p}_i \tag{5.204}$$

Then the following expression can be obtained by Eq. (5.203):

$$\sigma = \frac{1}{(2\pi)^2 (v_1 - v_2)} \int \mathrm{d}^3\vec{p}_1' \mathrm{d}^3\vec{p}_2' |\mathscr{M}|^2 \delta^4 (p_1' + p_2' - p_1 - p_2) \tag{5.205}$$

Since the particles 1 and 2 only move in the $z$-axis direction, using Eq. (5.165) we can obtain

$$v = |\vec{v}_1 - \vec{v}_2| = v_1 - v_2 = \frac{|\vec{p}_1|}{E_1} - \frac{|\vec{p}_2|}{E_2} = \frac{E_2|\vec{p}_1| - E_1|\vec{p}_2|}{E_1 E_2} \tag{5.206}$$

So Eq. (5.205) can be rewritten as

$$\sigma = \frac{E_1 E_2}{(2\pi)^2 \left(E_2|\vec{p}_1| - E_1|\vec{p}_2|\right)} \int \mathrm{d}^3\vec{p}_1' \mathrm{d}^3\vec{p}_2' |\mathscr{M}|^2 \delta^4 (p_1' + p_2' - p_1 - p_2) \tag{5.207}$$

Note that Eq. (5.205) or Eq. (5.207) gives an integral cross section that holds in any coordinate system of inertial motion along the $z$-axis.

Eq. (5.207) can be rewritten as

$$\sigma = \int \mathrm{d}^3\vec{p}_1' \mathrm{d}^3\vec{p}_2' \frac{d^6\sigma}{\mathrm{d}^3\vec{p}_1' \mathrm{d}^3\vec{p}_2'} \tag{5.208}$$

where

$$\frac{d^6\sigma}{\mathrm{d}^3\vec{p}_1' \mathrm{d}^3\vec{p}_2'} = \frac{E_1 E_2}{(2\pi)^2 \left(E_2|\vec{p}_1| - E_1|\vec{p}_2|\right)} |\mathscr{M}|^2 \delta^4 (p_1' + p_2' - p_1 - p_2) \tag{5.209}$$

is a 6-differential cross section of two emitted particles with definite $\vec{p}_1' \vec{p}_2'$. Assume that Eq. (5.209) is an expression given in the $S$ system, and it is assumed that the $S'$ system moves at a uniform speed $v$ in the $z$-axis direction relative to the $S$ system. We make convention that the relative velocity between the two coordinate systems is less than the velocity of the incident particle 1 in the laboratory system. And assume that the momentum-energy 4 vector of the particles in the $S'$ system is $\left(\vec{q}, ie\right)$. According to Eq. (5.209) the following expression can be written in the $S'$ system:

$$\frac{d^6\sigma'}{d^3\vec{q}_1'd^3\vec{q}_2'} = \frac{e_1e_2}{(2\pi)^2\left(e_2|\vec{q}_1| - e_1|\vec{q}_2|\right)}|\mathscr{M}|^2\delta^4(q_1' + q_2' - q_1 - q_2) \quad (5.210)$$

where $\mathscr{M}$ is a transition matrix element in the $S'$ system. There are the relations

$$q_{1x}' = p_{1x}', \quad q_{1y}' = p_{1y}', \quad q_{2x}' = p_{2x}', \quad q_{2y}' = p_{2y}' \quad (5.211)$$

And using the transformation relations given by Eq. (5.156), we can obtain

$$\begin{gathered} q_1 = \gamma p_1 - \beta\gamma E_1, \quad e_1 = -\beta\gamma p_1 + \gamma E_1, \quad q_2 = \gamma p_2 - \beta\gamma E_2, \quad e_2 = -\beta\gamma p_2 + \gamma E_2, \\ q_{1z}' = \gamma p_{1z}' - \beta\gamma E_1', \quad e_1' = -\beta\gamma p_{1z}' + \gamma E_1', \quad q_{2z}' = \gamma p_{2z}' - \beta\gamma E_2', \quad e_2' = -\beta\gamma p_{2z}' + \gamma E_2' \end{gathered} \quad (5.212)$$

Note that $q_1, q_2, p_1, p_2$ represent the momentum values of the particles in the above equations; they are not the momentum-energy 4 vector. First we find the factor

$$\begin{aligned} (e_2q_1 - e_1q_2)/\gamma^2 &= (-\beta p_2 + E_2)(p_1 - \beta E_1) - (-\beta p_1 + E_1)(p_2 - \beta E_2) \\ &= -\beta p_1p_2 + \beta^2 p_2E_1 + E_2p_1 - \beta E_1E_2 + \beta p_1p_2 - \beta^2 p_1E_2 \\ &\quad -E_1p_2 + \beta E_1E_2 = \left(1 - \beta^2\right)(E_2p_1 - E_1p_2) \end{aligned} \quad (5.213)$$

Then we can get

$$\frac{e_1e_2}{e_2q_1 - e_1q_2} = \frac{\gamma^2(E_1 - \beta p_1)(E_2 - \beta p_2)}{E_2p_1 - E_1p_2} \quad (5.214)$$

We have known

$$\begin{gathered} \delta\left(q_{1x}' + q_{2x}' - q_{1x} - q_{2x}\right) = \delta\left(p_{1x}' + p_{2x}' - p_{1x} - p_{2x}\right), \\ \delta\left(q_{1y}' + q_{2y}' - q_{1y} - q_{2y}\right) = \delta\left(p_{1y}' + p_{2y}' - p_{1y} - p_{2y}\right) \end{gathered} \quad (5.215)$$

Because the incident particle 1 and target 2 move only in the $z$ direction in the $S$ and $S'$ systems before the collision, therefore there are actually $p_{1x} = p_{2x} = p_{1y} = p_{2y} = 0$ and $q_{1x} = q_{2x} = q_{1y} = q_{2y} = 0$ in Eq. (5.215). Using Eq. (5.212) we can obtain

$$\delta\left(q'_{1z} + q'_{2z} - q_1 - q_2\right) = \frac{1}{\gamma}\delta\left[\left(p'_{1z} + p'_{2z} - p_1 - p_2\right) - \beta\left(E'_1 + E'_2 - E_1 - E_2\right)\right],$$
$$\delta\left(e'_1 + e'_2 - e_1 - e_2\right) = \frac{1}{\gamma}\delta\left[\left(E'_1 + E'_2 - E_1 - E_2\right) - \beta\left(p'_{1z} + p'_{2z} - p_1 - p_2\right)\right] \tag{5.216}$$

So we can get

$$\begin{aligned}&\delta\left(q'_{1z} + q'_{2z} - q_1 - q_2\right)\delta\left(e'_1 + e'_2 - e_1 - e_2\right)\\&= \frac{1}{\gamma^2}\delta\left(p'_{1z} + p'_{2z} - p_1 - p_2\right)\delta\left(E'_1 + E'_2 - E_1 - E_2\right)\end{aligned} \tag{5.217}$$

The solution of the positive energy plane wave of the Dirac equation has been given by Eq. (5.93) as

$$u_s\left(\vec{p}\right) = u\left(\vec{p}, +\right)\chi_s \tag{5.218}$$

where

$$u\left(\vec{p}, +\right) = \sqrt{\frac{E+m}{2m}}\begin{pmatrix}\hat{I}\\ \dfrac{\hat{\sigma}\cdot\vec{p}}{E+m}\end{pmatrix} \tag{5.219}$$

where $\chi_s$ is a nucleon spin wave function with a spin projection of $s$. In Eq. (5.218) the upper and lower components of the four-dimensional wave functions $u\left(\vec{p}, +\right)$ are permitted to act on the nucleon spin wave functions $\chi_s$, respectively. Let $\hat{t}_S$ and $\hat{t}_{S'}$ be the $t$ matrix operators of the two-body NN interaction of the $S$ system and the $S'$ system, respectively, and the nuclear force or $t$ matrix generally belongs to be the Lorentz invariant. The two-body transition matrix elements appearing in Eqs. (5.209) and (5.210) are as follows, respectively:

$$\mathscr{M} = \overline{u}_{s'_2}\left(\vec{p}'_2\right)\overline{u}_{s'_1}\left(\vec{p}'_1\right)\hat{t}_S u_{s_1}\left(\vec{p}_1\right)u_{s_2}\left(\vec{p}_2\right) \tag{5.220}$$

$$\mathscr{N} = \overline{u}_{s'_2}\left(\vec{q}'_2\right)\overline{u}_{s'_1}\left(\vec{q}'_1\right)\hat{t}_{S'} u_{s_1}\left(\vec{q}_1\right)u_{s_2}\left(\vec{q}_2\right) \tag{5.221}$$

where $\overline{u}_s = u_s^+\gamma^0$. Since the Dirac equation satisfies Lorentz invariance, its solutions satisfy the following relations:

$$\begin{aligned}&\overline{u}_{s'_1}\left(\vec{q}'_1\right) = \overline{u}_{s'_1}\left(\vec{p}'_1\right), \quad \overline{u}_{s'_2}\left(\vec{q}'_2\right) = \overline{u}_{s'_2}\left(\vec{p}'_2\right),\\&u_{s_1}\left(\vec{q}_1\right) = u_{s_1}\left(\vec{p}_1\right), \quad u_{s_2}\left(\vec{q}_2\right) = u_{s_2}\left(\vec{p}_2\right)\end{aligned} \tag{5.222}$$

And since there is $\hat{t}_{S'} = \hat{t}_S$, so we can get

$$\mathcal{N} = \mathcal{M} \tag{5.223}$$

The following expression can be obtained comparing Eq. (5.209) with Eq. (5.210) and using Eqs. (5.214), and (5.217), and (5.223):

$$\frac{\mathrm{d}^6\sigma'}{\mathrm{d}^3\vec{q}_1'\mathrm{d}^3\vec{q}_2'} = \frac{\left(E_1 - \beta|\vec{p}_1|\right)\left(E_2 - \beta|\vec{p}_2|\right)}{E_1 E_2}\frac{\mathrm{d}^6\sigma}{\mathrm{d}^3\vec{p}_1'\mathrm{d}^3\vec{p}_2'} \tag{5.224}$$

From the above formula it can be seen that when $v = \beta = 0$ the differential cross sections in the two coordinate systems are equal.

## 5.3 Relativistic Optical Model and Phenomenological Optical Potential

In 1979, Arnold and Clark [10, 11] developed a nucleon relativistic optical model based on the Dirac equation. And the elastic scattering data of the medium energy protons were analyzed with relativistic phenomenological optical potential (RPOP). In order to better meet the experimental data of the proton elastic scattering angular distribution $\frac{\mathrm{d}\sigma}{\mathrm{d}\Omega}$, the polarization analyzing power $A_y(\theta)$, and spin rotation function $Q(\theta)$ for some specific targets and energies, the expressions and parameters of RPOP were further studied [12–15]. Then the energy-related proton RPOP [16–19] and neutron RPOP [18, 19] for certain target nuclei were studied. Later the universal proton RPOP [20–22] and neutron RPOP [23, 24] that can be used in a variety of target nuclei and energy regions below 1000 MeV were also developed.

The neutrons and protons in the nucleus are all in the bound states, they all belong to the positive energy nucleons of total energy smaller than $M_\mathrm{t}c^2(\mathrm{t} = \mathrm{n}, \mathrm{p})$, $M_\mathrm{t}$ is the stationary mass of the nucleon t. The nuclear density is close to constant in the center region of the nucleus. If we look at the nucleus as an infinite medium without a surface area, we can introduce the hypothetical concept of nuclear matter. The nucleons in the nuclear matter are generally the positive energy nucleons of total energy smaller than $M_\mathrm{t}c^2$. If we study one of the nucleons in the nuclear matter, this nucleon must be in the potential field generated by other nucleons around it. This potential field is generated by this nucleon exchanging mesons with the surrounding nucleons in the nuclear matter. We call this potential field the nucleon self-energy $\Sigma_\mathrm{t}$, and we can also call $\Sigma_\mathrm{t}$ the mean field felt by this nucleon. The studied nucleon may be in the bound state of total energy smaller than $M_\mathrm{t}c^2$; and if it obtains enough kinetic energy, then it may also be in the scattering state with the energy greater than $M_\mathrm{t}c^2$. However, as long as it is in nuclear matter, it will still be affected by the self-energy $\Sigma_\mathrm{t}$. According to the local density approximation, if a nucleon is in the nuclear surface area with smaller nuclear density, the nucleon is considered to be in the nuclear matter with smaller nuclear density. Only when a nucleon is far from the nucleus and is no longer affected by the nuclear potential field, it does think that the nucleon has become a free nucleon. The total energy of a free nucleon must be greater than $M_t c^2$.

In order to solve the relativistic two-body motion equation, it is generally decomposed into a free motion equation that describes the system center of momentum motion without potential field and a one-body motion equation that uses the mean field method to describe the relative motion of two particles in the center of momentum system. Dirac equation is a one-body motion equation that describes the reaction of spin $\frac{1}{2}$ particle with the target nucleus in the center of momentum system. Under normal circumstances, there is no need to solve the motion equation of the system center of momentum. It is only necessary to solve the one-body equation describing the movement with the two-body interaction in the center of momentum system. If necessary, the calculation results can be transformed to the laboratory system.

In the $\hbar = c = 1$ unit system, the stationary Dirac equation for free nucleon can be written from Eqs. (5.20) and (5.19) as follows:

$$\left(\vec{\alpha} \cdot \vec{k}_{\rm t} + \beta M_{\rm t}\right) \psi_{\rm t} = E_{\rm t} \psi_{\rm t} \tag{5.225}$$

If the studied nucleon is in an infinite nuclear matter unrelated to space coordinate, it will feel the existence of self-energy (mean field) $\Sigma_{\rm t}(k_{\rm t})$, and Eq. (5.225) becomes [25]

$$\left[\vec{\alpha} \cdot \vec{k}_{\rm t} + \beta (M_{\rm t} + \Sigma_{\rm t}(k_{\rm t}))\right] \psi_{\rm t}(k_{\rm t}) = E_{\rm t} \psi_{\rm t}(k_{\rm t}) \tag{5.226}$$

In a static infinite nuclear matter that is independent of space coordinate, the self-energy $\Sigma_{\rm t}(k_{\rm t})$ should satisfy translational and rotational invariance, and it is assumed that it also satisfies the spatial reflection invariance.

As can be seen from Eq. (5.138) and Table 5.1, under the Pauli metric, the 16 linearly independent matrices $I$, $\mathrm{i}\gamma_5$, $\mathrm{i}\gamma_\mu$, $\mathrm{i}\gamma_5\gamma_\mu$ and $\sigma_{\mu\nu} = \frac{1}{2\mathrm{i}}\left(\gamma_\mu\gamma_\nu - \gamma_\nu\gamma_\mu\right)$ are scalar, pseudo-scalar, vector, axis-vector, and tensor matrices, respectively. Under the Bjorken-Drell (BD) metric, the following 16 linear independent matrices can be represented as $I$, $\gamma^5$, $\gamma^\mu$, $\gamma^5\gamma^\mu$ and $\sigma^{\mu\nu} = \frac{\mathrm{i}}{2}(\gamma^\mu\gamma^\nu - \gamma^\nu\gamma^\mu)$. Although known by Eq. (5.57), $\gamma_4 = \gamma^0$, $\gamma_i = -\mathrm{i}\gamma^i$, but known by Eqs. (5.32) and (5.56), $x_4 = \mathrm{i}x_0$, $x_i$(Pauli metric) $= -x_i$(BD metric), therefore it is available that $\gamma_\mu x_\mu$(Pauli metric) $= \mathrm{i}\gamma^\mu x_\mu$(BD metric). Notice $\sigma^{\mu\mu} = 0$, the most common form of self-energy $\Sigma_{\rm t}$ of the nucleon t under the BD metric is [2, 26, 27]

$$\Sigma_{\rm t} = \Sigma_{\rm t}^{\rm s} + \gamma^5 \Sigma_{\rm t}^{\rm ps} + \gamma^\mu \Sigma_{\rm t}^{\mu,{\rm v}} + \gamma^5 \gamma^\mu \Sigma_{\rm t}^{\mu,{\rm pv}} + \sigma^{0i} \Sigma_{\rm t}^{0i,{\rm T}} + \sigma^{ij} \Sigma_{\rm t}^{ij,{\rm T}} \tag{5.227}$$

It can be seen from Eq. (5.61) that under spatial reflection $k^{0'} = k^0$, $k^{i'} = -k^i$, and using Eq. (5.132) it can be seen that $k_{\rm t}^i \gamma^i$ remains unchanged under spatial reflection.

Note that there is a coefficient $\beta$ at the front of Eq. (5.226), and it can be seen from Eq. (5.132) that $\beta = \gamma_4 = \gamma^0$ is also unchanged under the spatial reflection transformation. Eq. (5.58) has given $\gamma^5 = -\gamma_5$, and according to Eq. (5.136) or Eq. (5.137) it can be seen that the sign of $\gamma^5$ will change under spatial reflection, so that the second term representing the pseudo-scalar (ps) and the fourth term representing the axis-vector (pv) in Eq. (5.227) do not satisfy the parity conservation, i.e., they cannot be selected. Note that the following covariant 4-vectors can be given by Eq. (5.61):

$$k = \begin{pmatrix} k_0 \\ k_1 \\ k_2 \\ k_3 \end{pmatrix} = \begin{pmatrix} E \\ -k_x \\ -k_y \\ -k_z \end{pmatrix}, \quad j = \begin{pmatrix} j_0 \\ j_1 \\ j_2 \\ j_3 \end{pmatrix} = \begin{pmatrix} \rho \\ -j_x \\ -j_y \\ -j_z \end{pmatrix} \tag{5.228}$$

where $\rho$ and $\vec{j}\left(j_x, j_y, j_z\right)$ represent the nucleon density and nucleon flow density, respectively. Since it is assumed that the nuclear matter has a uniform finite density and is in a stable state, so the nucleon flow density is 0, such that the tensor formed by $k_t^i j_t^j$ will not contribute, therefore the last item formed by $\sigma^{ij}$ in Eq. (5.227) does not contribute; however the term related to $k_t^i j_t^0 = k_t^i \rho_t$ has contribution to the penultimate item in Eq. (5.227). And let

$$\gamma^0 \Sigma_t^{0,\mathrm{v}} \equiv \gamma^0 \Sigma_t^0, \quad \gamma^i \Sigma_t^{i,\mathrm{v}} \equiv \gamma^i k_t^i \Sigma_t^{\mathrm{v}}, \quad \sigma^{0i} \Sigma_t^{0i,\mathrm{T}} \equiv \mathrm{i}\gamma^0 \gamma^i k_t^i \rho_t \overline{\Sigma}_t^{\mathrm{T}} \equiv \gamma^0 \gamma^i k_t^i \Sigma_t^{\mathrm{T}} \tag{5.229}$$

so Eq. (5.227) can be rewritten as

$$\Sigma_t = \Sigma_t^{\mathrm{s}} + \gamma^0 \Sigma_t^0 + \vec{\gamma} \cdot \vec{k}_t \Sigma_t^{\mathrm{v}} + \gamma^0 \vec{\gamma} \cdot \vec{k}_t \Sigma_t^{\mathrm{T}} \tag{5.230}$$

where $\vec{\gamma} \cdot \vec{k}_t = \gamma^i k_t^i$. $\Sigma_t^s$, $\Sigma_t^0$, $\Sigma_t^{\mathrm{v}}$, and $\Sigma_t^{\mathrm{T}}$ are the scalar, 4-vector time-like component, 4-vector space-like components, and tensor terms, respectively. For the sake of simplicity, we omit the tensor term, so Eq. (5.230) is simplified to

$$\Sigma_t = \Sigma_t^s + \gamma^0 \Sigma_t^0 + \vec{\gamma} \cdot \vec{k}_t \Sigma_t^{\mathrm{v}} \tag{5.231}$$

In the $\hbar = c = 1$ unit system, substituting Eq. (5.231) into Eq. (5.226) and using Eq. (5.57), as well as noting that $\alpha^i$ in the BD metric is equal to $\alpha_i$ in the Pauli metric, so we can obtain

$$\left[\left(1 + \Sigma_t^{\mathrm{v}}\right) \vec{\alpha} \cdot \vec{p}_t + \beta\left(M_t + \Sigma_t^{\mathrm{s}}\right) + \Sigma_t^0\right] \psi_t = E_t \psi_t \tag{5.232}$$

If we only study the on-shell nucleon, that is, $E_t = \frac{p_t^2}{2M_t}$, then the self-energy $\Sigma_t^{\mathrm{v}}$, $\Sigma_t^s$, $\Sigma_t^0$ and wave function $\psi_t$ are only related to $p_t$ and Fermi momentum $k_{t\mathrm{F}}$, where $k_{t\mathrm{F}}$ is determined by the density of the nuclear matter of the nucleon t. If $k_{n\mathrm{F}} \neq k_{p\mathrm{F}}$, it is an asymmetric nuclear matter that shows that the number of neutrons and the number of protons are not equal. In general, $k_{tF}$ is not explicitly included in the expression.

The various terms in Eq. (5.232) are divided by $\left(1+\Sigma_{t}^{v}\right)$ and note that

$$\begin{aligned}\frac{E_t}{1+\Sigma_{t}^{v}}\psi_{t} &= \frac{1+\Sigma_{t}^{v}}{1+\Sigma_{t}^{v}}E_{t}\psi_{t} - \frac{\Sigma_{t}^{v}}{1+\Sigma_{t}^{v}}E_{t}\psi_{t},\\ \frac{M_{t}+\Sigma_{t}^{s}}{1+\Sigma_{t}^{v}} &= \frac{\left(1+\Sigma_{t}^{v}\right)M_{t}+\Sigma_{t}^{s}-\Sigma_{t}^{v}M_{t}}{1+\Sigma_{t}^{v}}\end{aligned} \tag{5.233}$$

So we can obtain

$$\left[\vec{\alpha}\cdot\vec{p}+\beta(M+U^{s})+U^{0}\right]\psi = E\psi \tag{5.234}$$

where

$$U^{s} = \frac{\Sigma^{s}-\Sigma^{v}M}{1+\Sigma^{v}}, \qquad U^{0} = \frac{\Sigma^{0}+\Sigma^{v}E}{1+\Sigma^{v}} \tag{5.235}$$

For the sake of simplicity, in the above two formulas we omit the footnote t representing the nucleon type. Due to the propagation of electromagnetic fields, the Coulomb potential $V_{C}$ should be introduced. It is determined by the empirical charge distribution of the nucleus. Here we temporarily add the $V_{C}$ contribution according to the practice in the references, so Eq. (5.234) can be rewritten as [10]

$$\left[\vec{\alpha}\cdot\vec{p}+\beta(M+U^{s})+U^{0}+V_{C}\right]\psi = E\psi \tag{5.236}$$

Equation (5.38) has given

$$\vec{\alpha} = \begin{pmatrix}\hat{0} & \hat{\sigma}\\ \hat{\sigma} & \hat{0}\end{pmatrix}, \qquad \beta = \begin{pmatrix}\hat{I} & \hat{0}\\ \hat{0} & -\hat{I}\end{pmatrix} \tag{5.237}$$

and let

$$\psi = \begin{pmatrix}\psi_{u}\\ \psi_{d}\end{pmatrix} \tag{5.238}$$

so based on the discussion in Sect. 5.1, it can be seen that $\psi_{u}$ corresponds to the large upper component with positive energy and $\psi_{d}$ corresponds to the little lower component with positive energy. Then the following equation can be obtained by Eq. (5.236):

$$\begin{pmatrix} \left(\hat{\sigma}\cdot\vec{p}\right)\psi_{\mathrm{d}} \\ \left(\hat{\sigma}\cdot\vec{p}\right)\psi_{\mathrm{u}} \end{pmatrix} + \begin{pmatrix} \left(M+U^{\mathrm{s}}+U^{0}+V_{\mathrm{C}}\right)\psi_{\mathrm{u}} \\ -\left(M+U^{\mathrm{s}}-U^{0}-V_{\mathrm{C}}\right)\psi_{\mathrm{d}} \end{pmatrix} = E\begin{pmatrix} \psi_{\mathrm{u}} \\ \psi_{\mathrm{d}} \end{pmatrix} \tag{5.239}$$

$$\left(\hat{\sigma}\cdot\vec{p}\right)\psi_{\mathrm{d}} + \left(M+U^{s}+U^{0}+V_{\mathrm{C}}\right)\psi_{\mathrm{u}} = E\psi_{\mathrm{u}} \tag{5.240}$$

$$\left(\hat{\sigma}\cdot\vec{p}\right)\psi_{\mathrm{u}} - \left(M+U^{s}-U^{0}-V_{\mathrm{C}}\right)\psi_{\mathrm{d}} = E\psi_{\mathrm{d}} \tag{5.241}$$

$$\psi_{\mathrm{d}} = \frac{\hat{\sigma}\cdot\vec{p}}{D}\psi_{\mathrm{u}} \tag{5.242}$$

$$\left[\left(\hat{\sigma}\cdot\vec{p}\right)\frac{1}{D}\left(\hat{\sigma}\cdot\vec{p}\right) + \left(M+U^{s}+U^{0}+V_{\mathrm{C}}\right)\right]\psi_{\mathrm{u}} = E\psi_{\mathrm{u}} \tag{5.243}$$

where

$$D = E + M + U^{s} - U^{0} - V_{\mathrm{C}} \tag{5.244}$$

First let us analyze the following expressions:

$$A \equiv \left(\hat{\sigma}\cdot\vec{p}\right)\frac{1}{D}\left(\hat{\sigma}\cdot\vec{p}\right)\psi_{\mathrm{u}} = \frac{1}{D}\left(\hat{\sigma}\cdot\vec{p}\right)\left(\hat{\sigma}\cdot\vec{p}\right)\psi_{\mathrm{u}} + \left[\left(\hat{\sigma}\cdot\vec{p}\right)\frac{1}{D}\right]\left(\hat{\sigma}\cdot\vec{p}\right)\psi_{\mathrm{u}} \tag{5.245}$$

Here we only discuss the spherical nuclei, so $D$ is just a function of $r$, and there is

$$\vec{p} = -\mathrm{i}\vec{\nabla} = -\mathrm{i}\frac{\vec{r}}{r}\frac{\mathrm{d}}{\mathrm{d}r} \tag{5.246}$$

Then using Eq. (2.16) we can obtain

$$\begin{aligned} A &= \frac{1}{D(r)}p^{2}\psi_{\mathrm{u}} + \left\{-\mathrm{i}\frac{\hat{\sigma}\cdot\vec{r}}{r}\left[\frac{\mathrm{d}}{\mathrm{d}r}\left(\frac{1}{D(r)}\right)\right]\left(\hat{\sigma}\cdot\vec{p}\right)\right\}\psi_{\mathrm{u}} \\ &= \frac{1}{D(r)}p^{2}\psi_{\mathrm{u}} + \frac{\mathrm{i}}{D^{2}(r)}\left(\frac{\mathrm{d}D(r)}{\mathrm{d}r}\right)\left(\frac{\hat{\sigma}\cdot\vec{r}}{r}\right)\left(\hat{\sigma}\cdot\vec{p}\right)\psi_{\mathrm{u}} \end{aligned} \tag{5.247}$$

Equation (13.4.35) in Ref. [28] has given

$$\left(\hat{\sigma}\cdot\vec{a}\right)\left(\hat{\sigma}\cdot\vec{b}\right) = \vec{a}\cdot\vec{b} + \mathrm{i}\hat{\sigma}\cdot\left(\vec{a}\times\vec{b}\right) \tag{5.248}$$

So there is

$$\left(\hat{\sigma}\cdot\vec{r}\right)\left(\hat{\sigma}\cdot\vec{p}\right)=\vec{r}\cdot\vec{p}+\mathrm{i}\hat{\sigma}\cdot\left(\vec{r}\times\vec{p}\right)=\vec{r}\cdot\vec{p}+\mathrm{i}\hat{\sigma}\cdot\vec{L},\quad \vec{L}=\vec{r}\times\vec{p} \tag{5.249}$$

From Eq. (5.247) we can get

$$A=\frac{1}{D(r)}\left\{p^2+\mathrm{i}\left[\frac{\mathrm{d}}{\mathrm{d}r}(\ln D(r))\right]\frac{\vec{r}\cdot\vec{p}}{r}-\left[\frac{\mathrm{d}}{\mathrm{d}r}(\ln D(r))\right]\frac{\hat{\sigma}\cdot\vec{L}}{r}\right\}\psi_\mathrm{u} \tag{5.250}$$

Substituting Eq. (5.250) into Eq. (5.243), we can also get

$$\begin{aligned}&\frac{1}{D(r)}\left\{p^2+\mathrm{i}\left[\frac{\mathrm{d}}{\mathrm{d}r}(\ln D(r))\right]\frac{\vec{r}\cdot\vec{p}}{r}-\left[\frac{\mathrm{d}}{\mathrm{d}r}(\ln D(r))\right]\frac{\hat{\sigma}\cdot\vec{L}}{r}\right\}\psi_\mathrm{u}\\&\quad+\left(M+U^\mathrm{s}+U^0+V_\mathrm{C}\right)\psi_\mathrm{u}=E\psi_\mathrm{u}\end{aligned} \tag{5.251}$$

And using Eq. (5.244) the above formula can be rewritten as

$$\begin{aligned}&\left\{\frac{p^2}{2E}+\frac{\mathrm{i}}{2E}\left[\frac{\mathrm{d}}{\mathrm{d}r}(\ln D(r))\right]\frac{\vec{r}\cdot\vec{p}}{r}-\frac{1}{2E}\left[\frac{\mathrm{d}}{\mathrm{d}r}(\ln D(r))\right]\frac{\hat{\sigma}\cdot\vec{L}}{r}\right\}\psi_\mathrm{u}\\&\quad+\frac{1}{2E}\left(M+U^\mathrm{s}+U^0+V_\mathrm{C}\right)\left(E+M+U^\mathrm{s}-U^0-V_\mathrm{C}\right)\psi_\mathrm{u}\\&\quad=\frac{1}{2}\left(E+M+U^s-U^0-V_\mathrm{C}\right)\psi_\mathrm{u}\end{aligned} \tag{5.252}$$

Note that there is the following relation:

$$\begin{aligned}&\left(M+U^s+U^0+V_\mathrm{C}\right)\left(E+M+U^s-U^0-V_\mathrm{C}\right)\\&=E\left(M+U^s+U^0+V_\mathrm{C}\right)+M^2+2MU^s+(U^s)^2-\left(U^0+V_\mathrm{C}\right)^2\end{aligned} \tag{5.253}$$

Substituting above formula into Eq. (5.252) and moving the terms corresponding to the first two items at the right end of the above formula to the right side of the equal sign of Eq. (5.252), then we can obtain

$$\begin{aligned}&\left\{\frac{p^2}{2E}+\frac{\mathrm{i}}{2E}\left[\frac{\mathrm{d}}{\mathrm{d}r}(\ln D(r))\right]\frac{\vec{r}\cdot\vec{p}}{r}-\frac{1}{2E}\left[\frac{\mathrm{d}}{\mathrm{d}r}(\ln D(r))\right]\frac{\hat{\sigma}\cdot\vec{L}}{r}\right\}\psi_\mathrm{u}\\&\quad+\frac{1}{2E}\left[2MU^\mathrm{s}+(U^\mathrm{s})^2-\left(U^0+V_\mathrm{C}\right)^2\right]\psi_u=\frac{E^2-M^2-2E\left(U^0+V_\mathrm{C}\right)}{2E}\psi_\mathrm{u}\end{aligned} \tag{5.254}$$

The above formula can be further rewritten as

$$\begin{aligned}&\left\{\frac{p^2}{2E}+U^0+V_\mathrm{C}+\frac{\mathrm{i}}{2E}\left[\frac{1}{D(r)}\frac{\mathrm{d}D(r)}{\mathrm{d}r}\right]\frac{\vec{r}\cdot\vec{p}}{r}-\frac{1}{2ErD(r)}\frac{\mathrm{d}D(r)}{\mathrm{d}r}\hat{\sigma}\cdot\vec{L}\right.\\&\left.+\frac{1}{2E}\left[U^\mathrm{s}(2M+U^\mathrm{s})-\left(U^0+V_\mathrm{C}\right)^2\right]\right\}\psi_\mathrm{u}=\frac{E^2-M^2}{2E}\psi_\mathrm{u}\end{aligned}\tag{5.255}$$

Next transform the wave function as follows:

$$\psi_\mathrm{u}(r)=\sqrt{D(r)}\varphi(r)\tag{5.256}$$

One has known

$$\vec{p}=-\mathrm{i}\vec{\nabla}=-\mathrm{i}\frac{\vec{r}}{r}\frac{\mathrm{d}}{\mathrm{d}r},\quad \vec{r}\cdot\vec{p}=-\mathrm{i}r\frac{\mathrm{d}}{\mathrm{d}r},\quad p^2=-\vec{\nabla}^2=-\frac{1}{r^2}\frac{\mathrm{d}}{\mathrm{d}r}\left(r^2\frac{\mathrm{d}}{\mathrm{d}r}\right)\tag{5.257}$$

So we can obtain

$$\frac{\mathrm{d}}{\mathrm{d}r}\psi_\mathrm{u}=\sqrt{D}\frac{\mathrm{d}\varphi}{\mathrm{d}r}+\frac{1}{2}\sqrt{\frac{1}{D}}\frac{\mathrm{d}D}{\mathrm{d}r}\varphi\tag{5.258}$$

$$\begin{aligned}p^2\psi_\mathrm{u}&=-\nabla^2\psi_\mathrm{u}=-\frac{1}{r^2}\frac{\mathrm{d}}{\mathrm{d}r}\left[r^2\left(\sqrt{D}\frac{\mathrm{d}\varphi}{\mathrm{d}r}+\frac{1}{2}\sqrt{\frac{1}{D}}\frac{\mathrm{d}D}{\mathrm{d}r}\varphi\right)\right]\\&=-\sqrt{D}\frac{1}{r^2}\frac{\mathrm{d}}{\mathrm{d}r}\left(r^2\frac{\mathrm{d}\varphi}{\mathrm{d}r}\right)-\frac{1}{2}\sqrt{\frac{1}{D}}\frac{\mathrm{d}D}{\mathrm{d}r}\frac{\mathrm{d}\varphi}{\mathrm{d}r}-\frac{1}{2}\sqrt{\frac{1}{D}}\frac{\mathrm{d}D}{\mathrm{d}r}\frac{\mathrm{d}\varphi}{\mathrm{d}r}\\&\quad+\frac{1}{4}\sqrt{\frac{1}{D^3}}\left(\frac{\mathrm{d}D}{\mathrm{d}r}\right)^2\varphi-\frac{1}{2}\sqrt{\frac{1}{D}}\frac{1}{r^2}\frac{\mathrm{d}}{\mathrm{d}r}\left(r^2\frac{\mathrm{d}D}{\mathrm{d}r}\right)\varphi\end{aligned}\tag{5.259}$$

$$\mathrm{i}\frac{1}{D}\frac{\mathrm{d}D}{\mathrm{d}r}\frac{\vec{r}\cdot\vec{p}}{r}\psi_\mathrm{u}=\frac{1}{D}\frac{\mathrm{d}D}{\mathrm{d}r}\left(\sqrt{D}\frac{\mathrm{d}\varphi}{\mathrm{d}r}+\frac{1}{2}\sqrt{\frac{1}{D}}\frac{\mathrm{d}D}{\mathrm{d}r}\varphi\right)=\sqrt{\frac{1}{D}}\frac{\mathrm{d}D}{\mathrm{d}r}\frac{\mathrm{d}\varphi}{\mathrm{d}r}+\frac{1}{2}\sqrt{\frac{1}{D^3}}\left(\frac{\mathrm{d}D}{\mathrm{d}r}\right)^2\varphi\tag{5.260}$$

So we can get

$$\begin{aligned}\left(p^2+\mathrm{i}\frac{1}{D}\frac{\mathrm{d}D}{\mathrm{d}r}\frac{\vec{r}\cdot\vec{p}}{r}\right)\psi_\mathrm{u}&=-\sqrt{D}\frac{1}{r^2}\frac{\mathrm{d}}{\mathrm{d}r}\left(r^2\frac{\mathrm{d}\varphi}{\mathrm{d}r}\right)+\frac{3}{4}\sqrt{\frac{1}{D^3}}\left(\frac{\mathrm{d}D}{\mathrm{d}r}\right)^2\varphi\\&\quad-\frac{1}{2}\sqrt{\frac{1}{D}}\frac{1}{r^2}\frac{\mathrm{d}}{\mathrm{d}r}\left(r^2\frac{\mathrm{d}D}{\mathrm{d}r}\right)\varphi\end{aligned}\tag{5.261}$$

Substitute Eqs. (5.256) and (5.261) into Eq. (5.255) and eliminate $\sqrt{D}$, then we can obtain [10, 20, 23]

$$\left[\frac{p^2}{2E} + U_{\text{eff}}^{(\text{u})}(r) + V_{\text{C}}(r) + U_{\text{SO}}^{(\text{u})}(r)\,\vec{\sigma}\cdot\vec{L}\right]\varphi(r) = \frac{E^2 - M^2}{2E}\varphi(r) \tag{5.262}$$

where

$$U_{\text{eff}}^{(\text{u})}(r) = U^0(r) + \frac{1}{2E}\left[U^s(r)\,(2M + U^s(r)) - \left(U^0(r) + V_C(r)\right)^2\right] + U_{\text{D}}^{(\text{u})}(r) \tag{5.263}$$

$$U_{\text{D}}^{(\text{u})}(r) = \frac{1}{2E}\left[\frac{3}{4}\left(\frac{1}{\text{D}(r)}\frac{\text{d}D(r)}{dr}\right)^2 - \frac{1}{2r^2D(r)}\frac{\text{d}}{\text{d}r}\left(r^2\frac{\text{d}D(r)}{\text{d}r}\right)\right] \tag{5.264}$$

$$U_{\text{SO}}^{(\text{u})}(r) = -\frac{1}{2ErD(r)}\frac{\text{d}D(r)}{\text{d}r} \tag{5.265}$$

Since $D(r)$ always appears in both the molecule and the denominator of the formulas, Eq. (5.244) can be modified to [20]

$$D(r) = \left(E + M + U^s(r) - U^0(r) - V_{\text{C}}(r)\right)/(E + M) \tag{5.266}$$

where $U_{\text{D}}(r)$ is called the Darwin potential, and it is usually considered to be a small amount.

Equation (5.262) is an equation that describes the upper component of the positive energy nucleon wave function. The equation for the lower component of the positive energy nucleon wave function is derived below. From Eq. (5.240) we can obtain

$$\psi_{\text{u}} = \frac{\hat{\sigma}\cdot\vec{p}}{T}\psi_{\text{d}} \tag{5.267}$$

$$T = E - M - U^s - U^0 - V_{\text{C}} \tag{5.268}$$

$$\left[\left(\hat{\sigma}\cdot\vec{p}\right)\frac{1}{T}\left(\hat{\sigma}\cdot\vec{p}\right) - \left(M + U^s - U^0 - V_{\text{C}}\right)\right]\psi_{\text{d}} = E\psi_{\text{d}} \tag{5.269}$$

For spherical nuclei, refer to Eqs. (5.245) and (5.250); Eq. (5.269) can be rewritten as

$$\begin{aligned}&\frac{1}{T(r)}\left\{p^2 + \text{i}\left[\frac{\text{d}}{\text{d}r}(\ln T(r))\right]\frac{\vec{r}\cdot\vec{p}}{r} - \left[\frac{\text{d}}{\text{d}r}(\ln T(r))\right]\frac{\hat{\sigma}\cdot\vec{L}}{r}\right\}\psi_{\text{d}}\\&-\left(M + U^s - U^0 - V_C\right)\psi_{\text{d}} = E\psi_{\text{d}}\end{aligned} \tag{5.270}$$

Using Eq. (5.268) the above formula can be rewritten as

$$\begin{aligned}&\left\{\frac{p^2}{2E}+\frac{\mathrm{i}}{2E}\left[\frac{\mathrm{d}}{\mathrm{d}r}(\ln T(r))\right]\frac{\vec{r}\cdot\vec{p}}{r}-\frac{1}{2E}\left[\frac{\mathrm{d}}{\mathrm{d}r}(\ln T(r))\right]\frac{\hat{\sigma}\cdot\vec{L}}{r}\right\}\psi_{\mathrm{d}}\\&-\frac{1}{2E}\left(M+U^{s}-U^{0}-V_{\mathrm{C}}\right)\left(E-M-U^{s}-U^{0}-V_{\mathrm{C}}\right)\psi_{\mathrm{d}}\\&=\frac{1}{2}\left(E-M-U^{s}-U^{0}-V_{\mathrm{C}}\right)\psi_{\mathrm{d}}\end{aligned}\tag{5.271}$$

And there is the following relation:

$$\begin{aligned}&-\left(M+U^{s}-U^{0}-V_{\mathrm{C}}\right)\left(E-M-U^{s}-U^{0}-V_{\mathrm{C}}\right)\\&=-E\left(M+U^{s}-U^{0}-V_{\mathrm{C}}\right)+M^2+2MU^{s}+(U^{s})^2-\left(U^{0}+V_{\mathrm{C}}\right)^2\end{aligned}\tag{5.272}$$

Substituting Eq. (5.272) into Eq. (5.271) and moving the terms corresponding the first two terms of the right side of Eq. (5.272) to the right side of the equal sign of Eq. (5.271), we can obtain

$$\begin{aligned}&\left\{\frac{p^2}{2E}+\frac{\mathrm{i}}{2E}\left[\frac{\mathrm{d}}{\mathrm{d}r}(\ln T(r))\right]\frac{\vec{r}\cdot\vec{p}}{r}-\frac{1}{2E}\left[\frac{\mathrm{d}}{\mathrm{d}r}(\ln T(r))\right]\frac{\hat{\sigma}\cdot\vec{L}}{r}\right\}\psi_{\mathrm{d}}\\&+\frac{1}{2E}\left[2MU^{s}+(U^{s})^2-\left(U^{0}+V_{\mathrm{C}}\right)^2\right]\psi_{\mathrm{d}}=\frac{E^2-M^2-2E\left(U^{0}+V_{\mathrm{C}}\right)}{2E}\psi_{\mathrm{d}}\end{aligned}\tag{5.273}$$

The above formula can be further rewritten as

$$\begin{aligned}&\left\{\frac{p^2}{2E}+U^{0}+V_{\mathrm{C}}+\frac{\mathrm{i}}{2E}\left[\frac{1}{T(r)}\frac{\mathrm{d}T(r)}{\mathrm{d}r}\right]\frac{\vec{r}\cdot\vec{p}}{r}-\frac{1}{2ErT(r)}\frac{\mathrm{d}T(r)}{\mathrm{d}r}\hat{\sigma}\cdot\vec{L}\right.\\&\left.+\frac{1}{2E}\left[U^{s}(2M+U^{s})-\left(U^{0}+V_{\mathrm{C}}\right)^2\right]\right\}\psi_{\mathrm{d}}=\frac{E^2-M^2}{2E}\psi_{\mathrm{d}}\end{aligned}\tag{5.274}$$

The form of the above formula is exactly the same as Eq. (5.255) for the upper component. Then the following transform of the wave function is introduced:

$$\psi_{\mathrm{d}}(r)=\sqrt{T(r)}\phi(r)\tag{5.275}$$

The same derivation method as the upper component equation can be used to obtain

$$\left[\frac{p^2}{2E}+U_{\mathrm{eff}}^{(\mathrm{d})}(r)+V_{\mathrm{C}}(r)+U_{\mathrm{SO}}^{(\mathrm{d})}(r)\,\vec{\sigma}\cdot\vec{L}\right]\phi(r)=\frac{E^2-M^2}{2E}\phi(r)\tag{5.276}$$

where

$$U_{\text{eff}}^{(\text{d})}(r) = U^0(r) + \frac{1}{2E}\left[U^s(r)\,(2M + U^s(r)) - \left(U^0(r) + V_{\text{C}}(r)\right)^2\right] + U_{\text{D}}^{(\text{d})}(r) \tag{5.277}$$

$$U_{\text{D}}^{(\text{d})}(r) = \frac{1}{2E}\left[\frac{3}{4}\left(\frac{1}{\text{T}(r)}\frac{\text{d}T(r)}{dr}\right)^2 - \frac{1}{2r^2T(r)}\frac{\text{d}}{\text{d}r}\left(r^2\frac{\text{d}T(r)}{\text{d}r}\right)\right] \tag{5.278}$$

$$U_{\text{SO}}^{(\text{d})}(r) = -\frac{1}{2ErT(r)}\frac{\text{d}T(r)}{\text{d}r} \tag{5.279}$$

$$T(r) = \left(E - M - U^s(r) - U^0(r) - V_{\text{C}}(r)\right)/(E - M) \tag{5.280}$$

The form of Eq. (5.276) for the lower component is exactly the same as Eq. (5.262) for the upper component; their difference is that the definitions of $T(r)$ and $D(r)$ are different.

Noting that the $p$ in the first item at the left end of Eqs. (5.262) and (5.276) are operator, and there is $\vec{p} = -\text{i}\vec{\nabla}$. We know that $E^2 - M^2 = k^2$ by Eq. (5.3) or Eq. (5.160). If the nuclear potentials $U^s(r)$ and $U^0(r)$ as well as the Coulomb potential $V_{\text{C}}(r)$ are all equal to 0, Eq. (5.262) and Eq. (5.276) degenerate, respectively, into

$$\left(\nabla^2 + k^2\right)\varphi = 0, \qquad \left(\nabla^2 + k^2\right)\phi = 0 \tag{5.281}$$

The above two formulas are all spherical Bessel equations, so the spherical Bessel function $\text{j}_l(kr)$ and the spherical Neumann function $\text{n}_l(kr)$ are applicable to the upper component wave function and the lower component wave function in any positive energy situation.

In areas outside the nucleus, the Coulomb potential is

$$V_{\text{C}}(r) = \frac{zZe^2}{r} \tag{5.282}$$

where e is the electron charge and $z$ and $Z$ are the charge numbers carried by the incident particle and the target nucleus, respectively. When the nuclear potential is 0 and only the Coulomb potential exists, the motion equation satisfied by the charged particles in the $\hbar = c = 1$ unit system is

$$\left[-\frac{1}{2\mu}\nabla^2 + V_{\text{C}}(r)\right]\psi_{\text{C}}\left(\vec{r}\right) = \frac{k^2}{2\mu}\psi_{\text{C}}\left(\vec{r}\right) \tag{5.283}$$

where

$$\mu = \frac{MM_{\text{A}}}{M + M_{\text{A}}} \tag{5.284}$$

where $M$ and $M_A$ are the masses of the incident particle and target nucleus, respectively, and $\mu$ is called reduced mass. We can also rewrite Eq. (5.283) as

$$\left(\nabla^2 + k^2 - \frac{2\eta k}{r}\right)\psi_C\left(\vec{r}\right) = 0 \tag{5.285}$$

where

$$\eta = \frac{\mu z Z e^2}{k} \tag{5.286}$$

We refer to Eq. (5.283) and Eq. (5.285) as the Coulomb equation, which contains the charge numbers $z$ and $Z$ of the incident particle and the target, and the masses $M$ and $M_A$ and other physical quantities. The Coulomb equation is derived from the non-relativistic Schrodinger equation and is suitable for the lower energy region. When the energy is quite high, there will be $k^2 \gg \frac{2\eta k}{r}$ in the area outside the nucleus, so it can be seen from Eq. (5.285) that the Coulomb potential is no longer important for very high energy. In this case this equation basically becomes the spherical Bessel equation.

Previously, we added the Coulomb potential $V_C$ in the Dirac equation (5.236) using side-by-side with $U^0$, which is currently consistent method in the literature. But this approach will bring about the following two problems: Firstly, when we make both the nuclear potential $U^0$ and $U^s$ 0, because there is a non-constant $V_C(r)$ in $D(r)$ and $T(r)$, the nuclear potential $U_D^{(\rho)}(r)$ and $U_{SO}^{(\rho)}(r)(\rho = \mathrm{u, d})$ will not disappear, and secondly, even if we ignore the $V_C(r)$ contained in $D(r)$ and $T(r)$, at this time the terms $\left(V_C(r) - \frac{1}{2E}V_C^2(r)\right)$ are still retained in Eq. (5.262) and Eq. (5.276). If the second term of $\left(V_C(r) - \frac{1}{2E}V_C^2(r)\right)$ is ignored and Eqs. (5.262) and (5.276) are rewritten according to the form of the Coulomb equation (5.283), when compared them with the Coulomb equation (5.283), it will be found that there is one more coefficient $E/\mu$ in front of $V_C(r)$ in the newly obtained equation. This coefficient may be close to 1 only if the energy is very low and the target nucleus is very heavy. It can be seen that the current general method of the adding Coulomb potential $V_C$ in Dirac equation (5.236) is somewhat unreasonable in physics.

It can be seen from the positive energy nucleon plane waves given by Eqs. (5.80) and (5.93) that when the energy is very low, there is $p \ll E + M$; the contribution of the lower component in the incident plane wave is very small. We also know that the Coulomb potential only plays a significant role in low energy conditions, and the influence of the Coulomb potential on the high energy reaction is negligible. Therefore, we propose that the Coulomb potential $V_C$ is not added using side-by-side with $U^0$ in the Dirac equation (5.236). Instead, the Coulomb potential $\frac{\mu}{E}V_C$ is only added using side-by-side with $U_{\mathrm{eff}}^{(\mathrm{u})}$ in the upper component equation (5.262), and the influence of the Coulomb potential $V_C$ is ignored in the lower component equation (5.276). Since the Coulomb potential only works under low energy conditions, this method of adding Coulomb potential according to the non-relativistic

Schrodinger equation is reasonable. Thus Eqs. (5.262), (5.263), (5.266), (5.276), (5.277), and (5.280) will be replaced by the following equations based on our proposed approach, respectively:

$$\left[\frac{p^2}{2E} + U_{\text{eff}}^{(\text{u})}(r) + \frac{\mu}{E} V_{\text{C}}(r) + U_{\text{SO}}^{(\text{u})}(r)\, \vec{\sigma} \cdot \vec{L}\right] \varphi(r) = \frac{E^2 - M^2}{2E} \varphi(r) \tag{5.287}$$

$$U_{\text{eff}}^{(\text{u})}(r) = U^0(r) + \frac{1}{2E}\left[U^{\text{s}}(r)\,(2M + U^{\text{s}}(r)) - \left(U^0(r)\right)^2\right] + U_{\text{D}}^{(\text{u})}(r) \tag{5.288}$$

$$D(r) = \left(E + M + U^{\text{s}}(r) - U^0(r)\right)/(E + M) \tag{5.289}$$

$$\left[\frac{p^2}{2E} + U_{\text{eff}}^{(\text{d})}(r) + U_{\text{SO}}^{(\text{d})}(r)\, \vec{\sigma} \cdot \vec{L}\right] \phi(r) = \frac{E^2 - M^2}{2E} \phi(r) \tag{5.290}$$

$$U_{\text{eff}}^{(\text{d})}(r) = U^0(r) + \frac{1}{2E}\left[U^{\text{s}}(r)\,(2M + U^{\text{s}}(r)) - \left(U^0(r)\right)^2\right] + U_{\text{D}}^{(\text{d})}(r) \tag{5.291}$$

$$T(r) = \left(E - M - U^{\text{s}}(r) - U^0(r)\right)/(E - M) \tag{5.292}$$

where $U_{\text{D}}^{(\text{u})}(r)$ and $U_{\text{SO}}^{(\text{u})}(r)$ have been given by Eqs. (5.264) and (5.265), respectively; $U_{\text{D}}^{(\text{d})}(r)$ and $U_{\text{SO}}^{(\text{d})}(r)$ have been given by Eq. (5.278) and Eq. (5.279), respectively. In this way, when let the nuclear potential $U^0$ and $U^s$ be all 0, the equation (5.290) automatically degenerates into the spherical Bessel equation, and the equation (5.287) automatically degenerates into the Coulomb equation (5.283), and they can be applied to any target nuclei. In fact, when the energy is quite high, the equation (5.287) is also equivalent to the spherical Bessel equation.

If the following nuclear potential symbols are introduced

$$U_{\text{c}}^{(\rho)}(r) = \frac{E}{\mu} U_{\text{eff}}^{(\rho)}(r), \qquad U_{\text{so}}^{(\rho)}(r) = \frac{E}{\mu} U_{\text{SO}}^{(\rho)}(r), \qquad \rho = \text{u,d} \tag{5.293}$$

Equations (5.287) and (5.290) can be rewritten as

$$\left[-\frac{\nabla^2}{2\mu} + U_{\text{c}}^{(\text{u})}(r) + V_{\text{C}}(r) + U_{\text{so}}^{(\text{u})}(r)\, \vec{\sigma} \cdot \vec{L}\right] \varphi(r) = \frac{k^2}{2\mu} \varphi(r) \tag{5.294}$$

$$\left[-\frac{\nabla^2}{2\mu} + U_{\text{c}}^{(\text{d})}(r) + U_{\text{so}}^{(\text{d})}(r)\, \vec{\sigma} \cdot \vec{L}\right] \phi(r) = \frac{k^2}{2\mu} \phi(r) \tag{5.295}$$

Equations (5.287) and (5.290) are called the Schrodinger-like equations, whereas Eq. (5.294) and Eq. (5.295) have been completely transformed into the Schrodinger equation form, but the nuclear potentials defined by Eq. (5.293) need to be taken.

When the energy is not very high, it is only necessary to solve the upper component Eq. (5.287) or Eq. (5.294), so they are also called the spherical nucleus relativistic optical model equations. If assume that both $U^{\text{s}}(r)$ and $U^0(r)$ contain real and imaginary parts and phenomenally introduce their relations with the coordinates

$r$, the incident particle energy, and the $A$ and $Z$ of the target, then solve the relativistic optical model equation (5.287) or (5.294), and by adjusting the parameters to meet the experimental data, thus the relativistic phenomenological optical potential (RPOP) can be obtained.

If the nuclear bound state is described by Eq. (5.287) or Eq. (5.294), the mass-energy relation of the on-shell nucleons corresponding to Eq. (5.3) becomes $E^2 = -p^2 + M^2,\ 0 < E < M$. So the bound state nucleons still belong to the positive energy nucleons.

The $S$ matrix elements can be solved by boundary conditions of the spherical nucleus relativistic optical model Eq. (5.287) or (5.294) (discussed in the next section). The elastic scattering amplitude can be written according to Eqs. (2.70), (2.82), and (2.83) as follows:

$$\hat{F}(\theta, \varphi) = A(\theta)\hat{I} + B(\theta)\, \vec{n} \cdot \hat{\sigma} \tag{5.296}$$

$$A(\theta) = f_{\mathrm{C}}(\theta) + \frac{\mathrm{i}}{2k}\sum_l \left[(l+1)\left(1 - S_l^{l+\frac{1}{2}}\right) + l\left(1 - S_l^{l-\frac{1}{2}}\right)\right] \mathrm{e}^{2\mathrm{i}\sigma_l}\mathrm{P}_l(\cos\theta) \tag{5.297}$$

$$B(\theta) = \frac{1}{2k}\sum_l \left(S_l^{l+\frac{1}{2}} - S_l^{l-\frac{1}{2}}\right) \mathrm{e}^{2\mathrm{i}\sigma_l}\mathrm{P}_l^1(\cos\theta) \tag{5.298}$$

where $\vec{n}$ is a unit vector perpendicular to the reaction plane, generally taking the $y$-axis in the $\vec{n}$ direction. Moreover, the following formulas for the polarization analyzing power $A_y(\theta)$ and the spin rotation function $Q(\theta)$ have been given by Eqs. (2.131) and (2.132) as follows:

$$A_y(\theta) = \frac{2\,\mathrm{Re}\,(A(\theta)B^*(\theta))}{I_0(\theta)} \tag{5.299}$$

$$Q(\theta) = \frac{2\mathrm{Im}(A(\theta)B^*(\theta))}{I_0(\theta)} \tag{5.300}$$

$$I_0(\theta) = |A(\theta)|^2 + |B(\theta)|^2 \tag{5.301}$$

where $I_0(\theta)$ is the differential cross section of the unpolarized incident nucleons.

In the non-relativistic optical model, the spin-orbital coupling potential is added separately. But the spin-orbital coupling potential in the relativistic optical model equation is naturally generated by the Dirac equation, and it is closely related to the central potential $U_{\mathrm{eff}}(r)$. So it has obvious advantages to analyze the polarization observable quantities. Therefore, since the birth of the relativistic optical model, it has been used to analyze the elastic scattering angular distribution $\frac{\mathrm{d}\sigma}{\mathrm{d}\Omega}$, polarization analyzing power $A_y(\theta)$, and spin rotation function $Q(\theta)$ of the spherical nucleus at the same time, and the better results have been obtained [10–21]. From experimental data and theoretical calculation results, it can be seen that the polarization phenomena exist obviously until the incident nucleon energy 1000 MeV.

## 5.4 Dirac S Matrix Theory of the Relativistic Nuclear Reactions

The Dirac equation is only applicable to spin $\frac{1}{2}$ particles, so only the Dirac S matrix theory of the two-body nuclear reactions of the incident spin $\frac{1}{2}$ particles and various possible outgoing spin $\frac{1}{2}$ particles is studied here.

The two-body nuclear reaction can be expressed as a + A → b + B, a and b represent spin $\frac{1}{2}$ particles, and the reaction channels a + A and b + B are expressed by $\alpha$ and $\alpha'$, respectively. $m$ and $m'$ represent the stationary masses of the particles a and b; $M_A$ and $M_B$ represent the stationary masses of the target A and the residual nucleus B. $v$ and $v'$ represent the velocities of the incident particle a and the outgoing particle b. Next we let $\varepsilon_{\alpha n}$ and $\varepsilon_{\alpha' n'}$ represent the excitation energies for the $n$th exited state of the target A and the $n'$th exited state of the residual nucleus B, respectively. $I_nM_n$ and $I'_nM'_n$ represent their corresponding angular momentum and its $z$ component, $\Phi_{I_nM_n}(\xi)$ and $\Phi_{I'_nM'_n}(\xi')$ represent their corresponding spin wave functions. $\xi$ and $\xi'$ represent the internal freedom of A and B, respectively. We call the system, for which the sum of incident particle momentum and target momentum is equal to 0, as center of momentum (c.m.) system, and set $E$ as the total energy of the whole system in c.m. system. In the $\hbar = c = 1$ natural unit system we let

$$
\begin{aligned}
k &\equiv k_{\alpha n,\alpha n} = \sqrt{E^2 - m^2}\ , \\
k' &\equiv k_{\alpha' n',\alpha n} = \left[E^2 - m^2 + \left(M_A^2 + m^2 + \varepsilon_{\alpha n}^2 - M_B^2 - m'^2 - \varepsilon_{\alpha' n'}^2\right)\right]^{1/2}
\end{aligned}
\tag{5.302}
$$

Select the direction of incident particle as the $z$-axis, according to Eq. (5.80) and Eq. (3.1.22) given in Ref. [4], we can write the initial state total wave function of the positive energy stationary incident particles as

$$
\begin{aligned}
\Psi^{(\mathrm{i})}_{\alpha n \nu M_n} &= \frac{1}{\sqrt{v}} u\left(\vec{p}\right) \mathrm{e}^{\mathrm{i}kz} \chi_\nu \Phi_{I_nM_n}(\xi) \\
&= \sqrt{\frac{4\pi}{v}} u\left(\vec{p}\right) \sum_l \hat{l} \mathrm{j}_l(kr)\, \mathrm{i}^l \mathrm{Y}_{l0}(\theta, 0) \chi_\nu \Phi_{I_nM_n}(\xi)
\end{aligned}
\tag{5.303}
$$

where $\hat{l} \equiv \sqrt{2l+1}$. $v$ is the velocity of the incident particle relative to the target in the $n$th excited state, and $\chi_\nu$ is the incident particle spin wave function with a magnetic quantum number $\nu$. $\mathrm{j}_l$ is the spherical Bessel function. $u\left(\vec{p}\right)$ has been given by Eq. (5.93) as follows:

$$
u\left(\vec{p}\right) = \sqrt{\frac{E+m}{2m}} \begin{pmatrix} \hat{I} \\ \dfrac{\hat{\sigma} \cdot \vec{p}}{E+m} \end{pmatrix}
\tag{5.304}
$$

where $\hat{I}$ is the unit 2 × 2 matrix and $\vec{p}$ is the momentum operator. Eq. (5.303) denotes that the upper and lower components of $u\left(\vec{p}\right)$ act on the two-dimensional spin wave function $\chi_\nu$ at the same time. When the incident particle energy is very low, there are $p \ll (E+m)$ and $\sqrt{\frac{E+m}{2m}} \to 1$, then from Eq. (5.304) we can see that the contribution of the lower component in this case can be ignored, and Eq. (5.303) will automatically degenerate into non-relativistic theoretical formula. Only when the energy of the incident particle is quite high, the contribution of the lower component needs to be considered. Introduce the following symbols:

$$\mathbb{Y}_{lj\nu}(\Omega) = C_{l0\ \frac{1}{2}\nu}^{j\nu}\,\mathrm{i}^l\mathrm{Y}_{l0}(\theta,0)\chi_\nu, \quad \mathrm{i}^l\mathrm{Y}_{l0}(\theta,0)\chi_\nu = \sum_j C_{l0\ \frac{1}{2}\nu}^{j\nu}\,\mathbb{Y}_{lj\nu}(\Omega) \tag{5.305}$$

$$\begin{aligned} \varphi_{\alpha nlj}^{JM}(\Omega,\xi) &= \sum_{\nu M_n} C_{j\nu\ I_nM_n}^{J\ M}\,\mathbb{Y}_{lj\nu}(\Omega)\,\Phi_{I_nM_n}(\xi), \\ \mathbb{Y}_{lj\nu}(\Omega)\,\Phi_{I_nM_n}(\xi) &= \sum_{JM} C_{j\nu\ I_nM_n}^{J\ M}\,\varphi_{\alpha nlj}^{JM}(\Omega,\xi) \end{aligned} \tag{5.306}$$

So Eq. (5.303) can be rewritten as

$$\Psi_{\alpha n\nu M_n}^{(\mathrm{i})} = \sqrt{\frac{4\pi}{v}}\,u\left(\vec{p}\right)\sum_{ljJM}\hat{l}\,\mathrm{j}_l(kr)\,C_{l0\ \frac{1}{2}\nu}^{j\ \nu}\,C_{j\nu\ I_nM_n}^{J\ M}\,\varphi_{\alpha nlj}^{JM}(\Omega,\xi) \tag{5.307}$$

If the incident particle is a charged particle, there will be a long-range Coulomb potential $V_\mathrm{C}(r)$. Since the Coulomb potential can be ignored at very high energy, we only consider the Coulomb potential and Coulomb scattering in the upper component. The discussion in the 5.3 section has pointed out that the Coulomb functions obtained by Coulomb equation can be used directly in the case of relativity.

For the charged particles, the upper component of the initial state total wave function of the incident particles can be obtained from Eqs. (5.307) and (5.304) as

$$\Psi_{\alpha n\nu M_n}^{(\mathrm{i})(\mathrm{u})} = \sqrt{\frac{4\pi}{v}}\sqrt{\frac{E+m}{2m}}\sum_{ljJM}\hat{l}\,\mathrm{e}^{\mathrm{i}\sigma_l}\,\frac{F_l(kr)}{kr}\,C_{l0\ \frac{1}{2}\nu}^{j\ \nu}\,C_{j\nu\ I_nM_n}^{J\ M}\,\varphi_{\alpha nlj}^{JM}(\Omega,\xi) \tag{5.308}$$

where $F_l(kr)$ is Coulomb wave function. Further we can rewrite the above formula as

$$\Psi_{\alpha n\nu M_n}^{(\mathrm{i})(\mathrm{u})} = \frac{\mathrm{i}\sqrt{\pi}}{kr\sqrt{v}}\sqrt{\frac{E+m}{2m}}\sum_{ljJM}\hat{l}\,\mathrm{e}^{\mathrm{i}\sigma_l}\,C_{l0\ \frac{1}{2}\nu}^{j\ \nu}\,C_{j\nu\ I_nM_n}^{J\ M}\left[(G_l-\mathrm{i}F_l)-(G_l+\mathrm{i}F_l)\right]\varphi_{\alpha nlj}^{JM}(\Omega,\xi) \tag{5.309}$$

where $\frac{G_l-\mathrm{i}F_l}{r}$ represents the spherical incident wave and $\frac{G_l+\mathrm{i}F_l}{r}$ represents the spherical outgoing wave. If the incident particles do not undergo reaction, the strengths of the spherical incident wave and the spherical outgoing wave are equal to each other. If the incident particles undergone elastic scattering or other two-body reactions with the target, after the $S$ matrix elements are introduced into the upper component, the upper component of the total wave function of the system can be obtained as follows:

$$\Psi^{(t)(u)}_{\alpha n \nu M_n} = \frac{i\sqrt{\pi}}{kr}\sqrt{\frac{E+m}{2m}}\sum_{ljJM}\hat{l}e^{i\sigma_l}C^{j\ \nu}_{l0\ \frac{1}{2}\nu}C^{J\ M}_{j\nu\ I_nM_n}\sum_{\alpha'n'l'j'}\frac{1}{\sqrt{v'}} \left[(G_l - iF_l)\delta_{\alpha'n'l'j',\alpha nlj} - S^{(u)J}_{\alpha'n'l'j',\alpha nlj}(G_{l'} + iF_{l'})\right]\varphi^{JM}_{\alpha'n'l'j'}(\Omega,\xi) \tag{5.310}$$

There is a formula

$$i(G_c - iF_c)\,\delta_{c'c} - iS_{c'c}(G_{c'} + iF_{c'}) = 2F_c\delta_{c'c} + i(\delta_{c'c} - S_{c'c})(G_{c'} + iF_{c'}) \tag{5.311}$$

So Eq. (5.310) can be rewritten as

$$\Psi^{(t)(u)}_{\alpha n \nu M_n} = \frac{\sqrt{4\pi}}{kr}\sqrt{\frac{E+m}{2m}}\sum_{ljJM}\hat{l}e^{i\sigma_l}C^{j\ \nu}_{l0\ \frac{1}{2}\nu}C^{J\ M}_{j\nu\ I_nM_n} \left[\frac{1}{\sqrt{v}}F_l\,\varphi^{JM}_{\alpha nlj}(\Omega,\xi) + \sum_{\alpha'n'l'j'}\frac{i}{2\sqrt{v'}}\left(\delta_{\alpha'n'l'j',\alpha nlj} - S^{(u)J}_{\alpha'n'l'j',\alpha nlj}\right)(G_{l'} + iF_{l'})\varphi^{JM}_{\alpha'n'l'j'}(\Omega,\xi)\right] \tag{5.312}$$

The first term in the square bracket of the above formula is clearly the upper component of the initial state wave function of the incident particles given by Eq. (5.308). So according to the two terms in square bracket in Eq. (5.312), the upper component of the total wave function of the system given by Eq. (5.312) can be divided into two terms representing the initial state and the final state as

$$\Psi^{(t)(u)}_{\alpha n \nu M_n} = \Psi^{(i)(u)}_{\alpha n \nu M_n} + \Psi^{(f)(u)}_{\alpha n \nu M_n} \tag{5.313}$$

where

$$\Psi^{(f)(u)}_{\alpha n \nu M_n} = \frac{i\sqrt{\pi}}{kr}\sqrt{\frac{E+m}{2m}}\sum_{ljJM}\hat{l}e^{i\sigma_l}C^{j\ \nu}_{l0\ \frac{1}{2}\nu}C^{J\ M}_{j\nu\ I_nM_n}\sum_{\alpha'n'l'j'}\frac{1}{\sqrt{v'}} \left(\delta_{\alpha'n'l'j',\alpha nlj} - S^{(u)J}_{\alpha'n'l'j',\alpha nlj}\right)(G_{l'} + iF_{l'})\,\varphi^{JM}_{\alpha'n'l'j'}(\Omega,\xi) \tag{5.314}$$

Equation (3.8.5) in Ref. [4] has given

$$G_{l'} + \mathrm{i}F_{l'} \xrightarrow[r\to\infty]{} \mathrm{e}^{\mathrm{i}k'r}(-\mathrm{i})^{l'}\mathrm{e}^{\mathrm{i}\sigma_{l'}} \tag{5.315}$$

We can obtain the following expression substituting Eqs. (5.306), (5.305), and (5.315) into Eq. (5.314):

$$\begin{aligned}\Psi^{(\mathrm{f})(\mathrm{u})}_{\alpha n\nu M_n} \xrightarrow[r\to\infty]{} &\frac{\mathrm{i}\sqrt{\pi}}{k}\sqrt{\frac{E+m}{2m}}\sum_{ljJM}\hat{l}\mathrm{e}^{\mathrm{i}\sigma_l}C^{j\ \nu}_{l0\ \frac{1}{2}\nu}C^{J\ M}_{j\nu\ I_nM_n}\sum_{\alpha'n'l'j'}\frac{1}{\sqrt{v'}}\mathrm{e}^{\mathrm{i}\sigma_{l'}}\\ &\left(\delta_{\alpha'n'l'j',\alpha nlj} - S^{(\mathrm{u})J}_{\alpha'n'l'j',\alpha nlj}\right)\frac{\mathrm{e}^{\mathrm{i}k'r}}{r}\sum_{\nu'M'_nm'_l\,m'_j}C^{j'\ m'_j}_{l'm'_l\ \frac{1}{2}\nu'}C^{J\ M}_{j'm'_j\ I'_nM'_n}\mathrm{Y}_{l'm'_l}\chi_{\nu'}\Phi_{I'_nM'_n}\end{aligned} \tag{5.316}$$

And we can write the following formula according to Eqs. (5.312) and (5.313):

$$\Psi^{(\mathrm{i})(\mathrm{u})}_{\alpha n\nu M_n} = \frac{\sqrt{4\pi}}{kr}\sqrt{\frac{E+m}{2m}}\sum_{ljJM}\hat{l}\mathrm{e}^{\mathrm{i}\sigma_l}C^{j\ \nu}_{l0\ \frac{1}{2}\nu}C^{J\ M}_{j\nu\ I_nM_n}\frac{1}{\sqrt{v}}F_l\varphi^{JM}_{\alpha nlj}(\Omega,\xi) \tag{5.317}$$

Further the following formula can be obtained substituting Eqs. (5.306) and (5.305) into Eq. (5.317):

$$\Psi^{(\mathrm{i})(\mathrm{u})}_{\alpha n\nu M_n} = \sqrt{\frac{4\pi}{v}}\sqrt{\frac{E+m}{2m}}\sum_l\hat{l}\mathrm{e}^{\mathrm{i}\sigma_l}\frac{F_l(kr)}{kr}\mathrm{i}^l\mathrm{Y}_{l0}(\theta,0)\chi_\nu\Phi_{I_nM_n} \tag{5.318}$$

When $r \to \infty$ the following relation can be obtained when making comparison between Eq. (3.3.76) and Eq. (3.3.45) given in Ref. [4]:

$$\sqrt{4\pi}\sum_l\hat{l}\mathrm{e}^{\mathrm{i}\sigma_l}\frac{F_l(kr)}{kr}\mathrm{i}^l\mathrm{Y}_{l0}(\theta,0) \xrightarrow[r\to\infty]{} \mathrm{e}^{\mathrm{i}kz} + f_{\mathrm{C}}(\theta)\frac{\mathrm{e}^{\mathrm{i}kr}}{r} \tag{5.319}$$

where $f_{\mathrm{C}}(\theta)$ has been given by Eqs. (3.3.41), (3.3.3), and (3.3.42) in Ref. [4]. The first term on the right end of Eq. (5.319) represents the undistorted incident plane wave, and the second term represents the spherical outgoing wave of Coulomb scattering induced by Coulomb potential. In Eq. (5.318) the term corresponding to the second term of Eq. (5.319) is

$$\Psi^{(2)(\mathrm{u})}_{\alpha n\nu M_n} \xrightarrow[r\to\infty]{} \frac{1}{\sqrt{v}}\sqrt{\frac{E+m}{2m}}f_{\mathrm{C}}(\theta)\frac{\mathrm{e}^{\mathrm{i}kr}}{r}\chi_\nu\Phi_{I_nM_n} \tag{5.320}$$

The upper component of the total reaction wave function of the system can be obtained from Eqs. (5.316) and (5.320) as

$$
\begin{aligned}
\Psi^{(\mathrm{r})(\mathrm{u})}_{\alpha n\nu M_n} &= \Psi^{(\mathrm{f})(\mathrm{u})}_{\alpha n\nu M_n} + \Psi^{(2)(\mathrm{u})}_{\alpha n\nu M_n} \xrightarrow[r\to\infty]{} \sqrt{\frac{E+m}{2m}} \sum_{\alpha' n'\nu' M_n'} \frac{1}{\sqrt{v'}} \Bigg[ f_{\mathrm{C}}(\theta)\delta_{\alpha' n',\alpha n}\delta_{\nu' M_n',\nu M_n} \\
&+ \frac{\mathrm{i}\sqrt{\pi}}{k} \sum_{\substack{ljJM\\ l'j'm'_l m'_j}} \hat{l}\mathrm{e}^{\mathrm{i}(\sigma_l+\sigma_{l'})} \left( \delta_{\alpha' n' l' j',\alpha n l j} - S^{(\mathrm{u})J}_{\alpha' n' l' j',\alpha n l j} \right) \\
&\quad C^{j\ \nu}_{l0\ \frac{1}{2}\nu} C^{J\ M}_{j\nu\ I_n M_n} C^{j'\ m'_j}_{l' m'_l\ \frac{1}{2}\nu'} C^{J\ M}_{j'\ m'_j\ I'_n M'_n} \mathrm{Y}_{l' m'_l}(\theta,\varphi) \Bigg] \frac{\mathrm{e}^{\mathrm{i}k' r}}{r} \chi_{\nu'} \Phi_{I'_n M'_n}
\end{aligned}
\tag{5.321}
$$

From the above formula, the reaction amplitude of the upper component can be written as

$$
\begin{aligned}
f^{(\mathrm{u})}_{\alpha' n'\nu' M_n',\alpha n\nu M_n} &= \sqrt{\frac{E+m}{2m}} \Bigg[ f_{\mathrm{C}}(\theta)\delta_{\alpha' n',\alpha n}\delta_{\nu' M_n',\nu M_n} + \frac{\mathrm{i}\sqrt{\pi}}{k} \sum_{\substack{ljJM\\ l'j'm'_l m'_j}} \hat{l}\mathrm{e}^{\mathrm{i}(\sigma_l+\sigma_{l'})} \\
&\left( \delta_{\alpha' n' l' j',\alpha n l j} - S^{(\mathrm{u})J}_{\alpha' n' l' j',\alpha n l j} \right) C^{j\ \nu}_{l0\ \frac{1}{2}\nu} C^{J\ M}_{j\nu\ I_n M_n} C^{j'\ m'_j}_{l' m'_l\ \frac{1}{2}\nu'} C^{J\ M}_{j'\ m'_j\ I'_n M'_n} \mathrm{Y}_{l' m'_l}(\theta,\varphi) \Bigg]
\end{aligned}
\tag{5.322}
$$

The lower component of the initial state total wave function of the incident particles can be written from Eqs. (5.307) and (5.304) as

$$
\Psi^{(\mathrm{i})(\mathrm{d})}_{\alpha n\nu M_n} = \sqrt{\frac{4\pi}{v}} \sqrt{\frac{E+m}{2m}} \frac{\hat{\sigma}\cdot\vec{p}}{E+m} \sum_{ljJM} \hat{l}\mathrm{j}_l(kr) C^{j\ \nu}_{l0\ \frac{1}{2}\nu} C^{J\ M}_{j\nu\ I_n M_n} \varphi^{JM}_{\alpha n l j}(\Omega,\xi)
\tag{5.323}
$$

Further the above formula can be rewritten as

$$
\begin{aligned}
\Psi^{(\mathrm{i})(\mathrm{d})}_{\alpha n\nu M_n} &= \frac{\mathrm{i}\sqrt{\pi}}{\sqrt{v}} \sqrt{\frac{E+m}{2m}} \frac{\hat{\sigma}\cdot\vec{p}}{E+m} \sum_{ljJM} \hat{l} C^{j\ \nu}_{l0\ \frac{1}{2}\nu} C^{J\ M}_{j\nu\ I_n M_n} \\
&\left[ (-\mathrm{n}_l - \mathrm{i}\mathrm{j}_l) - (-\mathrm{n}_l + \mathrm{i}\mathrm{j}_l) \right] \varphi^{JM}_{\alpha n l j}(\Omega,\xi)
\end{aligned}
\tag{5.324}
$$

where $k(-\mathrm{n}_l - \mathrm{i}\mathrm{j}_l)$ represents spherical incident wave and $k(-\mathrm{n}_l + \mathrm{i}\mathrm{j}_l)$ represents spherical outgoing wave. If the incident particles undergo elastic scattering or other two-body reactions with the target, after the *S* matrix element is introduced into the lower component, the lower component of the total wave function of the system can be obtained as

$$\Psi_{\alpha n\nu M_n}^{(\mathrm{t})(\mathrm{d})} = \frac{\mathrm{i}\sqrt{\pi}}{k}\sqrt{\frac{E+m}{2m}}\frac{\hat{\sigma}\cdot\vec{p}}{E+m}\sum_{ljJM}\hat{l}C_{l0\ \frac{1}{2}\nu}^{j\ \nu}C_{j\nu\ I_nM_n}^{J\ M}\sum_{\alpha' n' l' j'}\frac{1}{\sqrt{v'}}$$
$$\left[k(-\mathrm{n}_l - \mathrm{i}\mathrm{j}_l)\,\delta_{\alpha' n' l' j',\alpha nlj} - S_{\alpha' n' l' j',\alpha nlj}^{(\mathrm{d})J}k'(-\mathrm{n}_{l'} + \mathrm{i}\mathrm{j}_{l'})\right]\varphi_{\alpha' n' l' j'}^{JM}(\Omega,\ \xi) \tag{5.325}$$

There is the formula

$$\mathrm{i}(-\mathrm{n}_c - \mathrm{i}\mathrm{j}_c)\,\delta_{c'c} - \mathrm{i}S_{c'c}(-\mathrm{n}_{c'} + \mathrm{i}\mathrm{j}_{c'}) = 2\mathrm{j}_c\delta_{c'c} + \mathrm{i}(\delta_{c'c} - S_{c'c})(-\mathrm{n}_{c'} + \mathrm{i}\mathrm{j}_{c'}) \tag{5.326}$$

So Eq. (5.325) can be rewritten as

$$\Psi_{\alpha n\nu M_n}^{(\mathrm{t})(\mathrm{d})} = \frac{\sqrt{4\pi}}{k}\sqrt{\frac{E+m}{2m}}\frac{\hat{\sigma}\cdot\vec{p}}{E+m}\sum_{ljJM}\hat{l}\,C_{l0\ \frac{1}{2}\nu}^{j\ \nu}C_{j\nu\ I_nM_n}^{J\ M}\left[\frac{k}{\sqrt{v}}\mathrm{j}_l\,\varphi_{\alpha nlj}^{JM}(\Omega,\ \xi)\right.$$
$$\left.+\sum_{\alpha' n' l' j'}\frac{\mathrm{i}}{2\sqrt{v'}}\left(\delta_{\alpha' n' l' j',\alpha nlj} - S_{\alpha' n' l' j',\alpha nlj}^{(\mathrm{d})J}\right)k'(-\mathrm{n}_{l'} + \mathrm{i}\mathrm{j}_{l'})\varphi_{\alpha' n' l' j'}^{JM}(\Omega,\ \xi)\right] \tag{5.327}$$

The first term in the square bracket of the above formula is the lower component of the initial state wave function of the incident particles given by Eq. (5.323). So according to the two terms in square bracket in Eq. (5.327), we can divide the lower component of the total wave function of the system into two terms representing the initial state and the final state as

$$\Psi_{\alpha n\nu M_n}^{(\mathrm{t})(\mathrm{d})} = \Psi_{\alpha n\nu M_n}^{(\mathrm{i})(\mathrm{d})} + \Psi_{\alpha n\nu M_n}^{(\mathrm{f})(\mathrm{d})} \tag{5.328}$$

where

$$\Psi_{\alpha n\nu M_n}^{(\mathrm{f})(\mathrm{d})} = \frac{\mathrm{i}\sqrt{\pi}}{k}\sqrt{\frac{E+m}{2m}}\frac{\hat{\sigma}\cdot\vec{p}}{E+m}\sum_{ljJM}\hat{l}C_{l0\ \frac{1}{2}\nu}^{j\ \nu}C_{j\nu\ I_nM_n}^{J\ M}\sum_{\alpha' n' l' j'}\frac{1}{\sqrt{v'}}$$
$$\left(\delta_{\alpha' n' l' j',\alpha nlj} - S_{\alpha' n' l' j',\alpha nlj}^{(\mathrm{d})J}\right)k'\,(-\mathrm{n}_{l'} + \mathrm{i}\mathrm{j}_{l'})\,\varphi_{\alpha' n' l' j'}^{JM}(\Omega,\ \xi) \tag{5.329}$$

Equation (3.1106) has given

$$-\mathrm{n}_l(kr) + \mathrm{i}\,\mathrm{j}_l(kr) \xrightarrow[r\to\infty]{} (-\mathrm{i})^l\frac{\mathrm{e}^{\mathrm{i}kr}}{kr} \tag{5.330}$$

Then the following expression can be obtained substituting Eqs. (5.306), (5.305), and (5.330) into Eq. (5.329):

$$\Psi^{(\mathrm{f})(\mathrm{d})}_{\alpha n \nu\, M_n} \xrightarrow[r\to\infty]{} \frac{\mathrm{i}\sqrt{\pi}}{k}\sqrt{\frac{E+m}{2m}}\frac{\hat{\sigma}\cdot\vec{p}}{E+m}\sum_{ljJM}\hat{l}C^{j\ \nu}_{l0\ \frac{1}{2}\nu}C^{J\ M}_{j\nu\ I_nM_n}\sum_{\alpha' n' l' j'}\frac{1}{\sqrt{v'}}$$
$$\left(\delta_{\alpha' n' l' j',\alpha n l j} - S^{(\mathrm{d})J}_{\alpha' n' l' j',\alpha n l j}\right)\frac{\mathrm{e}^{\mathrm{i}k' r}}{r}\sum_{\overline{\nu}' M_n' m_l' m_j'} C^{j'\ m_j'}_{l'm'_l\ \frac{1}{2}\overline{\nu}'}C^{J\ M}_{j'\ m_j'\ I_n' M_n'}\mathrm{Y}_{l'm'_l}\chi_{\overline{\nu}'}\Phi_{I_n' M_n'} \tag{5.331}$$

There are the following relations:

$$\vec{\nabla} = \vec{n}\frac{\partial}{\partial r} + \frac{1}{r}\vec{\nabla}_{\Omega}, \qquad \vec{n} = \frac{\vec{r}}{r} = \hat{r} = \vec{e}_r \tag{5.332}$$

$$\left(\vec{\nabla}_{\Omega}\right)_{\theta} = \frac{\partial}{\partial\theta}, \qquad \left(\vec{\nabla}_{\Omega}\right)_{\varphi} = \frac{1}{\sin\theta}\frac{\partial}{\partial\varphi} \tag{5.333}$$

$$\vec{l} = \vec{r}\times\vec{p}, \qquad \vec{p} = -\mathrm{i}\vec{\nabla} \tag{5.334}$$

It can be seen that $\vec{l}$ is perpendicular to $\vec{r}$, and we can obtain

$$l_r = 0, \qquad l_{\theta} = \frac{\mathrm{i}}{\sin\theta}\frac{\partial}{\partial\varphi}, \qquad l_{\varphi} = -\mathrm{i}\frac{\partial}{\partial\theta} \tag{5.335}$$

$$\left(\vec{\nabla}_{\Omega}\right)_{\theta} = \mathrm{i}l_{\varphi}, \qquad \left(\vec{\nabla}_{\Omega}\right)_{\varphi} = -\mathrm{i}l_{\theta} \tag{5.336}$$

$$\vec{\nabla} = \vec{n}\frac{\partial}{\partial r} - \frac{\mathrm{i}}{r}\left(\vec{\mathrm{e}}_{\varphi}l_{\theta} - \vec{\mathrm{e}}_{\theta}l_{\varphi}\right) \tag{5.337}$$

Note that the three spherical coordinate axes are perpendicular to each other and we can obtain

$$\vec{n}\times\vec{l} = \vec{\mathrm{e}}_{\varphi}\,l_{\theta} - \vec{\mathrm{e}}_{\theta}\,l_{\varphi}, \qquad \vec{\nabla}_{\Omega} = -\mathrm{i}\left(\vec{n}\times\vec{l}\right) \tag{5.338}$$

$$\vec{\nabla} = \vec{n}\frac{\partial}{\partial r} - \mathrm{i}\frac{1}{r}\left(\vec{n}\times\vec{l}\right) \tag{5.339}$$

Let $\vec{a} = \vec{n},\ \vec{b} = \vec{l}$ in Eq. (5.248), and apparently there is $\vec{n}\cdot\vec{l} = 0$, so we can get

$$\mathrm{i}\hat{\sigma}\cdot\left(\vec{n}\times\vec{l}\right) = \left(\hat{\sigma}\cdot\vec{n}\right)\left(\hat{\sigma}\cdot\vec{l}\right) \tag{5.340}$$

The following formula can be obtained using Eqs. (5.339) and (5.340):

$$\vec{\sigma} \cdot \vec{p} = -\mathrm{i}\vec{\sigma} \cdot \vec{\nabla} = -\mathrm{i}(\vec{\sigma} \cdot \hat{r})\left(\frac{\mathrm{d}}{\mathrm{d}r} - \frac{\hat{\sigma} \cdot \vec{l}}{r}\right) \tag{5.341}$$

The total angular momentum of the system is

$$\vec{J} = \vec{j} + \vec{I}_n = \vec{l} + \vec{s} + \vec{I}_n \tag{5.342}$$

$$J^2 = l^2 + s^2 + I_n^2 + 2\vec{l} \cdot \vec{s} + 2\vec{l} \cdot \vec{I}_n + 2\vec{s} \cdot \vec{I}_n \tag{5.343}$$

Ignore the last two terms of the above formula, and let the above formula act on its eigenfunction; the following result can be obtained:

$$K_{\alpha nl}^J \equiv 2\vec{l} \cdot \vec{s} = \hat{\sigma} \cdot \vec{l} = J(J+1) - l(l+1) - \frac{3}{4} - I_n(I_n+1) \tag{5.344}$$

So Eq. (5.341) can be rewritten as

$$\vec{\sigma} \cdot \vec{p} = -\mathrm{i}\vec{\sigma} \cdot \vec{\nabla} = -\mathrm{i}(\vec{\sigma} \cdot \hat{r})\left(\frac{d}{dr} - \frac{K_{\alpha nl}^J}{r}\right) \tag{5.345}$$

We will substitute Eq. (5.345) into Eq. (5.331). First note

$$\left(\frac{\mathrm{d}}{\mathrm{d}r} - \frac{K_{\alpha' n' l'}^J}{r}\right)\left(\frac{\mathrm{e}^{\mathrm{i}k' r}}{r}\right) = \left(\mathrm{i}k' - \frac{1}{r} - \frac{K_{\alpha' n' l'}^J}{r}\right)\left(\frac{\mathrm{e}^{\mathrm{i}k' r}}{r}\right) \xrightarrow[r \to \infty]{} \mathrm{i}k'\left(\frac{\mathrm{e}^{\mathrm{i}k' r}}{r}\right) \tag{5.346}$$

and it is known that

$$\hat{\sigma} \cdot \hat{r} = \frac{1}{r}\hat{\sigma} \cdot \vec{r} = \frac{1}{r}\sum_{\eta}(-1)^{\eta}\hat{\sigma}_{\eta} r_{-\eta} \tag{5.347}$$

Equation (2.21) has given

$$\hat{\sigma}_{\eta}\chi_{\nu} = \sqrt{2}(-1)^{\frac{1}{2}-\nu} C_{\frac{1}{2}\ \nu+\eta\ \ \frac{1}{2}\ -\nu}^{1\ \eta}\chi_{\nu+\eta} \tag{5.348}$$

So we can get

$$(\hat{\sigma} \cdot \hat{r})\chi_{\nu} = \frac{1}{r}\sum_{\eta}(-1)^{\eta} r_{-\eta}\sqrt{2}(-1)^{\frac{1}{2}-\nu} C_{\frac{1}{2}\ \nu+\eta\ \ \frac{1}{2}\ -\nu}^{1\ \eta}\chi_{\nu+\eta} \tag{5.349}$$

Equation (3.942) has also given

$$rY_{1\eta}(\Omega) = \sqrt{\frac{3}{4\pi}} r_\eta \tag{5.350}$$

so there is

$$(\hat{\sigma} \cdot \hat{r})\chi_\nu = \sqrt{\frac{8\pi}{3}} \sum_\eta (-1)^{\frac{1}{2}-\nu+\eta} C^{1\ \ \eta}_{\frac{1}{2}\ \nu+\eta\ \frac{1}{2}\ -\nu} Y_{1\ -\eta}(\Omega)\chi_{\nu+\eta} \tag{5.351}$$

From Eq. (3.943) we have

$$\begin{aligned} Y_{lm_l} Y_{1\ -\eta} &= \sqrt{\frac{3}{4\pi}} \hat{l} \sum_L \frac{1}{\hat{L}} C^{L\ \ m_l-\eta}_{l\ m_l\ 1\ -\eta} C^{L\ 0}_{l0\ \ 10} Y_{L\ \ m_l-\eta} \\ &= \sqrt{\frac{3}{4\pi}} (-1)^{1-\eta} \sum_L C^{l\ m_l}_{1\ \eta\ \ L\ \ m_l-\eta} C^{L\ 0}_{l0\ \ 10} Y_{L\ \ m_l-\eta} \end{aligned} \tag{5.352}$$

Then we can obtain the following formula substituting Eqs. (5.345), (5.346), (5.351), and (5.352) into (5.331):

$$\begin{aligned} &\Psi^{(\mathrm{f})(\mathrm{d})}_{\alpha n\nu M_n} \xrightarrow[r\to\infty]{} \frac{\mathrm{i}\sqrt{\pi}}{k} \sqrt{\frac{E+m}{2m}} \sum_{\alpha' n' \overline{\nu}' M'_n} \frac{k'}{E+m} \frac{1}{\sqrt{v'}} \sum_{\substack{ljJM \\ l'j'm'_l m'_j}} \hat{l}\left(\delta_{\alpha' n' l' j', \alpha nlj} - S^{(\mathrm{d})J}_{\alpha' n' l' j', \alpha nlj}\right) \\ &C^{j\ \nu}_{l0\ \frac{1}{2}\nu} C^{J\ M}_{j\nu\ I_n M_n} C^{j'\ \ m'_j}_{l' m'_l\ \frac{1}{2}\overline{\nu}'} C^{J\ M}_{j' m'_j\ I'_n M'_n} \sqrt{2} \sum_\eta (-1)^{\frac{1}{2}+\overline{\nu}'} C^{1\ \ \eta}_{\frac{1}{2}\ \overline{\nu}'+\eta\ \frac{1}{2}\ -\overline{\nu}'} \sum_L C^{l'\ \ m'_l}_{1\ \eta\ \ L\ m'_l-\eta} \\ &C^{L\ 0}_{l'\ 0\ 1\ 0} Y_{L\ m'_l-\eta}(\theta, \varphi) \frac{\mathrm{e}^{\mathrm{i}k' r}}{r} \chi_{\overline{\nu}'+\eta} \Phi_{I'_n M'_n} \end{aligned} \tag{5.353}$$

Let

$$\nu' = \overline{\nu}' + \eta, \qquad \overline{\nu}' = \nu' - \eta \tag{5.354}$$

then Eq. (5.353) can be rewritten as

$$\Psi^{(\mathrm{f})(\mathrm{d})}_{\alpha n\nu M_n} \xrightarrow[r\to\infty]{} \frac{\mathrm{i}\sqrt{\pi}}{k}\sqrt{\frac{E+m}{2m}} \sum_{\alpha' n'\nu' M'_n} \frac{\sqrt{2}\,k'}{E+m}\frac{1}{\sqrt{\nu'}} \sum_{\substack{ljJM \\ l'j'm'_l m'_j}} \hat{l}\left(\delta_{\alpha' n' l' j',\alpha n l j} - S^{(\mathrm{d})J}_{\alpha' n' l' j',\alpha n l j}\right)$$
$$C^{j\ \nu}_{l0\ \frac{1}{2}\nu} C^{J\ M}_{j\nu\ I_n M_n} C^{J\ M}_{j'\ m'_j\ I'_n M'_n} \sum_{\eta} (-1)^{\frac{1}{2}+\nu'-\eta} C^{j'\ m'_j}_{l' m'_l\ \frac{1}{2}\ \nu'-\eta} C^{\frac{1}{2}\ \ \eta}_{\frac{1}{2}\ \nu'\ \frac{1}{2}\ -\nu'+\eta}$$
$$\sum_L C^{l'\ m'_l}_{1\ \eta\ L\ m'_l-\eta} C^{L\ 0}_{l'\ 0\ 10} Y_{L\ m'_l-\eta}(\theta,\varphi)\frac{\mathrm{e}^{\mathrm{i}k' r}}{r}\chi_{\nu'}\Phi_{I'_n M'_n} \tag{5.355}$$

From the C-G coefficients in the above formula, we can see

$$\nu + M_n = m'_j + M'_n, \quad m'_j = m'_l + \nu' - \eta \tag{5.356}$$

and the following relations can be obtained:

$$\begin{aligned} m'_l - \eta = \nu + M_n - \nu' - M'_n, \quad \nu' - \eta = \nu + M_n - m'_l - M'_n, \\ \eta = m'_l + \nu' + M'_n - \nu - M_n \end{aligned} \tag{5.357}$$

So Eq. (5.355) can be rewritten as

$$\Psi^{(\mathrm{f})(\mathrm{d})}_{\alpha n\nu M_n} \xrightarrow[r\to\infty]{} \frac{\mathrm{i}\sqrt{\pi}}{k}\sqrt{\frac{E+m}{2m}} \sum_{\alpha' n'\nu' M'_n} \frac{\sqrt{2}k'}{E+m}\frac{1}{\sqrt{\nu'}} \sum_{\substack{ljJM \\ l'j'm'_l m'_j}} \hat{l}\left(\delta_{\alpha' n' l' j',\alpha n l j} - S^{(\mathrm{d})J}_{\alpha' n' l' j',\alpha n l j}\right)$$
$$C^{j\ \nu}_{l0\ \frac{1}{2}\nu} C^{J\ M}_{j\nu\ I_n M_n} C^{J\ M}_{j' m'_j\ I'_n M'_n} (-1)^{\frac{1}{2}+\nu+M_n-m'_l-M'_n} C^{j' m'_j}_{l' m'_l\ \frac{1}{2}\ \nu+M_n-m'_l-M'_n} C^{\frac{1}{2}\ \ m'_l+\nu'+M'_n-\nu-M_n}_{\frac{1}{2}\ \nu'\ \frac{1}{2}\ -\nu-M_n+m'_l+M'_n}$$
$$\sum_L C^{l'\ m'_l}_{1\ m'_l+\nu'+M'_n-\nu-M_n\ L\ \nu+M_n-\nu'-M'_n} C^{L\ 0}_{l' 0\ 10} Y_{L\ \nu+M_n-\nu'-M'_n}(\theta,\varphi)\frac{\mathrm{e}^{\mathrm{i}k' r}}{r}\chi_{\nu'}\Phi_{I'_n M'_n} \tag{5.358}$$

From the above formula the reaction amplitude of the lower component can be written as

$$f^{(\mathrm{d})}_{\alpha' n' \nu' M'_n,\ \alpha n \nu M_n} = \sqrt{\frac{E+m}{2m}}\frac{\sqrt{2}k'}{E+m}\frac{\mathrm{i}\sqrt{\pi}}{k}\sum_{\substack{ljJM\\ l'j'm'_l m'_j}} \hat{l}\left(\delta_{\alpha' n' l' j', \alpha n l j} - S^{(\mathrm{d})J}_{\alpha' n' l' j', \alpha n l j}\right) C^{j\ \nu}_{l0\ \frac{1}{2}\nu} C^{J\ M}_{j\nu\ I_n M_n}$$

$$C^{J\ M}_{j' m'_j\ I'_n M'_n}(-1)^{\frac{1}{2}+\nu+M_n-m'_l-M'_n} C^{j'\ m'_j}_{l' m'_l\ \frac{1}{2}\ \nu+M_n-m'_l-M'_n} C^{1\ \ m'_l+\nu'+M'_n-\nu-M_n}_{\frac{1}{2}\nu'\ \frac{1}{2}\ -\nu-M_n+m'_l+M'_n}$$

$$\sum_L C^{l'\ m'_l}_{1\ \ m'_l+\nu'+M'_n-\nu-M_n\ \ L\ \nu+M_n-\nu'-M'_n} C^{L\ 0}_{l'\ 0\ \ 1\ 0} \mathrm{Y}_{L\ \nu+M_n-\nu'-M'_n}(\theta, \varphi) \tag{5.359}$$

Equation (2.253) has given

$$\mathrm{Y}_{l' m'_l}(\theta, \varphi) = (-1)^{m'_l}\sqrt{\frac{2l'+1}{4\pi}}\sqrt{\frac{(l'-m'_l)!}{(l'+m'_l)!}}\mathrm{P}^{m'_l}_{l'}(\cos\theta)\mathrm{e}^{\mathrm{i}m'_l\varphi} \tag{5.360}$$

Introduce the following symbols for Eq. (5.322):

$$f^{(\mathrm{u})ll'J}_{\nu'\nu}(m'_l) = \hat{l}\hat{l}'\mathrm{e}^{\mathrm{i}(\sigma_l+\sigma_{l'})}\sqrt{\frac{(l'-m'_l)!}{(l'+m'_l)!}}\sum_{jj'} C^{j\ \nu}_{l0\ \frac{1}{2}\nu} C^{J\ M}_{j\nu\ I_n M_n}$$
$$C^{j'\ m'_j}_{l' m'_l\ \frac{1}{2}\nu'} C^{J\ M}_{j' m'_j\ I'_n M'_n}\left(\delta_{\alpha' n' l' j', \alpha n l j} - S^{(\mathrm{u})J}_{\alpha' n' l' j', \alpha n l j}\right) \tag{5.361}$$

and we can get the following relation based on the C-G coefficients in the above formula:

$$m'_l = \nu + M_n - \nu' - M'_n \tag{5.362}$$

So Eq. (5.322) can be rewritten as

$$f^{(\mathrm{u})}_{\alpha' n' \nu' M'_n, \alpha n \nu M_n} = \sqrt{\frac{E+m}{2m}}\Big[f_{\mathrm{C}}(\theta)\delta_{\alpha' n', \alpha n}\delta_{\nu' M'_n, \nu M_n}$$
$$+\frac{\mathrm{i}}{2k}(-1)^{m'_l}\sum_{ll'J} f^{(\mathrm{u})ll'J}_{\nu'\nu}(m'_l)\mathrm{P}^{m'_l}_{l'}(\cos\theta)\ \mathrm{e}^{\mathrm{i}m'_l\varphi}\Big] \tag{5.363}$$

Introduce the following symbols for Eq. (5.359):

$$f_{\nu'\nu}^{(\mathrm{d})ll'm_l'LJ}\left(\nu + M_n - \nu' - M_n'\right) = \hat{l}\hat{L}\sqrt{\frac{\left(L - \nu - M_n + \nu' + M_n'\right)!}{\left(L + \nu + M_n - \nu' - M_n'\right)!}}\sum_{jj'} C_{l0\ \frac{1}{2}\nu}^{j\ \nu} C_{j\nu\ I_n M_n}^{J\ M}$$
$$C_{j'm'_j\ I_n' M_n'}^{J\ M}\left(\delta_{\alpha' n' l' j', \alpha n l j} - S_{\alpha' n' l' j', \alpha n l j}^{(\mathrm{d})J}\right)(-1)^{\frac{1}{2}+\nu+M_n-m_l'-M_n'} C_{l'm'_l\ \frac{1}{2}\ \nu+M_n-m_l'-M_n'}^{j'\ m_j'}$$
$$C_{\frac{1}{2}\ \nu'\ \frac{1}{2}\ -\nu-M_n+m_l'+M_n'}^{1\ m_l'+\nu'+M_n'-\nu-M_n} C_{1\ m_l'+\nu'+M_n'-\nu-M_n\ L\ \nu+M_n-\nu'-M_n'}^{l'\ m_l'} C_{l'0\ 10}^{L\ 0} \tag{5.364}$$

So Eq. (5.359) can be rewritten as

$$f_{\alpha' n' \nu' M_n', \alpha n \nu M_n}^{(\mathrm{d})} = \sqrt{\frac{E+m}{2m}} \frac{\sqrt{2}\, k'}{E+m} \frac{\mathrm{i}}{2k} (-1)^{\nu+M_n-\nu'-M_n'}$$
$$\sum_{ll'm_l'LJ} f_{\nu'\nu}^{(\mathrm{d})ll'm_l'LJ}\left(\nu + M_n - \nu' - M_n'\right) \mathrm{P}_L^{\nu+M_n-\nu'-M_n'}(\cos\theta)\, \mathrm{e}^{\mathrm{i}\left(\nu+M_n-\nu'-M_n'\right)\varphi} \tag{5.365}$$

Referring to Eqs. (2.258)–(2.261), we introduce the following symbols according to Eqs. (5.363) and (5.365):

$$A_{M_n'M_n}^{(\mathrm{u})}(\theta) = \sqrt{\frac{E+m}{2m}}\left[f_{\mathrm{C}}(\theta)\, \delta_{\alpha' n', \alpha n}\delta_{M_n', M_n}\right.$$
$$\left. + \frac{\mathrm{i}}{2k}(-1)^{M_n-M_n'}\sum_{ll'J} f_{\frac{1}{2}\frac{1}{2}}^{(\mathrm{u})ll'J}\left(M_n - M_n'\right) \mathrm{P}_{l'}^{M_n-M_n'}(\cos\theta)\right] \tag{5.366}$$

$$B_{M_n'M_n}^{(\mathrm{u})}(\theta) = \sqrt{\frac{E+m}{2m}}\left[-\frac{1}{2k}(-1)^{M_n-M_n'-1}\right.$$
$$\left.\sum_{ll'J} f_{\frac{1}{2}\ -\frac{1}{2}}^{(\mathrm{u})ll'J}\left(M_n - M_n' - 1\right)\mathrm{P}_{l'}^{M_n-M_n'-1}(\cos\theta)\right] \tag{5.367}$$

$$C_{M_n'M_n}^{(\mathrm{u})}(\theta) = \sqrt{\frac{E+m}{2m}}\left[\frac{1}{2k}(-1)^{M_n-M_n'+1}\right.$$
$$\left.\sum_{ll'J} f_{-\frac{1}{2}\ \frac{1}{2}}^{(\mathrm{u})ll'J}\left(M_n - M_n' + 1\right)\mathrm{P}_{l'}^{M_n-M_n'+1}(\cos\theta)\right] \tag{5.368}$$

$$D_{M_n'M_n}^{(\mathrm{u})}(\theta) = \sqrt{\frac{E+m}{2m}}\left[f_{\mathrm{C}}(\theta)\delta_{\alpha' n', \alpha n}\delta_{M_n', M_n}\right.$$
$$\left. + \frac{\mathrm{i}}{2k}(-1)^{M_n-M_n'}\sum_{ll'J} f_{-\frac{1}{2}\ -\frac{1}{2}}^{(\mathrm{u})ll'J}\left(M_n - M_n'\right)\mathrm{P}_{l'}^{M_n-M_n'}(\cos\theta)\right] \tag{5.369}$$

$$A^{(\mathrm{d})}_{M_n' M_n}(\theta) = \sqrt{\frac{E+m}{2m}}\frac{\sqrt{2}k'}{E+m}\left[\frac{\mathrm{i}}{2k}(-1)^{M_n - M_n'} \sum_{ll'm_l'LJ} f^{(\mathrm{d})ll'm_l'LJ}_{\frac{1}{2}\frac{1}{2}}\left(M_n - M_n'\right) \mathrm{P}_L^{M_n - M_n'}(\cos\theta)\right] \tag{5.370}$$

$$B^{(\mathrm{d})}_{M_n' M_n}(\theta) = \sqrt{\frac{E+m}{2m}}\frac{\sqrt{2}k'}{E+m}\left[-\frac{1}{2k}(-1)^{M_n - M_n' - 1} \sum_{ll'm_l'LJ} f^{(\mathrm{d})ll'm_l'LJ}_{\frac{1}{2}\ -\frac{1}{2}}(M_n - M_n' - 1)\mathrm{P}_L^{M_n - M_n' - 1}(\cos\theta)\right] \tag{5.371}$$

$$C^{(\mathrm{d})}_{M_n' M_n}(\theta) = \sqrt{\frac{E+m}{2m}}\frac{\sqrt{2}k'}{E+m}\left[\frac{1}{2k}(-1)^{M_n - M_n' + 1} \sum_{ll'm_l'LJ} f^{(\mathrm{d})ll'm_l'LJ}_{-\frac{1}{2}\ \frac{1}{2}}(M_n - M_n' + 1)\mathrm{P}_L^{M_n - M_n' + 1}(\cos\theta)\right] \tag{5.372}$$

$$D^{(\mathrm{d})}_{M_n' M_n}(\theta) = \sqrt{\frac{E+m}{2m}}\frac{\sqrt{2}k'}{E+m}\left[\frac{\mathrm{i}}{2k}(-1)^{M_n - M_n'} \sum_{ll'm_l'LJ} f^{(\mathrm{d})ll'm_l'LJ}_{-\frac{1}{2}\ -\frac{1}{2}}(M_n - M''_n)\mathrm{P}_L^{M_n - M_n'}(\cos\theta)\right] \tag{5.373}$$

Note that there is a coefficient $\frac{\sqrt{2}k'}{E+m}$ in all matrix elements of the lower component of the reaction amplitude given by Eqs. (5.370)–(5.373). When the incident particle energy is very low, this coefficient is very small. So it can be seen that the contribution of the lower component can be ignored in very low energy case. Since when the incident particle energy is very low, the coefficient $\sqrt{\frac{E+m}{2m}} \to 1$, so the matrix elements of the upper component of the reaction amplitude given by Eqs. (5.366)–(5.369) will automatically degenerate into the non-relativistic reaction amplitude matrix elements given by Eqs. (2.258)–(2.261) in very low energy case.

It can be seen that the upper and lower components in the spin space are incoherent when determining the polarization physical quantities based on the expression form of the four-dimensional spin operator $\vec{S}$ given by Eq. (5.71). In the case of omitting the footnote $M_n{}'M_n$ and the argument $(\theta)$ or $(\theta, \varphi)$, starting from Eqs. (5.363) and (5.365), the following matrices of reaction amplitude with definite $M_n{}'M_n$ can be written for the upper component and the lower component, respectively:

$$\hat{F}^{(\mathrm{u})} = \begin{pmatrix} A^{(\mathrm{u})} & -\mathrm{i}B^{(\mathrm{u})}\mathrm{e}^{-\mathrm{i}\varphi} \\ \mathrm{i}C^{(\mathrm{u})}\mathrm{e}^{\mathrm{i}\varphi} & D^{(\mathrm{u})} \end{pmatrix} \mathrm{e}^{\mathrm{i}(M_n - M'_n)\varphi} \tag{5.374}$$

$$\hat{F}^{(\mathrm{d})} = \begin{pmatrix} A^{(\mathrm{d})} & -\mathrm{i}B^{(\mathrm{d})}e^{-\mathrm{i}\varphi} \\ iC^{(\mathrm{d})}e^{i\varphi} & D^{(\mathrm{d})} \end{pmatrix} \mathrm{e}^{\mathrm{i}(M_n - M'_n)\varphi} \tag{5.375}$$

If the $y$-axis is selected in the direction of the unit vector $\vec{n}$ perpendicular to the reaction plane, then there is $\varphi = 0$, and the above two formulas can be simplified to

$$\hat{F}^{(\mathrm{u})} = \begin{pmatrix} A^{(\mathrm{u})} & -\mathrm{i}B^{(\mathrm{u})} \\ \mathrm{i}C^{(\mathrm{u})} & D^{(\mathrm{u})} \end{pmatrix} \tag{5.376}$$

$$\hat{F}^{(\mathrm{d})} = \begin{pmatrix} A^{(\mathrm{d})} & -\mathrm{i}B^{(\mathrm{d})} \\ \mathrm{i}C^{(\mathrm{d})} & D^{(\mathrm{d})} \end{pmatrix} \tag{5.377}$$

The following relations can be proved by means of the method to prove Eqs. (2.295) and (2.296) in Sect. 2.7:

$$D^{(\mathrm{u})}_{M'_n M_n}(\theta) = (-1)^{\Pi + \Lambda + I_n + I'_n + M_n - M'_n} A^{(\mathrm{u})}_{-M'_n\ -M_n}(\theta) \tag{5.378}$$

$$C^{(\mathrm{u})}_{M'_n M_n}(\theta) = (-1)^{\Pi + \Lambda + I_n + I'_n + M_n - M'_n} B^{(\mathrm{u})}_{-M'_n\ -M_n}(\theta) \tag{5.379}$$

Note the following two parameters defined by Eqs. (2.290) and (2.291) are used in the above relations:

$$\Pi = \begin{cases} 0 & \text{when } \pi_i \pi_{I_n} = \pi'_i \pi'_{I_n} \\ 1 & \text{when } \pi_i \pi_{I_n} = -\pi'_i \pi'_{I_n} \end{cases} \tag{5.380}$$

$$\Lambda = \begin{cases} 0 & \text{when } I_n \text{ and } I'_n \text{ are all integers} \\ 1 & \text{when } I_n \text{ and } I'_n \text{ are all semi-odd numbers} \end{cases} \tag{5.381}$$

Below we analyze the C-G coefficients in the lower component of the reaction amplitude given by Eq. (5.359). First we get

$$C^{j\ \nu}_{l0\ \frac{1}{2}\nu} = (-1)^{l+\frac{1}{2}-j} C^{j\ -\nu}_{l0\ \frac{1}{2}\ -\nu}, \quad C^{J\ M}_{j\nu\ I_n M_n} = (-1)^{j+I_n-J} C^{J\ -M}_{j\ -\nu\ I_n\ -M_n},$$

$$C^{J\ M}_{j'\ m'_j\ I'_n M'_n} = (-1)^{j'+I'_n-J} C^{J\ -M}_{j'\ -m'_j\ I'_n\ -M'_n},$$

$$(-1)^{\frac{1}{2}+\nu+M_n-m'_l-M'_n} = (-1)^{2(\nu+M_n-m'_l-M'_n)}(-1)^{\frac{1}{2}-\nu-M_n+m'_l+M'_n} = -(-1)^{\frac{1}{2}-\nu-M_n+m'_l+M'_n}$$

$$C^{j'\ m'_j}_{l'm'_l\ \frac{1}{2}\ \nu+M_n-m'_l-M'_n} = (-1)^{l'+\frac{1}{2}-j'} C^{j'\ -m'_j}_{l'\ -m'_l\ \frac{1}{2}\ -\nu-M_n+m'_l+M'_n},$$

$$C^{1\ \ m'_l+\nu'+M'_n-\nu-M_n}_{\frac{1}{2}\ \nu'\ \frac{1}{2}\ -\nu-M_n+m'_l+M'_n} = C^{1\ \ -m'_l-\nu'-M'_n+\nu+M_n}_{\frac{1}{2}\ -\nu'\ \frac{1}{2}\ \nu+M_n-m'_l-M'_n},$$

$$C^{l'\ m'_l}_{1\ m'_l+\nu'+M'_n-\nu-M_n\ L\ \nu+M_n-\nu'-M'_n} = (-1)^{1+L-l'} C^{l'\ -m'_l}_{1\ -m'_l-\nu'-M'_n+\nu+M_n\ L\ -\nu-M_n+\nu'+M'_n}$$

$$Y_{L\ \nu+M_n-\nu'-M'_n}(\theta,0) = (-1)^{\nu+M_n-\nu'-M'_n} Y_{L\ -\nu-M_n+\nu'+M'_n}(\theta,0)$$

In this way we can see that if we change the sign of all the spin magnetic quantum numbers in Eq. (5.359) in $\varphi = 0$ case, what phase factor compared to the original formula is going to be present

$$\begin{aligned}&(-1)^{l+\frac{1}{2}-j+j+I_n-J+j'+I'_n-J+1+l'+\frac{1}{2}-j'+1+L-l'+\nu+M_n-\nu'-M'_n}\\ &= (-1)^{l+I_n-J+I'_n-J+1+L+\nu+M_n-\nu'-M'_n} = (-1)^{2J-1+l+I_n+I'_n+L+\nu+M_n-\nu'-M'_n}\end{aligned}$$

It can be seen that for the lower component, the two parameters defined by Eqs. (2.290) and (2.291) can also be introduced, but it should be noted that the orbital angular momentum of the emitted particles is $L$ instead of $l'$. Then the following relations can be obtained for the lower component:

$$D^{(\mathrm{d})}_{M'_nM_n}(\theta) = (-1)^{\Pi+\Lambda+I_n+I'_n+M_n-M'_n} A^{(\mathrm{d})}_{-M'_n\ -M_n}(\theta) \tag{5.382}$$

$$C^{(\mathrm{d})}_{M'_nM_n}(\theta) = (-1)^{\Pi+\Lambda+I_n+I'_n+M_n-M'_n} B^{(\mathrm{d})}_{-M'_n\ -M_n}(\theta) \tag{5.383}$$

They are similar with Eq. (5.378) and Eq. (5.379), respectively.

When the incident particle, target, and the residual nucleus are all unpolarized, refer to Eqs. (2.271), (2.278), and (2.279), the following differential cross section of the outgoing particle can be obtained by using Eqs. (5.378), (5.379), (5.382), and (5.383) as follows:

$$I^0_{\alpha'n',\alpha n} = \frac{1}{2I_n+1}\sum_{\rho M_nM'_n}\frac{1}{2}\mathrm{tr}\left\{\hat{F}^{(\rho)}_{M'_nM_n}\hat{F}^{(\rho)+}_{M'_nM_n}\right\} = \frac{1}{2I_n+1}\sum_{\rho M_nM'_n} I^{(\rho)}_{0,M'_nM_n} \tag{5.384}$$

$$I^{(\rho)}_{0,M'_nM_n} = \left|A^{(\rho)}_{M'_nM_n}\right|^2 + \left|B^{(\rho)}_{M'_nM_n}\right|^2, \quad \rho = \mathrm{u,d} \tag{5.385}$$

The component of the polarization vector of the outgoing particles corresponding to the unpolarized incident particles is

$$P^{0\ i}_{\alpha'n',\alpha\ n} = \frac{1}{2I_n+1}\sum_{\rho M_nM'_n}\frac{1}{2}\mathrm{tr}\left\{\hat{\sigma}_i\hat{F}^{(\rho)}_{M'_nM_n}\hat{F}^{(\rho)+}_{M'_nM_n}\right\}, \quad i = x,y,z \tag{5.386}$$

From Eq. (2.303) we obtain

$$P^{0x}_{\alpha' n', \alpha n} = 0, \quad P^{0y}_{\alpha' n', \alpha n} = P_2, \quad P^{0z}_{\alpha' n', \alpha n} = 0 \tag{5.387}$$

$$P_2 = \frac{1}{I^0_{\alpha' n', \alpha n}} \frac{2}{2I_n + 1} \sum_{\rho M_n M'_n} \mathrm{Re}\left( A^{(\rho)}_{M'_n M_n} C^{(\rho)*}_{M'_n M_n} \right) \tag{5.388}$$

In the case where the target and the residual nucleus are unpolarized and the incident particle is polarized, the analyzing power of the incident particles is

$$A^i_{\alpha' n', \alpha n} = \frac{1}{I^0_{\alpha' n', \alpha n}} \frac{1}{2I_0 + 1} \sum_{\rho M_n M'_n} \frac{1}{2} \mathrm{tr}\left\{ \hat{F}^{(\rho)}_{M'_n M_n} \hat{\sigma}_i \hat{F}^{(\rho)+}_{M'_n M_n} \right\}, \quad i = x,y,z \tag{5.389}$$

So from Eq. (2.301) we can obtain

$$A^x_{\alpha' n', \alpha n} = 0, \quad A^y_{\alpha' n', \alpha n} = P_1, \quad A^z_{\alpha' n', \alpha n} = 0 \tag{5.390}$$

$$P_1 = \frac{1}{I^0_{\alpha' n', \alpha n}} \frac{2}{2I_n + 1} \sum_{\rho M_n M'_n} \mathrm{Re}\left( A^{(\rho)}_{M'_n M_n} B^{(\rho)*}_{M'_n M_n} \right) \tag{5.391}$$

The differential cross section of the outgoing particles is

$$I_{\alpha' n', \alpha n} = I^0_{\alpha' n', \alpha n} \left( 1 + p_y P_1 \right) \tag{5.392}$$

where $p_i (i = x, y, z)$ is the component of the initial polarization vector of the incident particles.

The polarization transfer coefficient of the outgoing particles is defined as

$$K^j_i = \frac{1}{I^0_{\alpha' n', \alpha n}} \frac{1}{2I_0 + 1} \sum_{\rho M_n M'_n} \frac{1}{2} \mathrm{tr}\left\{ \hat{\sigma}_j \hat{F}^{(\rho)}_{M'_n M_n} \hat{\sigma}_i \hat{F}^{(\rho)+}_{M'_n M_n} \right\}, \quad i,j = x, y, z \tag{5.393}$$

According to the results given in Sect. 2.7 we have

$$\begin{aligned} &K^x_x = W_2, \quad &K^y_x = 0, \quad &K^z_x = -Q_1, \\ &K^x_y = 0, \quad &K^y_y = J_0, \quad &K^z_y = 0, \\ &K^x_z = Q_2, \quad &K^y_z = 0, \quad &K^z_z = W_1 \end{aligned} \tag{5.394}$$

$$Q_1 = \frac{1}{I^0_{\alpha' n', \alpha n}} \frac{2}{2I_n + 1} \sum_{\rho M_n M'_n} \mathrm{Im}\left( A^{(\rho)}_{M'_n M_n} B^{(\rho)*}_{M'_n M_n} \right) \tag{5.395}$$

$$Q_2 = \frac{1}{I^0_{\alpha' n', \alpha n}} \frac{2}{2I_n + 1} \sum_{\rho M_n M'_n} \mathrm{Im}\left( A^{(\rho)}_{M'_n M_n} C^{(\rho)*}_{M'_n M_n} \right) \tag{5.396}$$

$$W_1 = \frac{1}{I^0_{\alpha'n',\alpha n}} \frac{1}{2I_n+1} \sum_{\rho M_n M'_n} \left( \left| A^{(\rho)}_{M'_n M_n} \right|^2 - \left| B^{(\rho)}_{M'_n M_n} \right|^2 \right) \tag{5.397}$$

$$W_2 = \frac{1}{I^0_{\alpha'n',\alpha n}} \frac{1}{2I_n+1} \sum_{\rho M_n M'_n} \mathrm{Re} \left( A^{(\rho)}_{M'_n M_n} D^{(\rho)*}_{M'_n M_n} - B^{(\rho)}_{M'_n M_n} C^{(\rho)*}_{M'_n M_n} \right) \tag{5.398}$$

$$J_0 = \frac{1}{I^0_{\alpha'n',\alpha n}} \frac{1}{2I_n+1} \sum_{\rho M_n M'_n} \mathrm{Re} \left( A^{(\rho)}_{M'_n M_n} D^{(\rho)*}_{M'_n M_n} + B^{(\rho)}_{M'_n M_n} C^{(\rho)*}_{M'_n M_n} \right) \tag{5.399}$$

The components of the polarization vector of the outgoing particles are

$$P^x_{\alpha'n',\alpha n} = \frac{I^0_{\alpha'n',\alpha n}}{I_{\alpha'n',\alpha n}} (p_x W_2 + p_z Q_2) \tag{5.400}$$

$$P^y_{\alpha'n',\alpha n} = \frac{I^0_{\alpha'n',\alpha n}}{I_{\alpha'n',\alpha n}} (P_2 + p_y J_0) \tag{5.401}$$

$$P^z_{\alpha'n',\alpha n} = \frac{I^0_{\alpha'n',\alpha n}}{I_{\alpha'n',\alpha n}} (-p_x Q_1 + p_z W_1) \tag{5.402}$$

This section develops the Dirac $S$ matrix theory of the relativistic nuclear reactions. In the case of the unpolarized target and residual nucleus which spins are not equal to 0, this section also gives the formulas for calculating the differential cross section and the polarization physical quantities of the relativistic two-body nuclear reactions of spin $\frac{1}{2}$ incident particles and various possible spin $\frac{1}{2}$ emitted particles. The above relativistic theories and calculation formulas will automatically degenerate into corresponding non-relativistic theories and calculation formulas when the incident particle energy is very low.

## 5.5 Dirac Coupling Channel Theory Including Elastic Scattering and Collective Inelastic Scattering Channels

For collective deformation nuclei, refer to Eq. (5.234) not including the Coulomb potential $V_C$, the Dirac equation describing the nucleon scattering process can be written as [29–33]

$$\left[ \vec{\alpha} \cdot \vec{p} + \beta(m + U_s) + U_0 + H_T \right] \psi = E\psi \tag{5.403}$$

where $m$ is the nucleon stationary mass; $U_s$ and $U_0$ are the complex nuclear potentials; $H_T$ is a non-relativistic internal motion Hamiltonian of the target nucleus

$$H_{\mathrm{T}}\Phi_{I_nM_n} = \varepsilon_n\Phi_{I_nM_n} \tag{5.404}$$

The scalar potential $U_{\mathrm{s}}$ and the time-like component $U_0$ of the 4 vector potential appearing in Eq. (5.403) can take the phenomenological potential of the Woods-Saxon form as

$$U_j = V_j f\left(r, R_{j1}, a_{j1}\right) + \mathrm{i}\, W_j f\left(r, R_{j2}, a_{j2}\right), \quad j = \mathrm{s}, 0 \tag{5.405}$$

where

$$f\left(r, R_{jn}, a_{jn}\right) = \left[1 + \exp\left(\frac{r - R_{jn}}{a_{jn}}\right)\right]^{-1}, \quad j = \mathrm{s}, 0 \tag{5.406}$$

In the above formulas, $n = 1$ and 2 correspond to the optical potential real and imaginary parts, respectively. Of course, other forms of the phenomenological optical potential can also be included in Eq. (5.405), for example, the surface type optical potential obtained by $\frac{\mathrm{d}}{\mathrm{d}x}\left(\frac{1}{1+\mathrm{e}^x}\right)$ can also be included.

The optical potential of the nucleus is the potential field generated by the nucleus itself. If the density distribution of the nucleus deviates from the sphere, the corresponding optical potential will also deviate from the spherical symmetry. Here we assume that $U_{\mathrm{s}}$ and $U_0$ are deformed potentials.

The most common form of the deformed nucleus radius is

$$R_j \to R_j\left[1 + \sum_{\lambda\mu}\alpha_{\lambda\mu}\mathrm{Y}_{\lambda\mu}(\theta', \varphi')\right] \tag{5.407}$$

where $\theta'$ and $\varphi'$ are the direction angles in the coordinate system fixed on the target nucleus. Here we assume that the target nucleus is an axis-symmetric rotational nucleus, so from Eq. (5.407) we can get

$$R_{jn} \to R_{jn}\left[1 + \sum_{\lambda}\beta_\lambda \mathrm{Y}_{\lambda 0}(\theta', 0)\right], \quad n = 1, 2 \tag{5.408}$$

where $\beta_\lambda$ is the $\lambda$ order deformation parameter of the target nucleus. The nuclear potential appearing in the Dirac equation (5.234) or (5.403) is

$$U_{\mathrm{D}} = \gamma^0 U_{\mathrm{s}} + U_0 \tag{5.409}$$

When the target nucleus radius $R_{jn}$ appearing in Eq. (5.405) is transformed based on Eq. (5.408), from Eq. (5.409) we can obtain

$$U_{\mathrm{D}} \rightarrow \overline{U}_{\mathrm{D}} + \Delta U_{\mathrm{D}} \tag{5.410}$$

where $\overline{U}_{\mathrm{D}}$ represents the spherical potential that does not contain $\beta_\lambda$ items, $\Delta U_{\mathrm{D}}$ represents the nuclear potential contributed by the deformation items containing $\beta_\lambda$ in Eq. (5.408), and we can write its expression according to Eq. (5.405) as follows:

$$\Delta U_{\mathrm{D}} = \left(\sum_{\lambda} \beta_\lambda \mathrm{Y}_{\lambda 0}(\theta',0)\right)\left(\gamma^0 F_{\mathrm{s}}(r) + F_0(r)\right) \tag{5.411}$$

$$F_j(r) = V_j R_{j1} \frac{\partial}{\partial R_{j1}} f\left(r, R_{j1}, a_{j1}\right) + \mathrm{i}\, W_j R_{j2} \frac{\partial}{\partial R_{j2}} f\left(r, R_{j2}, a_{j2}\right), \quad j = \mathrm{s}, 0 \tag{5.412}$$

Due to the existence of the deformation potential energy term $\Delta U_{\mathrm{D}}$, it is possible for the incident nucleon to occur the inelastic scattering with the target nucleus. In the above theoretical formulas related to optical potential, the microscopic optical potential obtained from the theories of nuclear force or meson exchange can also be used.

Substituting Eq. (5.410) into Eq. (5.403) and replacing $\overline{U}$ with a spherical symmetric potential $U(r)$ we can obtain

$$\left[\vec{\alpha} \cdot \vec{p} + \beta(m + U_{\mathrm{s}}(r)) + U_0(r) + H_{\mathrm{T}} + \Delta U_{\mathrm{D}}\right] \psi = E\psi \tag{5.413}$$

Let

$$\psi = \begin{pmatrix} \psi_{\mathrm{u}} \\ \psi_{\mathrm{d}} \end{pmatrix} \tag{5.414}$$

then utilizing Eq. (5.38), Eq. (5.413) can be rewritten as

$$\begin{pmatrix} \left(\hat{\sigma} \cdot \vec{p}\right) \psi_{\mathrm{d}} \\ \left(\hat{\sigma} \cdot \vec{p}\right) \psi_{\mathrm{u}} \end{pmatrix} + \begin{pmatrix} \left[m + U_{\mathrm{s}} + U_0 + \frac{\mu}{E} V_{\mathrm{C}} + H_{\mathrm{T}} + g(F_0 + F_s)\right] \psi_{\mathrm{u}} \\ \left[-m - U_{\mathrm{s}} + U_0 + H_T + g(F_0 - F_s)\right] \psi_{\mathrm{d}} \end{pmatrix} = E \begin{pmatrix} \psi_{\mathrm{u}} \\ \psi_{\mathrm{d}} \end{pmatrix} \tag{5.415}$$

Note that in the upper component equation, refer to the discussion made on the Coulomb potential and Eq. (5.287) in the 5.3 section, the spherical nucleus Coulomb potential $\frac{\mu}{E} V_{\mathrm{C}}$ is added side-by-side with $U_0$, where $\mu = \frac{mM}{m+M}$, $m$ and $M$ are the masses of the incident particle and target nucleus, respectively, and $\mu$ is called the reduced mass. And $g$ appeared in Eq. (5.415) is

$$g = \sum_{\lambda} \beta_{\lambda} \mathrm{Y}_{\lambda 0}(\theta', 0) \tag{5.416}$$

Equation (5.341) has given

$$\hat{\sigma} \cdot \vec{p} = -\mathrm{i}(\hat{\sigma} \cdot \hat{r}) \left( \frac{\mathrm{d}}{\mathrm{d}r} - \frac{\hat{\sigma} \cdot \vec{l}}{r} \right) \tag{5.417}$$

where $\hat{r} = \vec{r}/r$. So from Eq. (5.415) we can obtain

$$\mathrm{i}(\hat{\sigma} \cdot \hat{r}) \left( \frac{\mathrm{d}}{\mathrm{d}r} - \frac{\hat{\sigma} \cdot \vec{l}}{r} \right) \psi_{\mathrm{d}} + \left( E - H_T - m - U_s - U_0 - \frac{\mu}{E} V_C \right) \psi_{\mathrm{u}} = (F_0 + F_s) g \psi_{\mathrm{u}} \tag{5.418}$$

$$\mathrm{i}(\hat{\sigma} \cdot \hat{r}) \left( \frac{\mathrm{d}}{\mathrm{d}r} - \frac{\hat{\sigma} \cdot \vec{l}}{r} \right) \psi_u + (E - H_T + m + U_s - U_0)\, \psi_{\mathrm{d}} = (F_0 - F_s) g \psi_{\mathrm{d}} \tag{5.419}$$

The wave functions $\psi_{\mathrm{u}}$ and $\psi_{\mathrm{d}}$ with the total angular momentum $JM$ can be expressed, respectively, as

$$\psi_{\mathrm{u},\, JM}\left(\vec{r}, \xi\right) = \frac{1}{r} \sum_{nlj} u^J_{nlj}(r)\, \varphi^{JM}_{nlj}(\Omega, \xi), \quad \psi_{\mathrm{d},\, JM}\left(\vec{r}, \xi\right) = \frac{1}{r} \sum_{nlj} d^J_{nlj}(r)\, \varphi^{JM}_{nlj}(\Omega, \xi) \tag{5.420}$$

where $n$ is the energy level symbol of the target nucleus and $\varphi^{JM}_{nlj}$ has been given by Eq. (5.306) and Eq. (5.305). The $\alpha' = \alpha$ scattering processes will be discussed in this section.

Using Eq. (5.305), Eq. (5.351), and the first equation of Eq. (5.352), we can obtain

$$\begin{aligned} \mathrm{i}(\hat{\sigma} \cdot \hat{r})\, \mathbb{Y}_{l'j'm'_j}(\Omega) = {} & \sqrt{2}\, \mathrm{i}^{l'+1} \hat{l}' \sum_{\nu' m'_l} C^{j'\ m'_j}_{l'm'_l\ \frac{1}{2}\nu'} \sum_{\eta} (-1)^{\frac{1}{2}-\nu'+\eta} \\ & C^{1\ \ \eta}_{\frac{1}{2}\ \nu'+\eta\ \frac{1}{2}\ -\nu'} \sum_{L} \frac{1}{\hat{L}} C^{L\ \ m'_l-\eta}_{l'm'_l\ 1\ -\eta} C^{L\,0}_{l'0\ 10} \mathrm{Y}_{L\ m'_l-\eta}(\Omega) \chi_{\nu'+\eta} \end{aligned} \tag{5.421}$$

Note that

$$\chi^+_{\nu} \chi_{\nu'+\eta} = \delta_{\nu\ \nu'+\eta} \tag{5.422}$$

$$\int \mathrm{Y}_{lm_l}^{*} \mathrm{Y}_{L\ m_l'-\eta} \mathrm{d}\Omega = \delta_{lL} \delta_{m_l\ m_l'-\eta} \tag{5.423}$$

The following relations can be obtained by using Eqs. (5.421)–(5.423):

$$\eta = \nu - \nu' = m_l' - m_l, \qquad m_j' = m_l' + \nu' = m_l + \nu = m_j \tag{5.424}$$

Then we get

$$\begin{aligned} X \equiv \int \mathbb{Y}_{ljm_l}^{+} \mathrm{i}(\hat{\sigma}\cdot\hat{r}) \mathbb{Y}_{l'j'm_j'} \mathrm{d}\Omega = \sqrt{2}\, \mathrm{i}^{l'-l+1} \frac{\hat{l'}}{\hat{l}} \sum_{\nu\, m_l} C_{lm_l\ \frac{1}{2}\nu}^{j\ m_j} \delta_{m_j m_j'} \sum_{\nu' m_l'} C_{l'm'_l\ \frac{1}{2}\nu'}^{j'\ m_j} \\ \sum_{\eta} (-1)^{\frac{1}{2}-\nu'+\eta} C_{\frac{1}{2}\nu\ \frac{1}{2}\ -\nu'}^{1\ \eta} C_{l'm_l'\ 1\ -\eta}^{l\ m_l'-\eta} C_{l'0\ 10}^{l\ 0} \\ = \sqrt{3}\, \mathrm{i}^{l'-l+1} \frac{\hat{l'}}{\hat{l}} C_{l'0\ 10}^{l\ 0} \delta_{m_j m_j'} \sum_{\nu\nu' m_l m_l'} C_{lm_l\ \frac{1}{2}\nu}^{j\ m_j} C_{l'm'_l\ \frac{1}{2}\nu'}^{j'\ m_j} \sum_{\eta} C_{\frac{1}{2}\nu'\ 1\ -\eta}^{\frac{1}{2}\ \nu'} C_{l'm'_l\ 1\ -\eta}^{l\ m_l} \end{aligned} \tag{5.425}$$

Utilizing Eq. (1.63) we can obtain

$$\sum_{\eta} C_{l'm'_l\ 1\ -\eta}^{l\ m_l} C_{\frac{1}{2}\nu\ 1\ -\eta}^{\frac{1}{2}\ \nu'} = (-1)^{l+m_l} \hat{l} \sum_{t\ m_t} \hat{t}\, W\left(ll'\frac{1}{2}\frac{1}{2};1t\right) C_{l'm'_l\ \frac{1}{2}\nu'}^{t\ m_t} C_{l\ -m_l\ t\ m_t}^{\frac{1}{2}\ \nu} \tag{5.426}$$

When deriving the above formula Eq. (5.424) is used. We can also obtain

$$\sum_{m_l'\nu'} C_{l'm'_l\ \frac{1}{2}\nu'}^{j'\ m_j} C_{l'm'_l\ \frac{1}{2}\nu'}^{t\ m_t} = \delta_{tj'} \delta_{m_t m_j} \tag{5.427}$$

$$\sum_{m_l\nu} (-1)^{l+m_l} C_{lm_l\ \frac{1}{2}\nu}^{j\ m_j} C_{l\ -m_l\ j'\ m_j}^{\frac{1}{2}\ \nu} = (-1)^{l+\frac{1}{2}-j} \frac{\sqrt{2}}{\hat{j}} \delta_{jj'} \tag{5.428}$$

Then we get

$$X = \sqrt{6}\, \mathrm{i}^{l'-l+1} (-1)^{l+\frac{1}{2}-j} \hat{l'} C_{l'0\ 10}^{l\ 0} W\left(ll'\frac{1}{2}\frac{1}{2};1j\right) \delta_{jj'} \delta_{m_j m_j'} \tag{5.429}$$

Using the orthogonal normalization properties of the wave functions, we can obtain

$$\begin{aligned} \int \varphi_{nlj}^{JM+}(\Omega,\xi)\,\psi_{u,JM}\left(\vec{r},\xi\right)\mathrm{d}\Omega\mathrm{d}\xi &= \frac{1}{r}u_{nlj}^{J}(r), \\ \int \varphi_{nlj}^{JM+}(\Omega,\xi)\,\psi_{d,JM}(\bar{r},\xi)\mathrm{d}\Omega\mathrm{d}\xi &= \frac{1}{r}d_{nlj}^{J}(r) \end{aligned} \tag{5.430}$$

The following equation can be obtained utilizing Eqs. (5.417), (5.420), (5.306), (5.425), (5.429), and (5.344):

$$\begin{aligned} &\int \varphi_{nlj}^{JM+}(\Omega,\xi)\,\mathrm{i}(\hat{\sigma}\cdot\hat{r})\left(\frac{\mathrm{d}}{\mathrm{d}r}-\frac{\hat{\sigma}\cdot\vec{l}}{r}\right)\psi_{u,JM}\left(\vec{r},\xi\right)\mathrm{d}\Omega\mathrm{d}\xi \\ &=\sum_{l'j'}\left(\frac{\mathrm{d}}{\mathrm{d}r}-\frac{K_{nl'}^{J}}{r}\right)\left(\frac{u_{nl'j'}^{J}(r)}{r}\right)\sum_{m_jM_n}C_{jm_j\ I_nM_n}^{J\ M}C_{j'm_j\ I_nM_n}^{J\ M}X \\ &=\sum_{l'}B_{ll'j}\left(\frac{\mathrm{d}}{\mathrm{d}r}-\frac{K_{nl'}^{J}}{r}\right)\left(\frac{u_{nl'j}^{J}(r)}{r}\right) \end{aligned} \tag{5.431}$$

where

$$B_{ll'j}=\sqrt{6}\,\mathrm{i}^{l'-l+1}(-1)^{l+\frac{1}{2}-j}\,\hat{l}'\,C_{l'0\ \ 10}^{l\ 0}W\left(ll'\frac{1}{2}\frac{1}{2};1j\right) \tag{5.432}$$

In the same way we can obtain

$$\begin{aligned} &\int \varphi_{nlj}^{JM+}(\Omega,\xi)\,\mathrm{i}(\hat{\sigma}\cdot\hat{r})\left(\frac{\mathrm{d}}{\mathrm{d}r}-\frac{\hat{\sigma}\cdot\vec{l}}{r}\right)\psi_{d,JM}\left(\vec{r},\xi\right)\mathrm{d}\Omega\mathrm{d}\xi \\ &=\sum_{l'}B_{ll'j}\left(\frac{\mathrm{d}}{\mathrm{d}r}-\frac{K_{nl'}^{J}}{r}\right)\left(\frac{d_{nl'j}^{J}(r)}{r}\right) \end{aligned} \tag{5.433}$$

Equations (3.1215) and (3.1220) have given, respectively

$$\mathrm{Y}_{\lambda 0}(\theta',0)=\sum_{m_\lambda}(-1)^{m_\lambda}D_{m_\lambda 0}^{\lambda *}(\alpha,\beta,\gamma)\mathrm{Y}_{\lambda\ -m_\lambda}(\Omega) \tag{5.434}$$

$$\int \Phi_{I_nM_n}^{*}D_{m_\lambda 0}^{\lambda *}\Phi_{I_n'M_n'}\mathrm{d}\xi=\frac{\hat{I}_n'}{\hat{I}_n}C_{I_n'M_n'\ \lambda m_\lambda}^{I_n\ M_n}C_{I_n'0\ \ \lambda 0}^{I_n0} \tag{5.435}$$

And we can obtain

$$\int \mathrm{Y}^*_{lm_l}\mathrm{Y}_{\lambda\ -m_\lambda}\mathrm{Y}_{l'm'_l}\mathrm{d}\Omega = \frac{\hat{\lambda}\,\hat{l}'}{\sqrt{4\pi}\hat{l}} C^{l\ m_l}_{l'm'_l\ \lambda\ -m_\lambda} C^{l\ 0}_{l'0\ \lambda 0} \tag{5.436}$$

Then the following equation can be obtained utilizing Eqs. (5.420), (5.306), (5.416), and (5.434)–(5.436):

$$\begin{aligned} &\int \varphi^{JM+}_{nlj}(\Omega,\xi)\, g\psi_{\mathrm{u},JM}(\hat{r},\xi)\mathrm{d}\Omega\mathrm{d}\xi \\ &= \sum_\lambda \beta_\lambda \sum_{n'l'j'} \frac{1}{r} u^J_{n'l'j'}(r)\, \mathrm{i}^{l'-l} \sum_{m_jM_n} C^{J\ M}_{jm_j\ I_nM_n} \sum_{m'_jM'_n} C^{J\ M}_{j'm'_j\ I'_nM'_n} \sum_{m_lm'_l\nu} C^{j\ m_j}_{lm_l\ \frac{1}{2}\nu} \\ &\quad C^{j'\ m'_j}_{l'm'_l\ \frac{1}{2}\nu} \sum_{m_\lambda}(-1)^{m_\lambda} \frac{\hat{I}'_n}{\hat{I}_n} C^{I_nM_n}_{I'_nM'_n\ \lambda m_\lambda} C^{I_n0}_{I'_n0\ \lambda 0} \frac{\hat{\lambda}\hat{l}'}{\sqrt{4\pi}\hat{l}} C^{l\ m_l}_{l'm'_l\ \lambda\ -m_\lambda} C^{l\ 0}_{l'0\ \lambda 0} \end{aligned} \tag{5.437}$$

First using Eq. (3.14) we find

$$\begin{aligned} &(-1)^{m_\lambda} \sum_{m_lm'_l\nu} C^{j\ m_j}_{lm_l\ \frac{1}{2}\nu} C^{j'\ m'_j}_{l'm'_l\ \frac{1}{2}\nu} C^{l\ m_l}_{l'm'_l\ \lambda\ -m_\lambda} \\ &= (-1)^{l+l'-\lambda-j'+1+m'_j+m_\lambda} \frac{\hat{j}}{\hat{\lambda}} \hat{l}\hat{j}' C^{\lambda\ m_\lambda}_{j\ -m_j\ j'm'_j} W\left(j\frac{1}{2}\lambda l'; lj'\right) \\ &= (-1)^{l+\frac{1}{2}+m_j} \frac{\hat{j}\hat{j}'\hat{l}}{\hat{\lambda}} C^{\lambda\ m_\lambda}_{j\ -m_j\ j'm'_j} W\left(jlj'l'; \frac{1}{2}\lambda\right) \end{aligned} \tag{5.438}$$

Then utilizing Eq. (1.63) we can get

$$\begin{aligned} &\sum_{m_\lambda} C^{\lambda\ m_\lambda}_{j\ -m_j\ j'm'_j} C^{I_n\ M_n}_{I'_nM'_n\ \lambda m_\lambda} \\ &= (-1)^{I'_n+\lambda-I_n} \sum_{J'M'} \hat{\lambda}\hat{J}'\ W(jj'I_nI'_n; \lambda J')\ C^{J'\ M'}_{j'm'_j\ I'_nM'_n} C^{I_n\ M_n}_{j\ -m_j\ J'M'} \end{aligned} \tag{5.439}$$

We can also obtain

$$\sum_{m'_jM'_n} C^{J\ M}_{j'm'_j\ I'_nM'_n} C^{J'\ M'}_{j'm'_j\ I'_nM'_n} = \delta_{JJ'}\delta_{MM'} \tag{5.440}$$

$$\sum_{m_jM_n} C^{J\ M}_{jm_j\ I_nM_n} (-1)^{m_j} C^{I_nM_n}_{j\ -m_j\ J'M'} = (-1)^{I_n-J} \frac{\hat{I}_n}{\hat{J}} \delta_{JJ'}\delta_{MM'} \tag{5.441}$$

Thus the following equation can be obtained substituting Eqs. (5.438)–(5.441) into Eq. (5.437):

$$\begin{aligned}
&\int \varphi_{nlj}^{JM+}(\Omega,\xi)\, g\psi_{\mathrm{u},JM}\left(\vec{r},\xi\right)\mathrm{d}\Omega\mathrm{d}\xi \\
&= \sum_{\lambda}\beta_\lambda \sum_{n'l'j'} \frac{1}{r}\, u_{n'l'j'}^{J}(r)\mathrm{i}^{l'-l}\frac{\hat{I}'_n}{\hat{I}_n} C_{I'_n 0\ \lambda 0}^{I_n\ 0}\frac{\hat{\lambda}\hat{l}'}{\sqrt{4\pi}\hat{l}} C_{l'0\ \lambda 0}^{l\ 0}(-1)^{l'+\frac{1}{2}} \\
&\quad \frac{\hat{j}\hat{j}'\hat{l}}{\hat{\lambda}} W\left(jlj'l';\frac{1}{2}\lambda\right)(-1)^{I'_n+\lambda-I_n}\hat{\lambda}\hat{J}W(jj'I_nI'_n;\lambda J)(-1)^{I_n-J}\frac{\hat{I}_n}{\hat{J}} \\
&= \frac{1}{\sqrt{4\pi}}\sum_{\lambda}\beta_\lambda \sum_{n'l'j'} \frac{1}{r}\, u_{n'l'j'}^{J}(r)\mathrm{i}^{l'-l}(-1)^{\lambda+l'+I'_n+\frac{1}{2}-J} \\
&\quad \hat{\lambda}\hat{l}'\hat{j}\hat{j}'\hat{I}'_n C_{l'0\ \lambda 0}^{l\ 0} C_{I'_n 0\ \lambda 0}^{I_n\ 0} W\left(jj'll';\lambda\frac{1}{2}\right)W(jj'I_nI'_n;\lambda J) \\
&= \sum_{\lambda}\beta_\lambda \sum_{n'l'j'} V_{nlj,n'l'j'}^{\lambda J}\frac{1}{r}u_{n'l'j'}^{J}(r)
\end{aligned} \tag{5.442}$$

where

$$\begin{aligned}
V_{nlj,n'l'j'}^{\lambda J} = &\frac{1}{\sqrt{4\pi}}\mathrm{i}^{l'-l}(-1)^{I_n+\frac{1}{2}-J}\hat{l}\hat{l}'\hat{j}\hat{j}'\hat{I}'_n C_{l'0\ \ l0}^{\lambda\ 0} C_{I'_n 0\ \ \lambda 0}^{I_n\ 0} \\
&W\left(jj'll';\lambda\frac{1}{2}\right)W(jj'I_nI'_n;\lambda J)
\end{aligned} \tag{5.443}$$

In the same way we can get

$$\int \varphi_{nlj}^{JM+}(\Omega,\xi)\, g\psi_{\mathrm{d},JM}\left(\vec{r},\xi\right)\mathrm{d}\Omega\mathrm{d}\xi = \sum_{\lambda}\beta_\lambda \sum_{n'l'j'} V_{nlj,n'l'j'}^{\lambda J}\frac{1}{r}d_{n'l'j'}^{J}(r) \tag{5.444}$$

From Eq. (5.418) and Eq. (5.419) we can also obtain the following equations utilizing Eqs. (5.430), (5.431), (5.433), (5.442), and (5.444):

$$\begin{aligned}
&\sum_{l'} B_{ll'j} r\left(\frac{\mathrm{d}}{\mathrm{d}r} - \frac{K_{nl'}^{J}}{r}\right)\left(\frac{d_{nl'j}^{J}(r)}{r}\right) + \left(E - \varepsilon_n - m - U_\mathrm{s} - U_0 - \frac{\mu}{E}V_\mathrm{C}\right)u_{nlj}^{J}(r) \\
&= (F_0 + F_\mathrm{s})\sum_{\lambda}\beta_\lambda \sum_{n'l'j'} V_{nlj,n'l'j'}^{\lambda J}u_{n'l'j'}^{J}(r)
\end{aligned} \tag{5.445}$$

$$\begin{aligned}
&\sum_{l'} B_{ll'j} r\left(\frac{\mathrm{d}}{\mathrm{d}r} - \frac{K_{nl'}^{J}}{r}\right)\left(\frac{u_{nl'j}^{J}(r)}{r}\right) + (E - \varepsilon_n + m + U_\mathrm{s} - U_0)\, d_{nlj}^{J}(r) \\
&= (F_0 - F_\mathrm{s})\sum_{\lambda}\beta_\lambda \sum_{n'l'j'} V_{nlj,n'l'j'}^{\lambda J}d_{n'l'j'}^{J}(r)
\end{aligned} \tag{5.446}$$

Let $r = r_m$ be the outer boundary to find numerical solution of the coupling channel equation of the above radial wave function, when $r > r_m$ the nuclear potential can be ignored. We stipulate that the symbol "′" represents the derivative for $k'r$. Notice here $\alpha' = \alpha$ is no longer marked for scattering process. According to Eq. (5.310), we can see that the radial wave function of the upper component satisfies the following boundary condition:

$$\left.\frac{\left(u^J_{n'l'j'}(r)\right)'}{u^J_{n'l'j'}(r)}\right|_{r_m} = \left.\frac{\left(G'_{l'}(k'r) - \mathrm{i}F'_{l'}(k'r)\right)\delta_{n'l'j',nlj} - S^{(\mathrm{u})J}_{n'l'j',nlj}\left(G'_{l'}(k'r) + \mathrm{i}F'_{l'}(k'r)\right)}{\left(G_{l'}(k'r) - \mathrm{i}F_{l'}(k'r)\right)\delta_{n'l'j',nlj} - S^{(\mathrm{u})J}_{n'l'j',nlj}\left(G_{l'}(k'r) + \mathrm{i}F_{l'}(k'r)\right)}\right|_{r_m} \tag{5.447}$$

According to Eq. (5.325), the radial wave function of the lower component satisfies the following boundary condition:

$$\left.\frac{\left(d^J_{n'l'j'}(r)\right)'}{d^J_{n'l'j'}(r)}\right|_{r_m} = \left.\frac{\left[-(k'r\mathrm{n}_{l'}(k'r))' - \mathrm{i}(k'r\mathrm{j}_{l'}(k'r))'\right]\delta_{n'l'j',nlj} - S^{(\mathrm{d})J}_{n'l'j',nlj}\left[-(k'r\mathrm{n}_{l'}(k'r))' + \mathrm{i}(k'r\mathrm{j}_{l'}(k'r))'\right]}{\left[-(k'r\mathrm{n}_{l'}(k'r)) - \mathrm{i}(k'r\mathrm{j}_{l'}(k'r))\right]\delta_{n'l'j',nlj} - S^{(\mathrm{d})J}_{n'l'j',nlj}\left[-(k'r\mathrm{n}_{l'}(k'r)) + \mathrm{i}(k'r\mathrm{j}_{l'}(k'r))\right]}\right|_{r_m} \tag{5.448}$$

From the similarity of the lower component equation (5.295) and the upper component equation (5.294), the similarity between the lower component boundary condition (5.448) and the upper component boundary condition (5.447) can be understood. Thus using the various $S$ matrix elements $S^{(\rho)J}_{n'l'j',nlj}(\rho = \mathrm{u}, \mathrm{d})$ obtained from the above boundary conditions and the calculation formulas given in the 5.4 section, the elastic and inelastic scattering cross sections, differential cross sections, and various polarization physical quantities of spin $\frac{1}{2}$ particles can be calculated.

## 5.6 Relativistic Collective Deformation RDWBA Method and Calculation of Nucleon Polarization Quantities

Equations (3.12.57), (3.12.56), and (7.2.40) in Ref. [4] give the following differential cross section formulas in the distorted wave Born approximation (DWBA) theory for a + A → b + B reaction:

$$\frac{\mathrm{d}\sigma}{\mathrm{d}\Omega} = \frac{1}{\hat{i}_a^2 \hat{I}_A^2}\sum_{m_a m_A m_b m_B}\left(\frac{\mathrm{d}\sigma}{\mathrm{d}\Omega}\right)_{\beta\alpha} \tag{5.449}$$

$$\left(\frac{\mathrm{d}\sigma}{\mathrm{d}\Omega}\right)_{\beta\alpha}=\frac{\mu_\alpha\mu_\beta}{\left(2\pi\hbar^2\right)^2}\frac{k_\beta}{k_\alpha}\left|\mathcal{T}_{\beta\alpha}\right|^2 \tag{5.450}$$

$$\begin{aligned}\mathcal{T}_{\beta\alpha}=&\int\chi_\beta^{(-)*}\left(\vec{k}_\beta,\vec{r}_\beta\right)\Phi_\beta^*(\xi_\beta)\\&\left[U_\beta\left(\vec{r}_\beta,\ \xi_\beta\right)-\overline{U}_\beta\left(\vec{r}_\beta\right)\right]\Phi_\alpha(\xi_\alpha)\chi_\alpha^{(+)}\left(\vec{k}_\alpha,\vec{r}_\alpha\right)\mathrm{d}\vec{r}_\beta\mathrm{d}\xi_\beta\end{aligned} \tag{5.451}$$

where $\alpha$ and $\beta$ represent the incident channel and the outgoing channel, respectively. For the scattering process the target nucleus is the same as the residual nucleus, so $\vec{r}_\alpha=\vec{r}_\beta=\vec{r}$ can be taking. For the inelastic scattering reaction of the rotational nucleus, $\xi_\alpha$ and $\xi_\beta$ are the intrinsic coordinate Euler angles, and there are $\mathrm{d}\xi=\ \sin\beta\mathrm{d}\beta\mathrm{d}\alpha\mathrm{d}\gamma$ and

$$U_\beta\left(\vec{r}_\beta,\xi_\beta\right)-\overline{U}_\beta\left(\vec{r}_\beta\right)=\Delta U_\mathrm{D} \tag{5.452}$$

For the axis-symmetric rotational nuclei, $\Delta U_\mathrm{D}$ has been given by Eq. (5.411). Assume that the target nucleus and the residual nucleus are in $n$ and $n'$ states of the ground state rotation band, respectively, their angular momentum are $I_n$ and $I'_{n'}$, respectively, then the corresponding rotational nucleus eigenfunctions are, respectively

$$\Phi_{I_nM_n}=\sqrt{\frac{2I_n+1}{8\pi^2}}D_{M_n0}^{I_n*} \tag{5.453}$$

$$\Phi_{I'_nM'_n}=\sqrt{\frac{2I'_n+1}{8\pi^2}}D_{M'_n0}^{I'_n*} \tag{5.454}$$

The above two formulas show that the theory discussed in this section only applies to even $A$ nuclei, which $I_n$ and $I'_n$ are integers. Eqs. (3.1215), (3.1219), and (3.1220) have given

$$\mathrm{Y}_{\lambda0}(\theta',0)=\sum_{M_\lambda}(-1)^{M_\lambda}D_{M_\lambda0}^{\lambda*}(\alpha,\beta,\gamma)\mathrm{Y}_{\lambda\ -M_\lambda}(\Omega_r) \tag{5.455}$$

$$\int D_{M'_n0}^{I'_n}D_{M_\lambda0}^{\lambda*}D_{M_n0}^{I_n*}\sin\beta\mathrm{d}\beta\mathrm{d}\alpha\mathrm{d}\gamma=\frac{8\pi^2}{\hat{I}_n'^2}C_{I_nM_n\ \ \lambda M_\lambda}^{I'_nM'_n}C_{I_n0\ \ \lambda0}^{I'_n0} \tag{5.456}$$

$$\int\Phi_{I'_nM'_n}^*D_{M_\lambda0}^{\lambda*}\Phi_{I_nM_n}\mathrm{d}\xi=\frac{\hat{I}_n}{\hat{I}'_n}C_{I_nM_n\ \ \lambda M_\lambda}^{I'_nM'_n}C_{I_n0\ \ \lambda0}^{I'_n0} \tag{5.457}$$

where $\chi_\alpha^{(+)}$ and $\chi_\beta^{(-)}$ in Eq. (5.451) take the solutions of a spherical relativistic optical model with a target nucleus with spin 0; they can be expressed using an upper component and a lower component, respectively, as follows:

$$\chi_\alpha^{(+)} \to \psi_\nu\left(\vec{k},\vec{r}\right) = \begin{pmatrix} u_\nu\left(\vec{k},\vec{r}\right) \\ d_\nu\left(\vec{k},\vec{r}\right) \end{pmatrix} \tag{5.458}$$

$$\chi_\beta^{(-)} \to \psi_{\nu'}\left(\vec{k}',\vec{r}\right) = \begin{pmatrix} u_{\nu'}\left(\vec{k}',\vec{r}\right) \\ d_{\nu'}\left(\vec{k}',\vec{r}\right) \end{pmatrix} \tag{5.459}$$

where $\nu$ and $\nu'$ are the spin magnetic quantum numbers of incident nucleon and outgoing nucleon, respectively; $\vec{k}$ and $\vec{k}'$ are the momentum wave vectors of incident nucleon and outgoing nucleon, respectively. According to the discussion in Sect. 5.3, $u_\nu(\vec{k},\vec{r})$ and $u_{\nu'}(\vec{k}',\vec{r})$ are the solutions of the upper component of the spherical relativistic optical model in the presence of the Coulomb potential; $d_\nu(\vec{k},\vec{r})$ and $d_{\nu'}(\vec{k}',\vec{r})$ are the solutions of the lower component of the spherical relativistic optical model in the absence of the Coulomb potential. Note that the incident nucleon is along the $z$-axis direction, then using Eqs. (5.411), (5.57), and (5.452)–(5.459), the transition matrix elements of the collective inelastic scattering can be obtained from Eq. (5.451) as follows:

$$\mathcal{T}^{(\rho)}_{n'\nu'M_n',n\nu M_n}(\theta) = \frac{\hat{I}_n}{\hat{I}_n'}\sum_\lambda \beta_\lambda C^{I_n'0}_{I_n0\ \lambda0}(-1)^{M_n'-M_n} C^{I_n'M_n'}_{I_nM_n\ \ \lambda\ \ M_n'-M_n} q^{(\rho)}_{n'\nu'M_n',n\nu M_n}, \quad \rho = \text{u, d} \tag{5.460}$$

$$q^{(\text{u})}_{n'\nu'M_n',n\nu M_n} = \int u^+_{\nu'}(\vec{k}',\vec{r})\,(F_0(r)+F_\text{s}(r))\text{Y}_{\lambda\ M_n-M_n'}(\Omega_r)u_\nu(\vec{k},\vec{r})\,r^2\text{d}r\text{d}\Omega_r \tag{5.461}$$

$$q^{(\text{d})}_{n'\nu'M_n',n\nu M_n} = \int d^+_{\nu'}(\vec{k}',\vec{r})\,(F_0(r)-F_\text{s}(r))\text{Y}_{\lambda\ M_n-M_n'}(\Omega_r)d_\nu(\vec{k},\vec{r})\,r^2\text{d}r\text{d}\Omega_r \tag{5.462}$$

The above formulas give the transition matrix elements contributed by the upper and lower components of the relativistic wave functions. Note that $(F_0(r) + F_\text{s}(r))$ appear in the formula of the upper component; and $(F_0(r) - F_\text{s}(r))$ appear in the formula of the lower component.

The following expressions can be obtained according to Eqs. (5.239), (5.241), and (5.446):

$$d_\nu(\vec{k},\vec{r}) = D_n(\hat{\sigma}\cdot\vec{p})\,u_\nu(\vec{k},\vec{r}) \tag{5.463}$$

$$D_n = \left(E - \varepsilon_n + m + \overline{U}_s - \overline{U}_0\right)^{-1} \tag{5.464}$$

where $m$ is the nucleon stationary mass. In the case of a spherical target nucleus with spin 0, based on Eq. (5.345) one can obtain

$$\vec{\sigma} \cdot \vec{p} = -\mathrm{i}\vec{\sigma} \cdot \vec{\nabla} = -\mathrm{i}(\vec{\sigma} \cdot \hat{r}) \left(\frac{\mathrm{d}}{\mathrm{d}r} - \frac{S_{lj}}{r}\right) \tag{5.465}$$

where

$$S_{lj} = j(j+1) - l(l+1) - \frac{3}{4} \tag{5.466}$$

When the direction of the wave vector $\vec{k}$ of the incident nucleon is selected as the $z$-axis and in the case of the target nucleus with spin 0, the general form of the coefficient of the front side of the spin wave function $\chi_\nu$ is as (see Eqs. (3.5.2) and (3.4.6) in Ref. [4]):

$$u_\nu\left(\vec{k}, \vec{r}\right) = \sqrt{4\pi}\sqrt{\frac{E+m}{2m}} \sum_{lj} \hat{l} C_{l0\,\frac{1}{2}\nu}^{j\,\nu} \mathrm{e}^{\mathrm{i}\sigma_l} \frac{F_{lj}(kr)}{kr} \mathbb{Y}_{lj\nu} \tag{5.467}$$

$$\mathbb{Y}_{lj\nu} = C_{l0\,\frac{1}{2}\nu}^{j\,\nu} \mathrm{i}^l \mathrm{Y}_{l0}(\Omega_r) \tag{5.468}$$

where $F_{lj}(kr)$ is a radial wave function obtained by relativistic optical model equation. For the Schrodinger-like equation (5.287), in the $\hbar = c = 1$ unit system, there is

$$k = p = \sqrt{E^2 - m^2} \tag{5.469}$$

Multiply the coefficient $\sqrt{\frac{E+m}{2m}}$ in Eq. (5.467) so that when there is no nuclear potential and Coulomb potential, this formula will automatically degenerate to a planar wave that is consistent with Eq. (5.93), and the coefficient is close to 1 when the energy is very low. In the DWBA theory, the outgoing wave is also a spherical distorted wave, but the $\vec{k}'$ direction is not along the $z$-axis, set $\theta$ to be the angle between $\vec{k}'$ and the $z$-axis (i.e., $\vec{k}$ direction), $\gamma$ is the angle between $\vec{k}'$ and $\vec{r}$, so the coefficient of the front side of the outgoing distorted wave $\chi_{\nu'}$ is

$$u_{\nu'}\left(\vec{k}', \vec{r}\right) = \sqrt{4\pi}\sqrt{\frac{E'+m}{2m}} \sum_{l'j'} \hat{l}' C_{l'0\,\frac{1}{2}\nu'}^{j'\,\nu'} \mathrm{e}^{\mathrm{i}\sigma_{l'}} \frac{F_{l'j'}(k'r)}{k'r} \mathbb{Y}_{l'j'\nu'} \tag{5.470}$$

$$\mathbb{Y}_{l'j'\nu'} = C^{j'\ \nu'}_{l'0\ \frac{1}{2}\nu'} \mathrm{i}^{l'} \mathrm{Y}_{l'0}(\gamma, 0) = C^{j'\ \nu'}_{l'0\ \frac{1}{2}\nu'} \mathrm{i}^{l'} \frac{\sqrt{4\pi}}{\hat{l}'} \sum_{m'} \mathrm{Y}_{l'm'}(\Omega_r)\, \mathrm{Y}^*_{l'm'}(\Omega_{k'}) \tag{5.471}$$

where $E' = \sqrt{E^2 + \varepsilon_n^2 - \varepsilon_{n'}^2}$. And note

$$\mathrm{Y}^*_{l'm'}(\Omega_r)\, \mathrm{Y}_{\lambda\ M_n - M_n'}(\Omega_r) = (-1)^{m'} \frac{1}{\sqrt{4\pi}} \sum_{L_1 M_1} \frac{\hat{l}'\hat{\lambda}}{\hat{L}_1} C^{L_1\ M_1}_{l'\ -m'\ \lambda\ M_n - M_n'} C^{L_1\ 0}_{l'0\ \lambda 0} \mathrm{Y}_{L_1 M_1}(\Omega_r) \tag{5.472}$$

$$\int \mathrm{Y}_{L_1 M_1}(\Omega_r)\, \mathrm{d}\varphi = 2\pi\, \mathrm{Y}_{L_1 0}(\theta, 0)\, \delta_{M_1 0} \tag{5.473}$$

From the above two formulas, $m' = M_n - M_n'$ can be got. Further we can obtain

$$\int \mathrm{Y}_{L_1 0}(\theta, 0) \mathrm{Y}_{l0}(\theta, 0) \mathrm{d}\cos\theta = \frac{1}{2\pi} \delta_{L_1 l} \tag{5.474}$$

Then we can get

$$\int \mathrm{Y}^*_{l'\ M_n - M_n'} \mathrm{Y}_{\lambda\ M_n - M_n'} \mathrm{Y}_{l0}\, \mathrm{d}\Omega_r = (-1)^{M_n - M_n'} \frac{\hat{l}'}{\sqrt{4\pi}} \frac{\hat{\lambda}}{\hat{l}} C^{l\ 0}_{l'\ -M_n + M_n'\ \lambda\ M_n - M_n'} C^{l\ 0}_{l'0\ \lambda 0} \tag{5.475}$$

And there is the relation

$$(\hat{\sigma} \cdot \hat{r})\, (\hat{\sigma} \cdot \hat{r}) = \hat{I} \tag{5.476}$$

Noting that $\vec{k}'$ is axis-symmetric relative to the $z$-axis, the following result can be obtained substituting Eqs. (5.463), (5.465), (5.467), and (5.470) into Eqs. (5.460)–(5.462) and using Eqs. (5.468), (5.471), (5.475), and (5.476):

$$\begin{aligned} \mathcal{T}^{(\rho)}_{n'\nu' M_n', n\nu M_n}(\theta) &= \frac{1}{kk'} \sqrt{\frac{(E+m)\,(E + \varepsilon_n - \varepsilon_{n'} + m)}{4m^2}} \frac{\hat{I}_n}{\hat{I}'_n} \sum_{lj l' j'} \hat{l}\ \mathrm{i}^{l-l'} \mathrm{e}^{\mathrm{i}(\sigma_l - \sigma_{l'})} \\ & C^{j\ \nu}_{l0\ \frac{1}{2}\nu} C^{j'\ \nu'}_{l'0\ \frac{1}{2}\nu'} \left( \sum_{\lambda} \beta_\lambda \hat{\lambda} C^{I_n' 0}_{I_n 0\ \lambda 0} C^{I_n' M_n'}_{I_n M_n\ \lambda\ M_n' - M_n} C^{l\ 0}_{l'0\ \lambda 0} C^{l\ 0}_{l'\ -M_n + M_n'\ \lambda\ M_n - M_n'} \right) \\ & \mathrm{Y}_{l'\ M_n - M_n'}(\theta, 0)\, T^{(\rho)}_{n'l'j', nlj}, \quad \rho = \mathrm{u,d} \end{aligned} \tag{5.477}$$

$$T^{(\mathrm{u})}_{n'l'j',nlj} = \int F_{l'j'}(k'r)(F_0(r) + F_s(r))F_{lj}(kr)\,\mathrm{d}r \tag{5.478}$$

$$T^{(\mathrm{d})}_{n'l'j',nlj} = \int \left[\left(\frac{\mathrm{d}}{\mathrm{d}r} - \frac{S_{l'j'}}{r}\right)\left(\frac{F_{l'j'}(k'r)}{r}\right)\right] D_{n'}(F_0(r) - F_s(r)) \\ D_n\left[\left(\frac{\mathrm{d}}{\mathrm{d}r} - \frac{S_{lj}}{r}\right)\left(\frac{F_{lj}(kr)}{r}\right)\right] r^2\mathrm{d}r \tag{5.479}$$

The definition of $D_n$ has been given by Eq. (5.464). And we know that when the energy of incident particles is very low, the contribution of the lower component can be ignored.

The theory described above is called the relativistic collective deformation RDWBA method. In the expression of the above transition operator, $I_n + \lambda - I'_n$ and $l' + \lambda - l$ are required to be even. When substituting the transition matrix elements of $\rho = \mathrm{u}$ and d given by Eq. (5.477) into Eqs. (5.450) and (5.449), the contributions of the upper component and the lower component to the collective excitation inelastic scattering differential cross section can be obtained. The optical potential parameters can be determined by fitting the elastic scattering experimental data, while the deformation parameters $\beta_\lambda$ can be determined by fitting the inelastic scattering experimental data. The $\lambda$ value takes the smallest one or two positive integers greater than or equal to 2 that satisfy the angular momentum coupling relation $\vec{I}_n + \vec{\lambda} - \vec{I}'_n$ as well as that $I_n + \lambda - I'_n$ is even number. In Refs. [34–40] the collective inelastic scattering problem of even-even target nuclei with spin 0 is studied.

It can be seen according to Eqs. (5.449) and (5.450) that the reaction amplitude corresponding to Eqs. (2.250) and (2.251) is as follows:

$$f^{(\rho)}_{n'\nu'M'_n,\,n\nu M_n}(\theta) = \frac{\mu}{2\pi\hbar^2}\sqrt{\frac{k'}{k}}T^{(\rho)}_{n'\nu'M'_n,\,n\nu M_n}(\theta), \qquad \rho = \mathrm{u},\,\mathrm{d} \tag{5.480}$$

where $\mu$ is the nucleon reduced mass. The RDWBA method is an approximation theory. The distorted wave Born approximation is that the wave function of the outgoing particle is replaced with the distorted wave function calculated by the spherical optical model. Of course, in this case the coupling of the incident nucleon angular momentum $\vec{j}$ and the target nucleus spin $\vec{I}_n$ is not considered. It can be seen that there is no C - G coefficient to represent the coupling of nucleon angular momentum $\vec{j}$ and target spin $\vec{I}_n$ in the transition matrix element expression given by Eq. (5.477). When the coupling between inelastic scattering channel and elastic scattering channel is relatively weak, the calculation of inelastic scattering reaction by RDWBA method still has a certain degree of credibility.

Refer to Eqs. (5.376) and (5.377), and let

$$
\begin{aligned}
&A^{(\rho)}_{M'_n M_n}(\theta) = f^{(\rho)}_{n'\frac{1}{2}M'_n,\, n\frac{1}{2}M_n}(\theta), && B^{(\rho)}_{M'_n M_n}(\theta) = \mathrm{i}f^{(\rho)}_{n'\frac{1}{2}M'_n, n\ -\frac{1}{2}M_n}(\theta), \\
&C^{(\rho)}_{M'_n M_n}(\theta) = -\mathrm{i}f^{(\rho)}_{n'\ -\frac{1}{2}M'_n,\, n\frac{1}{2}M_n}(\theta), && D^{(\rho)}_{M'_n M_n}(\theta) = f^{(\rho)}_{n'\ -\frac{1}{2}M'_n,\, n\ -\frac{1}{2}M_n}(\theta), \quad \rho = \mathrm{u,d}
\end{aligned} \tag{5.481}
$$

Thus the cross sections, differential cross sections, and various polarization physical quantities of the nucleon collective inelastic scattering with target can be calculated using the formulas given in Sect. 5.4.

## 5.7 Relativistic Impulse Approximation of Elastic Scattering and Calculation of Nucleon Polarization Quantities

If the incident nucleon energy is relatively high, its wavelength is smaller than the distance between the nucleons in the nucleus. The interaction between the incident nucleon and the nucleon in the nucleus can be regarded as quasi-free nucleon-nucleon (NN) scattering and can be handled by multiple scattering theories. At this time, we need to know the free NN scattering amplitude and the density distribution of the nucleus. When the incident nucleon energy is high enough, the correction of the NN scattering amplitude by the nuclear density is negligible. For the elastic scattering process, there is no need to consider changes in the internal states of the nucleus. So in 1983, the theory of relativistic impulse approximation (RIA) of the medium energy nucleon elastic scattering was proposed by McNeil et al. [41–43], and this theory was soon applied by others and further studied [44–47]. Subsequently, a number of studies on the nucleon elastic scattering with target using RIA theory were carried out [48–57].

In the stationary Dirac equation (5.226), the Bjurken-Drell metric is used, $p_0 = p^0 = E$ is taken, and the self-energy operator $\Sigma$ is expressed as an elastic scattering optical potential $\hat{U}\left(\vec{r}\right)$. And then the wave function $\psi^{(+)}_{\vec{k},s}\left(\vec{r}\right)$ describing the positive energy nucleon is used, where $s$ represents the nucleon spin magnetic quantum number. Thus in the $\hbar = c = 1$ unit system using Eq. (5.61) and Eq. (5.57), the following equation can be obtained by multiplying $\gamma^0$ from left side of Eq. (5.226) with definite energy

$$
\left[\gamma^\mu p_\mu - m - \hat{U}\left(\vec{r}\right)\right] \psi^{(+)}_{\vec{k},s}\left(\vec{r}\right) = 0 \tag{5.482}
$$

where $m$ represents the nucleon stationary mass, the on-shell mass-energy relation is $E = \sqrt{k^2 + m^2}$, and there are $p^\mu = \left(E, -\mathrm{i}\vec{\nabla}\right)$, $p_\mu = \left(E, \mathrm{i}\vec{\nabla}\right)$.

We use $\left|\vec{k}, s(+)\right\rangle$ to represent the positive energy free nucleon wave function in momentum space when the potential $\hat{U}\left(\vec{r}\right) = 0$, the momentum vector is $\vec{k}$, and the spin projection is $s$. Then from Eq. (5.482) we can obtain

$$\left(\gamma^{\mu} p_{\mu} - m\right) \left|\vec{k}, s(+)\right\rangle = 0 \tag{5.483}$$

And there is

$$\left|\vec{k}, s(+)\right\rangle = u\left(\vec{k}, +\right) \left|\chi_s\right\rangle \tag{5.484}$$

where $\left|\chi_s\right\rangle$ is the nucleon spin wave function. Equation (5.93) has given

$$u\left(\vec{k}, +\right) = \sqrt{\frac{E+m}{2m}} \begin{pmatrix} \hat{I} \\ \dfrac{\hat{\sigma} \cdot \vec{k}}{E+m} \end{pmatrix} \tag{5.485}$$

Equation (5.484) indicates that both the upper component and lower component of $u\left(\vec{k}, +\right)$ all act on the two-dimensional spin wave function $\left|\chi_s\right\rangle$, respectively. For the negative energy states there are

$$\left(\gamma^{\mu} p_{\mu} - m\right) \left|\vec{k}, s(-)\right\rangle = 0 \tag{5.486}$$

$$\left|\vec{k}, s(-)\right\rangle = u\left(\vec{k}, -\right) \left|\chi_s\right\rangle \tag{5.487}$$

Equation (5.94) has given

$$u\left(\vec{k}, -\right) = \sqrt{\frac{E+m}{2m}} \begin{pmatrix} \dfrac{\hat{\sigma} \cdot \vec{k}}{E+m} \\ \hat{I} \end{pmatrix} \tag{5.488}$$

When the potential $\hat{U}\left(\vec{r}\right) = 0$ and the momentum vector is $\vec{k}$, using Eqs. (5.484), (5.487), (5.80), and (5.86), the stationary wave function of the positive energy (+) and negative energy (−) free nucleons with the spin projection $s$ can be written as follows:

$$\varphi^{(\pm)}_{\vec{k},s}\left(\vec{r}\right) = \mathrm{e}^{\pm i\vec{k}\cdot\vec{r}} u\left(\vec{k}, \pm\right) \left|\chi_s\right\rangle \tag{5.489}$$

Their conjugate wave functions are

$$\varphi^{(\pm)+}_{\vec{k}',s'}\left(\vec{r}\right) = \langle\chi_{s'}|u^{+}\left(\vec{k}', \pm\right) e^{\mp i\vec{k}'\cdot\vec{r}} \tag{5.490}$$

From the above results it can be seen that $\varphi^{(+)}_{\vec{k},s}\left(\vec{r}\right)$ is the positive energy solution of Eq. (5.482) in $\hat{U}\left(\vec{r}\right) = 0$ case. It can also be seen that the general solution of Eq. (5.482) is

$$\psi^{(+)}_{\vec{k},s}\left(\vec{r}\right) = \varphi^{(+)}_{\vec{k},s}\left(\vec{r}\right) + \frac{1}{\gamma^{\mu}p_{\mu} - m + i\delta}\hat{U}\left(\vec{r}\right)\psi^{(+)}_{\vec{k},s}\left(\vec{r}\right) \tag{5.491}$$

The introduction of $i\delta$ is to deal with the poles and one finally makes $\delta \to 0+$. Here $(\gamma^{\mu}p_{\mu} - m + i\delta)^{-1}$ is a propagator.

Let

$$L \equiv \gamma^{\mu}p_{\mu} - m \tag{5.492}$$

so Eq. (5.491) can be rewritten as

$$\psi^{(+)}_{\vec{k},s}\left(\vec{r}\right) = \varphi^{(+)}_{\vec{k},s} + L^{-1}\hat{U}\left(\vec{r}\right)\psi^{(+)}_{\vec{k},s}\left(\vec{r}\right) \tag{5.493}$$

Since

$$\hat{U}\left(\vec{r}\right)\psi^{(+)}_{\vec{k},s}\left(\vec{r}\right) = \int \delta\left(\vec{r} - \vec{r}'\right)\hat{U}\left(\vec{r}'\right)\psi^{(+)}_{\vec{k},s}\left(\vec{r}'\right)d\vec{r}' \tag{5.494}$$

so there is

$$\begin{aligned} L^{-1}\hat{U}\left(\vec{r}\right)\psi^{(+)}_{\vec{k},s}\left(\vec{r}\right) &= \int L^{-1}\delta\left(\vec{r} - \vec{r}'\right)\hat{U}\left(\vec{r}'\right)\psi^{(+)}_{\vec{k},s}\left(\vec{r}'\right)d\vec{r}' \\ &\equiv \int G\left(\vec{r}, \vec{r}'\right)\hat{U}\left(\vec{r}'\right)\psi^{(+)}_{\vec{k},s}\left(\vec{r}'\right)d\vec{r}' \end{aligned} \tag{5.495}$$

where

$$G\left(\vec{r}, \vec{r}'\right) = L^{-1}\delta\left(\vec{r} - \vec{r}'\right), \quad LG\left(\vec{r}, \vec{r}'\right) = \delta\left(\vec{r} - \vec{r}'\right) \tag{5.496}$$

Note here we only study the stationary state problem. Substituting Eq. (5.495) into Eq. (5.493), one can get

$$\psi^{(+)}_{\vec{k},s}\left(\vec{r}\right) = \varphi^{(+)}_{\vec{k},s}\left(\vec{r}\right) + \int G\left(\vec{r}, \vec{r}'\right)\hat{U}\left(\vec{r}'\right)\psi^{(+)}_{\vec{k},s}\left(\vec{r}'\right)d\vec{r}' \tag{5.497}$$

There is the following δ function formula:

$$\delta\left(\vec{r}-\vec{r}'\right)=\frac{1}{(2\pi)^3}\int e^{i\,\vec{q}\cdot\left(\vec{r}-\vec{r}'\right)}d\vec{q} \tag{5.498}$$

Thus the following expression can be obtained from Eqs. (5.496), (5.492), and (5.498):

$$G\left(\vec{r},\vec{r}'\right)=\frac{1}{\gamma^\mu p_\mu-m}\frac{1}{(2\pi)^3}\int e^{i\,\vec{q}\cdot\left(\vec{r}-\vec{r}'\right)}d\vec{q} \tag{5.499}$$

We have

$$\frac{1}{\gamma^\mu p_\mu-m}=\frac{\gamma^\mu p_\mu+m}{\left(\gamma^\mu p_\mu\right)^2-m^2} \tag{5.500}$$

Substituting Eq. (5.500) into Eq. (5.499) and letting $(\gamma^\mu p_\mu)^2$ act on the function $e^{i\vec{q}\cdot\left(\vec{r}-\vec{r}'\right)-iE(t-t')}=e^{-iq_\mu(r-r')^\mu}$, and using Eqs. (5.59) and (5.53) as well as noting $p_\mu=\left(E,i\vec{\nabla}\right)$, then one can obtain

$$\left(\gamma^\mu p_\mu\right)^2=\gamma^\mu p_\mu\,\gamma^\nu p_\nu\rightarrow q_\mu q_\nu\gamma^\mu\gamma^\nu=\frac{1}{2}q_\mu q_\nu(\gamma^\mu\gamma^\nu+\gamma^\nu\gamma^\mu)=E^2-q^2$$

Further there is

$$\left(\gamma^\mu p_\mu\right)^2-m^2=E^2-m^2-q^2=-\left(q-\sqrt{E^2-m^2}\right)\left(q+\sqrt{E^2-m^2}\right) \tag{5.501}$$

The following expression can be obtained substituting Eqs. (5.500) and (5.501) into Eq. (5.499):

$$G\left(\vec{r},\vec{r}'\right)=-\frac{\gamma^\mu p_\mu+m}{(2\pi)^3}\int\frac{e^{i\vec{q}\cdot\left(\vec{r}-\vec{r}'\right)}}{\left(q-\sqrt{E^2-m^2}\right)\left(q+\sqrt{E^2-m^2}\right)}d\vec{q} \tag{5.502}$$

In spherical coordinate system $\vec{q}\cdot\left(\vec{r}-\vec{r}'\right)=q\left|\vec{r}-\vec{r}'\right|\cos\theta$ is hold, and there is

$$d\vec{q}=\sin\theta\,d\theta\,d\varphi\,q^2dq \tag{5.503}$$

Then when letting $x=\left|\vec{r}-\vec{r}'\right|$ the following expression can be obtained by Eq. (5.502):

$$
\begin{aligned}
G\left(\vec{r}, \vec{r}'\right) &= -\frac{\gamma^\mu p_\mu + m}{(2\pi)^3} \int \frac{\mathrm{e}^{iqx\cos\theta}}{\left(q - \sqrt{E^2 - m^2}\right)\left(q + \sqrt{E^2 - m^2}\right)} d\varphi \, \mathrm{d}\cos\theta \, q^2 \mathrm{d}q \\
&= -\frac{\gamma^\mu p_\mu + m}{4\pi^2 \mathrm{i}x} \int_0^\infty \frac{q}{\left(q - \sqrt{E^2 - m^2}\right)\left(q + \sqrt{E^2 - m^2}\right)} \left(\mathrm{e}^{\mathrm{i}qx} - \mathrm{e}^{-\mathrm{i}qx}\right) \mathrm{d}q
\end{aligned}
\tag{5.504}
$$

Therefore using the residual theorem, the following result can be obtained from Eq. (5.504) [4]:

$$
\begin{aligned}
G\left(\vec{r}, \vec{r}'\right) &= -\frac{\gamma^\mu p_\mu + m}{4\pi^2 \mathrm{i}x} \int_{-\infty}^\infty \frac{q\mathrm{e}^{\mathrm{i}qx}}{\left(q - \sqrt{E^2 - m^2}\right)\left(q + \sqrt{E^2 - m^2}\right)} \mathrm{d}q \\
&= -\frac{\gamma^\mu p_\mu + m}{4\pi} \frac{\mathrm{e}^{\mathrm{i}k\left|\vec{r} - \vec{r}'\right|}}{\left|\vec{r} - \vec{r}'\right|}
\end{aligned}
\tag{5.505}
$$

Here $k = \sqrt{E^2 - m^2}$ is the momentum wave vector value of the incident or outgoing nucleon.

We know that only when $r'$ is relatively smaller $\hat{U}\left(\vec{r}'\right)$ given in Eq. (5.497) is not 0, so when $r \to \infty$ there must be $r' \ll r$ in $\hat{U}\left(\vec{r}'\right) \neq 0$ region, then one can get

$$
\begin{aligned}
k\left|\vec{r} - \vec{r}'\right| &= k\left(r^2 - 2\vec{r} \cdot \vec{r}' + (r')^2\right)^{1/2} \approx kr\left(1 - \frac{\vec{r}}{r^2} \cdot \vec{r}'\right) \\
&= kr - \vec{k}' \cdot \vec{r}'
\end{aligned}
\tag{5.506}
$$

$$
\vec{k}' = k\frac{\vec{r}}{r}
\tag{5.507}
$$

If the $z$-axis is taken along the $\vec{k}'$ direction, because $\left|\vec{k}'\right| = k$ so there are $k'_x = k'_y = 0,\ k'_z = k$, then from Eqs. (5.484) and (5.485) we can obtain

$$
\begin{aligned}
&\sum_s \left|\vec{k}', s'(+)\right\rangle\left\langle\vec{k}', s'(+)\right|\gamma^0 \\
&= \frac{E+m}{2m}\left[\begin{pmatrix} 1 \\ 0 \\ \frac{k}{E+m} \\ 0 \end{pmatrix}\begin{pmatrix} 1 & 0 & -\frac{k}{E+m} & 0 \end{pmatrix} + \begin{pmatrix} 0 \\ 1 \\ 0 \\ -\frac{k}{E+m} \end{pmatrix}\begin{pmatrix} 0 & 1 & 0 & \frac{k}{E+m} \end{pmatrix}\right] \\
&= \frac{E+m}{2m}\left[\begin{pmatrix} 1 & 0 & -\frac{k}{E+m} & 0 \\ 0 & 0 & 0 & 0 \\ \frac{k}{E+m} & 0 & -\left(\frac{k}{E+m}\right)^2 & 0 \\ 0 & 0 & 0 & 0 \end{pmatrix} + \begin{pmatrix} 0 & 0 & 0 & 0 \\ 0 & 1 & 0 & \frac{k}{E+m} \\ 0 & 0 & 0 & 0 \\ 0 & -\frac{k}{E+m} & 0 & -\left(\frac{k}{E+m}\right)^2 \end{pmatrix}\right] \\
&= \frac{E+m}{2m}\begin{pmatrix} 1 & 0 & -\frac{k}{E+m} & 0 \\ 0 & 1 & 0 & \frac{k}{E+m} \\ \frac{k}{E+m} & 0 & -\left(\frac{k}{E+m}\right)^2 & 0 \\ 0 & -\frac{k}{E+m} & 0 & -\left(\frac{k}{E+m}\right)^2 \end{pmatrix}
\end{aligned} \tag{5.508}
$$

Utilizing Eq. (5.57) we can get

$$
E\gamma^0 - \gamma^i k_i' + m = \begin{pmatrix} (E+m)\hat{I} & \hat{0} \\ \hat{0} & -(E-m)\hat{I} \end{pmatrix} - \begin{pmatrix} \hat{0} & \hat{\sigma}_i k_i' \\ -\hat{\sigma}_i k_i' & \hat{0} \end{pmatrix} \tag{5.509}
$$

We can still get

$$
E - m = \frac{E^2 - m^2}{E+m} = \frac{k^2}{E+m} \tag{5.510}
$$

$$
\hat{\sigma}_i k_i' = \begin{pmatrix} 0 & 1 \\ 1 & 0 \end{pmatrix} k_x' + \begin{pmatrix} 0 & -\mathrm{i} \\ \mathrm{i} & 0 \end{pmatrix} k_y' + \begin{pmatrix} 1 & 0 \\ 0 & -1 \end{pmatrix} k_z' = \begin{pmatrix} k_z' & k_x' - \mathrm{i}k_y' \\ k_x' + \mathrm{i}k_y' & -k_z' \end{pmatrix} \tag{5.511}
$$

For a plane wave which selects the direction of the outgoing particle beam as the $z$-axis, so there are $k_z' = k,\ \ k_x' = k_y' = 0$. In this case substituting Eqs. (5.510) and (5.511) into Eq. (5.509) and comparing them with Eq. (5.508), we can obtain

$$E\gamma^0 - \vec{\gamma} \cdot \vec{k}' + m = 2m \sum_s \left| \vec{k}', s'(+) \right\rangle \overline{\left\langle \vec{k}', s'(+) \right|} \tag{5.512}$$

where

$$\overline{\left\langle \vec{k}', s'(+) \right|} = \left\langle \vec{k}', s'(+) \right| \gamma^0 \tag{5.513}$$

The above simplified proof of taking $\vec{k}'$ direction as $z$-axis does not lose its universality. The following result can be obtained substituting Eqs. (5.505), (5.512), and (5.489) into Eq. (5.497):

$$\begin{aligned} \psi^{(+)}_{\vec{k},s}\left(\vec{r}\right) = {} & \mathrm{e}^{\mathrm{i}\vec{k}\cdot\vec{r}} \left| \vec{k}, s(+) \right\rangle - \frac{m}{2\pi} \left( \sum_{s'} \left| \vec{k}', s'(+) \right\rangle \overline{\left\langle \vec{k}, s'(+) \right|} \right) \\ & \int \frac{\mathrm{e}^{\mathrm{i}k\left|\vec{r}-\vec{r}'\right|}}{\left|\vec{r}-\vec{r}'\right|} \hat{U}\left(\vec{r}'\right) \psi^{(+)}_{\vec{k},s}\left(\vec{r}'\right) \mathrm{d}\vec{r}' \end{aligned} \tag{5.514}$$

Using Eq. (5.506) and Eq. (5.507), Eq. (5.514) can be rewritten as follows:

$$\begin{aligned} \psi^{(+)}_{\vec{k},s}\left(\vec{r}\right) = {} & \mathrm{e}^{\mathrm{i}\vec{k}\cdot\vec{r}} \left| \vec{k}, s(+) \right\rangle - \frac{m}{2\pi} \left( \sum_{s'} |\vec{k}' s'(+)\rangle \overline{\left\langle \vec{k}', s'(+) \right|} \right) \\ & \left[ \int \mathrm{e}^{-\mathrm{i}\vec{k}'\cdot\vec{r}'} \hat{U}\left(\vec{r}'\right) \psi^{(+)}_{\vec{k},s}\left(\vec{r}'\right) \mathrm{d}\vec{r}' \right] \frac{\mathrm{e}^{\mathrm{i}kr}}{r} \end{aligned} \tag{5.515}$$

From the above formula, we can extract the scattering amplitude as [41]

$$F_{s',s}\left(\vec{k}', \vec{k}; E\right) = -\frac{m}{2\pi} \overline{\left\langle \vec{k}', s'(+) \right|} \int \mathrm{e}^{-\mathrm{i}\vec{k}'\cdot\vec{r}'} \hat{U}\left(\vec{r}'\right) \psi^{(+)}_{\vec{k},s}\left(\vec{r}'\right) \mathrm{d}\vec{r}' \tag{5.516}$$

We introduce the Dirac $T$ matrix in the following way:

$$\hat{U}\left(\vec{r}\right) \psi^{(+)}_{\vec{k},s}\left(\vec{r}\right) = \hat{T}\left(\vec{r}\right) \varphi^{(+)}_{\vec{k},s}\left(\vec{r}\right) \tag{5.517}$$

and introduce the following symbol:

$$\left\langle \vec{k}' \middle| \hat{T} \middle| \vec{k} \right\rangle = \int \mathrm{e}^{-\mathrm{i}\vec{k}'\cdot\vec{r}} \hat{T}\left(\vec{r}\right) \mathrm{e}^{\mathrm{i}\vec{k}\cdot\vec{r}} \mathrm{d}\vec{r} \tag{5.518}$$

So Eq. (5.516) can be rewritten as

$$F_{s',s}\left(\vec{k}',\vec{k};E\right)=-\frac{m}{2\pi}\overline{\left\langle\vec{k}',s'(+)\right|}\left\langle\vec{k}'\left|\hat{T}\right|\vec{k}\right\rangle\left|\vec{k},s(+)\right\rangle \tag{5.519}$$

If assumed that the incident nucleon energy is relatively high, it only interacts with each nucleon in the target nucleus once, i.e., ignoring the influence of the density of the nucleus on the NN scattering, then the $T$ matrix element can be approximated as follows:

$$\left\langle\vec{k}'\left|\hat{T}\right|\vec{k}\right\rangle\cong\left\langle\vec{k}'\left|\hat{T}^{(1)}\right|\vec{k}\right\rangle=\left\langle 0\left|\sum_{i=1}^{A}\left\langle\vec{k}'\left|\hat{t}_i\right|\vec{k}\right\rangle\right|0\right\rangle \tag{5.520}$$

where $|0\rangle$ represents the ground state of the spin saturation two-full shell even-even nucleus, $A$ is the mass number of the target nucleus, and $\hat{t}_i$ is the $t$ matrix of scattering of the incident nucleon with the $i$th nucleon in the target nucleus. From Eq. (5.518) it can be seen that $\left|\vec{k}\right\rangle=\mathrm{e}^{\mathrm{i}\vec{k}\cdot\vec{r}}$ is the non-relativistic plane wave, so $\left\langle\vec{k}'\left|\hat{t}_i\right|\vec{k}\right\rangle$ is the non-relativistic $t$ matrix element. Substituting Eq. (5.520) into Eq. (5.519) one can obtain

$$F_{s',s}\left(\vec{k}',\vec{k};E\right)=-\frac{m}{2\pi}\overline{\left\langle\vec{k}',s'(+)\right|}\left\langle 0\left|\sum_{i=1}^{A}\left\langle\vec{k}'\left|\hat{t}_i\right|\vec{k}\right\rangle\right|0\right\rangle\left|\vec{k},s(+)\right\rangle \tag{5.521}$$

We study in the NA center of momentum system composed of the incident nucleon and target nucleus. $\vec{k}$ and $\vec{k}'$ appearing in Eq. (5.521) are the wave vectors of the incident nucleon and the outgoing nucleon in the NA center of momentum system, respectively. And let

$$\vec{k}_a\equiv\frac{1}{2}\left(\vec{k}+\vec{k}'\right),\qquad\vec{q}\equiv\vec{k}-\vec{k}' \tag{5.522}$$

be the average momentum wave vector and momentum transfer wave vector of the incident nucleon and the outgoing nucleon, respectively. Apparently there are

$$\vec{k}=\vec{k}_a+\frac{1}{2}\vec{q},\qquad\vec{k}'=\vec{k}_a-\frac{1}{2}\vec{q} \tag{5.523}$$

And from $\vec{k}'=k\hat{r}$ given by Eq. (5.507) we know

$$\vec{k}_a\cdot\vec{q}=0 \tag{5.524}$$

For the sake of simplicity, we assume that each nucleon in the target nucleus is stationary relative to the center of the target nucleus, i.e., ignoring the nucleon Fermi movement in the nucleus [2, 42, 48, 49]. In the NA center of momentum system, the total momentum wave vectors of the target nucleus and the recoil nucleus should be $-\vec{k}$ and $-\vec{k}'$, respectively; the average momentum wave vectors of each nucleon of the target nucleus and the recoil nucleus are $-\vec{k}/A$ and $-\vec{k}'/A$, respectively. This coordinate system is usually called the Breit coordinate system. In RIA theory, assuming that the nucleons not collided are not affected, we set the momentum wave vector of the nucleon which will be collided before the colliding is $\vec{p} = -\vec{k}/A + \vec{x}$, and each nucleon momentum wave vector of the other $A - 1$ nucleons, which will not be collided, is $-\vec{k}/A - \vec{x}/(A-1)$. At this time, the total momentum wave vector of the target nucleus is still $-\vec{k}$, but $\vec{x}$ is to be determined. After the collision, the momentum wave vector of the emitted nucleon is $\vec{k}' = \vec{k} - \vec{q}$. The momentum wave vector of the nucleon, which has been collided, changed to be $\vec{p}' = -\vec{k}/A + \vec{x} + \vec{q}$. The momentum wave vectors of the other $A - 1$ nucleons, which were not collided, do not change. And the total momentum wave vector of the recoil nucleus is $-\vec{k} + \vec{q} = -\vec{k}'$. Since we study elastic scattering, it is required that the momentum value of the collided nucleon in the target nucleus should not change before and after the collision, that is,

$$\left|-\vec{k}/A + \vec{x}\right| = \left|-\vec{k}/A + \vec{x} + \vec{q}\right| \tag{5.525}$$

So we can get

$$\vec{q} \cdot \vec{q} + 2\left(-\vec{k}/A + \vec{x}\right) \cdot \vec{q} = 0 \tag{5.526}$$

Using Eq. (5.524) the above formula can be rewritten as

$$\vec{q} \cdot \vec{q} + 2\left(-\vec{k}/A + \vec{k}_a/A + \vec{x}\right) \cdot \vec{q} = 0 \tag{5.527}$$

Therefore we can obtain [2, 42, 48, 49]

$$\vec{x} = \vec{k}/A - \vec{k}_a/A - \vec{q}/2 \tag{5.528}$$

$$\vec{p} = -k_a/A - \vec{q}/2, \quad \vec{p}' = -\vec{k}_a/A + \vec{q}/2 \tag{5.529}$$

The total energies of the incident nucleon and the collided nucleon before being collided can be written according to Eqs. (5.523), (5.529), and (5.524) as follows, respectively:

$$E_{(1)} = \left(k_a^2 + q^2/4 + m^2\right)^{1/2} \tag{5.530}$$

$$E_{(2)} = \left[(k_a/A)^2 + q^2/4 + m^2\right]^{1/2} \tag{5.531}$$

From Eq. (5.523) and Eq. (5.524) we can get

$$k_a = \left(k^2 - q^2/4\right)^{1/2} \tag{5.532}$$

The physical quantity $s$ introduced according to Eq. (5.177) can be expressed using Eqs. (5.523) and (5.529) as follows:

$$s = \left(E_{(1)} + E_{(2)}\right)^2 - \left(\vec{k} + \vec{p}\right)^2 = \left(E_{(1)} + E_{(2)}\right)^2 - k_a^2(1 - 1/A)^2 \tag{5.533}$$

If $A = 1$, Eq. (5.533) degenerates into a formula for a two-nucleon system. It can be seen that $s$ is the square of the total energy of the two particle system in the center of momentum system and is the Lorentz invariant. And let

$$t = -q^2 \tag{5.534}$$

Here $t$ and $q$ are not Lorentz invariants. $k$ is determined by the kinetic energy $T_L$ of the incident nucleon in the laboratory system. If $T_L$ and $q^2$ are known, $s$ and $t$ can be calculated out. However, when $T_L$ is determined, the $k$ values obtained in different coordinate systems are different, so the $q$ values are also different.

The general form of NN scattering amplitude that satisfies Lorentz invariance is [41–49, 55]

$$\begin{aligned}\mathcal{F}_{NN} = {} & F_{\mathrm{S}} I(1)I(2) + F_{\mathrm{P}}\gamma^5(1)\gamma^{5+}(2) + F_{\mathrm{V}}\gamma^{\mu}(1)\gamma^{\mu+}(2) \\ & + F_{\mathrm{A}}\gamma^5(1)\gamma^{5+}(2)\gamma^{\mu}(1)\gamma^{\mu+}(2) + F_{\mathrm{T}}\sigma^{\mu\nu}(1)\sigma^{\mu\nu+}(2)\end{aligned} \tag{5.535}$$

where $F_{\mathrm{S}}$, $F_{\mathrm{P}}$, $F_{\mathrm{V}}$, $F_{\mathrm{A}}$, and $F_{\mathrm{T}}$ represent scalar, pseudo-scalar, vector, axial-vector, and anti-symmetric tensor items, respectively. Here (1) and (2) represent the incident nucleon and the being collided nucleon in the target nucleus, respectively. For the sake of simplicity, there is no distinction between neutrons and protons here.

In order to rewrite Eq. (5.535) into another form, firstly the following structural matrices are introduced [55, 58]:

$$\begin{aligned}\Gamma_1 = I = \begin{pmatrix} \hat{I} & \hat{0} \\ \hat{0} & \hat{I} \end{pmatrix}, \quad \Gamma_2 = \gamma^0 = \begin{pmatrix} \hat{I} & \hat{0} \\ \hat{0} & -\hat{I} \end{pmatrix}, \\ \Gamma_3 = \gamma^5 = \begin{pmatrix} \hat{0} & \hat{I} \\ \hat{I} & \hat{0} \end{pmatrix}, \quad \Gamma_4 = \gamma^0\gamma^5 = \begin{pmatrix} \hat{0} & \hat{I} \\ -\hat{I} & \hat{0} \end{pmatrix}\end{aligned} \tag{5.536}$$

It is easy to write the following relations using Eqs. (5.57)–(5.59):

$$\begin{aligned}
&\gamma^5(1)\gamma^{5+}(2) = \Gamma_3(1)\Gamma_3(2),\\
&\gamma^\mu(1)\gamma^{\mu+}(2) = \Gamma_2(1)\Gamma_2(2) - \Gamma_4(1)\Gamma_4(2)\vec{\Sigma}(1)\cdot\vec{\Sigma}(2),\\
&\gamma^5(1)\gamma^{5+}(2)\gamma^\mu(1)\gamma^{\mu+}(2) = \Gamma_4(1)\Gamma_4(2) - \Gamma_2(1)\Gamma_2(2)\vec{\Sigma}(1)\cdot\vec{\Sigma}(2)
\end{aligned} \tag{5.537}$$

The definition of the 4 × 4 matrix $\vec{\Sigma}$ has been given by Eq. (5.71). The tensor matrices are defined as

$$\sigma^{\mu\nu} = \frac{\mathrm{i}}{2}(\gamma^\mu\gamma^\nu - \gamma^\nu\gamma^\mu), \quad \sigma^{\mu\nu+} = -\frac{\mathrm{i}}{2}(\gamma^{\nu+}\gamma^{\mu+} - \gamma^{\mu+}\gamma^{\nu+}) \tag{5.538}$$

We can further find the following expression using $\hat{\sigma}_i\hat{\sigma}_j = \delta_{ij}\hat{I} + \mathrm{i}\varepsilon_{ijk}\hat{\sigma}_k$, $i, j, k = 1, 2, 3$

$$\begin{aligned}
\sigma^{\mu\nu}(1)\sigma^{\mu\nu+}(2) &= \frac{1}{4}[\gamma^\mu(1)\gamma^\nu(1)\gamma^{\nu+}(2)\gamma^{\mu+}(2) - \gamma^\nu(1)\gamma^\mu(1)\gamma^{\nu+}(2)\gamma^{\mu+}(2)\\
&\quad -\gamma^\mu(1)\gamma^\nu(1)\gamma^{\mu+}(2)\gamma^{\nu+}(2) + \gamma^\nu(1)\gamma^\mu(1)\gamma^{\mu+}(2)\gamma^{\nu+}(2)]\\
&= \frac{1}{2}[\gamma^\mu(1)\gamma^\nu(1)\gamma^{\nu+}(2)\gamma^{\mu+}(2) - \gamma^\nu(1)\gamma^\mu(1)\gamma^{\nu+}(2)\gamma^{\mu+}(2)]\\
&= \frac{1}{2}\{-[\gamma^0(1)\gamma^i(1)\gamma^i(2)\gamma^0(2) - \gamma^i(1)\gamma^0(1)\gamma^i(2)\gamma^0(2)\\
&\quad +\gamma^i(1)\gamma^0(1)\gamma^0(2)\gamma^i(2) - \gamma^0(1)\gamma^i(1)\gamma^0(2)\gamma^i(2)]\\
&\quad +[\gamma^i(1)\gamma^j(1)\gamma^j(2)\gamma^i(2) - \gamma^j(1)\gamma^i(1)\gamma^j(2)\gamma^i(2)]\}\\
&= 2\gamma^0(1)\gamma^0(2)\gamma^i(1)\gamma^i(2) - \gamma^i(1)\gamma^j(1)\gamma^i(2)\gamma^j(2)\,\big|_{i\neq j}
\end{aligned}$$

From $\gamma^i = \Gamma_4\Sigma_i$ and $\hat{\sigma}_i\hat{\sigma}_j\big|_{i\neq j} = \mathrm{i}\varepsilon_{ijk}\hat{\sigma}_k$ we can get

$$\begin{aligned}
-\gamma^i(1)\gamma^j(1)\gamma^i(2)\gamma^j(2)\big|_{i\neq j} &= -(\Gamma_4(1)\Gamma_4(2))^2\sum_{i\neq j}\Sigma_i(1)\Sigma_j(1)\Sigma_i(2)\Sigma_j(2)\\
&= 2(\Gamma_4(1)\Gamma_4(2))^2\vec{\Sigma}(1)\cdot\vec{\Sigma}(2)
\end{aligned}$$

Then we obtain

$$\begin{aligned}
&\sigma^{\mu\nu}(1)\,\sigma^{\mu\nu+}(2)\\
&= 2\Gamma_2(1)\Gamma_2(2)\Gamma_4(1)\Gamma_4(2)\vec{\Sigma}(1)\cdot\vec{\Sigma}(2) + 2(\Gamma_4(1))^2(\Gamma_4(2))^2\vec{\Sigma}(1)\cdot\vec{\Sigma}(2)\\
&= 2[\Gamma_1(1)\Gamma_1(2) + \Gamma_3(1)\Gamma_3(2)]\vec{\Sigma}(1)\cdot\vec{\Sigma}(2)
\end{aligned} \tag{5.539}$$

The following expression can be obtained substituting Eqs. (5.537) and (5.539) into Eq. (5.535):

$$\mathcal{F}_{NN} = \sum_{n=1}^{4} \left(a_n + b_n \vec{\Sigma}(1) \cdot \vec{\Sigma}(2)\right) \Gamma_n(1)\Gamma_n(2) \tag{5.540}$$

where

$$\begin{aligned} &a_1 = F_{\mathrm{S}}, \quad a_2 = F_{\mathrm{V}}, \quad a_3 = F_{\mathrm{P}}, \quad a_4 = F_{\mathrm{A}}, \\ &b_1 = 2F_{\mathrm{T}}, \quad b_2 = -F_{\mathrm{A}}, \quad b_3 = 2F_{\mathrm{T}}, \quad b_4 = -F_{\mathrm{V}} \end{aligned} \tag{5.541}$$

The coefficients $F_i$ for each item in Eq. (5.535) are all functions of $s$ and $t$ given by Eqs. (5.533) and (5.534), usually $s$ is not labeled, and they can be expressed only as $\mathcal{F}_{NN}(q)$.

Set $f_{\mathrm{NN}}$ as the non-relativistic NN scattering amplitude in the NA center of momentum system; its matrix element of the spin state is

$$M_{s_1' s_2', s_1 s_2} = \chi_{s_2'}^{+} \chi_{s_1'}^{+} f_{\mathrm{NN}} \chi_{s_1} \chi_{s_2} \tag{5.542}$$

Using the positive energy plane wave of the Dirac equation given by Eq. (5.218) and using Eq. (5.523) and Eq. (5.529), the matrix element of the plane wave state of the relativistic NN scattering amplitude $\mathcal{F}_{NN}$ can be written as follows [42]:

$$\widetilde{M}_{s_1' s_2',\, s_1 s_2} = \bar{u}_{s_2'}\left(\frac{1}{2}\vec{q} - \frac{1}{A}\vec{k}_a\right)\bar{u}_{s_1'}\left(\vec{k}_a - \frac{1}{2}\vec{q}\right)\mathcal{F}_{NN}\, u_{s_1}\left(\vec{k}_a + \frac{1}{2}\vec{q}\right)u_{s_2}\left(-\frac{1}{2}\vec{q} - \frac{1}{A}\vec{k}_a\right) \tag{5.543}$$

If we make a non-relativistic approximation, i.e., we ignore the contribution of the small lower component of the positive energy plane wave, at this time there is $u_s\left(\vec{k}\right) \sim \chi_s$, but the difference between the non-relativistic and relativistic cases is a normalized coefficient $\sqrt{\frac{E+m}{2m}}$, so we let

$$\begin{aligned} &B\left(\vec{k}_a, \vec{q}\right) \\ &= \frac{\left[\left(E\left(\frac{1}{2}\vec{q} - \frac{1}{A}\vec{k}_a\right) + m\right)\left(E\left(\vec{k}_a - \frac{1}{2}\vec{q}\right) + m\right)\left(E\left(\vec{k}_a + \frac{1}{2}\vec{q}\right) + m\right)\left(E\left(-\frac{1}{2}\vec{q} - \frac{1}{A}\vec{k}_a\right) + m\right)\right]^{1/2}}{4m^2} \end{aligned} \tag{5.544}$$

where $E\left(\vec{k}_a + \frac{1}{2}\vec{q}\right)$ is the energy corresponding to the nucleon momentum $\left(\vec{k}_a + \frac{1}{2}\vec{q}\right)$ and $\vec{k}_a$ and $\vec{q}$ have been given by Eq. (5.522). Since the impulse approximation is only applicable to medium-high energy nuclear reactions, the value of $B\left(\vec{k}_a, \vec{q}\right)$ given by the above formula should be significantly greater than 1. Here we give a convention

$$\widetilde{M}_{s_1' s_2', s_1 s_2} = B\left(\vec{k}_a, \vec{q}\right) M_{s_1' s_2', s_1 s_2} \tag{5.545a}$$

The corresponding convention given in the early literature of RIA theory [41–47] is as

$$\widetilde{M}_{s_1' s_2', s_1 s_2} = \frac{1}{2\mathrm{i}k} M_{s_1' s_2', s_1 s_2} \tag{5.545b}$$

This convention will make the dimensions of the two matrix elements in Eq. (5.545b) different. The nuclear force parameters corresponding to the above two kind of convention methods are different.

When the incident plane wave is taken as $\mathrm{e}^{\mathrm{i}\vec{k}\cdot\vec{r}_i}$ (there is no coefficient $1/(2\pi)^{3/2}$ in front), Eq. (3.12.50) of Ref. [4] and Eq. (3.76) of Ref. [49] all give the relation between the non-relativistic NN scattering amplitude $f_{\mathrm{NN}}$ and the $t$ matrix element as follows:

$$f_{\mathrm{NN}}(q) = -\frac{\mu_{\mathrm{NN}}}{2\pi\hbar^2}\left\langle \vec{k}' \middle| \hat{t}_{\mathrm{NN}} \middle| \vec{k} \right\rangle \tag{5.546}$$

where $\mu_{\mathrm{NN}}$ is the NN reduced mass and $\hat{t}_{\mathrm{NN}}$ is the $t$ matrix in the NA center of mass system. The usual nuclear force parameters are given in the NN center of mass system. In the case of relativity, since the general convention is that $t$ matrix is Lorentz invariant, so the $t$ matrix in the NN center of momentum system can be used directly in the NA center of momentum system.

Note the following relation:

$$\left\langle \vec{k}' \middle| \delta\left(\vec{r} - \vec{r}_i\right) \middle| \vec{k} \right\rangle = \int \delta\left(\vec{r} - \vec{r}_i\right) \mathrm{e}^{\mathrm{i}\vec{q}\cdot\vec{r}} \mathrm{d}\vec{r} = \mathrm{e}^{\mathrm{i}\vec{q}\cdot\vec{r}_i}$$

so under the zero-range force approximation and refer to Eq. (5.546), the NN scattering amplitudes in Eq. (5.521) for the above two convention methods can be expressed, respectively, as

$$-\frac{m}{2\pi}\left\langle \vec{k}'\middle|\hat{t}_i\middle|\vec{k}\right\rangle = \frac{1}{B\left(\vec{k}_a, \vec{q}\right)}\mathcal{F}_{NN}(q)\exp\left(\mathrm{i}\,\vec{q}\cdot\vec{r}_i\right) \tag{5.547a}$$

$$-\frac{m}{2\pi}\left\langle \vec{k}'\middle|\hat{t}_i\middle|\vec{k}\right\rangle = 2\mathrm{i}k\mathcal{F}_{NN}(q)\exp\left(\mathrm{i}\,\vec{q}\cdot\vec{r}_i\right) \tag{5.547b}$$

As mentioned earlier, $\left\langle \vec{k}'\middle|\hat{t}_i\middle|\vec{k}\right\rangle$ is the non-relativistic $t$ matrix element, and $\mathcal{F}_{NN}(q)$ is the relativistic scattering amplitude, so the convention corrections are introduced in the above two equations. And we can rewrite the above two formulas as

$$\left\langle \vec{k}'\middle|\hat{t}_i\middle|\vec{k}\right\rangle = -\frac{2\pi}{mB\left(\vec{k}_a, \vec{q}\right)}\mathcal{F}_{NN}(q)\exp\left(\mathrm{i}\,\vec{q}\cdot\vec{r}_i\right) \tag{5.548a}$$

$$\left\langle \vec{k}'\middle|\hat{t}_i\middle|\vec{k}\right\rangle = -\frac{4\pi\mathrm{i}k}{m}\mathcal{F}_{NN}(q)\exp\left(\mathrm{i}\,\vec{q}\cdot\vec{r}_i\right) \tag{5.548b}$$

We know that the optical potential is equivalent to the mass operator. The $\hat{U}\left(\vec{r}\right)$ appearing in the previous equations is the mass operator $\Sigma$ in the Dirac equation (5.226), and $\hat{U}\left(\vec{r}\right)$ is converted to $\hat{T}\left(\vec{r}\right)$ by using Eq. (5.517), so $\left\langle \vec{k}'\middle|\hat{T}\middle|\vec{k}\right\rangle$ given by Eq. (5.520) can be regarded as the optical potential $U_{\mathrm{opt}}(q)$ in the momentum space. Introduce a coefficient symbol

$$C(q) = \begin{cases} -\dfrac{2\pi}{mB\left(\vec{k}_a, \vec{q}\right)}, & \text{For convention given by Eq.(5.545a)} \\ -\dfrac{4\pi\mathrm{i}k}{m}, & \text{For convention given by Eq.(5.545b)} \end{cases} \tag{5.549}$$

So substituting Eq. (5.548a) and Eq. (5.548b) into Eq. (5.520), we can obtain

$$U_{\mathrm{opt}}(q) = C(q)\left\langle 0\middle|\sum_{i=1}^{A}\mathcal{F}_{NN}(q)\exp\left(\mathrm{i}\vec{q}\cdot\vec{r}_i\right)\middle|0\right\rangle \tag{5.550}$$

In the independent particle shell model, the anti-symmetric wave function of the double-full shell target nucleus can be written as [55]

$$|0\rangle = \frac{1}{\sqrt{A!}}\det\prod_{i=1}^{A}u_{nlj\mu}\left(\vec{r}_i\right) \tag{5.551}$$

where

$$u_{nlj\mu}\left(\vec{r}\right) = \begin{pmatrix} \phi_{nlj}(r) \\ -\mathrm{i}\hat{\sigma}\cdot\hat{r}\lambda_{nlj}(r) \end{pmatrix} Y_{lj}^{\mu}(\hat{r}) \tag{5.552}$$

We have the following expressions according to Eqs. (17.3.21) and (17.3.22) in Ref. [28]:

$$Y_{lj}^{\mu}(\hat{r}) = \sum_{\nu} C_{l\ \mu-\nu\ \frac{1}{2}\nu}^{j\ \mu}\,\mathrm{i}^{l}\mathrm{Y}_{l\ \mu-\nu}(\hat{r})\chi_{\frac{1}{2}\nu} \tag{5.553}$$

$$\phi_{nlj}(r) = u_{nlj}(r)/r \tag{5.554}$$

where $u_{nlj}(r)$ is a radial wave function of the nuclear bound state, and neutrons and protons are not distinguished. The Dirac equation can be written as

$$\left[\vec{\alpha}\cdot\vec{p} + \beta(m + U_{\mathrm{S}}) + U_0\right] u\left(\vec{r}\right) = Eu\left(\vec{r}\right) \tag{5.555}$$

And the following expression can be obtained from Eqs. (5.239), (5.241), and (5.345):

$$\lambda_{nlj}(r) = D\left(\frac{\mathrm{d}}{\mathrm{d}r} - \frac{S_{lj}}{r}\right)\phi_{nlj}(r) \tag{5.556}$$

where

$$S_{lj} = j(j+1) - l(l+1) - \frac{3}{4} \tag{5.557}$$

$$D = (E + m + U_{\mathrm{S}} - U_0)^{-1} \tag{5.558}$$

The effect of Coulomb potential is not considered in the lower component of the nucleon wave function. The following expression can be obtained from Eqs. (5.553), (5.351), and (5.352):

$$Z_{lj}^{\mu}(\hat{r}) \equiv -\mathrm{i}(\hat{\sigma}\cdot\hat{r})Y_{lj}^{\mu}(\hat{r}) = \sqrt{2}\,\mathrm{i}^{l+3}\hat{l}\sum_{\nu} C_{l\ \mu-\nu\ \frac{1}{2}\nu}^{j\ \mu}\sum_{\eta}(-1)^{\frac{1}{2}-\nu+\eta} C_{\frac{1}{2}\ \nu+\eta\ \frac{1}{2}\ -\nu}^{1\ \eta}$$
$$\sum_{L}\frac{1}{\hat{L}} C_{l\ \mu-\nu\ 1\ -\eta}^{L\ \mu-\nu-\eta}\, C_{l0\ 10}^{L\ 0}\mathrm{Y}_{L\ \mu-\nu-\eta}(\hat{r})\,\chi_{\nu+\eta} \tag{5.559}$$

So Eq. (5.552) can be rewritten as

$$u_{nlj\mu}\left(\vec{r}\right) = \begin{pmatrix} \phi_{nlj}(r)Y_{lj}^{\mu}(\hat{r}) \\ \lambda_{nlj}(r)Z_{lj}^{\mu}(\hat{r}) \end{pmatrix} \tag{5.560}$$

Introduce the following momentum distribution shape factors according to Eqs. (5.540) and (5.550):

$$\rho_n(q) = \left\langle 0 \left| \sum_i \Gamma_n(i)\, \mathrm{e}^{\mathrm{i}\vec{q}\cdot\vec{r}_i} \right| 0 \right\rangle \tag{5.561}$$

$$\vec{j}_n(q) = \left\langle 0 \left| \sum_i \Gamma_n(i) \vec{\Sigma}(i)\, \mathrm{e}^{\mathrm{i}\vec{q}\cdot\vec{r}_i} \right| 0 \right\rangle \tag{5.562}$$

where $\rho_n(q)$ and $\vec{j}_n(q)$ are determined by the wave function $|0\rangle$ of the target nucleus and have no relationship with the nuclear force parameters $a_n$ and $b_n$. So Eq. (5.550) can be rewritten as

$$U_{\text{opt}}(q) = C(q) \sum_{n=1}^{4} \Gamma_n \left( a_n(q)\rho_n(q) + b_n(q)\vec{j}_n(q) \cdot \vec{\Sigma} \right) \tag{5.563}$$

Using the Fourier transformation, the optical potential of the coordinate space can be obtained as [2, 46–48]

$$U_{\text{opt}}(r) = \frac{1}{(2\pi)^3} \int \mathrm{e}^{-\mathrm{i}\vec{q}\cdot\vec{r}}\, U_{\text{opt}}(q)\, \mathrm{d}\vec{q} \tag{5.564}$$

Substituting Eq. (5.563) into Eq. (5.564), we can obtain

$$U_{\text{opt}}(r) = \sum_{n=1}^{4} \Gamma_n \left( U_n(r) + \vec{W}_n(r) \cdot \vec{\Sigma} \right) \tag{5.565}$$

where

$$U_n(r) = \frac{1}{(2\pi)^3} \int \mathrm{e}^{-\mathrm{i}\vec{q}\cdot\vec{r}}\, C(q) a_n(q) \rho_n(q) \mathrm{d}\vec{q} \tag{5.566}$$

$$\vec{W}_n(r) = \frac{1}{(2\pi)^3} \int e^{-\mathrm{i}\vec{q}\cdot\vec{r}}\, C(q) b_n(q) \vec{j}_n(q) \mathrm{d}\vec{q} \tag{5.567}$$

Notice that $U_n(r)$ and $\vec{W}_n(r)$ in Eq. (5.565) correspond to $a_n$ and $b_n$ in Eq. (5.541), respectively, for example, $U_1(r)$ is equivalent to $U_S(r)$, $U_2(r)$ is equivalent to $U_V(r)$. It is also noted that the optical potential $U_{\text{opt}}(r)$ contains terms obviously appearing spin operator $\vec{\Sigma}$, which are obtained from the space-like components of the 4 vector terms and the 4 axis-vector terms as well as the tensor terms in the nuclear force. There are 8 items in the NN scattering amplitude given by Eq. (5.535), which correspond to scalar, pseudo-scalar, time-like, and space-like components of the

4-vector and the 4-axis-vector, and the corresponding tensors with $\sigma^{0i}$ and $\sigma^{ij}$. Due to the requirement to satisfy the conservation of parity, it is generally believed that the pseudo-scalar and axis-vector terms do not contribute. And it is believed that in the case of the stationary state of the spherical symmetric system, the 4-vector space-like component terms and tensor terms ($b_3$) of $\sigma^{0i}$ do not contribute [2, 46–48, 53, 57, 59]. In addition, the calculation results also show that the contribution of tensor items ($b_1$) of $\sigma^{ij}$ is also very small [46, 47]. Therefore it can be seen that the relativistic optical potential is mainly contributed by scalar term $U_S$ and 4-vector time-like component term $U_0$.

The spherical optical potential $U_{opt}(r)$ can be calculated using Eqs. (5.565)–(5.567), (5.561), (5.562), (5.551), (5.560), and (5.541) and nuclear force parameters. Then substituting $U_{opt}(r)$ into Eq. (5.482) and making a transformation similar to the 5.3 section, and introducing the Coulomb potential, the relativistic optical model calculation can be carried out. In the specific calculation, the selection of NN nuclear force should be considered. For example, someone used a simple Yukawa potential [43]; and some people use the RIA method to study the elastic scattering process of protons and non-double full shell nucleus (such as $^{13}$C) [55].

When the elastic scattering amplitudes are obtained, the differential cross section, analyzing power, and spin rotation function of the elastic scattering can be calculated using Eqs. (5.296)–(5.301) [2, 43–49, 51–57].

## 5.8 Relativistic Impulse Approximation of Inelastic Scattering and Calculation of Nucleon Polarization Quantities

### *5.8.1 Relativistic Distorted Wave Impulse Approximation of Single Particle State Inelastic Scattering [58, 60, 61]*

We consider the process by which the nucleus in the initial state $\Psi_{J_i M_i}$ is excited to the final state $\Psi_{J_f M_f}$. Its transition amplitude can be written as

$$T_{\mathrm{f\,i}} = \left\langle \sum_{j=1}^{A} \overline{\psi}^{(-)}_{\vec{k}',s'} \overline{\Psi}_{J_f M_f} \hat{t}(0,\ j)\, \psi^{(+)}_{\vec{k},s} \Psi_{J_i M_i} \right\rangle \tag{5.568}$$

where $\psi^{(+)}_{\vec{k},s}$ and $\psi^{(-)}_{\vec{k}',s'}$ are the wave functions representing the incident nucleon and the outgoing nucleon with the positive energy, respectively. The incident and outgoing nucleons have a label number of 0. $\Psi_{JM}$ is a wave function of the nucleus composed of $A$ nucleons ($j = 1, 2, \ldots, A$). $\hat{t}(0, j)$ is a $t$ matrix describing the interaction between the label 0 nucleon and the $j$th nucleon in the nucleus. And $\psi_{\vec{k},s}$ and $\Psi_{JM}$ satisfy the following Dirac equations, respectively:

$$\left(\vec{\alpha}\cdot\vec{p}+\beta m+V_0\right)\psi_{\vec{k},s}=E_k\psi_{\vec{k},s} \tag{5.569}$$

$$\left[\sum_{j=1}^{A}\left(\vec{\alpha}_j\cdot\vec{p}_j+\beta_j m\right)+V_{\mathrm{t}}\right]\Psi_{JM}=E\Psi_{JM} \tag{5.570}$$

where $V_{\mathrm{t}}$ in Eq. (5.570) is the potential felt by the nucleon in the target nucleus, it is selected to solve the bound state equation (5.570), so that $E_{\mathrm{i}}$ and $\Psi_{J_{\mathrm{i}}M_{\mathrm{i}}}$ for the initial state and $E_{\mathrm{f}}$ and $\Psi_{J_{\mathrm{f}}M_{\mathrm{f}}}$ for the final state can be obtained. The final state f is a single particle excited state. Equation (5.569) is an equation describing scattering state of nucleon labeled with 0. In higher energy condition, it can be assumed that the incident nucleon only interacts with one nucleon in the target nucleus and then runs out of the nucleus. Based on the nuclear wave function $\Psi_{J_{\mathrm{i}}M_{\mathrm{i}}}$ the optical potential $V_0$ can be obtained by the RIA method described in the 5.7 section, and substituting the obtained optical potential $V_0$ into Eq. (5.569), then the distorted wave $\psi^{(+)}_{\vec{k},s}$ can be solved. If starting from the nuclear wave function $\Psi_{J_{\mathrm{f}}M_{\mathrm{f}}}$, the distorted wave $\psi^{(-)}_{\vec{k}',s'}$ can be found by the same method. The $t$ matrix $\hat{t}(0,j)$ in Eq. (5.568) can be given with the method described in the 5.7 section. If all wave functions and $t$ matrices are given in the spatial coordinate, the transition amplitude $T_{\mathrm{f\,i}}$ can be obtained from Eq. (5.568). Then the differential cross section and various polarization physical quantities of the single particle state inelastic scattering can be calculated using Eq. (5.480) and Eq. (5.481) as well as the calculation formulas given in Sect. 5.4. The above method is called the relativistic distorted wave impulse approximation of the single particle state inelastic scattering.

### 5.8.2 *Relativistic Distorted Wave Impulse Approximation of Collective State Inelastic Scattering [62–64]*

We know that when calculating the target nucleus ground state wave function $|0\rangle$ appeared in Eqs. (5.550) and (5.551), we must first give the radius $R$ of the nucleus, so the RIA optical potential $U_{\mathrm{opt}}$ given by Eq. (5.565) is actually a function of the nuclear radius $R$. We assume that the nucleus is not spherical and its radius is deformed similar to Eq. (5.407), so we introduce the following deformation optical potential according to Eq. (5.565):

$$\begin{aligned}\Delta U=\sum_{n=1}^{4}\Gamma_n\Bigg[&\left(\sum_{\lambda}\beta_\lambda^{sn}\mathrm{Y}_{\lambda0}(\theta',0)\right)R_{sn}\frac{\partial U_n(r)}{\partial R_{sn}}\\&+\left(\sum_{\lambda}\beta_\lambda^{0n}\mathrm{Y}_{\lambda0}(\theta',0)\right)R_{0n}\frac{\partial \vec{W}_n(r)}{\partial R_{0n}}\cdot\vec{\Sigma}\Bigg]\end{aligned} \tag{5.571}$$

where $\beta_{\lambda}^{sn}$ and $\beta_{\lambda}^{0n}$ $(n = 1, 2, 3, 4)$ are the adjustable deformation parameters. And let

$$\Delta U_{\mathrm{D}} = \gamma^0 \Delta U \tag{5.572}$$

Then it is possible to perform various calculations using the methods described in Sect. 5.6. Since the optical potential used here is calculated using the RIA method, the theory is said to be a relativistic distorted wave impulse approximation of the collective state inelastic scattering.

## 5.9 Relativistic Impulse Approximation of (p, n) Reactions and Calculation of Nucleon Polarization Quantities

In the real part of the nucleon universal optical potential, it usually contains a term that is proportional to $\alpha \equiv \frac{N-Z}{A}$. The coefficient signs at the front edge of the $\frac{N-Z}{A}$ terms of the neutron and proton optical potentials are opposed; this form of optical potential indicates that the optical potential should be isospin related. Lane proposed the following form of optical potential [65]:

$$V = V_0 + \frac{\vec{t} \cdot \vec{T}}{A} V_1 \tag{5.573}$$

where $V_0$ and $V_1$ are the isospin scalar part and isospin vector part of the optical potential, respectively. $\vec{t}$ and $\vec{T}$ are the isospin operators of the incident nucleon and the target nucleus, respectively, and they are all vectors. Their components in the rectangular coordinate system are $(\hat{t}_x, \hat{t}_y, \hat{t}_z)$ and $(\hat{T}_x, \hat{T}_y, \hat{T}_z)$, respectively, and their components in the spherical basis coordinate system are $(\hat{t}_+, \hat{t}_0, \hat{t}_-)$ and $(\hat{T}_+, \hat{T}_0, \hat{T}_-)$, respectively, and there are $\hat{t}_0 = \hat{t}_z$, $\hat{T}_0 = \hat{T}_z$. The total isospin value of the nucleon is $t = \frac{1}{2}$, and its isospin z component value $\tau = \frac{1}{2}$ represents the neutron and $\tau = -\frac{1}{2}$ represents the proton. The total isospin value of the target nucleus is $T = \frac{A}{2}$, and if the nucleus is composed entirely of neutrons, there is $T_z = T_0 = \frac{A}{2}$; if the nucleus is composed entirely of protons, there is $T_z = T_0 = -\frac{A}{2}$, then there is $T_z = T_0 = \frac{N-Z}{2}$ for the nucleus composed of $A = N + Z$. If a proton in the nucleus becomes a neutron, $N$ increases by 1, $Z$ decreases by 1, and then $T_z = T_0$ increases by 1.

In spherical basis coordinate there is

$$\vec{t} \cdot \vec{T} = \sum_{\mu=-1}^{1} (-1)^{\mu} \hat{t}_{\mu} \hat{T}_{-\mu} \tag{5.574}$$

and there is the following angular momentum formula:

$$J_0\psi_{jm} = J_z\psi_{jm} = m\psi_{jm} \tag{5.575}$$

Introducing the angular momentum operators

$$J_\pm \equiv \mp\frac{1}{\sqrt{2}}\left(J_x \pm \mathrm{i}J_y\right) \tag{5.576}$$

and using the commutative relation $[J_i, J_k] = \mathrm{i}\varepsilon_{ikl}J_l$ of the angular momentum operators, it can be proved that

$$[J_z, J_\pm] = \mp\frac{1}{\sqrt{2}}\left\{\,\mathrm{i}J_y \pm \mathrm{i}(-\mathrm{i})J_x\right\} = -\frac{1}{\sqrt{2}}\left(J_x \pm \mathrm{i}J_y\right) = \pm J_\pm \tag{5.577}$$

Let

$$\varphi_{jm}^{(\pm)} = J_\pm\psi_{jm} \tag{5.578}$$

so we can obtain

$$J_z\left(J_\pm\psi_{jm}\right) = \{J_\pm J_z + [J_z,\ J_\pm]\,\}\,\psi_{jm} = J_\pm(J_z \pm 1)\,\psi_{jm} = (m \pm 1)\left(J_\pm\psi_{jm}\right) \tag{5.579}$$

and

$$J_z\varphi_{jm}^{(\pm)} = (m \pm 1)\,\varphi_{jm}^{(\pm)} \tag{5.580}$$

And let

$$J_\pm\psi_{jm} = \varphi_{jm}^{(\pm)} = \Gamma_\pm\psi_{j\ m\pm 1} \tag{5.581}$$

and from Eq. (5.576) we know

$$\left(J_\pm\right)^+ = -J_\mp \tag{5.582}$$

so from the above two formulas we can obtain

$$|\Gamma_\pm|^2 = \left\langle J_\pm\psi_{jm}, J_\pm\psi_{jm}\right\rangle = -\left\langle\psi_{jm}|J_\mp J_\pm|\psi_{jm}\right\rangle \tag{5.583}$$

We can also obtain

$$\begin{aligned} J_{\mp}J_{\pm} &= -\frac{1}{2}\left(J_x \mp \mathrm{i}\, J_y\right)\left(J_x \pm \mathrm{i}\, J_y\right) = -\frac{1}{2}\left(J_x^2 + J_y^2 \pm \mathrm{i}\, J_x J_y \mp \mathrm{i}\, J_y J_x\right) \\ &= -\frac{1}{2}\left(J_x^2 + J_y^2 \pm \mathrm{i}\left[J_x,\ J_y\right]\right) = -\frac{1}{2}\left[J^2 - J_z(J_z \pm 1)\right] \end{aligned} \tag{5.584}$$

$$|\Gamma_{\pm}|^2 = \frac{1}{2}[J(J+1) - m(m \pm 1)] = \frac{1}{2}(J \mp m)(J \pm m + 1) \tag{5.585}$$

Then we can get

$$J_{\pm}\psi_{jm} = \frac{1}{\sqrt{2}}[(j \mp m)(j \pm m + 1)]^{1/2}\psi_{j\ \ m\pm 1} \tag{5.586}$$

Let $\varphi_{t\tau}$ and $\Phi_{TT_0}$ be the isospin wave functions of the nucleon and the target nucleus, respectively, so that from Eq. (5.575) we can obtain

$$\hat{t}_0\varphi_{t\tau} = \tau\varphi_{t\tau}, \quad \hat{T}_0\Phi_{TT_0} = T_0\Phi_{TT_0} \tag{5.587}$$

From Eq. (5.586) we know

$$\hat{t}_{\pm}\varphi_{t\tau} = \frac{1}{\sqrt{2}}[(t \mp \tau)\ (t \pm \tau + 1)]^{1/2}\varphi_{t\ \ \tau\pm 1} \tag{5.588}$$

If $\tau = \frac{1}{2}$ (representing neutron), there will be

$$\hat{t}_{+}\varphi_{\frac{1}{2}\frac{1}{2}} = 0, \qquad \hat{t}_{-}\varphi_{\frac{1}{2}\frac{1}{2}} = \frac{1}{\sqrt{2}}\varphi_{\frac{1}{2}\ -\frac{1}{2}} \tag{5.589}$$

If $\tau = -\frac{1}{2}$ (representing proton), there will be

$$\hat{t}_{+}\varphi_{\frac{1}{2}\ -\frac{1}{2}} = \frac{1}{\sqrt{2}}\varphi_{\frac{1}{2}\ \frac{1}{2}}, \qquad \hat{t}_{-}\varphi_{\frac{1}{2}\ -\frac{1}{2}} = 0 \tag{5.590}$$

From Eq. (5.586) we can also write

$$\hat{T}_{\pm}\Phi_{TT_0} = \frac{1}{\sqrt{2}}[(T \mp T_0)\ (T \pm T_0 + 1)]^{1/2}\Phi_{T\ \ T_0\pm 1} \tag{5.591}$$

So we have

$$\hat{T}_{+}\Phi_{\frac{A}{2}\ \frac{N-Z}{2}} = \frac{1}{\sqrt{2}}\sqrt{Z(N+1)}\Phi_{\frac{A}{2}\ \frac{N-Z+2}{2}}\ , \quad \hat{T}_{-}\Phi_{\frac{A}{2}\ \frac{N-Z}{2}} = \frac{1}{\sqrt{2}}\sqrt{N(Z+1)}\Phi_{\frac{A}{2}\ \frac{N-Z-2}{2}} \tag{5.592}$$

Using the formulas given above, for the p + A reaction there is

$$\left(\vec{t}\cdot\vec{T}\right)\varphi_{\frac{1}{2}\,-\frac{1}{2}}\Phi_{\frac{A}{2}\,\frac{N-Z}{2}}=-\frac{N-Z}{4}\varphi_{\frac{1}{2}\,-\frac{1}{2}}\Phi_{\frac{A}{2}\,\frac{N-Z}{2}}-\frac{1}{2}\sqrt{N(Z+1)}\varphi_{\frac{1}{2}\,\frac{1}{2}}\Phi_{\frac{A}{2}\,\frac{N-Z-2}{2}} \tag{5.593}$$

The first item in above formula represents the (p, p) elastic scattering channel, and the second one represents the (p, n) reaction channel. For the n + A reaction there is

$$\left(\vec{t}\cdot\vec{T}\right)\varphi_{\frac{1}{2}\,\frac{1}{2}}\Phi_{\frac{A}{2}\,\frac{N-Z}{2}}=\frac{N-Z}{4}\varphi_{\frac{1}{2}\,\frac{1}{2}}\Phi_{\frac{A}{2}\,\frac{N-Z}{2}}-\frac{1}{2}\sqrt{Z(N+1)}\varphi_{\frac{1}{2}\,-\frac{1}{2}}\Phi_{\frac{A}{2}\,\frac{N-Z+2}{2}} \tag{5.594}$$

The first item in above formula represents the (n, n) elastic scattering channel, and the second one represents the (n, p) reaction channel.

It is assumed that the target nucleus is a spherical nucleus with spin 0, regardless of the internal freedom and excited state of the target nucleus, i.e., only its isospin state is considered. For incident and outgoing nucleons, their spatial coordinates and spins must be considered to be same as the spherical optical model. The isospin wave function of the target nucleus is still represented by $\Phi_{\frac{A}{2}\,\frac{N-Z}{2}}$, but the nucleon wave function is rewritten as

$$\varphi_{\frac{1}{2}\tau}=\sum_{l'j'm'_j}\frac{u^{\tau}_{l'j'}}{r}\mathcal{Y}^{m'_j}_{l'\frac{1}{2}j'}(\Omega)\chi_{\frac{1}{2}\tau} \tag{5.595}$$

where

$$\mathcal{Y}^{m'_j}_{l'\frac{1}{2}j'}(\Omega)=\sum_{m'_l\nu'}C^{j'm'_j}_{l'm'_l\;\frac{1}{2}\nu'}\,\mathrm{i}^{l'}Y_{l'm'_l}(\Omega)\chi_{\frac{1}{2}\nu'} \tag{5.596}$$

where $\chi_{\frac{1}{2}\tau}$ is a nucleon isospin wave function. Using the new $\varphi_{\frac{1}{2}\tau}$ given by Eq. (5.595), based on Eqs. (5.589) and (5.590), we agree

$$\hat{t}_{-}\varphi_{\frac{1}{2}\,\frac{1}{2}}=\sum_{l'j'm'_j}\frac{u^{-\frac{1}{2}}_{l'j'}(r)}{r}\mathcal{Y}^{m'_j}_{l'\frac{1}{2}j'}(\Omega)\frac{1}{\sqrt{2}}\chi_{\frac{1}{2}\,-\frac{1}{2}} \tag{5.597}$$

$$\hat{t}_{+}\varphi_{\frac{1}{2}\,-\frac{1}{2}}=\sum_{l'j'm'_j}\frac{u^{\frac{1}{2}}_{l'j'}(r)}{r}\mathcal{Y}^{m'_j}_{l'\frac{1}{2}j'}(\Omega)\frac{1}{\sqrt{2}}\chi_{\frac{1}{2}\,\frac{1}{2}} \tag{5.598}$$

This means that under the action of the isospin operator $\hat{t}_{\pm}$, when the proton is turned into a neutron or the neutron is turned into a proton, the corresponding nucleon space wave function is turned into a space wave function that represents the new nucleon. This is physically reasonable. This agreement is to ensure that $\tau$ in the radial wave function $u_{l'j'}^{\tau}(r)$ is consistent with $\tau$ in $\chi_{\frac{1}{2}\tau}$. The total wave function of the system can be expressed as

$$\psi_{\tau} = \varphi_{\frac{1}{2}\tau} \Phi_{\frac{A}{2}\ \frac{N-Z}{2}} \tag{5.599}$$

In the case of non-relativistic theory, the stationary Schrodinger equation satisfied by $\psi_{\tau}$ is

$$\left[ -\frac{\hbar^2}{2\mu_{\tau}} \vec{\nabla}^2 + \left( V_0 + \frac{\vec{t} \cdot \vec{T}}{A} V_1 \right) \right] \psi_{\tau} = E_{\tau} \psi_{\tau} \tag{5.600}$$

where $\mu_{\tau}$ and $E_{\tau}$ are the reduced mass and the energy of the nucleon $\tau$, respectively. For $\tau = -\frac{1}{2}$ protons, while substituting Eqs. (5.599), (5.595), (5.593), and (5.598) into Eq. (5.600) and multiplying $\sum_{\tau' T'_0} \Phi^{+}_{\frac{A}{2} T'_0} \chi^{+}_{\frac{1}{2}\tau'} r \int \mathrm{d}\Omega\, \mathbb{Y}^{m_j +}_{l\frac{1}{2}j}(\Omega)$ from the left side of the obtained expression, the following equation can be obtained:

$$\begin{aligned} &\left[ \frac{\hbar^2}{2\mu_{\mathrm{p}}} \left( \frac{\mathrm{d}^2}{\mathrm{d}r^2} - \frac{l(l+1)}{r^2} \right) - \left( V_0 + V_{\mathrm{C}} - \frac{N-Z}{4A} V_1 \right) + E_{\mathrm{p}} \right] u^{\mathrm{p}}_{lj}(r) \\ &+ \sqrt{\frac{N(Z+1)}{2A}}\, V_1\, u^{\mathrm{n}}_{lj}(r) = 0 \end{aligned} \tag{5.601}$$

In the above equation the Coulomb potential $V_{\mathrm{C}}$ is added. For $\tau = \frac{1}{2}$ neutrons, similar methods are used, but Eqs. (5.594) and (5.597) need to be used, and the following equation can be obtained:

$$\begin{aligned} &\left[ \frac{\hbar^2}{2\mu_{\mathrm{n}}} \left( \frac{\mathrm{d}^2}{\mathrm{d}r^2} - \frac{l(l+1)}{r^2} \right) - \left( V_0 + \frac{N-Z}{4A} V_1 \right) + E_{\mathrm{n}} \right] u^{\mathrm{n}}_{lj}(r) \\ &+ \sqrt{\frac{Z(N+1)}{2A}}\, V_1\, u^{\mathrm{p}}_{lj}(r) = 0 \end{aligned} \tag{5.602}$$

Whether proton is the incident particle or neutron is the incident particle, Eqs. (5.601) and (5.602) must be solved together. The $S$ matrix elements $S^{lj}_{\mathrm{p,p}}$, $S^{lj}_{\mathrm{n,p}}$, $S^{lj}_{\mathrm{n,n}}$, $S^{lj}_{\mathrm{p,n}}$ can be obtained through boundary conditions, and the relevant cross sections, differential cross sections, and polarization quantities can also be calculated. Through (p, n) or (n, p) reaction the target nucleus will become its own isospin analogous state (IAS).

In the case of relativity, the following form of optical potential can be taken [66]:

$$U_{\mathrm{opt}} = \left[ U^0_s + \gamma^0 U^0_0 \right] + \left[ U^1_s + \gamma^0 U^1_0 \right] \left( \vec{t} \cdot \vec{T} \right) / A \tag{5.603}$$

The Dirac equation given by Eq. (5.226) is

$$\left[\vec{\alpha} \cdot \vec{p} + \beta\left(m_\tau + U_{\text{opt}}\right) - E_\tau\right] \psi_\tau = 0 \tag{5.604}$$

Substituting Eq. (5.603) into Eq. (5.604) and using Eqs. (5.593), (5.594), (5.597), and (5.598), the following equations can be obtained:

$$\begin{aligned}&\left[\vec{\alpha} \cdot \vec{p} + \beta\left(m_{\text{p}} + U_{\text{s}}^0\right) + U_0^0 - \frac{N-Z}{4A}\left(\beta U_{\text{s}}^1 + U_0^1\right) - E_{\text{p}}\right]\psi_{\text{p}} \\ &\quad - \frac{\sqrt{N(Z+1)}}{2A}\left(\beta U_{\text{s}}^1 + U_0^1\right)\psi_{\text{n}} = 0\end{aligned} \tag{5.605}$$

$$\begin{aligned}&\left[\vec{\alpha} \cdot \vec{p} + \beta\left(m_{\text{n}} + U_{\text{s}}^0\right) + U_0^0 + \frac{N-Z}{4A}\left(\beta U_{\text{s}}^1 + U_0^1\right) - E_{\text{n}}\right]\psi_{\text{n}} \\ &\quad - \frac{\sqrt{Z(N+1)}}{2A}\left(\beta U_{\text{s}}^1 + U_0^1\right)\psi_{\text{p}} = 0\end{aligned} \tag{5.606}$$

where $\psi_{\text{p}}$ and $\psi_{\text{n}}$ are the 4-component wave functions and they can be divided into upper and lower components. Referring to Eq. (5.287) the proton upper component equation needs to be added to the Coulomb potential $\frac{\mu_{\text{p}}}{E_{\text{p}}} V_{\text{C}}$; $\mu_{\text{p}}$ is the proton reduced mass. Eqs. (5.605) and (5.606) should be solved simultaneously. The optical potential appeared in the equations may use the phenomenological potential, or the microscopic potential calculated by the RIA method; it means that the RIA method can be used to study (p, n) and (n, p) reactions. The above equations can also be solved by the coupling channel method described in Sect. 5.5. In the spherical nucleus case, the $S$ matrix elements $S_{\text{p,p}}^{(\rho)lj}, S_{\text{n,p}}^{(\rho)lj}, S_{\text{n,n}}^{(\rho)lj}, S_{\text{p,n}}^{(\rho)lj}$ are introduced, where $\rho = \text{u, d}$ represent the upper component and the lower component, respectively. Finally, the relevant cross sections, differential cross sections, and polarization quantities can be calculated by using the calculation formulas given in Sect. 5.4.

## 5.10 Relativistic Classical Field Theory and Lagrangian Density in Quantum Hadron Dynamics

### *5.10.1 Relativistic Classical Field Theory [2, 67, 68]*

Under the $\hbar = c = 1$ unit system and the Bjorken-Drell metric, the physical quantity of the field is set to be

$$\varphi_\lambda(x), \qquad \lambda = 1,2,\cdots,n \tag{5.607}$$

where

$$x^{\mu} = \left(x^0, x^1, x^2, x^3\right) = \left(t, \vec{r}\right), \quad x_{\mu} = (x_0, x_1, x_2, x_3) = \left(t, -\vec{r}\right) \tag{5.608}$$

and $n$ represents the number of physical quantities of the field at each space-time point. Assume that the Lagrangian density $\mathcal{L}$ is a function of $\varphi_{\lambda}$ and $\partial_{\mu}\varphi_{\lambda}$, $\partial_{\mu} \equiv \frac{\partial}{\partial x^{\mu}}$ has been given by Eq. (5.60). The Lagrange function is equal to

$$L = \int \mathrm{d}^4 x \mathcal{L}\left(\varphi_{\lambda}, \partial_{\mu}\varphi_{\lambda}\right) \tag{5.609}$$

Make a slight change for the physical quantity of the field as follows:

$$\varphi_{\lambda}(x) \rightarrow \varphi_{\lambda}(x) + \delta\varphi_{\lambda}(x) \tag{5.610}$$

and it is required that $\delta\varphi_{\lambda}(x)$ is equal to zero at the boundary of the Lagrange function integral; at the same time assume that the realistic field physical quantity $\varphi_{\lambda}$ makes the Lagrange function taking the extreme value, i.e., any $\delta\varphi_{\lambda}$ does not change the Lagrange function value. So the partial integral can be used to obtain

$$\begin{aligned} \delta L &= \int \mathrm{d}^4 x \left\{ \frac{\partial \mathcal{L}}{\partial \varphi_{\lambda}} \delta\varphi_{\lambda} + \frac{\partial \mathcal{L}}{\partial\left(\partial_{\mu}\varphi_{\lambda}\right)} \delta\left(\partial_{\mu}\varphi_{\lambda}\right) \right\} \\ &= \int \mathrm{d}^4 x \delta\varphi_{\lambda} \left\{ \frac{\partial \mathcal{L}}{\partial \varphi_{\lambda}} - \partial_{\mu}\left( \frac{\partial \mathcal{L}}{\partial\left(\partial_{\mu}\varphi_{\lambda}\right)} \right) \right\} = 0 \end{aligned} \tag{5.611}$$

Thus the Euler-Lagrange equation is obtained as

$$\frac{\partial \mathcal{L}}{\partial \varphi_{\lambda}} - \partial_{\mu}\left( \frac{\partial \mathcal{L}}{\partial\left(\partial_{\mu}\varphi_{\lambda}\right)} \right) = 0, \qquad \lambda = 1,\ 2,\ \cdots,\ n \tag{5.612}$$

The energy-momentum tensor is defined as

$$T^{\mu\nu} \equiv -g^{\mu\nu}\mathcal{L} + \frac{\partial \mathcal{L}}{\partial\left(\partial_{\mu}\varphi_{\lambda}\right)} \partial^{\nu}\varphi_{\lambda} \tag{5.613}$$

The metric matrix element $g^{\mu\nu}$ has been given by Eq. (5.53). The following expressions can be obtained using Eqs. (5.613) and (5.612):

$$\begin{aligned} \partial_{\mu}T^{\mu\nu} &= -g^{\mu\nu}\partial_{\mu}\mathcal{L} + \partial_{\mu}\left( \frac{\partial \mathcal{L}}{\partial\left(\partial_{\mu}\varphi_{\lambda}\right)} \right) \partial^{\nu}\varphi_{\lambda} + \frac{\partial \mathcal{L}}{\partial\left(\partial_{\mu}\varphi_{\lambda}\right)} \partial_{\mu}\partial^{\nu}\varphi_{\lambda} \\ &= -g^{\mu\nu}\partial_{\mu}\mathcal{L} + \frac{\partial \mathcal{L}}{\partial \varphi_{\lambda}} \partial^{\nu}\varphi_{\lambda} + \frac{\partial \mathcal{L}}{\partial\left(\partial_{\mu}\varphi_{\lambda}\right)} \partial_{\mu}\partial^{\nu}\varphi_{\lambda}, \end{aligned}$$

$$\partial_\mu \mathcal{L} = \frac{\partial \mathcal{L}}{\partial \varphi_\lambda}\partial_\mu \varphi_\lambda + \frac{\partial \mathcal{L}}{\partial(\partial_{\mu'}\varphi_\lambda)}\partial_\mu \partial_{\mu'}\varphi_\lambda,$$

$$-g^{\mu\nu}\partial_\mu \mathcal{L} = \delta_{\mu\nu}\begin{cases} -\dfrac{\partial \mathcal{L}}{\partial \varphi_\lambda}\partial^0 \varphi_\lambda - \dfrac{\partial \mathcal{L}}{\partial(\partial_{\mu'}\varphi_\lambda)}\partial_{\mu'}\partial^0 \varphi_\lambda, & \nu = 0 \\ -\dfrac{\partial \mathcal{L}}{\partial \varphi_\lambda}\partial^i \varphi_\lambda - \dfrac{\partial \mathcal{L}}{\partial(\partial_{\mu'}\varphi_\lambda)}\partial_{\mu'}\partial^i \varphi_\lambda, & \nu = i = 1,\ 2,\ 3 \end{cases}$$

Then we get the continuous equation as

$$\partial_\mu T^{\mu\nu} = 0 \tag{5.614}$$

The 4-momentum is defined as

$$P^\nu = \int \mathrm{d}\vec{r}\ T^{0\nu} \tag{5.615}$$

And the conjugate physical quantity $\pi_\lambda$ corresponding to the generalized coordinate $\varphi_\lambda$ is defined as

$$\pi_\lambda \equiv \frac{\partial \mathcal{L}}{\partial \dot{\varphi}_\lambda} \tag{5.616}$$

where $\dot{\varphi}_\lambda = \frac{\partial \varphi_\lambda}{\partial t}$, $\pi_\lambda$ is the generalized momentum. Then we introduce the Hamiltonian density $\mathcal{H}$ and from Eq. (5.613) we can get

$$\mathcal{H} \equiv \pi_\lambda \dot{\varphi}_\lambda - \mathcal{L} = T^{00} \tag{5.617}$$

The total energy of the field is

$$E = P^0 = H = \int \mathrm{d}\vec{r}\mathcal{H} \tag{5.618}$$

In Eq. (5.617), it is understood that $\mathcal{L}$ and $\mathcal{H}$ are the functions of $\varphi_\sigma$, $\pi_\sigma$, and $\dot{\varphi}_\sigma$; moreover $\dot{\varphi}_\lambda$ is a function of $\varphi_\sigma$ and $\pi_\sigma$. From Eq. (5.617) we can obtain

$$\frac{\partial \mathcal{H}}{\partial \pi_\sigma} = \dot{\varphi}_\sigma + \pi_\lambda \frac{\partial \dot{\varphi}_\lambda}{\partial \pi_\sigma} - \frac{\partial \mathcal{L}}{\partial \dot{\varphi}_\lambda}\frac{\partial \dot{\varphi}_\lambda}{\partial \pi_\sigma} = \dot{\varphi}_\sigma \tag{5.619}$$

Utilizing Eq. (5.612) we also have

$$\frac{\partial \mathcal{H}}{\partial \varphi_\sigma} = \pi_\lambda \frac{\partial \dot{\varphi}_\lambda}{\partial \varphi_\sigma} - \frac{\partial \mathcal{L}}{\partial \varphi_\sigma} - \frac{\partial \mathcal{L}}{\partial \dot{\varphi}_\lambda}\frac{\partial \dot{\varphi}_\lambda}{\partial \pi_\sigma} = -\frac{\partial \mathcal{L}}{\partial \varphi_\sigma}$$
$$= -\partial_\mu \left[\frac{\partial \mathcal{L}}{\partial(\partial_\mu \varphi_\sigma)}\right] = -\dot{\pi}_\sigma - \frac{\partial}{\partial x_i}\left[\frac{\partial \mathcal{L}}{\partial\left(\frac{\partial \varphi_\sigma}{\partial x_i}\right)}\right] \tag{5.620}$$

$$\frac{\partial \mathcal{H}}{\partial\left(\frac{\partial \varphi_\sigma}{\partial x_i}\right)} = \pi_\lambda \frac{\partial \dot{\varphi}_\lambda}{\partial\left(\frac{\partial \varphi_\sigma}{\partial x_i}\right)} - \frac{\partial \mathcal{L}}{\partial\left(\frac{\partial \varphi_\sigma}{\partial x_i}\right)} - \frac{\partial \mathcal{L}}{\partial \dot{\varphi}_\lambda}\frac{\partial \dot{\varphi}_\lambda}{\partial\left(\frac{\partial \varphi_\sigma}{\partial x_i}\right)} = -\frac{\partial \mathcal{L}}{\partial\left(\frac{\partial \varphi_\sigma}{\partial x_i}\right)} \tag{5.621}$$

Substituting Eq. (5.621) into Eq. (5.620) we can get

$$\dot{\pi}_\sigma = -\frac{\partial \mathcal{H}}{\partial \varphi_\sigma} + \frac{\partial}{\partial x_i}\left[\frac{\partial \mathcal{H}}{\partial\left(\frac{\partial \varphi_\sigma}{\partial x_i}\right)}\right] \tag{5.622}$$

Equations (5.619) and (5.622) are the Hamiltonian canonical motion equations of the field.

### *5.10.2 Lagrangian Density in Quantum Hadron Dynamics [2, 68–74]*

In order to study high-temperature and high-density systems, it is necessary to introduce strong interacting particles on the basis of relativistic field theory and develop quantum Hadron dynamics models. In the Walecka model [69] it is believed that nucleon-nucleon interaction is achieved by exchanging one boson, and there are many kinds of mesons and photons to be changed between the nucleons. Introduce the following Lagrangian density:

$$\mathcal{L} = \mathcal{L}_N + \mathcal{L}_M + \mathcal{L}_{\text{int}} \tag{5.623}$$

where $\mathcal{L}_N$ and $\mathcal{L}_M$ represent the Lagrangian densities of free nucleons and free mesons, respectively, and $\mathcal{L}_{\text{int}}$ represents the Lagrangian density of interaction between nucleons and mesons. The Lagrangian density of free nucleons is taken as

$$\mathcal{L}_N = \overline{\psi}_i\left(\mathrm{i}\gamma^\mu \partial_\mu - M\right)\psi_i \tag{5.624}$$

where $\psi_i$ is the Dirac spinor field describing the $i$th nucleon in the nucleus, and $M$ is the nucleon stationary mass. The sum of $i$ is included in Eq. (5.624). Using partial integral we have

$$\overline{\psi}_i \mathrm{i}\gamma^\mu \partial_\mu \psi_i \rightarrow -\mathrm{i}\gamma^\mu \left(\partial_\mu \overline{\psi}_i\right) \psi_i$$

In Eq. (5.612) when $\varphi_\lambda$ is replaced with $\overline{\psi}_i$ we can obtain

$$\frac{\partial \mathcal{L}_N}{\partial \overline{\psi}_i} = -M\psi_i \tag{5.625}$$

$$\partial_\mu \left[\frac{\partial \mathcal{L}_N}{\partial \left(\partial_\mu \overline{\psi}_i\right)}\right] = -\mathrm{i}\gamma^\mu \partial_\mu \psi_i \tag{5.626}$$

Using the Lagrange equation (5.612) we can get

$$\left(\mathrm{i}\gamma^\mu \partial_\mu - M\right) \psi_i = 0 \tag{5.627}$$

This is the Dirac equation (5.66) of free nucleons.

σ meson is a scalar meson whose spin is equal to zero in charge neutrality (isospin scalar) and is used to simulate the attraction between nucleons. However, the existence of σ mesons has not been confirmed experimentally so far. In Eq. (5.612) $\varphi_\lambda$ is replaced with $\sigma$, and the Lagrangian density of the free σ meson is taken as

$$\mathcal{L}_\sigma = \frac{1}{2}\partial_\mu \sigma\, \partial^\mu \sigma - U(\sigma) \tag{5.628}$$

$$U(\sigma) = \frac{1}{2} m_\sigma^2 \sigma^2 + \frac{1}{3} g_2 \sigma^3 + \frac{1}{4} g_3 \sigma^4 \tag{5.629}$$

where $m_\sigma$ is σ meson stationary mass, the second and third terms of $U(\sigma)$ are the non-linear potential of σ meson, and $g_2$ and $g_3$ are the non-linear term coupling constants of σ meson self-interaction. Substituting the scalar meson $\mathcal{L}_\sigma$ without the non-linear terms by taking $g_2 = g_3 = 0$ into Eq. (5.617), the following expression can be obtained:

$$\begin{aligned} \mathcal{H}_\sigma &= \pi_\sigma^2 - \frac{1}{2}\pi_\sigma^2 + \frac{1}{2}\left(\vec{\nabla}\sigma\right)^2 + \frac{1}{2} m_\sigma^2 \sigma^2 \\ &= \frac{1}{2}\left[\pi_\sigma^2 + \left(\vec{\nabla}\sigma\right)^2 + m_\sigma^2 \sigma^2\right] \end{aligned} \tag{5.630}$$

This proves that the Hamiltonian density of scalar mesons without non-linear terms is always positive. We know $\partial_\mu \partial^\mu = \partial^\mu \partial_\mu$ according to Eqs. (5.55) and (5.60), and then the following expressions can be obtained from Eqs. (5.628) and (5.629):

$$\frac{\partial \mathcal{L}_\sigma}{\partial \sigma} = -m_\sigma^2 \sigma - g_2 \sigma^2 - g_3 \sigma^3 \tag{5.631}$$

$$\partial_\mu \left[ \frac{\partial \mathcal{L}_\sigma}{\partial (\partial_\mu \sigma)} \right] = \partial_\mu \partial^\mu \sigma \tag{5.632}$$

Using the Lagrange equation (5.612), the free σ meson equation is obtained as

$$\left( \partial_\mu \partial^\mu + m_\sigma^2 \right) \sigma + g_2 \sigma^2 + g_3 \sigma^3 = 0 \tag{5.633}$$

Substituting Eq. (5.65) into Eq. (5.6) we can get the Klein-Gordon (KG) equation of free particles as

$$(\partial_\mu \partial^\mu + m_0^2) \psi = 0 \tag{5.634}$$

In $g_2 = g_3 = 0$ case, Eq. (5.633) is the same as Eq. (5.634).

The ω meson is a vector meson whose spin is equal to 1 in charge neutrality (isospin scalar), and the field physical quantity $\omega^\mu$ is 4-vector. The ω mesons are used to describe short-range repulsion between nucleons. The Lagrangian density of the free ω mesons is taken as

$$\mathcal{L}_\omega = -\frac{1}{4} \Omega_{\mu\nu} \Omega^{\mu\nu} + \frac{1}{2} m_\omega^2 \omega_\mu \omega^\mu + \frac{1}{4} C_3 \left( \omega_\mu \omega^\mu \right)^2 \tag{5.635}$$

The anti-symmetric field tensor of the ω mesons is

$$\Omega^{\mu\nu} = \partial^\mu \omega^\nu - \partial^\nu \omega^\mu \tag{5.636}$$

where $m_\omega$ is ω meson stationary mass and $C_3$ is a non-linear term constant. The following relations can be obtained from Eqs. (5.55) and (5.60):

$$\omega_\mu \omega^\mu = \omega^\mu \omega_\mu, \qquad \Omega_{\mu\nu} \Omega^{\mu\nu} = \Omega^{\mu\nu} \Omega_{\mu\nu} \tag{5.637}$$

We can get the following expressions from Eqs. (5.635) and (5.636):

$$\frac{\partial \mathcal{L}_\omega}{\partial \omega_\nu} = m_\omega^2 \omega^\nu + C_3 \left( \omega_\mu \omega^\mu \right) \omega^\nu \tag{5.638}$$

$$\partial_\mu \left[ \frac{\partial \mathcal{L}_\omega}{\partial (\partial_\mu \omega_\nu)} \right] = -\partial_\mu \Omega^{\mu\nu} \tag{5.639}$$

In the derivation of Eq. (5.639), a number 2 will come out when making the square of $\Omega_{\mu\nu}$, and the other 2 can be understood by the following formula:

$$\left[\frac{\partial \Omega_{\mu\nu}}{\partial(\partial_1\omega_2)}\right]\Omega^{\mu\nu}=\left[\frac{\partial(\partial_1\omega_2-\partial_2\omega_1)}{\partial(\partial_1\omega_2)}\right]\left(\partial^1\omega^2-\partial^2\omega^1\right)$$
$$+\left[\frac{\partial(\partial_2\omega_1-\partial_1\omega_2)}{\partial(\partial_1\omega_2)}\right]\left(\partial^2\omega^1-\partial^1\omega^2\right)=\left(\partial^1\omega^2-\partial^2\omega^1\right)-\left(\partial^2\omega^1-\partial^1\omega^2\right)=2\Omega^{12}$$

So the equation of the free ω vector meson obtained by using Eq. (5.612) is

$$\partial_\mu\Omega^{\mu\nu}+m_\omega^2\omega^\nu+C_3\left(\omega_\mu\omega^\mu\right)\omega^\nu=0 \tag{5.640}$$

If $C_3=0$ above formula becomes

$$\partial_\mu\Omega^{\mu\nu}+m_\omega^2\omega^\nu=0 \tag{5.641}$$

This formula is called the Proca equation of vector particles. Taking divergence of Eq. (5.641) we can obtain [9]

$$\partial_\nu\left(\partial_\mu\Omega^{\mu\nu}+m_\omega^2\omega^\nu\right)=\partial_\nu\partial_\mu\partial^\mu\omega^\nu-\partial_\nu\partial_\mu\partial^\nu\omega^\mu+m_\omega^2\partial_\nu\omega^\nu=0$$

Since the sums to $\mu$, $\nu$ should be done simultaneously, it can be proved that

$$\partial_\nu\partial_\mu\partial^\mu\omega^\nu-\partial_\nu\partial_\mu\partial^\nu\omega^\mu=\partial_\nu\partial_\mu\partial^\mu\omega^\nu-\partial_\mu\partial_\nu\partial^\mu\omega^\nu=0$$

So for the free vector particles with mass $m\neq 0$, there is the following relation:

$$\partial_\nu\omega^\nu=0 \tag{5.642}$$

This formula is called the Lorentz gauge of vector particles. This formula does not necessarily hold when there is an interaction potential. The following equation can be obtained substituting Eq. (5.642) into Eqs. (5.640) and (5.641):

$$\left(\partial_\mu\partial^\mu+m_\omega^2\right)\omega^\nu+C_3\left(\omega_\mu\omega^\mu\right)\omega^\nu=0 \tag{5.643}$$

$$\left(\partial_\mu\partial^\mu+m_\omega^2\right)\omega^\nu=0 \tag{5.644}$$

Equation (5.644) is the KG equation satisfied by the vector meson.

The ρ meson is a vector meson whose spin is equal to 1 with charge (isospin vector), its field physical quantity is represented by $\vec{\rho}^{\mu}$, where the arrow represents a three-dimensional vector in the isospin space, and it can be positively charged (+), uncharged (0), and negatively charged (−). The isospin space of the nucleon is two-dimensional, and the ρ meson isospin space is three-dimensional. The inclusion of ρ mesons in the Lagrangian density is to better describe the isospin effect. The Lagrangian density of the ρ mesons is taken as [2, 70–72]

$$\mathcal{L}_\rho = -\frac{1}{4}\vec{R}_{\mu\nu} \cdot \vec{R}^{\mu\nu} + \frac{1}{2}m_\rho^2 \vec{\rho}_\mu \cdot \vec{\rho}^{\mu} \tag{5.645}$$

where

$$\vec{R}^{\mu\nu} = \partial^\mu \vec{\rho}^{\nu} - \partial^\nu \vec{\rho}^{\mu} - g_\rho \left(\vec{\rho}^{\mu} \times \vec{\rho}^{\nu}\right) \tag{5.646}$$

and $m_\rho$ is ρ meson stationary mass. In order to make the ρ meson coupled with the isospin vector flow, a non-linear ρ meson interaction term is added in Eq. (5.646), and $g_\rho$ is a ρ meson coupling constant. If this item is not added, the field equation of the ρ meson can be derived by the same way with the derivation of the ω meson equation. Their difference is that the ρ meson is an isospin vector meson. Therefore let [2]

$$\vec{L}^{\mu\nu} = \partial^\mu \vec{\rho}^{\nu} - \partial^\nu \vec{\rho}^{\mu} \tag{5.647}$$

$$\mathcal{L}_\rho^0 = -\frac{1}{4}\vec{L}_{\mu\nu} \cdot \vec{L}^{\mu\nu} + \frac{1}{2}m_\rho^2 \vec{\rho}_\mu \cdot \vec{\rho}^{\mu} \tag{5.648}$$

then we can obtain

$$\begin{aligned} \Delta\mathcal{L}_\rho &\equiv \mathcal{L}_\rho - \mathcal{L}_\rho^0 \\ &= \frac{1}{4}g_\rho\left[\left(\vec{\rho}_\mu \times \vec{\rho}_\nu\right) \cdot \vec{L}^{\mu\nu} + \left(\vec{\rho}^{\mu} \times \vec{\rho}^{\nu}\right) \cdot \vec{L}_{\mu\nu} - g_\rho\left(\vec{\rho}_\mu \times \vec{\rho}_\nu\right) \cdot \left(\vec{\rho}^{\mu} \times \vec{\rho}^{\nu}\right)\right] \\ &= \frac{1}{4}g_\rho\left[-\left(\vec{\rho}_\nu \times \vec{\rho}_\mu\right) \cdot \vec{R}^{\mu\nu} + \left(\vec{\rho}^{\mu} \times \vec{\rho}^{\nu}\right) \cdot \vec{L}_{\mu\nu}\right] \end{aligned} \tag{5.649}$$

Using the vector formula $\left(\vec{a} \times \vec{b}\right) \cdot \vec{c} = \vec{a} \cdot \left(\vec{b} \times \vec{c}\right)$, from above formula the following expression can be obtained:

$$\Delta\mathcal{L}_\rho = \frac{1}{4}g_\rho\left[-\vec{\rho}_\nu \cdot \left(\vec{\rho}_\mu \times \vec{R}^{\mu\nu}\right) + \left(\vec{\rho}^{\mu} \times \vec{\rho}^{\nu}\right) \cdot \vec{L}_{\mu\nu}\right] \tag{5.650}$$

When substituting $\Delta\mathcal{L}_\rho$ into the Lagrange equation (5.612), we ignore the contributions of the following items:

$$\left[\frac{\partial(\vec{\rho}^{\mu'} \times \vec{\rho}^{\nu'})}{\partial \vec{\rho}_\nu}\right] \cdot \vec{L}_{\mu'\nu'}, \quad \vec{\rho}_{\nu'} \cdot \left[\frac{\partial(\vec{\rho}_{\mu'} \times \vec{R}^{\mu'\nu'})}{\partial \vec{\rho}_\nu}\right], \quad \vec{\rho}_{\nu'} \cdot \left[\frac{\partial(\vec{\rho}_{\mu'} \times \vec{R}^{\mu'\nu'})}{\partial\left(\partial_\mu \vec{\rho}_\nu\right)}\right]$$

So we can get

$$\frac{\partial \Delta \mathcal{L}_\rho}{\partial \vec{\rho}_\nu} = -\frac{1}{4} g_\rho \left( \vec{\rho}_\mu \times \vec{R}^{\mu\nu} \right) \tag{5.651}$$

$$\frac{\partial \Delta \mathcal{L}_\rho}{\partial \left( \partial_\mu \vec{\rho}_\nu \right)} = \frac{1}{2} g_\rho \left( \vec{\rho}^\mu \times \vec{\rho}^\nu \right) \tag{5.652}$$

Referring to Eq. (5.640) the field equation of the ρ meson can be written as

$$\partial_\mu \vec{L}^{\mu\nu} + m_\rho^2 \vec{\rho}^\nu = g_\rho \left[ \frac{1}{4} \left( \vec{\rho}_\mu \times \vec{R}^{\mu\nu} \right) + \frac{1}{2} \partial_\mu \left( \vec{\rho}^\mu \times \vec{\rho}^\nu \right) \right] \tag{5.653}$$

If the non-linear term is not included in Eq. (5.646), i.e., let $g_\rho = 0$ in Eq. (5.653), and using the Lorentz gauge $\partial_\nu \vec{\rho}^\nu = 0$ of the vector particle, Eq. (5.653) can be rewritten as

$$\left( \partial_\mu \partial^\mu + m_\rho^2 \right) \vec{\rho}^\nu = 0 \tag{5.654}$$

This is the KG equation satisfied by ρ meson.

π meson is a pseudo-scalar meson whose spin is equal to zero with charge (isospin vector); the field physical quantity is represented by symbol $\vec{\pi}$, where the arrow indicates that π meson is a three-dimensional vector in the isospin space. Similarly, in order to consider the coupling problem with the isospin vector flow, the following form of Lagrangian density is introduced for π meson [2, 70, 71, 74]:

$$\mathcal{L}_\pi = \frac{1}{2} \vec{P}_\mu \cdot \vec{P}^\mu - \frac{1}{2} m_\pi^2 \vec{\pi} \cdot \vec{\pi} \tag{5.655}$$

$$\vec{P}^\mu = \partial^\mu \vec{\pi} - g_\rho \vec{\rho}^\mu \times \vec{\pi} \tag{5.656}$$

Let

$$\mathcal{L}_\pi^0 = \frac{1}{2} \partial_\mu \vec{\pi} \cdot \partial^\mu \vec{\pi} - \frac{1}{2} m_\pi^2 \vec{\pi} \cdot \vec{\pi} \tag{5.657}$$

then we can obtain

$$\begin{aligned} \Delta \mathcal{L}_\pi &\equiv \mathcal{L}_\pi - \mathcal{L}_\pi^0 \\ &= -\frac{1}{2} g_\rho \left[ \left( \vec{\rho}_\mu \times \vec{\pi} \right) \cdot \partial^\mu \vec{\pi} + \left( \vec{\rho}^\mu \times \vec{\pi} \right) \cdot \partial_\mu \vec{\pi} - g_\rho \left( \vec{\rho}_\mu \times \vec{\pi} \right) \cdot \left( \vec{\rho}^\mu \times \vec{\pi} \right) \right] \\ &= -\frac{1}{2} g_\rho \left[ - \left( \vec{\pi} \times \vec{\rho}_\mu \right) \cdot \vec{P}^\mu + \left( \vec{\rho}^\mu \times \vec{\pi} \right) \cdot \partial_\mu \vec{\pi} \right] \\ &= -\frac{1}{2} g_\rho \left[ - \vec{\pi} \cdot \left( \vec{\rho}_\mu \times \vec{P}^\mu \right) + \left( \vec{\rho}^\mu \times \vec{\pi} \right) \cdot \partial_\mu \vec{\pi} \right] \end{aligned} \tag{5.658}$$

When substituting $\Delta\mathcal{L}_\pi$ into the Lagrange equation (5.612), we ignore the contributions of the following items:

$$\left[\frac{\partial\left(\vec{\rho}^{\mu} \times \vec{\pi}\right)}{\partial\vec{\pi}}\right] \cdot \partial_\mu \vec{\pi}, \qquad \vec{\pi} \cdot \left[\frac{\partial\left(\vec{\rho}_{\mu'} \times \vec{P}^{\mu'}\right)}{\partial\left(\partial_\mu \vec{\pi}\right)}\right]$$

So we can get

$$\frac{\partial \Delta\mathcal{L}_\pi}{\partial\vec{\pi}} = \frac{1}{2} g_\rho \left(\vec{\rho}_\mu \times \vec{P}^{\mu}\right) \tag{5.659}$$

$$\frac{\partial \Delta\mathcal{L}_\pi}{\partial\left(\partial_\mu \vec{\pi}\right)} = -\frac{1}{2} g_\rho \left(\vec{\rho}^{\mu} \times \vec{\pi}\right) \tag{5.660}$$

Refer to Eqs. (5.634) and (5.653) and noting the difference between the $m^2$ term signs in Eq. (5.648) and Eq. (5.657), the field equation of the $\pi$ meson can be written as

$$\left(\partial_\mu \partial^\mu + m_\pi^2\right) \vec{\pi} = g_\rho \left[\frac{1}{2}\left(\vec{\rho}_\mu \times \vec{P}^{\mu}\right) + \frac{1}{2}\partial_\mu\left(\vec{\rho}^{\mu} \times \vec{\pi}\right)\right] \tag{5.661}$$

If the non-linear term is not included in Eq. (5.656), i.e., let $g_\rho = 0$ in Eq. (5.661), and then Eq. (5.661) can be rewritten as

$$\left(\partial_\mu \partial^\mu + m_\pi^2\right) \vec{\pi} = 0 \tag{5.662}$$

This is the KG equation satisfied by $\pi$ meson.

The electromagnetic field is a vector field with a spin equal to 1 and a mass equal to zero, and it is expressed as $A^\mu$, $A = (A^0, A^1, A^2, A^3) = \left(\phi, \vec{A}\right)$, $\vec{A}$ is a three-dimensional vector potential and $\phi$ represents a scalar potential. The Lagrangian density of the free electromagnetic field is taken as

$$\mathcal{L}_\mathrm{A} = -\frac{1}{4} F_{\mu\nu} F^{\mu\nu} \tag{5.663}$$

$$F^{\mu\nu} = \partial^\mu A^\nu - \partial^\nu A^\mu \tag{5.664}$$

From the Lagrange equation (5.612) we can obtain

$$\partial_\mu F^{\mu\nu} = 0 \tag{5.665}$$

The free electromagnetic field satisfies the following Lorentz gauge given by Eq. (5.642):

$$\partial_\nu A^\nu = 0 \tag{5.666}$$

So Eq. (5.665) can be rewritten as

$$\partial_\mu \partial^\mu A^\nu = 0 \tag{5.667}$$

The Lagrangian densities of the interaction between the nucleon and σ, ω, ρ mesons can be taken as

$$\mathcal{L}_{\text{int}, \sigma} = -g_\sigma \overline{\Psi}_i \sigma \psi_i \tag{5.668}$$

$$\mathcal{L}_{\text{int}, \omega} = -g_\omega \overline{\Psi}_i \gamma^\mu \omega_\mu \psi_i - \frac{f_\omega}{4M} \overline{\Psi}_i \sigma^{\mu\nu} \Omega_{\mu\nu} \psi_i \tag{5.669}$$

$$\mathcal{L}_{\text{int}, \rho} = -g_\rho \overline{\Psi}_i \gamma^\mu \vec{\tau} \cdot \vec{\rho}_\mu \psi_i - \frac{f_\rho}{4M} \overline{\Psi}_i \sigma^{\mu\nu} \vec{\tau} \cdot \vec{L}_{\mu\nu} \psi_i \tag{5.670}$$

where $\sigma^{\mu\nu}$ has been given by Eq. (5.538) and $\Omega_{\mu\nu}$ and $\vec{L}_{\mu\nu}$ have been given by Eq. (5.636) and Eq. (5.647), respectively. $\vec{\tau}$ is a nucleon Pauli isospin matrix. The first items of Eq. (5.669) and Eq. (5.670) represent vector coupling, and the second items represent tensor coupling and $f_\omega$ and $f_\rho$ are tensor coupling constants. For π meson there is

$$\mathcal{L}_{\text{int}, \pi} = -\mathrm{i} g_\pi \overline{\Psi}_i \gamma^5 \vec{\tau} \cdot \vec{\pi} \psi_i - \frac{f_\pi}{m_\pi} \overline{\Psi}_i \gamma^5 \gamma^\mu \left( \vec{\tau} \cdot \partial_\mu \vec{\pi} \right) \psi_i \tag{5.671}$$

In the above formula the first term represents the pseudo-scalar (ps) coupling, and the second term represents the axis-vector (pv) coupling. For the electromagnetic field there is

$$\mathcal{L}_{\text{int}, \mathrm{A}} = -\mathrm{e} \overline{\psi}_i \gamma^\mu \frac{1 - \tau_3}{2} A_\mu \psi_i \tag{5.672}$$

In the above formula $\tau_3 = 1$ represents the neutrons, and $\tau_3 = -1$ represents the protons. And e is a proton charge.

### 5.10.3 Relativistic Mean Field Equations

Substituting the interaction Lagrangian density introduced earlier into the Lagrange equation (5.612) and merging the obtained result into the corresponding free field equation, the nucleon Dirac equation (5.627) becomes

$$\left[ \gamma^\mu \left( \mathrm{i} \partial_\mu - V_\mu \right) - (M + S) - \sigma^{\mu\nu} T_{\mu\nu} \right] \psi_i = 0 \tag{5.673}$$

where

$$S(x) = g_\sigma \sigma(x) + \mathrm{i} g_\pi \gamma^5 \vec{\tau} \cdot \vec{\pi}(x) \tag{5.674}$$

The first item on the right end of the above formula is the scalar field, and the second item is the pseudo-scalar field related to the isospin; $x$ represents the position of the space-time vertex where the interaction occurs.

$$V_\mu(x) = g_\omega \omega_\mu(x) + g_\rho \vec{\tau} \cdot \vec{\rho}_\mu(x) + \frac{f_\pi}{m_\pi} \gamma^5 \vec{\tau} \cdot \partial_\mu \vec{\pi}(x) + \mathrm{e} \frac{1 - \tau_3}{2} A_\mu(x) \tag{5.675}$$

The first item at the right end of the above formula is the vector term, the second item is the vector field related to the isospin, the third item is the axis-vector field related to the isospin, and the fourth item is the Coulomb potential.

$$T_{\mu\nu}(x) = \frac{f_\omega}{4M} G_{\mu\nu}(x) + \frac{f_\rho}{4M} \vec{\tau} \cdot L_{\mu\nu}(x) \tag{5.676}$$

The first item on the right end of the above formula is the tensor field, and the second item is the tensor field related to the isospin.

The σ meson equation (5.633) becomes

$$\left(\partial_\mu \partial^\mu + m_\sigma^2\right) \sigma = -g_\sigma \rho_\mathrm{S} - g_2 \sigma^2 - g_3 \sigma^3 \tag{5.677}$$

where

$$\rho_\mathrm{S}(x) = \sum_{i=1}^{A} \overline{\psi}_i(x) \, \psi_i(x) \tag{5.678}$$

$\rho_\mathrm{S}(x)$ is a nucleon scalar density.

The ω meson equation (5.640) becomes

$$\partial_\mu \Omega^{\mu\nu} + m_\omega^2 \omega^\nu = g_\omega j^\nu - \frac{f_\omega}{2M} t^\nu - C_3 (m_\omega \omega^\mu) \omega^\nu \tag{5.679}$$

where

$$j^\nu(x) = \sum_{i=1}^{A} \overline{\psi}_i(x) \gamma^\nu \psi_i(x) \tag{5.680}$$

$$t^\nu(x) = \sum_{i=1}^{A} \partial_\mu \left(\overline{\psi}_i(x) \sigma^{\mu\nu} \psi_i(x)\right) \tag{5.681}$$

Here $j^{\nu}$ is the nucleon flow density, and there is $j = (j^0, j^1, j^2, j^3) = \left(\rho_B, \vec{j}\right)$, where

$$\rho_B(x) = \sum_{i=1}^{A} \psi_i^+(x)\,\psi_i(x) \tag{5.682}$$

$\rho_B$ is the usual nucleon density, which is different from the nucleon scalar density $\rho_S$ given by Eq. (5.678). $t^{\nu}$ is called nucleon tensor flow density.

The ρ meson equation (5.653) becomes

$$\partial_\mu \vec{L}^{\mu\nu} + m_\rho^2 \vec{\rho}^{\nu} = g_\rho\left[\vec{j}_\tau^{\nu} + \frac{1}{4}\left(\vec{\rho}_\mu \times \vec{R}^{\mu\nu}\right) + \frac{1}{2}\partial_\mu\left(\vec{\rho}^{\mu} \times \vec{\rho}^{\nu}\right)\right] - \frac{f_\rho}{2M}\vec{t}_\tau^{\nu} \tag{5.683}$$

where

$$\vec{j}_\tau^{\nu}(x) = \sum_{i=1}^{A} \overline{\psi}_i(x)\gamma^{\nu}\vec{\tau}\psi_i(x) \tag{5.684}$$

$$\vec{t}_\tau^{\nu}(x) = \sum_{i=1}^{A} \partial_\mu\left(\overline{\psi}_i(x)\sigma^{\mu\nu}\vec{\tau}\psi_i(x)\right) \tag{5.685}$$

Here $\vec{j}_\tau^{\nu}$ is a nucleon isospin vector flow density vector, and $\vec{t}_\tau^{\nu}$ is a nucleon isospin tensor flow density vector. In the case of taking $f_\rho = 0$ and not considering π mesons, if the non-linear term related to $g_\rho$ is not added in Eq. (5.646), the second and third items related to $g_\rho$ will not appear at the right end of Eq. (5.683). If the initial state isospin vector flow $\vec{j}_\tau^{\nu}$ is 0, at this time, a ρ meson field $\vec{\rho}^{\nu} = 0$ is obtained. Then in the nucleon equations (5.673) and (5.675), the $g_\rho$ term also has no contribution, so it cannot cause the nucleons to produce an isospin vector flow, which is equivalent to that the ρ meson field does not work. This is why the non-linear term related to $g_\rho$ is added in Eq. (5.646).

The π meson equation (5.661) becomes

$$\left(\partial_\mu\partial^\mu + m_\pi^2\right)\vec{\pi} = -\mathrm{i}g_\pi\vec{\rho}_{\mathrm{ps}} + \frac{f_\pi}{m_\pi}\vec{\rho}_{\mathrm{pv}} + g_\rho\left[\frac{1}{2}\left(\vec{\rho}_\mu \times \vec{P}^{\mu}\right) + \frac{1}{2}\partial_\mu\left(\vec{\rho}^{\mu} \times \vec{\pi}\right)\right] \tag{5.686}$$

where

$$\vec{\rho}_{\mathrm{ps}}(x) = \sum_{i=1}^{A} \overline{\psi}_i(x)\gamma^5\vec{\tau}\psi_i(x) \tag{5.687}$$

$$\vec{\rho}_{\text{pv}}(x) = \sum_{i=1}^{A} \partial_\mu \left( \overline{\psi}_i(x) \gamma^5 \gamma^\mu \vec{\tau} \psi_i(x) \right) \tag{5.688}$$

Here $\vec{\rho}_{\text{ps}}$ is a nucleon pseudo-scalar density vector, and $\vec{\rho}_{\text{pv}}$ is a nucleon axis-vector density vector. The $g_\rho$ terms in Eq. (5.686) are generated by non-linear terms, and it can be seen that π mesons and ρ mesons are coupled each other.

The equation of electromagnetic field (5.665) becomes

$$\partial_\mu F^{\mu\nu} = e j_p^\nu \tag{5.689}$$

where

$$j_{\text{p}}^\nu(x) = \sum_{i=1}^{A} \overline{\psi}_i(x) \gamma^\nu \frac{1-\tau_3}{2} \psi_i(x) \tag{5.690}$$

here $j_{\text{p}}^\nu$ is a proton flow density.

In general, we only study the static properties of the nucleus. It can be assumed that the meson field and the photon field are static classical fields, and the nucleons perform independent movements in the classical field, then the relativistic mean field theory can be formed.

The $A$ nucleons in the nucleus are all the positive nucleons above the Dirac sea, so it can be said that the above theory was established without considering the negative energy nucleons in the Dirac sea. If we want to consider the Dirac sea effect, i.e., considering vacuum polarization, the theory must be renormalized. When considering vacuum polarization, Lagrangian density will change somewhat [74–77]. However, the introduced theory previously is an effective theory containing multiple parameters. When determining the theoretical parameters by meeting the experimental data, most of the vacuum polarization effects can be considered. Therefore, the relativistic mean field theory by using the approximation without considering the Dirac sea effect still has considerable practical value.

## 5.11 Relativistic Green Function Theory at Zero Temperature

Let $|\Psi_0\rangle$ represent the ground state of a non-interacting fermions system at finite density. The ground state $|\Psi_0\rangle$ contains positive energy baryon levels filled to some Fermi wave number $k_{\text{F}}$ and no antiparticles and free mesons.

### 5.11.1 *The Propagator of Neutral Scalar Bosons with Spinless [2, 67]*

The Klein-Gorden equation satisfied by the wave function $\sigma\left(\vec{x}, t\right)$ of the free scalar meson σ can be obtained by substituting Eq. (5.65) into Eq. (5.6) or by Eq. (5.634) as follows:

$$\left(\partial_\mu \partial^\mu + m_\sigma^2\right) \sigma\left(\vec{x}, t\right) = 0 \tag{5.691}$$

In $g_2 = g_3 = 0$ case, from Eq. (5.628) and (5.629), we can obtain

$$\mathcal{L}_\sigma = \frac{1}{2} \partial_\mu \sigma \, \partial^\mu \sigma - \frac{1}{2} m_\sigma^2 \sigma^2 \tag{5.692}$$

where $\sigma$ is called the generalized coordinate. The generalized momentum can be obtained by Eq. (5.616).

$$\pi \equiv \frac{\partial \mathcal{L}_\sigma}{\partial \dot{\sigma}} = \dot{\sigma} \tag{5.693}$$

Using Eq. (5.617) and Eq. (5.692), we can write

$$\mathcal{H}_\sigma = \pi^2 - \mathcal{L}_\sigma = \pi^2 - \frac{1}{2} \partial_\mu \sigma \, \partial^\mu \sigma + \frac{1}{2} m_\sigma^2 \sigma^2 = \frac{1}{2} \left[ \left(\vec{\nabla} \sigma\right) \cdot \left(\vec{\nabla} \sigma\right) + \pi^2 + m_\sigma^2 \sigma^2 \right] \tag{5.694}$$

The following canonical motion equations can be obtained according to the Eqs. (5.619) and (5.622):

$$\frac{\partial \sigma}{\partial t} = \pi, \qquad \frac{\partial \pi}{\partial t} = \vec{\nabla}^2 \sigma - m_\sigma^2 \sigma \tag{5.695}$$

In $t = t'$ case, introduce the following equal-time commutation relation:

$$\left[\sigma\left(\vec{x}, t\right), \pi\left(\vec{x}', t\right)\right] = \mathrm{i}\delta\left(\vec{x} - \vec{x}'\right) \tag{5.696}$$

This formula is similar to Eq. (4.175), which is the usual boson quantification rule.

To facilitate the discussion, we assume that the field is limited to a large cubic box of volume $V = L^3$ and impose the following periodic boundary conditions:

$$k_i = \frac{2\pi \, n_i}{L}, \qquad i = 1, 2, 3; \qquad n_i = 0, \pm 1, \pm 2, \cdots \tag{5.697}$$

The simplest eigenfunction of Eq. (5.691) is the plane wave. We can write $\sigma$ and $\pi$ as the following superposed forms of the plane waves:

$$\sigma = \frac{1}{\sqrt{V}} \sum_{\vec{k}} q_{\vec{k}} e^{i\vec{k}\cdot\vec{x}}, \qquad \pi = \frac{1}{\sqrt{V}} \sum_{\vec{k}} p_{\vec{k}} e^{-i\vec{k}\cdot\vec{x}} \tag{5.698}$$

Since $\sigma$ and $\pi$ are the Hermitian operators, so there are

$$q_{-\vec{k}} = q_{\vec{k}}^{+}, \qquad p_{-\vec{k}} = p_{\vec{k}}^{+} \tag{5.699}$$

where $q_{\vec{k}}$ and $p_{\vec{k}}$ are all time functions. We can rewrite Eq. (5.698) as

$$q_{\vec{k}} = \frac{1}{\sqrt{V}} \int d\vec{x}\ \sigma\, e^{-i\vec{k}\cdot\vec{x}}, \qquad p_{\vec{k}} = \frac{1}{\sqrt{V}} \int d\vec{x}\ \pi\, e^{i\vec{k}\cdot\vec{x}} \tag{5.700}$$

Then we can obtain the following commutation relation from Eqs. (5.696) and (5.700):

$$\left[q_{\vec{k}}, p_{\vec{k}'}\right] = i\delta_{\vec{k}\vec{k}'} \tag{5.701}$$

Substituting Eq. (5.698) into Eq. (5.695), then comparing it with Eq. (5.698), and noting that $\sigma$ and $\pi$ are all real numbers, thus we can obtain

$$\frac{dq_{\vec{k}}}{dt} = p_{\vec{k}}^{+}, \qquad \frac{dp_{\vec{k}}}{dt} = -\omega_k^2 q_{\vec{k}}^{+} \tag{5.702}$$

$$\omega_k^2 \equiv m_\sigma^2 + k^2 \tag{5.703}$$

Equation (5.703) means that the σ meson is on-shell. It can also be seen that the movements of different degrees of freedom marked with different $\vec{k}$ can be completely separated.

From Eq. (5.702) we can see that $q_{\vec{k}}$, $q_{\vec{k}}^{+}$, $p_{\vec{k}}$, $p_{\vec{k}}^{+}$ are all the superpositions of $e^{-i\omega_k t}$ and $e^{i\omega_k t}$. We will quantify $q_{\vec{k}}$ in the following way:

$$q_{\vec{k}} = \frac{1}{\sqrt{2\omega_k}} \left(c_{\vec{k}} e^{-i\omega_k t} + c_{-\vec{k}}^{+} e^{i\omega_k t}\right) \tag{5.704}$$

where $c_{-\vec{k}}^{+}$ and $c_{\vec{k}}$ are the σ meson creation operator and annihilation operator, respectively. The following expression can be obtained by Eqs. (5.702), (5.699), and (5.704):

$$p_{\vec{k}} = -\mathrm{i}\sqrt{\frac{\omega_k}{2}}\left(c_{-\vec{k}}\mathrm{e}^{-\mathrm{i}\omega_k t} - c^{+}_{\vec{k}}\mathrm{e}^{\mathrm{i}\omega_k t}\right) \tag{5.705}$$

Further we can obtain the following relations from Eqs. (5.704) and (5.705):

$$c_{\vec{k}} = \left(\sqrt{\frac{\omega_k}{2}}q_{\vec{k}} + \mathrm{i}\frac{1}{\sqrt{2\omega_k}}p_{-\vec{k}}\right)\mathrm{e}^{\mathrm{i}\omega_k t}, \quad c^{+}_{\vec{k}} = \left(\sqrt{\frac{\omega_k}{2}}q_{-\vec{k}} - \mathrm{i}\frac{1}{\sqrt{2\omega_k}}p_{\vec{k}}\right)\mathrm{e}^{-\mathrm{i}\omega_k t} \tag{5.706}$$

Substituting Eq. (5.706) into Eq. (5.701), we can obtain the following commutation relation:

$$\left[c_{\vec{k}}, c^{+}_{\vec{k}}\right] = \delta_{\vec{k}\vec{k}'} \tag{5.707}$$

The remaining pairs of operators are all commutated. The selection of the front edge coefficient of the right end of Eq. (5.704) ensures that the commutative relation Eq. (5.707) can be obtained.

Since $\sigma$ is a real number, from Eqs. (5.698) and (5.704) we can obtain

$$\begin{aligned}\sigma(x) &= \frac{1}{\sqrt{V}}\sum_{\vec{k}}\frac{1}{\sqrt{2\omega_k}}\left(c_{\vec{k}}\mathrm{e}^{-\mathrm{i}\omega_k t} + c^{+}_{-\vec{k}}\mathrm{e}^{\mathrm{i}\omega_k t}\right)e^{\mathrm{i}\vec{k}\cdot\vec{x}} \\ &= \frac{1}{\sqrt{V}}\sum_{\vec{k}}\frac{1}{\sqrt{2\omega_k}}\left(c_{\vec{k}}\mathrm{e}^{-\mathrm{i}k_\mu x^\mu} + c^{+}_{\vec{k}}\mathrm{e}^{\mathrm{i}k_\mu x^\mu}\right)\end{aligned} \tag{5.708}$$

where $k_\mu x^\mu = k_0 t - \vec{k}\cdot\vec{x}$. In the on-shell case there is $k_0 = \omega_k \equiv \left(k^2 + m_\sigma^2\right)^{1/2}$.

Note that $|\Psi_0\rangle$ is the vacuum state of the $\sigma$ meson, and the zero order Green function of the neutral free scalar meson $\sigma$ is defined as

$$\mathrm{i}\Delta^0(x, x') = \langle\Psi_0|\mathrm{T}\{\sigma(x)\sigma(x')\}\,|\Psi_0\rangle \tag{5.709}$$

where $\mathrm{T}\{\cdots\}$ represents the time ordered product of $\{\cdots\}$. So there is

$$\mathrm{i}\Delta^0(x, x') = \langle\Psi_0|\sigma(x)\sigma(x')|\Psi_0\rangle\theta(t-t') + \langle\Psi_0|\sigma(x')\sigma(x)|\Psi_0\rangle\theta(t'-t) \tag{5.710}$$

where $\theta(t - t')$ is the Heaviside step function. Note that for all $\vec{k}$ there are $c_{\vec{k}}|\Psi_0\rangle = 0$ and $\langle\Psi_0|c^{+}_{\vec{k}} = 0$; therefore we can obtain the following expression using Eq. (5.708):

$$
\begin{aligned}
\langle \Psi_0 | \sigma(x)\sigma(x') | \Psi_0 \rangle &= \frac{1}{V} \sum_{\vec{k}\,\vec{k}'} \frac{1}{2\sqrt{\omega_k \omega_{k'}}} \langle \Psi_0 | \left( c_{\vec{k}}\, \mathrm{e}^{-\mathrm{i}k_\mu x^\mu} + c_{\vec{k}}^{+} \mathrm{e}^{\mathrm{i}k_\mu x^\mu} \right) \\
&\left( c_{\vec{k}'}\, \mathrm{e}^{-\mathrm{i}k'_\mu x'^\mu} + c_{\vec{k}'}^{+} \mathrm{e}^{\mathrm{i}k'_\mu x'^\mu} \right) | \Psi_0 \rangle = \frac{1}{V} \sum_{\vec{k}} \frac{1}{2\omega_k} \mathrm{e}^{-\mathrm{i}k_\mu (x-x')^\mu}
\end{aligned}
\tag{5.711}
$$

The following expression can also be obtained in the same way:

$$
\langle \Psi_0 | \sigma(x')\sigma(x) | \Psi_0 \rangle = \frac{1}{V} \sum_{\vec{k}} \frac{1}{2\omega_k} \mathrm{e}^{\mathrm{i}k_\mu (x-x')^\mu}
\tag{5.712}
$$

From Eq. (5.697) we know

$$
\Delta k_i = \frac{2\pi}{L} \Delta n_i, \qquad \Delta n_i = 1; \qquad i = 1, 2, 3
\tag{5.713}
$$

So there is

$$
\frac{1}{V} \sum_{\vec{k}} = \frac{1}{V} \sum_{\vec{k}} \Delta n_1 \Delta n_2 \Delta n_3 = \frac{L^3}{V} \frac{1}{(2\pi)^3} \sum_{\vec{k}} \Delta k_1 \Delta k_2 \Delta k_3
\tag{5.714}
$$

In the limit of an infinitely large volume, that is, $V \to \infty$ and $\Delta k_i \to 0$, there is [9]

$$
\frac{1}{V} \sum_{\vec{k}} \to \frac{1}{(2\pi)^3} \int \mathrm{d}\vec{k}
\tag{5.715}
$$

Then the following expression can be obtained substituting Eqs. (5.711), (5.712), and (5.715) into (5.710):

$$
\begin{aligned}
\mathrm{i}\Delta^0(x, x') &= \frac{1}{V} \sum_{\vec{k}} \frac{1}{2\omega_k} \left[ \mathrm{e}^{-\mathrm{i}k_\mu (x-x')^\mu} \theta(t-t') + \mathrm{e}^{\mathrm{i}k_\mu (x-x')^\mu} \theta(t'-t) \right] \\
&= \frac{1}{(2\pi)^3} \int \mathrm{d}\vec{k}\, \mathrm{e}^{\mathrm{i}\vec{k}\cdot(\vec{x}-\vec{x}')} \frac{1}{2\omega_k} \left[ \mathrm{e}^{-\mathrm{i}\omega_k (t-t')} \theta(t-t') + e^{\mathrm{i}\omega_k (t-t')} \theta(t'-t) \right]
\end{aligned}
\tag{5.716}
$$

And because

$$
\theta(t-t') + \theta(t'-t) = 1
\tag{5.717}
$$

thus we can define temporarily the following quantity using Eq. (5.703):

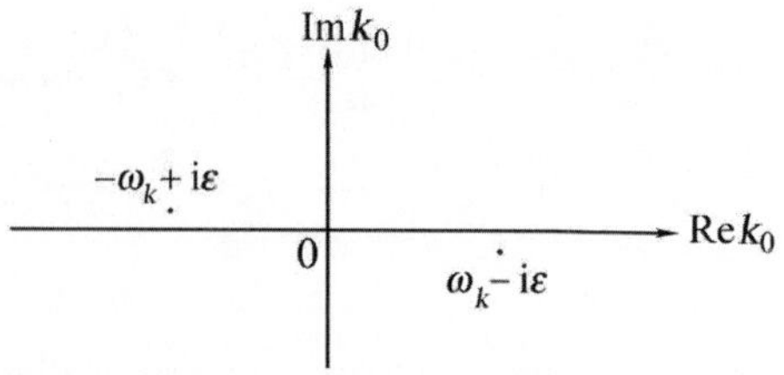


**Fig. 5.3** Integral poles

$$\begin{aligned} I &\equiv \frac{1}{2\pi i}\int \frac{e^{-ik_0(t-t')}}{k_\mu k^\mu - m_\sigma^2 + i\varepsilon} dk_0 = \frac{1}{2\pi i}\int \frac{e^{-ik_0(t-t')}}{k_0^2 - \omega_k^2 + i\varepsilon} dk_0 \\ &= \frac{1}{2\pi i}\int \frac{e^{-ik_0(t-t')}}{(k_0 - \omega_k + i\varepsilon)(k_0 + \omega_k - i\varepsilon)} [\theta(t-t') + \theta(t'-t)] dk_0 \end{aligned} \tag{5.718}$$

where $k_0$ is equivalent to off-shell energy and $\varepsilon$ is a positive infinitesimal.

In the integral given by Eq. (5.718), there is a pole in both upper and lower half planes (see Fig. 5.3). The integrals in Eq. (5.718) for $\theta(t - t')$ and $\theta(t' - t)$ terms are doing using the residual theorem in the lower half plane and the upper half plane, respectively, and one obtains

$$I = -\frac{1}{2\omega_k}\left[e^{-i\omega_k(t-t')}\theta(t-t') + e^{i\omega_k(t-t')}\theta(t'-t)\right] \tag{5.719}$$

Equation (5.716) can be rewritten as follows using Eqs. (5.718) and (5.719):

$$i\Delta^0(x, x') = i\int \frac{d^4k}{(2\pi)^4} \frac{e^{-ik_\mu(x-x')^\mu}}{k_\mu k^\mu - m_\sigma^2 + i\varepsilon} \tag{5.720}$$

In the above formula, only the integral on $k_0$ has an extreme value. From the Fourier transform relation, it can be seen that

$$\Delta^0(k) = \frac{1}{k_\mu k^\mu - m_\sigma^2 + i\varepsilon} \tag{5.721}$$

This is the propagator of a neutral scalar boson with spinless in the momentum space.

### 5.11.2 *The Propagator of Fermions with Spin $\frac{1}{2}$ [2, 6–8]*

Equation (5.66) or (5.627) has given the Dirac equation of the free spin $\frac{1}{2}$ particle as

$$\left(i\gamma^\mu \partial_\mu - M\right)\psi = 0 \tag{5.722}$$

where $M$ is the spin $\frac{1}{2}$ particle stationary mass. The particles with semi-odd number spin are called fermions, and the particles with integer spin are called bosons. Baryons are the heavier strong interaction fermions. The nucleon belongs to baryon. All mesons are bosons. The hadron is the collective name for all the particles involved in the strong interaction, such as baryon and meson. Eq. (5.624) has given the corresponding Lagrangian density as

$$\mathcal{L} = \overline{\psi}\left(\mathrm{i}\gamma^{\mu}\partial_{\mu} - M\right)\psi \tag{5.723}$$

Since the spinor field $\psi$ is a complex, we may consider $\psi(x)$ and $\overline{\psi}(x)$ as independent variables. Taking $\psi$ as the generalized coordinates, then the generalized momentum can be obtained by Eq. (5.616) as

$$\Pi = \frac{\partial \mathcal{L}}{\partial \dot{\psi}} = \mathrm{i}\psi^{+} \tag{5.724}$$

Utilizing Eq. (5.617) the Hamiltonian density can be obtained as

$$\mathcal{H} = \psi^{+}\left(-\mathrm{i}\vec{\alpha}\cdot\vec{\nabla} + \beta M\right)\psi = \psi^{+}\mathrm{i}\frac{\partial}{\partial t}\psi \tag{5.725}$$

In $t = t'$ case, we introduce the following equal-time anti-commutation relation:

$$\left\{\psi_{\alpha}\left(\vec{x},\, t\right),\psi_{\beta}^{+}\left(\vec{x}',\, t\right)\right\} = \delta_{\alpha\beta}\delta\left(\vec{x} - \vec{x}'\right) \tag{5.726}$$

and the other pairs of spinor fields are all anti-commutated each other.

Both the positive- and negative-energy solutions of the free nucleon Dirac equation can be written as follows by Eqs. (5.80), (5.86), (5.93), and (5.94):

$$\psi^{(+)}(x) = \mathrm{e}^{-\mathrm{i}k_{\mu}x^{\mu}}u(k) \tag{5.727}$$

$$u(k) = \sqrt{\frac{E+M}{2M}}\begin{pmatrix} \chi \\ \dfrac{\hat{\sigma}\cdot\vec{k}}{E+M}\chi \end{pmatrix} \tag{5.728}$$

$$\psi^{(-)}(x) = \mathrm{e}^{\mathrm{i}k_{\mu}x^{\mu}}v(k) \tag{5.729}$$

$$v(k) = \sqrt{\frac{E+M}{2M}}\begin{pmatrix} \dfrac{\hat{\sigma}\cdot\vec{k}}{E+M}\chi \\ \chi \end{pmatrix} \tag{5.730}$$

where $\chi_1 = \begin{pmatrix} 1 \\ 0 \end{pmatrix}$, $\chi_2 = \begin{pmatrix} 0 \\ 1 \end{pmatrix}$. $u(k)$ and $v(k)$ are all four-component vectors and satisfy the orthonormalization relations given by Eq. (5.91)

$$\begin{aligned} &\overline{u}^{(s)}(k)\,u^{(s')}(k) = \delta_{ss'}, \qquad \overline{v}^{(s)}(k)v^{(s')}(k) = -\delta_{ss'}, \\ &\overline{v}^{(s)}(k)\,u^{(s')}(k) = \overline{u}^{(s)}(k)v^{(s')}(k) = 0 \end{aligned} \tag{5.731}$$

where $s$ and $s'$ stand for nucleon spins. Substituting Eq. (5.727) into Eq.(5.722) there is

$$\left[\,\mathrm{i}\gamma^0(-\mathrm{i}E) + \mathrm{i}\gamma^j(\mathrm{i}k_j) - M\right]u(k) = \left[\gamma^0 E - \vec{\gamma}\cdot\vec{k} - M\right]u(k) = 0$$

Then from the above formula we can get

$$\left(\gamma^\mu \overline{k}_\mu - M\right)u(k) = 0, \quad \gamma^\mu \overline{k}_\mu \equiv \gamma^0 E - \vec{\gamma}\cdot\vec{k} \tag{5.732}$$

Note that $\overline{k}_\mu$ is a newly introduced symbol. Substituting Eq. (5.729) into Eq. (5.722) there is

$$\left(\gamma^\mu \overline{k}_\mu + M\right)v(k) = 0 \tag{5.733}$$

The following relation can be obtained by Eqs. (5.57) and (5.59):

$$(\gamma^\mu)^+ = \gamma^0\gamma^\mu\gamma^0 \tag{5.734}$$

Taking complex conjugate to Eq. (5.732) and using Eq. (5.734), we can obtain

$$u^+(k)\left(\gamma^0\gamma^\mu\gamma^0\overline{k}_\mu - \gamma^0 M\gamma^0\right) = 0$$

That is,

$$\overline{u}(k)\left(\gamma^\mu \overline{k}_\mu - M\right) = 0 \tag{5.735}$$

In the same way we can get

$$\overline{v}(k)\left(\gamma^\mu \overline{k}_\mu + M\right) = 0 \tag{5.736}$$

In the particle stationary system, there is $\overline{k}_\mu = (M, 0)$; thus Eqs. (5.732) and (5.733) are converted into [8]

$$\left(\gamma^0 - 1\right)u(M, 0) = 0, \qquad \left(\gamma^0 + 1\right)v(M, 0) = 0 \tag{5.737}$$

We can directly get the following relations from Eqs. (5.85), (5.90), and (5.92):

$$u^{(1)}(M,0)=\begin{pmatrix}1\\0\\0\\0\end{pmatrix},\quad u^{(2)}(M,0)=\begin{pmatrix}0\\1\\0\\0\end{pmatrix},$$
$$v^{(1)}(M,0)=\begin{pmatrix}0\\0\\1\\0\end{pmatrix},\quad v^{(2)}(M,0)=\begin{pmatrix}0\\0\\0\\1\end{pmatrix} \tag{5.738}$$

Substituting Eq. (5.738) into Eq. (5.737), we can confirm that Eq. (5.738) is indeed the solution of Eq. (5.737). And let

$$\varphi^{(1)}(M,0)=\begin{pmatrix}1\\0\end{pmatrix},\quad \varphi^{(2)}(M,0)=\begin{pmatrix}0\\1\end{pmatrix},$$
$$\phi^{(1)}(M,0)=\begin{pmatrix}1\\0\end{pmatrix},\quad \phi^{(2)}(M,0)=\begin{pmatrix}0\\1\end{pmatrix} \tag{5.739}$$

in this case, Eqs. (5.728) and (5.730) can be rewritten as

$$u^{(s)}(k)=\sqrt{\frac{E+M}{2M}}\begin{pmatrix}\varphi^{(s)}(M,\ 0)\\ \dfrac{\hat{\sigma}\cdot\vec{k}}{E+M}\varphi^{(s)}(M,\ 0)\end{pmatrix},$$
$$v^{(s)}(k)=\sqrt{\frac{E+M}{2M}}\begin{pmatrix}\dfrac{\hat{\sigma}\cdot\vec{k}}{E+M}\phi^{(s)}(M,\ 0)\\ \phi^{(s)}(M,\ 0)\end{pmatrix} \tag{5.740}$$

We can also obtain

$$\gamma^{\mu}\overline{k}_{\mu}+M=\gamma^{0}E-\gamma^{i}k^{i}+M=\begin{pmatrix}E+M & -\hat{\sigma}\cdot\vec{k}\\ \hat{\sigma}\cdot\vec{k} & -(E-M)\end{pmatrix} \tag{5.741}$$

$$\frac{\gamma^{\mu}\overline{k}_{\mu}+M}{[2M(E+M)]^{1/2}}u^{(s)}(M,0)=\sqrt{\frac{E+M}{2M}}\begin{pmatrix}\varphi^{(s)}(M,\ 0)\\ \dfrac{\hat{\sigma}\cdot\vec{k}}{E+M}\varphi^{(s)}(M,\ 0)\end{pmatrix}=u^{(s)}(k) \tag{5.742}$$

$$-\gamma^{\mu}\overline{k}_{\mu}+M=-\gamma^{0}E+\gamma^{i}k^{i}+M=\begin{pmatrix}-(E-M) & \hat{\sigma}\cdot\vec{k}\\ -\hat{\sigma}\cdot\vec{k} & E+M\end{pmatrix} \tag{5.743}$$

$$\frac{-\gamma^{\mu}\overline{k}_{\mu}+M}{[2M(E+M)]^{1/2}}v^{(s)}(M,0)=\sqrt{\frac{E+M}{2M}}\begin{pmatrix}\dfrac{\hat{\sigma}\cdot\vec{k}}{E+M}\phi^{(s)}(M,\ 0)\\ \phi^{(s)}(M,\ 0)\end{pmatrix}=v^{(s)}(k) \tag{5.744}$$

We have known $\overline{u}^{(s)}(k)=u^{(s)+}(k)\gamma^{0}$, $\overline{v}^{(s)}(k)=v^{(s)+}(k)\gamma^{0}$; by using Eqs. (5.57) and (5.59), we can obtain the following relations from Eqs. (5.742) and (5.744):

$$\overline{u}^{(s)}(k)=\overline{u}^{(s)}(M,0)\frac{\gamma^{\mu}\overline{k}_{\mu}+M}{[2M(E+M)]^{1/2}} \tag{5.745}$$

$$\overline{v}^{(s)}(k)=\overline{v}^{(s)}(M,0)\frac{-\gamma^{\mu}\overline{k}_{\mu}+M}{[2M(E+M)]^{1/2}} \tag{5.746}$$

First the following formula can be found:

$$\begin{aligned}\sum_{s=1,2}u^{(s)}(M,0)\overline{u}^{(s)}(M,0)&=\left[\begin{pmatrix}1\\0\\0\\0\end{pmatrix}(1\ 0\ 0\ 0)+\begin{pmatrix}0\\1\\0\\0\end{pmatrix}(0\ 1\ 0\ 0)\right]\\ &=\begin{pmatrix}1&0&0&0\\0&1&0&0\\0&0&0&0\\0&0&0&0\end{pmatrix}=\begin{pmatrix}\hat{I}&\hat{0}\\ \hat{0}&\hat{0}\end{pmatrix}=\frac{1+\gamma^{0}}{2}\end{aligned} \tag{5.747}$$

So there is

$$\begin{aligned}\Lambda_{+}(k)&\equiv\sum_{s=1,2}u^{(s)}(k)\,\overline{u}^{(s)}(k)\\ &=\frac{1}{2M(E+M)}(\gamma^{\mu}\overline{k}_{\mu}+M)\left[\sum_{s=1,\ 2}u^{(s)}(M,\ 0)\,\overline{u}^{(s)}(M,\ 0)\right](\gamma^{\mu}\overline{k}_{\mu}+M)\\ &=\frac{1}{2M(E+M)}(\gamma^{\mu}\overline{k}_{\mu}+M)\frac{1+\gamma^{0}}{2}(\gamma^{\mu}\overline{k}_{\mu}+M)\end{aligned} \tag{5.748}$$

Then the following relation can be obtained by using Eqs. (5.728) and (5.731):

$$\overline{u}^{(s)}(k)u^{(s)}(k) = \frac{E+M}{2M}\begin{pmatrix}\chi^{(s)+} & \frac{\hat{\sigma}\cdot\vec{k}}{E+M}\chi^{(s)+}\end{pmatrix}\begin{pmatrix}\hat{I} & \hat{0}\\ \hat{0} & -\hat{I}\end{pmatrix}\begin{pmatrix}\chi^{(s)}\\ \frac{\hat{\sigma}\cdot\vec{k}}{E+M}\chi^{(s)}\end{pmatrix}$$
$$= \frac{E+M}{2M}\left[1-\frac{\left(\hat{\sigma}\cdot\vec{k}\right)^2}{(E+M)^2}\right] = \frac{1}{2M(E+M)}\left[(E+M)^2-\left(\hat{\sigma}\cdot\vec{k}\right)^2\right] = 1$$

Further we can get

$$\left(\hat{\sigma}\cdot\vec{k}\right)^2 = (E+M)^2 - 2M(E+M) = (E+M)(E-M) \tag{5.749}$$

And we can obtain the following formula using Eqs. (5.741) and (5.749):

$$\begin{aligned}
&\left(\gamma^{\mu}\overline{k}_{\mu}+M\right)\frac{1+\gamma^0}{2}\left(\gamma^{\mu}\overline{k}_{\mu}+M\right)\\
&= \begin{pmatrix}E+M & -\hat{\sigma}\cdot\vec{k}\\ \hat{\sigma}\cdot\vec{k} & -(E-M)\end{pmatrix}\begin{pmatrix}\hat{I} & \hat{0}\\ \hat{0} & \hat{0}\end{pmatrix}\begin{pmatrix}E+M & -\hat{\sigma}\cdot\vec{k}\\ \hat{\sigma}\cdot\vec{k} & -(E-M)\end{pmatrix}\\
&= \begin{pmatrix}E+M & -\hat{\sigma}\cdot\vec{k}\\ \hat{\sigma}\cdot\vec{k} & -(E-M)\end{pmatrix}\begin{pmatrix}E+M & -\hat{\sigma}\cdot\vec{k}\\ \hat{0} & \hat{0}\end{pmatrix}\\
&= \begin{pmatrix}(E+M)^2 & -(E+M)\hat{\sigma}\cdot\vec{k}\\ (E+M)\hat{\sigma}\cdot\vec{k} & -\left(\hat{\sigma}\cdot\vec{k}\right)^2\end{pmatrix} = (E+M)\begin{pmatrix}E+M & -\hat{\sigma}\cdot\vec{k}\\ \hat{\sigma}\cdot\vec{k} & -(E-M)\end{pmatrix}\\
&= (E+M)\left(\gamma^{\mu}\overline{k}_{\mu}+M\right)
\end{aligned} \tag{5.750}$$

Then the following result can be obtained by Eq. (5.748):

$$\Lambda_{+}(k) \equiv \sum_{s=1,2} u^{(s)}(k)\,\overline{u}^{(s)}(k) = \frac{\gamma^{\mu}\overline{k}_{\mu}+M}{2M} \tag{5.751}$$

In the same way we can prove

$$\Lambda_{-}(k) \equiv -\sum_{s=1,2} v^{(s)}(k)\,\overline{v}^{(s)}(k) = \frac{-\gamma^{\mu}\overline{k}_{\mu}+M}{2M} \tag{5.752}$$

So from the above two formulas, the following relation can be seen

$$\Lambda_+(k) + \Lambda_-(k) = I \tag{5.753}$$

Therefore we call $\Lambda_+(k)$ and $\Lambda_-(k)$ as the projection operators of the positive- and negative-energy states, respectively.

We still use the periodic boundary conditions given by Eq. (5.697), so the spinor fields appeared in Eqs. (5.722), (5.723), and (5.726) can be expanded, respectively [2, 6, 7]:

$$\begin{aligned}\psi\left(\vec{x}, t\right) &= \frac{1}{\sqrt{V}}\sum_{\vec{k}s}\sqrt{\frac{M}{E}}\left[a_{\vec{k}s} u^{(s)}(k)\,\mathrm{e}^{-\mathrm{i}k_\mu x^\mu} + b^+_{\vec{k}s} v^{(s)}(k)\,\mathrm{e}^{\mathrm{i}k_\mu x^\mu}\right],\\ \overline{\psi}\left(\vec{x}, t\right) &= \frac{1}{\sqrt{V}}\sum_{\vec{k}s}\sqrt{\frac{M}{E}}\left[a^+_{\vec{k}s} \overline{u}^{(s)}(k)\,\mathrm{e}^{\mathrm{i}k_\mu x^\mu} + b_{\vec{k}s} \overline{v}^{(s)}(k)\,\mathrm{e}^{-\mathrm{i}k_\mu x^\mu}\right]\end{aligned} \tag{5.754}$$

In above formulas, $a^+$ and $a$, $b^+$ and $b$ represent the creation and annihilation operators of nucleon and anti-nucleon, respectively. Next we will prove that the expansion coefficients selected in Eq. (5.754) will make the introduced creation and annihilation operators satisfy the following anti-commutation relations:

$$\left\{a_{\vec{k}s}, a^+_{\vec{k}'s'}\right\} = \delta_{ss'}\delta\left(\vec{k} - \vec{k}'\right), \qquad \left\{b_{\vec{k}s}, b^+_{\vec{k}'s'}\right\} = \delta_{ss'}\delta\left(\vec{k} - \vec{k}'\right) \tag{5.755}$$

The other pairs of these operators are all anti-commutated each other. First we assume that Eq. (5.755) and other anti-commutated relations are valid, and then substituting Eq. (5.754) into Eq. (5.726) and using Eqs. (5.751), (5.752), and (5.715), as well as noting $t' = t$, thus we can obtain [6, 7]

$$\begin{aligned}&\left\{\psi_\alpha\left(\vec{x}, t\right), \psi^+_\beta\left(\vec{x}', t\right)\right\} = \frac{1}{V}\sum_{\vec{k}\vec{k}'}\frac{M}{\sqrt{E_k E_{k'}}}\sum_{ss'}\delta_{ss'}\delta\left(\vec{k} - \vec{k}'\right)\\ &\quad\left[u^{(s)}_\alpha(k)\overline{u}^{(s')}_\beta(k')\gamma^0 \mathrm{e}^{\mathrm{i}\vec{k}\cdot\left(\vec{x}-\vec{x}'\right)} + v^{(s)}_\alpha(k)\overline{v}^{(s')}_\beta(k')\gamma^0 \mathrm{e}^{-\mathrm{i}\vec{k}\cdot\left(\vec{x}-\vec{x}'\right)}\right]\\ &= \frac{1}{V}\sum_{\vec{k}}\frac{1}{2E}\mathrm{e}^{\mathrm{i}\vec{k}\cdot\left(\vec{x}-\vec{x}'\right)}\left\{\left[\left(\gamma^0 k_0 - \vec{\gamma}\cdot\vec{k} + M\right) - \left(-\gamma^0 k_0 - \vec{\gamma}\cdot\vec{k} + M\right)\right]\gamma^0\right\}_{\alpha\beta}\\ &= \frac{1}{(2\pi)^3}\int \mathrm{d}\vec{k}\,\mathrm{e}^{\mathrm{i}\vec{k}\cdot\left(\vec{x}-\vec{x}'\right)}\delta_{\alpha\beta} = \delta_{\alpha\beta}\delta\left(\vec{k} - \vec{k}'\right)\end{aligned} \tag{5.756}$$

Note that in the on-shell case $k_0 = E$. This proves that $\psi_\alpha\left(\vec{x}, t\right)$ and $\overline{\psi}\left(\vec{x}', t\right)$ expressed by Eq. (5.754) may satisfy Eq. (5.726) under the condition that Eq. (5.755) is satisfied.

Since the nucleon is fermion, the zero order Green function of the nucleon is defined as

$$
\begin{aligned}
iG^0_{\alpha\beta}(x-x') &= \langle \Psi_0 | \mathrm{T}\{\psi_\alpha(x)\,\overline{\psi}_\beta(x')\}|\Psi_0\rangle \\
&= \langle \Psi_0|\psi_\alpha(x)\,\overline{\psi}_\beta(x')|\Psi_0\rangle \theta(t-t') - \langle \Psi_0|\overline{\psi}_\beta(x')\,\psi_\alpha(x)|\Psi_0\rangle\theta(t'-t)
\end{aligned}
\tag{5.757}
$$

Note the following relations:

$$
\begin{gathered}
b_{\vec{k}s}|\Psi_0\rangle = 0, \quad \text{for all } \vec{k}; \quad a^+_{\vec{k}s}|\Psi_0\rangle = 0, \quad \left|\vec{k}\right| < k_\mathrm{F}; \\
a_{\vec{k}s}|\Psi_0\rangle = 0, \quad \left|\vec{k}\right| > k_\mathrm{F}
\end{gathered}
\tag{5.758}
$$

Also note that only the positive energy nucleon can be created above the Fermi surface; using Eqs. (5.754), (5.758), and (5.715), we can get firstly

$$
\begin{aligned}
\langle \Psi_0|\psi_\alpha(x)\,\overline{\psi}_\beta(x')|\Psi_0\rangle &= \frac{1}{V}\sum_{\vec{k}\vec{k}'}\frac{M}{\sqrt{E_k E_{k'}}}\sum_{ss'}\left\langle \Psi_0\middle|a_{\vec{k}s}a^+_{\vec{k}'s'}\middle|\Psi_0\right\rangle \\
&\quad \left[u^{(s)}(k)\overline{u}^{(s')}(k')\right]_{\alpha\beta} e^{\mathrm{i}\left(\vec{k}\cdot\vec{x}-\vec{k}'\cdot\vec{x}'\right)}e^{-\mathrm{i}(E_k t - E_{k'}t')} \\
&= \frac{1}{(2\pi)^3}\int \mathrm{d}\vec{k}\,\frac{1}{2E}\left(\gamma^\mu \overline{k}_\mu + M\right)_{\alpha\beta}\left[1-\theta\left(k_F - \left|\vec{k}\right|\right)\right] e^{i\vec{k}\cdot\left(\vec{x}-\vec{x}'\right)}e^{-\mathrm{i}E(t-t')}
\end{aligned}
\tag{5.759}
$$

There are the relations

$$
\pm\theta(\pm t) = \mathrm{i}\int_{-\infty}^{\infty}\frac{\mathrm{d}\omega}{2\pi}\frac{e^{-\mathrm{i}\omega t}}{\omega \pm \mathrm{i}\varepsilon}
\tag{5.760}
$$

$$
e^{-\mathrm{i}E(t-t')}\theta(t-t') = \mathrm{i}\int\frac{\mathrm{d}\omega}{2\pi}\frac{e^{-\mathrm{i}(\omega+E)(t-t')}}{\omega+\mathrm{i}\varepsilon}
\tag{5.761}
$$

Let $k_0 = \omega + E$, $\omega = k_0 - E$, $\mathrm{d}\omega = \mathrm{d}k_0$, $E$ is the on-shell energy, and $k_0$ is equivalent to the off-shell energy. Then Eq. (5.761) may be rewritten as

$$
e^{-\mathrm{i}E(t-t')}\theta(t-t') = \mathrm{i}\int\frac{\mathrm{d}k_0}{2\pi}\frac{e^{-ik_0(t-t')}}{k_0 - E + \mathrm{i}\varepsilon}
\tag{5.762}
$$

So the following expression can be obtained utilizing Eqs. (5.759) and (5.762):

$$
\begin{aligned}
&\langle \Psi_0|\psi_\alpha(x)\,\overline{\psi}_\beta(x')\,|\Psi_0\rangle\theta(t-t') \\
&= \mathrm{i}\int\frac{\mathrm{d}^4k}{(2\pi)^4}\frac{1}{2E}\left(\gamma^\mu\overline{k}_\mu + M\right)_{\alpha\beta}\frac{1-\theta\left(k_F - \left|\vec{k}\right|\right)}{k_0 - E + \mathrm{i}\varepsilon}e^{-ik_\mu(x-x')^\mu}
\end{aligned}
\tag{5.763}
$$

And we can write the following expression by using Eqs. (5.754) and (5.758):

$$
\begin{aligned}
&\langle \Psi_0|\overline{\psi}_\beta(x')\psi_\alpha(x)|\Psi_0\rangle \\
&= \frac{1}{V}\sum_{\vec{k}\vec{k}'}\frac{M}{\sqrt{E_k E_{k'}}}\sum_{ss'}\langle \Psi_0|\left(a^+_{\vec{k}'s'}\overline{u}^{(s')}_\beta(k')\mathrm{e}^{\mathrm{i}k'_\mu x'^\mu}+b_{\vec{k}'s'}\overline{v}^{(s')}_\beta(k')\mathrm{e}^{-\mathrm{i}k'_\mu x'^\mu}\right) \\
&\qquad \left(a_{\vec{k}s}u^{(s)}_\alpha(k)\mathrm{e}^{-\mathrm{i}k_\mu x^\mu}+b^+_{\vec{k}s}v^{(s)}_\alpha(k)\mathrm{e}^{\mathrm{i}k_\mu x^\mu}\right)|\Psi_0\rangle \equiv I + II
\end{aligned} \tag{5.764}
$$

where

$$
\begin{aligned}
I &= \frac{1}{V}\sum_{\vec{k}\vec{k}'}\frac{M}{\sqrt{E_k E_{k'}}}\sum_{ss'}\langle \Psi_0|a^+_{\vec{k}'s'}a_{\vec{k}s}|\Psi_0\rangle\left(\overline{u}^{(s')}_\beta(k)u^{(s)}_\alpha(k)\right)\mathrm{e}^{\mathrm{i}\left(\vec{k}\cdot\vec{x}-\vec{k}'\cdot\vec{x}'\right)}\mathrm{e}^{-\mathrm{i}(E_k t-E_{k'}t')} \\
&= \frac{1}{V}\sum_{\vec{k}}\frac{M}{E}\left(\sum_s \overline{u}^{(s)}_\beta(k)u^{(s)}_\alpha(k)\right)\theta\left(k_F-\left|\vec{k}\right|\right)\mathrm{e}^{\mathrm{i}\vec{k}\cdot\left(\vec{x}-\vec{x}'\right)}\mathrm{e}^{-\mathrm{i}E(t-t')}
\end{aligned} \tag{5.765}
$$

Since $\overline{u}_\beta$ and $u_\alpha$ are only the components of the vector, there are

$$
\overline{u}_\beta u_\alpha = u_\alpha \overline{u}_\beta = (u\overline{u})_{\alpha\beta} \tag{5.766}
$$

We may see the following expression from Eqs. (5.760)–(5.762):

$$
-\mathrm{e}^{-\mathrm{i}E(t-t')}\theta(t'-t) = \mathrm{i}\int\frac{\mathrm{d}\omega}{2\pi}\frac{\mathrm{e}^{-\mathrm{i}(\omega+E)(t-t')}}{\omega-\mathrm{i}\varepsilon} = \mathrm{i}\int\frac{\mathrm{d}k_0}{2\pi}\frac{\mathrm{e}^{-\mathrm{i}k_0(t-t')}}{k_0-E-\mathrm{i}\varepsilon} \tag{5.767}
$$

Then from Eq. (5.765) and Eq. (5.767), the following expression can be obtained using Eqs. (5.715) and (5.751):

$$
-I\times\theta(t'-t) = \mathrm{i}\int\frac{\mathrm{d}^4k}{(2\pi)^4}\frac{1}{2E}\left(\gamma^\mu\overline{k}_\mu+M\right)_{\alpha\beta}\frac{\theta\left(k_F-\left|\vec{k}\right|\right)}{k_0-E-\mathrm{i}\varepsilon}\mathrm{e}^{-\mathrm{i}k_\mu(x-x')^\mu} \tag{5.768}
$$

According to Eq. (5.764), in the similar way we can prove

$$
\begin{aligned}
II &= \frac{1}{V}\sum_{\vec{k}\vec{k}'}\frac{M}{\sqrt{E_k E_{k'}}}\sum_{ss'}\langle \Psi_0|b_{\vec{k}'s'}b^+_{\vec{k}s}|\Psi_0\rangle\left(\overline{v}^{(s')}_\beta(k)v^{(s)}_\alpha(k)\right)\mathrm{e}^{-\mathrm{i}\left(\vec{k}\cdot\vec{x}-\vec{k}'\cdot\vec{x}'\right)}\mathrm{e}^{\mathrm{i}(E_k t-E_{k'}t')} \\
&= \frac{1}{V}\sum_{\vec{k}}\frac{M}{E}\left(\sum_s \overline{v}^{(s)}_\beta(k)v^{(s)}_\alpha(k)\right)\mathrm{e}^{-\mathrm{i}\vec{k}\cdot\left(\vec{x}-\vec{x}'\right)}\mathrm{e}^{\mathrm{i}E(t-t')} \\
&= -\frac{1}{V}\sum_{\vec{k}}\frac{1}{2E}\left(-\gamma^\mu\overline{k}_\mu+M\right)_{\alpha\beta}\mathrm{e}^{-\mathrm{i}\vec{k}\cdot\left(\vec{x}-\vec{x}'\right)}\mathrm{e}^{\mathrm{i}E(t-t')}
\end{aligned} \tag{5.769}
$$

This term represents the generation of the negative-energy nucleon in vacuum. Introduce the new symbol

$$\gamma^\mu \widetilde{k}_\mu = -\gamma^0 E - \vec{\gamma} \cdot \vec{k} \tag{5.770}$$

Letting $\vec{k} \to -\vec{k}$ after the summation symbol of Eq. (5.769), we can get

$$II = -\frac{1}{V}\sum_{\vec{k}} \frac{1}{2E}\left(\gamma^\mu \widetilde{k}_\mu + M\right)_{\alpha\beta} e^{iE(t-t')} e^{i\vec{k}\cdot\left(\vec{x}-\vec{x}'\right)} \tag{5.771}$$

From Eq. (5.760)–(5.762) we know

$$-e^{iE(t-t')}\theta(t'-t) = i\int \frac{d\omega}{2\pi}\frac{e^{-i(\omega-E)(t-t')}}{\omega - i\varepsilon} = i\int \frac{dk_0}{2\pi}\frac{e^{-ik_0(t-t')}}{k_0 + E - i\varepsilon} \tag{5.772}$$

$$-II \times \theta(t'-t) = -i\int \frac{d^4k}{(2\pi)^4}\frac{1}{2E}\left(\gamma^\mu \widetilde{k}_\mu + M\right)_{\alpha\beta}\frac{1}{k_0 + E - i\varepsilon} e^{-ik_\mu(x-x')^\mu} \tag{5.773}$$

It can be seen that the zero order Green function (propagator) of the free nucleon in momentum space is, according to Eqs. (5.757), (5.763), (5.764), (5.768), and (5.773), as follows:

$$\begin{aligned} G^0_{\alpha\beta}(k) = \frac{1}{2E}\Bigg\{ &\left(\gamma^\mu \overline{k}_\mu + M\right)_{\alpha\beta}\left[\frac{1-\theta\left(k_F - \left|\vec{k}\right|\right)}{k_0 - E + i\varepsilon} + \frac{\theta\left(k_F - \left|\vec{k}\right|\right)}{k_0 - E - i\varepsilon}\right] \\ &- \left(\gamma^\mu \widetilde{k}_\mu + M\right)_{\alpha\beta}\frac{1}{k_0 + E - i\varepsilon}\Bigg\} \end{aligned} \tag{5.774}$$

where

$$\gamma^\mu \overline{k}_\mu = \gamma^0 E - \vec{\gamma} \cdot \vec{k}, \qquad \gamma^\mu \widetilde{k}_\mu = -\gamma^0 E - \vec{\gamma} \cdot \vec{k} \tag{5.775}$$

The first item in the first line of the right end of Eq. (5.774) represents the positive energy nucleon propagator above the Fermi surface, the second item represents the positive energy nucleon-hole propagator below the Fermi surface, and the last item appeared in the second line on the right end represents the propagator of negative-energy nucleon-hole (called anti-nucleon) in the infinite Dirac sea. In the non-relativistic (NR) approximation, $\left|\vec{k}\right| \ll M,\ E \cong M + k^2/2M$, the lower component of the positive energy nucleon will tend to have no contribution. Notice $(\gamma^0 E + M)_{\alpha\beta} \xrightarrow[\text{NR}]{} 2M\delta_{\alpha\beta}$ and ignore the last item representing the anti-nucleon in Eq. (5.774), we obtain

$$G^0_{\alpha\beta}(k) \xrightarrow[\text{NR}]{} \delta_{\alpha\beta}\left[\frac{1-\theta\left(k_F-\left|\vec{k}\right|\right)}{k_0-E+\mathrm{i}\varepsilon}+\frac{\theta\left(k_F-\left|\vec{k}\right|\right)}{k_0-E-\mathrm{i}\varepsilon}\right] \tag{5.776}$$

This is just the non-relativistic nucleon zero order Green function [28].

Equation (5.774) can be rewritten as

$$\begin{aligned} 2EG^0_{\alpha\beta}(k) &= \left(\gamma^\mu \overline{k}_\mu + M\right)_{\alpha\beta}\left[\frac{1}{k_0-E-\mathrm{i}\varepsilon}-\frac{1}{k_0-E+\mathrm{i}\varepsilon}\right]\theta\left(k_F-\left|\vec{k}\right|\right) \\ &+\left[\frac{\left(\gamma^\mu \overline{k}_\mu + M\right)_{\alpha\beta}}{k_0-E+\mathrm{i}\varepsilon}-\frac{\left(\gamma^\mu \widetilde{k}_\mu + M\right)_{\alpha\beta}}{k_0+E-\mathrm{i}\varepsilon}\right] \end{aligned} \tag{5.777}$$

Using the principal value integral formula

$$\frac{1}{k_0-E\pm\mathrm{i}\varepsilon}=\mathcal{P}\left(\frac{1}{k_0-E}\right)\mp\mathrm{i}\pi\delta(k_0-E) \tag{5.778}$$

in the first item containing $\theta\left(k_{\mathrm{F}}-\left|\vec{k}\right|\right)$ of Eq. (5.777), the real part of the principal value integral is just eliminated. The corresponding usual symbol is

$$\gamma^\mu k_\mu = \gamma^0 k_0 - \vec{\gamma}\cdot\vec{k} \tag{5.779}$$

then we can get the relation

$$\gamma^\mu k_\mu \delta(k_0-E) = \gamma^\mu \overline{k}_\mu \delta(k_0-E) \tag{5.780}$$

In order to derive the second item of Eq. (5.777), we first seek

$$\begin{aligned} &(k_0-E+\mathrm{i}\varepsilon)(k_0+E-\mathrm{i}\varepsilon) = k_0^2 - E^2 + \mathrm{i}\varepsilon = k_\mu k^\mu - M^2 + \mathrm{i}\varepsilon, \\ &(k_0+E)\gamma^\mu \overline{k}_\mu - (k_0-E)\gamma^\mu \widetilde{k}_\mu = 2E\left(\gamma^0 k_0 - \vec{\gamma}\cdot\vec{k}\right) = 2E\gamma^\mu k_\mu, \\ &(k_0+E)M - (k_0-E)M = 2EM \end{aligned}$$

Then by Eq. (5.777) we can obtain

$$\begin{aligned} G^0_{\alpha\beta}(k) &= \left(\gamma^\mu k_\mu + M\right)_{\alpha\beta}\left\{\frac{1}{k_\mu k^\mu - M^2 + \mathrm{i}\varepsilon}+\frac{\mathrm{i}\pi}{E}\delta(k_0-E)\,\theta\left(k_{\mathrm{F}}-\left|\vec{k}\right|\right)\right\} \\ &\equiv G^0_{\mathrm{F}}(k)_{\alpha\beta} + G^0_{\mathrm{D}}(k)_{\alpha\beta} \end{aligned} \tag{5.781}$$

The two propagators appeared in the second line of the above formula correspond to the first term and the second term in brace $\{\ldots\}$ of the above formula, respectively.

Dirac propagator $G_D^0$ does not contribute when $k_F \to 0$; $\delta(k_0 - E)$ represents that $G_D^0$ is on-shell. In the case of $k_F > 0$ finite nuclear density, one should carry out the study in the ground state coordinate system in which total three-dimensional momentum is equal to 0. $G_F^0(k)$ is called the Feynman propagator of the free nucleon and anti-nucleon without interaction in the vacuum state. It is off-shell. One can also derive $G_F^0(k)$ in another way. The equation satisfied by the propagator of a free nucleus in vacuum is [6, 7]

$$\left(\mathrm{i}\gamma^\mu \partial_\mu - M\right) G_F(x, x') = \delta^4(x - x') \tag{5.782}$$

We can see from the above equation that $G_F(x, x')$ is only related to $(x - x')$, and we may obtain the following expression:

$$G_F(x, x') = G_F(x - x') = \int \frac{\mathrm{d}^4 k}{(2\pi)^4} \frac{\mathrm{e}^{-\mathrm{i}k_\mu (x-x')^\mu}}{\gamma^\mu k_\mu - M + \mathrm{i}\varepsilon} \tag{5.783}$$

Then we get

$$G_F(k) = \frac{1}{\gamma^\mu k_\mu - M + \mathrm{i}\varepsilon} = \frac{\gamma^\mu k_\mu + M}{k_\mu k^\mu - M^2 + \mathrm{i}\varepsilon} \tag{5.784}$$

### 5.11.3 The Propagator of Neutral Vector Bosons with Spin 1

When the nonlinear term constant $C_3 = 0$, the Lagrangian density of the free vector meson given by Eq. (5.635) is

$$\mathcal{L} = -\frac{1}{4}\Omega_{\mu\nu}\Omega^{\mu\nu} + \frac{1}{2}m^2 V_\mu V^\mu \tag{5.785}$$

where

$$\Omega^{\mu\nu} = \partial^\mu V^\nu - \partial^\nu V^\mu \tag{5.786}$$

where $m$ is meson stationary mass. The Proca equation corresponding with $\mathcal{L}$ has been given by Eq. (5.641)

$$\partial_\mu \Omega^{\mu\nu} + m^2 V^\nu = 0 \tag{5.787}$$

Free vector meson satisfies the Lorentz gauge given by Eq. (5.642):

$$\partial_\nu V^\nu = 0 \tag{5.788}$$

Then the vector meson satisfies the Klein-Gorden equation given by Eq. (5.644):

$$\left(\partial_\mu \partial^\mu + m^2\right) V^\nu = \left(\Box + m^2\right) V^\nu = 0 \tag{5.789}$$

Since the 4-dimensional vector $V^\nu$ satisfies the Lorentz gauge condition given by Eq. (5.788), one of the four degrees of freedom is not independent. Since $\Omega^{\mu\nu}$ is anti-symmetric for $\mu\nu$ and noting Eq. (5.60), so we can write

$$\Omega^{0j} = \dot{V}^j + \nabla_j V^0 = -\Omega^{j0} \tag{5.790}$$

Note $\partial_j = -\partial^j, \quad V_j = -V^j$, we can also write

$$\Omega_{0j} = \dot{V}_j - \nabla_j V_0 = -\Omega_{j0} \tag{5.791}$$

$$\Omega^{0j} = -\Omega^{j0} = -\Omega_{0j} = \Omega_{j0} \tag{5.792}$$

It can be seen that $\dot{V}^0 = \dot{V}_0$ is not included in $\mathcal{L}$; therefore $V^0 = V_0$ is not an independent variable [9]. And we can also write

$$\begin{gathered} \Omega^{ij} = -\nabla_i V^j + \nabla_j V^i = -\Omega^{ji}, \qquad \Omega_{ij} = \nabla_i V_j - \nabla_j V_i = -\Omega_{ji}, \\ \Omega^{ij} = -\Omega^{ji} = \Omega_{ij} = -\Omega_{ji} \end{gathered} \tag{5.793}$$

Noting $V_j = -V^j$, so we can get

$$-\frac{1}{4}\Omega_{\mu\nu}\Omega^{\mu\nu} = \frac{1}{2}\left(\dot{V}^j + \nabla_j V^0\right)\left(\dot{V}^j + \nabla_j V^0\right) - \frac{1}{2}\left(\vec{\nabla} \times \vec{V}\right)^2 \tag{5.794}$$

The conjugate momentum can be obtained by using Eqs. (5.616) and (5.794):

$$\pi^j = \frac{\partial \mathcal{L}}{\partial \dot{V}^j} = \dot{V}^j + \nabla_j V^0 \tag{5.795}$$

Since $V^0$ is not an independent variable, we take $V^0 = 0$ here, and then from Eq. (5.795) we get

$$\pi^j = \dot{V}^j \tag{5.796}$$

For the certain $\vec{k}$, we take the unit vectors as

$$\vec{e}_{\vec{k}3} = \frac{\vec{k}}{\left|\vec{k}\right|}, \qquad \vec{e}_{\vec{k}i} \cdot \vec{e}_{\vec{k}j} = \delta_{ij}\ , \qquad i, j = 1, 2, 3 \tag{5.797}$$

In above formulas, $i = 1, 2$ represents a transverse wave, and $i = 3$ represents a longitudinal wave. Since $\vec{V}$ is the Hermitian operator, so $\vec{V}$ can be expanded with the plane waves and quantized as follows:

$$\vec{V}\left(\vec{x}, t\right) = \frac{1}{\sqrt{V}}\sum_{\vec{k}}\frac{1}{\sqrt{2\omega_k}}\sum_{i=1}^{3}\vec{e}_{\vec{k}i}\left[a_{\vec{k}i}e^{-i\omega_k t + i\vec{k}\cdot\vec{x}} + \text{h.c.}\right] \tag{5.798}$$

where h.c. represents the complex conjugate term of the previous term and $a$ and $a^+$ are the particle annihilation and creation operators, respectively. Substituting $V^i$ given by Eq. (5.798) into Eq. (5.789), we can obtain

$$\omega_k^2 = k^2 + m^2 \tag{5.799}$$

We can obtain the following expression according to Eqs. (5.796) and (5.798):

$$\pi\left(\vec{x}, t\right) = \frac{1}{\sqrt{V}}\sum_{\vec{k}}(-i)\sqrt{\frac{\omega_k}{2}}\sum_{i=1}^{3}\vec{e}_{\vec{k}i}\left[a_{\vec{k}i}e^{-i\omega_k t + i\vec{k}\cdot\vec{x}} - \text{h.c.}\right] \tag{5.800}$$

Assume that the annihilation operator and the creation operator satisfy the following commutation relations:

$$\left[a_{\vec{k}i}, a^+_{\vec{k}'j}\right] = \delta_{ij}\delta_{\vec{k}\vec{k}'}, \qquad \left[a_{\vec{k}i}, a_{\vec{k}'j}\right] = \left[a^+_{\vec{k}i}, a^+_{\vec{k}'j}\right] = 0 \tag{5.801}$$

We can prove the following equal-time commutation relation by use of Eqs. (5.798), (5.800), and (5.801):

$$\begin{aligned}\left[V^i\left(\vec{x}, t\right), \pi^j\left(\vec{x}', t\right)\right] &= i\delta_{ij}\frac{1}{V}\sum_{\vec{k}}e^{i\vec{k}\cdot\left(\vec{x}-\vec{x}'\right)} \\ &= i\delta_{ij}\frac{1}{(2\pi)^3}\int d\vec{k}e^{i\vec{k}\cdot\left(\vec{x}-\vec{x}'\right)} = i\delta_{ij}\delta\left(\vec{x} - \vec{x}'\right)\end{aligned} \tag{5.802}$$

And it can also be proved that

$$\left[V^i\left(\vec{x}, t\right), V^j\left(\vec{x}', t\right)\right] = \left[\pi^i\left(\vec{x}, t\right), \pi^j\left(\vec{x}', t\right)\right] = 0 \tag{5.803}$$

These are just the equal-time commutation relations required by the quantization rule.

Set $\vec{e}_{\vec{k}0}$ be the unit vector along the time axis and require to satisfy the following conditions in the 4-dimensional space-time coordinate system:

$$\vec{e}_{\vec{k}\mu} \cdot \vec{e}_{\vec{k}\nu} = \delta_{\mu\nu}, \qquad \mu,\nu = 0,1,2,3 \tag{5.804}$$

Be careful not to attempt to express the four unit vector axes of the 4-dimensional space-time coordinate in 3-dimensional space. Since $V^0$ is not an independent variable, refer to Eq. (5.798), assume that we can expand the field quantity $V$ in the 4-dimensional space-time coordinate as follows:

$$V\left(\vec{x}, t\right) = \frac{1}{\sqrt{V}} \sum_{\vec{k}} \frac{1}{\sqrt{2\omega_k}} \sum_{\mu=0}^{3} \vec{e}_{\vec{k}\mu} \left[a_{\vec{k}\mu} e^{-ik_\rho x^\rho} + \text{h.c.}\right] \tag{5.805}$$

And assume that there are the following communication relations:

$$\left[a_{\vec{k}\mu}, a^+_{\vec{k}'\nu}\right] = \delta_{\mu\nu}\delta_{\vec{k}\vec{k}'}, \qquad \left[a_{\vec{k}\mu}, a_{\vec{k}\nu}\right] = \left[a^+_{\vec{k}\mu}, a^+_{\vec{k}\nu}\right] = 0 \tag{5.806}$$

The zero order Green function of a free vector meson is defined as

$$\mathrm{i}D^0_{\mu\nu}(x, x') = -\langle \Psi_0 | \mathrm{T}\{ V_\mu(x)\, V^\nu(x')\} | \Psi_0 \rangle \tag{5.807}$$

When the study work is done only in 3-dimensional coordinate space, the above formula is reduced to

$$\mathrm{i}D^0_{ij}(x, x') = \langle \Psi_0 | \mathrm{T}\{ V_i(x)\, V_j(x')\} | \Psi_0 \rangle \tag{5.808}$$

This is the definition of the usual propagator. Eq. (5.807) can be rewritten as

$$\mathrm{i}D^0_{\mu\nu}(x, x') = -\langle \Psi_0 | V_\mu(x)\, V^\nu(x')\, | \Psi_0 \rangle \theta(t - t') - \langle \Psi_0 | V^\nu(x')\, V_\mu(x)\, | \Psi_0 \rangle \theta(t' - t) \tag{5.809}$$

Noting $V_i = -V^i$, then we obtain

$$\begin{aligned} \mathrm{i}D^0_{ij}(x, x') &= \frac{-g_{\mu\nu}}{V} \sum_{\vec{k}} \frac{1}{2\omega_k} \left[e^{-ik_\rho(x-x')^\rho} \theta(t - t') + e^{ik_\rho(x-x')^\rho} \theta(t' - t)\right] \\ &= \frac{-g_{\mu\nu}}{(2\pi)^3} \int \mathrm{d}\vec{k}\, e^{i\vec{k}\cdot\left(\vec{x}-\vec{x}'\right)} \frac{1}{2\omega_k} \left[e^{-i\omega_k(t-t')} \theta(t - t') + e^{i\omega_k(t-t')} \theta(t' - t)\right] \end{aligned} \tag{5.810}$$

Using the derivation method of Eq. (5.720), we can obtain

$$\mathrm{i}D^0_{\mu\nu}(x, x') = \mathrm{i}\frac{-g_{\mu\nu}}{(2\pi)^4} \int \mathrm{d}^4k \frac{e^{-ik_\rho(x-x')^\rho}}{k_\lambda k^\lambda - m^2 + i\varepsilon} \tag{5.811}$$

So the propagator of the free vector mesons in momentum space is

$$D^0_{\mu\nu}(k) = \frac{-g_{\mu\nu}}{k_\lambda k^\lambda - m^2 + \mathrm{i}\varepsilon} \tag{5.812}$$

Another derivation method of Eq. (5.812) is given below. In the Eq. (5.795), no longer taking non-independent variable $V^0 = 0$, but in Eq. (5.787) taking $\nu = 0$, then by using Eqs. (5.790)–(5.792) and (5.795) we can obtain

$$V^0 = \frac{1}{m^2}\vec{\nabla}\cdot\vec{\pi} \tag{5.813}$$

Substituting Eq. (5.813) into Eq. (5.795), we can also obtain [9]

$$\vec{\pi} = \dot{\vec{V}} + \frac{1}{m^2}\vec{\nabla}\left(\vec{\nabla}\cdot\vec{\pi}\right) \tag{5.814}$$

Since $\vec{V}$ and $\vec{\pi}$ are the Hermitian operators, the following plane wave expansion can be done:

$$\vec{V} = \frac{1}{\sqrt{V}}\sum_{\vec{k}}\frac{1}{\sqrt{2\omega_k}}\sum_{i=1}^{3}\left(\vec{q}_{\vec{k}i}\mathrm{e}^{-\mathrm{i}\omega_k t + \mathrm{i}\vec{k}\cdot\vec{x}} + \text{h.c.}\right) \tag{5.815}$$

$$\vec{\pi} = \frac{1}{\sqrt{V}}\sum_{\vec{k}}\frac{1}{\sqrt{2\omega_k}}\sum_{i=1}^{3}\left(\vec{p}_{\vec{k}i}\mathrm{e}^{-\mathrm{i}\omega_k t + \mathrm{i}\vec{k}\cdot\vec{x}} + \text{h.c.}\right) \tag{5.816}$$

Further from Eqs. (5.815) and (5.816), we can obtain

$$\dot{\vec{V}} = \frac{1}{\sqrt{V}}\sum_{\vec{k}}\frac{1}{\sqrt{2\omega_k}}\sum_{i=1}^{3}\left[(-\mathrm{i}\omega_k)\,\vec{q}_{\vec{k}i}\mathrm{e}^{-\mathrm{i}\omega_k t + \mathrm{i}\vec{k}\cdot\vec{x}} + \text{h.c.}\right] \tag{5.817}$$

$$\vec{\nabla}\cdot\vec{\pi} = \frac{1}{\sqrt{V}}\sum_{\vec{k}}\frac{1}{\sqrt{2\omega_k}}\sum_{i=1}^{3}\left[\mathrm{i}k_i\left(\vec{e}_{\vec{k}i}\cdot\vec{p}_{\vec{k}i}\right)\mathrm{e}^{-\mathrm{i}\omega_k t + \mathrm{i}\vec{k}\cdot\vec{x}} + \text{h.c.}\right] \tag{5.818}$$

$$\begin{aligned}\frac{1}{m^2}\vec{\nabla}\left(\vec{\nabla}\cdot\vec{\pi}\right) = {} & \frac{1}{\sqrt{V}}\sum_{\vec{k}}\frac{1}{\sqrt{2\omega_k}}\sum_{i=1}^{3}\vec{\mathrm{e}}_{\vec{k}i}\frac{1}{m^2} \\ & \left[(\mathrm{i}k_i)\sum_{j=1}^{3}\mathrm{i}k_j\left(\vec{\mathrm{e}}_{\vec{k}j}\cdot\vec{p}_{\vec{k}j}\right)\mathrm{e}^{-\mathrm{i}\omega_k t + \mathrm{i}\vec{k}\cdot\vec{x}} + \text{h.c.}\right]\end{aligned} \tag{5.819}$$

The following relations can be obtained substituting Eqs. (5.816), (5.817), and (5.819) into Eq. (5.814):

$$\begin{aligned} \vec{p}_{\vec{k}i} &= -\mathrm{i}\omega_k \vec{q}_{\vec{k}i} - \vec{\mathrm{e}}_{\vec{k}i}\frac{k_i}{m^2}\sum_{j=1}^{3} k_j\left(\vec{\mathrm{e}}_{\vec{k}j}\cdot\vec{p}_{\vec{k}j}\right), \\ \vec{p}^{\,+}_{\vec{k}i} &= \mathrm{i}\omega_k \vec{q}^{\,+}_{\vec{k}i} - \vec{\mathrm{e}}_{\vec{k}i}\frac{k_i}{m^2}\sum_{j=1}^{3} k_j\left(\vec{\mathrm{e}}_{\vec{k}j}\cdot\vec{p}^{\,+}_{\vec{k}j}\right) \end{aligned} \tag{5.820}$$

Then rewrite the above formulas as

$$\begin{aligned} \vec{q}_{\vec{k}i} &= \frac{\mathrm{i}}{\omega_k}\left[\vec{p}_{\vec{k}i} + \vec{\mathrm{e}}_{\vec{k}i}\frac{k_i}{m^2}\sum_{j=1}^{3} k_j\left(\vec{\mathrm{e}}_{\vec{k}j}\cdot\vec{p}_{\vec{k}j}\right)\right], \\ \vec{q}^{\,+}_{\vec{k}i} &= \frac{-\mathrm{i}}{\omega_k}\left[\vec{p}^{\,+}_{\vec{k}i} + \vec{\mathrm{e}}_{\vec{k}i}\frac{k_i}{m^2}\sum_{j=1}^{3} k_j\left(\vec{\mathrm{e}}_{\vec{k}j}\cdot\vec{p}^{\,+}_{\vec{k}j}\right)\right] \end{aligned} \tag{5.821}$$

Introduce the quantization in the following way:

$$\vec{p}_{\vec{k}i} = \vec{\mathrm{e}}_{\vec{k}i}(-\mathrm{i}\omega_k)a_{\vec{k}i}, \qquad \vec{p}^{\,+}_{\vec{k}i} = \vec{\mathrm{e}}_{\vec{k}i}(\mathrm{i}\omega_k)a^{+}_{\vec{k}i} \tag{5.822}$$

Einstein's convention is still used below. In the KD metric, two identical Greek letter subscripts represent sums from 0 to 3, and two identical English letter subscripts represent sums from 1 to 3. So the Eq. (5.821) can be rewritten as

$$\vec{q}_{\vec{k}i} = \vec{\mathrm{e}}_{\vec{k}i}\left(a_{\vec{k}i} + \frac{k_ik_j}{m^2}a_{\vec{k}j}\right), \qquad \vec{q}^{\,+}_{\vec{k}i} = \vec{\mathrm{e}}_{\vec{k}i}\left(a^{+}_{\vec{k}i} + \frac{k_ik_j}{m^2}a^{+}_{\vec{k}j}\right) \tag{5.823}$$

Substituting Eq. (5.823) and Eq. (5.822) into Eq. (5.815) and Eq. (5.816), respectively, we can get the following relations:

$$\vec{V} = \frac{1}{\sqrt{V}}\sum_{\vec{k}}\frac{1}{\sqrt{2\omega_k}}\sum_{i=1}^{3}\vec{\mathrm{e}}_{\vec{k}i}\left[\left(a_{\vec{k}i} + \frac{k_ik_j}{m^2}a_{\vec{k}j}\right)\mathrm{e}^{-\mathrm{i}\omega_k t + \mathrm{i}\vec{k}\cdot\vec{x}} + \mathrm{h.c.}\right] \tag{5.824}$$

$$\vec{\pi} = \frac{1}{\sqrt{V}}\sum_{\vec{k}}(-\mathrm{i})\sqrt{\frac{\omega_k}{2}}\sum_{i=1}^{3}\vec{\mathrm{e}}_{\vec{k}i}\left[a_{\vec{k}i}\mathrm{e}^{-\mathrm{i}\omega_k t + \mathrm{i}\vec{k}\cdot\vec{x}} - \mathrm{h.c.}\right] \tag{5.825}$$

The equal-time commutation relation can be obtained, by using Eqs. (5.824), (5.825), and Eq. (5.801), as follows:

$$\left[V^i\left(\vec{x},\ t\right),\ \pi^j\left(\vec{x}',\ t\right)\right] = \mathrm{i}\left(\delta_{ij} + \frac{k_i k_l}{m^2}\delta_{lj}\right)\frac{1}{V}\sum_{\vec{k}} \mathrm{e}^{\mathrm{i}\vec{k}\cdot\left(\vec{x}-\vec{x}'\right)}$$
$$= \mathrm{i}\left(\delta_{ij} + \frac{k_i k_l}{m^2}\delta_{lj}\right)\delta\left(\vec{x} - \vec{x}'\right) \tag{5.826}$$

This formula is more the term $\frac{k_i k_l}{m^2}\delta_{lj}$ than Eq. (5.802). Of course it can be prove that

$$\left[V^i\left(\vec{x},\ t\right),\ V^j\left(\vec{x}',\ t\right)\right] = \left[\pi^i\left(\vec{x},\ t\right),\ \pi^j\left(\vec{x}',\ t\right)\right] = 0$$

Equation (5.805) can be rewritten as follows referring to Eq. (5.824):

$$V\left(\vec{x}, t\right) = \frac{1}{\sqrt{V}}\sum_{\vec{k}}\frac{1}{\sqrt{2\omega_k}}\sum_{\mu=0}^{3}\vec{e}_{\vec{k}\mu}\left[\left(a_{\vec{k}\mu} + \frac{k_\mu k_\nu}{m^2}a_{\vec{k}\nu}\right)\mathrm{e}^{-\mathrm{i}k_\rho x^\rho} + \mathrm{h.c.}\right] \tag{5.827}$$

Using the similar method, from Eq. (5.807) the following results can be obtained by using Eq. (5.827):

$$\mathrm{i}D^0_{\mu\nu}(x, x') = \mathrm{i}\int \frac{\mathrm{d}^4k}{(2\pi)^4}\frac{-g^{\mu\nu} + \dfrac{k_\mu k_\nu}{m^2}}{k_\lambda k^\lambda - m^2 + \mathrm{i}\varepsilon}\mathrm{e}^{-\mathrm{i}k_\rho(x-x')^\rho} \tag{5.828}$$

$$D^0_{\mu\nu}(k) = \frac{-g^{\mu\nu} + \dfrac{k_\mu k_\nu}{m^2}}{k_\lambda k^\lambda - m^2 + \mathrm{i}\varepsilon} \tag{5.829}$$

This is exactly the result given in Refs. [8, 26, 78].

Since the vector mesons must be coupled with the conserved nucleon current in practical applications, the conserved nucleon current $j^\nu(x)$ must satisfy $\partial_\nu j^\nu(x) = 0$; therefore $k_\nu j^\nu(k) = 0$ must be satisfied in momentum space. When $D^0_{\mu\nu}(k)$ given by Eq. (5.829) acting on $j^\nu(k)$, the second term will give $k_\mu k_\nu j^\nu = 0$, so the $k_\mu k_\nu$ term of $D^0_{\mu\nu}(k)$ given by Eq. (5.829) does not contribute to the physical quantity [8, 69, 79]. At the same time, the term $\frac{k_i k_l}{m^2}\delta_{lj}$ appeared in Eq. (5.826) violates the usual quantization rule, which also indicates that the $k_\mu k_\nu$ term should not be retained in Eq. (5.829). Therefore, in practical applications, the $D^0_{\mu\nu}(k)$ given by Eq. (5.812) is used.

Although the time-like component $V^0$ of the free vector meson field is not an independent variable, but the zero order propagators obtained here are generally recognized results. In addition, this problem was studied by introducing Lagrange multiplier $\lambda$ in the Refs. [8, 79]. The selection of $\lambda$ value is uncertain, but in the end it is recommended to use $D^0_{\mu\nu}(k)$ given by Eq. (5.812).

### 5.11.4 Feynman Rules of Nucleon-Meson Interaction in Momentum Representation

The Feynman rules of the non-relativistic many-body theory are summarized by introducing secondary quantification, Wick theorem, expressions of various Green functions, corresponding Feynman diagrams, and so on in Ref. [28]. In the relativistic many-body theory, it is also necessary to go through a series of theoretical preparations, in particular to use the Wick theorem of bosons and fermions, and then to summarize the Feynman rules of nucleon-meson interaction through a large number of examples [2, 8, 67, 78]. The relativistic many-body theory will automatically degenerate to the non-relativistic many-body theory under very low energy condition.

Let $|\Psi\rangle$ represent the ground state of a system of interacting many nucleons at finite density ($|\Psi\rangle$ is vacuum state for mesons); the corresponding nucleon propagator is defined as

$$\mathrm{i}G_{\alpha\beta}(x, x') = \langle\Psi|\mathrm{T}\{\,\psi_\alpha(x)\,\overline{\psi}_\beta(x')\}\;|\Psi\rangle \tag{5.830}$$

where $\psi(x)$ is the nucleon field in Heisenberg representation. $G_{\alpha\beta}$ is a matrix element and $G$ represents a matrix. The interacting nucleon propagator $G$ and the non-interacting nucleon propagator $G^0$ satisfy the following Dyson equation:

$$G(k) = G^0(k) + G^0(k)\Sigma(k)G(k) \tag{5.831}$$

where $\Sigma$ represents the self-energy operator. The above Dyson equation, the Hartree-Fock equation of the self-energy operator $\Sigma$, and the interacting meson propagator can be represented by Fig. 5.4 when only $\sigma$ mesons exist.

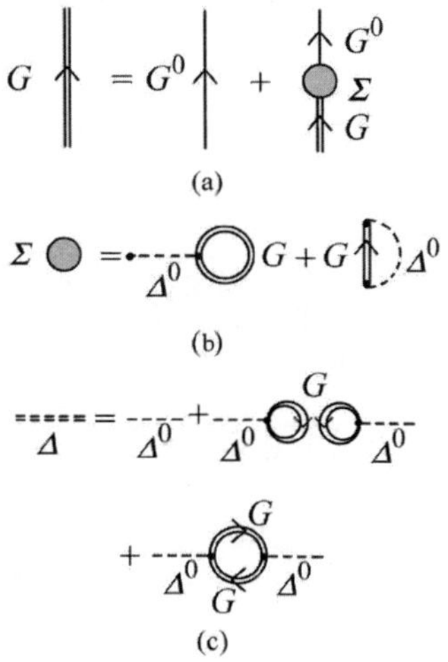


**Fig. 5.4** Graphical representation of the self-consistent relativistic Hartree-Fock equation (**a**) Nucleon Dyson equation; (**b**) Hartree-Fock equation of $\Sigma$; (**c**) Interacting $\sigma$ meson propagator

In the non-relativistic theory, the nucleon-nucleon interaction occurs simultaneously for two nucleons, whereas in the relativistic theory, the nucleon-nucleon interaction occurs by exchanging mesons. Since the propagation of mesons needs time, so the relativistic nucleon-nucleon interaction does not occur at the same time for two nucleons. In the Feynman diagram, the position point where two nucleon lines meet a meson line is called vertex, a meson has two vertexes and the number of the vertexes is even in a Feynman diagram. Therefore, the non-relativistic first order graph corresponds to relativistic second order graph, and the non-relativistic second order graph corresponds to relativistic fourth order graph.

The main advantage of calculating the interacting propagators is that they allow one to evaluate observable quantities. For example, using the definition of the nucleon Green function given by Eq. (5.830) and the anti-commutation property of the nucleon wave function, the expectation value of the operator $\Gamma$ can be written as

$$\begin{aligned}\langle\Psi|\overline{\psi}(x)\Gamma\psi(x)\,|\Psi\rangle &= -\lim_{\vec{y}\to\vec{x}}\lim_{y^0\to x^{0+}}\langle\Psi|\mathrm{T}\left[\psi_\alpha(x)\,\overline{\psi}_\beta(y)\right]|\Psi\rangle\Gamma_{\beta\alpha}\\ &= -\lim_{\vec{y}\to\vec{x}}\lim_{y^0\to x^{0+}}\mathrm{tr}[\Gamma\,\mathrm{i}G(x,\ y)]\end{aligned} \tag{5.832}$$

where the relation $\mathrm{tr}\left(\hat{A}\hat{B}\right) = \sum_{\alpha\beta} A_{\beta\alpha}B_{\alpha\beta}$ is used, $y^0 \to x^{0+}$ represents that $y^0$ is less than $x^{0+}$ but infinitely closer to $x^{0+}$. For a two-line closed ring graph (loop) of the operator $G$ interacting with $\sigma$ meson, in this case there is $x' = x$, refer to above formula we can write

$$\begin{aligned}\langle\Psi|\overline{\psi}(x)\,\psi(x)|\Psi\rangle &= -\lim_{\vec{y}\to\vec{x}}\lim_{y^0\to x^{0+}}\langle\Psi|\mathrm{T}\{\,\psi_\alpha(x)\,\overline{\psi}_\beta(y)\}\,|\Psi\rangle\delta_{\beta\alpha}\\ &= -\lim_{\vec{y}\to\vec{x}}\lim_{y^0\to x^{0+}}\mathrm{tr}[\mathrm{i}G(x,\ y)]\end{aligned} \tag{5.833}$$

First the physical process of the interaction between the nucleon and the meson is expressed by the connected Feynman diagram with two external propagator lines and $n$ vertexes connected by internal propagator lines. Assign a direction to each line. The rule that points out what each part of the diagram corresponds to what each part in the physical formula is called the Feynman rule. We will summarize some of the main elements of the Feynman rule of the nucleon-meson interaction in the momentum representation using the Bjorken-Drell metric as follows:

1. Each directed line gets a factor, such as:

   A single solid line represents a free nucleon propagator $\to \mathrm{i}G^0$.
   The double solid lines represent interacting nucleon propagator $\to \mathrm{i}G$.
   The dotted line represents the free $\sigma$ meson propagator $\to \mathrm{i}\Delta^0$.
   The ripple line represents the free ω meson propagator $\to \mathrm{i}D^0_{\mu\nu}$.
   Similar methods can be used for isospin vector meson ρ and π.

2. The vertexes of σ and ω mesons are assigned a factor $-ig_\sigma$ and $-ig_\omega\gamma^\mu$, respectively. Conserve the energy and momentum at each vertex. In general, spin and isospin are summed at the vertex, but if the spin and isospin of the two nucleons at the vertex are the same, there is no need to sum them. Similar methods can be used for ρ and π mesons.
3. Sum over all repeated indices.
4. Integrate $\int \frac{d^4q}{(2\pi)^4}$ over all independent four-momentum $q$.
5. Include a factor of $(-1)$ for each closed fermions loop, and trace the corresponding physical quantity, such as the following correspondence:

$-\mathrm{tr}(iG)$

$-\mathrm{tr}[(-ig_\omega\gamma^\nu)iG]$

Require placing the contribution of the vertex of ω meson line and the nucleon loop in the trace.

$-\mathrm{tr}(iGiG)$

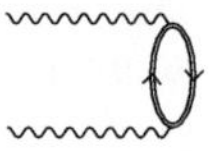

$-\mathrm{tr}[(-ig_\omega\gamma^\nu)iG(-ig_\omega\gamma^\eta)iG]$

Require placing the contributions of the two vertexes of ω meson lines and the nucleon loop in the trace. If the contribution of the vertex located on the nucleon loop contains the isospin matrix, it should also be placed in the trace. Then we should make tracing of the two kinds of matrices, respectively. Other cases could be dealt with in a similar manner.

## 5.12 Real Part of Nucleon Relativistic Microscopic Optical Potential and Relativistic Nuclear Matter Properties

### *5.12.1 Self-Energy Operator and Green Function in Nuclear Matter [26, 80]*

When the tensor term is omitted, the self-energy operator in nuclear matter has been given by Eq. (5.231) as

$$\Sigma(k) = \Sigma^s(k) + \gamma^0\Sigma^0(k) + \vec{\gamma}\cdot\vec{k}\Sigma^v(k) \tag{5.834}$$

For the no-interacting state $|\Psi_0\rangle$, refer to Eqs. (5.66) and (5.62), we can write the definition of the zero order Green function $G^0$ as

$$\left(\gamma^{\mu}k_{\mu} - M\right) G^{0} = I \tag{5.835}$$

The above formula can be formally written as

$$\left(G^{0}\right)^{-1} = \gamma^{\mu}k_{\mu} - M \tag{5.836}$$

The Dyson equation (5.831) can also be rewritten as

$$\left(1 - G^{0}\Sigma\right) G = G^{0} \tag{5.837}$$

For the above equation, multiplying $(G^{0})^{-1}$ from left end and $G^{-1}$ from right end, and using Eqs. (5.836) and (5.834), we can obtain

$$G^{-1} = \left(G^{0}\right)^{-1} - \Sigma = \gamma^{\mu}k_{\mu} - M - \Sigma^{\mathrm{s}} - \gamma^{0}\Sigma^{0} - \vec{\gamma}\cdot\vec{k}\Sigma^{\mathrm{v}} \tag{5.838}$$

If we let

$$k_0^{*} = k_0 - \Sigma^{0}, \quad \vec{k}^{*} = \vec{k}(1 + \Sigma^{\mathrm{v}}), \quad M_k^{*} = M + \Sigma^{\mathrm{s}} \tag{5.839}$$

Equation (5.838) can be rewritten as

$$G^{-1} = \gamma^{\mu}k_{\mu}^{*} - M_k^{*} \tag{5.840}$$

We also define

$$E_k^{*} = \left(k^{*2} + M_k^{*2}\right)^{1/2} \tag{5.841}$$

Formally $E_k^{*}$ is equivalent to on-shell energy when existence of interactions. We know that in no-interacting case the on-shell single particle energy should be $k_0 = E_k = (k^2 + M^2)^{1/2}$, and $k_0 = E_k$ corresponds to $k_0^{*} = E_k^{*}$, i.e., when $k_0 = E_k$ there is $k_0^{*} = E_k^{*}$. In $\Sigma \to 0$ case we can easily see that the above statement is correct. So when $k_0 = E_k$ from the first equation of Eq. (5.839) and Eq. (5.841), we can obtain

$$\begin{aligned} E_k = E_k^{*} + \Sigma^{0} = &\left\{k^2\left[1 + \Sigma^{\mathrm{v}}\left(\left|\vec{k}\right|, \Sigma_k\right)\right]^2 + \left[M + \Sigma^{\mathrm{s}}\left(\left|\vec{k}\right|, E_k\right)\right]^2\right\}^{1/2} \\ &+ \Sigma^{0}\left(\left|\vec{k}\right|, E_k\right) \end{aligned} \tag{5.842}$$

In the above formula, $E_k$ is a transcendent function that requires to be solved by self-consistent iteration way. Obviously $E_k$ is only related to $\left|\vec{k}\right|$ and nuclear matter density expressed by $k_{\mathrm{F}}$. And from Eq. (5.842) we can see $E_k \underset{\Sigma\to 0}{\longrightarrow} \left(k^2 + M^2\right)^{1/2}$.

Since Eq. (5.840) and Eq. (5.836) are the same in form, so refer to Eq. (5.781), we can write

$$G(k) = G_{\mathrm{F}}(k) + G_{\mathrm{D}}(k) \tag{5.843}$$

$$G_{\mathrm{F}}(k) = \left(\gamma^{\mu} k_{\mu}^{*} + M_{k}^{*}\right) \frac{1}{k_{\mu}^{*} k^{\mu *} - M_{k}^{*2} + \mathrm{i}\varepsilon} \tag{5.844}$$

$$G_{\mathrm{D}}(k) = \left(\gamma^{\mu} k_{\mu}^{*} + M_{k}^{*}\right) \frac{\mathrm{i}\pi}{E_{k}^{*}} \delta(k_0 - E_k)\, \theta\left(k_F - \left|\vec{k}\right|\right) \tag{5.845}$$

where $E_k$ in Eq. (5.845) is the on-shell single particle energy given by Eq. (5.842).

## *5.12.2 Real Part of Nucleon Relativistic Microscopic Optical Potential*

1. σ Meson

Figure 5.5 shows the Feynman diagram of Hartree-Fock equation of exchange σ mesons. According to the Feynman rules described in the 5.11 section, the expression corresponding to the abovementioned Feynman diagram can be written as

$$\begin{aligned} &\mathrm{i}G^0(k)\Sigma_\sigma(k)G(k) \\ &\quad = \mathrm{i}G^0(k)(-\mathrm{i}g_\sigma)(-\mathrm{i}g_\sigma)\, \mathrm{i}\Delta^0(0) \int \frac{\mathrm{d}^4 q}{(2\pi)^4} \mathrm{tr}(-\mathrm{i}G(q))\, \mathrm{i}G(k) \\ &\quad\quad + \mathrm{i}G^0(k)(-\mathrm{i}g_\sigma)(-\mathrm{i}g_\sigma) \int \frac{\mathrm{d}^4 q}{(2\pi)^4} \mathrm{i}\Delta^0(k-q)\mathrm{i}G(q)\, \mathrm{i}G(k) \end{aligned} \tag{5.846}$$

where $G^0\Sigma_\sigma G$ on the left side of the above equation is regarded as a total propagator, so only one imaginary unit i is multiplied in front. Or according to the Feynman rules, interaction $V$, effective interaction $T$, and self-energy $\Sigma$ contribute $-\mathrm{i}V$, $-\mathrm{i}T$, $-\mathrm{i}\Sigma$, respectively, and there is $(-\mathrm{i})^3 = \mathrm{i}$. Then noting Eq. (5.721) we can obtain

$$\begin{aligned} \Sigma_\sigma(k) &= -\mathrm{i}g_\sigma^2 \Delta^0(0) \int \frac{\mathrm{d}^4 q}{(2\pi)^4} \mathrm{tr}G(q) + \mathrm{i}g_\sigma^2 \int \frac{\mathrm{d}^4 q}{(2\pi)^4} \Delta^0(k-q) G(q) \\ &= \mathrm{i}g_\sigma^2 \int \frac{\mathrm{d}^4 q}{(2\pi)^4} \left\{ \frac{1}{m_\sigma^2} \mathrm{tr}G(q) + \frac{G(q)}{(k-q)_\mu (k-q)^\mu - m_\sigma^2 + \mathrm{i}\varepsilon} \right\} \end{aligned} \tag{5.847}$$

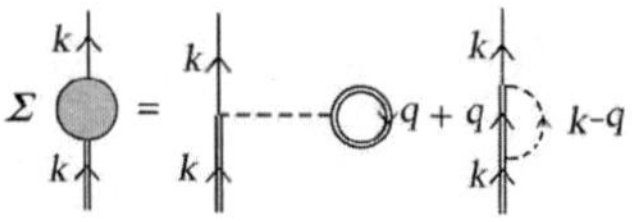


**Fig. 5.5** Feynman diagram of Hartree-Fock equation of exchange σ mesons

where the first term is the Hartree term, which is independent with momentum $k$ ; and the second term is the Fock term, which is related to momentum $k$.

We have known $G = G_F + G_D$, where the Feynman propagator $G_F$ represents the contribution of virtual nucleons and anti-nucleons in vacuum, and the $G_F$ term in the integral given by Eq. (5.847) are divergent. In principle, through studying vacuum fluctuation correction, the contribution of the $G_F$ item can be studied using the renormalization method [75]. However, this is a subject that needs to be further studied. Here we only study the real nucleons in the Fermi sea, i.e., we neglect the contribution of the $G_F$ item.

When taking $G = G_D$, using Eq. (5.845) we can rewrite Eq. (5.847) as

$$\Sigma_\sigma(k) = \mathrm{i}g_\sigma^2 \int \frac{\mathrm{d}^4 q}{(2\pi)^4} \left\{ \frac{1}{m_\sigma^2} \mathrm{tr}\left(\gamma^\mu q_\mu^* + M_q^*\right) + \frac{\gamma^\mu q_\mu^* + M_q^*}{(k-q)_\mu (k-q)^\mu - m_\sigma^2 + \mathrm{i}\varepsilon} \right\} \frac{\mathrm{i}\pi}{E_q^*} \delta(q_0 - E_q)\, \theta\left(k_F - |\vec{q}|\right) = \Sigma_\sigma^H(k) + \Sigma_\sigma^F(k) \tag{5.848}$$

Here we assume that there is only one kind of nucleon in the nuclear matter, i.e., without considering the nucleon isospin. Then we can find the Hartre term as

$$\Sigma_\sigma^H(k) = -g_\sigma^2 \frac{\pi}{(2\pi)^4} \int \mathrm{d}\vec{q}\, \frac{4M_q^*}{m_\sigma^2 E_q^*} \theta\left(k_F - |\vec{q}|\right) = -\left(\frac{g_\sigma}{m_\sigma}\right)^2 \frac{1}{\pi^2} \int_0^{k_F} \frac{M_q^*}{E_q^*} q^2 \mathrm{d}q \tag{5.849}$$

It can be seen that the Hartree item is independent with momentum $k$. In the following we will find the Fock item by using Eq. (5.848). Noting when $q_0 = E_q$ there is $q_0^* = E_q^*$, then we can obtain

$$\begin{aligned}
\Sigma_\sigma^F(k) &= \mathrm{i}g_\sigma^2 \int \frac{\mathrm{d}^4 q}{(2\pi)^4} \frac{\gamma^\mu q_\mu^* + M_q^*}{(k-q)_\mu (k-q)^\mu - m_\sigma^2 + \mathrm{i}\varepsilon} \frac{\mathrm{i}\pi}{\mathrm{E}_\mathrm{q}^*} \delta(q_0 - E_q) \theta\left(k_F - |\vec{q}|\right) \\
&= -\frac{\pi g_\sigma^2}{(2\pi)^4} \int \mathrm{d}\vec{q}\, \frac{\gamma^\mu q_\mu^* + M_q^*}{(k-q)_\mu (k-q)^\mu - m_\sigma^2} \frac{\theta\left(k_F - |\vec{q}|\right)}{E_q^*} \Big|_{q_0 = E_q} \\
&= -\frac{\pi g_\sigma^2}{(2\pi)^4} \int \mathrm{d}\vec{q} \left( \frac{M_q^*}{E_q^*} + \gamma^0 - \frac{\vec{\gamma} \cdot \vec{q}^{\,*}}{E_q^*} \right) \frac{\theta\left(k_F - |\vec{q}|\right)}{(k-q)_\mu (k-q)^\mu - m_\sigma^2} \Big|_{q_0 = E_q}
\end{aligned} \tag{5.850}$$

Taking $z$-axis along the $\vec{k}$ direction and setting that the angle between $\vec{\gamma}$ and $\vec{k}$ is $\theta_1$ in the $\varphi = 0$ plane, i.e., $\vec{\gamma} = (\gamma, \theta_1, 0)$, and letting $\vec{q} = (q, \theta, \varphi)$, then in the rectangular coordinate system we can write

$$q^1 = q\sin\theta\cos\varphi \qquad \gamma^1 = \gamma\sin\theta_1$$
$$q^2 = q\sin\theta\sin\varphi \qquad \gamma^2 = 0$$
$$q^3 = q\cos\theta \qquad \gamma^3 = \gamma\cos\theta_1$$

From Eq. (5.839) we know $\vec{q}^{\,*} = (1+\Sigma^{v})\,\vec{q}$, and we can obtain

$$\vec{\gamma}\cdot\vec{q}^{\,*} = \gamma\sin\theta_1 q^*\sin\theta\cos\varphi + \gamma\cos\theta_1 q^*\cos\theta \tag{5.851}$$

Substituting this formula into Eq. (5.850), when making integral to $\varphi$, we can know that the first item in Eq. (5.851) does not contribute. When only retaining the items that have contributions to the integral, from Eq. (5.851) we can get

$$\vec{\gamma}\cdot\vec{q}^{\,*} = \left(\vec{\gamma}\cdot\vec{k}\right)\left(\vec{k}\cdot\vec{q}^{\,*}\right)/k^2 \tag{5.852}$$

We can also obtain

$$\begin{aligned}(k-q)_\mu(k-q)^\mu|_{q_0=E_q} &= \left(E-E_q\right)^2 - \left(\vec{k}-\vec{q}\right)^2 \\ &= \left(E-E_q\right)^2 - k^2 - q^2 + 2kq\cos\theta\end{aligned} \tag{5.853}$$

Let

$$A_i(k,q) = k^2 + q^2 + m_i^2 - \left(E-E_q\right)^2, \qquad i = \sigma,\ \omega,\ \pi,\ \rho \tag{5.854}$$

so there is

$$\frac{1}{(k-q)_\mu(k-q)^\mu - m_i^2}|_{q_0=E_q} = \frac{1}{-A_i(k,q) + 2kq\cos\theta} \tag{5.855}$$

And there are the following indefinite integral formulas:

$$\int\frac{du}{a+bu} = \frac{1}{b}\ln(a+bu) + C, \qquad \int\frac{u\,du}{a+bu} = \frac{1}{b^2}\left[a+bu-a\ln(a+bu)\right] + C \tag{5.856}$$

Let us find the following integral:

$$\begin{aligned}2\pi\int_{-1}^{1}\frac{1}{-A_i(k,q)+2kq\cos\theta}\,\mathrm{d}\cos\theta &= -\frac{2\pi}{2kq}\ln\left|\frac{-A_i(k,\ q)-2kq}{-A_i(k,\ q)+2kq}\right| \\ &= -\frac{\pi}{kq}\ln\left|\frac{A_i(k,\ q)+2kq}{A_i(k,\ q)-2kq}\right| = -\frac{\pi}{kq}\Theta_i(k,q)\end{aligned} \tag{5.857}$$

where

$$\Theta_i(k,\ q) = \ln\left|\frac{A_i(k,\ q) + 2kq}{A_i(k,\ q) - 2kq}\right|, \qquad i = \sigma,\omega,\pi,\rho \tag{5.858}$$

We can also get the following integral:

$$\begin{aligned} 2\pi\int_{-1}^{1}\frac{kq\cos\theta}{-A_i(k,\ q) + 2kq\cos\theta}\mathrm{d}\cos\theta &= \frac{2\pi kq}{4k^2q^2}\left[4kq - A_i(k,\ q)\ \Theta_i(k,\ q)\right] \\ &= 2\pi\left[1 - \frac{A_i(k,\ q)}{4kq}\Theta_i(k,\ q)\right] = -2\pi\Phi_i(k,\ q) \end{aligned} \tag{5.859}$$

where

$$\Phi_i(k,\ q) = \frac{A_i(k,\ q)\ \Theta_i(k,\ q)}{4kq} - 1, \qquad i = \sigma,\omega,\pi,\rho \tag{5.860}$$

Therefore we can obtain the following result according to Eqs. (5.850), (5.852), (5.855), (5.857), and (5.859):

$$\begin{aligned} \Sigma_\sigma^{\mathrm{F}}(k,\ E) &= \frac{\pi g_\sigma^2}{(2\pi)^4}\left\{\frac{\pi}{k}\int_0^{k_{\mathrm{F}}}\frac{M_q^*}{E_q^*}\Theta_\sigma(k,q)q\mathrm{d}q\right. \\ &\left.+\gamma^0\frac{\pi}{k}\int_0^{k_{\mathrm{F}}}\Theta_\sigma(k,q)q\mathrm{d}q - \vec{\gamma}\cdot\vec{k}\frac{2\pi}{k^2}\int_0^{k_{\mathrm{F}}}\frac{q^*}{E_q^*}\Phi_\sigma(k,q)q\mathrm{d}q\right\} \\ &= \frac{g_\sigma^2}{16\pi^2 k}\left\{\int_0^{k_{\mathrm{F}}}\frac{M_q^*}{E_q^*}\Theta_\sigma(k,q)q\mathrm{d}q + \gamma^0\int_0^{k_{\mathrm{F}}}\Theta_\sigma(k,q)q\mathrm{d}q\right. \\ &\left.-\frac{2\vec{\gamma}\cdot\vec{k}}{k}\int_0^{k_{\mathrm{F}}}\frac{q^*}{E_q^*}\Phi_\sigma(k,q)q\mathrm{d}q\right\} \end{aligned} \tag{5.861}$$

The coefficients of the constant term which is not related to $\gamma$ matrix, $\gamma^0$ term, and $\vec{\gamma}\cdot\vec{k}$ term in the above equation are $V_\sigma^{\mathrm{F-s}}(k)$, $V_\sigma^{\mathrm{F-0}}(k)$, and $V_\sigma^{\mathrm{F-v}}(k)$, respectively.

2. ω Meson

In Fig. 5.5 we replace the dotted line with a corrugated line to represent the exchange of the ω meson. According to Feynman rules, we can write

$$\begin{aligned} \mathrm{i}G^0(k)\Sigma_\omega(k)\ G(k) &= \mathrm{i}G^0(k)(-\mathrm{i}g_\omega\gamma^\mu)\ \mathrm{i}D_{\mu\nu}^0(0) \\ &\int\frac{\mathrm{d}^4q}{(2\pi)^4}\mathrm{tr}[(-\mathrm{i}g_\omega\gamma^\nu)(-\mathrm{i}G(q))]\ \mathrm{i}G(k) + \mathrm{i}G^0(k)(-\mathrm{i}g_\omega\gamma^\mu) \\ &\int\frac{\mathrm{d}^4q}{(2\pi)^4}\ \mathrm{i}D_{\mu\nu}^0(k-q)\ \mathrm{i}G(q)(-\mathrm{i}g_\omega\gamma^\nu)\ \mathrm{i}G(k) \end{aligned} \tag{5.862}$$

And using Eq. (5.812) we can further obtain

$$\begin{aligned}\Sigma_\omega(k) &= -\mathrm{i}g_\omega^2\gamma^\mu \frac{g_{\mu\nu}}{m_\omega^2}\int \frac{\mathrm{d}^4q}{(2\pi)^4}\mathrm{tr}(\gamma^\nu G(q)) + \mathrm{i}g_\omega^2\gamma^\mu \int \frac{\mathrm{d}^4q}{(2\pi)^4} D^0_{\mu\nu}(k-q)G(q)\gamma^\nu \\ &= -\mathrm{i}g_\omega^2\gamma^\mu g_{\mu\mu}\int \frac{\mathrm{d}^4q}{(2\pi)^4}\left[\frac{\mathrm{tr}(\gamma^\mu G(q))}{m_\omega^2} + \frac{G(q)\gamma^\mu}{(k-q)_\lambda(k-q)^\lambda - m_\omega^2 + \mathrm{i}\varepsilon}\right] \\ &= \Sigma_\omega^{\mathrm{H}}(k) + \Sigma_\omega^{\mathrm{F}}(k)\end{aligned} \tag{5.863}$$

The above formula implicit summation to $\mu$. It can be seen that when the sum to $\mu$ is not carried out, according Eq. (5.149) we know

$$g_{\mu\mu}\mathrm{tr}\{\gamma^\mu(\gamma^\nu q_\nu^* + M_k^*)\} = 4q_\mu^* \tag{5.864}$$

Still taking $G = G_{\mathrm{D}}$, so using Eqs. (5.845) and (5.864), we can obtain the following formula from Eq. (5.863):

$$\Sigma_\omega^{\mathrm{H}}(k) = -\mathrm{i}g_\omega^2\gamma^\mu \int \frac{\mathrm{d}^4q}{(2\pi)^4}\frac{4q_\mu^*}{m_\omega^2}\frac{\mathrm{i}\pi}{E_q^*}\delta(q_0 - E_q)\,\theta\left(k_{\mathrm{F}} - |\vec{q}|\right) \tag{5.865}$$

Utilizing Eq. (5.852) we also have

$$\gamma^\mu g_\mu^*\Big|_{q_0=E_q} = \gamma^0 E_q^* - \vec{\gamma}\cdot\vec{k}\,\frac{\vec{k}\cdot\vec{q}^{\,*}}{k^2} \tag{5.866}$$

Since there is $\vec{k}\cdot\vec{q}^{\,*} = kq^*\cos\theta$, when making integral to $\cos\theta$ it can be known that the second item in Eq. (5.866) has no contribution, so the following result can be obtained by Eq. (5.865).

$$\Sigma_\omega^{\mathrm{H}}(k) = \gamma^0\frac{g_\omega^2}{\pi^2 m_\omega^2}\int_0^{k_{\mathrm{F}}} q^2\mathrm{d}q = \gamma^0\left(\frac{g_\omega}{m_\omega}\right)^2\frac{k_{\mathrm{F}}^3}{3\pi^2} \tag{5.867}$$

And the following formula can be obtained according to Eqs. (5.863) and (5.845):

$$\Sigma_\omega^{\mathrm{F}}(k) = -\mathrm{i}g_\omega^2\int \frac{\mathrm{d}^4q}{(2\pi)^4}\frac{g_{\mu\mu}\gamma^\mu\left(\gamma^\nu q_\nu^* + M_q^*\right)\gamma^\mu}{(k-q)_\lambda(k-q)^\lambda - m_\omega^2 + \mathrm{i}\varepsilon}\frac{\mathrm{i}\pi}{E_q^*}\delta(q_0 - E_q)\,\theta\left(k_{\mathrm{F}} - |\vec{q}|\right) \tag{5.868}$$

Notice

$$\sum_{\mu} g_{\mu\mu}\gamma^{\mu}\gamma^{\nu}\gamma^{\mu} = \sum_{\mu} g_{\mu\mu}\gamma^{\mu}(-\gamma^{\mu}\gamma^{\nu} + 2g^{\mu\nu}) = -4\gamma^{\nu} + 2\gamma^{\nu} = -2\gamma^{\nu} \tag{5.869}$$

so there is

$$\sum_{\mu} g_{\mu\mu}\gamma^{\mu}\left(\gamma^{\nu}q_{\nu}^{*} + M_{q}^{*}\right)\gamma^{\mu} = -2\gamma^{\nu}q_{\nu}^{*} + 4M_{q}^{*} \tag{5.870}$$

Then Eq. (5.868) can be rewritten as

$$\Sigma_{\omega}^{\mathrm{F}}(k) = -\pi g_{\omega}^{2}\int\frac{\mathrm{d}^{4}q}{(2\pi)^{4}}\frac{2\gamma^{\nu}q_{\nu}^{*} - 4M_{q}^{*}}{(k-q)_{\lambda}(k-q)^{\lambda} - m_{\omega}^{2} + \mathrm{i}\varepsilon}\frac{1}{E_{q}^{*}}\delta\left(q_{0} - E_{q}\right)\theta\left(k_{\mathrm{F}} - |\vec{q}|\right) \tag{5.871}$$

Comparing this formula with the first equation of Eq. (5.850) and then according to Eq. (5.861), we can obtain

$$\begin{aligned} \Sigma_{\omega}^{\mathrm{F}}(k,\,E) = \frac{g_{\omega}^{2}}{16\pi^{2}k}\Bigg\{ & -4\int_{0}^{k_{\mathrm{F}}}\frac{M_{q}^{*}}{E_{q}^{*}}\Theta_{\omega}(k,\ q)q\mathrm{d}q \\ & +2\gamma^{0}\int_{0}^{k_{\mathrm{F}}}\Theta_{\omega}(k,\,q)\,q\mathrm{d}q - \frac{4\vec{\gamma}\cdot\vec{k}}{k}\int_{0}^{k_{\mathrm{F}}}\frac{q^{*}}{E_{q}^{*}}\Phi_{\omega}(k,\ q)q\mathrm{d}q\Bigg\} \end{aligned} \tag{5.872}$$

The contribution of ω meson tensor term is not considered here.

3. π Meson

Equation (5.671) gives the interaction Lagrangian density of the π meson with the nucleon. The first and second terms represent the pseudo-scalar (ps) and axis-vector (pv) coupling, respectively. Replace the σ meson with the π meson in Fig. 5.5, and compare Eq. (5.671) with Eq. (5.669) describing the ω meson; according to the Feynman rules we can write

$$\begin{aligned} & \mathrm{i}G^{0}(k)\Sigma_{\pi}^{\mathrm{ps}}(k)G(k) = \mathrm{i}G^{0}(k)g_{\pi}\gamma^{5}\tau_{i}\mathrm{i}\Delta_{ij}^{0}(0) \\ & \int\frac{\mathrm{d}^{4}q}{(2\pi)^{4}}\mathrm{tr}\left[\left(g_{\pi}\gamma^{5}\tau_{j}\right)\left(-\mathrm{i}G(q)\right)\right]\mathrm{i}G(k) + \mathrm{i}G^{0}(k)g_{\pi}\gamma^{5}\tau_{i} \\ & \int\frac{\mathrm{d}^{4}q}{(2\pi)^{4}}\mathrm{i}\Delta_{ij}^{0}(k-q)\mathrm{i}G(q)g_{\pi}\gamma^{5}\tau_{j}\mathrm{i}G(k) \end{aligned} \tag{5.873}$$

And then get

$$\Sigma_{\pi}^{\mathrm{ps}}(k) = \mathrm{i}g_{\pi}^{2} \int \frac{\mathrm{d}^{4}q}{(2\pi)^{4}} \left[\gamma^{5}\tau_{i}\Delta_{ij}^{0}(0)\,\mathrm{tr}\left(\gamma^{5}\tau_{j}G(g)\right) - \gamma^{5}\tau_{i}\Delta_{ij}^{0}(k-q)G(q)\gamma^{5}\tau_{j}\right] \quad (5.874)$$

In the above formula the first item is Hartree one, and the second item is Fock one. $\pi$ meson is a pseudo-scalar meson, whose propagator similar to $\sigma$ meson is taken as

$$\Delta_{ij}^{0}(q) = \frac{\delta_{ij}}{q_{\mu}q^{\mu} - m_{\pi}^{2} + \mathrm{i}\varepsilon} \quad (5.875)$$

where $i$ and $j$ are isospin footnotes and $\delta_{ij}$ means that the charge exchange between the $\pi$ meson and the nucleon is not considered, i.e., the nuclear force is approximately independent of the charge. When $G(q)$ in the first item of Eq. (5.874) is replaced by $G_{\mathrm{D}}(q)$ given by Eq. (5.845), because there are $\mathrm{tr}\gamma^{5} = 0$ and $\mathrm{tr}(\gamma^{5}\gamma^{\mu}) = 0$, so we can get

$$\Sigma_{\pi}^{\mathrm{ps-H}}(k) = 0 \quad (5.876)$$

That is, the Hartree item of $\Sigma_{\pi}^{\mathrm{ps}}$ has no contribution. Noting $\tau_{i}\tau_{i} = 3$ and $\{\gamma^{5}, \gamma^{\mu}\} = 0$, the following result can be obtained from Eqs. (5.874) and (5.875):

$$\begin{aligned}\Sigma_{\pi}^{\mathrm{ps}}(k) = \Sigma_{\pi}^{\mathrm{ps-F}}(k) &= -\mathrm{i}g_{\pi}^{2} \int \frac{\mathrm{d}^{4}q}{(2\pi)^{4}} \frac{\gamma^{5}\tau_{i}G(q)\gamma^{5}\tau_{i}}{(k-q)_{\mu}(k-q)^{\mu} - m_{\pi}^{2} + \mathrm{i}\varepsilon} \\ &= \mathrm{i}g_{\pi}^{2} \int \frac{\mathrm{d}^{4}q}{(2\pi)^{4}} \frac{3\left(\gamma^{\mu}q_{\mu}^{*} - M_{q}^{*}\right)}{(k-q)_{\mu}(k-q)^{\mu} - m_{\pi}^{2} + \mathrm{i}\varepsilon} \frac{\mathrm{i}\pi}{E_{\mu}^{*}} \delta\left(q_{0} - E_{q}\right) \theta\left(k_{\mathrm{F}} - |\vec{q}|\right)\end{aligned} \quad (5.877)$$

Comparing this formula with $\Sigma_{\sigma}^{\mathrm{F}}(k)$ in Eq. (5.848), here an extra 3 is multiplied, and in the front of $M_{q}^{*}$ there is a negative sign, so according to Eq. (5.861) we can write

$$\Sigma_{\pi}^{\mathrm{ps}}(k, E) = \frac{3g_{\pi}^{2}}{16\pi^{2}k} \int_{0}^{k_{\mathrm{F}}} q\mathrm{d}q \left[ -\frac{M_{q}^{*}}{E_{q}^{*}} \Theta_{\pi}(k,q) + \gamma^{0}\Theta_{\pi}(k,q) - \frac{2\vec{\gamma}\cdot\vec{k}}{k} \frac{q^{*}}{E_{q}^{*}} \Phi_{\pi}(k,q) \right] \quad (5.878)$$

It is generally believed that the pseudo-scalar coupling of $\pi$ mesons cannot give reasonable results, and the pseudo-scalar coupling of $\pi$ mesons can be converted into axis-vector coupling, so the contribution of $\Sigma_{\pi}^{\mathrm{ps}}$ term is generally not considered [26, 80].

Referring to Eq. (5.708) the $\pi$ meson wave function of the isospin $i$ component can be written as

$$\pi_i(x) = \frac{1}{\sqrt{V}} \sum_{\vec{k}} \frac{1}{\sqrt{2\omega_k}} \left( a_{\vec{k}i}^{-} e^{-ik_\mu x^\mu} + a_{\vec{k}i}^{+} e^{ik_\mu x^\mu} \right) \tag{5.879}$$

And we can also write the following expression according to Eqs. (5.720) and (5.721):

$$i\Delta_{ij}^0(x, x') = \langle \Psi_0 | T\{\pi_i(x)\pi_j(x')\} | \Psi_0 \rangle = i \int \frac{d^4k}{(2\pi)^4} \Delta_{ij}^0(k) \, e^{-ik_\mu (x-x')^\mu} \tag{5.880}$$

In the Fock term at the right end of Fig. 5.5, there are two vertices of the upper annihilation meson $x$ and the lower generation meson $x'$. Since $\partial_\mu \equiv \frac{\partial}{\partial x^\mu}$, according to Eq. (5.879) it is known that $\partial_{\mu x} \vec{\pi}$ contributes $-ik_\mu$ and $\partial_{\mu x'} \vec{\pi}$ contributes $ik_\mu$ in Eq. (5.671). According to Eq. (5.671), in Eq. (5.873) let the corresponding vertex $x$ term $g_\pi \gamma^5 \to -i\frac{f_\pi}{m_\pi}\gamma^5\gamma^\mu(-ip_\mu)$ and let the corresponding vertex $x'$ term $g_\pi\gamma^5 \to -i\frac{f_\pi}{m_\pi}\gamma^5\gamma^\nu(ip_\nu)$, where $p = 0^+$ or $k - q$, so for the axis-vector (pv) coupling term we can get

$$\begin{aligned}
iG^0(k)\Sigma_\pi^{pv}(k)G(k) = {} & iG^0(k)\left(-i\frac{f_\pi}{m_\pi}\right)\gamma^5\gamma^\mu\left(-i0_\mu^+\right) \tau_i i\Delta_{ij}^0(0) \\
& \int \frac{d^4q}{(2\pi)^4} \text{tr}\left[\left(-i\frac{f_\pi}{m_\pi}\right)\gamma^5\gamma^\nu (i0_\nu^+)\tau_j(-iG(q))\right] iG(k) \\
& + iG^0(k)\left(-i\frac{f_\pi}{m_\pi}\right)\gamma^5 \int \frac{d^4q}{(2\pi)^4}\gamma^\mu\left(-i(k-q)_\mu\right) \tau_i i\Delta_{ij}^0(k-q) iG(q) \\
& \left(-i\frac{f_\pi}{m_\pi}\right)\gamma^5\gamma^\nu i(k-q)_\nu \tau_j iG(k)
\end{aligned} \tag{5.881}$$

According to the above formula, $\Sigma_\pi^{pv}$ is divided into Hartree and Fock terms:

$$\Sigma_\pi^{pv}(k) = \Sigma_\pi^{pv-H}(k) + \Sigma_\pi^{pv-F}(k) \tag{5.882}$$

In Eq. (5.881) $G(q)$ is replaced with $G_D(q)$. By using $\text{tr}\gamma^5 = 0$, $\text{tr}(\gamma^5\gamma^\nu) = 0$, tr $(\gamma^5\gamma^\nu\gamma^\mu) = 0$ or by $0_\nu^+ \to 0$ we can always get

$$\Sigma_\pi^{pv-H}(k) = 0 \tag{5.883}$$

So we can obtain

$$
\begin{aligned}
\Sigma_\pi^{\mathrm{pv}}(k) &= \Sigma_\pi^{\mathrm{pv-F}}(k) \\
&= \mathrm{i}\left(\frac{f_\pi}{m_\pi}\right)^2 \int \frac{\mathrm{d}^4 q}{(2\pi)^4} \gamma^5 \gamma^\mu (k-q)_\mu \tau_i \Delta_{ij}^0(k-q)\, G(q)\, \gamma^5 \gamma^\nu (k-q)_\nu \tau_j \\
&= 3\mathrm{i}\left(\frac{f_\pi}{m_\pi}\right)^2 \int \frac{\mathrm{d}^4 q}{(2\pi)^4} \frac{\gamma^5 \gamma^\mu (k-q)_\mu G(q) \gamma^5 \gamma^\nu (k-q)_\nu}{(k-q)_\eta (k-q)^\eta - m_\pi^2 + \mathrm{i}\varepsilon} \\
&= 3\mathrm{i}\left(\frac{f_\pi}{m_\pi}\right)^2 \int \frac{\mathrm{d}^4 q}{(2\pi)^4} \frac{\gamma^5 \gamma^\mu (k-q)_\mu \left(\gamma^\lambda q_\lambda^* + M_q^*\right) \gamma^5 \gamma^\nu (k-q)_\nu}{(k-q)_\eta (k-q)^\eta - m_\pi^2 + \mathrm{i}\varepsilon} \\
&\quad \frac{\mathrm{i}\pi}{E_q^*} \delta(q_0 - E_q)\, \theta\left(k_{\mathrm{F}} - |\vec{q}|\right)
\end{aligned} \tag{5.884}
$$

Note that $\gamma^\lambda \gamma^\nu = -\gamma^\nu \gamma^\lambda + 2g^{\lambda\nu}$, $g^{\lambda\nu}(k-q)_\nu = \delta_{\lambda\nu}(k-q)^\nu$, and when the $\gamma^\mu(k-q)_\mu \gamma^\nu(k-q)_\nu$ item is making summation to $\mu$, $\nu$, the $\mu \neq \nu$ items will cancel each other, and there is $\gamma^5\gamma^5 = 1$. When the sum to $\mu$ is not carried out, $\gamma^\mu\gamma^\mu = g^{\mu\mu}$ can be obtained by Eq. (5.59). Then using Eq. (5.852) we can obtain

$$
\begin{aligned}
I &\equiv \gamma^5 \gamma^\mu (k-q)_\mu \left(\gamma^\lambda q_\lambda^* + M_q^*\right) \gamma^5 \gamma^\nu (k-q)_\nu \delta(q_0 - E_q) \\
&= \gamma^\mu (k-q)_\mu \left(\gamma^\lambda q_\lambda^* - M_q^*\right) \gamma^\nu (k-q)_\nu \delta(q_0 - E_q) \\
&= \left[-(k-q)_\mu (k-q)^\mu \left(\gamma^\lambda q_\lambda^* + M_q^*\right) + 2\gamma^\mu (k-q)_\mu q_\nu^* (k-q)^\nu\right] \delta(q_0 - E_q) \\
&= \Bigg\{ \left[-(E_k - E_q)^2 + \left(\vec{k} - \vec{q}\right)^2\right] M_q^* \\
&\quad + \gamma^0 \left[(E_k - E_q)^2 E_q^* + \left(\vec{k} - \vec{q}\right)^2 E_q^* - 2(E_k - E_q)\vec{q}^{\,*} \cdot \left(\vec{k} - \vec{q}\right)\right] \\
&\quad + \vec{\gamma} \cdot \vec{k} \Bigg[\frac{\vec{k} \cdot \vec{q}^{\,*}}{k^2} \left((E_k - E_q)^2 - \left(\vec{k} - \vec{q}\right)^2\right) \\
&\quad - 2\left(1 - \frac{\vec{k} \cdot \vec{q}}{k^2}\right)\left(E_q^*(E_k - E_q) - \vec{q}^{\,*} \cdot \left(\vec{k} - \vec{q}\right)\right)\Bigg]\Bigg\} \delta(q_0 - E_q)
\end{aligned} \tag{5.885}
$$

Simplify some of the items in the above formula as follows:

$$
-\left(\vec{k} - \vec{q}\right)^2 \frac{\vec{k} \cdot \vec{q}^{\,*}}{k^2} + 2\left(1 - \frac{\vec{k} \cdot \vec{q}}{k^2}\right) \vec{q}^{\,*} \cdot \left(\vec{k} - \vec{q}\right) = \vec{k} \cdot \vec{q}^{\,*} \left(1 + \frac{\vec{q}^{\,2}}{k^2}\right) - 2qq^* \tag{5.886}
$$

So Eq. (5.885) can be rewritten as

$$
\begin{aligned}
I = \Big\{ & \left[ -(E_k - E_q)^2 + k^2 + q^2 - 2\vec{k} \cdot \vec{q} \right] M_q^* + \gamma^0 \Big[ E_q^* (E_k - E_q)^2 \\
& + \left( k^2 + q^2 - 2\vec{k} \cdot \vec{q} \right) E_q^* + 2qq^* (E_k - E_q) - 2\vec{k} \cdot \vec{q}^{\,*} (E_k - E_q) \Big] \\
& + \vec{\gamma} \cdot \vec{k} \Bigg[ -2(E_k - E_q) E_q^* + \frac{2\vec{k} \cdot \vec{q}}{k^2} (E_k - E_q) E_q^* + \frac{\vec{k} \cdot \vec{q}^{\,*}}{k^2} (E_k - E_q)^2 \\
& + \vec{k} \cdot \vec{q}^{\,*} \left( 1 + \frac{q^2}{k^2} \right) - 2qq^* \Bigg] \Big\} \delta(q_0 - E_q)
\end{aligned}
\tag{5.887}
$$

The following expression can be got utilizing Eqs. (5.732) and (5.852):

$$
\gamma^\mu \overline{q}_\mu^* + M_q^* = M_q^* + \gamma^0 E_q^* - \vec{\gamma} \cdot \vec{k} \frac{\vec{k} \cdot \vec{q}^{\,*}}{k^2}
\tag{5.888}
$$

Comparing the first equation of Eq. (5.850) with Eq. (5.861) and noting Eq. (5.888), it can be seen that after removing the same coefficient, 1 corresponds to $\Theta_\sigma$, and $\vec{k} \cdot \vec{q}^{\,*}$ corresponds to $2kq^* \Phi_\sigma$. And comparing the first equation of Eq. (5.850) with Eq. (5.884) and using the result given by Eq. (5.887), referring to Eq. (5.861), we can obtain

$$
\begin{aligned}
\Sigma_\pi^{\text{pv}}(k) = & \frac{3}{16\pi^2 k} \left( \frac{f_\pi}{m_\pi} \right)^2 \int_0^{k_F} q \mathrm{d}q \Bigg\{ \frac{M_q^*}{E_q^*} \Big[ \left( -(E_k - E_q)^2 + k^2 + q^2 \right) \Theta_\pi(k, q) \\
& - 4kq \Phi_\pi(k, q) \Big] + \gamma^0 \Bigg[ \left( (E_k - E_q)^2 + k^2 + q^2 + \frac{2qq^* (E_k - E_q)}{E_q^*} \right) \Theta_\pi(k, q) \\
& - \left( 4kq + \frac{4kq^* (E_k - E_q)}{E_q^*} \right) \Phi_\pi(k, q) \Bigg] - \vec{\gamma} \cdot \vec{k} \Bigg[ \left( 2(E_k - E_q) + \frac{2qq^*}{E_q^*} \right) \Theta_\pi(k, q) \\
& - \left( \frac{2kq^*}{E_q^*} \left( 1 + \frac{q^2}{k^2} \right) + \frac{4kq}{k^2} (E_k - E_q) + \frac{2kq^*}{k^2 E_q^*} (E_k - E_q)^2 \right) \Phi_\pi(k, q) \Bigg] \Bigg\}
\end{aligned}
\tag{5.889}
$$

Using Eq. (5.860) and Eq. (5.854) to simplify the first item on the right side of the above formula as follows

$$
\left[ -(E_k - E_q)^2 + k^2 + q^2 \right] \Theta_\pi - 4kq \Phi_\pi = 4kq - m_\pi^2 \Theta_\pi
\tag{5.890}
$$

then simplify some of the items in the second term on the right side of Eq. (5.889) as follows:

$$\left(k^2+q^2\right)\Theta_\pi-4kq\Phi_\pi=\left[\left(E_k-E_q\right)^2-m_\pi^2\right]\Theta_\pi+4kq \tag{5.891}$$

So Eq. (5.889) can be rewritten as

$$\begin{aligned}\Sigma_\pi^{\mathrm{pv}}(k,E)=&\frac{3}{16\pi^2k}\left(\frac{f_\pi}{m_\pi}\right)^2\int_0^{k_\mathrm{F}}q\mathrm{d}q\left\{\frac{M_q^*}{E_q^*}\left[4kq-m_\pi^2\Theta_\pi(k,\ q)\right]\right.\\&+\gamma^0\left[4kq+\left(\left(E_k-E_q\right)^2-m_\pi^2+\frac{2qq^*\left(E_k-E_q\right)}{E_q^*}\right)\Theta_\pi(k,\ q)\right.\\&\left.-\frac{4kq^*\left(E_k-E_q\right)}{E_q^*}\Phi_\pi(k,\ q)\right]-\frac{2\vec{\gamma}\cdot\vec{k}}{k}\left[k\left(E_k-E_q+\frac{qq^*}{E_q^*}\right)\Theta_\pi(k,\ q)\right.\\&\left.\left.-\left(\frac{q^*}{E_q^*}\left(k^2+q^2+\left(E_k-E_q\right)^2\right)+2q\left(E_k-E_q\right)\right)\Phi_\pi(k,\ q)\right]\right\}\end{aligned} \tag{5.892}$$

Because the contribution of the $\pi$ meson pseudo-scalar terms is generally not considered, the coupling terms with one vertex as the pseudo-scalar and the other vertex as the axis-vector are not studied here.

4. ρ Meson

Equation (5.670) gives the interaction Lagrangian density of the ρ meson with the nucleon. The first and second terms represent vector coupling and tensor coupling, respectively. Referring to Eqs. (5.812) and (5.875), we can write the propagators of the ρ mesons without interaction as follows:

$$D^0_{\mu\nu,ab}(q)=\frac{-g_{\mu\nu}}{q_\lambda q^\lambda-m_\rho^2+\mathrm{i}\varepsilon}\delta_{ab} \tag{5.893}$$

where $\delta_{ab}$ represents that the charge exchange between the ρ meson and the nucleon is not considered. First studying the vector items, referring to Eq. (5.862) we can write

$$\begin{aligned}\mathrm{i}G^0(k)\Sigma_\rho^{\mathrm{V}}(k)G(k)=&\,\mathrm{i}G^0(k)\left(-\mathrm{i}g_\rho\gamma^\mu\right)\tau_a\mathrm{i}D^0_{\mu\nu,ab}(0)\\&\int\frac{\mathrm{d}^4q}{(2\pi)^4}\mathrm{tr}\left[\tau_b\left(-\mathrm{i}g_\rho\gamma^\nu\right)\left(-\mathrm{i}G(q)\right)\right]\mathrm{i}G(k)+\mathrm{i}G^0(k)\left(-\mathrm{i}g_\rho\gamma^\mu\right)\tau_a\\&\int\frac{\mathrm{d}^4q}{(2\pi)^4}\,\mathrm{i}D^0_{\mu\nu,ab}(k-q)\,\mathrm{i}G(q)\left(-\mathrm{i}g_\rho\gamma^\nu\right)\tau_a\mathrm{i}G(k)\end{aligned} \tag{5.894}$$

and get

$$
\begin{aligned}
\Sigma_{\rho}^{\mathrm{V}}(k) &= -\mathrm{i}g_{\rho}^{2}\gamma^{\mu}\tau_{a}\int\frac{\mathrm{d}^{4}q}{(2\pi)^{4}}D_{\mu\nu,ab}^{0}(0)\ \mathrm{tr}(\tau_{b}\gamma^{\nu}G(q)) \\
&\quad +\mathrm{i}g_{\rho}^{2}\gamma^{\mu}\tau_{a}\int\frac{\mathrm{d}^{4}q}{(2\pi)^{4}}D_{\mu\nu,ab}^{0}(k-q)\ G(q)\gamma^{\nu}\tau_{b} \\
&= -\mathrm{i}g_{\rho}^{2}\gamma^{\mu}\tau_{a}\int\frac{\mathrm{d}^{4}q}{(2\pi)^{4}}g_{\mu\mu}\left[\frac{\mathrm{tr}(\tau_{a}\gamma^{\mu}G(q))}{m_{\rho}^{2}}+\frac{G(q)\ \gamma^{\mu}\tau_{a}}{(k-q)_{\lambda}(k-q)^{\lambda}-m_{\rho}^{2}+\mathrm{i}\varepsilon}\right]
\end{aligned}
\tag{5.895}
$$

Notice $\mathrm{tr}\tau_a = 0\ \ (a = 1, 2, 3)$, so we can get

$$
\Sigma_{\rho}^{\mathrm{V-H}}(k) = 0 \tag{5.896}
$$

Then using Eq. (5.845) and noting $\tau_a\tau_a = 3$, from Eq. (5.895) we can obtain

$$
\begin{aligned}
\Sigma_{\rho}^{\mathrm{V}}(k) &= \Sigma_{\rho}^{\mathrm{V-F}}(k) \\
&= -\mathrm{i}g_{\rho}^{2}\int\frac{\mathrm{d}^{4}q}{(2\pi)^{4}}\frac{3g_{\mu\mu}\gamma^{\mu}\left(\gamma^{\nu}q_{\nu}^{*}+M_{q}^{*}\right)\gamma^{\mu}}{(k-q)_{\lambda}(k-q)^{\lambda}-m_{\rho}^{2}+\mathrm{i}\varepsilon}\frac{\mathrm{i}\pi}{E_{q}^{*}}\delta\left(q_{0}-E_{q}\right)\theta\left(k_{\mathrm{F}}-\left|\vec{q}\right|\right)
\end{aligned}
\tag{5.897}
$$

When including summation to $\mu$ there is $g_{\mu\mu}\gamma^{\mu}\gamma^{\mu} = (g_{\mu\mu})^2 = 4$, so we get

$$
\begin{aligned}
g_{\mu\mu}\gamma^{\mu}\left(\gamma^{\nu}q_{\nu}^{*}+M_{q}^{*}\right)\gamma^{\mu} &= -\gamma^{\nu}q_{\nu}^{*}g_{\mu\mu}\gamma^{\mu}\gamma^{\mu}+2g_{\mu\mu}g^{\mu\nu}q_{\nu}^{*}\gamma^{\mu}+M_{q}^{*}g_{\mu\mu}\gamma^{\mu}\gamma^{\mu} \\
&= 2\left(-\gamma^{\nu}q_{\nu}^{*}+2M_{q}^{*}\right)
\end{aligned}
\tag{5.898}
$$

Comparing the first equation of Eq. (5.850) with Eq. (5.897) and using the result given by Eq. (5.861), and comparing Eq. (5.888) with Eq. (5.898), as well as using the conclusion obtained through the analyzing after Eq. (5.888): 1 corresponds to $\Theta$ and $\vec{k}\cdot\vec{q}^{\,*}$ corresponds to $2kq^*\Phi$, then we can obtain

$$
\Sigma_{\rho}^{\mathrm{V}}(k, E) = \frac{3g_{\rho}^{2}}{8\pi^{2}k}\int_{0}^{k_{\mathrm{F}}}q\mathrm{d}q\left[-\frac{2M_{q}^{*}}{E_{q}^{*}}\Theta_{\rho}(k,\ q)+\gamma^{0}\Theta_{\rho}(k,\ q)-\frac{2\vec{\gamma}\cdot\vec{k}}{k}\frac{q^{*}}{E_{q}^{*}}\Phi_{\rho}(k,\ q)\right]
\tag{5.899}
$$

In the tensor term of the Lagrangian density of the interaction between ρ meson and nucleon given by Eq. (5.670), $\sigma^{\mu\nu}$ and $\vec{L}_{\mu\nu}$ have been given by Eq. (5.538) and Eq. (5.647), respectively. And noting that $\partial_{\mu x}$ contributes $-\mathrm{i}k_{\mu}$, $\partial_{\mu x'}$ contributes $\mathrm{i}k_{\mu}$, then at the vertex $x$ we can find

$$\begin{aligned}\sigma^{\mu\nu}\vec{L}_{\mu\nu} &= \frac{\mathrm{i}}{2}\left(\gamma^{\mu}\gamma^{\nu}-\gamma^{\nu}\gamma^{\mu}\right)\left(\partial_{\mu}\vec{\rho}_{\nu}-\partial_{\nu}\vec{\rho}_{\mu}\right)\\ &= \frac{\mathrm{i}}{2}\left(\gamma^{\mu}\gamma^{\nu}\partial_{\mu}\vec{\rho}_{\nu}-\gamma^{\nu}\gamma^{\mu}\partial_{\mu}\vec{\rho}_{\nu}-\gamma^{\mu}\gamma^{\nu}\partial_{\nu}\vec{\rho}_{\mu}+\gamma^{\nu}\gamma^{\mu}\partial_{\nu}\vec{\rho}_{\mu}\right)\\ &= \mathrm{i}\left(\gamma^{\mu}\gamma^{\nu}\partial_{\mu}\vec{\rho}_{\nu}-\gamma^{\nu}\gamma^{\mu}\partial_{\mu}\vec{\rho}_{\nu}\right)=\mathrm{i}\left(2\gamma^{\mu}\gamma^{\nu}\partial_{\mu}\vec{\rho}_{\nu}-2g^{\mu\nu}\partial_{\mu}\vec{\rho}_{\nu}\right)\\ &= 2\mathrm{i}\left(\gamma^{\mu}\gamma^{\nu}\partial_{\mu}\vec{\rho}_{\nu}-\partial^{\nu}\vec{\rho}_{\nu}\right)=2\left(\gamma^{\mu}q_{\mu}\gamma^{\nu}\vec{\rho}_{\nu}-q^{\nu}\vec{\rho}_{\nu}\right)\end{aligned} \tag{5.900}$$

where $\sigma^{\mu\nu}\vec{L}_{\mu\nu}$ changes its sign at the vertex $x'$. The momentum propagated in the Hartree term is 0, i.e., in Eq. (5.900), there are $q_{\mu}\to 0$ and $q^{\mu}\to 0$, so we can get

$$\Sigma_{\rho}^{\mathrm{T-H}}(k)=0 \tag{5.901}$$

Comparing Eq. (5.670) with the second item of Eq. (5.898), noting that the multiplication of two imaginary unit i from two $\sigma^{\mu\nu}$ will come out $(-1)$, and using the result given by Eq. (5.900), referring to the first formula of Eq. (5.884), we can write

$$\begin{aligned}&\Sigma_{\rho}^{\mathrm{T}}(k)=\Sigma_{\rho}^{\mathrm{T-F}}(k)=4\mathrm{i}\left(\frac{f_{\rho}}{4M}\right)^{2}\int\frac{\mathrm{d}^{4}q}{(2\pi)^{4}}\left[\left(\gamma^{\xi}(k-q)_{\xi}\gamma^{\mu}-(k-q)^{\mu}\right)\tau_{a}\right.\\ &\left.D^{0}_{\mu\nu,ab}(k-q)G(q)\left(-\gamma^{\zeta}(k-q)_{\zeta}\gamma^{\nu}+(k-q)^{\nu}\right)\tau_{b}\right]=12\mathrm{i}\left(\frac{f_{\rho}}{4M}\right)^{2}\int\frac{\mathrm{d}^{4}q}{(2\pi)^{4}}\\ &\frac{\left[-\gamma^{\xi}(k-q)_{\xi}\gamma^{\mu}+(k-q)^{\mu}\right]\left(\gamma^{\eta}q^{*}_{\eta}+M^{*}_{q}\right)\left[-\gamma^{\zeta}(k-q)_{\zeta}g_{\mu\mu}\gamma^{\mu}+(k-q)_{\mu}\right]}{(k-q)_{\lambda}(k-q)^{\lambda}-m_{\rho}^{2}+\mathrm{i}\varepsilon}\\ &\frac{\mathrm{i}\pi}{E^{*}_{q}}\delta\left(q_{0}-E_{q}\right)\theta\left(k_{\mathrm{F}}-|\vec{q}|\right)\end{aligned} \tag{5.902}$$

And we can find

$$\begin{aligned}I\equiv&\left[-\gamma^{\xi}(k-q)_{\xi}\gamma^{\mu}+(k-q)^{\mu}\right]\left(\gamma^{\eta}q^{*}_{\eta}+M^{*}_{q}\right)\left[-\gamma^{\zeta}(k-q)_{\zeta}g_{\mu\mu}\gamma^{\mu}+(k-q)_{\mu}\right]\\ =&\left[-\gamma^{\xi}(k-q)_{\xi}\gamma^{\mu}+(k-q)^{\mu}\right]\left[-\gamma^{\eta}q^{*}_{\eta}\gamma^{\zeta}(k-q)_{\zeta}g_{\mu\mu}\gamma^{\mu}+\gamma^{\eta}q^{*}_{\eta}(k-q)_{\mu}\right.\\ &\left.-\gamma^{\zeta}(k-q)_{\zeta}g_{\mu\mu}\gamma^{\mu}M^{*}_{q}+(k-q)_{\mu}M^{*}_{q}\right]\\ =&\,\gamma^{\xi}(k-q)_{\xi}\gamma^{\mu}\gamma^{\eta}q^{*}_{\eta}\gamma^{\zeta}(k-q)_{\zeta}g_{\mu\mu}\gamma^{\mu}-\gamma^{\xi}(k-q)_{\xi}\gamma^{\mu}\gamma^{\eta}q^{*}_{\eta}(k-q)_{\mu}\\ &+\gamma^{\xi}(k-q)_{\xi}\gamma^{\mu}\gamma^{\zeta}(k-q)_{\zeta}g_{\mu\mu}\gamma^{\mu}M^{*}_{q}-\gamma^{\xi}(k-q)_{\xi}\gamma^{\mu}(k-q)_{\mu}M^{*}_{q}\\ &-\gamma^{\eta}q^{*}_{\eta}\gamma^{\zeta}(k-q)_{\zeta}g_{\mu\mu}\gamma^{\mu}(k-q)^{\mu}+\gamma^{\eta}q^{*}_{\eta}(k-q)_{\mu}(k-q)^{\mu}\\ &-\gamma^{\zeta}(k-q)_{\zeta}g_{\mu\mu}\gamma^{\mu}(k-q)^{\mu}M^{*}_{q}+(k-q)_{\mu}(k-q)^{\mu}M^{*}_{q}\end{aligned} \tag{5.903}$$

Simplify some of the items in the above formula as follows:

$$\begin{aligned}
&\gamma^{\xi}(k-q)_{\xi}\gamma^{\mu}\gamma^{\eta}q_{\eta}^{*}\gamma^{\zeta}(k-q)_{\zeta}g_{\mu\mu}\gamma^{\mu} \\
&\quad = -\gamma^{\xi}(k-q)_{\xi}\gamma^{\eta}q_{\eta}^{*}\gamma^{\mu}\gamma^{\zeta}(k-q)_{\zeta}g_{\mu\mu}\gamma^{\mu} + 2\gamma^{\xi}(k-q)_{\xi}g^{\mu\eta}q_{\eta}^{*}\gamma^{\zeta}(k-q)_{\zeta}g_{\mu\mu}\gamma^{\mu} \\
&\quad = \gamma^{\xi}(k-q)_{\xi}\gamma^{\eta}q_{\eta}^{*}\gamma^{\zeta}(k-q)_{\zeta}\gamma^{\mu}g_{\mu\mu}\gamma^{\mu} - 2\gamma^{\xi}(k-q)_{\xi}\gamma^{\eta}q_{\eta}^{*}g^{\mu\zeta}(k-q)_{\zeta}g_{\mu\mu}\gamma^{\mu} \\
&\qquad +2\gamma^{\xi}(k-q)_{\xi}\gamma^{\zeta}(k-q)_{\zeta}\gamma^{\eta}q_{\eta}^{*} \\
&\quad = 2\gamma^{\xi}(k-q)_{\xi}\gamma^{\eta}q_{\eta}^{*}\gamma^{\zeta}(k-q)_{\zeta} + 2\gamma^{\xi}(k-q)_{\xi}\gamma^{\zeta}(k-q)_{\zeta}\gamma^{\eta}q_{\eta}^{*} \\
&\quad = -2\gamma^{\xi}(k-q)_{\xi}\gamma^{\zeta}(k-q)_{\zeta}\gamma^{\eta}q_{\eta}^{*} + 4\gamma^{\xi}(k-q)_{\xi}q_{\eta}^{*}g^{\eta\zeta}(k-q)_{\zeta} \\
&\qquad +2\gamma^{\xi}(k-q)_{\xi}\gamma^{\zeta}(k-q)_{\zeta}\gamma^{\eta}q_{\eta}^{*} \\
&\quad = 4\gamma^{\xi}(k-q)_{\xi}q_{\eta}^{*}(k-q)^{\eta}
\end{aligned} \tag{5.904}$$

$$\begin{aligned}
&\gamma^{\xi}(k-q)_{\xi}\gamma^{\mu}\gamma^{\zeta}(k-q)_{\zeta}g_{\mu\mu}\gamma^{\mu}M_q^{*} \\
&\quad = -\gamma^{\xi}(k-q)_{\xi}\gamma^{\zeta}(k-q)_{\zeta}\gamma^{\mu}g_{\mu\mu}\gamma^{\mu}M_q^{*} + 2\gamma^{\xi}(k-q)_{\xi}g^{\mu\zeta}(k-q)_{\zeta}g_{\mu\mu}\gamma^{\mu}M_q^{*} \\
&\quad = -2\gamma^{\xi}(k-q)_{\xi}\gamma^{\zeta}(k-q)_{\zeta}M_q^{*}
\end{aligned} \tag{5.905}$$

$$\begin{aligned}
&-\gamma^{\eta}q_{\eta}^{*}\gamma^{\zeta}(k-q)_{\zeta}g_{\mu\mu}\gamma^{\mu}(k-q)^{\mu} = -\gamma^{\eta}q_{\eta}^{*}\gamma^{\zeta}(k-q)_{\zeta}\gamma^{\mu}(k-q)_{\mu} \\
&\quad = \gamma^{\zeta}(k-q)_{\zeta}\gamma^{\eta}q_{\eta}^{*}\gamma^{\mu}(k-q)_{\mu} - 2g^{\eta\zeta}q_{\eta}^{*}(k-q)_{\zeta}\gamma^{\mu}(k-q)_{\mu} \\
&\quad = -\gamma^{\zeta}(k-q)_{\zeta}\gamma^{\mu}(k-q)_{\mu}\gamma^{\eta}q_{\eta}^{*} + 2\gamma^{\zeta}(k-q)_{\zeta}g^{\eta\mu}q_{\eta}^{*}(k-q)_{\mu} \\
&\qquad -2q_{\eta}^{*}(k-q)^{\eta}\gamma^{\mu}(k-q)_{\mu} \\
&\quad = -\gamma^{\zeta}(k-q)_{\zeta}\gamma^{\mu}(k-q)_{\mu}\gamma^{\eta}q_{\eta}^{*}
\end{aligned} \tag{5.906}$$

And notice

$$\begin{aligned}
\gamma^{\xi}(k-q)_{\xi}\gamma^{\zeta}(k-q)_{\zeta} &= -(k-q)_{\xi}\gamma^{\zeta}\gamma^{\xi}(k-q)_{\zeta} + 2g^{\xi\zeta}(k-q)_{\xi}(k-q)_{\zeta} \\
&= (k-q)_{\xi}(k-q)^{\xi}
\end{aligned} \tag{5.907}$$

Also note that the $\xi \neq \zeta$ items in the first term after the first equal sign of the above formula will disappear when the sum to $\xi$ and $\zeta$ is proceeded. From Eq. (5.903) the following result can be obtained by using Eqs. (5.904)–(5.907), (5.854), and (5.852):

$$
\begin{aligned}
I_{|q_0=E_q} &= 4\gamma^{\xi}(k-q)_{\xi}q^*_{\eta}(k-q)^{\eta} - \gamma^{\xi}(k-q)_{\xi}\gamma^{\mu}(k-q)_{\mu}\gamma^{\eta}q^*_{\eta} \\
&\quad -2\gamma^{\xi}(k-q)_{\xi}\gamma^{\zeta}(k-q)_{\zeta}M^*_q - \gamma^{\xi}(k-q)_{\xi}\gamma^{\mu}(k-q)_{\mu}M^*_q \\
&\quad -\gamma^{\zeta}(k-q)_{\zeta}\gamma^{\mu}(k-q)_{\mu}\gamma^{\eta}q^*_{\eta} + (k-q)_{\mu}(k-q)^{\mu}\gamma^{\eta}q^*_{\eta} \\
&\quad -\gamma^{\zeta}(k-q)_{\zeta}\gamma^{\mu}(k-q)_{\mu}M^*_q + (k-q)_{\mu}(k-q)^{\mu}M^*_q \\
&= -(k-q)_{\mu}(k-q)^{\mu}\gamma^{\eta}q^*_{\eta} - 3M^*_q(k-q)_{\mu}(k-q)^{\mu} + 4\gamma^{\xi}(k-q)_{\xi}q^*_{\eta}(k-q)^{\eta} \\
&= -\left(\gamma^0E^*_q - \vec{\gamma}\cdot\vec{q}^{\,*}\right)\left[\left(E_k-E_q\right)^2 - \left(\vec{k}-\vec{q}\right)^2\right] \\
&\quad -3M^*_q\left[\left(E_k-E_q\right)^2 - \left(\vec{k}-\vec{q}\right)^2\right] \\
&\quad +4\left[\gamma^0\left(E_k-E_q\right) - \vec{\gamma}\cdot\left(\vec{k}-\vec{q}\right)\right]\left[E^*_q\left(E_k-E_q\right) - \vec{q}^{\,*}\cdot\left(\vec{k}-\vec{q}\right)\right] \\
&= -\gamma^0E^*_q\left(m^2_{\rho} - A_{\rho} + 2\vec{k}\cdot\vec{q}\right) + \vec{\gamma}\cdot\vec{k}\,\frac{\vec{k}\cdot\vec{q}^{\,*}}{k^2}\left(m^2_{\rho} - A_{\rho} + 2\vec{k}\cdot\vec{q}\right) \\
&\quad -3M^*_q\left(m^2_{\rho} - A_{\rho} + 2\vec{k}\cdot\vec{q}\right) \\
&\quad +4\gamma^0\left[E^*_q\left(E_k-E_q\right)^2 - \vec{k}\cdot\vec{q}^{\,*}\left(E_k-E_q\right) + qq^*\left(E_k-E_q\right)\right] \\
&\quad -4\vec{\gamma}\cdot\vec{k}\left(1-\frac{\vec{k}\cdot\vec{q}}{k^2}\right)\left[E^*_q\left(E_k-E_q\right) - \vec{k}\cdot\vec{q}^{\,*} + qq^*\right] \\
&= -3M^*_q\left(m^2_{\rho} - A_{\rho} + 2\vec{k}\cdot\vec{q}\right) - \gamma^0\left[E^*_q\left(m^2_{\rho} - A_{\rho} + 2\vec{k}\cdot\vec{q}\right) - 4E^*_q\left(E_k-E_q\right)^2\right. \\
&\quad \left. +4\vec{k}\cdot\vec{q}^{\,*}\left(E_k-E_q\right) - 4qq^*\left(E_k-E_q\right)\right] \\
&\quad +\vec{\gamma}\cdot\vec{k}\left[\frac{\vec{k}\cdot\vec{q}^{\,*}}{k^2}\left(m^2_{\rho} - A_{\rho}\right) + 2qq^*\cos^2\theta - 4E^*_q\left(E_k-E_q\right)\right. \\
&\quad \left. +4\frac{\vec{k}\cdot\vec{q}}{k^2}E^*_q\left(E_k-E_q\right) + 4\vec{k}\cdot\vec{q}^{\,*} - 4qq^* - 4qq^*\cos^2\theta + 4\vec{k}\cdot\vec{q}\,\frac{qq^*}{k^2}\right]
\end{aligned}
\tag{5.908}
$$

There is an indefinite integral formula as

$$
\int \frac{u^2\mathrm{d}u}{a+bu} = \frac{1}{b^3}\left[\frac{1}{2}(a+bu)^2 - 2a(a+bu) + a^2\log{(a+bu)}\right] + C \tag{5.909}
$$

Referring to Eq. (5.859) and using Eqs. (5.858) and (5.860), we can obtain

$$
\begin{aligned}
&2\pi\int_{-1}^{1}\frac{k^2q^2\cos^2\theta}{-A_i+2kq\cos\theta}\,\mathrm{d}\cos\theta \\
&\quad=\frac{2\pi k^2q^2}{(2kq)^3}\left\{\frac{1}{2}\left[(2kq-A_i)^2-(2kq+A_i)^2\right]+8kqA_i-A_i^2\Theta_i\right\} \qquad (5.910)\\
&\quad=\frac{\pi}{4kq}\left(-4kqA_i+8kqA_i-A_i^2\Theta_i\right)=\pi A_i\left(1-\frac{A_i\Theta_i}{4kq}\right)=-\pi A_i\Phi_i
\end{aligned}
$$

The previous summary of the correspondence is that 1 corresponds to $\Theta_i$ and $\vec{k}\cdot\vec{q}^{\,*}$ corresponds to $2kq^*\Phi_i$. The comparison between Eq. (5.910) and Eq. (5.859) shows that $\left(\vec{k}\cdot\vec{q}\right)^2=k^2q^2\cos^2\theta$ should correspond to $\frac{2kq\Phi_i}{2\pi}\pi A_i=kqA_i\Phi_i$. Referring to Eq. (5.889) and using Eq. (5.908), from Eq. (5.902) we can obtain

$$
\begin{aligned}
\Sigma_\rho^{\mathrm{T}}(k)=&\frac{12}{16\pi^2k}\left(\frac{f_\rho}{4M}\right)^2\int_0^{k_{\mathrm{F}}}q\mathrm{d}q\Bigg\{-\frac{3M_q^*}{E_q^*}\left[\left(m_\rho^2-A_\rho\right)\Theta_\rho+4kq\Phi_\rho\right]\\
&-\gamma^0\Bigg[\left(m_\rho^2-A_\rho\right)\Theta_\rho+4kq\Phi_\rho-4\left(E_k-E_q\right)^2\Theta_\rho+\frac{8kq^*}{E_q^*}\left(E_k-E_q\right)\Phi_\rho\\
&-\frac{4qq^*\left(E_k-E_q\right)}{E_q^*}\Theta_\rho\Bigg]+\vec{\gamma}\cdot\vec{k}\Bigg[\frac{2q^*}{kE_q^*}\left(m_\rho^2-A_\rho\right)\Phi_\rho-4\left(E_k-E_q\right)\Theta_\rho\\
&+\frac{8q}{k}\left(E_k-E_q\right)\Phi_\rho+\frac{8kq^*}{E_q^*}\Phi_\rho-\frac{4qq^*}{E_q^*}\Theta_\rho+\frac{8q^2q^*}{kE_q^*}\Phi_\rho-\frac{2q^*}{kE_q^*}A_\rho\Phi_\rho\Bigg]\Bigg\}
\end{aligned}
\tag{5.911}
$$

From Eq. (5.859) we know

$$
4kq\Phi_\rho=A_\rho\Theta_\rho-4kq \tag{5.912}
$$

Merge the coefficients of the containing $\frac{4q^*}{kE_q^*}\Phi_\rho$ items of $\vec{\gamma}\cdot\vec{k}$ items in Eq. (5.911) as follows:

$$
2k^2+2q^2+\frac{1}{2}m_\rho^2-A_\rho=k^2+q^2+\left(E_k-E_q\right)^2-\frac{1}{2}m_\rho^2 \tag{5.913}
$$

So Eq. (5.911) can be rewritten as

$$\Sigma_{\rho}^{\mathrm{T}}(k,E)=\frac{3}{4\pi^2 k}\left(\frac{f_\rho}{4M}\right)^2\int_0^{k_{\mathrm{F}}} q\mathrm{d}q\Bigg\{-\frac{3M_q^*}{E_q^*}\left(m_\rho^2\Theta_\rho-4kq\right)-\gamma^0\Big[m_\rho^2\Theta_\rho$$
$$-4kq-4\left(E_k-E_q\right)^2\Theta_\rho-\frac{4qq^*\left(E_k-E_q\right)}{E_q^*}\Theta_\rho+\frac{8kq^*\left(E_k-E_q\right)}{E_q^*}\Phi_\rho\Big]$$
$$+4\vec{\gamma}\cdot\vec{k}\Big[-\frac{qq^*}{E_q^*}\Theta_\rho-\left(E_k-E_q\right)\Theta_\rho+\frac{2q}{k}\left(E_k-E_q\right)\Phi_\rho$$
$$+\frac{q^*}{kE_q^*}\left(k^2+q^2+\left(E_k-E_q\right)^2-\frac{1}{2}m_\rho^2\right)\Phi_\rho\Big]\Bigg\} \tag{5.914}$$

For ρ mesons, if the tensor coupling exists at the vertex $x$ and the vector coupling exists at the vertex $x'$; or the vector coupling exists at the vertex $x$ and the tensor coupling exists at the vertex $x'$, so the vector-tensor coupling terms will appear. From the previous discussion it can be seen that the vertex on the loop of the Hartree item does not contribute for either vector or tensor coupling, so we can get

$$\Sigma_\rho^{\mathrm{VT-H}}(k)=0 \tag{5.915}$$

Also note that the negative sign needs to be added when $\sigma^{\mu\nu}\vec{L}_{\mu\nu}$ given by Eq. (5.900) acting on the vertex $x'$. Referring to the second term of the first equation of Eq. (5.895) and the first equation of Eq. (5.902), we can write

$$\Sigma_\rho^{\mathrm{VT}}(k)=\Sigma_\rho^{\mathrm{VT-F}}(k)=2\mathrm{i}g_\rho\frac{f_\rho}{4M}\int\frac{\mathrm{d}^4q}{(2\pi)^4}\Big\{\left[\gamma^\xi(k-q)_\xi\gamma^\mu-(k-q)^\mu\right]$$
$$\tau_a D^0_{\mu\nu,ab}(k-q)G(q)\gamma^\nu\tau_b-\gamma^\mu\tau_a D^0_{\mu\nu,ab}(k-q)G(q)\left[\gamma^\xi(k-q)_\xi\gamma^\nu-(k-q)^\nu\right]\tau_b\Big\}$$
$$=-6\mathrm{i}g_\rho\frac{f_\rho}{4M}\int\frac{\mathrm{d}^4q}{(2\pi)^4}\left\{\frac{\left[\gamma^\xi(k-q)_\xi\gamma^\mu-(k-q)^\mu\right]g_{\mu\nu}G(q)\gamma^\nu}{(k-q)_\lambda(k-q)^\lambda-m_\rho^2+\mathrm{i}\varepsilon}\right.$$
$$\left.-\frac{\gamma^\mu g_{\mu\nu}G(q)\left[\gamma^\xi(k-q)_\xi\gamma^\nu-(k-q)^\nu\right]}{(k-q)_\lambda(k-q)^\lambda-m_\rho^2+\mathrm{i}\varepsilon}\right\}$$
$$=-6\mathrm{i}g_\rho\frac{f_\rho}{4M}\int\frac{\mathrm{d}^4q}{(2\pi)^4}\left\{\frac{\left[\gamma^\xi(k-q)_\xi\gamma^\mu-(k-q)^\mu\right]\left(\gamma^\eta q_\eta^*+M_q^*\right)g_{\mu\mu}\gamma^\mu}{(k-q)_\lambda(k-q)^\lambda-m_\rho^2+\mathrm{i}\varepsilon}\right.$$
$$\left.-\frac{\gamma^\mu g_{\mu\mu}\left(\gamma^\eta q_\eta^*+M_q^*\right)\left[\gamma^\xi(k-q)_\xi\gamma^\mu-(k-q)^\mu\right]}{(k-q)_\lambda(k-q)^\lambda-m_\rho^2+\mathrm{i}\varepsilon}\right\}\frac{\mathrm{i}\pi}{E_q^*}\delta\left(q_0-E_q\right)\theta\left(k_{\mathrm{F}}-|\vec{q}|\right) \tag{5.916}$$

We can also find

$$
\begin{aligned}
I \equiv & \left[\gamma^{\xi}(k-q)_{\xi}\gamma^{\mu}-(k-q)^{\mu}\right]\left(\gamma^{\eta}q_{\eta}^{*}+M_{q}^{*}\right)g_{\mu\mu}\gamma^{\mu} \\
& -\gamma^{\mu}g_{\mu\mu}\left(\gamma^{\eta}q_{\eta}^{*}+M_{q}^{*}\right)\left[\gamma^{\xi}(k-q)_{\xi}\gamma^{\mu}-(k-q)^{\mu}\right] \\
= & \gamma^{\xi}(k-q)_{\xi}\gamma^{\mu}\gamma^{\eta}q_{\eta}^{*}g_{\mu\mu}\gamma^{\mu}+\gamma^{\xi}(k-q)_{\xi}\gamma^{\mu}g_{\mu\mu}\gamma^{\mu}M_{q}^{*} \\
& -(k-q)^{\mu}\gamma^{\eta}q_{\eta}^{*}g_{\mu\mu}\gamma^{\mu}-(k-q)^{\mu}g_{\mu\mu}\gamma^{\mu}M_{q}^{*} \\
& -\gamma^{\mu}g_{\mu\mu}\gamma^{\eta}q_{\eta}^{*}\gamma^{\xi}(k-q)_{\xi}\gamma^{\mu}+\gamma^{\mu}g_{\mu\mu}\gamma^{\eta}q_{\eta}^{*}(k-q)^{\mu} \\
& -\gamma^{\mu}g_{\mu\mu}\gamma^{\xi}(k-q)_{\xi}\gamma^{\mu}M_{q}^{*}+\gamma^{\mu}g_{\mu\mu}(k-q)^{\mu}M_{q}^{*} \\
= & -\gamma^{\xi}(k-q)_{\xi}\gamma^{\eta}q_{\eta}^{*}\gamma^{\mu}g_{\mu\mu}\gamma^{\mu}+2\gamma^{\xi}(k-q)_{\xi}g^{\mu\eta}q_{\eta}^{*}g_{\mu\mu}\gamma^{\mu} \\
& +4\gamma^{\xi}(k-q)_{\xi}M_{q}^{*}-\gamma^{\eta}q_{\eta}^{*}\gamma^{\mu}(k-q)_{\mu}+\gamma^{\mu}g_{\mu\mu}\gamma^{\eta}q_{\eta}^{*}\gamma^{\mu}\gamma^{\xi}(k-q)_{\xi} \\
& -2\gamma^{\mu}g_{\mu\mu}\gamma^{\eta}q_{\eta}^{*}g^{\mu\xi}(k-q)_{\xi}+\gamma^{\mu}(k-q)_{\mu}\gamma^{\eta}q_{\eta}^{*} \\
& +\gamma^{\mu}g_{\mu\mu}\gamma^{\mu}\gamma^{\xi}(k-q)_{\xi}M_{q}^{*}-2\gamma^{\mu}g_{\mu\mu}g^{\mu\xi}(k-q)_{\xi}M_{q}^{*} \\
= & -4\gamma^{\xi}(k-q)_{\xi}\gamma^{\eta}q_{\eta}^{*}+2\gamma^{\xi}(k-q)_{\xi}\gamma^{\eta}q_{\eta}^{*}+4\gamma^{\xi}(k-q)_{\xi}M_{q}^{*} \\
& +\gamma^{\mu}(k-q)_{\mu}\gamma^{\eta}q_{\eta}^{*}-2q_{\eta}^{*}g^{\eta\mu}(k-q)_{\mu}-\gamma^{\mu}g_{\mu\mu}\gamma^{\mu}\gamma^{\eta}q_{\eta}^{*}\gamma^{\xi}(k-q)_{\xi} \\
& +2\gamma^{\mu}g_{\mu\mu}g^{\mu\eta}q_{\eta}^{*}\gamma^{\xi}(k-q)_{\xi}-2\gamma^{\mu}(k-q)_{\mu}\gamma^{\eta}q_{\eta}^{*} \\
& +\gamma^{\mu}(k-q)_{\mu}\gamma^{\eta}q_{\eta}^{*}+4\gamma^{\xi}(k-q)_{\xi}M_{q}^{*}-2\gamma^{\xi}(k-q)_{\xi}M_{q}^{*} \\
= & -2\gamma^{\xi}(k-q)_{\xi}\gamma^{\eta}q_{\eta}^{*}-2(k-q)^{\mu}q_{\mu}^{*}-2\gamma^{\eta}q_{\eta}^{*}\gamma^{\xi}(k-q)_{\xi}+6\gamma^{\xi}(k-q)_{\xi}M_{q}^{*} \\
= & -4g^{\xi\eta}(k-q)_{\xi}q_{\eta}^{*}-2(k-q)^{\mu}q_{\mu}^{*}+6\gamma^{\xi}(k-q)_{\xi}M_{q}^{*} \\
= & 6\left[-(k-q)^{\mu}q_{\mu}^{*}+\gamma^{\xi}(k-q)_{\xi}M_{q}^{*}\right]
\end{aligned}
\tag{5.917}
$$

Further from above formula we can get

$$
\begin{aligned}
I_{|q_0=E_q} &= 6\left[-(E_k-E_q)E_q^{*}+\left(\vec{k}-\vec{q}\right)\cdot\vec{q}^{\,*}+\gamma^{0}(E_k-E_q)M_q^{*}-\vec{\gamma}\cdot\left(\vec{k}-\vec{q}\right)M_q^{*}\right] \\
&= 6\left[-(E_k-E_q)E_q^{*}+\vec{k}\cdot\vec{q}^{\,*}-qq^{*}+\gamma^{0}(E_k-E_q)M_q^{*}-\vec{\gamma}\cdot\vec{k}\left(1-\frac{\vec{k}\cdot\vec{q}}{k^2}\right)M_q^{*}\right]
\end{aligned}
\tag{5.918}
$$

Then the following expression can be obtained by Eqs. (5.916) and (5.918):

$$
\begin{aligned}
\Sigma_{\rho}^{\mathrm{VT}}(k,E) &= -36\mathrm{i}g_{\rho}\frac{f_{\rho}}{4M}\int\frac{\mathrm{d}^4q}{(2\pi)^4}\frac{1}{(k-q)_{\lambda}(k-q)^{\lambda}-m_{\rho}^2+\mathrm{i}\varepsilon} \\
&\quad\Big[-\big(E_k-E_q\big)E_q^*+\vec{k}\cdot\vec{q}^{\,*}-qq^*+\gamma^0\big(E_k-E_q\big)M_q^* \\
&\quad-\vec{\gamma}\cdot\vec{k}\left(1-\frac{\vec{k}\cdot\vec{q}}{k^2}\right)M_q^*\Big]\frac{\mathrm{i}\pi}{E_q^*}\delta\big(q_0-E_q\big)\,\theta\big(k_{\mathrm{F}}-|\vec{q}|\big) \\
&= \frac{9}{4\pi^2k}g_{\rho}\frac{f_{\rho}}{4M}\int_0^{k_F}q\mathrm{d}q\Big\{\big(E_k-E_q\big)\Theta_{\rho}+\frac{qq^*}{E_q^*}\Theta_{\rho}-\frac{2kq^*}{E_q^*}\Phi_{\rho} \\
&\quad-\gamma^0\frac{\big(E_k-E_q\big)M_q^*}{E_q^*}\Theta_{\rho}+\vec{\gamma}\cdot\vec{k}\left(\frac{M_q^*}{E_q^*}\Theta_{\rho}-\frac{2qM_q^*}{kE_q^*}\Phi_{\rho}\right)\Big\}
\end{aligned}
\tag{5.919}
$$

For a nuclear matter containing only neutrons or protons, the previously obtained self-energy formula can be used directly. For $N = Z$ symmetric nuclear matter, it is necessary to multiply the isospin number $\lambda = 2$ in front of the self-energy formula and using $\rho = \frac{2}{3\pi^2}k_{\mathrm{F}}^3$ to determine Fermi momentum. For $N \neq Z$ asymmetric nuclear matter, neutrons and protons have Fermi momentum $k_{\mathrm{Fn}}$ and $k_{\mathrm{Fp}}$, respectively, so their self-energies obtained separately need to be added. Finally the optical potential of the finite nucleus is obtained by the local density approximation.

For the nucleons describing the Feynman diagram intermediate state, because Dirac propagator $G_D$ containing δ function is used, so the intermediate state nucleons are on-shell. In the $A_i$ expression given by Eq. (5.854), if $k$ and $E$ are not related, i.e., both are independent variables, then the real part of the relativistic microscopic potential given above is off-shell; if taking $E = \left(k^{*2} + M_k^{*2}\right)^{1/2} + \Sigma^0(k)$ or $E = k^2 + M^2$, it is on-shell.

### 5.12.3 *Relativistic Nuclear Matter Properties*

The Dirac equation for free nucleon is

$$
\left(\vec{\alpha}\cdot\vec{k}+\beta M\right)u\left(\vec{k},\xi\right)=E_k u\left(\vec{k},\xi\right) \tag{5.920}
$$

where $\xi$ represents spin. When there is an interaction in the nuclear matter, Eq. (5.920) can be rewritten as

$$
\left(\vec{\alpha}\cdot\vec{k}^{\,*}+\beta M_k^*\right)u\left(\vec{k},\xi\right)=E_k^* u\left(\vec{k},\xi\right) \tag{5.921}
$$

And referring to Eq. (5.93), the plane wave solution of Eq. (5.921) can be written as

$$u\left(\vec{k},\xi\right)=\sqrt{\frac{E_k^*+M_k^*}{2M_k^*}}\begin{pmatrix}\chi_\xi\\ \dfrac{\hat{\sigma}\cdot\vec{k}^*}{E_k^*+M_k^*}\chi_\xi\end{pmatrix}\tag{5.922}$$

Moreover utilizing Eq. (5.839), Eq. (5.921) can be rewritten as

$$\left[\vec{\alpha}\cdot\vec{k}+\beta M+\beta\Sigma^{\mathrm{s}}+\Sigma^{0}+\vec{\alpha}\cdot\vec{k}\Sigma^{\mathrm{v}}\right]u\left(\vec{k},\xi\right)=E_k u\left(\vec{k},\xi\right)\tag{5.923}$$

In the above formula, $\vec{\alpha}\cdot\vec{k}+\beta M$ are the kinetic energy terms, and the sum of the three terms related to self-energy $\Sigma$ are potential energy terms.

In the 5.11 section, the nucleon wave function $\psi\left(\vec{x},t\right)$ given by Eq. (5.754) and Eq. (5.755) has been proved to satisfy the anti-commutative relation given by Eq. (5.726). Therefore, in the case with interaction, the plane wave function of the positive energy nucleons is taken as

$$\psi_{\vec{k}\xi T}\left(\vec{x},t\right)=\frac{1}{\sqrt{V}}\sqrt{\frac{M_k^*}{E_k^*}}u\left(\vec{k},\xi\right)\mathrm{e}^{-\mathrm{i}k_\mu^* x^\mu}\chi_T\tag{5.924}$$

where $T$ stands for isospin. The wave function so defined is orthogonal normalized by using $\psi^+\psi$. In the $N=Z$ symmetric nuclear matter, using Eq. (5.678) a scalar density can be given as

$$\rho_{\mathrm{S}}=\sum_{\xi T}\int^{k_{\mathrm{F}}}\frac{V\mathrm{d}\vec{k}}{(2\pi)^3}\overline{\psi}_{\vec{k}\xi T}\psi_{\vec{k}\xi T}\tag{5.925}$$

where $\int^{k_{\mathrm{F}}}\frac{V\mathrm{d}\vec{k}}{(2\pi)^3}$ is the number of states in the phase space. Since $\overline{u}\left(\vec{k},\xi\right)u\left(\vec{k},\xi\right)=1$, using Eq. (5.715) the following expression can be obtained from Eqs. (5.924) and (5.925) as

$$\rho_{\mathrm{S}}=4\int^{k_{\mathrm{F}}}\frac{\mathrm{d}\vec{k}}{(2\pi)^3}\frac{M_k^*}{E_k^*}=\frac{2}{\pi^2}\int_0^{k_{\mathrm{F}}}\frac{M_k^*}{E_k^*}k^2\mathrm{d}k\tag{5.926}$$

where $E_k^*$ has been given by Eq. (5.841). And using Eq. (5.682) the usual nucleon density can also be given as

$$\rho_{\mathrm{B}}=\sum_{\xi T}\int^{k_{\mathrm{F}}}\frac{V\mathrm{d}\vec{k}}{(2\pi)^3}\psi^+_{\vec{k}\xi T}\psi_{\vec{k}\xi T}\tag{5.927}$$

Since $\frac{M_k^*}{E_k^*}u^+\left(\vec{k},\xi\right)u\left(\vec{k},\xi\right)=1$, so we have

$$\rho_{\mathrm{B}} = \frac{2}{3\pi^2} k_{\mathrm{F}}^3 \tag{5.928}$$

Equation (5.617) gives the Hamiltonian density as

$$\mathcal{H} = \pi_\lambda \dot{\varphi}_\lambda - \mathcal{L} = T^{00} \tag{5.929}$$

where $\dot{\varphi}_\lambda = \frac{\partial \varphi_\lambda}{\partial t}, \quad \pi_\lambda = \frac{\partial \mathcal{L}}{\partial \dot{\varphi}_\lambda}$. The energy density of the system is

$$\varepsilon = \langle \Psi | T^{00} | \Psi \rangle \tag{5.930}$$

Since $\mathcal{L}$ contains the contributions of nucleon and various mesons, the energy density calculated from the above formula must also be related to nucleon and various mesons. Some people have studied Eq. (5.930) when only considering σ and ω mesons [26, 69]. It is believed that in $N = Z$ symmetric nuclear matter, only the smaller items need to be ignored to write Eq. (5.930) in the following familiar form:

$$\begin{aligned} \varepsilon &= 2\lambda \int^{k_{\mathrm{F}}} \frac{\mathrm{d}\vec{k}}{(2\pi)^3} T(k) + 2\lambda \int^{k_{\mathrm{F}}} \frac{\mathrm{d}\vec{k}}{(2\pi)^3} \frac{1}{2} V(k) \\ &= \frac{\lambda}{\pi^2} \int_0^{k_{\mathrm{F}}} T(k)\, k^2 \mathrm{d}k + \frac{\lambda}{2\pi^2} \int_0^{k_{\mathrm{F}}} V(k)\, k^2 \mathrm{d}k \end{aligned} \tag{5.931}$$

where $T(k)$ and $V(k)$ are the distributions of nucleon kinetic energy density and nucleon potential energy density, respectively. The 2 in $2\lambda$ represents two spins, and $\lambda = 2$ taken here represents two isospins.

The kinetic energy density distribution is

$$\begin{aligned} T(k) &= \frac{M_k^*}{E_k^*} u^+\left(\vec{k}, \xi\right) \left(\vec{\alpha} \cdot \vec{k} + \beta M\right) u\left(\vec{k}, \xi\right) \\ &= \frac{E_k^* + M_k^*}{2E_k^*} \chi_\xi^+ \begin{pmatrix} I & \dfrac{\hat{\sigma} \cdot \vec{k}^*}{E_k^* + M_k^*} \end{pmatrix} \begin{pmatrix} M & \hat{\sigma} \cdot \vec{k} \\ \hat{\sigma} \cdot \vec{k} & -M \end{pmatrix} \begin{pmatrix} I \\ \dfrac{\hat{\sigma} \cdot \vec{k}^*}{E_k^* + M_k^*} \end{pmatrix} \chi_\xi \\ &= \frac{E_k^* + M_k^*}{2E_k^*} \chi_\xi^+ \begin{pmatrix} I & \dfrac{\hat{\sigma} \cdot \vec{k}^*}{E_k^* + M_k^*} \end{pmatrix} \begin{pmatrix} M + \dfrac{kk^*}{E_k^* + M_k^*} \\ \hat{\sigma} \cdot \vec{k}^* - M \dfrac{\hat{\sigma} \cdot \vec{k}^*}{E_k^* + M_k^*} \end{pmatrix} \chi_\xi \\ &= \frac{E_k^* + M_k^*}{2E_k^*} \left( M + \frac{kk^*}{E_k^* + M_k^*} + \frac{kk^*}{E_k^* + M_k^*} - M \frac{E_k^* - M_k^*}{E_k^* + M_k^*} \right) \\ &= \frac{MM_k^* + kk^*}{E_k^*} \end{aligned} \tag{5.932}$$

The distribution of potential energy density is

$$\begin{aligned}
V(k) &= \frac{M_k^*}{E_k^*} u^+\left(\vec{k}, \xi\right)\left(\beta\Sigma^{\mathrm{s}} + \Sigma^0 + \vec{\alpha}\cdot\vec{k}\Sigma^{\mathrm{v}}\right) u\left(\vec{k}, \xi\right) \\
&= \frac{E_k^* + M_k^*}{2E_k^*}\chi_\xi^+ \begin{pmatrix} I & \dfrac{\hat{\sigma}\cdot\vec{k}^*}{E_k^* + M_k^*}\end{pmatrix} \begin{pmatrix} \Sigma^{\mathrm{s}} + \Sigma^0 & \hat{\sigma}\cdot\vec{k}\Sigma^{\mathrm{v}} \\ \hat{\sigma}\cdot\vec{k}\Sigma^{\mathrm{v}} & -\Sigma^{\mathrm{s}} + \Sigma^0 \end{pmatrix} \begin{pmatrix} I \\ \dfrac{\hat{\sigma}\cdot\vec{k}^*}{E_k^* + M_k^*}\end{pmatrix}\chi_\xi \\
&= \frac{E_k^* + M_k^*}{2E_k^*}\chi_\xi^+ \begin{pmatrix} I & \dfrac{\hat{\sigma}\cdot\vec{k}^*}{E_k^* + M_k^*}\end{pmatrix} \begin{pmatrix} \Sigma^{\mathrm{s}} + \Sigma^0 + \dfrac{kk^*\Sigma^{\mathrm{v}}}{E_k^* + M_k^*} \\ \hat{\sigma}\cdot\vec{k}\Sigma^{\mathrm{v}} - \left(\Sigma^{\mathrm{s}} - \Sigma^0\right)\dfrac{\hat{\sigma}\cdot\vec{k}^*}{E_k^* + M_k^*} \end{pmatrix}\chi_\xi \\
&= \frac{E_k^* + M_k^*}{2E_k^*}\left(\Sigma^{\mathrm{s}} + \Sigma^0 + \frac{kk^*\Sigma^{\mathrm{v}}}{E_k^* + M_k^*} + \frac{kk^*\Sigma^{\mathrm{v}}}{E_k^* + M_k^*} - \left(\Sigma^{\mathrm{s}} - \Sigma^0\right)\frac{E_k^* - M_k^*}{E_k^* + M_k^*}\right) \\
&= \frac{\Sigma^{\mathrm{s}}M_k^* + \Sigma^{\mathrm{v}}kk^*}{E_k^*} + \Sigma^0
\end{aligned} \tag{5.933}$$

If only σ and ω mesons are considered, the energy density is formally divided into kinetic energy terms, Hartree potential energy terms, and Fock potential energy terms, namely,

$$\varepsilon = \varepsilon_{\mathrm{T}} + \varepsilon_{\mathrm{H}} + \varepsilon_{\mathrm{F}} \tag{5.934}$$

For $N = Z$ symmetric nuclear matter, from Eq. (5.931) and Eq. (5.932), we can obtain

$$\varepsilon_{\mathrm{T}} = \frac{2}{\pi^2}\int_0^{k_{\mathrm{F}}} \frac{MM_k^* + kk^*}{E_k^*} k^2 \mathrm{d}k \tag{5.935}$$

In fact the contribution of the self-energy operator $\Sigma$ is also included in the $\varepsilon_T$ expression given above. And there is

$$\varepsilon_{\mathrm{T}} \xrightarrow[\Sigma\to 0 \ \text{ or } \rho_{\mathrm{B}}\to 0]{} \frac{2}{\pi^2}\int_0^{k_{\mathrm{F}}} E_k k^2 \mathrm{d}k \tag{5.936}$$

Equations (5.849), (5.861), (5.867), and (5.872) give the $\lambda = 1$ self-energy expressions, respectively, as follows:

$$\Sigma^{\mathrm{s}}_{\mathrm{H}\sigma} = -\left(\frac{g_\sigma}{m_\sigma}\right)^2 \frac{1}{\pi^2}\int_0^{k_\mathrm{F}} \frac{M_q^*}{E_q^*} q^2 \mathrm{d}q \tag{5.937}$$

$$\Sigma^{\mathrm{s}}_{\mathrm{F}\sigma} = \frac{g_\sigma^2}{16\pi^2 k}\int_0^{k_\mathrm{F}} \frac{M_q^*}{E_q^*} \Theta_\sigma q \mathrm{d}q \tag{5.938}$$

$$\Sigma^{0}_{\mathrm{F}\sigma} = \frac{g_\sigma^2}{16\pi^2 k}\int_0^{k_\mathrm{F}} \Theta_\sigma q \mathrm{d}q \tag{5.939}$$

$$\Sigma^{\mathrm{v}}_{\mathrm{F}\sigma} = -\frac{g_\sigma^2}{8\pi^2 k^2}\int_0^{k_\mathrm{F}} \frac{q^*}{E_q^*} \Phi_\sigma q \mathrm{d}q \tag{5.940}$$

$$\Sigma^{0}_{\mathrm{H}\omega} = \left(\frac{g_\omega}{m_\omega}\right)^2 \frac{k_\mathrm{F}^3}{3\pi^2} \tag{5.941}$$

$$\Sigma^{\mathrm{s}}_{\mathrm{F}\omega} = -\frac{g_\omega^2}{4\pi^2 k}\int_0^{k_\mathrm{F}} \frac{M_q^*}{E_q^*} \Theta_\omega q \mathrm{d}q \tag{5.942}$$

$$\Sigma^{0}_{\mathrm{F}\omega} = \frac{g_\omega^2}{8\pi^2 k}\int_0^{k_\mathrm{F}} \Theta_\omega q \mathrm{d}q \tag{5.943}$$

$$\Sigma^{\mathrm{v}}_{\mathrm{F}\omega} = -\frac{g_\omega^2}{4\pi^2 k^2}\int_0^{k_\mathrm{F}} \frac{q^*}{E_q^*} \Phi_\omega q \mathrm{d}q \tag{5.944}$$

When $\lambda = 2$ the following expression can be found using Eqs. (5.931), (5.933), (5.937), and (5.941):

$$\begin{aligned}\varepsilon_\mathrm{H} &= \frac{\lambda}{2\pi^2}\int_0^{k_\mathrm{F}} \frac{M_k^*}{E_k^*}\left[-\left(\frac{g_\sigma}{m_\sigma}\right)^2 \frac{\lambda}{\pi^2}\int_0^{k_\mathrm{F}} \frac{M_q^*}{E_q^*} q^2 \mathrm{d}q\right] k^2 \mathrm{d}k \\ &+ \frac{\lambda}{2\pi^2}\int_0^{k_\mathrm{F}} \left[\left(\frac{g_\omega}{m_\omega}\right)^2 \frac{\lambda k_\mathrm{F}^3}{3\pi^2}\right] k^2 \mathrm{d}k = -\frac{1}{2}\left(\frac{g_\sigma}{m_\sigma}\right)^2 \rho_\mathrm{S}^2 + \frac{1}{2}\left(\frac{g_\omega}{m_\omega}\right)^2 \rho_\mathrm{B}^2\end{aligned} \tag{5.945}$$

We can also find

$$\begin{aligned}\varepsilon_\mathrm{F} &= \frac{\lambda}{2\pi^2}\int_0^{k_\mathrm{F}} \left\{ \frac{\lambda}{E_k^*}\left[\left(\Sigma^{\mathrm{s}}_{\mathrm{F}\sigma} + \Sigma^{\mathrm{s}}_{\mathrm{F}\omega}\right) M_k^* + \left(\Sigma^{\mathrm{v}}_{\mathrm{F}\sigma} + \Sigma^{\mathrm{v}}_{\mathrm{F}\omega}\right) k k^*\right] \right. \\ &\quad \left. + \lambda\left(\Sigma^{0}_{\mathrm{F}\sigma} + \Sigma^{0}_{\mathrm{F}\omega}\right) \right\} k^2 \mathrm{d}k \\ &= \frac{2}{\pi^2}\int_0^{k_\mathrm{F}} \left\{ \frac{1}{E_k^*}\left[\left(\Sigma^{\mathrm{s}}_{\mathrm{F}\sigma} + \Sigma^{\mathrm{s}}_{\mathrm{F}\omega}\right) M_k^* + \left(\Sigma^{\mathrm{v}}_{\mathrm{F}\sigma} + \Sigma^{\mathrm{v}}_{\mathrm{F}\omega}\right) k k^*\right] \right. \\ &\quad \left. + \left(\Sigma^{0}_{\mathrm{F}\sigma} + \Sigma^{0}_{\mathrm{F}\omega}\right) \right\} k^2 \mathrm{d}k\end{aligned} \tag{5.946}$$

The binding energy of a single nucleon in nuclear matter is

$$\frac{\varepsilon}{\rho_{\mathrm{B}}} - M = -a, \qquad a \sim 16\,\mathrm{MeV} \tag{5.947}$$

According to the law of thermodynamics, the formula for calculating the pressure $p$ is

$$p = \rho_{\mathrm{B}}^2 \frac{\partial}{\partial \rho_{\mathrm{B}}}\left(\frac{\varepsilon}{\rho_{\mathrm{B}}}\right) = -\varepsilon + \rho_{\mathrm{B}} \frac{\partial \varepsilon}{\partial \rho_{\mathrm{B}}} \tag{5.948}$$

Let $p = 0$ in Eq. (5.948) then $\rho_0 \equiv \rho_{\mathrm{B}}$ can be solved. At this time, the system is the most stable, $p = 0$ is called the saturation condition, and $\rho_0$ is called the saturation density. Generally $\rho_0 \approx 0.15$~$0.16\ \mathrm{fm}^{-3}$, $k_{\mathrm{F}}$ can be obtained by $\rho_0$. The chemical potential under zero temperature is defined as

$$\mu = \left(\frac{\partial E}{\partial N}\right)_{\mathrm{V}} = \left(\frac{\partial \varepsilon}{\partial \rho_{\mathrm{B}}}\right)_{\mathrm{V}} \tag{5.949}$$

For the ideal gas there is $\mu = E_{\mathrm{F}}$. Fermi energy $E_{\mathrm{F}}$ is the solution of Eq. (5.842) when $\left|\vec{k}\right| = k_{\mathrm{F}}$. So the saturation condition can be written as

$$p = -\varepsilon + E_{\mathrm{F}}\rho_0 = 0 \tag{5.950}$$

If the effective nuclear force has only two free parameters $g_\sigma$ and $g_\omega$, they can be determined by the above nuclear matter property parameters $a$ and $\rho_0$.

The incompressible coefficient is [28]

$$\begin{aligned} K &= 9\rho_{\mathrm{B}}^2 \frac{\partial^2(\varepsilon/\rho_{\mathrm{B}})}{\partial \rho_{\mathrm{B}}^2}\bigg|_{\rho_{\mathrm{B}}=\rho_0} = 9\rho_{\mathrm{B}}^2 \left[\frac{\partial}{\partial \rho_{\mathrm{B}}}\left(-\frac{\varepsilon}{\rho_{\mathrm{B}}^2} + \frac{1}{\rho_{\mathrm{B}}}\frac{\partial \varepsilon}{\partial \rho_{\mathrm{B}}}\right)\right]\bigg|_{\rho_{\mathrm{B}}=\rho_0} \\ &= 9\rho_{\mathrm{B}}\left[-\frac{2}{\rho_{\mathrm{B}}^2}\left(-\varepsilon + \rho_{\mathrm{B}}\frac{\partial \varepsilon}{\partial \rho_{\mathrm{B}}}\right) + \frac{\partial^2 \varepsilon}{\partial \rho_{\mathrm{B}}^2}\right]\bigg|_{\rho_{\mathrm{B}}=\rho_0} = 9\rho_{\mathrm{B}}\frac{\partial^2 \varepsilon}{\partial \rho_{\mathrm{B}}^2}\bigg|_{\rho_{\mathrm{B}}=\rho_0} \end{aligned} \tag{5.951}$$

The saturation condition is used in the above formula derivation.

If the effective nuclear force has more free parameters, they can also be constrained by other properties of nuclear matter such as the incompressible coefficient $K$ and the nucleon effective mass $M_k^*/M$. For the asymmetric nuclear matter, the symmetric energy $E_{\mathrm{sym}}$ and symmetric energy slope $L$ can also be used to constrain the effective nuclear force parameters. If the theory is used for finite nuclei, they can also be constrained by experimental data on the binding energies and charge radii of some specific nuclei.

## 5.13 Imaginary Part of Nucleon Relativistic Microscopic Optical Potential

We only study the lowest order Feynman diagrams that have contributions to the imaginary part of the self-energy $\Sigma$ [25, 80]. For the case where the incident nucleon energy is above the Fermi surface, only the 2p-1h polarization diagrams are considered for the time being. When only σ and ω mesons are included, Fig. 5.6 gives the fourth order Feynman diagrams that need to be considered.

Each graph in Fig. 5.6 can be regarded as a second order correction graph of meson propagators. Figure 5.7 shows the Feynman diagram corresponding to the second order propagators of σ mesons.

### *5.13.1 Contribution of σ − σ Meson Exchange Processes to the Imaginary Part of the Optical Potential*

Using the Feynman rules, the expression corresponding to Fig. 5.7 describing the exchange σ − σ mesons can be written as

$$\begin{aligned} \mathrm{i}\Delta^{(2)}(q) &= \mathrm{i}\Delta^0(q)(-\mathrm{i}q_\sigma)\int\frac{\mathrm{d}^4p}{(2\pi)^4}\mathrm{tr}[-\mathrm{i}G(p)\,\mathrm{i}G(p+q)](-\mathrm{i}q_\sigma)\,\mathrm{i}\Delta^0(q) \\ &\equiv \mathrm{i}\Delta^0(q)\,\pi_\sigma(q)\Delta^0(q) \end{aligned} \tag{5.952}$$

And we can get

$$\pi_\sigma(q) = -\mathrm{i}g_\sigma^2\int\frac{\mathrm{d}^4p}{(2\pi)^4}\mathrm{tr}[G(p)G(p+q)] \tag{5.953}$$

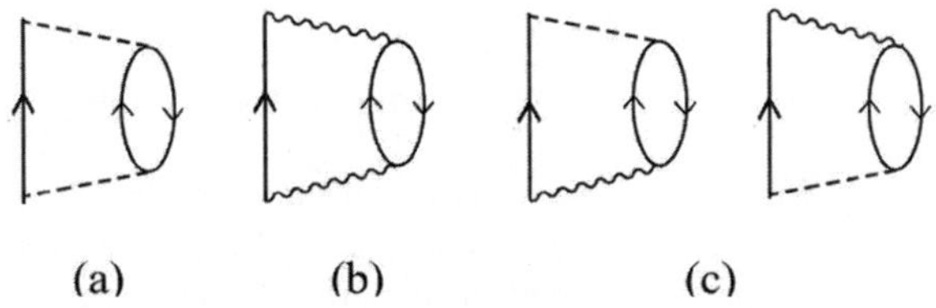

**Fig. 5.6** σ and ω mesons forth order polarization diagrams (------------ σ meson, ∼∼∼∼∼∼∼ ω meson) (**a**) exchange σ − σ mesons (**b**) exchange ω − ω mesons (**c**) exchange σ − ω mesons

**Fig. 5.7** Feynman diagram corresponding to second order propagators of σ mesons

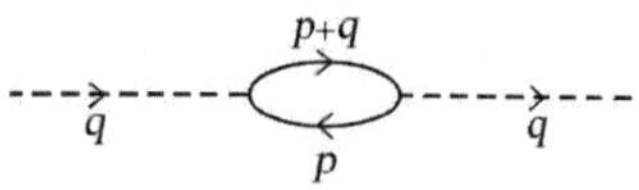

Note that $\pi_\sigma(q)$ is in the nuclear matter, so we use $G$ instead of $G^0$. Eq. (5.721) has given

$$\Delta^0(q) = \frac{-1}{q^2 - q_0^2 + m_\sigma^2} \tag{5.954}$$

Let us change the Fock term self-energy operator diagram on the right side of Fig. 5.5 to the following form:

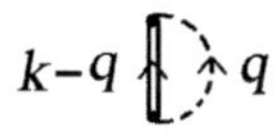


In the above figure, $\uparrow q$ represents a positive energy $\sigma$ meson. According to the second term of the first equation of Eq. (5.847), one can write the expression corresponding to the above graph as

$$\Sigma_\sigma^0(k) = \mathrm{i}g_\sigma^2 \int \frac{\mathrm{d}^4 q}{(2\pi)^4} \Delta^0(q)\, G(k-q) \tag{5.955}$$

If the second order propagator given in Fig. 5.7 is embedded in the above figure, it becomes

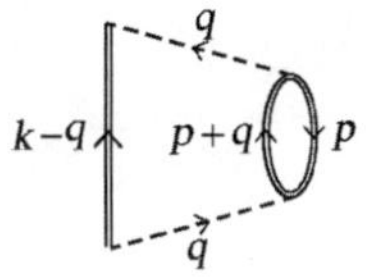


As long as the replacement $\Delta^0(q)$ with $\Delta^0(q)\,\pi_\sigma(q)\Delta^0(q)$ in Eq. (5.955) is used, the expression corresponding to the above figure can be obtained as

$$\Sigma_\sigma(k) = \mathrm{i}g_\sigma^2 \int \frac{\mathrm{d}^4 q}{(2\pi)^4} \left(\Delta^0(q)\right)^2 \pi_\sigma(q)\, G(k-q) \tag{5.956}$$

where $\pi_\sigma(q)$ has been given by Eq. (5.953).

The self-energy operator given by Eq. (5.834) is

$$\Sigma(k, E) = \Sigma^{\mathrm{s}}(k, E) + \gamma^0 \Sigma^0(k, E) + \vec{\gamma} \cdot \vec{k} \Sigma^{\mathrm{v}}(k, E) \tag{5.957}$$

And there are

$$\begin{gathered} \Sigma^{\mathrm{s}}(k, E) = V_{\mathrm{s}}(k, E) + \mathrm{i}W_{\mathrm{s}}(k, E), \quad \Sigma^0(k, E) = V_0(k, E) + \mathrm{i}W_0(k, E), \\ \Sigma^{\mathrm{v}}(k, E) = V_{\mathrm{v}}(k, E) + \mathrm{i}W_{\mathrm{v}}(k, E) \end{gathered} \tag{5.958}$$

where $k$ and $E$ are the momentum and energy of the nucleon, respectively, and in the on-shell case there is $k_0 = E$.

The following expression can be written according to Eqs. (5.843)–(5.845) as

$$\begin{aligned} G(k-q) = \left[\gamma^\mu (k^* - q)_\mu + M^*_{k-q}\right] \Bigg\{ &\frac{1}{(k^* - q)_\mu (k^* - q)^\mu - M^{*2}_{k-q} + \mathrm{i}\varepsilon} \\ &+ \frac{\mathrm{i}\pi}{E^*_{k-q}} \delta\left(k_0^* - q_0 - E^*_{k-q}\right) \theta\left(k_\mathrm{F} - \left|\vec{k} - \vec{q}\right|\right) \Bigg\} \end{aligned} \tag{5.959}$$

The practical Hartree-Fock calculation shows that the $V_\mathrm{v}$ value is very small and can be ignored. From the discussion in the 5.12 section, it can be seen that for $V_\mathrm{s}$ and $V_0$ their Hartree terms are independent with momentum. Since $V_\mathrm{s}$ and $V_0$ values are of the same order of magnitude as the nucleon stationary mass $M$, the total calculation result of Hartree-Fock term is also relatively weak in relation with momentum, so here we assume that $V_\mathrm{s}$ and $V_0$ are constants. Then when only the real part of $\Sigma$ is taken in Eq. (5.839), one can obtain

$$k_0^* = k_0 - V_0 = E - V_0, \qquad \vec{k}^* = \vec{k}, \qquad M^* = M + V_\mathrm{s} \tag{5.960}$$

Define and find

$$E^2_{k-q} \equiv \left(\vec{k} - \vec{q}\right)^2 + M^{*2} = \left(\vec{k}^* - \vec{q}\right)^2 + M^{*2} = E^{*2}_{k-q} \tag{5.961}$$

so Eq. (5.959) can be rewritten as

$$\begin{aligned} G(k-q) = \left[\gamma^\mu (k^* - q)_\mu + M^*\right] \Bigg\{ &\frac{1}{(k^* - q)_\mu (k^* - q)^\mu - M^{*2} + \mathrm{i}\varepsilon} \\ &+ \frac{\mathrm{i}\pi}{E_{k-q}} \delta\left(k_0 - V_0 - q_0 - E_{k-q}\right) \theta\left(k_\mathrm{F} - \left|\vec{k} - \vec{q}\right|\right) \Bigg\} \end{aligned} \tag{5.962}$$

Substitute Eq. (5.962) into Eq. (5.956), and we first make the integral to $q_0$. The second item in Eq. (5.962) has only imaginary part, and for the first term when using the residual theorem, only the imaginary part is retained. Firstly we analyze the following expression:

$$\begin{aligned} (k^* - q)_\mu (k^* - q)^\mu - M^{*2} + \mathrm{i}\varepsilon &= (k_0 - V_0 - q_0)^2 - \left(\vec{k} - \vec{q}\right)^2 - M^{*2} + \mathrm{i}\varepsilon \\ &= \left[q_0 - \left(k_0 - V_0 + E_{k-q}\right) + \mathrm{i}\varepsilon\right]\left[q_0 - \left(k_0 - V_0 - E_{k-q}\right) - \mathrm{i}\varepsilon\right] \end{aligned} \tag{5.963}$$

The pole in the second medium bracket [ ] part of the above equation is in the upper half plane. This pole requires

$$q_0 = k_0 - V_0 - E_{k-q} \tag{5.964}$$

This formula is the same as the requirement of the δ function in the second item of Eq. (5.962). The residual theorem can be used to obtain

$$\frac{2\pi \mathrm{i}}{k_0 - V_0 - E_{k-q} - \left(k_0 - V_0 + E_{k-q}\right)} = -\frac{\pi \mathrm{i}}{E_{k-q}}$$

And noting that $1 - \theta\left(k_{\mathrm{F}} - \left|\vec{k} - \vec{q}\right|\right) = \theta\left(\left|\vec{k} - \vec{q}\right| - k_{\mathrm{F}}\right)$, then the imaginary part of the optical potential contributed by the $\sigma$ meson can be obtained by Eq. (5.956) as

$$W_\sigma(k) = g_\sigma^2 \int \frac{\mathrm{d}\vec{q}}{(2\pi)^3} \left(\Delta^0(q)\right)^2 \frac{\gamma^\mu (k^* - q)_\mu + M^*}{2E_{k-q}} \mathrm{Im}\pi_\sigma(q) \\ \theta\left(\left|\vec{k} - \vec{q}\right| - k_{\mathrm{F}}\right)\Big|_{q_0 = k_0 - V_0 - E_{k-q}} \tag{5.965}$$

where $k_0 - V_0 - q_0 = E_{k-q}$ represents that $W_\sigma(k)$ is on-shell. Moreover, the contribution of the pole of meson propagator to the imaginary part of the optical potential is ignored in the above formula.

Select the $z$-axis along the $\vec{k}$ direction; the entire reaction system is axis-symmetric relative to the $z$-axis. The following relation can be got from Eqs. (5.964) and (5.961):

$$q_0 = E - V_0 - \left[k^2 + q^2 - 2kq\cos\theta + M^{*2}\right]^{1/2} \tag{5.966}$$

We know $\mathrm{d}\vec{q} = q^2 \mathrm{d}q \mathrm{d}\cos\theta\, \mathrm{d}\varphi$, cos$\theta$ can be replaced with $q_0$ in integral variables, so the following relations can be obtained by Eq. (5.966):

$$\mathrm{d}q_0 = \frac{kq}{E_{k-q}} \mathrm{d}\cos\theta, \qquad \mathrm{d}\cos\theta = \frac{E_{k-q}}{kq} \mathrm{d}q_0 \tag{5.967}$$

$$\mathrm{d}\vec{q} = \frac{E_{k-q}}{k} \mathrm{d}q_0 q \mathrm{d}q \mathrm{d}\varphi \tag{5.968}$$

In Eq. (5.965) there is $\int \mathrm{d}\varphi = 2\pi$. $\theta\left(\left|\vec{k} - \vec{q}\right| - k_{\mathrm{F}}\right)$ requests $\left|\vec{k} - \vec{q}\right| \geq k_{\mathrm{F}}$, and using Eq. (5.966) we can get

$$(q_0)_{\max} = E - V_0 - \left(k_F^2 + M^{*2}\right)^{1/2} \tag{5.969}$$

Define

$$\widetilde{E}_F = \left(k_F^2 + M^{*2}\right)^{1/2} + V_0 \tag{5.970}$$

then there are

$$(q_0)_{\max} = E - \widetilde{E}_F, \qquad (q_0)_{\min} = 0 \tag{5.971}$$

From Eq. (5.964) and Eq. (5.961) we can get

$$q_0 = E - V_0 - \left(\left|\vec{k} - \vec{q}\right|^2 + M^{*2}\right)^{1/2}, \quad \left|\vec{k} - \vec{q}\right| = \sqrt{(E - V_0 - q_0)^2 - M^{*2}} \equiv P \tag{5.972}$$

The condition that the second equation in the above equations has a solution is

$$(E - V_0 - q_0)^2 \geq M^{*2} \tag{5.973}$$

It can be seen from Eq. (5.969) that this condition is satisfied. Then from Eq. (5.966) we can obtain

$$\cos\theta = \frac{k^2 + q^2 + M^{*2} - (E - V_0 - q_0)^2}{2kq} \tag{5.974}$$

Since $|\cos\theta| \leq 1$ we can also obtain

$$\begin{aligned} X &\equiv \left[k^2 + q^2 + M^{*2} - (E - V_0 - q_0)^2\right]^2 - 4k^2q^2 \\ &= q^4 - 2\left[(E - V_0 - q_0)^2 + k^2 - M^{*2}\right]q^2 + \left[(E - V_0 - q_0)^2 - (k^2 + M^{*2})\right]^2 \leq 0 \end{aligned} \tag{5.975}$$

When $X = 0$ the solution of $q^2$ obtained from the above formula and Eq. (5.972) is

$$\begin{aligned} q_r^2 &\equiv (E - V_0 - q_0)^2 + k^2 - M^{*2} \pm 2k\sqrt{(E - V_0 - q_0)^2 - M^{*2}} \\ &= k^2 + P^2 \pm 2kP \end{aligned} \tag{5.976}$$

The definition of $P$ is given by Eq. (5.972). The two solutions obtained from the above formula are

$$q_{r1} = k + P, \qquad q_{r2} = |k - P| \tag{5.977}$$

Because $X \to \infty$ when $q \to \infty$, so the area satisfying $X \leq 0$ is

$$q_{r2} \leq q \leq q_{r1} \tag{5.978}$$

Then we get

$$q_{\max} = k + P, \qquad q_{\min} = |k - P| \tag{5.979}$$

According to Eq. (5.957) we can write

$$W = W^{\mathrm{s}} + \gamma^0 W^0 + \vec{\gamma} \cdot \vec{k} W^{\mathrm{v}} \tag{5.980}$$

Noting $\vec{k} \cdot \vec{k} = k^i k^i = k_i k_i = k^2$, $\vec{\gamma} \cdot \vec{k} = \gamma^i k^i$, then using Eq. (5.149) we can obtain the following results by Eq. (5.980) [81]:

$$W^{\mathrm{s}} = \frac{1}{4}\mathrm{tr}W, \quad W^0 = \frac{1}{4}\mathrm{tr}\left(\gamma^0 W\right), \quad W^{\mathrm{v}} = -\frac{1}{4k^2}\mathrm{tr}\left(\vec{\gamma} \cdot \vec{k} W\right) \tag{5.981}$$

And notice

$$\frac{1}{4}\mathrm{tr}\left[\gamma^0\gamma^0\left(k_0^* - q_0\right)\right] = E - V_0 - q_0 = E_{k-q} \tag{5.982}$$

$$\frac{1}{4}\mathrm{tr}\left\{\vec{\gamma} \cdot \vec{k}\left[\gamma^\mu (k^* - q)_\mu\right]\right\} = k^2 - \vec{k} \cdot \vec{q} \tag{5.983}$$

so from Eq. (5.966) and Eq. (5.972) we find

$$\begin{aligned} 2\vec{k} \cdot \vec{q} &= k^2 + q^2 - (E - V_0 - q_0)^2 + M^{*2} = k^2 + q^2 - P^2, \\ k^2 - \vec{k} \cdot \vec{q} &= \frac{1}{2}\left(k^2 - q^2 + P^2\right) \end{aligned} \tag{5.984}$$

Here we introduce an abbreviated symbol representing a double definite integral:

$$\oiint \equiv \int_0^{E-\tilde{E}_{\mathrm{F}}} \mathrm{d}q_0 \int_{|k-P|}^{k+P} q\mathrm{d}q \tag{5.985}$$

We can write the following expressions according to Eqs. (5.965), (5.968), (5.971), (5.979), and (5.981)–(5.984):

$$W_\sigma^{\mathrm{s}}(k, E) = \frac{g_\sigma^2 M^*}{8\pi^2 k} \oiint \left(\Delta^0(q)\right)^2 \mathrm{Im}\pi_\sigma(q) \tag{5.986}$$

$$W_{\sigma}^{0}(k,E)=\frac{g_{\sigma}^{2}}{8\pi^{2}k}\oiint\left(E-V_{0}-q_{0}\right)\left(\Delta^{0}(q)\right)^{2}\operatorname{Im}\pi_{\sigma}(q) \tag{5.987}$$

$$W_{\sigma}^{\mathrm{v}}(k,E)=-\frac{g_{\sigma}^{2}}{16\pi^{2}k^{3}}\oiint\left(k^{2}-q^{2}+P^{2}\right)\left(\Delta^{0}(q)\right)^{2}\operatorname{Im}\pi_{\sigma}(q) \tag{5.988}$$

When carrying out the practical calculation, pay attention to whether $(\Delta^0(q))^2$ has $\rightarrow\infty$ singularities.

We use the form of the Green function given by Eq. (5.774), the last item of which describes the propagation of anti-nucleons with negative energy. Due to the very high energy required to generate anti-nucleon particle-hole pair, the contribution of this item can be ignored in the energy areas interested by us. So $G(p)$ can be expressed in nuclear matter as

$$G(p)=\frac{\gamma^{\mu}\overline{p}_{\mu}^{*}+M^{*}}{2E_{p}}\left[\frac{\theta\left(|\vec{p}|-k_{\mathrm{F}}\right)}{p_{0}-V_{0}-E_{p}+\mathrm{i}\varepsilon}+\frac{\theta\left(k_{\mathrm{F}}-|\vec{p}|\right)}{p_{0}-V_{0}-E_{p}-\mathrm{i}\varepsilon}\right] \tag{5.989}$$

where

$$E_{p}^{2}=p^{2}+M^{*2} \tag{5.990}$$

We know $E_p^{*2}=p^{*2}+M^{*2}=E_p^2$, so according to Eq. (5.732) we have

$$\gamma^{\mu}\overline{p}_{\mu}^{*}=\gamma^{0}E_{p}-\vec{\gamma}\cdot\vec{p} \tag{5.991}$$

And we can write

$$G(p+q)=\frac{\gamma^{\mu}(\overline{p}^{*}+q)_{\mu}+M^{*}}{2E_{p+q}}\left[\frac{\theta\left(|\vec{p}+\vec{q}|-k_{\mathrm{F}}\right)}{p_{0}-V_{0}+q_{0}-E_{p+q}+\mathrm{i}\varepsilon}+\frac{\theta\left(k_{\mathrm{F}}-|\vec{p}+\vec{q}|\right)}{p_{0}-V_{0}+q_{0}-E_{p+q}-\mathrm{i}\varepsilon}\right] \tag{5.992}$$

where

$$E_{p+q}^{2}=\left(\vec{p}+\vec{q}\right)^{2}+M^{*2} \tag{5.993}$$

$$\gamma^{\mu}(\overline{p}^{*}+q)_{\mu}=\gamma^{0}\left(E_{p}+q_{0}\right)-\vec{\gamma}\cdot\left(\vec{p}+\vec{q}\right) \tag{5.994}$$

Substituting Eq. (5.989) and Eq. (5.992) into Eq. (5.953), and we first study the integral to $p_0$. Note that for functions with two poles, if both poles are in the upper half plane or both poles in the lower half plane, the contribution of the loop integration of the half plane without pole is zero, so we first remove these items without contribution. Then the loop integration for lower half plane or upper half plane can give

$$\begin{aligned}
\pi_\sigma(q) = & -\mathrm{i}g_\sigma^2 \int \frac{\mathrm{d}\vec{p}}{(2\pi)^3} \int \frac{\mathrm{d}p_0}{2\pi} \frac{\mathrm{tr}\left[\left(\gamma^\mu \overline{p}_\mu^* + M^*\right)\left(\gamma^\nu (\overline{p}^* + q)_\nu + M^*\right)\right]}{4E_p E_{p+q}} \\
& \left\{ \frac{\theta\left(k_\mathrm{F} - |\vec{p}|\right) \theta\left(|\vec{p} + \vec{q}| - k_\mathrm{F}\right)}{(p_0 - V_0 - E_p - \mathrm{i}\varepsilon)(p_0 - V_0 + q_0 - E_{p+q} + \mathrm{i}\varepsilon)} \right. \\
& \left. + \frac{\theta\left(|\vec{p}| - k_\mathrm{F}\right) \theta\left(k_\mathrm{F} - |\vec{p} + \vec{q}|\right)}{(p_0 - V_0 - E_p + \mathrm{i}\varepsilon)(p_0 - V_0 + q_0 - E_{p+q} - \mathrm{i}\varepsilon)} \right\} \\
= & -g_\sigma^2 \int \frac{\mathrm{d}\vec{p}}{(2\pi)^3} \frac{\mathrm{tr}\left[\left(\gamma^\mu \overline{p}_\mu^* + M^*\right)\left(\gamma^\nu (\overline{p}^* + q)_\nu + M^*\right)\right]}{4E_p E_{p+q}} \\
& \left\{ \frac{\theta\left(k_\mathrm{F} - |\vec{p}|\right) \theta\left(|\vec{p} + \vec{q}| - k_\mathrm{F}\right)}{E_{p+q} - E_p - q_0 - \mathrm{i}\varepsilon} - \frac{\theta\left(|\vec{p}| - k_\mathrm{F}\right) \theta\left(k_\mathrm{F} - |\vec{p} + \vec{q}|\right)}{E_{p+q} - E_p - q_0 + \mathrm{i}\varepsilon} \right\}
\end{aligned} \tag{5.995}$$

Select $z$-axis along $\vec{q}$ direction, $\mathrm{d}\vec{p} = p^2 \mathrm{d}p\,\mathrm{d}\cos\theta\,\mathrm{d}\varphi$, and $2\pi$ can be obtained from the integral to $\mathrm{d}\varphi$. Using Eq. (5.990) and Eq. (5.993), we can obtain

$$\begin{aligned}
\mathrm{d}E_p^2 \mathrm{d}E_{p+q}^2 &= \begin{vmatrix} \dfrac{\partial E_p^2}{\partial p} & \dfrac{\partial E_p^2}{\partial \cos\theta} \\ \dfrac{\partial E_{p+q}^2}{\partial p} & \dfrac{\partial E_{p+q}^2}{\partial \cos\theta} \end{vmatrix} \mathrm{d}p\,\mathrm{d}\cos\theta \\
&= \begin{vmatrix} 2p & 0 \\ 2p + 2q\cos\theta & 2pq \end{vmatrix} \mathrm{d}p\,\mathrm{d}\cos\theta = 4p^2 q\,\mathrm{d}p\,\mathrm{d}\cos\theta
\end{aligned}$$

So we get

$$p^2 \mathrm{d}p\,\mathrm{d}\cos\theta = \frac{E_p E_{p+q}}{q} \mathrm{d}E_p \mathrm{d}E_{p+q} \tag{5.996}$$

Using the following principal value integral formula

$$\mathrm{Im}\frac{1}{E_{p+q}-E_p-q_0\pm \mathrm{i}\varepsilon}=\mp\pi\delta\left(E_{p+q}-E_p-q_0\right) \tag{5.997}$$

from Eq. (5.996) we can obtain

$$\begin{aligned}\mathrm{Im}\pi_\sigma(q)=&-\frac{g_\sigma^2}{16\pi q}\int \mathrm{d}E_p\mathrm{d}E_{p+q}\mathrm{tr}\left[\left(\gamma^\mu\overline{p}_\mu^*+M^*\right)\left(\gamma^\nu(\overline{p}^*+q)_\nu+M^*\right)\right]\\&\left[\theta\left(k_\mathrm{F}-|\vec{p}|\right)\theta\left(|\vec{p}+\vec{q}|-k_\mathrm{F}\right)+\theta\left(|\vec{p}|-k_\mathrm{F}\right)\theta\left(k_\mathrm{F}-|\vec{p}+\vec{q}|\right)\right]\\&\delta\left(E_{p+q}-E_p-q_0\right)\end{aligned} \tag{5.998}$$

And there is

$$\begin{aligned}\mathrm{tr}\left[\left(\gamma^\mu\overline{p}_\mu^*+M^*\right)\left(\gamma^\nu(\overline{p}^*+q)_\nu+M^*\right)\right]&=4\lambda\left(E_p^2+E_pq_0-p^2-\vec{p}\cdot\vec{q}+M^{*2}\right)\\&=4\lambda\left(E_pq_0-\vec{p}\cdot\vec{q}+2M^{*2}\right)\end{aligned} \tag{5.999}$$

where $\lambda$ is a nucleon isospin degeneracy. From $\delta(E_{p+q}-E_p-q_0)$ we can obtain

$$\begin{aligned}p^2+q^2+2\vec{p}\cdot\vec{q}+M^{*2}&=p^2+M^{*2}+q_0^2+2E_pq_0,\\E_pq_0-\vec{p}\cdot\vec{q}&=-\frac{1}{2}q_\mu q^\mu\end{aligned} \tag{5.1000}$$

and then get

$$\mathrm{tr}\left[\left(\gamma^\mu\overline{p}_\mu^*+M^*\right)\left(\gamma^\nu(\overline{p}^*+q)_\nu+M^*\right)\right]=8\lambda\left(M^{*2}-\frac{1}{4}q_\mu q^\mu\right) \tag{5.1001}$$

So Eq. (5.998) can be rewritten as

$$\begin{aligned}\mathrm{Im}\pi_\sigma(q)=&-\frac{\lambda g_\sigma^2}{2\pi q}\left(M^{*2}-\frac{1}{4}q_\mu q^\mu\right)\int \mathrm{d}E_p\mathrm{d}E_{p+q}\left[\theta\left(k_\mathrm{F}-|\vec{p}|\right)\theta\left(|\vec{p}+\vec{q}|-k_\mathrm{F}\right)\right.\\&\left.+\theta\left(|\vec{p}|-k_\mathrm{F}\right)\theta\left(k_\mathrm{F}-|\vec{p}+\vec{q}|\right)\right]\delta\left(E_{p+q}-E_p-q_0\right)\end{aligned} \tag{5.1002}$$

Since

$$E_{p+q}^2 = p^2 + q^2 + 2pq\cos\theta + M^{*2} = E_p^2 + q^2 + 2pq\cos\theta$$

so there is the following requirement:

$$|\cos\theta| = \left|\frac{E_{p+q}^2 - E_p^2 - q^2}{2pq}\right| \le 1 \tag{5.1003}$$

In Eq. (5.1002) we can use the δ function to finish the integral to $dE_{p+q}$, but a ladder function θ(*X*) must be multiplied, where

$$\begin{aligned} X &\equiv 4p^2q^2 - \left(E_{p+q}^2 - E_p^2 - q^2\right)^2 \\ &= 4\left(E_p^2 - M^{*2}\right)q^2 - \left(E_p^2 + q_0^2 + 2E_pq_0 - E_p^2 - q^2\right)^2 \\ &= 4E_p^2q^2 - 4M^{*2}q^2 - 4E_p^2q_0^2 - \left(q_\mu q^\mu\right)^2 - 4E_pq_0q_\mu q^\mu \\ &= -4E_p^2q_\mu q^\mu - 4E_pq_0q_\mu q^\mu - 4M^{*2}q^2 - \left(q_\mu q^\mu\right)^2 \end{aligned} \tag{5.1004}$$

Let $X = 0$, then from the above formula we can obtain

$$q_\mu q^\mu E_p^2 + q_0q_\mu q^\mu E_p + M^{*2}q^2 + \frac{1}{4}\left(q_\mu q^\mu\right)^2 = 0$$

When $q_\mu q^\mu \neq 0$ the above formula can be rewritten as

$$E_p^2 + q_0E_p + \frac{M^{*2}q^2}{q_\mu q^\mu} + \frac{1}{4}q_\mu q^\mu = 0 \tag{5.1005}$$

The solution of $E_p$ of the above equation, when taking the positive sign for its second term, is

$$E_r = -\frac{1}{2}q_0 + \frac{1}{2}q\sqrt{1 - 4M^{*2}/\left(q_\mu q^\mu\right)} \tag{5.1006}$$

So θ(*X*) can be replaced with θ($E_p - E_r$). Define

$$E_{\mathrm{F}} = \left(k_{\mathrm{F}}^2 + M^{*2}\right)^{1/2} \tag{5.1007}$$

and use the on-shell relation of the $\vec{p}$ particles and $\vec{p} + \vec{q}$ particles, then Eq. (5.1002) can be rewritten as

$$
\begin{aligned}
\mathrm{Im}\pi_\sigma(q) = -\frac{\lambda g_\sigma^2}{2\pi q}\left(M^{*2} - \frac{1}{4}q_\mu q^\mu\right)\int \mathrm{d}E_p \theta(E_p - E_r) \\
\left[\theta(E_\mathrm{F} - E_p)\,\theta(E_p + q_0 - E_\mathrm{F}) + \theta(E_p - E_\mathrm{F})\,\theta(E_\mathrm{F} - E_p - q_0)\right]
\end{aligned}
\tag{5.1008}
$$

In Eq. (5.1008), both items in the above formula all require $E_p \geq E_r$. The first item in the medium bracket also requires $E_p \leq E_\mathrm{F}$, $E_p \geq E_\mathrm{F} - q_0$, $q_0 \geq E_\mathrm{F} - E_p \geq 0$, and it means that the meson belongs to a positive energy one; the second item also requires $E_p \geq E_\mathrm{F}$, $E_p \leq E_\mathrm{F} - q_0$, $q_0 \leq E_\mathrm{F} - E_p \leq 0$, and it means that the meson belongs to a negative energy one. As can be seen from Eq. (5.971), the second item should not be considered here. Then from Eq. (5.1008) we obtain

$$
\mathrm{Im}\pi_\sigma(q) = -\frac{\lambda g_\sigma^2}{2\pi q}\left(M^{*2} - \frac{1}{4}q_\mu q^\mu\right)(E_\mathrm{F} - E_x) \tag{5.1009}
$$

$$
E_x = \min\left(E_\mathrm{F}, E_{\max}\right), \quad E_{\max} = \max\left(E_r, E_\mathrm{F} - q_0, M^*\right) \tag{5.1010}
$$

### 5.13.2 Contribution of $\omega - \omega$ Meson Exchange Processes to the Imaginary Part of the Optical Potential

In Fig. 5.7, ω mesons are used instead of σ mesons. The Feynman rules can be used to write the corresponding expression as

$$
\begin{aligned}
\mathrm{i}D_{\mu\nu}^{(2)}(q) &= \mathrm{i}D_{\mu\alpha}^0(q)(-\mathrm{i}g_\omega)\int \frac{\mathrm{d}^4p}{(2\pi)^4}\mathrm{tr}\left[-\gamma^\alpha \mathrm{i}G(p)\gamma^\beta \mathrm{i}G(p+q)\right](-\mathrm{i}g_\omega)\,\mathrm{i}D_{\beta\nu}^0(q) \\
&\equiv \mathrm{i}D_{\mu\alpha}^0(q)\pi_\omega^{\alpha\beta}(q)D_{\beta\nu}^0(q)
\end{aligned}
\tag{5.1011}
$$

and get

$$
\pi_\omega^{\alpha\beta}(q) = -\mathrm{i}g_\omega^2\int \frac{\mathrm{d}^4p}{(2\pi)^4}\mathrm{tr}\left[\gamma^\alpha G(p)\gamma^\beta G(p+q)\right] \tag{5.1012}
$$

When q is exchanged with $k - q$, the second item of the first equation of Eq. (5.863) becomes

$$
\Sigma_\omega^0(k) = \mathrm{i}g_\omega^2\gamma^\mu\int \frac{\mathrm{d}^4q}{(2\pi)^4}D_{\mu\nu}^0(q)\,G(k-q)\,\gamma^\nu \tag{5.1013}
$$

When $D_{\mu\nu}^0(q)$ is replaced with $D_{\mu\alpha}^0(q)\pi_\omega^{\alpha\beta}(q)D_{\beta\nu}^0(q)$, the above formula becomes

$$\Sigma_\omega(k) = \mathrm{i}g_\omega^2\gamma^\mu \int \frac{\mathrm{d}^4q}{(2\pi)^4} D_{\mu\alpha}^0(q)\pi_\omega^{\alpha\beta}(q)D_{\beta\nu}^0(q)\, G(k-q)\gamma^\nu \tag{5.1014}$$

Equation (5.812) has given

$$D_{\mu\nu}^0(q) = \frac{-g_{\mu\nu}}{q_\lambda q^\lambda - m_\omega^2} = D^0(q)g_{\mu\nu} \tag{5.1015}$$

where

$$D^0(q) = \frac{1}{q^2 - q_0^2 + m_\omega^2} \tag{5.1016}$$

So Eq. (5.1014) can be rewritten as

$$\Sigma_\omega(k) = \mathrm{i}g_\omega^2\gamma^\mu \int \frac{\mathrm{d}^4q}{(2\pi)^4} \left(D^0(q)\right)^2 g_{\mu\mu}\pi_\omega^{\mu\nu}(q)g_{\nu\nu}G(k-q)\,\gamma^\nu \tag{5.1017}$$

Caring out the derivation similar to σ mesons, the following expressions similar to Eq. (5.965) can be obtained as

$$\begin{aligned} W_\omega(k) = g_\omega^2 \int \frac{\mathrm{d}\vec{q}}{(2\pi)^3} \left(D^0(q)\right)^2 \gamma^\mu \frac{\gamma^\lambda(k^* - q)_\lambda + M^*}{2E_{k-q}} \\ g_{\mu\mu}\ \mathrm{Im}\pi_\omega^{\mu\nu}(q)g_{\nu\nu}\gamma^\nu\, \theta\left(\left|\vec{k} - \vec{q}\right| - k_\mathrm{F}\right)\Big|_{q_0 = k_0 - V_0 - E_{k-q}} \end{aligned} \tag{5.1018}$$

Utilizing the analyses by Eqs (5.965)–(5.979) and using Eqs. (5.149), (5.150), and (5.981)–(5.984), the following results can be obtained from Eq. (5.1018):

$$W_\omega^\mathrm{s}(k,E) = \frac{g_\omega^2 M^*}{8\pi^2 k} \oiint \left(D^0(q)\right)^2 g_{\mu\mu}\mathrm{Im}\pi_\omega^{\mu\mu}(q) \tag{5.1019}$$

$$\begin{aligned} W_\omega^0(k,E) = -\frac{g_\omega^2}{8\pi^2 k} \oiint \left(D^0(q)\right)^2 \Big[(E - V_0 - q_0)g_{\mu\mu}\mathrm{Im}\pi_\omega^{\mu\mu}(q) \\ -\mathrm{Im}\pi_\omega^{0\nu}(q)(k^* - q)_\nu - (k^* - q)_\mu \mathrm{Im}\pi_\omega^{\mu 0}(q)\Big] \end{aligned} \tag{5.1020}$$

$$\begin{aligned} W_\omega^\mathrm{v}(k,E) = \frac{g_\omega^2}{8\pi^2 k^3} \oiint \left(D^0(q)\right)^2 \Big[\frac{1}{2}\left(k^2 - q^2 + P^2\right) g_{\mu\mu}\mathrm{Im}\pi_\omega^{\mu\mu}(q) \\ -k^i\mathrm{Im}\pi_\omega^{i\nu}(q)\,(k^* - q)_\nu - k^i(k^* - q)_\mu\mathrm{Im}\pi_\omega^{\mu i}(q)\Big] \end{aligned} \tag{5.1021}$$

In the above three formulas, there are the summations to $\mu$, $\nu$, $i$.

Comparing Eq. (5.953) and Eq. (5.1012), referring to Eq. (5.998), we can write

$$\begin{aligned}\mathrm{Im}\pi_{\omega}^{\mu\nu}(q) = &-\frac{g_{\omega}^{2}}{16\pi q}\int \mathrm{d}E_{p}\mathrm{d}E_{p+q}\mathrm{tr}\left[\gamma^{\mu}(\gamma^{\lambda}\overline{p}_{\lambda}^{*}+M^{*})\gamma^{\nu}\left(\gamma^{\eta}(\overline{p}^{*}+q)_{\eta}+M^{*}\right)\right]\\ &\left[\theta\left(k_{\mathrm{F}}-|\vec{p}|\right)\theta\left(|\vec{p}+\vec{q}|-k_{\mathrm{F}}\right)\right.\\ &\left.+\theta\left(|\vec{p}|-k_{\mathrm{F}}\right)\theta\left(k_{\mathrm{F}}-|\vec{p}+\vec{q}|\right)\right]\delta(E_{p+q}-E_{p}-q_{0})\end{aligned} \tag{5.1022}$$

Using Eq. (5.149) and Eq. (5.150) and taking into account the nucleon isospin degeneracy $\lambda$, we can obtain

$$\begin{aligned}t_{\omega}^{\mu\nu} &\equiv \mathrm{tr}\left[\gamma^{\mu}(\gamma^{\lambda}\overline{p}_{\lambda}^{*}+M^{*})\gamma^{\nu}\left(\gamma^{\eta}(\overline{p}^{*}+q)_{\eta}+M^{*}\right)\right]\\ &= 4\lambda\left[\left(M^{*2}-\overline{p}_{\eta}^{*}(\overline{p}^{*}+q)^{\eta}\right)g^{\mu\nu}+g^{\mu\mu}\overline{p}_{\mu}^{*}(\overline{p}^{*}+q)^{\nu}+g^{\nu\nu}\overline{p}_{\nu}^{*}(\overline{p}^{*}+q)^{\mu}\right]\end{aligned} \tag{5.1023}$$

And utilizing Eq. (5.1000) we can also obtain

$$\overline{p}_{\eta}^{*}(\overline{p}^{*}+q)^{\eta} = E_{p}^{2}+E_{p}q_{0}-p^{2}-\vec{p}\cdot\vec{q} = M^{*2}-\frac{1}{2}q_{\eta}q^{\eta} \tag{5.1024}$$

Using the analyses from Eq. (5.1002) to Eq. (5.1010), the following expression containing summation to $\mu$ can be obtained from Eq. (5.1022)–(5.1024):

$$\mathrm{Im}\pi_{\omega}^{\Lambda}(q) \equiv \frac{1}{4}\sum_{\mu}g_{\mu\mu}\mathrm{Im}\pi_{\omega}^{\mu\mu}(q) = -\frac{\lambda g_{\omega}^{2}}{2\pi q}\left(M^{*2}-\frac{1}{4}q_{\eta}q^{\eta}\right)(E_{\mathrm{F}}-E_{x}) \tag{5.1025}$$

Utilizing Eq. (5.1023) and Eq. (5.1024), we can get

$$t_{\omega}^{00} = 8\lambda\left(\frac{1}{4}q_{\eta}q^{\eta}+q_{0}E_{p}+E_{p}^{2}\right) \tag{5.1026}$$

$$t_{\omega}^{0i} = t_{\omega}^{i0} = 4\lambda\left[E_{p}(p+q)^{i}+p^{i}(E_{p}+q_{0})\right] = 4\lambda\left[(q_{0}+2E_{p})p^{i}+E_{p}q^{i}\right] \tag{5.1027}$$

$$t_{\omega}^{ij} = 4\lambda\left(-\frac{1}{2}q_{\eta}q^{\eta}\delta_{ij}+p^{i}(p+q)^{j}+p^{j}(p+q)^{i}\right) \tag{5.1028}$$

Using Eq. (5.1026), from Eq. (5.1022) we can obtain

$$\mathrm{Im}\pi_{\omega}^{00}(q) = -\frac{\lambda g_{\omega}^{2}}{2\pi q}\left[\frac{1}{4}q_{\eta}q^{\eta}(E_{\mathrm{F}}-E_{x})+\frac{1}{2}q_{0}(E_{\mathrm{F}}^{2}-E_{x}^{2})+\frac{1}{3}(E_{\mathrm{F}}^{3}-E_{x}^{3})\right] \tag{5.1029}$$

By using Eq. (5.984) and Eq. (5.1000), the following expressions can be obtained, respectively:

$$\vec{k} \cdot \vec{q} = \frac{1}{2}\left(k^2 + q^2 - P^2\right) \tag{5.1030}$$

$$\vec{p} \cdot \vec{q} = \frac{1}{2} q_\eta q^\eta + q_0 E_p \tag{5.1031}$$

Previously $\vec{k}$ has been selected as the $z$-axis; here let $(\theta_q, \varphi_q)$ and $(\theta_p, \varphi_p)$ be the polar and azimuthal angles of $\vec{q}$ and $\vec{p}$, respectively, and $\theta$ is the angle between $\vec{q}$ and $\vec{p}$. When $\vec{k} \cdot \vec{p}$ appears only in a linear form, since $\int \cos(\varphi_p - \varphi_q)\, \mathrm{d}\varphi_p = 0$, using the additive angle formula, we can obtain

$$k^2 \vec{p} \cdot \vec{q} = \vec{k} \cdot \vec{q}\, \vec{k} \cdot \vec{p} \tag{5.1032}$$

$$\vec{k} \cdot \vec{p} = 2k^2 \left(\frac{1}{2} q_\eta q^\eta + q_0 E_p\right) / \left(k^2 + q^2 - P^2\right) \tag{5.1033}$$

And the following expressions can be obtained using Eqs. (5.1027), (5.1028), (5.1030)–(5.1033):

$$\begin{aligned}
k^i t_\omega^{0i} &= k^i t_\omega^{i0} = 4\lambda\left[\left(q_0 + 2E_p\right) \vec{k} \cdot \vec{p} + E_p \vec{k} \cdot \vec{q}\right] \\
&= \frac{2\lambda}{k^2 + q^2 - P^2}\left[\left(q_0 + 2E_p\right) 4k^2 \left(\frac{1}{2} q_0^2 - \frac{1}{2} q^2 + q_0 E_p\right) + \left(k^2 + q^2 - P^2\right)^2 E_p\right] \\
&= \frac{2\lambda}{k^2 + q^2 - P^2}\left\{4k^2\ \frac{1}{2} q_0 q_\eta q^\eta + \left(2q_0^2 - q^2\right) E_p + 2q_0 E_p^2\right] + \left(k^2 + q^2 - P^2\right)^2 E_p\right\} \\
&= \frac{2\lambda}{k^2 + q^2 - P^2}\left\{2k^2 q_0 q_\eta q^\eta + \left[4k^2\left(2q_0^2 - q^2\right) + \left(k^2 + q^2 - P^2\right)^2\right] E_p + 8k^2 q_0 E_p^2\right\}
\end{aligned} \tag{5.1034}$$

$$\begin{aligned}
(k - q)^i t_\omega^{0i} &= (k - q)^i t_\omega^{i0} = 4\lambda\left[\left(q_0 + 2E_p\right)\left(\vec{k} \cdot \vec{p} - \vec{p} \cdot \vec{q}\right) + \left(\vec{k} \cdot \vec{p} - q^2\right) E_p\right] \\
&= \frac{2\lambda}{k^2 + q^2 - P^2}\left\{\left(q_0 + 2E_p\right)\left[4k^2 - 2\left(k^2 + q^2 - P^2\right)\right]\left(\frac{1}{2} q_\eta q^\eta + q_0 E_p\right)\right. \\
&\quad \left. + \left(k^2 + q^2 - P^2\right)\left(k^2 + q^2 - P^2 - 2q^2\right) E_p\right\} \\
&= \frac{2\lambda}{k^2 + q^2 - P^2}\left\{2\left(k^2 - q^2 + P^2\right)\left(q_0 + 2E_p\right)\left(\frac{1}{2} q_0^2 - \frac{1}{2} q^2 + q_0 E_p\right)\right. \\
&\quad \left. + \left[\left(k^2 - P^2\right)^2 - q^4\right] E_p\right\} \\
&= \frac{2\lambda}{k^2 + q^2 - P^2}\left\{\left(k^2 - q^2 + P^2\right) q_0 q_\eta q^\eta + \left[2\left(k^2 - q^2 + P^2\right)\left(2q_0^2 - q^2\right)\right.\right. \\
&\quad \left.\left. + \left(\left(k^2 - P^2\right)^2 - q^4\right)\right] E_p + 4\left(k^2 - q^2 + P^2\right) q_0 E_p^2\right\}
\end{aligned} \tag{5.1035}$$

First we find

$$\left(\frac{1}{2}q_\eta q^\eta + q_0 E_p\right)^2 = \frac{1}{4}\left(q_\eta q^\eta\right)^2 + q_0 q_\eta q^\eta E_p + q_0^2 E_p^2 \tag{5.1036}$$

Further we can find

$$\begin{aligned}
k^i(k-q)^j t^{ij}_\omega &= 4\lambda\Bigg[-\frac{1}{2}q_\eta q^\eta\left(k^2 - \vec{k}\cdot\vec{q}\right) + \vec{k}\cdot\vec{p}\left(\vec{k}\cdot\vec{p} + \vec{k}\cdot\vec{q} - \vec{p}\cdot\vec{q} - q^2\right) \\
&\quad + \left(\vec{k}\cdot\vec{p} + \vec{k}\cdot\vec{q}\right)\left(\vec{k}\cdot\vec{p} - \vec{p}\cdot\vec{q}\right)\Bigg] \\
&= 4\lambda\Bigg[-\frac{1}{2}q_\eta q^\eta\left(k^2 - \vec{k}\cdot\vec{q}\right) + 2\left(\vec{k}\cdot\vec{p}\right)\left(\vec{k}\cdot\vec{p} + \vec{k}\cdot\vec{q} - \vec{p}\cdot\vec{q}\right) \\
&\quad - \left(q^2\vec{k}\cdot\vec{p} + \vec{k}\cdot\vec{q}\vec{p}\cdot\vec{q}\right)\Bigg] \\
&= \frac{4\lambda}{\left(k^2+q^2-P^2\right)^2}\Bigg\{-\frac{1}{4}q_\eta q^\eta\left(k^2 - q^2 + P^2\right)\left(k^2 + q^2 - P^2\right)^2 \\
&\quad + 8k^4\left(\frac{1}{2}q_\eta q^\eta + q_0 E_p\right)^2 + 2k^2\left(k^2+q^2-P^2\right)^2\left(\frac{1}{2}q_\eta q^\eta + q_0 E_p\right) \\
&\quad - 4k^2\left(\frac{1}{2}q_\eta q^\eta + q_0 E_p\right)^2\left(k^2+q^2-P^2\right) - 2k^2q^2\left(\frac{1}{2}q_\eta q^\eta + q_0 E_p\right) \\
&\quad \left(k^2+q^2-P^2\right) - \frac{1}{2}\left(k^2+q^2-P^2\right)^3\left(\frac{1}{2}q_\eta q^\eta + q_0 E_p\right)\Bigg\} \\
&\equiv \frac{4\lambda}{\left(k^2+q^2-P^2\right)^2}\left[A_0(k,q) + A_1(k,q)E_p + A_2(k,q)E_p^2\right]
\end{aligned} \tag{5.1037}$$

From the above formula we can obtain

$$\begin{aligned}
A_0(k,q) &= -\frac{1}{4}q_\eta q^\eta\left(k^2-q^2+P^2\right)\left(k^2+q^2-P^2\right)^2 + 2k^4\left(q_\eta q^\eta\right)^2 \\
&\quad + k^2 q_\eta q^\eta\left(k^2+q^2-P^2\right)^2 - k^2\left(q_\eta q^\eta\right)^2\left(k^2+q^2-P^2\right) \\
&\quad - k^2q^2q_\eta q^\eta\left(k^2+q^2-P^2\right) - \frac{1}{4}q_\eta q^\eta\left(k^2+q^2-P^2\right)^3 \\
&= k^2\left(q_\eta q^\eta\right)^2\left(k^2-q^2+P^2\right) \\
&\quad + q_\eta q^\eta\left(k^2+q^2-P^2\right)\Big[-\frac{1}{4}\left(k^2-q^2+P^2\right)\left(k^2+q^2-P^2\right) \\
&\quad + k^2\left(k^2+q^2-P^2\right) - k^2q^2 - \frac{1}{4}\left(k^2+q^2-P^2\right)^2\Big] \\
&= k^2\left(q_\eta q^\eta\right)^2\left(k^2-q^2+P^2\right) + \frac{1}{2}k^2q_\eta q^\eta\left(k^2+q^2-P^2\right)\left(k^2-q^2-P^2\right) \\
&= \frac{1}{2}k^2q_\eta q^\eta\left[2q_\eta q^\eta\left(k^2-q^2+P^2\right) + \left(\left(k^2-P^2\right)^2 - q^4\right)\right]
\end{aligned} \tag{5.1038}$$

$$\begin{aligned}A_1(k,q)&=8k^4q_0q_\eta q^\eta+2k^2\left(k^2+q^2-P^2\right)^2q_0-4k^2\left(k^2+q^2-P^2\right)q_0q_\eta q^\eta\\&\quad-2k^2\left(k^2+q^2-P^2\right)q_0q^2-\frac{1}{2}q_0\left(k^2+q^2-P^2\right)^3\\&=q_0\Big[8k^4q_\eta q^\eta-2k^2\left(k^2+q^2-P^2\right)\left(2q_0^2-q^2\right)\\&\quad+\frac{1}{2}\left(k^2+q^2-P^2\right)^2\left(3k^2-q^2+P^2\right)\Big]\end{aligned}\tag{5.1039}$$

$$A_2(k,q)=8k^4q_0^2-4k^2q_0^2\left(k^2+q^2-P^2\right)=4k^2q_0^2\left(k^2-q^2+P^2\right)\tag{5.1040}$$

So the following results can be obtained using the similar method as before:

$$\begin{aligned}\mathrm{Im}\ \pi_\omega^\alpha(q)&\equiv(k-q)^i\mathrm{Im}\ \pi_\omega^{0i}(q)=(k-q)^i\mathrm{Im}\ \pi_\omega^{i0}(q)\\&=-\frac{\lambda g_\omega^2}{8\pi q\left(k^2+q^2-P^2\right)}\Big\{\left(k^2-q^2+P^2\right)q_0q_\eta q^\eta(E_\mathrm{F}-E_x)\\&\quad+\frac{1}{2}\left[2\left(k^2-q^2+P^2\right)\left(2q_0^2-q^2\right)+\left(\left(k^2-P^2\right)^2-q^4\right)\right]\left(E_\mathrm{F}^2-E_x^2\right)\\&\quad+\frac{4}{3}\left(k^2-q^2+P^2\right)q_0\left(E_\mathrm{F}^3-E_x^3\right)\Big\}\end{aligned}\tag{5.1041}$$

$$\begin{aligned}\mathrm{Im}\pi_\omega^\beta(q)&\equiv k^i\mathrm{Im}\pi_\omega^{0i}(q)=k^i\mathrm{Im}\pi_\omega^{i0}(q)\\&=-\frac{\lambda g_\omega^2}{8\pi q\left(k^2+q^2-P^2\right)}\Big\{2k^2q_0q_\eta q^\eta(E_\mathrm{F}-E_x)\\&\quad+\frac{1}{2}\left[4k^2\left(2q_0^2-q^2\right)+\left(k^2+q^2-P^2\right)^2\right]\left(E_\mathrm{F}^2-E_x^2\right)+\frac{8}{3}k^2q_0\left(E_\mathrm{F}^3-E_x^3\right)\Big\}\end{aligned}\tag{5.1042}$$

$$\begin{aligned}\mathrm{Im}\pi_\omega^\gamma(q)&\equiv k^i(k-q)^j\mathrm{Im}\pi_\omega^{ij}(q)=k^i(k-q)^j\ \mathrm{Im}\pi_\omega^{ji}(q)\\&=-\frac{\lambda g_\omega^2}{4\pi q\left(k^2+q^2-P^2\right)^2}\Big\{A_0(k,q)(E_\mathrm{F}-E_x)\\&\quad+\frac{1}{2}A_1(k,\ q)\left(E_\mathrm{F}^2-E_x^2\right)+\frac{1}{3}A_3(k,\ q)\left(E_\mathrm{F}^3-E_x^3\right)\Big\}\end{aligned}\tag{5.1043}$$

Equations (5.1019)–(5.1021) can be rewritten, using the above results, as

$$W_\omega^\mathrm{s}(k,E)=\frac{g_\omega^2M^*}{2\pi^2k}\oiint\left(D^0(q)\right)^2\mathrm{Im}\pi_\omega^\Lambda(q)\tag{5.1044}$$

$$\begin{aligned}W_\omega^0(k,E)=-\frac{g_\omega^2}{4\pi^2k}\oiint\left(D^0(q)\right)^2\big[&2(E-V_0-q_0)\mathrm{Im}\pi_\omega^\Lambda(q)\\&-\mathrm{Im}\pi_\omega^{00}(q)+\mathrm{Im}\pi_\omega^\alpha(q)\big]\end{aligned}\tag{5.1045}$$

$$W^{\mathrm{v}}_{\omega}(k,E)=\frac{g_{\omega}^{2}}{4\pi^{2}k^{3}}\oiint\left(D^{0}(q)\right)^{2}\left[\left(k^{2}-q^{2}+P^{2}\right)\mathrm{Im}\pi_{\omega}^{\Lambda}(q)\right.$$
$$\left.-(E-V_{0}-q_{0})\mathrm{Im}\pi_{\omega}^{\beta}(q)+\mathrm{Im}\pi_{\omega}^{\gamma}(q)\right] \tag{5.1046}$$

## 5.13.3 *Contribution of σ − ω Meson Exchange Processes to the Imaginary Part of the Optical potential*

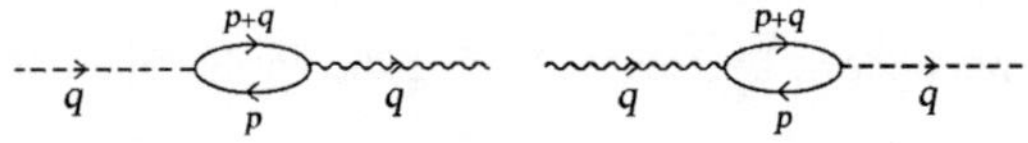


We consider the two second order propagators shown in the above figures at the same time. Using Eq. (5.1015) and Eq. (5.1016), the expressions corresponding to the above figures can be obtained as

$$\begin{aligned}\mathrm{i}D_{\mu}^{(2)}(q)&=\mathrm{i}D_{\mu\alpha}^{0}(q)(-\mathrm{i}g_{\omega})\int\frac{\mathrm{d}^{4}p}{(2\pi)^{4}}\mathrm{tr}\left[-\gamma^{\alpha}\mathrm{i}G(p)\,\mathrm{i}G(p+q)\right](-\mathrm{i}g_{\sigma})\,\mathrm{i}\Delta^{0}(q)\\&\quad+\mathrm{i}\Delta^{0}(q)(-\mathrm{i}g_{\sigma})\int\frac{\mathrm{d}^{4}p}{(2\pi)^{4}}\mathrm{tr}\left[-\mathrm{i}G(p)\gamma^{\alpha}\mathrm{i}G(p+q)\right](-\mathrm{i}g_{\omega})\,\mathrm{i}D_{\alpha\mu}^{0}(q)\\&\equiv\mathrm{i}D^{0}(q)\pi_{\mu}^{\omega\sigma}(q)\Delta^{0}(q)+\mathrm{i}\Delta^{0}(q)\pi_{\mu}^{\sigma\omega}(q)D^{0}(q)\end{aligned} \tag{5.1047}$$

where

$$\begin{aligned}\pi_{\mu}^{\omega\sigma}(q)&=-\mathrm{i}g_{\sigma}g_{\omega}g_{\mu\mu}\int\frac{\mathrm{d}^{4}p}{(2\pi)^{4}}\mathrm{tr}\left[\gamma^{\mu}G(p)G(p+q)\right],\\\pi_{\mu}^{\sigma\omega}(q)&=-\mathrm{i}g_{\sigma}g_{\omega}g_{\mu\mu}\int\frac{\mathrm{d}^{4}p}{(2\pi)^{4}}\mathrm{tr}\left[G(p)\gamma^{\mu}G(p+q)\right]\end{aligned} \tag{5.1048}$$

First using Eq. (5.149) we can obtain

$$\begin{aligned}&g_{\mu\mu}\mathrm{tr}\left\{\gamma^{\mu}\left(\gamma^{\lambda}\overline{p}_{\lambda}^{*}+M^{*}\right)\left[\gamma^{\eta}\left(\overline{p}^{*}+q\right)_{\eta}+M^{*}\right]\right\}=4\lambda M^{*}\left(2\overline{p}_{\mu}^{*}+q_{\mu}\right),\\&g_{\mu\mu}\mathrm{tr}\left\{\left(\gamma^{\lambda}\overline{p}_{\lambda}^{*}+M^{*}\right)\gamma^{\mu}\left[\gamma^{\eta}\left(\overline{p}^{*}+q\right)_{\eta}+M^{*}\right]\right\}=4\lambda M^{*}\left(2\overline{p}_{\mu}^{*}+q_{\mu}\right)\end{aligned} \tag{5.1049}$$

So according to Eq. (5.1048), Eq. (5.989), and Eq. (5.149) and that $g^{\mu\nu}$ is not equal to 0 only when $\mu=\nu$, we can get

$$\pi_\mu^{\omega\sigma}(q) = \pi_\mu^{\sigma\omega}(q) \equiv \frac{1}{2}\pi_\mu^{\mathrm{M}}(q) \tag{5.1050}$$

Referring to Eq. (5.1017) we can write

$$\begin{aligned}
\Sigma_{\omega\sigma}(k) &= \mathrm{i}g_\sigma g_\omega \int \frac{\mathrm{d}^4q}{(2\pi)^4}\Delta^0(q)D^0(q)\gamma^\mu \frac{1}{2}\pi_\mu^{\mathrm{M}}(q)G(k-q), \\
\Sigma_{\sigma\omega}(k) &= \mathrm{i}g_\sigma g_\omega \int \frac{\mathrm{d}^4q}{(2\pi)^4}\Delta^0(q)D^0(q)\frac{1}{2}\pi_\mu^{\mathrm{M}}(q)G(k-q)\gamma^\mu
\end{aligned} \tag{5.1051}$$

The following expressions similar to Eq. (5.965) or Eq. (5.1018) can be obtained from Eq. (5.1051) as

$$\begin{aligned}
W_{\omega\sigma}(k) &= g_\sigma g_\omega \int \frac{\mathrm{d}\vec{q}}{(2\pi)^3}\Delta^0(q)D^0(q)\gamma^\mu \frac{\gamma^\lambda(k^*-q)_\lambda + M^*}{2E_{k-q}} \\
&\quad \frac{1}{2}\mathrm{Im}\pi_\mu^{\mathrm{M}}(q)\,\theta\left(\left|\vec{k}-\vec{q}\right| - k_{\mathrm{F}}\right)\Big|_{q_0=k_0-V_0-E_{k-q}} \\
W_{\sigma\omega}(k) &= g_\sigma g_\omega \int \frac{\mathrm{d}\vec{q}}{(2\pi)^3}\Delta^0(q)D^0(q)\frac{\gamma^\lambda(k^*-q)_\lambda + M^*}{2E_{k-q}}\gamma^\mu \\
&\quad \frac{1}{2}\mathrm{Im}\pi_\mu^{\mathrm{M}}(q)\,\theta\left(\left|\vec{k}-\vec{q}\right| - k_{\mathrm{F}}\right)\Big|_{q_0=k_0-V_0-E_{k-q}}
\end{aligned} \tag{5.1052}$$

Using Eq. (5.149) we can prove to have the following relations:

$$\begin{aligned}
\mathrm{tr}\left\{\gamma^\mu\left[\gamma^\lambda(k^*-q)_\lambda + M^*\right]\right\} &= \mathrm{tr}\left\{\left[\gamma^\lambda(k^*-q)_\lambda + M^*\right]\gamma^\mu\right\}, \\
\mathrm{tr}\left\{\gamma^0\gamma^\mu\left[\gamma^\lambda(k^*-q)_\lambda + M^*\right]\right\} &= \mathrm{tr}\left\{\gamma^0\left[\gamma^\lambda(k^*-q)_\lambda + M^*\right]\gamma^\mu\right\}, \\
\mathrm{tr}\left\{\gamma^i\gamma^\mu\left[\gamma^\lambda(k^*-q)_\lambda + M^*\right]\right\} &= \mathrm{tr}\left\{\gamma^i\left[\gamma^\lambda(k^*-q)_\lambda + M^*\right]\gamma^\mu\right\}
\end{aligned} \tag{5.1053}$$

$\omega\sigma$ and $\sigma\omega$ two items overall contributions in Eq. (5.1052) are, using Eqs. (5.1053) and (5.981)–(5.984) and caring out the similar discussions, as follows:

$$W_{\mathrm{M}}^{\mathrm{s}}(k,E) = \frac{g_\sigma g_\omega}{8\pi^2 k}\oiint \Delta^0(q)D^0(q)\left[(E - V_0 - q_0)\mathrm{Im}\pi_0^{\mathrm{M}}(q) - (k-q)_i\mathrm{Im}\pi_i^{\mathrm{M}}(q)\right] \tag{5.1054}$$

$$W_{\mathrm{M}}^{0}(k,E) = \frac{g_\sigma g_\omega M^*}{8\pi^2 k}\oiint \Delta^0(q)D^0(q)\mathrm{Im}\pi_0^{\mathrm{M}}(q) \tag{5.1055}$$

$$W_{\mathrm{M}}^{\mathrm{v}}(k,E) = -\frac{g_\sigma g_\omega M^*}{8\pi^2 k^3}\oiint \Delta^0(q)D^0(q)\,k_i\mathrm{Im}\pi_i^{\mathrm{M}}(q) \tag{5.1056}$$

Comparing Eq. (5.1012) and Eq. (5.1048) and using Eqs. (5.1049) and (5.1050), referring to Eq. (5.1022), we can write

$$
\begin{aligned}
\mathrm{Im}\pi_{\mu}^{\mathrm{M}}(q) = & -\frac{g_{\sigma}g_{\omega}}{16\pi q}\int \mathrm{d}E_{p}\mathrm{d}E_{p+q}8\lambda M^{*}(2\overline{p}^{*}+q)_{\mu} \\
& \left[\theta\left(k_{\mathrm{F}}-|\vec{p}|\right)\theta\left(|\vec{p}+\vec{q}|-k_{\mathrm{F}}\right)+\theta\left(|\vec{p}|-k_{\mathrm{F}}\right)\theta\left(k_{\mathrm{F}}-|\vec{p}+\vec{q}|\right)\right] \\
& \delta\left(E_{p+q}-E_{p}-q_{0}\right)
\end{aligned}
\tag{5.1057}
$$

So we can get

$$
\mathrm{Im}\pi_{0}^{\mathrm{M}}(q) = -\frac{\lambda g_{\sigma}g_{\omega}M^{*}}{2\pi q}(E_{\mathrm{F}}+E_{x}+q_{0})(E_{\mathrm{F}}-E_{x}) \tag{5.1058}
$$

And let

$$
t_{i}^{\mathrm{M}} = -8\lambda M^{*}(2p+q)_{i} \tag{5.1059}
$$

so we can get the following expressions utilizing Eqs. (5.1030)–(5.1033) as

$$
\begin{aligned}
k_{i}t_{i}^{\mathrm{M}} &= -8\lambda M^{*}\left(2\vec{k}\cdot\vec{p}+\vec{k}\cdot\vec{q}\right) \\
&= -\frac{8\lambda M^{*}}{k^{2}+q^{2}-P^{2}}\left[4k^{2}\left(\frac{1}{2}q_{\eta}q^{\eta}+q_{0}E_{p}\right)+\frac{1}{2}\left(k^{2}+q^{2}-P^{2}\right)^{2}\right] \\
&= -\frac{4\lambda M^{*}}{k^{2}+q^{2}-P^{2}}\left[4k^{2}q_{\eta}q^{\eta}+\left(k^{2}+q^{2}-P^{2}\right)^{2}+8k^{2}q_{0}E_{p}\right]
\end{aligned}
\tag{5.1060}
$$

$$
\begin{aligned}
(k-q)_{i}t_{i}^{\mathrm{M}} &= -8\lambda M^{*}\left(2\vec{k}\cdot\vec{p}-2\vec{p}\cdot\vec{q}+\vec{k}\cdot\vec{q}-q^{2}\right) \\
&= -\frac{8\lambda M^{*}}{k^{2}+q^{2}-P^{2}}\left[4k^{2}\left(\frac{1}{2}q_{\eta}q^{\eta}+q_{0}E_{p}\right)-2\left(\frac{1}{2}q_{\eta}q^{\eta}+q_{0}E_{p}\right)\left(k^{2}+q^{2}-P^{2}\right)\right. \\
&\quad \left.+\frac{1}{2}\left(k^{2}+q^{2}-P^{2}\right)^{2}-q^{2}\left(k^{2}+q^{2}-P^{2}\right)\right] \\
&= -\frac{4\lambda M^{*}}{k^{2}+q^{2}-P^{2}}\left\{\left[\left(k^{2}-P^{2}\right)^{2}-q^{4}\right]+4\left(k^{2}-q^{2}+P^{2}\right)\left(\frac{1}{2}q_{\eta}q^{\eta}+q_{0}E_{p}\right)\right\}
\end{aligned}
\tag{5.1061}
$$

And then we can find

$$
\begin{aligned}
\mathrm{Im}\pi_{\alpha}^{\mathrm{M}}(q) &\equiv (k-q)_{i}\mathrm{Im}\pi_{i}^{\mathrm{M}}(q) \\
&= \frac{\lambda g_{\sigma}g_{\omega}M^{*}}{4\pi q\left(k^{2}+q^{2}-P^{2}\right)}\left\{\left[\left(k^{2}-P^{2}\right)^{2}-q^{4}\right](E_{\mathrm{F}}-E_{x})\right. \\
&\quad \left.+2\left(k^{2}-q^{2}+P^{2}\right)\left[q_{\eta}q^{\eta}(E_{\mathrm{F}}-E_{x})+q_{0}\left(E_{\mathrm{F}}^{2}-E_{x}^{2}\right)\right]\right\}
\end{aligned}
\tag{5.1062}
$$

$$\begin{aligned}\mathrm{Im}\pi_{\beta}^{\mathrm{M}}(q) &\equiv k_i \mathrm{Im}\pi_i^{\mathrm{M}}(q)\\ &= \frac{\lambda g_\sigma g_\omega M^*}{4\pi q\left(k^2+q^2-P^2\right)}\left\{\left[4k^2 q_\eta q^\eta + \left(k^2+q^2-P^2\right)^2\right](E_{\mathrm{F}}-E_x)\right.\\ &\quad \left. +4k^2 q_0\left(E_{\mathrm{F}}^2-E_x^2\right)\right\}\end{aligned} \tag{5.1063}$$

So Eq. (5.1054) and Eq. (5.1056) can be rewritten as

$$W_{\mathrm{M}}^{\mathrm{s}}(k,E)=\frac{g_\sigma g_\omega}{8\pi^2 k}\oiint \Delta^0(q)\,D^0(q)\,\left[(E-V_0-q_0)\mathrm{Im}\pi_0^{\mathrm{M}}(q)-\mathrm{Im}\pi_\alpha^{\mathrm{M}}(q)\right] \tag{5.1064}$$

$$W_{\mathrm{M}}^{\mathrm{v}}(k,E)=-\frac{g_\sigma g_\omega M^*}{8\pi^2 k^3}\oiint \Delta^0(q)\,D^0(q)\mathrm{Im}\pi_\beta^M(q) \tag{5.1065}$$

## 5.13.4 Contribution of $\pi^{ps}-\pi^{ps}$ Meson Exchange Processes to the Imaginary Part of Optical Potential

Comparing Eq. (5.669) with Eq. (5.671), the vertex contribution of the ω meson is $-\mathrm{i}g_\omega$; thus the vertex of the π meson ps term should contribute $g_\pi\gamma^5\tau_i$. Therefore the expression of the second order propagator of the exchange $\pi^{\mathrm{ps}}-\pi^{\mathrm{ps}}$ mesons can be written as

$$\begin{aligned}\mathrm{i}\Delta_{ij}^{(2)\mathrm{ps}}(q) &= \mathrm{i}\Delta_{il}^0(q)g_\pi^2\int\frac{\mathrm{d}^4p}{(2\pi)^4}\mathrm{tr}\left[-\mathrm{i}\gamma^5\tau_l G(p)\,\mathrm{i}\gamma^5\tau_m G(p+q)\right]\mathrm{i}\Delta_{mj}^0(q)\\ &= \mathrm{i}\Delta_{il}^0(q)\pi_{lm}^{\mathrm{ps}}\Delta_{mj}^0(q)\end{aligned} \tag{5.1066}$$

where

$$\pi_{lm}^{\mathrm{ps}}(q)=\mathrm{i}g_\pi^2\int\frac{\mathrm{d}^4p}{(2\pi)^4}\mathrm{tr}\left[\gamma^5\tau_l G(p)\gamma^5\tau_m G(p+q)\right] \tag{5.1067}$$

After $q$ is exchanged with $k-q$, the second item of Eq. (5.874) becomes

$$\Sigma_\pi^{0\ \mathrm{ps}}(k)=-\mathrm{i}g_\pi^2\gamma^5\tau_i\int\frac{\mathrm{d}^4q}{(2\pi)^4}\Delta_{ij}^0(q)G(k-q)\gamma^5\tau_j \tag{5.1068}$$

When replacing $\Delta_{ij}^0(q)$ with $\Delta_{il}^0(q)\pi_{lm}^{\mathrm{ps}}(q)\Delta_{mj}^0(q)$ in the above formula, we get

$$\Sigma_{\pi}^{\mathrm{ps}}(k) = -\mathrm{i}g_{\pi}^{2}\gamma^{5}\tau_{i}\int\frac{\mathrm{d}^{4}q}{(2\pi)^{4}}\Delta_{il}^{0}(q)\pi_{lm}^{\mathrm{ps}}(q)\Delta_{mj}^{0}(q)G(k-q)\gamma^{5}\tau_{j} \tag{5.1069}$$

From Eq. (5.875) we can write

$$\Delta_{ij}^{0}(q) = \frac{-\delta_{ij}}{q^{2}-q_{0}^{2}+m_{\pi}^{2}} = \Delta_{\pi}^{0}(q)\,\delta_{ij} \tag{5.1070}$$

$$\Delta_{\pi}^{0}(q) = \frac{-1}{q^{2}-q_{0}^{2}+m_{\pi}^{2}} \tag{5.1071}$$

So Eq. (5.1069) can be rewritten as

$$\Sigma_{\pi}^{\mathrm{ps}}(k) = -\mathrm{i}g_{\pi}^{2}\gamma^{5}\tau_{i}\int\frac{\mathrm{d}^{4}q}{(2\pi)^{4}}\left(\Delta_{\pi}^{0}(q)\right)^{2}\pi_{ij}^{\mathrm{ps}}(q)G(k-q)\tau_{j} \tag{5.1072}$$

In the expression of $\pi_{ij}^{\mathrm{ps}}(q)$ given by Eq. (5.1067), there is $\mathrm{tr}\{\tau_i\tau_j\} = 3\delta_{ij}$, and then we have

$$\pi_{ij}^{\mathrm{ps}}(q) = 3\delta_{ij}\pi^{\mathrm{ps}}(q) \tag{5.1073}$$

$$\pi^{\mathrm{ps}}(q) = \mathrm{i}g_{\pi}^{2}\int\frac{\mathrm{d}^{4}q}{(2\pi)^{4}}\mathrm{tr}\left[\gamma^{5}G(p)\gamma^{5}G(p+q)\right] \tag{5.1074}$$

Making a similar derivation and noting $\tau_i\tau_i = 3$, an expression similar to Eq. (5.965) or Eq. (5.1018) can be obtained as follows:

$$\begin{aligned} W_{\pi}^{\mathrm{ps}}(k) = {} & -9g_{\pi}^{2}\gamma^{5}\int\frac{\mathrm{d}\vec{q}}{(2\pi)^{3}}\left(\Delta_{\pi}^{0}(q)\right)^{2}\frac{\gamma^{\lambda}(k^{*}-q)_{\lambda}+M^{*}}{2E_{k-q}} \\ & \mathrm{Im}\pi^{\mathrm{ps}}(q)\gamma^{5}\,\theta\left(\left|\vec{k}-\vec{q}\right|-k_{\mathrm{F}}\right)\Big|_{q_0=k_0-V_0-E_{k-q}} \end{aligned} \tag{5.1075}$$

Noting $\gamma^5\gamma^\lambda\gamma^5 = -\gamma^\lambda$ and using the other relations such as Eq. (5.981), the following results can be obtained by Eq. (5.1075):

$$W_{\pi}^{\mathrm{ps-s}}(k,E) = -\frac{9g_{\pi}^{2}M^{*}}{8\pi^{2}k}\oiint\left(\Delta_{\pi}^{0}(q)\right)^{2}\,\mathrm{Im}\pi^{\mathrm{ps}}(q) \tag{5.1076}$$

$$W_{\pi}^{\mathrm{ps-0}}(k,E) = \frac{9g_{\pi}^{2}}{8\pi^{2}k}\oiint(E-V_{0}-q_{0})\left(\Delta^{0}(q)\right)^{2}\,\mathrm{Im}\pi^{\mathrm{ps}}(q) \tag{5.1077}$$

$$W_{\pi}^{\mathrm{ps-v}}(k,E) = -\frac{9g_{\pi}^{2}}{16\pi^{2}k^{3}}\oiint\left(k^{2}-q^{2}+P^{2}\right)\left(\Delta_{\pi}^{0}(q)\right)^{2}\,\mathrm{Im}\pi^{\mathrm{ps}}(q) \tag{5.1078}$$

Using Eq. (5.999) and Eq. (5.1000) we can get

$$\begin{aligned}&\mathrm{tr}\left[\gamma^5\left(\gamma^\mu \overline{p}^*_\mu + M^*\right)\gamma^5\left(\gamma^\nu (\overline{p}^* + q)_\nu + M^*\right)\right]\\&\quad = -4\lambda\left(E_p^2 + E_p q_0 - p^2 - \vec{p}\cdot\vec{q} - M^{*2}\right) = 2\lambda q_\mu q^\mu\end{aligned} \tag{5.1079}$$

Then from Eq. (5.1074) we obtain

$$\mathrm{Im}\pi^{\mathrm{ps}}(q) = \frac{\lambda g_\pi^2}{8\pi q} q_\mu q^\mu (E_{\mathrm{F}} - E_x) \tag{5.1080}$$

Similar to the real part of the optical potential, the contribution of the $\pi$ meson pseudo-scalar term is generally not considered for the imaginary part of the optical potential.

### 5.13.5 Contribution of $\pi^{pv} - \pi^{pv}$ Meson Exchange Processes to the Imaginary Part of the Optical Potential

Through the analyses before Eq. (5.881), in the case of exchange $\pi^{\mathrm{pv}} - \pi^{\mathrm{pv}}$ mesons, the vertex $x$ corresponds to $-\mathrm{i}\frac{f_\pi}{m_\pi}\gamma^5\gamma^\mu(-\mathrm{i}q_\mu)$, and the vertex $x'$ corresponds to $-\mathrm{i}\frac{f_\pi}{m_\pi}\gamma^5\gamma^\nu(\mathrm{i}q_\nu)$. Referring to Eq. (5.1066) we can write the expression of the corresponding second order propagator as

$$\begin{aligned}\mathrm{i}\Delta_{ij}^{(2)\mathrm{pv}}(q) &= \mathrm{i}\Delta_{il}^0(q)\left(-\mathrm{i}\frac{f_\pi}{m_\pi}\right)^2\int\frac{\mathrm{d}^4p}{(2\pi)^4}\mathrm{tr}\{-\mathrm{i}\gamma^5\gamma^\mu(-\mathrm{i}q_\mu)\tau_l G(p)\,\mathrm{i}\gamma^5\gamma^\nu(\mathrm{i}q_\nu)\tau_m G(p+q)\}\ \mathrm{i}\Delta_{mj}^0(q)\\&\equiv \mathrm{i}\Delta_{il}^0(q)\pi_{lm}^{\mathrm{pv}}(q)\Delta_{mj}^0(q)\end{aligned} \tag{5.1081}$$

where

$$\pi_{lm}^{\mathrm{pv}}(q) = -\mathrm{i}\left(\frac{f_\pi}{m_\pi}\right)^2\int\frac{\mathrm{d}^4p}{(2\pi)^4}\mathrm{tr}\{\gamma^5\gamma^\mu q_\mu\tau_l G(p)\gamma^5\gamma^\nu q_\nu\tau_m G(p+q)\} \tag{5.1082}$$

Noting $\mathrm{tr}\{\tau_l\tau_m\} = 3\delta_{lm}$ the following expressions can be obtained by the above formula:

$$\pi_{lm}^{\mathrm{pv}}(q) = 3\delta_{lm}\pi^{\mathrm{pv}}(q) \tag{5.1083}$$

$$\pi^{\mathrm{pv}}(q) = -\mathrm{i}\left(\frac{f_\pi}{m_\pi}\right)^2\int\frac{\mathrm{d}^4p}{(2\pi)^4}\mathrm{tr}\{\gamma^5\gamma^\mu q_\mu G(p)\gamma^5\gamma^\nu q_\nu G(p+q)\} \tag{5.1084}$$

When $q$ is exchanged with $k - q$, the first equation of Eq. (5.884) becomes

$$\Sigma_{\pi}^{0\mathrm{pv}}(k)=\mathrm{i}\left(\frac{f_{\pi}}{m_{\pi}}\right)^{2}\int\frac{\mathrm{d}^{4}q}{(2\pi)^{4}}\gamma^{5}\gamma^{\mu}q_{\mu}\tau_{i}\Delta_{ij}^{0}(q)G(k-q)\gamma^{5}\gamma^{\nu}q_{\nu}\tau_{j} \tag{5.1085}$$

Replacing $\Delta_{ij}^{0}$ with $\Delta_{il}^{0}(q)\pi_{lm}^{\mathrm{pv}}(q)\Delta_{mj}^{0}(q)$ in the above formula we can get

$$\begin{aligned}\Sigma_{\pi}^{\mathrm{pv}}(k)&=\mathrm{i}\left(\frac{f_{\pi}}{m_{\pi}}\right)^{2}\int\frac{\mathrm{d}^{4}q}{(2\pi)^{4}}\gamma^{5}\gamma^{\mu}q_{\mu}\tau_{i}\Delta_{il}^{0}(q)\pi_{lm}^{\mathrm{pv}}(q)\Delta_{mj}^{0}(q)G(k-q)\gamma^{5}\gamma^{\nu}q_{\nu}\tau_{j}\\&=9\mathrm{i}\left(\frac{f_{\pi}}{m_{\pi}}\right)^{2}\int\frac{\mathrm{d}^{4}q}{(2\pi)^{4}}\left(\Delta_{\pi}^{0}(q)\right)^{2}\gamma^{5}\gamma^{\mu}q_{\mu}\pi^{\mathrm{pv}}(q)G(k-q)\gamma^{5}\gamma^{\nu}q_{\nu}\end{aligned} \tag{5.1086}$$

and then get

$$\begin{aligned}W_{\pi}^{\mathrm{pv}}(k)=9\left(\frac{f_{\pi}}{m_{\pi}}\right)^{2}\int\frac{\mathrm{d}\vec{q}}{(2\pi)^{3}}\left(\Delta_{\pi}^{0}(q)\right)^{2}\gamma^{5}\gamma^{\mu}q_{\mu}\frac{\gamma^{\lambda}(k^{*}-q)_{\lambda}+M^{*}}{2E_{k-q}}\\\mathrm{Im}\pi^{\mathrm{pv}}(q)\,\gamma^{5}\gamma^{\nu}q_{\nu}\,\theta\left(\left|\vec{k}-\vec{q}\right|-k_{\mathrm{F}}\right)\Big|_{q_{0}=k_{0}-V_{0}-E_{k-q}}\end{aligned} \tag{5.1087}$$

Noting $\gamma^{5}\gamma^{\mu}=-\gamma^{\mu}\gamma^{5}$ and

$$\begin{aligned}\mathrm{tr}\left[\gamma^{0}\gamma^{\mu}q_{\mu}\gamma^{\lambda}(k^{*}-q)_{\lambda}\gamma^{\nu}q_{\nu}\right]&=4\left[q^{0}(k^{*}-q)^{\mu}q_{\mu}+q^{0}(k^{*}-q)^{\mu}q_{\mu}-(k^{*}-q)^{0}q_{\mu}q^{\mu}\right]\\&=4\left[2q^{0}q_{\mu}(k^{*}-q)^{\mu}-(k^{*}-q)^{0}q_{\mu}q^{\mu}\right]\end{aligned} \tag{5.1088}$$

$$k^{i}\mathrm{tr}\left[\gamma^{i}\gamma^{\mu}q_{\mu}\gamma^{\lambda}(k^{*}-q)_{\lambda}\gamma^{\nu}q_{\nu}\right]=4k^{i}\left[2q^{i}q_{\mu}(k^{*}-q)^{\mu}-(k^{*}-q)^{i}q_{\mu}q^{\mu}\right] \tag{5.1089}$$

and using Eqs. (5.981)–(5.984), (5.1030), from Eq. (5.1087) we can obtain

$$W_{\pi}^{\mathrm{pv-s}}(k,E)=-\left(\frac{f_{\pi}}{m_{\pi}}\right)^{2}\frac{9M^{*}}{8\pi^{2}k}\oint\!\!\oint\left(\Delta_{\pi}^{0}(q)\right)^{2}q_{\mu}q^{\mu}\mathrm{Im}\pi^{\mathrm{pv}}(q) \tag{5.1090}$$

$$\begin{aligned}W_{\pi}^{pv-0}(k,E)=-\left(\frac{f_{\pi}}{m_{\pi}}\right)^{2}\frac{9}{8\pi^{2}k}\oint\!\!\oint\left(\Delta_{\pi}^{0}(q)\right)^{2}\Big\{(E-V_{0}-q_{0})q_{\mu}q^{\mu}\\-2q_{0}\left[(E-V_{0}-q_{0})q_{0}-\frac{1}{2}(k^{2}-q^{2}-P^{2})\right]\Big\}\mathrm{Im}\pi^{\mathrm{pv}}(q)\end{aligned} \tag{5.1091}$$

$$\begin{aligned}W_{\pi}^{\mathrm{pv-v}}(k,E)=\left(\frac{f_{\pi}}{m_{\pi}}\right)^{2}\frac{9}{8\pi^{2}k^{3}}\oint\!\!\oint\left(\Delta_{\pi}^{0}(q)\right)^{2}\Big\{\frac{1}{2}(k^{2}-q^{2}+P^{2})q_{\mu}q^{\mu}\\-(k^{2}+q^{2}-P^{2})\left[(E-V_{0}-q_{0})q_{0}-\frac{1}{2}(k^{2}-q^{2}-P^{2})\right]\Big\}\mathrm{Im}\pi^{\mathrm{pv}}(q)\end{aligned} \tag{5.1092}$$

Utilizing Eq. (5.1000) we can get

$$q_\eta \overline{p}^{*\eta} = -\frac{1}{2} q_\eta q^\eta, \qquad q_\eta (\overline{p}^* + q)^\eta = \frac{1}{2} q_\eta q^\eta \tag{5.1093}$$

Referring to Eq. (5.1023) and using Eqs. (5.1024) and (5.1093), we can find

$$\begin{aligned}
&\mathrm{tr}\left\{\gamma^5\gamma^\mu q_\mu (\gamma^\lambda \overline{p}^*_\lambda + M^*)\gamma^5\gamma^\nu q_\nu \left[\gamma^\eta (\overline{p}^* + q)_\eta + M^*\right]\right\} \\
&= \mathrm{tr}\left\{\gamma^\mu q_\mu (\gamma^\lambda \overline{p}^*_\lambda - M^*)\gamma^\nu q_\nu \left[\gamma^\eta (\overline{p}^* + q)_\eta + M^*\right]\right\} \\
&= -4\lambda \left\{\left[M^{*2} + \overline{p}^*_\eta (\overline{p}^* + q)^\eta\right] g^{\mu\nu} - g^{\mu\mu}\overline{p}^*_\mu (\overline{p}^* + q)^\nu - g^{\nu\nu}\overline{p}^*_\nu (\overline{p}^* + q)^\mu\right\} q_\mu q_\nu \\
&= -4\lambda\left\{\left[M^{*2} + \overline{p}^*_\eta (\overline{p}^* + q)^\eta\right] q_\mu q^\mu - q_\mu \overline{p}^{*\mu} q_\nu (\overline{p}^* + q)^\nu - q_\nu \overline{p}^{*\nu} q_\mu (\overline{p}^* + q)^\mu\right\} \\
&= -4\lambda\left[2\left(M^{*2} - \frac{1}{4} q_\eta q^\eta\right) q_\mu q^\mu + \frac{1}{2}\left(q_\mu q^\mu\right)^2\right] \\
&= -8\lambda M^{*2} q_\mu q^\mu
\end{aligned} \tag{5.1094}$$

Therefore refer to and compared with Eqs. (5.994), (5.998), (5.1001), (5.1002), (5.1008), (5.1009), etc, we can find the following result from Eqs. (5.1084) and (5.1094):

$$\pi^{\mathrm{PV}}(q) = \left(\frac{f_\pi}{m_\pi}\right)^2 \frac{\lambda M^{*2}}{2\pi q} q_\mu q^\mu (E_{\mathrm{F}} - E_x) \tag{5.1095}$$

### 5.13.6 Contribution of $\rho^{\mathrm{V}} - \rho^{\mathrm{V}}$ Meson Exchange Processes to the Imaginary Part of the Optical Potential

The expression of the second order propagator that exchanges two ρ mesons with vector interactions is

$$\begin{aligned}
\mathrm{i}D^{(2)\mathrm{V}}_{\mu\nu,ij}(q) &= \mathrm{i}D^0_{\mu\alpha,il}(q) g^2_\rho \int \frac{\mathrm{d}^4 p}{(2\pi)^4} \mathrm{tr}\left\{\left(-\mathrm{i}\gamma^\alpha \tau_l\right)\left(-\mathrm{i}G(p)\right)\left(-\mathrm{i}\gamma^\beta \tau_m\right) \mathrm{i}G(p+q)\right\} \mathrm{i}D^0_{\beta\nu,mj}(q) \\
&\equiv \mathrm{i}D^0_{\mu\alpha,il}(q) \Pi^{\alpha\beta,lm}_{\mathrm{V}}(q) \Delta^0_{\beta\nu,mj}(q)
\end{aligned} \tag{5.1096}$$

where

$$\Pi_{\mathrm{V}}^{\alpha\beta,lm}(q) = -\mathrm{i}g_{\rho}^2 \int \frac{\mathrm{d}^4 p}{(2\pi)^4} \mathrm{tr}\left[\gamma^{\alpha}\tau_l G(p)\gamma^{\beta}\tau_m G(p+q)\right] = 3\delta_{lm}\Pi_{\mathrm{V}}^{\alpha\beta}(q) \tag{5.1097}$$

$$\Pi_{\mathrm{V}}^{\alpha\beta}(q) = -\mathrm{i}g_{\rho}^2 \int \frac{\mathrm{d}^4 p}{(2\pi)^4} \mathrm{tr}\left[\gamma^{\alpha}G(p)\gamma^{\beta}G(p+q)\right] \tag{5.1098}$$

First Eq. (5.893) is rewritten as

$$D^0_{\mu\nu,ij}(q) = g_{\mu\nu}\delta_{ij}D^0_{\rho}(q) \tag{5.1099}$$

$$D^0_{\rho}(q) = \frac{1}{q^2 - q_0^2 + m_{\rho}^2} \tag{5.1100}$$

After the exchanging $q$ with $k-q$ the second item of the first equation of Eq. (5.1013) becomes

$$\Sigma_{\rho}^{0\mathrm{V}}(k) = \mathrm{i}g_{\rho}^2 \int \frac{\mathrm{d}^4 q}{(2\pi)^4} \gamma^{\mu}\tau_i D^0_{\mu\nu,ij}(q)\, G(k-q)\gamma^{\nu}\tau_j \tag{5.1101}$$

When $D^0_{\mu\nu,ij}(q)$ in above formula is replaced by $D_{\mu\alpha,il}(q)\Pi_{\mathrm{V}}^{\alpha\beta,lm}(q)D^0_{\beta\nu,mj}(q)$, using Eq. (5.1099) we can obtain

$$\Sigma_{\rho}^{\mathrm{V}}(k) = 9\mathrm{i}g_{\rho}^2 \int \frac{\mathrm{d}^4 q}{(2\pi)^4} \left(D^0_{\rho}(q)\right)^2 \gamma^{\mu} g_{\mu\mu}\Pi_{\mathrm{V}}^{\mu\nu}(q) g_{\nu\nu} G(k-q)\gamma^{\nu} \tag{5.1102}$$

Comparing Eq. (5.1102) and Eq. (5.1017), referring to Eq. (5.1018), we get

$$\begin{aligned} W_{\rho}^{\mathrm{V}}(k) = 9g_{\rho}^2 \int \frac{\mathrm{d}\vec{q}}{(2\pi)^3} \left(D^0_{\rho}(q)\right)^2 \gamma^{\mu} \frac{\gamma^{\lambda}(k^* - q)_{\lambda} + M^*}{2E_{k-q}} \\ g_{\mu\mu}\mathrm{Im}\Pi_{\mathrm{V}}^{\mu\nu}(q) g_{\nu\nu}\gamma^{\nu}\, \theta\left(\left|\vec{k}-\vec{q}\right| - k_{\mathrm{F}}\right)\Big|_{q_0 = k_0 - V_0 - E_{k-q}} \end{aligned} \tag{5.1103}$$

Comparing Eq. (5.1098) with Eq. (5.1012), we can know that only replacing ω with ρ in Eqs. (5.1025), (5.1029), (5.1041)–(5.1043), $\mathrm{Im}\Pi^{\Lambda}_{\rho\mathrm{V}}(q)$, $\mathrm{Im}\Pi^{00}_{\rho\mathrm{V}}(q)$, $\mathrm{Im}\Pi^{\alpha}_{\rho\mathrm{V}}(q)$, $\mathrm{Im}\Pi^{\beta}_{\rho\mathrm{V}}(q)$, $\mathrm{Im}\Pi^{\gamma}_{\rho\mathrm{V}}(q)$ can be obtained. And comparing Eq. (5.1103) with Eq. (5.1018), we can know that only making the following replaces: $\pi_{\omega}$ with $\Pi_{\rho\mathrm{V}}$, $D^0(q)$ with $D^0_{\rho}(q)$, $g_{\omega}^2$ with $9g_{\rho}^2$ in Eqs. (5.1044), (5.1045), and (5.1046), then $W^{\mathrm{s}}_{\rho\mathrm{V}}(k,E)$, $W^{0}_{\rho\mathrm{V}}(k,E)$, $W^{\mathrm{v}}_{\rho\mathrm{V}}(k,E)$ can also be obtained.

### 5.13.7 Contribution of $\rho^T - \rho^T$ Meson Exchange Processes to the Imaginary Part of Optical Potential

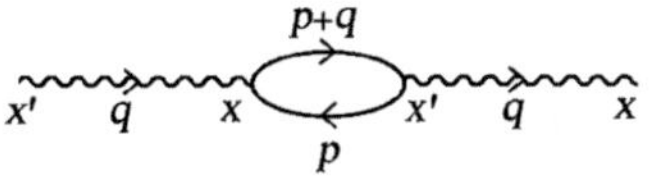


From Eqs. (5.670) and (5.900), it can be seen that the contribution of the ρ meson tensor term at the vertex $x$ is $-\mathrm{i}\left(\frac{f_\rho}{2M}\right)(\gamma^\mu q_\mu \gamma^\nu - q^\nu)\tau_i$; the contribution at the vertex $x'$ need to change its sign. The expression of the second order propagator for exchanging two ρ mesons with tensor interactions shown in the above figure is

$$\begin{aligned}\mathrm{i}D^{(2)\mathrm{T}}_{\mu\nu,ij}(q) &= \mathrm{i}D^0_{\mu\alpha,il}(q)\left(\frac{f_\rho}{2M}\right)^2\int\frac{\mathrm{d}^4p}{(2\pi)^4}\\ &\mathrm{tr}\left\{\mathrm{i}(\gamma^\xi q_\xi\gamma^\alpha - q^\alpha)\,\tau_l(-\mathrm{i}G(p))\left[-\mathrm{i}(\gamma^\zeta q_\zeta\gamma^\beta - q^\beta)\right]\tau_m\mathrm{i}G(p+q)\right\}\mathrm{i}D^0_{\beta\nu,mj}(q)\\ &\equiv \mathrm{i}D^0_{\mu\alpha,il}(q)\Pi^{\alpha\beta,lm}_{\mathrm{T}}(q)D^0_{\beta\nu,mj}(q)\end{aligned} \tag{5.1104}$$

where

$$\begin{aligned}\Pi^{\alpha\beta,lm}_{\mathrm{T}}(q) &= \mathrm{i}\left(\frac{f_\rho}{2M}\right)^2\int\frac{\mathrm{d}^4p}{(2\pi)^4}\mathrm{tr}\left[(\gamma^\xi q_\xi\gamma^\alpha - q^\alpha)\,\tau_lG(p)\right.\\ &\left.(\gamma^\zeta q_\zeta\gamma^\beta - q^\beta)\,\tau_mG(p+q)\right] = 3\delta_{lm}\Pi^{\alpha\beta}_{\mathrm{T}}(q)\end{aligned} \tag{5.1105}$$

$$\begin{aligned}\Pi^{\alpha\beta}_{\mathrm{T}}(q) &= \mathrm{i}\left(\frac{f_\rho}{2M}\right)^2\int\frac{\mathrm{d}^4p}{(2\pi)^4}\mathrm{tr}\left[(\gamma^\xi q_\xi\gamma^\alpha - q^\alpha)G(p)\right.\\ &\left.(\gamma^\zeta q_\zeta\gamma^\beta - q^\beta)G(p+q)\right]\end{aligned} \tag{5.1106}$$

After exchanging $q$ and $k-q$, the second equation of Eq. (5.1020) becomes

$$\begin{aligned}\Sigma^{0\mathrm{T}}_\rho(k) &= \mathrm{i}\left(\frac{f_\rho}{2M}\right)^2\int\frac{\mathrm{d}^4q}{(2\pi)^4}(\gamma^\xi q_\xi\gamma^\mu - q^\mu)\,\tau_iD^0_{\mu\nu,ij}(q)\\ &G(k-q)(-\gamma^\zeta q_\zeta\gamma^\nu + q^\nu)\,\tau_j\end{aligned} \tag{5.1107}$$

Similar methods can be used to obtain

$$\Sigma_{\rho}^{\mathrm{T}}(k)=9\mathrm{i}\left(\frac{f_{\rho}}{2M}\right)^{2}\int\frac{\mathrm{d}^{4}q}{(2\pi)^{4}}\left(D_{\rho}^{0}(q)\right)^{2}\left(\gamma^{\xi}q_{\xi}\gamma^{\mu}-q^{\mu}\right)$$
$$g_{\mu\mu}\Pi_{\mathrm{T}}^{\mu\nu}(q)g_{\nu\nu}G(k-q)\left(-\gamma^{\zeta}q_{\zeta}\gamma^{\nu}+q^{\nu}\right) \tag{5.1108}$$

Comparing Eq. (5.1108) and Eq. (5.956) and referring to (5.965), we get

$$W_{\rho}^{\mathrm{T}}(k)=9\left(\frac{f_{\rho}}{2M}\right)^{2}\int\frac{\mathrm{d}\vec{q}}{(2\pi)^{3}}\left(D_{\rho}^{0}(q)\right)^{2}\left(\gamma^{\xi}q_{\xi}\gamma^{\mu}-q^{\mu}\right)\frac{\gamma^{\lambda}(k^{*}-q)_{\lambda}+M^{*}}{2E_{k-q}}$$
$$g_{\mu\mu}\mathrm{Im}\Pi_{T}^{\mu\nu}(q)g_{\nu\nu}\left(-\gamma^{\zeta}q_{\zeta}\gamma^{\nu}+q^{\nu}\right)\theta\left(\left|\vec{k}-\vec{q}\right|-k_{\mathrm{F}}\right)\Big|_{q_{0}=k_{0}-V_{0}-E_{k-q}} \tag{5.1109}$$

In the following we make trace

$$\mathrm{tr}\left[\left(\gamma^{\xi}q_{\xi}\gamma^{\mu}-q^{\mu}\right)\left(-\gamma^{\zeta}q_{\zeta}\gamma^{\nu}+q^{\nu}\right)\right]=4\left(g^{\mu\nu}q_{\xi}q^{\xi}-q^{\mu}q^{\nu}\right) \tag{5.1110}$$

Referring to the process of the derivation of Eq. (5.986), the following result can be obtained from the first formula of Eq. (5.981):

$$W_{\rho\mathrm{T}}^{\mathrm{s}}(k,E)=\left(\frac{f_{\rho}}{2M}\right)^{2}\frac{9M^{*}}{8\pi^{2}k}\oiint\left(D_{\rho}^{0}(q)\right)^{2}\left[g_{\mu\mu}\mathrm{Im}\Pi_{\mathrm{T}}^{\mu\mu}(q)q_{\xi}q^{\xi}-q_{\mu}\mathrm{Im}\Pi_{\mathrm{T}}^{\mu\nu}(q)q_{\nu}\right] \tag{5.1111}$$

Using Eq. (5.152) we can find

$$\begin{aligned}
&\mathrm{tr}\{\gamma^{0}\gamma^{\xi}q_{\xi}\gamma^{\mu}\gamma^{\lambda}(k^{*}-q)_{\lambda}\gamma^{\zeta}q_{\zeta}\gamma^{\nu}\}\\
&=4\Big\{q^{0}\left[(k^{*}-q)^{\mu}q^{\nu}+g^{\mu\nu}(k^{*}-q)^{\lambda}q_{\lambda}-q^{\mu}(k^{*}-q)^{\nu}\right]\\
&\quad-q^{0\mu}\left[(k^{*}-q)^{\lambda}q_{\lambda}q^{\nu}+q^{\nu}(k^{*}-q)^{\lambda}q_{\lambda}-q_{\xi}q^{\xi}(k^{*}-q)^{\nu}\right]\\
&\quad+(k^{*}-q)^{0}\left[q^{\mu}q^{\nu}+q^{\mu}q^{\nu}-g^{\mu\nu}q_{\xi}q^{\xi}\right]\\
&\quad-q^{0}\left[q^{\mu}(k^{*}-q)^{\nu}+q^{\nu}(k^{*}-q)^{\mu}-g^{\mu\nu}q_{\lambda}(k^{*}-q)^{\lambda}\right]\\
&\quad+q^{0\nu}\left[q^{\mu}q_{\lambda}(k^{*}-q)^{\lambda}+q_{\xi}q^{\xi}(k^{*}-q)^{\mu}-q_{\lambda}(k^{*}-q)^{\lambda}q^{\mu}\right]\Big\}\\
&=4\Big\{q^{0}\left[2g^{\mu\nu}q_{\lambda}(k^{*}-q)^{\lambda}-2q^{\mu}(k^{*}-q)^{\nu}\right]\\
&\quad+(k^{*}-q)^{0}\left(2q^{\mu}q^{\nu}-g^{\mu\nu}q_{\xi}q^{\xi}\right)-2g^{0\mu}q^{\nu}q_{\lambda}(k^{*}-q)^{\lambda}\\
&\quad+\left[g^{0\mu}(k^{*}-q)^{\nu}+g^{0\nu}(k^{*}-q)^{\mu}\right]q_{\xi}q^{\xi}\Big\}
\end{aligned} \tag{5.1112}$$

and then get

$$
\begin{aligned}
&\operatorname{tr}\left\{\gamma^0\left(\gamma^\xi q_\xi \gamma^\mu - q^\mu\right)\left(\gamma^\lambda (k^* - q)_\lambda + M^*\right)\left(-\gamma^\zeta q_\zeta \gamma^\nu + q^\nu\right)\right\} \\
&= -4\Big\{q^\mu q^\nu (k^* - q)^0 - q^\mu\Big[(k^* - q)^0 q^\nu + g^{0\nu} q_\lambda (k^* - q)^\lambda - q^0 (k^* - q)^\nu\Big] \\
&\quad -q^\nu\Big[q^0 (k^* - q)^\mu + (k^* - q)^0 q^\mu - g^{0\mu} q_\lambda (k^* - q)^\lambda\Big] \\
&\quad +q^0\Big[2g^{\mu\nu} q_\lambda (k^* - q)^\lambda - 2q^\mu (k^* - q)^\nu\Big] \\
&\quad +(k^* - q)^0\left(2q^\mu q^\nu - g^{\mu\nu} q_\xi q^\xi\right) - 2g^{0\mu} q^\nu q_\lambda (k^* - q)^\lambda \\
&\quad +\left[g^{0\mu} (k^* - q)^\nu + g^{0\nu} (k^* - q)^\mu\right] q_\xi q^\xi\Big\} \\
&= -4\Big\{ q^0\Big[2g^{\mu\nu} q_\lambda (k^* - q)^\lambda - q^\mu (k^* - q)^\nu - q^\nu (k^* - q)^\mu\Big] \\
&\quad +(k^* - q)^0\left(q^\mu q^\nu - g^{\mu\nu} q_\xi q^\xi\right) - \left(g^{0\mu} q^\nu + g^{0\nu} q^\mu\right) q_\lambda (k^* - q)^\lambda \\
&\quad +\left[g^{0\mu} (k^* - q)^\nu + g^{0\nu} (k^* - q)^\mu\right] q_\xi q^\xi\Big\}
\end{aligned}
\tag{5.1113}
$$

So according to Eq. (5.1113), etc., we obtain

$$
\begin{aligned}
W^0_{\rho \mathrm{T}}(k, E) = &-\left(\frac{f_\rho}{2M}\right)^2 \frac{9}{8\pi^2 k} \oint\!\!\!\oint \left(D^0_\rho(q)\right)^2 \Big\{\Big[2g_{\mu\mu} \mathrm{Im}\Pi^{\mu\mu}_{\mathrm{T}}(q) q_\lambda (k^* - q)^\lambda \\
&-q_\mu \mathrm{Im}\Pi^{\mu\nu}_{\mathrm{T}}(q)(k^* - q)_\nu - (k^* - q)_\mu \mathrm{Im}\Pi^{\mu\nu}_{\mathrm{T}}(q) q_\nu\Big] q_0 \\
&-(E - V_0 - q_0)\left[g_{\mu\mu} \mathrm{Im}\Pi^{\mu\mu}_{\mathrm{T}}(q) q_\xi q^\xi - q_\mu \mathrm{Im}\Pi^{\mu\nu}_{\mathrm{T}}(q) q_\nu\right] \\
&-\Big[\mathrm{Im}\Pi^{0\nu}_{\mathrm{T}}(q) q_\nu + q_\mu \mathrm{Im}\Pi^{\mu 0}_{\mathrm{T}}(q)\Big] q_\lambda (k^* - q)^\lambda \\
&+\Big[\mathrm{Im}\Pi^{0\nu}_{\mathrm{T}}(q)(k^* - q)_\nu + (k^* - q)_\mu \mathrm{Im}\Pi^{\mu 0}_{\mathrm{T}}(q)\Big] q_\xi q^\xi\Big\} \\
= &-\left(\frac{f_\rho}{2M}\right)^2 \frac{9}{8\pi^2 k} \oint\!\!\!\oint \left(D^0_\rho(q)\right)^2 \Big\{ \Big[2q_0 q_\lambda (k^* - q)^\lambda - (E - V_0 - q_0) q_\xi q^\xi\Big] g_{\mu\mu} \mathrm{Im}\Pi^{\mu\mu}_{\mathrm{T}}(q) \\
&+\Big[(E - V_0 - q_0) q_\mu q_\nu - q_0 q_\mu (k^* - q)_\nu - q_0 (k^* - q)_\mu q_\nu\Big] \mathrm{Im}\Pi^{\mu\nu}_{\mathrm{T}}(q) \\
&+\Big[(k^* - q)_\nu q_\xi q^\xi - q_\nu q_\lambda (k^* - q)^\lambda\Big] \mathrm{Im}\Pi^{0\nu}_{\mathrm{T}}(q) \\
&+\Big[(k^* - q)_\mu q_\xi q^\xi - q_\mu q_\lambda (k^* - q)^\lambda\Big] \mathrm{Im}\Pi^{\mu 0}_{T}(q)\Big\}
\end{aligned}
\tag{5.1114}
$$

Since the properties of $g^{00} = 1$ and $q^0 = q_0$ are not used in the derivation of Eqs. (5.1112) and (5.1113), from Eq. (5.1113), we can obtain

$$\begin{aligned}&\mathrm{tr}\{k^i\gamma^i(\gamma^\xi q_\xi\gamma^\mu - q^\mu)\ [\gamma^\lambda(k^* - q)_\lambda + M^*](-\gamma^\zeta q_\zeta\gamma^\nu + q^\nu)\}\\&\quad = -4k^i\Big\{q^i\Big[2g^{\mu\nu}q_\lambda(k^* - q)^\lambda - q^\mu(k^* - q)^\nu - q^\nu(k^* - q)^\mu\Big]\\&\qquad +(k^* - q)^i(q^\mu q^\nu - g^{\mu\nu}q_\xi q^\xi) - (g^{i\mu}q^\nu + g^{i\nu}q^\mu)q_\lambda(k^* - q)^\lambda\\&\qquad +[g^{i\mu}(k^* - q)^\nu + g^{i\nu}(k^* - q)^\mu]q_\xi q^\xi\Big\}\end{aligned} \tag{5.1115}$$

and then get

$$\begin{aligned}W^{\mathrm{v}}_{\rho\mathrm{T}}(k, E) &= \left(\frac{f_\rho}{2M}\right)^2\frac{9}{8\pi^2k^3}\oiint\left(D^0_\rho(q)\right)^2k^i\Big\{\Big[2g_{\mu\mu}\mathrm{Im}\Pi^{\mu\mu}_{\mathrm{T}}(q)q_\lambda(k^* - q)^\lambda\\&\quad -q_\mu\mathrm{Im}\Pi^{\mu\nu}_{\mathrm{T}}(q)(k^* - q)_\nu - (k^* - q)_\mu\mathrm{Im}\Pi^{\mu\nu}_{\mathrm{T}}(q)q_\nu\Big]q^i\\&\quad -(k - q)^i\left[g_{\mu\mu}\mathrm{Im}\Pi^{\mu\mu}_{\mathrm{T}}(q)q_\xi q^\xi - q_\mu\mathrm{Im}\Pi^{\mu\nu}_{\mathrm{T}}(q)q_\nu\right]\\&\quad -\left[\mathrm{Im}\Pi^{i\nu}_{\mathrm{T}}(q)q_\nu + q_\mu\mathrm{Im}\Pi^{\mu\, i}_{\mathrm{T}}(q)\right]q_\lambda(k^* - q)^\lambda\\&\quad +\Big[\mathrm{Im}\Pi^{i\nu}_{\mathrm{T}}(q)(k^* - q)_\nu + (k^* - q)_\mu\mathrm{Im}\Pi^{\mu\, i}_{\mathrm{T}}(q)\Big]q_\xi q^\xi\Big\}\\&= \left(\frac{f_\rho}{2M}\right)^2\frac{9}{8\pi^2k^3}\oiint\left(D^0_\rho(q)\right)^2k^i\Big\{\Big[2q^iq_\lambda(k^* - q)^\lambda - (k - q)^iq_\xi q^\xi\Big]g_{\mu\mu}\mathrm{Im}\Pi^{\mu\mu}_{\mathrm{T}}(q)\\&\quad +\Big[(k - q)^iq_\mu q_\nu - q^iq_\mu(k^* - q)_\nu - q^i(k^* - q)_\mu q_\nu\Big]\mathrm{Im}\Pi^{\mu\nu}_{\mathrm{T}}(q)\\&\quad +\Big[(k^* - q)_\nu q_\xi q^\xi - q_\nu q_\lambda(k^* - q)^\lambda\Big]\mathrm{Im}\Pi^{i\nu}_{\mathrm{T}}(q)\\&\quad +\Big[(k^* - q)_\mu q_\xi q^\xi - q_\mu q_\lambda(k^* - q)^\lambda\Big]\mathrm{Im}\Pi^{\mu\, i}_{\mathrm{T}}(q)\Big\}\\&= \left(\frac{f_\rho}{2M}\right)^2\frac{9}{8\pi^2k^3}\oiint\left(D^0_\rho(q)\right)^2\Big\{\Big[2\vec{k}\cdot\vec{q}q_\lambda(k^* - q)^\lambda - \left(k^2 - \vec{k}\cdot\vec{q}\right)q_\xi q^\xi\Big]g_{\mu\mu}\mathrm{Im}\Pi^{\mu\mu}_{\mathrm{T}}(q)\\&\quad +\Big[\left(k^2 - \vec{k}\cdot\vec{q}\right)q_\mu q_\nu - \vec{k}\cdot\vec{q}\left(q_\mu(k^* - q)_\nu + (k^* - q)_\mu q_\nu\right)\Big]\mathrm{Im}\Pi^{\mu\nu}_{\mathrm{T}}(q)\\&\quad +\Big[q_\xi q^\xi(k^* - q)_\nu - q_\lambda(k^* - q)^\lambda q_\nu\Big]k^i\mathrm{Im}\Pi^{i\nu}_{\mathrm{T}}(q)\\&\quad +\Big[q_\xi q^\xi(k^* - q)_\mu - q_\lambda(k^* - q)^\lambda q_\mu\Big]k^i\mathrm{Im}\Pi^{\mu\, i}_{\mathrm{T}}(q)\Big\}\end{aligned} \tag{5.1116}$$

Using Eq. (5.152) we can also obtain

$$
\begin{aligned}
&\mathrm{tr}\left\{\gamma^{\xi}q_{\xi}\gamma^{\mu}\gamma^{\lambda}\overline{p}_{\lambda}^{*}\gamma^{\zeta}q_{\zeta}\gamma^{\nu}\gamma^{\eta}(\overline{p}^{*}+q)_{\eta}\right\}\\
&=4\lambda\Big\{q^{\mu}\left[\overline{p}_{\xi}^{*}q^{\xi}(\overline{p}^{*}+q)^{\nu}+q^{\nu}\overline{p}_{\lambda}^{*}(\overline{p}^{*}+q)^{\lambda}-\overline{p}^{*\nu}q_{\lambda}(\overline{p}^{*}+q)^{\lambda}\right]\\
&-q^{\xi}\overline{p}_{\xi}^{*}\left[q^{\mu}(\overline{p}^{*}+q)^{\nu}+q^{\nu}(\overline{p}^{*}+q)^{\mu}-g^{\mu\nu}q_{\lambda}(\overline{p}^{*}+q)^{\lambda}\right]\\
&+q_{\xi}q^{\xi}\left[\overline{p}^{*\mu}(\overline{p}^{*}+q)^{\nu}+\overline{p}^{*\nu}(\overline{p}^{*}+q)^{\mu}-g^{\mu\nu}\overline{p}_{\lambda}^{*}(\overline{p}^{*}+q)^{\lambda}\right]\\
&-q^{\nu}\left[\overline{p}^{*\mu}q_{\lambda}(\overline{p}^{*}+q)^{\lambda}+\overline{p}_{\xi}^{*}q^{\xi}(\overline{p}^{*}+q)^{\mu}-q^{\mu}\overline{p}_{\lambda}^{*}(\overline{p}^{*}+q)^{\lambda}\right]\\
&+q_{\lambda}(\overline{p}^{*}+q)^{\lambda}\left[\overline{p}^{*\mu}q^{\nu}+g^{\mu\nu}\overline{p}_{\xi}^{*}q^{\xi}-q^{\mu}\overline{p}^{*\nu}\right]\Big\}\\
&=4\lambda\Big\{q^{\mu}\left[2q^{\nu}\overline{p}_{\lambda}^{*}(\overline{p}^{*}+q)^{\lambda}-2\overline{p}^{*\nu}q_{\lambda}(\overline{p}^{*}+q)^{\lambda}\right]\\
&-q^{\xi}\overline{p}_{\xi}^{*}\left[2q^{\nu}(\overline{p}^{*}+q)^{\mu}-2g^{\mu\nu}q_{\lambda}(\overline{p}^{*}+q)^{\lambda}\right]\\
&+q_{\xi}q^{\xi}\left[\overline{p}^{*\mu}(\overline{p}^{*}+q)^{\nu}+\overline{p}^{*\nu}(\overline{p}^{*}+q)^{\mu}-g^{\mu\nu}\overline{p}_{\lambda}^{*}(\overline{p}^{*}+q)^{\lambda}\right]\Big\}
\end{aligned}
\tag{5.1117}
$$

and then get

$$
\begin{aligned}
t_{\mathrm{T}}^{\mu\nu}&\equiv\mathrm{tr}\left\{\left(\gamma^{\xi}q_{\xi}\gamma^{\mu}-q^{\mu}\right)\left(\gamma^{\lambda}\overline{p}_{\lambda}^{*}+M^{*}\right)\left(\gamma^{\zeta}q_{\zeta}\gamma^{\nu}-q^{\nu}\right)\left(\gamma^{\eta}(\overline{p}^{*}+q)_{\eta}+M^{*}\right)\right\}\\
&=4\lambda\Big\{M^{*2}\left(q^{\mu}q^{\nu}-q^{\mu}q^{\nu}-q^{\nu}q^{\mu}+q^{\mu}q^{\nu}+q^{\mu}q^{\nu}-g^{\mu\nu}q_{\xi}q^{\xi}\right)+q^{\mu}q^{\nu}\overline{p}_{\lambda}^{*}(\overline{p}^{*}+q)^{\lambda}\\
&-q^{\mu}\left[\overline{p}_{\xi}^{*}q^{\xi}(\overline{p}^{*}+q)^{\nu}+q^{\nu}\overline{p}_{\lambda}^{*}(\overline{p}^{*}+q)^{\lambda}-\overline{p}^{*\nu}q_{\lambda}(\overline{p}^{*}+q)^{\lambda}\right]\\
&-q^{\nu}\left[q^{\mu}\overline{p}_{\lambda}^{*}(\overline{p}^{*}+q)^{\lambda}+\overline{p}^{*\mu}q_{\lambda}(\overline{p}^{*}+q)^{\lambda}-q^{\xi}\overline{p}_{\xi}^{*}(\overline{p}^{*}+q)^{\mu}\right]\\
&+q^{\mu}\left[2q^{\nu}\overline{p}_{\lambda}^{*}(\overline{p}^{*}+q)^{\lambda}-2\overline{p}^{*\nu}q_{\lambda}(\overline{p}^{*}+q)^{\lambda}\right]\\
&-q^{\xi}\overline{p}_{\xi}^{*}\left[2q^{\nu}(\overline{p}^{*}+q)^{\mu}-2g^{\mu\nu}q_{\lambda}(\overline{p}^{*}+q)^{\lambda}\right]\\
&+q_{\xi}q^{\xi}\left[\overline{p}^{*\mu}(\overline{p}^{*}+q)^{\nu}+\overline{p}^{*\nu}(\overline{p}^{*}+q)^{\mu}-g^{\mu\nu}\overline{p}_{\lambda}^{*}(\overline{p}^{*}+q)^{\lambda}\right]\Big\}\\
&=4\lambda\Big\{M^{*2}\left(q^{\mu}q^{\nu}-g^{\mu\nu}q_{\xi}q^{\xi}\right)+q^{\mu}q^{\nu}\overline{p}_{\lambda}^{*}(\overline{p}^{*}+q)^{\lambda}\\
&-\left[q^{\mu}(\overline{p}^{*}+q)^{\nu}+q^{\nu}(\overline{p}^{*}+q)^{\mu}-2g^{\mu\nu}q_{\lambda}(\overline{p}^{*}+q)^{\lambda}\right]q^{\xi}\overline{p}_{\xi}^{*}\\
&-(q^{\mu}\overline{p}^{*\nu}+q^{\nu}\overline{p}^{*\mu})q_{\lambda}(\overline{p}^{*}+q)^{\lambda}\\
&+\left[\overline{p}^{*\mu}(\overline{p}^{*}+q)^{\nu}+\overline{p}^{*\nu}(\overline{p}^{*}+q)^{\mu}-g^{\mu\nu}\overline{p}_{\lambda}^{*}(\overline{p}^{*}+q)^{\lambda}\right]q_{\xi}q^{\xi}\Big\}
\end{aligned}
\tag{5.1118}
$$

From the above formula it can be seen that $\Pi_{\mathrm{T}}^{\mu\nu}(q)$ is symmetrical for $\mu$, $\nu$.

Using the similar method the following expressions containing summation for $\mu$ can be obtained from Eqs. (5.1106), (5.1118), (5.1093), and (5.1024):

$$\begin{aligned}\mathrm{Im}\Pi_{\mathrm{T}}^{\Lambda}(q) &\equiv \sum_{\mu} g_{\mu\mu}\mathrm{Im}\Pi_{\mathrm{T}}^{\mu\mu}(q) \\ &= \left(\frac{f_{\rho}}{2M}\right)^{2}\frac{\lambda}{4\pi q}\left[-3M^{*2}q_{\xi}q^{\xi} + q_{\xi}q^{\xi}\left(M^{*2} - \frac{1}{2}q_{\eta}q^{\eta}\right) + \frac{1}{2}q_{\xi}q^{\xi}(q_{\eta}q^{\eta})\right. \\ &\quad - \frac{1}{2}q_{\xi}q^{\xi}4q_{\eta}q^{\eta} + \frac{1}{2}q_{\xi}q^{\xi}q_{\eta}q^{\eta} + q_{\xi}q^{\xi}2\left(M^{*2} - \frac{1}{2}q_{\eta}q^{\eta}\right) \\ &\quad \left.-4q_{\xi}q^{\xi}\left(M^{*2} - \frac{1}{2}q_{\eta}q^{\eta}\right)\right](E_{\mathrm{F}} - E_{x}) \\ &= -\left(\frac{f_{\rho}}{2M}\right)^{2}\frac{\lambda}{\pi q}\left(M^{*2} + \frac{1}{8}q_{\xi}q^{\xi}\right)q_{\eta}q^{\eta}(E_{\mathrm{F}} - E_{x})\end{aligned} \tag{5.1119}$$

We can also obtain

$$\begin{aligned}\mathrm{Im}\Pi_{\mathrm{T}}^{00}(q) &= \left(\frac{f_{\rho}}{2M}\right)^{2}\frac{\lambda}{4\pi q}\int_{E_x}^{E_{\mathrm{F}}}\left[M^{*2}(q_0^2 - q_{\xi}q^{\xi}) + q_0^2\left(M^{*2} - \frac{1}{2}q_{\eta}q^{\eta}\right)\right. \\ &\quad + q_{\xi}q^{\xi}(E_p + q_0)q_0 - q_{\xi}q^{\xi}\frac{1}{2}q_{\eta}q^{\eta} - \frac{1}{2}q_{\xi}q^{\xi}2q_0E_p + 2q_{\xi}q^{\xi}E_p(E_p + q_0) \\ &\quad \left.-q_{\xi}q^{\xi}\left(M^{*2} - \frac{1}{2}q_{\eta}q^{\eta}\right)\right]dE_p \\ &= \left(\frac{f_{\rho}}{2M}\right)^{2}\frac{\lambda}{4\pi q}\left[\left(2M^{*2}q^2 + \frac{1}{2}q_0^2q_{\xi}q^{\xi}\right)(E_{\mathrm{F}} - E_x) + q_0q_{\xi}q^{\xi}\left(E_{\mathrm{F}}^2 - E_x^2\right)\right. \\ &\quad \left.+\frac{2}{3}q_{\xi}q^{\xi}\left(E_{\mathrm{F}}^3 - E_x^3\right)\right]\end{aligned} \tag{5.1120}$$

According to Eq. (5.1118) and using Eqs. (5.1024) and (5.1093), we can obtain

$$\begin{aligned}t_{\mathrm{T}}^{0i} &= 4\lambda\Big\{\left[M^{*2} + \overline{p}_{\lambda}^{*}(\overline{p}^{*} + q)^{\lambda}\right]q_0q^i - q_{\xi}\overline{p}^{*\xi}\left[q_0p^i + (2q_0 + E_p)q^i\right] \\ &\quad - q_{\lambda}(\overline{p}^{*} + q)^{\lambda}(q_0p^i + E_pq^i) + q_{\xi}q^{\xi}\left[(q_0 + 2E_p)p^i + E_pq^i\right]\Big\} \\ &= 4\lambda\bigg(\Big\{\left[M^{*2} + \overline{p}_{\lambda}^{*}(\overline{p}^{*} + q)^{\lambda}\right]q_0 - q_{\xi}\overline{p}^{*\xi}(2q_0 + E_p) - q_{\lambda}(\overline{p}^{*} + q)^{\lambda}E_p + q_{\xi}q^{\xi}E_p\Big\}q^i \\ &\quad - \left[q_{\xi}\overline{p}^{*\xi}q_0 + q_{\lambda}(\overline{p}^{*} + q)^{\lambda}q_0 - q_{\xi}q^{\xi}(q_0 + 2E_p)p^i\right]\bigg) \\ &= 4\lambda\Big\{\left[2\left(M^{*2} + \frac{1}{4}q_{\xi}q^{\xi}\right)q_0 + q_{\xi}q^{\xi}E_p\right]q^i + q_{\xi}q^{\xi}(q_0 + 2E_p)p^i\Big\}\end{aligned} \tag{5.1121}$$

$$
\begin{aligned}
t_{\mathrm{T}}^{ij} = 4\lambda\Big\{ & M^{*2}\left(q^i q^j + q_\xi q^\xi \delta_{ij}\right) + \overline{p}_\lambda^*\left(\overline{p}^* + q\right)^\lambda q^i q^j - q_\xi \overline{p}^{*\xi} \\
& \left[q^i (p+q)^j + q^j (p+q)^i + 2q_\lambda\left(\overline{p}^* + q\right)^\lambda \delta_{ij}\right] - q_\lambda\left(\overline{p}^* + q\right)^\lambda\left(q^i p^j + q^j p^i\right) \\
& + q_\xi q^\xi\left[p^i (p+q)^j + p^j (p+q)^i + \overline{p}_\lambda^*\left(\overline{p}^* + q\right)^\lambda \delta_{ij}\right] \Big\} \\
= 4\lambda\Big\{ & M^{*2}\left(q^i q^j + q_\xi q^\xi \delta_{ij}\right) + \left(M^{*2} - \frac{1}{2} q_\xi q^\xi\right) q^i q^j \\
& + \frac{1}{2} q_\xi q^\xi\left[q^i (p+q)^j + q^j (p+q)^i + q_\eta q^\eta \delta_{ij}\right] - \frac{1}{2} q_\eta q^\eta\left(q^i p^j + q^j p^i\right) \\
& + q_\xi q^\xi\left[p^i (p+q)^j + p^j (p+q)^i + \left(M^{*2} - \frac{1}{2} q_\eta q^\eta\right) \delta_{ij}\right] \Big\} \\
= 8\lambda\Big\{ & M^{*2} q_\xi q^\xi \delta_{ij} + \left(M^{*2} + \frac{1}{4} q_\xi q^\xi\right) q^i q^j \\
& + \frac{1}{2} q_\xi q^\xi\left[p^i (p+q)^j + p^j (p+q)^i\right] \Big\}
\end{aligned}
\tag{5.1122}
$$

From Eq. (5.1121) and Eq. (5.1122), we can further obtain the following expressions using Eqs. (5.1030)–(5.1033):

$$
\begin{aligned}
q^i t_{\mathrm{T}}^{0i} &= 4\lambda\Big\{q^2\left[2\left(M^{*2} + \frac{1}{4} q_\xi q^\xi\right) q_0 + q_\xi q^\xi E_p\right] \\
&\quad + q_\xi q^\xi\left[\frac{1}{2} q_0 q_\eta q^\eta + \left(2q_0^2 - q^2\right) E_p + 2q_0 E_p^2\right]\Big\} \\
&= 4\lambda q_0\left[2M^{*2} q^2 + \frac{1}{2} q_0^2 q_\xi q^\xi + 2q_\xi q^\xi\left(q_0 E_p + E_p^2\right)\right]
\end{aligned}
\tag{5.1123}
$$

$$
\begin{aligned}
k^i t_{\mathrm{T}}^{0i} &= 4\lambda\left\{\frac{1}{2}\left(k^2 + q^2 - P^2\right)\left[2\left(M^{*2} + \frac{1}{4} q_\xi q^\xi\right) q_0 + q_\xi q^\xi E_p\right]\right. \\
&\quad \left. + \frac{2k^2 q_\xi q^\xi}{k^2 + q^2 - P^2}\left[\frac{1}{2} q_0 q_\eta q^\eta + \left(2q_0^2 - q^2\right) E_p + 2q_0 E_p^2\right]\right\} \\
&= \frac{4\lambda}{k^2 + q^2 - P^2}\left(\left[\left(k^2 + q^2 - P^2\right)^2\left(M^{*2} + \frac{1}{4} q_\xi q^\xi\right) + k^2\left(q_\xi q^\xi\right)^2\right] q_0\right. \\
&\quad \left. + q_\xi q^\xi\left\{\left[\frac{1}{2}\left(k^2 + q^2 - P^2\right)^2 + 2k^2\left(2q_0^2 - q^2\right)\right] E_p + 4k^2 q_0 E_p^2\right\}\right)
\end{aligned}
\tag{5.1124}
$$

$$
\begin{aligned}
q^i q^j t_{\mathrm{T}}^{ij} &= 8\lambda\left\{M^{*2} q_\xi q^\xi q^2 + \left(M^{*2} + \frac{1}{4} q_\xi q^\xi\right) q^4 + q_\xi q^\xi \vec{p} \cdot \vec{q}\left(\vec{p} \cdot \vec{q} + q^2\right)\right\} \\
&= 8\lambda\left\{\left(M^{*2} q_0^2 + \frac{1}{4} q_\xi q^\xi q^2\right) q^2 + q_\xi q^\xi\left[\left(\tfrac{1}{2} q_\eta q^\eta + q_0 E_p\right)^2 + \left(\frac{1}{2} q_\eta q^\eta + q_0 E_p\right) q^2\right]\right\} \\
&= 8\lambda q_0^2\left[M^{*2} q^2 + \frac{1}{4} q_0^2 q_\xi q^\xi + q_\xi q^\xi\left(q_0 E_p + E_p^2\right)\right]
\end{aligned}
\tag{5.1125}
$$

$$
\begin{aligned}
q^i k^j t_{\mathrm{T}}^{ij} &= 8\lambda\Big\{M^{*2} q_\xi q^\xi \vec{k}\cdot\vec{q} + \Big(M^{*2} + \frac{1}{4} q_\xi q^\xi\Big) q^2 \vec{k}\cdot\vec{q} \\
&\quad + \frac{1}{2} q_\xi q^\xi \Big[\vec{p}\cdot\vec{q}\Big(\vec{k}\cdot\vec{p} + \vec{k}\cdot\vec{q}\Big) + \vec{k}\cdot\vec{p}\Big(\vec{p}\cdot\vec{q} + q^2\Big)\Big]\Big\} \\
&= 8\lambda\Bigg(\frac{1}{2}(k^2 + q^2 - P^2)\Big(M^{*2} q_0^2 + \frac{1}{4} q^2 q_\xi q^\xi\Big) + \frac{q_\xi q^\xi}{2(k^2 + q^2 - P^2)} \\
&\quad \Big\{4k^2\big(\tfrac{1}{2} q_\eta q^\eta + q_0 E_p\big)^2 + \Big(\frac{1}{2} q_\eta q^\eta + q_0 E_p\Big)\Big[\frac{1}{2}(k^2 + q^2 - P^2)^2 + 2k^2 q^2\Big]\Big\}\Bigg) \\
&= \frac{4\lambda}{k^2 + q^2 - P^2}\Big(\Big\{(k^2 + q^2 - P^2)^2\Big(M^{*2} q_0^2 + \frac{1}{4} q^2 q_\xi q^\xi\Big) + \frac{1}{4}\Big[(k^2 + q^2 - P^2)^2 \\
&\quad + 4k^2 q_0^2\Big](q_\xi q^\xi)^2\Big\} + 2 q_0 q_\xi q^\xi \Big\{\Big[k^2(2q_0^2 - q^2) + \frac{1}{4}(k^2 + q^2 - P^2)^2\Big] E_p \\
&\quad + 2k^2 q_0 E_p^2\Big\}\Big)
\end{aligned}
\tag{5.1126}
$$

In the following we will make an approximation; it is considered that the contribution of the $\cos(\varphi_p - \varphi_q)$ term can be ignored when using the additive angle formula to $\left(\vec{k}\cdot\vec{p}\right)^2$, i.e., Eq. (5.1032) can still be used here. So from Eq. (5.1122) we can obtain

$$
\begin{aligned}
k^i k^j t_{\mathrm{T}}^{ij} &= 8\lambda\Big\{M^{*2} q_\xi q^\xi k^2 + \Big(M^{*2} + \frac{1}{4} q_\xi q^\xi\Big)\Big(\vec{k}\cdot\vec{q}\Big)^2 + q_\xi q^\xi \vec{k}\cdot\vec{p}\,\Big(\vec{k}\cdot\vec{p} + \vec{k}\cdot\vec{q}\Big)\Big\} \\
&= 8\lambda\Big\{M^{*2} k^2 q_\xi q^\xi + \frac{1}{4}(k^2 + q^2 - P^2)^2\Big(M^{*2} + \frac{1}{4} q_\xi q^\xi\Big) + \frac{2k^2 q_\xi q^\xi}{(k^2 + q^2 - P^2)^2} \\
&\quad \Big[2k^2\Big(\frac{1}{2} q_\eta q^\eta + q_0 E_p\Big)^2 + \frac{1}{2}(k^2 + q^2 - P^2)^2\Big(\frac{1}{2} q_\eta q^\eta + q_0 E_p\Big)\Big]\Big\} \\
&= \frac{8\lambda}{(k^2 + q^2 - P^2)^2}\Big(\Big\{(k^2 + q^2 - P^2)^2\Big[M^{*2} k^2 q_\xi q^\xi + \frac{1}{4}(k^2 + q^2 - P^2)^2 \\
&\quad \Big(M^{*2} + \frac{1}{4} q_\xi q^\xi\Big)\Big] + k^2 (q_\xi q^\xi)^2\Big[k^2 q_\eta q^\eta + \frac{1}{2}(k^2 + q^2 - P^2)^2\Big]\Big\} \\
&\quad + 4k^2 q_0 q_\xi q^\xi\Big\{\Big[k^2 q_\eta q^\eta + \frac{1}{4}(k^2 + q^2 - P^2)^2\Big] E_p + k^2 q_0 E_p^2\Big\}\Big)
\end{aligned}
\tag{5.1127}
$$

Introduce and find the following physical quantities using Eqs. (5.1123)–(5.1127), and (5.1106):

$$
\begin{aligned}
\mathrm{Im}\Pi_T^1(q) \equiv q^i \mathrm{Im}\Pi_{\mathrm{T}}^{0i}(q) &= \left(\frac{f_\rho}{2M}\right)^2 \frac{\lambda q_0}{4\pi q}\Big\{\Big(2M^{*2} q^2 + \frac{1}{2} q_0^2 q_\xi q^\xi\Big)(E_{\mathrm{F}} - E_x) \\
&\quad + 2 q_\xi q^\xi \Big[\frac{1}{2} q_0 (E_{\mathrm{F}}^2 - E_x^2) + \frac{1}{3}(E_{\mathrm{F}}^3 - E_x^3)\Big]\Big\}
\end{aligned}
\tag{5.1128}
$$

$$\mathrm{Im}\Pi_{\mathrm{T}}^{2}(q) \equiv k^{i}\mathrm{Im}\Pi_{T}^{0i}(q) = \left(\frac{f_{\rho}}{2M}\right)^{2} \frac{\lambda}{4\pi q\left(k^{2}+q^{2}-P^{2}\right)}$$
$$\left(\left[\left(k^{2}+q^{2}-P^{2}\right)^{2}\left(M^{*2}+\frac{1}{4}q_{\xi}q^{\xi}\right)+k^{2}\left(q_{\xi}q^{\xi}\right)^{2}\right] q_{0}\left(E_{\mathrm{F}}-E_{x}\right)\right.$$
$$\left.+q_{\xi}q^{\xi}\left\{\frac{1}{2}\left[\frac{1}{2}\left(k^{2}+q^{2}-P^{2}\right)^{2}+2k^{2}\left(2q_{0}^{2}-q^{2}\right)\right]\left(E_{\mathrm{F}}^{2}-E_{x}^{2}\right)+\frac{4}{3}k^{2}q_{0}\left(E_{\mathrm{F}}^{3}-E_{x}^{3}\right)\right\}\right) \tag{5.1129}$$

$$\mathrm{Im}\Pi_{\mathrm{T}}^{3}(q) \equiv q^{i}q^{j}\mathrm{Im}\Pi_{\mathrm{T}}^{ij}(q) = \left(\frac{f_{\rho}}{2M}\right)^{2} \frac{\lambda q_{0}^{2}}{2\pi q}\left\{\left(M^{*2}q^{2}+\frac{1}{4}q_{0}^{2}q_{\xi}q^{\xi}\right)\left(E_{\mathrm{F}}-E_{x}\right)\right.$$
$$\left.+q_{\xi}q^{\xi}\left[\frac{1}{2}q_{0}\left(E_{\mathrm{F}}^{2}-E_{x}^{2}\right)+\frac{1}{3}\left(E_{\mathrm{F}}^{3}-E_{x}^{3}\right)\right]\right\} \tag{5.1130}$$

$$\mathrm{Im}\Pi_{\mathrm{T}}^{4}(q) \equiv q^{i}k^{j}\mathrm{Im}\Pi_{\mathrm{T}}^{ij}(q)$$
$$= \left(\frac{f_{\rho}}{2M}\right)^{2} \frac{\lambda}{4\pi q\left(k^{2}+q^{2}-P^{2}\right)}\left(\left\{\left(k^{2}+q^{2}-P^{2}\right)^{2}\left(M^{*2}q_{o}^{2}+\frac{1}{4}q^{2}q_{\xi}q^{\xi}\right)\right.\right.$$
$$\left.+\frac{1}{4}\left[\left(k^{2}+q^{2}-P^{2}\right)^{2}+4k^{2}q_{0}^{2}\right]\left(q_{\xi}q^{\xi}\right)^{2}\right\}\left(E_{\mathrm{F}}-E_{x}\right)+2q_{0}q_{\xi}q^{\xi}$$
$$\left.\left\{\frac{1}{2}\left[k^{2}\left(2q_{0}^{2}-q^{2}\right)+\frac{1}{4}\left(k^{2}+q^{2}-P^{2}\right)^{2}\right]\left(E_{\mathrm{F}}^{2}-E_{x}^{2}\right)+\frac{2}{3}k^{2}q_{0}\left(E_{\mathrm{F}}^{3}-E_{x}^{3}\right)\right\}\right) \tag{5.1131}$$

$$\mathrm{Im}\Pi_{\mathrm{T}}^{5}(q) \equiv k^{i}k^{j}\mathrm{Im}\Pi_{\mathrm{T}}^{ij}(q) = \left(\frac{f_{\rho}}{2M}\right)^{2} \frac{\lambda}{2\pi q\left(k^{2}+q^{2}-P^{2}\right)^{2}}\left(\left\{\left(k^{2}+q^{2}-P^{2}\right)^{2}\right.\right.$$
$$\left[M^{*2}k^{2}q_{\xi}q^{\xi}+\frac{1}{4}\left(k^{2}+q^{2}-P^{2}\right)^{2}\left(M^{*2}+\frac{1}{4}q_{\xi}q^{\xi}\right)\right]+k^{2}\left(q_{\xi}q^{\xi}\right)^{2}$$
$$\left.\left[k^{2}q_{\eta}q^{\eta}+\frac{1}{2}\left(k^{2}+q^{2}-P^{2}\right)^{2}\right]\right\}\left(E_{\mathrm{F}}-E_{x}\right)+4k^{2}q_{0}q_{\xi}q^{\xi}$$
$$\left.\left\{\frac{1}{2}\left[k^{2}q_{\eta}q^{\eta}+\frac{1}{4}\left(k^{2}+q^{2}-P^{2}\right)^{2}\right]\left(E_{\mathrm{F}}^{2}-E_{x}^{2}\right)+\frac{1}{3}k^{2}q_{0}\left(E_{\mathrm{F}}^{3}-E_{x}^{3}\right)\right\}\right) \tag{5.1132}$$

Based on Eq. (5.1030) we can also obtain

$$q_{\lambda}\left(k^{*}-q\right)^{\lambda} = \left(E-V_{0}-q_{0}\right)q_{0}-\frac{1}{2}\left(k^{2}-q^{2}-P^{2}\right) \tag{5.1133}$$

Equations (5.1111), (5.1114), and (5.1116) can be rewritten, using Eqs. (5.984), (5.1030), and (5.1133) and the symbols introduced by Eqs. (5.1119), (5.1120), (5.1128)–(5.1132), as follows:

$$W_{\rho\mathrm{T}}^{\mathrm{s}}(k,E) = \left(\frac{f_{\rho}}{2M}\right)^{2} \frac{9M^{*}}{8\pi^{2}k}\oiint\left(D_{\rho}^{0}(q)\right)^{2}$$
$$\left[q_{\xi}q^{\xi}\mathrm{Im}\Pi_{\mathrm{T}}^{\Lambda}(q)-q_{0}^{2}\mathrm{Im}\Pi_{\mathrm{T}}^{00}(q)+2q_{0}\mathrm{Im}\Pi_{\mathrm{T}}^{1}(q)-\mathrm{Im}\Pi_{\mathrm{T}}^{3}(q)\right] \tag{5.1134}$$

$$
\begin{aligned}
&W^{0}_{\rho\mathrm{T}}(k,E)\\
&=-\left(\tfrac{f_\rho}{2M}\right)^2\frac{9}{8\pi^2 k}\oiint\left(D^0_\rho(q)\right)^2\{[(E-V_0-q_0)(q_0^2+q^2)-q_0(k^2-q^2-P^2)]\mathrm{Im}\Pi^\Lambda_\mathrm{T}(q)\\
&\quad-[(E-V_0-q_0)(q_0^2+2q^2)-q_0(k^2-q^2-P^2)]\mathrm{Im}\Pi^{00}_\mathrm{T}(q)\\
&\quad+[2(E-V_0-q_0)q_0-(k^2+q^2-P^2)]\mathrm{Im}\Pi^1_\mathrm{T}(q)+2q^2\mathrm{Im}\Pi^2_\mathrm{T}(q)\\
&\quad+(E-V_0+q_0)\mathrm{Im}\Pi^3_\mathrm{T}(q)-2q_0\mathrm{Im}\Pi^4_\mathrm{T}(q)\}
\end{aligned}
\tag{5.1135}
$$

$$
\begin{aligned}
W^{\mathrm{V}}_{\rho\mathrm{T}}(k,E)&=\left(\frac{f_\rho}{2M}\right)^2\frac{9}{8\pi^2 k^3}\oiint\left(D^0_\rho(q)\right)^2\Big(\Big\{\left(k^2+q^2-P^2\right)\Big[(E-V_0-q_0)q_0\\
&\quad-\frac{1}{2}\left(k^2-q^2-P^2\right)\Big]-\frac{1}{2}\left(k^2-q^2+P^2\right)q_\xi q^\xi\Big\}\mathrm{Im}\Pi^\Lambda_\mathrm{T}(q)\\
&\quad-q_0\Big[(E-V_0-q_0)\left(k^2+q^2-P^2\right)-\frac{1}{2}q_0\left(k^2-q^2+P^2\right)\Big]\mathrm{Im}\Pi^{00}_\mathrm{T}(q)\\
&\quad-\Big[\left(k^2-q^2+P^2\right)q_0-\left(k^2+q^2-P^2\right)(E-V_0-2q_0)\Big]\mathrm{Im}\Pi^1_\mathrm{T}(q)\\
&\quad-2\Big[(E-V_0-q_0)q^2-\left(k^2-P^2\right)q_0\Big]\mathrm{Im}\Pi^2_\mathrm{T}(q)+\frac{1}{2}\left(3k^2+q^2-P^2\right)\mathrm{Im}\Pi^3_\mathrm{T}(q)\\
&\quad-\frac{1}{2}\Big[2(E-V_0-3q_0)q_0+k^2+7q^2-P^2\Big]\mathrm{Im}\Pi^4_\mathrm{T}(q)-2q_\xi q^\xi\mathrm{Im}\Pi^5_\mathrm{T}(q)\Big)
\end{aligned}
\tag{5.1136}
$$

### 5.13.8 Contribution of $\rho^V-\rho^T$ Meson Exchange Processes to the Imaginary Part of Optical Potential

The expression of the second order propagator, which contains the interactions of ρ meson vector and tensor simultaneously, is

$$
\begin{aligned}
\mathrm{i}D^{(2)\,\mathrm{VT}}_{\mu\nu,ij}(q)&=\mathrm{i}D^0_{\mu\alpha,il}(q)g_\rho\left(\frac{f_\rho}{2M}\right)\int\frac{\mathrm{d}^4p}{(2\pi)^4}\mathrm{tr}\left[\mathrm{i}\left(\gamma^\xi q_\xi\gamma^\alpha-q^\alpha\right)\tau_l(-\mathrm{i}G(p))\left(-\mathrm{i}\gamma^\beta\right)\right.\\
&\quad\left.\tau_m\mathrm{i}G(p+q)\right]\mathrm{i}D^0_{\beta\nu,mj}(q)+\mathrm{i}D^0_{\mu\alpha,il}(q)g_\rho\left(\frac{f_\rho}{2M}\right)\\
&\quad\int\frac{\mathrm{d}^4p}{(2\pi)^4}\mathrm{tr}\left[\left(-\mathrm{i}\gamma^\alpha\right)\tau_l(-\mathrm{i}G(p))(-\mathrm{i})\left(\gamma^\xi q_\xi\gamma^\beta-q^\beta\right)\tau_m\mathrm{i}G(p+q)\right]\mathrm{i}D^0_{\beta\nu,mj}(q)\\
&\equiv\mathrm{i}D^0_{\mu\alpha,il}(q)\left(\Pi^{\alpha\beta,lm}_{\mathrm{TV}}(q)+\Pi^{\alpha\beta,lm}_{\mathrm{VT}}(q)\right)D^0_{\beta\nu,mj}(q)
\end{aligned}
\tag{5.1137}
$$

where

$$
\begin{aligned}
\Pi^{\alpha\beta,lm}_{\mathrm{TV}}(q)&=\mathrm{i}g_\rho\left(\frac{f_\rho}{2M}\right)\int\frac{\mathrm{d}^4p}{(2\pi)^4}\mathrm{tr}\left[\left(\gamma^\xi q_\xi\gamma^\alpha-q^\alpha\right)\tau_lG(p)\gamma^\beta\tau_mG(p+q)\right]\\
&=3\delta_{lm}\Pi^{\alpha\beta}_{\mathrm{TV}}(q)
\end{aligned}
\tag{5.1138}
$$

$$\Pi_{\mathrm{TV}}^{\alpha\beta}(q) = \mathrm{i}g_\rho\left(\frac{f_\rho}{2M}\right)\int\frac{\mathrm{d}^4p}{(2\pi)^4}\mathrm{tr}\left[(\gamma^\xi q_\xi\gamma^\alpha - q^\alpha)G(p)\gamma^\beta G(p+q)\right] \tag{5.1139}$$

$$\begin{aligned}\Pi_{\mathrm{VT}}^{\alpha\beta,lm}(q) &= -\mathrm{i}g_\rho\left(\frac{f_\rho}{2M}\right)\int\frac{\mathrm{d}^4p}{(2\pi)^4}\mathrm{tr}\left[\gamma^\alpha\tau_l G(p)\left(\gamma^\xi q_\xi\gamma^\beta - q^\beta\right)\tau_m G(p+q)\right]\\ &= 3\delta_{lm}\Pi_{\mathrm{VT}}^{\alpha\beta}(q)\end{aligned} \tag{5.1140}$$

$$\Pi_{\mathrm{VT}}^{\alpha\beta}(q) = -\mathrm{i}g_\rho\left(\frac{f_\rho}{2M}\right)\int\frac{\mathrm{d}^4p}{(2\pi)^4}\mathrm{tr}\left[\gamma^\alpha G(p)\left(\gamma^\xi q_\xi\gamma^\beta - q^\beta\right)G(p+q)\right] \tag{5.1141}$$

We can find

$$\begin{aligned}&\mathrm{tr}\left\{\left(\gamma^\xi q_\xi\gamma^\mu - q^\mu\right)\left(\gamma^\lambda\overline{p}_\lambda^* + M^*\right)\gamma^\nu\left(\gamma^\eta(\overline{p}^* + q)_\eta + M^*\right)\right\}\\ &= M^*\mathrm{tr}\left\{\left(\gamma^\xi q_\xi\gamma^\mu - q^\mu\right)\left[\gamma^\lambda\overline{p}_\lambda^*\gamma^\nu + \gamma^\nu\gamma^\eta(\overline{p}^* + q)_\eta\right]\right\}\\ &= 4\lambda M^*\left\{-q^\mu[\overline{p}^{*\nu} + (\overline{p}^* + q)^\nu] + q^\mu\overline{p}^{*\nu} + q^\nu\overline{p}^{*\mu} - g^{\mu\nu}q_\lambda\overline{p}^{*\lambda}\right.\\ &\quad\left. + q^\mu(\overline{p}^* + q)^\nu + g^{\mu\nu}q_\lambda(\overline{p}^* + q)^\lambda - q^\nu(\overline{p}^* + q)^\mu\right\}\\ &= 4\lambda M^*\left(g^{\mu\nu}q_\xi q^\xi - q^\mu q^\nu\right)\end{aligned} \tag{5.1142}$$

$$\begin{aligned}&\mathrm{tr}\left\{\gamma^\mu\left(\gamma^\lambda\overline{p}_\lambda^* + M^*\right)\left(\gamma^\zeta q_\zeta\gamma^\nu - q^\nu\right)\left[\gamma^\eta(\overline{p}^* + q)_\eta + M^*\right]\right\}\\ &= M^*\mathrm{tr}\left[\gamma^\mu\left(\gamma^\zeta q_\zeta\gamma^\nu - q^\nu\right)\gamma^\eta(\overline{p}^* + q)_\eta + \gamma^\mu\gamma^\lambda\overline{p}_\lambda^*\left(\gamma^\zeta q_\zeta\gamma^\nu - q^\nu\right)\right]\\ &= 4\lambda M^*\left\{-q^\nu[(\overline{p}^* + q)^\mu + \overline{p}^{*\mu}] + q^\mu(\overline{p}^* + q)^\nu + q^\nu(\overline{p}^* + q)^\mu\right.\\ &\quad\left. - g^{\mu\nu}q_\lambda(\overline{p}^* + q)^\lambda + \overline{p}^{*\mu}q^\nu + g^{\mu\nu}q_\lambda\overline{p}^{*\lambda} - q^\mu\overline{p}^{*\nu}\right\}\\ &= -4\lambda M^*\left(g^{\mu\nu}q_\xi q^\xi - q^\mu q^\nu\right)\end{aligned} \tag{5.1143}$$

Therefore it is known that according to Eqs. (5.1139), (5.1141), (5.989), and (5.992), there is $\Pi_{\mathrm{TV}}^{\mu\nu}(q) = \Pi_{\mathrm{VT}}^{\mu\nu}(q)$, and let

$$\Pi_{\rho\mathrm{M}}^{\mu\nu}(q) = g_{\mu\mu}\left(\Pi_{\mathrm{TV}}^{\mu\nu}(q) + \Pi_{\mathrm{VT}}^{\mu\nu}(q)\right)g_{\nu\nu} \tag{5.1144}$$

Refer to Eq. (5.1109) we can write

$$\begin{aligned}W_\rho^{\mathrm{TV}}(k) = 9g_\rho\left(\frac{f_\rho}{2M}\right)\int\frac{\mathrm{d}\vec{q}}{(2\pi)^3}\left(D_\rho^0(q)\right)^2\left(\gamma^\xi q_\xi\gamma^\mu - q^\mu\right)\frac{\gamma^\lambda(k^* - q)_\lambda + M^*}{2E_{k-q}}\\ \frac{1}{2}\mathrm{Im}\Pi_{\rho\mathrm{M}}^{\mu\nu}(q)\gamma^\nu\theta\left(\left|\vec{k} - \vec{q}\right| - k_\mathrm{F}\right)\Big|_{q_0 = k_0 - V_0 - E_{k-q}}\end{aligned} \tag{5.1145}$$

$$W_\rho^{\mathrm{VT}}(k) = 9g_\rho\left(\frac{f_\rho}{2M}\right)\int\frac{\mathrm{d}\vec{q}}{(2\pi)^3}\left(D_\rho^0(q)\right)^2\gamma^\mu\frac{\gamma^\lambda(k^*-q)_\lambda+M^*}{2E_{k-q}}$$
$$\frac{1}{2}\mathrm{Im}\Pi_{\rho\mathrm{M}}^{\mu\nu}(q)\left(-\gamma^\zeta q_\zeta\gamma^\nu+q^\nu\right)\theta\left(\left|\vec{k}-\vec{q}\right|-k_\mathrm{F}\right)\Big|_{q_0=k_0-V_0-E_{k-q}} \tag{5.1146}$$

and we can find

$$\begin{aligned}
&\mathrm{tr}\left[\left(\gamma^\xi q_\xi\gamma^\mu-q^\mu\right)\left(\gamma^\lambda(k^*-q)_\lambda+M^*\right)\gamma^\nu\right]\\
&\quad=4\left[-q^\mu(k^*-q)^\nu+q^\mu(k^*-q)^\nu+q^\nu(k^*-q)^\mu-g^{\mu\nu}q_\lambda(k^*-q)^\lambda\right]\\
&\quad=-4\left[g^{\mu\nu}q_\lambda(k^*-q)^\lambda-(k^*-q)^\mu q^\nu\right],\\
&\mathrm{tr}\left\{\gamma^\mu\left[\gamma^\lambda(k^*-q)_\lambda+M^*\right]\left(-\gamma^\zeta q_\zeta\gamma^\nu+q^\nu\right)\right\}\\
&\quad=4\left[q^\nu(k^*-q)^\mu-q^\nu(k^*-q)^\mu-g^{\mu\nu}q_\lambda(k^*-q)^\lambda+q^\mu(k^*-q)^\nu\right]\\
&\quad=-4\left[g^{\mu\nu}q_\lambda(k^*-q)^\lambda-q^\mu(k^*-q)^\nu\right]
\end{aligned} \tag{5.1147}$$

So we can obtain

$$\begin{aligned}
W_{\rho\mathrm{M}}^{\mathrm{s}}(k,E)&\equiv W_\rho^{\mathrm{TVs}}(k,E)+W_\rho^{\mathrm{VTs}}(k,E)\\
&=-g_\rho\left(\frac{f_\rho}{2M}\right)\frac{9}{8\pi^2k}\iint\left(D_\rho^0(q)\right)^2\left\{g^{\mu\mu}q_\lambda(k^*-q)^\lambda\mathrm{Im}\Pi_{\rho\mathrm{M}}^{\mu\mu}(q)\right.\\
&\quad\left.-\frac{1}{2}\left[q^\mu(k^*-q)^\nu+q^\nu(k^*-q)^\mu\right]\mathrm{Im}\Pi_{\rho\mathrm{M}}^{\mu\nu}(q)\right\}
\end{aligned} \tag{5.1148}$$

We can still get

$$\begin{aligned}
&\mathrm{tr}\left\{\gamma^0\left(\gamma^\xi q_\xi\gamma^\mu-q^\mu\right)\left[\gamma^\lambda(k^*-q)_\lambda+M^*\right]\gamma^\nu\right\}=4M^*\left(g^{\mu\nu}q^0-g^{0\mu}q^\nu\right),\\
&\mathrm{tr}\left\{\gamma^0\gamma^\mu\left[\gamma^\lambda(k^*-q)_\lambda+M^*\right]\left(-\gamma^\zeta q_\zeta\gamma^\nu+q^\nu\right)\right\}=4M^*\left(g^{\mu\nu}q^0-g^{0\nu}q^\mu\right)
\end{aligned} \tag{5.1149}$$

So we can obtain

$$\begin{aligned}
&W_{\rho\mathrm{M}}^0(k,E)\equiv W_\rho^{\mathrm{TV0}}(k,E)+W_\rho^{\mathrm{VT0}}(k,E)\\
&=g_\rho\left(\frac{f_\rho}{2M}\right)\frac{9}{8\pi^2k}\iint\left(D_\rho^0(q)\right)^2\left[q_0g^{\mu\mu}\mathrm{Im}\Pi_{\rho\mathrm{M}}^{\mu\mu}(q)-\frac{1}{2}\left(q^\mu\mathrm{Im}\Pi_{\rho\mathrm{M}}^{\mu0}(q)+q^\nu\mathrm{Im}\Pi_{\rho\mathrm{M}}^{0\nu}(q)\right)\right]
\end{aligned} \tag{5.1150}$$

Further we can also get

$$\begin{aligned}&\mathrm{tr}\{\gamma^i(\gamma^\xi q_\xi\gamma^\mu - q^\mu)\left[\gamma^\lambda(k^* - q)_\lambda + M^*\right]\gamma^\nu\} = 4M^*(g^{\mu\nu}q^i - g^{i\mu}q^\nu),\\&\mathrm{tr}\{\gamma^i\gamma^\mu\left[\gamma^\lambda(k^* - q)_\lambda + M^*\right](-\gamma^\xi q_\xi\gamma^\nu + q^\nu)\} = 4M^*(g^{\mu\nu}q^i - g^{i\nu}q^\mu)\end{aligned} \tag{5.1151}$$

So we can obtain

$$\begin{aligned}W^{\mathrm{v}}_{\rho\mathrm{M}}(k, E) &\equiv W^{\mathrm{TVv}}_{\rho}(k, E) + W^{\mathrm{VTv}}_{\rho}(k, E)\\&= -g_\rho\left(\frac{f_\rho}{2M}\right)\frac{9M^*}{8\pi^2k^3}\oint\!\!\oint\left(D^0_\rho(q)\right)^2 k^i\left[q^i g^{\mu\mu}\mathrm{Im}\Pi^{\mu\mu}_{\rho\mathrm{M}}(q)\right.\\&\quad\left.+\frac{1}{2}\left(q^\mu\mathrm{Im}\Pi^{\mu i}_{\rho\mathrm{M}}(q) + q^\nu\mathrm{Im}\Pi^{i\nu}_{\rho\mathrm{M}}(q)\right)\right]\end{aligned} \tag{5.1152}$$

Then the following expression can be obtained using Eqs. (5.1139) and (5.1141)–(5.1144):

$$\mathrm{Im}\Pi^{\mu\nu}_{\rho\mathrm{M}}(q) = g_\rho\left(\frac{f_\rho}{2M}\right)\frac{\lambda}{2\pi q}(g^{\mu\nu}q_\xi q^\xi - q^\mu q^\nu)(E_\mathrm{F} - E_x) \tag{5.1153}$$

And use the above formula to introduce the following symbols:

$$\mathrm{Im}\Pi^{\Lambda}_{\rho\mathrm{M}}(q) \equiv \sum_\mu g^{\mu\mu}\mathrm{Im}\Pi^{\mu\mu}_{\rho\mathrm{M}}(q) = g_\rho\left(\frac{f_\rho}{2M}\right)\frac{3\lambda}{2\pi q}q_\xi q^\xi(E_\mathrm{F} - E_x) \tag{5.1154}$$

$$\mathrm{Im}\Pi^{00}_{\rho\mathrm{M}}(q) = -g_\rho\left(\frac{f_\rho}{2M}\right)\frac{\lambda q}{2\pi}(E_\mathrm{F} - E_x) \tag{5.1155}$$

at the same time using Eq. (5.1030) we can also obtain

$$\mathrm{Im}\Pi^{1}_{\rho\mathrm{M}}(q) \equiv q^i\mathrm{Im}\Pi^{0i}_{\rho\mathrm{M}}(q) = g_\rho\left(\frac{f_\rho}{2M}\right)\frac{\lambda q_0 q}{2\pi}(E_\mathrm{F} - E_x) \tag{5.1156}$$

$$\mathrm{Im}\Pi^{2}_{\rho\mathrm{M}}(q) \equiv k^i\mathrm{Im}\Pi^{0i}_{\rho\mathrm{M}}(q) = g_\rho\left(\frac{f_\rho}{2M}\right)\frac{\lambda q_0}{4\pi q}(k^2 + q^2 - P^2)(E_\mathrm{F} - E_x) \tag{5.1157}$$

$$\mathrm{Im}\Pi^{3}_{\rho\mathrm{M}}(q) \equiv q^i q^j\mathrm{Im}\Pi^{ij}_{\rho\mathrm{M}}(q) = g_\rho\left(\frac{f_\rho}{2M}\right)\frac{\lambda q}{2\pi}(q_0^2 - 2q^2)(E_\mathrm{F} - E_x) \tag{5.1158}$$

$$\begin{aligned}\mathrm{Im}\Pi^{4}_{\rho\mathrm{M}}(q) &\equiv q^i k^j\mathrm{Im}\Pi^{ij}_{\rho\mathrm{M}}(q)\\&= g_\rho\left(\frac{f_\rho}{2M}\right)\frac{\lambda}{4\pi q}(k^2 + q^2 - P^2)(q_0^2 - 2q^2)(E_\mathrm{F} - E_x)\end{aligned} \tag{5.1159}$$

Equations (5.1148), (5.1150), and (5.1152) can be rewritten, using the previously introduced symbols, as follows:

$$\begin{aligned} W^{\mathrm{s}}_{\rho\mathrm{M}}(k,E) = &-g_\rho\left(\frac{f_\rho}{2M}\right)\frac{9}{8\pi^2 k}\oint\!\!\!\oint\left(D^0_\rho(q)\right)^2\left\{\left[(E-V_0-q_0)q_0\right.\right.\\ &\left.-\frac{1}{2}\left(k^2-q^2-P^2\right)\right]\mathrm{Im}\Pi^{\Lambda}_{\rho\mathrm{M}}(q)-(E-V_0-q_0)q_0\mathrm{Im}\Pi^{00}_{\rho\mathrm{M}}(q)\\ &-(E-V_0-2q_0)\mathrm{Im}\Pi^{1}_{\rho\mathrm{M}}(q)-q_0\mathrm{Im}\Pi^{2}_{\rho\mathrm{M}}(q)+\mathrm{Im}\Pi^{3}_{\rho\mathrm{M}}(q)\\ &\left.-\mathrm{Im}\Pi^{4}_{\rho\mathrm{M}}(q)\right\} \end{aligned} \tag{5.1160}$$

$$W^{0}_{\rho\mathrm{M}}(k,E) = g_\rho\left(\frac{f_\rho}{2M}\right)\frac{9M^*}{8\pi^2 k}\oint\!\!\!\oint\left(D^0_\rho(q)\right)^2\left[q_0\mathrm{Im}\Pi^{\Lambda}_{\rho\mathrm{M}}(q)-q_0\mathrm{Im}\Pi^{00}_{\rho\mathrm{M}}(q)-\mathrm{Im}\Pi^{1}_{\rho\mathrm{M}}(q)\right] \tag{5.1161}$$

$$\begin{aligned} W^{\mathrm{v}}_{\rho\mathrm{M}}(k,E) = &-g_\rho\left(\frac{f_\rho}{2M}\right)\frac{9M^*}{8\pi^2 k^3}\oint\!\!\!\oint\left(D^0_\rho(q)\right)^2\left[\frac{1}{2}\left(k^2+q^2-P^2\right)\mathrm{Im}\Pi^{\Lambda}_{\rho\mathrm{M}}(q)\right.\\ &\left.+q_0\mathrm{Im}\Pi^{2}_{\rho\mathrm{M}}(q)+\mathrm{Im}\Pi^{4}_{\rho\mathrm{M}}(q)\right] \end{aligned} \tag{5.1162}$$

In the Feynman diagrams discussed in this section, for the three internal nucleon lines, since Dirac propagators $G_{\mathrm{D}}$ containing δ function are used, they must be on-shell. But the poles of the meson propagators are not used in the formula derivation, $q_0$ and $\vec{q}$ are independent of each other, i.e., the meson is off-shell. The energy conservation at the vertices is ensured by $q_0 = E - V_0 - E_{k-q}$ given by Eq. (5.964). And Eq. (5.966) gives another form of this formula as

$$q_0 = E - V_0 - \left[k^2 + q^2 - 2kq\cos\theta + M^{*2}\right]^{1/2}$$

For the definite $E$, $V_0$, $M^*$, $k$, we know that $q$ and cos$\theta$ are not independent, and it has been given by Eq. (5.978) that cos$\theta$ of $-1 \rightarrow 1$ corresponds to $q$ of $q_{r2} \rightarrow q_{r1}$, so the integral over $q$ appeared in the formulas is essentially an integral over cos$\theta$. In the above formulas, $E$ and $k$ are independent variables, so $E$ and $k$, $q_0$ and $q$ can be off-shell, at this time, the imaginary part of the relativistic microscopic optical potential given by the formulas in this section is also off-shell. If taking

$$E = \left(k^2 + M^{*2}\right)^{1/2} + V_0 \tag{5.1163}$$

the corresponding imaginary part of the relativistic microscopic optical potential is on-shell.

In Ref. [80] the nucleon elastic scattering angular distribution $I(\theta)$, analyzing power $A_y(\theta)$, and spin rotation function $Q(\theta)$ were calculated by using the relativistic microscopic optical potential and local density approximation.

## 5.14 Contribution to Imaginary Part of Nucleon Relativistic Microscopic Optical Potential by Fourth Order Exchange Diagrams

Figure 5.8a, b represent, respectively, Hartree and Fock terms of the second order Feynman diagrams of the exchange mesons between particles discussed in Sect. 5.12; and Fig. 5.8c, d represent, respectively, the corresponding Hartree and Fock terms of the second order Feynman diagrams of the exchange mesons between the hole states under the Fermi surface.

Figure 5.9a is a fourth order particle polarization diagram with an intermediate state of 2plh discussed in the 5.13 section. Figure 5.9b is a corresponding fourth order particle exchange diagram, which has also the contribution for nucleon relativistic microscopic optical potential [81] that will be discussed in this section. Figure 5.9c, d are the fourth order hole correlation diagrams corresponding to Fig. 5.9a, b, respectively, which can be used to find a single particle potential imaginary part of the hole state below the Fermi surface with an intermediate state of 1p2h [82–84].

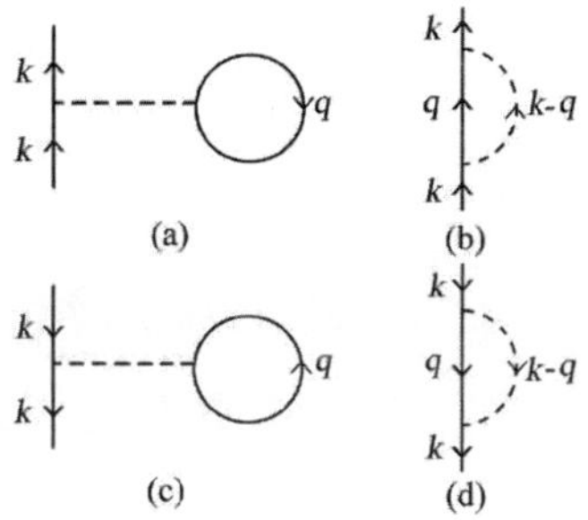


**Fig. 5.8** Second order Feynman diagrams of exchange σ mesons (**a**) Particle Hartree term (**b**) Particle Fock term (**c**) Hole Hartree term (**d**) Hole Fock term

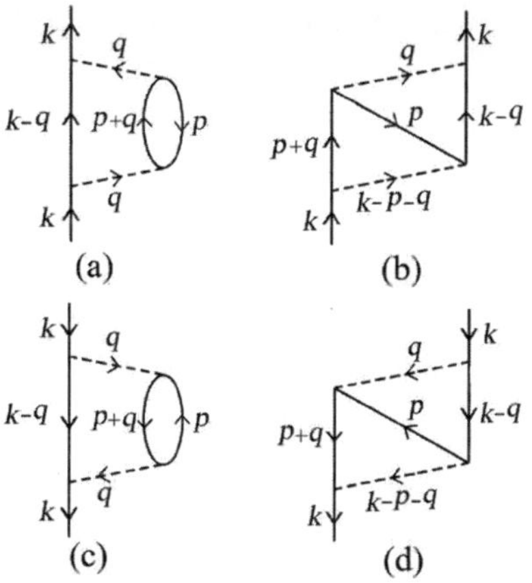


**Fig. 5.9** Fourth order Feynman diagrams for the exchange two σ mesons (**a**) Particle direct polarization graph (**b**) Particle exchange polarization graph (**c**) Hole direct correlation graph (**d**) Hole exchange correlation graph

1. *Exchange σ - σ mesons*

According to Feynman rules, the expression corresponding to Fig. 5.9b of the exchange σ - σ mesons is

$$\begin{aligned} \mathrm{i}G^0(k)\Sigma_\sigma(k)G(k) = \mathrm{i}G^0(k)\,(-\mathrm{i}g_\sigma)^2 \int \frac{\mathrm{d}^4q}{(2\pi)^4}\mathrm{i}\Delta^0(q)\,\mathrm{i}G(k-q)(-\mathrm{i}g_\sigma)^2 \\ \int \frac{\mathrm{d}^4p}{(2\pi)^4}\mathrm{i}G(p)\,\mathrm{i}\Delta^0(k-p-q)\,\mathrm{i}G(p+q)\,\mathrm{i}G(k) \end{aligned} \tag{5.1164}$$

Define the following symbols containing the $\gamma$ matrix:

$$\pi_\sigma(k,q) = \mathrm{i}g_\sigma^2 \int \frac{\mathrm{d}^4p}{(2\pi)^4} G(p)G(p+q)\Delta^0(k-p-q) \tag{5.1165}$$

then we can get

$$\Sigma_\sigma(k) = \mathrm{i}g_\sigma^2 \int \frac{\mathrm{d}^4q}{(2\pi)^4}\Delta^0(q)G(k-q)\pi_\sigma(k,q) \tag{5.1166}$$

Comparing Eq. (5.1166) and Eq. (5.956), based on Eq. (5.965), we can write

$$\begin{aligned} W_\sigma(k) = g_\sigma^2 \int \frac{\mathrm{d}\vec{q}}{(2\pi)^3}\Delta^0(q)\frac{\gamma^\mu(k^*-q)_\mu + M^*}{2E_{k-q}}\mathrm{Im}\pi_\sigma(k,q) \\ \theta\left(\left|\vec{k}-\vec{q}\right| - k_\mathrm{F}\right)\Big|_{q_0=E-V_0-E_{k-q}} \end{aligned} \tag{5.1167}$$

From Fig. 5.7b it can be seen that $p+q$ is a particle state and $p$ is a hole state, so it can be known from Eqs. (5.989) and (5.992) that

$$G(p) = \frac{\gamma^\nu \overline{p}_\nu^* + M^*}{2E_p}\,\frac{\theta\left(k_\mathrm{F} - |\vec{p}|\right)}{p_0 - V_0 - E_p - \mathrm{i}\varepsilon} \tag{5.1168}$$

$$G(p+q) = \frac{\gamma^\eta(\overline{p}^*+q)_\eta + M^*}{2E_{p+q}}\,\frac{\theta\left(|\vec{p}+\vec{q}| - k_\mathrm{F}\right)}{p_0 - V_0 + q_0 - E_{p+q} + \mathrm{i}\varepsilon} \tag{5.1169}$$

According to Eq. (5.721) we can write

$$\Delta^0(k-p-q) = \frac{1}{(k^*-p^*-q)_\xi(k^*-p^*-q)^\xi - m_\sigma^2} \tag{5.1170}$$

Comparing Eq. (5.1165) and Eq. (5.953) and referring to Eqs. (5.1001), (5.1002), (5.1008), and (5.1009), we can write

$$\mathrm{Im}\pi_\sigma(k,q)=\frac{g_\sigma^2}{16\pi q}\int_{E_x}^{E_F}\mathrm{d}E_p\left(\gamma^\nu\overline{p}_\nu^*+M^*\right)\left[\gamma^\eta(\overline{p}^*+q)_\eta+M^*\right]\Delta^0(k-p-q) \tag{5.1171}$$

The following expression can be obtained using Eqs. (5.1030), (5.1031), and (5.1033):

$$\begin{aligned}\overline{p}_\xi^*(k^*-q)^\xi &= (E-V_0-q_0)E_p-\vec{k}\cdot\vec{p}+\vec{p}\cdot\vec{q}=(E-V_0-q_0)E_p\\ &\quad+\frac{1}{k^2+q^2-P^2}\left[(k^2+q^2-P^2)-2k^2\right]\left(\frac{1}{2}q_\eta q^\eta+q_0E_p\right)\\ &=-\frac{1}{2}F(q)q_\eta q^\eta+\left[(E-V_0-q_0)-F(q)q_0\right]E_p\end{aligned} \tag{5.1172}$$

where

$$F(q)=\frac{k^2-q^2+P^2}{k^2+q^2-P^2} \tag{5.1173}$$

We can obtain the following tracing formula using Eqs. (5.149), (5.1172), (5.1133), and (5.1024):

$$\begin{aligned}&\mathrm{tr}\left\{\left[\gamma^\mu(k^*-q)_\mu+M^*\right]\left(\gamma^\nu\overline{p}_\nu^*+M^*\right)\left[\gamma^\eta(\overline{p}^*+q)_\eta+M^*\right]\right\}\\ &=4\lambda M^*\left[\overline{p}_\xi^*(k^*-q)^\xi+(\overline{p}^*+q)_\xi(k^*-q)^\xi+\overline{p}_\xi^*(\overline{p}^*+q)^\xi+M^{*2}\right]\\ &=4\lambda M^*\Big\{-F(q)q_\eta q^\eta+2\left[(E-V_0-q_0)-F(q)q_0\right]E_p\\ &\quad+(E-V_0-q_0)q_0-\frac{1}{2}(k^2-q^2-P^2)+2M^{*2}-\frac{1}{2}q_\eta q^\eta\Big\}\\ &\equiv 4\lambda\left[Q_\sigma^{\mathrm{s}}(q)+R_\sigma^{\mathrm{s}}(q)E_p\right]\end{aligned} \tag{5.1174}$$

where

$$Q_\sigma^{\mathrm{s}}(q)=M^*\left[2M^{*2}+(E-V_0-q_0)q_0-\frac{1}{2}(k^2-q^2-P^2)-\left(F(q)+\frac{1}{2}\right)q_\eta q^\eta\right] \tag{5.1175}$$

$$R_\sigma^{\mathrm{s}}(q)=2M^*\left[(E-V_0-q_0)-F(q)q_0\right] \tag{5.1176}$$

We can also find

$$
\begin{aligned}
&\mathrm{tr}\Big\{\gamma^0\left[\gamma^\mu(k^*-q)_\mu+M^*\right]\left(\gamma^\nu\overline{p}^*_\nu+M^*\right)\left[\gamma^\eta(\overline{p}^*+q)_\eta+M^*\right]\Big\}\\
&=4\lambda\Big\{M^{*2}\Big[(k^*-q)^0+\overline{p}^{*0}+(\overline{p}^*+q)^0\Big]+(k^*-q)^0\overline{p}^*_\xi(\overline{p}^*+q)^\xi\\
&\quad+(\overline{p}^*+q)^0\overline{p}^*_\xi(k^*-q)^\xi-\overline{p}^{*0}(\overline{p}^*+q)_\xi(k^*-q)^\xi\Big\}\\
&=4\lambda\Big(M^{*2}(E-V_0+2E_p)+(E-V_0-q_0)\Big(M^{*2}-\frac{1}{2}q_\eta q^\eta\Big)\\
&\quad+q_0\Big\{-\frac{1}{2}F(q)q_\eta q^\eta+[(E-V_0-q_0)-F(q)q_0]E_p\Big\}\\
&\quad-\Big[(E-V_0-q_0)q_0-\frac{1}{2}(k^2-q^2-P^2)\Big]E_p\Big)\\
&\equiv 4\lambda\left[Q^0_\sigma(q)+R^0_\sigma(q)E_p\right]
\end{aligned}
\tag{5.1177}
$$

where

$$
Q^0_\sigma(q)=2M^{*2}\Big(E-V_0-\frac{1}{2}q_0\Big)-\frac{1}{2}[(E-V_0-q_0)+F(q)q_0]q_\eta q^\eta \tag{5.1178}
$$

$$
R^0_\sigma(q)=2M^{*2}+\frac{1}{2}(k^2-q^2-P^2)-F(q)q_0^2 \tag{5.1179}
$$

And we can obtain the following expressions using Eqs. (5.1030), (5.1031), and (5.1033):

$$
\begin{aligned}
&k^i\mathrm{tr}\Big\{\gamma^i\left[\gamma^\mu(k^*-q)_\mu+M^*\right]\left(\gamma^\nu\overline{p}^*_\nu+M^*\right)\left[\gamma^\eta(\overline{p}^*+q)_\eta+M^*\right]\Big\}\\
&=4\lambda k^i\Big\{M^{*2}\left[(k-q)^i+p^i+(p+q)^i\right]+(k^*-q)^i\overline{p}^*_\xi(\overline{p}^*+q)^\xi\\
&\quad+(p+q)^i\overline{p}^*_\xi(k^*-q)^\xi-p^i(\overline{p}^*+q)_\xi(k^*-q)^\xi\Big\}\\
&=4\lambda\Big[M^{*2}\Big(k^2+2\vec{k}\cdot\vec{p}\Big)+\Big(k^2-\vec{k}\cdot\vec{q}\Big)\overline{p}^*_\xi(\overline{p}^*+q)^\xi\\
&\quad+\vec{k}\cdot\vec{q}\,\overline{p}^*_\xi(k^*-q)^\xi-\vec{k}\cdot\vec{p}\,q_\xi(k^*-q)^\xi\Big]\\
&=4\lambda\Big(M^{*2}k^2+\frac{1}{2}(k^2-q^2+P^2)\Big(M^{*2}-\frac{1}{2}q_\eta q^\eta\Big)\\
&\quad+\frac{1}{2}(k^2+q^2-P^2)\Big\{-\frac{1}{2}F(q)q_\eta q^\eta+[(E-V_0-q_0)-F(q)q_0]E_p\Big\}\\
&\quad+\frac{2k^2}{k^2+q^2-P^2}\Big(\frac{1}{2}q_\eta q^\eta+q_0E_p\Big)\Big[2M^{*2}-(E-V_0-q_0)q_0\\
&\quad+\frac{1}{2}(k^2-q^2-P^2)\Big]\Big)\equiv 4\lambda\left[Q^{\mathrm{v}}_\sigma(q)+R^{\mathrm{v}}_\sigma(q)E_p\right]
\end{aligned}
\tag{5.1180}
$$

where

$$Q^{\mathrm{v}}_{\sigma}(q) = \frac{1}{2}\left(3k^2 - q^2 + P^2\right) M^{*2} - \frac{1}{2}\left(k^2 - q^2 + P^2\right) q_{\eta} q^{\eta} + Z_0(q) q_{\eta} q^{\eta} \quad (5.1181)$$

$$R^{\mathrm{v}}_{\sigma}(q) = \frac{1}{2}\left(k^2 + q^2 - P^2\right) \left[(E - V_0 - q_0) - F(q) q_0\right] + 2Z_0(q) q_0 \quad (5.1182)$$

$$Z_0(q) = \frac{k^2}{k^2 + q^2 - P^2}\left[2M^{*2} - (E - V_0 - q_0) q_0 + \frac{1}{2}\left(k^2 - q^2 - P^2\right)\right] \quad (5.1183)$$

We can also obtain the following expression using Eqs. (5.1030), (5.1031), (5.1033), and (5.972):

$$\begin{aligned}
&(k^* - p^* - q)_{\xi}(k^* - p^* - q)^{\xi} \\
&\quad = \left(E - V_0 - q_0 - E_p\right)^2 - k^2 - p^2 - q^2 + 2\vec{k} \cdot \vec{p} + 2\vec{k} \cdot \vec{q} - 2\vec{p} \cdot \vec{q} \\
&\quad = P^2 + M^{*2} - 2(E - V_0 - q_0)E_p + E_p^2 - k^2 - E_p^2 + M^{*2} - q^2 + k^2 + q^2 - P^2 \\
&\qquad + \frac{2}{k^2 + q^2 - P^2}\left[2k^2 - \left(k^2 + q^2 - P^2\right)\right]\left(\frac{1}{2} q_{\eta} q^{\eta} + q_0 E_p\right) \\
&\quad = 2\left[M^{*2} - (E - V_0 - q_0)E_p + F(q)\left(\frac{1}{2} q_{\eta} q^{\eta} + q_0 E_p\right)\right]
\end{aligned} \quad (5.1184)$$

Then from Eq. (5.1170) we can obtain

$$\Delta^0(k - p - q) = \frac{1}{S_{\sigma}(q) + T(q) E_p} \quad (5.1185)$$

where

$$S_i(q) = 2M^{*2} + F(q) q_{\eta} q^{\eta} - m_i^2, \; i = \sigma, \omega \quad (5.1186)$$

$$T(q) = -2(E - V_0 - q_0) + 2F(q) q_0 \quad (5.1187)$$

There are the indefinite integral formulas:

$$\int \frac{\mathrm{d}x}{a + bx} = \frac{1}{b} \ln (a + bx) + C \quad (5.1188)$$

$$\int \frac{x\mathrm{d}x}{a + bx} = \frac{1}{b^2}\left[a + bx - a \ln (a + bx)\right] + C \quad (5.1189)$$

And we define the following quantities:

$$I_i(q) = \int_{E_x}^{E_F} \frac{dE_p}{S_i(q) + T(q)E_p} = \frac{1}{T(q)} \ln\left(\frac{S_i(q) + T(q)E_F}{S_i(q) + T(q)E_x}\right), \qquad i = \sigma,\omega \tag{5.1190}$$

$$\begin{aligned} J_i(q) &= \int_{E_x}^{E_F} \frac{E_p dE_p}{S_i(q) + T(q)E_p} \\ &= \frac{1}{(T(q))^2}\left[T(q)(E_F - E_x) - S_i(q)\ln\left(\frac{S_i(q) + T(q)E_F}{S_i(q) + T(q)E_x}\right)\right], \qquad i = \sigma,\omega \end{aligned} \tag{5.1191}$$

Comparing Eq. (5.1167) and Eq. (5.965) and according to Eqs. (5.986)–(5.988), the imaginary part of the relativistic microscopic optical potential contributed by the fourth order exchange graph of the σ meson can be written as

$$W_\sigma^s(k, E) = \frac{g_\sigma^2}{8\pi^2 k} \oint\!\!\oint \Delta^0(q) \mathrm{Im}\pi_\sigma^s(k, q) \tag{5.1192}$$

$$W_\sigma^0(k, E) = \frac{g_\sigma^2}{8\pi^2 k} \oint\!\!\oint \Delta^0(q) \mathrm{Im}\pi_\sigma^0(k, q) \tag{5.1193}$$

$$W_\sigma^v(k, E) = -\frac{g_\sigma^2}{8\pi^2 k^3} \oint\!\!\oint \Delta^0(q) \mathrm{Im}\pi_\sigma^v(k, q) \tag{5.1194}$$

It is noted that the contributions of $[\gamma^\mu(k^* - q)_\mu + M^*]$ in Eq. (5.1167) have been put in $\mathrm{Im}\pi_\sigma^l$ $(l = \mathrm{s}, 0, \mathrm{v})$. The definition of the symbol $\oint\!\!\oint$ has been given by Eq. (5.985).

Multiplying $[\gamma^\mu(k^* - q)_\mu + M^*]$ on the left end of Eq. (5.1171) and making trace according to Eq. (5.981), and noting that the negative sign of $W^v$ item has been placed in Eq. (5.1194), then using the results given by Eq. (5.1174)–(5.1198) we can obtain

$$\mathrm{Im}\pi_\sigma^l(k, q) = \frac{\lambda g_\sigma^2}{4\pi q}\left(Q_\sigma^l(q) I_\sigma(q) + R_\sigma^l(q) J_\sigma(q)\right), \quad l = \mathrm{s,0,v} \tag{5.1195}$$

2. *Exchange ω − ω mesons*

The expression corresponding to Fig. 5.9b of the exchange ω − ω mesons is

$$
\begin{aligned}
&\mathrm{i}G^0(k)\Sigma_\omega(k)G(k) \\
&= \mathrm{i}G^0(k)\,(-\mathrm{i}g_\omega)^2\gamma^\mu\int\frac{\mathrm{d}^4q}{(2\pi)^4}\mathrm{i}D^0_{\mu\alpha}(q)\,\mathrm{i}G(k-q)\,(-\mathrm{i}g_\omega)^2 \\
&\int\frac{\mathrm{d}^4p}{(2\pi)^4}\gamma^\alpha\mathrm{i}G(p)\gamma^\beta\mathrm{i}D^0_{\beta\nu}(k-p-q)\,\mathrm{i}G(p+q)\gamma^\nu\mathrm{i}G(k)
\end{aligned}
\tag{5.1196}
$$

Utilize Eq. (5.1015) and define

$$
\pi^{\mu\nu}_\omega(k,q) = \mathrm{i}g^2_\omega\int\frac{\mathrm{d}^4p}{(2\pi)^4}\gamma^\mu G(p)\gamma^\nu G(p+q)D^0(k-p-q) \tag{5.1197}
$$

then we can get

$$
\Sigma_\omega(k) = \mathrm{i}g^2_\omega\int\frac{\mathrm{d}^4q}{(2\pi)^4}D^0(q)\gamma^\mu G(k-q)g_{\mu\mu}\pi^{\mu\nu}_\omega(k,q)g_{\nu\nu}\gamma^\nu \tag{5.1198}
$$

Comparing Eq. (5.1198) with Eq. (5.1166) and according to Eq. (5.1167), we can write

$$
\begin{aligned}
W_\omega(k) = g^2_\omega\int\frac{\mathrm{d}^4q}{(2\pi)^3}D^0(q)\,\gamma^\mu\frac{\gamma^\eta(k^*-q)_\eta+M^*}{2E_{k-q}} \\
g_{\mu\mu}\mathrm{Im}\pi^{\mu\nu}_\omega(k,q)g_{\nu\nu}\gamma^\nu\,\theta\left(\left|\vec{k}-\vec{q}\right|-k_\mathrm{F}\right)\Big|_{q_0=E-V_0-E_{k-q}}
\end{aligned}
\tag{5.1199}
$$

Then comparing Eq. (5.1197) with Eq. (5.1165) and according to Eq. (5.1171), we can write

$$
\mathrm{Im}\pi^{\mu\nu}_\omega(k,q) = \frac{g^2_\omega}{16\pi q}\int_{E_x}^{E_\mathrm{F}}\mathrm{d}E_p\gamma^\mu\left(\gamma^\xi\overline{p}^*_\xi+M^*\right)\gamma^\nu\left[\gamma^\zeta\left(\overline{p}^*+q\right)_\zeta+M^*\right]D^0(k-p-q) \tag{5.1200}
$$

The following expression can be written according to Eqs. (5.1016), (5.1170), and (5.1185):

$$
D^0(k-p-q) = -\frac{1}{S_\omega(q)+T(q)E_p} \tag{5.1201}
$$

The following tracing formula can be obtained using Eqs. (5.870), (5.149)–(5.152), (5.1172), (5.1133), and (5.1024):

$$\begin{aligned}
&\operatorname{tr}\Big\{ g_{\mu\mu}\gamma^{\mu}\Big[\gamma^{\eta}(k^* - q)_{\eta} + M^*\Big]\gamma^{\mu}\Big(\gamma^{\xi}\overline{p}_{\xi}^{*}+M^*\Big)g_{\nu\nu}\gamma^{\nu}\Big[\gamma^{\zeta}(\overline{p}^* + q)_{\zeta} + M^*\Big)\Big]\gamma^{\nu}\Big\} \\
&= 16\operatorname{tr}\Big\{\Big[-\frac{1}{2}\gamma^{\eta}(k^* - q)_{\eta} + M^*\Big]\Big(\gamma^{\xi}\overline{p}_{\xi}^{*}+M^*\Big)\ \Big[-\frac{1}{2}\gamma^{\zeta}(\overline{p}^* + q)_{\zeta} + M^*\Big]\Big\} \\
&= 16\lambda M^*\Big[-2\overline{p}_{\xi}^{*}(k^* - q)^{\xi} + (\overline{p}^* + q)_{\xi}(k^* - q)^{\xi} - 2\overline{p}_{\xi}^{*}(\overline{p}^* + q)^{\xi}\Big] \\
&= -16\lambda M^*\Big[\overline{p}_{\xi}^{*}(k^* - q)^{\xi} - q_{\xi}(k^* - q)^{\xi} + 2\overline{p}_{\xi}^{*}(\overline{p}^* + q)^{\xi}\Big] \\
&= -16\lambda M^*\Big\{-\frac{1}{2}F(q)q_{\eta}q^{\eta} + [(E - V_0 - q_0) - F(q)q_0]E_p \\
&\quad -(E - V_0 - q_0)q_0 + \frac{1}{2}(k^2 - q^2 - P^2) + 2\Big(M^{*2} - \frac{1}{2}q_{\eta}q^{\eta}\Big)\Big\} \\
&= -16\lambda\big[Q_{\omega}^{s}(q) + R_{\omega}^{s}(q)E_p\big]
\end{aligned} \tag{5.1202}$$

where

$$Q_{\omega}^{s}(q) = M^*\Big[2M^{*2} - (E - V_0 - q_0)q_0 + \frac{1}{2}(k^2 - q^2 - P^2) - \frac{1}{2}(F(q) + 2)q_{\eta}q^{\eta}\Big] \tag{5.1203}$$

$$R_{\omega}^{s}(q) = M^*[(E - V_0 - q_0) - F(q)q_0] \tag{5.1204}$$

We can also obtain

$$\begin{aligned}
&\operatorname{tr}\Big\{\gamma^0 g_{\mu\mu}\gamma^{\mu}\Big[\gamma^{\eta}(k^* - q)_{\eta} + M^*\Big]\gamma^{\mu}\Big(\gamma^{\xi}\overline{p}_{\xi}^{*}+M^*\Big)g_{\nu\nu}\gamma^{\nu}\Big[\gamma^{\zeta}(\overline{p}^* + q)_{\zeta} + M^*\Big]\gamma^{\nu}\Big\} \\
&= 16\operatorname{tr}\Big\{\gamma^0\Big[-\frac{1}{2}\gamma^{\eta}(k^* - q)_{\eta} + M^*\Big]\Big(\gamma^{\xi}\overline{p}_{\xi}^{*}+M^*\Big)\ \Big[-\frac{1}{2}\gamma^{\zeta}(\overline{p}^* + q)_{\zeta} + M^*\Big]\Big\} \\
&= 16\lambda\Big\{M^{*2}\Big[-2(k^* - q)^0 + 4\overline{p}^{*0} - 2(\overline{p}^* + q)^0\Big] + (k^* - q)^0\overline{p}_{\xi}^{*}(\overline{p}^* + q)^{\xi} \\
&\quad +(\overline{p}^* + q)^0\overline{p}_{\xi}^{*}(k^* - q)^{\xi} - \overline{p}^{*0}(\overline{p}^* + q)_{\xi}(k^* - q)^{\xi}\Big\} \\
&= -16\lambda\Big\{2M^{*2}(E - V_0 - E_p) - (E - V_0 - q_0)\Big(M^{*2} - \frac{1}{2}q_{\eta}q^{\eta}\Big) + \frac{1}{2}F(q)q_0q_{\eta}q^{\eta} \\
&\quad -[(E - V_0 - q_0) - F(q)q_0]q_0E_p + \Big[(E - V_0 - q_0)q_0 - \frac{1}{2}(k^2 - q^2 - P^2)\Big]E_p\Big\} \\
&\equiv -16\lambda\big[Q_{\omega}^{0}(q) + R_{\omega}^{0}(q)E_p\big]
\end{aligned} \tag{5.1205}$$

where

$$Q_{\omega}^{0}(q) = M^{*2}(E - V_0 + q_0) + \frac{1}{2}[(E - V_0 - q_0) + F(q)q_0]q_\eta q^\eta \tag{5.1206}$$

$$R_{\omega}^{0}(q) = -2M^{*2} - \frac{1}{2}\left(k^2 - q^2 - P^2\right) + F(q)q_0^2 \tag{5.1207}$$

And the following expression can also be obtained:

$$\begin{aligned}
&k^i \mathrm{tr}\left\{\gamma^i g_{\mu\mu}\gamma^\mu\left[\gamma^\eta(k^* - q)_\eta + M^*\right]\gamma^\mu\left(\gamma^\xi \overline{p}_\xi^* + M^*\right)g_{\nu\nu}\gamma^\nu\left[\gamma^\zeta(\overline{p}^* + q)_\zeta + M^*\right]\gamma^\nu\right\}\\
&= 16k^i \mathrm{tr}\left\{\gamma^i\left[-\frac{1}{2}\gamma^\eta(k^* - q)_\eta + M^*\right]\left(\gamma^\xi \overline{p}_\xi^* + M^*\right)\left[-\frac{1}{2}\gamma^\zeta(\overline{p}^* + q)_\zeta + M^*\right]\right\}\\
&= 16\lambda k^i\left\{M^{*2}\left[-2(k-q)^i + 4p^i - 2(p+q)^i\right] + (k-q)^i \overline{p}_\xi^*(\overline{p}^* + q)^\xi\right.\\
&\quad \left.+(p+q)^i \overline{p}_\xi^*(k^* - q)^\xi - p^i(\overline{p}^* + q)_\xi(k^* - q)^\xi\right\}\\
&= -16\lambda\left[2M^{*2}\left(k^2 - \vec{k}\cdot\vec{p}\right) - \left(k^2 - \vec{k}\cdot\vec{q}\right)\overline{p}_\xi^*(\overline{p}^* + q)^\xi\right.\\
&\quad \left.-\vec{k}\cdot\vec{q}\,\overline{p}_\xi^*(k^* - q)^\xi + \vec{k}\cdot\vec{p}\,q_\xi(k^* - q)^\xi\right]\\
&= -16\lambda\left\{2M^{*2}k^2 - \frac{1}{2}\left(k^2 - q^2 + P^2\right)\left(M^{*2} - \frac{1}{2}q_\eta q^\eta\right)\right.\\
&\quad -\frac{1}{2}\left(k^2 + q^2 - P^2\right)\left\{-\frac{1}{2}F(q)q_\eta q^\eta + [(E - V_0 - q_0) - F(q)q_0]E_p\right\}\\
&\quad \left.-\frac{2k^2}{k^2 + q^2 - P^2}\left(\frac{1}{2}q_\eta q^\eta + q_0 E_p\right)\left[2M^{*2} - (E - V_0 - q_0)q_0 + \frac{1}{2}\left(k^2 - q^2 - P^2\right)\right]\right\}\\
&= -16\lambda\left[Q_{\omega}^{\mathrm{v}}(q) + R_{\omega}^{\mathrm{v}}(q)E_p\right]
\end{aligned} \tag{5.1208}$$

where

$$Q_{\omega}^{\mathrm{v}}(R) = \frac{1}{2}M^{*2}\left(3k^2 + q^2 - P^2\right) + \frac{1}{2}\left(k^2 - q^2 + P^2\right)q_\eta q^\eta - Z_0(q)q_\eta q^\eta \tag{5.1209}$$

$$R_{\omega}^{\mathrm{v}}(R) = -\frac{1}{2}\left(k^2 + q^2 - P^2\right)\left[(E - V_0 - q_0) - F(q)q_0\right] - 2Z_0(q)q_0 \tag{5.1210}$$

Referring to Eqs. (5.1192)–(5.1194) we can obtain

$$W_{\omega}^{\mathrm{s}}(k, E) = \frac{g_\omega^2}{8\pi^2 k}\oiint D^0(q)\mathrm{Im}\pi_{\omega}^{\mathrm{s}}(k, q) \tag{5.1211}$$

$$W_{\omega}^{0}(k, E) = \frac{g_\omega^2}{8\pi^2 k}\oiint D^0(q)\mathrm{Im}\pi_{\omega}^{0}(k, q) \tag{5.1212}$$

$$W_{\omega}^{\mathrm{v}}(k,E)=-\frac{g_{\omega}^{2}}{8\pi^{2}k^{3}}\oiint D^{0}(q)\mathrm{Im}\pi_{\omega}^{\mathrm{v}}(k,q) \tag{5.1213}$$

Referring to Eq. (5.1195) we can also obtain the following result comparing Eqs. (5.1199)–(5.1202) with Eqs. (5.1167), (5.1171), (5.1185), and (5.1174):

$$\mathrm{Im}\pi_{\omega}^{l}(k,q)=\frac{\lambda g_{\omega}^{2}}{\pi q}\left(Q_{\omega}^{l}(q)I_{\omega}(q)+R_{\omega}^{l}(q)J_{\omega}(q)\right),\qquad l=\mathrm{s,0,v} \tag{5.1214}$$

3. *Exchange σ − ω mesons*

The expression corresponding to Fig. 5.9b of the exchange σ − ω and σ − ω mesons is

$$\begin{aligned}
\mathrm{i}G^{0}(k)\Sigma_{\sigma\omega}(k)G(k)=\mathrm{i}G^{0}(k)&\Bigg[(-\mathrm{i}g_{\sigma})^{2}\int\frac{\mathrm{d}^{4}q}{(2\pi)^{4}}\,\mathrm{i}\Delta^{0}(q)\,\mathrm{i}G(k-q)\\
&(-\mathrm{i}g_{\omega})^{2}\int\frac{\mathrm{d}^{4}p}{(2\pi)^{4}}\mathrm{i}G(p)\gamma^{\mu}\mathrm{i}D_{\mu\nu}^{0}(k-p-q)\,\mathrm{i}G(p+q)\gamma^{\nu}\\
&+(-\mathrm{i}g_{\omega})^{2}\gamma^{\mu}\int\frac{\mathrm{d}^{4}q}{(2\pi)^{4}}\,\mathrm{i}D_{\mu\nu}^{0}(q)\,\mathrm{i}G(k-q)\,(-\mathrm{i}g_{\sigma})^{2}\\
&\int\frac{\mathrm{d}^{4}p}{(2\pi)^{4}}\gamma^{\nu}\mathrm{i}G(p)\,\mathrm{i}\Delta^{0}(k-p-q)\,\mathrm{i}G(p+q)\Bigg]\mathrm{i}G(k)
\end{aligned} \tag{5.1215}$$

Utilize Eq. (5.1015) and define

$$\begin{aligned}
\pi_{\sigma\omega}^{\mu}(k,q)&=\mathrm{i}g_{\sigma}g_{\omega}\int\frac{\mathrm{d}^{4}p}{(2\pi)^{4}}G(p)\gamma^{\mu}G(p+q)D^{0}(k-p-q)\\
\pi_{\omega\sigma}^{\mu}(k,q)&=\mathrm{i}g_{\sigma}g_{\omega}\int\frac{\mathrm{d}^{4}p}{(2\pi)^{4}}\gamma^{\mu}G(p)G(p+q)\Delta^{0}(k-p-q)
\end{aligned} \tag{5.1216}$$

then we can get

$$\begin{aligned}
\Sigma_{\sigma\omega}(k)=\mathrm{i}g_{\sigma}g_{\omega}\int\frac{\mathrm{d}^{4}q}{(2\pi)^{4}}&\left(\Delta^{0}(q)G(k-q)g_{\mu\mu}\pi_{\sigma\omega}^{\mu}(k,q)\gamma^{\mu}\right.\\
&\left.+D^{0}(q)\gamma^{\mu}G(k-q)g_{\mu\mu}\pi_{\omega\sigma}^{\mu}(k,q)\right)
\end{aligned} \tag{5.1217}$$

Do the following tracing derivation:

$$\begin{aligned}
&\mathrm{tr}\Big\{\Big[\gamma^\eta(k^*-q)_\eta+M^*\Big](\gamma^\xi\overline{p}_\xi{}^*+M^*)g_{\mu\mu}\gamma^\mu\Big[\gamma^\zeta(\overline{p}^*+q)_\zeta+M^*\Big]\gamma^\mu\Big\}\\
&=4\mathrm{tr}\Big\{\Big[\gamma^\eta(k^*-q)_\eta+M^*\Big](\gamma^\xi\overline{p}_\xi{}^*+M^*)\Big[-\frac{1}{2}\gamma^\zeta(\overline{p}^*+q)_\zeta+M^*\Big]\Big\}\\
&=8\lambda M^*\Big[2\overline{p}^*_\xi(k^*-q)^\xi-(\overline{p}^*+q)_\xi(k^*-q)^\xi-\overline{p}^*_\xi(\overline{p}^*+q)^\xi+2M^{*2}\Big]\\
&=-8\lambda M^*\Big\{\frac{1}{2}F(q)q_\eta q^\eta-[(E-V_0-q_0)-F(q)q_0]E_p\\
&\quad+(E-V_0-q_0)q_0-\frac{1}{2}\left(k^2-q^2-P^2\right)+3M^{*2}-\frac{1}{2}q_\eta q^\eta\Big\}\\
&\equiv-8\lambda\big[Q^{\mathrm{s}}_{\sigma\omega}(q)+R^{\mathrm{s}}_{\sigma\omega}(q)E_p\big]
\end{aligned}\tag{5.1218}$$

where

$$Q^{\mathrm{s}}_{\sigma\omega}(q)=M^*\Big[3M^{*2}+(E-V_0-q_0)q_0-\frac{1}{2}\left(k^2-q^2-P^2\right)+\frac{1}{2}(F(q)-1)q_\eta q^\eta\Big]\tag{5.1219}$$

$$R^{\mathrm{s}}_{\sigma\omega}(q)=-M^*[(E-V_0-q_0)-F(q)q_0]\tag{5.1220}$$

$$\begin{aligned}
&\mathrm{tr}\Big\{g_{\mu\mu}\gamma^\mu\Big[\gamma^\eta(k^*-q)_\eta+M^*\Big]\gamma^\mu\Big(\gamma^\xi\overline{p}^*_\xi+M^*\Big)\Big[\gamma^\zeta(\overline{p}^*+q)_\zeta+M^*\Big]\Big\}\\
&=4\mathrm{tr}\Big\{\Big[-\frac{1}{2}\gamma^\eta(k^*-q)_\eta+M^*\Big]\Big(\gamma^\xi\overline{p}^*_\xi+M^*\Big)\Big[\gamma^\zeta(\overline{p}^*+q)_\zeta+M^*\Big]\Big\}\\
&=8\lambda M^*\Big[-\overline{p}^*_\xi(k^*-q)^\xi-(\overline{p}^*+q)_\xi(k^*-q)^\xi+2\overline{p}^*_\xi(\overline{p}^*+q)^\xi+2M^{*2}\Big]\\
&=8\lambda M^*\Big\{F(q)q_\eta q^\eta-2[(E-V_0-q_0)-F(q)q_0]E_p\\
&\quad-(E-V_0-q_0)q_0+\frac{1}{2}\left(k^2-q^2-P^2\right)+4M^{*2}-q_\eta q^\eta\Big\}\\
&\equiv 8\lambda\big[Q^{\mathrm{s}}_{\omega\sigma}(q)+R^{\mathrm{s}}_{\omega\sigma}(q)E_p\big]
\end{aligned}\tag{5.1221}$$

where

$$Q^{\mathrm{s}}_{\omega\sigma}(q)=M^*\big[4M^{*2}-(E-V_0-q_0)q_0+\frac{1}{2}\left(k^2-q^2-P^2\right)+(F(q)-1)q_\eta q^\eta\big]\tag{5.1222}$$

$$R^{\mathrm{s}}_{\omega\sigma}(q)=-2M^*[(E-V_0-q_0)-F(q)q_0]\tag{5.1223}$$

$$
\begin{aligned}
&\mathrm{tr}\Big\{\gamma^0\Big[\gamma^\eta(k^*-q)_\eta+M^*\Big]\Big(\gamma^\xi\overline{p}^*_\xi+M^*\Big)g_{\mu\mu}\gamma^\mu\Big[\gamma^\zeta(\overline{p}^*+q)_\zeta+M^*\Big]\gamma^\mu\Big\}\\
&\quad=4\mathrm{tr}\Big\{\gamma^0\Big[\gamma^\eta(k^*-q)_\eta+M^*\Big]\Big(\gamma^\xi\overline{p}^*_\xi+M^*\Big)\Big[-\frac{1}{2}\gamma^\zeta(\overline{p}^*+q)_\zeta+M^*\Big]\Big\}\\
&\quad=8\lambda\Big\{M^{*2}\Big[2(k^*-q)^0+2\overline{p}^{*0}-(\overline{p}^*+q)^0\Big]-(k^*-q)^0\overline{p}^*_\xi(\overline{p}^*+q)^\xi\\
&\qquad-(\overline{p}^*+q)^0\overline{p}^*_\xi(k^*-q)^\xi+\overline{p}^{*0}(\overline{p}^*+q)_\xi(k^*-q)^\xi\Big\}\\
&\quad=-8\lambda\Big(-M^{*2}\big(2E-2V_0-3q_0+E_p\big)+(E-V_0-q_0)\Big(M^{*2}-\frac{1}{2}q_\eta q^\eta\Big)\\
&\qquad+q_0\Big\{-\frac{1}{2}F(q)q_\eta q^\eta+[(E-V_0-q_0)-F(q)q_0]E_p\Big\}\\
&\qquad-E_p\Big[(E-V_0-q_0)q_0-\frac{1}{2}\big(k^2-q^2-P^2\big)\Big]\Big)\\
&\quad\equiv-8\lambda\big[Q^0_{\sigma\omega}(q)+R^0_{\sigma\omega}(q)E_p\big]
\end{aligned}
\tag{5.1224}
$$

where

$$
Q^0_{\sigma\omega}(q)=-M^{*2}(E-V_0-2q_0)-\frac{1}{2}[(E-V_0-q_0)+F(q)q_0]q_\eta q^\eta \tag{5.1225}
$$

$$
R^0_{\sigma\omega}(q)=-M^{*2}+\frac{1}{2}\big(k^2-q^2-P^2\big)-F(q)q_0^2 \tag{5.1226}
$$

Referring to the above three formulas, we can obtain

$$
\begin{aligned}
&\mathrm{tr}\Big\{\gamma^0g_{\mu\mu}\gamma^\mu\Big[\gamma^\eta(k^*-q)_\eta+M^*\Big]\gamma^\mu\Big(\gamma^\xi\overline{p}^*_\xi+M^*\Big)\Big[\gamma^\zeta(\overline{p}^*+q)_\zeta+M^*\Big]\Big\}\\
&\quad=4\mathrm{tr}\Big\{\gamma^0\Big[-\frac{1}{2}\gamma^\eta(k^*-q)_\eta+M^*\Big]\Big(\gamma^\xi\overline{p}^*_\xi+M^*\Big)\Big[\gamma^\zeta(\overline{p}^*+q)_\zeta+M^*\Big]\Big\}\\
&\quad=8\lambda\Big\{M^{*2}\Big[-(k^*-q)^0+2\overline{p}^{*0}+2(\overline{p}^*+q)^0\Big]-(k^*-q)^0\overline{p}^*_\xi(\overline{p}^*+q)^\xi\\
&\qquad-(\overline{p}^*+q)^0\overline{p}^*_\xi(k^*-q)^\xi+\overline{p}^{*0}(\overline{p}^*+q)_\xi(k^*-q)^\xi\Big\}\\
&\quad\equiv8\lambda\big[Q^0_{\omega\sigma}(q)+R^0_{\omega\sigma}(q)E_p\big]
\end{aligned}
\tag{5.1227}
$$

where

$$
Q^0_{\omega\sigma}(q)=-2M^{*2}(E-V_0-2q_0)+\frac{1}{2}[(E-V_0-q_0)+F(q)q_0]q_\eta q^\eta \tag{5.1228}
$$

$$
R^0_{\omega\sigma}(q)=4M^{*2}-\frac{1}{2}\big(k^2-q^2-P^2\big)+F(q)q_0^2 \tag{5.1229}
$$

Then perform the following tracing derivation:

$$
\begin{aligned}
& k^i \mathrm{tr}\Big\{ \gamma^i \Big[\gamma^\eta (k^* - q)_\eta + M^*\Big] \Big(\gamma^\xi \overline{p}^*_\xi + M^*\Big) g_{\mu\mu} \gamma^\mu \Big[\gamma^\zeta (\overline{p}^* + q)_\zeta + M^*\Big] \gamma^\mu \Big\} \\
& = 4k^i \mathrm{tr}\Big\{ \gamma^i \Big[\gamma^\eta (k^* - q)_\eta + M^*\Big] \Big(\gamma^\xi \overline{p}^*_\xi + M^*\Big) \Big[-\frac{1}{2}\gamma^\zeta (\overline{p}^* + q)_\zeta + M^*\Big]\Big\} \\
& = 8\lambda k^i \Big\{ M^{*2}\big[2(k-q)^i + 2p^i - (p+q)^i\big] - (k-q)^i \overline{p}^*_\xi (\overline{p}^* + q)^\xi \\
& \quad - (p+q)^i \overline{p}^*_\xi (k^* - q)^\xi + p^i (\overline{p}^* + q)_\xi (k^* - q)^\xi \Big\} \\
& = -8\lambda \Big[-M^{*2}\Big(2k^2 - 3\vec{k}\cdot\vec{q} + \vec{k}\cdot\vec{p}\Big) + \Big(k^2 - \vec{k}\cdot\vec{q}\Big)\overline{p}^*_\xi (\overline{p}^* + q)^\xi \\
& \quad + \vec{k}\cdot\vec{q}\, \overline{p}^*_\xi (k^* - q)^\xi - \vec{k}\cdot\vec{p}\, q_\xi (k^* - q)^\xi \Big] \\
& = -8\lambda \Big(-\frac{1}{2} M^{*2}(k^2 - 3q^2 + 3P^2) + \frac{1}{2}(k^2 - q^2 + P^2)\Big(M^{*2} - \frac{1}{2} q_\eta q^\eta\Big) \\
& \quad + \frac{1}{2}(k^2 + q^2 - P^2)\Big\{-\frac{1}{2}F(q) q_\eta q^\eta + [(E - V_0 - q_0) - F(q) q_0] E_p\Big\} \\
& \quad - \frac{2k^2}{k^2 + q^2 - P^2}\Big(\frac{1}{2} q_\eta q^\eta + q_0 E_p\Big)\Big[M^{*2} + (E - V_0 - q_0) q_0 - \frac{1}{2}(k^2 - q^2 - P^2)\Big]\Big) \\
& \equiv -8\lambda \big[Q^{\mathrm{v}}_{\sigma\omega}(q) + R^{\mathrm{v}}_{\sigma\omega}(q) E_p\big]
\end{aligned}
\tag{5.1230}
$$

where

$$
Q^{\mathrm{v}}_{\sigma\omega}(q) = M^{*2}(q^2 - P^2) - \frac{1}{2}(k^2 - q^2 + P^2) q_\eta q^\eta - Z_1(q) q_\eta q^\eta \tag{5.1231}
$$

$$
R^{\mathrm{v}}_{\sigma\omega}(q) = \frac{1}{2}(k^2 + q^2 - P^2)[(E - V_0 - q_0) - F(q) q_0] - 2Z_1(q) q_0 \tag{5.1232}
$$

$$
Z_1(q) = \frac{k^2}{k^2 + q^2 - P^2}\Big[M^{*2} + (E - V_0 - q_0) q_0 - \frac{1}{2}(k^2 - q^2 - P^2)\Big] \tag{5.1233}
$$

Referring to the above four formulas, we can also obtain

$$
\begin{aligned}
& k^i \mathrm{tr}\Big\{ \gamma^i g_{\mu\mu} \gamma^\mu \Big[\gamma^\eta (k^* - q)_\eta + M^*\Big] \gamma^\mu \Big(\gamma^\xi \overline{p}^*_\xi + M^*\Big) \Big[\gamma^\zeta (\overline{p}^* + q)_\zeta + M^*\Big] \Big\} \\
& = 4k^i \mathrm{tr}\Big\{\gamma^i \Big[-\frac{1}{2}\gamma^\eta (k^* - q)_\eta + M^*\Big] \Big(\gamma^\xi \overline{p}^*_\xi + M^*\Big) \Big[\gamma^\zeta (\overline{p}^* + q)_\zeta + M^*\Big] \Big\} \\
& = 8\lambda k^i \Big\{ M^{*2}\big[-(k-q)^i + 2p^i + 2(p+q)^i\big] - (k-q)^i \overline{p}^*_\xi (\overline{p}^* + q)^\xi \\
& \quad - (p+q)^i \overline{p}^*_\xi (k^* - q)^\xi + p^i (\overline{p}^* + q)_\xi (k^* - q)^\xi \Big\} \\
& \equiv 8\lambda \big[Q^{\mathrm{v}}_{\omega\sigma}(q) + R^{\mathrm{v}}_{\omega\sigma}(q) E_p\big]
\end{aligned}
\tag{5.1234}
$$

where

$$Q^{\mathrm{v}}_{\omega\sigma}(q) = 2M^{*2}\left(q^2 - P^2\right) + \frac{1}{2}\left(k^2 - q^2 + P^2\right)q_\eta q^\eta + Z_2(q)q_\eta q^\eta \tag{5.1235}$$

$$R^{\mathrm{v}}_{\omega\sigma}(q) = -\frac{1}{2}\left(k^2 - q^2 - P^2\right)\left[(E - V_0 - q_0) - F(q)q_0\right] + 2Z_2(q)q_0 \tag{5.1236}$$

$$Z_2(q) = \frac{k^2}{k^2 + q^2 - P^2}\left[4M^{*2} + (E - V_0 - q_0)q_0 - \frac{1}{2}\left(k^2 - q^2 - P^2\right)\right] \tag{5.1237}$$

Referring to Eqs. (5.1192)–(5.1194) and (5.1211)–(5.1213), we can obtain

$$W^{\mathrm{s}}_{\sigma\omega}(k, E) = \frac{g_\sigma g_\omega}{8\pi^2 k}\oiint\left[\Delta^0(q)\mathrm{Im}\pi^{\mathrm{s}}_{\sigma\omega}(k,q) + D^0(q)\mathrm{Im}\pi^{\mathrm{s}}_{\omega\sigma}(k,q)\right] \tag{5.1238}$$

$$W^{0}_{\sigma\omega}(k, E) = \frac{g_\sigma g_\omega}{8\pi^2 k}\oiint\left[\Delta^0(q)\mathrm{Im}\pi^{0}_{\sigma\omega}(k,q) + D^0(q)\mathrm{Im}\pi^{0}_{\omega\sigma}(k,q)\right] \tag{5.1239}$$

$$W^{\mathrm{v}}_{\sigma\omega}(k, E) = -\frac{g_\sigma g_\omega}{8\pi^2 k^3}\oiint\left[\Delta^0(q)\mathrm{Im}\pi^{\mathrm{v}}_{\sigma\omega}(k,q) + D^0(q)\mathrm{Im}\pi^{\mathrm{v}}_{\omega\sigma}(k,q)\right] \tag{5.1240}$$

Referring to Eqs. (5.1195) and (5.1214), we can also obtain

$$\begin{aligned}
\mathrm{Im}\pi^{l}_{\sigma\omega}(k, q) &= \frac{\lambda g_\sigma g_\omega}{2\pi q}\left[Q^{l}_{\sigma\omega}(q)I_\omega(q) + R^{l}_{\sigma\omega}(q)J_\omega(q)\right],\\
\mathrm{Im}\pi^{l}_{\omega\sigma}(k, q) &= \frac{\lambda g_\sigma g_\omega}{2\pi q}\left[Q^{l}_{\omega\sigma}(q)I_\sigma(q) + R^{l}_{\omega\sigma}(q)J_\sigma(q)\right], \quad l = \mathrm{s,0,v}
\end{aligned} \tag{5.1241}$$

For the sake of simplicity, this section only discusses the contributions of σ and ω mesons.

## 5.15 Relativistic Bethe-Salpeter Equation

The contents of the previous few sections belong to the relativistic mean field theory. The relativistic mean field theory can be used to study the properties of nuclear matter or finite nuclei; to study spherical nuclei or deformed nuclei; to study the ground state or the excited state of the nucleus. For example, the relativistic Hartree-Fock-Bogoliubov equation, the relativistic RPA giant resonance theory, and so on all belong to the relativistic mean field theory. These theories can be used to study the nuclear pair effects, shell effects, deformation effects, collective excitation, microscopic fission barrier, etc, and the time-related relativistic nuclear problems. These theories mentioned above are mainly used to study the nuclear structure problem, so this book will not include these contents.

The relativistic mean field theory belongs to the relativistic one-body theory. The relativistic theory based on the two-body interaction (two-body realistic nuclear force) belongs to the relativistic two-body theory.

Let the 4-vector momentum before and after the interaction of the two nucleons be $p_1$,$p_2$ and $p_1'$,$p_2'$, respectively, and let

$$P = p_1 + p_2 = p_1' + p_2', \quad q = \frac{1}{2}(p_1 - p_2), \quad q' = \frac{1}{2}(p_1' - p_2') \tag{5.1242}$$

where $P$ is the total 4-vector momentum of the two nucleon system, which is the conservation quantity, $q$ and $q'$ are the 4-vector momentum of the relative motion of the initial state and the final state, respectively. Set $v(p_1', p_2'; p_1, p_2) = v(q', q; P)$ as the interaction potential between two free nucleons. In nuclear medium, the interaction between two nucleons is affected by the strong interaction of other nucleons and the Pauli incompatibility principle. In the many-body theory, the interaction $v$ includes the sum of all connected two-body irreducible graphs, but the ladder approximations are usually used, i.e., only one order graphs are retained, so the interaction between two nucleons in the nuclear medium can be expressed by the effective interaction or the reaction matrix $t(p_1', p_2'; p_1, p_2) = t(q', q; P)$.

Figure 5.10 is a schematic diagram of the two nucleon reaction in the nuclear medium under ladder approximation, where $k$ is the relative motion 4-vector momentum in the intermediate state. According to Feynman rules, the $v$ line and the $t$ line correspond to $-iv$ and $-it$, respectively, and the double nucleon Green function contributes $i^2 = -1$, so the expression corresponding to this graph is

$$\mathrm{t}(q', q; P) = v(q', q; P) + \int \frac{d^4k}{(2\pi)^4} v(q', k; P)\, iG(k, P)\, t(k, q; P) \tag{5.1243}$$

This is the relativistic Bethe-Salpeter (BS) equation [85–89]. Since the BS equation is written out according to the Feynman rules, the BS equation satisfies the Lorentz covariant. For the sake of simplicity, the spin and isospin of the nucleon are not marked in the above formula. Because the influence of the other nucleons in the medium is considered through iteration method in Eq. (5.1243), $G(k, P)$ can be considered to represent the propagator of two free nucleons. At this time, only the

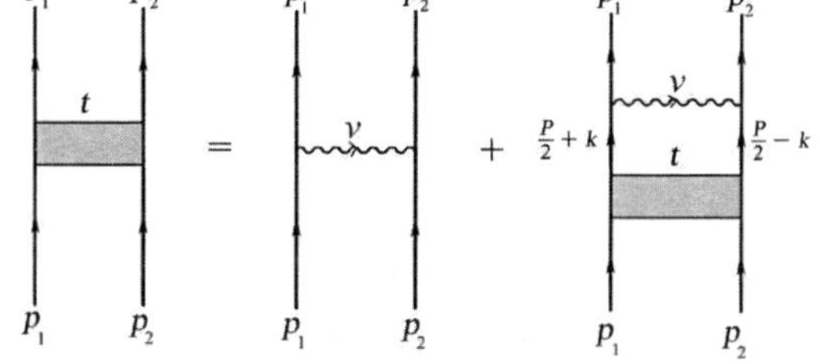


**Fig. 5.10** A schematic diagram of two nucleon reaction in the nuclear medium under ladder approximation

Fermi propagator $G_F^0$ item can be retained in the single nucleon propagator given by Eq. (5.781), so we can write

$$G(k,P)=\frac{\gamma^\mu\left(\frac{1}{2}P+k\right)_\mu+M}{\left(\frac{1}{2}P+k\right)_\mu\left(\frac{1}{2}P+k\right)^\mu-M^2+\mathrm{i}\varepsilon}\;\frac{\gamma^\mu\left(\frac{1}{2}P-k\right)_\mu+M}{\left(\frac{1}{2}P-k\right)_\mu\left(\frac{1}{2}P-k\right)^\mu-M^2+\mathrm{i}\varepsilon} \tag{5.1244}$$

where $M$ is a nucleon stationary mass and $G(k, P)$ is of-shell.

First define

$$E_{k\pm}\equiv E_{\frac{1}{2}P\pm k}=\left[M^2+\left(\frac{1}{2}\vec{P}\pm\vec{k}\right)^2\right]^{1/2} \tag{5.1245}$$

and let us do the following analyses:

$$\begin{aligned}\left(\frac{1}{2}P+k\right)_\mu\left(\frac{1}{2}P+k\right)^\mu-M^2+\mathrm{i}\varepsilon&=\left(\frac{1}{2}P_0+k_0\right)^2-\left(\frac{1}{2}\vec{P}+\vec{k}\right)^2\\-M^2+\mathrm{i}\varepsilon&=\left(\frac{1}{2}P_0+k_0-E_{k+}-\mathrm{i}\varepsilon\right)\left(\frac{1}{2}P_0+k_0+E_{k+}+\mathrm{i}\varepsilon\right)\end{aligned} \tag{5.1246}$$

$$\left(\frac{1}{2}P-k\right)_\mu\left(\frac{1}{2}P-k\right)^\mu-M^2+\mathrm{i}\varepsilon=\left(\frac{1}{2}P_0-k_0-E_{k-}-\mathrm{i}\varepsilon\right)\left(\frac{1}{2}P_0-k_0+E_{k-}+\mathrm{i}\varepsilon\right) \tag{5.1247}$$

There is a principal integral formula

$$\begin{aligned}\frac{1}{(x_1-\mathrm{i}\varepsilon)(x_2-\mathrm{i}\varepsilon)}&=\left(\mathcal{P}\frac{1}{x_1}+\mathrm{i}\pi\delta(x_1)\right)\frac{1}{x_2-\mathrm{i}\varepsilon}+\frac{1}{x_1-\mathrm{i}\varepsilon}\left(\mathcal{P}\frac{1}{x_2}+\mathrm{i}\pi\delta(x_2)\right)\\&=2\left(\mathcal{P}\frac{1}{x_1}+\mathrm{i}\pi\delta(x_1)\right)\left(\mathcal{P}\frac{1}{x_2}+\mathrm{i}\pi\delta(x_2)\right)\end{aligned} \tag{5.1248}$$

When making the numerical integrals, since the contributions to the real part represented by the $\mathcal{P}$ item on both sides of the pole will strongly cancel each other, therefore the contribution of $\mathcal{P}$ item can be approximately ignored. So we can obtain

$$\frac{1}{(x_1-\mathrm{i}\varepsilon)(x_2-\mathrm{i}\varepsilon)}\approx-2\pi^2\delta(x_1)\,\delta(x_2) \tag{5.1249}$$

Substituting Eqs. (5.1246) and (5.1247) into Eq. (5.1244) and using the principal integral formula Eq. (5.1249) for the first item taking positive energy in Eqs. (5.1246) and (5.1247), so we can get

$$G(k, P) = -\frac{\pi^2}{2E_{k+}E_{k-}}\left[\gamma^\mu\left(\frac{1}{2}P + k\right)_\mu + M\right]\left[\gamma^\mu\left(\frac{1}{2}P - k\right)_\mu + M\right]$$
$$\delta\left(\frac{1}{2}P_0 + k_0 - E_{k+}\right)\delta\left(\frac{1}{2}P_0 - k_0 - E_{k-}\right) \tag{5.1250}$$

In the case of nuclear matter and on-shell, the positive energy state projection operator $\Lambda_+$ given by Eq. (5.751) can be used, and from $\frac{1}{2}P_0 + k_0 - E_{k+} = 0$ and $\frac{1}{2}P_0 - k_0 - E_{k-} = 0$we can obtain $k_0 - \frac{1}{2}(E_{k+} - E_{k-}) = 0$and $P_0 - (E_{k+} + E_{k-}) = 0$, so Eq. (5.1250) can be rewritten as

$$G(k, P) \equiv G(k, P_0) = -\frac{2\pi^2 M^2}{E_{k+}E_{k-}}\Lambda_+^{(1)}\left(\frac{1}{2}\vec{P} + \vec{k}\right)\Lambda_+^{(2)}\left(\frac{1}{2}\vec{P} - \vec{k}\right)$$
$$\delta\left[k_0 - \frac{1}{2}(E_{k+} - E_{k-})\right]\delta[P_0 - (E_{k+} + E_{k-})] \tag{5.1251}$$

where (1) and (2) represent the particle labels. And we have

$$\Lambda_+\left(\vec{p}\right) = \frac{\gamma^0 E_p - \vec{\gamma}\cdot\vec{p} + M}{2M} = \sum_{\lambda=\pm\frac{1}{2}} u\left(\vec{p}, \lambda\right)\overline{u}\left(\vec{p}, \lambda\right) \tag{5.1252}$$

where $u\left(\vec{p}, \lambda\right)$ is the positive energy Dirac spinor for the momentum $\vec{p}$ and the spin projection $\lambda$. From Eq. (5.93) or Eq. (5.728) we can write

$$u\left(\vec{p}, \lambda\right) = \sqrt{\frac{E+M}{2M}}\begin{pmatrix} \chi_\lambda \\ \frac{\hat{\sigma}\cdot\vec{p}}{E+M}\chi_\lambda \end{pmatrix} \tag{5.1253}$$

where $\chi_{\frac{1}{2}} = \begin{pmatrix} 1 \\ 0 \end{pmatrix}, \chi_{-\frac{1}{2}} = \begin{pmatrix} 0 \\ 1 \end{pmatrix}$.

First let

$$E = E_{k+} + E_{k-} \tag{5.1254}$$

and from Eqs. (5.1242) and (5.177) we know

$$s = P_\mu P^\mu = P_0^2 - \vec{P}^2 \tag{5.1255}$$

Using δ function formula

$$\delta\left(x^2 - a^2\right) = \frac{\delta(x-a) + \delta(x+a)}{2|a|} \tag{5.1256}$$

we can write

$$2E\,\delta\left(P_0^2 - E^2\right) = \delta(P_0 - E) + \delta(P_0 + E) \tag{5.1257}$$

Due to $P_0 > 0$, using Eq. (5.1255) we can obtain

$$\delta(P_0 - E) = 2E\,\delta\left(s - E^2 + \vec{P}^2\right) \tag{5.1258}$$

and Eq. (5.1251) can be rewritten as

$$G(k, P) \equiv G(k, s) = -4\pi^2 M^2 \frac{E_{k+} + E_{k-}}{E_{k+}E_{k-}} \Lambda_+^{(1)}\left(\frac{1}{2}\vec{P} + \vec{k}\right)\Lambda_+^{(2)}\left(\frac{1}{2}\vec{P} - \vec{k}\right)$$
$$\delta\left[k_0 - \frac{1}{2}(E_{k+} - E_{k-})\right]\delta\left[s - (E_{k+} + E_{k-})^2 + \vec{P}^2\right] \tag{5.1259}$$

Let $z - \alpha = \rho\,e^{i\varphi}$, there is the following loop integral on the full plane

$$\oint \frac{dz}{z - \alpha} = \oint \frac{d\left(\alpha + \rho\,e^{i\varphi}\right)}{\rho\,e^{i\varphi}} = \int_0^{2\pi} \frac{i\rho\,e^{i\varphi}d\varphi}{\rho\,e^{i\varphi}} = 2\pi i$$

If we make a half-loop integration of the half plane corresponding to the positive real axis, and there is no singularity on the longitudinal complex axis, so there is $\oint \frac{dz}{z-\alpha} = \pi i$. Then for the function $f(s)$ using the residual theorem of the positive real axis half plane, we can write

$$f(s) = \frac{1}{\pi i}\int_{4M^2}^{\infty} \frac{f(s')}{s' - s - i\varepsilon} ds' \tag{5.1260}$$

In the above formula, $4M^2$ is the lower limit of $s$. Replace $f(s)$ with $G(k, s)$ in Eq. (5.1260) and use Eq. (5.1259), then we can obtain

$$iG(k, P) = iG(k, s) = -2\pi \frac{2M^2(E_{k+} + E_{k-})}{E_{k+}E_{k-}}$$
$$\frac{\Lambda_+^{(1)}\left(\frac{1}{2}\vec{P} + \vec{k}\right)\Lambda_+^{(2)}\left(\frac{1}{2}\vec{P} - \vec{k}\right)}{(E_{k+} + E_{k-})^2 - \vec{P}^2 - s - i\varepsilon}\delta\left[k_0 - \frac{1}{2}(E_{k+} - E_{k-})\right] \tag{5.1261}$$

Define

$$E_{q\pm} = E_{\frac{1}{2}P\pm q} = \left[M^2 + \left(\frac{1}{2}\vec{P} \pm \vec{q}\right)^2\right]^{1/2} \tag{5.1262}$$

and since $s$ is a conservation quantity, in Eq. (5.1261) $s$ can be taken as

$$s = \left(E_{q+} + E_{q-}\right)^2 - \vec{P}^2 \tag{5.1263}$$

So Eq. (5.1261) can be rewritten as

$$iG(k, P) = 2\pi \frac{2M^2(E_{k+} + E_{k-})}{E_{k+}E_{k-}} \frac{\Lambda_+^{(1)}\left(\frac{1}{2}\vec{P} + \vec{k}\right)\Lambda_+^{(2)}\left(\frac{1}{2}\vec{P} - \vec{k}\right)}{\left(E_{q+} + E_{q-}\right)^2 - (E_{k+} + E_{k-})^2 + i\varepsilon} \delta\left[k_0 - \frac{1}{2}(E_{k+} - E_{k-})\right] \tag{5.1264}$$

Substituting Eq. (5.1264) into the BS equation given by Eq. (5.1243) and noting that there is the δ function containing $k_0$ in Eq. (5.1264), we can obtain

$$t\left(\vec{q}', \vec{q}; \vec{P}\right) = v\left(\vec{q}', \vec{q}; \vec{P}\right) + \int \frac{d\vec{k}}{(2\pi)^3} v\left(\vec{q}', \vec{k}; \vec{P}\right) g\left(\vec{k}, \vec{P}\right) t\left(\vec{k}, \vec{q}; \vec{P}\right) \tag{5.1265}$$

where

$$g\left(\vec{k}, \vec{P}\right) = \frac{2M^2(E_{k+} + E_{k-})}{E_{k+}E_{k-}} \frac{\Lambda_+^{(1)}\left(\frac{1}{2}\vec{P} + \vec{k}\right)\Lambda_+^{(2)}\left(\frac{1}{2}\vec{P} - \vec{k}\right)}{\left(E_{q+} + E_{q-}\right)^2 - (E_{k+} + E_{k-})^2 + i\varepsilon} \tag{5.1266}$$

Then, $G(k, P)$ was transformed into a three-dimensional propagator $g\left(\vec{k}, \vec{P}\right)$; we call Eq. (5.1260) and Eq. (5.1264) as Blankenbecler-Sugar (BbS) selection [90]. It can be seen from Eq. (5.1264) that the relative momentum time-like component $k_0$ of the intermediate state is not an independent variable. Since the two-nucleon interaction also contains elastic scattering process, it is possible to infer that the time-like components $q_0$ and $q_0'$ of the initial and final states are not independent variables, so Eqs. (5.1265) and (5.1266) are equations that only contain three-dimensional momentum $\vec{q}$, $\vec{k}$, $\vec{q}'$, and $\vec{P}$.

Using the mean approximation of angles, there are $\left(\frac{1}{2}\vec{P} + \vec{k}\right)^2 \approx \frac{1}{4}\vec{P}^2 + \vec{k}^2$ and $\left(\frac{1}{2}\vec{P} - \vec{k}\right)^2 \approx \frac{1}{4}\vec{P}^2 + \vec{k}^2$, and then Eq. (5.1264) and Eq. (5.1266) can be simplified to

$$\mathrm{i}G(k, P) = 2\pi \frac{M^2}{E_{k+}} \frac{\Lambda_+^{(1)}\left(\frac{1}{2}\vec{P} + \vec{k}\right)\Lambda_+^{(2)}\left(\frac{1}{2}\vec{P} - \vec{k}\right)}{E_{q+}^2 - E_{k+}^2 + \mathrm{i}\varepsilon} \delta(k_0) \tag{5.1267}$$

$$g\left(\vec{k}, \vec{P}\right) = \frac{M^2}{E_{k+}} \frac{\Lambda_+^{(1)}\left(\frac{1}{2}\vec{P} + \vec{k}\right)\Lambda_+^{(2)}\left(\frac{1}{2}\vec{P} - \vec{k}\right)}{E_{q+}^2 - E_{k+}^2 + \mathrm{i}\varepsilon} \tag{5.1268}$$

It can be seen from $\delta(k_0)$ that in this case there is $k_0 = q_0 = q_0' = 0$. The positive energy Dirac spinor that appears in Eq. (5.1252) can be expressed by $u\left(\vec{p}\right)$ when without considering nucleon spin and isospin. Using $\overline{u}_1\left(\frac{1}{2}\vec{P} + \vec{q}'\right)\overline{u}_2\left(\frac{1}{2}\vec{P} - \vec{q}'\right)$ and $u_1\left(\frac{1}{2}\vec{P} + \vec{q}\right)u_2\left(\frac{1}{2}\vec{P} - \vec{q}\right)$ to act on Eq. (5.1265) from left and right sides, respectively, since $u\left(\vec{p}\right)$ and $v\left(\vec{q}', \vec{q}; \vec{P}\right)$ are both Lorentz invariants, so we can define

$$\begin{aligned} &\overline{u}_1\left(\frac{1}{2}\vec{P} + \vec{q}'\right)\overline{u}_2\left(\frac{1}{2}\vec{P} - \vec{q}'\right)v\left(\vec{q}', \vec{q}; \vec{P}\right)u_1\left(\frac{1}{2}\vec{P} + \vec{q}\right)u_2\left(\frac{1}{2}\vec{P} - \vec{q}\right) \\ &\quad = \overline{u}_1\left(\vec{q}'\right)\overline{u}_2\left(-\vec{q}'\right)\, v\left(\vec{q}', \vec{q}\right)\, u_1\left(\vec{q}\right)u_2\left(-\vec{q}\right) \equiv V\left(\vec{q}', \vec{q}\right) \end{aligned} \tag{5.1269}$$

$$\begin{aligned} &\overline{u}_1\left(\frac{1}{2}\vec{P} + \vec{q}'\right)\overline{u}_2\left(\frac{1}{2}\vec{P} - \vec{q}'\right)t\left(\vec{q}', \vec{q}; \vec{P}\right)\, u_1\left(\frac{1}{2}\vec{P} + \vec{q}\right)u_2\left(\frac{1}{2}\vec{P} - \vec{q}\right) \\ &\quad = \overline{u}_1\left(\vec{q}'\right)\overline{u}_2\left(-\vec{q}'\right)\, t\left(\vec{q}', \vec{q}; \vec{P}\right)u_1\left(\vec{q}\right)u_2\left(-\vec{q}\right) \equiv T\left(\vec{q}', \vec{q}; \vec{P}\right) \end{aligned} \tag{5.1270}$$

Strictly speaking, the $t$ matrix $t\left(\vec{q}', \vec{q}; \vec{P}\right)$ is not a Lorentz invariant. Using Eq. (5.1268) and the projection operator expression without summation to $\lambda$ given by Eq. (5.1252), we can obtain

$$T\left(\vec{q}', \vec{q}; \vec{P}\right) = V\left(\vec{q}', \vec{q}\right) + \int \frac{\mathrm{d}\vec{k}}{(2\pi)^3} V\left(\vec{q}', \vec{k}\right) \frac{M^2}{E_{k+}} \frac{1}{E_{q+}^2 - E_{k+}^2 + \mathrm{i}\varepsilon} T\left(\vec{k}, \vec{q}; \vec{P}\right) \tag{5.1271}$$

The equation composed of Eqs. (5.1265) and (5.1266) and the equation given by Eq. (5.1271) are all called BbS equation.

In the two-nucleon center of momentum system ($\vec{P} = 0$), Eqs. (5.1267) and (5.1271) are simplified to

$$\mathrm{i}G(k, s) = 2\pi \frac{M^2}{E_{k+}} \frac{\Lambda_+^{(1)}\left(\vec{k}\right)\Lambda_+^{(2)}\left(-\vec{k}\right)}{\frac{s}{4} - E_k^2 + \mathrm{i}\varepsilon} \delta(k_0) \tag{5.1272}$$

$$T\left(\vec{q}', \vec{q}\right) = V\left(\vec{q}', \vec{q}\right) + \int \frac{\mathrm{d}\vec{k}}{(2\pi)^3} V\left(\vec{q}', \vec{k}\right) \frac{M^2}{E_k} \frac{1}{\vec{q}^{\,2} - \vec{k}^{\,2} + \mathrm{i}\varepsilon} T\left(\vec{k}, \vec{q}\right) \tag{5.1273}$$

where

$$E_k = \left[M^2 + \vec{k}^{\,2}\right]^{1/2} \tag{5.1274}$$

According to the residual theorem of the positive real axis half plane, it can be written that

$$f(P_0) = \frac{1}{\pi \mathrm{i}} \int_{2M}^{\infty} \frac{f\left(P_0'\right)}{P_0' - P_0 - \mathrm{i}\varepsilon} \mathrm{d}P_0' \tag{5.1275}$$

Utilizing $G(k, P_0)$ to replace $f(P_0)$ in Eq. (5.1275) and using Eq. (5.1251), we can obtain

$$\begin{aligned} \mathrm{i}G(k, P) = \mathrm{i}G(k, P_0) &= -2\pi \frac{M^2}{E_{k+}E_{k-}} \frac{\Lambda_+^{(1)}\left(\frac{1}{2}\vec{P} + \vec{k}\right)\Lambda_+^{(2)}\left(\frac{1}{2}\vec{P} - \vec{k}\right)}{(E_{k+} + E_{k-}) - P_0 - \mathrm{i}\varepsilon} \\ &\quad \delta\left[k_0 - \frac{1}{2}(E_{k+} - E_{k-})\right] \\ &= 2\pi \frac{M^2}{E_{k+}E_{k-}} \frac{\Lambda_+^{(1)}\left(\frac{1}{2}\vec{P} + \vec{k}\right)\Lambda_+^{(2)}\left(\frac{1}{2}\vec{P} - \vec{k}\right)}{E_{q+} + E_{q-} - E_{k+} - E_{k-} + \mathrm{i}\varepsilon} \delta\left[k_0 - \frac{1}{2}(E_{k+} - E_{k-})\right] \end{aligned} \tag{5.1276}$$

And substituting Eq. (5.1276) into Eq. (5.1243), we can obtain Eq. (5.1265), where

$$g\left(\vec{k}, \vec{P}\right) = \frac{M^2}{E_{k+}E_{k-}} \frac{\Lambda_+^{(1)}\left(\frac{1}{2}\vec{P} + \vec{k}\right)\Lambda_+^{(2)}\left(\frac{1}{2}\vec{P} - \vec{k}\right)}{E_{q+} + E_{q-} - E_{k+} - E_{k-} + \mathrm{i}\varepsilon} \tag{5.1277}$$

As before, we use the mean approximation of angles and the matrix element of the positive energy Dirac spinor, the following equation can be obtained:

$$T\left(\vec{q}', \vec{q}; \vec{P}\right) = V\left(\vec{q}', \vec{q}\right) + \int \frac{\mathrm{d}\vec{k}}{(2\pi)^3} V\left(\vec{q}', \vec{k}\right) \frac{M^2}{E_{k+}^2} \frac{1}{2E_{q+} - 2E_{k+} + \mathrm{i}\varepsilon} T\left(\vec{k}, \vec{q}; \vec{P}\right) \tag{5.1278}$$

The equation composed of Eq. (5.1265) and Eq. (5.1277) and the equation given by Eq. (5.1278) are all called Thompson equation [91]. In the two-nucleon center of momentum system ($\vec{P} = 0$), Eq. (5.1278) is simplified to

$$T\left(\vec{q}', \vec{q}\right) = V\left(\vec{q}', \vec{q}\right) + \int \frac{\mathrm{d}\vec{k}}{(2\pi)^3} V\left(\vec{q}', \vec{k}\right) \frac{M^2}{E_k^2} \frac{1}{2E_q - 2E_k + \mathrm{i}\varepsilon} T\left(\vec{k}, \vec{q}\right) \tag{5.1279}$$

Since the BbS equation and the Thompson equation are derived from the BS equation but a certain approximation was made, it can only be said that they approximately satisfy Lorentz invariance, and at the same time it can be assumed that the $t$ matrix $t\left(\vec{q}', \vec{q}; \vec{P}\right)$ also approximately satisfies Lorentz invariance.

## 5.16 Bonn One Boson Exchange Potential

For the description of strong interacting nuclear forces, the meson theory based on perturbation theory is usually applied. The contribution to the lowest order of NN scattering is the one boson exchange graph. This section introduces the Bonn one boson exchange potential (OBEP) [92–94].

When the distance between the two nucleons is very short, the meson exchange graph will no longer be applicable due to the internal structure of the hadron. The usual method is to introduce the vertex shape factor phenomenally. Since it is a repulsive force in a very short range, it is equivalent to establishing a repulsion wall.

Use $(J^\pi, I)$ to represent the spin, parity, and isospin of the meson. In the one boson exchange (OBE), three isospin scalar mesons $\sigma(0^+, 0)$, $\omega(1^-, 0)$, and $\eta(0^-, 0)$ and three isospin vector mesons $\delta(0^+, 1)$, $\rho(1^-, 1)$, and $\pi(0^-, 1)$ are included. Since the existence of $\sigma$ meson has not been confirmed experimentally, people usually regard $\sigma$ exchange as an effective $2\pi$ mesons exchange.

Refer to Eqs. (5.668)–(5.671), we can write the Lagrangian densities of the scalar s($\sigma$), vector v($\omega$), pseudo-scalar ps($\eta$), and axial vector pv($\eta$) mesons, which all belong to the isospin scalar mesons coupled to the nucleons, as follows [89, 92–94]:

$$\mathcal{L}_\mathrm{s} = -g_\mathrm{s}\overline{\psi}\psi\varphi^{(\mathrm{s})} \tag{5.1280}$$

$$\mathcal{L}_{\mathrm{v}} = -g_{\mathrm{v}}\overline{\psi}\gamma^{\mu}\psi\varphi_{\mu}^{(\mathrm{v})} - \frac{f_{\mathrm{v}}}{4M}\overline{\psi}\sigma^{\mu\nu}\psi\left(\partial_{\mu}\varphi_{\nu}^{(\mathrm{v})} - \partial_{\nu}\varphi_{\mu}^{(\mathrm{v})}\right) \tag{5.1281}$$

$$\mathcal{L}_{\mathrm{ps}} = -g_{\mathrm{ps}}\overline{\psi}\, i\gamma^{5}\psi\varphi^{(\mathrm{ps})} \tag{5.1282}$$

$$\mathcal{L}_{\mathrm{pv}} = -\frac{f_{\mathrm{ps}}}{m_{\mathrm{ps}}}\overline{\psi}\gamma^{5}\gamma^{\mu}\psi\ \partial_{\mu}\varphi^{(\mathrm{ps})} \tag{5.1283}$$

where $\psi$ is a nucleon field and $\varphi_{(\mu)}^{(\alpha)}$ is a meson field. For the corresponding isospin vector meson δ, ρ, π, we use $\vec{\tau}\cdot\vec{\varphi}_{(\mu)}^{(\alpha)}$ instead of $\varphi_{(\mu)}^{(\alpha)}$.

The study of the interaction between two bare nucleons is usually carried out in the two-nucleon center of momentum (c.m.) system. Figure 5.11 gives a Feynman diagram of the contribution of one boson exchange to NN scattering in the c.m. system. From Eq. (5.721) we can write the scalar meson propagator corresponding to this graph as

$$\Delta_{\mathrm{s}}^{0}(q'-q) = \frac{1}{\left(q_0'-q_0\right)^2 - \left(\vec{q}\,'-\vec{q}\right)^2 - m_{\mathrm{s}}^2} \tag{5.1284}$$

Equation (5.1267) indicates that the two-nucleon propagator contains $\delta(k_0)$ in the case by using the mean approximation of angles; therefore there is $k_0 = q_0 = q_0' = 0$, so the following three-dimensional static propagator can be used [92]:

$$\Delta_{\mathrm{s}}^{0}(q'-q) = -\frac{1}{\left(\vec{q}\,'-\vec{q}\right)^2 + m_{\mathrm{s}}^2} \tag{5.1285}$$

OBEP is defined as

$$v_{\mathrm{OBEP}} = \sum_{\alpha=\sigma,\delta,\omega,\rho,\eta,\pi} v_{\alpha}^{\mathrm{OBE}} \tag{5.1286}$$

For the isospin vector mesons δ, ρ, π, above formula needs to multiply by a factor $\vec{\tau}_1\cdot\vec{\tau}_2$. The on-shell positive energy free nucleon plane wave spinor is

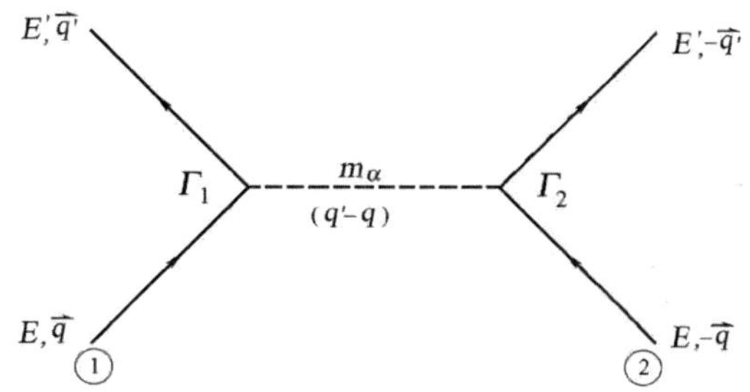


**Fig. 5.11** Feynman diagram of the contribution of one boson exchange to NN scattering in c. m. system

$$u\left(\vec{q}, \lambda\right) = \sqrt{\frac{E+M}{2M}} \begin{pmatrix} \chi_\lambda \\ \frac{\hat{\sigma} \cdot \vec{q}}{E+M} \chi_\lambda \end{pmatrix} \tag{5.1287}$$

where $E = \left(\vec{q}^{\,2} + M^2\right)^{1/2}$. In addition to the Feynman rules described in Sect. 5.11, there are the following Feynman rules:

1. The initial state positive energy nucleon outside line is expressed by $u\left(\vec{q}, \lambda\right)$, and the final state positive energy nucleon outside line is expressed by $\overline{u}\left(\vec{q}, \lambda\right)$.
2. A $v$ line describing nucleon-nucleon interaction is expressed by $-\mathrm{i}v$.
3. In c.m. system, the two-nucleon initial state is expressed by $\left|\vec{q}\lambda_1\lambda_2\right\rangle$, and the corresponding final state is expressed by $\left\langle\vec{q}'\lambda_1'\lambda_2'\right|$.

Using the Feynman rules, we can first write the OBE amplitude of the scalar meson as

$$\left\langle\vec{q}'\lambda_1'\lambda_2'\left|v_{\mathrm{s}}^{\mathrm{OBE}}\right|\vec{q}\lambda_1\lambda_2\right\rangle = -\frac{g_{\mathrm{s}}^2\overline{u}\left(\vec{q}', \lambda_1'\right)u\left(\vec{q}, \lambda_1\right)\overline{u}\left(-\vec{q}', \lambda_2'\right)u\left(-\vec{q}, \lambda_2\right)}{\left(\vec{q}' - \vec{q}\right)^2 + m_{\mathrm{s}}^2} \tag{5.1288}$$

According to Eq. (5.812) the static propagator of the vector meson is

$$D_{\mu\nu}^0(q' - q) = \frac{g_{\mu\nu}}{\left(\vec{q}' - \vec{q}\right)^2 + m_{\mathrm{v}}^2} \tag{5.1289}$$

Here introduce the following $\gamma$ matrix expressions with the lower footnote in the BD metric:

$$\gamma_\mu \equiv g_{\mu\mu}\gamma^\mu \tag{5.1290}$$

$$\sigma_{\mu\nu} \equiv \frac{\mathrm{i}}{2}\left(\gamma_\mu\gamma_\nu - \gamma_\nu\gamma_\mu\right) = \frac{\mathrm{i}}{2} g_{\mu\mu}g_{\nu\nu}(\gamma^\mu\gamma^\nu - \gamma^\nu\gamma^\mu) = g_{\mu\mu}g_{\nu\nu}\sigma^{\mu\nu} \tag{5.1291}$$

Note that the $\gamma$ matrices with the lower footnote when using the BD metric in the above two formulas are not the same as the $\gamma$ matrices with the lower footnote when using the Pauli metric introduced by Eq. (5.26). From Eq. (5.880) we can see that the meson propagator in space-time coordinate is

$$\mathrm{i}\Delta^0(x, x') = \mathrm{i}\int \frac{\mathrm{d}^4 k}{(2\pi)^4}\Delta^0(k)\,\mathrm{e}^{-\mathrm{i}k_\mu(x-x')^\mu} \tag{5.1292}$$

The contribution of $\partial_{\mu x}$ at the vertex $x$ is $-\mathrm{i}k_\mu$; the contribution of $\partial_{\mu x'}$ at the vertex $x'$ is $\mathrm{i}k_\mu$. In the second item of Eq. (5.1281), $\sigma^{\mu\nu}$ is anti-symmetric; therefore $\partial_\mu\varphi_\nu^{(\mathrm{v})}$ and $-\partial_\nu\varphi_\mu^{(\mathrm{v})}$ have the same contribution. So we take the second item and multiply it by 2, and make the change $g_{\mu\nu}\to g_{\mu\mu}$ in Eq. (5.1289) to be used to the first vertex; also note that in Fig. 5.11 the 4-momentum of the meson propagator is $q'-q$, and then the OBE amplitude of the vector meson can be obtained as

$$\begin{aligned}&\left\langle \vec{q}'\lambda_1'\lambda_2'\middle| v_\mu^{\mathrm{OBE}}\middle|\vec{q}\lambda_1\lambda_2\right\rangle\\ &= \left[g_{\mathrm{v}}\bar{u}\left(\vec{q}',\lambda_1'\right)\gamma_\mu u\left(\vec{q},\lambda_1\right)+\frac{f_{\mathrm{v}}}{2M}\bar{u}\left(\vec{q}',\lambda_1'\right)\sigma_{\mu\nu}\mathrm{i}(q'-q)^\nu u\left(\vec{q},\lambda_1\right)\right]\\ &\quad\left[g_{\mathrm{v}}\bar{u}\left(-\vec{q}',\lambda_2'\right)\gamma^\mu u\left(-\vec{q},\lambda_2\right)-\frac{f_{\mathrm{v}}}{2M}\bar{u}\left(-\vec{q}',\lambda_2'\right)\sigma^{\mu\nu}\mathrm{i}(q'-q)_\nu u\left(-\vec{q}',\lambda_2\right)\right]\\ &\quad\frac{1}{\left(\vec{q}'-\vec{q}\right)^2+m_{\mathrm{v}}^2}\end{aligned} \tag{5.1293}$$

We can also write

$$\left\langle \vec{q}'\lambda_1'\lambda_2'\middle| v_{\mathrm{ps}}^{\mathrm{OBE}}\middle|\vec{q}\lambda_1\lambda_2\right\rangle = \frac{-g_{\mathrm{ps}}^2\bar{u}\left(\vec{q}',\lambda_1'\right)\,\mathrm{i}\gamma^5 u\left(\vec{q},\lambda_1\right)\bar{u}\left(-\vec{q}',\lambda_2'\right)\,\mathrm{i}\gamma^5 u\left(-\vec{q},\lambda_2\right)}{\left(\vec{q}'-\vec{q}\right)^2+m_{\mathrm{ps}}^2} \tag{5.1294}$$

$$\begin{aligned}\left\langle \vec{q}'\lambda_1'\lambda_2'\middle| v_{\mathrm{pv}}^{\mathrm{OBE}}\middle|\vec{q}\lambda_1\lambda_2\right\rangle &= \frac{f_{\mathrm{ps}}^2}{m_{\mathrm{ps}}^2}\bar{u}\left(\vec{q}',\lambda_1'\right)\gamma^5\gamma^\mu\mathrm{i}(q'-q)_\mu u\left(\vec{q},\lambda_1\right)\bar{u}\left(-\vec{q}',\lambda_2'\right)\\ &\quad\gamma^5\gamma^\mu\,\mathrm{i}(q'-q)_\mu u\left(-\vec{q},\lambda_2\right)\frac{1}{\left(\vec{q}'-\vec{q}\right)^2+m_{\mathrm{ps}}^2}\end{aligned} \tag{5.1295}$$

The shape factor used for each meson-nucleon vertex is

$$F_\alpha\left[\left(\vec{q}'-\vec{q}\right)^2\right] = \left[\frac{\Lambda_\alpha^2-m_\alpha^2}{\Lambda_\alpha^2+\left(\vec{q}'-\vec{q}\right)^2}\right]^{n_\alpha} \tag{5.1296}$$

where $\Lambda_\alpha$ is called truncation mass.

Since the multiple meson exchanges occur between two bare nucleons, it is necessary to use Brueckner theory to obtain an effective interaction that considers the effect of multiple meson exchanges. Now use symbol $v$ instead of $v_{\mathrm{OBEP}}$, and let

$$V\left(\vec{q}'\lambda_1'\lambda_2', \vec{q}\lambda_1\lambda_2\right) = \left\langle \vec{q}'\lambda_1'\lambda_2' | v | \vec{q}\lambda_1\lambda_2 \right\rangle \tag{5.1297}$$

so the Thompson equation in the c.m. system given by Eq. (5.1279) can be rewritten as

$$T\left(\vec{q}'\lambda_1'\lambda_2', \vec{q}\lambda_1\lambda_2\right) = V\left(\vec{q}'\lambda_1'\lambda_2', \vec{q}\lambda_1\lambda_2\right) + \sum_{\tilde{\lambda}_1\tilde{\lambda}_2} \int \frac{\mathrm{d}\vec{k}}{(2\pi)^3} V\left(\vec{q}'\lambda_1'\lambda_2', \vec{k}\tilde{\lambda}_1\tilde{\lambda}_2\right)$$
$$\frac{M^2}{E_k^2} \frac{1}{2E_q - 2E_k} T\left(\vec{k}\tilde{\lambda}_1\tilde{\lambda}_2, \vec{q}\lambda_1\lambda_2\right) \tag{5.1298}$$

In the c.m. system we choose the direction of the initial state incident particles as the $z$-axis, and $\vec{q}'$ of the final state particle is located on the $xz$ plane (see Fig. 5.12); it is equivalent that $\vec{q}'$ is obtained by rotating $\vec{q}$ an angle $\theta$ around the $y$-axis.

A point $P(x, y, z)$ in the original coordinate system becomes $P(x', y', z')$ in the new coordinate system after rotation (see Fig. 5.13), and there are

$$\begin{aligned} x' &= x\cos\theta - z\sin\theta \\ y' &= y \\ z' &= x\sin\theta + z\cos\theta \end{aligned} \tag{5.1299}$$

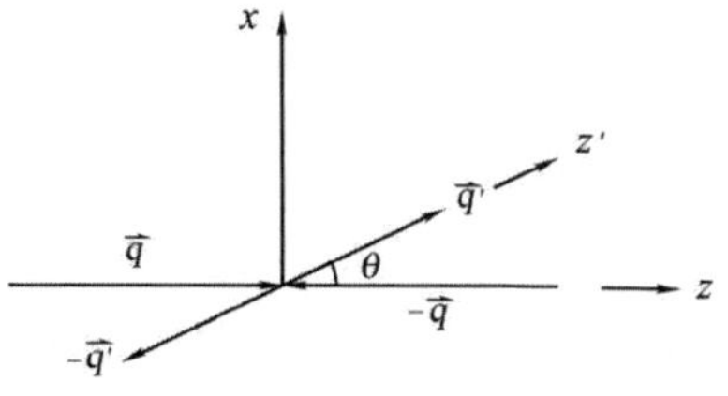


**Fig. 5.12** Scattering schematic diagram of two particles in c.m. system

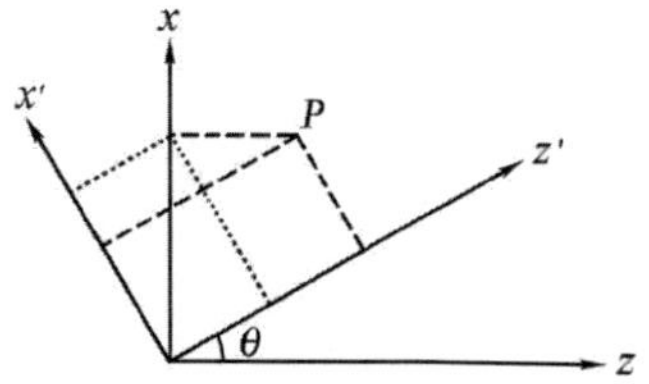


**Fig. 5.13** Two coordinate systems that one of them rotates an angle $\theta$ around the $y$-axis

According to the above formula, we can find

$$\left.\frac{\partial}{\partial\theta}\right|_{\theta=0} = \left.\left(\frac{\partial x'}{\partial\theta}\frac{\partial}{\partial x'} + \frac{\partial z'}{\partial\theta}\frac{\partial}{\partial z'}\right)\right|_{\theta=0} = -z\frac{\partial}{\partial x} + x\frac{\partial}{\partial z}$$

The wave function $\psi\left(\vec{r}'\right)$ can be written as

$$\psi\left(\vec{r}'\right) = \hat{R}_y\psi\left(\vec{r}\right) = \psi'\left(\vec{r}\right) \tag{5.1300}$$

The operator $\hat{R}_y$ can be obtained from the following expansion of $\psi\left(\vec{r}'\right)$

$$\begin{aligned}
\psi\left(\vec{r}'\right) &= \psi(x\cos\theta - z\sin\theta,\ y,\ x\sin\theta + z\cos\theta) \\
&= \psi(x, y, z) + \theta\left(\tfrac{\partial}{\partial\theta}\psi\left(\vec{r}'\right)\right)_{\theta=0} + \frac{\theta^2}{2!}\left(\frac{\partial^2}{\partial\theta^2}\psi\left(\vec{r}'\right)\right)_{\theta=0} + \cdots\cdots \\
&= \left[1 - \theta\left(z\frac{\partial}{\partial x} - x\frac{\partial}{\partial z}\right) + \frac{\theta^2}{2!}\left(z\frac{\partial}{\partial x} - x\frac{\partial}{\partial z}\right)^2 + \cdots\cdots\right]\psi\left(\vec{r}\right) \\
&= \mathrm{e}^{-\theta\left(z\frac{\partial}{\partial x} - x\frac{\partial}{\partial z}\right)}\psi(\vec{r})
\end{aligned} \tag{5.1301}$$

$$\hat{R}_y = \mathrm{e}^{-\mathrm{i}\theta\,\hat{L}_y} \tag{5.1302}$$

Let $\chi_{\lambda_1}, \chi_{\lambda_2}, \chi_{\lambda_1'}, \chi_{\lambda_2'}$ be the two-dimensional helicity wave functions of the four particles taking their own motion direction as the axis, and $|\lambda_1\rangle$, $|\lambda_2\rangle$, $|\lambda_1'\rangle$, $|\lambda_2'\rangle$ be their two-dimensional spin wave functions with the previously selected common $z$-axis as the axis; also note that $-\vec{q}$ of the $\lambda_2$ particle is in the opposite direction with $\vec{q}$ of the $\lambda_1$ particle, and $\vec{q}'$ is rotated relative to $\vec{q}$, so we can write

$$|\lambda_1\rangle = \chi_{\lambda_1}, \quad |\lambda_2\rangle = \chi_{-\lambda_2} \tag{5.1303}$$

$$|\lambda_1'\rangle = \mathrm{e}^{-\frac{\mathrm{i}}{2}\theta\,\hat{\sigma}_y}\chi_{\lambda_1'}, \quad |\lambda_2'\rangle = \mathrm{e}^{-\frac{\mathrm{i}}{2}\theta\,\hat{\sigma}_y}\chi_{-\lambda_2'} \tag{5.1304}$$

where $|\lambda_i^{(')}\rangle$ $(i = 1, 2)$ is the eigenstate of the helicity operator $\frac{1}{2}\hat{\sigma}_i \cdot \hat{p}_i$ $\left(= \frac{1}{2}\hat{\sigma}_{iz^{(')}}\right)$ of the $i$th nucleon with unit momentum $\hat{p}_i$, (′) represents whether there is a apostrophe or no apostrophe, so there is

$$\frac{1}{2}\hat{\sigma}_i \cdot \hat{p}_i\left|\lambda_i^{(')}\right\rangle = \lambda_i^{(')}\left|\lambda_i^{(')}\right\rangle, \quad i = 1,2 \tag{5.1305}$$

where $\lambda_i^{(')}$ represents the nucleon helicity value. The definition of a reduced rotation matrix element is

$$d^J_{\lambda'\lambda}(\theta) = \left\langle J\lambda' \middle| e^{-i\theta \hat{J}_y} \middle| J\lambda \right\rangle \tag{5.1306}$$

And it has the following properties:

$$d^J_{\lambda'\lambda}(\theta) = d^J_{-\lambda\ -\lambda'}(\theta) = (-1)^{\lambda'-\lambda} d^J_{\lambda\lambda'}(\theta) \tag{5.1307}$$

$$\int d^J_{\lambda'\lambda}(\theta) d^{J'}_{\lambda'\lambda}(\theta)\, \mathrm{d}\cos\theta = \frac{2}{2J+1}\delta_{JJ'} \tag{5.1308}$$

$$d^J_{00}(\theta) = \mathrm{P}_J(\cos\theta) \tag{5.1309}$$

Let

$$|\lambda_1\lambda_2\rangle = \chi_{\lambda_1}\chi_{-\lambda_2}, \quad |\lambda_1'\lambda_2'\rangle = \chi_{\lambda_1'}\chi_{-\lambda_2'} \tag{5.1310}$$

then we can write the following two-nucleon spin coupled state wave functions as

$$\varphi_{\lambda_1\lambda_2} = \sum_S C^{S\ \ \lambda_1-\lambda_2}_{\frac{1}{2}\lambda_1\ \frac{1}{2}\ -\lambda_2} |\lambda_1\lambda_2\rangle, \quad \varphi_{\lambda_1'\lambda_2'} = \sum_{S'} C^{S'\ \ \lambda_1'-\lambda_2'}_{\frac{1}{2}\lambda_1'\ \frac{1}{2}\ -\lambda_2'} |\lambda_1'\lambda_2'\rangle \tag{5.1311}$$

Because the total spin of the two nucleon system is conserved, $S' = S$ should be maintained when the above two formulas are expanded at the same time; their desirable values are 0 and 1. Therefore, according to Eq. (5.1306) we can write the following expansion [93–95]:

$$\langle \vec{q}'\lambda_1'\lambda_2'|v|\vec{q}\lambda_1\lambda_2\rangle = \frac{1}{4\pi}\sum_S (2S+1)\, d^S_{\lambda_1'-\lambda_2'\ \ \lambda_1-\lambda_2}(\theta)\ \langle \lambda_1'\lambda_2'|v^S(q',q)|\lambda_1\lambda_2\rangle \tag{5.1312}$$

Utilizing Eq. (5.1308), from above formula we can get

$$\langle \lambda_1'\lambda_2'|v^S(q',q)|\lambda_1\lambda_2\rangle = 2\pi \int_{-1}^{1} d^S_{\lambda_1'-\lambda_2'\ \ \lambda_1-\lambda_2}(\theta)\ \langle \vec{q}'\lambda_1'\lambda_2'|v|\vec{q}\lambda_1\lambda_2\rangle \mathrm{d}\cos\theta \tag{5.1313}$$

Then one can carry out the following expansion:

$$\begin{aligned} T\left(\vec{q}'\lambda_1'\lambda', \vec{q}\lambda_1\lambda_2\right) &\equiv \langle \lambda_1'\lambda_2'|T(\vec{q}', \vec{q})\ |\lambda_1\lambda_2\rangle \\ &= \frac{1}{2}\sum_{L'} (2L'+1)\ T^{L'}\left(q'\lambda_1'\lambda_2', q\lambda_1\lambda_2\right)\ \mathrm{P}_{L'}(\cos\theta) \end{aligned} \tag{5.1314}$$

Substituting Eq. (5.1312) and Eq. (5.1314) into Eq. (5.1298), multiplying $\mathrm{P}_L(\cos\theta)$ from the left and integral over $\cos\theta$, and noting that $\vec{q}' \cdot \vec{k}$ and $\vec{q}' \cdot \vec{q}$ with $\vec{k} \cdot \vec{q}$ satisfy the additive angle formula, so it can be converted into an integral equation containing only variable $k$; each item of these is a $4 \times 4$ matrix. Taking into account that the parity is conserved, time inversion is not changed, and the total spin is conserved, so only the following 6 of the 16 matrix elements $\langle \lambda_1' \lambda_2' | T(\vec{q}', \vec{q}) | \lambda_1 \lambda_2 \rangle$ are independent [95–98]:

$$\langle ++|T|++\rangle, \quad \langle ++|T|--\rangle, \quad \langle +-|T|+-\rangle$$
$$\langle +-|T|-+\rangle, \quad \langle ++|T|+-\rangle, \quad \langle +-|T|++\rangle$$

where + and − represent the helicity value $\frac{1}{2}$ and $-\frac{1}{2}$. In the on-shell case, from the time inversion invariant, we can get $\langle ++|T|+-\rangle = -\langle +-|T|++\rangle$, so only 5 matrix elements are independent.

It is also necessary to consider the nucleon isospin, distinguish between neutrons and protons, and include Coulomb interaction in p-p interactions.

There is a relation between the $S$ matrix and the $T$ matrix as [92, 93]

$$\langle p_1' p_2' | S | p_1 p_2 \rangle = \langle p_1' p_2' | p_1 p_2 \rangle - 2\pi \mathrm{i}\, \delta^{(4)}(p_1' + p_2' - p_1 - p_2) \langle p_1' p_2' | T | p_1 p_2 \rangle \tag{5.1315}$$

After the $S$ matrix elements are obtained, the cross sections, angular distributions, and polarization data of the NN scattering can be calculated. At the same time, OBEP can also be used to solve the homogeneous equation of the deuterium bound state to meet the properties of the deuterium ground state [92–95].

In the Ref. [92] three groups of OBEP (Table 5.2) of the relativistic momentum space obtained by the Thompson equation and the pv coupling of π, η mesons are given, where $g_{\text{ps}} = f_{\text{ps}} \frac{2M}{m_{\text{ps}}}$, $M$ is the average nucleon mass. The above Bonn potential parameters are obtained by the Thompson equation calculation through fitting the NN scattering experimental data and the properties of the deuteron ground state. But the Thompson equation only allows the projection of the intermediate state to the positive energy scattering state, i.e., in the process of determining the nuclear force parameters by fitting to the experimental data, the influence of the intermediate state

**Table 5.2** The relativistic momentum space OBEP obtained by Thompson equation and pv coupling of π,η mesons [92]

| | $m_\alpha$ | Bonn A | | Bonn B | | Bonn C | |
|---|---|---|---|---|---|---|---|
| $\alpha$ | (MeV) | $g_\alpha^2/4\pi$ | $\Lambda_\alpha$(GeV) | $g_\alpha^2/4\pi$ | $\Lambda_\alpha$(GeV) | $g_\alpha^2/4\pi$ | $\Lambda_\alpha$(GeV) |
| σ | 550 | 8.3143 | 2.0 | 8.0769 | 2.0 | 8.0279 | 1.8 |
| δ | 983 | 0.7709 | 2.0 | 3.1155 | 1.5 | 5.0742 | 1.5 |
| ω | 782.6 | 20 | 1.5 | 20 | 1.5 | 20 | 1.5 |
| ρ | 769 | 0.99 | 1.3 | 0.95 | 1.3 | 0.95 | 1.3 |
| η | 548.8 | 7 | 1.5 | 5 | 1.5 | 3 | 1.5 |
| π | 138.03 | 14.9 | 1.05 | 14.6 | 1.2 | 14.6 | 1.3 |

Note: $g_{\text{ps}} = f_{\text{ps}} \frac{2M}{m_{\text{ps}}}$, $f_\omega/g_\omega = 0$, $f_\rho/g_\rho = 6.1$, $n_\alpha = 1$

as a negative energy state has been ignored. In fact, this kind of determining nuclear force parameters is equivalent to considering the vacuum polarization effect to a certain extent. That is, in the nuclear matter calculation by using the Thompson equation or the BbS equation, the intermediate state as a negative energy state is not considered, it is consistent with that of determining the nuclear force parameter, and it is also equivalent to considering the vacuum polarization effect to a certain extent.

In recent years, a new breakthrough has been made in the study of nuclear forces with the chiral effective field theory [99]. This is currently the forefront of the study of nuclear force.

## 5.17 Nucleon Relativistic Microscopic Optical Potential Based on the Dirac-Brueckner-Hartree-Fock Theory

### *5.17.1 Relativistic Brueckner Theory*

The Dirac equation satisfied by a nucleon with momentum $k$ in nuclear medium has been given by Eq. (5.482) as

$$\left(\gamma^{\mu} k_{\mu} - M - \Sigma(k)\right) u\left(\vec{k}, \lambda\right) = 0 \tag{5.1316}$$

where $\lambda$ represents the nucleon spin projection; self-energy $\Sigma(k)$ is equivalent to the mean field. Eq. (5.158) has given

$$\Sigma(k) = \Sigma^{\mathrm{s}}(k) + \gamma^0 \Sigma^0(k) + \vec{\gamma} \cdot \vec{k} \Sigma^{\mathrm{v}}(k) \tag{5.1317}$$

The components of self-energy can be written according to Eq. (5.981) as

$$\Sigma^{\mathrm{s}}(k) = \frac{1}{4} \mathrm{tr} \Sigma(k), \quad \Sigma^0(k) = \frac{1}{4} \mathrm{tr}\left(\gamma^0 \Sigma(k)\right), \quad \Sigma^{\mathrm{v}}(k) = -\frac{1}{4k^2} \mathrm{tr}\left(\vec{\gamma} \cdot \vec{k} \Sigma(k)\right) \tag{5.1318}$$

$\Sigma(k)$ and its three components are all complex numbers. Substituting Eq. (5.1317) into Eq. (5.1316) we can obtain

$$\left[\gamma^0 \left(k_0 - \Sigma^0(k)\right) - \vec{\gamma} \cdot \vec{k} (1 + \Sigma^{\mathrm{v}}(k)) - (M + \Sigma^{\mathrm{s}}(k))\right] u\left(\vec{k}, \lambda\right) = 0 \tag{5.1319}$$

Let

$$M^*(k) = \frac{M + \Sigma^{\mathrm{s}}(k)}{1 + \Sigma^{\mathrm{v}}(k)}, \quad k_0^* = \frac{k_0 - \Sigma^0(k)}{1 + \Sigma^{\mathrm{v}}(k)}, \quad \vec{k}^* = \vec{k} \tag{5.1320}$$

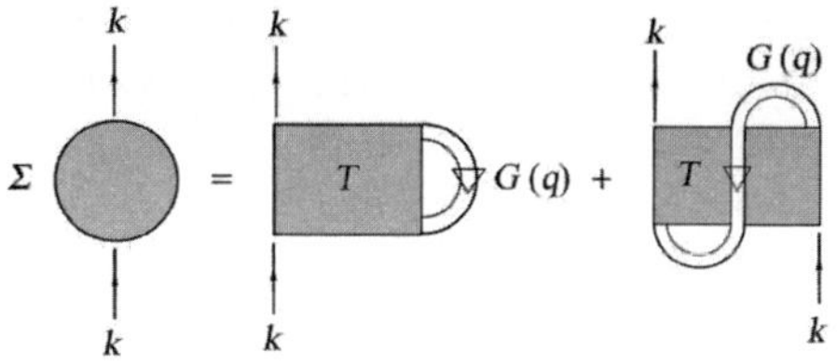


**Fig. 5.14** Relationship between valence nucleon self-energy $\Sigma$, effective interaction $T$, and self-consistent propagator $G$

The definition here is somewhat different from Eq. (5.839). Dividing the right end of Eq. (5.839) by $[1 + \Sigma^{\mathrm{v}}(k)]$ will give Eq. (5.1320). The advantage of Eq. (5.1320) is $\vec{k}^* = \vec{k}$. So Eq. (5.1319) can be rewritten as

$$\left[\gamma^{\mu} k_{\mu}^{*} - M^{*}(k)\right] u\left(\vec{k}, \lambda\right) = 0 \tag{5.1321}$$

The above formula is formally equivalent to the Dirac equation of a free nucleon. We call this kind of nucleons an equivalent free nucleon in nuclear matter, or simply a valence nucleon.

Let $T(k)$ be the effective interaction or reaction matrix of the two nucleons in nuclear matter discussed in the 5.16 section. The relationship between the valence nucleon self-energy $\Sigma(k)$, effective interaction $T(k)$, and self-consistent propagator $G$ can be expressed by Fig. 5.14 [87], where the first item on the right represents the Hartree direct item, and the second item represents the Fock exchange item. In Feynman rules, the interaction line $V$ or $T$ is expressed by $-\mathrm{i}V$ or $-\mathrm{i}T$, so from Fig. 5.14 we can write

$$\begin{aligned} \mathrm{i}G(k)\Sigma(k)G(k) = \mathrm{i}G(k) \int \frac{\mathrm{d}^4 q}{(2\pi)^4} \left[(-\mathrm{i}G(q))(-\mathrm{i}T(kq,kq))\right] \mathrm{i}G(k) \\ +\mathrm{i}G(k) \int \frac{\mathrm{d}^4 q}{(2\pi)^4} \mathrm{i}G(q)(-\mathrm{i}T(kq,qk))\, \mathrm{i}G(k) \end{aligned} \tag{5.1322}$$

and get [100]

$$\Sigma(k) = -\mathrm{i} \int \frac{\mathrm{d}^4 q}{(2\pi)^4} \{\mathrm{tr}[G(q)T(kq,kq)] - G(q)T(kq,qk)\} \tag{5.1323}$$

The specific numerical calculation of the relativistic Hartree-Fock (HF) theory in the nuclear matter (n.m.) stationary coordinate system shows that the relation of $\Sigma^s$ and $\Sigma^0$ with momentum in the interesting nuclear matter density range is very weak, and there is $\Sigma^{\mathrm{v}} \ll 1$, so the following approximation is used for Eq. (5.1317):

$$\Sigma(k) \cong (M^* - M) + \gamma^0 \Sigma^0 \tag{5.1324}$$

where $M^* \cong M + \Sigma^s$ is consistent with Eq. (5.1320). In the specific calculation, we first take $M^*$ and $\Sigma^0$ as the real constants, i.e., ignoring their imaginary parts, it is called the quasi particle approximation. This is similar to the fact that the mean field contains the momentum independent Lorentz scalar $\Sigma^s$ and time-like vector $\Sigma^0$. In the calculation process, the new $M^*$ and $\Sigma^0$ are obtained, and their values can be determined by repeated iterations. The approximation expression (5.1324) is obtained in the relativistic HF calculation in the n.m. system, and Eq. (5.1324) can be generalized in any coordinate system as

$$\Sigma(k) \cong (M^* - M) + \gamma^\mu \Sigma_\mu \tag{5.1325}$$

Note in the n.m. system $\Sigma^\mu = \delta_{\mu 0}\Sigma^0$, $\vec{\Sigma} = 0$.

In the n.m. system, using the symbol introduced by Eq. (5.1320) and letting $E_k^* = E^*(k)$, according to Eq. (5.781) the propagator of the valence nucleon in the nuclear matter can be written as

$$G(k) = \left(\gamma^\mu k_\mu^* + M^*\right)\left[\frac{1}{k_\nu^* k^{\nu *} - M^{*2} + \mathrm{i}\varepsilon} + \frac{\mathrm{i}\pi}{E_k^*}\delta\left(k_0^* - E_k^*\right)\theta\left(k_\mathrm{F} - \underline{k}\right)\right] \tag{5.1326}$$

where $\underline{k} = \left|\vec{k}\right|$. The δ function in the above formula indicates that the valence nucleon is on-shell, and there is

$$E_k^* = \left(\underline{k}^2 + M^{*2}\right)^{1/2} \tag{5.1327}$$

$E_k^*$ is obtained through the nuclear matter correction to the definition of $E_k$. Utilizing Eq. (5.1325) we can obtain

$$\frac{1}{2E_k^*}\delta\left(k_0^* - E_k^*\right) = \delta\left(k_\nu^* k^{\nu *} - M^{*2}\right)\theta\left(k_0^*\right) \equiv \delta^+\left(k_\nu^* k^{\nu *} - M^{*2}\right) \tag{5.1328}$$

The above formula has given the definition of the $\delta^+$ function, and it is a Lorentz scalar.

According to Eq. (5.93), the plane wave spinor of the positive energy valence nucleon in the nuclear matter satisfying Eq. (5.1321) can be written as

$$u\left(\vec{k}, \lambda\right) = \sqrt{\frac{E_k^* + M^*}{2M^*}}\begin{pmatrix} \chi_\lambda \\ \dfrac{\hat{\sigma}\cdot\vec{k}}{E_k^* + M^*}\chi_\lambda \end{pmatrix} \tag{5.1329}$$

The baryon flow density 4-vector is defined as

$$B^{\mu} = \overline{\psi}\gamma^{\mu}\psi \tag{5.1330}$$

where $\psi$ is a baryon Dirac spinor. The baryon density is

$$\rho_B = \psi^{+}\psi = \overline{\psi}\gamma^{0}\psi = B^{0} \tag{5.1331}$$

where $\vec{B}\left(B^1, B^2, B^3\right)$ is a baryon flow density. In the n.m. system (baryons do not flow) and in the case of zero temperature, there is

$$B_{\mu}B^{\mu} = \left(B^{0}\right)^{2} = \rho_B^2 \tag{5.1332}$$

For the symmetric nuclear matter Eq. (5.928) has given $\rho_{\mathrm{B}} = \frac{2}{3\pi^2}k_{\mathrm{F}}^3$. When we do observing in any new coordinate system, which moves at a constant speed $\vec{v}$ relative to n.m. system, baryons are flowing at a speed $-\vec{v}$. Therefore it can be seen that the baryon velocity in the new observation system is

$$\vec{\beta} \equiv \frac{\vec{v}}{c} = -\frac{\vec{B}}{B^0}, \qquad \vec{B} = -\vec{\beta}B^0 \tag{5.1333}$$

Since $B_{\mu}B^{\mu}$ is a Lorentz invariant, using Eqs. (5.1332) and (5.1333) in any coordinate system, we can obtain

$$B_{\mu}B^{\mu} = \left(B^{0}\right)^{2} - \vec{B}^{2} = \left(1-\beta^2\right)\left(B^{0}\right)^{2} = \rho_B^2 \tag{5.1334}$$

$$\gamma = \left(1-\beta^2\right)^{-1/2} = \frac{B^0}{\rho_B} \tag{5.1335}$$

In the n.m. system $\vec{\beta} = \frac{\vec{v}}{c} = 0$, there will be a baryon flow density $\vec{B} = 0$ and a self-energy flow density $\vec{\Sigma} = 0$. When the total three-dimensional momentum of the nuclear matter is 0, i.e., in the n.m. system, for the symmetric nuclear matter the distribution of the momentum density can be represented by $\theta(k_{\mathrm{F}} - \underline{k})$, which is called the Fermi sphere. Here $\vec{k}$ represents the speed with which the nucleons do irregular Fermi movement in nuclear matter; by paying attention $\vec{k}^{*} = \vec{k}$ we can obtain

$$\begin{aligned}
k_{\mathrm{F}}^2 + k_{\nu}^{*}k^{*\nu} - \frac{\left(k_{\nu}^{*}B^{\nu}\right)^2}{B_{\mu}B^{\mu}} &= k_{\mathrm{F}}^2 + k_0^{*2} - \underline{k}^2 - \frac{\left(k_0^{*}B_0 - \vec{k}\cdot\vec{B}\right)^2}{\rho_B^2} \\
&= k_{\mathrm{F}}^2 + k_0^{*2} - \underline{k}^2 - \left(k_0^{*2} + \underline{k}^2\beta^2 + 2\beta\, k_0^{*}\underline{k}\right) \\
&\xrightarrow[\beta\to 0]{} k_{\mathrm{F}}^2 - \underline{k}^2 = \left(k_{\mathrm{F}} - \underline{k}\right)\left(k_{\mathrm{F}} + \underline{k}\right)
\end{aligned} \tag{5.1336}$$

Because of $\theta(k_F + \underline{k}) = 1$, the Lorentz invariant $n(k^*; B)$ representing the distribution of nucleon momentum density can be taken as

$$n(k^*; B) = \theta\left[k_F^2 + k_\nu^* k^{\nu *} - \frac{(k_\nu^* B^\nu)^2}{(B_\mu B^\mu)}\right] \xrightarrow[\text{n.m.system}]{} \theta(k_F - \underline{k}) \tag{5.1337}$$

Equation (5.1326) can be rewritten, using Eqs. (5.1328) and (5.1337), as follows:

$$\begin{aligned} G(k; B) &= (\gamma^\mu k_\mu^* + M^*)\left[\frac{1}{k_\nu^* k^{\nu *} - M^{*2} + i\varepsilon} + 2\pi i\, \delta^+(k_\nu^* k^{\nu *} - M^{*2})\, n^*(k^*; B)\right] \\ &= G_F(k; B) + G_D(k; B) \end{aligned} \tag{5.1338}$$

At this time $G(k, B)$ becomes Lorentz invariant. Since $G_F$ represents the propagation between virtual nucleon and anti-nucleon with effective mass $M^*$, if only the actual valence nucleons are considered when calculating self-energy $\Sigma$, so while making calculation for the expression of $\Sigma(k)$ given by Eq. (5.1323), $G(q)$ can be replaced by $G_D(q, B)$ [87, 100]. Then we get

$$\Sigma(k) = -i\int \frac{d^4 q}{(2\pi)^4}\{\mathrm{tr}[G_D(q; B)T(kq, kq) - G_D(q; B)T(kq, qk)]\} \tag{5.1339}$$

## 5.17.2 Relativistic Pauli Incompatibility Operators

The two intermediate state nucleons shown in Fig. 5.10 can be represented by $\frac{1}{2}P^* + k$ and $\frac{1}{2}P^* - k$ in nuclear matter. According to Eq. (5.1337), their density distributions are $n\left(\frac{1}{2}P^* + k; B\right)$ and $n\left(\frac{1}{2}P^* - k; B\right)$, respectively. From the discussion in the 5.16 section, we can see that in the case of angle averaging, there is $k_0 = 0$. Since it is required that one of the two intermediate state nucleons cannot be scattered to the occupying state of another nucleon, the following Pauli incompatibility operator can be introduced in the n.m. system as

$$Q(k, P^*; B) = \left[1 - n\left(\frac{1}{2}P^* + k; B\right)\right]\left[1 - n\left(\frac{1}{2}P^* - k; B\right)\right] \tag{5.1340}$$

where $Q(k, P^*; B)$ represents the part that is not occupied by any one of the two intermediate state nucleons. Note that the above Pauli incompatibility operator does not exclude the occupying state of the nucleons in the nuclear matter. Therefore it is not convenient for practical utilization.

Note that there is a total invariant in any coordinate system as

$$s^* = P_\mu^* P^{*\mu} = P_0^{*2} - \vec{P}^{*2} \tag{5.1341}$$

In the n.m. system there is $\vec{B} = 0$, i.e., the medium does not flow. In the two particle center of momentum (c.m.) system, there is $\vec{P}_c^* = 0$, i.e., the coordinate origin is fixed at the center of momentum of two interacting particles, and the entire nuclear matter is flowing relative to the center of momentum of the two particles. It can be seen from Eq. (5.165) that the velocity of baryon flow is

$$\vec{\beta} = \frac{\vec{P}^*}{P_0^*} \tag{5.1342}$$

where $\vec{\beta}$ represents the velocity of the center of momentum of the two particles relative to the n.m. system, and when making observation in the c.m. system the velocity of the baryon flow is $-\vec{\beta}$. And there is

$$\gamma = \left(1 - \beta^2\right)^{-1/2} = \frac{P_0^*}{\sqrt{s^*}} \tag{5.1343}$$

The Lorentz matrix of the Pauli metric given by Eq. (5.102) or Eq. (5.450) should become in the BD metric as follows:

$$\alpha = \begin{pmatrix} \gamma & 0 & 0 & -\frac{\gamma}{c}\beta \\ 0 & 1 & 0 & 0 \\ 0 & 0 & 1 & 0 \\ -c\gamma\beta & 0 & 0 & \gamma \end{pmatrix} \tag{5.1344}$$

For space-time coordinate $\left(t, \vec{r}\right)$ using the transformation given by the above matrix, the result given by Eq. (5.104) can also be obtained. If the components of the speed $\vec{v}$ on $(x, y, z)$ three axes are all non-zero, in the $\hbar = c = 1$ unit system, the space-time coordinate $\left(a^0, \vec{a}\right)$ in the original coordinate system can be changed to $\left(a_c^0, \vec{a}_c\right)$ in the new coordinate system represented by footnote $c$ by using the following formulas [87]:

$$a_c^0 = \gamma\left(a^0 - \vec{\beta} \cdot \vec{a}\right) \tag{5.1345}$$

$$\vec{a}_c = \vec{a} + \vec{\beta}\gamma\left(\frac{\gamma}{\gamma+1}\vec{\beta}\cdot\vec{a} - a^0\right) \tag{5.1346}$$

From above two formulas, the corresponding Lorentz transformation matrix can be written as

$$\alpha = \begin{pmatrix} \gamma & -\gamma\beta_x & -\gamma\beta_y & -\gamma\beta_z \\ -\gamma\beta_x & 1+\frac{\gamma^2\beta_x^2}{\gamma+1} & \frac{\gamma^2\beta_x\beta_y}{\gamma+1} & \frac{\gamma^2\beta_x\beta_z}{\gamma+1} \\ -\gamma\beta_y & \frac{\gamma^2\beta_y\beta_x}{\gamma+1} & 1+\frac{\gamma^2\beta_y^2}{\gamma+1} & \frac{\gamma^2\beta_y\beta_z}{\gamma+1} \\ -\gamma\beta_z & \frac{\gamma^2\beta_z\beta_x}{\gamma+1} & \frac{\gamma^2\beta_z\beta_y}{\gamma+1} & 1+\frac{\gamma^2\beta_z^2}{\gamma+1} \end{pmatrix} \tag{5.1347}$$

And notice

$$\frac{\gamma^2\beta^2}{\gamma+1} = \frac{\gamma^2\beta^2(\gamma-1)}{\gamma^2-1} = \frac{\gamma^2\beta^2(\gamma-1)}{\gamma^2(1-\gamma^{-2})} = \gamma - 1$$

then when $\beta_x = \beta_y = 0$ and $\beta_z = \beta$, Eq. (5.1347) can be automatically reduced to Eq. (5.1344).

We take n.m. system as the original coordinate system, and take c.m. system with $\vec{P}_c^* = 0$ as the new coordinate system $c$, so there are

$$\left(\vec{p}_1^*\right)_c = \left(\frac{1}{2}\vec{P}^* + \vec{k}\right)_c = \vec{k}, \quad \left(\vec{p}_2^*\right)_c = \left(\frac{1}{2}\vec{P}^* - \vec{k}\right)_c = -\vec{k} \tag{5.1348}$$

In the following, $k$ will be used to represent the $\underline{k} = \left|\vec{k}\right|$ used previously. In the n.m. system the particles are on-shell, there is $\left(p_1^*\right)^0 = E^*(p_1)$, in the c.m. system, there is $\left(p_1^*\right)_c^0 = E^*(p_{1c}) = E^*(k)$, and we can obtain

$$\begin{aligned} P_0^{*2} - \vec{P}^{*2} &= s^* = \left(P_0^*\right)_c^2 = \left[E^*(p_{1c}) + E^*(p_{2c})\right]^2 \\ &= \left[E^*(k) + E^*(-k)\right]^2 = 4\left(k^2 + M^{*2}\right) \end{aligned} \tag{5.1349}$$

and

$$k = \left(\frac{1}{4}s^* - M^{*2}\right)^{1/2} \tag{5.1350}$$

In the following we study the Pauli incompatibility operator in the c.m. system. First let

$$E_{\mathrm{F}}^{*}=\left(k_{\mathrm{F}}^{2}+M^{*2}\right)^{1/2} \tag{5.1351}$$

where $k_{\mathrm{F}}$ is defined in the n.m. system. From Eq. (5.1336) the following expression can be obtained in the c.m. system by using Eqs. (5.1349)–(5.1351) and (5.1334):

$$\begin{aligned} n(k^{*};B) &= \theta\left[E_{\mathrm{F}}^{*2}-\frac{\left(k_{\nu}^{*}B^{\nu}\right)^{2}}{\rho_{B}^{2}}\right] \\ &= \theta\left[\left(E_{\mathrm{F}}^{*}-\frac{k_{\nu}^{*}B^{\nu}}{\rho_{B}}\right)\left(E_{\mathrm{F}}^{*}+\frac{k_{\nu}^{*}B^{\nu}}{\rho_{B}}\right)\right]=\theta\left(E_{\mathrm{F}}^{*}-\frac{k_{\nu}^{*}B^{\nu}}{\rho_{B}}\right) \end{aligned} \tag{5.1352}$$

Since $\dfrac{k_{\nu}^{*}B^{\nu}}{\rho_{B}}=\dfrac{1}{\rho_{B}}\left(k_{0}^{*}B^{0}-\vec{k}\cdot\vec{B}\right)$, the observed $\vec{B}/\rho_{B}$ is negative number flow in the c.m. system, so the second factor after the second equal sign of Eq. (5.1352) is always positive. Note that the term after the last equal sign in Eq. (5.1352) contains $\vec{k}\cdot\vec{B}=kB\cos\theta_{k}$ item, so that the θ function given by this formula forms a Fermi ellipsoid. Using Eq. (5.1333) and Eq. (5.1335), from Eq. (5.1352) we can see that in the c.m. system the condition, which guarantee that the intermediate state particles are outside the Fermi ellipsoid formed by the on-shell particle occupancy states, is as follows:

$$E_{\mathrm{F}}^{*}\leq\frac{\left(E_{k}^{*}B^{0}-\vec{k}\cdot\vec{B}\right)}{\rho_{B}}=\gamma\left(E_{k}^{*}+\vec{k}\cdot\vec{\beta}\right) \tag{5.1353}$$

Then it can be obtained that the Pauli incompatibility operator satisfied by the two intermediate state particles with $\vec{k}$ and $-\vec{k}$ in the c.m. system is

$$Q(k^{*};B)=\theta\left[\gamma\left(E_{k}^{*}+\vec{k}\cdot\vec{\beta}\right)-E_{\mathrm{F}}^{*}\right]\theta\left[\gamma\left(E_{k}^{*}-\vec{k}\cdot\vec{\beta}\right)-E_{\mathrm{F}}^{*}\right] \tag{5.1354}$$

In the above formula the multiplied two terms on the right side of the equal sign correspond to $\vec{k}$ and $-\vec{k}$ two intermediate state particles, respectively. We choose the $\vec{\beta}$ direction as the $z$-axis and the angle between $\vec{k}$ and $\vec{\beta}$ as $\theta_{k}$. From Eq. (5.1353) we can see that the equation satisfied by the above Fermi ellipsoid surface is

$$\gamma\left(E_{k}^{*}+k\beta\cos\theta_{k}\right)-E_{\mathrm{F}}^{*}=0 \tag{5.1355}$$

Taking $\cos\theta_{k}=0$ in Eq. (5.1355), so we can get

$$\gamma^{2}\left(k^{2}+M^{*2}\right)=k_{\mathrm{F}}^{2}+M^{*2},\qquad k^{2}-k_{\mathrm{F}}^{2}+\beta^{2}E_{F}^{*2}=0 \tag{5.1356}$$

Let its solution be

$$k_{\mathrm{u}} = \sqrt{k_{\mathrm{F}}^2 - \beta^2 E_{\mathrm{F}}^{*2}} \tag{5.1357}$$

Taking $\cos\theta_k = 1$ in Eq. (5.1355), so we can get

$$\begin{aligned} &E_k^* = \frac{1}{\gamma} E_{\mathrm{F}}^* - k\beta, \quad k^2 = k_{\mathrm{F}}^2 - \beta^2 E_{\mathrm{F}}^{*2} + k^2\beta^2 - \frac{2k\beta}{\gamma} E_{\mathrm{F}}^*, \\ &\frac{1}{\gamma^2} k^2 + \frac{2k\beta}{\gamma} E_{\mathrm{F}}^* + \beta^2 E_{\mathrm{F}}^{*2} = k_{\mathrm{F}}^2, \quad \frac{k}{\gamma} + \beta E_{\mathrm{F}}^* = k_{\mathrm{F}} \end{aligned} \tag{5.1358}$$

Let its solution be

$$k_{\mathrm{v}} = \gamma \left(k_{\mathrm{F}} - \beta E_{\mathrm{F}}^*\right) \tag{5.1359}$$

Taking $\cos\theta_k = -1$ in Eq. (5.1355), then we can get

$$\frac{k}{\gamma} - \beta E_{\mathrm{F}}^* = k_{\mathrm{F}} \tag{5.1360}$$

Let its solution be

$$k_+ = \gamma \left(k_{\mathrm{F}} + \beta E_{\mathrm{F}}^*\right) \tag{5.1361}$$

From Eq. (5.1359) and Eq. (5.1361), it can be seen that the total length of the long axis of the Fermi ellipsoid is $2\gamma k_{\mathrm{F}}$. Since the short half axis of the Fermi ellipsoid is perpendicular to the $\vec{\beta}$ direction, the short half axis corresponding to the center of the major long axis is still to be $k_{\mathrm{F}}$, but for the short radius $k_{\mathrm{u}}$ corresponding to the center of momentum there is $k_{\mathrm{u}} < k_{\mathrm{F}}$. And let

$$k_- = \min\left(k_{\mathrm{u}}, k_{\mathrm{v}}\right) \tag{5.1362}$$

$$\cos\alpha = \frac{\gamma\, E_k^* - E_{\mathrm{F}}^*}{\gamma\beta k} \tag{5.1363}$$

so Eq. (5.1354) can be rewritten as

$$Q(k^*; B) = \begin{cases} 0 & k < k_- \\ \theta(\cos\alpha - \cos\theta_k)\, \theta(\cos\theta_k + \cos\alpha), & k_- \le k \le k_+ \\ 1 & k > k_+ \end{cases} \tag{5.1364}$$

The relativistic Pauli incompatibility operator after averaging for the direction angle of $\vec{k}$ can be obtained by Eq. (5.1364) as

$$\overline{Q}(k^*;B) = \frac{1}{4\pi}\int d\Omega_k Q(k^*;B) \tag{5.1365}$$

In the $k_- \leq k \leq k_+$ case of Eq. (5.1364), the condition, which guarantee that the two θ functions are equal to 1 at the same time, is

$$-\cos\alpha \leq \cos\theta_k \leq \cos\alpha \tag{5.1366}$$

So the following expression can be got from Eqs. (5.1363)–(5.1366):

$$\overline{Q}(k^*;B) = \begin{cases} 0 & k < k_- \\ \dfrac{\gamma E_k^* - E_F^*}{\gamma\beta k} & k_- \leq k \leq k_+ \\ 1 & k > k_+ \end{cases} \tag{5.1367}$$

In Refs. [87, 101] the more detailed derivations of the above relativistic Pauli incompatibility operator in the c.m. system are given, and Refs. [88, 102–104] also quoted its final result.

In the following we study the Pauli incompatibility operator in the n.m. system. According to Eq. (5.1337) we can write

$$n\left(\frac{1}{2}P^* \pm k;B\right) = \theta\left\{k_F^2 + \left(\frac{1}{2}P^* \pm k\right)_\nu\left(\frac{1}{2}P^* \pm k\right)^\nu - \left[\left(\frac{1}{2}P^* \pm k\right)_\nu B^\nu\right]^2/(B_\mu B^\mu)\right\} \tag{5.1368}$$

The following expressions can be obtained using Eqs. (5.1333)–(5.1335), and (5.1342):

$$\left(\frac{1}{2}P^* \pm k\right)_\nu\left(\frac{1}{2}P^* \pm k\right)^\nu = \left(\frac{1}{2}P_0^* \pm E_k^*\right)^2 - \left(\frac{1}{2}P_0^*\vec{\beta} \pm \vec{k}\right)^2 \tag{5.1369}$$

$$\begin{aligned}\left[\left(\frac{1}{2}P^* \pm k\right)_\nu B^\nu\right]^2/(B_\mu B^\mu) &= \left[\left(\frac{1}{2}P_0^* \pm E_k^*\right)B^0 - \left(\frac{1}{2}P_0^*\vec{\beta} \pm \vec{k}\right)\cdot\vec{B}\right]^2/\rho_B^2 \\ &= \gamma^2\left[\left(\frac{1}{2}P_0^* \pm E_k^*\right) + \left(\frac{1}{2}P_0^*\beta^2 \pm \vec{k}\cdot\vec{\beta}\right)\right]^2\end{aligned} \tag{5.1370}$$

Let

$$D_k^{(\pm)} = \frac{1}{2}P_0^* \pm E_k^* \tag{5.1371}$$

$$R_k^{(\pm)} = D_k^{(\pm)} + \frac{1}{2}P_0^*\beta^2 = \frac{1}{2}P_0^*(1+\beta^2) \pm E_k^* \tag{5.1372}$$

From Eq. (5.1368) we can obtain that the equation satisfied by the Fermi ellipsoid surface in the n.m. system is

$$
\begin{aligned}
& k_{\mathrm{F}}^2+\left(\frac{1}{2} P^* \pm k\right)_\nu\left(\frac{1}{2} P^* \pm k\right)^\nu-\left[\left(\frac{1}{2} P^* \pm k\right)_\nu B^\nu\right]^2 /\left(B_\mu B^\mu\right) \\
& =k_F^2+D_k^{(\pm) 2}-\left(\frac{1}{2} P_0^* \vec{\beta} \pm \vec{k}\right)^2-\gamma^2\left(R_k^{(\pm)} \pm \vec{k} \cdot \vec{\beta}\right)^2 \\
& =k_{\mathrm{F}}^2+D_k^{(\pm) 2}-\frac{1}{4} P_0^{* 2} \beta^2-k^2 \mp P_0^* \beta k \cos \theta_k \\
& -\gamma^2\left[R_k^{(\pm) 2}+k^2 \beta^2 \cos ^2 \theta_k \pm 2 k \beta R_k^{(\pm)} \cos \theta_k\right]=0
\end{aligned}
\tag{5.1373}
$$

Here introduce the following symbols:

$$
F_k^{(\pm)}=\gamma R_k^{(\pm)}+\frac{1}{2 \gamma} P_0^* \tag{5.1374}
$$

$$
C_k^{(\pm)}=\gamma^2 R_k^{(\pm) 2}-D_k^{(\pm) 2}+\frac{1}{4} \beta^2 P_0^{* 2}+k^2-k_{\mathrm{F}}^2 \tag{5.1375}
$$

so it can be seen that when $k \geq k_{\mathrm{F}}$, $F_k^{(\pm)}$ and $C_k^{(\pm)}$ are all positive. For $\frac{1}{2} P^* \pm k$ from Eq. (5.1373) we can get the following equations, respectively:

$$
\gamma^2 k^2 \beta^2 \cos ^2 \theta_k+2 \gamma k \beta F_k^{(+)} \cos \theta_k+C_k^{(+)}=0 \tag{5.1376}
$$

$$
\gamma^2 k^2 \beta^2 \cos ^2 \theta_k-2 \gamma k \beta F_k^{(-)} \cos \theta_k+C_k^{(-)}=0 \tag{5.1377}
$$

The above two formulas are equivalent to Eq. (5.1373) to be multiplied by (-1), so that if the left end of the above two formulas is greater than 0, it means that it is outside the Fermi ellipsoid.

The solution of Eq. (5.1376) is

$$
\cos \theta_{k \pm}^{(+)}=\frac{1}{\gamma^2 k^2 \beta^2}\left\{-\gamma k \beta F_k^{(+)} \pm\left[\left(\gamma k \beta F_k^{(+)}\right)^2-\gamma^2 k^2 \beta^2 C_k^{(+)}\right]^{1 / 2}\right\} \equiv \frac{k_{\pm}^{(+)}}{k} \tag{5.1378}
$$

where

$$
k_{\pm}^{(+)}=\frac{1}{\gamma \beta}\left[-F_k^{(+)} \pm\left(F_k^{(+) 2}-C_k^{(+)}\right)^{1 / 2}\right] \tag{5.1379}
$$

The solution of Eq. (5.1377) is

$$\cos\theta_{k\pm}^{(-)} \equiv \frac{k_{\pm}^{(-)}}{k} \tag{5.1380}$$

where

$$k_{\pm}^{(-)} = \frac{1}{\gamma\beta}\left[F_k^{(-)} \pm \left(F_k^{(-)2} - C_k^{(-)}\right)^{1/2}\right] \tag{5.1381}$$

From Eq. (5.1374) and Eq. (5.1375), we can obtain

$$\begin{aligned} H_k^{(\pm)} &\equiv F_k^{(\pm)2} - C_k^{(\pm)} = \frac{1}{4\gamma^2}P_0^{*2} + P_0^* R_k^{(\pm)} + D_k^{(\pm)2} - \frac{1}{4}\beta^2 P_0^{*2} - k^2 + k_{\mathrm{F}}^2 \\ &= P_0^* R_k^{(\pm)} + D_k^{(\pm)2} + \frac{1}{4}\left(1 - 2\beta^2\right)P_0^{*2} - k^2 + k_{\mathrm{F}}^2 \end{aligned} \tag{5.1382}$$

It can be seen that $H_k^{(\pm)} \geq 0$ are the conditions that Eqs. (5.1376) and (5.1377) have solutions, respectively, and $H_k^{(\pm)} < 0$ represent that Eqs. (5.1376) and (5.1377) have no solutions, respectively. That is, if the left end of the corresponding equation is greater than 0, it is outside the Fermi ellipsoid. Also notice $k_{-}^{(+)} < 0$.

We can express the Pauli incompatibility operator in the n.m. system as

$$Q(k^*, P^*; B) = Q^{(+)}(k^*, P^*; B)Q^{(-)}(k^*, P^*; B) \tag{5.1383}$$

where

$$Q^{(\pm)}(k^*, P^*; B) = \begin{cases} 0, & k < k_{+}^{(\pm)} \text{ and } k < -k_{-}^{(\pm)} \text{ and } H_k^{(\pm)} \geq 0 \\ \theta\left(\cos\theta_k - \dfrac{k_{+}^{(\pm)}}{k}\right), & k \geq \max\left(k_{+}^{(\pm)},\ -k_{+}^{(\pm)}\right) \text{ and } k < -k_{-}^{(\pm)} \text{ and } H_k^{(\pm)} \geq 0 \\ \theta\left(\dfrac{k_{-}^{(\pm)}}{k} - \cos\theta_k\right), & k < k_{+}^{(\pm)} \text{ and } k \geq \max\left(k_{-}^{(\pm)},\ -k_{-}^{(\pm)}\right) \text{ and } H_k^{(\pm)} \geq 0 \\ 1, & \left(k < -k_{+}^{(\pm)} \text{ and } k < k_{-}^{(\pm)} \text{ and } H_k^{(\pm)} \geq 0\right) \text{ or } H_k^{(\pm)} < 0 \end{cases} \tag{5.1384}$$

Note $\cos\theta_k \geq \cos\theta_{k_+}^{(\pm)}$ and $\cos\theta_k \leq \cos\theta_{k_-}^{(\pm)}$ cannot be satisfied at the same time. The relativistic Pauli incompatibility operator after the angle averaging in the n.m. system is

$$\overline{Q}(k^*, P^*; B) = \frac{1}{2}\int_{-1}^{1} Q(k^*, P^*; B)\, \mathrm{d}\cos\theta_k \tag{5.1385}$$

Referring to the Pauli incompatibility operator in the c.m. system described previously, according to Eqs. (5.1383)–(5.1385), their analytical expressions may be further found.

### 5.17.3 Calculation Formulas for T Matrix Elements and Nuclear Matter Properties in Symmetric Nuclear Matter

According to Eq. (5.1279) the Thompson equation in nuclear matter can be written in the c.m. system as

$$T\left(\vec{q}', \vec{q}\right) = V\left(\vec{q}', \vec{q}\right) + \int \frac{\mathrm{d}\vec{k}}{(2\pi)^3} V\left(\vec{q}', \vec{k}\right) \frac{M^{*2}}{E_k^{*2}} \frac{\overline{Q}(k^*; B)}{2E_q^* - 2E_k^*} T\left(\vec{k}, \vec{q}\right) \tag{5.1386}$$

where $\overline{Q}(k^*; B)$ is the angle averaging Pauli incompatibility operator. $T\left(\vec{q}', \vec{q}\right)$ is the $T$ matrix element in the c.m. system. The speed $\vec{\beta}$ of the center of momentum relative to the n.m. system has been given by Eq. (5.1342). Using the $\vec{J} = \vec{L} + \vec{S}$ angular momentum coupling mode, the method used previously to study the Bonn one meson exchange potential can also be used to convert Eq. (5.1386) into a one-dimensional equation of relative momentum $k$, and then we can use the inverse matrix method to solve the obtained equation. Let $I = 0$, 1 be the total isospin of the two-nucleon system, and the anti-symmetry requires

$$(-1)^{L+S+I} = -1 \tag{5.1387}$$

where $L$ is the orbital angular momentum.

As mentioned previously, for symmetric nuclear matter and in the on-shell case, there are only five linear independent helicity $T$ matrix elements, so that the $T$ matrix elements can be expanded with five linearly independent covariables. The following 5 Fermi covariables are usually used [87, 103, 105, 106]:

$$S = 1, \quad V = \gamma_1^\mu \gamma_{2\mu}, \quad T = \sigma_1^{\mu\nu} \sigma_{2\mu\nu}, \quad A = \gamma_1^5 \gamma_1^\mu \gamma_2^5 \gamma_{2\mu}, \quad P = \gamma_1^5 \gamma_2^5 \tag{5.1388}$$

We also define the following five exchange Fermi covariables:

$$\widetilde{S} = \widetilde{S}S, \quad \widetilde{V} = \widetilde{S}V, \quad \widetilde{T} = \widetilde{S}T, \quad \widetilde{A} = \widetilde{S}A, \quad \widetilde{P} = \widetilde{S}P \tag{5.1389}$$

where $\widetilde{S}$ is the exchange covariable of $S$, whose role is to exchange the Dirac footmarks of particles 1 and 2, for example, $\widetilde{S}u(1)_\sigma u(2)_\tau = u(1)_\tau u(2)_\sigma$. The Fierz transform relations satisfied between 5 exchange Fermi covariables and 5 original Fermi covariables are as follows [105–107]:

$$\begin{pmatrix} \widetilde{S} \\ \widetilde{V} \\ \widetilde{T} \\ \widetilde{A} \\ \widetilde{P} \end{pmatrix} = \mathcal{F} \begin{pmatrix} S \\ V \\ T \\ A \\ P \end{pmatrix} = \frac{1}{4} \begin{pmatrix} 1 & 1 & \frac{1}{2} & -1 & 1 \\ 4 & -2 & 0 & -2 & -4 \\ 12 & 0 & -2 & 0 & 12 \\ -4 & -2 & 0 & -2 & 4 \\ 1 & -1 & \frac{1}{2} & 1 & 1 \end{pmatrix} \begin{pmatrix} S \\ V \\ T \\ A \\ P \end{pmatrix} \tag{5.1390}$$

Equation (5.1388) has given the specific form of the Fermi covariant operators $\Gamma_{\mathrm{i}} = \{S, V, T, A, P\}$, and Eq. (5.1389) can be verified from Eq. (5.1390) using the $\gamma$ matrix commutative relations.

According to the Hartree-Fock theory, the $t$ matrix whose isospin is $I$ should be written as

$$t^I(q, \theta) = t^{I,\mathrm{dir}}(q, \theta) - t^{I,\mathrm{exc}}(q, \theta) \tag{5.1391}$$

where $\theta$ represents the angle between $\vec{q}$ and $\vec{q}\,'$. The direct item can be expanded as follows:

$$t^{I,\mathrm{dir}}(q,\theta) = \frac{1}{2}\left[F_S^I(q,\theta)S + F_V^I(q,\theta)\,V + F_T^I(q,\theta)T + F_A^I(q,\theta)A + F_P^I(q,\theta)P\right] \tag{5.1392}$$

If $\vec{q}$ corresponds to an angle $\theta$, then $-\vec{q}$ corresponds to an angle $\pi - \theta$, and the two-nucleon system with isospin $I = 1$ is symmetric and that with isospin $I = 0$ is anti-symmetric. We can write [103, 106]

$$\begin{aligned} t^{I,\mathrm{exc}}(q, \theta) = (-1)^{I+1}\frac{1}{2}\Big[&F_S^I(q,\ \pi-\theta)\widetilde{S} + F_V^I(q, \pi-\theta)\,\widetilde{V} \\ &+F_T^I(q,\ \pi-\theta)\widetilde{T} + F_A^I(q,\ \pi-\theta)\widetilde{A} + F_P^I(q,\ \pi-\theta)\widetilde{P}\Big] \end{aligned} \tag{5.1393}$$

The five Fermi covariables given by Eq. (5.1388) are not unique. For example, $P$ is a pseudo-scalar quantity, but $P$ can also be replaced by a normal axis vector or a complete axis vector form [103, 106].

In the actual calculation, it was found that if all meson exchange potentials $V$ were converted to effective interaction $T$, one could not obtain the satisfactory results, so it was proposed to divide $T$ into two parts [96, 97]:

$$T = V + \Delta T \tag{5.1394}$$

where $V$ is the bare nucleon-nucleon interaction. Since the one $\pi$ exchange plays a major role in the NN scattering process, in order to obtain satisfactory calculation results, the subtracted $T$ matrix method is proposed [103, 106]. For the $\pi$ and $\eta$ mesons the bare NN interaction represented by a complete pseudo-vector is used, while for other mesons the effective interaction represented by a pseudo-scalar is used, namely,

$$T = T_{\text{sub}} + V_{\pi,\eta} \tag{5.1395}$$

The Pauli operator $Q$ is related to speed $\beta$, so the $T$ matrix element derived from Eq. (5.1386) is also related to speed $\beta$. Section 5.15 has clearly pointed out that the $T$ matrix approximately satisfies the Lorentz invariance. Therefore, in the projection method discussed earlier in this section, the expansion of the $T$ matrix element with five linear independent covariables can only be regarded as an approximate method. Strictly speaking, the results obtained by this method may lose the non-Lorentz invariant component of the $T$ matrix element, maybe which will cause the wrong calculation result to some extent. The above method for solving the $T$ matrix is performed in the c.m. system. The bare nucleon-nucleon interaction $V$ obtained by the method described in Sect. 5.16 is the Lorentz invariant, but the entire process from the interaction $V$ to the $T$ matrix cannot be guaranteed to be Lorentz invariant. Therefore, in general the $T$ matrix found in the c.m. system is not applicable directly in the n.m. system. The coordinate system transformation of the calculation results can be considered according to the method described in Sect. 5.2. Of course, it can also be done that the Thompson equation is solved directly in the n.m. system like Ref. [89], but the Pauli incompatibility operator of the n.m. system given earlier in this section should be used.

After the $T$ matrix is solved in the nuclear matter, the nucleon self energy $\Sigma(k)$ in the nuclear matter can be calculated by Eqs. (5.1337)–(5.1339). Note that we use the $\bar{u}u = 1$ plane wave normalization method, the nucleon spinor field given by Eq. (5.754) or (5.924) contains the factor $\sqrt{\frac{M}{E}}$, so the relativistic single particle potential in nuclear matter is

$$U(k) = \frac{M^*}{E_k^*} \text{Re}\,\langle k|\Sigma(k)|k\rangle = \sum_{q \le k_{\text{F}}} \frac{M^{*2}}{E_k^* E_q^*} \text{Re}\,\langle kq|T|kq - qk\rangle \tag{5.1396}$$

If $\Sigma^{\text{v}}(k)$ is ignored, there is $\Sigma(k) \cong \Sigma^{\text{s}}(k) + \gamma^0 \Sigma^0(k)$; in this case the $\Sigma^0(k)$ item corresponds to the $u^+u = 1$ normalization method, so from Eq. (5.1396) we can obtain

$$U(k) = \frac{M^*}{E_k^*}\Sigma^{\mathrm{s}}(k) + \Sigma^0(k) \tag{5.1397}$$

The energy density in nuclear matter comes from kinetic energy and half potential energy, and there is

$$\varepsilon = \sum_{k\lambda} \frac{M^*}{E_k^*} \bar{u}\left(\vec{k}, \lambda\right) \left[\vec{\gamma} \cdot \vec{k} + M + \frac{1}{2}\Sigma(k)\right] u\left(\vec{k}, \lambda\right) \tag{5.1398}$$

The energy of each nucleon in nuclear matter is called binding energy, and its expression is

$$E_b = \frac{E}{A} - M = \frac{\varepsilon}{\rho_{\mathrm{B}}} - M \tag{5.1399}$$

In the on-shell case, the single particle energy without static mass contribution in nuclear matter can be obtained by Eq. (5.1320) as

$$\varepsilon_{\mathrm{sp}} = (1 + \Sigma^{\mathrm{v}}(k)) \left[k^2 + M^*(k)\right]^{1/2} + \Sigma^0(k) - M \tag{5.1400}$$

At the same time, the pressure $p$ can be calculated using Eq. (5.948), and the incompressible coefficient $K$ can be calculated using Eq. (5.951).

### *5.17.4 Nucleon Self-Energy in Asymmetric Nuclear Matter [103, 104]*

The Fermi momentum of neutrons and protons in symmetric nuclear matter are equal. However, in asymmetric nuclear matter, neutrons and protons occupy the Fermi spheres of different sizes. Therefore they also have different Pauli incompatibility operators and correspond to different effective masses and self-energies. For asymmetric nuclear matter, it is necessary to distinguish between nn, pp, np three reaction channels. For the np reaction channel, taking into account the anti-symmetry effect, the following decompositions on the interaction $V$ and effective interaction $T$ can be done:

$$V_{\mathrm{np}} = V_{\mathrm{np}}^{\mathrm{dir}} - V_{\mathrm{np}}^{\mathrm{exc}}, \qquad T_{\mathrm{np}} = T_{\mathrm{np}}^{\mathrm{dir}} - T_{\mathrm{np}}^{\mathrm{exc}} \tag{5.1401}$$

where

$$V_{\mathrm{np}}^{\mathrm{dir}} = \langle \mathrm{np}|v|\mathrm{np}\rangle, \qquad V_{\mathrm{np}}^{\mathrm{exc}} = \langle \mathrm{np}|v|\mathrm{pn}\rangle \tag{5.1402}$$

and there are the similar expressions for $T$.

In the BS equation of symmetric nuclear matter given by Eq. (5.1243), i$G$ represents the propagator of two free nucleons, which can be formally represented by i$GG$. For nuclear matter, it is also necessary to add a Pauli operator $Q$, $v$ and $t$ both need to take their helicity matrix elements, and then Eq. (5.1243) can be written in a simplified form as

$$T = V + \mathrm{i}\int VQGGT \tag{5.1403}$$

For asymmetric nuclear matter, the following BS equations need to be processed simultaneously:

$$T_{\mathrm{nn}} = V_{\mathrm{nn}} + \mathrm{i}\int V_{\mathrm{nn}}Q_{\mathrm{nn}}G_{\mathrm{n}}G_{\mathrm{n}}T_{\mathrm{nn}} \tag{5.1404}$$

$$T_{\mathrm{pp}} = V_{\mathrm{pp}} + \mathrm{i}\int V_{\mathrm{pp}}Q_{\mathrm{pp}}G_{\mathrm{p}}G_{\mathrm{p}}T_{\mathrm{pp}} \tag{5.1405}$$

$$T_{\mathrm{np}}^{\mathrm{dir}} = V_{\mathrm{np}}^{\mathrm{dir}} + \mathrm{i}\int V_{\mathrm{np}}^{\mathrm{dir}}Q_{\mathrm{np}}G_{\mathrm{n}}G_{\mathrm{p}}T_{\mathrm{np}}^{\mathrm{dir}} + \mathrm{i}\int V_{\mathrm{np}}^{\mathrm{exc}}Q_{\mathrm{pn}}G_{\mathrm{p}}G_{\mathrm{n}}T_{\mathrm{np}}^{\mathrm{exc}} \tag{5.1406}$$

$$T_{\mathrm{np}}^{\mathrm{exc}} = V_{\mathrm{np}}^{\mathrm{exc}} + \mathrm{i}\int V_{\mathrm{np}}^{\mathrm{exc}}Q_{\mathrm{pn}}G_{\mathrm{p}}G_{\mathrm{n}}T_{\mathrm{np}}^{\mathrm{dir}} + \mathrm{i}\int V_{\mathrm{np}}^{\mathrm{dir}}Q_{\mathrm{np}}G_{\mathrm{n}}G_{\mathrm{p}}T_{\mathrm{np}}^{\mathrm{exc}} \tag{5.1407}$$

where $V_{ij}(i,j = \mathrm{n},\mathrm{p})$ represent the one boson exchange potential between bare nucleons.

Assuming that in the np reaction channel neutrons and protons have the same average effective mass, from Eq. (5.1393) it can be seen that the relations of $T_{\mathrm{np}}^{\mathrm{dir}}$, $V_{\mathrm{np}}^{\mathrm{dir}}$ and $T_{\mathrm{np}}^{\mathrm{exc}}$, $V_{\mathrm{np}}^{\mathrm{exc}}$ are established through the Fierz transformation given by Eq. (5.1390), so Eq. (5.1406) and Eq. (5.1407) can be reduced to

$$T_{\mathrm{np}} = V_{\mathrm{np}} + \mathrm{i}\int V_{\mathrm{np}}Q_{\mathrm{np}}G_{\mathrm{n}}G_{\mathrm{p}}T_{\mathrm{np}} \tag{5.1408}$$

According to Eq. (5.1279) it can be written that the effective Thompson propagator in nuclear matter is

$$\mathrm{i}G_iG_j = \frac{M_i^*}{E_i^*}\frac{M_j^*}{E_j^*}\frac{1}{\sqrt{s^*} - E_i^* - E_j^* + \mathrm{i}\varepsilon} \tag{5.1409}$$

The Fermi momentum $k_{\mathrm{F_n}}$ and $k_{\mathrm{F_p}}$ of neutrons and protons in asymmetric nuclear matter are different, so two Fermi ellipsoids of different sizes are considered for the Pauli incompatibility operator of the np reaction channel.

For neutrons and protons, their baryon flow speed $\vec{\beta}$ and the corresponding $\gamma$ are the same, and it is also noted that the effective masses of neutrons and protons are assumed to be equal for the np reaction channel. So the following relations can be introduced for $i = \mathrm{n, p}$ according to Eqs. (5.1357), (5.1359), (5.1361)–(5.1364):

$$k_{\mathrm{u}i} = \sqrt{k_{\mathrm{F}_i}^2 - \beta^2 E_{\mathrm{F}_i}^{*2}} \tag{5.1410}$$

$$k_{\mathrm{v}i} = \gamma\left(k_{\mathrm{F}_i} - \beta E_{\mathrm{F}_i}^*\right) \tag{5.1411}$$

$$k_{+i} = \gamma\left(k_{\mathrm{F}_i} + \beta E_{\mathrm{F}_i}^*\right) \tag{5.1412}$$

$$k_{-i} = \min\left(k_{\mathrm{u}i}, k_{\mathrm{v}i}\right) \tag{5.1413}$$

$$\cos\alpha_i = \frac{\gamma E_k^* - E_{\mathrm{F}_i}^*}{\gamma\beta k} \tag{5.1414}$$

$$Q_i(k^*; B) = \begin{cases} 0, & k < k_{-i} \\ \Theta(\cos\alpha_i - \cos\theta_k)\,\Theta(\cos\theta_k + \cos\alpha_i), & k_{-i} \le k \le k_{+i} \\ 1, & k > k_{+i} \end{cases} \tag{5.1415}$$

The angle averaging Pauli incompatibility operator of the np reaction channel in the c.m. system can be approximated as

$$\overline{Q}_{\mathrm{np}}(k^*; B) = \frac{1}{2}\int_{-1}^{1} Q_{\mathrm{n}}(k^*; B) Q_{\mathrm{p}}(k^*; B) \mathrm{d}\cos\theta_k \tag{5.1416}$$

According to Eq. (5.1323), the neutron self-energy and proton self-energy can be formally written as follows, respectively:

$$\Sigma_{\mathrm{n}} = -\mathrm{i}\int_{\mathrm{F_n}} \left[\mathrm{tr}(G_{\mathrm{n}} T_{\mathrm{nn}}) - G_{\mathrm{n}} T_{\mathrm{nn}}\right] - \mathrm{i}\int_{\mathrm{F_p}} \left[\mathrm{tr}\left(G_{\mathrm{p}} T_{\mathrm{np}}\right) - G_{\mathrm{p}} T_{\mathrm{np}}\right] \tag{5.1417}$$

$$\Sigma_{\mathrm{p}} = -\mathrm{i}\int_{\mathrm{F_p}} \left[\mathrm{tr}\left(G_{\mathrm{p}} T_{\mathrm{pp}}\right) - G_{\mathrm{p}} T_{\mathrm{pp}}\right] - \mathrm{i}\int_{\mathrm{F_n}} \left[\mathrm{tr}\left(G_{\mathrm{n}} T_{\mathrm{np}}\right) - G_{\mathrm{n}} T_{\mathrm{np}}\right] \tag{5.1418}$$

The asymmetry degree of nuclear matter is defined as

$$\beta = \frac{\rho_{\mathrm{n}} - \rho_{\mathrm{p}}}{\rho} \tag{5.1419}$$

where $\rho_{\mathrm{n}}$, $\rho_{\mathrm{p}}$, $\rho$ are neutron, proton, and total nuclear density in nuclear matter, respectively. The binding energy $\mathrm{E_b}(\rho, \beta)$ is a function of $\rho$ and $\beta$, and it can be expanded as

$$E_{\mathrm{b}}(\rho, \beta) = E_{\mathrm{b}}(\rho) + E_{\mathrm{sym}}(\rho)\beta^2 + O\left(\beta^4\right) \tag{5.1420}$$

From the above formula, the symmetric energy can be obtained as

$$E_{\text{sym}}(\rho) = \frac{1}{2}\frac{\partial^2 E_{\text{b}}(\rho,\ \beta)}{\partial \beta^2}\bigg|_{\beta=0} \tag{5.1421}$$

If the $\beta^4$ term and the more high order terms of $\beta$ are ignored in Eq. (5.1420), we can obtain

$$E_{\text{sym}}(b) \cong E_{\text{b}}(\rho, \beta = 1) - E_{\text{b}}(\rho, \beta = 0) \tag{5.1422}$$

### 5.17.5 Nucleon Relativistic Microscopic Optical Potential Based on DBHF Theory

The Dirac-Brueckner-Hartree-Fock (DBHF) theory described above can be used to calculate the properties of nuclear matter and the nucleon self-energy in nuclear matter. The real and imaginary parts of the nucleon self-energy given by Eq. (5.1318) correspond to the real and imaginary parts of the nucleon optical potential in the nuclear matter. The nuclear matter saturation density $\rho_0$~0.16 fm$^{-3}$, the DBHF numerical calculation at low density will encounter divergence problems, in Ref. [103] only the calculation results of the $\rho \geq 0.08$ fm$^{-3}$ range are given. If we want to use the local density approximation to calculate the microscopic optical potential of finite nuclei, we will encounter the problem of how to select the self-energy of the low-density region in nuclear matter. In order to solve this problem, people usually use a so-called relativistic mean field method with density-related coupling constants [108–114]. The general approach is to adjust the coupling constants of the σ, ω, etc. mesons in the relativistic mean field theory to fit the self-energy obtained by the microscopic DBHF calculation at this density. Then using the density-related coupling constants obtained above and the local density approximation, the relativistic mean field method can be used to calculate the ground state properties [108–113] or the microscopic optical potential [114] of the finite nucleus. After the density- related meson coupling constants are determined, the finite nucleus equation can be solved directly in the relativistic mean field theory. In the Ref. [114] the density-related meson coupling constants are obtained only from the real part of the microscopic optical potential calculated by DBHF theory. In fact, the density-related meson coupling constants can be obtained simultaneously from the real and imaginary parts of the microscopic optical potential in the asymmetric nuclear matter. How to solve finite nucleus problems directly by DBHF theory is a theory to be further developed in the future. In addition, in Ref. [115] the off-shell self-energy of asymmetric nuclear matter was also studied using DBHF theory.

## 5.18 Proca Relativistic Dynamics Equation of Spin 1 Particle and Its Application in Elastic Scattering Calculations

Deuteron is a weakly bound state vector particle with spin 1 consisting of neutron and proton. In order to calculate the cross sections, differential cross sections, and various polarization physical quantities of the elastic scattering and other reactions induced by the deuterons, we first need to study the relativistic dynamics equation of the particles with spin 1. We do not consider the limited size and internal structure of the deuterons for the time being; we only think of the deuteron as a point particle. In addition, the W boson is also a particle with spin 1 and non-zero mass, while the photon is a particle with spin 1 and mass 0. This section discusses the Proca relativistic dynamics equation that describes the particles with spin 1.

### *5.18.1 Proca Equation of Free Particles*

The physical quantity of the vector particle field with spin 1 can be described by $\psi^\mu(x)$, and the corresponding free vector particle Lagrangian density can be expected to be [116, 117]

$$\mathcal{L} = -\frac{1}{4}F_{\mu\nu}F^{\mu\nu} + \frac{1}{2}m^2\psi_\mu\psi^\mu \tag{5.1423}$$

where the field strength tensor is

$$F^{\mu\nu} = \partial^\mu\psi^\nu - \partial^\nu\psi^\mu \tag{5.1424}$$

Substituting Eq. (5.1423) into the Lagrange equation (5.612), we can get

$$\partial_\mu F^{\mu\nu} + m^2\psi^\nu = 0 \tag{5.1425}$$

This is the Proca equation satisfied by the free particles with spin 1 [116–118]. In order to eliminate the additional components of the particle field, the following Lorentz gauge of the vector particles given by Eq. (5.642) can be used:

$$\partial_\nu\psi^\nu = 0 \tag{5.1426}$$

At this time, Eq. (5.1425) becomes the following Klein-Gordon equation:

$$\left(\partial_\mu\partial^\mu + m^2\right)\psi^\nu = 0 \tag{5.1427}$$

However, when there is an interaction potential, Eq. (5.1426) and Eq. (5.1427) do not necessarily hold.

### *5.18.2 Proca Equation with Interaction Potential*

We assume that the interaction potential of the point deuteron and the nucleus consists of the scalar potential $S$ and the time-like component $V$ of the vector potential, and the interaction potential is added in the following form [118]:

$$\partial_\mu \to D_\mu = \partial_\mu + \mathrm{i}V\delta_{\mu 0} \tag{5.1428}$$

$$m \to \widetilde{m} = m + S \tag{5.1429}$$

Equation (5.5) has given $-\mathrm{i}E \to \frac{\partial}{\partial t}$, so Eq. (5.1428) is equivalent to $E \to E - V$. In this way, the Proca equation (5.1425) can be rewritten as

$$D_\mu(D^\mu \psi^\nu - D^\nu \psi^\mu) + \widetilde{m}^2 \psi^\nu = 0 \tag{5.1430}$$

Referring to the electromagnetic field theory, the wave function $\psi$ can be expressed as

$$\psi = \begin{pmatrix} \varphi \\ \vec{A} \end{pmatrix} \tag{5.1431}$$

where $\varphi$ and $\vec{A}$ are the scalar component and vector component of the electromagnetic field, respectively. For $\nu = 0$ there is $\psi^0 = \varphi$, and note that the $\mu = \nu$ item in Eq. (5.1430) has no contribution. Let us assume that $S$ and $V$ are both spherical symmetrical, and note that

$$-D_i D^0 \psi^i = -\partial_i\left(\frac{\partial}{\partial t} + \mathrm{i}V\right)\psi^i = \mathrm{i}\partial_i(E - V)\,\psi^i = -i\frac{\mathrm{d}V}{\mathrm{d}r}\hat{r}\cdot\vec{A} + \mathrm{i}(E - V)\vec{\nabla}\cdot\vec{A}$$

Let

$$\omega = E - V \tag{5.1432}$$

$$\Theta = \frac{1}{\omega}\frac{\mathrm{d}V}{\mathrm{d}r} \tag{5.1433}$$

then from Eq. (5.1430) we can get

$$\left(\vec{\nabla}^2 - \widetilde{m}^2\right)\varphi = -\mathrm{i}\omega\left(\Theta\,\hat{r}\cdot\vec{A} - \vec{\nabla}\cdot\vec{A}\right) \tag{5.1434}$$

For $\nu = i = 1, 2, 3$ items, first note

$$D_\mu D^\mu \psi^i\big|_{\mu\neq i} = \left(\partial_\mu + \mathrm{i}V\delta_{\mu 0}\right)\left(\partial^\mu + \mathrm{i}V\delta_{\mu 0}\right)\psi^i\big|_{\mu\neq i} = \left(-\omega^2 - \vec{\nabla}^2\right)\psi^i + \vec{\nabla}_i^2\psi^i,$$

$$-D_\mu D^i \psi^\mu\big|_{\mu\neq i} = -\mathrm{i}\omega\,\partial_i\phi + \sum_{j\neq i}\vec{\nabla}_j\vec{\nabla}_i\psi^j,$$

$$\sum_i\left(\vec{\nabla}_i^2\psi^i + \sum_{j\neq i}\vec{\nabla}_j\vec{\nabla}_i\psi^j\right) = \sum_{ij}\vec{\nabla}_i\vec{\nabla}_j\psi^j = \vec{\nabla}\left(\vec{\nabla}\cdot\vec{A}\right)$$

then from Eq. (5.1430) we can obtain

$$\left(\vec{\nabla}^2 + \omega^2 - \widetilde{m}^2\right)\vec{A} = \vec{\nabla}\left(\vec{\nabla}\cdot\vec{A}\right) - \mathrm{i}\omega\vec{\nabla}\varphi \tag{5.1435}$$

In order to eliminate the $\hat{r}\cdot\vec{A}$ item in Eq. (5.1434), let

$$\vec{A} = \frac{E}{\omega}\vec{A}_1 \tag{5.1436}$$

then we can obtain

$$\vec{\nabla}\cdot\vec{A} = \frac{E}{\omega}\vec{\nabla}\cdot\vec{A}_1 + \frac{E}{\omega^2}\frac{\mathrm{d}V}{\mathrm{d}r}\hat{r}\cdot\vec{A}_1 = \frac{E}{\omega}\vec{\nabla}\cdot\vec{A}_1 + \frac{E}{\omega}\Theta\hat{r}\cdot\vec{A}_1,$$
$$\vec{\nabla}\vec{A} = \frac{E}{\omega}\vec{\nabla}\vec{A}_1 + \frac{E}{\omega^2}\frac{\mathrm{d}V}{\mathrm{d}r}\hat{r}\vec{A}_1 = \frac{E}{\omega}\vec{\nabla}\vec{A}_1 + \frac{E}{\omega}\Theta\hat{r}\vec{A}_1$$

So Eq. (5.1434) can be rewritten as

$$\left(\vec{\nabla}^2 - \widetilde{m}^2\right)\varphi = \mathrm{i}E\,\vec{\nabla}\cdot\vec{A}_1 \tag{5.1437}$$

There are the following formulas:

$$\vec{\nabla}\left(\frac{1}{r}\right) = -\frac{\vec{r}}{r^3}, \qquad \vec{\nabla}\cdot\vec{r} = 3 \tag{5.1438}$$

so we can get

$$\vec{\nabla}\cdot\hat{r} = \vec{\nabla}\cdot\left(\frac{\vec{r}}{r}\right) = \vec{\nabla}\left(\frac{1}{r}\right)\cdot\vec{r} + \frac{1}{r}\vec{\nabla}\cdot\vec{r} = \frac{2}{r} \tag{5.1439}$$

Notice $\vec{\nabla}^2 = \vec{\nabla}\cdot\vec{\nabla}$; we can also find

$$\vec{\nabla}^2\vec{A} = \vec{\nabla}\cdot\left(\vec{\nabla}\vec{A}\right) = \vec{\nabla}\cdot\left(\frac{E}{\omega}\vec{\nabla}\vec{A}_1 + \frac{E}{\omega}\Theta\hat{r}\vec{A}_1\right) = \frac{E}{\omega}\vec{\nabla}^2\vec{A}_1 + \frac{E}{\omega}\Theta\hat{r}\cdot\vec{\nabla}\vec{A}_1$$
$$+\frac{E}{\omega}\Theta^2\vec{A}_1 + \frac{E}{\omega}\Theta'\vec{A}_1 + \frac{E}{\omega}\frac{2}{r}\Theta\vec{A}_1 + \frac{E}{\omega}\Theta\hat{r}\cdot\vec{\nabla}\vec{A}_1 \tag{5.1440}$$

where

$$\Theta' = \frac{\mathrm{d}\Theta}{\mathrm{d}r} \tag{5.1441}$$

And notice

$$\vec{\nabla}\left(\hat{r}\cdot\vec{A}_1\right) = \vec{\nabla}\left(\frac{\vec{r}\cdot\vec{A}_1}{r}\right) = -\frac{\vec{r}}{r^3}\left(\vec{r}\cdot\vec{A}_1\right) + \frac{1}{r}\vec{\nabla}\left(\vec{r}\cdot\vec{A}_1\right) \tag{5.1442}$$

so we can obtain

$$\vec{\nabla}\left(\vec{\nabla}\cdot\vec{A}\right) = \vec{\nabla}\left(\frac{E}{\omega}\vec{\nabla}\cdot\vec{A}_1 + \frac{E}{\omega}\Theta\hat{r}\cdot\vec{A}_1\right) = \frac{E}{\omega}\vec{\nabla}\left(\vec{\nabla}\cdot\vec{A}_1\right) + \frac{E}{\omega}\Theta\hat{r}\left(\vec{\nabla}\cdot\vec{A}_1\right)$$
$$+\frac{E}{\omega}\Theta^2\hat{r}\left(\hat{r}\cdot\vec{A}_1\right) + \frac{E}{\omega}\Theta'\hat{r}\left(\hat{r}\cdot\vec{A}_1\right) - \frac{E}{\omega}\Theta\frac{1}{r}\hat{r}\left(\hat{r}\cdot\vec{A}_1\right) + \frac{E}{\omega}\Theta\vec{\nabla}\left(\hat{r}\cdot\vec{A}_1\right) \tag{5.1443}$$

Substituting Eq. (5.1440) and Eq. (5.1443) into Eq. (5.1435), we can obtain

$$\left(\vec{\nabla}^2 + \omega^2 - \widetilde{m}^2\right)\vec{A}_1 = -\left(\Theta' + \Theta^2 + \frac{2}{r}\Theta\right)\vec{A}_1 - 2\Theta\left(\hat{r}\cdot\vec{\nabla}\right)\vec{A}_1 + \vec{\nabla}\left(\vec{\nabla}\cdot\vec{A}_1\right)$$
$$+\frac{\Theta}{r}\left(\vec{r}\vec{\nabla}\cdot\vec{A}_1 + \vec{\nabla}\left(\vec{r}\cdot\vec{A}_1\right)\right) + \left(\Theta' + \Theta^2 - \frac{\Theta}{r}\right)\hat{r}\left(\hat{r}\cdot\vec{A}_1\right) + \vec{R} \tag{5.1444}$$

$$\vec{R} = -\mathrm{i}\frac{\omega^2}{E}\vec{\nabla}\varphi \tag{5.1445}$$

If $\vec{A}_1$ is known, $\varphi$ can be solved from Eq. (5.1437), and $\vec{R}$ can be obtained using Eq. (5.1445). Introduce the following operator:

$$\hat{Z} \equiv \vec{\nabla}^2 + \omega^2 - \widetilde{m}^2 \tag{5.1446}$$

Since in the above formula, $E^2$ contained in $\omega^2$ and $m^2$ contained in $\widetilde{m}^2$ will offset each other, $\hat{Z}$ can be approximated as much less than $\omega^2$ when the energy is not very high. Eq. (5.1445) can be formally rewritten, according to Eq. (5.1437), as

$$\vec{R} = -\omega^2 \vec{\nabla}\left(\frac{1}{\omega^2 - \hat{Z}} \vec{\nabla} \cdot \vec{A}_1\right) \tag{5.1447}$$

### 5.18.3 *Some Expressions Related to S=1 Spin Operators [119]*

Equation (3.51) has given that in the rectangular basis representation, the matrix element of the spin operator of particle with spin 1 in the rectangular coordinate system is

$$\left[\hat{S}_k\right]_{ij} = -\mathrm{i}\varepsilon_{kij} = -\mathrm{i}\varepsilon_{ijk} \tag{5.1448}$$

where $\varepsilon_{ijk}$ is a three-dimensional anti-symmetry unit tensor given by Eq. (1.7). $\hat{S}$ is a $3 \times 3$ matrix, and $\vec{r}$ and $\vec{A}$ are three-dimensional vectors; therefore $\left(\hat{S} \cdot \vec{r}\right)\left(\hat{S} \cdot \vec{r}\right)\vec{A}$ is also a three-dimensional vector, and using Eq. (5.1448) we can write its vector element as

$$\left[\left(\hat{S} \cdot \vec{r}\right)\left(\hat{S} \cdot \vec{r}\right)\vec{A}\right]_i = -\varepsilon_{ijk}\varepsilon_{jlm} r_k r_m A_l \tag{5.1449}$$

There is the following equation [120]:

$$\varepsilon_{rmn}\varepsilon_{rst} = \delta_{ms}\delta_{nt} - \delta_{mt}\delta_{ns} \tag{5.1450}$$

The left end of the above formula implies summation to $r$. So Eq. (5.1449) can be rewritten as

$$\left[\left(\hat{S} \cdot \vec{r}\right)\left(\hat{S} \cdot \vec{r}\right)\vec{A}\right]_i = r_k r_k A_i - r_k r_i A_k \tag{5.1451}$$

Turn the above formula into vector form as

$$\vec{r}\left(\vec{r} \cdot \vec{A}\right) = r^2\vec{A} - \left(\hat{S} \cdot \vec{r}\right)\left(\hat{S} \cdot \vec{r}\right)\vec{A} \tag{5.1452}$$

There is also the following vector cross multiplication formula:

$$\left(\vec{C} \times \vec{D}\right)_i = \varepsilon_{ijk} C_j D_k \tag{5.1453}$$

Using the above formula we can write

$$\left[\vec{\nabla} \times \left(\vec{r} \times \vec{A}\right)\right]_i = \varepsilon_{ijk} \nabla_j \varepsilon_{klm} r_l A_m \tag{5.1454}$$

Noting $p_j = -\mathrm{i}\nabla_j$ and using Eq. (5.1448), the right end of the above formula can be rewritten as

$$\varepsilon_{ijk} \nabla_j \varepsilon_{lkm} r_l A_m = -\mathrm{i}\left[S_j\right]_{ik} p_j \left[S_l\right]_{km} r_l A_m$$

Substituting above formula into Eq. (5.1454), the following vector relation can be obtained:

$$\vec{\nabla} \times \left(\vec{r} \times \vec{A}\right) = -\mathrm{i}\left(\hat{S} \cdot \vec{p}\right)\left(\hat{S} \cdot \vec{r}\right)\vec{A} \tag{5.1455}$$

Using the relation $p_i x_j = x_j p_i - \mathrm{i}\delta_{ij}$, we can find

$$\vec{p}\,\vec{r} = p_i \vec{e}_i x_j \vec{e}_j = \left(x_j p_i - \mathrm{i}\delta_{ij}\right) \vec{e}_i \vec{e}_j = \vec{r}\,\vec{p} - \mathrm{i}\vec{e}_i \vec{e}_i$$

And noting $\hat{S}^2 = 2$, we can write

$$\left(\hat{S} \cdot \vec{p}\right)\left(\hat{S} \cdot \vec{r}\right) = \left(\hat{S} \cdot \vec{r}\right)\left(\hat{S} \cdot \vec{p}\right) - 2\mathrm{i} \tag{5.1456}$$

In this way, Eq. (5.1455) can be rewritten as

$$\vec{\nabla} \times \left(\vec{r} \times \vec{A}\right) = -\mathrm{i}\left(\hat{S} \cdot \vec{r}\right)\left(\hat{S} \cdot \vec{p}\right)\vec{A} - 2\vec{A} \tag{5.1457}$$

Noting the orbital angular momentum $\vec{L} = -\mathrm{i}\left(\vec{r} \times \vec{\nabla}\right)$, so we can obtain

$$\left(\vec{r} \times \vec{\nabla}\right) \times \vec{A} = \mathrm{i}\vec{L} \times \vec{A} \tag{5.1458}$$

and we can write

$$\mathrm{i}\left(\vec{L} \times \vec{A}\right)_i = \mathrm{i}\varepsilon_{ijk} L_j A_k = -\mathrm{i}\varepsilon_{jik} L_j A_k \tag{5.1459}$$

So the following relation can be obtained according to Eqs. (5.1448), (5.1458), and (5.1459):

$$\left(\vec{r} \times \vec{\nabla}\right) \times \vec{A} = \left(\hat{S} \cdot \vec{L}\right)\vec{A} \tag{5.1460}$$

There is the following vector identity:

$$\vec{C}\left(\vec{\nabla} \cdot \vec{D}\right) = \left(\vec{C} \cdot \vec{\nabla}\right)\vec{D} + \vec{\nabla} \times \left(\vec{C} \times \vec{D}\right) - \left(\vec{D} \cdot \vec{\nabla}\right)\vec{C} + \vec{D}\left(\vec{\nabla} \cdot \vec{C}\right) \tag{5.1461}$$

And notice

$$\left(\vec{A} \cdot \vec{\nabla}\right) \vec{r} = A_i \vec{e}_i \cdot \vec{e}_j \frac{\mathrm{d}}{\mathrm{d}x_j} x_k \vec{e}_k = A_i \vec{e}_i \cdot \vec{e}_j \vec{e}_j = \vec{A} \tag{5.1462}$$

using the above formula from Eq. (5.1461), we can get

$$\vec{r}\left(\vec{\nabla} \cdot \vec{A}\right) = 2\vec{A} + \left(\vec{r} \cdot \vec{\nabla}\right)\vec{A} + \vec{\nabla} \times \left(\vec{r} \times \vec{A}\right) \tag{5.1463}$$

Substituting Eq. (5.1457) into the above formula, we have

$$\vec{r}\left(\vec{\nabla} \cdot \vec{A}\right) = \left(\vec{r} \cdot \vec{\nabla}\right)\vec{A} - \mathrm{i}\left(\hat{S} \cdot \vec{r}\right)\left(\hat{S} \cdot \vec{p}\right)\vec{A} \tag{5.1464}$$

From Eq. (5.1453) we can write

$$\left[\vec{r} \times \left(\vec{\nabla} \times \vec{A}\right)\right]_i = \varepsilon_{ijk} r_j \varepsilon_{klm} \nabla_l A_m = \varepsilon_{kij} \varepsilon_{klm} r_j \nabla_l A_m \tag{5.1465}$$

Using Eq. (5.1450) we can also write

$$\varepsilon_{kij}\varepsilon_{klm} = \delta_{il}\delta_{jm} - \delta_{im}\delta_{jl} + \delta_{ml}\delta_{ij} - \delta_{ml}\delta_{ij} = \varepsilon_{ikm}\varepsilon_{kjl} + \delta_{ml}\delta_{ij} - \delta_{im}\delta_{jl}$$

So Eq. (5.1465) can be rewritten as

$$\left[\vec{r} \times \left(\vec{\nabla} \times \vec{A}\right)\right]_i = \left(\varepsilon_{kjl}\varepsilon_{ikm} + \delta_{ij}\delta_{ml} - \delta_{im}\delta_{jl}\right) r_j \nabla_l A_m \tag{5.1466}$$

Write the above formula by vector form as

$$\vec{r} \times \left(\vec{\nabla} \times \vec{A}\right) = \left(\vec{r} \times \vec{\nabla}\right) \times \vec{A} + \vec{r}\left(\vec{\nabla} \cdot \vec{A}\right) - \left(\vec{r} \cdot \vec{\nabla}\right)\vec{A} \tag{5.1467}$$

Utilizing Eq. (5.1460) and Eq. (5.1464), from Eq. (5.1467), we can get

$$\vec{r} \times \left(\vec{\nabla} \times \vec{A}\right) = \left(\hat{S} \cdot \vec{L}\right)\vec{A} - \mathrm{i}\left(\hat{S} \cdot \vec{r}\right)\left(\hat{S} \cdot \vec{p}\right)\vec{A} \tag{5.1468}$$

There is also the following vector identity:

$$\vec{\nabla}\left(\vec{C} \cdot \vec{D}\right) = \left(\vec{D} \cdot \vec{\nabla}\right)\vec{C} + \left(\vec{C} \cdot \vec{\nabla}\right)\vec{D} + \vec{D} \times \left(\vec{\nabla} \times \vec{C}\right) + \vec{C} \times \left(\vec{\nabla} \times \vec{D}\right) \tag{5.1469}$$

Noting there is $\vec{\nabla} \times \vec{r} = 0$, from the above formula we can write

$$\vec{\nabla}\left(\vec{r} \cdot \vec{A}\right) = \left(\vec{A} \cdot \vec{\nabla}\right)\vec{r} + \left(\vec{r} \cdot \vec{\nabla}\right)\vec{A} + \vec{r} \times \left(\vec{\nabla} \times \vec{A}\right) \tag{5.1470}$$

From the above formula, the following relation can be obtained using Eqs. (5.1462), (5.1467), and (5.1463) as

$$\vec{\nabla}\left(\vec{r} \cdot \vec{A}\right) = 3\vec{A} + \left(\vec{r} \cdot \vec{\nabla}\right)\vec{A} + \left(\vec{r} \times \vec{\nabla}\right) \times \vec{A} + \vec{\nabla} \times \left(\vec{r} \times \vec{A}\right) \tag{5.1471}$$

Substituting Eq. (5.1457) and Eq. (5.1460) into the above formula, we can obtain

$$\vec{\nabla}\left(\vec{r} \cdot \vec{A}\right) = \vec{A} + \left(\vec{r} \cdot \vec{\nabla}\right)\vec{A} + \left(\hat{S} \cdot \vec{L}\right)\vec{A} - \mathrm{i}\left(\hat{S} \cdot \vec{r}\right)\left(\hat{S} \cdot \vec{p}\right)\vec{A} \tag{5.1472}$$

### 5.18.4 Proca Equation in the Form of Schrodinger-like Equation

Equation (3.927) gives the following expression:

$$T_2\left(\vec{a}, \vec{b}\right) \cdot T_2\left(\vec{c}, \vec{d}\right) = \left(\vec{a} \cdot \vec{c}\right)\left(\vec{b} \cdot \vec{d}\right) - \frac{1}{2}\left(\vec{a} \times \vec{b}\right) \cdot \left(\vec{c} \times \vec{d}\right) - \frac{1}{3}\left(\vec{a} \cdot \vec{b}\right)\left(\vec{c} \cdot \vec{d}\right) \tag{5.1473}$$

If let $\vec{a} = \vec{b} = \hat{S},\ \vec{c} = \vec{A},\ \vec{d} = \vec{B}$ in the above equation, we can get

$$T_{\mathrm{AB}} = \left(\hat{S} \cdot \vec{A}\right)\left(\hat{S} \cdot \vec{B}\right) - \frac{\mathrm{i}}{2}\hat{S} \cdot \left(\vec{A} \times \vec{B}\right) - \frac{2}{3}\left(\vec{A} \cdot \vec{B}\right) \tag{5.1474}$$

If taking $\vec{A} = \vec{B} = \vec{r}$, we can define the following spatial-spatial tensor operator:

$$T_{\mathrm{RR}} = r^2\left[\left(\hat{S}\cdot\hat{r}\right)^2 - \frac{2}{3}\right] \tag{5.1475}$$

If taking $\vec{A} = \vec{r},\ \vec{B} = \vec{p}$, we can define the following spatial-momentum tensor operator:

$$\begin{aligned} T_{\mathrm{RP}} &= \left(\hat{S}\cdot\vec{r}\right)\left(\hat{S}\cdot\vec{p}\right) - \frac{\mathrm{i}}{2}\hat{S}\cdot\vec{L} - \frac{2}{3}\left(\vec{r}\cdot\vec{p}\right) \\ &= \left(\hat{S}\cdot\vec{r}\right)\left(\hat{S}\cdot\vec{p}\right) - \frac{\mathrm{i}}{2}\hat{S}\cdot\vec{L} + \frac{2\mathrm{i}}{3}\left(\vec{r}\cdot\vec{\nabla}\right) \end{aligned} \tag{5.1476}$$

Let

$$k^2 = E^2 - m^2 \tag{5.1477}$$

and note $\vec{\nabla}^2 = -p^2$; dividing Eq. (5.1444) by $(-2E)$ we can obtain

$$\begin{aligned} &\left\{\frac{p^2}{2E} + \frac{1}{2E}\left(-\omega^2 + \widetilde{m}^2 + k^2 - \left[\Theta' + \Theta\left(\Theta + \frac{2}{r}\right)\right]\right)\right\}\vec{A}_1 \\ &- \frac{1}{2E}\frac{2\Theta}{r}\left(\vec{r}\cdot\vec{\nabla}\right)\vec{A}_1 + \frac{1}{2E}\vec{\nabla}\left(\vec{\nabla}\cdot\vec{A}_1\right) + \frac{1}{2E}\frac{\Theta}{r}\left[\vec{r}\vec{\nabla}\cdot\vec{A}_1 + \vec{\nabla}\left(\vec{r}\cdot\vec{A}_1\right)\right] \\ &+ \frac{1}{2E}\left[\Theta' + \Theta\left(\Theta - \frac{1}{r}\right)\right]\hat{r}\left(\hat{r}\cdot\vec{A}_1\right) + \frac{\vec{R}}{2E} = \frac{k^2}{2E}\vec{A}_1 \end{aligned} \tag{5.1478}$$

When the energy is very low, $\frac{k^2}{2E}$ is equivalent to non-relativistic kinetic energy. The following three items appeared in Eq. (5.1478) are first combined by using Eqs. (5.1464), (5.1472), and (5.1476):

$$\begin{aligned} &\vec{r}\vec{\nabla}\cdot\vec{A}_1 + \vec{\nabla}\left(\vec{r}\cdot\vec{A}_1\right) - 2\left(\vec{r}\cdot\vec{\nabla}\right)\vec{A}_1 = \left(\vec{r}\cdot\vec{\nabla}\right)\vec{A}_1 - \mathrm{i}\left(\hat{S}\cdot\vec{r}\right)\left(\hat{S}\cdot\vec{p}\right)\vec{A}_1 + \vec{A}_1 \\ &\quad + \left(\vec{r}\cdot\vec{\nabla}\right)\vec{A}_1 + \left(\hat{S}\cdot\vec{L}\right)\vec{A}_1 - \mathrm{i}\left(\hat{S}\cdot\vec{r}\right)\left(\hat{S}\cdot\vec{p}\right)\vec{A}_1 - 2\left(\vec{r}\cdot\vec{\nabla}\right)\vec{A}_1 \\ &= \vec{A}_1 + \left(\hat{S}\cdot\vec{L}\right)\vec{A}_1 - 2\mathrm{i}\left(\hat{S}\cdot\vec{r}\right)\left(\hat{S}\cdot\vec{p}\right)\vec{A}_1 \\ &= \vec{A}_1 + \left(\hat{S}\cdot\vec{L}\right)\vec{A}_1 - 2\mathrm{i}\left[T_{\mathrm{RP}} + \frac{\mathrm{i}}{2}\left(\hat{S}\cdot\vec{L}\right) - \frac{2\mathrm{i}}{3}\left(\vec{r}\cdot\vec{\nabla}\right)\right]\vec{A}_1 \\ &= \vec{A}_1 + 2\left(\hat{S}\cdot\vec{L}\right)\vec{A}_1 - \frac{4}{3}\left(\vec{r}\cdot\vec{\nabla}\right)\vec{A}_1 - 2\mathrm{i}T_{\mathrm{RP}} \end{aligned} \tag{5.1479}$$

According to Eq. (5.1452) and Eq. (5.1475), we have

$$\hat{r}\left(\hat{r}\cdot\vec{A}_1\right) = \left[1-\left(\hat{S}\cdot\hat{r}\right)^2\right]\vec{A}_1 = -\frac{T_{\mathrm{RR}}}{r^2}\vec{A}_1 + \frac{1}{3}\vec{A}_1 \tag{5.1480}$$

Then merge some of the central potentials appeared previously as follows:

$$-\left[\Theta' + \Theta\left(\Theta + \frac{2}{r}\right)\right] + \frac{\Theta}{r} + \frac{1}{3}\left[\Theta' + \Theta\left(\Theta - \frac{1}{r}\right)\right] = -\frac{2}{3}\left[\Theta' + \Theta\left(\Theta + \frac{2}{r}\right)\right] \tag{5.1481}$$

Using the result given by Eq. (5.1479)–(5.1481), Eq. (5.1478) can be rewritten as

$$\begin{aligned}&\left[\frac{p^2}{2E} + U_{\mathrm{c}} + U_{\mathrm{SO}}\hat{S}\cdot\vec{L} + U_{\mathrm{D}}\left(\vec{r}\cdot\vec{\nabla}\right) + U_{\mathrm{RR}}\left(\frac{T_{\mathrm{RR}}}{r^2}\right) + \mathrm{i}U_{\mathrm{RP}}T_{\mathrm{RP}}\right]\vec{A}_1\\ &+\frac{1}{2E}\left(\vec{Q}+\vec{R}\right) = \frac{k^2}{2E}\vec{A}_1\end{aligned} \tag{5.1482}$$

where $U_{\mathrm{c}}$, $U_{\mathrm{SO}}$, $U_{\mathrm{D}}$, $U_{\mathrm{RR}}$, $U_{\mathrm{RP}}$ are the central potential, spin-orbit coupling potential, Darwin potential, spatial-spatial tensor potential, and spatial-momentum tensor potential, respectively. Their expressions are as follows:

$$U_{\mathrm{c}} = \frac{1}{2E}\left\{-\omega^2 + \widetilde{m}^2 + k^2 - \frac{2}{3}\left[\Theta' + \Theta\left(\Theta + \frac{2}{r}\right)\right]\right\} \tag{5.1483}$$

$$U_{\mathrm{SO}} = \frac{1}{2E}\frac{2\Theta}{r} \tag{5.1484}$$

$$U_{\mathrm{D}} = -\frac{1}{2E}\frac{4\Theta}{3r} \tag{5.1485}$$

$$U_{\mathrm{RR}} = -\frac{1}{2E}\left[\Theta' + \Theta\left(\Theta - \frac{1}{r}\right)\right] \tag{5.1486}$$

$$U_{\mathrm{RP}} = -\frac{1}{2E}\frac{2\Theta}{r} \tag{5.1487}$$

And there is

$$\vec{Q} = \vec{\nabla}\left(\vec{\nabla}\cdot\vec{A}_1\right) \tag{5.1488}$$

where $\vec{R}$ has been given by Eq. (5.1445). Equation (5.1482) is called the Proca equation in the form of Schrodinger-like equation. In addition, it is also necessary to artificially add Coulomb potential. The deuteron Coulomb potential $V_{\mathrm{C}}$ can also be discussed like the discussion to the spin $\frac{1}{2}$ particle Coulomb potential in Sect. 5.3. Then one can let $U_{\mathrm{c}} \to U_{\mathrm{c}} + \frac{\mu}{E}V_{\mathrm{C}}$ in equation (5.1482), where $\mu$ is the deuteron reduced mass. If the other components of the optical potential are added in addition to the scalar potential $S$ and the time-like component $V$ of the vector potential in the nuclear potential, the resulting Proca equation also needs to be changed accordingly.

### *5.18.5 Discussion on the Application of the Proca Equation to Calculate the Elastic Scatterings Between Deuteron and Nucleus*

When the Proca equation is used to calculate the elastic scattering of the deuteron and the nucleus, the deuteron phenomenological optical potential can be directly used, and the deuteron optical potential can also be obtained from the nucleon-nucleon interaction potential using the folding model. The contribution of the deuteron breakup channel can be or cannot be considered in the folding model. It is also possible to use the deuteron microscopic optical potential obtained from effective nuclear force, realistic nuclear force, or meson exchange coupling constants.

Equation (5.1482) and Eq. (5.1437) constitute a set of simultaneous equations for the 4-vector field $\left(\varphi, \vec{A}_1\right)$. If we make $\vec{Q} = \vec{R} = 0$ in Eq. (5.1482), it is equivalent to ignore $\vec{\nabla}\left(\vec{\nabla} \cdot \vec{A}_1\right)$ items and $\varphi$ items. In fact, as long as we make $\vec{\nabla} \cdot \vec{A}_1 = 0$, we will get $\vec{Q} = \vec{R} = 0$ simultaneously. At this time, Eq. (5.1482) gives three equations of the three components of $\vec{A}_1$; they are not coupled to each other. If the incident particle direction is selected as the $z$-axis, the entire system is axis-symmetric, so only the two equations of the radial component and the polar angle component of $\vec{A}_1$ need to be considered in the spherical coordinate system. For the two variables we can use mathematical methods such as transforming wave functions and separating variables as we solve the Schrodinger equation in the non-relativistic case. When solving the radial equation, the $S$ matrix can be introduced to give the scattering amplitude, and then the differential cross sections and various polarization physical quantities of the elastic scattering of the deuteron with nucleus can be calculated. Sometimes it is thought that in Eq. (5.1482) the Darwin term and two tensor terms contribute less and can be ignored. In the Ref. [118] the differential cross section $\frac{d\sigma}{d\Omega}$, the vector analyzing power $A_y$, and the tensor analyzing power $A_{yy}$ of the elastic scattering of the 400 MeV d + $^{58}$Ni and 700 MeV d + $^{40}$Ca in the laboratory system were calculated by using Proca equation. $^{58}$Ni and $^{40}$Ca are the target nuclei with spinless. According to the theory introduced in Chap. 3 of this book, the target nucleus with non-zero spin can also be calculated.

When $\vec{\nabla} \cdot \vec{A}_1 \neq 0$, it can be seen from Eqs. (5.1482) and (5.1437) that the three components of $\vec{A}_1$ and the scalar wave function $\varphi$ satisfy the coupling equations. In the axis-symmetric system, there are also three wave function components to be coupled and solved. And there is the radial variable in each wave function component. In this case, how to solve the numerical problems and how to calculate the differential cross sections and various polarization physical quantities of the nuclear reaction are the topics that need to be further studied.

In the Proca equation, the wave function describing the field physical quantity is represented by a 4-vector field consisting of a scalar field $\varphi$ and a vector field $\vec{A}$. There is also a Kemmer-Duffin-Petiau (KDP) equation [118, 119, 121] that can be used to describe spin 1 particles, where the wave function describing the field physical quantity is represented by a 10-dimensional vector formed by a column constituted by $\left(\varphi, \vec{A}, \vec{E}, -\vec{B}\right)$. We know that in electromagnetic field theory $\vec{E}$ and $\vec{B}$ can be derived from $\varphi$ and $\vec{A}$. According to convention, all kinds of physical quantities should be able to be derived from the wave function, so putting $\vec{E}$ and $\vec{B}$ belong to the derived quantities into the wave function as a basic physical quantity is not consistent with the convention. In fact, when solving the KDP equation, it will be ultimately converted into the equations of $\varphi$ and $\vec{A}$ to be solved.

## 5.19 Weinberg Relativistic Dynamics Equation with Spin 1 Particles and Discussion on its Application

### *5.19.1 Weinberg Equation of Free Particles*

The spin operators of the spin $\frac{1}{2}$ nucleon are represented by $2 \times 2$ matrices. In the Dirac equation, the wave function is divided into upper and lower components, so the entire spin space is described by $4 \times 4$ $\gamma$ matrices. And in the case of low energy approximation, the contribution of the lower component can be ignored. Then the obtained equation will automatically degenerate into $2 \times 2$ spin space with only the upper component. The spin operators of the spin 1 particle are represented by $3 \times 3$ matrices. In the Weinberg equation [122, 123] describing the spin 1 particles, the wave function is also expressed as

$$\psi = \begin{pmatrix} \vec{A} \\ \vec{B} \end{pmatrix} \tag{5.1489}$$

where $\vec{A}$ and $\vec{B}$ are the upper and lower components of the wave function, and both are three-dimensional vectors. The Weinberg equation can be written as [116–118]

$$\left(p_\mu p_\nu \gamma^{\mu\nu} - m^2\right)\psi = 0 \tag{5.1490}$$

where $m$ is the particle stationary mass. $\gamma^{\mu\nu}$ is $6 \times 6$ matrix and their specific expressions are

$$\gamma^{00} = \begin{pmatrix} \hat{I}_3 & \hat{0} \\ \hat{0} & \hat{I}_3 \end{pmatrix} \tag{5.1491}$$

$$\gamma^{0i} = \gamma^{i0} = \begin{pmatrix} \hat{0} & \hat{S}_i \\ -\hat{S}_i & \hat{0} \end{pmatrix} \tag{5.1492}$$

$$\gamma^{ij} = \begin{pmatrix} \hat{S}_i\hat{S}_j + \hat{S}_j\hat{S}_i - \delta_{ij}\hat{I}_3 & \hat{0} \\ \hat{0} & -\left(\hat{S}_i\hat{S}_j + \hat{S}_j\hat{S}_i - \delta_{ij}\hat{I}_3\right) \end{pmatrix} \tag{5.1493}$$

where $\hat{I}_k$ is a unit $k \times k$ matrix, $\hat{S}_i$ is spin operator of spin 1 particle.

Let us do the following matrix decomposition:

$$p_\mu p_\nu \gamma^{\mu\nu} = \begin{pmatrix} \Gamma_{11} & \Gamma_{12} \\ \Gamma_{21} & \Gamma_{22} \end{pmatrix} \tag{5.1494}$$

where $\Gamma_{ij}(i,j = 1,2)$ is $3 \times 3$ matrix. For the stationary systems, according to Eq. (5.61) we can see

$$(p_0, p_1, p_2, p_3) = \left(E, -\vec{p}\right) \tag{5.1495}$$

The following expressions can be obtained using Eqs. (5.1491)–(5.1495):

$$\Gamma_{11} = E^2 + 2\left(\hat{S}\cdot\vec{p}\right)^2 - \vec{p}^{\,2} = 2\left(\hat{S}\cdot\vec{p}\right)^2 + m^2 \tag{5.1496}$$

$$\Gamma_{12} = -2E\,\hat{S}\cdot\vec{p} \tag{5.1497}$$

$$\Gamma_{21} = 2E\,\hat{S}\cdot\vec{p} \tag{5.1498}$$

$$\Gamma_{22} = -\left[E^2 + 2\left(\hat{S}\cdot\vec{p}\right)^2 - \vec{p}^{\,2}\right] = -\left[2\left(\hat{S}\cdot\vec{p}\right)^2 + m^2\right] \tag{5.1499}$$

Substituting Eq. (5.1494) into Eq. (5.1490) we can get

$$\begin{pmatrix} \Gamma_{11} - m^2 & \Gamma_{12} \\ \Gamma_{21} & \Gamma_{22} - m^2 \end{pmatrix} \begin{pmatrix} \vec{A} \\ \vec{B} \end{pmatrix} = 0 \tag{5.1500}$$

We can also rewrite the above formula as

$$\frac{1}{2}\begin{pmatrix} -(\Gamma_{11}-m^2) & -\Gamma_{12} \\ \Gamma_{21} & \Gamma_{22}-m^2 \end{pmatrix}\begin{pmatrix} \vec{A} \\ \vec{B} \end{pmatrix}=0 \tag{5.1501}$$

Then the following equation can be obtained substituting Eqs. (5.1496)–(5.1499) into Eq. (5.1501):

$$\begin{pmatrix} -\left(\hat{S}\cdot\vec{p}\right)^2 & E\left(\hat{S}\cdot\vec{p}\right) \\ E\left(\hat{S}\cdot\vec{p}\right) & -\left[\left(\hat{S}\cdot\vec{p}\right)^2+m^2\right] \end{pmatrix}\begin{pmatrix} \vec{A} \\ \vec{B} \end{pmatrix}=0 \tag{5.1502}$$

This is the Weinberg equation for free particles with spin 1.

### 5.19.2 Weinberg Equation with Interaction Potential

We assume that the particle with spin 1 only has the scalar potential $S$ and the time-like component $V$ of the vector potential, and the interaction potentials are still introduced by the method given Eq. (5.1428) and Eq. (5.1429) [118]. And noting Eq. (5.62) we can rewrite Eq. (5.1490) as

$$\left[\left(\mathrm{i}\partial_\mu - V\delta_{\mu 0}\right)\left(\mathrm{i}\partial_\nu - V\delta_{\nu 0}\right)\gamma^{\mu\nu}-\widetilde{m}^2\right]\psi=0 \tag{5.1503}$$

Let us do the following matrix decomposition:

$$\left(\mathrm{i}\partial_\mu - V\delta_{\mu 0}\right)\left(\mathrm{i}\partial_\nu - V\delta_{\nu 0}\right)\gamma^{\mu\nu}=\begin{pmatrix} \Lambda_{11} & \Lambda_{12} \\ \Lambda_{21} & \Lambda_{22} \end{pmatrix} \tag{5.1504}$$

where $\Lambda_{ij}(i,j=1,2)$ is 3×3 matrix. For the stationary systems, using $\omega=E-V$ introduced by Eq. (5.1432), the following expressions can be obtained using Eqs. (5.1491)–(5.1493) and (5.1504):

$$\Lambda_{11}=\omega^2+\vec{\nabla}^2+2\left(\hat{S}\cdot\vec{p}\right)^2 \tag{5.1505}$$

$$\Lambda_{12}=-2\omega^2\left(\hat{S}\cdot\vec{p}\right) \tag{5.1506}$$

$$\Lambda_{21}=2\omega\left(\hat{S}\cdot\vec{p}\right) \tag{5.1507}$$

$$\Lambda_{22}=-\left[\omega^2+\vec{\nabla}^2+2\left(\hat{S}\cdot\vec{p}\right)^2\right] \tag{5.1508}$$

So the following equation can be obtained using Eqs. (5.1503)–(5.1508):

$$\begin{pmatrix} \omega^2 - \widetilde{m}^2 + \vec{\nabla}^2 + 2\left(\hat{S}\cdot\vec{p}\right)^2 & -2\omega\left(\hat{S}\cdot\vec{p}\right) \\ 2\omega\left(\hat{S}\cdot\vec{p}\right) & -\left(\omega^2 + \widetilde{m}^2 + \vec{\nabla}^2 + 2\left(\hat{S}\cdot\vec{p}\right)^2\right) \end{pmatrix} \begin{pmatrix} \vec{A} \\ \vec{B} \end{pmatrix} = 0 \tag{5.1509}$$

And we can further get

$$\left[\vec{\nabla}^2 + 2\left(\hat{S}\cdot\vec{p}\right)^2 + \omega^2 - \widetilde{m}^2\right]\vec{A} - 2\omega\left(\hat{S}\cdot\vec{p}\right)\vec{B} = 0 \tag{5.1510}$$

$$2\omega\left(\hat{S}\cdot\vec{p}\right)\vec{A} - \left[\vec{\nabla}^2 + 2\left(\hat{S}\cdot\vec{p}\right)^2 + \omega^2 + \widetilde{m}^2\right]\vec{B} = 0 \tag{5.1511}$$

Equation (5.1511) can be formally rewritten as

$$\vec{B} = \frac{2\omega\left(\hat{S}\cdot\vec{p}\right)}{\vec{\nabla}^2 + 2\left(\hat{S}\cdot\vec{p}\right)^2 + \omega^2 + \widetilde{m}^2}\vec{A} \tag{5.1512}$$

Since $\omega = E - V$, the molecule of Eq. (5.1512) ~$E$, and the denominator ~$E^2$, so when the incident particle energy is quite low there is $p << E$, in this case the contribution of the lower component $\vec{B}$ item can be ignored, which is similar to the result of the Dirac equation. However, when the energy is relatively high, the contribution of the $\vec{B}$ item cannot be ignored. Then Eq. (5.1510) and Eq. (5.1511) can be rewritten as

$$\left[\vec{\nabla}^2 - 2\left(\hat{S}\cdot\vec{\nabla}\right)^2 + \omega^2 - \widetilde{m}^2\right]\vec{A} + 2i\omega\left(\hat{S}\cdot\vec{\nabla}\right)\vec{B} = 0 \tag{5.1513}$$

$$\left[\vec{\nabla}^2 - 2\left(\hat{S}\cdot\vec{\nabla}\right)^2 + \omega^2 + \widetilde{m}^2\right]\vec{B} + 2i\omega\left(\hat{S}\cdot\vec{\nabla}\right)\vec{A} = 0 \tag{5.1514}$$

In the above equation the Coulomb potential can be introduced utilizing the method used in Sect. 5.3 or 5.18. If the other components of the optical potential are added in addition to the scalar potential $S$ and the time-like component $V$ of the vector potential in the nuclear potential, the form of the equation will also change accordingly.

### 5.19.3 Discussion on the Application of the Weinberg Equation to Calculate the Elastic Scatterings Between Deuteron and Nucleus

It can be seen that there is no coupling between the three components of the $\vec{A}$ and $\vec{B}$ in Eqs. (5.1513) and (5.1514). In the axis-symmetric system, only two components of $r$ and $\theta$ of $\vec{A}$ and $\vec{B}$ are required to be studied, and in the specific calculation only the radial components of $\vec{A}$ and $\vec{B}$ are required to be used. For the radial components of $\vec{A}$ and $\vec{B}$, the form of the separation variable solution can be given, but in Eqs. (5.1513) and (5.1514) there all are $\vec{\nabla}^2$, $\left(\hat{S}\cdot\vec{\nabla}\right)^2$, and $\left(\hat{S}\cdot\vec{\nabla}\right)$ terms, so it needs to find a transformation method for the wave functions to get the equations to be independently satisfied by the radial wave functions of the radial components of $\vec{A}$ and $\vec{B}$. In this way, the commonly used method for solving the coupled channel equations can be used, and one can introduce the upper and lower component $S$ matrix elements through the boundary conditions. According to the method introduced in Sect. 5.4, considering the upper and lower components at the same time, the $S$ matrix theory of the elastic scattering of the spin 1 particles can be developed, and then the differential cross sections and various polarization physical quantities of deuteron elastic scattering can also be calculated.

## 5.20 The Relativistic Nuclear Reaction Theory Considering the Internal Structure of the Incident Deuterons

The theory of relativistic nuclear reactions induced by deuterons introduced in the previous two sections regards the deuteron as a point particle, regardless of the internal structure or internal freedom of the deuteron. The calculation results show that the strength of the spin-orbital coupling potential obtained by the above theories is too weak, and the calculated vector analyzing power $A_y$ and tensor analyzing power $A_{yy}$ are also significantly smaller than the experimental value [118]. The above results show that it is necessary to consider the influence of the internal structure of the weakly bound state deuteron on the nuclear reaction process when studying the relativistic nuclear reaction induced by deuterons.

The multiple scattering processes should also be considered in the deuteron induced nuclear reaction theory, and ladder approximations can also be used. Let the total 4-momentum of the two nucleon in the deuterium nucleus be $P$, and the relative motion 4-momentum of the two nucleons before and after the reaction be $p$ and $p'$, respectively. Refer to the BS equation given by Eq. (5.1243), the transition matrix of the d − A reaction can be written as [124–126]

$$T_{\mathrm{d}}(P,p',p)=V_{\mathrm{d}}(P,p',p)+\int\frac{\mathrm{d}^4P'}{(2\pi)^4}V_{\mathrm{d}}(P',p',p)\,\mathrm{i}G(P',p',p)T_{\mathrm{d}}(P',p',p) \tag{5.1515}$$

where

$$V_{\mathrm{d}}(P,p',p)=V_1\left(\frac{P}{2}+p',\frac{P}{2}+p\right)+V_2\left(\frac{P}{2}-p',\frac{P}{2}-p\right) \tag{5.1516}$$

$V_1$ and $V_1$ represent the interaction potentials of the first nucleon and the second nucleon in the deuterium nucleus with the target nucleus, respectively, and $V_{\mathrm{d}}$ does not include the interaction potential between the two nucleons in the deuterium nucleus. But after the two nucleons interact with the target nucleus respectively, the two nucleon relative motion 4-momentum $p$ has changed to be $p'$. Since the $p \to p'$ change may be caused by various total 4-momentum $P'$, so there is an equation (5.1515). In Eq. (5.1515), $G$ is the two-nucleon Green function of the deuterium two-nucleon system. In order to obtain the deuteron optical potential, $G$ is divided into two parts contributed by the deuterium center of mass movement and the deuterium bound state [124–126]:

$$\mathrm{i}G(P,p',p)=\frac{R(P,p',p)}{P_\mu P^\mu-M_{\mathrm{d}}^2+\mathrm{i}\varepsilon}+M(P,p',p) \tag{5.1517}$$

The first term in the above formula is similar to the nucleon Feynman propagator and describes the propagation of the deuteron and anti-deuteron as point particles, $R(P', p',p)$ is the residue when $P_\mu P^\mu = M_{\mathrm{d}}^2$. The second term $M(P,p',p)$ describes the contribution of the internal structure of the deuterium nucleus to $G$. Using Eq. (5.1517), Eq. (5.1515) can be decomposed into the following two equations:

$$T_{\mathrm{d}}(P,p',p)=U_{\mathrm{d}}(P,p',p)+\int\frac{\mathrm{d}^4P'}{(2\pi)^4}U_{\mathrm{d}}(P',p',p)\frac{R(P',p',p)}{P'_\mu P'^\mu-M_{\mathrm{d}}^2+\mathrm{i}\varepsilon}T_d(P',p',p) \tag{5.1518}$$

$$U_{\mathrm{d}}(P,p',p)=V_{\mathrm{d}}(P,p',p)+\int\frac{\mathrm{d}^4P'}{(2\pi)^4}V_{\mathrm{d}}(P',p',p)M(P',p',p)\,U_{\mathrm{d}}(P',p',p) \tag{5.1519}$$

From Eq. (5.1519) it can be seen that $U_{\mathrm{d}}$ only contains the influence of the internal structure of the deuterium nucleus on the deuteron optical potential, so this equation can be called a relativistic folding model describing the medium energy d-A scattering [127]. Eq. (5.1518) describes the effect of multiple scattering of the center of mass of the deuterium nucleus.

Deuterium wave function is composed of the coupled two nucleon positive energy Dirac spinors

$$\chi_{P,m} = \psi(p) \sum_{\sigma_1 \sigma_2} C^{1m}_{\frac{1}{2}\sigma_1 \ \frac{1}{2}\sigma_2} u^{(1)}_{\sigma_1} \left(\frac{1}{2}P + p\right) u^{(2)}_{\sigma_2} \left(\frac{1}{2}P - p\right) \tag{5.1520}$$

where $\psi(p)$ is the normalization S state probability amplitude of deuterium nucleus. So the relativistic folding optical potential of the deuterium nucleus is [127]

$$\langle P', m' | U_{\text{d}} | P, m \rangle = \langle \chi_{P',m'} | (V_1 + V_2) | \chi_{P,m} \rangle \tag{5.1521}$$

where

$$V_i = V_i^{\text{s}} + \gamma^0 V_i^0, \quad i = 1,2 \tag{5.1522}$$

$V_i^{\text{s}}$ and $V_i^0$ may use the phenomenological potentials. The contribution of the negative energy nucleon state is ignored in the above expression. And from Eq. (5.1518) we can obtain

$$\begin{aligned} \langle P', m' | T_{\text{d}} | P, m \rangle &= \langle P', m' | U_{\text{d}} | P, m \rangle \\ &+ \sum_{m''} \int \frac{\text{d}^4 P''}{(2\pi)^4} \frac{\langle P', m' | U_{\text{d}} R | P'', m'' \rangle}{P''_\mu P''^\mu - M_{\text{d}}^2 + \text{i}\varepsilon} \langle P'', m'' | T_{\text{d}} | P, m \rangle \end{aligned} \tag{5.1523}$$

Using the abovementioned positive energy impulse approximation method developed by Santos et al., the differential cross sections, vector analyzing power $A_y$, and tensor analyzing power $A_{yy}$ of the d+$^{40}$Ca and d+$^{58}$Ni reactions induced by 400 MeV and 700 MeV deuterons are calculated, and the obtained results are better fitting with the experimental data [124–126].

If we want to study the reaction process, which includes both the scattering channel and the breakup channel, induced by the medium energy deuterons, we can develop the relativistic continuous discretization coupling channel (RCDCC) method [128].

The N + d reaction can also be regarded as a d + N reaction, so the relativistic three-body Bethe-Salpeter equation, which studies the medium energy N + d reaction, also belongs to the deuteron induced relativistic nuclear reaction theory.

The basic idea of the Faddeev equation is to write the three-body transition operator $T$ as

$$T = T^{(1)} + T^{(2)} + T^{(3)} \tag{5.1524}$$

where $T^{(1)}(E)$ represents the transition operator of the interaction between particles 2 and 3 after all other possible three-body interaction processes (including processes that do not interact) have occurred; it can be represented by a graph

$T^{(1)} =$ 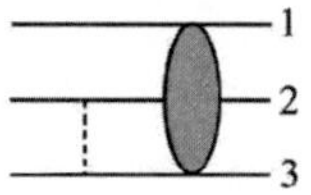

(5.1525)

The shaded liquid drop diagram shows all the interaction processes that occur between the three particles. In the same way we define

$T^{(2)} =$ 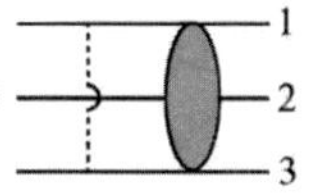

(5.1526)

$T^{(3)} =$ 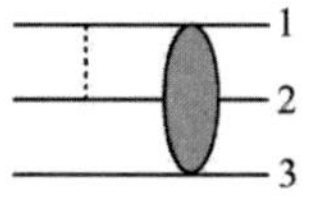

(5.1527)

Since the sum of $T^{(1)}$, $T^{(2)}$, and $T^{(3)}$ clearly includes all possible perturbation graphs of the three-body system, the decomposition of Eq. (5.1524) is correct.

We break $T^{(1)}$ down into two parts: in the first part, particle 1 never interacts with particle 2 or 3, the sum of these graphs is exactly the two-body transition operator $t_1$, that is,

$t_1 =$ 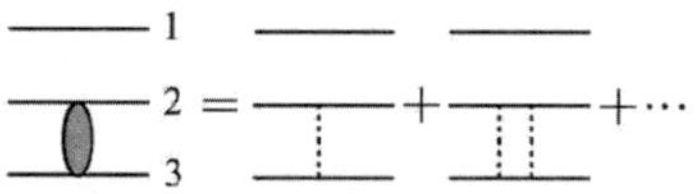

(5.1528)

In the second part, after the interaction between the three particles occurs in any order, the interaction of particles 1 and 3 or particles 1 and 2 occurs again, and then particles 2 and 3 undergo one or more interactions. The sum of these contributions can be expressed as

$\widetilde{T}^{(1)} =$ 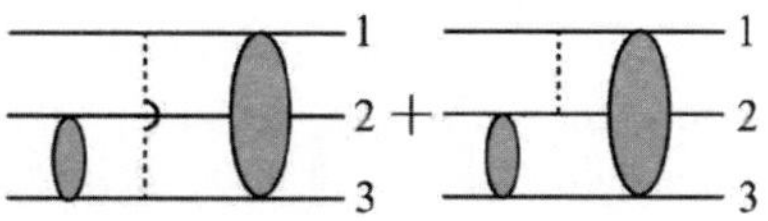

(5.1529)

Using $G^i$ to represent the Green function of the $i$th nucleon, according to the Feynman rules, the expression of the two-nucleon Green function propagated by nucleon 2 and nucleon 3 is

$$\mathrm{i}G^2\mathrm{i}G^3 = -G^2G^3 \equiv -G_1 \tag{5.1530}$$

The following expression can be known according to Eqs. (5.1526)–(5.1530):

$$\widetilde{T}^{(1)} = -t_1G_1T^{(2)} - t_1G_1T^{(3)} \tag{5.1531}$$

The graph with a dotted line between nucleon 2 and nucleon 3 cannot be included in Eq. (5.1529) because it is already included in the left liquid drop diagram. The following relation can be seen according to Eqs. (5.1525), (5.1528), (5.1529), and (5.1531):

$$T^{(1)} = t_1 + \widetilde{T}^{(1)} = t_1 - t_1 G_1 T^{(2)} - t_1 G_1 T^{(3)} \tag{5.1532}$$

In the same way we can get

$$T^{(2)} = t_2 - t_2 G_2 T^{(3)} - t_2 G_2 T^{(1)} \tag{5.1533}$$

$$T^{(3)} = t_3 - t_3 G_3 T^{(1)} - t_3 G_3 T^{(2)} \tag{5.1534}$$

Write the above three formulas in matrix form as [129, 130]

$$\begin{pmatrix} T^{(1)} \\ T^{(2)} \\ T^{(3)} \end{pmatrix} = \begin{pmatrix} t_1 \\ t_2 \\ t_3 \end{pmatrix} - \begin{pmatrix} 0 & t_1 G_1 & t_1 G_1 \\ t_2 G_2 & 0 & t_2 G_2 \\ t_3 G_3 & t_3 G_3 & 0 \end{pmatrix} \begin{pmatrix} T^{(1)} \\ T^{(2)} \\ T^{(3)} \end{pmatrix} \tag{5.1535}$$

Equation (5.1535) can be called the relativistic Faddeev equation. The two-body $t$ matrix satisfies the BS equation given by Eq. (5.1243) and can be expressed formally as

$$t = v + \mathrm{i} \int vGt \tag{5.1536}$$

where $G$ is the two-nucleon Green function defined by Eq. (5.1530). Therefore, Eq. (5.1535) can also be called a three-body BS equation. In the specific calculation, the four-dimensional BS equation can be converted into a three-dimensional Thompson equation or BbS equation. The two-body interaction potential $v$ can use a simplified separation potential [129, 130].

Using the above three-body BS equation, $\pi$–d elastic scattering and three-body breakup reaction can also be studied, and the vector analyzing power $\mathrm{i}T_{11}$ and the tensor analyzing power $T_{20}$, $T_{21}$, $T_{22}$ of $\pi$–d elastic scattering can be calculated [131, 132]. Some people also discussed the weak interaction process of $\beta$ decay of polarized nucleon and negative $\pi$ meson decay [133].

## References

1. Ma ZY, Zhang JS. Advanced quantum mechanics. Harbin: Harbin University of Industry Press; 2013. p. 120. (in Chinese)
2. Serot BD, Walecka JD. The relativistic nuclear many-body problem. In: Negele JW, Vogt E, editors. Advances in nuclear physics, vol. 16. New York/London: Plenum Press; 1986. p. 16, 23, 28, 97, 106, 126, 238.

3. Yukawa H. On the interaction of elementary particles. I. Proc Phys Maths Soc (Japan). 1935;17:48.
4. Shen QB. Nuclear reaction theory of low and medium energy (upper volume). Beijing: Science Press; 2005. (in Chinese)
5. Dirac PAM. The quantum theory of the electron, I & II. Proc R Soc (London). 1928;A117:610, A118:351.
6. Bjorken JD, Drell SD. Relativistic quantum mechanics. New York: McGraw-Hill Book Company; 1964. p. 25, 28, 93.
7. Bjorken JD, Drell SD. Relativistic quantum fields. New York: McGraw-Hill Book Company; 1965. p. 56, 387.
8. Itzykson C, Zuber JB. Quantum field theory. New York: McGraw-Hill Book Company; 1980. p. 55, 127, 217.
9. Li ZD. Introduction to particle physics and field theory (upper volume). Beijing: Science Press; 1984. p. 72. (in Chinese)
10. Arnold LG, Clark BC, Mercer RL. Relativistic optical model analysis of medium energy p-$^{4}$He elastic scattering experiments. Phys Rev. 1979;C19:917.
11. Arnold LG, Clark BC. Relativistic nucleon-nucleus optical model. Phys Lett. 1979;B84:46.
12. Arnold LG, Clark BC, Mercer RL, Schwandt P. Dirac optical model analysis of p-$^{40}$Ca elastic scattering at 180 MeV and the wine-bottle-bottom shape. Phys Rev. 1949;1981:C23.
13. Clark B C, Mercer R L, and Schwandt P. Prediction of the spin-rotation function $Q(\theta)$ using a Dirac equation based optical model. Phys Lett, 1983, B122:211
14. Kobos AM, Cooper ED, Johansson JI. Phenomenological study of relativistic optical model potentials in proton elastic scattering. Nucl Phys. 1985;A445:605.
15. Kozack R, Madland DG. Parameter correlations and ambiguities in Dirac phenomenology. Nucl Phys. 1993;A552:469.
16. Cooper ED, Clark BC, Kozack R. Global optical potentials for elastic p+$^{40}$Ca scattering using the Dirac equation. Phys Rev. 1987;C36:2170.
17. Cooper ED, Clark BC, Hama S, Mercer RL. Dirac-global fits to calcium elastic scattering data in the range 21-200 MeV. Phys Lett. 1988;B206:588.
18. Kozack R, Madland DG. Dirac optical potentials for nucleon scattering by $^{208}$Pb at intermediate energies. Phys Rev. 1989;C39:1461.
19. Kozack R, Madland DG. Prediction of intermediate- energy neutron scattering observables from a Dirac optical potential. Nucl Phys. 1990;A509:664.
20. Hama S, Clark BC, Cooper ED, Sherif ED, Mercer RL. Global Dirac optical potentials for elastic proton scattering from heavy nuclei. Phys Rev. 1990;C41:2737.
21. Cooper ED, Hama S, Clark BC, Mercer RL. Global Dirac phenomenology for proton-nucleus elastic scattering. Phys Rev. 1993;C47:297.
22. Typel S, Riedl O, Wolter HH. Elastic proton-nucleus scattering and the optical potential in a relativistic mean field model. Nucl Phys. 2002;A709:299.
23. Shen QB, Feng DC, Zhuo YZ. Neutron relativistic phenomenological and microscopic optical potential. Phys Rev. 1991;C43:2773.
24. Shen QB, Feng DC, Zhuo YZ. Neutron relativistic phenomenological optical potential and microscopic optical potential. High Energy Phys Nucl Phys. 1991;15:1033. (in Chinese)
25. Horowitz CJ. The Lorentz structure of the imaginary optical potential. Nucl Phys. 1984;A412: 228.
26. Horowitz CJ, Serot BD. Properties of nuclear and neutron matter in a relativistic Hartree-Fock theory. Nucl Phys. 1983;A399:529.
27. Jaminon M. Local and nonlocal space-like fields in the relativistic description of nucleon-nucleus scattering. Nucl Phys. 1983;A402:366.
28. Shen QB. Nuclear reaction theory of low and medium energy (middle volume). Beijing: Science Press; 2012. (in Chinese)
29. Raynal J. Ambiguity on the imaginary potentials in the Dirac formalism for the elastic and the inelastic scattering of nucleons. Phys Lett. 1987;196:7.

30. Swiniarski RD, Pham DL, Raynal J. Dirac coupled-channels analysis of inelastic scattering of 800 MeV polarized protons from $^{16}O$, $^{24}Mg$ and $^{26}Mg$. Phys Lett. 1988;213:247.
31. Shim S, Clark BC, Hama S, et al. Dirac coupled channels calculation for p+$^{40}Ca$ inelastic scattering using the relativistic impulse approximation. Phys Rev. 1968;1988:C38.
32. Kurth L, Clark BC, Cooper ED, et al. Dirac coupled channel calculations for proton inelastic scattering from spherically symmetric nuclei for projectile energies of 362, 500, and 800 MeV. Phys Rev. 1994;C49:2086.
33. Shim S, Kim MW, Clark BC, Kerr K. Dirac coupled channel analyses of 800 MeV proton inelastic scattering from $^{24}Mg$. Phys Rev. 1999;C59:317.
34. Sherif HS, Cooper ED, Sawafta RI. Relativistic DWBA calculations for polarization transfer in proton inelastic scattering. Phys Lett. 1985;B158:193.
35. Sherif H S, Sawafta R I, and Cooper E D. Proton inelastic scattering at intermediate energies and Dirac- equation-based optical potentials. Nucl Phys, 1986, A449:709
36. Johansson J I, Cooper E D Sherif H S. Relativistic DWBA calculations for proton inelastic scattering. Nucl Phys, 1988, A476:663
37. Lisantti J, Horen DJ, Bertrand FE, et al. Collective model distorted-wave Born approximation analysis of 500-MeV proton scattering from $^{40}Ca$. Phys Rev. 1989;C39:568.
38. Lisantti J, Mcdaniels DK, Tang Z, et al. Elastic and inelastic scattering of 280 and 489 MeV protons from $^{58}Ni$. Nucl Phys. 1990;A511:643.
39. Raynal J. Ambiguity on the imaginary potentials in the Dirac formalism for the elastic and the inelastic scattering of nucleons. Phys Lett. 1987;B196:7.
40. Frekers D, Wong S S M, Azuma R E, et al. Elastic and inelastic scattering of 362 MeV polarized proton from $^{40}Ca$. Phys Rev, 1987, C35:2236
41. McNeil JA, Shepard JR, Wallace SJ. Impulse- approximation Dirac optical potential. Phys Rev Lett. 1983;50:1439.
42. McNeil JA, Ray L, Wallace SJ. Impulse- approximation Dirac optical potential. Phys Rev. 1983;C27:2123.
43. Shepard JR, McNeil JA, Wallace SJ. Relativistic impulse approximation for p-nucleus elastic scattering. Phys Rev Lett. 1983;50:1443.
44. Clark B C, Hama S, Mercer R L, et al. Dirac-equation impulse approximation for intermediate-energy nucleon- nucleus scattering. Phys Rev Lett, 1983, 50:1644
45. Clark BC, Hama S, Mercer RL, et al. Energy dependence of the relativistic impulse approximation for proton-nucleus elastic scattering. Phys Rev. 1983;C28:1421.
46. Miller LD. Comment on nucleon-nucleus impulse- approximation optical-model potentials for the Dirac equation. Phys Rev Lett. 1807;1983:51.
47. Clark BC, Hama S, McNeil JA, et al. Respond. Phys Rev Lett. 1983;51:1808; Errata. Phys Rev Lett. 1984;53:302.
48. Ray L, Hoffmann GW. Relativistic and nonrelativistic impulse approximation descriptions of 300-1000 MeV proton + nucleus elastic scattering. Phys Rev. 1985;C31:538, 1986;C34:2353 (E).
49. Ray L, Hoffmann GW, Coker WK. Nonrelativistic and relativistic descriptions of proton–nucleus scattering. Phys Rep. 1992;212:223.
50. Horowitz CJ. Relativistic Love-Franey model: covariant representation of the NN interaction for N-nucleus scattering. Phys Rev. 1985;C31:1340.
51. Hynes MV, Picklesimer A, Tandy PC, Thaler RM. Relativistic (Dirac equation) effects in microscopic elastic scattering calculations. Phys Rev. 1985;C31:1438.
52. Fergerson RW, Barlett ML, Hoffmann GW, et al. Spin-rotation parameter Q for 800 MeV proton elastic scattering from $^{16}O$, $^{40}Ca$, and $^{208}Pb$. Phys Rev. 1986;C33:239.
53. Murdock DP, Horowitz CJ. Microscopic relativistic description of proton-nucleus scattering. Phys Rev. 1987;C35:1442.
54. Tjon JA, Wallace SJ. Generalized impulse approximation for relativistic proton scattering. Phys Rev. 1987;C36:1085.

55. Ray L, Hoffmann GW, Barlett ML, et al. Relativistic impulse approximation description of polarized proton elastic scattering from polarized $^{13}C$. Phys Rev. 1988;C37:1169.
56. Hoffmann GW, Barlett ML, Ciskowski D, et al. Cross sections, analyzing powers, and spin-rotation- depolarization observables for 500 MeV proton elastic scattering from $^{12}C$ and $^{13}C$. Phys Rev. 1990;C41:1651.
57. Kaki K, Toki H. Relativistic impulse approach for proton elastic scattering with $^{58}Ni$ and $^{120}Sn$ at $E_p$=200,300 and 400 MeV. Nucl Phys. 2001;A696:453.
58. Shepard JR, Rost E, Piekarewicz J. Microscopic relativistic nucleon-nucleus inelastic scattering. Phys Rev. 1984;C30:1604.
59. Schwandt P. The single-particle potential in a relativistic mean field model: theoretical basis and application to nuclear structure and reactions. Lectures presented at the 1983 RCNP-Kikuchi summer school held at Kyoto (Japan), May 23–27.
60. Sparrow DA, Piekarewicz J, Rost E, et al. Relativistic impulse approximation, nuclear currents, and the spin- difference function. Phys Rev Lett. 1985;54:2207.
61. Shepard JR, Rost E, McNeil JA. Relativistic plane-wave impulse approximation for nuclear inelastic scattering of proton and electrons. Phys Rev. 1986;C33:634.
62. Piekarewicz J, Amado RD, Sparrow DA, McNeil JA. Dirac theory of nucleon-nucleus collective excitation. Phys Rev. 1983;C28:2392.
63. Rost E, Shepard J R, Siciliano E R McNeil J A. Impulse approximation Dirac theory of inelastic proton nucleus collective excitation. Phys Rev, 1984, C29:209
64. Shim S, Clark BC, Cooper ED, et al. Comparison of relativistic and nonrelativistic approaches to the collective treatment of p+$^{40}$Ca inelastic scattering. Phys Rev. 1990;C42:1592.
65. Lane AM. New term in the nuclear optical potential: Implications for (p,n) mirror state reactions. Phys Rev Lett. 1962;8:171; Isobaric spin dependence of the optical potential and quasi-elastic (p,n)reactions. Nucl Phys. 1962;35:676.
66. Clark BC, Hama S, Sugarbaker E, et al. Relativistic description of (p,n) reaction to the isobaric analog state. Phys Rev. 1984;C30:314.
67. Zhu HY. Quantum field theory. Beijing: Science Press; 1960. p. 12, 19, 51, 75. (in Chinese)
68. Ring P. Relativistic mean field theory in finite nuclei. Prog Part Nucl Phys. 1996;37:193.
69. Walecka JA. The theory of highly condensed matter. Ann Phys (NY). 1974;83:491.
70. Horowitz CJ, Serot BD. Self-consistent Hartree description of finite nuclei in a relativistic quantum field theory. Nucl Phys. 1981;A368:503.
71. Price CE, Walker GE. Self-consistent Hartree description of deformed nuclei in a relativistic quantum field theory. Phys Rev. 1987;C36:354.
72. Gambhir YK, Ring P, Thimet A. Relativistic mean field theory for finite nuclei. Ann Phys (NY). 1990;198:132.
73. Ebran JP, Khan E, Arteaga DP, Vretenar D. Relativistic Hartree-Fock-Bogoliubov model for deformed nuclei. Phys Rev. 2011;C83:064323.
74. Yao HB, Wu SS. Study of various charged ρ-meson masses in asymmetric nuclear matter. Chin Phys. 2009;C33:842.
75. Chin SA. A relativistic many-body theory of high density matter. Ann Phys (NY). 1977;108: 301.
76. Furnstahl RJ, Price CE. Vacuum polarization currents in finite nuclei. Phys Rev. 1990;C41: 1792.
77. Haga A, Tamenaga S, Toki H, Horikawa Y. Relativistic Hartree approach with exact treatment of vacuum polarization for finite nuclei. Phys Rev. 2004;C70:064322.
78. Bogoliubov NN, Shirkov DV. Quantum fields. Moscow: Benjamin/Cummings Publishing Company; 1983. p. 167, 336.
79. Zhou BR. Quantum field theory. Beijing: Higher Education Press; 2007. p. 45, 72. (in Chinese)
80. Ma ZY, Zhu P, Gu YQ, Zhuo YZ. Optical potentials in relativistic meson-nucleon model. Nucl Phys. 1988;A490:619.

81. Han YL, Shen QB, Zhuo YZ. Tianming G. Contribution of the fourth order exchange diagram to relativistic microscopic optical potential. High Energy Phys Nucl Phys. 1993;17:751. (in Chinese)
82. Steffani MH, Betz M, Maris AJ. Widths of nuclear hole states in a relativistic model. Nucl Phys. 1989;A493:493.
83. Ma ZY, Sun ZY, Zhuo YZ. Second order correction of relativistic optical potential. Commun Theor Phys. 1990;13:327.
84. Dieperink AEL, Piekarewicz J, Wehrberger K. Imaginary part of the nucleon self-energy in a relativistic field theory. Phys Rev. 1990;C41:R2479.
85. Salpeter EE, Bethe HA. A relativistic equation for bound-state problems. Phys Rev. 1951;84: 1232.
86. Woloshyn RM, Jackson AD. Comparison of three-dimensional relativistic scattering equations. Nucl Phys. 1973;B64:269.
87. Horowitz CJ, Serot BD. The relativistic two-nucleon problem in nuclear matter. Nucl Phys. 1987;A464:613.
88. Haar BT, Malfliet R. Nucleons, mesons and deltas in nuclear matter. A relativistic Dirac-Brueckner approach. Phys Rep. 1987;149:207.
89. Brockmann R, Machleidt. Relativistic nuclear structure. I. Nuclear matter. Phys Rev. 1965;1990:C42.
90. Blankenbecler R, Sugar R. Linear integral equations for relativistic multichannel scattering. Phys Rev. 1966;142:1051.
91. Thompson RH. Three-dimensional Bethe-Salpeter equation applied to the nucleon-nucleon interaction. Phys Rev. 1970;D1:110.
92. Machleidt R. The meson theory of nuclear forces and nuclear structure. Adv Nucl Phys. 1989;19(189)
93. Machleidt R, Holinde K, Elster C. The Bonn meson-exchange model for the nucleon-nucleon interaction. Phys Rep. 1987;149:1.
94. Machleidt R. High-precision, charge-dependent Bonn nucleon-nucleon potential. Phys Rev. 2001;C63:024001.
95. Erkelenz K. Current status of the relativistic two-nucleon one boson exchange potential. Phys Rep. 1974;13:191.
96. Schiller E, Muther H. Correlations and the Dirac structure of the nucleon self-energy. Eur Phys J. 2001;A11:15.
97. Elsenhans H, Muther H, Machleidt R. Parametrization of the relativistic effective interaction in nuclear matter. Nucl Phys. 1990;A515:715.
98. Brown GE, Jackson AD. The nucleon-nucleon interaction. Amsterdam: North-Holland; 1976.
99. Machleidt R. Chiral effective field theory for nuclear forces: achievements and challenges. EPJ Web Conf. 2014;66:01011.
100. Horowitz CJ, Serot BD. Two-nucleon correlations in a relativistic theory of nuclear matter. Phys Lett. 1984;B137:287.
101. Sehn L, Faessler A, Fuchs C. Pauli operator for colliding nuclear matter. J Phys G Nucl Part Phys. 1998;24:135.
102. Sehn L, Fuchs C, Faessler A. Nucleon self-energy in the relativistic Brueckner approach. Phys Rev. 1997;C56:216.
103. van Dalen ENE, Fuchs C, Faessler A. The relativistic Dirac-Brueckner approach to asymmetric nuclear matter. Nucl Phys. 2004;A744:227.
104. van Dalen ENE, Fuchs C, Faessler A. Dirac-Brueckner-Hartree-Fock calculations for isospin asymmetric nuclear matter based on improved approximation schemes. Eur Phys J. 2007;A31: 29.
105. Tjon JA, Wallace SJ. Meson theoretical basis for Dirac impulse approximation. Phys Rev. 1985;C32:267.
106. Gross-boelting T, Fuchs C, Faessler A. Covariant representations of the relativistic Brueckner T-matrix and the nuclear matter problem. Nucl Phys. 1999;A648:105.

107. Goldberger ML, Grisaru MT, MacDowell SW, Wong DY. Theory of low-energy nucleon-nucleon scattering. Phys Rev. 1960;120:2250.
108. Brockmann R, Toki H. Relativistic density-dependent Hartree approach for finite nuclei. Phys Rev Lett. 1992;68:3408.
109. Boersma HF, Malfliet R. From nuclear matter to finite nuclei. I. Parametrization of the Dirac-Brueckner G matrix. Phys Rev. 1994;C49:233.
110. Shen H, Sugahara Y, Toki H. Relativistic mean field approach with density dependent couplings for finite nuclei. Phys Rev. 1997;C55:1211.
111. Ulrych S, Muther H. Relativistic structure of the nucleon self-energy in asymmetric nuclei. Phys Rev. 1997;C56:1788.
112. Muther H, Machleidt R, Brockmann R. Relativistic nuclear structure. II. Finite nuclei. Phys Rev. 1981;1990:C42.
113. Ma ZY, Liu L. Effective Dirac Brueckner-Hartree-Fock for asymmetric nuclear matter and finite nuclei. Phys Rev. 2002;C66:024321.
114. Rong J, Ma ZY, Giai NV. Isospin-dependent optical potentials in Dirac-Brueckner-Hartree-Fock approach. Phys Rev. 2006;C73:014614.
115. van Dalen ENE, Muther H. Off-shell behavior of nucleon self-energy in asymmetric nuclear matter. Phys Rev. 2010;C82:014319.
116. Santos FD, van Dam H. Relativistic dynamics of spin-one particles and deuteron-nucleus scattering. Phys Rev. 1986;C34:250.
117. Santos FD. Relativistic spin-1 dynamics and deuteron-nucleus elastic scattering. Phys Lett. 1986;B175:110.
118. Mishra VK, Hama S, Clark BC, et al. Implications of various spin-one relativistic wave equations for intermediate-energy deuteron-nucleus scattering. Phys Rev. 1991;C43:801.
119. Kozack RE, Clark BC, Hama S, et al. Spin-one Kemmer-Duffin-Petiau equations and intermediate-energy deuteron- nucleus scattering. Phys Rev. 1989:C40, 2181.
120. Varshalovich DA, Moskalev AN, Khersonskii VK. Quantum theory of angular momentum. Singapore/New Jersey/Hong Kong: World Scientific; 1988.
121. Kozack RE, Clark BC, Hama S, et al. Relativistic deuteron-nucleus scattering in the Kemmer-Duffin-Petiau formalism. Phys Rev. 1988;C37:2898.
122. Weinberg S. Feynman rules for any spin. Phys Rev. 1964;133:B1318.
123. Weinberg S. Feynman rules for any spin. II. Massless particles. Phys Rev. 1964;134:B882.
124. Amorim A, Santos FD. From Dirac phenomenology to deuteron-nucleus elastic scattering at intermediate energies. Phys Rev. 1991;C44:2100.
125. Amorim A, Santos FD. Relativistic impulse approximation for deuteron scattering. Phys Lett. 1992;B297:31.
126. Pinto JP, Amorim A, Santos FD. Microscopic relativistic model for deuteron-nucleus scattering. Phys Rev. 1996;C53:2376.
127. Santos FD, Amorim A. Relativistic folding model for intermediate energy deuteron-nucleus scattering. Phys Rev. 1988;C37:1183.
128. Bertulani CA. Relativistic continuum-continuum coupling in the dissociation of halo nuclei. Phys Rev Lett. 2005;94:072701.
129. Rupp G, Tjon JA. Bethe-Salpeter calculation of three-nucleon observables with rank-one separable potentials. Phys Rev. 1988;C37:1729.
130. Adam J Jr, Van Orden JW. Comprehensive treatment of electromagnetic interactions and three-body spectator equations. Phys Rev. 2005;C71:034003.
131. Garcilazo H. Relativistic Faddeev theory of the$\pi$NN system with application to$\pi$ d scattering. Phys Rev. 1987;C35:1804.
132. Garcilazo H. N$\Delta$ interaction in $\pi$d breakup. Phys Rev. 1990;C42:2334.
133. Robson BA. The theory of polarization phenomena. Oxford: Clarendon Press; 1974.

# Chapter 6
# Basis of the Polarized Particle Transport Theory

**Abstract** This chapter proposes the polarized nucleus transport theory and introduces the polarization theories of $\overrightarrow{\frac{1}{2}} + \mathrm{A} \rightarrow \overrightarrow{\frac{1}{2}} + \mathrm{B}$, $\vec{1} + \mathrm{A} \rightarrow \vec{1} + \mathrm{B}$, $\vec{1} + \mathrm{A} \rightarrow \overrightarrow{\frac{1}{2}} + \mathrm{B}$, and $\overrightarrow{\frac{1}{2}} + \mathrm{A} \rightarrow \vec{1} + \mathrm{B}$ reactions related to the polar angle $\theta$ and azimuthal angle $\varphi$ at the same time, and puts forward the specific concept of establishing a polarization nuclear database. It is pointed out that polarization quantities that are only related to the polar angles $\theta$ should be stored in the database. The Monte-Carlo methods for simulating the reactions including polarization channels are given, and the method of the coordinate system rotation and the transformation of the polarization rates of transported particles are presented. The transport equation of polarized particles with spin $\frac{1}{2}$ or 1 is also proposed and developed. Finally, the prospective problems of the future development of the polarized particle transport theory are discussed.

**Keywords** Analyzing power · Azimuthal angle · Elastic scattering angular distribution · Monte-Carlo method · Polarization nuclear database · Polarized particle transport equation · Reaction amplitude

## 6.1 Introduction to the Polarized Particle Transport Theory

The particle transport theory is a mathematical theory that studies the process of particle transport. The transport process refers to the process in which various physical quantities change with time and space due to changes in the position, momentum, and other characteristic quantities of each particle when a large number of particles move in space or in a certain medium. In particle transport theory, particles are considered as point particles, so their location, kinetic energy, and flight direction can be accurately described.

In the usual particle transport theory, the polarization of the particles is not considered, and all particles are considered to be unpolarized. In fact, when a particle moves under the action of a potential field, not only its coordinate position and momentum are

 
Q.-B. Shen, *Polarization Theory of Nuclear Reactions*,
https://doi.org/10.1007/978-3-031-11878-4_6

constantly changing, but the particle itself will also perform a self-rotation motion called spin. Due to the quantum effect, the spin of the particle can only be an integer or a semi-odd number. For example, the spin of electrons, neutrons, and protons is $\frac{1}{2}$, the spin of the deuteron is 1, and the spin of the $\alpha$ particle is 0. Both experiments and theories have proved that for particles with spin $\frac{1}{2}$, there is a spin-orbital coupling potential, and for particles with spin 1, there is a tensor potential in addition to the spin-orbital coupling potential. Therefore, the forces felt by the particles with different spin magnetic quantum numbers are different, so their motion behavior will also be different; this is the polarization phenomenon. In order to describe the physically more realistic polarization particle transport process that exists objectively, the polarized particle transport theory should be established and developed.

The transfer process of spin polarized electrons in ferromagnets and superconductors has become a widely studied field [1–10]. The spin $\frac{1}{2}$ electrons experience electromagnetic interactions and need to study the effects of spin-orbital coupling potential during polarized electron transport process. The corresponding theory is called polarized electron transport theory. The content to be discussed in this chapter can be called polarized nucleus transport theory, but this chapter is limited to study the transport theory of polarized spin $\frac{1}{2}$ or 1 nucleon and light complex particles. The interaction of nucleon and light complex particles with the nucleus is a strong interaction.

Whether particle transport theory can be used to accurately describe the transport behavior of a large number of particles in the medium depends on whether there are accurate microscopic nuclear data of nuclear reactions between particles and medium nuclei. Therefore, it can be said that the improvement and refinement of microscopic nuclear data will directly promote the improvement of the accuracy and credibility of the calculation results of particle transport theory. For polarized nucleus transport theory, not only the usual microscopic nuclear data but also the microscopic polarization nuclear data are required. The above few chapters of this book have discussed in detail that if the incident particles are polarized, the differential cross sections obtained by considering the polarization effect and without considering the polarization effect are different. Obviously, considering the polarization effect can more accurately and objectively reflect the real physical process; without considering the polarization effect is only an approximate description of the real physical process. If the microscopic nuclear data containing the polarization effect is used in the nuclear transport theory, the polarized nucleus transport theory is formed. However, due to a variety of reasons, so far all current nucleon transport calculations do not consider the polarization effect.

The direction of the incident particle is selected as the $z$-axis, and the $y$-axis is selected in the unit vector $\vec{n}$ direction perpendicular to the reaction plane. Equation (2.116) has given that when the target nucleus spin is 0, the nucleon elastic scattering differential cross section is

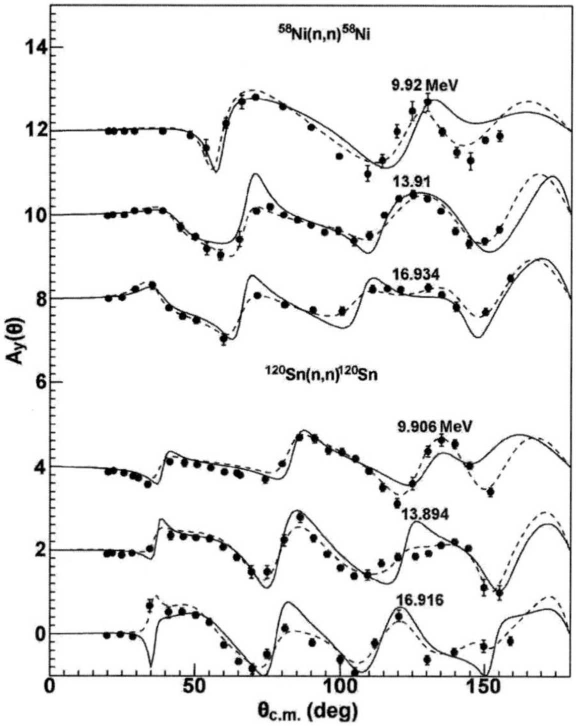


**Fig. 6.1** Calculated analyzing powers $A_y(\theta)$ using the microscopic optical potential obtained by Skyrme force parameter SkC (solid line) and the phenomenological optical potential given by Koning-Delaroche (dashed line) compared with experimental data (black symbols) for n + $^{58}$Ni and n + $^{120}$Sn elastic scatterings. The value shown in the figure is the incident neutron energy in MeV. The curves and data points at the bottom represent true values, while the others are added by 2, 4, 8, 10, 12, respectively

$$\frac{d\sigma}{d\Omega} = \frac{d\sigma^0}{d\Omega}\left[1 + p_y A_y(\theta)\right] \tag{6.1}$$

where $\frac{d\sigma^0}{d\Omega}$ is the unpolarization differential cross section, $p_y$ is the polarization rate of the incident particles, and $|p_y| \le 1$, $A_y(\theta)$ is the analyzing power of the incident particles (see Fig. 6.1 and Fig. 6.2) [11]. The elastic scattering integral cross section corresponding to Eq. (6.1) is

$$\sigma = \int \frac{d\sigma}{d\Omega} d\Omega = \sigma^0 + \sigma^{\mathrm{p}} \tag{6.2}$$

where

$$\sigma^0 = \int \frac{d\sigma^0}{d\Omega} d\Omega \tag{6.3}$$

$$\sigma^{\mathrm{p}} = p_y \int \frac{d\sigma^0}{d\Omega} A_y(\theta)\, d\Omega \tag{6.4}$$

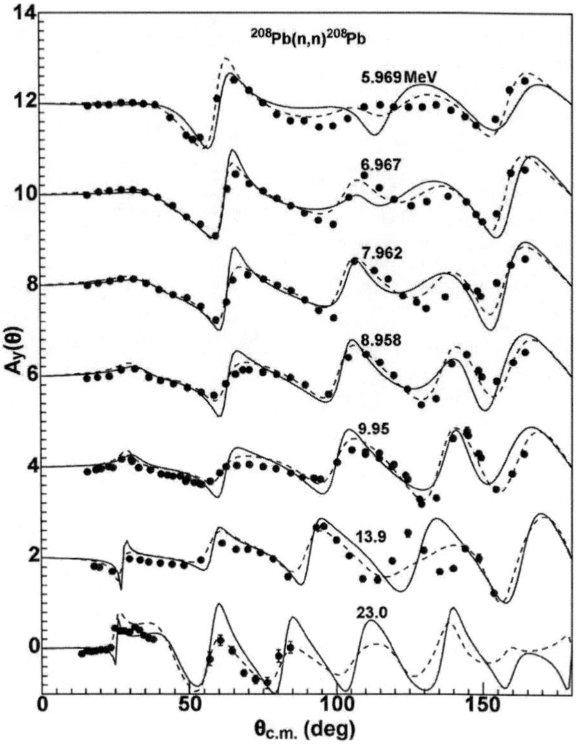


**Fig. 6.2** The same as Fig. 6.1, but for target $^{208}$Pb and while the others are added by 2, 4, 6, 8, 10, 12, respectively

where $\sigma^0$ is the unpolarization elastic scattering integral cross section and $\sigma^{\rm P}$ is the elastic scattering integral cross section contributed by the polarized component of the incident particles. For spin 1 particles, we also need to consider tensor analyzing power. For spin 1 particles and target nuclei with non-zero spin, there are the similar or more complex results compared with the above expressions.

The specific numerical results show that when the neutron energy $E_n \leq 1\,\text{keV}$, $\frac{d\sigma^0}{d\Omega}$ of the heavier nuclei are very close to isotropy, $|A_y(\theta)|_{\max} << 1$, so the polarization effect cannot be considered for the thermal neutron reactor. The average neutron energy of the fast neutron reactors is ~0.1 MeV. For heavy and medium heavy nuclei, when the incident neutron energy is 0.1 MeV, the difference of the elastic scattering angular distribution with isotropy is not yet obvious, and $|A_y(\theta)|_{\max}$ is still significantly less than 1. But in the fast neutron reactor, there is a certain proportion of the neutrons that energy can achieve 0.5 – 1.0 MeV. In this case, the neutron elastic scattering angular distribution has a clear anisotropy trend. And the value of $|A_y(\theta)|_{\max}$ can already compare with 1. Especially for $^{23}$Na and other more light nuclei, even for 0.1 MeV incident neutrons, the elastic scattering angular distribution has already shown obvious anisotropy, and the difference of $|A_y(\theta)|_{\max}$ from 1 is not much; moreover there is that about half of the neutrons energy $E_n > 0.1$ MeV.

Of course, if the light medium in the fast neutron reactor is only in the outer region where the neutrons have been greatly slowed down, the polarization effect will not be considered. In short, further research is needed on whether fast neutron reactors need to consider the polarization effects.

For nuclear reactions between nucleons or light complex particles, even if the energy is quite low, the angular distribution also presents anisotropy, so the polarization effect should also be considered for fusion reactions that belong to low energy category. In particular, some people are currently studying whether fusion reactions using polarized charged particles as fuel may increase the fusion reaction rate [12–17], and others are exploring the possibility of building accelerator-driven fusion reactors [18]. Although the main concern at present is the influence of polarization effects in primary fusion reactions, both the usual concept fusion reactors that people are exploring and the study of nuclear fusion using polarized particles as fuels all hope to achieve long-term stable operation. Therefore, it is possible to need discussion on the polarization effects of secondary reactions and even the transport problem of the polarized particles in the future.

In an accelerator-driven sub-critical device, the proton energy is ~GeV, and the neutrons generated by the fragmentation reaction are also mainly in MeV and GeVenergy area in the central area of the device. For nuclear devices with nucleon energy in MeV and GeV energy zone, the differential cross section and integral cross section of the unpolarized nucleons produced by elastic scattering and direct reaction maybe have a degree of difference compared with the differential cross section and integral cross section that consider the polarization effect. Moreover, even if the incident particles are unpolarized, the secondary particles generated by the nuclear reaction are also polarized. In some nuclear devices, sometimes there are the processes of transportation of the polarized deuterons and other light complex particles.

Due to the existence of spin-related nuclear force components in strong interactions, the particles will be polarized in the nuclear reaction, which is an objective physical process. Therefore, it is necessary to use polarized nucleus transport theory to simulate the corresponding polarization particle transport process for nuclear devices with nucleon energy in Mev and GeV energy zone. However, considering that the spin-orbit coupling potential is small relative to the center potential, and that there are positive and negative countervailing phenomena in the polarization quantities such as the analyzing power $A_y(\theta)$, for the question that how much difference between the calculation results of considering and without considering polarization effect is there, only when the polarization theory of nuclear reaction has been relatively perfect, the polarization nuclear data has accumulated to a certain extent, and more calculations of the polarized nucleus transport theory have been carried out, it will be possible to give a relatively clear view and answer. Even if the polarization effect is proved to have little effect on particle transport in the future, it is of great scientific value to study the physical process of objective existence. And in the process of studying the polarization problem, it is possible to find that the nuclear polarization effect has some special application value and development prospects. The in-depth development of basic research is more conducive to the transformation of basic theory into application.

In the polarized nucleus transport theory discussed in this chapter, we assume that both the target nucleus and the residual nucleus are unpolarized, and spin $S \geq \frac{3}{2}$ particles are also unpolarized, the pre-equilibrium reactions and equilibrium reactions, three-body or multi-body reactions are also unpolarized. In other words, we only study the polarized nucleus transport problems of spin $\frac{1}{2}$ and 1 particles in shape (potential) elastic scattering and two-body direct reactions.

## 6.2 Polarized Particle Transport Theory for $\vec{\frac{1}{2}} + \mathrm{A} \rightarrow \vec{\frac{1}{2}} + \mathrm{B}$ Reactions

### 6.2.1 *Polarization Theory of Spin $\frac{1}{2}$ Particles Related to Azimuthal Angle*

In $\vec{\frac{1}{2}} + \mathrm{A} \rightarrow \vec{\frac{1}{2}} + \mathrm{B}$ reaction, both the incident and outgoing particles are polarized particles of the spin $\frac{1}{2}$, and the spins of the target A and the residual nucleus B are $I$ and $I'$, respectively, and they are both unpolarized. Equation (2.267) has given that the reaction amplitude with a definite spin magnetic quantum numbers $M_I' M_I$ is

$$\hat{\mathcal{F}}_{M_I' M_I}(\theta, \varphi) = \begin{pmatrix} A_{M_I' M_I}(\theta) & -\mathrm{i}B_{M_I' M_I}(\theta)\, e^{-\mathrm{i}\varphi} \\ \mathrm{i}C_{M_I' M_I}(\theta)\, e^{\mathrm{i}\varphi} & D_{M_I' M_I}(\theta) \end{pmatrix} e^{\mathrm{i}(M_I - M_I')\varphi} \tag{6.5}$$

where $A_{M_I' M_I}, B_{M_I' M_I}, C_{M_I' M_I}, D_{M_I' M_I}$ have been given by Eqs. (2.258)–(2.251). Because in the calculations of the polarization quantities, when both the target and the residual nucleus are unpolarized, $\hat{\mathcal{F}}_{M_I' M_I}$ and $\hat{\mathcal{F}}^{+}_{M_I' M_I}$ always appear at the same time, so the factor $\mathrm{e}^{\mathrm{i}(M_I - M_I')\varphi}$ must be eliminated. Then we may introduce

$$\hat{F} = \begin{pmatrix} A & -\mathrm{i}B\mathrm{e}^{-\mathrm{i}\varphi} \\ \mathrm{i}C\mathrm{e}^{\mathrm{i}\varphi} & D \end{pmatrix}, \quad \hat{F}^{+} = \begin{pmatrix} A^{*} & -\mathrm{i}C^{*}\mathrm{e}^{-\mathrm{i}\varphi} \\ \mathrm{i}B^{*}\mathrm{e}^{\mathrm{i}\varphi} & D^{*} \end{pmatrix} \tag{6.6}$$

The footnotes $M_I' M_I$ and the argument ($\theta$) of the symbols $A$, $B$, $C$, and $D$ are omitted, and the footnotes $M_I' M_I$ and the arguments ($\theta$, $\varphi$) of the reaction amplitude matrices $\hat{F}$ and $\hat{F}^{+}$ are also omitted. From Eq. (6.6) one can get

$$\hat{F}\hat{F}^{+} = \begin{pmatrix} |A|^2 + |B|^2 & -\mathrm{i}(AC^* + BD^*)\,\mathrm{e}^{-\mathrm{i}\varphi} \\ \mathrm{i}(CA^* + DB^*)\,\mathrm{e}^{\mathrm{i}\varphi} & |C|^2 + |D|^2 \end{pmatrix} \tag{6.7}$$

We use $\alpha'$, $\alpha$ and $n'$, $n$ to represent the type of the outgoing particles and the incident particles and the excited state label of the residual nucleus and target, respectively. Next a simplification summarization symbol is introduced as

$$\widetilde{\Sigma} \equiv \frac{1}{2I+1} \sum_{M_I M_I'} \tag{6.8}$$

The differential cross section of the unpolarized projectiles can be obtained by using Eqs. (2.295) and (2.296) as

$$I_0(\theta) = \widetilde{\Sigma}\left( |A|^2 + |B|^2 \right) \tag{6.9}$$

where the footnotes $\alpha' n'$, $\alpha\, n$ of $I_0(\theta)$ for outgoing channels and incident channels were dropped.

Using Eqs (2.6) and (6.6) one can obtain

$$\hat{\sigma}_x \hat{F}^{+} = \begin{pmatrix} \mathrm{i}B^*\mathrm{e}^{\mathrm{i}\varphi} & D^* \\ A^* & -\mathrm{i}C^*\mathrm{e}^{-\mathrm{i}\varphi} \end{pmatrix},\ \hat{\sigma}_y \hat{F}^{+} = \begin{pmatrix} B^*\mathrm{e}^{\mathrm{i}\varphi} & -\mathrm{i}D^* \\ \mathrm{i}A^* & C^*\mathrm{e}^{-\mathrm{i}\varphi} \end{pmatrix},\ \hat{\sigma}_z \hat{F}^{+} = \begin{pmatrix} A^* & -\mathrm{i}C^*\mathrm{e}^{-\mathrm{i}\varphi} \\ -\mathrm{i}B^*\mathrm{e}^{\mathrm{i}\varphi} & -D^* \end{pmatrix} \tag{6.10}$$

$$\begin{aligned} \hat{F}\,\hat{\sigma}_x \hat{F}^{+} &= \begin{pmatrix} \mathrm{i}AB^*\mathrm{e}^{\mathrm{i}\varphi} - \mathrm{i}BA^*\mathrm{e}^{-\mathrm{i}\varphi} & AD^* - BC^*\mathrm{e}^{-2\mathrm{i}\varphi} \\ -CB^*\mathrm{e}^{2\mathrm{i}\varphi} + DA^* & \mathrm{i}CD^*\mathrm{e}^{\mathrm{i}\varphi} - \mathrm{i}DC^*\mathrm{e}^{-\mathrm{i}\varphi} \end{pmatrix}, \\ \hat{F}\,\hat{\sigma}_y \hat{F}^{+} &= \begin{pmatrix} AB^*\mathrm{e}^{\mathrm{i}\varphi} + BA^*\mathrm{e}^{-\mathrm{i}\varphi} & -\mathrm{i}AD^* - \mathrm{i}BC^*\mathrm{e}^{-2\mathrm{i}\varphi} \\ \mathrm{i}CB^*\mathrm{e}^{2\mathrm{i}\varphi} + \mathrm{i}DA^* & CD^*\mathrm{e}^{\mathrm{i}\varphi} + DC^*e^{-\mathrm{i}\varphi} \end{pmatrix}, \\ \hat{F}\,\hat{\sigma}_z \hat{F}^{+} &= \begin{pmatrix} |A|^2 - |B|^2 & -\mathrm{i}(AC^* - BD^*)\,\mathrm{e}^{-\mathrm{i}\varphi} \\ \mathrm{i}(CA^* - DB^*)\,\mathrm{e}^{\mathrm{i}\varphi} & |C|^2 - |D|^2 \end{pmatrix} \end{aligned} \tag{6.11}$$

We can get the following relations by using Eqs. (2.295) and (2.296):

$$\widetilde{\Sigma}\mathrm{Im}(AB^* + CD^*) = \widetilde{\Sigma}\mathrm{Im}(AB^* + A^*B) = 0 \tag{6.12}$$

$$\widetilde{\Sigma}\,\mathrm{Re}\,(AB^* + CD^*) = 2\widetilde{\Sigma}\,\mathrm{Re}\,(AB^*) \tag{6.13}$$

Then, by using Eq. (6.11) the following expressions can be obtained:

$$\begin{aligned} \frac{1}{2}\widetilde{\Sigma}\mathrm{tr}\left(\hat{F}\,\hat{\sigma}_x \hat{F}^{+}\right) &= -\widetilde{\Sigma}\mathrm{Im}[(AB^* + CD^*)\,e^{\mathrm{i}\varphi}] \\ &= -\widetilde{\Sigma}\,\mathrm{Re}\,(AB^* + CD^*)\sin\varphi = -2\widetilde{\Sigma}\,\mathrm{Re}\,(AB^*)\sin\varphi \end{aligned} \tag{6.14}$$

$$\frac{1}{2}\widetilde{\Sigma}\mathrm{tr}\left(\hat{F}\,\hat{\sigma}_y\hat{F}^+\right)=\widetilde{\Sigma}\,\mathrm{Re}\left[(AB^* + CD^*)\,e^{i\varphi}\right]$$
$$=\widetilde{\Sigma}\,\mathrm{Re}\,(AB^* + CD^*)\cos\varphi=2\widetilde{\Sigma}\,\mathrm{Re}\,(AB^*)\cos\varphi \tag{6.15}$$

When the $y$-axis is selected in the direction $\vec{n}$ perpendicular to the reaction plan, the following analyzing power only related to the angle $\theta$ is obtained:

$$A_y(\theta)=\frac{2}{I_0(\theta)}\widetilde{\Sigma}\,\mathrm{Re}\,(AB^*) \tag{6.16}$$

In the general case, the analyzing power related to the angles $\theta$, $\varphi$ is defined as

$$\overline{A}_i(\theta,\,\varphi)=\frac{1}{2I_0(\theta)}\widetilde{\Sigma}\mathrm{tr}(\hat{F}\,\hat{\sigma}_i\hat{F}^+),\qquad i=x,y,z \tag{6.17}$$

Then the following results can be obtained, from Eqs. (6.11), (6.14)–(6.17), (2.295), and (2.296), as follows:

$$\overline{A}_x(\theta,\,\varphi)=-A_y(\theta)\sin\varphi,\quad \overline{A}_y(\theta,\,\varphi)=A_y(\theta)\cos\varphi,\quad \overline{A}_z(\theta,\,\varphi)=0 \tag{6.18}$$

The differential cross section of the outgoing particle corresponding to the polarized incident particle is

$$I(\theta,\varphi)=I_0(\theta)\left[1+\sum_i p_i\overline{A}_i(\theta,\varphi)\right]=I_0(\theta)\left[1+p_x\overline{A}_x(\theta,\varphi)+p_y\overline{A}_y(\theta,\varphi)\right] \tag{6.19}$$

where $p_x$, $p_y$, $p_z$ are the incident particle polarization rates along the $x$-, $y$-, and $z$-axis, respectively.

We can obtain the following expressions from Eqs. (2.6) and (6.7):

$$\begin{aligned}
\frac{1}{2}\mathrm{tr}\left(\hat{\sigma}_x\hat{F}\hat{F}^+\right)&=\frac{1}{2}\left[\mathrm{i}(CA^*+DB^*)\,\mathrm{e}^{\mathrm{i}\varphi}-\mathrm{i}(AC^*+BD^*)\,\mathrm{e}^{-\mathrm{i}\varphi}\right]\\
&=\mathrm{Im}\left[(AC^*+BD^*)\,\mathrm{e}^{-\mathrm{i}\varphi}\right]\\
\frac{1}{2}\mathrm{tr}\left(\hat{\sigma}_y\hat{F}\hat{F}^+\right)&=\frac{1}{2}\left[(CA^*+DB^*)\,\mathrm{e}^{\mathrm{i}\varphi}+(AC^*+BD^*)\,\mathrm{e}^{-\mathrm{i}\varphi}\right]\\
&=\mathrm{Re}\left[(AC^*+BD^*)\,\mathrm{e}^{-\mathrm{i}\varphi}\right]\\
\frac{1}{2}\mathrm{tr}\left(\hat{\sigma}_z\hat{F}\hat{F}^+\right)&=\frac{1}{2}\left(|A|^2+|B|^2-|C|^2-|D|^2\right)
\end{aligned} \tag{6.20}$$

We can also get the following relations by using the similar method to derive Eqs. (6.12) and (6.13):

$$\widetilde{\Sigma}\mathrm{Im}(AC^* + BD^*) = \widetilde{\Sigma}\mathrm{Im}(AC^* + CA^*) = 0 \tag{6.21}$$

$$\widetilde{\Sigma}\,\mathrm{Re}\,(AC^* + BD^*) = 2\widetilde{\Sigma}\,\mathrm{Re}\,(AC^*) \tag{6.22}$$

When the $y$-axis is selected in the direction $\vec{n}$ perpendicular to the reaction plan, the polarization rates, which are only related to angle $\theta$, of the outgoing particles corresponding to the unpolarized incident particles are as follows:

$$P^y(\theta) = \frac{2}{I_0(\theta)}\widetilde{\Sigma}\,\mathrm{Re}\,(AC^*) \tag{6.23}$$

In the general cases, the polarization vectors, which are related to the angles $\theta$, $\varphi$, of the outgoing particles corresponding to the unpolarized incident particles can be written as

$$\overline{P}^i(\theta, \varphi) = \frac{1}{2I_0(\theta)}\widetilde{\Sigma}\mathrm{tr}(\hat{\sigma}_i\hat{F}\hat{F}^+), \qquad i = x,y,z \tag{6.24}$$

Then we can obtain the following results from Eqs. (6.20), (6.23), (6.24), (2.295), and (2.296):

$$\overline{P}^x(\theta, \varphi) = -P^y(\theta)\sin\varphi, \quad \overline{P}^y(\theta, \varphi) = P^y(\theta)\cos\varphi, \quad \overline{P}^z(\theta, \varphi) = 0 \tag{6.25}$$

And we can obtain the following expressions from Eqs. (2.6) and (6.11):

$$\begin{aligned}
&\frac{1}{2}\mathrm{tr}\left(\hat{\sigma}_x\hat{F}\,\hat{\sigma}_x\hat{F}^+\right) = \mathrm{Re}\,\left(AD^* - BC^*\mathrm{e}^{-2\mathrm{i}\varphi}\right),\\
&\frac{1}{2}\mathrm{tr}\left(\hat{\sigma}_y\hat{F}\,\hat{\sigma}_x\hat{F}^+\right) = -\mathrm{Im}\left(AD^* - BC^*\mathrm{e}^{-2\mathrm{i}\varphi}\right),\\
&\frac{1}{2}\mathrm{tr}\left(\hat{\sigma}_z\hat{F}\,\hat{\sigma}_x\hat{F}^+\right) = -\mathrm{Im}\left[(AB^* - CD^*)\,\mathrm{e}^{\mathrm{i}\varphi}\right],\\
&\frac{1}{2}\mathrm{tr}\left(\hat{\sigma}_x\hat{F}\,\hat{\sigma}_y\hat{F}^+\right) = \mathrm{Im}\left(AD^* + BC^*\mathrm{e}^{-2\mathrm{i}\varphi}\right),\\
&\frac{1}{2}\mathrm{tr}\left(\hat{\sigma}_y\hat{F}\,\hat{\sigma}_y\hat{F}^+\right) = \mathrm{Re}\,\left(AD^* + BC^*\mathrm{e}^{-2\mathrm{i}\varphi}\right),\\
&\frac{1}{2}\mathrm{tr}\left(\hat{\sigma}_z\hat{F}\,\hat{\sigma}_y\hat{F}^+\right) = \mathrm{Re}\,\left(AB^* - CD^*\mathrm{e}^{\mathrm{i}\varphi}\right),\\
&\frac{1}{2}\mathrm{tr}\left(\hat{\sigma}_x\hat{F}\,\hat{\sigma}_z\hat{F}^+\right) = \mathrm{Im}\left[(AC^* - BD^*)\,\mathrm{e}^{-\mathrm{i}\varphi}\right],\\
&\frac{1}{2}\mathrm{tr}\left(\hat{\sigma}_y\hat{F}\,\hat{\sigma}_z\hat{F}^+\right) = \mathrm{Re}\,\left[(AC^* - BD^*)\,\mathrm{e}^{-\mathrm{i}\varphi}\right],\\
&\frac{1}{2}\mathrm{tr}\left(\hat{\sigma}_z\hat{F}\,\hat{\sigma}_z\hat{F}^+\right) = \frac{1}{2}\left(|A|^2 - |B|^2 - |C|^2 + |D|^2\right)
\end{aligned} \tag{6.26}$$

We can also get the following relations by using Eqs. (2.295) and (2.296):

$$\widetilde{\Sigma}\,\mathrm{Re}\,(AC^* - BD^*) = 0, \qquad \widetilde{\Sigma}\,\mathrm{Re}\,(AB^* - CD^*) = 0 \tag{6.27}$$

$$\widetilde{\Sigma}\mathrm{Im}(AC^* - BD^*) = 2\widetilde{\Sigma}\mathrm{Im}(AC^*), \quad \widetilde{\Sigma}\mathrm{Im}(AB^* - CD^*) = 2\widetilde{\Sigma}\mathrm{Im}(AB^*) \tag{6.28}$$

Equation (2.298) has proved that $\sum\limits_{M_I M_I'} A_{M_I' M_I} D^*_{M_I' M_I}$ and $\sum\limits_{M_I M_I'} B_{M_I' M_I} C^*_{M_I' M_I}$ are both the real numbers, so we have

$$\widetilde{\Sigma}\mathrm{Im}(AD^*) = 0, \qquad \widetilde{\Sigma}\mathrm{Im}(BC^*) = 0 \tag{6.29}$$

Let us introduce the following polarization transfer coefficient components related only to the angle $\theta$:

$$\begin{aligned} R_1(\theta) &= \frac{1}{I_0(\theta)}\widetilde{\Sigma}\,\mathrm{Re}\,(AD^*), \qquad R_2(\theta) = \frac{1}{I_0(\theta)}\widetilde{\Sigma}\,\mathrm{Re}\,(BC^*), \\ Q_1(\theta) &= \frac{2}{I_0(\theta)}\widetilde{\Sigma}\mathrm{Im}(AC^*), \qquad Q_2(\theta) = \frac{2}{I_0(\theta)}\widetilde{\Sigma}\mathrm{Im}(AB^*), \\ W(\theta) &= \frac{1}{I_0(\theta)}\widetilde{\Sigma}\left(|A|^2 - |B|^2\right) \end{aligned} \tag{6.30}$$

The polarization transfer coefficient related simultaneously to the angles $\theta$ and $\varphi$ is defined as

$$K_i^j(\theta, \varphi) = \frac{1}{2I_0(\theta)}\widetilde{\Sigma}\mathrm{tr}(\hat{\sigma}_j \hat{F}\,\hat{\sigma}_i \hat{F}^+), \quad i,j = x,y,z \tag{6.31}$$

where $i$ and $j$ represent the incident and outgoing particle footnote, respectively. Then we can obtain the following results according to Eqs. (6.26)–(6.31):

$$\begin{aligned} &K_x^x(\theta, \varphi) = R_1(\theta) - R_2(\theta)\cos(2\varphi), \quad K_y^x(\theta, \varphi) = -R_2(\theta)\sin(2\varphi), \\ &K_z^x(\theta, \varphi) = Q_1(\theta)\cos\varphi, \quad K_x^y(\theta, \varphi) = -R_2(\theta)\sin(2\varphi), \\ &K_y^y(\theta, \varphi) = R_1(\theta) + R_2(\theta)\cos(2\varphi), \quad K_z^y(\theta, \varphi) = Q_1(\theta)\sin\varphi, \\ &K_x^z(\theta, \varphi) = -Q_2(\theta)\cos\varphi, \quad K_y^z(\theta, \varphi) = -Q_2(\theta)\sin\varphi, \\ &K_z^z(\theta, \varphi) = W(\theta) \end{aligned} \tag{6.32}$$

The components of the polarization vector of the outgoing particle corresponding to the polarized incident particle are

$$P_j(\theta,\varphi) = \frac{I_0(\theta)}{I(\theta,\varphi)}\left[\overline{P}^j(\theta,\varphi) + \sum_{i=x,y,z} p_i K_i^j(\theta,\varphi)\right], \quad j = x,y,z \tag{6.33}$$

where $p_i$ $(i = x, y, z)$ are the incident particle polarization rates.

### 6.2.2 Polarization Nuclear Database of Spin $\frac{1}{2}$ Particles

The polarization nuclear database of spin $\frac{1}{2}$ particles to be discussed in this section contains only the following eight differential physical quantities related to the incident particle energy and polar angle $\theta$ for the shape elastic scattering and two-body direct reaction channels:

1. The differential cross section corresponding to unpolarized incident particles $I_0(\theta)$
2. The analyzing power of the incident particle $A_y(\theta)$
3. The polarization rate of the emitted particle corresponding to the unpolarized incident particle $P^y(\theta)$
4. The polarization transfer coefficient components $R_1(\theta)$, $R_2(\theta)$, $Q_1(\theta)$, $Q_2(\theta)$, $W(\theta)$

The expressions of the above physical quantities are given in Sect. 6.2.1.

For elastic scattering, due to time reversal invariance, $P^y(\theta) = A_y(\theta)$ is still satisfied when the spins of the target and the residual nucleus are not equal to 0. This relation has been verified in numerical calculations. For the case where the spins of the target and the residual nucleus are equal to 0, we have $P^y(\theta) = A_y(\theta) = P$, $Q_1(\theta) = Q_2(\theta) = Q$, $W(\theta) = W$. $P$, $Q$, and $W$ as $\theta$ function have been given by Eqs. (2.131)–(2.133), respectively. $Q$ is called a spin rotation function, and $W$ is called a spin continuation function. And in the case where the spins of the target and the residual nucleus are equal to 0, we have $R_1(\theta) + R_2(\theta) = 1$, $R_1(\theta) - R_2(\theta) = W$.

The experimental measurements can be carried out for $I_0(\theta)$ and $A_y(\theta)$, and nowadays a number of experimental data for them have been accumulated. For the shape elastic scattering we have $P^y(\theta) = A_y(\theta)$, but it is still to be studied how to measure directly or indirectly the $P^y(\theta)$ for two-body direct reaction and the five polarization transfer coefficient components for the shape elastic scattering and two-body direct reaction.

We specify that the direction of incident particles is $z$-axis and there is $\varphi = 0$ on the $xz$ plane, so that the angle $\varphi$ does not change when the coordinate system is transformed between the laboratory system and the center-of-mass (c.m.) system. It can be seen from Eq. (6.19) that $I(\theta, \varphi)$ and $I_0(\theta)$ are transformed according to the angular distribution transformation formula given by Eq. (2.152), and there are

$$\overline{A}_i(\theta_L, \varphi) = \overline{A}_i(\theta_C, \varphi), \qquad i = x, y, z \tag{6.34}$$

Multiply $I(\theta, \varphi)$ on the two ends of Eq. (6.33) from the left, and perform a coordinate system transformation to the just obtained formula, and then we can obtain

$$P_j(\theta_L, \varphi) = P_j(\theta_C, \varphi), \quad \overline{P}^j(\theta_L, \varphi) = \overline{P}^j(\theta_C, \varphi), \\ K_i^j(\theta_L, \varphi) = K_i^j(\theta_C, \varphi), \quad i, j = x, y, z \tag{6.35}$$

The transformation formula between $\theta_L$ and $\theta_C$ has been given by Eq. (2.157). Note that for a + A → b + B reaction, there is

$$\gamma = \left[\frac{m_a m_b (m_b + M_B)}{M_A M_B (m_a + M_A)} \frac{E_C}{E_C + Q}\right]^{1/2} \tag{6.36}$$

where $Q$ is the reaction energy. In the case of the elastic scattering, there is $\gamma = \frac{m}{M}$. In general, the theoretical calculations for the differential cross sections and the polarization quantities are doing in c.m. system. But for the convenience of users, it is recommended to preserve the polarization nuclear data in the laboratory system in the polarization nuclear database.

### 6.2.3 *Main Points of Polarized Particle Transport Theory for Spin $\frac{1}{2}$ Particles*

Here we will discuss the Monte-Carlo method to simulate the polarized particle transport process. For the sake of simplicity, it is assumed here that only $\vec{\frac{1}{2}} +$ A $\rightarrow \vec{\frac{1}{2}} +$ B reaction occurs in the device. This reaction can be a shape elastic scattering or a two-body direct reaction. All other reaction channels are omitted temporarily.

We choose studying the polarized particle transport process in the laboratory system, and choose the initial incident particle direction as the $z$-axis. The components $p_{0x}, p_{0y}, p_{0z}$ of the initial incident particle polarization vector are determined by experiment or artificial hypothesis. Take the position where the initial reaction (also known as the first reaction) $\vec{\frac{1}{2}} +$ A occurs as the coordinate origin $O_0$. When the incident particle energy $E_0$ in the laboratory system is determined, the following eight laboratory system differential data, $I_0(\theta)$, $A_y(\theta)$, $P^y(\theta)$, $R_1(\theta)$, $R_2(\theta)$, $Q_1(\theta)$, $Q_2(\theta)$, $W(\theta)$ can be obtained from the polarization nuclear database. Then, the various polarization quantities related to angles $(\theta, \varphi)$ can be calculated by using Eqs. (6.18), (6.25), and (6.32).

When the polarization rates $p_{0x}$ and $p_{0y}$ of the initial incident particles are known, the differential cross section corresponding to energy $E_0$ and any angle $(\theta, \varphi)$ of the outgoing particles can be obtained from Eq. (6.19). Regardless of the other reaction channels, after sampling, assuming that at the direction $(\theta_1, \varphi_1)$ and position $O_1$ with the distance $L_1$ from the origin $O_0$, an outgoing particle of the first reaction reacts with the target nucleus again, called the second reaction. The corresponding incident particle laboratory system energy of the second reaction is $E_1$. The Stokes vector expression of the initial incident particle of the first reaction at the second reaction space point $O_1$ in the $O_0$ system is [19]

$$S_1^{(\mathrm{i})}(\theta_1) = I_0(\theta_1)\begin{pmatrix} 1 \\ p_{0x} \\ p_{0y} \\ p_{0z} \end{pmatrix} \tag{6.37}$$

The Stokes vector expression of the corresponding outgoing particles is

$$S_1^{(\mathrm{f})}(\theta_1, \varphi_1) = I(\theta_1, \varphi_1)\begin{pmatrix} 1 \\ P_{1x}(\theta_1, \varphi_1) \\ P_{1y}(\theta_1, \varphi_1) \\ P_{1z}(\theta_1, \varphi_1) \end{pmatrix} \tag{6.38}$$

It can be seen that the above two vectors satisfy the following equation according to Eqs. (6.19) and (6.33):

$$S_1^{(\mathrm{f})} = Z_1 S_1^{(\mathrm{i})} \tag{6.39}$$

where

$$Z_1 = \begin{pmatrix} 1 & \overline{A}_x & \overline{A}_y & 0 \\ \overline{P}^x & K_x^x & K_y^x & K_z^x \\ \overline{P}^y & K_x^y & K_y^y & K_z^y \\ 0 & K_x^z & K_y^z & K_z^z \end{pmatrix}_1 \tag{6.40}$$

The footnote 1 in the above expression corresponds to $(\theta_1, \varphi_1)$ and energy $E_0$. $Z_1$ is called Mueller matrix [19]. After the first reaction at origin $O_0$, a relation similar to Eq. (6.39) can be established at any angles $(\theta, \varphi)$ in $O_0$ system; of course it is also true at the point $(L_1, \theta_1, \varphi_1)$, i.e., the relation (6.39) is determined by the first reaction.

The Cartesian coordinate components of the point $O_1$ in the $O_0$ system are

$$X_1 = L_1 \sin\theta_1 \cos\varphi_1, \quad Y_1 = L_1 \sin\theta_1 \sin\varphi_1, \quad Z_1 = L_1 \cos\theta_1 \tag{6.41}$$

Equation (2.203) has given the $D$ matrix of the rotating angles $(\theta, \varphi)$ in the three-dimensional Cartesian coordinate system as

$$\hat{B}(\theta, \varphi) = \begin{pmatrix} \cos\theta\cos\varphi & -\sin\varphi & \sin\theta\cos\varphi \\ \cos\theta\sin\varphi & \cos\varphi & \sin\theta\sin\varphi \\ -\sin\theta & 0 & \cos\theta \end{pmatrix} \tag{6.42}$$

Equation (2.204) gives the corresponding inverse matrix as

$$\hat{b}(\theta, \varphi) = \begin{pmatrix} \cos\theta\cos\varphi & \cos\theta\sin\varphi & -\sin\theta \\ -\sin\varphi & \cos\varphi & 0 \\ \sin\theta\cos\varphi & \sin\theta\sin\varphi & \cos\theta \end{pmatrix} \tag{6.43}$$

From the discussion in Section 2.5, we know that when we rotate the coordinate system and the vector of the object, we can use the following two formulas, respectively:

$$\vec{p}' = \hat{b}(\theta, \varphi)\vec{p}, \quad \vec{p}' = \hat{B}(\theta, \varphi)\vec{p} \tag{6.44}$$

Turn the three Cartesian coordinate axes in the $O_0$ system by angles $(\theta_1, \varphi_1)$ to be three new coordinate axes of the $O_1$ system, and then the coordinate origin carrying the three new coordinate axes are moved flatly from $O_0$ to $O_1$; the $O_0O_1$ is just the direction of the new $z$-axis.

The polarization vector of the emitted particles at the point $O_1$ induced by the first reaction at the point $O_0$ in the $O_0$ system is $\vec{P}_1(\theta_1, \varphi_1)$; it is also the polarization vector $\vec{p}_1$ of the incident particle of the second reaction in the $O_1$ system. This situation is equivalent rotating angles $(\theta_1, \varphi_1)$ to the coordinate system, and the object is not rotating, so there is

$$\vec{p}_1 = \hat{b}(\theta_1, \varphi_1)\vec{P}_1(\theta_1, \varphi_1) \tag{6.45}$$

That is, $\vec{P}_1$ observed in the $O_0$ system is changed to $\vec{p}_1$ observed in the $O_1$ system. Using Eq. (6.43) the following expressions can be derived from Eq. (6.45):

$$\begin{aligned} p_{1x} &= \cos\theta_1\cos\varphi_1 P_{1x}(\theta_1, \varphi_1) + \cos\theta_1\sin\varphi_1 P_{1y}(\theta_1, \varphi_1) - \sin\theta_1 P_{1z}(\theta_1, \varphi_1) \\ p_{1y} &= -\sin\varphi_1 P_{1x}(\theta_1, \varphi_1) + \cos\varphi_1 P_{1y}(\theta_1, \varphi_1) \\ p_{1z} &= \sin\theta_1\cos\varphi_1 P_{1x}(\theta_1, \varphi_1) + \sin\theta_1\sin\varphi_1 P_{1y}(\theta_1, \varphi_1) + \cos\theta_1 P_{1z}(\theta_1, \varphi_1) \end{aligned} \tag{6.46}$$

In the Monte-Carlo simulation calculations, if we determine by sampling that a nuclear reaction will take place somewhere, we consider that there is a 100 percent probability that a nuclear reaction will take place there. So the unpolarization differential cross section $I_0(\theta)$ corresponding to the energy $E_1$ in the $O_1$ system can still be found in the polarization nuclear database. The energy of the second reaction incident particles along the new $z$-axis at the point $O_1$ is $E_1$, and the corresponding incident particle polarization rates can be found from Eq. (6.45) or Eq. (6.46). Using the polarization nuclear database and Eq. (6.19), the differential cross section for any angles $(\theta, \varphi)$ of the emitted particles after the second reaction at the point $O_1$ can be obtained. Without taking into account other reaction channels, after sampling, it is assumed that at the direction $(\theta_2, \varphi_2)$ and the position $O_2$ with the distance $L_2$ from the origin $O_1$, an outgoing particle of the second reaction reacts with the target nucleus again, called the third reaction. The coordinate components of the origin $O_2$ at the $O_1$ system are

$$x_2 = L_2 \sin\theta_2 \cos\varphi_2, \quad y_2 = L_2 \sin\theta_2 \sin\varphi_2, \quad z_2 = L_2 \cos\theta_2 \tag{6.47}$$

The coordinate axes of the $O_0$ system compared with the coordinate axes of the $O_1$ system are equivalent to reverse rotating angles $(\theta_1, \varphi_1)$, so in order to change the coordinates of the point $O_2$ to be observed in the $O_0$ system but still using $O_1$ as the coordinate origin, the following equation can be used:

$$\begin{pmatrix} X_2 \\ Y_2 \\ Z_2 \end{pmatrix} = \hat{B}(\theta_1, \varphi_1) \begin{pmatrix} x_2 \\ y_2 \\ z_2 \end{pmatrix} \tag{6.48}$$

The expression of $\hat{B}$ has been given by Eq. (6.42). The coordinate components in the laboratory coordinate system $O_0$ of the emitted particle from the second reaction at the point $O_2$ are

$$X = X_1 + X_2, \quad Y = Y_1 + Y_2, \quad Z = Z_1 + Z_2 \tag{6.49}$$

In the $O_1$ system, the Stokes vector of the incident particle at the point $O_2$ is

$$S_2^{(\mathrm{i})}(\theta_2) = I_0(\theta_2) \begin{pmatrix} 1 \\ p_{1x} \\ p_{1y} \\ p_{1z} \end{pmatrix} \tag{6.50}$$

The corresponding Stokes vector of the outgoing particles at the point $O_2$ is

$$S_2^{(\mathrm{f})}(\theta_2, \varphi_2) = I(\theta_2, \varphi_2) \begin{pmatrix} 1 \\ P_{2x}(\theta_2, \varphi_2) \\ P_{2y}(\theta_2, \varphi_2) \\ P_{2z}(\theta_2, \varphi_2) \end{pmatrix} \tag{6.51}$$

The above two formulas satisfy the following relation:

$$S_2^{(\mathrm{f})} = Z_2 S_2^{(\mathrm{i})} \tag{6.52}$$

The footnote 2 corresponds to $(\theta_2, \varphi_2)$ and energy $E_1$. Equation (6.52) is determined by the second reaction.

The above method can be extended to the third, fourth, and so on reaction processes. Note that the coordinate $(X, Y, Z)$ and other physical quantities of the last emitted particle in the laboratory system $O_0$ must always be recorded. It can be seen that the main purpose of the above Monte-Carlo simulation calculation is to calculate the polarization rates of the outgoing particles after the collision. The polarization analyzing power, polarization transfer coefficient, the differential cross section, and the corresponding outgoing particle polarization rates of the unpolarized incident particles can be directly detected from the polarization nuclear database. However, note that the real Monte-Carlo simulation process must contain all possible reaction channels at the same time.

## 6.3 Polarized Particle Transport Theory for $\vec{1} + \mathrm{A} \to \vec{1} + \mathrm{B}$ Reactions

### *6.3.1 Polarization Quantities of Spin 1 Particles Related to Azimuthal Angle*

In the $\vec{1} + \mathrm{A} \to \vec{1} + \mathrm{B}$ reaction, it is assumed that both the incident particles and the outgoing particles are polarized particles of spin 1, and the spins of the target nucleus A and the residual nucleus B are $I$ and $I'$, respectively, and both are unpolarized. The reaction amplitude with a definite spin magnetic quantum number $M_I' M_I$ has been given by Eq. (3.374) as

$$\hat{\mathcal{F}}_{M_I' M_I}(\theta, \varphi) = \begin{pmatrix} B_{M_I' M_I}(\theta) & -\mathrm{i}C_{M_I' M_I}(\theta)\, e^{-\mathrm{i}\varphi} & -E_{M_I' M_I}(\theta)\, e^{-2\mathrm{i}\varphi} \\ \mathrm{i}O_{M_I' M_I}(\theta)\, e^{\mathrm{i}\varphi} & A_{M_I' M_I}(\theta) & -\mathrm{i}D_{M_I' M_I}(\theta)\, e^{-\mathrm{i}\varphi} \\ -F_{M_I' M_I}(\theta)\, e^{2\mathrm{i}\varphi} & \mathrm{i}G_{M_I' M_I}(\theta)\, e^{\mathrm{i}\varphi} & H_{M_I' M_I}(\theta) \end{pmatrix} e^{\mathrm{i}(M_I - M_I')\varphi} \tag{6.53}$$

where $A_{M_I' M_I}$,$B_{M_I' M_I}$ etc. have been given by Eqs. (3.355)–(3.363). Because in the calculation of the polarization quantities, when both the target nucleus and the residual nucleus are unpolarized, $\hat{\mathcal{F}}_{M_I' M_I}$ and $\hat{\mathcal{F}}^{+}_{M_I' M_I}$ always appear at the same time, the factor $\mathrm{e}^{\mathrm{i}(M_I - M_I')\varphi}$ must be eliminated, so we introduce the following symbols:

$$\hat{F}=\begin{pmatrix} B & -\mathrm{i}C\mathrm{e}^{-\mathrm{i}\varphi} & -E\mathrm{e}^{-2\mathrm{i}\varphi} \\ \mathrm{i}O\mathrm{e}^{\mathrm{i}\varphi} & A & -\mathrm{i}D\mathrm{e}^{-\mathrm{i}\varphi} \\ -F\mathrm{e}^{2\mathrm{i}\varphi} & \mathrm{i}G\mathrm{e}^{\mathrm{i}\varphi} & H \end{pmatrix}, \quad \hat{F}^{+}=\begin{pmatrix} B^* & -\mathrm{i}O^*\mathrm{e}^{-\mathrm{i}\varphi} & -F^*\mathrm{e}^{-2\mathrm{i}\varphi} \\ \mathrm{i}C^*\mathrm{e}^{\mathrm{i}\varphi} & A^* & -\mathrm{i}G^*\mathrm{e}^{-\mathrm{i}\varphi} \\ -E^*\mathrm{e}^{2\mathrm{i}\varphi} & \mathrm{i}D^*\mathrm{e}^{\mathrm{i}\varphi} & H^* \end{pmatrix} \tag{6.54}$$

where the footnote $M_I'M_I$ and the argument $(\theta)$ of the symbols $A$, $B$, etc. are omitted and the footnote $M_I'M_I$ and argument $(\theta,\varphi)$ of $\hat{F}$ and $\hat{F}^{+}$ are also omitted. Equation (3.377) has given

$$\hat{F}\hat{F}^{+} = \begin{pmatrix} |B|^2+|C|^2+|E|^2 & -\mathrm{i}(BO^*+CA^*+ED^*)\mathrm{e}^{-\mathrm{i}\varphi} & -(BF^*+CG^*+EH^*)\mathrm{e}^{-2\mathrm{i}\varphi} \\ \mathrm{i}(OB^*+AC^*+DE^*)\mathrm{e}^{\mathrm{i}\varphi} & |O|^2+|A|^2+|D|^2 & -\mathrm{i}(OF^*+AG^*+DH^*)\mathrm{e}^{-\mathrm{i}\varphi} \\ -(FB^*+GC^*+HE^*)\,\mathrm{e}^{2\mathrm{i}\varphi} & \mathrm{i}(FO^*+GA^*+HD^*)\,\mathrm{e}^{\mathrm{i}\varphi} & |F|^2+|G|^2+|H|^2 \end{pmatrix} \tag{6.55}$$

Also using the symbol introduced by Eq. (6.8), the differential cross section of the unpolarized incident particles can be obtained by using Eqs. (3.493)–(3.497) as

$$I_0(\theta)=\frac{1}{3}\widetilde{\Sigma}\left[\,|A|^2+2\left(\,|B|^2+|C|^2+|D|^2+|E|^2\right)\right] \tag{6.56}$$

where the footnotes $\alpha'n'$, $\alpha n$ of the outgoing and incident channels of $I_0(\theta)$ are omitted.

Equations (3.33) and (3.36) have given the expressions of the spin vector operator and the spin tensor operator of the spin 1 particle in the rectangular coordinate system of the spherical basis representation, respectively. For the tensor operator $\hat{Q}_{ij}$, we take the five independent tensor operators with footnote $xy$, $xz$, $yz$, $xx-yy$, $zz$. Utilizing Eq. (3.33) and Eq. (6.54), we can obtain

$$\hat{S}_x\hat{F}^{+}=\frac{1}{\sqrt{2}}\begin{pmatrix} \mathrm{i}C^*\mathrm{e}^{\mathrm{i}\varphi} & A^* & -\mathrm{i}G^*\mathrm{e}^{-\mathrm{i}\varphi} \\ B^*-E^*\mathrm{e}^{2\mathrm{i}\varphi} & -\mathrm{i}O^*\mathrm{e}^{-\mathrm{i}\varphi}+\mathrm{i}D^*\mathrm{e}^{\mathrm{i}\varphi} & -F^*\mathrm{e}^{-2\mathrm{i}\varphi}+H^* \\ \mathrm{i}C^*\mathrm{e}^{\mathrm{i}\varphi} & A^* & -\mathrm{i}G^*\mathrm{e}^{-\mathrm{i}\varphi} \end{pmatrix},$$

$$\hat{S}_y\hat{F}^{+}=\frac{\mathrm{i}}{\sqrt{2}}\begin{pmatrix} -\mathrm{i}C^*\mathrm{e}^{\mathrm{i}\varphi} & -A^* & \mathrm{i}G^*\mathrm{e}^{-\mathrm{i}\varphi} \\ B^*+E^*\mathrm{e}^{2\mathrm{i}\varphi} & -\mathrm{i}O^*\mathrm{e}^{-\mathrm{i}\varphi}-\mathrm{i}D^*\mathrm{e}^{\mathrm{i}\varphi} & -F^*\mathrm{e}^{-2\mathrm{i}\varphi}-H^* \\ \mathrm{i}C^*\mathrm{e}^{\mathrm{i}\varphi} & A^* & -\mathrm{i}G^*\mathrm{e}^{-\mathrm{i}\varphi} \end{pmatrix}, \tag{6.57}$$

$$\hat{S}_z\hat{F}^{+}=\begin{pmatrix} B^* & -\mathrm{i}O^*\mathrm{e}^{-\mathrm{i}\varphi} & -F^*\mathrm{e}^{-2\mathrm{i}\varphi} \\ 0 & 0 & 0 \\ E^*\mathrm{e}^{2\mathrm{i}\varphi} & -\mathrm{i}D^*\mathrm{e}^{-\mathrm{i}\varphi} & -H^* \end{pmatrix}$$

$$\hat{F}\hat{S}_x\hat{F}^+ = \frac{1}{\sqrt{2}}\begin{pmatrix} \mathrm{i}\left(BC^*\mathrm{e}^{\mathrm{i}\varphi} - CB^*\mathrm{e}^{-\mathrm{i}\varphi} + CE^*\mathrm{e}^{\mathrm{i}\varphi} - EC^*\mathrm{e}^{-\mathrm{i}\varphi}\right) & BA^* - CO^*\mathrm{e}^{-2\mathrm{i}\varphi} + CD^* - EA^*\mathrm{e}^{-2\mathrm{i}\varphi} & \mathrm{i}\left(-BG^*\mathrm{e}^{-\mathrm{i}\varphi} + CF^*\mathrm{e}^{-3\mathrm{i}\varphi} - CH^*\mathrm{e}^{-\mathrm{i}\varphi} + EG^*\mathrm{e}^{-3\mathrm{i}\varphi}\right) \\ -OC^*\mathrm{e}^{2\mathrm{i}\varphi} + AB^* - AE^*\mathrm{e}^{2\mathrm{i}\varphi} + DC^* & \mathrm{i}\left(OA^*\mathrm{e}^{\mathrm{i}\varphi} - AO^*\mathrm{e}^{-\mathrm{i}\varphi} + AD^*\mathrm{e}^{\mathrm{i}\varphi} - DA^*\mathrm{e}^{-\mathrm{i}\varphi}\right) & OG^* - AF^*\mathrm{e}^{-2\mathrm{i}\varphi} + AH^* - DG^*\mathrm{e}^{-2\mathrm{i}\varphi} \\ \mathrm{i}\left(-FC^*\mathrm{e}^{3\mathrm{i}\varphi} + GB^*\mathrm{e}^{\mathrm{i}\varphi} - GE^*\mathrm{e}^{3\mathrm{i}\varphi} + HC^*\mathrm{e}^{\mathrm{i}\varphi}\right) & -FA^*\mathrm{e}^{-2\mathrm{i}\varphi} + GO^* - GD^*\mathrm{e}^{2\mathrm{i}\varphi} + HA^* & \mathrm{i}\left(FG^*\mathrm{e}^{\mathrm{i}\varphi} - GF^*\mathrm{e}^{-\mathrm{i}\varphi} + GH^*\mathrm{e}^{\mathrm{i}\varphi} - HG^*\mathrm{e}^{-\mathrm{i}\varphi}\right) \end{pmatrix},$$

$$\hat{F}\hat{S}_y\hat{F}^+ = \frac{\mathrm{i}}{\sqrt{2}}\begin{pmatrix} -\mathrm{i}\left(BC^*\mathrm{e}^{\mathrm{i}\varphi} + CB^*\mathrm{e}^{-\mathrm{i}\varphi} + CE^*\mathrm{e}^{\mathrm{i}\varphi} + EC^*\mathrm{e}^{-\mathrm{i}\varphi}\right) & -BA^* - CO^*\mathrm{e}^{-2\mathrm{i}\varphi} - CD^* - EA^*\mathrm{e}^{-2\mathrm{i}\varphi} & \mathrm{i}\left(BG^*\mathrm{e}^{-\mathrm{i}\varphi} + CF^*\mathrm{e}^{-3\mathrm{i}\varphi} + CH^*\mathrm{e}^{-\mathrm{i}\varphi} + EG^*\mathrm{e}^{-3\mathrm{i}\varphi}\right) \\ OC^*\mathrm{e}^{2\mathrm{i}\varphi} + AB^* + AE^*\mathrm{e}^{2\mathrm{i}\varphi} + DC^* & -\mathrm{i}\left(OA^*\mathrm{e}^{\mathrm{i}\varphi} + AO^*\mathrm{e}^{-\mathrm{i}\varphi} + AD^*\mathrm{e}^{\mathrm{i}\varphi} + DA^*\mathrm{e}^{-\mathrm{i}\varphi}\right) & -OG^* - AF^*\mathrm{e}^{-2\mathrm{i}\varphi} - AH^* - DG^*\mathrm{e}^{-2\mathrm{i}\varphi} \\ \mathrm{i}\left(FC^*\mathrm{e}^{3\mathrm{i}\varphi} + GB^*\mathrm{e}^{\mathrm{i}\varphi} + GE^*\mathrm{e}^{3\mathrm{i}\varphi} + HC^*\mathrm{e}^{\mathrm{i}\varphi}\right) & FA^*\mathrm{e}^{2\mathrm{i}\varphi} + GO^* + GD^*\mathrm{e}^{2\mathrm{i}\varphi} + HA^* & -\mathrm{i}\left(FG^*\mathrm{e}^{\mathrm{i}\varphi} + GF^*\mathrm{e}^{-\mathrm{i}\varphi} + GH^*\mathrm{e}^{\mathrm{i}\varphi} + HG^*\mathrm{e}^{-\mathrm{i}\varphi}\right) \end{pmatrix},$$

$$\hat{F}\hat{S}_z\hat{F}^+ = \begin{pmatrix} |B|^2 - |E|^2 & -\mathrm{i}\left(BO^*\mathrm{e}^{-\mathrm{i}\varphi} - ED^*\mathrm{e}^{-\mathrm{i}\varphi}\right) & -BF^*\mathrm{e}^{-2\mathrm{i}\varphi} + EH^*\mathrm{e}^{-2\mathrm{i}\varphi} \\ \mathrm{i}\left(OB^*\mathrm{e}^{\mathrm{i}\varphi} - DE^*\mathrm{e}^{\mathrm{i}\varphi}\right) & |O|^2 - |D|^2 & -\mathrm{i}\left(OF^*\mathrm{e}^{-\mathrm{i}\varphi} - DH^*\mathrm{e}^{-\mathrm{i}\varphi}\right) \\ -FB^*\mathrm{e}^{2\mathrm{i}\varphi} + HE^*\mathrm{e}^{2\mathrm{i}\varphi} & \mathrm{i}\left(FO^*\mathrm{e}^{\mathrm{i}\varphi} - HD^*\mathrm{e}^{\mathrm{i}\varphi}\right) & |F|^2 - |H|^2 \end{pmatrix} \tag{6.58}$$

From Eq. (6.58) we can get

$$\begin{aligned} \frac{1}{\sqrt{2}}\mathrm{tr}\left(\hat{F}\hat{S}_x\hat{F}^+\right) &= -\mathrm{Im}\left[\left(BC^* + CE^* + OA^* + AD^* + FG^* + GH^*\right)\mathrm{e}^{\mathrm{i}\varphi}\right], \\ \frac{1}{\sqrt{2}}\mathrm{tr}\left(\hat{F}\hat{S}_y\hat{F}^+\right) &= \mathrm{Re}\left[\left(BC^* + CE^* + OA^* + AD^* + FG^* + GH^*\right)\mathrm{e}^{\mathrm{i}\varphi}\right], \\ \mathrm{tr}\left(\hat{F}\hat{S}_z\hat{F}^+\right) &= |B|^2 - |E|^2 + |O|^2 - |D|^2 + |F|^2 - |H|^2 \end{aligned} \tag{6.59}$$

The following expressions can be obtained from Eq. (6.59) and using Eqs. (3.493)–(3.499)

$$\begin{aligned} \frac{1}{\sqrt{2}}\widetilde{\Sigma}\,\mathrm{tr}\left(\hat{F}\hat{S}_x\hat{F}^+\right) &= -2\widetilde{\Sigma}\,\mathrm{Re}\left(BC^* + CE^* + OA^*\right)\sin\varphi, \\ \frac{1}{\sqrt{2}}\widetilde{\Sigma}\,\mathrm{tr}\left(\hat{F}\hat{S}_y\hat{F}^+\right) &= 2\widetilde{\Sigma}\,\mathrm{Re}\left(BC^* + CE^* + OA^*\right)\cos\varphi, \\ \widetilde{\Sigma}\,\mathrm{tr}\left(\hat{F}\hat{S}_z\hat{F}^+\right) &= 0 \end{aligned} \tag{6.60}$$

Define the following vector analyzing power only related to angle $\theta$

$$A_y(\theta) = \frac{2\sqrt{2}}{3I_0}\tilde{\Sigma}\ \mathrm{Re}\,(BC^* + CE^* + OA^*) \tag{6.61}$$

In the general case, the vector analyzing power related to the angle $\theta$, $\varphi$ is

$$\overline{A}_i(\theta,\varphi) = \frac{1}{3I_0}\tilde{\Sigma}\ \mathrm{tr}(\hat{F}\hat{S}_i\hat{F}^+), \quad i = x,y,z \tag{6.62}$$

Then the following results can be got according to Eqs. (6.60)–(6.62):

$$\overline{A}_x(\theta,\varphi) = -A_y(\theta)\sin\varphi, \quad \overline{A}_y(\theta,\varphi) = A_y(\theta)\cos\varphi, \quad \overline{A}_z(\theta,\varphi) = 0 \tag{6.63}$$

Utilizing Eq. (3.36) and Eq. (6.54) we can obtain

$$\hat{Q}_{xy}\hat{F}^+ = \frac{3\mathrm{i}}{2}\begin{pmatrix} E^*\mathrm{e}^{2\mathrm{i}\varphi} & -\mathrm{i}D^*\mathrm{e}^{\mathrm{i}\varphi} & -H^* \\ 0 & 0 & 0 \\ B^* & -\mathrm{i}O^*\mathrm{e}^{-\mathrm{i}\varphi} & -F^*\mathrm{e}^{-2\mathrm{i}\varphi} \end{pmatrix},$$
$$\hat{Q}_{xz}\hat{F}^+ = \frac{3}{2\sqrt{2}}\begin{pmatrix} \mathrm{i}C^*\mathrm{e}^{2\mathrm{i}\varphi} & A^* & -\mathrm{i}G^*\mathrm{e}^{-\mathrm{i}\varphi} \\ B^* + E^*\mathrm{e}^{2\mathrm{i}\varphi} & -\mathrm{i}O^*\mathrm{e}^{-\mathrm{i}\varphi} - \mathrm{i}D^*\mathrm{e}^{\mathrm{i}\varphi} & -F^*\mathrm{e}^{-2\mathrm{i}\varphi} - H^* \\ -\mathrm{i}C^*\mathrm{e}^{\mathrm{i}\varphi} & -A^* & \mathrm{i}G^*\mathrm{e}^{-\mathrm{i}\varphi} \end{pmatrix},$$
$$\hat{Q}_{yz}\hat{F}^+ = \frac{3\mathrm{i}}{2\sqrt{2}}\begin{pmatrix} -\mathrm{i}C^*\mathrm{e}^{\mathrm{i}\varphi} & -A^* & \mathrm{i}G^*\mathrm{e}^{-\mathrm{i}\varphi} \\ B^* - E^*\mathrm{e}^{2\mathrm{i}\varphi} & -\mathrm{i}O^*\mathrm{e}^{-\mathrm{i}\varphi} + \mathrm{i}D^*\mathrm{e}^{\mathrm{i}\varphi} & -F^*\mathrm{e}^{-2\mathrm{i}\varphi} + H^* \\ -\mathrm{i}C^*\mathrm{e}^{\mathrm{i}\varphi} & -A^* & \mathrm{i}G^*\mathrm{e}^{-\mathrm{i}\varphi} \end{pmatrix},$$
$$\hat{Q}_{xx-yy}\hat{F}^+ = 3\begin{pmatrix} -E^*\mathrm{e}^{2\mathrm{i}\varphi} & \mathrm{i}D^*\mathrm{e}^{\mathrm{i}\varphi} & H^* \\ 0 & 0 & 0 \\ B^* & -\mathrm{i}O^*\mathrm{e}^{-\mathrm{i}\varphi} & -F^*\mathrm{e}^{-2\mathrm{i}\varphi} \end{pmatrix},$$
$$\hat{Q}_{zz}\hat{F}^+ = \begin{pmatrix} B^* & -\mathrm{i}O^*\mathrm{e}^{-\mathrm{i}\varphi} & -F^*\mathrm{e}^{-2\mathrm{i}\varphi} \\ -2\mathrm{i}C^*\mathrm{e}^{\mathrm{i}\varphi} & -2A^* & 2\mathrm{i}G^*\mathrm{e}^{-\mathrm{i}\varphi} \\ -E^*\mathrm{e}^{2\mathrm{i}\varphi} & \mathrm{i}D^*\mathrm{e}^{\mathrm{i}\varphi} & H^* \end{pmatrix} \tag{6.64}$$

$$\hat{F}\hat{Q}_{xy}\hat{F}^+ = \frac{3\mathrm{i}}{2}\begin{pmatrix} BE^*\mathrm{e}^{2\mathrm{i}\varphi} - EB^*\mathrm{e}^{-2\mathrm{i}\varphi} & -\mathrm{i}(BD^*\mathrm{e}^{\mathrm{i}\varphi} - EO^*\mathrm{e}^{-3\mathrm{i}\varphi}) & -BH^* + EF^*e^{-4\mathrm{i}\varphi} \\ \mathrm{i}(OE^*\mathrm{e}^{3\mathrm{i}\varphi} - DB^*\mathrm{e}^{-\mathrm{i}\varphi}) & OD^*\mathrm{e}^{2\mathrm{i}\varphi} - DO^*\mathrm{e}^{-2\mathrm{i}\varphi} & -\mathrm{i}(OH^*\mathrm{e}^{\mathrm{i}\varphi} - DF^*\mathrm{e}^{-3\mathrm{i}\varphi}) \\ -FE^*\mathrm{e}^{4\mathrm{i}\varphi} + HB^* & \mathrm{i}(FD^*\mathrm{e}^{3\mathrm{i}\varphi} - HO^*\mathrm{e}^{-\mathrm{i}\varphi}) & FH^*\mathrm{e}^{2\mathrm{i}\varphi} - HF^*\mathrm{e}^{-2\mathrm{i}\varphi} \end{pmatrix},$$

$$\hat{F}\hat{Q}_{xz}\hat{F}^+ = \frac{3}{2\sqrt{2}}\begin{pmatrix} \mathrm{i}(BC^*\mathrm{e}^{\mathrm{i}\varphi} - CB^*\mathrm{e}^{-\mathrm{i}\varphi} - CE^*\mathrm{e}^{\mathrm{i}\varphi} + EC^*\mathrm{e}^{-\mathrm{i}\varphi}) & BA^* - CO^*\mathrm{e}^{-2\mathrm{i}\varphi} - CD^* + EA^*\mathrm{e}^{-2\mathrm{i}\varphi} & -\mathrm{i}(BG^*\mathrm{e}^{-\mathrm{i}\varphi} - CF^*\mathrm{e}^{-3\mathrm{i}\varphi} - CH^*\mathrm{e}^{-\mathrm{i}\varphi} + EG^*\mathrm{e}^{-3\mathrm{i}\varphi}) \\ -OC^*\mathrm{e}^{2\mathrm{i}\varphi} + AB^* + AE^*\mathrm{e}^{2\mathrm{i}\varphi} - DC^* & \mathrm{i}(OA^*\mathrm{e}^{\mathrm{i}\varphi} - AO^*\mathrm{e}^{-\mathrm{i}\varphi} - AD^*\mathrm{e}^{\mathrm{i}\varphi} + DA^*e^{-\mathrm{i}\varphi}) & OG^* - AF^*\mathrm{e}^{-2\mathrm{i}\varphi} - AH^* + DG^*\mathrm{e}^{-2\mathrm{i}\varphi} \\ -\mathrm{i}(FC^*\mathrm{e}^{3\mathrm{i}\varphi} - GB^*\mathrm{e}^{\mathrm{i}\varphi} - GE^*\mathrm{e}^{3\mathrm{i}\varphi} + HC^*\mathrm{e}^{\mathrm{i}\varphi}) & -FA^*\mathrm{e}^{2\mathrm{i}\varphi} + GO^* + GD^*\mathrm{e}^{2\mathrm{i}\varphi} - HA^* & \mathrm{i}(FG^*\mathrm{e}^{\mathrm{i}\varphi} - GF^*\mathrm{e}^{-\mathrm{i}\varphi} - GH^*\mathrm{e}^{\mathrm{i}\varphi} + HG^*\mathrm{e}^{-\mathrm{i}\varphi}) \end{pmatrix},$$

$$
\hat{F}\hat{Q}_{yz}\hat{F}^{+}=\frac{3\mathrm{i}}{2\sqrt{2}}\begin{pmatrix}
-\mathrm{i}\left(BC^{*}\mathrm{e}^{\mathrm{i}\varphi}+CB^{*}\mathrm{e}^{-\mathrm{i}\varphi}-CE^{*}e^{\mathrm{i}\varphi}-EC^{*}\mathrm{e}^{-\mathrm{i}\varphi}\right) & -BA^{*}-CO^{*}\mathrm{e}^{-2\mathrm{i}\varphi}+CD^{*}+EA^{*}\mathrm{e}^{-2\mathrm{i}\varphi} & \mathrm{i}\left(BG^{*}\mathrm{e}^{-\mathrm{i}\varphi}+CF^{*}e^{-3\mathrm{i}\varphi}-CH^{*}\mathrm{e}^{-\mathrm{i}\varphi}-EG^{*}\mathrm{e}^{-3\mathrm{i}\varphi}\right)\\
OC^{*}\mathrm{e}^{2\mathrm{i}\varphi}+AB^{*}-AE^{*}\mathrm{e}^{2\mathrm{i}\varphi}-DC^{*} & -\mathrm{i}\left(OA^{*}\mathrm{e}^{\mathrm{i}\varphi}+AO^{*}\mathrm{e}^{-\mathrm{i}\varphi}-AD^{*}\mathrm{e}^{\mathrm{i}\varphi}-DA^{*}\mathrm{e}^{-\mathrm{i}\varphi}\right) & -OG^{*}-AF^{*}\mathrm{e}^{-2\mathrm{i}\varphi}+AH^{*}+DG^{*}\mathrm{e}^{-2\mathrm{i}\varphi}\\
\mathrm{i}\left(FC^{*}\mathrm{e}^{3\mathrm{i}\varphi}+GB^{*}\mathrm{e}^{\mathrm{i}\varphi}-GE^{*}\mathrm{e}^{3\mathrm{i}\varphi}-HC^{*}\mathrm{e}^{\mathrm{i}\varphi}\right) & FA^{*}\mathrm{e}^{2\mathrm{i}\varphi}+GO^{*}-GD^{*}\mathrm{e}^{2\mathrm{i}\varphi}-HA^{*} & -\mathrm{i}\left(FG^{*}\mathrm{e}^{\mathrm{i}\varphi}+GF^{*}\mathrm{e}^{-\mathrm{i}\varphi}-GH^{*}\mathrm{e}^{\mathrm{i}\varphi}-HG^{*}\mathrm{e}^{-\mathrm{i}\varphi}\right)
\end{pmatrix},
$$

$$
\hat{F}\hat{Q}_{xx-yy}\hat{F}^{+}=3\begin{pmatrix}
-BE^{*}\mathrm{e}^{2\mathrm{i}\varphi}-EB^{*}\mathrm{e}^{-2\mathrm{i}\varphi} & \mathrm{i}\left(BD^{*}\mathrm{e}^{\mathrm{i}\varphi}+EO^{*}\mathrm{e}^{-3\mathrm{i}\varphi}\right) & BH^{*}+EF^{*}\mathrm{e}^{-4\mathrm{i}\varphi}\\
-\mathrm{i}\left(OE^{*}\mathrm{e}^{3\mathrm{i}\varphi}+DB^{*}\mathrm{e}^{-\mathrm{i}\varphi}\right) & -OD^{*}\mathrm{e}^{2\mathrm{i}\varphi}-DO^{*}\mathrm{e}^{-2\mathrm{i}\varphi} & \mathrm{i}\left(OH^{*}\mathrm{e}^{\mathrm{i}\varphi}+DF^{*}\mathrm{e}^{-3\mathrm{i}\varphi}\right)\\
FE^{*}\mathrm{e}^{4\mathrm{i}\varphi}+HB^{*} & -\mathrm{i}\left(FD^{*}\mathrm{e}^{3\mathrm{i}\varphi}+HO^{*}\mathrm{e}^{-\mathrm{i}\varphi}\right) & -FH^{*}\mathrm{e}^{2\mathrm{i}\varphi}-HF^{*}\mathrm{e}^{-2\mathrm{i}\varphi}
\end{pmatrix},
$$

$$
\hat{F}\hat{Q}_{zz}\hat{F}^{+}=\begin{pmatrix}
|B|^{2}-2|C|^{2}+|E|^{2} & -\mathrm{i}(BO^{*}-2CA^{*}+ED^{*})\mathrm{e}^{-\mathrm{i}\varphi} & -(BF^{*}-2CG^{*}+EH^{*})\mathrm{e}^{-2\mathrm{i}\varphi}\\
\mathrm{i}(OB^{*}-2AC^{*}+DE^{*})\mathrm{e}^{\mathrm{i}\varphi} & |O|^{2}-2|A|^{2}+|D|^{2} & -\mathrm{i}(OF^{*}-2AG^{*}+DH^{*})\mathrm{e}^{-\mathrm{i}\varphi}\\
-(FB^{*}-2GC^{*}+HE^{*})\mathrm{e}^{2\mathrm{i}\varphi} & \mathrm{i}(FO^{*}-2GA^{*}+HD^{*})\mathrm{e}^{\mathrm{i}\varphi} & |F|^{2}-2|G|^{2}+|H|^{2}
\end{pmatrix}
\tag{6.65}
$$

From Eq. (6.65) we can also obtain

$$
\begin{aligned}
\mathrm{tr}\left(\hat{F}\hat{Q}_{xy}\hat{F}^{+}\right)&=-3\mathrm{Im}\left[(BE^{*}+OD^{*}+FH^{*})\,\mathrm{e}^{2\mathrm{i}\varphi}\right],\\
\mathrm{tr}\left(\hat{F}\hat{Q}_{xz}\hat{F}^{+}\right)&=-\frac{3}{\sqrt{2}}\mathrm{Im}\left[(BC^{*}-CE^{*}+OA^{*}-AD^{*}+FG^{*}-GH^{*})\,\mathrm{e}^{\mathrm{i}\varphi}\right],\\
\mathrm{tr}\left(\hat{F}\hat{Q}_{yz}\hat{F}^{+}\right)&=\frac{3}{\sqrt{2}}\mathrm{Re}\left[(BC^{*}-CE^{*}+OA^{*}-AD^{*}+FG^{*}-GH^{*})\,\mathrm{e}^{\mathrm{i}\varphi}\right],\\
\mathrm{tr}\left(\hat{F}\hat{Q}_{xx-yy}\hat{F}^{+}\right)&=-6\mathrm{Re}\left[(BE^{*}+OD^{*}+FH^{*})\,\mathrm{e}^{2\mathrm{i}\varphi}\right],\\
\mathrm{tr}\left(\hat{F}\hat{Q}_{zz}\hat{F}^{+}\right)&=|B|^{2}+|E|^{2}+|O|^{2}+|D|^{2}+|F|^{2}+|H|^{2}-2\left(|C|^{2}+|A|^{2}+|G|^{2}\right)
\end{aligned}
\tag{6.66}
$$

So the following expressions can be obtained from Eq. (6.66) and using Eqs. (3.493)–(3.499)

$$
\begin{aligned}
\widetilde{\Sigma}\mathrm{tr}\left(\hat{F}\hat{Q}_{xy}\hat{F}^{+}\right)&=-3\widetilde{\Sigma}\,\mathrm{Re}\,(2BE^{*}+OD^{*})\ \sin(2\varphi),\\
\widetilde{\Sigma}\mathrm{tr}\left(\hat{F}\hat{Q}_{xz}\hat{F}^{+}\right)&=-3\sqrt{2}\,\widetilde{\Sigma}\mathrm{Im}(BC^{*}+OA^{*}+FG^{*})\ \cos\varphi,\\
\widetilde{\Sigma}\mathrm{tr}\left(\hat{F}\hat{Q}_{yz}\hat{F}^{+}\right)&=-3\sqrt{2}\,\widetilde{\Sigma}\mathrm{Im}(BC^{*}+OA^{*}+FG^{*})\ \sin\varphi,\\
\widetilde{\Sigma}\mathrm{tr}\left(\hat{F}\hat{Q}_{xx-yy}\hat{F}^{+}\right)&=-6\widetilde{\Sigma}\,\mathrm{Re}\,(2BE^{*}+OD^{*})\ \cos(2\varphi),\\
\widetilde{\Sigma}\mathrm{tr}\left(\hat{F}\hat{Q}_{zz}\hat{F}^{+}\right)&=2\left(|B|^{2}+|D|^{2}+|E|^{2}-|A|^{2}-2|C|^{2}\right)
\end{aligned}
\tag{6.67}
$$

The definition of tensor analyzing power related to angles $\theta$, $\varphi$ is

$$\overline{A}_i(\theta, \varphi) = \frac{1}{3I_0} \widetilde{\Sigma} \, \mathrm{tr}(\hat{F}\hat{Q}_i\hat{F}^+), \qquad i = xy, xz, yz, xx - yy, zz \tag{6.68}$$

We define the following tensor analyzing powers that are only related to angle $\theta$:

$$\begin{aligned} A_1(\theta) &= \frac{1}{I_0} \widetilde{\Sigma} \, \mathrm{Re}\,(2BE^* + OD^*), \quad A_2(\theta) = \frac{\sqrt{2}}{I_0} \widetilde{\Sigma} \mathrm{Im}(BC^* + OA^* + FG^*), \\ A_3(\theta) &= \frac{2}{3I_0} \widetilde{\Sigma} \left( |B|^2 + |D|^2 + |E|^2 - |A|^2 - 2|C|^2 \right) \end{aligned} \tag{6.69}$$

So the following results can be obtained from Eqs. (6.67)–(6.69):

$$\begin{aligned} &\overline{A}_{xy}(\theta, \varphi) = -A_1(\theta) \sin(2\varphi), \quad \overline{A}_{xz}(\theta, \varphi) = -A_2(\theta) \cos\varphi, \\ &\overline{A}_{yz}(\theta, \varphi) = -A_2(\theta) \sin\varphi, \qquad \overline{A}_{xx-yy}(\theta, \varphi) = -2A_1(\theta) \cos(2\varphi), \\ &\overline{A}_{zz}(\theta, \varphi) = A_3(\theta) \end{aligned} \tag{6.70}$$

Referring to Eq. (3.214), we can write the differential cross section of the outgoing particle corresponding to the polarized incident particle as

$$\begin{aligned} I(\theta, \varphi) = I_0(\theta) \Big[ 1 &+ \frac{3}{2} (p_x \overline{A}_x(\theta, \varphi) + p_y \overline{A}_y(\theta, \varphi)) \\ &+ \frac{2}{3} (p_{xy} \overline{A}_{xy}(\theta, \varphi) + p_{xz} \overline{A}_{xz}(\theta, \varphi) + p_{yz} \overline{A}_{yz}(\theta, \varphi)) \\ &+ \frac{1}{6} p_{xx-yy} \overline{A}_{xx-yy}(\theta, \varphi) + \frac{1}{2} p_{zz} \overline{A}_{zz}(\theta, \varphi) \Big] \end{aligned} \tag{6.71}$$

where $p_{xx-yy} \equiv p_{xx} - p_{yy}$, $p_i$ $(i = x, y, xy, xz, yz, xx - yy, zz)$ are the polarization rates of the incident particles.

From Eq. (3.33) and Eq. (6.55), we can obtain

$$\begin{aligned} \frac{1}{\sqrt{2}} \mathrm{tr}\left(\hat{S}_x \hat{F} \hat{F}^+\right) &= \frac{\mathrm{i}}{2} \big[ (OB^* + AC^* + DE^*) \, \mathrm{e}^{\mathrm{i}\varphi} - (BO^* + CA^* + ED^*) \, \mathrm{e}^{-\mathrm{i}\varphi} \\ &\quad + (FO^* + GA^* + HD^*) \, \mathrm{e}^{\mathrm{i}\varphi} - (OF^* + AG^* + DH^*) \, \mathrm{e}^{-\mathrm{i}\varphi} \big] \\ &= -\mathrm{Im} \left[ (OB^* + AC^* + DE^* + FO^* + GA^* + HD^*) \, \mathrm{e}^{\mathrm{i}\varphi} \right], \\ \frac{1}{\sqrt{2}} \mathrm{tr}\left(\hat{S}_y \hat{F} \hat{F}^+\right) &= \frac{1}{2} \big[ (OB^* + AC^* + DE^*) \, \mathrm{e}^{\mathrm{i}\varphi} + (BO^* + CA^* + ED^*) \, \mathrm{e}^{-\mathrm{i}\varphi} \\ &\quad + (FO^* + GA^* + HD^*) \, \mathrm{e}^{\mathrm{i}\varphi} + (OF^* + AG^* + DH^*) \, \mathrm{e}^{-\mathrm{i}\varphi} \big] \\ &= \mathrm{Re} \left[ (OB^* + AC^* + DE^* + FO^* + GA^* + DH^*) \, \mathrm{e}^{\mathrm{i}\varphi} \right], \\ \mathrm{tr}\left(\hat{S}_z \hat{F} \hat{F}^+\right) &= |B|^2 + |C|^2 + |E|^2 - |F|^2 - |G|^2 - |H|^2 \end{aligned} \tag{6.72}$$

The following expressions can be obtained from Eq. (6.72) and using Eqs. (3.493)–(3.499):

$$\begin{aligned}\frac{1}{\sqrt{2}}\widetilde{\Sigma}\mathrm{tr}\left(\hat{S}_x\hat{F}\hat{F}^+\right) &= -2\widetilde{\Sigma}\,\mathrm{Re}\,(OB^* + AC^* + DE^*)\sin\varphi,\\ \frac{1}{\sqrt{2}}\widetilde{\Sigma}\mathrm{tr}\left(\hat{S}_y\hat{F}\hat{F}^+\right) &= 2\widetilde{\Sigma}\,\mathrm{Re}\,(OB^* + AC^* + DE^*)\cos\varphi,\\ \widetilde{\Sigma}\mathrm{tr}\left(\hat{S}_z\hat{F}\hat{F}^+\right) &= 0\end{aligned} \tag{6.73}$$

Define

$$P^y(\theta) = \frac{2\sqrt{2}}{3I_0}\widetilde{\Sigma}\,\mathrm{Re}\,(OB^* + AC^* + DE^*) \tag{6.74}$$

and

$$\overline{P}^i(\theta,\varphi) = \frac{1}{3I_0}\widetilde{\Sigma}\mathrm{tr}\left(\hat{S}_i\hat{F}\hat{F}^+\right), \quad i = x,y,z \tag{6.75}$$

Then the following results can be obtained according to Eqs. (6.73)–(6.75):

$$\overline{P}^x(\theta,\varphi) = -P^y(\theta)\sin\varphi, \quad \overline{P}^y(\theta,\varphi) = P^y(\theta)\cos\varphi, \quad \overline{P}^z(\theta,\varphi) = 0 \tag{6.76}$$

From Eq. (3.36) and Eq. (6.55) we can get

$$\begin{aligned}\mathrm{tr}\left(\hat{Q}_{xy}\hat{F}\hat{F}^+\right) &= -3\mathrm{Im}\left[(FB^* + GC^* + HE^*)\,\mathrm{e}^{2\mathrm{i}\varphi}\right],\\ \mathrm{tr}\left(\hat{Q}_{xz}\hat{F}\hat{F}^+\right) &= -\frac{3}{\sqrt{2}}\mathrm{Im}\left[(OB^* + AC^* + DE^* - FO^* - GA^* - HD^*)\,\mathrm{e}^{\mathrm{i}\varphi}\right],\\ \mathrm{tr}\left(\hat{Q}_{yz}\hat{F}\hat{F}^+\right) &= \frac{3}{\sqrt{2}}\mathrm{Re}\left[(OB^* + AC^* + DE^* - FO^* - GA^* - HD^*)\,\mathrm{e}^{\mathrm{i}\varphi}\right],\\ \mathrm{tr}\left(\hat{Q}_{xx-yy}\hat{F}\hat{F}^+\right) &= -6\,\mathrm{Re}\left[(FB^* + GC^* + HE^*)\,\mathrm{e}^{2\mathrm{i}\varphi}\right],\\ \mathrm{tr}\left(\hat{Q}_{zz}\hat{F}\hat{F}^+\right) &= |B|^2 + |C|^2 + |E|^2 + |F|^2 + |G|^2 + |H|^2 - 2\left(|O|^2 + |A|^2 + |D|^2\right)\end{aligned} \tag{6.77}$$

So the following expressions can be obtained from Eq. (6.77) and using Eqs. (3.493)–(3.499):

$$\begin{aligned}
\frac{1}{3}\widetilde{\Sigma}\mathrm{tr}\left(\hat{Q}_{xy}\hat{F}\hat{F}^{+}\right) &= -\widetilde{\Sigma}\,\mathrm{Re}\,(2FB^{*}+GC^{*})\sin(2\varphi),\\
\frac{1}{3}\widetilde{\Sigma}\mathrm{tr}\left(\hat{Q}_{xz}\hat{F}\hat{F}^{+}\right) &= -\sqrt{2}\,\widetilde{\Sigma}\mathrm{Im}(OB^{*}+AC^{*}+DE^{*})\cos\varphi,\\
\frac{1}{3}\widetilde{\Sigma}\mathrm{tr}\left(\hat{Q}_{yz}\hat{F}\hat{F}^{+}\right) &= -\sqrt{2}\,\widetilde{\Sigma}\mathrm{Im}(OB^{*}+AC^{*}+DE^{*})\sin\varphi,\\
\frac{1}{3}\widetilde{\Sigma}\mathrm{tr}\left(\hat{Q}_{xx-yy}\hat{F}\hat{F}^{+}\right) &= -2\widetilde{\Sigma}\,\mathrm{Re}\,(2FB^{*}+GC^{*})\cos(2\varphi),\\
\frac{1}{3}\widetilde{\Sigma}\mathrm{tr}\left(\hat{Q}_{zz}\hat{F}\hat{F}^{+}\right) &= \frac{2}{3}\widetilde{\Sigma}\left(|B|^{2}+|C|^{2}+|E|^{2}-|A|^{2}-2|D|^{2}\right)
\end{aligned}\tag{6.78}$$

Define

$$\begin{aligned}
P^{1}(\theta) &= \frac{1}{I_{0}}\widetilde{\Sigma}\,\mathrm{Re}\,(2FB^{*}+GC^{*}),\\
P^{2}(\theta) &= \frac{\sqrt{2}}{I_{0}}\widetilde{\Sigma}\mathrm{Im}(OB^{*}+AC^{*}+DE^{*}),\\
P^{3}(\theta) &= \frac{2}{3I_{0}}\widetilde{\Sigma}\left(|B|^{2}+|C|^{2}+|E|^{2}-|A|^{2}-2|D|^{2}\right)
\end{aligned}\tag{6.79}$$

and

$$\overline{P}^{i}(\theta,\varphi) = \frac{1}{3I_{0}}\widetilde{\Sigma}\mathrm{tr}\left(\hat{Q}_{i}\hat{F}\hat{F}^{+}\right),\qquad i=\gamma \tag{6.80}$$

where the coordinate symbol set $\gamma$ has been given by Eq. (3.824). Then the following results can be obtained from Eqs. (6.78)–(6.80):

$$\begin{aligned}
&\overline{P}^{xy}(\theta,\varphi) = -P^{1}(\theta)\sin(2\varphi), \quad \overline{P}^{xz}(\theta,\varphi) = -P^{2}(\theta)\cos\varphi,\\
&\overline{P}^{yz}(\theta,\varphi) = -P^{2}(\theta)\sin\varphi, \qquad \overline{P}^{xx-yy}(\theta,\varphi) = -2P^{1}(\theta)\cos(2\varphi),\\
&\overline{P}^{zz}(\theta,\varphi) = P^{3}(\theta)
\end{aligned}\tag{6.81}$$

From Eq. (3.33) and Eq. (6.58) we can find

$$\begin{aligned}
&\mathrm{tr}\left(\hat{S}_x\hat{F}\hat{S}_x\hat{F}^+\right)=\mathrm{Re}\left[(AB^*+DC^*+GO^*+HA^*)-(OC^*+AE^*+FA^*+GD^*)\,\mathrm{e}^{2\mathrm{i}\varphi}\right],\\
&\mathrm{tr}\left(\hat{S}_y\hat{F}\hat{S}_x\hat{F}^+\right)=\mathrm{Im}\left[(AB^*+DC^*+GO^*+HA^*)-(OC^*+AE^*+FA^*+GD^*)\,\mathrm{e}^{2\mathrm{i}\varphi}\right],\\
&\mathrm{tr}\left(\hat{S}_z\hat{F}\hat{S}_x\hat{F}^+\right)=-\sqrt{2}\mathrm{Im}\left[(BC^*+CE^*-FG^*-GH^*)\,\mathrm{e}^{\mathrm{i}\varphi}\right],\\
&\mathrm{tr}\left(\hat{S}_x\hat{F}\hat{S}_y\hat{F}^+\right)=-\mathrm{Im}\left[(AB^*+DC^*+GO^*+HA^*)+(OC^*+AE^*+FA^*+GD^*)\,\mathrm{e}^{2\mathrm{i}\varphi}\right],\\
&\mathrm{tr}\left(\hat{S}_y\hat{F}\hat{S}_y\hat{F}^+\right)=\mathrm{Re}\left[(AB^*+DC^*+GO^*+HA^*)+(OC^*+AE^*+FA^*+GD^*)\,\mathrm{e}^{2\mathrm{i}\varphi}\right],\\
&\mathrm{tr}\left(\hat{S}_z\hat{F}\hat{S}_y\hat{F}^+\right)=\sqrt{2}\mathrm{Re}\left[(BC^*+CE^*-FG^*-GH^*)\,\mathrm{e}^{\mathrm{i}\varphi}\right],\\
&\mathrm{tr}\left(\hat{S}_x\hat{F}\hat{S}_z\hat{F}^+\right)=-\sqrt{2}\mathrm{Im}\left[(OB^*-DE^*+FO^*-HD^*)\,\mathrm{e}^{\mathrm{i}\varphi}\right],\\
&\mathrm{tr}\left(\hat{S}_y\hat{F}\hat{S}_z\hat{F}^+\right)=\sqrt{2}\mathrm{Re}\left[(OB^*-DE^*+FO^*-HD^*)\,\mathrm{e}^{\mathrm{i}\varphi}\right],\\
&\mathrm{tr}\left(\hat{S}_z\hat{F}\hat{S}_z\hat{F}^+\right)=|B|^2-|E|^2-|F|^2+|H|^2
\end{aligned} \tag{6.82}$$

Using the similar method mentioned above, from Eq. (6.82) we can obtain

$$\begin{aligned}
&\widetilde{\Sigma}\mathrm{tr}\left(\hat{S}_x\hat{F}\hat{S}_x\hat{F}^+\right)=2\widetilde{\Sigma}\,\mathrm{Re}\left[(AB^*+DC^*)-(OC^*+AE^*)\cos(2\varphi)\right],\\
&\widetilde{\Sigma}\mathrm{tr}\left(\hat{S}_y\hat{F}\hat{S}_x\hat{F}^+\right)=-2\widetilde{\Sigma}\,\mathrm{Re}\,(OC^*+AE^*)\sin(2\varphi),\\
&\widetilde{\Sigma}\mathrm{tr}\left(\hat{S}_z\hat{F}\hat{S}_x\hat{F}^+\right)=-2\sqrt{2}\,\widetilde{\Sigma}\mathrm{Im}(BC^*+CE^*)\cos\varphi,\\
&\widetilde{\Sigma}\mathrm{tr}\left(\hat{S}_x\hat{F}\hat{S}_y\hat{F}^+\right)=-2\widetilde{\Sigma}\,\mathrm{Re}\,(OC^*+AE^*)\sin(2\varphi),\\
&\widetilde{\Sigma}\mathrm{tr}\left(\hat{S}_y\hat{F}\hat{S}_y\hat{F}^+\right)=2\widetilde{\Sigma}\,\mathrm{Re}\left[(AB^*+DC^*)+(OC^*+AE^*)\cos(2\varphi)\right]\\
&\widetilde{\Sigma}\mathrm{tr}\left(\hat{S}_z\hat{F}\hat{S}_y\hat{F}^+\right)=-2\sqrt{2}\,\widetilde{\Sigma}\mathrm{Im}(BC^*+CE^*)\sin\varphi,\\
&\widetilde{\Sigma}\mathrm{tr}\left(\hat{S}_x\hat{F}\hat{S}_z\hat{F}^+\right)=-2\sqrt{2}\,\widetilde{\Sigma}\mathrm{Im}(OB^*-DE^*)\cos\varphi,\\
&\widetilde{\Sigma}\mathrm{tr}\left(\hat{S}_y\hat{F}\hat{S}_z\hat{F}^+\right)=-2\sqrt{2}\,\widetilde{\Sigma}\mathrm{Im}(OB^*-DE^*)\sin\varphi,\\
&\widetilde{\Sigma}\mathrm{tr}\left(\hat{S}_z\hat{F}\hat{S}_z\hat{F}^+\right)=2\left(|B|^2-|E|^2\right)
\end{aligned} \tag{6.83}$$

Define

$$R_1(\theta) = \frac{2}{3I_0}\tilde{\Sigma}\,\mathrm{Re}\,(AB^* + DC^*), \quad R_2(\theta) = \frac{2}{3I_0}\tilde{\Sigma}\,\mathrm{Re}\,(OC^* + AE^*),$$
$$Q_1(\theta) = \frac{2\sqrt{2}}{3I_0}\tilde{\Sigma}\mathrm{Im}(BC^* + CE^*), \quad Q_2(\theta) = \frac{2\sqrt{2}}{3I_0}\tilde{\Sigma}\mathrm{Im}(OB^* - DE^*) \tag{6.84}$$
$$W_1(\theta) = \frac{2}{3I_0}\tilde{\Sigma}\left(|B|^2 - |E|^2\right)$$

and the following polarization transfer coefficients related to angles $\theta$, $\varphi$

$$K_i^j(\theta, \varphi) = \frac{1}{3I_0}\tilde{\Sigma}\mathrm{tr}\left(\hat{S}_j\hat{F}\hat{S}_i\hat{F}^+\right), \quad i, j = x, y, z \tag{6.85}$$

Then the following results can be obtained from Eqs. (6.83)–(6.85):

$$\begin{aligned}
&K_x^x(\theta, \varphi) = R_1(\theta) - R_2(\theta)\cos(2\varphi), && K_x^y(\theta, \varphi) = -R_2(\theta)\sin(2\varphi),\\
&K_x^z(\theta, \varphi) = -Q_1(\theta)\cos\varphi, && K_y^x(\theta, \varphi) = -R_2(\theta)\sin(2\varphi),\\
&K_y^y(\theta, \varphi) = R_1(\theta) + R_2(\theta)\cos(2\varphi), && K_y^z(\theta, \varphi) = -Q_1(\theta)\sin\varphi,\\
&K_z^x(\theta, \varphi) = -Q_2(\theta)\cos\varphi, && K_z^y(\theta, \varphi) = -Q_2(\theta)\sin\varphi\\
&K_z^z(\theta, \varphi) = W_1(\theta)
\end{aligned} \tag{6.86}$$

From Eq. (3.36) and Eq. (6.58), we can also obtain

$$\mathrm{tr}\left(\hat{Q}_{xy}\hat{F}\hat{S}_x\hat{F}^+\right) = \frac{3}{\sqrt{2}}\mathrm{Re}\left[(GB^* + HC^*)\,\mathrm{e}^{\mathrm{i}\varphi} - (FC^* + GE^*)\,\mathrm{e}^{3\mathrm{i}\varphi}\right],$$
$$\mathrm{tr}\left(\hat{Q}_{xz}\hat{F}\hat{S}_x\hat{F}^+\right) = \frac{3}{2}\mathrm{Re}\left[(AB^* + DC^* - GO^* - HA^*) - (OC^* + AE^* - FA^* - GD^*)\,\mathrm{e}^{2\mathrm{i}\varphi}\right],$$
$$\mathrm{tr}\left(\hat{Q}_{yz}\hat{F}\hat{S}_x\hat{F}^+\right) = \frac{3}{2}\mathrm{Im}\left[(AB^* + DC^* - GO^* - HA^*) - (OC^* + AE^* - FA^* - GD^*)\,\mathrm{e}^{2\mathrm{i}\varphi}\right],$$
$$\mathrm{tr}\left(\hat{Q}_{xx-yy}\hat{F}\hat{S}_x\hat{F}^+\right) = -3\sqrt{2}\mathrm{Im}\left[(GB^* + HC^*)\,\mathrm{e}^{\mathrm{i}\varphi} - (FC^* + GE^*)\,\mathrm{e}^{3\mathrm{i}\varphi}\right],$$
$$\mathrm{tr}\left(\hat{Q}_{zz}\hat{F}\hat{S}_x\hat{F}^+\right) = -\sqrt{2}\mathrm{Im}\left\{\left[(BC^* + CE^* + FG^* + GH^*) - 2(OA^* + AD^*)\right]\mathrm{e}^{\mathrm{i}\varphi}\right\},$$
$$\mathrm{tr}\left(\hat{Q}_{xy}\hat{F}\hat{S}_y\hat{F}^+\right) = -\frac{3}{\sqrt{2}}\mathrm{Im}\left[(GB^* + HC^*)\,\mathrm{e}^{\mathrm{i}\varphi} + (FC^* + GE^*)\,\mathrm{e}^{3\mathrm{i}\varphi}\right],$$
$$\mathrm{tr}\left(\hat{Q}_{xz}\hat{F}\hat{S}_y\hat{F}^+\right) = -\frac{3}{2}\mathrm{Im}\left[(AB^* + DC^* - GO^* - HA^*) + (OC^* + AE^* - FA^* - GD^*)\,\mathrm{e}^{2\mathrm{i}\varphi}\right],$$
$$\mathrm{tr}\left(\hat{Q}_{yz}\hat{F}\hat{S}_y\hat{F}^+\right) = \frac{3}{2}\mathrm{Re}\left[(AB^* + DC^* - GO^* - HA^*) + (OC^* + AE^* - FA^* - GD^*)\,\mathrm{e}^{2\mathrm{i}\varphi}\right]$$
$$\mathrm{tr}\left(\hat{Q}_{xx-yy}\hat{F}\hat{S}_y\hat{F}^+\right) = -3\sqrt{2}\mathrm{Re}\left[(GB^* + HC^*)\,\mathrm{e}^{\mathrm{i}\varphi} + (FC^* + GE^*)\,\mathrm{e}^{3\mathrm{i}\varphi}\right],$$

$$
\begin{aligned}
\mathrm{tr}\left(\hat{Q}_{zz}\hat{F}\hat{S}_y\hat{F}^+\right) &= \sqrt{2}\mathrm{Re}\left\{ [(BC^* + CE^* + FG^* + GH^*) - 2(OA^* + AD^*)]\,\mathrm{e}^{\mathrm{i}\varphi}\right\},\\
\mathrm{tr}\left(\hat{Q}_{xy}\hat{F}\hat{S}_z\hat{F}^+\right) &= -3\mathrm{Im}\left[(FB^* - HE^*)\,\mathrm{e}^{2\mathrm{i}\varphi}\right],\\
\mathrm{tr}\left(\hat{Q}_{xz}\hat{F}\hat{S}_z\hat{F}^+\right) &= -\frac{3}{\sqrt{2}}\mathrm{Im}\left[(OB^* - DE^* - FO^* + HD^*)\,\mathrm{e}^{\mathrm{i}\varphi}\right],\\
\mathrm{tr}\left(\hat{Q}_{yz}\hat{F}\hat{S}_z\hat{F}^+\right) &= \frac{3}{\sqrt{2}}\mathrm{Re}\left[(OB^* - DE^* - FO^* + HD^*)\,\mathrm{e}^{\mathrm{i}\varphi}\right],\\
\mathrm{tr}\left(\hat{Q}_{xx-yy}\hat{F}\hat{S}_z\hat{F}^+\right) &= -6\mathrm{Re}\left[(FB^* - HE^*)\,\mathrm{e}^{2\mathrm{i}\varphi}\right],\\
\mathrm{tr}\left(\hat{Q}_{zz}\hat{F}\hat{S}_z\hat{F}^+\right) &= |B|^2 - |E|^2 + |F|^2 - |H|^2 - 2\left(|O|^2 - |D|^2\right)
\end{aligned}
\tag{6.87}
$$

By using the similar method mentioned above, from Eq. (6.87) we can find

$$
\begin{aligned}
\widetilde{\Sigma}\mathrm{tr}\left(\hat{Q}_{xy}\hat{F}\hat{S}_x\hat{F}^+\right) &= 3\sqrt{2}\,\widetilde{\Sigma}[\mathrm{Re}\,(GB^*)\cos\varphi - \mathrm{Re}\,(FC^*)\cos(3\varphi)],\\
\widetilde{\Sigma}\mathrm{tr}\left(\hat{Q}_{xz}\hat{F}\hat{S}_x\hat{F}^+\right) &= 3\widetilde{\Sigma}\mathrm{Im}(OC^* + AE^*)\sin(2\varphi),\\
\widetilde{\Sigma}\mathrm{tr}\left(\hat{Q}_{yz}\hat{F}\hat{S}_x\hat{F}^+\right) &= 3\widetilde{\Sigma}[\mathrm{Im}(AB^* + DC^*) - \mathrm{Im}(OC^* + AE^*)\cos(2\varphi)],\\
\widetilde{\Sigma}\mathrm{tr}\left(\hat{Q}_{xx-yy}\hat{F}\hat{S}_x\hat{F}^+\right) &= -6\sqrt{2}\,\widetilde{\Sigma}[\mathrm{Re}\,(GB^*)\sin\varphi - \mathrm{Re}\,(FC^*)\sin(3\varphi)],\\
\widetilde{\Sigma}\mathrm{tr}\left(\hat{Q}_{zz}\hat{F}\hat{S}_x\hat{F}^+\right) &= -2\sqrt{2}\,\widetilde{\Sigma}\mathrm{Re}\,(BC^* + CE^* - 2OA^*)\sin\varphi,\\
\widetilde{\Sigma}\mathrm{tr}\left(\hat{Q}_{xy}\hat{F}\hat{S}_y\hat{F}^+\right) &= -3\sqrt{2}\,\widetilde{\Sigma}[\mathrm{Re}\,(GB^*)\sin\varphi + \mathrm{Re}\,(FC^*)\sin(3\varphi)],\\
\widetilde{\Sigma}\mathrm{tr}\left(\hat{Q}_{xz}\hat{F}\hat{S}_y\hat{F}^+\right) &= -3\widetilde{\Sigma}[\mathrm{Im}(AB^* + DC^*) + \mathrm{Im}(OC^* + AE^*)\cos(2\varphi)],\\
\widetilde{\Sigma}\mathrm{tr}\left(\hat{Q}_{yz}\hat{F}\hat{S}_y\hat{F}^+\right) &= -3\widetilde{\Sigma}\mathrm{Im}(OC^* + AE^*)\sin(2\varphi),\\
\widetilde{\Sigma}\mathrm{tr}\left(\hat{Q}_{xx-yy}\hat{F}\hat{S}_y\hat{F}^+\right) &= -6\sqrt{2}\,\widetilde{\Sigma}[\mathrm{Re}\,(GB^*)\cos\varphi + \mathrm{Re}\,(FC^*)\cos(3\varphi)],\\
\widetilde{\Sigma}\mathrm{tr}\left(\hat{Q}_{zz}\hat{F}\hat{S}_y\hat{F}^+\right) &= 2\sqrt{2}\,\widetilde{\Sigma}[\mathrm{Re}\,(BC^* + CE^* - 2OA^*)]\cos\varphi,\\
\widetilde{\Sigma}\mathrm{tr}\left(\hat{Q}_{xy}\hat{F}\hat{S}_z\hat{F}^+\right) &= -6\widetilde{\Sigma}\mathrm{Im}(FB^*)\cos(2\varphi),\\
\widetilde{\Sigma}\mathrm{tr}\left(\hat{Q}_{xz}\hat{F}\hat{S}_z\hat{F}^+\right) &= -3\sqrt{2}\,\widetilde{\Sigma}\mathrm{Re}\,(OB^* - DE^*)\sin\varphi,\\
\widetilde{\Sigma}\mathrm{tr}\left(\hat{Q}_{yz}\hat{F}\hat{S}_z\hat{F}^+\right) &= 3\sqrt{2}\,\widetilde{\Sigma}\mathrm{Re}\,(OB^* - DE^*)\cos\varphi,\\
\widetilde{\Sigma}\mathrm{tr}\left(\hat{Q}_{xx-yy}\hat{F}\hat{S}_z\hat{F}^+\right) &= 12\widetilde{\Sigma}\mathrm{Im}(FB^*)\sin(2\varphi),\\
\widetilde{\Sigma}\mathrm{tr}\left(\hat{Q}_{zz}\hat{F}\hat{S}_z\hat{F}^+\right) &= 0
\end{aligned}
\tag{6.88}
$$

Define

$$
\begin{aligned}
&R_3(\theta) = \frac{\sqrt{2}}{I_0}\widetilde{\Sigma}\,\mathrm{Re}\,(GB^*), && R_4(\theta) = \frac{\sqrt{2}}{I_0}\widetilde{\Sigma}\,\mathrm{Re}\,(FC^*),\\
&R_5(\theta) = \frac{2\sqrt{2}}{3I_0}\widetilde{\Sigma}\,\mathrm{Re}\,(BC^* + CE^* - 2OA^*), && R_6(\theta) = \frac{\sqrt{2}}{I_0}\widetilde{\Sigma}\,\mathrm{Re}\,(OB^* - DE^*),\\
&Q_3(\theta) = \frac{1}{I_0}\widetilde{\Sigma}\mathrm{Im}(OC^* + AE^*), && Q_4(\theta) = \frac{1}{I_0}\widetilde{\Sigma}\mathrm{Im}(AB^* + DC^*),\\
&Q_5(\theta) = \frac{2}{I_0}\widetilde{\Sigma}\mathrm{Im}(FB^*)
\end{aligned}
\tag{6.89}
$$

and

$$
K_i^j(\theta,\varphi) = \frac{1}{3I_0}\widetilde{\Sigma}\mathrm{tr}\left(\hat{Q}_j\hat{F}\hat{S}_i\hat{F}^+\right), \qquad i = x, y, z; \quad j = \gamma \tag{6.90}
$$

Then the following results can be obtained from Eqs. (6.88)–(6.90):

$$
\begin{aligned}
K_x^{xy}(\theta,\varphi) &= R_3(\theta)\cos\varphi - R_4(\theta)\cos(3\varphi),\\
K_x^{xz}(\theta,\varphi) &= Q_3(\theta)\sin(2\varphi),\\
K_x^{yz}(\theta,\varphi) &= Q_4(\theta) - Q_3(\theta)\cos(2\varphi),\\
K_x^{xx-yy}(\theta,\varphi) &= -2R_3(\theta)\sin\varphi + 2R_4(\theta)\sin(3\varphi),\\
K_x^{zz}(\theta,\varphi) &= -R_5(\theta)\sin\varphi,\\
K_y^{xy}(\theta,\varphi) &= -R_3(\theta)\sin\varphi - R_4(\theta)\sin(3\varphi),\\
K_y^{xz}(\theta,\varphi) &= -Q_4(\theta) - Q_3(\theta)\cos(2\varphi),\\
K_y^{yz}(\theta,\varphi) &= -Q_3(\theta)\sin(2\varphi),\\
K_y^{xx-yy}(\theta,\varphi) &= -2R_3(\theta)\cos\varphi - 2R_4(\theta)\cos(3\varphi),\\
K_y^{zz}(\theta,\varphi) &= R_5(\theta)\cos\varphi,\\
K_z^{xy}(\theta,\varphi) &= -Q_5(\theta)\cos(2\varphi),\\
K_z^{xz}(\theta,\varphi) &= -R_6(\theta)\sin\varphi,\\
K_z^{yz}(\theta,\varphi) &= R_6(\theta)\cos\varphi,\\
K_z^{xx-yy}(\theta,\varphi) &= 2Q_5(\theta)\sin(2\varphi),\\
K_z^{zz}(\theta,\varphi) &= 0
\end{aligned}
\tag{6.91}
$$

From Eq. (3.33) and Eq. (6.65) we can get

$$
\begin{aligned}
&\mathrm{tr}\left(\hat{S}_x\hat{F}\hat{Q}_{xy}\hat{F}^+\right)=\frac{3}{\sqrt{2}}\mathrm{Re}\left[(BD^*+OH^*)\,\mathrm{e}^{\mathrm{i}\varphi}-(OE^*+FD^*)\,\mathrm{e}^{3\mathrm{i}\varphi}\right],\\
&\mathrm{tr}\left(\hat{S}_y\hat{F}\hat{Q}_{xy}\hat{F}^+\right)=-\frac{3}{\sqrt{2}}\mathrm{Im}\left[(BD^*+OH^*)\,\mathrm{e}^{\mathrm{i}\varphi}+(OE^*+FD^*)\,\mathrm{e}^{3\mathrm{i}\varphi}\right],\\
&\mathrm{tr}\left(\hat{S}_z\hat{F}\hat{Q}_{xy}\hat{F}^+\right)=-3\mathrm{Im}\left[(BE^*-FH^*)\,\mathrm{e}^{2\mathrm{i}\varphi}\right],\\
&\mathrm{tr}\left(\hat{S}_x\hat{F}\hat{Q}_{xz}\hat{F}^+\right)=\frac{3}{2}\mathrm{Re}\left[(AB^*-DC^*+GO^*-HA^*)-(OC^*-AE^*+FA^*-GD^*)\,\mathrm{e}^{2\mathrm{i}\varphi}\right],\\
&\mathrm{tr}\left(\hat{S}_y\hat{F}\hat{Q}_{xz}\hat{F}^+\right)=\frac{3}{2}\mathrm{Im}\left[(AB^*-DC^*+GO^*-HA^*)-(OC^*-AE^*+FA^*-GD^*)\,\mathrm{e}^{2\mathrm{i}\varphi}\right],\\
&\mathrm{tr}\left(\hat{S}_z\hat{F}\hat{Q}_{xz}\hat{F}^+\right)=-\frac{3}{\sqrt{2}}\mathrm{Im}\left[(BC^*-CE^*-FG^*+GH^*)\,\mathrm{e}^{\mathrm{i}\varphi}\right],\\
&\mathrm{tr}\left(\hat{S}_x\hat{F}\hat{Q}_{yz}\hat{F}^+\right)=-\frac{3}{2}\mathrm{Im}\left[(AB^*-DC^*+GO^*-HA^*)+(OC^*-AE^*+FA^*-GD^*)\,\mathrm{e}^{2\mathrm{i}\varphi}\right],\\
&\mathrm{tr}\left(\hat{S}_y\hat{F}\hat{Q}_{yz}\hat{F}^+\right)=\frac{3}{2}\mathrm{Re}\left[(AB^*-DC^*+GO^*-HA^*)+(OC^*-AE^*+FA^*-GD^*)\,\mathrm{e}^{2\mathrm{i}\varphi}\right],\\
&\mathrm{tr}\left(\hat{S}_z\hat{F}\hat{Q}_{yz}\hat{F}^+\right)=\frac{3}{\sqrt{2}}\mathrm{Re}\left[(BC^*-CE^*-FG^*+GH^*)\,\mathrm{e}^{\mathrm{i}\varphi}\right],\\
&\mathrm{tr}\left(\hat{S}_x\hat{F}\hat{Q}_{xx-yy}\hat{F}^+\right)=-3\sqrt{2}\mathrm{Im}\left[(BD^*+OH^*)\,\mathrm{e}^{\mathrm{i}\varphi}-(OE^*+FD^*)\,\mathrm{e}^{3\mathrm{i}\varphi}\right],\\
&\mathrm{tr}\left(\hat{S}_y\hat{F}\hat{Q}_{xx-yy}\hat{F}^+\right)=-3\sqrt{2}\mathrm{Re}\left[(BD^*+OH^*)\,\mathrm{e}^{\mathrm{i}\varphi}+(OE^*+FD^*)\,\mathrm{e}^{3\mathrm{i}\varphi}\right],\\
&\mathrm{tr}\left(\hat{S}_z\hat{F}\hat{Q}_{xx-yy}\hat{F}^+\right)=-6\mathrm{Re}\left[(BE^*-FH^*)\,\mathrm{e}^{2\mathrm{i}\varphi}\right],\\
&\mathrm{tr}\left(\hat{S}_x\hat{F}\hat{Q}_{zz}\hat{F}^+\right)=-\sqrt{2}\mathrm{Im}\left[(OB^*-2AC^*+DE^*+FO^*-2GA^*+HD^*)\,\mathrm{e}^{\mathrm{i}\varphi}\right],\\
&\mathrm{tr}\left(\hat{S}_y\hat{F}\hat{Q}_{zz}\hat{F}^+\right)=\sqrt{2}\mathrm{Re}\left[(OB^*-2AC^*+DE^*+FO^*-2GA^*+HD^*)\,\mathrm{e}^{\mathrm{i}\varphi}\right],\\
&\mathrm{tr}\left(\hat{S}_z\hat{F}\hat{Q}_{zz}\hat{F}^+\right)=|B|^2-2|C|^2+|E|^2-|F|^2+2|G|^2-|H|^2
\end{aligned}
\tag{6.92}
$$

By using the similar method mentioned above, from Eq. (6.92) we can find

$$
\begin{aligned}
&\widetilde{\Sigma}\mathrm{tr}\left(\hat{S}_x\hat{F}\hat{Q}_{xy}\hat{F}^+\right)=3\sqrt{2}\widetilde{\Sigma}[\,\mathrm{Re}\,(BD^*)\cos\varphi-\,\mathrm{Re}\,(OE^*)\cos(3\varphi)],\\
&\widetilde{\Sigma}\mathrm{tr}\left(\hat{S}_y\hat{F}\hat{Q}_{xy}\hat{F}^+\right)=-3\sqrt{2}\,\widetilde{\Sigma}[\,\mathrm{Re}\,(BD^*)\sin\varphi+\,\mathrm{Re}\,(OE^*)\sin(3\varphi)],\\
&\widetilde{\Sigma}\mathrm{tr}\left(\hat{S}_z\hat{F}\hat{Q}_{xy}\hat{F}^+\right)=-6\widetilde{\Sigma}\mathrm{Im}(BE^*)\cos(2\varphi),\\
&\widetilde{\Sigma}\mathrm{tr}\left(\hat{S}_x\hat{F}\hat{Q}_{xz}\hat{F}^+\right)=3\widetilde{\Sigma}\mathrm{Im}(OC^*-AE^*)\sin(2\varphi),\\
&\widetilde{\Sigma}\mathrm{tr}\left(\hat{S}_y\hat{F}\hat{Q}_{xz}\hat{F}^+\right)=3\widetilde{\Sigma}[\mathrm{Im}(AB^*-DC^*)-\mathrm{Im}(OC^*-AE^*)\cos(2\varphi)],\\
&\widetilde{\Sigma}\mathrm{tr}\left(\hat{S}_z\hat{F}\hat{Q}_{xz}\hat{F}^+\right)=-3\sqrt{2}\,\widetilde{\Sigma}\,\mathrm{Re}\,(BC^*-CE^*)\sin\varphi,\\
&\widetilde{\Sigma}\mathrm{tr}\left(\hat{S}_x\hat{F}\hat{Q}_{yz}\hat{F}^+\right)=-3\widetilde{\Sigma}[\mathrm{Im}(AB^*-DC^*)+\mathrm{Im}(OC^*-AE^*)\cos(2\varphi)],\\
&\widetilde{\Sigma}\mathrm{tr}\left(\hat{S}_y\hat{F}\hat{Q}_{yz}\hat{F}^+\right)=-3\widetilde{\Sigma}\mathrm{Im}(OC^*-AE^*)\sin(2\varphi),\\
&\widetilde{\Sigma}\mathrm{tr}\left(\hat{S}_z\hat{F}\hat{Q}_{yz}\hat{F}^+\right)=3\sqrt{2}\,\widetilde{\Sigma}\,\mathrm{Re}\,(BC^*-CE^*)\cos\varphi,
\end{aligned}
$$

$$
\begin{aligned}
&\widetilde{\Sigma}\mathrm{tr}\left(\hat{S}_x\hat{F}\hat{Q}_{xx-yy}\hat{F}^+\right) = -6\sqrt{2}\,\widetilde{\Sigma}[\,\mathrm{Re}\,(BD^*)\sin\varphi - \mathrm{Re}\,(OE^*)\sin(3\varphi)],\\
&\widetilde{\Sigma}\mathrm{tr}\left(\hat{S}_y\hat{F}\hat{Q}_{xx-yy}\hat{F}^+\right) = -6\sqrt{2}\,\widetilde{\Sigma}[\,\mathrm{Re}\,(BD^*)\cos\varphi + \mathrm{Re}\,(OE^*)\cos(3\varphi)],\\
&\widetilde{\Sigma}\mathrm{tr}\left(\hat{S}_z\hat{F}\hat{Q}_{xx-yy}\hat{F}^+\right) = 12\widetilde{\Sigma}\mathrm{Im}(BE^*)\sin(2\varphi),\\
&\widetilde{\Sigma}\mathrm{tr}\left(\hat{S}_x\hat{F}\hat{Q}_{zz}\hat{F}^+\right) = -2\sqrt{2}\,\widetilde{\Sigma}\,\mathrm{Re}\,(OB^* - 2AC^* + DE^*)\sin\varphi,\\
&\widetilde{\Sigma}\mathrm{tr}\left(\hat{S}_y\hat{F}\hat{Q}_{zz}\hat{F}^+\right) = 2\sqrt{2}\,\widetilde{\Sigma}\,\mathrm{Re}\,(OB^* - 2AC^* + DE^*)\cos\varphi,\\
&\widetilde{\Sigma}\mathrm{tr}\left(\hat{S}_z\hat{F}\hat{Q}_{zz}\hat{F}^+\right) = 0
\end{aligned}
\tag{6.93}
$$

Define

$$
\begin{aligned}
&R_7(\theta) = \frac{\sqrt{2}}{I_0}\widetilde{\Sigma}\,\mathrm{Re}\,(BD^*), && R_8(\theta) = \frac{\sqrt{2}}{I_0}\widetilde{\Sigma}\,\mathrm{Re}\,(OE^*),\\
&R_9(\theta) = \frac{\sqrt{2}}{I_0}\widetilde{\Sigma}\,\mathrm{Re}\,(BC^* - CE^*), && R_{10}(\theta) = \frac{2\sqrt{2}}{3I_0}\widetilde{\Sigma}\,\mathrm{Re}\,(OB^* - 2AC^* + DE^*),\\
&Q_6(\theta) = \frac{2}{I_0}\widetilde{\Sigma}\mathrm{Im}(BE^*), && Q_7(\theta) = \frac{1}{I_0}\widetilde{\Sigma}\mathrm{Im}(OC^* - AE^*),\\
&Q_8(\theta) = \frac{1}{I_0}\widetilde{\Sigma}\mathrm{Im}(AB^* - DC^*)
\end{aligned}
\tag{6.94}
$$

and

$$
K_i^j(\theta,\varphi) = \frac{1}{3I_0}\widetilde{\Sigma}\mathrm{tr}\left(\hat{S}_j\hat{F}\hat{Q}_i\hat{F}^+\right), \quad i=\gamma;\quad j=x,y,z
\tag{6.95}
$$

Then the following results can be obtained by using Eqs. (6.93)–(6.95):

$$
\begin{aligned}
&K_{xy}^x(\theta,\varphi) = R_7(\theta)\cos\varphi - R_8(\theta)\cos(3\varphi),\\
&K_{xy}^y(\theta,\varphi) = -R_7(\theta)\sin\varphi - R_8(\theta)\sin(3\varphi),\\
&K_{xy}^z(\theta,\varphi) = -Q_6(\theta)\cos(2\varphi),\\
&K_{xz}^x(\theta,\varphi) = Q_7(\theta)\sin(2\varphi),\\
&K_{xz}^y(\theta,\varphi) = Q_8(\theta) - Q_7(\theta)\cos(2\varphi),\\
&K_{xz}^z(\theta,\varphi) = -R_9(\theta)\sin\varphi,\\
&K_{yz}^x(\theta,\varphi) = -Q_8(\theta) - Q_7(\theta)\cos(2\varphi),\\
&K_{yz}^y(\theta,\varphi) = -Q_7(\theta)\sin(2\varphi),\\
&K_{yz}^z(\theta,\varphi) = R_9(\theta)\cos\varphi,
\end{aligned}
$$

$$
\begin{aligned}
K^x_{xx-yy}(\theta,\varphi) &= -2R_7(\theta)\sin\varphi + 2R_8(\theta)\sin(3\varphi),\\
K^y_{xx-yy}(\theta,\varphi) &= -2R_7(\theta)\cos\varphi - 2R_8(\theta)\cos(3\varphi),\\
K^z_{xx-yy}(\theta,\varphi) &= 2Q_6(\theta)\sin(2\varphi),\\
K^x_{zz}(\theta,\varphi) &= -R_{10}(\theta)\sin\varphi,\\
K^y_{zz}(\theta,\varphi) &= R_{10}(\theta)\cos\varphi,\\
K^z_{zz}(\theta,\varphi) &= 0
\end{aligned}
\tag{6.96}
$$

From Eq. (3.36) and Eq. (6.65), we can also get

$$
\begin{aligned}
&\mathrm{tr}\left(\hat{Q}_{xy}\hat{F}\hat{Q}_{xy}\hat{F}^+\right)=\frac{9}{2}\mathrm{Re}\left(HB^*-FE^*\mathrm{e}^{4\mathrm{i}\varphi}\right),\\
&\mathrm{tr}\left(\hat{Q}_{xz}\hat{F}\hat{Q}_{xy}\hat{F}^+\right)=\frac{9}{2\sqrt{2}}\mathrm{Re}\left[(BD^*-OH^*)\,\mathrm{e}^{\mathrm{i}\varphi}-(OE^*-FD^*)\,\mathrm{e}^{3\mathrm{i}\varphi}\right],\\
&\mathrm{tr}\left(\hat{Q}_{yz}\hat{F}\hat{Q}_{xy}\hat{F}^+\right)=-\frac{9}{2\sqrt{2}}\mathrm{Im}\left[(BD^*-OH^*)\,\mathrm{e}^{\mathrm{i}\varphi}+(OE^*-FD^*)\,\mathrm{e}^{3\mathrm{i}\varphi}\right],\\
&\mathrm{tr}\left(\hat{Q}_{xx-yy}\hat{F}\hat{Q}_{xy}\hat{F}^+\right)=-9\mathrm{Im}\left(HB^*-FE^*\mathrm{e}^{4\mathrm{i}\varphi}\right),\\
&\mathrm{tr}\left(\hat{Q}_{zz}\hat{F}\hat{Q}_{xy}\hat{F}^+\right)=-3\mathrm{Im}\left[(BE^*+FH^*-2OD^*)\,\mathrm{e}^{2\mathrm{i}\varphi}\right],\\
&\mathrm{tr}\left(\hat{Q}_{xy}\hat{F}\hat{Q}_{xz}\hat{F}^+\right)=\frac{9}{2\sqrt{2}}\mathrm{Re}\left[(GB^*-HC^*)\,\mathrm{e}^{\mathrm{i}\varphi}-(FC^*-GE^*)\,\mathrm{e}^{3\mathrm{i}\varphi}\right],\\
&\mathrm{tr}\left(\hat{Q}_{xz}\hat{F}\hat{Q}_{xz}\hat{F}^+\right)=\frac{9}{4}\mathrm{Re}\left[(AB^*-DC^*-GO^*+HA^*)-(OC^*-AE^*-FA^*+GD^*)\,\mathrm{e}^{2\mathrm{i}\varphi}\right],\\
&\mathrm{tr}\left(\hat{Q}_{yz}\hat{F}\hat{Q}_{xz}\hat{F}^+\right)=\frac{9}{4}\mathrm{Im}\left[(AB^*-DC^*-GO^*+HA^*)-(OC^*-AE^*-FA^*+GD^*)\,\mathrm{e}^{2\mathrm{i}\varphi}\right],\\
&\mathrm{tr}\left(\hat{Q}_{xx-yy}\hat{F}\hat{Q}_{xz}\hat{F}^+\right)=-\frac{9}{\sqrt{2}}\mathrm{Im}\left[(GB^*-HC^*)\,\mathrm{e}^{\mathrm{i}\varphi}-(FC^*-GE^*)\,\mathrm{e}^{3\mathrm{i}\varphi}\right],\\
&\mathrm{tr}\left(\hat{Q}_{zz}\hat{F}\hat{Q}_{xz}\hat{F}^+\right)=-\frac{3}{\sqrt{2}}\mathrm{Im}\left\{\left[BC^*-CE^*+FG^*-GH^*-2(OA^*-AD^*)\right]\mathrm{e}^{\mathrm{i}\varphi}\right\},\\
&\mathrm{tr}\left(\hat{Q}_{xy}\hat{F}\hat{Q}_{yz}\hat{F}^+\right)=-\frac{9}{2\sqrt{2}}\mathrm{Im}\left[(GB^*-HC^*)\,\mathrm{e}^{\mathrm{i}\varphi}+(FC^*-GE^*)\,\mathrm{e}^{3\mathrm{i}\varphi}\right],\\
&\mathrm{tr}\left(\hat{Q}_{xz}\hat{F}\hat{Q}_{yz}\hat{F}^+\right)=-\frac{9}{4}\mathrm{Im}\left[(AB^*-DC^*-GO^*+HA^*)+(OC^*-AE^*-FA^*+GD^*)\,\mathrm{e}^{2\mathrm{i}\varphi}\right],\\
&\mathrm{tr}\left(\hat{Q}_{yz}\hat{F}\hat{Q}_{yz}\hat{F}^+\right)=\frac{9}{4}\mathrm{Re}\left[(AB^*-DC^*-GO^*+HA^*)+(OC^*-AE^*-FA^*+GD^*)\,\mathrm{e}^{2\mathrm{i}\varphi}\right],\\
&\mathrm{tr}\left(\hat{Q}_{xx-yy}\hat{F}\hat{Q}_{yz}\hat{F}^+\right)=-\frac{9}{\sqrt{2}}\mathrm{Re}\left[(GB^*-HC^*)\,e^{\mathrm{i}\varphi}+(FC^*-GE^*)\,\mathrm{e}^{3\mathrm{i}\varphi}\right],\\
&\mathrm{tr}\left(\hat{Q}_{zz}\hat{F}\hat{Q}_{yz}\hat{F}^+\right)=\frac{3}{\sqrt{2}}\mathrm{Re}\left\{\left[BC^*-CE^*+FG^*-GH^*-2(OA^*-AD^*)\right]\mathrm{e}^{\mathrm{i}\varphi}\right\},\\
&\mathrm{tr}\left(\hat{Q}_{xy}\hat{F}\hat{Q}_{xx-yy}\hat{F}^+\right)=9\mathrm{Im}\left(HB^*+FE^*\mathrm{e}^{4\mathrm{i}\varphi}\right),\\
&\mathrm{tr}\left(\hat{Q}_{xz}\hat{F}\hat{Q}_{xx-yy}\hat{F}^+\right)=-\frac{9}{\sqrt{2}}\mathrm{Im}\left[(BD^*-OH^*)\,\mathrm{e}^{\mathrm{i}\varphi}-(OE^*-FD^*)\,\mathrm{e}^{3\mathrm{i}\varphi}\right],
\end{aligned}
$$

$$\begin{aligned}
\mathrm{tr}\left(\hat{Q}_{yz}\hat{F}\hat{Q}_{xx-yy}\hat{F}^{+}\right)&=-\frac{9}{\sqrt{2}}\mathrm{Re}\left[(BD^{*}-OH^{*})\,\mathrm{e}^{\mathrm{i}\varphi}+(OE^{*}-FD^{*})\,\mathrm{e}^{3\mathrm{i}\varphi}\right],\\
\mathrm{tr}\left(\hat{Q}_{xx-yy}\hat{F}\hat{Q}_{xx-yy}\hat{F}^{+}\right)&=18\,\mathrm{Re}\left(HB^{*}+FE^{*}\mathrm{e}^{4\mathrm{i}\varphi}\right),\\
\mathrm{tr}\left(\hat{Q}_{zz}\hat{F}\hat{Q}_{xx-yy}\hat{F}^{+}\right)&=-6\,\mathrm{Re}\left[(BE^{*}+FH^{*}-2OD^{*})\,\mathrm{e}^{2\mathrm{i}\varphi}\right],\\
\mathrm{tr}\left(\hat{Q}_{xy}\hat{F}\hat{Q}_{zz}\hat{F}^{+}\right)&=-3\,\mathrm{Im}\left[(FB^{*}-2GC^{*}+HE^{*})\,\mathrm{e}^{2\mathrm{i}\varphi}\right],\\
\mathrm{tr}\left(\hat{Q}_{xz}\hat{F}\hat{Q}_{zz}\hat{F}^{+}\right)&=-\frac{3}{\sqrt{2}}\mathrm{Im}\left[(OB^{*}-2AC^{*}+DE^{*}-FO^{*}+2GA^{*}-HD^{*})\,\mathrm{e}^{\mathrm{i}\varphi}\right],\\
\mathrm{tr}\left(\hat{Q}_{yz}\hat{F}\hat{Q}_{zz}\hat{F}^{+}\right)&=\frac{3}{\sqrt{2}}\mathrm{Re}\left[(OB^{*}-2AC^{*}+DE^{*}-FO^{*}+2GA^{*}-HD^{*})\,\mathrm{e}^{\mathrm{i}\varphi}\right],\\
\mathrm{tr}\left(\hat{Q}_{xx-yy}\hat{F}\hat{Q}_{zz}\hat{F}^{+}\right)&=-6\,\mathrm{Re}\left[(FB^{*}-2GC^{*}+HE^{*})\,\mathrm{e}^{2\mathrm{i}\varphi}\right],\\
\mathrm{tr}\left(\hat{Q}_{zz}\hat{F}\hat{Q}_{zz}\hat{F}^{+}\right)&=|B|^{2}+|E|^{2}+|F|^{2}+|H|^{2}-2\left(|C|^{2}+|G|^{2}+|O|^{2}+|D|^{2}\right)+4|A|^{2}
\end{aligned}\tag{6.97}$$

By using the similar method mentioned above, from Eq. (6.97) we can find

$$\begin{aligned}
\tilde{\Sigma}\mathrm{tr}\left(\hat{Q}_{xy}\hat{F}\hat{Q}_{xy}\hat{F}^{+}\right)&=\frac{9}{2}\tilde{\Sigma}\,\mathrm{Re}\left[HB^{*}-FE^{*}\cos(4\varphi)\right],\\
\tilde{\Sigma}\mathrm{tr}\left(\hat{Q}_{xz}\hat{F}\hat{Q}_{xy}\hat{F}^{+}\right)&=-\frac{9}{\sqrt{2}}\tilde{\Sigma}\left[\mathrm{Im}(BD^{*})\sin\varphi-\mathrm{Im}(OE^{*})\sin(3\varphi)\right],\\
\tilde{\Sigma}\mathrm{tr}\left(\hat{Q}_{yz}\hat{F}\hat{Q}_{xy}\hat{F}^{+}\right)&=-\frac{9}{\sqrt{2}}\tilde{\Sigma}\left[\mathrm{Im}(BD^{*})\cos\varphi+\mathrm{Im}(OE^{*})\cos(3\varphi)\right],\\
\tilde{\Sigma}\mathrm{tr}\left(\hat{Q}_{xx-yy}\hat{F}\hat{Q}_{xy}\hat{F}^{+}\right)&=9\tilde{\Sigma}\,\mathrm{Re}\left(FE^{*}\right)\sin(4\varphi),\\
\tilde{\Sigma}\mathrm{tr}\left(\hat{Q}_{zz}\hat{F}\hat{Q}_{xy}\hat{F}^{+}\right)&=-6\tilde{\Sigma}\,\mathrm{Re}\left(BE^{*}-OD^{*}\right)\sin(2\varphi),\\
\tilde{\Sigma}\mathrm{tr}\left(\hat{Q}_{xy}\hat{F}\hat{Q}_{xz}\hat{F}^{+}\right)&=-\frac{9}{\sqrt{2}}\tilde{\Sigma}\left[\mathrm{Im}(GB^{*})\sin\varphi-\mathrm{Im}(FC^{*})\sin(3\varphi)\right],\\
\tilde{\Sigma}\mathrm{tr}\left(\hat{Q}_{xz}\hat{F}\hat{Q}_{xz}\hat{F}^{+}\right)&=\frac{9}{2}\tilde{\Sigma}\left[\mathrm{Re}\left(AB^{*}-DC^{*}\right)-\mathrm{Re}\left(OC^{*}-AE^{*}\right)\cos(2\varphi)\right],\\
\tilde{\Sigma}\mathrm{tr}\left(\hat{Q}_{yz}\hat{F}\hat{Q}_{xz}\hat{F}^{+}\right)&=-\frac{9}{2}\tilde{\Sigma}\,\mathrm{Re}\left(OC^{*}-AE^{*}\right)\sin(2\varphi),\\
\tilde{\Sigma}\mathrm{tr}\left(\hat{Q}_{xx-yy}\hat{F}\hat{Q}_{xz}\hat{F}^{+}\right)&=-9\sqrt{2}\,\tilde{\Sigma}\left[\mathrm{Im}(GB^{*})\cos\varphi-\mathrm{Im}(FC^{*})\cos(3\varphi)\right],\\
\tilde{\Sigma}\mathrm{tr}\left(\hat{Q}_{zz}\hat{F}\hat{Q}_{xz}\hat{F}^{+}\right)&=-3\sqrt{2}\,\tilde{\Sigma}\left[\mathrm{Im}(BC^{*}-CE^{*})-2\mathrm{Im}(OA^{*})\cos\varphi\right],\\
\tilde{\Sigma}\mathrm{tr}\left(\hat{Q}_{xy}\hat{F}\hat{Q}_{yz}\hat{F}^{+}\right)&=-\frac{9}{\sqrt{2}}\tilde{\Sigma}\left[\mathrm{Im}(GB^{*})\cos\varphi+\mathrm{Im}(FC^{*})\cos(3\varphi)\right],\\
\tilde{\Sigma}\mathrm{tr}\left(\hat{Q}_{xz}\hat{F}\hat{Q}_{yz}\hat{F}^{+}\right)&=-\frac{9}{2}\tilde{\Sigma}\,\mathrm{Re}\left(OC^{*}-AE^{*}\right)\sin(2\varphi),\\
\tilde{\Sigma}\mathrm{tr}\left(\hat{Q}_{yz}\hat{F}\hat{Q}_{yz}\hat{F}^{+}\right)&=\frac{9}{2}\tilde{\Sigma}\left[\mathrm{Re}\left(AB^{*}-DC^{*}\right)+\mathrm{Re}\left(OC^{*}-AE^{*}\right)\cos(2\varphi)\right],
\end{aligned}$$

$$\widetilde{\Sigma}\mathrm{tr}\left(\hat{Q}_{xx-yy}\hat{F}\hat{Q}_{yz}\hat{F}^{+}\right)=9\sqrt{2}\,\widetilde{\Sigma}[\mathrm{Im}(GB^{*})\sin\varphi+\mathrm{Im}(FC^{*})\sin(3\varphi)],$$
$$\widetilde{\Sigma}\mathrm{tr}\left(\hat{Q}_{zz}\hat{F}\hat{Q}_{yz}\hat{F}^{+}\right)=-3\sqrt{2}\,\widetilde{\Sigma}\mathrm{Im}(BC^{*}-CE^{*})\sin\varphi,$$
$$\widetilde{\Sigma}\mathrm{tr}\left(\hat{Q}_{xy}\hat{F}\hat{Q}_{xx-yy}\hat{F}^{+}\right)=9\widetilde{\Sigma}\,\mathrm{Re}\,(FE^{*})\sin(4\varphi),$$
$$\widetilde{\Sigma}\mathrm{tr}\left(\hat{Q}_{xz}\hat{F}\hat{Q}_{xx-yy}\hat{F}^{+}\right)=-9\sqrt{2}\,\widetilde{\Sigma}[\mathrm{Im}(BD^{*})\cos\varphi-\mathrm{Im}(OE^{*})\cos(3\varphi)],$$
$$\widetilde{\Sigma}\mathrm{tr}\left(\hat{Q}_{yz}\hat{F}\hat{Q}_{xx-yy}\hat{F}^{+}\right)=9\sqrt{2}\,\widetilde{\Sigma}[\mathrm{Im}(BD^{*})\sin\varphi+\mathrm{Im}(OE^{*})\sin(3\varphi)],$$
$$\widetilde{\Sigma}\mathrm{tr}\left(\hat{Q}_{xx-yy}\hat{F}\hat{Q}_{xx-yy}\hat{F}^{+}\right)=18\widetilde{\Sigma}[\,\mathrm{Re}\,(HB^{*})+\,\mathrm{Re}\,(FE^{*})\cos(4\varphi)],$$
$$\widetilde{\Sigma}\mathrm{tr}\left(\hat{Q}_{zz}\hat{F}\hat{Q}_{xx-yy}\hat{F}^{+}\right)=-12\widetilde{\Sigma}\,\mathrm{Re}\,(BE^{*}-OD^{*})\cos(2\varphi),$$
$$\widetilde{\Sigma}\mathrm{tr}\left(\hat{Q}_{xy}\hat{F}\hat{Q}_{zz}\hat{F}^{+}\right)=-6\widetilde{\Sigma}\,\mathrm{Re}\,(FB^{*}-GC^{*})\sin(2\varphi),$$
$$\widetilde{\Sigma}\mathrm{tr}\left(\hat{Q}_{xz}\hat{F}\hat{Q}_{zz}\hat{F}^{+}\right)=-3\sqrt{2}\,\widetilde{\Sigma}\mathrm{Im}(OB^{*}-2AC^{*}+DE^{*})\cos\varphi,$$
$$\widetilde{\Sigma}\mathrm{tr}\left(\hat{Q}_{yz}\hat{F}\hat{Q}_{zz}\hat{F}^{+}\right)=-3\sqrt{2}\,\widetilde{\Sigma}\mathrm{Im}(OB^{*}-2AC^{*}+DE^{*})\sin\varphi,$$
$$\widetilde{\Sigma}\mathrm{tr}\left(\hat{Q}_{xx-yy}\hat{F}\hat{Q}_{zz}\hat{F}^{+}\right)=-12\widetilde{\Sigma}\,\mathrm{Re}\,(FB^{*}-GC^{*})\cos(2\varphi),$$
$$\widetilde{\Sigma}\mathrm{tr}\left(\hat{Q}_{zz}\hat{F}\hat{Q}_{zz}\hat{F}^{+}\right)=2\widetilde{\Sigma}\left[\,|B|^{2}+|E|^{2}-2\left(\,|C|^{2}+|D|^{2}-|A|^{2}\right)\right] \tag{6.98}$$

Define

$$R_{11}(\theta)=\frac{3}{2I_0}\widetilde{\Sigma}\,\mathrm{Re}\,(HB^{*}),\qquad R_{12}(\theta)=\frac{3}{2I_0}\widetilde{\Sigma}\,\mathrm{Re}\,(FE^{*}),$$
$$R_{13}(\theta)=\frac{2}{I_0}\widetilde{\Sigma}\,\mathrm{Re}\,(BE^{*}-OD^{*}),\qquad R_{14}(\theta)=\frac{3}{2I_0}\widetilde{\Sigma}\,\mathrm{Re}\,(AB^{*}-DC^{*}),$$
$$R_{15}(\theta)=\frac{3}{2I_0}\widetilde{\Sigma}\,\mathrm{Re}\,(OC^{*}-AE^{*}),\qquad R_{16}(\theta)=\frac{2}{I_0}\widetilde{\Sigma}\,\mathrm{Re}\,(FB^{*}-GC^{*}),$$
$$Q_{9}(\theta)=\frac{3}{\sqrt{2}I_0}\widetilde{\Sigma}\mathrm{Im}(BD^{*}),\qquad Q_{10}(\theta)=\frac{3}{\sqrt{2}I_0}\widetilde{\Sigma}\mathrm{Im}(OE^{*}),$$
$$Q_{11}(\theta)=\frac{3}{\sqrt{2}I_0}\widetilde{\Sigma}\mathrm{Im}(GB^{*}),\qquad Q_{12}(\theta)=\frac{3}{\sqrt{2}I_0}\widetilde{\Sigma}\mathrm{Im}(FC^{*}),$$
$$Q_{13}(\theta)=\frac{\sqrt{2}}{I_0}\widetilde{\Sigma}\mathrm{Im}(BC^{*}-CE^{*}),\qquad Q_{14}(\theta)=\frac{2\sqrt{2}}{I_0}\widetilde{\Sigma}\mathrm{Im}(OA^{*}),$$
$$Q_{15}(\theta)=\frac{\sqrt{2}}{I_0}\widetilde{\Sigma}\mathrm{Im}(OB^{*}-2AC^{*}+DE^{*}),$$
$$W_{2}(\theta)=\frac{2}{3I_0}\widetilde{\Sigma}\left[\,|B|^{2}+|E|^{2}-2\left(\,|C|^{2}+|D|^{2}-|A|^{2}\right)\right] \tag{6.99}$$

and

$$K_i^j(\theta,\varphi)=\frac{1}{3I_0}\widetilde{\Sigma}\mathrm{tr}\left(\hat{Q}_j\hat{F}\hat{Q}_i\hat{F}^+\right),\qquad i,j=\gamma \tag{6.100}$$

Then the following results can be obtained from Eqs. (6.98)–(6.100):

$$\begin{aligned}
&K_{xy}^{xy}(\theta,\varphi)=R_{11}(\theta)-R_{12}(\theta)\cos(4\varphi),\\
&K_{xy}^{xz}(\theta,\varphi)=-Q_9(\theta)\sin\varphi+Q_{10}(\theta)\sin(3\varphi),\\
&K_{xy}^{yz}(\theta,\varphi)=-Q_9(\theta)\cos\varphi-Q_{10}(\theta)\cos(3\varphi),\\
&K_{xy}^{xx-yy}(\theta,\varphi)=2R_{12}(\theta)\sin(4\varphi),\\
&K_{xy}^{zz}(\theta,\varphi)=-R_{13}(\theta)\sin(2\varphi),\\
&K_{xz}^{xy}(\theta,\varphi)=-Q_{11}(\theta)\sin\varphi+Q_{12}(\theta)\sin(3\varphi),\\
&K_{xz}^{xz}(\theta,\varphi)=R_{14}(\theta)-R_{15}(\theta)\cos(2\varphi),\\
&K_{xz}^{yz}(\theta,\varphi)=-R_{15}(\theta)\sin(2\varphi),\\
&K_{xz}^{xx-yy}(\theta,\varphi)=-2Q_{11}(\theta)\cos\varphi+2Q_{12}(\theta)\cos(3\varphi),\\
&K_{xz}^{zz}(\theta,\varphi)=-Q_{13}(\theta)+Q_{14}(\theta)\cos\varphi,\\
&K_{yz}^{xy}(\theta,\varphi)=-Q_{11}(\theta)\cos\varphi-Q_{12}(\theta)\cos(3\varphi),\\
&K_{yz}^{xz}(\theta,\varphi)=-R_{15}(\theta)\sin(2\varphi)=K_{xz}^{yz}(\theta,\varphi),\\
&K_{yz}^{yz}(\theta,\varphi)=R_{14}(\theta)+R_{15}(\theta)\cos(2\varphi),\\
&K_{yz}^{xx-yy}(\theta,\varphi)=2Q_{11}(\theta)\sin\varphi+2Q_{12}(\theta)\sin(3\varphi),\\
&K_{yz}^{zz}(\theta,\varphi)=-Q_{13}(\theta)\sin\varphi,\\
&K_{xx-yy}^{xy}(\theta,\varphi)=2R_{12}(\theta)\sin(4\varphi)=K_{xy}^{xx-yy}(\theta,\varphi),\\
&K_{xx-yy}^{xz}(\theta,\varphi)=-2Q_9(\theta)\cos\varphi+2Q_{10}(\theta)\cos(3\varphi),\\
&K_{xx-yy}^{yz}(\theta,\varphi)=2Q_9(\theta)\sin\varphi+2Q_{10}(\theta)\sin(3\varphi),\\
&K_{xx-yy}^{xx-yy}(\theta,\varphi)=4R_{11}(\theta)+4R_{12}(\theta)\cos(4\varphi),\\
&K_{xx-yy}^{zz}(\theta,\varphi)=-2R_{13}(\theta)\cos(2\varphi),\\
&K_{zz}^{xy}(\theta,\varphi)=-R_{16}(\theta)\sin(2\varphi),\\
&K_{zz}^{xz}(\theta,\varphi)=-Q_{15}(\theta)\cos\varphi,\\
&K_{zz}^{yz}(\theta,\varphi)=-Q_{15}(\theta)\sin\varphi,\\
&K_{zz}^{xx-yy}(\theta,\varphi)=-2R_{16}(\theta)\cos(2\varphi),\\
&K_{zz}^{zz}(\theta,\varphi)=W_2(\theta)
\end{aligned}\tag{6.101}$$

Referring to Eqs. (3.214), (6.71), and (6.33), the polarization rates related to the angles $\theta$, $\varphi$ simultaneously of the spin 1 outgoing particles corresponding to the spin 1 polarized incident particles can be written as

$$P_j(\theta, \varphi) = \frac{I_0(\theta)}{I(\theta, \varphi)} \left[ \overline{P}^j(\theta, \varphi) + \sum_{i=\varepsilon} \overline{p}_i K_i^j(\theta, \varphi) \right], \quad j = \varepsilon \tag{6.102}$$

The definition of the subscript symbol set $\varepsilon = x, y, z, xy, xz, yz, xx - yy, zz$ has been given by Eq. (3.476). Similar to Eq. (3.898), here we define

$$\overline{p}_i = \begin{cases} \frac{3}{2} p_i, & i = x, y, z \\ \frac{2}{3} p_i, & i = xy, xz, yz \\ \frac{1}{6} p_i, & i = xx - yy \\ \frac{1}{2} p_i, & i = zz \end{cases} \tag{6.103}$$

Note if we make $\varphi = 0$, the results obtained in Sect. 6.3.1 will automatically degenerate to the results obtained in Sect. 3.8.

### 6.3.2 Polarization Nuclear Database of Spin 1 Particles

According to the previous discussion, the polarization nuclear database of spin 1 particles should contain the following differential physical quantities of the polar angle $\theta$ for different incident particle energies of the shape elastic scattering and the two-body direct reaction channels:

1. Differential cross section corresponding to unpolarized incident particles $I_0(\theta)$.
2. The incident particle vector analyzing power $A_y(\theta)$; tensor analyzing powers $A_i(\theta)$ $(i = 1\sim3)$.
3. Corresponding to the unpolarized incident particles, the outgoing particle vector polarization rate $P^y(\theta)$; tensor polarization rates $P^i(\theta)$ $(i = 1\sim3)$.
4. Polarization transfer coefficient components $R_i(\theta)$ $(i = 1\sim16)$, $Q_i(\theta)$ $(i = 1\sim15)$, $W_i(\theta)$ $(i = 1, 2)$.

The expressions of the above physical quantities are given in Sect. 6.3.1. Among them, $I_0(\theta)$ and the polarization analyzing powers can be experimentally measured, and the physical quantities that cannot be measured experimentally can be obtained through theoretical calculations.

The coordinate system transformation expressions given by Eqs. (6.34)–(6.36) and the related discussion are also applicable in this section, and it is still recommended to preserve the polarization nuclear data of the laboratory system in the polarization nuclear database.

### 6.3.3 *Main Points of Polarized Particle Transport Theory for Spin 1 Particles*

For the sake of simplicity, assuming that only $\vec{1}$ + A → $\vec{1}$ + B reaction occurs in the device, the reaction can be a shape elastic scattering or a two-body direct reaction, and all other reaction channels are omitted for the time being.

We choose to study in the laboratory system and choose the initial incident particle direction as the $z$-axis, and the initial incident particle polarization rates $p_{0i}( i = \varepsilon \equiv x, y, z, xy, xz, yz, xx - yy, zz)$ are determined by experiment or artificial hypothesis. Take the position where the initial reaction (also known as the first reaction) occurs as the origin $O_0$ of the coordinate system. After the incident particle energy $E_0$ in the laboratory system is determined, the 42 kinds of the differential physical quantities in the laboratory system listed in Sect. 6.3.2 can be obtained from the polarization nuclear database of the $\vec{1}$ + A → $\vec{1}$ + B reaction. Then the various polarization quantities related to angles $(\theta, \varphi)$ can be calculated according to Eqs. (6.63), (6.70), (6.71), (6.76), (6.81), (6.86), (6.91), (6.96), and (6.101).

When the polarization rates $p_{0i}(i = \varepsilon)$ of the initial incident particles are known, the differential cross section corresponding to energy $E_0$ and any angle $(\theta, \varphi)$ of the outgoing particle can be obtained from Eq. (6.71). Regardless of other reaction channels, after sampling, assume that the reaction of the emitted particle of the first reaction with target nucleus occurs at the position $O_1$ with the direction of $(\theta_1, \varphi_1)$ and the distance $L_1$ from the origin $O_0$, which is called the second reaction. The incident particle energy corresponding to the second reaction is $E_1$. In $O_0$ system, the Stokes vector expressions of the initial incident particle and the outgoing particle of the first reaction at the second reaction space point $O_1$ are as follows, respectively:

$$S_1^{(i)}(\theta_1) = I_0(\theta_1)\begin{pmatrix} 1 \\ \overline{p}_{0x} \\ \overline{p}_{0y} \\ \overline{p}_{0z} \\ \overline{p}_{0xy} \\ \overline{p}_{0xz} \\ \overline{p}_{0yz} \\ \overline{p}_{0\,xx-yy} \\ \overline{p}_{0zz} \end{pmatrix}, \quad S_1^{(f)}(\theta_1, \varphi_1) = I(\theta_1, \varphi_1)\begin{pmatrix} 1 \\ P_{1x}(\theta_1, \varphi_1) \\ P_{1y}(\theta_1, \varphi_1) \\ P_{1z}(\theta_1, \varphi_1) \\ P_{1xy}(\theta_1, \varphi_1) \\ P_{1xz}(\theta_1, \varphi_1) \\ P_{1yz}(\theta_1, \varphi_1) \\ P_{1\,xx-yy}(\theta_1, \varphi_1) \\ P_{1zz}(\theta_1, \varphi_1) \end{pmatrix} \tag{6.104}$$

The definition of $\overline{p}_{0i}$ is given by Eq. (6.103). The above two vectors satisfy the following relation:

$$S_1^{(f)} = Z_1 S_1^{(i)} \tag{6.105}$$

where

$$Z_1 = \begin{pmatrix} 1 & \overline{A}_x & \overline{A}_y & 0 & \overline{A}_{xy} & \overline{A}_{xz} & \overline{A}_{yz} & \overline{A}_{xx-yy} & \overline{A}_{zz} \\ \overline{P}^x & K_x^x & K_y^x & K_z^x & K_{xy}^x & K_{xz}^x & K_{yz}^x & K_{xx-yy}^x & K_{zz}^x \\ \overline{P}^y & K_x^y & K_y^y & K_z^y & K_{xy}^y & K_{xz}^y & K_{yz}^y & K_{xx-yy}^y & K_{zz}^y \\ 0 & K_x^z & K_y^z & K_z^z & K_{xy}^z & K_{xz}^z & K_{yz}^z & K_{xx-yy}^z & 0 \\ \overline{P}^{xy} & K_x^{xy} & K_y^{xy} & K_z^{xy} & K_{xy}^{xy} & K_{xz}^{xy} & K_{yz}^{xy} & K_{xx-yy}^{xy} & K_{zz}^{xy} \\ \overline{P}^{xz} & K_x^{xz} & K_y^{xz} & K_z^{xz} & K_{xy}^{xz} & K_{xz}^{xz} & K_{yz}^{xz} & K_{xx-yy}^{xz} & K_{zz}^{xz} \\ \overline{P}^{yz} & K_x^{yz} & K_y^{yz} & K_z^{yz} & K_{xy}^{yz} & K_{xz}^{yz} & K_{yz}^{yz} & K_{xx-yy}^{yz} & K_{zz}^{yz} \\ \overline{P}^{xx-yy} & K_x^{xx-yy} & K_y^{xx-yy} & K_z^{xx-yy} & K_{xy}^{xx-yy} & K_{xz}^{xx-yy} & K_{yz}^{xx-yy} & K_{xx-yy}^{xx-yy} & K_{zz}^{xx-yy} \\ \overline{P}^{zz} & K_x^{zz} & K_y^{zz} & 0 & K_{xy}^{zz} & K_{xz}^{zz} & K_{yz}^{zz} & K_{xx-yy}^{zz} & K_{zz}^{zz} \end{pmatrix}_1 \tag{6.106}$$

where footnote 1 corresponds to $(\theta_1, \varphi_1)$ and energy $E_0$. Equation (6.105) is a combination of Eq. (6.71) and Eq. (6.102). After the first reaction occurs at the point $O_0$, the equation similar to Eq. (6.105) at any angles $(\theta, \varphi)$ holds, and of course it also holds at the angles $(\theta_1, \varphi_1)$, that is, the relation (6.105) is determined by the first reaction.

The Cartesian coordinates of the point $O_1$ in $O_0$ system are

$$X_1 = L_1 \sin\theta_1 \cos\varphi_1, \quad Y_1 = L_1 \sin\theta_1 \sin\varphi_1, \quad Z_1 = L_1 \cos\varphi_1 \tag{6.107}$$

In the $O_0$ coordinate system, the three Cartesian coordinate system axes after rotating the angles $(\theta_1, \varphi_1)$ become the three new coordinate axes of the new coordinate system, and then the coordinate origin of the new coordinate system is moved from $O_0$ to $O_1$; the $O_0O_1$ direction is just the new $z$-axis direction in $O_1$ system.

The rank-2 tensor $\hat{g}_2$ with 5 components is rotated by angles $(\theta, \varphi)$ to be changed to a new rank-2 tensor $\hat{T}_2$, and its transformation relation is known from Eq. (1.27) as

$$\hat{T}_{2M'} = \sum_{M=-2}^{2} \hat{g}_{2M} D_{MM'}^2(\varphi, \theta, 0) \tag{6.108}$$

The expression of the rotation function $D_{MM'}^2(\alpha, \beta, \gamma)$ of the Euler angles is

$$D_{MM'}^2(\alpha, \beta, \gamma) = e^{-\mathrm{i}M\alpha} d_{MM'}^2(\beta) e^{-\mathrm{i}M'\gamma} \tag{6.109}$$

where $d_{MM'}^2(\beta)$ have been provided in Table 6.1. From Eq. (6.109) we know

**Table 6.1** Expressions of $d^2_{MM'}(\beta)$[20]

| $M$ | $M'$ | | | | |
|---|---|---|---|---|---|
| | 2 | 1 | 0 | −1 | −2 |
| 2 | $\frac{(1+\cos\beta)^2}{4}$ | $-\frac{\sin\beta(1+\cos\beta)}{2}$ | $\frac{1}{2}\sqrt{\frac{3}{2}}\sin^2\beta$ | $-\frac{\sin\beta(1-\cos\beta)}{2}$ | $\frac{(1-\cos\beta)^2}{4}$ |
| 1 | $\frac{\sin\beta(1+\cos\beta)}{2}$ | $\frac{2\cos^2\beta+\cos\beta-1}{2}$ | $-\sqrt{\frac{3}{2}}\sin\beta\cos\beta$ | $-\frac{2\cos^2\beta-\cos\beta-1}{2}$ | $-\frac{\sin\beta(1-\cos\beta)}{2}$ |
| 0 | $\frac{1}{2}\sqrt{\frac{3}{2}}\sin^2\beta$ | $\sqrt{\frac{3}{2}}\sin\beta\cos\beta$ | $\frac{3\cos^2\beta-1}{2}$ | $-\sqrt{\frac{3}{2}}\sin\beta\cos\beta$ | $\frac{1}{2}\sqrt{\frac{3}{2}}\sin^2\beta$ |
| −1 | $\frac{\sin\beta(1-\cos\beta)}{2}$ | $-\frac{2\cos^2\beta-\cos\beta-1}{2}$ | $\sqrt{\frac{3}{2}}\sin\beta\cos\beta$ | $\frac{2\cos^2\beta+\cos\beta-1}{2}$ | $-\frac{\sin\beta(1+\cos\beta)}{2}$ |
| −2 | $\frac{(1-\cos\beta)^2}{4}$ | $\frac{\sin\beta(1-\cos\beta)}{2}$ | $\frac{1}{2}\sqrt{\frac{3}{2}}\sin^2\beta$ | $\frac{\sin\beta(1+\cos\beta)}{2}$ | $\frac{(1+\cos\beta)^2}{4}$ |

$$D^2_{MM'}(\varphi, \theta, 0) = d^2_{MM'}(\theta)\, \mathrm{e}^{-\mathrm{i}M\varphi} \tag{6.110}$$

When we study the spin 1 particles rotating in the five dimensional Cartesian coordinate system, we can take the following expressions given by Eq. (3.142) as the basis functions of the polarization tensor of the five-dimensional orthogonal Cartesian coordinate system

$$\left(\hat{H}_{xx-yy}, \hat{H}_{xy}, \hat{H}_{xz}, \hat{H}_{yz}, \hat{H}_{zz}\right) = \left(\sqrt{\frac{1}{6}}\hat{Q}_{xx-yy}, \sqrt{\frac{2}{3}}\hat{Q}_{xy}, \sqrt{\frac{2}{3}}\hat{Q}_{xz}, \sqrt{\frac{2}{3}}\hat{Q}_{yz}, \sqrt{\frac{1}{2}}\hat{Q}_{zz}\right) \tag{6.111}$$

So the expression (3.22) can be rewritten as

$$\hat{T}_{2\ \pm 2} = \frac{1}{\sqrt{2}}\left(\hat{H}_{xx-yy} \pm \mathrm{i}\hat{H}_{xy}\right),\quad \hat{T}_{2\ \pm 1} = \mp\frac{1}{\sqrt{2}}\left(\hat{H}_{xz} \pm \mathrm{i}\hat{H}_{yz}\right),\quad \hat{T}_{2\ 0} = \hat{H}_{zz} \tag{6.112}$$

And the following relations can be obtained by Eq. (6.112) or Eq. (3.21):

$$\begin{aligned} &\hat{H}_{xx-yy} = \frac{1}{\sqrt{2}}\left(\hat{T}_{2\ -2} + \hat{T}_{2\ 2}\right),\quad \hat{H}_{xy} = \frac{\mathrm{i}}{\sqrt{2}}\left(\hat{T}_{2\ -2} - \hat{T}_{2\ 2}\right),\\ &\hat{H}_{xz} = \frac{1}{\sqrt{2}}\left(\hat{T}_{2\ -1} - \hat{T}_{2\ 1}\right),\qquad \hat{H}_{yz} = \frac{\mathrm{i}}{\sqrt{2}}\left(\hat{T}_{2\ -1} + \hat{T}_{2\ 1}\right).\\ &\hat{H}_{zz} = \hat{T}_{2\ 0} \end{aligned} \tag{6.113}$$

That is exactly the result given by Eq. (3.142). It is visible that Eq. (6.112) and Eq. (6.113) are the unitary transformation relations between the spherical basis coordinate system and the orthogonal Cartesian coordinate system of the five-dimensional rank-2 tensor basis functions.

We use Eq. (6.112) to describe the transformation between $\hat{g}_{2M}$ and $\hat{H}_i$ in Eq. (6.108), where $M = \pm 2, \pm 1, 0$; $i = xx - yy, xy, xz, yz, zz$. Thus, according to Eq. (6.108), Eq. (6.110), and Table 6.1, the following results can be derived from Eq. (6.112):

$$
\begin{aligned}
\hat{T}_{22} = & \frac{(1+\cos\theta)^2}{4}\frac{1}{\sqrt{2}}\left(\hat{H}_{xx-yy}+\mathrm{i}\hat{H}_{xy}\right)\mathrm{e}^{-2\mathrm{i}\varphi} \\
& -\frac{\sin\theta(1+\cos\theta)}{2}\frac{1}{\sqrt{2}}\left(\hat{H}_{xz}+\mathrm{i}\hat{H}_{yz}\right)\mathrm{e}^{-\mathrm{i}\varphi}+\frac{1}{2}\sqrt{\frac{3}{2}}\sin^2\theta\,\hat{H}_{zz} \\
& +\frac{\sin\theta(1-\cos\theta)}{2}\frac{1}{\sqrt{2}}\left(\hat{H}_{xz}-\mathrm{i}\hat{H}_{yz}\right)\mathrm{e}^{\mathrm{i}\varphi} \\
& +\frac{(1-\cos\theta)^2}{4}\frac{1}{\sqrt{2}}\left(\hat{H}_{xx-yy}-\mathrm{i}\hat{H}_{xy}\right)\mathrm{e}^{2\mathrm{i}\varphi} \\
= & \frac{1}{\sqrt{2}}\left\{\left[\frac{1+\cos^2\theta}{2}\hat{H}_{xx-yy}+\cos\theta\left(\mathrm{i}\hat{H}_{xy}\right)\right]\cos(2\varphi)\right. \\
& -\left[\cos\theta\,\hat{H}_{xx-yy}+\frac{1+\cos^2\theta}{2}\left(\mathrm{i}\hat{H}_{xy}\right)\right]\mathrm{i}\sin(2\varphi) \\
& -\left[\sin\theta\cos\theta\hat{H}_{xz}+\sin\theta\left(\mathrm{i}\hat{H}_{yz}\right)\right]\cos\varphi \\
& \left.+\left[\sin\theta\hat{H}_{xz}+\sin\theta\cos\theta\left(\mathrm{i}\hat{H}_{yz}\right)\right]\mathrm{i}\sin\varphi+\frac{\sqrt{3}}{2}\sin^2\theta\,\hat{H}_{zz}\right\} \\
= & \frac{1}{\sqrt{2}}\left\{\left[\frac{1+\cos^2\theta}{2}\left(\cos(2\varphi)\hat{H}_{xx-yy}+\sin(2\varphi)\hat{H}_{xy}\right)\right.\right. \\
& \left.-\frac{1}{2}\sin(2\theta)\left(\cos\varphi\hat{H}_{xz}+\sin\varphi\hat{H}_{yz}\right)+\frac{\sqrt{3}}{2}\sin^2\theta\,\hat{H}_{zz}\right] \\
& +\mathrm{i}\left[\cos\theta\left(\cos(2\varphi)\hat{H}_{xy}-\sin(2\varphi)\hat{H}_{xx-yy}\right)\right. \\
& \left.\left.-\sin\theta\left(\cos\varphi\hat{H}_{yz}-\sin\varphi\hat{H}_{xz}\right)\right]\right\}
\end{aligned}
$$

$$
\begin{aligned}
\hat{T}_{21} = & -\frac{\sin\theta(1+\cos\theta)}{2}\frac{1}{\sqrt{2}}\left(\hat{H}_{xx-yy}+\mathrm{i}\hat{H}_{xy}\right)\mathrm{e}^{-2\mathrm{i}\varphi} \\
& -\frac{2\cos^2\theta+\cos\theta-1}{2}\frac{1}{\sqrt{2}}\left(\hat{H}_{xz}+\mathrm{i}\hat{H}_{yz}\right)\mathrm{e}^{-\mathrm{i}\varphi}+\sqrt{\frac{3}{2}}\sin\theta\cos\theta\hat{H}_{zz} \\
& -\frac{2\cos^2\theta-\cos\theta-1}{2}\frac{1}{\sqrt{2}}\left(\hat{H}_{xz}-\mathrm{i}\hat{H}_{yz}\right)\mathrm{e}^{\mathrm{i}\varphi} \\
& +\frac{\sin\theta(1-\cos\theta)}{2}\frac{1}{\sqrt{2}}\left(\hat{H}_{xx-yy}-\mathrm{i}\hat{H}_{xy}\right)\mathrm{e}^{2\mathrm{i}\varphi} \\
= & -\frac{1}{\sqrt{2}}\left\{\left[\sin\theta\cos\theta\hat{H}_{xx-yy}+\sin\theta\left(\mathrm{i}\hat{H}_{xy}\right)\right]\cos(2\varphi)\right. \\
& -\left[\sin\theta\hat{H}_{xx-yy}+\sin\theta\cos\theta\left(\mathrm{i}\hat{H}_{xy}\right)\right]\mathrm{i}\sin(2\varphi) \\
& +\left[\left(2\cos^2\theta-1\right)\hat{H}_{xz}+\cos\theta\left(\mathrm{i}\hat{H}_{yz}\right)\right]\cos\varphi \\
& \left.-\left[\cos\theta\hat{H}_{xz}+\left(2\cos^2\theta-1\right)\left(\mathrm{i}\hat{H}_{yz}\right)\right]\mathrm{i}\sin\varphi-\frac{\sqrt{3}}{2}\sin(2\theta)\hat{H}_{zz}\right\}
\end{aligned}
$$

$$= -\frac{1}{\sqrt{2}}\Big\{\Big[\frac{1}{2}\sin(2\theta)\big(\cos(2\varphi)\hat{H}_{xx-yy} + \sin(2\varphi)\hat{H}_{xy}\big)$$
$$+\cos(2\theta)\big(\cos\varphi\hat{H}_{xz} + \sin\varphi\hat{H}_{yz}\big) - \frac{\sqrt{3}}{2}\sin(2\theta)\hat{H}_{zz}\Big]$$
$$+\mathrm{i}\,\big[\sin\theta\big(\cos(2\varphi)\hat{H}_{xy} - \sin(2\varphi)\hat{H}_{xx-yy}\big)$$
$$+\cos\theta\big(\cos\varphi\hat{H}_{yz} - \sin\varphi\hat{H}_{xz}\big)\big]\Big\}$$

$$\hat{T}_{20} = \frac{1}{2}\sqrt{\frac{3}{2}}\sin^2\theta\frac{1}{\sqrt{2}}\big(\hat{H}_{xx-yy} + \mathrm{i}\hat{H}_{xy}\big)\,\mathrm{e}^{-2i\varphi}$$
$$+\sqrt{\frac{3}{2}}\sin\theta\cos\theta\frac{1}{\sqrt{2}}\big(\hat{H}_{xz} + \mathrm{i}\hat{H}_{yz}\big)\,\mathrm{e}^{-\mathrm{i}\varphi} + \frac{3\cos^2\theta - 1}{2}\hat{H}_{zz}$$
$$+\sqrt{\frac{3}{2}}\sin\theta\cos\theta\frac{1}{\sqrt{2}}\big(\hat{H}_{xz} - \mathrm{i}\hat{H}_{yz}\big)\,\mathrm{e}^{\mathrm{i}\varphi}$$
$$+\frac{1}{2}\sqrt{\frac{3}{2}}\sin^2\theta\frac{1}{\sqrt{2}}\big(\hat{H}_{xx-yy} - \mathrm{i}\hat{H}_{xy}\big)\,\mathrm{e}^{2\mathrm{i}\varphi}$$
$$= \frac{\sqrt{3}}{2}\Big\{\sin^2\theta\big[\cos(2\varphi)\hat{H}_{xx-yy} + \sin(2\varphi)\hat{H}_{xy}\big]$$
$$+\sin(2\theta)\big[\cos\varphi\hat{H}_{xz} + \sin\varphi\hat{H}_{yz}\big] + \frac{1}{\sqrt{3}}\big(3\cos^2\theta - 1\big)\hat{H}_{zz}\Big\},$$

$$\hat{T}_{2\ -1} = -\frac{\sin\theta(1-\cos\theta)}{2}\frac{1}{\sqrt{2}}\big(\hat{H}_{xx-yy} + \mathrm{i}\hat{H}_{xy}\big)\,\mathrm{e}^{-2\mathrm{i}\varphi}$$
$$+\frac{2\cos^2\theta - \cos\theta - 1}{2}\frac{1}{\sqrt{2}}\big(\hat{H}_{xz} + \mathrm{i}\hat{H}_{yz}\big)\,\mathrm{e}^{-\mathrm{i}\varphi} - \sqrt{\frac{3}{2}}\sin\theta\cos\theta\hat{H}_{zz}$$
$$+\frac{2\cos^2\theta + \cos\theta - 1}{2}\frac{1}{\sqrt{2}}\big(\hat{H}_{xz} - \mathrm{i}\hat{H}_{yz}\big)\,\mathrm{e}^{\mathrm{i}\varphi}$$
$$+\frac{\sin\theta(1+\cos\theta)}{2}\frac{1}{\sqrt{2}}\big(\hat{H}_{xx-yy} - \mathrm{i}\hat{H}_{xy}\big)\,\mathrm{e}^{2\mathrm{i}\varphi}$$
$$= \frac{1}{\sqrt{2}}\Big\{\big[\sin\theta\cos\theta\hat{H}_{xx-yy} - \sin\theta\big(\mathrm{i}\hat{H}_{xy}\big)\big]\cos(2\varphi)$$
$$+\big[\sin\theta\hat{H}_{xx-yy} - \sin\theta\cos\theta\big(\mathrm{i}\hat{H}_{xy}\big)\big]\,\mathrm{i}\sin(2\varphi)$$
$$+\big[\big(2\cos^2\theta - 1\big)\hat{H}_{xz} - \cos\theta\big(\mathrm{i}\hat{H}_{yz}\big)\big]\,\cos\varphi$$
$$+\big[\cos\theta\hat{H}_{xz} - \big(2\cos^2\theta - 1\big)\big(\mathrm{i}\hat{H}_{yz}\big)\big]\,\mathrm{i}\sin\varphi - \frac{\sqrt{3}}{2}\sin(2\theta)\hat{H}_{zz}\Big\}$$

$$
\begin{aligned}
&= \frac{1}{\sqrt{2}}\left\{\left[\frac{1}{2}\sin(2\theta)\left(\cos(2\varphi)\hat{H}_{xx-yy} + \sin(2\varphi)\hat{H}_{xy}\right)\right.\right.\\
&\quad\left. + \cos(2\theta)\left(\cos\varphi\hat{H}_{xz} + \sin\varphi\hat{H}_{yz}\right) - \frac{\sqrt{3}}{2}\sin(2\theta)\hat{H}_{zz}\right]\\
&\quad - \mathrm{i}\left[\sin\theta\left(\cos(2\varphi)\hat{H}_{xy} - \sin(2\varphi)\hat{H}_{xx-yy}\right)\right.\\
&\quad\left.\left. + \cos\theta\left(\cos\varphi\hat{H}_{yz} - \sin\varphi\hat{H}_{xz}\right)\right]\right\},
\end{aligned}
$$

$$
\begin{aligned}
\hat{T}_{2\ -2} &= \frac{(1-\cos\theta)^2}{4}\frac{1}{\sqrt{2}}\left(\hat{H}_{xx-yy} + \mathrm{i}\hat{H}_{xy}\right)\mathrm{e}^{-2\mathrm{i}\varphi}\\
&\quad + \frac{\sin\theta(1-\cos\theta)}{2}\frac{1}{\sqrt{2}}\left(\hat{H}_{xz} + \mathrm{i}\hat{H}_{yz}\right)\mathrm{e}^{-\mathrm{i}\varphi} + \frac{1}{2}\sqrt{\frac{3}{2}}\sin^2\theta\hat{H}_{zz}\\
&\quad - \frac{\sin\theta(1+\cos\theta)}{2}\frac{1}{\sqrt{2}}\left(\hat{H}_{xz} - \mathrm{i}\hat{H}_{yz}\right)\mathrm{e}^{\mathrm{i}\varphi}\\
&\quad + \frac{(1+\cos\theta)^2}{4}\frac{1}{\sqrt{2}}\left(\hat{H}_{xx-yy} - \mathrm{i}\hat{H}_{xy}\right)\mathrm{e}^{2\mathrm{i}\varphi}\\
&= \frac{1}{\sqrt{2}}\left\{\left[\frac{1+\cos^2\theta}{2}\hat{H}_{xx-yy} - \cos\theta\left(\mathrm{i}\hat{H}_{xy}\right)\right]\cos(2\varphi)\right.\\
&\quad + \left[\cos\theta\hat{H}_{xx-yy} - \frac{1+\cos^2\theta}{2}\left(\mathrm{i}\hat{H}_{xy}\right)\right]\mathrm{i}\sin(2\varphi)\\
&\quad - \left[\sin\theta\cos\theta\,\hat{H}_{xz} - \sin\theta\left(\mathrm{i}\hat{H}_{yz}\right)\right]\cos\varphi\\
&\quad\left. - \left[\sin\theta\,\hat{H}_{xz} - \sin\theta\cos\theta\left(\mathrm{i}\hat{H}_{yz}\right)\right]\mathrm{i}\sin\varphi + \frac{\sqrt{3}}{2}\sin^2\theta\hat{H}_{zz}\right\}\\
&= \frac{1}{\sqrt{2}}\left\{\left[\frac{1+\cos^2\theta}{2}\left(\cos(2\varphi)\hat{H}_{xx-yy} + \sin(2\varphi)\hat{H}_{xy}\right)\right.\right.\\
&\quad\left. - \frac{1}{2}\sin(2\theta)\left(\cos\varphi\,\hat{H}_{xz} + \sin\varphi\,\hat{H}_{yz}\right) + \frac{\sqrt{3}}{2}\sin^2\theta\,\hat{H}_{zz}\right]\\
&\quad - \mathrm{i}\left[\cos\theta\left(\cos(2\varphi)\hat{H}_{xy} - \sin(2\varphi)\hat{H}_{xx-yy}\right)\right.\\
&\quad\left.\left. - \sin\theta\left(\cos\varphi\,\hat{H}_{yz} - \sin\varphi\,\hat{H}_{xz}\right)\right]\right\}
\end{aligned}
\tag{6.114}
$$

By using Eq. (6.113) and from Eq. (6.114), we can obtain

$$\begin{aligned}
\hat{H}'_{xx-yy} &= \frac{1}{\sqrt{2}}\left(\hat{T}_{2\ -2} + \hat{T}_{22}\right) \\
&= \frac{1+\cos^2\theta}{2}\left(\cos(2\varphi)\hat{H}_{xx-yy} + \sin(2\varphi)\hat{H}_{xy}\right) \\
&\quad - \frac{1}{2}\sin(2\theta)\left(\cos\varphi\hat{H}_{xz} + \sin\varphi\hat{H}_{yz}\right) + \frac{\sqrt{3}}{2}\sin^2\theta\,\hat{H}_{zz} \\
&= \frac{1+\cos^2\theta}{2}\cos(2\varphi)\hat{H}_{xx-yy} + \frac{1+\cos^2\theta}{2}\sin(2\varphi)\hat{H}_{xy} \\
&\quad - \frac{1}{2}\sin(2\theta)\cos\varphi\hat{H}_{xz} - \frac{1}{2}\sin(2\theta)\sin\varphi\hat{H}_{yz} + \frac{\sqrt{3}}{2}\sin^2\theta\,\hat{H}_{zz}
\end{aligned}$$

$$\begin{aligned}
\hat{H}'_{xy} &= \frac{\mathrm{i}}{\sqrt{2}}\left(\hat{T}_{2\ -2} - \hat{T}_{22}\right) \\
&= \cos\theta\left(\cos(2\varphi)\hat{H}_{xy} - \sin(2\varphi)\hat{H}_{xx-yy}\right) - \sin\theta\left(\cos\varphi\hat{H}_{yz} - \sin\varphi\hat{H}_{xz}\right) \\
&= -\cos\theta\sin(2\varphi)\hat{H}_{xx-yy} + \cos\theta\cos(2\varphi)\hat{H}_{xy} + \sin\theta\sin\varphi\hat{H}_{xz} - \sin\theta\cos\varphi\hat{H}_{yz}
\end{aligned}$$

$$\begin{aligned}
\hat{H}'_{xz} &= \frac{1}{\sqrt{2}}\left(\hat{T}_{2\ -1} - \hat{T}_{21}\right) \\
&= \frac{1}{2}\sin(2\theta)\left(\cos(2\varphi)\hat{H}_{xx-yy} + \sin(2\varphi)\hat{H}_{xy}\right) \\
&\quad + \cos(2\theta)\left(\cos\varphi\hat{H}_{xz} + \sin\varphi\hat{H}_{yz}\right) - \frac{\sqrt{3}}{2}\sin(2\theta)\hat{H}_{zz} \\
&= \frac{1}{2}\sin(2\theta)\cos(2\varphi)\hat{H}_{xx-yy} + \frac{1}{2}\sin(2\theta)\sin(2\varphi)\hat{H}_{xy} \\
&\quad + \cos(2\theta)\cos\varphi\hat{H}_{xz} + \cos(2\theta)\sin\varphi\hat{H}_{yz} - \frac{\sqrt{3}}{2}\sin(2\theta)\hat{H}_{zz}
\end{aligned}$$

$$\begin{aligned}
\hat{H}'_{yz} &= \frac{\mathrm{i}}{\sqrt{2}}\left(\hat{T}_{2\ -1} + \hat{T}_{21}\right) \\
&= \sin\theta\left(\cos(2\varphi)\hat{H}_{xy} - \sin(2\varphi)\hat{H}_{xx-yy}\right) + \cos\theta\left(\cos\varphi\hat{H}_{yz} - \sin\varphi\hat{H}_{xz}\right) \\
&= -\sin\theta\sin(2\varphi)\hat{H}_{xx-yy} + \sin\theta\cos(2\varphi)\hat{H}_{xy} - \cos\theta\sin\varphi\hat{H}_{xz} + \cos\theta\cos\varphi\hat{H}_{yz}
\end{aligned}$$

$$\begin{aligned}
\hat{H}'_{zz} &= \hat{T}_{20} \\
&= \frac{\sqrt{3}}{2}\Big\{\sin^2\theta\left[\cos(2\varphi)\hat{H}_{xx-yy} + \sin(2\varphi)\hat{H}_{xy}\right] \\
&\quad + \sin(2\theta)\left[\cos\varphi\hat{H}_{xz} + \sin\varphi\hat{H}_{yz}\right] + \frac{1}{\sqrt{3}}\left(3\cos^2\theta - 1\right)\hat{H}_{zz}\Big\} \\
&= \frac{\sqrt{3}}{2}\sin^2\theta\cos(2\varphi)\hat{H}_{xx-yy} + \frac{\sqrt{3}}{2}\sin^2\theta\sin(2\varphi)\hat{H}_{xy} \\
&\quad + \frac{\sqrt{3}}{2}\sin(2\theta)\cos\varphi\hat{H}_{xz} + \frac{\sqrt{3}}{2}\sin(2\theta)\sin\varphi\hat{H}_{yz} + \frac{1}{2}\left(3\cos^2\theta - 1\right)\hat{H}_{zz}
\end{aligned} \tag{6.115}$$

According to the original definition of the rotation function $D$ given by Eq. (1.27), we define the rotation matrix $\hat{\Gamma}(\theta, \varphi)$ of the five-dimensional orthogonal Cartesian coordinate system based on Eq. (6.115) as follows:

$$\hat{H}_j'(\theta,\varphi)=\sum_i \hat{H}_i \Gamma_{ij}(\theta,\varphi),\qquad i,j=xx-yy,xy,xz,yz,zz \tag{6.116}$$

Then, according to Eq. (6.115), we have the following matrix $\hat{\Gamma}(\theta,\varphi)$

$$\hat{\Gamma}(\theta,\varphi)=\begin{pmatrix} \frac{1+\cos^2\theta}{2}\cos(2\varphi) & -\cos\theta\sin(2\varphi) & \frac{\sin(2\theta)\cos(2\varphi)}{2} & -\sin\theta\sin(2\varphi) & \frac{\sqrt{3}}{2}\sin^2\theta\cos(2\varphi) \\ \frac{1+\cos^2\theta}{2}\sin(2\varphi) & \cos\theta\cos(2\varphi) & \frac{\sin(2\theta)\sin(2\varphi)}{2} & \sin\theta\cos(2\varphi) & \frac{\sqrt{3}}{2}\sin^2\theta\sin(2\varphi) \\ -\frac{1}{2}\sin(2\theta)\cos\varphi & \sin\theta\sin\varphi & \cos(2\theta)\cos\varphi & -\cos\theta\sin\varphi & \frac{\sqrt{3}}{2}\sin(2\theta)\cos\varphi \\ -\frac{1}{2}\sin(2\theta)\sin\varphi & -\sin\theta\cos\varphi & \cos(2\theta)\sin\varphi & \cos\theta\cos\varphi & \frac{\sqrt{3}}{2}\sin(2\theta)\sin\varphi \\ \frac{\sqrt{3}}{2}\sin^2\theta & 0 & -\frac{\sqrt{3}}{2}\sin(2\theta) & 0 & \frac{1}{2}(3\cos^2\theta-1) \end{pmatrix} \tag{6.117}$$

Let $\hat{\gamma}(\theta,\varphi)$ be the inverse matrix of $\hat{\Gamma}(\theta,\varphi)$. Since the inverse matrix of the real matrix is its transpose matrix, then we can get the matrix $\hat{\gamma}(\theta,\varphi)$ from Eq. (6.117) as follows:

$$\hat{\gamma}(\theta,\varphi)=\begin{pmatrix} \frac{1+\cos^2\theta}{2}\cos(2\varphi) & \frac{1+\cos^2\theta}{2}\sin(2\varphi) & -\frac{1}{2}\sin(2\theta)\cos\varphi & -\frac{1}{2}\sin(2\theta)\sin\varphi & \frac{\sqrt{3}}{2}\sin^2\theta \\ -\cos\theta\sin(2\varphi) & \cos\theta\cos(2\varphi) & \sin\theta\sin\varphi & -\sin\theta\cos\varphi & 0 \\ \frac{\sin(2\theta)\cos(2\varphi)}{2} & \frac{\sin(2\theta)\sin(2\varphi)}{2} & \cos(2\theta)\cos\varphi & \cos(2\theta)\sin\varphi & -\frac{\sqrt{3}}{2}\sin(2\theta) \\ -\sin\theta\sin(2\varphi) & \sin\theta\cos(2\varphi) & -\cos\theta\sin\varphi & \cos\theta\cos\varphi & 0 \\ \frac{\sqrt{3}}{2}\sin^2\theta\cos(2\varphi) & \frac{\sqrt{3}}{2}\sin^2\theta\sin(2\varphi) & \frac{\sqrt{3}}{2}\sin(2\theta)\cos\varphi & \frac{\sqrt{3}}{2}\sin(2\theta)\sin\varphi & \frac{1}{2}(3\cos^2\theta-1) \end{pmatrix} \tag{6.118}$$

We can see that at $\theta=\varphi=0$, $\hat{\Gamma}$ and $\hat{\gamma}$ are all unit matrices.

Equation (6.116) describing five-dimensional Cartesian coordinate system transformation can be written as a matrix form

$$\hat{H}'(\theta,\varphi)=\hat{\gamma}(\theta,\varphi)\hat{H} \tag{6.119}$$

where $\hat{H}$ is a five-dimensional tensor in space coordinates, and each of its components is a $3\times 3$ matrix in the spin space. Similar to Eq. (6.111), the following particle polarization tensor components in the five-dimensional orthogonal Cartesian coordinate system are introduced as

$$\left(h_{xx-yy}, h_{xy}, h_{xz}, h_{yz}, h_{zz}\right) \equiv \left(\sqrt{\frac{1}{6}}p_{xx-yy}, \sqrt{\frac{2}{3}}p_{xy}, \sqrt{\frac{2}{3}}p_{xz}, \sqrt{\frac{2}{3}}p_{yz}, \sqrt{\frac{1}{2}}p_{zz}\right) \tag{6.120}$$

The spin 1 particle polarization tensor belongs to an object property, when the coordinate system does not rotate but only the object rotates, which can be written out according to Eq. (2.205) as

$$\vec{h}'(\theta, \varphi) = \hat{\Gamma}(\theta, \varphi)\vec{h} \tag{6.121}$$

For example, suppose that the components of the orthogonal polarization tensor of a particle prior to rotation are $h_{xx-yy} = h_{xy} = h_{xz} = h_{yz} = 0$, $h_{zz} = h_{ZZ}$, and then the orthogonal polarization tensor of the particle is rotated by angles $(\theta, \varphi)$; or that in the Cartesian coordinate system *XYZ* taking the polarization direction as *Z*-axis, there is only one component $h_{ZZ}$ of the orthogonal polarization tensor of the particle is not zero, and then the *XYZ* coordinate system is rotated inversely by angles $(\theta, \varphi)$, for the both cases the following equation can be written, according to Eq. (6.121), as

$$\begin{pmatrix} h_{xx-yy} \\ h_{xy} \\ h_{xz} \\ h_{yz} \\ h_{zz} \end{pmatrix} = \hat{\Gamma}(\theta, \varphi) \begin{pmatrix} 0 \\ 0 \\ 0 \\ 0 \\ h_{ZZ} \end{pmatrix} \tag{6.122}$$

Then using Eq. (6.117) and from Eq. (6.122), we can obtain

$$\begin{aligned} &h_{xx-yy} = \frac{\sqrt{3}}{2}\sin^2\theta\cos(2\varphi)h_{ZZ}, \quad h_{xy} = \frac{\sqrt{3}}{2}\sin^2\theta\sin(2\varphi)h_{ZZ} \\ &h_{xz} = \frac{\sqrt{3}}{2}\sin(2\theta)\cos\varphi h_{ZZ}, \quad h_{yz} = \frac{\sqrt{3}}{2}\sin(2\theta)\sin\varphi h_{ZZ} \\ &h_{zz} = \frac{1}{2}\left(3\cos^2\theta - 1\right)h_{ZZ} \end{aligned} \tag{6.123}$$

And applying Eq. (6.120) the above formulas can be rewritten as

$$\begin{aligned} &p_{xx-yy} = \frac{3}{2}\sin^2\theta\cos(2\varphi)p_{ZZ}, \quad p_{xy} = \frac{3}{4}\sin^2\theta\sin(2\varphi)p_{ZZ} \\ &p_{xz} = \frac{3}{4}\sin(2\theta)\cos\varphi p_{ZZ}, \quad p_{yz} = \frac{3}{4}\sin(2\theta)\sin\varphi p_{ZZ} \\ &p_{zz} = \frac{1}{2}\left(3\cos^2\theta - 1\right)p_{ZZ} \end{aligned} \tag{6.124}$$

These formulas are identical to the formulas Eqs. (3.176) and (3.177) obtained by other means.

Based on Eq. (1.55), the following generalized Pauli matrix has been introduced from Eq. (4.73):

$$\hat{\Sigma}_M(S) = \hat{T}_{1M}(S) = \sqrt{\frac{3}{S(S+1)}}\,\hat{S}_M(S), \quad M = \pm 1, 0 \tag{6.125}$$

where $\hat{S}_M(S)$ is the spin operator of the spin $S$ particle. When $S = 1$ from Eq. (6.125) we can obtain

$$\hat{\Sigma}_M(1) = \sqrt{\frac{3}{2}}\,\hat{S}_M(1), \quad M = \pm 1, 0 \tag{6.126}$$

According to Eq. (1.22) we can get

$$\hat{\Sigma}_x = \frac{1}{\sqrt{2}}\left(\hat{\Sigma}_{-1} - \hat{\Sigma}_1\right), \quad \hat{\Sigma}_y = \frac{\mathrm{i}}{\sqrt{2}}\left(\hat{\Sigma}_{-1} + \hat{\Sigma}_1\right), \quad \hat{\Sigma}_z = \hat{\Sigma}_0 \tag{6.127}$$

The following spin $S = 1$ particle polarization vector components in three- dimensional orthogonal Cartesian coordinate system are introduced as

$$h_i = \sqrt{\frac{3}{2}}\,p_i, \quad i = x, y, z \tag{6.128}$$

If throughout the $S = 1$ particle polarization theory, the basis polarization tensor functions and the particle polarization tensor components in the five-dimensional Cartesian coordinate system introduced by Eqs. (6.111) and (6.120) are used, and the basis spin functions and the particle polarization vector components in the three-dimensional Cartesian coordinate system introduced by Eqs. (6.127) and (6.128) are used, the spin 1 particle density matrix given by Eq. (3.214) becomes

$$\hat{\rho}_{\text{in}} = \frac{1}{3}\left(\hat{I} + \sum_{i=x,\,y,\,z} h_i\hat{\Sigma}_i + \sum_{\substack{i=xx-yy,\,xy,\\ xz,\,yz,\,zz}} h_i\hat{H}_i\right) \tag{6.129}$$

It can be seen that the use of orthogonal Cartesian coordinate system in the derivation of theoretical formulas will reduce some of the trouble in coefficient processing.

The outgoing particle polarization rates $P_{1i}(i = x, y, z, xy, xz, yz, xx - yy, zz)$ at the point $O_1$ after the first reaction can be obtained from Eq. (6.105); these are the calculation results in $O_0$ system. The corresponding polarization rates $p_{1i}$ of the incident particles for the second reaction in $O_1$ system are

$$\begin{pmatrix} p_{1x} \\ p_{1y} \\ p_{1z} \end{pmatrix} = \hat{b}(\theta_1, \varphi_1) \begin{pmatrix} P_{1x} \\ P_{1y} \\ P_{1z} \end{pmatrix}, \quad \begin{pmatrix} \sqrt{\frac{1}{6}} p_{1\,xx-yy} \\ \sqrt{\frac{2}{3}} p_{1xy} \\ \sqrt{\frac{2}{3}} p_{1xz} \\ \sqrt{\frac{2}{3}} p_{1yz} \\ \sqrt{\frac{1}{2}} p_{1zz} \end{pmatrix} = \hat{\gamma}(\theta_1, \varphi_1) \begin{pmatrix} \sqrt{\frac{1}{6}} P_{1\,xx-yy} \\ \sqrt{\frac{2}{3}} P_{1xy} \\ \sqrt{\frac{2}{3}} P_{1xz} \\ \sqrt{\frac{2}{3}} P_{1yz} \\ \sqrt{\frac{1}{2}} P_{1zz} \end{pmatrix} \tag{6.130}$$

The expressions of $\hat{b}$ and $\hat{\gamma}$ have been given by Eq. (6.43) and Eq. (6.118), respectively. The energy of the emitted particles appeared at the point $O_1$ after the first reaction is $E_1$, and then using the polarization nuclear database of the $\vec{1}$ + A→ $\vec{1}$ + B reaction and Eq. (6.71), the differential cross section of the emitted particles after the reaction occurred at point $O_1$ can be obtained. In the case without considering the other reaction channels, after sampling, assume that in the $O_1$ system the third reaction occurs at the point $O_2$ where is at the direction $(\theta_2, \varphi_2)$ and the distance $L_2$ from the point $O_1$. The coordinates of the point $O_2$ in the $O_1$ system are

$$x_2 = L_2 \sin\theta_2 \cos\varphi_2, \quad y_2 = L_2 \sin\theta_2 \sin\varphi_2, \quad z_2 = L_2 \cos\theta_2 \tag{6.131}$$

When transforming the coordinates of the point $O_2$ to the $O_0$ system, but still taking $O_1$ as the coordinate origin, there is

$$\begin{pmatrix} X_2 \\ Y_2 \\ Z_2 \end{pmatrix} = \hat{B}(\theta_1, \varphi_1) \begin{pmatrix} x_2 \\ y_2 \\ z_2 \end{pmatrix} \tag{6.132}$$

The matrix $\hat{B}$ has been given by Eq. (2.203) or Eq. (6.42). The particle coordinates at the point $O_2$ in the $O_0$ laboratory system are

$$X = X_1 + X_2, \quad Y = Y_1 + Y_2, \quad Z = Z_1 + Z_2 \tag{6.133}$$

In the $O_1$ system, the Stokes vectors of the incident particles and the outgoing particles at the place $O_2$ are as follows, respectively:

$$S_2^{(\mathrm{i})}(\theta_2) = I_0(\theta_2)\begin{pmatrix} 1 \\ \overline{p}_{1x} \\ \overline{p}_{1y} \\ \overline{p}_{1z} \\ \overline{p}_{1xy} \\ \overline{p}_{1xz} \\ \overline{p}_{1yz} \\ \overline{p}_{1\,xx-yy} \\ \overline{p}_{1zz} \end{pmatrix}, \quad S_2^{(\mathrm{f})}(\theta_2,\varphi_2) = I(\theta_2,\varphi_2)\begin{pmatrix} 1 \\ P_{2x}(\theta_2,\varphi_2) \\ P_{2y}(\theta_2,\varphi_2) \\ P_{2z}(\theta_2,\varphi_2) \\ P_{2xy}(\theta_2,\varphi_2) \\ P_{2xz}(\theta_2,\varphi_2) \\ P_{2yz}(\theta_2,\varphi_2) \\ P_{2\,xx-yy}(\theta_2,\varphi_2) \\ P_{2zz}(\theta_2,\varphi_2) \end{pmatrix} \tag{6.134}$$

The definition of $\overline{p}_{1i}$ is given by Eq. (6.103). The above two vectors satisfy the following relation:

$$S_2^{(\mathrm{f})} = Z_2 S_2^{(\mathrm{i})} \tag{6.135}$$

The expression of $Z_2$ can be obtained by changing $(\theta_1, \varphi_1)$ appeared in $Z_1$ given by Eq. (6.106) to $(\theta_2, \varphi_2)$, but the corresponding energy is $E_1$. Equation (6.135) is determined by the second reaction.

The above method can be extended to the third, fourth, etc. reaction processes. Note that the coordinates $(X, Y, Z)$ and other physical quantities of the last outgoing particle in the $O_0$ laboratory system must always be tracked.

## 6.4 Polarized Particle Transport Theory for $\vec{1} + \mathrm{A} \to \vec{\frac{1}{2}} + \mathrm{B}$ and $\vec{\frac{1}{2}} + \mathrm{A} \to \vec{1} + \mathrm{B}$ Reactions

### *6.4.1 Polarization Quantities for* $\vec{1} + \mathrm{A} \to \vec{\frac{1}{2}} + \mathrm{B}$ *Reactions Related to Azimuthal Angle*

In the $\vec{1} + \mathrm{A} \to \vec{\frac{1}{2}} + \mathrm{B}$ reaction, the spin of the incident particle is 1, and the spin of the outgoing particle is $\frac{1}{2}$. The spin of the target nucleus A and the residual nucleus B is $I$ and $I'$, respectively, and both are unpolarized. Equation (3.567) has given the reaction amplitude with definite spin magnetic quantum number $M'_I M_I$ as

$$\hat{\mathcal{F}}_{M_I'M_I}(\theta,\varphi)=\begin{pmatrix} A_{M_I'M_I}(\theta)\,e^{\frac{i}{2}\varphi} & -\mathrm{i}C_{M_I'M_I}(\theta)\,e^{-\frac{i}{2}\varphi} & -E_{M_I'M_I}(\theta)\,e^{-\frac{3i}{2}\varphi} \\ -F_{M_I'M_I}(\theta)\,e^{\frac{3i}{2}\varphi} & \mathrm{i}D_{M_I'M_I}(\theta)\,e^{\frac{i}{2}\varphi} & B_{M_I'M_I}(\theta)\,e^{-\frac{i}{2}\varphi} \end{pmatrix} e^{\mathrm{i}(M_I-M_I')\varphi} \tag{6.136}$$

where $A_{M_I'M_I}$,$B_{M_I'M_I}$, etc. have been given by Eqs. (3.554)–(3.559). Because in the calculation of the polarization quantities, when both the target nucleus and the residual nucleus are unpolarized, $\hat{\mathcal{F}}_{M_I'M_I}$ and $\hat{\mathcal{F}}^{+}_{M_I'M_I}$ always appear at the same time, the factor $\mathrm{e}^{\mathrm{i}(M_I-M_I')\varphi}$ must be eliminated, so we introduce

$$\hat{F}=\begin{pmatrix} A\mathrm{e}^{\frac{i}{2}\varphi} & -\mathrm{i}C\mathrm{e}^{-\frac{i}{2}\varphi} & -E\mathrm{e}^{-\frac{3i}{2}\varphi} \\ -F\mathrm{e}^{\frac{3i}{2}\varphi} & \mathrm{i}D\mathrm{e}^{\frac{i}{2}\varphi} & B\mathrm{e}^{-\frac{i}{2}\varphi} \end{pmatrix},\quad \hat{F}^{+}=\begin{pmatrix} A^*\mathrm{e}^{-\frac{i}{2}\varphi} & -F^*\mathrm{e}^{-\frac{3i}{2}\varphi} \\ \mathrm{i}C^*\mathrm{e}^{\frac{i}{2}\varphi} & -\mathrm{i}D^*\mathrm{e}^{-\frac{i}{2}\varphi} \\ -E^*\mathrm{e}^{\frac{3i}{2}\varphi} & B^*\mathrm{e}^{\frac{i}{2}\varphi} \end{pmatrix} \tag{6.137}$$

where the footnote $M_I'M_I$ and the argument $(\theta)$ of the symbols $A$, $B$, etc. are omitted and the footnote $M_I'M_I$ and argument $(\theta,\varphi)$ of $\hat{F}$ and $\hat{F}^{+}$ are also omitted. From Eq. (6.137) we can obtain

$$\hat{F}\hat{F}^{+}=\begin{pmatrix} |A|^2+|C|^2+|E|^2 & -(AF^*+CD^*+EB^*)\,\mathrm{e}^{-\mathrm{i}\varphi} \\ -(FA^*+DC^*+BE^*)\,\mathrm{e}^{\mathrm{i}\varphi} & |F|^2+|D|^2+|B|^2 \end{pmatrix} \tag{6.138}$$

Also using the symbol introduced by Eq. (6.8), the differential cross section of the unpolarized incident particles can be obtained by using Eqs. (3.641)–(3.643) as

$$I_0(\theta)=\frac{1}{3}\mathrm{tr}\left(\hat{F}\hat{F}^{+}\right)=\frac{2}{3}\tilde{\Sigma}\left(|A|^2+|C|^2+|E|^2\right) \tag{6.139}$$

where the footnotes $\alpha'n'$, $\alpha n$ of the outgoing and incident channels of $I_0(\theta)$ are omitted.

Utilizing Eq. (3.33) and Eq. (6.137) we can obtain

$$\hat{S}_x\hat{F}^{+}=\frac{1}{\sqrt{2}}\begin{pmatrix} \mathrm{i}C^*\mathrm{e}^{\frac{i}{2}\varphi} & -\mathrm{i}D^*\mathrm{e}^{-\frac{i}{2}\varphi} \\ A^*\mathrm{e}^{-\frac{i}{2}\varphi}-E^*\mathrm{e}^{\frac{3i}{2}\varphi} & -F^*\mathrm{e}^{-\frac{3i}{2}\varphi}+B^*\mathrm{e}^{\frac{i}{2}\varphi} \\ \mathrm{i}C^*\mathrm{e}^{\frac{i}{2}\varphi} & -\mathrm{i}D^*e^{-\frac{i}{2}\varphi} \end{pmatrix},$$

$$\hat{S}_y\hat{F}^{+}=\frac{\mathrm{i}}{\sqrt{2}}\begin{pmatrix} -\mathrm{i}C^*\mathrm{e}^{\frac{i}{2}\varphi} & \mathrm{i}D^*\mathrm{e}^{-\frac{i}{2}\varphi} \\ A^*\mathrm{e}^{-\frac{i}{2}\varphi}+E^*\mathrm{e}^{\frac{3i}{2}\varphi} & -F^*\mathrm{e}^{-\frac{3i}{2}\varphi}-B^*\mathrm{e}^{\frac{i}{2}\varphi} \\ \mathrm{i}C^*\mathrm{e}^{\frac{i}{2}\varphi} & -\mathrm{i}D^*\mathrm{e}^{-\frac{i}{2}\varphi} \end{pmatrix}, \tag{6.140}$$

$$\hat{S}_z\hat{F}^{+}=\begin{pmatrix} A^*\mathrm{e}^{-\frac{i}{2}\varphi} & -F^*\mathrm{e}^{-\frac{3i}{2}\varphi} \\ 0 & 0 \\ E^*\mathrm{e}^{\frac{3i}{2}\varphi} & -B^*\mathrm{e}^{\frac{i}{2}\varphi} \end{pmatrix}$$

$$\hat{F}\hat{S}_x\hat{F}^+ = \frac{\mathrm{i}}{\sqrt{2}}$$
$$\begin{pmatrix} AC^*\mathrm{e}^{\mathrm{i}\varphi} - CA^*\mathrm{e}^{-\mathrm{i}\varphi} + CE^*\mathrm{e}^{\mathrm{i}\varphi} - EC^*\mathrm{e}^{-\mathrm{i}\varphi} & -AD^* + CF^*\mathrm{e}^{-2\mathrm{i}\varphi} - CB^* + ED^*\mathrm{e}^{-2\mathrm{i}\varphi} \\ -FC^*\mathrm{e}^{2\mathrm{i}\varphi} + DA^* - DE^*\mathrm{e}^{2\mathrm{i}\varphi} + BC^* & FD^*\mathrm{e}^{\mathrm{i}\varphi} - DF^*\mathrm{e}^{-\mathrm{i}\varphi} + DB^*\mathrm{e}^{\mathrm{i}\varphi} - BD^*\mathrm{e}^{-\mathrm{i}\varphi} \end{pmatrix},$$
$$\hat{F}\hat{S}_y\hat{F}^+ = \frac{1}{\sqrt{2}}$$
$$\begin{pmatrix} AC^*\mathrm{e}^{\mathrm{i}\varphi} + CA^*\mathrm{e}^{-\mathrm{i}\varphi} + CE^*\mathrm{e}^{\mathrm{i}\varphi} + EC^*\mathrm{e}^{-\mathrm{i}\varphi} & -AD^* - CF^*\mathrm{e}^{-2\mathrm{i}\varphi} - CB^* - ED^*\mathrm{e}^{-2\mathrm{i}\varphi} \\ -FC^*\mathrm{e}^{2\mathrm{i}\varphi} - DA^* - DE^*\mathrm{e}^{2\mathrm{i}\varphi} - BC^* & FD^*\mathrm{e}^{\mathrm{i}\varphi} + DF^*\mathrm{e}^{-\mathrm{i}\varphi} + DB^*\mathrm{e}^{\mathrm{i}\varphi} + BD^*\mathrm{e}^{-\mathrm{i}\varphi} \end{pmatrix},$$
$$\hat{F}\hat{S}_z\hat{F}^+ = \begin{pmatrix} |A|^2 - |E|^2 & -(AF^* - EB^*)\,\mathrm{e}^{-\mathrm{i}\varphi} \\ -(FA^* - BE^*)\,\mathrm{e}^{\mathrm{i}\varphi} & |F|^2 - |B|^2 \end{pmatrix} \tag{6.141}$$

Then from Eq. (6.141) we can find

$$\begin{aligned} \frac{1}{\sqrt{2}}\mathrm{tr}\left(\hat{F}\hat{S}_x\hat{F}^+\right) &= -\,\mathrm{Im}\left[(AC^* + CE^* + FD^* + DB^*)\,\mathrm{e}^{\mathrm{i}\varphi}\right], \\ \frac{1}{\sqrt{2}}\mathrm{tr}\left(\hat{F}\hat{S}_y\hat{F}^+\right) &= \mathrm{Re}\left[(AC^* + CE^* + FD^* + DB^*)\,\mathrm{e}^{\mathrm{i}\varphi}\right], \\ \mathrm{tr}\left(\hat{F}\hat{S}_z\hat{F}^+\right) &= |A|^2 - |E|^2 + |F|^2 - |B|^2 \end{aligned} \tag{6.142}$$

The following results can be obtained from Eq. (6.142) and using Eqs. (3.638)–(3.642):

$$\begin{aligned} \frac{1}{\sqrt{2}}\widetilde{\Sigma}\mathrm{tr}\left(\hat{F}\hat{S}_x\hat{F}^+\right) &= -\,2\widetilde{\Sigma}\,\mathrm{Re}\,(AC^* + CE^*)\sin\varphi, \\ \frac{1}{\sqrt{2}}\widetilde{\Sigma}\mathrm{tr}\left(\hat{F}\hat{S}_y\hat{F}^+\right) &= 2\widetilde{\Sigma}\,\mathrm{Re}\,(AC^* + CE^*)\cos\varphi, \\ \widetilde{\Sigma}\,\mathrm{tr}\left(\hat{F}\hat{S}_z\hat{F}^+\right) &= 0 \end{aligned} \tag{6.143}$$

Define

$$A_y(\theta) = \frac{2\sqrt{2}}{3I_0}\widetilde{\Sigma}\,\mathrm{Re}\,(AC^* + CE^*) \tag{6.144}$$

and

$$\overline{A}_i(\theta, \varphi) = \frac{1}{3I_0}\widetilde{\Sigma}\mathrm{tr}\left(\hat{F}\hat{S}_i\hat{F}^+\right), \qquad i = x,y,z \tag{6.145}$$

Then the following results can be obtained from Eqs. (6.143)–(6.145):

$$\overline{A}_x(\theta,\varphi) = -A_y(\theta)\sin\varphi, \quad \overline{A}_y(\theta,\varphi) = A_y(\theta)\cos\varphi, \quad \overline{A}_z(\theta,\varphi) = 0 \tag{6.146}$$

Utilizing Eq. (3.36) and Eq. (6.137) we can find

$$\hat{Q}_{xy}\hat{F}^+ = \frac{3\mathrm{i}}{2}\begin{pmatrix} E^*\mathrm{e}^{\frac{3\mathrm{i}}{2}\varphi} & -B^*\mathrm{e}^{\frac{\mathrm{i}}{2}\varphi} \\ 0 & 0 \\ A^*\mathrm{e}^{-\frac{\mathrm{i}}{2}\varphi} & -F^*\mathrm{e}^{-\frac{3\mathrm{i}}{2}\varphi} \end{pmatrix},$$
$$\hat{Q}_{xz}\hat{F}^+ = \frac{3}{2\sqrt{2}}\begin{pmatrix} \mathrm{i}C^*\mathrm{e}^{\frac{\mathrm{i}}{2}\varphi} & -\mathrm{i}D^*\mathrm{e}^{-\frac{\mathrm{i}}{2}\varphi} \\ A^*\mathrm{e}^{-\frac{\mathrm{i}}{2}\varphi} + E^*\mathrm{e}^{\frac{3\mathrm{i}}{2}\varphi} & -F^*\mathrm{e}^{-\frac{3\mathrm{i}}{2}\varphi} - B^*\mathrm{e}^{\frac{\mathrm{i}}{2}\varphi} \\ -\mathrm{i}C^*\mathrm{e}^{\frac{\mathrm{i}}{2}\varphi} & \mathrm{i}D^*\mathrm{e}^{-\frac{\mathrm{i}}{2}\varphi} \end{pmatrix},$$
$$\hat{Q}_{yz}\hat{F}^+ = \frac{3\mathrm{i}}{2\sqrt{2}}\begin{pmatrix} -\mathrm{i}C^*\mathrm{e}^{\frac{\mathrm{i}}{2}\varphi} & \mathrm{i}D^*\mathrm{e}^{-\frac{\mathrm{i}}{2}\varphi} \\ A^*\mathrm{e}^{-\frac{\mathrm{i}}{2}\varphi} - E^*\mathrm{e}^{\frac{3\mathrm{i}}{2}\varphi} & -F^*\mathrm{e}^{-\frac{3\mathrm{i}}{2}\varphi} + B^*\mathrm{e}^{\frac{\mathrm{i}}{2}\varphi} \\ -\mathrm{i}C^*\mathrm{e}^{\frac{\mathrm{i}}{2}\varphi} & \mathrm{i}D^*\mathrm{e}^{-\frac{\mathrm{i}}{2}\varphi} \end{pmatrix},$$
$$\hat{Q}_{xx-yy}\hat{F}^+ = 3\begin{pmatrix} -E^*\mathrm{e}^{\frac{3\mathrm{i}}{2}\varphi} & B^*\mathrm{e}^{\frac{\mathrm{i}}{2}\varphi} \\ 0 & 0 \\ A^*\mathrm{e}^{-\frac{\mathrm{i}}{2}\varphi} & -F^*\mathrm{e}^{-\frac{3\mathrm{i}}{2}\varphi} \end{pmatrix}, \quad \hat{Q}_{zz}\hat{F}^+ = \begin{pmatrix} A^*\mathrm{e}^{-\frac{\mathrm{i}}{2}\varphi} & -F^*\mathrm{e}^{-\frac{3\mathrm{i}}{2}\varphi} \\ -2\mathrm{i}C^*\mathrm{e}^{\frac{\mathrm{i}}{2}\varphi} & 2\mathrm{i}D^*\mathrm{e}^{-\frac{\mathrm{i}}{2}\varphi} \\ -E^*\mathrm{e}^{\frac{3\mathrm{i}}{2}\varphi} & B^*\mathrm{e}^{\frac{\mathrm{i}}{2}\varphi} \end{pmatrix} \tag{6.147}$$

$$\hat{F}\hat{Q}_{xy}\hat{F}^+ = \frac{3\mathrm{i}}{2}\begin{pmatrix} AE^*\mathrm{e}^{2\mathrm{i}\varphi} - EA^*\mathrm{e}^{-2\mathrm{i}\varphi} & -AB^*\mathrm{e}^{\mathrm{i}\varphi} + EF^*\mathrm{e}^{-3\mathrm{i}\varphi} \\ -FE^*\mathrm{e}^{3\mathrm{i}\varphi} + BA^*\mathrm{e}^{-\mathrm{i}\varphi} & FB^*\mathrm{e}^{2\mathrm{i}\varphi} - BF^*\mathrm{e}^{-2\mathrm{i}\varphi} \end{pmatrix},$$
$$\hat{F}\hat{Q}_{xz}\hat{F}^+ = \frac{3\mathrm{i}}{2\sqrt{2}}$$
$$\begin{pmatrix} AC^*\mathrm{e}^{\mathrm{i}\varphi} - CA^*\mathrm{e}^{-\mathrm{i}\varphi} - CE^*\mathrm{e}^{\mathrm{i}\varphi} + EC^*\mathrm{e}^{-\mathrm{i}\varphi} & -\left(AD^* - CF^*\mathrm{e}^{-2\mathrm{i}\varphi} - CB^* + ED^*\mathrm{e}^{-2\mathrm{i}\varphi}\right) \\ -\left(FC^*\mathrm{e}^{2\mathrm{i}\varphi} - DA^* - DE^*\mathrm{e}^{2\mathrm{i}\varphi} + BC^*\right) & FD^*\mathrm{e}^{\mathrm{i}\varphi} - DF^*\mathrm{e}^{-\mathrm{i}\varphi} - DB^*\mathrm{e}^{\mathrm{i}\varphi} + BD^*\mathrm{e}^{-\mathrm{i}\varphi} \end{pmatrix},$$
$$\hat{F}\hat{Q}_{yz}\hat{F}^+ = \frac{3}{2\sqrt{2}}$$
$$\begin{pmatrix} AC^*\mathrm{e}^{\mathrm{i}\varphi} + CA^*\mathrm{e}^{-\mathrm{i}\varphi} - CE^*\mathrm{e}^{\mathrm{i}\varphi} - EC^*\mathrm{e}^{-\mathrm{i}\varphi} & -AD^* - CF^*\mathrm{e}^{-2\mathrm{i}\varphi} + CB^* + ED^*\mathrm{e}^{-2\mathrm{i}\varphi} \\ -FC^*\mathrm{e}^{2\mathrm{i}\varphi} - DA^* + DE^*\mathrm{e}^{2\mathrm{i}\varphi} + BC^* & FD^*\mathrm{e}^{\mathrm{i}\varphi} + DF^*\mathrm{e}^{-\mathrm{i}\varphi} - DB^*\mathrm{e}^{\mathrm{i}\varphi} - BD^*\mathrm{e}^{-\mathrm{i}\varphi} \end{pmatrix},$$
$$\hat{F}\hat{Q}_{xx-yy}\hat{F}^+ = 3\begin{pmatrix} -AE^*\mathrm{e}^{2\mathrm{i}\varphi} - EA^*\mathrm{e}^{-2\mathrm{i}\varphi} & AB^*\mathrm{e}^{\mathrm{i}\varphi} + EF^*\mathrm{e}^{-3\mathrm{i}\varphi} \\ FE^*\mathrm{e}^{3\mathrm{i}\varphi} + BA^*\mathrm{e}^{-\mathrm{i}\varphi} & -FB^*\mathrm{e}^{2\mathrm{i}\varphi} - BF^*\mathrm{e}^{-2\mathrm{i}\varphi} \end{pmatrix},$$
$$\hat{F}\hat{Q}_{zz}\hat{F}^+ = \begin{pmatrix} |A|^2 - 2|C|^2 + |E|^2 & -(AF^* - 2CD^* + EB^*)\,\mathrm{e}^{-\mathrm{i}\varphi} \\ -(FA^* - 2DC^* + BE^*)\,\mathrm{e}^{\mathrm{i}\varphi} & |F|^2 - 2|D|^2 + |B|^2 \end{pmatrix} \tag{6.148}$$

From Eq. (6.148) we can further find

$$\begin{aligned}
\mathrm{tr}\left(\hat{F}\hat{Q}_{xy}\hat{F}^{+}\right) &= -3\mathrm{Im}\left[(AE^{*}+FB^{*})\,\mathrm{e}^{2\mathrm{i}\varphi}\right],\\
\mathrm{tr}\left(\hat{F}\hat{Q}_{xz}\hat{F}^{+}\right) &= -\frac{3}{\sqrt{2}}\mathrm{Im}\left[(AC^{*}-CE^{*}+FD^{*}-DB^{*})\,\mathrm{e}^{\mathrm{i}\varphi}\right],\\
\mathrm{tr}\left(\hat{F}\hat{Q}_{yz}\hat{F}^{+}\right) &= \frac{3}{\sqrt{2}}\,\mathrm{Re}\left[(AC^{*}-CE^{*}+FD^{*}-DB^{*})\,\mathrm{e}^{\mathrm{i}\varphi}\right],\\
\mathrm{tr}\left(\hat{F}\hat{Q}_{xx-yy}\hat{F}^{+}\right) &= -6\,\mathrm{Re}\left[(AE^{*}+FB^{*})\,\mathrm{e}^{2\mathrm{i}\varphi}\right],\\
\mathrm{tr}\left(\hat{F}\hat{Q}_{zz}\hat{F}^{+}\right) &= |A|^{2}+|E|^{2}+|F|^{2}+|B|^{2}-2\left(|C|^{2}+|D|^{2}\right)
\end{aligned} \tag{6.149}$$

The following results can be obtained from Eq. (6.149) and using Eqs. (3.638)–(3.642):

$$\begin{aligned}
\widetilde{\Sigma}\mathrm{tr}\left(\hat{F}\hat{Q}_{xy}\hat{F}^{+}\right) &= -6\widetilde{\Sigma}\,\mathrm{Re}\,(AE^{*})\sin(2\varphi),\\
\widetilde{\Sigma}\mathrm{tr}\left(\hat{F}\hat{Q}_{xz}\hat{F}^{+}\right) &= -3\sqrt{2}\,\widetilde{\Sigma}\mathrm{Im}(AC^{*}-CE^{*})\cos\varphi,\\
\widetilde{\Sigma}\mathrm{tr}\left(\hat{F}\hat{Q}_{yz}\hat{F}^{+}\right) &= -3\sqrt{2}\,\widetilde{\Sigma}\mathrm{Im}(AC^{*}-CE^{*})\sin\varphi,\\
\widetilde{\Sigma}\mathrm{tr}\left(\hat{F}\hat{Q}_{xx-yy}\hat{F}^{+}\right) &= -12\widetilde{\Sigma}\,\mathrm{Re}\,(AE^{*})\cos(2\varphi),\\
\widetilde{\Sigma}\mathrm{tr}\left(\hat{F}\hat{Q}_{zz}\hat{F}^{+}\right) &= 2\widetilde{\Sigma}\left(|A|^{2}+|E|^{2}-2|C|^{2}\right)
\end{aligned} \tag{6.150}$$

Define

$$\begin{aligned}
&A_{1}(\theta)=\frac{2}{I_{0}}\widetilde{\Sigma}\,\mathrm{Re}\,(AE^{*}),\qquad A_{2}(\theta)=\frac{\sqrt{2}}{I_{0}}\widetilde{\Sigma}\mathrm{Im}(AC^{*}-CE^{*}),\\
&A_{3}(\theta)=\frac{2}{3I_{0}}\widetilde{\Sigma}\left(|A|^{2}+|E|^{2}-2|C|^{2}\right)
\end{aligned} \tag{6.151}$$

and

$$\overline{A}_{i}(\theta,\varphi)=\frac{1}{3I_{0}}\widetilde{\Sigma}\mathrm{tr}\left(\hat{F}\hat{Q}_{i}\hat{F}^{+}\right),\qquad i=\gamma \tag{6.152}$$

Then we can obtain the following results from Eqs. (6.150)–(6.152):

$$\begin{aligned}
&\overline{A}_{xy}(\theta,\varphi)=-A_{1}(\theta)\sin(2\varphi),\quad \overline{A}_{xz}(\theta,\varphi)=-A_{2}(\theta)\cos\varphi,\\
&\overline{A}_{yz}(\theta,\varphi)=-A_{2}(\theta)\sin\varphi,\qquad \overline{A}_{xx-yy}(\theta,\varphi)=-2A_{1}(\theta)\cos(2\varphi),\\
&\overline{A}_{zz}(\theta,\varphi)=A_{3}(\theta)
\end{aligned} \tag{6.153}$$

Referring to Eq. (3.214), we can write the differential cross section of the outgoing particles corresponding to the polarized incident particles as

$$\begin{aligned}I(\theta, \varphi) = I_0(\theta)\Big\{1 &+ \frac{3}{2}\left[p_x\overline{A}_x(\theta, \varphi) + p_y\overline{A}_y(\theta, \varphi)\right] \\ &+ \frac{2}{3}\left[p_{xy}\overline{A}_{xy}(\theta, \varphi) + p_{xz}\overline{A}_{xz}(\theta, \varphi) + p_{yz}\overline{A}_{yz}(\theta, \varphi)\right] \\ &+ \frac{1}{6}p_{xx-yy}\overline{A}_{xx-yy}(\theta, \varphi) + \frac{1}{2}p_{zz}\overline{A}_{zz}(\theta, \varphi)\Big\}\end{aligned} \tag{6.154}$$

From Eq. (2.6) and Eq. (6.138) we can obtain

$$\begin{aligned}\mathrm{tr}\left(\hat{\sigma}_x\hat{F}\hat{F}^+\right) &= -2\,\mathrm{Re}\left[(FA^* + DC^* + BE^*)\,\mathrm{e}^{\mathrm{i}\varphi}\right], \\ \mathrm{tr}\left(\hat{\sigma}_y\hat{F}\hat{F}^+\right) &= -2\mathrm{Im}\left[(FA^* + DC^* + BE^*)\,\mathrm{e}^{\mathrm{i}\varphi}\right], \\ \mathrm{tr}\left(\hat{\sigma}_z\hat{F}\hat{F}^+\right) &= |A|^2 + |C|^2 + |E|^2 - |F|^2 - |D|^2 - |B|^2\end{aligned} \tag{6.155}$$

Using the similar method mentioned above, the following results can be obtained from Eq. (6.155)

$$\begin{aligned}\widetilde{\Sigma}\mathrm{tr}\left(\hat{\sigma}_x\hat{F}\hat{F}^+\right) &= -2\widetilde{\Sigma}\,\mathrm{Re}\,(2FA^* + DC^*)\cos\varphi, \\ \widetilde{\Sigma}\mathrm{tr}\left(\hat{\sigma}_y\hat{F}\hat{F}^+\right) &= -2\widetilde{\Sigma}\,\mathrm{Re}\,(2FA^* + DC^*)\sin\varphi, \\ \widetilde{\Sigma}\mathrm{tr}\left(\hat{\sigma}_Z\hat{F}\hat{F}^+\right) &= 0\end{aligned} \tag{6.156}$$

Define

$$P^x(\theta) = \frac{2}{3I_0}\widetilde{\Sigma}\,\mathrm{Re}\,(2FA^* + DC^*) \tag{6.157}$$

and

$$\overline{P}^i(\theta, \varphi) = \frac{1}{3I_0}\widetilde{\Sigma}\mathrm{tr}\left(\hat{\sigma}_i\hat{F}\hat{F}^+\right), \qquad i = x,y,z \tag{6.158}$$

Then we can obtain the following results from Eqs. (6.155)–(6.158):

$$\overline{P}^x(\theta, \varphi) = -P^x(\theta)\cos\varphi, \quad \overline{P}^y(\theta, \varphi) = -P^x(\theta)\sin\varphi, \quad \overline{P}^z(\theta, \varphi) = 0 \tag{6.159}$$

From Eq. (2.6) and Eq. (6.141), we can also find

$$
\begin{aligned}
\mathrm{tr}\left(\hat{\sigma}_x\hat{F}\hat{S}_x\hat{F}^+\right) &= -\frac{\mathrm{i}}{\sqrt{2}}\left(FC^*\mathrm{e}^{2\mathrm{i}\varphi} - DA^* + DE^*\mathrm{e}^{2\mathrm{i}\varphi} - BC^*\right.\\
&\quad \left. + AD^* - CF^*\mathrm{e}^{-2\mathrm{i}\varphi} + CB^* - ED^*\mathrm{e}^{-2\mathrm{i}\varphi}\right)\\
&= \sqrt{2}\mathrm{Im}\left[(AD^* + CB^*) + (FC^* + DE^*)\,\mathrm{e}^{2\mathrm{i}\varphi}\right],\\
\mathrm{tr}\left(\hat{\sigma}_y\hat{F}\hat{S}_x\hat{F}^+\right) &= \frac{1}{\sqrt{2}}\left(-FC^*\mathrm{e}^{2\mathrm{i}\varphi} + DA^* - DE^*\mathrm{e}^{2\mathrm{i}\varphi} + BC^*\right.\\
&\quad \left. + AD^* - CF^*\mathrm{e}^{-2\mathrm{i}\varphi} + CB^* - ED^*\mathrm{e}^{-2\mathrm{i}\varphi}\right)\\
&= \sqrt{2}\,\mathrm{Re}\left[(AD^* + CB^*) - (FC^* + DE^*)\,\mathrm{e}^{2\mathrm{i}\varphi}\right],\\
\mathrm{tr}\left(\hat{\sigma}_z\hat{F}\hat{S}_x\hat{F}^+\right) &= -\sqrt{2}\mathrm{Im}\left[(AC^* + CE^* - FD^* - DB^*)\,\mathrm{e}^{\mathrm{i}\varphi}\right],\\
\mathrm{tr}\left(\hat{\sigma}_x\hat{F}\hat{S}_y\hat{F}^+\right) &= -\sqrt{2}\,\mathrm{Re}\left[(AD^* + CB^*) + (FC^* + DE^*)\,\mathrm{e}^{2\mathrm{i}\varphi}\right],\\
\mathrm{tr}\left(\hat{\sigma}_y\hat{F}\hat{S}_y\hat{F}^+\right) &= \sqrt{2}\mathrm{Im}\left[(AD^* + CB^*) - (FC^* + DE^*)\,\mathrm{e}^{2\mathrm{i}\varphi}\right]\\
\mathrm{tr}\left(\hat{\sigma}_z\hat{F}\hat{S}_y\hat{F}^+\right) &= \sqrt{2}\,\mathrm{Re}\left[(AC^* + CE^* - FD^* - DB^*)\,\mathrm{e}^{\mathrm{i}\varphi}\right]\\
\mathrm{tr}\left(\hat{\sigma}_x\hat{F}\hat{S}_z\hat{F}^+\right) &= -2\,\mathrm{Re}\left[(FA^* - BE^*)\,\mathrm{e}^{\mathrm{i}\varphi}\right]\\
\mathrm{tr}\left(\hat{\sigma}_y\hat{F}\hat{S}_z\hat{F}^+\right) &= -2\mathrm{Im}\left[(FA^* - BE^*)\,\mathrm{e}^{\mathrm{i}\varphi}\right]\\
\mathrm{tr}\left(\hat{\sigma}_z\hat{F}\hat{S}_z\hat{F}^+\right) &= |A|^2 - |E|^2 - |F|^2 + |B|^2
\end{aligned}
\tag{6.160}
$$

Using the similar method, the following results can be obtained from Eq. (6.160):

$$
\begin{aligned}
\widetilde{\Sigma}\mathrm{tr}\left(\hat{\sigma}_x\hat{F}\hat{S}_x\hat{F}^+\right) &= 2\sqrt{2}\,\widetilde{\Sigma}\,\mathrm{Re}\,(FC^*)\sin(2\varphi),\\
\widetilde{\Sigma}\mathrm{tr}\left(\hat{\sigma}_y\hat{F}\hat{S}_x\hat{F}^+\right) &= 2\sqrt{2}\,\widetilde{\Sigma}[\,\mathrm{Re}\,(AD^*) - \mathrm{Re}\,(FC^*)\cos(2\varphi)],\\
\widetilde{\Sigma}\mathrm{tr}\left(\hat{\sigma}_z\hat{F}\hat{S}_x\hat{F}^+\right) &= -2\sqrt{2}\,\widetilde{\Sigma}\mathrm{Im}(AC^* + CE^*)\cos\varphi,\\
\widetilde{\Sigma}\mathrm{tr}\left(\hat{\sigma}_x\hat{F}\hat{S}_y\hat{F}^+\right) &= -2\sqrt{2}\,\widetilde{\Sigma}[\,\mathrm{Re}\,(AD^*) + \mathrm{Re}\,(FC^*)\cos(2\varphi)],\\
\widetilde{\Sigma}\mathrm{tr}\left(\hat{\sigma}_y\hat{F}\hat{S}_y\hat{F}^+\right) &= -2\sqrt{2}\,\widetilde{\Sigma}\,\mathrm{Re}\,(FC^*)\sin(2\varphi),\\
\widetilde{\Sigma}\mathrm{tr}\left(\hat{\sigma}_z\hat{F}\hat{S}_y\hat{F}^+\right) &= -2\sqrt{2}\,\widetilde{\Sigma}\mathrm{Im}(AC^* + CE^*)\sin\varphi,\\
\widetilde{\Sigma}\mathrm{tr}\left(\hat{\sigma}_x\hat{F}\hat{S}_z\hat{F}^+\right) &= 4\widetilde{\Sigma}\mathrm{Im}(FA^*)\sin\varphi,\\
\widetilde{\Sigma}\mathrm{tr}\left(\hat{\sigma}_y\hat{F}\hat{S}_z\hat{F}^+\right) &= -4\widetilde{\Sigma}\mathrm{Im}(FA^*)\cos\varphi,\\
\widetilde{\Sigma}\mathrm{tr}\left(\hat{\sigma}_z\hat{F}\hat{S}_z\hat{F}^+\right) &= 2\widetilde{\Sigma}\left(|A|^2 - |E|^2\right)
\end{aligned}
\tag{6.161}
$$

Define

$$\begin{aligned} R_1(\theta) &= \frac{2\sqrt{2}}{3I_0}\widetilde{\Sigma}\,\mathrm{Re}\,(FC^*), & R_2(\theta) &= \frac{2\sqrt{2}}{3I_0}\widetilde{\Sigma}\,\mathrm{Re}\,(AD^*), \\ Q_1(\theta) &= \frac{2\sqrt{2}}{3I_0}\widetilde{\Sigma}\mathrm{Im}(AC^* + CE^*), & Q_2(\theta) &= \frac{4}{3I_0}\widetilde{\Sigma}\mathrm{Im}(FA^*), \\ W_1(\theta) &= \frac{2}{3I_0}\widetilde{\Sigma}\left(|A|^2 - |E|^2\right) \end{aligned} \tag{6.162}$$

and

$$K_i^j(\theta,\varphi) = \frac{1}{3I_0}\widetilde{\Sigma}\mathrm{tr}\left(\hat{\sigma}_j\hat{F}\hat{S}_i\hat{F}^+\right), \qquad i,j = x,y,z \tag{6.163}$$

Then we can obtain the following results from Eqs. (6.161)–(6.163):

$$\begin{aligned} &K_x^x(\theta,\varphi) = R_1(\theta)\sin(2\varphi), && K_x^y(\theta,\varphi) = R_2(\theta) - R_1(\theta)\cos(2\varphi), \\ &K_x^z(\theta,\varphi) = -Q_1(\theta)\cos\varphi, && K_y^x(\theta,\varphi) = -R_2(\theta) - R_1(\theta)\cos(2\varphi), \\ &K_y^y(\theta,\varphi) = -R_1(\theta)\sin(2\varphi) = -K_x^x(\theta,\varphi), && \\ &K_y^z(\theta,\varphi) = -Q_1(\theta)\sin\varphi, && K_z^x(\theta,\varphi) = Q_2(\theta)\sin\varphi, \\ &K_z^y(\theta,\varphi) = -Q_2(\theta)\cos\varphi, && K_z^z(\theta,\varphi) = W_1(\theta) \end{aligned} \tag{6.164}$$

From Eq. (2.6) and Eq. (6.148), we can also find

$$\begin{aligned} \mathrm{tr}\left(\hat{\sigma}_x\hat{F}\hat{Q}_{xy}\hat{F}^+\right) &= 3\mathrm{Im}\left(AB^*\mathrm{e}^{i\varphi} + FE^*\mathrm{e}^{3i\varphi}\right), \\ \mathrm{tr}\left(\hat{\sigma}_y\hat{F}\hat{Q}_{xy}\hat{F}^+\right) &= 3\,\mathrm{Re}\left(AB^*\mathrm{e}^{i\varphi} - FE^*\mathrm{e}^{3i\varphi}\right), \\ \mathrm{tr}\left(\hat{\sigma}_z\hat{F}\hat{Q}_{xy}\hat{F}^+\right) &= -3\mathrm{Im}\left[(AE^* - FB^*)\,\mathrm{e}^{2i\varphi}\right], \\ \mathrm{tr}\left(\hat{\sigma}_x\hat{F}\hat{Q}_{xz}\hat{F}^+\right) &= \frac{3}{\sqrt{2}}\mathrm{Im}\left[(AD^* - CB^*) + (FC^* - DE^*)\,\mathrm{e}^{2i\varphi}\right], \\ \mathrm{tr}\left(\hat{\sigma}_y\hat{F}\hat{Q}_{xz}\hat{F}^+\right) &= \frac{3}{\sqrt{2}}\,\mathrm{Re}\left[(AD^* - CB^*) - (FC^* - DE^*)\,\mathrm{e}^{2i\varphi}\right], \\ \mathrm{tr}\left(\hat{\sigma}_z\hat{F}\hat{Q}_{xz}\hat{F}^+\right) &= -\frac{3}{\sqrt{2}}\mathrm{Im}\left[(AC^* - CE^* - FD^* + DB^*)\,\mathrm{e}^{i\varphi}\right], \\ \mathrm{tr}\left(\hat{\sigma}_x\hat{F}\hat{Q}_{yz}\hat{F}^+\right) &= -\frac{3}{\sqrt{2}}\,\mathrm{Re}\left[(AD^* - CB^*) + (FC^* - DE^*)\,\mathrm{e}^{2i\varphi}\right], \\ \mathrm{tr}\left(\hat{\sigma}_y\hat{F}\hat{Q}_{yz}\hat{F}^+\right) &= \frac{3}{\sqrt{2}}\mathrm{Im}\left[(AD^* - CB^*) - (FC^* - DE^*)\,\mathrm{e}^{2i\varphi}\right], \\ \mathrm{tr}\left(\hat{\sigma}_z\hat{F}\hat{Q}_{yz}\hat{F}^+\right) &= \frac{3}{\sqrt{2}}\,\mathrm{Re}\left[(AC^* - CE^* - FD^* + DB^*)\,\mathrm{e}^{i\varphi}\right], \end{aligned}$$

$$
\begin{aligned}
&\mathrm{tr}\left(\hat{\sigma}_x\hat{F}\hat{Q}_{xx-yy}\hat{F}^+\right)=6\,\mathrm{Re}\left(AB^*\mathrm{e}^{\mathrm{i}\varphi}+FE^*\mathrm{e}^{3\mathrm{i}\varphi}\right),\\
&\mathrm{tr}\left(\hat{\sigma}_y\hat{F}\hat{Q}_{xx-yy}\hat{F}^+\right)=-6\mathrm{Im}\left(AB^*\mathrm{e}^{\mathrm{i}\varphi}-FE^*\mathrm{e}^{3\mathrm{i}\varphi}\right),\\
&\mathrm{tr}\left(\hat{\sigma}_z\hat{F}\hat{Q}_{xx-yy}\hat{F}^+\right)=-6\,\mathrm{Re}\left[(AE^*-FB^*)\,\mathrm{e}^{2\mathrm{i}\varphi}\right],\\
&\mathrm{tr}\left(\hat{\sigma}_x\hat{F}\hat{Q}_{zz}\hat{F}^+\right)=-2\,\mathrm{Re}\left[(FA^*-2DC^*+BE^*)\,\mathrm{e}^{\mathrm{i}\varphi}\right],\\
&\mathrm{tr}\left(\hat{\sigma}_y\hat{F}\hat{Q}_{zz}\hat{F}^+\right)=-2\mathrm{Im}\left[(FA^*-2DC^*+BE^*)\,\mathrm{e}^{\mathrm{i}\varphi}\right],\\
&\mathrm{tr}\left(\hat{\sigma}_z\hat{F}\hat{Q}_{zz}\hat{F}^+\right)=|A|^2+|E|^2-|F|^2-|B|^2-2\left(|C|^2-|D|^2\right)
\end{aligned}
\tag{6.165}
$$

Using the similar method the following results can be obtained from Eq. (6.165):

$$
\begin{aligned}
&\widetilde{\Sigma}\mathrm{tr}\left(\hat{\sigma}_x\hat{F}\hat{Q}_{xy}\hat{F}^+\right)=3\widetilde{\Sigma}[\,\mathrm{Re}\,(AB^*)\sin\varphi+\,\mathrm{Re}\,(FE^*)\sin(3\varphi)],\\
&\widetilde{\Sigma}\mathrm{tr}\left(\hat{\sigma}_y\hat{F}\hat{Q}_{xy}\hat{F}^+\right)=3\widetilde{\Sigma}[\,\mathrm{Re}\,(AB^*)\cos\varphi-\,\mathrm{Re}\,(FE^*)\cos(3\varphi)],\\
&\widetilde{\Sigma}\mathrm{tr}\left(\hat{\sigma}_z\hat{F}\hat{Q}_{xy}\hat{F}^+\right)=-6\widetilde{\Sigma}\mathrm{Im}(AE^*)\cos(2\varphi),\\
&\widetilde{\Sigma}\mathrm{tr}\left(\hat{\sigma}_x\hat{F}\hat{Q}_{xz}\hat{F}^+\right)=3\sqrt{2}\widetilde{\Sigma}[\mathrm{Im}(AD^*)+\mathrm{Im}(FC^*)\cos(2\varphi)],\\
&\widetilde{\Sigma}\mathrm{tr}\left(\hat{\sigma}_y\hat{F}\hat{Q}_{xz}\hat{F}^+\right)=3\sqrt{2}\widetilde{\Sigma}\mathrm{Im}(FC^*)\sin(2\varphi),\\
&\widetilde{\Sigma}\mathrm{tr}\left(\hat{\sigma}_z\hat{F}\hat{Q}_{xz}\hat{F}^+\right)=-3\sqrt{2}\widetilde{\Sigma}\,\mathrm{Re}\,(AC^*-CE^*)\sin\varphi,\\
&\widetilde{\Sigma}\mathrm{tr}\left(\hat{\sigma}_x\hat{F}\hat{Q}_{yz}\hat{F}^+\right)=3\sqrt{2}\widetilde{\Sigma}\mathrm{Im}(FC^*)\sin(2\varphi),\\
&\widetilde{\Sigma}\mathrm{tr}\left(\hat{\sigma}_y\hat{F}\hat{Q}_{yz}\hat{F}^+\right)=3\sqrt{2}\widetilde{\Sigma}[\mathrm{Im}(AD^*)-\mathrm{Im}(FC^*)\cos(2\varphi)],\\
&\widetilde{\Sigma}\mathrm{tr}\left(\hat{\sigma}_z\hat{F}\hat{Q}_{yz}\hat{F}^+\right)=3\sqrt{2}\widetilde{\Sigma}\,\mathrm{Re}\,(AC^*-CE^*)\cos\varphi,\\
&\widetilde{\Sigma}\mathrm{tr}\left(\hat{\sigma}_x\hat{F}\hat{Q}_{xx-yy}\hat{F}^+\right)=6\widetilde{\Sigma}[\,\mathrm{Re}\,(AB^*)\cos\varphi+\,\mathrm{Re}\,(FE^*)\cos(3\varphi)],\\
&\widetilde{\Sigma}\mathrm{tr}\left(\hat{\sigma}_y\hat{F}\hat{Q}_{xx-yy}\hat{F}^+\right)=-6\widetilde{\Sigma}[\,\mathrm{Re}\,(AB^*)\sin\varphi-\,\mathrm{Re}\,(FE^*)\sin(3\varphi)],\\
&\widetilde{\Sigma}\mathrm{tr}\left(\hat{\sigma}_z\hat{F}\hat{Q}_{xx-yy}\hat{F}^+\right)=12\widetilde{\Sigma}(AE^*)\sin(2\varphi),\\
&\widetilde{\Sigma}\mathrm{tr}\left(\hat{\sigma}_x\hat{F}\hat{Q}_{zz}\hat{F}^+\right)=-4\widetilde{\Sigma}\,\mathrm{Re}\,(FA^*-DC^*)\cos\varphi,\\
&\widetilde{\Sigma}\mathrm{tr}\left(\hat{\sigma}_y\hat{F}\hat{Q}_{zz}\hat{F}^+\right)=-4\widetilde{\Sigma}\,\mathrm{Re}\,(FA^*-DC^*)\sin\varphi,\\
&\widetilde{\Sigma}\mathrm{tr}\left(\hat{\sigma}_z\hat{F}\hat{Q}_{zz}\hat{F}^+\right)=0
\end{aligned}
\tag{6.166}
$$

Define

$$\begin{aligned}
&R_3(\theta)=\frac{1}{I_0}\widetilde{\Sigma}\,\mathrm{Re}\,(AB^*), && R_4(\theta)=\frac{1}{I_0}\widetilde{\Sigma}\,\mathrm{Re}\,(FE^*),\\
&R_5(\theta)=\frac{\sqrt{2}}{I_0}\widetilde{\Sigma}\,\mathrm{Re}\,(AC^*-CE^*), && R_6(\theta)=\frac{4}{3I_0}\widetilde{\Sigma}\,\mathrm{Re}\,(FA^*-DC^*),\\
&Q_3(\theta)=\frac{2}{I_0}\widetilde{\Sigma}\mathrm{Im}(AE^*), && Q_4(\theta)=\frac{\sqrt{2}}{I_0}\widetilde{\Sigma}\mathrm{Im}(AD^*),\\
&Q_5(\theta)=\frac{\sqrt{2}}{I_0}\widetilde{\Sigma}\mathrm{Im}(FC^*)
\end{aligned} \tag{6.167}$$

and

$$K_i^j(\theta,\varphi)=\frac{1}{3I_0}\widetilde{\Sigma}\mathrm{tr}\left(\hat{\sigma}_j\hat{F}\hat{Q}_i\hat{F}^+\right),\qquad i=\gamma;\ \ j=x,y,z \tag{6.168}$$

Then we can obtain the following results from Eqs. (6.166)–(6.168):

$$\begin{aligned}
K_{xy}^x(\theta,\varphi)&=R_3(\theta)\sin\varphi+R_4(\theta)\sin(3\varphi),\\
K_{xy}^y(\theta,\varphi)&=R_3(\theta)\cos\varphi-R_4(\theta)\cos(3\varphi),\\
K_{xy}^z(\theta,\varphi)&=-Q_3(\theta)\cos(2\varphi),\\
K_{xz}^x(\theta,\varphi)&=Q_4(\theta)+Q_5(\theta)\cos(2\varphi),\\
K_{xz}^y(\theta,\varphi)&=Q_5(\theta)\sin(2\varphi),\\
K_{xz}^z(\theta,\varphi)&=-R_5(\theta)\sin\varphi,\\
K_{yz}^x(\theta,\varphi)&=Q_5(\theta)\sin(2\varphi),\\
K_{yz}^y(\theta,\varphi)&=Q_4(\theta)-Q_5(\theta)\cos(2\varphi),\\
K_{yz}^z(\theta,\varphi)&=R_5(\theta)\sin(2\varphi),\\
K_{xx-yy}^x(\theta,\varphi)&=2R_3(\theta)\cos\varphi+2R_4(\theta)\cos(3\varphi),\\
K_{xx-yy}^y(\theta,\varphi)&=-2R_3(\theta)\sin\varphi+2R_4(\theta)\sin(3\varphi),\\
K_{xx-yy}^z(\theta,\varphi)&=2Q_3(\theta)\sin(2\varphi),\\
K_{zz}^x(\theta,\varphi)&=-R_6(\theta)\cos\varphi,\\
K_{zz}^y(\theta,\varphi)&=-R_6(\theta)\sin\varphi,\\
K_{zz}^z(\theta,\varphi)&=0
\end{aligned} \tag{6.169}$$

Referring to Eqs. (3.214), (6.154), and (6.33), we can write the polarization rates related to the angles $\theta$, $\varphi$ simultaneously of the spin $\frac{1}{2}$ outgoing particles corresponding to the $\vec{1}+\mathrm{A}\rightarrow\vec{\frac{1}{2}}+\mathrm{B}$ reaction as

$$P_j(\theta,\varphi)=\frac{I_0(\theta)}{I(\theta,\varphi)}\left[\overline{P}^j(\theta,\ \varphi)+\sum_{i=\varepsilon}\overline{p}_iK_i^j(\theta,\ \varphi)\right],\qquad j=x,y,z \tag{6.170}$$

The definition of the subscript symbol set $\varepsilon$ has been given by Eq. (3.476), and the definition of $\overline{p}_i$ has been given by Eq. (6.103).

Note that if we make $\varphi = 0$, the results obtained in Sect. 6.4.1 will automatically degenerate to the results obtained in Sect. 3.9.

### 6.4.2 Polarization Nuclear Database for $\vec{1} + \mathrm{A} \rightarrow \overrightarrow{\frac{1}{2}} + \mathrm{B}$ Reactions

According to the previous discussion, the polarization nuclear database for $\vec{1} + \mathrm{A} \rightarrow \overrightarrow{\frac{1}{2}} + \mathrm{B}$ reactions should contain the following differential physical quantities of the polar angle $\theta$ for different incident particle energies of the two-body direct reaction channels:

1. Differential cross section corresponding to unpolarized incident particles $I_0(\theta)$.
2. The incident particle vector analyzing power $A_y(\theta)$; tensor analyzing powers $A_i(\theta)$ $(i = 1\sim3)$.
3. The outgoing particle vector polarization rates corresponding to the unpolarized incident particle $P^x(\theta)$.
4. Polarization transfer coefficient components $R_i(\theta)$ $(i = 1\sim6)$, $Q_i(\theta)$ $(i = 1 - 5)$, $W_1(\theta)$ .

The expressions of the above physical quantities are given in Sect. 6.4.1. Among them, $I_0(\theta)$ and the polarization analyzing powers can be experimentally measured, and the physical quantities that cannot be measured experimentally can be obtained through theoretical calculations.

The coordinate system transformation expressions given by Eqs. (6.34)–(6.36) and the related discussion are also applicable in this section, and it is still recommended to preserve the polarization nuclear data of the laboratory system in the polarization nuclear database.

### 6.4.3 Polarization Quantities for $\overrightarrow{\frac{1}{2}} + \mathrm{A} \rightarrow \vec{1} + \mathrm{B}$ Reactions Related to Azimuthal Angle

In the $\overrightarrow{\frac{1}{2}} + \mathrm{A} \rightarrow \vec{1} + \mathrm{B}$ reaction, the spin of the incident particle is $\frac{1}{2}$, and the spin of the outgoing particle is 1. The spin of the target nucleus and the residual nucleus is $I$ and $I'$, respectively, and both are unpolarized. Equation (3.685) has given the reaction amplitude with definite spin magnetic quantum number $M_I' M_I$ as

$$\hat{\mathcal{F}}_{M_I'M_I}(\theta,\varphi)=\begin{pmatrix} A_{M_I'M_I}(\theta)\,e^{-\frac{i}{2}\varphi} & -F_{M_I'M_I}(\theta)\,e^{-\frac{3i}{2}\varphi} \\ -iC_{M_I'M_I}(\theta)\,e^{\frac{i}{2}\varphi} & iD_{M_I'M_I}(\theta)\,e^{-\frac{i}{2}\varphi} \\ -E_{M_I'M_I}(\theta)\,e^{\frac{3i}{2}\varphi} & B_{M_I'M_I}(\theta)\,e^{\frac{i}{2}\varphi} \end{pmatrix} e^{i(M_I-M_I')\varphi} \tag{6.171}$$

where $A_{M_I'M_I}$,$B_{M_I'M_I}$, etc. have been given by Eqs. (3.673)–(3.678). Because in the calculation of the polarization quantities, when both the target nucleus and the residual nucleus are unpolarized, $\hat{\mathcal{F}}_{M_I'M_I}$ and $\hat{\mathcal{F}}^{+}_{M_I'M_I}$ always appear at the same time, the factor $e^{i(M_I-M_I')\varphi}$ must be eliminated, so we introduce

$$\hat{F}=\begin{pmatrix} Ae^{-\frac{i}{2}\varphi} & -Fe^{-\frac{3i}{2}\varphi} \\ -iCe^{\frac{i}{2}\varphi} & iDe^{-\frac{i}{2}\varphi} \\ -Ee^{\frac{3i}{2}\varphi} & Be^{\frac{i}{2}\varphi} \end{pmatrix},\quad \hat{F}^{+}=\begin{pmatrix} A^*e^{\frac{i}{2}\varphi} & iC^*e^{-\frac{i}{2}\varphi} & -E^*e^{-\frac{3i}{2}\varphi} \\ -F^*e^{\frac{3i}{2}\varphi} & -iD^*e^{\frac{i}{2}\varphi} & B^*e^{-\frac{i}{2}\varphi} \end{pmatrix} \tag{6.172}$$

where the footnote $M_I'M_I$ and the argument $(\theta)$ of the symbols $A$, $B$, etc. are omitted and the footnote $M_I'M_I$ and argument $(\theta,\varphi)$ of $\hat{F}$ and $\hat{F}^{+}$ are also omitted. From Eq. (6.172) we can obtain

$$\hat{F}\hat{F}^{+}=\begin{pmatrix} |A|^2+|F|^2 & i(AC^*+DE^*)e^{-i\varphi} & -(AE^*+FB^*)\,e^{-2i\varphi} \\ -i(CA^*+DF^*)\,e^{i\varphi} & |C|^2+|D|^2 & i(CE^*+DB^*)\,e^{-i\varphi} \\ -(EA^*+BF^*)\,e^{2i\varphi} & -i(EC^*+BD^*)\,e^{i\varphi} & |E|^2+|B|^2 \end{pmatrix} \tag{6.173}$$

Also using the symbol introduced by Eq. (6.8), the differential cross section of the unpolarized incident particles can be obtained by using Eqs. (3.751)–(3.753) as

$$I_0(\theta)=\frac{1}{2}\mathrm{tr}\left(\hat{F}\hat{F}^{+}\right)=\tilde{\Sigma}\left(|A|^2+|C|^2+|E|^2\right) \tag{6.174}$$

where the footnotes $\alpha'n'$, $\alpha n$ of the outgoing and incident channels of $I_0(\theta)$ were omitted.

Utilizing Eq. (2.6) and Eq. (6.172), we can obtain

$$\hat{\sigma}_x\hat{F}^{+}=\begin{pmatrix} -F^*e^{\frac{3i}{2}\varphi} & -iD^*e^{\frac{i}{2}\varphi} & B^*e^{-\frac{i}{2}\varphi} \\ A^*e^{\frac{i}{2}\varphi} & iC^*e^{-\frac{i}{2}\varphi} & -E^*e^{-\frac{3i}{2}\varphi} \end{pmatrix},\quad \hat{\sigma}_y\hat{F}^{+}=\begin{pmatrix} iF^*e^{\frac{3i}{2}\varphi} & -D^*e^{\frac{i}{2}\varphi} & -iB^*e^{-\frac{i}{2}\varphi} \\ iA^*e^{\frac{i}{2}\varphi} & -C^*e^{-\frac{i}{2}\varphi} & -iE^*e^{-\frac{3i}{2}\varphi} \end{pmatrix},$$
$$\hat{\sigma}_z\hat{F}^{+}=\begin{pmatrix} A^*e^{\frac{i}{2}\varphi} & iC^*e^{-\frac{i}{2}\varphi} & -E^*e^{-\frac{3i}{2}\varphi} \\ F^*e^{\frac{3i}{2}\varphi} & iD^*e^{\frac{i}{2}\varphi} & B^*e^{-\frac{i}{2}\varphi} \end{pmatrix} \tag{6.175}$$

$$\hat{F}\hat{\sigma}_x\hat{F}^+ = \begin{pmatrix} -\left(AF^*e^{i\varphi} + FA^*e^{-i\varphi}\right) & -i\left(AD^* + FC^*e^{-2i\varphi}\right) & AB^*e^{-i\varphi} + FE^*e^{-3i\varphi} \\ i\left(CF^*e^{2i\varphi} + DA^*\right) & -\left(CD^*e^{i\varphi} + DC^*e^{-i\varphi}\right) & -i\left(CB^* + DE^*e^{-2i\varphi}\right) \\ EF^*e^{3i\varphi} + BA^*e^{i\varphi} & i\left(ED^*e^{2i\varphi} + BC^*\right) & -\left(EB^*e^{i\varphi} + BE^*e^{-i\varphi}\right) \end{pmatrix},$$

$$\hat{F}\hat{\sigma}_y\hat{F}^+ = \begin{pmatrix} i\left(AF^*e^{i\varphi} - FA^*e^{-i\varphi}\right) & -AD^* + FC^*e^{-2i\varphi} & -i\left(AB^*e^{-i\varphi} - FE^*e^{-3i\varphi}\right) \\ CF^*e^{2i\varphi} - DA^* & i\left(CD^*e^{i\varphi} - DC^*e^{-i\varphi}\right) & -CB^* + DE^*e^{-2i\varphi} \\ -i\left(EF^*e^{3i\varphi} - BA^*e^{i\varphi}\right) & ED^*e^{2i\varphi} - BC^* & i\left(EB^*e^{i\varphi} - BE^*e^{-i\varphi}\right) \end{pmatrix},$$

$$\hat{F}\hat{\sigma}_z\hat{F}^+ = \begin{pmatrix} |A|^2 - |F|^2 & i(AC^* - FD^*)e^{-i\varphi} & -(AE^* - FB^*)\,e^{-2i\varphi} \\ -i(CA^* - DF^*)\,e^{i\varphi} & |C|^2 - |D|^2 & i(CE^* - DB^*)e^{-i\varphi} \\ -(EA^* - BF^*)\,e^{2i\varphi} & -i(EC^* - BD^*)\,e^{i\varphi} & |E|^2 - |B|^2 \end{pmatrix} \tag{6.176}$$

From Eq. (6.176) we can find

$$\begin{aligned} \mathrm{tr}\left(\hat{F}\hat{\sigma}_x\hat{F}^+\right) &= -2\,\mathrm{Re}\left[(AF^* + CD^* + EB^*)\,e^{i\varphi}\right], \\ \mathrm{tr}\left(\hat{F}\hat{\sigma}_y\hat{F}^+\right) &= -2\mathrm{Im}\left[(AF^* + CD^* + EB^*)\,e^{i\varphi}\right], \\ \mathrm{tr}\left(\hat{F}\hat{\sigma}_z\hat{F}^+\right) &= |A|^2 + |C|^2 + |E|^2 - |F|^2 - |D|^2 - |B|^2 \end{aligned} \tag{6.177}$$

The following results can be obtained from Eq. (6.177) and using Eqs. (3.751)–(3.753), (3.641), and (3.642):

$$\begin{aligned} \widetilde{\Sigma}\mathrm{tr}\left(\hat{F}\hat{\sigma}_x\hat{F}^+\right) &= -2\widetilde{\Sigma}\,\mathrm{Re}\,(2AF^* + CD^*)\cos\varphi, \\ \widetilde{\Sigma}\mathrm{tr}\left(\hat{F}\hat{\sigma}_y\hat{F}^+\right) &= -2\widetilde{\Sigma}\,\mathrm{Re}\,(2AF^* + CD^*)\sin\varphi, \\ \widetilde{\Sigma}\mathrm{tr}\left(\hat{F}\hat{\sigma}_z\hat{F}^+\right) &= 0 \end{aligned} \tag{6.178}$$

Define

$$A_x(\theta) = \frac{1}{I_0}\widetilde{\Sigma}\,\mathrm{Re}\,(2AF^* + CD^*) \tag{6.179}$$

and

$$\overline{A}_i(\theta, \varphi) = \frac{1}{2I_0}\widetilde{\Sigma}\mathrm{tr}\left(\hat{F}\hat{\sigma}_i\hat{F}^+\right), \quad i = x, y, z \tag{6.180}$$

Then we can obtain the following results from Eqs. (6.178)–(6.180):

$$\overline{A}_x(\theta, \varphi) = -A_x(\theta)\cos\varphi, \quad \overline{A}_y(\theta, \varphi) = -A_x(\theta)\sin\varphi, \quad \overline{A}_z(\theta, \varphi) = 0 \tag{6.181}$$

The differential cross section of the outgoing particles corresponding to the polarization incident particles is

$$I(\theta, \varphi) = I_0(\theta)\left[1 + p_x\overline{A}_x(\theta, \varphi) + p_y\overline{A}_y(\theta, \varphi)\right] \tag{6.182}$$

From Eq. (3.33) and Eq. (6.173), we can find

$$\begin{aligned}
\mathrm{tr}\left(\hat{S}_x\hat{F}\hat{F}^+\right) &= -\frac{\mathrm{i}}{\sqrt{2}}[(CA^* + DF^*)\,\mathrm{e}^{\mathrm{i}\varphi} - (AC^* + FD^*)\,\mathrm{e}^{-\mathrm{i}\varphi} \\
&\quad + (EC^* + BD^*)\,\mathrm{e}^{\mathrm{i}\varphi} - (CE^* + DB^*)\,\mathrm{e}^{-\mathrm{i}\varphi}] \\
&= \sqrt{2}\mathrm{Im}\left[(CA^* + DF^* + EC^* + BD^*)\,\mathrm{e}^{\mathrm{i}\varphi}\right], \\
\mathrm{tr}\left(\hat{S}_y\hat{F}\hat{F}^+\right) &= -\frac{1}{\sqrt{2}}[(CA^* + DF^*)\,\mathrm{e}^{\mathrm{i}\varphi} + (AC^* + FD^*)\,\mathrm{e}^{-\mathrm{i}\varphi} \\
&\quad + (EC^* + BD^*)\,\mathrm{e}^{\mathrm{i}\varphi} + (CE^* + DB^*)\,\mathrm{e}^{-\mathrm{i}\varphi}] \\
&= -\sqrt{2}\,\mathrm{Re}\left[(CA^* + DF^* + EC^* + BD^*)\,\mathrm{e}^{\mathrm{i}\varphi}\right], \\
\mathrm{tr}\left(\hat{S}_z\hat{F}\hat{F}^+\right) &= |A|^2 + |F|^2 - |E|^2 - |B|^2
\end{aligned} \tag{6.183}$$

Using the method to find Eq. (6.178), the following results can be obtained by Eq. (6.183):

$$\begin{aligned}
\widetilde{\Sigma}\mathrm{tr}\left(\hat{S}_x\hat{F}\hat{F}^+\right) &= 2\sqrt{2}\,\widetilde{\Sigma}\,\mathrm{Re}\,(CA^* + DF^*)\sin\varphi, \\
\widetilde{\Sigma}\mathrm{tr}\left(\hat{S}_y\hat{F}\hat{F}^+\right) &= -2\sqrt{2}\,\widetilde{\Sigma}\,\mathrm{Re}\,(CA^* + DF^*)\cos\varphi, \\
\widetilde{\Sigma}\mathrm{tr}\left(\hat{S}_z\hat{F}\hat{F}^+\right) &= 0
\end{aligned} \tag{6.184}$$

Define

$$P^y(\theta) = \frac{\sqrt{2}}{I_0}\widetilde{\Sigma}\,\mathrm{Re}\,(CA^* + DF^*) \tag{6.185}$$

and

$$\overline{P}^i(\theta, \varphi) = \frac{1}{2I_0}\widetilde{\Sigma}\mathrm{tr}\left(\hat{S}_i\hat{F}\hat{F}^+\right), \quad i = x, y, z \tag{6.186}$$

We can obtain the following results from Eqs. (6.184)–(6.186):

$$\overline{P}^x(\theta, \varphi) = P^y(\theta)\sin\varphi, \quad \overline{P}^y(\theta, \varphi) = -P^y(\theta)\cos\varphi, \quad \overline{P}^z(\theta, \varphi) = 0 \tag{6.187}$$

From Eq. (3.36) and Eq. (6.173), we can also find

$$\begin{aligned}
\mathrm{tr}\left(\hat{Q}_{xy}\hat{F}\hat{F}^{+}\right) &= -3\mathrm{Im}\left[(EA^{*}+BF^{*})\,\mathrm{e}^{2\mathrm{i}\varphi}\right],\\
\mathrm{tr}\left(\hat{Q}_{xz}\hat{F}\hat{F}^{+}\right) &= \frac{3}{\sqrt{2}}\mathrm{Im}\left[(CA^{*}+DF^{*}-EC^{*}-BD^{*})\,\mathrm{e}^{\mathrm{i}\varphi}\right]\\
\mathrm{tr}\left(\hat{Q}_{yz}\hat{F}\hat{F}^{+}\right) &= -\frac{3}{\sqrt{2}}\,\mathrm{Re}\left[(CA^{*}+DF^{*}-EC^{*}-BD^{*})\,\mathrm{e}^{\mathrm{i}\varphi}\right]\\
\mathrm{tr}\left(\hat{Q}_{xx-yy}\hat{F}\hat{F}^{+}\right) &= -6\,\mathrm{Re}\left[(EA^{*}+BF^{*})\,\mathrm{e}^{2\mathrm{i}\varphi}\right]\\
tr\left(\hat{Q}_{zz}\hat{F}\hat{F}^{+}\right) &= =|A|^2+|F|^2+|E|^2+|B|^2-2\left(|C|^2+|D|^2\right)
\end{aligned} \tag{6.188}$$

Using the method to find Eq. (6.178), the following results can be obtained by Eq. (6.188):

$$\begin{aligned}
\widetilde{\Sigma}\mathrm{tr}\left(\hat{Q}_{xy}\hat{F}\hat{F}^{+}\right) &= -6\widetilde{\Sigma}\,\mathrm{Re}\,(EA^{*})\sin(2\varphi),\\
\widetilde{\Sigma}\mathrm{tr}\left(\hat{Q}_{xz}\hat{F}\hat{F}^{+}\right) &= 3\sqrt{2}\,\widetilde{\Sigma}\mathrm{Im}(CA^{*}+DF^{*})\cos\varphi,\\
\widetilde{\Sigma}\mathrm{tr}\left(\hat{Q}_{yz}\hat{F}\hat{F}^{+}\right) &= 3\sqrt{2}\,\widetilde{\Sigma}\mathrm{Im}(CA^{*}+DF^{*})\sin\varphi,\\
\widetilde{\Sigma}\mathrm{tr}\left(\hat{Q}_{xx-yy}\hat{F}\hat{F}^{+}\right) &= -12\widetilde{\Sigma}\,\mathrm{Re}\,(EA^{*})\cos(2\varphi),\\
\widetilde{\Sigma}\mathrm{tr}\left(\hat{Q}_{zz}\hat{F}\hat{F}^{+}\right) &= 2\widetilde{\Sigma}\left(|A|^2+|E|^2-2|C|^2\right)
\end{aligned} \tag{6.189}$$

Define

$$\begin{aligned}
&P^1(\theta)=\frac{3}{I_0}\widetilde{\Sigma}\,\mathrm{Re}\,(EA^{*}), \qquad P^2(\theta)=\frac{3}{\sqrt{2}I_0}\widetilde{\Sigma}\mathrm{Im}(CA^{*}+DF^{*}),\\
&P^3(\theta)=\frac{1}{I_0}\widetilde{\Sigma}\left(|A|^2+|E|^2-2|C|^2\right)
\end{aligned} \tag{6.190}$$

and

$$\overline{P}^i(\theta,\varphi)=\frac{1}{2I_0}\widetilde{\Sigma}\mathrm{tr}\left(\hat{Q}_i\hat{F}\hat{F}^{+}\right),\quad i=\gamma \tag{6.191}$$

We can obtain the following results from Eqs. (6.189)–(6.191):

$$\begin{aligned}
&\overline{P}^{xy}(\theta,\varphi)=-P^1(\theta)\sin(2\varphi), \quad \overline{P}^{xz}(\theta,\varphi)=P^2(\theta)\cos\varphi,\\
&\overline{P}^{yz}(\theta,\varphi)=P^2(\theta)\sin\varphi, \qquad \overline{P}^{xx-yy}(\theta)=-2P^1(\theta)\cos(2\varphi),\\
&\overline{P}^{zz}(\theta,\varphi)=P^3(\theta)
\end{aligned} \tag{6.192}$$

By use of Eq. (3.33) and Eq. (6.176), we can also find

$$
\begin{aligned}
\mathrm{tr}\left(\hat{S}_x\hat{F}\,\hat{\sigma}_x\hat{F}^+\right) &= -\sqrt{2}\mathrm{Im}\left[(DA^*+BC^*)+(CF^*+ED^*)\,\mathrm{e}^{2\mathrm{i}\varphi}\right],\\
\mathrm{tr}\left(\hat{S}_y\hat{F}\,\hat{\sigma}_x\hat{F}^+\right) &= \sqrt{2}\,\mathrm{Re}\left[(DA^*+BC^*)+(CF^*+ED^*)\,\mathrm{e}^{2\mathrm{i}\varphi}\right],\\
\mathrm{tr}\left(\hat{S}_z\hat{F}\,\hat{\sigma}_x\hat{F}^+\right) &= -2\,\mathrm{Re}\left[(AF^*-EB^*)\,\mathrm{e}^{\mathrm{i}\varphi}\right]\\
\mathrm{tr}\left(\hat{S}_x\hat{F}\,\hat{\sigma}_y\hat{F}^+\right) &= -\sqrt{2}\,\mathrm{Re}\left[(DA^*+BC^*)-(CF^*+ED^*)\,\mathrm{e}^{2\mathrm{i}\varphi}\right],\\
\mathrm{tr}\left(\hat{S}_y\hat{F}\,\hat{\sigma}_y\hat{F}^+\right) &= -\sqrt{2}\mathrm{Im}\left[(DA^*+BC^*)-(CF^*+ED^*)\,\mathrm{e}^{2\mathrm{i}\varphi}\right],\\
\mathrm{tr}\left(\hat{S}_z\hat{F}\,\hat{\sigma}_y\hat{F}^+\right) &= -2\mathrm{Im}\left[(AF^*-EB^*)\,\mathrm{e}^{\mathrm{i}\varphi}\right],\\
\mathrm{tr}\left(\hat{S}_x\hat{F}\,\hat{\sigma}_z\hat{F}^+\right) &= \sqrt{2}\mathrm{Im}\left[(CA^*-DF^*+EC^*-BD^*)\,\mathrm{e}^{\mathrm{i}\varphi}\right],\\
\mathrm{tr}\left(\hat{S}_y\hat{F}\,\hat{\sigma}_z\hat{F}^+\right) &= -\sqrt{2}\,\mathrm{Re}\left[(CA^*-DF^*+EC^*-BD^*)\,\mathrm{e}^{\mathrm{i}\varphi}\right],\\
\mathrm{tr}\left(\hat{S}_z\hat{F}\,\hat{\sigma}_z\hat{F}^+\right) &= |A|^2-|F|^2-|E|^2+|B|^2
\end{aligned}
\tag{6.193}
$$

Using the method to find Eq. (6.178), the following results can be obtained by Eq. (6.193):

$$
\begin{aligned}
\widetilde{\Sigma}\mathrm{tr}\left(\hat{S}_x\hat{F}\,\hat{\sigma}_x\hat{F}^+\right) &= -2\sqrt{2}\,\widetilde{\Sigma}\,\mathrm{Re}\,(CF^*)\sin(2\varphi),\\
\widetilde{\Sigma}\mathrm{tr}\left(\hat{S}_y\hat{F}\,\hat{\sigma}_x\hat{F}^+\right) &= 2\sqrt{2}\,\widetilde{\Sigma}[\mathrm{Re}\,(DA^*)+\mathrm{Re}\,(CF^*)\cos(2\varphi)],\\
\widetilde{\Sigma}\mathrm{tr}\left(\hat{S}_z\hat{F}\,\hat{\sigma}_x\hat{F}^+\right) &= 4\widetilde{\Sigma}\mathrm{Im}(AF^*)\sin\varphi,\\
\widetilde{\Sigma}\mathrm{tr}\left(\hat{S}_x\hat{F}\,\hat{\sigma}_y\hat{F}^+\right) &= -2\sqrt{2}\,\widetilde{\Sigma}[\mathrm{Re}\,(DA^*)-\mathrm{Re}\,(CF^*)\cos(2\varphi)],\\
\widetilde{\Sigma}\mathrm{tr}\left(\hat{S}_y\hat{F}\,\hat{\sigma}_y\hat{F}^+\right) &= 2\sqrt{2}\,\widetilde{\Sigma}\,\mathrm{Re}\,(CF^*)\sin(2\varphi),\\
\widetilde{\Sigma}\mathrm{tr}\left(\hat{S}_z\hat{F}\,\hat{\sigma}_y\hat{F}^+\right) &= -4\widetilde{\Sigma}\mathrm{Im}(AF^*)\cos\varphi,\\
\widetilde{\Sigma}\mathrm{tr}\left(\hat{S}_x\hat{F}\,\hat{\sigma}_z\hat{F}^+\right) &= 2\sqrt{2}\,\widetilde{\Sigma}\mathrm{Im}(CA^*-DF^*)\cos\varphi,\\
\widetilde{\Sigma}\mathrm{tr}\left(\hat{S}_y\hat{F}\,\hat{\sigma}_z\hat{F}^+\right) &= 2\sqrt{2}\,\widetilde{\Sigma}\mathrm{Im}(CA^*-DF^*)\sin\varphi,\\
\widetilde{\Sigma}\mathrm{tr}\left(\hat{S}_z\hat{F}\,\hat{\sigma}_z\hat{F}^+\right) &= 2\widetilde{\Sigma}\left(|A|^2-|F|^2\right)
\end{aligned}
\tag{6.194}
$$

Define

$$
\begin{aligned}
R_1(\theta) &= \frac{\sqrt{2}}{I_0}\widetilde{\Sigma}\,\mathrm{Re}\,(DA^*), \qquad R_2(\theta)=\frac{\sqrt{2}}{I_0}\widetilde{\Sigma}\,\mathrm{Re}\,(CF^*),\\
Q_1(\theta) &= \frac{2}{I_0}\widetilde{\Sigma}\mathrm{Im}(AF^*), \qquad Q_2(\theta)=\frac{\sqrt{2}}{I_0}\widetilde{\Sigma}\mathrm{Im}(CA^*-DF^*),\\
W_1(\theta) &= \frac{1}{I_0}\widetilde{\Sigma}\left(|A|^2-|F|^2\right)
\end{aligned}
\tag{6.195}
$$

and

$$K_i^j(\theta,\varphi)=\frac{1}{2I_0}\widetilde{\Sigma}\mathrm{tr}\left(\hat{S}_j\hat{F}\,\hat{\sigma}_i\hat{F}^+\right),\qquad i,j=x,y,z \tag{6.196}$$

We can obtain the following results from Eqs. (6.194)–(6.196):

$$\begin{aligned}
&K_x^x(\theta,\varphi)=-R_2(\theta)\sin(2\varphi), && K_x^y(\theta,\varphi)=R_1(\theta)+R_2(\theta)\cos(2\varphi),\\
&K_x^z(\theta,\varphi)=Q_1(\theta)\sin\varphi, && K_y^x(\theta,\varphi)=-R_1(\theta)+R_2(\theta)\cos(2\varphi),\\
&K_y^y(\theta,\varphi)=R_2(\theta)\sin(2\varphi), && K_y^z(\theta,\varphi)=-Q_1(\theta)\cos\varphi,\\
&K_z^x(\theta,\varphi)=Q_2(\theta)\cos\varphi, && K_z^y(\theta,\varphi)=Q_2(\theta)\sin\varphi\\
&K_z^z(\theta,\varphi)=W_1(\theta)
\end{aligned} \tag{6.197}$$

By use of Eq. (3.36) and Eq. (6.176), we can also find

$$\begin{aligned}
\mathrm{tr}\left(\hat{Q}_{xy}\hat{F}\,\hat{\sigma}_x\hat{F}^+\right)&=3\mathrm{Im}\left(BA^*\mathrm{e}^{\mathrm{i}\varphi}+EF^*\mathrm{e}^{3\mathrm{i}\varphi}\right),\\
\mathrm{tr}\left(\hat{Q}_{xz}\hat{F}\,\hat{\sigma}_x\hat{F}^+\right)&=-\frac{3}{\sqrt{2}}\mathrm{Im}\left[(DA^*-BC^*)+(CF^*-ED^*)\,\mathrm{e}^{2\mathrm{i}\varphi}\right],\\
\mathrm{tr}\left(\hat{Q}_{yz}\hat{F}\,\hat{\sigma}_x\hat{F}^+\right)&=\frac{3}{\sqrt{2}}\mathrm{Re}\left[(DA^*-BC^*)+(CF^*-ED^*)\,\mathrm{e}^{2\mathrm{i}\varphi}\right],\\
\mathrm{tr}\left(\hat{Q}_{xx-yy}\hat{F}\,\hat{\sigma}_x\hat{F}^+\right)&=6\,\mathrm{Re}\left(BA^*\mathrm{e}^{\mathrm{i}\varphi}+EF^*\mathrm{e}^{3\mathrm{i}\varphi}\right),\\
\mathrm{tr}\left(\hat{Q}_{zz}\hat{F}\,\hat{\sigma}_x\hat{F}^+\right)&=-2\,\mathrm{Re}\left[(AF^*+EB^*-2CD^*)\,\mathrm{e}^{\mathrm{i}\varphi}\right],\\
\mathrm{tr}\left(\hat{Q}_{xy}\hat{F}\,\hat{\sigma}_y\hat{F}^+\right)&=3\,\mathrm{Re}\left(BA^*\mathrm{e}^{\mathrm{i}\varphi}-EF^*\mathrm{e}^{3\mathrm{i}\varphi}\right),\\
\mathrm{tr}\left(\hat{Q}_{xz}\hat{F}\,\hat{\sigma}_y\hat{F}^+\right)&=-\frac{3}{\sqrt{2}}\mathrm{Re}\left[(DA^*-BC^*)-(CF^*-ED^*)\,\mathrm{e}^{2\mathrm{i}\varphi}\right],\\
\mathrm{tr}\left(\hat{Q}_{yz}\hat{F}\,\hat{\sigma}_y\hat{F}^+\right)&=-\frac{3}{\sqrt{2}}\mathrm{Im}\left[(DA^*-BC^*)-(CF^*-ED^*)\,\mathrm{e}^{2\mathrm{i}\varphi}\right],\\
\mathrm{tr}\left(\hat{Q}_{xx-yy}\hat{F}\,\hat{\sigma}_y\hat{F}^+\right)&=-6\mathrm{Im}\left(BA^*e^{\mathrm{i}\varphi}-EF^*\mathrm{e}^{3\mathrm{i}\varphi}\right),\\
\mathrm{tr}\left(\hat{Q}_{zz}\hat{F}\,\hat{\sigma}_y\hat{F}^+\right)&=-2\mathrm{Im}\left[(AF^*+EB^*-2CD^*)\,\mathrm{e}^{\mathrm{i}\varphi}\right],\\
\mathrm{tr}\left(\hat{Q}_{xy}\hat{F}\,\hat{\sigma}_z\hat{F}^+\right)&=-3\mathrm{Im}\left[(EA^*-BF^*)\,\mathrm{e}^{2\mathrm{i}\varphi}\right],\\
\mathrm{tr}\left(\hat{Q}_{xz}\hat{F}\,\hat{\sigma}_z\hat{F}^+\right)&=\frac{3}{\sqrt{2}}\mathrm{Im}\left[(CA^*-DF^*-EC^*+BD^*)\,\mathrm{e}^{\mathrm{i}\varphi}\right],\\
\mathrm{tr}\left(\hat{Q}_{yz}\hat{F}\,\hat{\sigma}_z\hat{F}^+\right)&=-\frac{3}{\sqrt{2}}\mathrm{Re}\left[(CA^*-DF^*-EC^*+BD^*)\,\mathrm{e}^{\mathrm{i}\varphi}\right],\\
\mathrm{tr}\left(\hat{Q}_{xx-yy}\hat{F}\,\hat{\sigma}_z\hat{F}^+\right)&=-6\,\mathrm{Re}\left[(EA^*-BF^*)\,\mathrm{e}^{2\mathrm{i}\varphi}\right],\\
\mathrm{tr}\left(\hat{Q}_{zz}\hat{F}\,\hat{\sigma}_z\hat{F}^+\right)&=|A|^2-|F|^2+|E|^2-|B|^2-2|C|^2+2|D|^2
\end{aligned} \tag{6.198}$$

Using the method to find Eq. (6.178), the following results can be obtained by Eq. (6.198):

$$\begin{aligned}
&\widetilde{\Sigma}\mathrm{tr}\left(\hat{Q}_{xy}\hat{F}\,\hat{\sigma}_x\hat{F}^+\right) = 3\widetilde{\Sigma}[\,\mathrm{Re}\,(BA^*)\sin\varphi + \mathrm{Re}\,(EF^*)\sin(3\varphi)],\\
&\widetilde{\Sigma}\mathrm{tr}\left(\hat{Q}_{xz}\hat{F}\,\hat{\sigma}_x\hat{F}^+\right) = -3\sqrt{2}\,\widetilde{\Sigma}[\mathrm{Im}(DA^*) + \mathrm{Im}(CF^*)\cos(2\varphi)],\\
&\widetilde{\Sigma}\mathrm{tr}\left(\hat{Q}_{yz}\hat{F}\,\hat{\sigma}_x\hat{F}^+\right) = -3\sqrt{2}\,\widetilde{\Sigma}\mathrm{Im}(CF^*)\sin(2\varphi),\\
&\widetilde{\Sigma}\mathrm{tr}\left(\hat{Q}_{xx-yy}\hat{F}\,\hat{\sigma}_x\hat{F}^+\right) = 6\widetilde{\Sigma}[\,\mathrm{Re}\,(BA^*)\cos\varphi + \mathrm{Re}\,(EF^*)\cos(3\varphi)],\\
&\widetilde{\Sigma}\mathrm{tr}\left(\hat{Q}_{zz}\hat{F}\,\hat{\sigma}_x\hat{F}^+\right) = -4\widetilde{\Sigma}\,\mathrm{Re}\,(AF^* - CD^*)\cos\varphi,\\
&\widetilde{\Sigma}\mathrm{tr}\left(\hat{Q}_{xy}\hat{F}\,\hat{\sigma}_y\hat{F}^+\right) = 3\widetilde{\Sigma}[\,\mathrm{Re}\,(BA^*)\cos\varphi - \mathrm{Re}\,(EF^*)\cos(3\varphi)],\\
&\widetilde{\Sigma}\mathrm{tr}\left(\hat{Q}_{xz}\hat{F}\,\hat{\sigma}_y\hat{F}^+\right) = -3\sqrt{2}\,\widetilde{\Sigma}\mathrm{Im}(CF^*)\sin(2\varphi),\\
&\widetilde{\Sigma}\mathrm{tr}\left(\hat{Q}_{yz}\hat{F}\,\hat{\sigma}_y\hat{F}^+\right) = -3\sqrt{2}\,\widetilde{\Sigma}[\mathrm{Im}(DA^*) - \mathrm{Im}(CF^*)\cos(2\varphi)],\\
&\widetilde{\Sigma}\mathrm{tr}\left(\hat{Q}_{xx-yy}\hat{F}\,\hat{\sigma}_y\hat{F}^+\right) = -6\widetilde{\Sigma}[\,\mathrm{Re}\,(BA^*)\sin\varphi - \mathrm{Re}\,(EF^*)\sin(3\varphi)],\\
&\widetilde{\Sigma}\mathrm{tr}\left(\hat{Q}_{zz}\hat{F}\,\hat{\sigma}_y\hat{F}^+\right) = -4\widetilde{\Sigma}\,\mathrm{Re}\,(AF^* - CD^*)\sin\varphi,\\
&\widetilde{\Sigma}\mathrm{tr}\left(\hat{Q}_{xy}\hat{F}\,\hat{\sigma}_z\hat{F}^+\right) = -6\widetilde{\Sigma}\mathrm{Im}(EA^*)\cos(2\varphi),\\
&\widetilde{\Sigma}\mathrm{tr}\left(\hat{Q}_{xz}\hat{F}\,\hat{\sigma}_z\hat{F}^+\right) = 3\sqrt{2}\,\widetilde{\Sigma}\,\mathrm{Re}\,(CA^* - DF^*)\sin\varphi,\\
&\widetilde{\Sigma}\mathrm{tr}\left(\hat{Q}_{yz}\hat{F}\,\hat{\sigma}_z\hat{F}^+\right) = -3\sqrt{2}\,\widetilde{\Sigma}\,\mathrm{Re}\,(CA^* - DF^*)\cos\varphi,\\
&\widetilde{\Sigma}\mathrm{tr}\left(\hat{Q}_{xx-yy}\hat{F}\,\hat{\sigma}_z\hat{F}^+\right) = 12\widetilde{\Sigma}\mathrm{Im}(EA^*)\sin(2\varphi),\\
&\widetilde{\Sigma}\mathrm{tr}\left(\hat{Q}_{zz}\hat{F}\,\hat{\sigma}_z\hat{F}^+\right) = 0
\end{aligned} \tag{6.199}$$

Define

$$\begin{aligned}
&R_3(\theta) = \frac{3}{2I_0}\widetilde{\Sigma}\,\mathrm{Re}\,(BA^*), && R_4(\theta) = \frac{3}{2I_0}\widetilde{\Sigma}\,\mathrm{Re}\,(EF^*),\\
&R_5(\theta) = \frac{2}{I_0}\widetilde{\Sigma}\,\mathrm{Re}\,(AF^* - CD^*), && R_6(\theta) = \frac{3}{\sqrt{2}I_0}\widetilde{\Sigma}\,\mathrm{Re}\,(CA^* - DF^*),\\
&Q_3(\theta) = \frac{3}{\sqrt{2}I_0}\widetilde{\Sigma}\mathrm{Im}(DA^*), && Q_4(\theta) = \frac{3}{\sqrt{2}I_0}\widetilde{\Sigma}\mathrm{Im}(CF^*),\\
&Q_5(\theta) = \frac{3}{I_0}\widetilde{\Sigma}\mathrm{Im}(EA^*)
\end{aligned} \tag{6.200}$$

and

$$K_i^j(\theta,\varphi) = \frac{1}{2I_0}\widetilde{\Sigma}\mathrm{tr}\left(\hat{Q}_j\hat{F}\,\hat{\sigma}_i\hat{F}^+\right), \qquad i = x,y,z; \quad j = \gamma \tag{6.201}$$

We can obtain the following results from Eqs. (6.199)–(6.201):

$$\begin{aligned}
&K_x^{xy}(\theta,\varphi)=R_3(\theta)\sin\varphi+R_4(\theta)\sin(3\varphi),\\
&K_x^{xz}(\theta,\varphi)=-Q_3(\theta)-Q_4(\theta)\cos(2\varphi),\\
&K_x^{yz}(\theta,\varphi)=-Q_4(\theta)\sin(2\varphi),\\
&K_x^{xx-yy}(\theta,\varphi)=2R_3(\theta)\cos\varphi+2R_4(\theta)\cos(3\varphi),\\
&K_x^{zz}(\theta,\varphi)=-R_5(\theta)\cos\varphi,\\
&K_y^{xy}(\theta,\varphi)=R_3(\theta)\cos\varphi-R_4(\theta)\cos(3\varphi),\\
&K_y^{xz}(\theta,\varphi)=-Q_4(\theta)\sin(2\varphi),\\
&K_y^{yz}(\theta,\varphi)=-Q_3(\theta)+Q_4(\theta)\cos(2\varphi),\\
&K_y^{xx-yy}(\theta,\varphi)=-2R_3(\theta)\sin\varphi+2R_4(\theta)\sin(3\varphi),\\
&K_y^{zz}(\theta,\varphi)=-R_5(\theta)\sin\varphi,\\
&K_z^{xy}(\theta,\varphi)=-Q_5(\theta)\cos(2\varphi),\\
&K_z^{xz}(\theta,\varphi)=R_6(\theta)\sin\varphi,\\
&K_z^{yz}(\theta,\varphi)=-R_6(\theta)\cos\varphi,\\
&K_z^{xx-yy}(\theta,\varphi)=2Q_5(\theta)\sin(2\varphi),\\
&K_z^{zz}(\theta,\varphi)=0
\end{aligned} \tag{6.202}$$

Referring to Eq. (6.33), we can write the polarization rates related to the angles $\theta$, $\varphi$ simultaneously of the spin 1 outgoing particle corresponding to the $\vec{\frac{1}{2}}+\mathrm{A}\rightarrow\vec{1}+\mathrm{B}$ reaction as

$$P_j(\theta,\varphi)=\frac{I_0(\theta)}{I(\theta,\varphi)}\left[\overline{P}^j(\theta,\varphi)+\sum_{i=x,\,y,\,z}p_iK_i^j(\theta,\varphi)\right],\qquad j=\varepsilon \tag{6.203}$$

The definition of the subscript symbol set $\varepsilon$ has been given by Eq. (3.476).

Note that if we make $\varphi=0$, the results obtained in Sect. 6.4.3 will automatically degenerate to the results obtained in Sect. 3.10.

### 6.4.4 Polarization Nuclear Database for $\vec{\frac{1}{2}}+\mathrm{A}\rightarrow\vec{1}+\mathrm{B}$ Reactions

According to the previous discussion, the polarization nuclear database for $\vec{\frac{1}{2}}+A\rightarrow\vec{1}+B$ reactions should contain the following differential physical quantities of the polar angle $\theta$ for different incident particle energies of the two-body direct reaction channels:

1. Differential cross section corresponding to unpolarized incident particles $I_0(\theta)$.
2. The incident particle vector analyzing power $A_x(\theta)$.
3. Corresponding to the unpolarized incident particles, the outgoing particle vector polarization rates $P^y(\theta)$; tensor polarization rates $P^i(\theta)$ $(i = 1 \sim 3)$.
4. Polarization transfer coefficient components $R_i(\theta)$ $(i = 1 \sim 6)$, $Q_i(\theta)$ $(i = 1 \sim 5)$, $W_1(\theta)$ .

The expressions of the above physical quantities are given in Sect. 6.4.3.

The coordinate system transformation expressions given by Eqs. (6.34)–(6.36) and the related discussion are also applicable in this section.

### 6.4.5 Main Points of Polarized Particle Transport Theory for $\vec{1} + \mathrm{A} \rightarrow \overrightarrow{\frac{1}{2}} + \mathrm{B}$ and $\overrightarrow{\frac{1}{2}} + \mathrm{A} \rightarrow \vec{1} + \mathrm{B}$ Reactions

For the sake of simplicity, we only consider the two-body direct reaction of $\vec{1} + \mathrm{A} \rightarrow \overrightarrow{\frac{1}{2}} + \mathrm{B}$ and $\overrightarrow{\frac{1}{2}} + \mathrm{A} \rightarrow \vec{1} + \mathrm{B}$ reactions and omit all other reaction channels.

The position, where the initial reaction (also known as the first reaction) $\vec{1} + \mathrm{A}$ takes place, is the origin $O_0$ of the coordinate system, and the direction of the incident particles is along the $z$-axis. After the laboratory system energy $E_0$ of the incident particle is determined, the 18 kinds of the laboratory system differential physical quantities listed in Sect. 6.4.2 can be obtained from the polarization nuclear database of the $\vec{1} + \mathrm{A} \rightarrow \overrightarrow{\frac{1}{2}} + \mathrm{B}$ reaction. Thus, the various polarization quantities related to the angles $(\theta, \varphi)$ can be calculated from Eqs. (6.146), (6.153), (6.154), (6.159), (6.164), and (6.169).

When the initial polarization rates $p_{0i}(i = \varepsilon)$ of the incident particles are known, the differential cross section of the outgoing particles can be obtained from Eq. (6.154). Regardless of other reaction channels, after sampling, assume that the reaction of the emitted particle of the first reaction with target nucleus occurs at the position $O_1$ with the direction of $(\theta_1, \varphi_1)$ and the distance $L_1$ from the origin $O_0$, which is called the second reaction. The incident particle energy corresponding to the second reaction is $E_1$. In $O_0$ system, the Stokes vector expression of the initial incident particles and the outgoing particles of the first reaction at the second reaction point $O_1$ are as follows, respectively:

$$S_1^{(\mathrm{i})}(\theta_1) = I_0(\theta_1)\begin{pmatrix} 1 \\ \overline{p}_{0x} \\ \overline{p}_{0y} \\ \overline{p}_{0z} \\ \overline{p}_{0xy} \\ \overline{p}_{0xz} \\ \overline{p}_{0yz} \\ \overline{p}_{0\,xx-yy} \\ \overline{p}_{0zz} \end{pmatrix}, \qquad S_1^{(\mathrm{f})}(\theta_1, \varphi_1) = I(\theta_1, \varphi_1)\begin{pmatrix} 1 \\ P_{1x}(\theta_1, \varphi_1) \\ P_{1y}(\theta_1, \varphi_1) \\ P_{1z}(\theta_1, \varphi_1) \end{pmatrix} \tag{6.204}$$

The definition of $\overline{p}_{0i}$ is given by Eq. (6.103). The above two vectors satisfy the following relation:

$$S_1^{(\mathrm{f})} = Z_1 S_1^{(\mathrm{i})} \tag{6.205}$$

where

$$Z_1 = \begin{pmatrix} 1 & \overline{A}_x & \overline{A}_y & 0 & \overline{A}_{xy} & \overline{A}_{xz} & \overline{A}_{yz} & \overline{A}_{xx-yy} & \overline{A}_{zz} \\ \overline{P}^x & K_x^x & K_y^x & K_z^x & K_{xy}^x & K_{xz}^x & K_{yz}^x & K_{xx-yy}^x & K_{zz}^x \\ \overline{P}^y & K_x^y & K_y^y & K_z^y & K_{xy}^y & K_{xz}^y & K_{yz}^y & K_{xx-yy}^y & K_{zz}^y \\ 0 & K_x^z & K_y^z & K_z^z & K_{xy}^z & K_{xz}^z & K_{yz}^z & K_{xx-yy}^z & 0 \end{pmatrix}_1 \tag{6.206}$$

The footnote 1 corresponds to $(\theta_1, \varphi_1)$ and energy $E_0$. Equation (6.205) is a combination of Eq. (6.154) and Eq. (6.170). Note that Eq. (6.205) is determined by the first reaction.

In the $O_0$ coordinate system, the three Cartesian coordinate system axes after rotating the angles $(\theta_1, \varphi_1)$ become the three new coordinate axes of the new coordinate system, then the coordinate origin of the new coordinate system is moved from $O_0$ to $O_1$, the $O_0O_1$ direction is just the new $z$-axis direction in $O_1$ system.

The polarization rates $P_{1_i}(i = x, y, z)$ of the spin $\frac{1}{2}$ particles can be obtained by Eq. (6.205). These are the calculation results in the $O_0$ system. The corresponding polarization rates $p_{1_i}$ of the second reaction incident particles in the $O_1$ system are

$$\begin{pmatrix} p_{1x} \\ p_{1y} \\ p_{1z} \end{pmatrix} = \hat{b}(\theta_1, \varphi_1)\begin{pmatrix} P_{1x}(\theta_1, \varphi_1) \\ P_{1y}(\theta_1, \varphi_1) \\ P_{1z}(\theta_1, \varphi_1) \end{pmatrix} \tag{6.207}$$

where the expression of $\hat{b}$ has been given by Eq. (2.204) or Eq. (6.43). The energy of the emitted particles appeared at point $O_1$ after the first reaction is $E_1$, and then using the polarization nuclear database of the $\overrightarrow{\frac{1}{2}} + \mathrm{A} \rightarrow \overrightarrow{1} + \mathrm{B}$ reaction and Eq. (6.182), the outgoing particle differential cross section for any angles $(\theta, \varphi)$ in the $O_1$ system after the reaction occurred at point $O_1$ can be obtained. In the case without considering other reaction channels, after sampling, assume that in the $O_1$ system the third reaction occurs at the point $O_2$, where is at the direction $(\theta_2, \varphi_2)$ and the distance $L_2$ from the point $O_1$. In the $O_1$ system, the Stokes vectors of the incident particles and the outgoing particles at the point $O_2$ are as follows, respectively:

$$S_2^{(i)}(\theta_2) = I_0(\theta_2)\begin{pmatrix} 1 \\ p_{1x} \\ p_{1y} \\ p_{1z} \end{pmatrix}, \qquad S_2^{(f)} = I(\theta_2, \varphi_2)\begin{pmatrix} 1 \\ P_{2x}(\theta_2, \varphi_2) \\ P_{2y}(\theta_2, \varphi_2) \\ P_{2z}(\theta_2, \varphi_2) \\ P_{2xy}(\theta_2, \varphi_2) \\ P_{2xz}(\theta_2, \varphi_2) \\ P_{2yz}(\theta_2, \varphi_2) \\ P_{2xx-yy}(\theta_2, \varphi_2) \\ P_{2zz}(\theta_2, \varphi_2) \end{pmatrix} \tag{6.208}$$

The above two vectors satisfy the following relation:

$$S_2^{(\mathrm{f})} = Z_2 S_2^{(\mathrm{i})} \tag{6.209}$$

where

$$Z_2 = \begin{pmatrix} 1 & \overline{A}_x & \overline{A}_y & 0 \\ \overline{P}^x & K_x^x & K_y^x & K_z^x \\ \overline{P}^y & K_x^y & K_y^y & K_z^y \\ 0 & K_x^z & K_y^z & K_z^z \\ \overline{P}^{xy} & K_x^{xy} & K_y^{xy} & K_z^{xy} \\ \overline{P}^{xz} & K_x^{xz} & K_y^{xz} & K_z^{xz} \\ \overline{P}^{yz} & K_x^{yz} & K_y^{yz} & K_z^{yz} \\ \overline{P}^{xx-yy} & K_x^{xx-yy} & K_y^{xx-yy} & K_z^{xx-yy} \\ \overline{P}^{zz} & K_x^{zz} & K_y^{zz} & 0 \end{pmatrix}_2 \tag{6.210}$$

were the footnote 2 corresponds the angles $(\theta_2, \varphi_2)$ and energy $E_1$. The physical quantities for the energy $E_1$ contained in the $Z_2$ matrix can be obtained from the polarization nuclear database according to Eqs. (6.181), (6.187), (6.192), (6.197), and (6.202).

In the $O_1$ coordinate system, the three Cartesian coordinate axes after rotating the angles $(\theta_2, \varphi_2)$ become the three new coordinate axes of the new coordinate system, then the coordinate origin of the new coordinate system is moved from $O_1$ to $O_2$, the $O_1O_2$ direction is just the new z-axis direction in $O_2$ system. Through sampling, the third reaction will occur at the position $(L_3, \theta_3, \varphi_3)$ in the $O_2$ system.

The polarization rates $P_{2i}(i = \varepsilon)$ of the spin 1 particles can be obtained by Eq. (6.209). These are the calculation results in the $O_1$ system. The polarization rates $p_{2i}$ of the incident particles corresponding to the third reaction in the $O_2$ system are

$$\begin{pmatrix} p_{2x} \\ p_{2y} \\ p_{2z} \end{pmatrix} = \hat{b}(\theta_2, \varphi_2) \begin{pmatrix} P_{2x}(\theta_2, \varphi_2) \\ P_{2y}(\theta_2, \varphi_2) \\ P_{2z}(\theta_2, \varphi_2) \end{pmatrix},$$

$$\begin{pmatrix} \sqrt{\frac{1}{6}} p_{2\,xx-yy} \\ \sqrt{\frac{2}{3}} p_{2xy} \\ \sqrt{\frac{2}{3}} p_{2xz} \\ \sqrt{\frac{2}{3}} p_{2yz} \\ \sqrt{\frac{1}{2}} p_{2zz} \end{pmatrix} = \hat{\gamma}(\theta_2, \varphi_2) \begin{pmatrix} \sqrt{\frac{1}{6}} P_{2\,xx-yy}(\theta_2, \varphi_2) \\ \sqrt{\frac{2}{3}} P_{2xy}(\theta_2, \varphi_2) \\ \sqrt{\frac{2}{3}} P_{2xz}(\theta_2, \varphi_2) \\ \sqrt{\frac{2}{3}} P_{2yz}(\theta_2, \varphi_2) \\ \sqrt{\frac{1}{2}} P_{2zz}(\theta_2, \varphi_2) \end{pmatrix} \tag{6.211}$$

where the expression of $\hat{\gamma}$ has been given by Eq. (6.118).

Equations (6.41), (6.47)–(6.49) are all applicable in this section.

The above method can be extended to the third, fourth, etc. reactions. Note that the coordinates $(X, Y, Z)$ and the other physical quantities of the last outgoing particle in the $O_0$ laboratory system must be tracked.

The introduced reaction process previously is that the $\vec{1} + \mathrm{A} \to \vec{\frac{1}{2}} + \mathrm{B}$ reaction occurs firstly and then the $\vec{\frac{1}{2}} + \mathrm{A} \to \vec{1} + \mathrm{B}$ reaction is followed. In the same way, we can study the reaction process that the $\vec{\frac{1}{2}} + \mathrm{A} \to \vec{1} + \mathrm{B}$ reaction occurs firstly and then the $\vec{1} + \mathrm{A} \to \vec{\frac{1}{2}} + \mathrm{B}$ reaction is followed. The two kinds of reactions are usually solved simultaneously with $\vec{\frac{1}{2}} + \mathrm{A} \to \vec{\frac{1}{2}} + \mathrm{B}$ and $\vec{1} + \mathrm{A} \to \vec{1} + \mathrm{B}$ reactions.

## 6.5 Polarized Particle Transport Equation

The usual particle transport equation is

$$\begin{aligned}
&\frac{1}{v}\frac{\partial\psi\left(\vec{r},E,\vec{\Omega},t\right)}{\partial t}+\vec{\Omega}\cdot\vec{\nabla}\psi\left(\vec{r},E,\vec{\Omega},t\right)+\Sigma_t\left(\vec{r},E,t\right)\psi\left(\vec{r},E,\vec{\Omega},t\right)\\
&=\iint\Sigma\left(\vec{r},t;E',\vec{\Omega}'\to E,\vec{\Omega}\right)\psi\left(\vec{r},E',\vec{\Omega}',t\right)\mathrm{d}E'\mathrm{d}\vec{\Omega}'+Q\left(\vec{r},E,\vec{\Omega},t\right)
\end{aligned}\tag{6.212}$$

Assume that we are working in a laboratory system. The particle velocity is $v=\sqrt{\frac{2E}{m}}$, where $m$ and $E$ represent the mass and energy of the particle, respectively. The $z$-axis of the Cartesian coordinate system $xyz$ is normally chosen in the initial incident particle direction, and the coordinate origin $O_0$ is usually chosen at the center of the nuclear device. $\vec{\Omega}$ is solid angle of emission, it represents the motion direction of the described particle relative to the $O_0$ Cartesian coordinate system, and the length of $\vec{\Omega}$ is 1. $\psi$ is the particle flux function with seven degrees of freedom. The relation between the macroscopic cross section $\Sigma$ and the microscopic cross section $\sigma$ is $\Sigma=\rho\sigma$, where $\rho$ is the density of the nuclear medium. If the density of the nuclear medium is independent of time $t$, $\rho=\rho\left(\vec{r}\right)$, then the particle transport equation can be solved separately; if the density of the nuclear medium is dependent of time $t$, $\rho=\rho\left(\vec{r},t\right)$, the particle transport equation is then solved in conjunction with the fluid mechanics equation, which describes how the density of the nuclear medium changes with time and space. Although this section chooses the case of $\rho=\rho\left(\vec{r},t\right)$, it does not give the equation to solve $\rho\left(\vec{r},t\right)$. In equation (6.212), $\Sigma_t$ is the macroscopic total cross section, and $\Sigma\left(\vec{r},t;E',\vec{\Omega}'\to E,\vec{\Omega}\right)$ represents the sum of the macroscopic double differential cross sections multiplied by the secondary particle number of the various reaction channels for all nuclides in the medium. $Q$ is the source term.

The polarized particle transport equation becomes

$$\begin{aligned}
&\frac{1}{v}\frac{\partial\psi\left(\vec{r},E,\vec{\Omega},s,t\right)}{\partial t}+\vec{\Omega}\cdot\vec{\nabla}\psi\left(\vec{r},E,\vec{\Omega},s,t\right)+\Sigma_t\left(\vec{r},E,t\right)\psi\left(\vec{r},E,\vec{\Omega},s,t\right)\\
&=\iint\Sigma_0\left(\vec{r},t;E',\vec{\Omega}'\to E,\vec{\Omega}\right)\psi\left(\vec{r},E',\vec{\Omega}',s,t\right)\mathrm{d}E'\mathrm{d}\vec{\Omega}'\\
&+\iint\sum_{s'}\left[\Sigma_{\mathrm{P}}\left(\vec{r},\ t;E',\ \vec{\Omega}',\ s'\to E,\ \vec{\Omega},\ s\right)\psi\left(\vec{r},E',\vec{\Omega}',s',t\right)\right]\mathrm{d}E'\mathrm{d}\vec{\Omega}'\\
&+Q\left(\vec{r},E,\vec{\Omega},s,t\right)
\end{aligned}\tag{6.213}$$

Let $S$ be the particle spin, and $s$ represents the particle spin magnetic quantum number, for the nucleons $S = \frac{1}{2}$, $s = \frac{1}{2}, -\frac{1}{2}$; for the deuterons $S = 1$, $s = 1, 0, -1$. In this case, $\psi$ is the particle flux function with eight degrees of freedom. In the above equation, $\Sigma_0$ corresponds to the unpolarization reaction channel part of $\Sigma$, which does not change the $s$ value; $\Sigma_\text{p}$ corresponds to the polarization reaction channel part of $\Sigma$, which may change the $s$ value.

The $\Sigma\left(\vec{r}, t; E', \vec{\Omega}' \to E, \vec{\Omega}\right)\psi\left(\vec{r}, E', \vec{\Omega}', t\right)$ term in Eq. (6.212) and the $\Sigma_0\left(\vec{r}, t; E', \vec{\Omega}' \to E, \vec{\Omega}\right)\psi\left(\vec{r}, E', \vec{\Omega}', s, t\right)$ term in Eq. (6.213) represent that the incident particles with various energies $E'$ and motion directions $\vec{\Omega}'$ occur nuclear reactions with the target at time $t$ and space point $\vec{r}$, and all of which can produce the outgoing particles with the energy $E$ and motion direction $\vec{\Omega}$. The $\Sigma_\text{p}$ term in Eq. (6.213) represents the contribution of the incident particles with spin magnetic quantum numbers $s'$ to the outgoing particles with spin magnetic quantum number $s$, where $s'$ may or may not be the same as $s$. For the integral of $\mathrm{d}E'\mathrm{d}\vec{\Omega}'$, $E'$ and $\vec{\Omega}'$ can be discretized, then this integral can be replaced by summation.

$\vec{\Omega}'$ is the direction angle $(\theta', \varphi')$ in the $O_0$ system. The second Cartesian coordinate system is set below, its origin $O_1$ is chosen at $\vec{r}$, and its new $z$-axis is chosen as $z_1$-axis in the direction of $\vec{\Omega}'$, that is, the $(\theta', \varphi')$ direction. Note that the $z_1$-axis is not in the direction of $\vec{r}$. The three new axes of $O_1$ system are equivalent to rotate angles $(\theta', \varphi')$ relative to the three axes of $O_0$ system. We know that the angles $(\theta', \varphi')$ and $(\theta, \varphi)$ are all relative to $O_0$ system, and the angle rotation relation between the axes of $O_1$ system and the axes of $O_0$ system is independent of the space coordinate $\vec{r}$. And because the nuclear reactions occur at the point $\vec{r}$ and the incident particles and the outgoing particles are also present at the point $\vec{r}$, in order to make it easier to analyze we move the origin $O_0$ to the point $O_1$, but we do not make any rotation for $O_0$ system axes. The three components of the unit vector $\vec{\Omega}$ in $O_0$ system are

$$\begin{aligned} \Omega_x &= \sin\theta\cos\varphi \\ \Omega_y &= \sin\theta\sin\varphi \\ \Omega_z &= \cos\theta \end{aligned} \tag{6.214}$$

However the three components of the same unit vector $\vec{\Omega}$ in $O_1$ system become

$$\begin{aligned} \Omega_{x_1} &= \sin\theta_1\cos\varphi_1 \\ \Omega_{y_1} &= \sin\theta_1\sin\varphi_1 \\ \Omega_{z_1} &= \cos\theta_1 \end{aligned} \tag{6.215}$$

$(\theta_1, \varphi_1)$ and $(\theta, \varphi)$ determined in this way have one-to-one correspondence. Using the matrix $\hat{b}(\theta', \varphi')$ given by Eq. (6.43), the unit vector $\vec{\Omega}$ given by Eq. (6.214) can be transformed into a representation in $O_1$ system, that is,

$$\begin{pmatrix} \Omega_{x_1} \\ \Omega_{y_1} \\ \Omega_{z_1} \end{pmatrix} = \hat{b}(\theta', \varphi') \begin{pmatrix} \Omega_x \\ \Omega_y \\ \Omega_z \end{pmatrix} \tag{6.216}$$

Using Eq. (6.43) and Eq. (6.214), the following expressions can be obtained from the above formula:

$$\begin{aligned} \Omega_{x_1} &= \cos\theta' \cos\varphi' \ \sin\theta \cos\varphi + \cos\theta' \sin\varphi' \ \sin\theta \sin\varphi - \sin\theta' \cos\theta \\ \Omega_{y_1} &= - \sin\varphi' \ \sin\theta \cos\varphi + \cos\varphi' \ \sin\theta \sin\varphi \\ \Omega_{z_1} &= \sin\theta' \cos\varphi' \ \sin\theta \cos\varphi + \sin\theta' \sin\varphi' \ \sin\theta \sin\varphi + \cos\theta' \cos\theta \end{aligned} \tag{6.217}$$

Comparing the third equation of Eq. (6.215) with the third equation of Eq. (6.217), we get

$$\begin{aligned} \cos\theta_1 &= \cos\theta' \cos\theta + \sin\theta' \sin\theta( \cos\varphi' \ \cos\varphi + \sin\varphi' \sin\varphi) \\ &= \cos\theta' \cos\theta + \sin\theta' \sin\theta \cos(\varphi - \varphi') \end{aligned} \tag{6.218}$$

Dividing the first equation by the second equation of Eq. (6.215) and Eq. (6.217), we can also get

$$\cot\varphi_1 = \cos\theta' \cot(\varphi - \varphi') - \frac{\sin\theta' \cot\theta}{\sin(\varphi - \varphi')} \tag{6.219}$$

The above two equations are consistent with Eq. (1.28). For the shape elastic scattering and two-body direct reaction channels which need to consider the polarization effect, because of the coordinate change between the laboratory and the center of mass systems, in the laboratory system there is a one-to-one correspondence between the energy $E$ of the outgoing particle and the angle $\theta_1$ of the outgoing particle. First of all, we can rewrite Eq. (6.218) as

$$\cos(\varphi - \varphi') = \frac{1}{\sin\theta' \sin\theta}( \cos\theta_1 - \cos\theta' \cos\theta) \tag{6.220}$$

When the outgoing particle energy $E$ is determined, $\cos\theta_1$ is also determined. So when making the integral of $d\Omega^{'}$ in the second term on the right side of Eq. (6.213), once $\theta^{'}$ is determined, $\varphi'$ can be calculated from Eq. (6.220), which is equivalent to reduce the double integrals of $d\vec{\Omega}'$ to single integral of d cos $\theta'$.

We shall consider the neutron transport as an example, and consider only one polarization reaction channel. At the time $t$ and the space point $\vec{r}$, the direction of the incident neutron with energy $E'$ in $O_0$ system is $\vec{\Omega}'$. Now let us study this problem in $O_1$ system; in this case the incident neutron direction is along the $z_1$-axis. When the laboratory system energy $E'$ of the incident neutrons is determined, the following eight differential components in the laboratory system can be obtained from the polarization nuclear database: $I_0(\theta_1)$, $A_y(\theta_1)$, $P^y(\theta_1)$, $R_1(\theta_1)$, $R_2(\theta_1)$, $Q_1(\theta_1)$, $Q_2(\theta_1)$, $W(\theta_1)$. Then we can calculate the various associated polarization quantities related to the angle $(\theta_1, \varphi_1)$ in $O_1$ system based on Eqs. (6.18), (6.25), and (6.32).

Let $P_i\left(\vec{r}, E', \vec{\Omega}', s', t\right)$, $i=x,y,z$ be the neutron polarization vector components in $O_0$ system with energy $E'$, motion direction $\vec{\Omega}'$, and spin magnetic quantum number $s'$. The three polarization rate functions are solved in conjunction with the neutron flux function $\psi\left(\vec{r}, E, \vec{\Omega}, s, t\right)$. Note that the particle motion direction and polarization direction are two different things. If we look at $P_i\left(\vec{r}, E', \vec{\Omega}', s', t\right)$, $i=x,y,z$ in $O_1$ system, they become

$$\vec{p}_1\left(\vec{r}, E', \vec{\Omega}', s', t\right) = \hat{b}(\theta', \varphi')\vec{P}\left(\vec{r}, E', \vec{\Omega}', s', t\right) \tag{6.221}$$

The matrix $\hat{b}(\theta', \varphi')$ has been given by Eq. (2.204) or Eq. (6.43); thus from the above formula, we can get

$$\begin{aligned} p_{1x} &= \cos\theta' \cos\varphi' P_x + \cos\theta' \sin\varphi' P_y - \sin\theta' P_z \\ p_{1y} &= -\sin\varphi' P_x + \cos\varphi' P_y \\ p_{1z} &= \sin\theta' \cos\varphi' P_x + \sin\theta' \sin\varphi' P_y + \cos\theta' P_z \end{aligned} \tag{6.222}$$

where $p_{1i} = p_{1i}\left(\vec{r}, E', \vec{\Omega}', s', t\right)$, $i=x, y, z$. So in $O_1$ system we can write the Stokes vector of the incident neutrons as

$$S^{(i)}\left(\vec{r}, E', E, \theta_1, \vec{\Omega}', s', t\right) = I_0(E', E, \theta_1)\begin{pmatrix} 1 \\ p_{1x}\left(\vec{r}, E', \vec{\Omega}', s', t\right) \\ p_{1y}\left(\vec{r}, E', \vec{\Omega}', s', t\right) \\ p_{1z}\left(\vec{r}, E', \vec{\Omega}', s', t\right) \end{pmatrix} \tag{6.223}$$

The Stokes vector of the outgoing neutrons is expressed as

$$\begin{aligned} &S^{(f)}\left(\vec{r}, E', E, \theta_1, \varphi_1, \vec{\Omega}', s', t\right) \\ &= I\left(\vec{r}, E', E, \theta_1, \varphi_1, \vec{\Omega}', s', t\right)\begin{pmatrix} 1 \\ P_{1x}\left(\vec{r}, E', E, \theta_1, \varphi_1, \vec{\Omega}', s', t\right) \\ P_{1y}\left(\vec{r}, E', E, \theta_1, \varphi_1, \vec{\Omega}'', s', t\right) \\ P_{1z}\left(\vec{r}, E', E, \theta_1, \varphi_1, \vec{\Omega}', s', t\right) \end{pmatrix} \end{aligned} \tag{6.224}$$

The above two formulas satisfy the following relation:

$$S^{(\mathrm{f})} = Z_1 S^{(\mathrm{i})} \tag{6.225}$$

where the expression of the Mueller matrix $Z_1$ has been given by Eq. (6.40) as

$$Z_1 = \begin{pmatrix} 1 & \overline{A}_x & \overline{A}_y & 0 \\ \overline{P}^x & K_x^x & K_y^x & K_z^x \\ \overline{P}^y & K_x^y & K_y^y & K_z^y \\ 0 & K_x^z & K_y^z & K_z^z \end{pmatrix}_1 \tag{6.226}$$

Here, the subscript 1 represents the energy $E'$ and the angles $(\theta_1, \varphi_1)$. The function $I\left(\vec{r}, E', E, \theta_1, \varphi_1, \Omega', s', t\right)$ is the microscopic differential cross section of the outgoing neutrons. Based on Eq. (6.218) and Eq. (6.219), we can get $\theta_1(\theta', \varphi', \theta, \varphi)$ and $\varphi_1(\theta', \varphi', \theta, \varphi)$, so $I\left(\vec{r}, E', E, \theta_1, \varphi_1, \Omega', s', t\right)$ is also a function of $(\theta', \varphi', \theta, \varphi)$. Let the spin magnetic quantum number of the outgoing neutrons be $s$, $I\left(\vec{r}, E', E, \theta_1, \varphi_1, \Omega', s', t\right)$ in Eq. (6.224) contains the contributions of $s = \pm\frac{1}{2}$ two parts.

The three components $(P_{1x}, P_{1y}, P_{1z})$ of the polarization vector $\vec{P}_1$ of the outgoing neutrons in Eq. (6.224) are given in $O_1$ system. If we look at this vector in $O_0$ system, it becomes

$$\vec{p} = \hat{B}(\theta', \varphi')\vec{P}_1 \tag{6.227}$$

The matrix $\hat{B}(\theta', \varphi')$ has been given by Eq. (2.203) or Eq. (6.42), and then from the above formula we can obtain

$$\begin{aligned} p_x &= \cos\theta' \cos\varphi' P_{1x} - \sin\varphi' P_{1y} + \sin\theta' \cos\varphi' P_{1z} \\ p_y &= \cos\theta' \sin\varphi' P_{1x} + \cos\varphi' P_{1y} + \sin\theta' \sin\varphi' P_{1z} \\ p_z &= -\sin\theta' P_{1x} + \cos\theta' P_{1z} \end{aligned} \tag{6.228}$$

where $p_i = p_i\left(\vec{r}, E', E, \theta_1, \varphi_1, \vec{\Omega}', s', t\right)$, $i = x,y,z$. The absolute value of the outgoing neutron polarization vector $\vec{p}$ in $O_0$ system can be obtained from the above equations as

$$p_0 = \sqrt{p_x^2 + p_y^2 + p_z^2} \tag{6.229}$$

The radial component of a vector $\vec{p}$ in spherical coordinates system is just $p = p_0$.

We know that the spin $\frac{1}{2}$ particle polarization degree is defined as

$$p = N_{\frac{1}{2}} - N_{-\frac{1}{2}} \tag{6.230}$$

Here we limit $0 \le N_s \le 1$, $N_{\frac{1}{2}} \ge N_{-\frac{1}{2}}$, $p \ge 0$. That is, we stipulate that the direction, in which the share $N_s$ of the spin projection is larger, is the particle positive polarization direction. Their normalization condition requires

$$N_{\frac{1}{2}} + N_{-\frac{1}{2}} = 1 \tag{6.231}$$

Thus, from the above two formulas we can derive

$$N_{\frac{1}{2}} = \frac{1}{2}(1+p), \quad N_{-\frac{1}{2}} = \frac{1}{2}(1-p) \tag{6.232}$$

Here we take $p = p_0$, $p_0$ has been given by Eq. (6.229). So $N_s$ obtained from above equations is also a function of $\left(\vec{r}, E', E, \theta_1, \varphi_1, \vec{\Omega}', s', t\right)$.

Using the above results, we can obtain that the polarization macroscopic double differential cross section in Eq. (6.213) is

$$\begin{aligned} &\Sigma_{\mathrm{p}}\left(\vec{r}, t; E', \vec{\Omega}', s' \to E, \vec{\Omega}, s\right) \\ &= \rho\left(\vec{r}, t\right) N_s\left(\vec{r}, E', E, \theta_1, \varphi_1, \vec{\Omega}', s', t\right) I\left(\vec{r}, E', E, \theta_1, \varphi_1, \vec{\Omega}', s', t\right) \end{aligned} \tag{6.233}$$

where $\rho\left(\vec{r}, t\right)$ is the density of the nuclear medium, $N_s$ has been given by Eq. (6.232), and $I\left(\vec{r}, E', E, \theta_1, \varphi_1, \vec{\Omega}', s', t\right)$ appearing in Eq. (6.224) can be found by Eq. (6.225). And then we can also find that the outgoing neutron polarization vector components in $O_0$ system with energy $E$, motion direction $\vec{\Omega}$, and spin magnetic quantum number $s$ are

$$\begin{aligned} &P_i\left(\vec{r}, E, \vec{\Omega}, s, t\right) \\ &= d_s \iint \sum_{s'} \left[N_s\left(\vec{r}, E', E, \theta_1, \varphi_1, \vec{\Omega}', s', t\right) p_i\left(\vec{r}, E', E, \theta_1, \varphi_1, \vec{\Omega}', s', t\right)\right] \mathrm{d}E' \mathrm{d}\vec{\Omega}' \\ &\qquad s = \frac{1}{2}, -\frac{1}{2}; \qquad i = x, y, z \end{aligned} \tag{6.234}$$

where $p_i = p_i\left(\vec{r}, E', E, \theta_1, \varphi_1, \vec{\Omega}', s', t\right)$, $i = x,y,z$ have been given by Eq. (6.228). The definition of the direction factor $d_s$ is

$$d_s = \begin{cases} 1, & \text{when } s = \dfrac{1}{2} \\ -1, & \text{when } s = -\dfrac{1}{2} \end{cases} \tag{6.235}$$

In the previous expressions, the relations between $(\theta_1, \varphi_1)$ and $(\theta', \varphi', \theta, \varphi)$ can be derived from Eq. (6.218) and Eq. (6.219). So the function related to angle $\left(\theta_1, \varphi_1, \vec{\Omega}'\right)$ after being integrated to angle $\vec{\Omega}'$ becomes just related to angle $\vec{\Omega}$. The neutron polarization vector components given by Eq. (6.234) are just that used in Eq. (6.221) or Eq. (6.222) in $O_0$ system. When the initial conditions and boundary conditions of the four functions to be solved are determined, after substituting Eq. (6.233) into Eq. (6.213), the obtained equation combined with Eqs. (6.234) can be solved jointly. In fact, three neutron polarization vector components can be calculated conveniently in the process of calculating neutron flux function. From the point of view of calculation mathematics, the calculation of neutron polarization vector components will not bring much difficulty. It can be predicted that when the required neutron normal and polarization microscopic nuclear data are available, the method of solving the usual neutron transport equation can be extended to solve the polarized neutron transport equation.

Next we study the polarized deuteron transport equation, and consider only one polarized reaction channel. At the time $t$ and the space point $\vec{r}$, the direction of the incident deuterons with energy $E'$ in $O_0$ system is $\vec{\Omega}'$. Now we are going to study in $O_1$ system; in this case the incident deuterons are along the $z_1$-axis direction. After the incident deuteron energy $E'$ in the laboratory system is determined, the 42 kinds of the differential physical quantities in the laboratory system $I_0(\theta_1)$, $A_y(\theta_1)$, $A_i(\theta_1)$ $(i = 1\sim3)$, $P^y(\theta_1)$, $P^i(\theta_1)$ $(i = 1\sim3)$, $R_i(\theta_1)$ $(i = 1\sim16)$, $Q_i(\theta_1)(i = 1\sim15)$, $W_i(\theta_1)$ $(i = 1, 2)$ can be obtained from the polarization nuclear database for the $\vec{1} +$ $\mathrm{A} \rightarrow \vec{1} + \mathrm{B}$ reaction. Then the various polarization quantities related to angles $(\theta_1, \varphi_1)$ can be calculated according to Eqs. (6.63), (6.70), (6.71), (6.76), (6.81), (6.86), (6.91), (6.96), and (6.101).

Let $P_i^{\mathrm{v}}\left(\vec{r}, E', \vec{\Omega}', s', t\right)$, $i = x, y, z$ be the deuteron polarization vector components in $O_0$ system with energy $E'$, motion direction $\vec{\Omega}'$, and spin magnetic quantum number $s'$, and let $P_i^{\mathrm{t}}\left(\vec{r}, E', \vec{\Omega}', s', t\right), i = xy,\ xz,\ yz,\ xx - yy,\ zz$ be the deuteron polarization tensor components in $O_0$ system with energy $E'$, motion direction $\vec{\Omega}'$, and spin magnetic quantum number $s'$. The eight polarization rate functions are solved in conjunction with the deuteron flux function $\psi\left(\vec{r}, E, \vec{\Omega}, s, t\right)$. If we look at $P_i^{\mathrm{v}}\left(\vec{r}, E', \vec{\Omega}', s', t\right), i = x, y, z$ in $O_1$ system, they become

$$\vec{p}_1^{\mathrm{v}}\left(\vec{r}, E', \vec{\Omega}', s', t\right) = \hat{b}(\theta', \varphi')\vec{P}^{\mathrm{v}}\left(\vec{r}, E', \vec{\Omega}', s', t\right) \tag{6.236}$$

where the matrix $\hat{b}(\theta', \varphi')$ has been given by Eq. (2.204) or Eq. (6.43). If we look at $P_i^{\mathrm{t}}\left(\vec{r}, E', \vec{\Omega}', s', t\right), i = xy, xz, yz, xx - yy, zz$ in $O_1$ system, they become

$$\begin{pmatrix} \sqrt{\frac{1}{6}}p^{\mathrm{t}}_{1\,xx-yy}\left(\vec{r},E',\vec{\Omega}',s',t\right) \\ \sqrt{\frac{2}{3}}p^{\mathrm{t}}_{1\,xy}\left(\vec{r},E',\vec{\Omega}',s',t\right) \\ \sqrt{\frac{2}{3}}p^{\mathrm{t}}_{1\,xz}\left(\vec{r},E',\vec{\Omega}',s',t\right) \\ \sqrt{\frac{2}{3}}p^{\mathrm{t}}_{1\,yz}\left(\vec{r},E',\vec{\Omega}',s',t\right) \\ \sqrt{\frac{1}{2}}p^{\mathrm{t}}_{1\,zz}\left(\vec{r},E',\vec{\Omega}',s',t\right) \end{pmatrix} = \hat{\gamma}(\theta',\varphi') \begin{pmatrix} \sqrt{\frac{1}{6}}P^{\mathrm{t}}_{xx-yy}\left(\vec{r},E',\vec{\Omega}',s',t\right) \\ \sqrt{\frac{2}{3}}P^{\mathrm{t}}_{xy}\left(\vec{r},E',\vec{\Omega}',s',t\right) \\ \sqrt{\frac{2}{3}}P^{\mathrm{t}}_{xz}\left(\vec{r},E',\vec{\Omega}',s',t\right) \\ \sqrt{\frac{2}{3}}P^{\mathrm{t}}_{yz}\left(\vec{r},E',\vec{\Omega}',s',t\right) \\ \sqrt{\frac{1}{2}}P^{\mathrm{t}}_{zz}\left(\vec{r},E',\vec{\Omega}',s',t\right) \end{pmatrix} \tag{6.237}$$

where the matrix $\hat{\gamma}(\theta', \varphi')$ has been given by Eq. (6.118). And according to Eq. (6.103) we define

$$\begin{aligned} &\bar{p}^{\mathrm{v}}_{1i}\left(\vec{r}, E', \vec{\Omega}', s', t\right) = \frac{3}{2}p^{\mathrm{v}}_{1i}\left(\vec{r}, E', \vec{\Omega}', s', t\right), \qquad i=x,y,z \\ &\bar{p}^{\mathrm{t}}_{1i}\left(\vec{r}, E', \vec{\Omega}', s', t\right) = \begin{cases} \frac{2}{3}p^{\mathrm{t}}_{1i}\left(\vec{r},E',\vec{\Omega}',s',t\right), & i=xy,\ xz,\ yz \\ \frac{1}{6}p^{\mathrm{t}}_{1i}\left(\vec{r},E',\vec{\Omega}',s',t\right), & i=xx-yy \\ \frac{1}{2}p^{\mathrm{t}}_{1i}\left(\vec{r},E',\vec{\Omega}',s',t\right), & i=zz \end{cases} \end{aligned} \tag{6.238}$$

Then the Stokes vector of the incident deuterons in $O_1$ system can be written out as

$$S^{(\mathrm{i})}\left(\vec{r}, E', E, \theta_1, \vec{\Omega}', s', t\right) = I_0(E', E, \theta_1) \begin{pmatrix} 1 \\ \bar{p}^{\mathrm{v}}_{1x}\left(\vec{r},E',\vec{\Omega}',s',t\right) \\ \bar{p}^{\mathrm{v}}_{1y}\left(\vec{r},E',\vec{\Omega}',s',t\right) \\ \bar{p}^{\mathrm{v}}_{1z}\left(\vec{r},E',\vec{\Omega}',s',t\right) \\ \bar{p}^{\mathrm{t}}_{1xy}\left(\vec{r},E',\vec{\Omega}',s',t\right) \\ \bar{p}^{\mathrm{t}}_{1xz}\left(\vec{r},E',\vec{\Omega}',s',t\right) \\ \bar{p}^{\mathrm{t}}_{1yz}\left(\vec{r},E',\vec{\Omega}',s',t\right) \\ \bar{p}^{\mathrm{t}}_{1\,xx-yy}\left(\vec{r},E',\vec{\Omega}',s',t\right) \\ \bar{p}^{\mathrm{t}}_{1zz}\left(\vec{r},E',\vec{\Omega}',s',t\right) \end{pmatrix} \tag{6.239}$$

The expression of Stokes vector for outgoing deuterons is

$$S^{(\mathrm{f})}\left(\vec{r}, E', E, \theta_1, \varphi_1, \vec{\Omega}', s', t\right)$$
$$= I\left(\vec{r}, E', E, \theta_1, \varphi_1, \vec{\Omega}', s', t\right)\begin{pmatrix} 1 \\ P^{\mathrm{v}}_{1x}\left(\vec{r}, E', E, \theta_1, \varphi_1, \vec{\Omega}', s', t\right) \\ P^{\mathrm{v}}_{1y}\left(\vec{r}, E', E, \theta_1, \varphi_1, \vec{\Omega}', s', t\right) \\ P^{\mathrm{v}}_{1z}\left(\vec{r}, E', E, \theta_1, \varphi_1, \vec{\Omega}', s', t\right) \\ P^{\mathrm{t}}_{1xy}\left(\vec{r}, E', E, \theta_1, \varphi_1, \vec{\Omega}', s', t\right) \\ P^{\mathrm{t}}_{1xz}\left(\vec{r}, E', E, \theta_1, \varphi_1, \vec{\Omega}', s', t\right) \\ P^{\mathrm{t}}_{1yz}\left(\vec{r}, E', E, \theta_1, \varphi_1, \vec{\Omega}', s', t\right) \\ P^{\mathrm{t}}_{1\,xx-yy}\left(\vec{r}, E', E, \theta_1, \varphi_1, \vec{\Omega}', s', t\right) \\ P^{\mathrm{t}}_{1zz}\left(\vec{r}, E', E, \theta_1, \varphi_1, \vec{\Omega}', s', t\right) \end{pmatrix} \tag{6.240}$$

The above two formulas satisfy the following relation:

$$S^{(\mathrm{f})} = Z_1 S^{(\mathrm{i})} \tag{6.241}$$

where the expression of the Mueller matrix $Z_1$ has been given by Eq. (6.106) as

$$Z_1 = \begin{pmatrix} 1 & \overline{A}_x & \overline{A}_y & 0 & \overline{A}_{xy} & \overline{A}_{xz} & \overline{A}_{yz} & \overline{A}_{xx-yy} & \overline{A}_{zz} \\ \overline{P}^x & K^x_x & K^x_y & K^x_z & K^x_{xy} & K^x_{xz} & K^x_{yz} & K^x_{xx-yy} & K^x_{zz} \\ \overline{P}^y & K^y_x & K^y_y & K^y_z & K^y_{xy} & K^y_{xz} & K^y_{yz} & K^y_{xx-yy} & K^y_{zz} \\ 0 & K^z_x & K^z_y & K^z_z & K^z_{xy} & K^z_{xz} & K^z_{yz} & K^z_{xx-yy} & 0 \\ \overline{P}^{xy} & K^{xy}_x & K^{xy}_y & K^{xy}_z & K^{xy}_{xy} & K^{xy}_{xz} & K^{xy}_{yz} & K^{xy}_{xx-yy} & K^{xy}_{zz} \\ \overline{P}^{xz} & K^{xz}_x & K^{xz}_y & K^{xz}_z & K^{xz}_{xy} & K^{xz}_{xz} & K^{xz}_{yz} & K^{xz}_{xx-yy} & K^{xz}_{zz} \\ \overline{P}^{yz} & K^{yz}_x & K^{yz}_y & K^{yz}_z & K^{yz}_{xy} & K^{yz}_{xz} & K^{yz}_{yz} & K^{yz}_{xx-yy} & K^{yz}_{zz} \\ \overline{P}^{xx-yy} & K^{xx-yy}_x & K^{xx-yy}_y & K^{xx-yy}_z & K^{xx-yy}_{xy} & K^{xx-yy}_{xz} & K^{xx-yy}_{yz} & K^{xx-yy}_{xx-yy} & K^{xx-yy}_{zz} \\ \overline{P}^{zz} & K^{zz}_x & K^{zz}_y & 0 & K^{zz}_{xy} & K^{zz}_{xz} & K^{zz}_{yz} & K^{zz}_{xx-yy} & K^{zz}_{zz} \end{pmatrix}_1 \tag{6.242}$$

Here the subscript 1 represents the energy $E'$ and the angles $(\theta_1, \varphi_1)$. The function $I\left(\vec{r}, E', E, \theta_1, \varphi_1, \Omega', s', t\right)$ is the microscopic differential cross section of the outgoing deuterons, and it is also a function of $(\theta', \varphi', \theta, \varphi)$ according to

Eq. (6.218) and Eq. (6.219). Let the spin magnetic quantum number of the outgoing deuterons be s; the function $I\left(\vec{r}, E', E, \theta_1, \varphi_1, \Omega', s', t\right)$ in Eq. (6.240) contains the contributions of three parts $s = \pm 1, 0$.

The three components $\left(P^{\mathrm{v}}_{1x}, P^{\mathrm{v}}_{1y}, P^{\mathrm{v}}_{1z}\right)$ of the polarization vector $\vec{P}^{\mathrm{v}}_1$ and the five components $\left(P^{\mathrm{t}}_{1xy}, P^{\mathrm{t}}_{1xz}, P^{\mathrm{t}}_{1yz}, P^{\mathrm{t}}_{1\,xx-yy}, P^{\mathrm{t}}_{1zz}\right)$ of the polarization tensor $\vec{P}^{\mathrm{t}}_1$ of the outgoing deuterons in Eq. (6.240) are given in $O_1$ system. If we look at them in $O_0$ system, $\vec{P}^{\mathrm{v}}_1$ becomes

$$\vec{p}^{\mathrm{v}} = \hat{B}(\theta', \varphi')\vec{P}^{\mathrm{v}}_1 \tag{6.243}$$

where matrix $\hat{B}(\theta', \varphi')$ has been given by Eq. (2.203) or Eq. (6.42); $\vec{P}^{\mathrm{t}}_1$ becomes

$$\begin{pmatrix} \sqrt{\frac{1}{6}}p^{\mathrm{t}}_{xx-yy} \\ \sqrt{\frac{2}{3}}p^{\mathrm{t}}_{xy} \\ \sqrt{\frac{2}{3}}p^{\mathrm{t}}_{xz} \\ \sqrt{\frac{2}{3}}p^{\mathrm{t}}_{yz} \\ \sqrt{\frac{1}{2}}p^{\mathrm{t}}_{zz} \end{pmatrix} = \hat{\Gamma}(\theta', \varphi') \begin{pmatrix} \sqrt{\frac{1}{6}}P^{\mathrm{t}}_{1\,xx-yy} \\ \sqrt{\frac{2}{3}}P^{\mathrm{t}}_{1\,xy} \\ \sqrt{\frac{2}{3}}P^{\mathrm{t}}_{1\,xz} \\ \sqrt{\frac{2}{3}}P^{\mathrm{t}}_{1\,yz} \\ \sqrt{\frac{1}{2}}P^{\mathrm{t}}_{1\,zz} \end{pmatrix} \tag{6.244}$$

where matrix $\hat{\Gamma}(\theta', \varphi')$ has been given by Eq. (6.117). The polarization vector $\vec{p}^{\mathrm{v}}$ and the polarization tensor $\vec{p}^{\mathrm{t}}$ of the outgoing deuterons in $O_0$ system are along the deuteron polarization direction. In the Cartesian coordinate system $XYZ$ taking the deuteron polarization direction as the $Z$-axis, only the component $p^{\mathrm{v}}_Z$ of the polarization vector $\vec{p}^{\mathrm{v}}$ and the component $p^{\mathrm{t}}_{ZZ}$ of the polarization tensor are not equal to 0. The absolute value of the outgoing deuteron polarization vector $\vec{p}^{\mathrm{v}}$ can be found to be

$$p^{\mathrm{v}}_Z = \sqrt{\left(p^{\mathrm{v}}_x\right)^2 + \left(p^{\mathrm{v}}_y\right)^2 + \left(p^{\mathrm{v}}_z\right)^2} \tag{6.245}$$

Assume that the polarization direction of the outgoing deuterons is $(\theta_2, \varphi_2)$ in $O_0$ system, and then the relations between the components of $\vec{p}^{\mathrm{t}}$ and $p^{\mathrm{t}}_{ZZ}$ can be given by Eq. (6.124) as follows:

$$p^{\mathrm{t}}_{xx-yy}=\frac{3}{2}\sin^2\theta_2\cos(2\varphi_2)p^{\mathrm{t}}_{ZZ},\quad p^{\mathrm{t}}_{xy}=\frac{3}{4}\sin^2\theta_2\sin(2\varphi_2)p^{\mathrm{t}}_{ZZ}$$
$$p^{\mathrm{t}}_{xz}=\frac{3}{4}\sin(2\theta_2)\cos\varphi_2 p^{\mathrm{t}}_{ZZ},\quad p^{\mathrm{t}}_{yz}=\frac{3}{4}\sin(2\theta_2)\sin\varphi_2 p^{\mathrm{t}}_{ZZ} \tag{6.246}$$
$$p^{\mathrm{t}}_{zz}=\frac{1}{2}(3\cos^2\theta_2-1)p^{\mathrm{t}}_{ZZ}$$

It can be derived from the above formulas that

$$\left(p^{\mathrm{t}}_{xx-yy}\right)^2+4\left(p^{\mathrm{t}}_{xy}\right)^2=\frac{9}{4}\sin^4(\theta_2)\left(p^{\mathrm{t}}_{ZZ}\right)^2$$
$$\left(p^{\mathrm{t}}_{xz}\right)^2+\left(p^{\mathrm{t}}_{yz}\right)^2=\frac{9}{4}\sin^2(\theta_2)\cos^2(\theta_2)\left(p^{\mathrm{t}}_{ZZ}\right)^2 \tag{6.247}$$
$$\left(p^{\mathrm{t}}_{zz}\right)^2=\frac{9}{4}\left(\cos^4\theta_2-\frac{2}{3}\cos^2\theta_2+\frac{1}{9}\right)\left(p^{\mathrm{t}}_{ZZ}\right)^2$$

$$\frac{1}{3}\left[\left(p^{\mathrm{t}}_{xx-yy}\right)^2+4\left(p^{\mathrm{t}}_{xy}\right)^2\right]+\frac{4}{3}\left[\left(p^{\mathrm{t}}_{xz}\right)^2+\left(p^{\mathrm{t}}_{yz}\right)^2\right]+\left(p^{\mathrm{t}}_{zz}\right)^2=\left(p^{\mathrm{t}}_{ZZ}\right)^2 \tag{6.248}$$

$$p^{\mathrm{t}}_{ZZ}=\left\{\frac{1}{3}\left[\left(p^{\mathrm{t}}_{xx-yy}\right)^2+4\left(p^{\mathrm{t}}_{xy}\right)^2\right]+\frac{4}{3}\left[\left(p^{\mathrm{t}}_{xz}\right)^2+\left(p^{\mathrm{t}}_{yz}\right)^2\right]+\left(p^{\mathrm{t}}_{zz}\right)^2\right\}^{1/2} \tag{6.249}$$

In fact, in the five-dimensional orthogonal Cartesian coordinate system, using the notations introduced by Eq. (6.120), we can write the following relation directly:

$$\left(h^{\mathrm{t}}_{xx-yy}\right)^2+\left(h^{\mathrm{t}}_{xy}\right)^2+\left(h^{\mathrm{t}}_{xz}\right)^2+\left(h^{\mathrm{t}}_{yz}\right)^2+\left(h^{\mathrm{t}}_{zz}\right)^2=\left(h^{\mathrm{t}}_{ZZ}\right)^2 \tag{6.250}$$

According to Eq. (6.120), we can get Eq. (6.248) from Eq. (6.250) immediately. This clearly shows that the use of five-dimensional orthogonal Cartesian coordinate system will bring greater convenience.

The components of the polarization vector $\vec{p}^{\mathrm{v}}$ and the polarization tensor $\vec{p}^{\mathrm{t}}$ of the outgoing deuterons in $O_0$ system are calculated results numerically, and $p^{\mathrm{v}}_Z$ and $p^{\mathrm{t}}_{ZZ}$ can be obtained by Eq. (6.245) and Eq. (6.249), respectively. The shares $N_1$, $N_0$, $N_{-1}$ of the three spin projections of the deuterons along the polarization direction satisfy the following relations given by Eqs. (3.163) and (3.165):

$$N_1+N_0+N_{-1}=1,\qquad p^{\mathrm{v}}_Z=N_1-N_{-1},\qquad p^{\mathrm{t}}_{ZZ}=N_1-2N_0+N_{-1} \tag{6.251}$$

From the above three formulas we can obtain

$$N_1=\frac{1}{6}\left(2+3p^{\mathrm{v}}_Z+p^{\mathrm{t}}_{ZZ}\right),\qquad N_0=\frac{1}{3}\left(1-p^{\mathrm{t}}_{ZZ}\right),$$
$$N_{-1}=\frac{1}{6}\left(2-3p^{\mathrm{v}}_Z+p^{\mathrm{t}}_{ZZ}\right) \tag{6.252}$$

Since $p_Z^{\mathrm{v}}$ and $p_{ZZ}^{\mathrm{t}}$ are the functions of $\left(\vec{r}, E', E, \theta_1, \varphi_1, \vec{\Omega}', s', t\right)$, the $N_s(s = \pm 1, 0)$ derived from the above equations are also the functions of $\left(\vec{r}, E', E, \theta_1, \varphi_1, \vec{\Omega}', s', t\right)$.

Using the above results we can obtain that the polarization macroscopic double differential cross section in Eq. (6.213) is

$$\begin{aligned} &\Sigma_{\mathrm{p}}\left(\vec{r}, t; E', \vec{\Omega}', s' \rightarrow E, \vec{\Omega}, s\right) \\ &= \rho\left(\vec{r}, t\right) N_s\left(\vec{r}, E', E, \theta_1, \varphi_1, \vec{\Omega}', s', t\right) I\left(\vec{r}, E', E, \theta_1, \varphi_1, \vec{\Omega}', s', t\right) \end{aligned} \tag{6.253}$$

where $\rho\left(\vec{r}, t\right)$ is the density of the nuclear medium, $N_s$ has been given by Eq. (6.252), and $I\left(\vec{r}, E', E, \theta_1, \varphi_1, \vec{\Omega}', s', t\right)$ appearing in Eq. (6.240) can be solved by Eq. (6.241). Furthermore, the polarization vector components of the outgoing deuterons with energy $E$, motion direction $\vec{\Omega}$, and spin quantum number $s$ in $O_0$ system can also be obtained as

$$\begin{aligned} &P_i^{\mathrm{v}}\left(\vec{r}, E, \vec{\Omega}, s, t\right) \\ &= d_s^{\mathrm{v}} \iint \sum_{s'} \left[N_s\left(\vec{r}, E', E, \theta_1, \varphi_1, \vec{\Omega}', s', t\right) p_i^{\mathrm{v}}\left(\vec{r}, E', E, \theta_1, \varphi_1, \vec{\Omega}', s', t\right)\right] \mathrm{d}E' \mathrm{d}\vec{\Omega}' \\ &\qquad s = \pm 1, 0; \qquad i = x, y, z \end{aligned} \tag{6.254}$$

where $p_i^{\mathrm{v}}\left(\vec{r}, E', E, \theta_1, \varphi_1, \vec{\Omega}', s', t\right)$, $i = x, y, z$ have been given by Eq. (6.243). In $O_0$ system the polarization tensor components of the outgoing deuterons with energy $E$, motion direction $\vec{\Omega}$, and spin quantum number $s$ are

$$\begin{aligned} &P_i^{\mathrm{t}}\left(\vec{r}, E, \vec{\Omega}, s, t\right) \\ &= d_s^{\mathrm{t}} \iint \sum_{s'} \left[N_s\left(\vec{r}, E', E, \theta_1, \varphi_1, \vec{\Omega}', s', t\right) p_i^{\mathrm{t}}\left(\vec{r}, E', E, \theta_1, \varphi_1, \vec{\Omega}', s', t\right)\right] \mathrm{d}E' \mathrm{d}\vec{\Omega}' \\ &\qquad s = \pm 1, 0; \qquad i = xx - yy, xy, xz, yz, zz \end{aligned} \tag{6.255}$$

where $p_i^{\mathrm{t}}\left(\vec{r}, E', E, \theta_1, \varphi_1, \vec{\Omega}', s', t\right)$, $i = xx - yy, xy, xz, yz, zz$ have been given by Eq. (6.244). The deuteron vector direction factor $d_s^{\mathrm{v}}$ is defined as

$$d_s^{\mathrm{v}} = \begin{cases} 1, & \text{when} \quad s = 1 \\ 0, & \text{when} \quad s = 0 \\ -1, & \text{when} \quad s = -1 \end{cases} \tag{6.256}$$

The deuteron tensor direction factor $d_s^{\mathrm{t}}$ is defined as

$$d_s^{\mathrm{t}} = \begin{cases} 1, & \text{when} \quad s = 1 \\ -2, & \text{when} \quad s = 0 \\ 1, & \text{when} \quad s = -1 \end{cases} \tag{6.257}$$

The components of the polarization vector and the polarization tensor of deuterons given by Eq. (6.254) and Eq. (6.255) are just that used in Eq. (6.236) and Eq. (6.237) in $O_0$ system, respectively. When the initial conditions and boundary conditions of the nine functions to be solved are determined, after substituting Eq. (6.253) into Eq. (6.213), the obtained equation combined with Eqs. (6.254) and (6.255) can be solved jointly.

For the transport problem of the spin 1 deuterons, in addition that the ionization loss effect needs to be considered, due to the presence of (d, n), (n, d), and so on reactions, it is necessary to study the conjunction transport problem of $\vec{1} + \mathrm{A} \rightarrow \vec{1} + \mathrm{B}$, $\vec{1} + \mathrm{A} \rightarrow \vec{\frac{1}{2}} + \mathrm{B}$, $\vec{\frac{1}{2}} + \mathrm{A} \rightarrow \vec{1} + \mathrm{B}$, $\vec{\frac{1}{2}} + \mathrm{A} \rightarrow \vec{\frac{1}{2}} + \mathrm{B}$ reaction channels. In other words, it is necessary to solve the polarized particle transport equation (6.213) by using the source term Q to combine the neutron transport with the charged particle transport. So this problem becomes quite complicated.

## 6.6 Prospect of the Polarized Particle Transport Theory

### *6.6.1 To Perfect the Polarization Theory of Nuclear Reactions*

The theory of calculating various polarization physical quantities from the nuclear reaction amplitude is called polarization theory of nuclear reactions, and it still belongs to the category of nuclear reaction theory. *S* matrix theory is the basic theory describing the nuclear reactions. The expressions of the reaction amplitude given by *S* matrix theory satisfy the basic physical requirements, such as angular momentum conservation and parity conservation. The polarization theory of the two-body reaction of the spin $\frac{1}{2}$ or 1 incident particle with a target and residual nucleus with arbitrary spins was studied in this book. Some internal relations of the matrix elements of the nuclear reaction amplitude described by the *S* matrix elements were found. The clear expressions of various polarization quantities of the corresponding nuclear reactions were obtained. The optical model, coupled channel optical model, distorted wave Born approximation, $R$ matrix theory, phase-shift analysis method, and even some microscopic nuclear reaction theory can all be used to calculate the $S$ matrix elements of the nuclear reactions. The above theories belong to the usual nuclear reaction theory. Therefore, the publication of this book makes that the polarization theory of the incident particles of spin $\frac{1}{2}$ or 1 and the unpolarized targets and residual nuclei with arbitrary spins have reached the stage where the actual calculations can be done. The particle n, p, t, $^3$Heof $S = \frac{1}{2}$ and d of $S = 1$ are the most concerned particles in the particle transport theory. In the future, it is necessary to study the universal optical potential of deuteron containing tensor term, and the polarization experimental data should be included in the process of adjusting the theoretical model parameters to fit the experimental data.

The spin of the particle α is equal to 0, and there is no polarization problem for itself. However, there are the polarization problems in the incident channel or outgoing channel corresponding to (n, α), (d, α), (α, n), (α, d) and other nuclear reactions indicated by $\overrightarrow{\frac{1}{2}} + \mathrm{A} \to 0 + \mathrm{B}$, $\vec{1} + \mathrm{A} \to 0 + \mathrm{B}$, $0 + \mathrm{A} \to \overrightarrow{\frac{1}{2}} + \mathrm{B}$, $0 + \mathrm{A} \to \vec{1} + \mathrm{B}$. It is relatively easy to study the polarization phenomena of the two-body direct reaction where only the incident channel or only the outgoing channel is polarized.

The spin of the particles such as $^5$He, $^7$Li, $^7$Be, $^9$Be and so on is $\frac{3}{2}$; these particles are relatively heavy. And for the sake of the integrity of the nuclear reaction polarization theory, the spin $\frac{3}{2}$ particle polarization theory should also be studied. This book begins by studying the polarization theory of the spin $\frac{3}{2}$ particles in spherical basis coordinates, then based on the studying on the five-dimensional rank-2 tensors required for spin 1 particles, the unitary transformation relations of the rank-2 and higher-rank tensors between the spherical basis coordinates system and the orthogonal Cartesian coordinate system are found, and the theoretical methods for studying the spin $\geq \frac{3}{2}$ particle polarization phenomena are given.

It is necessary to study the polarization effect of nuclear reactions in the energy region with the anisotropic differential cross sections. For the few-body nuclear reaction, there is a clear anisotropy phenomenon even in the keV energy region. If the target is a medium-weight nuclide or a heavy nuclide, the anisotropy is only apparent when the incident particle energy approaches the MeV energy region.

This book also studied some nuclear reactions for which both the incident particle and the target are polarized at the same time. They can be expressed as symbols $\overrightarrow{\frac{1}{2}} + \overrightarrow{\frac{1}{2}}$, $\overrightarrow{\frac{1}{2}} + \vec{1}$, $\vec{1} + \vec{1}$. Of course, their theoretical formulas are different corresponding to different outgoing channels. This book only studies part of the outgoing channels for above reactions. For example, we have studied $\vec{\mathrm{n}} + \vec{\mathrm{d}} \to \mathrm{n} + \mathrm{d}$ reaction, whereas we have not specifically studied $\vec{\mathrm{d}} + \vec{\mathrm{t}} \to \mathrm{n} + \alpha$ reaction. But it is not very difficult to study it using the methods described in this book. This book also studied $\vec{d} + \vec{d} \rightarrow p + t$ reaction, but because there are too many items in the formulas, we recommend using numerical calculation method including trace to deal with this problem.

For the case where the target and the residual nucleus are a medium-weight nuclide or a heavy nuclide, it is usually assumed that the target and the residual nucleus are unpolarized. For the target and the residual nucleus which are also polarized, when the polarization rates of the target and/or residual nucleus with arbitrary spin are known, the methods described in Sect. 2.8, 3.7, 3.9, 3.10 sections can also be used to calculate various polarization quantities, but this method cannot calculate the polarization rates of the residual nucleus. If the spins of the heavier polarized target and the residual nucleus are $\frac{1}{2}$ or 1, various polarization physical quantities can also be calculated using the methods described in Sects. 2.9, 3.11, and 3.12. Although the theory itself is relatively complex, however, compared with the previous method in Sects. 2.8, 3.7, 3.9, 3.10, it is not necessary to know the polarization rates of the residual nucleus in advance; conversely the polarization rates of the residual nucleus can be given through theoretical calculations.

In this book the light beam polarization theory has only made preliminary exploration, and the further research work on it still needs to be carried out. In addition, the polarization theory of the three-body reaction also needs to be further explored. In fact, some research works have been carried out on the polarization problem of three-body reaction in some papers studying the few-body reactions [21–26].

Since the pre-equilibrium and equilibrium compound nuclear reactions belong to the nuclear reaction process in which many nucleons are involved. Even if the incident nucleon is polarized, after the reaction process in which many nucleons participate, taking into account the average effect, the outgoing nucleons should be unpolarized.

Since the spin-orbit potential can be automatically generated in the relativistic Dirac equation, it is also a hot topic to study the polarization phenomena using relativistic nuclear reaction theory. In the relativistic optical model, the Dirac equation is converted into a Schrodinger-like equation, and naturally it becomes that using non-relativistic formulas to calculate the corresponding polarization physical quantities. This book develops the Dirac $S$ matrix theory of relativistic nuclear reactions. When the energy is quite low, this theory will automatically degenerate into a non-relativistic $S$ matrix theory for spin $\frac{1}{2}$ particles. In this way, as long as the relativistic reaction amplitude is solved, using the Dirac equation, the Dirac $S$ matrix theory can be used to calculate various physical quantities of spin $\frac{1}{2}$ particles, including an amount of polarization physical quantities. The relativistic polarization theory of the deuteron induced reactions needs to be studied further.

When the incident nucleon energy is high enough to need to be described by relativistic theory, the intra-nuclear cascade reaction becomes one of the very important reaction mechanisms. The quantum molecular dynamics (QMD) model is just a theory to describe the intra-nuclear cascade reaction. In the intra-nuclear cascade process, it is possible to impact the nucleon going outside the nucleus after only a few collisions, so the intra-nuclear cascade reaction belongs to a nuclear reaction in which a small number of nucleons participate. Assume that the incident nucleon is polarized and the nucleons in the target are unpolarized, so the first collision belongs to $\overrightarrow{\frac{1}{2}} + \frac{1}{2} \rightarrow \overrightarrow{\frac{1}{2}} + \overrightarrow{\frac{1}{2}}$ reaction. The following collision generally also belongs to this kind of reaction, but the collision between the two nucleons, which have been collided before, belongs to $\overrightarrow{\frac{1}{2}} + \overrightarrow{\frac{1}{2}} \rightarrow \overrightarrow{\frac{1}{2}} + \overrightarrow{\frac{1}{2}}$ reaction. The probability of such a collision is relatively small. In the QMD model, after simulating a large number of collision events, considering the axial symmetry of the system, the average double differential cross sections of the neutrons and protons being collided out by an incident nucleon can finally be obtained as $N_n(E, \theta)$ and $N_p(E, \theta)$. Making the integrals to them on $E$ and $\theta$ we can get the numbers of the neutrons and protons which are collided out. If the nucleon polarization rates are also passed during the nucleon-nucleon collision in the nucleus, the average nucleon polarization rates $P_{\mathrm{n}i}(E, \theta)$ and $P_{\mathrm{p}i}(E, \theta)$ $(i = x, y, z)$ of the neutrons and protons being collided out will be obtained. In order to realize the above idea, it is necessary to develop a polarization QMD model.

### 6.6.2 To Establish a Polarization Nuclear Database

The aim of the establishment of polarization nuclear database is to provide the basic nuclear data for the calculation of polarized nucleus transport theory. At present, we only consider the unpolarized target and residual nucleus.

With regard to polarization nuclear databases, we propose to use the classification based on the kinds of the incident particles, such as neutron polarization nuclear databases, proton polarization nuclear databases, deuteron polarization nuclear databases, and so on. In each particle polarization nuclear database, there are many kinds of targets, and for each target there are many possible reaction channels. For example, for the neutron induced nuclear reactions, the elastic scattering channel only needs to contain the shape elastic scattering part, the discrete inelastic scattering channel only needs to contain the corresponding direct inelastic scattering part, and the direct reaction parts of $(\mathrm{n}, \mathrm{p})$, $(\mathrm{n}, \mathrm{d})$, $(\mathrm{n}, \alpha)$ and so on reactions can also be included. The polarization problem is not considered for the parts contributed by the pre-equilibrium and equilibrium compound nuclear reaction, as well as for the parts contributed by the second emission and the third emission. For each reaction channel to consider the polarization effect, many sets of polarization differential physical quantities only related to angle $\theta$ for multi-energy points of the laboratory system are contained. It has been pointed out in 6.2–6.4 sections that for $\overrightarrow{\frac{1}{2}} +$

$A \rightarrow \overrightarrow{\frac{1}{2}} + B$ reaction there are 8 sets of polarization differential physical quantities, for $\vec{1} + A \rightarrow \overrightarrow{\frac{1}{2}} + B$ and $\overrightarrow{\frac{1}{2}} + A \rightarrow \vec{1} + B$ reactions there are 18 sets, respectively, and for $\vec{1} + A \rightarrow \vec{1} + B$ reaction there are 42 sets. It can also be inferred that there are 2 sets for $\overrightarrow{\frac{1}{2}} + A \rightarrow 0 + B$ and $0 + A \rightarrow \overrightarrow{\frac{1}{2}} + B$ reactions and 5 sets for $\vec{1} + A \rightarrow 0 + B$ and $0 + A \rightarrow \vec{1} + B$ reactions, respectively.

The establishment of a polarization nuclear database should be combined with the calculation and evaluation of a complete set of nuclear data, including cross sections, angular distributions, double differential cross sections, and polarization data. Only when the used theories and model parameters can better fit the multiple experimental data of the reaction at the same time, the polarization nuclear data predicted theoretically will possess a certain degree of credibility.

The establishment of the polarization nuclear database is a gradual accumulation process. For example, at the beginning it can only contain few targets and neutron induced data, and can only contain the data of the shape elastic scattering and the direct inelastic scattering of one or two discrete levels. In the initial stage, it needs only that we can give the basic data needed to carry out the research on the polarized nucleus transport theory for the individual macroscopic experiment inspection device.

In this book, for the polarization theory of spin 1 particles, the five- dimensional independent Cartesian coordinate system taking $\hat{Q}_i(i = xy, xz, yz, xx - yy, zz)$ given by Eq. (3.21) as the basis functions is used, which is not yet a five-dimensional orthogonal Cartesian coordinate system. In the future, if someone wishes using the five dimensional orthogonal Cartesian coordinate system taking $\hat{H}_i(i = xx - yy, xy, xz, yz, zz)$ given by Eq. (3.142), or Eq. (6.111), or Eq. (6.113) as the basis functions and using the generalized Pauli matrix introduced in this book to derive a set of spin 1 particle polarization theory formulas, so that the order and coefficients of the terms contained in the new formulas will be different from those given in this book. But I suggest that you still use the polarization physical quantities, which are only relevant to polar angles, in the polarization nuclear database as defined in this book; thus two sets of spin 1 particle polarization theoretical formulas can use the same spin 1 particle polarization nuclear database. In addition, the rotation matrices $\hat{\Gamma}$ and $\hat{\gamma}$ given by Eq. (6.117) and Eq. (6.118) should be used only for five-dimensional orthogonal Cartesian coordinate system, so for the spin 1 theoretical formulas given in this book, when making rotation using these matrices, appropriate changes should be made to the relevant physical quantities, such as doing in Eq. (6.130) (or Eq. (6.211), or Eq. (6.237), or Eq. (6.244)).

### 6.6.3 To Research the Polarized Particle Transport Theory

In the initial stage, we can choose to explore the polarized nucleus transport theory for macroscopic experiment inspection devices with incident neutron energy in the MeV region and containing only one or two nuclides in the nuclear medium. The Monte-Carlo method or the polarized particle transport equation method can all be used. It is recommended that the Cartesian coordinate system of the laboratory system is used and the initial incident neutron direction is taken as the $z$-axis. Then assume that the initial incident neutrons, targets, and residual nuclei are all unpolarized.

When doing the calculation, the usual nuclear database and polarization nuclear database are used at the time. First, in the usual nuclear database, we deduct the reaction channels that are already included in the polarization nuclear database. For example, in the polarization nuclear database, there are the shape elastic scattering channel and the first two level direct inelastic scattering channels, and then in the normal nuclear database, the elastic scattering channel can only include the contribution of the compound elastic scattering; in the first two level inelastic scattering channels, the contribution of direct inelastic scattering should be deducted. The above approach is equivalent to divide the original reaction channel into two reaction channels in form. The reaction channel, for which the polarization effect is not considered, is still placed in the normal nuclear database, and the reaction channel, for which the polarization effect is considered, is placed in the polarization nuclear database. When using the Monte-Carlo method, assume that the initial incident neutrons are unpolarized, after sampling, the obtained secondary neutrons generated through the unpolarization channel are still unpolarized, and secondary neutrons generated through the polarization channel are polarized. The polarization channels of the second reaction can process for the unpolarized and polarized secondary neutrons of the first reaction, and their corresponding outgoing particles are all polarized. For unpolarization channels, polarized incident neutrons are still treated as unpolarized neutrons, and the corresponding outgoing particles are also unpolarized. The ionization loss effect should be considered for the charged particle transport process.

Generally, in MeV energy region, the contribution of the elastic scattering channels basically comes from the shape elastic scattering; the main contribution of the low-energy discrete level inelastic scattering channels comes from the direct inelastic scattering. Although the polarization effect is not considered in the continuous inelastic scattering and (n, 2n), (n, 3n), and so on reaction channels, the contribution given by the reaction channels considering the polarization effect are still accounts for a considerable share. In the energy region that needs to be studied with relativistic theory, for the elastic scattering channel the polarization effect can be considered. At this time, since the incident particle energy is much larger than the heights of the discrete energy levels, the discrete level direct inelastic scattering and other direct reaction channels are not very important, but the intra-nuclear cascade mechanism does have a greater contribution. Therefore, the development of the polarization QMD model is very valuable for the study of the polarization problem of the relativistic nuclear reactions.

Both the experiments and the theories have proved that the nuclear potential of the spin $\frac{1}{2}$ nucleon contains spin-orbital potential, and the nuclear potential of the spin 1 deuteron contains tensor potential in addition to the spin-orbital potential. Therefore, the nuclear forces felt by the particles with different spin magnetic quantum numbers when moving in the nuclear potential are different. Therefore their motion behavior will also be different. This is the polarization phenomenon of nuclear reaction that exists objectively in nature. The polarization phenomenon is not considered in the existing transport theory that describes the movement of the heavy particles in nuclear media. It means that all particles are unpolarized. From a microscopic physical point of view, this is only an approximate approach. In order to be able to describe the physical process of the natural world in a more realistic way and to study the influence of polarization phenomenon, the research on polarized nucleus transport theory should be carried out. This shows that it is necessary to carry out research on polarized nucleus transport theory.

We know that the spins of most nuclei are not equal to 0. However, before writing this book, there was no polarization theory system that described the nuclear reaction of spin $\frac{1}{2}$ and 1 particle with target and residual nucleus with non-zero spin, so that a full set of polarization nuclear data could not be obtained. Of course, it was impossible to carry out polarization nucleus transport theory research in this case. However, in this book, the theory formulas describing the polarization reactions of the incident particles of spin $\frac{1}{2}$ and 1 with the unpolarized target and residual nucleus with arbitrary spins are clearly given. So it is feasible to establish a polarization nuclear database and carry out research on polarized nucleus transport theory.

For the first time, this book puts forward the research topic of establishing polarization nuclear database and carrying out polarized nucleus transport theory. I think that this topic will encounter many difficulties and challenges in the future development. In the future, how the subject will develop, and what will be the results and conclusions, these questions can only be answered by scientific researchers through their future scientific research work.

## References


1. Xiao XB, Li XM, Zhou GH. Properties of electron spin polarization transport of the quantum lines under the electromagnetic wave irradiation. J Phys. 2007;56:1649. (in Chinese)
2. Johnson M. Spin accumulation in gold films. Phys Rev Lett. 1993;70:2142.
3. Hershfield S, Zhao HL. Charge and spin transport through a metallic ferromagnetic-paramagnetic-ferromagnetic junction. Phys Rev. 1997;B56:3296.
4. Moroz AV, Barnes CHW. Effect of the spin-orbit interaction on the band structure and conductance of quasi-one-dimensional systems. Phys Rev. 1999;B60:14272.

5. Brataas A, Nazarov YV, bauer G E W. Finite-element theory of transport in ferromagnet-normal metal systems. Phys Rev Lett. 2000;84:2481.
6. Moroz AV, Samokhin KV, Barnes CHW. Spin-orbit coupling in interacting quasi-one-dimensional electron systems. Phys Rev Lett. 2000;84:4164.
7. Nadgorny B, Soulen RJ, Osofsky MS, et al. Transport spin polarization of $Ni_xFe_{1-x}$: Electronic kinematics and band structure. Phys Rev. 2000;B61:R3788.
8. Tang HX, Monzon FG, Lifshitz R, et al. Ballistic spin transport in a two-dimensional electron gas. Phys Rev. 2000;B61:4437.
9. Mireles F, Kirczenow G. Ballistic spin-polarized transport and Rashba spin precession in semiconductor nanowires. Phys Rev. 2001;B64:024426.
10. Wang XF. Spin transport of electrons through quantum wires with a spatially modulated Rashba spin-orbit interaction. Phys Rev. 2004;B69:035302.
11. Xu YL, Guo HR, Han YL, et al. New Skyrme interaction parameters for a unified description of the nuclear properties. J Phys G Nucl Part Phys. 2014;41:015101.
12. Kulsrud RM, Furth HP, Valeo EJ, Goldhaber M. Fusion reactor plasmas with polarized nuclei. Phys Rev Lett. 1982;49:1248.
13. More RM. Nuclear spin-polarized fuel in inertial fusion. Phys Rev Lett. 1983;51:396.
14. Fletcher KA, Ayer Z, Black TC, et al. Tensor analyzing power for $^2H(d,p)^3H$ and $^2H(d,n)^3He$ at deuteron energies of 25,40, and 80 keV. Phys Rev. 1994;C49:2305.
15. Zhang JS, Liu KF, Shuy GW. Neutron suppression in polarized dd fusion reaction. Phys Rev. 1999;C60:054614.
16. Schieck H, Paetz gen. The status of "polarized fusion". Eur Phys J. 2010;A44:321.
17. Schieck H, Paetz gen. "Polarized fusion": new aspects of an old project. Few-Body Syst. 2013;54:2159.
18. Liu KF, Chao AW. Accelerator based fusion reactor. Nucl Fusion. 2017;57:084002.
19. Robson BA. The theory of polarization phenomena. Oxford: Clarendon Press; 1974.
20. Varshalovich DA, Moskalev AN, Khersonskii VK. Quantum theory of angular momentum. Singapore/New Jersey/Hong Kong: World Scientific; 1988.
21. Deltuva A. Spin Observables in three-body direct nuclear reactions. Nucl Phys. 2009;A821:72.
22. Felsher PD, Howell CR, Tornow W, et al. Analyzing power measurement for the $\vec{d} + d \rightarrow d + p + n$ breakup reaction at 12 MeV. Phys Rev. 1997;C56:38.
23. Kievsky A, Viviani M, Rosati S. Cross section, polarization observables, and phase-shift parameters in p-d and n-d elastic scattering. Phys Rev. 1995;C52:R15.
24. Kievsky A, Rosati S, Viviani M. Proton-deuteron elastic scattering above the deuteron breakup. Phys Rev Lett. 1999;82:3759.
25. Kievsky A, Viviani M, Rosati S. Polarization observables in p-d scattering below 30 MeV. Phys Rev. 2001;C64:024002.
26. Kievsky A. Polarization observables in p-d and p-$^3$He scattering. Nucl Phys. 2001;A689:361c.

# Appendix: Clebsch-Gordan Coefficients, Racah Coefficients, and 9$j$ Symbols

In the Clebsch-Gordan (C-G) coefficient $C_{j_1m_1\ j_2m_2}^{j_3m_3}$, the angular momentum magnetic quantum numbers satisfy the relation expression $m_1 + m_2 = m_3$. Only $a$, $b$, $c$ are all integers and $a + b + c$ is even; the C-G coefficient $C_{a0\ b0}^{c0}$ with the magnetic quantum numbers to be zero may not be zero. The Racah coefficient is also known as the W coefficient. The following abbreviated symbols are introduced:

$$\hat{j} \equiv \sqrt{2j+1}\ ,\quad \hat{a} \equiv \sqrt{2a+1},\quad \cdots\cdots \tag{A1}$$

Here are some common formulas related to the C-G coefficients, the Racah coefficients, and the 9$j$ symbols [1, 2]:

$$\begin{aligned} C_{j_1m_1\ j_2m_2}^{j_3m_3} &= (-1)^{j_1+j_2-j_3} C_{j_1\ -m_1\ \ j_2\ -m_2}^{j_3\ -m_3} \\ &= (-1)^{j_1+j_2-j_3} C_{j_2\ m_2\ \ j_1\ m_1}^{j_3\ m_3} \end{aligned} \tag{A2}$$

$$\begin{aligned} C_{j_1m_1\ j_2m_2}^{j_3m_3} &= (-1)^{j_1-m_1} \frac{\hat{j}_3}{\hat{j}_2} C_{j_1\ m_1\ \ j_3\ -m_3}^{j_2\ -m_2} \\ &= (-1)^{j_1-m_1} \frac{\hat{j}_3}{\hat{j}_2} C_{j_3\ m_3\ \ j_1\ -m_1}^{j_2\ m_2} \\ &= (-1)^{j_2+m_2} \frac{\hat{j}_3}{\hat{j}_1} C_{j_3\ -m_3\ \ j_2\ m_2}^{j_1\ -m_1} \\ &= (-1)^{j_2+m_2} \frac{\hat{j}_3}{\hat{j}_1} C_{j_2\ -m_2\ \ j_3\ m_3}^{j_1\ m_1} \end{aligned} \tag{A3}$$

$$C_{j_1m_1\ 00}^{j_2m_2} = \delta_{j_1j_2}\delta_{m_1m_2} \tag{A4}$$

 
Q.-B. Shen, *Polarization Theory of Nuclear Reactions*,
https://doi.org/10.1007/978-3-031-11878-4

$$\sum_{m_1 m_2} C^{jm}_{j_1 m_1 \; j_2 m_2} C^{j'm'}_{j_1 m_1 \; j_2 m_2} = \delta_{jj'} \delta_{mm'} \tag{A5}$$

$$\sum_{jm} C^{jm}_{j_1 m_1 \; j_2 m_2} C^{jm}_{j_1 m'_1 \; j_2 m'_2} = \delta_{m_1 m'_1} \delta_{m_2 m'_2} \tag{A6}$$

$$\begin{aligned} W(abcd;ef) &= W(badc;ef) = W(cdab;ef) = W(dcba;ef) \\ &= W(acbd;fe) \end{aligned} \tag{A7}$$

$$W(abcd;ef) = (-1)^{e+f-a-d} W(ebcf;ad) = (-1)^{e+f-b-c} W(aefd;bc) \tag{A8}$$

$$\sum_e \hat{e}^2 \hat{f}^2 W(abcd;ef) W(abcd;eg) = \delta_{fg} \tag{A9}$$

$$\sum_e \hat{e}^2 (-)^{a+b-e} W(abcd;ef) W(bacd;eg) = W(afgb;cd) \tag{A10}$$

$$\sum_{\alpha\beta\delta\varepsilon\varphi} C^{e\varepsilon}_{a\alpha \; b\beta} C^{c\gamma}_{e\varepsilon \; d\delta} C^{f\varphi}_{b\beta \; d\delta} C^{c'\gamma'}_{a\alpha \; f\varphi} = \delta_{cc'} \delta_{\gamma\gamma'} \, \hat{e}\hat{f} \; W(abcd;ef) \tag{A11}$$

$$\sum_{\beta\delta\varepsilon} C^{e\varepsilon}_{a\alpha \; b\beta} C^{c\gamma}_{e\varepsilon \; d\delta} C^{f\varphi}_{b\beta \; d\delta} = \hat{e}\hat{f} \; C^{c\gamma}_{a\alpha \; f\varphi} \, W(abcd;ef) \tag{A12}$$

$$\sum_{\varepsilon} C^{e\varepsilon}_{a\alpha \; b\beta} C^{c\gamma}_{e\varepsilon \; d\delta} = \sum_{f\varphi} \hat{e}\hat{f} \; W(abcd;ef) C^{f\varphi}_{b\beta \; d\delta} C^{c\gamma}_{a\alpha \; f\varphi} \tag{A13}$$

$$W(abcd;0f) = \frac{(-)^{f-b-d} \delta_{ab} \delta_{cd}}{\hat{b}\hat{d}} \tag{A14}$$

$$C^{e0}_{a0 \; b0} W\left(abcd; e\frac{1}{2}\right) = -\frac{(-)^{c-d-a+b}}{\hat{a}\hat{b}} C^{e\,0}_{c\,-\frac{1}{2} \; d\frac{1}{2}} \tag{A15}$$

$$\begin{Bmatrix} a & b & c \\ d & e & f \\ g & h & i \end{Bmatrix} = \sum_k \hat{k}^2 W(aidh;kg) W(bfhd;ke) W(aibf;kc) \tag{A16}$$

$$\sum_c \hat{c}^2 W(aibf;kc) \begin{Bmatrix} a & b & c \\ d & e & f \\ g & h & i \end{Bmatrix} = W(aidh;kg) W(bfhd;ke) \tag{A17}$$

$$\delta_{cc'} \delta_{\gamma\gamma'} \begin{Bmatrix} a & b & c \\ d & e & f \\ g & h & i \end{Bmatrix} = \frac{\hat{c}}{\hat{f}\hat{g}\hat{h}\hat{i}^2} \sum_{\substack{\alpha\beta\delta\varepsilon \\ \varphi\rho\eta\nu}} C^{c\gamma}_{a\alpha \; b\beta} C^{h\eta}_{b\beta \; e\varepsilon} C^{i\nu}_{c'\gamma' \; f\varphi} C^{g\rho}_{a\alpha \; d\delta} C^{f\varphi}_{d\delta \; e\varepsilon} C^{i\nu}_{g\rho \; h\eta} \tag{A18}$$

$$C^{c\,\gamma}_{a\alpha\;\;b\beta}\begin{Bmatrix} a & b & c \\ d & e & f \\ g & h & i \end{Bmatrix} = \frac{\hat{c}}{\hat{f}\hat{g}\hat{h}\hat{i}^2}\sum_{\substack{\delta\varepsilon\varphi \\ \rho\eta\nu}} C^{h\,\eta}_{b\beta\;\;e\varepsilon} C^{i\,\nu}_{c\gamma\;f\varphi} C^{g\,\rho}_{a\alpha\;\;d\delta} C^{f\,\varphi}_{d\delta\;\;e\varepsilon} C^{i\,\nu}_{g\rho\;\;h\eta} \tag{A19}$$

$$\sum_{\delta\varphi\rho\nu} C^{i\nu}_{c\gamma\;\;f\varphi} C^{g\rho}_{a\alpha\;\;d\delta} C^{f\varphi}_{d\delta\;\;e\varepsilon} C^{i\nu}_{g\rho\;\;h\eta} = \sum_{b\beta} \frac{\hat{b}^2\hat{f}\hat{g}\hat{i}^2}{\hat{c}\hat{h}} C^{c\gamma}_{a\alpha\;\;b\beta} C^{h\eta}_{b\beta\;\;e\varepsilon}\begin{Bmatrix} a & b & c \\ d & e & f \\ g & h & i \end{Bmatrix} \tag{A20}$$

$$\sum_{\delta\rho} C^{g\rho}_{a\alpha\;\;d\delta} C^{f\varphi}_{d\delta\;\;e\varepsilon} C^{i\nu}_{g\rho\;\;h\eta} = \sum_{b\beta c\gamma} \frac{\hat{b}^2\hat{c}\hat{f}\hat{g}}{\hat{h}} C^{c\gamma}_{a\alpha\;\;b\beta} C^{h\eta}_{b\beta\;\;e\varepsilon} C^{i\nu}_{c\gamma\;f\varphi}\begin{Bmatrix} a & b & c \\ d & e & f \\ g & h & i \end{Bmatrix} \tag{A21}$$

$$\begin{Bmatrix} a & b & c \\ d & e & f \\ g & h & 0 \end{Bmatrix} = \frac{(-1)^{c+g-a-e}\delta_{cf}\delta_{gh}}{\hat{c}\hat{g}} W(abde;cg) \tag{A22}$$

$$\begin{Bmatrix} a & b & c \\ d & e & c \\ g & g & 1 \end{Bmatrix} = (-1)^{c+g-a-e}\frac{a(a+1)-b(b+1)-d(d+1)+e(e+1)}{\sqrt{4c(c+1)g(g+1)}\,\hat{c}\hat{g}} W(abde;cg) \tag{A23}$$

$$C^{e0}_{c0\;\;d0}\begin{Bmatrix} a & b & c \\ d & e & c \\ \frac{1}{2} & \frac{1}{2} & 1 \end{Bmatrix} = \frac{(-1)^{a-b-c+d+1}}{\sqrt{6}\,\hat{c}^2\hat{d}} C^{c1}_{a\frac{1}{2}\;\;b\frac{1}{2}} \tag{A24}$$

$$C^{e\;0}_{c+1\;\;0\;\;d0}\begin{Bmatrix} a & b & c \\ d & e & c+1 \\ \frac{1}{2} & \frac{1}{2} & 1 \end{Bmatrix} = (-1)^{a-\frac{1}{2}-c+d+e}\frac{(d-a)\hat{a}^2+(e-b)\hat{b}^2+c+1}{\sqrt{6(c+1)(2c+3)}\,\hat{c}^2\hat{d}} C^{c\;0}_{a\,\frac{1}{2}\;\;b\;\;-\frac{1}{2}} \tag{A25}$$

$$C^{e\;0}_{c-1\;\;0\;\;d0}\begin{Bmatrix} a & b & c \\ d & e & c-1 \\ \frac{1}{2} & \frac{1}{2} & 1 \end{Bmatrix} = (-1)^{a-\frac{1}{2}-c+d+e}\frac{(d-a)\hat{a}^2+(e-b)\hat{b}^2-c}{\sqrt{6c(2c-1)}\,\hat{c}^2\hat{d}} C^{c\;0}_{a\,\frac{1}{2}\;\;b\;\;-\frac{1}{2}} \tag{A26}$$

$$\sum_{cf} \hat{c}^2 \hat{f}^2 \hat{g}^2 \hat{h}^2 \begin{Bmatrix} a & b & c \\ d & e & f \\ g & h & i \end{Bmatrix} \begin{Bmatrix} a & b & c \\ d & e & f \\ j & k & i \end{Bmatrix} = \delta_{gj} \delta_{hk} \tag{A27}$$

$$\sum_{jk} (-1)^{2e+k-f-h} \hat{j}^2 \hat{k}^2 \begin{Bmatrix} a & b & c \\ e & d & f \\ j & k & i \end{Bmatrix} \begin{Bmatrix} a & e & j \\ d & b & k \\ g & h & i \end{Bmatrix} = \begin{Bmatrix} a & b & c \\ d & e & f \\ g & h & i \end{Bmatrix} \tag{A28}$$

In addition, the $3j$ symbol is

$$\begin{pmatrix} a & b & c \\ \alpha & \beta & \gamma \end{pmatrix} = \frac{(-1)^{b-a+\gamma}}{\hat{c}} C^{c\ -\gamma}_{a\alpha\ b\beta} \tag{A29}$$

The $6j$ symbol is

$$\begin{Bmatrix} a & b & e \\ d & c & f \end{Bmatrix} = (-1)^{a+b+c+d} W(abcd; ef) \tag{A30}$$

The $9j$ symbol $\begin{Bmatrix} a & b & c \\ d & e & f \\ g & h & i \end{Bmatrix}$ is invariant under interchange of rows and columns (reflection about a diagonal) and is multiplied by $(-1)^p$ (where $p = a + b + c + d + e + f + g + h + i$ ) upon interchange of two adjacent rows or columns.

## References

1. Rose M E. Elementary Theory of Angular Momentum. New York: John Wiley and Sons, Inc., 1957
2. Brink D M and Satchler G R. Angular Momentum. Oxford: Clarendon Press, 1962.

# Index




Q.-B. Shen, *Polarization Theory of Nuclear Reactions*,
https://doi.org/10.1007/978-3-031-11878-4

## D



## E

## L



## M



## N



## O

**P**



**Q**



**R**